Simple Models of Complex Nuclei

Contemporary Concepts in Physics

Volume 1, *Particle Physics and Introduction to Field Theory* (Revised and Updated First Edition) **T. D. Lee**

Volume 2, *Particle Physics: The Quest for the Substance of Substance*
L. B. Okun

Volume 3, *Gauge Fields and Strings*
A. M. Polyakov

Volume 4, *Nonlinear Physics: From the Pendulum to Turbulence and Chaos*
R. Z. Sagdeev, D. A. Usikov and **G. M. Zaslavsky**

Volume 5, *Particle Physics and Inflationary Cosmology*
Andrei Linde

Volume 6, *Algebraic Approaches to Nuclear Structure: Interacting Boson and Fermion Models*
Richard F. Casten, Pertti O. Lipas, David. D. Warner, Takaharu Otsuka, Kris Heyde and **Jerry P. Draayer,** edited by **Richard F. Casten**

Volume 7, *Simple Models of Complex Nuclei: The Shell Model and Interacting Boson Model*
Igal Talmi

Simple Models of Complex Nuclei

The Shell Model and Interacting Boson Model

Igal Talmi

Weizmann Institute of Science
Rehovot, Israel

harwood academic publishers
Switzerland Australia Belgium France Germany Great Britain India Japan Malaysia Netherlands Russia Singapore USA

Harwood Academic Publishers

Private Bag 8
Camberwell, Victoria 3124
Australia

58, rue Lhomond
75005 Paris
France

Glinkastrasse 13–15
O-1086 Berlin
Germany

Post Office Box 90
Reading, Berkshire RG1 8JL
Great Britain

3-14-9, Okubo
Shinjuku-ku, Tokyo 169
Japan

Emmaplein 5
1075 AW Amsterdam
Netherlands

820 Town Center Drive
Langhorne, Pennsylvania 19047
United States of America

Library of Congress Cataloging-in-Publication Data
Talmi, Igal, 1925–
Simple models of complex nuclei : the shell model and interacting boson model / Igal Talmi.
p. cm.—(Contemporary concepts in physics; v. 7)
Includes bibliographical references and index.
ISBN 3-7186-0551-1.—ISBN 3-7186-0550-3 (pbk.)
1. Nuclear shell model. 2. Interacting boson models. 3. Atomic structure. I. Title. II. Series.
QC173.T26 1993
539.7'43—dc20 92-12901
CIP

In memory of
Giulio (Yoel) Racah, teacher, colleague and friend,
and
Amos de-Shalit, colleague and friend

Contents

Preface to the Series **xvii**

Preface **xix**

1. *Introduction. Effective Interactions* **1**

Nuclear constituents. The shell model. Order of single nucleon levels and magic numbers. Single nucleon and single hole states. Use of effective interactions determined from experiment. The ^{38}Cl and ^{40}K case. Limitations of the approach. Plan of the book.

2. *Historical Survey* **21**

Early attempts in the 1930s. The Elsasser papers. The summary of Bethe and Bacher. The criticism by Niels Bohr. Emergence of magic numbers in heavy nuclei. The Feenberg–Hammack and Nordheim schemes. Single nucleon magnetic moments and nuclear structure. Introduction of spin-orbit interaction by M. G. Mayer and by Jensen et al. establishing the jj-coupling shell model.

3. *Single Particle in a Central Potential Well* **33**

Non-relativistic Schrödinger equation and its solutions. Spherical harmonics. Spin-orbit interaction and wave functions for $j = l \pm \frac{1}{2}$. Single nucleon magnetic moments.

4. *Harmonic Oscillator Potential* **45**

Harmonic oscillator Hamiltonian and eigenfunctions. Virial theorem and average of $\mathbf{r}^2$. Many particle wave functions, center-of-mass motion and translational invariance. Spurious states. Particles interacting by a two-body harmonic potential.

5. *Transformation under Rotations* **55**

Physical tensor fields and their transformation properties. Active interpretation of rotations of scalar fields and orbital angular momentum. Irreducible tensorial sets. Two component spinors and transformation laws. Intrinsic spin.

6. *Infinitesimal Rotations and Angular Momentum* **71**

Infinitesimal rotations and generators. Spinor fields, intrinsic spin and total angular momentum. Transformations of irreducible tensor operators under rotations. Finite rotations, an example. Lie algebras, the $O(3)$ and the $SU(2)$ groups.

7. *Coupling of Angular Momenta. $3j$-Symbols* **91**

Reduction of the external product of two irreducible tensorial sets. Clebsch–Gordan coefficients and their properties. Reduction formulae. Wigner's $3j$-symbols and their orthogonality and symmetry properties. Reduction of products of two spherical harmonics.

8. *The Wigner–Eckart Theorem* **107**

Matrix elements of irreducible tensor operators. The Wigner–Eckart theorem. Reduced matrix elements. Hermitian tensors. Magnetic moments of a combined system (Lande formula). Transition rates.

9. *Two Nucleon Wave Functions. $9j$-Symbols* **121**

Symmetric and antisymmetric wave functions. Isospin. Matrix elements of allowed β-decays. jj-coupling and LS-coupling and the transformation between them. Change of coupling transformations

and $9j$-symbols. Properties of $9j$-symbols. Relations with $3j$-symbols. Tensor products and their matrix elements.

10. *Matrix Elements of Two Nucleon Interactions. $6j$-Symbols* **145**

The Slater expansion. Direct and exchange terms. Matrix elements of scalar products and $6j$-symbols. Properties of $6j$-symbols and relations with $9j$-symbols and $3j$-symbols. Expansion of direct and exchange terms. Reduced matrix elements of spherical harmonics. Rates of electric and magnetic multipole transitions. Spin dependent and non-central interactions. Matrix elements in LS-coupling.

11. *Short Range Potentials—The δ-Interaction* **179**

Eigenstates and eigenvalues in LS-coupling and in jj-coupling for isospin $T = 1$ and $T = 0$ states. Eigenvalues of the $(\mathbf{s}_1 \cdot \mathbf{s}_2)\delta(|\mathbf{r}_1 - \mathbf{r}_2|)$ interaction. Comparison of level spacings calculated with zero range interactions and experimental data. Hamiltonian matrices in l^2 configurations, the LS-coupling and jj-coupling limits.

12. *The Pairing Interaction and the Surface Delta Interaction* **205**

General tensor expansion of interactions. Odd tensor interactions including the δ-potential in $T = 1$ states. The pairing interaction in j^n configurations, expansion in odd and even tensors. Mixing of several two nucleon configurations. The surface delta interaction. Eigenstates for $T = 1$ and $T = 0$ of mixed configurations and their eigenvalues.

13. *Two Nucleons in a Harmonic Oscillator Potential Well* **229**

Transformation of the Hamiltonian to the center-of-mass and relative coordinates. Expansion of two nucleon wave functions in functions of $\mathbf{R}$ and $\mathbf{r}$. Matrix elements of interactions. Harmonic oscillator integrals.

14. *Determinantal Many Nucleon Wave Functions* **237**

The m-scheme. Slater determinants. Matrix elements of single nucleon and two-nucleon operators. Closed shells, single nucleon and single hole states. Nucleon-hole configurations, the Pandya relation.

29. ***LS-Coupling of Protons and Neutrons. The SU(4) Scheme*** **599**

Symmetry properties of spin-isospin states. The $SU(4)$ group, its irreducible representations and quantum numbers. Wigner supermultiplets. The Majorana space exchange operator, its eigenvalues and relation to the $SU(4)$ Casimir operator. Simple binding energy formula, symmetry energy and pairing energy. l^n configurations. The $U(2l+1)$ group and its $O(2l+1)$ subgroup. Casimir operators. Seniority in l^n configurations with isospin. Binding energy formula. Favored and unfavored β-decay.

30. ***Special Proton Neutron Mixed Configurations. The SU(3) Scheme*** **621**

Degeneracies in the harmonic oscillator potential. The $SU(3)$ group and its irreducible representations. The Casimir operator and its eigenvalues. Two-nucleon interactions with $SU(3)$ symmetry and eigenvalues of $SU(3)$ Hamiltonians. Irreducible representations of $O(3)$ contained in $SU(3)$ irreducible representations. The K quantum number and rotational bands. The $1p$ shell and the $2s, 1d$ shell. Difficulties due to strong spin-orbit interaction.

31. ***Valence Protons and Neutrons in Different Orbits*** **645**

Mutual interaction energies. Weak coupling and center-of-mass theorem. Non-diagonal matrix elements. Special case with $j = \frac{1}{2}$ of one of the orbits. Examples of exact diagonalization—the ^{42}K spectrum, levels of ^{93}Mo. Spin gaps—Yrast traps. Other examples, ^{211}Po and ^{212}Po. Average attraction between protons and neutrons. Its effect on level order in ^{11}Be. Average repulsion of identical nucleons in different orbits. Implications for nuclear structure. General expressions with isospin. Matrix elements of single nucleon operators. Matrix elements for pick-up and stripping reactions. General expressions with isospin. A simple example of E2 transitions in $g^n_{9/2}p^{n'}_{1/2}$ mixed configurations.

32. ***Configuration Mixing. Effective Operators*** **681**

Matrix elements of interactions between various configurations differing by orbits of *two* nucleons. Second-order corrections to energies due to such configuration mixing, effective two-body interactions. Matrix elements of interactions between various configurations differing by the orbit of *one* nucleon. Change in radial functions of single

Preface to the Series

The series of volumes, *Contemporary Concepts in Physics*, is addressed to the professional physicist and to the serious graduate student of physics. The subjects to be covered will include those at the forefront of current research. It is anticipated that the various volumes in the series will be rigorous and complete in their treatment, supplying the intellectual tools necessary for the appreciation of the present status of the areas under consideration and providing the framework upon which future developments may be based.

Preface

Nuclear Shell Theory, which I co-authored with Amos de-Shalit, was published in 1963. Over the years I have been approached by colleagues who have urged me to revise and update that book. It was clear, however, that a simple revision would not be sufficient; a new book was needed. *Simple Models of Complex Nuclei: The Shell Model and Interacting Boson Model* is intended to satisfy that need.

One aim of *Nuclear Shell Theory* was to serve as an introduction to tensor algebra, a subject many nuclear physicists were still unfamiliar with. In that work, the use of group theory was limited, and the use of second quantization avoided entirely; neither was in wide use in the community at the time of publication. Much has changed since then. Tensor algebra and group theory are now commonly used. Tensor algebra is therefore presented here in less detail, while group theoretical methods and second quantization occupy large parts of the book. The use of group theory plays a paramount role in the interacting boson model, which is also presented here.

The present book may serve as an introduction to the use of group theory in spectroscopy. The emphasis is on general principles; no attempt is made at mathematical rigor or detailed description of group theory. Many textbooks are available now which specifically cover these areas.

As in *Nuclear Shell Theory*, the term *shell model* means here the *spherical* shell model. States of nucleons moving in a *deformed* potential well, not considered in this book, are simple only in an intrinsic frame of reference. States with definite angular momenta which should be projected from them are extremely complicated. The interacting boson model is included in this book because it is treated by the same methods developed for the spherical shell model. A possible connection between the shell model and the interacting boson model is discussed in detail.

No attempt has been made to consider all phenomena in nuclei that can be described by shell model wave functions. The experimental material included in this book is presented merely to illustrate the agreement between some predictions of the shell model and experimental results. The formalism developed may be used for complicated calculations, but such calculations are not carried out here. While they are interesting and important, such calculations, in general, are not quantitatively significant due to their complexity.

Unlike *Nuclear Shell Theory*, the present volume includes references. It was not possible to give proper credit to all papers that have been written on the subjects covered here. I tried to refer to original papers, but could not mention all the works that followed them. I would like to apologize to those authors, including good friends, whose papers do not appear in the list of references.

In this volume it has been necessary to shorten the discussion of some topics covered in *Nuclear Shell Theory*, and to omit others entirely. I believe, however, that the most important material found in that book is included here. In particular, the useful appendix appears here in an expanded version. My hope is that this book will be useful to as many physicists as was *Nuclear Shell Theory*.

I have learned much from collaborations and discussions with many colleagues. Although they are too numerous to permit individual mention here, I am very grateful to all of those colleagues, Akito Arima and Franco Iachello in particular. I wish to express my thanks to Mrs. Naomi Cohen for her skillful typing of the manuscript. Finally, I would like to thank my wife Chana; without her warm, strong and continuous support this book could not have been written.

Igal Talmi

1

Introduction. Effective Interactions

Atomic nuclei are very complex systems including large numbers of strongly interacting protons and neutrons. It is now recognized that these nucleons are not fundamental building blocks with some meson clouds around them. Nucleons are composite particles, each including three valence quarks. Quantum chromodynamics (QCD) of quarks and gluons is accepted as the fundamental theory describing, among other phenomena, the structure and interactions of nucleons. Unfortunately, there are great difficulties in obtaining a reasonable solution of QCD to low energy phenomena like the structure of the nucleon. The interaction between two nucleons is even more complicated. In high energy experiments, the composite nature of nucleons is clearly manifest. For low energy phenomena, it is usually assumed that it is possible to ignore the internal structure of nucleons. Nucleons may then be considered as basic constituents of nuclei and their complicated interactions may be replaced by some interaction which is a function of their coordinates, momenta and spins. The latter may be determined by scattering of nucleons by nucleons and by properties of the deuteron.

These simplifying assumptions are analogous to the treatment of ions in molecular and solid state physics. The complex interaction between ions, arising from the electrostatic and magnetic interactions between nuclei and electrons, are often replaced by local potentials.

Those are determined empirically from experimental data. A simple example is offered by the Morse potential which yields good description of certain states of diatomic molecules.

In practice, scattering of free nucleons does not determine completely the nuclear interaction. The deuteron has only one bound state and its properties do not provide sufficient information for constructing a potential interaction. Simple nuclear interactions were constructed with guidance from some theory. As a result, there are several such interactions available, all of which fit the two nucleon data. However, none of the experimental data from scattering experiments and the deuteron can determine the importance of three body interactions of nucleons. Theoretical considerations lead to such interactions but they are inconclusive about the magnitude of three body forces.

Effects due to the structure of nucleons, as well as explicit effects of meson exchanges, may not be successfully replaced by a simple interaction even at low energies. In particular, manifestations of such effects may depend on the density of nucleons in nuclei. Such density dependence leads naturally to three-body interactions between nucleons in nuclei. Still, there are no reliable estimates of the importance of such effects.

Considering nuclear structure at low excitation energies as due to nucleons interacting by two-body potential interactions is a very great simplification. There are certainly phenomena, specially at high energies, where this picture cannot provide a complete description. Yet, even with this simplification we are faced with the task of solving the non-relativistic Schrödinger equation for a many body system with strong interactions. Such a problem cannot be treated *exactly* by many body theory. Not even useful approximation procedures have been developed. To make progress in understanding nuclear structure, *models* have been suggested which are based on empirical evidence.

A similar situation is encountered in condensed matter physics. The interactions there are known precisely but in many cases, no exact solution is possible. Much progress was made by introducing and using successfully many models. Models are simplified or idealized physical systems which can be treated either exactly or, at least, by good approximation methods. In suggesting a model, an attempt is made to isolate the most important degrees of freedom and deal with them explicitly. Other degrees of freedom contribute only implicitly, their effect is limited to renormalization of the parameters of the model. A model must be internally consistent. It must reproduce to a sufficient accuracy the experimental data observed in the physical system which it represents. Successful models should be simple enough to be accessible to analysis and yet detailed enough to provide a good description of the relevant physical situation.

Nuclei are attractive objects for research since, in spite of their complexity, they exhibit remarkably simple regularities. Models of nuclear structure aim at describing correctly these regular features. In the shell model, degrees of freedom of individual nucleons are displayed. Another model, the collective model, very successfully describes collective motions of nucleons. It should not be forgotten that, unlike exact theories, models may be useful for describing only certain groups of phenomena. Beyond the domain in which a certain model is successful, other degrees of freedom must be taken explicitly into account. Still, a good model sometimes works well even where apparently it is not valid. This is in contrast to bad models which are useless even where they are supposed to work.

In the shell model, the complicated motion of nucleons in nuclei due to their mutual interactions is replaced by motion of *independent nucleons* in a spherical and static potential well. This is analogous to the motion of electrons in atoms where a major part of the central potential is due to the positively charged and massive nucleus. Similar to noble gas atoms, where electron shells are completely filled, nuclei with certain numbers of protons or neutrons ("magic numbers") turn out to be particularly stable. The shell model can account for such magic numbers as due to completely filled proton or neutron orbits provided the appropriate order of single nucleon levels is assumed. Almost all magic numbers were known in the early thirties but no satisfactory scheme of single nucleon orbits was discovered until 1949. The modern version of the shell model is based on a strong spin orbit interaction which yields large splittings between energies of the $j = l + \frac{1}{2}$ and $j' = l - \frac{1}{2}$ orbits. It predicts correctly all magic numbers and has been very successful in describing a whole host of experimental data.

It is rather difficult to reconcile the shell model with the strong and singular interaction between free nucleons. This interaction has strong repulsion at very short distance followed by a very strong short range attraction. Such an interaction should give rise to strong short range correlations between nucleons which are absent from the simple independent nucleon wave functions of the shell model. Starting from a simple shell model wave function, like that of closed shells, the strong short range interaction admixes into it many shell model states which lie at high excitation energies. This difficulty was recognized already in the thirties and still has no satisfactory solution. Many body theory has done much to elucidate this problem and supplies a framework within which the shell model may be formulated. It may be possible to use simple shell model wave functions to calculate nuclear energies and other observables if all effects of short range correlations can be transformed onto the Hamiltonian and other operators. Two body interactions become then strongly *renormalized*. They become rather regular at short distances and do not admix into simple shell model wave functions states of very high lying configurations.

It may be possible to adopt this approach and use simple shell model wave functions to calculate energies (and perhaps other observables) provided renormalized or *effective interactions* (or other effective operators) are used. In principle, such effective interactions could be calculated by many body theory starting from the interaction between free nucleons. Many attempts have been made to carry out this program but they met only modest success. Many calculations were performed on nuclear matter which is not directly accessible to experimental study. For finite nuclei, no reliable way was found to carry out the complicated renormalizations involved in deriving matrix elements of the effective interaction. No clear demonstration was ever given that the shell model is a good approximation nor a characterization of properties of the interactions for which this is the case. Today, as in 1949, the best proof for the validity of the shell model is the good agreement of its predictions with experiment. Let us now look at some of the predictions for which this good agreement is obtained in spite of our inability to obtain satisfactory solutions to the nuclear many body problem.

The magic numbers were explained when the order and spacings of single nucleon energies in the various orbits were determined. Energies of nuclear states in which a single nucleon occupies certain orbits cannot be reliably calculated. In the shell model it is *assumed* that energies of single nucleon orbits correspond roughly to levels of a single nucleon in simple potentials like the harmonic oscillator or a square well potential. They may be determined in many cases from experiment. The new important ingredient introduced in 1949 to the shell model is the strong spin-orbit interaction. It determines the levels which constitute major shells whose closures yield the magic numbers. The splitting between the $j = l + \frac{1}{2}$ and $j' = l - \frac{1}{2}$ levels for the state with highest l-value in a major oscillator shell is very large. As a result, the $j = l + \frac{1}{2}$ orbit is lowered into the lower oscillator major shell. Thus, the $1f_{7/2}$ orbit joins the $2s, 1d$ orbits to form the shell whose closure is at proton or neutron number 28. The $1g_{9/2}$ orbit joins the $2p, 1f_{5/2}$ orbits to form the major shell between 28 and 50. In this way also the magic numbers 82 and 126 are reproduced. The resulting order of single nucleon orbits is shown schematically in Fig. 1.1. The letters $s, p, d, f, g, \ldots$ stand for orbital angular momenta $l = 0, 1, 2, 3, 4, \ldots$. The numbers on the left of these letters define the order of orbits with the same value of l in the potential well. For instance, $1s$ is the lowest $l = 0$ orbit, $2s$ is the next $l = 0$ orbit, etc. The total angular momentum j is written on the right of the letter.

In the left column (a) of Fig. 1.1 the order of single nucleon levels is that of the harmonic oscillator potential well. In the middle column (b) the oscillator degeneracy is removed so as to yield, upon inclusion of the spin-orbit splitting, the level order in the right column (c). That schematic level order roughly corresponds to the observed order. Single nucleon orbits whose energies are close form major shells. It should be kept in mind that the shape of the central potential well which determines the order of single nucleon orbits, depends rather strongly on the occupation of orbits. In particular, as will be discussed in Section 38, the order of neutron orbits strongly depends on the occupation of proton orbits and *vice versa*.

To obtain magic numbers from the scheme of single nucleon orbits, shell model wave functions should be constructed according to the jj-coupling scheme. The orbits which nucleons occupy are characterized not only by the principal quantum number n and orbital angular momentum l. The total spin j of each nucleon is a good quantum number. States of several nucleons are then obtained by coupling the individual j spins to a total spin J. It is not sufficient to have the $j = l + \frac{1}{2}$ orbit lying lower than the $j' = l - \frac{1}{2}$ orbit. The observed magic numbers are reproduced if in states of several l-nucleons, up to $2l + 2$, each nucleon is occupying the $j = l + \frac{1}{2}$ orbit. The jj-coupling scheme is discussed in great detail in this volume.

The most impressive success of the shell model is the correct reproduction of the magic numbers. Single nucleon orbits are bunched into major shells and the number of identical nucleons, protons or neutrons, in completely filled shells is a magic number. Excitation of closed shells involves raising a nucleon to an orbit in a higher shell which requires relatively high energy. If a nucleon is added to a magic nucleus its separation energy is much smaller than the energy of removing a nucleon from closed shells. The extra nucleon must enter an unoccupied orbit where its kinetic energy is higher and the cen-

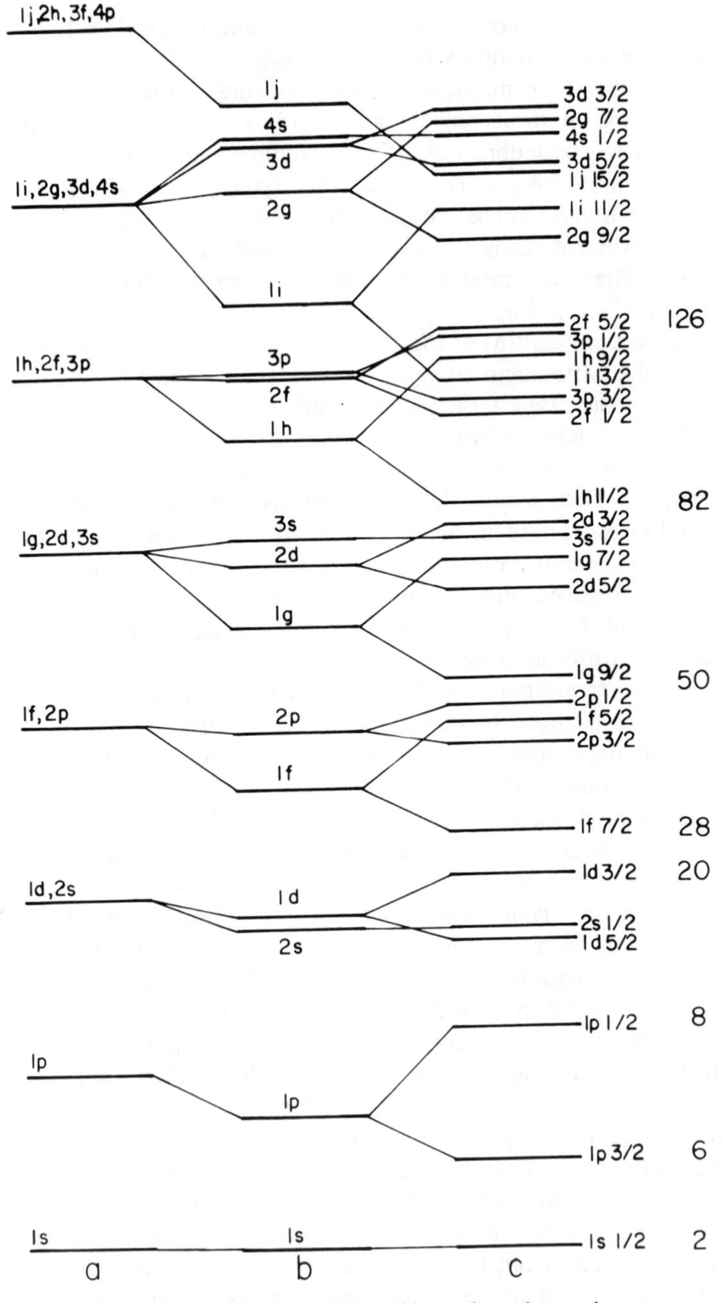

FIGURE 1.1. *Single nucleon orbits and magic numbers.*

tral potential less attractive. This makes magic nuclei more stable and abundant than their neighbors.

Spacings of single nucleon levels within a shell are smaller than spacings between orbits in different shells. Hence, lowest excited states of a single nucleon outside closed shells are obtained by the extra nucleon occupying the various orbits in its shell. The same occurs for nuclei in which a single nucleon is *missing* from closed shells. Such single nucleon states, as well as single *hole* states, are observed in actual nuclei lending strong support for the shell model. Such a spectrum is displayed in Fig. 1.2. All levels observed experimentally in the ^{209}Pb nucleus up to 2 MeV excitation energy, can be interpreted as due to the valence neutron occupying the various orbits in the major shell beyond the magic neutron number $N = 126$.

States of nuclei with one nucleon outside closed shells or one nucleon hole, are represented by shell model wave functions which are well defined. They may be used for calculating various observables of those nuclei. These include magnetic and quadrupole moments, electromagnetic transition probabilities, β-decay and cross-sections for various reactions. Results of such calculations, even in more complicated cases, are in fair agreement with experimental data. In some cases, there are clear indications that the operators whose matrix elements are calculated should be renormalized. It should be realized that energies of excitations of nucleons from closed shells are not very high compared to spacings of orbits within a major shell. At a few MeV above ground states of nuclei we see very many levels due to such excitations. The effects of high energy states admixed by short range correlations may be replaced by using an effective interaction. Still, it is expected that states whose energies are very close will be admixed by the effective interaction. Thus, simple shell model states, like single nucleon or single hole states, if they lie at sufficiently high energies, are not pure. They may be admixed with some of the states lying near them. Single nucleon properties are thus fragmented which means shared by other states. Since states which may be admixed have the same spins and parities, such fragmentation may take place only if nearby levels have the same spin and parity as the simple shell model state. The energy where states with given spin and parity become numerous, strongly depends on their spin. The higher the spin, the higher the energy where this occurs. The few MeV mentioned above should be measured from the energy of the lowest state with a given spin (*Yrast* state). High spin shell model states may be rather pure even if they lie several MeV above ground states.

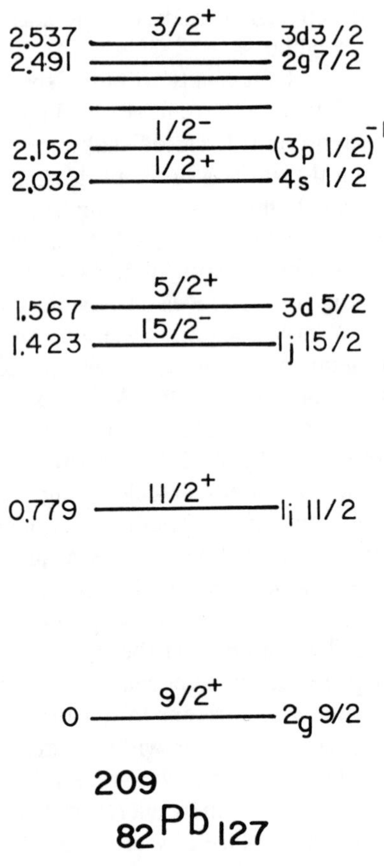

FIGURE 1.2. *Single nucleon levels of* ^{209}Pb.

As mentioned above, states of single nucleons at sufficiently high excitation may not be well described in terms of single nucleon wave functions. Similarly, single hole states of deep lying orbits have high excitations and may no longer be accurately described in terms of single hole wave functions. the deeper the state of the missing nucleon, the higher the excitation energy of the hole state.

The fragmentation of single nucleon or single hole states at higher excitations imposes practical limitations to the simple shell model picture. The statement has been made that the notion of simple nucleon orbits is valid only near the Fermi surface. This does not mean that deep lying orbits do not have a precise meaning. It is true that high lying single hole states are not exact eigenstates of the nuclear Hamiltonian. Still the description of low lying states in terms of fully occupied orbits may be a good approximation. For example, removing a nucleon from the lowest $1s$ orbit leaves the nucleus in such high excitation that the resulting state is strongly admixed with very many states with complicated shell model structure. In the ground state of a medium weight nucleus, however, it is expected that the lowest $1s$-orbit is fully occupied by two protons and two neutrons.

The success of the shell model in predicting magic numbers is based on the level scheme of a central potential well including spin-orbit interaction. The single nucleon orbits in a major shell are determined by this picture. Experimental observation of single nucleon levels and their properties lends strong support for the shell model. There is, however, no way available to *calculate* the energies of nuclei with closed shells. In order to use shell model wave functions we need to know the effective interaction between nucleons in the various orbits. As explained above, it has not been possible to calculate it from the interaction between free nucleons. This is true also of energies of nuclei in which there is one nucleon outside closed shells. We consider the total energy of such a nucleus in states where the extra (or *valence*) nucleon occupies various orbits. We may now subtract from them the total energy of the closed shell nucleus (the core) obtained by removing the valence nucleon. In the approximation that nucleon states in the core are the same as in the nucleus considered, these separation energies are equal to the expectation value of the kinetic energy of the valence nucleon and its interaction with the closed shells. These separation energies may be considered as energies of single nucleon orbits in a potential well. The potential well in this case depends on the occupation of orbits in closed shells. These single nucleon energies have not been calculated so far and we must determine them from experiment using observed levels like those in Fig. 1.2.

To obtain *quantitative* agreement between shell model calculation of energies and experimental data we should turn to nuclei with *several* nucleons outside closed shells. In such cases, there are, in general, several shell model states in any given *configuration*, i.e. given occupation numbers of valence nucleons in the various orbits. States of several nucleons in a j-orbit are degenerate if only the central potential is considered. The degeneracy is removed and eigenstates of the shell model Hamiltonian are determined if the mutual interactions of the valence nucleons are taken into account. To carry out the calculation of energy eigenvalues and determine the corresponding eigenstates, we should know the matrix elements of the effective interaction between states of valence nucleons. It should be considerably easier to obtain them from the interaction between free nucleons than to calculate total energies. Still, no reliable way of performing such calculations has been developed. It was mentioned above that in adopting the shell model we are relying on experimental facts. Also for the determination of matrix elements of the effective interaction we must use experimental data.

In the early days of the shell model, the singular features of the interaction between free nucleons were not yet known. Rather regular and simple potential interactions, with certain exchange properties, were constructed which gave a fair description of low energy scattering between free nucleons, as well as of properties of the deuteron. There was no problem in calculating matrix elements of these interactions between shell model states and they were used for calculations of nuclear energies and wave functions. Many papers reporting results of such calculations have been published but they are not reviewed in the present volume. The simple interactions used can be described only as simple prescriptions for the effective interaction. In many cases, their use led to mixing of several shell model configurations and this introduced parameters which were not determined in a consistent way. The good agreement with experiment which was often obtained did not have quantitative significance. An exception can be found in the work of Kurath (1956, 1957) in which an attempt was made to consider several nuclei with the valence $1p$-shell using a rather simple phenomenological interaction.

In order to determine matrix elements of the effective interaction from experimental energies, the number of the former should be smaller than the number of the latter. The number of matrix elements to be determined should therefore be minimized. One way to do it is to assume that those matrix elements are the same for all states of a given nucleus. It also helps to assume that these matrix elements are fairly constant, or change in a simple way, in a group of neighboring nuclei. Even if this happens to be the case, these assumptions do not reduce sufficiently the number of matrix elements of the effective interaction. The latter, as obtained formally from many body theory, includes two-nucleon terms, three-nucleon terms, four-nucleon terms etc. in its expansion. This feature is present even if the interaction between free nucleons is only a two-body interaction. There is no reliable theoretical method for estimating the relative importance of the various many-nucleon terms. If all of them are considered, they may still be determined by experimental energies. There will not be, in that case, any check for consistency of the procedure nor will it be possible to make any prediction.

If we assume that the important effective interaction has only two-nucleon terms, the number of matrix elements is drastically reduced. In this case, *matrix elements of the effective interaction in configurations with several valence nucleons are linear combinations of matrix elements between states of two nucleon configurations*. In a given subspace of the shell model, defined by the valence orbits, there is a well defined set of such matrix elements. The smaller the subspace, the smaller the number of these theoretical parameters. If enough experimental energies are known, they may be equated to the theoretical expressions in which the two-nucleon matrix elements appear as unknown parameters. The values of these parameters are then determined by a least squares fit of the theoretical expressions to the experimental energies. If good agreement is obtained between energies calculated by using matrix elements determined in this way and experimental energies, the procedure is consistent. The matrix elements thus obtained may be used for the calculation and prediction of energies of other states within the shell model subspace considered.

In order to illustrate the approach outlined above, let us consider a simple case. This is the relation between levels of ^{38}Cl and ^{40}K. It was the first successful application of extracting matrix elements of the effective interaction from experiment in a consistent way. It is still one of the finest examples of simple and successful shell model calculations.

The $^{38}_{17}\mathrm{Cl}_{21}$ nucleus has one neutron outside the closed shells of neutrons 20 and 9 protons in the $2s, 1d$ shell outside the closed proton shell of $Z = 8$. To make a *simple* analysis we make the drastic assumption that the $1d_{5/2}$ and $2s_{1/2}$ proton orbits are completely filled. The valence proton is thus in the $1d_{3/2}$ orbit. In ^{37}Cl the $J = \frac{1}{2}$, positive parity $(\frac{1}{2})^+$ state, interpreted as an excitation from the $2s_{1/2}$ proton orbit, lies at 1.73 MeV above the $(\frac{3}{2})^+$ ground state. The $(\frac{5}{2})^+$ level, due to proton excitation from the $1d_{5/2}$ orbit, is at 3.09 MeV and a $(\frac{7}{2})^-$ level, indicating excitation into the $1f_{7/2}$ orbit, is at 3.10 MeV. This situation does not contradict the simplifying assumption about the occupation of single proton orbits.

In $^{38}_{17}\mathrm{Cl}_{21}$ the lowest four levels have negative parities and spins 2^- (g.s.) 5^- (at .67 MeV), 3^- (at .76 MeV) and 4^- (at 1.31 MeV). We interpret them as due to the coupling between the valence $1d_{3/2}$ proton and the valence $1f_{7/2}$ neutron to form states with definite total J. As repeatedly emphasized, there is no known method to calculate energy spacings between these levels nor can it be proved that it makes any sense to assign them simple shell model configurations. Nevertheless, we proceed from the assumption that the observed energies are indeed equal to matrix-elements of the effective two-body interaction in the simple shell model configuration of a $1d_{3/2}$ proton and a $1f_{7/2}$ neutron.

Levels of ${}^{40}_{19}\mathrm{K}_{21}$ have been known several years before the ${}^{38}\mathrm{Cl}$ spectrum was measured. The ground state has spin 4^-, .03 MeV above it there is a 3^- state and the next states have spins 2^- (.80 MeV) and 5^- (.89 MeV). The spins (and parities) of these levels are expected from the configuration of three $1d_{3/2}$ protons and one $1f_{7/2}$ neutron. Since there can be at most four protons in the $1d_{3/2}$ orbit, there is only one state, with $J_\pi = \frac{3}{2}$, allowed by the Pauli principle in the $d^3_{3/2}$ configuration. This $J_\pi = \frac{3}{2}$ coupled with $j_\nu = \frac{7}{2}$ of the neutron yields states with $J = 2, 3, 4, 5$. In all these four states the interaction energy (and single particle energies) of the three $d_{3/2}$ protons is the same as in the $d^3_{3/2} J_\pi = \frac{3}{2}$ state. The energy eigenvalues in the $J = 2, 3, 4, 5$ states are different due to the interaction between the $d_{3/2}$ protons and $f_{7/2}$ neutron. There are three such interaction terms, each of which is a matrix element of the two body interaction between one $d_{3/2}$ proton and the $f_{7/2}$ neutron.

In the following sections we shall see several ways to express the interaction energy in these ${}^{40}K$ states in terms of two-body matrix elements. Let us quote here the results, expressed in terms of the matrix elements $V_J = \langle d_{3/2} f_{7/2} JM | V | d_{3/2} f_{7/2} JM \rangle$ of the effective interaction

$$\langle d^3_{3/2} f_{7/2} J = 4 | \sum_{i=1}^{3} V_{i4} | d^3_{3/2} f_{7/2} J = 4 \rangle = \tfrac{25}{28} V_2 + \tfrac{1}{6} V_3 + \tfrac{23}{35} V_4 + \tfrac{77}{60} V_5$$

$$\langle d^3_{3/2} f_{7/2} J = 3 | \sum_{i=1}^{3} V_{i4} | d^3_{3/2} f_{7/2} J = 3 \rangle = \tfrac{15}{28} V_2 + \tfrac{4}{3} V_3 + \tfrac{3}{14} V_4 + \tfrac{11}{12} V_5$$

$$\langle d^3_{3/2} f_{7/2} J = 2 | \sum_{i=1}^{3} V_{i4} | d^3_{3/2} f_{7/2} J = 2 \rangle = \tfrac{9}{14} V_2 + \tfrac{3}{4} V_3 + \tfrac{45}{28} V_4$$

$$\langle d^3_{3/2} f_{7/2} J = 5 | \sum_{i=1}^{3} V_{i4} | d^3_{3/2} f_{7/2} J = 5 \rangle = \tfrac{7}{12} V_3 + \tfrac{21}{20} V_4 + \tfrac{41}{30} V_5 \qquad (1.1)$$

A simple check on the numerical coefficients in (1.1) is obtained by putting all V_J equal. The coefficient of this matrix element is the sum of the coefficients in every row and it turns out to be 3. This is the number of interactions considered here. Subtracting the first equation in (1.1) from the others, we obtain the calculated energy *spacings* in ^{40}K in terms of $V_3 - V_2$, $V_4 - V_2$ and $V_5 - V_2$. Hence, from the spacings in ^{38}Cl we can thus calculate spacings of ^{40}K levels. The relation between the two sets of energies is linear and can be reversed. Using the known level spacings in ^{40}K, the spectrum of ^{38}Cl can be deduced. The calculated levels are 2^- g.s., 5^- at .70, 3^- at .75 and 4^- at 1.32 MeV. In 1955, the levels of ^{38}Cl were found to agree very well with the calculated ones (Goldstein and Talmi 1956 and independently Pandya 1956). The calculated and experimental levels of ^{38}Cl are shown in Fig. 1.3.

The interaction energy in the ground state of ^{40}K can also be calculated from (1.1) and some observed binding energies. This type of calculations is carried out in a subsequent section. Magnetic moments of ^{40}K levels can be simply related to those of ^{39}K and ^{41}Ca. Also this is discussed in the following. At this stage we can be just impressed by the excellent agreement displayed in Fig. 1.3. It is instructive to fully understand the meaning of this agreement. It does not imply that the *real* wave functions of ^{38}Cl or ^{40}K are those given by the simple shell model. The short range correlations introduced by the real nucleon-nucleon interaction admix into any shell model wave function many high lying states. Still, to calculate energies and perhaps other observables, it is possible to use simple shell model wave functions provided an effective interaction is adopted. The effect of the many admixed configurations seems to be just a drastic renormalization of the two body interaction between free nucleons.

On the practical level we observe that matrix elements taken from one nucleus can be used in calculations of a neighboring one. There is no need for considerable three-body terms nor for strong dependence of two-body matrix elements on the number of nucleons. This latter effect, had it existed, would have amounted to effective three-body interaction terms. We shall later see that in some cases also magnetic moments of actual nuclei can be well described in terms of effective or renormalized magnetic moments of individual nucleons. It is worth while to repeat that there is no *a priori* reason for the success of this approach. Matrix elements of the effective interaction could have changed drastically from one nucleus to the next one. The restrictions to simple configurations could have resulted in a horrendous effective

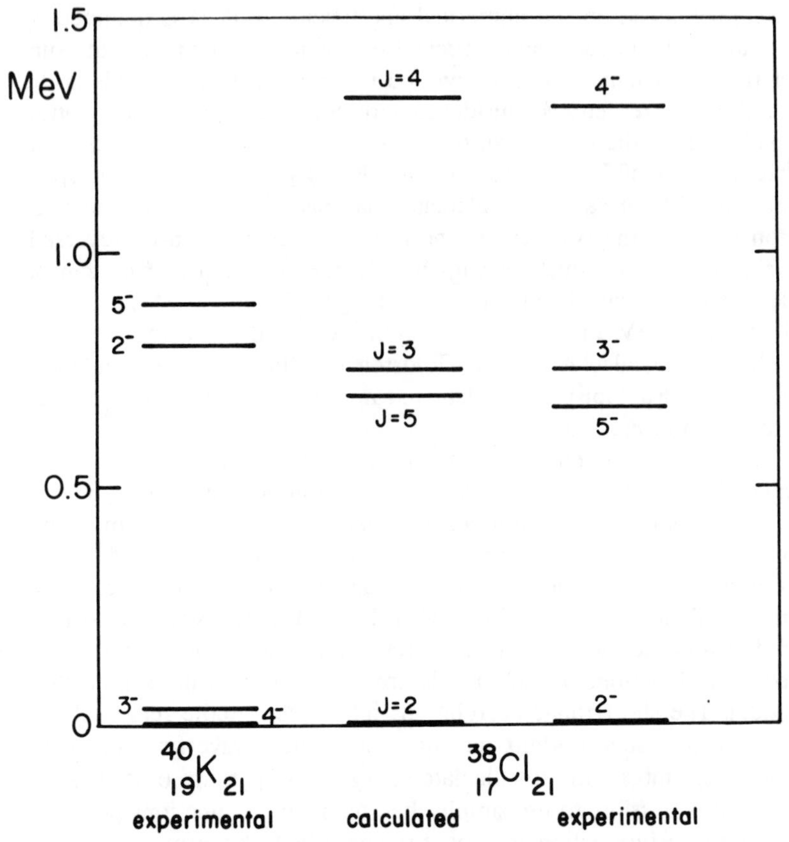

FIGURE 1.3. *Calculated and experimental levels of* ^{38}Cl.

interaction with many nucleon terms playing an important role. The approach presented here has been very encouraging in the search for a quantitative description of nuclear energies within the framework of the shell model.

In the following sections other examples are presented to demonstrate how matrix elements of the effective interactions can be determined from experimental data in a consistent way. Experimental level schemes may be found in the book by C. M. Lederer and V. S. Shirley, *Table of Isotopes*, 7th Edition, John Wiley and Sons, New York (1978) and in current issues of *Nuclear Data Sheets*, J. K. Tuli, Editor, Academic Press, New York. Ground state (binding) energies of nuclei are given by A. H. Wapstra and G. Audi in *Nuclear Physics* **A432** (1985) 1 and more recent values by A. H. Wapstra, G. Audi and R. Hoekstra in *Atomic Data and Nuclear Data Tables* **39** (1988) 281.

It is interesting to note that the need for effective interaction arises also in the atomic shell model. In atoms, the interaction between electrons is the well known Coulomb repulsion. Most of the atomic mass is carried by the nucleus which serves as a source of a strong and attractive potential well. Still, mixing of configurations leads to effective interactions between electrons which are different from the pure Coulomb repulsion. Some evidence for such renormalization of the two-electron interaction is presented in Section 16.

The renormalization leading to effective interactions need not be due only to many nucleon effects. At short distances, meson exchange corrections and quark effects may become very important. The correct description of the wave function of the two nucleons may require explicit introduction of mesons and quarks. Still, as long as such phenomena are largely independent of the presence of other nucleons, their effects may be incorporated in the matrix elements of the two-body effective interaction.

If the procedure described above yields good agreement between calculated and experimental energies, the deduced values of the two nucleon matrix elements may be trusted for other calculations. These values provide information about properties of the effective interaction. These are the values which the nuclear many body theory should eventually be able to calculate. If such shell model calculations are successful, the implication is that the renormalization of the nuclear interaction due to short range correlations affects just the two-body interaction. Three-nucleon terms seem then to be rather unimportant. As mentioned above, the modification of the two nucleon interaction may also be due to various quark effects at very short distances which are too complex to be included in the ordinary interaction between nucleons. The goals of the procedure described above are rather limited to calculation of energies of a group of nuclei with the same valence shell. The impressive feature of it is the possibility to obtain in this way *quantitative* agreement between shell model calculations and experimental energies. The relevance to nuclear many body theory is in realizing that two-nucleon effective interactions are all that is necessary to calculate nuclear energies. The two nucleon matrix elements obtained by this procedure, determine the structure of nuclei in their ground states and at low excitations. The information about a large number of two-nucleon matrix elements can point out general features of the effective interaction and the resulting structure of nuclei. These general features will be discussed in the following sections.

The use of effective interactions is based on mixing of shell model configuration by the interaction between free nucleons. The effect of that mixing, including high lying configurations, on low-lying levels, may be replaced by renormalization. The resulting effective interaction may be used with simple or pure shell model configurations to calculate energies of low-lying levels. A nice example was described above. In other cases, energies cannot be reproduced by using effective two-body interactions with pure configurations. If single nucleon orbits lie sufficiently close, it may be necessary to consider mixing of configurations in order to calculate energy levels. The effective interaction needed in this case includes *diagonal* matrix elements between states in the same configuration as well as *non-diagonal* matrix elements between states in different configurations. It should be noted that matrix elements of the effective interaction in a given nucleus depend on the choice of shell model configurations considered. If a given pure configuration is adopted, the effective interaction is due to *all* other configurations. If several configurations are considered, the effect of each of them on the others is treated explicitly. As a result, matrix elements of the effective interaction in the given configuration may well be changed if other configurations are taken into account.

Calculation of nuclear energies by the procedure described above has practical limitations. It works best if used with simple shell model configurations where nucleons occupy one or two orbits. If there are several configurations mixed, the detailed application of the method described above becomes very difficult. The number of matrix elements of the effective interaction which must be determined from experiment, increases rather rapidly with the number of valence orbits occupied. These are diagonal matrix elements between states of the same two-nucleon configuration as well as non-diagonal ones between states of different two nucleon configurations. In fact, even if all relevant matrix elements of the effective interaction were known, the problem in many cases would still have been beyond solution. The number of states which may be admixed and have to be taken into account, may become gigantic. This occurs especially in nuclei with both valence protons and neutrons outside closed shells. A concrete example is given at the end of Section 24 and the beginning of Section 38.

Use of effective interactions determined from experimental data is limited by the availability of a sufficient number of experimental energies compared to the number of matrix elements. In several cases, some of which are described in this volume, the procedure described above could be carried out even for heavier nuclei. Simple configurations could be used specially in nuclei with identical valence nucleons. In lighter nuclei, the number of valence orbits is rather small and there are relatively more experimental energies available for analysis. Systematic shell model calculations of nuclear energies and other observables in complicated cases were carried out in the $1p$ shell by Amit and Katz (1964) and in more detail by Cohen and Kurath (1965). In this shell, there are two orbits, the $1p_{3/2}$ orbit and the higher $1p_{1/2}$ orbit and configurations with nucleons in these orbits are mixed. The next higher shell, the $2s, 1d$ shell, includes the $1d_{5/2}$, $2s_{1/2}$ and $1d_{3/2}$ orbits. Extensive and successful shell model calculations, including mixing of configurations of nucleons in these three orbits, were carried out by Wildenthal et al. (1984). There are in this case 63 matrix elements of the effective interaction between two nucleon states and 3 single nucleon energies. In spite of the good agreement obtained between calculated energies (and other observables) and experimental data, not all of those matrix elements could have been accurately determined. These calculations will not be described in this volume and the interested reader is referred to the forthcoming book in this series by Wildenthal and Brown.

In spite of the complexity of the shell model approach, spectra of many nuclei with valence protons and neutrons show remarkably simple regularities. Levels may be grouped into "rotational bands" whose energies, within the band, are roughly proportional to $J(J+1)$. There are enhanced intra-band electromagnetic transitions and weaker ones between states in different bands. Such spectra are successfully described by the collective model of the nucleus. The collective model has no simple relationship to the spherical shell model and its detailed description is beyond the scope of the present volume. It can be approximately related to a model of the nucleus in which nucleons move independently in a *deformed* potential well. Eigenstates of nuclei, with well defined angular momenta must be projected from wave functions of such a system. The resulting wave functions, expressed in terms of spherical shell model states must be very very complicated.

A possible simple description of collective states of nuclei, including rotational spectra is offered by the *interacting boson model*. Building blocks of that model are bosons with $J = 0$ (s-bosons) and $J = 2$ (d-bosons) whose number N is fixed for a given nucleus. Such boson states may be related to states constructed from $J = 0$ and $J = 2$ nucleon pairs. A description of the interacting boson model and its relationship with the spherical shell model occupies a considerable part of the present volume.

In the next section a short review is given of the history of the shell model from 1932 to 1950. In Sections 3 to 10 a short description is presented of single nucleon and two nucleon wave functions, operators and tensor algebra. Results of tensor algebra are developed for actual use in the problems considered in this volume. No comprehensive treatment of this important field is attempted. Tensor algebra was developed by Racah (1942, 1943) who applied it in his work on atomic spectroscopy. It was independently developed by Wigner (1940) who did not publish his theory. Only about 10 years later a manuscript entitled "On the matrices which reduce the Kronecker product of representations of S. R. groups," was privately circulated. It was eventually published in *Quantum Theory of Angular Momentum*, L. C. Biedenharn and H. van Dam, Eds., Academic Press, New York, NY (1965). Detailed and extensive descriptions of tensor algebra may be found in several books. Two of the older ones are by A. R. Edmonds, entitled *Angular Momentum in Quantum Mechanics*, Princeton University Press, Princeton, NJ (1957) and by U. Fano and G. Racah, entitled *Irreducible Tensorial Sets*, Academic Press, New York, NY (1959). More recent books by L. C. Biedenharn and J. D. Louck were published in the series of Encyclopedia of Mathematics and Its Applications, Vol. 8, entitled *Angular Momentum in Quantum Physics, Theory and Applications*, and Vol. 9, entitled *The Racah–Wigner Algebra in Quantum Theory*, Addison-Wesley, Reading, MA (1981). Still more recent is a book by D. A. Varshalovich, A. N. Moskalev and V. K. Khersonskii, entitled *Quantum Theory of Angular Momentum*, World Scientific, Singapore (1988).

Two nucleon matrix elements are considered in Sections 11 to 13 and specific examples are given. Sections 14 to 18 deal with many nucleon wave functions and states and operators in the formalism of second quantization. Nucleon pairing, on which the concept of seniority is based, occupies a central position in this book. Seniority and its generalizations are discussed in detail in Sections 19 to 24 and also in Sections 25 to 28. In Section 31, matrix elements of two-body interactions and of single nucleon operators for nucleons in different orbits are considered. Interactions between different configurations are described in Section 32.

In Sections 25 to 30 the use of group theory in spectroscopy is described. Groups play an important role in physics and particularly in spectroscopy. If the Hamiltonian has a certain symmetry some simplification of the physical problem arises. The symmetry implies that the Hamiltonian is invariant under certain transformations. Such transformations form a group and hence group theoretical methods may be used to exploit the given symmetry and obtain powerful and elegant descriptions of the physical system. In this volume an introduction to the use of group theory in spectroscopy is given. No attempt is made to present a detailed description of group theoretical methods. Only the principles and some applications are described. As far as possible, most necessary results are derived by elementary methods.

Since nuclei are very complex systems, the nuclear Hamiltonian does not have many exact symmetries. Still, group theory can provide a simple scheme in which the Hamiltonian may be expressed in a simple way. The actual physical states may in some cases be rather close to states of good symmetry. In other cases, eigenstates may be expressed as linear combinations of states with good symmetry. Care must be taken when applying group theory to nuclei. It is necessary to find out in which cases the group theoretical description is a good approximation and where it fails. In nuclear physics, the emphasis should be on experimental data and not on the mathematical formulae, elegant and attractive as they may be. One should not be carried away by the beauty of the formalism, away from the solid ground of experiment.

Use of group theory is particularly important for the interacting boson model. Sections 33 to 36 contain a description of the interacting boson model with one kind of s and d bosons (IBA-1). The description of the interacting boson model presented in this volume does not provide many important details. The emphasis is on the physical basis of the model. Detailed and comprehensive theoretical derivations and formulae can be found in the book by F. Iachello and A. Arima, entitled *The Interacting Boson Model*, Cambridge University Press, Cambridge (1987) and in the book by D. Bonatsos, entitled *Interacting Boson Models of Nuclear Structure*, Oxford University Press, Oxford (1988). The shell model basis of this model is considered in Sections 37 and 38. The interacting boson model arising from such considerations has two kinds of s and d bosons, proton bosons and neutron bosons (IBA-2). It is described in Sections 39 and 40. In the last section there is a description of a particle (or several particles) strongly coupled to a rotating core. This restricted problem, which arises in the collective model and in interacting boson models, can be treated by methods developed in this volume.

2

Historical Survey

The shell model was applied to nuclei immediately after the realization that their building blocks are protons and neutrons. Theoretical physicists at that time had before their eyes a very successful shell model. The picture of electrons moving independently in a central potential well, due to the positively charged nucleus and obeying the Pauli principle, accounted very well for atomic structure. The filling of electron shells in atoms reproduced remarkably well the periodic table of elements. Without real knowledge of nuclear interactions, physicists tried to apply the shell model also to nuclei.

In the paper which marks the beginning of nuclear structure physics, Heisenberg (1932) presented the idea that nuclei are composed of protons and neutrons. Until then, protons and electrons, as well as α-particles, were suggested as building blocks of nuclei. In the same year, Bartlett considered the stability of various nuclei. On the basis of the rather meager experimental information he states (Bartlett 1932a) that "If an analogy with the external electronic system subsists, then the α-particle may represent a closed s-shell, with two neutrons and two protons, while O16 is obtained by adding on a closed p-shell, with six neutrons and six protons".

Bartlett goes on and considers the next d-shell and in another paper he tries to extend the scheme to heavier nuclei. There (Bartlett 1932b) he writes: "For some time there has been speculation as to whether or not the atomic nucleus can be regarded as consisting of shells of protons, just as the external structure is known to consist of shells of electrons. The writer has recently pointed out that the experimental evidence seems to demand a modification of this view, in that s, p, d (etc.) shells do exist, but that a closed shell of azimuthal quantum number l consists of $2l + 1$ protons and $2l + 1$ neutrons. It was shown that the facts are well represented for elements of mass number (M) less than 36. It is of interest to inquire if this scheme is capable of extension to elements of higher mass number". Looking at the experimental evidence he concludes that "The above evidence seems to indicate quite clearly that, for $M < 144$, the picture of closed shells of protons and neutrons is in agreement with the facts .... Furthermore, it seems that Russel-Saunders coupling holds ... ". This last conclusion remained unchallenged until 1949.

A much more serious and detailed study of the shell model for nuclei was taken up by Elsasser (1933). The title of his first paper and the two that follow is "On the Pauli principle in nuclei." This principle, together with the assumption of a central potential well, leads to shell structure in nuclei. He explained that the field due to $N - 1$ nucleons acting on the Nth nucleon has probably spherical symmetry leading to the analogy with electron shells in atoms. Since in the nucleus there is no massive center of force like in the atom, the order of single particle levels in nuclei may be different and should be determined from experiment. He concluded that the potential well ("pot") must have a rather flat bottom. Due to the analogy with electron shells in atoms, he believes that nuclei with closed shells should have "specially large binding energies". This belief, unfortunately, is still shared by some authors today. It will be shown in the following that it is based on an incorrect analogy with atomic energies.

In his next paper, Elsasser (1934a) considered heavier nuclei and refers to closed shells with proton or neutron numbers 50 and 82. He considered there the number of isotopes for given Z as well as the abundance of various isotopes. The single nucleon orbits that he adopts are shown in Table 2.1. In order to reproduce the order of

TABLE 2.1 *Order of single nucleon levels (Elsasser)*

orbit	$1s$	$1p$	$1d$	$1f$	$1g$	$2d$	$1h$
$2(2l+1)$	2	6	10	14	18	10	22
$\sum 2(2l+1)$	2	8	18	32	50	60	82

single nucleon levels he added repulsion at the center of the potential well which looks as "in a bottle".

In the third paper of the series, Elsasser (1934b) considers α-decay energies of heavier nuclei whose experimental levels became then available. His analysis clearly demonstrates the independence of proton and neutron shell closures. He finds a clear discontinuity at neutron number 126, which number is not simply reached from 82. The orbit $1i$ may contain 26 neutrons, hence, the next shell closure should be at 108 but in his analysis, 98 and 116 are candidates for shell closures. He then suggests that perhaps the $1d$ orbit becomes higher in energy and is empty at those mass numbers. Thus, $108 - 10 = 98$ and the number $116 - 98 = 18$ is the maximum number of neutrons in the $2g$ orbit. Beyond 116 the $2d$ orbit ("interior orbit") may be filled. The effect is similar to the filling of electron shells in atoms. These possible rearrangements may be also necessary for explaining the situation in Ca isotopes. If the $2s$ orbit is filled beyond proton or neutron number 18, there is shell closure at 20. For higher mass numbers Elsasser assumes that the $2s$ orbit becomes higher due to the repulsion at the center and therefore empty in accordance with the order of levels in Table 2.1.

In another paper, Elsasser (1935) considers information on binding energies obtained from radioactive nuclei. He notices the discontinuities in separation energies of proton pairs at $Z = 82$ and of neutron pairs at $N = 126$, and attributes them to shell closures. It is difficult to understand, in retrospect, how the solid evidence presented by him was ignored for about 13 years.

The situation in nuclear physics at that time was presented in a famous review article by Bethe and Bacher (1936). The shell model is described there by the following words: "The opposite extreme to the assumption of α-particles as nuclear subunits is that of independent motion of the individual protons and neutrons". In what is described as the zero-order approximation, only the single nucleon energies are considered. This leads "to the prediction of periodicities". The authors discuss the orbits in several potential wells ("holes"). These are the oscillator potential, the infinite square well and the finite well. They clearly state that "Whenever a shell is completed, we should expect a nucleus of particular stability. When a new shell is begun, the binding energy of the newly added particles should be less than that of the preceding particles which serve to complete the preceding shell. We should thus expect that the 3rd, 9th, 21st, etc. neutron or proton is less strongly bound than the 2nd, 8th, 20th".

Bethe and Bacher look at the experimental evidence for this effect. They expect shell closures at 2, 8 and 20 (where the $1d + 2s$ shell is filled). The situation for ^{40}Ca is not clear since "Unfortunately no exact data about nuclear masses are available for such high atomic weights ... ". In the case of the $1p$ shell, however, they find that the binding of a nucleon to ^{16}O is less strong than to ^{12}C. They state that this "is exactly what we must expect if ^{16}O marks the completion of a neutron and proton 'shell'". They conclude that this fact "constitutes very strong evidence for the shell structure indeed".

Before considering heavier nuclei, the authors find it "necessary to give a strong warning against taking the neutron and proton shells too literally. This has been done very frequently in the past with the effect of discrediting the whole concept of neutron and proton shells among physicists". The reason is that the zero-order approximation, of single nucleon energies must be complemented by considering the mutual interactions of the nucleons and possibly also configuration mixing. "This fact alone shows that the effects connected with the completion of a shell cannot be too well marked, and it seems reasonable to expect them to be the less well marked the greater the number of particles already at the nucleus". From this they reach a more specific warning. "Therefore, apparent *deviations from the simple shell structure* expected should of course be *attributed* to the *crude approximation* used. *Under no circumstances* do such deviations justify far reaching *ad-hoc assumptions* ... ". These words are still very appropriate today as they were then.

From the single particle levels in an infinite square well, Bethe and Bacher expect shell closures at proton or neutron number 2, 8, 20, 34, 40, 58, 92, 132. From experimental data "there is only one which can readily be explained on the grounds of these 'closed shells' viz, the case of ^{39}K. This nucleus constitutes actually quite a strong piece of evidence for the completion of a neutron shell with 20 neutrons". The fact that ^{39}K is stable rather than the expected ^{39}Ar, is attributed to the 21st neutron going into a new shell. The authors do not find evidence for the other shell closures they expect. On the other hand, they reject schemes of Bartlett, Elsasser and Guggenheimer who "have left out the shells $2s$, $3p$ and $3s$, without giving any reason for such a procedure. According to them, the $5g$ shell should be filled when there are 50 particles, the $4d, 6h$ shell with 82 particles. These numbers would agree with experiment, but they lack theoretical foundation".

Bethe and Bacher conclude "that the naive theory of neutron-proton shells fails for higher atomic numbers". They even offer a tentative explanation. The mutual interactions lower the ground state energy relative to the average energy of the configuration. The effect "being largest when the outermost shells of protons and neutrons are just half filled, because this state of affairs corresponds to the largest number of levels in the system". This effect becomes larger for heavier nuclei (more levels). "Thus it may happen that 48 rather than 58 protons and 82 rather than 92 neutrons, correspond to minimum energy".

In their review article, Bethe and Bacher describe various calculations of binding energies and nuclear energy levels. In those calculations, carried out by several authors, simple (direct and exchange) interactions have been used with single nucleon wave functions of a simple potential well. One conclusion about the coupling scheme in nuclei is interesting. "It seems reasonable to assume Russel-Saunders coupling to hold at least approximately in the nucleus ...."

The discussion of the shell model by Bethe and Bacher is rather balanced. They do not think that the numbers 50, 82 and 126 represent shell closures. Still, they reserve judgement: "In conclusion, we want to emphasize again that *reliable* conclusions about shell-structure of nuclei can only be drawn when atomic weight determinations will be available which are guaranteed to be accurate to at least three decimals". In the same year, however, a devastating criticism of the shell model was published by a most eminent physicist.

Niels Bohr (1936) in a lecture at the Royal Danish Academy, presented the compound nucleus model. Based on experimental data he explained that when a neutron hits a nucleus the "energy of the incident neutron will be rapidly divided among all nuclear particles with the result that for some time afterwards no single particle will possess sufficient kinetic energy to leave the nucleus". On the basis of this picture Bohr reaches the following conclusion.

"Quite apart from the problem of the nature of the nuclear constituents themselves, which is not of direct importance for the present discussion, it is, at any rate, clear that the nuclear models hitherto treated in detail are unsuited to account for the typical properties of nuclei for which, as we have seen, energy exchanges between the individual nuclear particles is a decisive factor. In fact, in these models it is, for the sake of simplicity, assumed that the state of motion of each particle in the nucleus can, in the first approximation, be treated as taking place in a conservative field of force, and can therefore be characterized by quantum numbers in a similar way to the motion of an electron in an ordinary atom. In the atom and in the nucleus we have indeed to do with two extreme cases of mechanical many-body problems for which a procedure of approximation resting on a combination of one-body problems, so effective in the former case, loses any validity in the latter ....”

Bohr’s criticism had a profound effect on the development of the nuclear shell model. His strong objections discouraged theoretical physicists from using it. Giulio Racah who started to work on nuclear spectroscopy was convinced that the shell model was indeed not valid for nuclei. He then applied the methods he developed to atomic spectroscopy. Calculations of nuclear energies were still carried out by Wigner and Feenberg, Hund, Jahn and some others. Most of their work, however, was concerned with light nuclei. Only some general statements were made about nuclei heavier than oxygen. The general attitude towards the shell model is clearly seen in the comprehensive book of Rosenfeld (1948). There, the “quasi-atomic model” plays a minor role, less important than the α-particle model. It is described only for nuclei up to ^{40}Ca, the magic numbers 50, 82 and 126 are not even mentioned.

The renaissance of the nuclear shell model began by a paper of Maria Goeppert-Mayer (1948). She presented in that paper strong experimental evidence for the reality of the “magic numbers” 20, 50, 82 and 126. Experimental data at that time were much more abundant than those available to Elsasser. She based her conclusions not only on binding energies (including delayed emissions of the 51st and 83rd neutron) but also on isotopic abundance, number of isotopes and isotones and also on neutron absorption cross sections. She observed that the change in nucleon separation energies beyond a magic number is “of order 30 percent” of the average binding energy but her conclusion is that “Nevertheless, the effect of closed shells in the nuclei seems very pronounced”. Her paper drew the attention of nuclear physicists to the existence of magic numbers in heavy nuclei. Some of them tried to obtain shell closures at the numbers 50, 82 and 126.

The paper of M. G. Mayer was published in August 1948. On December 27, 1948 two manuscripts were received by the *Physical Review.* Both presented single nucleon level schemes which reproduced the magic numbers as well as spins, magnetic moments, electromagnetic transitions and beta decay. Feenberg and Hammack (1949), following Feenberg (1949), adopted the level scheme of Elsasser with a minor modification. The “wine bottle potential” of Elsasser was introduced to push up the $2s$ level. Feenberg and Hammack, who attribute the repulsion at the center of the potential well to the Coulomb repulsion of the protons, suggest that it becomes effective only for heavier nuclei. The $2s$ orbit is fully occupied in the shell closure at 20 but it lies higher than the $1f$ and $1g$ orbits which are fully occupied at magic number 50 and the $1h$ and $2d$ orbits which are fully occupied at 82.

The other paper, by Nordheim (1949), presents a very different level scheme. In his model, the magic numbers 2, 8 and 20 are due to successive closure of the $1s, 2p$ and $2s, 1d$ orbits. The higher orbits which are also filled at 50 are $1f, 2p$ and $2d$ whereas 82 is obtained by closure of also the $1g$ and $2f$ orbits. Nordheim, as well as Feenberg and Hammack, do not specify exactly the orbits which close at magic number 126 since there are several possibilities and not enough experimental evidence. It is interesting to note that Nordheim's level scheme is based on the *same* experimental data used by Feenberg and Hammack. In both papers much attention is paid to magnetic moments of nuclei which provided for a long time strong evidence for shell structure of nuclei.

Lande (1933), in a short paper, and in an expanded version (Lande 1934), considered magnetic moments as due to a single nucleon. "The basic idea is this: *One* particle only, one proton or one neutron, is responsible for the total spin and the magnetic properties of the whole nucleus, the rest of it forming closed shells in general." The nature of those closed shells was not explained but the idea of single nucleon magnetic moments is very clearly stated. So sure was Lande of his picture that he tried to determine the magnetic moments of the proton and the neutron from the analysis of measured magnetic moments. As will be derived in Section 3 (he actually used the Lande formula (8.24)), the magnetic moments of a single nucleon with orbital angular momentum l are given, in nuclear magnetons, by

$$\mu = gj = g_l l + \tfrac{1}{2} g_s \qquad \text{for} \quad j = l + \tfrac{1}{2} \tag{2.1}$$

$$\mu = gj = \frac{j}{j+1}\left(g_l(l+1) - \tfrac{1}{2} g_s\right) \qquad \text{for} \quad j = l - \tfrac{1}{2} \tag{2.2}$$

In (2.1) and (2.2), g_l and g_s are the g-factors of the orbital angular momentum and intrinsic spin respectively. The best fit to the few experimental data known then, assuming $g_l = 1$ for protons and $g_l = 0$ for neutrons, was obtained by using the values $g_s = \frac{1}{2}\mu_\pi = \frac{1}{2}2$ for protons and $g_s = \frac{1}{2}\mu_\nu = \frac{1}{2}(-.6)$ for neutrons. The neutron magnetic moment has not been measured yet, and the proton moment has just been determined to be 2.5 or 3 nuclear magnetons. Lande was wondering about the discrepancy of his value of μ_π, which could be called the *effective* magnetic moment of the proton, and the one measured directly.

Schmidt (1937) plotted magnetic moments of nuclei as a function of spin. He noticed that for either odd proton nuclei or odd neutron nuclei, they fall into two groups lying roughly on two parallel lines. The slope of the lines for odd proton nuclei is ~ 1 and for odd neutron nuclei, the lines are roughly parallel to the abscissa. He noticed that these slopes agree with the model put forward by Lande (1934) and draws the lines through the values of the moments according to this picture, using the better measured values of the proton and neutron magnetic moments ("Schmidt lines"). Modern versions of those plots are presented in Fig. 2.1 and Fig. 2.2. He concludes that there must be corrections to the simple pictures which should account for the deviations of measured magnetic moments from the calculated values.

Margenau and Wigner (1940) offered another explanation of the fact that magnetic moments lie roughly on two lines. According to the super-multiplet theory of Wigner (considered in Section 29) the lowest levels of nuclei should have a total intrinsic spin $S = \frac{1}{2}$. This should be coupled to the total orbital angular momentum L to yield the total J. The magnetic moment of the nucleus turns out to be given by (2.1) or (2.2) where l is replaced by L and $\frac{1}{2}g_s$ is the magnetic moment of the proton (or neutron) for odd proton (or neutron) nuclei. The difference between the Margenau-Wigner magnetic moments and those of Lande-Schmidt lies in the value of g_l. If L is due to the orbital motion of *all* nucleons then g_l is expected to have the value Z/A. The fact that the slopes of the lines through the observed magnetic moments are close to 1 and 0 respectively, was taken as a strong indication that many nucleons form closed shells.

It is interesting to note that the behavior of magnetic moments was considered to be consistent with Russel-Saunders coupling (LS-coupling). Magnetic moments close to the Schmidt values served as evidence for *one* value of L to which $S = \frac{1}{2}$ is coupled. The admixtures of states with different L and S values have thus been shown to be small, lending support for the validity of LS-coupling.

Nordheim as well as Feenberg and Hammack quote values of magnetic moments as evidence for their level schemes. In the cases of a single nucleon outside closed shells or missing from them, the orbital angular momentum could be determined from the magnetic moment. For instance, Nordheim (1949) writes "It is remarkable, however, how near orbits appear after passing through the closed shell numbers ....

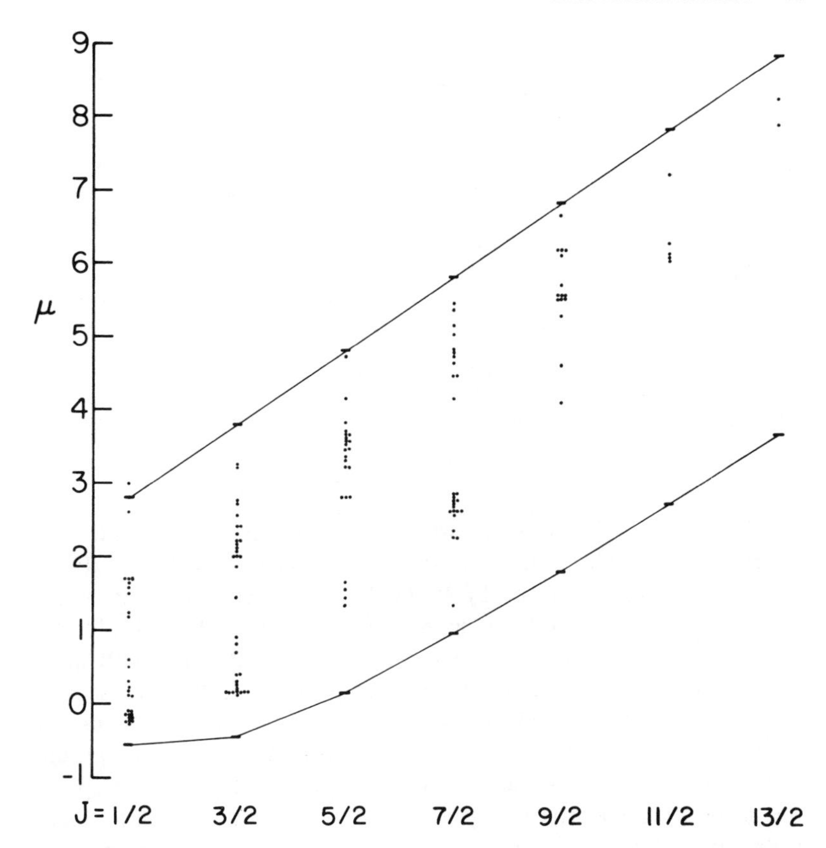

FIGURE 2.1. *Magnetic moments of odd proton nuclei (in nuclear magnetons).*

$^{19}_{9}F$ shows that here the $2s$ orbit is lower than the $1d$, $^{45}_{21}Sc$ brings the first $1f$. The $1g$'s appear fairly solidly with $^{123}_{51}Sb$, and $^{209}_{83}Bi$ gives beautiful evidence for a $1h$ orbit". The fact that these moments of a single g and h-nucleon are closer to the value for $j = l - \frac{1}{2}$ is not even mentioned. Whereas these magnetic moments agree with Nordheim's level scheme, Feenberg and Hammack have some difficulty with ^{209}Bi. According to their scheme the only $j = \frac{9}{2}$ orbit available for the 83rd proton is the $2g$ orbit which they believe should be higher. "The situation is not improved by the fact that the magnetic moment favors a predominantly $^{2}H_{9/2}$ state". They realize that the orbital angular momentum should be $L = 5$ but in their scheme the $1h$ orbit is completely filled at $Z = 82$.

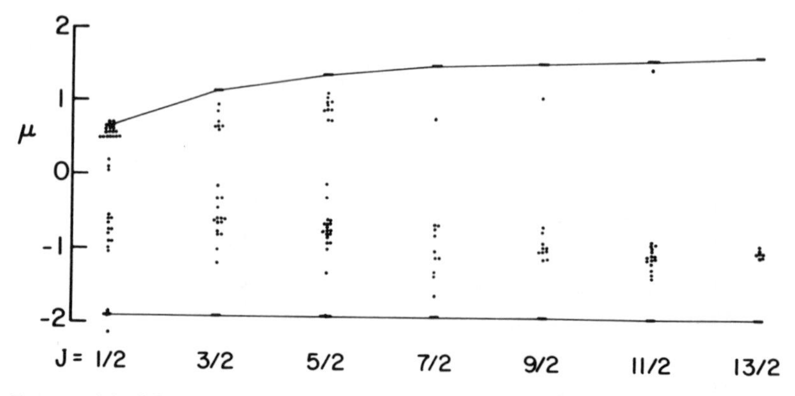

FIGURE 2.2. *Magnetic moments of odd neutron nuclei (in nuclear magnetons).*

The spell of LS-coupling which made it difficult to find level schemes with shell closures at 50, 82 and 126, was broken by Fermi who, after a seminar given by M. G. Mayer, asked her "Is there any indication of spin-orbit coupling in nuclei?" She thought and said "Yes and it explains everything". In her paper M.G. Mayer (1949) introduced strong spin-orbit interaction which gave rise to the level scheme shown in Fig. 1.1 reproducing all known magic numbers. Spins and magnetic moments followed naturally from that scheme once ground state spins of odd-even nuclei were assumed to be equal to the spin of the single nucleon orbit being filled, $J = j$. This assumption is clearly stated in a detailed paper (Mayer 1950a) where many experimental data are explained on the basis of the jj-coupling shell model. These include not only spins and magnetic moments of ground states of nuclei but also electromagnetic transitions and in particular the "islands of isomerism" near shell closures where single nucleon states may have widely different j-values. There were a few cases in which the rule $J = j$ could not be applied like the $J = \frac{5}{2}$ ground state of $^{55}_{25}$Mn which Mayer attributed to the $(f_{7/2})^5$ configuration. In a following paper (Mayer 1950b) M. G. Mayer carried out calculations with an attractive δ-potential for energy levels of some simple j^n configurations of identical nucleons for small values of l and j up to $l = 3$, $j = \frac{7}{2}$. She found that indeed the ground states of such configurations, for n even, have spin $J = 0$ and for n odd they have the spin $J = j$. She also found that the binding energy is proportional to $n/2$ for n even and to $(n-1)/2$ for odd values of n, thus reproducing the odd even variation of binding energies.

The single nucleon level scheme based on strong spin-orbit interaction was discovered independently by Haxel, Jensen and Suess. They wrote a series of short notes which were published in the journal *Die Naturwissenschaften*. The first note by Haxel, Jensen and Suess (1949a) was received by the editor on February 12, 1949. The second, by Suess, Haxel and Jensen (1949) reached the journal on March 15, 1949 and the third, by Jensen, Suess and Haxel (1949) was received by the editor on April 6, 1949. Then they sent a short paper to the *Physical Review* (Haxel, Jensen and Suess 1949b) in which they write "A simple explanation of the 'magic numbers' 14, 28, 50, 82 and 126 follows at once from the oscillator model of the nucleus, if one assumes that the spin-orbit coupling ... leads to a strong splitting of a term with angular momentum l into two distinct terms $j = l \pm \frac{1}{2}$". Their paper was published as a Letter to the Editor of the *Physical Review* in Vol. **75** (1949) 1766 even earlier than Mayer's 1949 paper in *Physical Review* **75** (1949) 1969. Her manuscript was received by the editor on February 4, 1949 whereas the Haxel, Jensen and Suess manuscript was received on April 18, 1949. Apparently, Mayer's paper was delayed since the editor had asked Feenberg, Hammack and Nordheim to write a note comparing the various models (1949). A paper following their Letter to the Editor was published by Haxel, Jensen and Suess (1950).

Since the trail blazing papers of M. G. Mayer and Jensen et al., the shell model made enormous impact on almost all aspects of nuclear physics. There is no way to present here all successes of the shell model. The shell model was accepted even though it still "lacks theoretical foundation". The main argument for the validity of the shell model is the agreement of its predictions with experiment. We should still keep in mind the warning of Bethe and Bacher (1936) that "apparent *deviations from the simple shell structure* expected should of course be *attributed* to the *crude approximation* used".

3

Single Particle in a Central Potential Well

In this section a brief review is given of single nucleon wave functions in a central (spherical) potential well. Only nuclear states in which there is one nucleon outside (or missing from) closed shells can be fully described by single nucleon wave functions. Other states are more complicated and will be described in subsequent sections. The importance of single nucleon wave functions lies in the fact that they are the building blocks of all shell model wave functions. In the following the basic concepts are presented for the sake of completeness and for establishing the notation. There will be no discussion of the various potential wells which are commonly used. An important exception is the harmonic oscillator potential. It is explicitly included (Section 4) because it appears in some subsequent discussions (Sections 13 and 30). It also offers a convenient opportunity to discuss the motion of the center of mass of all nucleons bound to a fixed point in space.

The Hamiltonian of a particle with mass m moving in a central potential is

$$H_0 = \frac{1}{2m}\mathbf{p}^2 + V(r) \qquad (3.1)$$

In the classical orbits due to such a Hamiltonian the orbital angular momentum is a constant of the motion. Its magnitude and direction in space are constant as the particle moves in its orbit. In quantum mechanics the corresponding feature holds. The Hamiltonian (3.1) is a function of the scalars $\mathbf{p}^2$ and $\mathbf{r}^2$ and hence it commutes with all components of the orbital angular momentum vector

$$\hbar \boldsymbol{l} = \mathbf{r} \times \mathbf{p} \tag{3.2}$$

The components of $\boldsymbol{l}$ satisfy commutation relations which follow directly from those of the components of $\mathbf{p}$ and $\mathbf{r}$. These are

$$\boxed{[l_x, l_y] = il_z \qquad [l_z, l_x] = il_y \qquad [l_y, l_z] = il_x} \tag{3.3}$$

Due to the fact that the Hamiltonian (3.1) commutes with all components of $\boldsymbol{l}$ we can characterize the eigenstates by the eigenvalues of $\boldsymbol{l}^2$ and l_z.

We first express the Hamiltonian (3.1) by using the operator $\boldsymbol{l}^2$. A direct calculation yields the relation

$$(\mathbf{r} \times \mathbf{p})^2 + (\mathbf{r} \cdot \mathbf{p})^2 = r^2 p^2 + i\hbar(\mathbf{r} \cdot \mathbf{p}) \tag{3.4}$$

Since the Hamiltonian (3.1) has spherical symmetry, it is convenient to use the polar coordinates r, θ, ϕ instead of x, y, z. The derivative with respect to r is

$$\frac{\partial}{\partial r} = \frac{1}{r}(\mathbf{r} \cdot \boldsymbol{\nabla}) \tag{3.5}$$

Using (3.4) and replacing $\mathbf{p}$ by $(\hbar/i)\boldsymbol{\nabla}$ we obtain for the Hamiltonian (3.1) the expression

$$\boxed{H_0 = -\frac{\hbar^2}{2m}\Delta + V(r) = -\frac{\hbar^2}{2m}\left(\frac{\partial^2}{\partial r^2} + \frac{2}{r}\frac{\partial}{\partial r} - \frac{1}{r^2}\boldsymbol{l}^2\right) + V(r)} \tag{3.6}$$

The discrete (bound) eigenstates of the corresponding Schrödinger equation

$$H_0 \phi(r, \theta, \phi) = E \phi(r, \theta, \phi)$$

can be expressed as

$$\boxed{\phi_{nlm}(r,\theta,\phi) = \frac{1}{r}R_{nl}(r)Y_{lm}(\theta,\phi)} \tag{3.7}$$

The radial function in (3.7) satisfies the differential equation

$$\boxed{\frac{d^2R_{nl}(r)}{dr^2} - \frac{l(l+1)}{r^2}R_{nl}(r) + \frac{2m}{\hbar^2}(E_{nl} - V(r))R_{nl}(r) = 0} \tag{3.8}$$

The eigenvalues E_{nl} are determined by the boundary conditions

$$R_{nl}(r) = 0 \quad \text{at} \quad r = 0 \qquad \text{and} \qquad R_{nl} \to 0 \quad \text{at} \quad r \to \infty \tag{3.9}$$

The number n is equal to the number of nodes (points for which $R_{nl}(r) = 0$) including the one at $r = 0$ but not the point at infinity. The radial functions are taken to be normalized

$$\int_0^\infty R_{nl}(r)R_{n'l}(r)dr = \delta_{nn'}$$

The angular functions $Y_{lm}(\theta,\phi)$ are the spherical harmonics of order l. They are defined in terms of $2l+1$ independent homogeneous polynominals of order l in x, y, z which written as $r^l f(\theta\phi)$, satisfy the equation

$$\Delta r^l f(\theta,\phi) = 0$$

The spherical harmonics are thus defined by $\Delta r^l Y_{lm}(\theta,\phi) = 0$ from which follows the condition

$$\boxed{l^2 Y_{lm}(\theta,\phi) = l(l+1)Y_{lm}(\theta,\phi)} \tag{3.10}$$

as well as by the more special condition

$$l_z Y_{lm}(\theta,\phi) = mY_{lm}(\theta,\phi) \tag{3.11}$$

Due to (3.10) the eigenvalues of the Hamiltonian (3.6) are independent of m. The operator $\boldsymbol{l}^2$ is a function of only θ, ϕ and $\partial/\partial\theta$, $\partial/\partial\phi$. It is equal to

$$\boxed{\boldsymbol{l}^2 = -\frac{1}{\sin\theta}\frac{\partial}{\partial\theta}\sin\theta\frac{\partial}{\partial\theta} - \frac{1}{\sin^2\theta}\frac{\partial^2}{\partial\phi^2}} \tag{3.12}$$

The l_z component is given by

$$l_z = \frac{1}{i}\left(x\frac{\partial}{\partial y} - y\frac{\partial}{\partial x}\right) = \frac{1}{i}\frac{\partial}{\partial\phi} \tag{3.13}$$

Hence, the dependence of $Y_{lm}(\theta, \phi)$ on ϕ is given by $e^{im\phi}$. The eigenfunctions (3.7) should be single valued functions of the space coordinates like the homogenous polynomials in x, y, z. This requirement implies that m must be an integer. From (3.13) follows that the maximum value of $|m|$ is l, which is the order of the homogenous polynomial. It also follows from (3.13) that the polynomials in x, y, z which are equal to $r^l Y_{lm}(\theta, \phi)$ have complex coefficients. In fact, they are polynomials in the linear combinations

$$x \pm iy = r\sin\theta e^{\pm i\phi} \qquad z = r\cos\theta \tag{3.14}$$

The behavior under complex conjugation of the spherical harmonics is given by

$$Y^*_{lm}(\theta, \phi) = (-1)^m Y_{l,-m}(\theta, \phi)$$

The spherical harmonics obey the orthogonality and normalization condition

$$\int_0^{\pi} \sin\theta d\theta \int_0^{2\pi} d\phi Y^*_{lm}(\theta, \pi) Y_{l'm'}(\theta, \phi) = \delta_{ll'}\,\delta_{mm'}$$

Some spherical harmonics are listed in the Appendix.

It is convenient to use spherical components of $\boldsymbol{l}$ in analogy with (3.14). These are defined by

$$\boxed{l^{\pm} = l_x \pm i l_y \qquad l^0 = l_z} \tag{3.15}$$

From the commutation relations (3.3) follows that l^+ and l^- are raising and lowering operators respectively. The meaning of these terms is that these operators satisfy the following relations

$$\boxed{l^{\pm}Y_{lm} = \sqrt{l(l+1) - m(m \pm 1)}Y_{lm\pm1} \qquad l^0 Y_{lm} = mY_{lm}} \tag{3.16}$$

Spherical harmonics play an important role in spectroscopy and their applications occupy many of the following pages.

The successful version of the nuclear shell model is based on a strong spin orbit interaction which will be now introduced. The intrinsic spin vector of a non-relativistic nucleon is given by

$$\begin{aligned} s_x &= \tfrac{1}{2}\sigma_x = \tfrac{1}{2}\begin{pmatrix} 0 & 1 \\ 1 & 0 \end{pmatrix} \\ s_y &= \tfrac{1}{2}\sigma_y = \tfrac{1}{2}\begin{pmatrix} 0 & -i \\ i & 0 \end{pmatrix} \\ s_z &= \tfrac{1}{2}\sigma_z = \tfrac{1}{2}\begin{pmatrix} 1 & 0 \\ 0 & 1 \end{pmatrix} \end{aligned} \tag{3.17}$$

The operator $\mathrm{s}^2 = s_x^2 + s_y^2 + s_z^2$ is proportional to the unit matrix. The proportionality factor is the eigenvalue $s(s+1)$ with $s = \frac{1}{2}$. The matrix of s_z in (3.17) is diagonal and has the eigenvalues $m_s = +\frac{1}{2}$ and $m_s = -\frac{1}{2}$. The eigenstates corresponding to these eigenvalues are the columns

$$\begin{aligned} \chi_{1/2} &= \begin{pmatrix} 1 \\ 0 \end{pmatrix} \qquad m_s = +\tfrac{1}{2} \text{ (spin up)} \qquad \text{and} \\ \chi_{-1/2} &= \begin{pmatrix} 0 \\ 1 \end{pmatrix} \qquad m_s = -\tfrac{1}{2} \text{ (spin down)} \end{aligned} \tag{3.18}$$

The eigenvalues of s_x and s_y are also $\frac{1}{2}$ and $-\frac{1}{2}$. The s_x and s_y matrices cannot be diagonal if s_z is diagonal. This follows from the commutation relations satisfied by the components (3.17), namely

$$\boxed{[s_x, s_y] = is_z \qquad [s_z, s_x] = is_y \qquad [s_y, s_z] = is_x} \tag{3.19}$$

These commutation relations are the same as those of the orbital angular momentum (3.3).

The orbital angular momentum $\boldsymbol{l}$ operates on the space coordinates of the nucleon whereas the intrinsic spin $\mathbf{s}$ operates on a two dimensional space whose basis is given by the columns (3.18). Hence, any component of $\mathbf{s}$ commutes with any component of $\boldsymbol{l}$. As a result, the components of the total angular momentum of the nucleon

$$\mathbf{j} = \mathbf{s} + \boldsymbol{l} \tag{3.20}$$

also satisfy the commutation relations (3.3). The total angular momentum vector $\mathbf{j}$ is often called the total spin of the nucleon or simply its spin.

As long as there is no term in the Hamiltonian which couples $\mathbf{s}$ and $\boldsymbol{l}$, the energy eigenvalue of a nucleon in a given n,l orbit (E_{nl}) is independent of the orientation of $\mathbf{s}$. The eigenstates of the Hamiltonian (3.1) for a nucleon with $s = \frac{1}{2}$ can be taken as the products $\phi_{nlm_l}\chi_{1/2}$ and $\phi_{nlm_l}\chi_{-1/2}$. This degeneracy of eigenvalues is removed when a spin orbit interaction is added to the Hamiltonian (3.1). Thus, we consider instead of (3.1) the Hamiltonian

$$H_o + V_{so}(r)(\mathbf{s}\cdot\boldsymbol{l}) \tag{3.21}$$

The potential $V_{so}(r)$ can be derived for electrons moving in an electrostatic potential. Its effects there are usually small. In nuclei, however, the spin orbit term is important and determines shell closures. The origin of the spin orbit potential $V_{so}(r)$ in nuclei is not well understood and when it is considered explicitly, its form and strength are determined phenomenologically.

The components of $\boldsymbol{l}$ which appear in $(\mathbf{s}\cdot\boldsymbol{l})$ in (3.21) have non-vanishing matrix elements between states with different values of m_l according to (3.16). They do not change, however, the value of l. Also $V_{so}(r)$ has non-vanishing matrix elements only between states with the same l. The spin orbit term in (3.21) is diagonal in the scheme in which $\mathbf{j}^2$ is diagonal. This follows directly from the operator identity

$$\mathbf{j}^2 = (\mathbf{s} + \boldsymbol{l})^2 = \mathbf{s}^2 + \boldsymbol{l}^2 + 2(\mathbf{s}\cdot\boldsymbol{l}) \tag{3.22}$$

Since $\boldsymbol{l}^2$ is diagonal, with eigenvalue $l(l+1)$, and $\mathbf{s}^2$ is always diagonal, the spin orbit interaction is diagonal if $\mathbf{j}^2$ is diagonal. The term $2(\mathbf{s}\cdot\boldsymbol{l})$ has then the eigenvalues

$$j(j+1) - l(l+1) - \tfrac{3}{4} \tag{3.23}$$

The eigenvalues of $\mathbf{j}^2$, denoted by $j(j+1)$ in (3.23), can be readily obtained. The maximum eigenvalue of j_z, for given l, is obtained from the definition (3.20) to be $m_{\max} = l + \frac{1}{2}$. Hence, for given l, the maximum value of j is $l + \frac{1}{2}$. For $l > 0$ there is another possible value of j. To any value of m satisfying $l - \frac{1}{2} \geq m \geq -l + \frac{1}{2}$ correspond *two* independent states, $\phi_{nl,m-1/2}\chi_{1/2}$ and $\phi_{nl,m+1/2}\chi_{-1/2}$. One linear combination of these is the state with given m and $j = l + \frac{1}{2}$ and the orthogonal combination is the state with $j = l - \frac{1}{2}$ and m.

Substituting these two possible values of j into (3.23) we obtain the eigenvalues of $2(\mathbf{s} \cdot \boldsymbol{l})$ as

$$\begin{aligned} (l+\tfrac{1}{2})(l+\tfrac{3}{2}) - l(l+1) - \tfrac{3}{4} &= l \qquad \text{for} \quad j = l + \tfrac{1}{2} \\ (l-\tfrac{1}{2})(l+\tfrac{1}{2}) - l(l+1) - \tfrac{3}{4} &= -(l+1) \qquad \text{for} \quad j = l - \tfrac{1}{2} \end{aligned} \tag{3.24}$$

The eigenvalues of the states with $j = l + \frac{1}{2}$ and $j' = l - \frac{1}{2}$ are no longer the same. For the state with $j = l + \frac{1}{2}$ the eigenvalue is given by the solution of the radial equation

$$\begin{aligned} &\frac{d^2 R_{nlj}(r)}{dr^2} - \frac{l(l+1)}{r^2} R_{nlj}(r) \\ &\quad + \frac{2m}{\hbar^2}\left(E_{nlj} - V(r) - \frac{l}{2} V_{so}(r)\right) R_{nlj}(r) = 0 \\ &\qquad\qquad j = l + \tfrac{1}{2} \end{aligned} \tag{3.25}$$

On the other hand, the eigenvalue of the state with $j = l - \frac{1}{2}$ is given by the solution of the equation

$$\begin{aligned} &\frac{d^2 R_{nlj}(r)}{dr^2} - \frac{l(l+1)}{r^2} R_{nlj}(r) \\ &\quad + \frac{2m}{\hbar^2}\left(E_{nlj} - V(r) + \frac{l+1}{2} V_{so}(r)\right) R_{nlj}(r) = 0 \\ &\qquad\qquad j = l - \tfrac{1}{2} \end{aligned} \tag{3.26}$$

In general, the radial functions of the states with $j = l + \frac{1}{2}$ and $j = l - \frac{1}{2}$ are different. An approximation which is usually made, is to assume that the spin orbit term is considerably smaller than the central potential $V(r)$. In that case, the radial functions can be approximated by those obtained from the solution of (3.8). These radial functions depend only on n and l. To first order in perturbation theory the contribution of the spin-orbit interaction is given by

$$(\mathbf{s}\cdot\boldsymbol{l})\int_0^\infty R_{nl}^2(r)V_{so}(r)dr = 2a_{nl}(\mathbf{s}\cdot\boldsymbol{l})$$

The eigenvalues which belong to the states with $j = l + \frac{1}{2}$ and $j = l - \frac{1}{2}$ are then given, in view of (3.24), by

$$\boxed{\begin{array}{ll} E_{nlj} = E_{nl} + a_{nl}l & j = l + \frac{1}{2} \\ E_{nlj} = E_{nl} - a_{nl}(l+1) & j = l - \frac{1}{2} \end{array}} \qquad (3.27)$$

In the case of electrons in atoms the potential $V_{so}(r)$ is repulsive and the single electron state with $j = l - \frac{1}{2}$ lies lower than the $j = l + \frac{1}{2}$ state. In nuclei, the spin orbit potential is attractive, hence $a_{nl} < 0$ and the state with $j = l + \frac{1}{2}$ is lower than the state with $j = l - \frac{1}{2}$. The only way that the actual values a_{nl} can be determined is from experiment. Phenomenologically, the absolute value of a_{nl} in a given shell is largest for the orbit with the largest l and $n = 1$ which leads to the observed magic numbers 28, 50, 82 and 126. Due to this feature, the only value of a_{nl} for $n = 1$ which can be obtained from experiment is for the $1p$ orbit. This value, as well as values of a_{nl} with $n > 1$ are listed in Table 3.1 for proton (π) orbits and neutron (ν) orbits. It is based on energy differences between levels with $j = l + \frac{1}{2}$ and $j = l - \frac{1}{2}$ in nuclei with one nucleon outside closed shells or one nucleon missing from closed shells (a hole state).

We can make use of the eigenvalues (3.27) to calculate the eigenstates with $j = l + \frac{1}{2}$ and $j = l - \frac{1}{2}$. To do this, we have to diagonalize the matrix of the spin orbit interaction in (3.21). We write the scalar product $2(\mathbf{s}\cdot\boldsymbol{l})$ as

$$2(\mathbf{s}\cdot\boldsymbol{l}) = s^+l^- + s^-l^+ + 2s_zl_z \qquad (3.28)$$

where s^+ and s^- are defined as l^+ and l^- are defined by (3.15). For any given eigenvalue $m < l + \frac{1}{2}$ of j_z there are two eigenstates which are linear combinations of $\phi_{nl,m-1/2}\chi_{1/2}$ and $\phi_{nl,m+1/2}\chi_{-1/2}$. We write the 2×2 matrix (3.28) in the scheme of these states. Due to the relations (3.16) which also hold for the components of $\mathbf{s}$, and the explicit form of the spin matrices (3.17) we obtain the matrix of $2(\mathbf{s}\cdot\boldsymbol{l})$

TABLE 3.1 *Spin-orbit interaction in nuclei (in MeV)*

Nucleus	Orbit	ΔE	$-a_{nl}$
$^{5}_{2}$He	$\nu\, 1p$	~ 4	~ 1.3
$^{15}_{8}$O	$\nu\, 1p$	6.18	2.06
$^{17}_{8}$O	$\nu\, 1d$	> 5	> 1
$^{39}_{19}$K	$\pi\, 1d$	> 4	$> .8$
$^{49}_{20}$Ca	$\nu\, 2p$	2.02	.67
$^{57}_{28}$Ni	$\nu\, 2p$	1.11	.37
$^{91}_{40}$Zr	$\nu\, 2d$	2.04	.41
$^{207}_{81}$Tl	$\pi\, 2d$	1.32	.26
$^{207}_{82}$Pb	$\nu\, 3p$	.90	.30
	$\nu\, 2f$	1.77	.25
$^{209}_{83}$Bi	$\pi\, 2f$	1.90	.27
$^{209}_{82}$Pb	$\nu\, 2g$	2.49	.28
	$\nu\, 3d$	1.00	.20

in the form

$$\begin{pmatrix} m - \frac{1}{2} & \sqrt{l(l+1) - (m - \frac{1}{2})(m + \frac{1}{2})} \\ \sqrt{l(l+1) - (m + \frac{1}{2})(m - \frac{1}{2})} & -m + \frac{1}{2} \end{pmatrix} \tag{3.29}$$

The normalized eigenstate corresponding to the eigenvalue l of the matrix (3.27) is found to be

$$\begin{aligned} \psi_{nl\,j=l+1/2,m} &= \begin{pmatrix} \sqrt{\dfrac{l + \frac{1}{2} + m}{2l+1}}\,\phi_{nlm-1/2} \\ \sqrt{\dfrac{l + \frac{1}{2} - m}{2l+1}}\,\phi_{nlm+1/2} \end{pmatrix} \\ &= \sqrt{\frac{l + \frac{1}{2} + m}{2l+1}}\,\phi_{nlm-1/2}\chi_{1/2} + \sqrt{\frac{l + \frac{1}{2} - m}{2l+1}}\,\phi_{nlm+1/2}\chi_{-1/2} \end{aligned} \tag{3.30}$$

For $m = l + \frac{1}{2}$ the eigenstate is just $\phi_{nlm}\chi_{1/2}$. The accepted phase convention which will be described later, is to take in (3.30) the positive value of the square root. The normalized eigenstate corresponding to the eigenvalue $-(l+1)$ of (3.29) is orthogonal to (3.30) and is given by

$$\boxed{\begin{aligned}\psi_{nl\,j=l-1/2,m} &= \begin{pmatrix} \sqrt{\dfrac{l+\frac{1}{2}-m}{2l+1}}\phi_{nlm-1/2} \\ -\sqrt{\dfrac{l+\frac{1}{2}+m}{2l+1}}\phi_{nlm+1/2} \end{pmatrix} \\ &= \sqrt{\frac{l+\frac{1}{2}-m}{2l+1}}\phi_{nlm-1/2}\chi_{1/2} \\ &\quad -\sqrt{\frac{l+\frac{1}{2}+m}{2l+1}}\phi_{nlm+1/2}\chi_{-1/2}\end{aligned}} \tag{3.31}$$

As we shall see later, the expressions (3.30) and (3.31) give correctly also the eigenstates of (3.25) and (3.26) provided we replace the radial parts of ϕ_{nlm_l} by $R_{nl,l+1/2}(r)$ in (3.30) and by $R_{nl,l-1/2}(r)$ in (3.31).

The wave function ψ_{nljm} in (3.30) or in (3.31) stands for a column with two rows. Similarly, ψ^*_{nljm} is a row obtained by transposing ψ_{nljm} and replacing the functions ψ_{nlm_l} by their complex conjugates. Operators should be considered as 2×2 matrices whose elements may well be operators acting on the space coordinates. For simplicity we will not introduce a special notation. The orthogonality and normalization of (3.30) and (3.31) will be written as

$$\int \psi^*_{nljm}\psi_{n'l'j'm'} = \delta_{nn'}\,\delta_{ll'}\,\delta_{jj'}\,\delta_{mm'}$$

We can use the states (3.30) and (3.31) to calculate magnetic moments of nuclei which are due to a single nucleon. Such is the case whenever there is one nucleon outside closed shells or in nuclei where there is one nucleon missing from closed shells. The operator from which the magnetic moment is calculated is

$$g_l \mathbf{l} + g_s \mathbf{s} \tag{3.32}$$

In (3.32) the g-factors are

$$g_l = \mu_N \quad \text{for protons} \qquad g_l = 0 \quad \text{for neutrons}$$

and

$$g_s = 2\mu_\pi = 2 \times 2.79\mu_N \quad \text{for protons,}$$

$$g_s = 2\mu_\nu = -2 \times 1.91\mu_N \quad \text{for neutrons}$$

The units of these g-factors are *nuclear magnetons* which are defined by

$$\mu_N = \frac{e\hbar}{2mc} \tag{3.33}$$

where m is the proton mass, and e its charge.

In states with definite given j, matrix elements of the operator (3.32) are equal to those of $g\mathbf{j}$ where g is a constant factor. This point will be discussed later in more detail. The *magnetic moment* is defined to be $\mu = gj$. Hence, to calculate μ we have to calculate the expectation value of the z-component of the operator (3.32) in the state with $m = j$. For $j = l + \frac{1}{2}$ the state with $m = j$ is simply $\phi_{nll}\chi_{1/2}$ and the expectation value of $g_l l_z + g_s s_z$ is

$$\boxed{\mu = g_l l + \tfrac{1}{2}g_s \qquad \text{for} \quad j = l + \tfrac{1}{2}} \tag{3.34}$$

According to (3.31) the state with $j = l - \frac{1}{2}$ and $m = l - \frac{1}{2}$ is

$$\frac{1}{\sqrt{2l+1}}\phi_{nl,l-1}\chi_{1/2} - \sqrt{\frac{2l}{2l+1}}\phi_{nll}\chi_{-1/2} \tag{3.35}$$

To calculate μ we apply the operator $g_l l_z + g_s s_z$ to the state (3.35) obtaining

$$\frac{1}{\sqrt{2l+1}}(g_l(l-1) + \tfrac{1}{2}g_s)\phi_{nl,l-1}\chi_{1/2}$$

$$- \sqrt{\frac{2l}{2l+1}}\left(g_l l - \tfrac{1}{2}g_s\right)\phi_{nll}\chi_{-1/2} \tag{3.36}$$

The expectation value of $g_l l_z + g_s s_z$ is obtained by taking the overlap of (3.36) with (3.35) yielding

$$\frac{1}{2l+1}\left(g_l(l-1)+\tfrac{1}{2}g_s\right)+\frac{2l}{2l+1}\left(g_l l-\tfrac{1}{2}g_s\right)$$
$$=\frac{2l-1}{2l+1}\left(g_l(l+1)-\tfrac{1}{2}g_s\right)$$

Thus, the magnetic moment of the $j = l - \frac{1}{2}$ state is given by

$$\boxed{\mu = \frac{j}{j+1}\left(g_l(l+1)-\tfrac{1}{2}g_s\right) \qquad \text{for} \quad j = l - \tfrac{1}{2}} \tag{3.37}$$

The expressions (3.34) and (3.37) are called the Schmidt values of magnetic moments. In the paper of Schmidt (1937), figures like Fig. 2.1 appear following the model and formulae suggested by Lande (1933, 1934). Looking at the available measured magnetic moments of nuclei, Lande concluded that they seem to be due to "one particle only, one proton or one neutron, [which] is responsible for the total spin and the magnetic properties of the whole nucleus, the rest of it forming closed shells in general."

4

Harmonic Oscillator Potential

Among all central potential wells the harmonic oscillator potential occupies a special position. As mentioned in the Introduction, it yields the order of single nucleon orbits which, together with a strong spin orbit interaction, give rise to the observed shells in nuclei. The harmonic oscillator potential, being proportional to r^2 does not go to zero for $r \to \infty$. The asymptotic form of the eigenfunctions is not given by

$$\exp(-\sqrt{2m|E_{nl}|}r/\hbar)$$

Still, harmonic oscillator wave functions may be reasonable approximations for low lying (bound) orbits. If beyond a certain value of r the potential no longer rises but levels off smoothly, wave functions of low energy orbits may not be affected too much. There are attractive and unique features of harmonic oscillator wave functions. Some of these are considered in the present section. Others will be taken up in subsequent sections.

The harmonic oscillator Hamiltonian for a single nucleon is

$$\boxed{H_0 = \frac{1}{2m}\mathbf{p}^2 + \frac{1}{2}m\omega^2\mathbf{r}^2} \tag{4.1}$$

The Schrödinger equation is accordingly

$$\left(-\frac{\hbar^2}{2m}\Delta + \frac{1}{2}m\omega^2 r^2\right)\phi(r,\theta,\phi) = E\phi(r,\theta,\phi) \tag{4.2}$$

The radial functions of the eigenstates (3.7) of equation (4.2) are given by

$$\boxed{R_{nl}(r) = \sqrt{\frac{(2\nu)^{l+3/2}2^{n+l+1}(n-1)!}{\sqrt{\pi}(2n+2l-1)!!}}\, r^{l+1}e^{-\nu r^2}L_{n-1}^{l+1/2}(2\nu r^2)} \tag{4.3}$$

The $L_{n-1}^{l+1/2}$ are associated Laguerre polynomials (see e.g. Magnus and Oberhettinger (1949)) and ν is defined by

$$\boxed{\nu = \frac{m\omega}{2\hbar}} \tag{4.4}$$

The energy eigenvalues of (4.2) which belong to the wave functions (4.3) are

$$E = \hbar\omega(2(n-1) + l + \tfrac{3}{2}) = \hbar\omega(N + \tfrac{3}{2}) \tag{4.5}$$

As explained in Section 3, n is the number of nodes of the radial function (including the point $r = 0$ and excluding $r = \infty$). It is the number indicated for the various orbits in Fig. 1.1. The number N can be called the principal quantum number. It characterizes the major shells in the oscillator potential well ($N = 0, 1, 2, \ldots$). The explicit expression of the radial function is given by

$$\boxed{\begin{aligned} R_{nl}(r) &= \sqrt{\frac{2^{l-n+3}(2\nu)^{l+3/2}(2l+2n-1)!!}{\sqrt{\pi}(n-1)!((2l+1)!!)^2}}\, r^{l+1}e^{-\nu r^2} \times \\ &\times \sum_{k=0}^{n-1}(-1)^k 2^k \binom{n-1}{k}\frac{(2l+1)!!}{(2l+1+2k)!!}(2\nu r^2)^k \end{aligned}} \tag{4.6}$$

For states with $n = 1$, the expression (4.6) simplifies into

$$\boxed{R_{n=1,l}(r) = \sqrt{\frac{2^{l+2}(2\nu)^{l+3/2}}{\sqrt{\pi}(2l+1)!!}}\, r^{l+1} e^{-\nu r^2}} \tag{4.7}$$

The distribution of nucleons in space is usually measured by the root mean square radius. The r.m.s. radius is defined as the square root of the expectation value of r^2 in the orbit considered. For harmonic oscillator wave functions this expectation value can be directly found without explicit calculations. The virial theorem for the harmonic oscillator Hamiltonian is obtained from the relation

$$[H_0, (\mathbf{r}\cdot\mathbf{p})] = 2\frac{\hbar}{i}\left(\frac{1}{2m}\mathbf{p}^2 - \frac{1}{2}m\omega^2\mathbf{r}^2\right) \tag{4.8}$$

by taking expectation values in eigenstates of the Hamiltonian. The expectation values of the commutator in (4.8) in such states vanish. Hence, the expectation value of the kinetic energy is equal to that of the potential energy of the oscillator Hamiltonian. Thus, the expectation value of $\frac{1}{2}m\omega^2 r^2$ is equal to one half the energy eigenvalue (4.5) in the state considered. As a result we obtain

$$\tfrac{1}{2}m\omega^2\langle Nl|r^2|Nl\rangle = \tfrac{1}{2}\hbar\omega(N + \tfrac{3}{2}) \tag{4.9}$$

From (4.9) follows that $\langle r^2\rangle$ depends only on N and hence it has the same expectation value in all orbits in an oscillator major shell. Its expectation value is given according to (4.9) by

$$\boxed{\langle Nl|r^2|Nl\rangle = \frac{\hbar}{m\omega}(N + \tfrac{3}{2}) = \frac{1}{2\nu}(N + \tfrac{3}{2})} \tag{4.10}$$

The r.m.s. of the charge distribution of a nucleus is given by the square root of the expectation value of $(1/Z)\sum_{i=1}^{Z} r_i^2$ where the summation is carried over all protons. The charge distribution can be measured experimentally and its r.m.s. radius may determine the oscillator constant ν. If all proton orbits up to the N_0-th oscillator shell are completely filled, the calculation of the r.m.s. radius is given by

$$\left\langle \frac{\sum r_i^2}{Z} \right\rangle = \langle r^2 \rangle^{1/2} = \frac{\left[\frac{1}{2\nu}\sum_{N=0}^{N_0}(N+\frac{3}{2})2\sum_{l}^{N}(2l+1)\right]^{1/2}}{\left[\sum_{N=0}^{N_0}2\sum_{l}^{N}(2l+1)\right]^{1/2}}$$

$$= \frac{1}{\sqrt{\nu}} \frac{\left[\sum_{N=0}^{N_0}(N+\frac{3}{2})\frac{1}{2}(N+1)(N+2)\right]^{1/2}}{\left[\sum_{N=0}^{N_0}(N+1)(N+2)\right]^{1/2}}$$

$$= \frac{1}{\sqrt{\nu}} \frac{\left[\frac{1}{8}(12+28N_0+23N_0^2+8N_0^3+N_0^4)\right]^{1/2}}{\left[\frac{1}{3}(6+11N_0+6N_0^2+N_0^3)\right]^{1/2}} \tag{4.11}$$

The summation over l in (4.11) is over orbits in the N-th shell, namely over values of $l = N, N-2, \ldots, 1$ or 0.

States of several nucleons will be considered in detail in subsequent sections. Nevertheless, we shall consider here an important aspect of the shell model. This is the lack of translational invariance of the shell model Hamiltonian. The system of A nucleons interacting among themselves must have eigenstates which are translationally invariant. These eigenstates are due to a Hamiltonian, containing kinetic energies of the nucleons and translationally invariant interactions between nucleons. By adopting the shell model we seem to be giving up translational invariance, since the nucleons move in a potential well which is fixed in space. This is a very minor problem in the atomic shell model since the atom has the nucleus at its center. Since the nucleus is very much heavier than the electrons, it is a good approximation to put it at a fixed point in space. Translational invariance of the atom can be restored by letting the wave function of the coordinates of the nucleus be a plane wave, with definite momentum. There is no such central mass (with independent coordinates) in the nucleus. If nucleons move independently in a potential well fixed in space, the center of mass of the nucleus is bound in some way to the potential well. This is in contrast with translational invariance for which states of the center of mass coordinate should be those of a free particle. We shall presently see that in the case of harmonic oscillator wave functions this difficulty can be overcome.

For A nucleons in the oscillator potential well the Hamiltonian is

$$H_0 = \frac{1}{2m}\sum_{i=1}^{A}\mathbf{p}_i^2 + \frac{1}{2}m\omega^2\sum_{i=1}^{A}\mathbf{r}_i^2 \tag{4.12}$$

The kinetic energy can be transformed by separating the total momentum of the system

$$\mathbf{P} = \sum_{i=1}^{A}\mathbf{p}_i \tag{4.13}$$

and defining nucleon momenta relative to the average momentum

$$\mathbf{p}_i' = \mathbf{p}_i - \frac{1}{A}\mathbf{P} \tag{4.14}$$

Thus, we obtain

$$\begin{aligned}\frac{1}{2m}\sum\mathbf{p}_i'^2 = \frac{1}{2m}\sum\left(\mathbf{p}_i - \frac{1}{A}\mathbf{P}\right)^2 &= \frac{1}{2m}\sum\mathbf{p}_i^2 \\ &\quad + \frac{1}{2mA}\mathbf{P}^2 - \frac{2}{2mA}\sum\mathbf{p}_i\cdot\mathbf{P} \\ &= \frac{1}{2m}\sum\mathbf{p}_i^2 - \frac{1}{2mA}\mathbf{P}^2\end{aligned} \tag{4.15}$$

Hence the kinetic energy in (4.12) becomes equal to

$$\frac{1}{2m}\sum_{i=1}^{A}\mathbf{p}_i'^2 + \frac{1}{2mA}\mathbf{P}^2 \tag{4.16}$$

The dependence on $\mathbf{r}_i^2$ in (4.12) is the same as the dependence on $\mathbf{p}_i^2$. Hence, we can obtain a similar separation of the harmonic oscillator potential energy. We define the center of mass coordinate by

$$\mathbf{R} = \frac{1}{A}\sum_{i=1}^{A}\mathbf{r}_i \tag{4.17}$$

The coordinate **R** is canonically conjugate to **P** as can be verified from the commutation relation. We now define relative coordinates by

$$\mathbf{r}'_i = \mathbf{r}_i - \mathbf{R} \tag{4.18}$$

The A vectors $\mathbf{r}'_i$ are not independent, their sum being equal to zero. We can define A-1 independent linear combinations of the vectors $\mathbf{r}'_i$ but we need not do it explicitly. In analogy with (4.15) we obtain the relation

$$\sum \mathbf{r}'^2_i = \sum (\mathbf{r}_i - \mathbf{R})^2 = \sum \mathbf{r}^2_i + A\mathbf{R}^2 - 2\sum \mathbf{r}_i \cdot \mathbf{R} = \sum \mathbf{r}^2_i - A\mathbf{R}^2 \tag{4.19}$$

Thus, the potential energy in (4.12) becomes equal to

$$\frac{1}{2}m\omega^2 \sum_{i=1}^{A} \mathbf{r}'^2_i + \frac{1}{2}mA\omega^2\mathbf{R}^2 \tag{4.20}$$

Combining the kinetic energy (4.16) and the potential energy (4.20) we obtain the oscillator Hamiltonian (4.12) in the form

$$\boxed{H_0 = \frac{1}{2m}\sum_{i=1}^{A} \mathbf{p}'^2_i + \frac{1}{2}m\omega^2 \sum_{i=1}^{A} \mathbf{r}'^2_i + \frac{1}{2mA}\mathbf{P}^2 + \frac{1}{2}mA\omega^2\mathbf{R}^2} \tag{4.21}$$

The Hamiltonian H_0 is thus a sum of an *intrinsic* part and an oscillator Hamiltonian for the center of mass. These two Hamiltonians are independent of each other. Their commutator vanishes due to the following commutation relations

$$[\mathbf{P}, \mathbf{r}'_i] = 0 \qquad [\mathbf{p}'_i, \mathbf{R}] = 0 \tag{4.22}$$

The mass which appears in the oscillator Hamiltonian of the center of mass is the total mass of the nucleus mA. The eigenvalues of that Hamiltonian are given as in (4.5) by

$$\hbar\omega(\mathcal{N} + \tfrac{3}{2}) = \hbar\omega(2(n_A - 1) + \mathcal{L} + \tfrac{3}{2}) \tag{4.23}$$

where $\mathcal{N}$ is the principal quantum number, $\mathcal{L}$ is the orbital angular momentum and n_A is the number of nodes in the wave function of **R**.

We need not discuss the intrinsic Hamiltonian in detail. As mentioned above, the A vectors $\mathbf{r}'_i$ are not independent. The momenta $\mathbf{p}'_i$ are not canonically conjugate to the coordinate $\mathbf{r}'_i$. An important feature is that relative momenta and coordinates of two nucleons are given by

$$\mathbf{p}_i - \mathbf{p}_j = \mathbf{p}'_i - \mathbf{p}'_j \qquad \mathbf{r}_i - \mathbf{r}_j = \mathbf{r}'_i - \mathbf{r}'_j \tag{4.24}$$

Let us now consider the wave functions of n nucleons in the harmonic oscillator potential well. A better approximation to actual nuclei is obtained by including in the Hamiltonian some residual interactions between nucleons. These will be discussed later in great detail. Some of them may be very complicated but here we consider only interactions which depend on the relative coordinates and momenta of the nucleons. Due to (4.24) such interactions when added to H_0 can be combined with the intrinsic part of (4.21) to form an intrinsic Hamiltonian. This more complicated Hamiltonian still commutes with $\mathbf{P}$ and $\mathbf{R}$. Any eigenstate of the system can then be taken as a product of an intrinsic state and a wave function $\Phi_{\mathcal{N}\mathcal{L}\mathcal{M}}(\mathbf{R})$ where $\mathcal{M}$ is the z-projection of $\mathcal{L}$. In particular, the lowest eigenstate of the Hamiltonian (4.21), even if interactions involving relative coordinates and momenta are added to it, is a product of an intrinsic state and the function $\Phi_{\mathcal{N}=0,\mathcal{L}=0}(\mathbf{R})$.

Given any such eigenstate, it is possible to restore the translational invariance of the true Hamiltonian of the nucleus containing kinetic energies and translationally invariant residual interactions. We may simply remove $\Phi_{\mathcal{N}\mathcal{L}\mathcal{M}}(\mathbf{R})$ and replace it by $\exp(i\mathbf{K}\cdot\mathbf{R})$. Thereby we obtain a state of the nucleus in which the intrinsic state is the same and the total momentum $\mathbf{P}$ has a definite eigenvalue $\hbar\mathbf{K}$.

In actual applications of the shell model it is impossible to obtain eigenstates of the many-body problem. Therefore, the residual interactions between nucleons are usually considered as first order perturbations. As a result, it is not always possible to deal with the center of mass motion in the simple manner described above. In many interesting cases, however, this is still possible. To approach the problem we should consider eigenstates of H_0.

The wave functions of the ground state configuration are of particular interest in this connection. The states of this configuration correspond to the lowest states of H_0 in which all nucleons occupy the lowest orbits according to the Pauli principle. The wave functions are antisymmetric in the variables of the nucleons. The center of mass coordinate is *symmetric* in all nucleon coordinates and so is any function of **R**. Exchange of nucleon coordinates $\mathbf{r}_i$ and $\mathbf{r}_j$ is, according to (4.18) equivalent to exchanging $\mathbf{r}'_i$ and $\mathbf{r}'_j$. Hence, the symmetry properties of shell model states are the symmetry properties of the intrinsic states. This is certainly true for the spin variables of the nucleons which are unaffected by the transformations (4.14) and (4.18)

There are no restrictions on the center of mass states due to antisymmetry. Hence, if all nucleons are in their lowest orbit, i.e. in the lowest state of H_0, the center of mass *must* be in *its* lowest state in which $\mathcal{N} = 0$, $\mathcal{L} = 0$. The nucleons may occupy any of the degenerate orbits with the same value of N. This observation was made by Bethe and Rose (1937). Translational invariance may be restored for all such eigenstates of H_0 as described above.

Thus, in the ground state configuration, the wave functions are products of the various intrinsic states with $\Phi_{\mathcal{N}=0,\mathcal{L}=0}(\mathbf{R})$. In the case of closed shells there is only one intrinsic state. In other cases, there are several shell model states due to different occupation of orbits in the oscillator major shell as well as to different angular momentum couplings. The main effect of the residual interactions in first order of perturbation theory is to yield eigenstates which are various admixtures (linear combinations) of such states. Still, as long as the nucleons occupy the same oscillator major shell, the center of mass is in the same $\mathcal{N} = 0$, $\mathcal{L} = 0$ state. The quantum numbers that characterize the various shell model states characterize also the corresponding intrinsic states. Moreover, energy differences between intrinsic states of the ground state configuration are determined by energy differences of the corresponding shell model states. The center of mass Hamiltonian contributes to all these shell model states the same energy $\hbar\omega\frac{3}{2}$. Magnetic moments and many other observables are also determined by the shell model wave functions due to the spherical symmetry of the $\Phi_{\mathcal{N}=0,\mathcal{L}=0}$ wave function. It should be kept in mind, however, that in calculating charge or mass distributions and any observable which depends on them, the center of mass motion should be taken into account. This can be conveniently done since the center of mass is in the well defined state $\Phi_{\mathcal{N}=0,\mathcal{L}=0}$.

The conclusion of this discussion is that for most applications which involve low lying states, the center of mass motion can be ignored. This statement is exact for states in the ground configuration and harmonic oscillator wave functions. When excited configurations are considered, the situation is no longer simple. Consider, for instance, states obtained by raising one nucleon to the next (higher) oscillator shell.

There is a linear combination of all these states which can be obtained from a wave function of the ground configuration by multiplying it by a component of **R**. Such a shell model state is the product of the *same* intrinsic state of the ground state configuration and a state of the center of mass with $\mathcal{N} = 1$, $\mathcal{L} = 1$. It has the same energy due to the intrinsic part of H_0, including interactions to first order in perturbation theory, as in the ground state configuration and its total energy is higher by $\hbar\omega$ due to center of mass motion. Certainly such a state should not be counted as an independent state of the nucleus. It is a *spurious state* due to the fact that the nucleus is bound to a fixed point in space (Elliott and Skyrme 1955). Unfortunately, it is difficult to remove spurious states even if harmonic oscillator wave functions are used. Due to the various residual interactions, spurious states have non-vanishing amplitudes in many shell model states of excited configurations. Care must be taken to remove such components from shell model states. In the following pages, however, we shall not encounter such complications.

The discussion above should not create the impression that there is a simple cure to the problem of center of mass motion even for the ground state configuration and harmonic oscillator wave functions. The large spin orbit splitting observed in nuclei leads to major shells which, beyond the $N = 2$ shell, are not identical with the oscillator shells. These major shells include one orbit, with the largest j, from the next oscillator shell. Hence states of nucleons in a major shell may have some spurious components in them. Moreover, it is only in the case of the harmonic oscillator potential that eigenstates can be written as products of an intrinsic function and a function of center of mass coordinates. In the case of all other potential wells, shell model wave functions of even the ground state configuration are complicated combinations of such products. Removing the contribution of the center of mass motion can be done only approximately. In states of the ground state configuration, as well as in some excited configurations, the corrections contributed by the center of mass motion turn out to be of order $1/A$. We will not discuss such corrections in detail. It is still important to realize that such corrections are the price we pay for the great convenience of using shell model wave functions of nucleons moving independently in a potential well.

Before leaving the subject of center of mass let us return to (4.21). From it follows that if we subtract $\frac{1}{2}m\omega^2 A\mathbf{R}^2$ from H_0 we obtain a translationary invarian Hamiltonian. The potential energy in this case becomes equal to

$$\frac{1}{2}m\omega^2\sum_i \mathbf{r}_i^2 - \frac{1}{2}mA\omega^2\mathbf{R}^2 = \frac{1}{2}m\omega^2\frac{1}{A}\sum_{i<j}(\mathbf{r}_i - \mathbf{r}_j)^2 \tag{4.25}$$

Hence, the potential energy of particles interacting by a two body harmonic potential is equal to the sum of potential energies of independent particles in an oscillator potential well and the potential energy of their center of mass (Post 1953). Adding to the Hamiltonian of such a system the term $\frac{1}{2}mA\omega^2\mathbf{R}^2$, the eigenstates are products of single particle wave functions. To obtain the correct eigenstates of the system, the center of mass motion should be treated as explained above.

5

Transformation Under Rotations

The shell model Hamiltonian is invariant under spatial rotations. This fact has profound implications for its eigenstates. Other physical operators which are not invariant may have well defined transformation properties under rotations. Much of the following is related to the behavior of various physical entities, operators as well as wave functions, under rotations in three-dimensional space. In the following pages a simple introduction to this subject is presented. First, single particle states and operators will be discussed. Subsequent sections will deal with transformation properties of more general systems of several nucleons.

The solutions of the non-relativistic Schrödinger equation, like the one obtained from (3.6), are wave functions of spinless particles. These wave functions are examples of a *physical scalar field*. To clarify this concept let us look at the time dependent Schrödinger equation

$$H(\mathbf{p},\mathbf{x})\phi(\mathbf{x},t) = i\hbar\frac{\partial}{\partial t}\phi(\mathbf{x},t) \tag{5.1}$$

The Hamiltonian, like the one in (3.6), is constructed from the coordinates at a point **P** in space and their conjugate momenta. The coordinates and momenta **x** and **p** at the point **P** are defined by a certain cartesian frame of reference $\sum$. If we rotate the axes of $\sum$ around the origin we obtain a new frame of reference $\sum'$. The coordinates and momenta in the rotated frame $\sum'$ of the *same* point in space **P** are $\mathbf{x}'$ and $\mathbf{p}'$. The new coordinates are obtained from the old ones by the transformation

$$\boxed{x_i' = \sum_k a_{ik} x_k} \tag{5.2}$$

where x_i, $i = 1,2,3$ stand for x, y, z. The matrix of the coefficients a_{ik} is a 3×3 real orthogonal matrix whose elements satisfy the relations

$$\boxed{\sum a_{ik} a_{jk} = \delta_{ij} \qquad \sum a_{ki} a_{kj} = \delta_{ij}} \tag{5.3}$$

The determinant of the matrix whose elements are a_{ik} is equal to $+1$. The components p_i' are given in terms of the p_i by the same transformation (5.2). From (5.3) follows that $x'^2 + y'^2 + z'^2 = x^2 + y^2 + z^2$ and $p_x'^2 + p_y'^2 + p_z'^2 = p_x^2 + p_y^2 + p_z^2$.

The Hamiltonian (3.1) which is constructed from $\mathbf{p}^2$ and $\mathbf{r}^2$ has the same form in the new coordinates and momenta

$$\frac{1}{2m}\mathbf{p}'^2 + V(|\mathbf{r}'|) = \frac{1}{2m}\mathbf{p}^2 + V(|\mathbf{r}|) \tag{5.4}$$

The relation (5.4) is the mathematical expression of the statement that the Hamiltonian (3.1) is invariant under rotations. The inverse transformation to (5.2) is given, due to (5.3) by the expression

$$\boxed{x_i = \sum x_k' a_{ki}} \tag{5.5}$$

Using this transformation we can express in the Schrödinger equation (5.1), **p** and **x** in terms of $\mathbf{x}'$ and $\mathbf{p}'$. We thus obtain

$$H(\mathbf{p}(\mathbf{p}'), \mathbf{x}(\mathbf{x}'))\phi(\mathbf{x}(\mathbf{x}'), t) = H(\mathbf{p}', \mathbf{x}')\phi(\mathbf{x}(\mathbf{x}'), t) = i\hbar\frac{\partial}{\partial t}\phi(\mathbf{x}(\mathbf{x}'), t) \tag{5.6}$$

The physical law expressed by (5.1) must be the *same law* in every frame of reference. Hence, the solution at the point $\mathbf{P}, \phi(\mathbf{P}, t)$ of (5.1) is equal to the solution of (5.6) in the rotated frame. In the frame $\sum$, the coordinates of $\mathbf{P}$ are $\mathbf{x}$ and the solution $\phi(\mathbf{P}, t)$ is given by $\phi(\mathbf{x}, t)$. In the rotated frame $\sum'$, where $\mathbf{P}$ has the coordinates $\mathbf{x}'$, the *same* solution $\phi(\mathbf{P}, t)$ is expressed by $\phi(\mathbf{x}(\mathbf{x}'), t) = \phi'(\mathbf{x}', t)$.

In general, the dependence of the function $\phi(\mathbf{x}(\mathbf{x}'), t)$ on $\mathbf{x}'$ is very different from the dependence of $\phi(\mathbf{x}, t)$ on $\mathbf{x}$. For example, the wave function of a particle in a d-orbit ($l = 2$) with $m = 0$ is proportional, according to (3.7), to

$$r^2 Y_{20}(\theta, \phi) = \sqrt{\frac{5}{16\pi}} r^2 (3\cos^2\theta - 1) = \sqrt{\frac{5}{16\pi}} (3z^2 - r^2) \qquad (5.7)$$

Any rotation around the z-axis leaves the z-coordinate of any point P unchanged $z' = z$. Since $r'^2 = r^2$ for any rotation, the functional form of (5.7) does not change under rotations around the z-axis. If, however, we consider, for instance, a clockwise rotation of the frame of reference by $\pi/2$ around the x-axis of $\sum$, the new coordinates of the point $\mathbf{P}$ are given by

$$x' = x \qquad y' = -z \qquad z' = y$$

Hence, the same physical state described by (5.7) in the frame $\sum$ is described in the new coordinates in the frame $\sum'$ by

$$\sqrt{\frac{5}{16\pi}} (3y'^2 - r^2).$$

Looking at the transformation properties of the *same* physical entity, when the frame of reference is rotated, is referred to as the *passive interpretation* of the rotation. It yields information about the physical nature of the field considered. We saw above an example of a scalar field and now we consider a different kind of field. Consider one of Maxwell's equations

$$\nabla \cdot \mathbf{E}(\mathbf{x}) = \sum_i \frac{\partial}{\partial x_i} E_i(\mathbf{x}) = 4\pi\rho(\mathbf{x}) \qquad (5.8)$$

This equation relates the divergence of the electric field to the charge density. As above, we consider the equation (5.8) at a point $\mathbf{P}$ in the frame of reference $\sum$ and express it in terms of the coordinated $\mathbf{x}'$

of $\mathbf{P}$ in a rotated frame $\sum'$. Using the transformation (5.5) we obtain from (5.8) the equation

$$\sum_i \frac{\partial}{\partial x_i} E_i(\mathbf{x}) = \sum_{i,k} a_{ki} \frac{\partial}{\partial x'_k} E_i(\mathbf{x}(\mathbf{x}')) = 4\pi\rho(\mathbf{x}(\mathbf{x}')) = 4\pi\rho'(\mathbf{x}') \tag{5.9}$$

The physical law expressed by Maxwell's equations should be the same in all frames of reference. The charge density is a physical scalar with respect to rotations in three dimensional space which means that the charge density $\rho'(\mathbf{x}')$ at the point $\mathbf{P}$ is equal to $\rho(\mathbf{x})$ at that point. Hence, the equation (5.9) will have the same form as the original equation (5.8) if the components of the electric field at the point $\mathbf{P}$ in the rotated frame $\sum'$ will be related to those in the frame $\sum$ by the relation

$$\sum_i a_{ki} E_i(\mathbf{x}(\mathbf{x}')) = E'_k(\mathbf{x}') \tag{5.10}$$

The relation (5.10) shows that the transformation law of the electric field under rotations is the same as that of the vector $\mathbf{x}$ in (5.2). The fact that the electric field behaves like a vector under rotations guarantees the validity of (5.8) in any (rotated) frame of reference.

It is possible to construct more general sets of components which transform linearly among themselves under rotations. We shall refer to the values of the components at a given point $\mathbf{P}$ in space and hence not indicate their dependence on the coordinates of $\mathbf{P}$. We start with two vectors $\mathbf{A}$ and $\mathbf{B}$ which transform under rotations like the electric field according to (5.10). Thus

$$A'_i = \sum a_{ik} A_k \qquad B'_i = \sum a_{ik} B_k \tag{5.11}$$

Consider now the nine products $A_i B_j$. When rotating the frame $\sum$ into the frame $\sum'$ these products transform linearly

$$A'_i B'_j = \sum a_{ik} a_{jl} A_k B_l \tag{5.12}$$

The products $A_i B_j$ are thus components of a *cartesian tensor* with two indices. The set of nine products is, however, not irreducible under rotations. This means that there are smaller sets of linear combinations of these products so that members of each set transform among themselves under rotations. The combination

$$\mathbf{A} \cdot \mathbf{B} = \sum_i A_i B_i = \sum_i A'_i B'_i \tag{5.13}$$

is the scalar product of the two vectors which is invariant under rotations. The three antisymmetric products

$$A_i B_j - A_j B_i = \epsilon_{ijk} C_k \tag{5.14}$$

are components of the vector product of the two vectors. Under rotations, the three components C_k transform irreducibly according to (5.2). If we take $\mathbf{B}$ to be the momentum $\mathbf{p}$ of a particle and $\mathbf{A}$ its coordinate vector $\mathbf{r}$, the vector $\mathbf{C}$ is the orbital angular momentum vector $\hbar \mathbf{l}$.

Consider now the six symmetrized products

$$A_i B_j + A_j B_i \tag{5.15}$$

Under rotations, the symmetric products (5.15) transform into linear combinations of symmetric products. In other words, the products $A'_i B'_j + A'_j B'_i$ are equal to linear combinations of the products $A_i B_j + A_j B_i$. The scalar product (5.13) can be removed by defining six symmetric combinations as follows

$$T_{ij} = A_i B_j + A_j B_i - \tfrac{1}{3}(\mathbf{A} \cdot \mathbf{B})\delta_{ij} \tag{5.16}$$

Also the components T_{ij} transform among themselves under rotations since $\mathbf{A} \cdot \mathbf{B}$ is transformed into itself. There are only five independent components among the six terms T_{ij} since they obey the condition

$$\sum T_{ii} = 0$$

If five independent linear combinations of the T_{ij}, are chosen, it will be shown later that they transform irreducibly under rotations. No smaller set of linear combination of them can be found whose members transform among themselves. The five independent terms are components of an *irreducible tensor* of rank 2. The relation of the rank and the number of components will now be explained.

If we take both vectors $\mathbf{A}$ and $\mathbf{B}$ to be the coordinate vector $\mathbf{r}$ of a particle, we can choose five independent linear combinations of the terms $x_i x_j + x_j x_i - \frac{1}{3}r^2\delta_{ij}$ as follows

$$3z^2 - r^2, \quad -z(x+iy), \quad z(x-iy), \quad (x+iy)^2, \quad (x-iy)^2$$

These terms are proportional to the spherical harmonics with rank 2, namely

$$r^2 Y_{20}(\theta,\phi), \quad r^2 Y_{21}(\theta,\phi), \quad r^2 Y_{2,-1}(\theta,\phi), \quad r^2 Y_{22}(\theta,\phi), \quad r^2 Y_{2,-2}(\theta,\phi).$$

This feature will be soon generalized. At this point it is worthwhile to show a concrete example of such a second rank irreducible tensor. The *quadrupole operator* of a nucleon is defined by

$$\boxed{Q_\kappa^{(2)} = \sqrt{\frac{16\pi}{5}} e r^2 Y_{2\kappa}(\theta,\phi)} \tag{5.17}$$

From k vectors $\mathbf{A}^{(j)}$ we can construct a cartesian tensor with k indices by taking products of components

$$T_{i_1 i_2 \ldots i_k} = A_{i_1}^{(1)} A_{i_2}^{(2)} \ldots A_{i_k}^{(k)} \tag{5.18}$$

The components of this tensor transform linearly under rotations. Since each vector $\mathbf{A}^{(j)}$ is transformed according to (5.10) (which is identical to (5.2)) we obtain

$$T'_{i_1 i_2 \ldots i_k} = A_{i_1}^{(1)'} A_{i_2}^{(2)'} \ldots A_{i_k}^{(k)'} = \sum a_{i_1 j_1} a_{i_2 j_2} \ldots a_{i_k j_k} A_{j_1}^{(1)} A_{j_2}^{(2)} \ldots A_{j_k}^{(k)} \tag{5.19}$$

The set of components (5.18) does not transform irreducibly under rotations. This was shown above in the case $k = 2$. It is possible to define smaller sets of linear combinations of the components (5.18) which transform irreducibly among themselves. In a subsequent section we will see how to carry out such a reduction in a systematic way. Here we will consider only the largest set of linear combinations which transform irreducibly under rotations.

To identify this largest set we put all vectors $\mathbf{A}^{(i)}$ equal to the coordinate vector $\mathbf{x}$ of a particle. The components (5.18) become then homogeneous polynomials $P_k(x,y,z)$ of order k. Very many of them become equal. Of the original 3^k components (5.18) there are only $\sum_{i=0}^{k}(k+1-i) = (k+1)(k+2)/2$ different ones. We impose the condition $\Delta P_k(x,y,z) = 0$ which amounts to obtaining a homogeneous polynomial of order $k-2$, all of whose coefficients vanish. There are thus $(k+1)(k+2)/2 - (k-1)k/2 = 2k+1$ independent polynomials satisfying the Laplace equation. As is well known, linear combinations of these polynomials can be chosen so that they can be expressed in terms of the $2k+1$ spherical harmonics of order k as $r^k Y_{k\kappa}(\theta,\phi)$. The set of spherical harmonics of order k transforms irreducibly under rotations. This will be shown in the next section. The fact that in the general case the largest set of linear combinations of the components which transforms irreducibly under rotations is indeed the set corresponding to the polynomials $P_k(x,y,z)$ will be shown in a subsequent section.

To obtain the $2k+1$ linear combinations of (5.18) which transform irreducibly under rotations we first fully symmetrize the components (5.18) as follows.

$$\tilde{T}_{i_1 i_2 \dots i_k} = \sum P A_{i_1}^{(1)} A_{i_2}^{(2)} \dots A_{i_k}^{(k)} \tag{5.20}$$

The summation in (5.20) is over all permutations of the vectors $\mathbf{A}^{(j)}$. Each term contains the product of components i_1 to i_k of each *one* of the vectors $\mathbf{A}^{(j)}$ in any possible order. The set (5.20) is in one to one correspondence with the set of components

$$x_{i_1} x_{i_2} \dots x_{i_k} \tag{5.21}$$

As we saw, there are $(k+1)(k+2)/2$ different terms (5.21). Linear combinations of the $\tilde{T}_{i_1 i_2} \dots x_{i_k}$ should now be taken with the same coefficients as those of the products (5.21) in the homogenous polynomials satisfying the Laplace equation $\Delta P_k(x,y,z) = 0$.

The $2k+1$ components which transform irreducibly under rotations are the components of an *irreducible tensor* of rank k. The transformations law of the general irreducible tensor can be inferred from the transformation law of the special tensor constructed from the components x_i. Thus, the standard basis for components of any irreducible tensor of rank k is the set which has the same transformations law as the spherical harmonics of order k, the $Y_{k\kappa}(\theta,\phi)$. In the frame of reference $\sum'$ the components of (5.21) are given by $x'_{i_1} x'_{i_2} \dots x'_{i_k}$ where x'_i is given in terms of the components of $\mathbf{x}$ in the frame $\sum$ by (5.2). The polar angles of the vector $\mathbf{x}$ are θ,ϕ whereas those of $\mathbf{x}'$ are θ',ϕ'. A rotation R of the frame of reference $\sum$ into the new frame $\sum'$, transforms the angles θ,ϕ into $\theta',\phi'(r'=r)$. The transformation properties of the spherical harmonics are given by the well known formula

$$Y_{k\kappa}(\theta',\phi') = \sum_{\kappa'} Y_{k\kappa'}(\theta,\phi) D^{(k)}_{\kappa'\kappa}(R) \tag{5.22}$$

Accordingly we define any set of $2k+1$ terms to be the components of an irreducible physical tensor of rank k if in the rotated frame $\sum'$ they are related to those in the frame $\sum$ by

$$\boxed{T^{(k)\prime}_{\kappa}(\mathbf{x}') = \sum_{\kappa'} T^{(k)}_{\kappa'}(\mathbf{x}) D^{(k)}_{\kappa'\kappa}(R)} \tag{5.23}$$

The vectors $\mathbf{x}'$ and $\mathbf{x}$ are the coordinates of the point $\mathbf{P}$ in space in the frames $\sum'$ and $\sum$ respectively.

The coordinate vector $\mathbf{x}$ and any other vector whose cartesian components transform according to (5.2) is an irreducible tensor of rank 1. The spherical components of $\mathbf{x}$ are proportional to $Y_{k=1,\kappa}$ and those of any vector $\mathbf{v}$ are

$$v_0 = v_z \qquad v_{+1} = -\frac{v_x + iv_y}{\sqrt{2}} \qquad v_{-1} = \frac{v_x - iv_y}{\sqrt{2}}$$

Under rotations these spherical components transform according to (5.23) with $D^{(1)}_{\kappa\kappa'}(R)$.

The spherical harmonics play here a role which seems different from their role in Section 3. There, a physical *scalar* field was considered and they were introduced as the angular parts of single nucleon wave functions which have a definite l-value. Here they are used for obtaining the transformation laws under rotations of irreducible *tensors* including tensor operators like (5.17). The connection between these two roles will become clearer later in this section and will be fully explained in the next section.

The notation used so far is of *one* coordinate vector $\mathbf{x}$. Many results which were derived above hold equally well for systems whose description involves *several* vectors $\mathbf{x}^{(i)}$. In particular, the definition of components which transform irreducibly among themselves according to (5.23) is very general. We shall not reformulate here the derivations for the general case since this will be done explicitly in the next section.

The transformation coefficients $D^k_{\kappa\kappa'}(R)$ depend on the rotation R defined by (5.2). The nine real coefficients a_{ik} obey six independent conditions in (5.3). Thus, there are only three independent real parameters which uniquely characterize any given rotation. Any rotation in the three-dimensional space has a fixed axis. Hence, two parameters can be chosen as the angles specifying the direction of the axis. The third can then be chosen as the angle of rotation around that axis. A more useful common choice is to define the rotation in terms of the three Euler angles. The transformation coefficients $D^{(k)}_{\kappa\kappa'}$ are the elements of a matrix of order $2k + 1$, called the Wigner D-matrix. All D-matrices with given k form a group of matrices which is an *irreducible representation* of the group $O(3)$ (or R_3) of three-dimensional rotations. The properties of the D-matrix are discussed in many books (e.g. Edmonds 1957) and will not be presented here.

As emphasized above, the passive interpretation does not provide information about the dependence of $\phi(\mathbf{x},t)$ in (5.6) on the coordinates $\mathbf{x}$. The same is generally true of tensor fields of higher rank. An important exception is offered by irreducible tensors constructed from the coordinates $\mathbf{x}$ of a particle as described above. These are all proportional to $r^k Y_{k\kappa}(\theta,\phi)$, a fact which has been used in the definition (5.23) of a general irreducible tensor of rank k. The spherical harmonics as well as irreducible tensors constructed from components of $\mathbf{x}$ and of the corresponding derivatives $(\partial/\partial x, \partial/\partial y, \partial/\partial z)$, form a special case of (5.23). Comparing (5.22) to the general definition (5.23) we see that $Y'_{k\kappa}$ is the *same* function of $\mathbf{x}'$ as $Y_{k\kappa}$ is of $\mathbf{x}$. This is, however, not true in general. We can look at the example given above of the electric field which is a physical tensor field of rank $k = 1$ generated by a static charge distribution. The transformation law of $\mathbf{E}$, or tensors constructed from its components at any point, is given by (5.23). The dependence of $\mathbf{E}$ on the coordinates in any given frame $\sum$ strongly depends on the form of the charge distribution. It could be very complicated if the charge distribution has some irregular shape.

On the other hand, there are cases in which the *active interpretation* of the rotation, to be described now, makes it possible to obtain further information. This is the case if we consider eigenvalue equations. Consider the stationary (bound) solutions of the Schrödinger equation (5.1) in the rotated frame given as in (5.6) by the equation

$$H(\mathbf{p}',\mathbf{x}')\phi(\mathbf{x}(\mathbf{x}')) = E\phi(\mathbf{x}(\mathbf{x}')) \tag{5.24}$$

The solution $\phi(\mathbf{x}(\mathbf{x}')) = \phi'(\mathbf{x}')$ may be considered as *another* solution of (5.24), with the *same* eigenvalue E. It is obtained by a "rigid rotation" of the solution $\phi(\mathbf{x})$ so that the field at the point $\mathbf{P}$ with coordinates $\mathbf{x}$ is carried over to a point $\mathbf{P}'$ with coordinates $\mathbf{x}'$ in the *same* frame of reference $\sum$.

Any Hamiltonian of physical interest has only a finite number of bound independent eigenstates which belong to the same eigenvalue. We can adopt a basis set for all eigenstates with the eigenvalue $\mathbf{E}$ in (5.24). We may include the state $\phi(\mathbf{x})$ of (5.24) in the basis set, so that $\phi(\mathbf{x})$ will be one of the set $\phi_m(\mathbf{x})$. Since $\phi(\mathbf{x}(\mathbf{x}'))$, as a function of $\mathbf{x}'$, is according to (5.24) an eigenstate with the eigenvalue E, it can be expanded as a linear combination of the $\phi_m(\mathbf{x}')$. We thus obtain

$$\phi_m(\mathbf{x}(\mathbf{x}')) = \sum \phi_{m'}(\mathbf{x}')D_{m'm}(R) \tag{5.25}$$

The coefficients $D_{m'm}$ depend on the rotation R which moves the field ϕ at the point $\mathbf{P}$ into the point $\mathbf{P}'$.

A rotationally invariant Hamiltonian (3.1) may have *accidental degeneracies* of its eigenvalues. A simple example is offered by the harmonic oscillator Hamiltonian (4.1). In such cases, the transformation (5.25) of all eigenstates is not irreducible. In such cases it is possible to consider among all eigenstates with the same eigenvalue of H sets which are also eigenstates of $\mathbf{l}^2$ with definite eigenvalue $l(l+1)$. We saw in Section 3 that such sets of eigenstates can be chosen according to (3.7) to be the $2l+1$ eigenfunctions

$$\frac{1}{r}R_{nl}(r)Y_{lm}(\theta,\phi)$$

Under rotations, the radial functions remain unchanged ($r' = r$) whereas the Y_{lm} transform irreducibly among themselves. The transformation (5.25) can be now written more explicitly as

$$\boxed{Y_{lm}(\theta(\theta',\phi'),\phi(\theta',\phi')) = \sum Y_{lm'}(\theta',\phi')D^{(l)}_{m'm}(R)} \tag{5.26}$$

Comparing (5.26) with the transformation (5.22) obtained above, we see that under rotations the $Y_{lm}(\theta,\phi)$, and hence $\phi_{nlm}(r,\theta,\phi)$ which are states of a *physical scalar field*, transform like the components of a *tensor field* of rank l.

To avoid confusion between the primed and unprimed angles in (5.22) and (5.26) let us make sure that these two relations are indeed equivalent. A rotation R of the frame of reference in which the coordinates of the point $\mathbf{P}$ change from $\mathbf{x}$ to $\mathbf{x}'$ is around a certain axis by a certain angle. On the other hand, in (5.26) the frame of reference is fixed and the field Y_{lm} is rigidly rotated so that its value at the point $\mathbf{P}$ with coordinates $\mathbf{x}$ is carried over to another point $\mathbf{P}'$ with coordinates $\mathbf{x}'$ (in the same frame $\sum$). Such a rotation must be around the same axis by an angle of rotation which has the same absolute value as in R but in the *opposite* direction. Hence, the rotation in (5.26) is the *inverse* of the rotation in (5.22). The D-matrices form an irreducible representation of the group $O(3)$ of three-dimensional rotations and hence, $D^{(l)}(R^{-1}) = [D^{(l)}(R)]^{-1}$. Thus, the relation (5.26) can be obtained from (5.22) and *vice versa.*

The transformation properties of states of several nucleons follow from (5.26). A product of several single nucleon wave functions is linearly transformed under rotations with coefficients which are products of elements of D-matrices. Such a transformation is usually not irreducible. In a subsequent section we deal with the problem of forming linear combinations of product wave functions which transform irreducibly under rotations.

In view of the identical mathematical expressions of (5.23) and (5.26) it is important to emphasize the difference in their physical meaning. The relation (5.23) is the transformation under rotations of the $2k + 1$ components of a physical tensor field of rank k. Even when considered at a given point, this is the law of transformation of the components when the same physical situation described in one frame of reference $\sum$ is changed to the description in a rotated frame $\sum'$. According to (5.23) the components are transformed without any reference to their possible functional dependence on the coordinates.

It was pointed out above that tensors constructed from components of the coordinate vector $\mathbf{x}$ form a special case. Any such irreducible tensor is proportional to $Y_{k\kappa}(\mathbf{x})$. This is also true of tensors constructed from components of $\mathbf{x}$ and derivatives with respect to them. The simplest example is $\boldsymbol{l}$ defined in (3.2). It is an irreducible tensor operator of rank $k = 1$ constructed from the two vectors $\mathbf{x}$ and $\boldsymbol{\nabla}$ (it is proportional to their vector product). The transformation law (5.23) of such tensors is a direct result of the transformation law (5.2) of $\mathbf{x}$ and the vector of derivatives $\boldsymbol{\nabla} \equiv (\partial/\partial x, \partial/\partial y, \partial/\partial z)$. Unlike the case of general physical tensor fields, the transformation properties of these $T_\kappa^{(k)}$ are fully determined by their functional dependence on the components of $\mathbf{x}$ and ∇. This holds also for irreducible tensors constructed from vectors $\mathbf{x}^{(1)},\mathbf{x}^{(2)}\ldots\nabla^{(1)},\nabla^{(2)},\ldots$ of several particles. In the following, the irreducible tensorial sets of operators to be considered are exclusively of this kind. In the rest of this section this class of operators will be extended to include also those of spin $\frac{1}{2}$ particles.

The components of the physical tensor field $T^k_\kappa(x)$ may be operators like the quadrupole operator (5.17). They may also describe particles which need $2k+1$ fields to characterize their states. In Section 3 we saw that nucleons with spin $\frac{1}{2}$ need *two* fields for their description. In the next section we will see that fields with $2k+1$ components may describe particles with *intrinsic* angular momentum k. On the other hand, any $Y_{lm}(\theta,\phi)$ which appears in (5.26) is a physical scalar field which is an eigenstate of a rotationally invariant operator. In such states the *orbital* angular momentum has a well-defined l-value. If an external force acts on the particle, it is still described by a scalar field but it may no longer have a definite value of l.

Another aspect of the situation should be pointed out. The Hamiltonian (3.1) is invariant under rotations, i.e. transforms under rotations as a $l=0$ scalar. Its solutions, however, may transform according to (5.28) as tensors of rank l. We deduced the transformation properties of the spherical harmonics Y_{lm} from the $2l+1$ fold degeneracy of the Schrödinger equation following from (3.6) or from the simpler eigenvalue equation (3.10). Conversely, starting from the transformation properties (5.26) we can deduce that there is a $2l+1$ fold degeneracy of eigenvalues of the equations satisfied by the Y_{lm} or ϕ_{lm}. This is a very simple manifestation of a general property. Whenever the symmetry of an eigenstate of a given Hamiltonian is *lower* than the symmetry of the Hamiltonian itself, the corresponding eigenvalue is degenerate.

It is important to realize that from the mathematical point of view, all *tensorial sets* with a given number, $2l+1$, of components are equivalent. They all transform under rotations according to the irreducible representations of $O(3)$ in $2l+1$ dimensions. It does not matter whether l is an *orbital* angular momentum or *intrinsic* spin or the rank of an irreducible tensor *operator*. We should be aware of the physical nature of the tensorial sets considered but, as will be shown later in detail, they all behave according to the same algebra of angular momentum.

The physical fields considered so far in this section, scalar, vector and irreducible tensor fields of higher ranks l, have all odd numbers of components, i.e. $2l+1$. There are however other sets with even numbers of components which transform irreducibly under rotations. The simplest of these, introduced in Section 3, is the two component field of a non relativistic spin $\frac{1}{2}$ particle. We shall now consider the transformation properties under rotation of this well known system.

In Section 3 we considered the Hamiltonian of a non-relativistic nucleon with spin $\frac{1}{2}$ in a central potential well. According to (3.21) that Hamiltonian is

$$H = H_0 + V_{so}(r)(\mathbf{s}\cdot\boldsymbol{l}) = \frac{1}{2m}\mathbf{p}^2 + V(r) + \tfrac{1}{2}V_{so}(r)(\boldsymbol{\sigma}\cdot\boldsymbol{l}) \quad (5.27)$$

with the *numerical* matrices defined by (3.17). To make it clear that we deal here with given numerical matrices, let us denote then by $\sigma_1, \sigma_2, \sigma_3$ and accordingly $\boldsymbol{\sigma}\cdot\boldsymbol{l}$ simply means $\sigma_1 l_x + \sigma_2 l_y + \sigma_3 l_z$. As explained in Section 3, the spinor field ψ of the nucleon is a column with two rows

$$\psi(\mathbf{x},t) = \begin{pmatrix} \phi_1(\mathbf{x},t) \\ \phi_2(\mathbf{x},t) \end{pmatrix} \quad (5.28)$$

where ϕ_1 and ϕ_2 are scalar fields of $\mathbf{x}$ and t. We consider again the time dependent Schrödinger equation

$$H(\mathbf{p},\mathbf{x})\psi(\mathbf{x},t) = i\hbar\frac{\partial}{\partial t}\psi(\mathbf{x},t) \quad (5.29)$$

at a given point $\mathbf{P}$ in space whose coordinates in the frame of reference $\sum$ are given by $\mathbf{x}$. The same equation (5.29) can be expressed in a rotated frame $\sum'$ in which the coordinates of $\mathbf{P}$ are given by $\mathbf{x}'$. Since H_0 is invariant under rotations, from (5.29) we obtain

$$\left[H_0(\mathbf{p}',\mathbf{x}') + \tfrac{1}{2}V_{so}(r)\sum_{i,k}\sigma_i a_{ki} l'_k\right]\begin{pmatrix} \phi_1(\mathbf{x}(\mathbf{x}'),t) \\ \phi_2(\mathbf{x}(\mathbf{x}'),t) \end{pmatrix}$$

$$= i\hbar\frac{\partial}{\partial t}\begin{pmatrix} \phi_1(\mathbf{x}(\mathbf{x}'),t) \\ \phi_2(\mathbf{x}(\mathbf{x}'),t) \end{pmatrix} \quad (5.30)$$

In deriving (5.30) we used (5.5) for expressing x_i in terms of x'_i and l_i in terms of the components l'_i in the rotated frame $\sum'$.

The numerical matrices multiplying l'_k in (5.30) are thus equal to

$$\sum a_{ki}\sigma_i \quad (5.31)$$

which may be very different from the numerical matrices σ_k which multiply l_k in (5.29). The spin orbit interaction in (5.30) does not have the same form as in (5.29). The physical situation must not change when we use instead of the frame of reference $\sum$, the rotated frame $\sum'$. This requirement implies that the components of ψ must change when going from $\sum$ to $\sum'$. Like the components of the electric field **E** they must transform under rotations but their transformation law is different.

A linear transformation of the components of (5.28) can be written as a unitary 2×2 matrix U which multiplies ψ. Multiplying (5.30) on the left by U we obtain

$$\left[H_0(\mathbf{p}',\mathbf{x}') + \tfrac{1}{2}V_{so}(r)\sum_{ik} a_{ki} U\sigma_i U^\dagger l'_k\right] U \begin{pmatrix} \phi_1(\mathbf{x}(\mathbf{x}'),t) \\ \phi_2(\mathbf{x}(\mathbf{x}'),t) \end{pmatrix}$$

$$= i\hbar \frac{\partial}{\partial t} U \begin{pmatrix} \phi_1(\mathbf{x}(\mathbf{x}'),t) \\ \psi_1(\mathbf{x}(\mathbf{x}'),t) \end{pmatrix} \tag{5.32}$$

In deriving (5.32) we used the fact that H_0 commutes with U and the unitarity of U, i.e. $U^\dagger U = 1$. We now see that the Hamiltonian in (5.32) will have, in the frame $\sum'$, the same form as (5.27) in $\sum$ if the transformation law

$$\begin{pmatrix} \phi'_1(\mathbf{x}',t) \\ \phi'_2(\mathbf{x}',t) \end{pmatrix} = U \begin{pmatrix} \phi_1(\mathbf{x}(\mathbf{x}'),t) \\ \phi_2(\mathbf{x}(\mathbf{x}'),t) \end{pmatrix} \tag{5.33}$$

is due to a matrix U which satisfies the condition

$$\boxed{\sum_i a_{ki} U\sigma_i U^\dagger = \sigma_k} \tag{5.34}$$

For every rotation defined by (5.2) or (5.5), a matrix U must be found which satisfies the condition (5.34). Due to the orthogonality of the matrices with elements a_{ik}, the numerical matrices (5.31) with $k = 1, 2, 3$ satisfy the same commutation and anticommutation relations as the matrices σ_k in (3.17). From this follows that there is a unitary transformation which transforms one set into the other, as we shall presently see.

The eigenvalues of the matrices (5.31) are also $+1$ and -1. Hence, there is a unitary transformation that brings (5.31) with $k = 3$ into the form of σ_3

$$U_1 \sum a_{3i}\sigma_i U_1^\dagger = \sigma_3 \tag{5.35}$$

From the commutation relations of σ_3 with $U_1(\sum a_{1i}\sigma_i \pm i \sum a_{2i}\sigma_i)U_1^\dagger$ follows that the matrix of $U_1(\sum a_{1i}\sigma_i)U_1^\dagger$ must be equal to

$$\begin{pmatrix} 0 & e^{i\alpha} \\ e^{-i\alpha} & 0 \end{pmatrix} \tag{5.36}$$

with real α. The transformation with the matrix

$$U_2 = \begin{pmatrix} e^{-(i/2)\alpha} & 0 \\ 0 & e^{(i/2)\alpha} \end{pmatrix}$$

brings (5.36) into the standard form of σ_1 and then $\sigma_2 = -i/2[\sigma_3, \sigma_1]$ must have the standard form given by (3.17). Thus, the matrix which brings the matrices (5.31) into the form (3.17), as required by the relation (5.34), is $U = U_2 U_1$. The U-matrices form a group which is a representation of the group of three dimensional rotations in the two-dimensional space of the states (5.28). Recalling the definition (5.23) of the D-matrices we may identify the U-matrices with the matrices $D^{(1/2)}_{mm'}(R)$.

As described above, the matrices σ_1, σ_2 and σ_3 are taken to be the numerical matrices (3.17) which are the *same* in all frames of references. As numerical matrices, the σ_i do not transform under their own power. The Schrödinger equation is invariant due to the transformation laws of the spinor fields. In fact, we can make use of these transformation properties to form a spin *vector*. If we take two spinors ψ_1 and ψ_2 we can construct three components

$$\psi_1^\dagger \sigma_i \psi_2 \tag{5.37}$$

In a rotated frame we obtain for (5.37) the following transformation laws

$$\boxed{\psi_1'^\dagger \sigma_i \psi_2' = \psi_1^\dagger U^\dagger \sigma_i U \psi_2 = \sum a_{ik} \psi_1^\dagger \sigma_k \psi_2} \tag{5.38}$$

The last equality is obtained from (5.34) due to the orthogonality conditions (5.3) or the unitarity of U. Thus, the components (5.37) transform under rotations like the components of a vector. It is in this sense that $\mathbf{s} \cdot \boldsymbol{l}$ can be considered as a scalar product of two vectors which is invariant under rotations. This justifies the notation $\sigma_x, \sigma_y, \sigma_z$ introduced intuitively in Section 3. To simplify matters we shall use this notation also in the following sections.

The two components spinor field describes a non-relativistic particle with spin $\frac{1}{2}$. Such a particle may well have also orbital angular momentum which should be coupled to the intrinsic spin. This was actually carried out in Section 3. Here we review this point by considering spatial rotations. We saw above that the orbital angular momentum appears in this approach by adopting the active representation. We should take up now this interpretation for the spinor field. In order to simplify the discussion we shall carry out this program in the next section after the introduction of *infinitesimal* rotations.

6

Infinitesimal Rotations and Angular Momentum

Instead of considering the transformation properties of irreducible tensors under finite rotations, we will now consider transformations under *infinitesimal* rotations. Those rotations can be characterized by choosing a definite axis and a rotation around it by an infinitesimal angle $\delta\phi$. Consider rotations around the z-axis ($z' = z$). Under a finite anticlockwise rotation of the frame of reference by an angle ϕ, the coordinates x', y' are given in terms of x, y by

$$\begin{aligned} x' &= x\cos\phi + y\sin\phi \qquad & x &= x'\cos\phi - y'\sin\phi \\ y' &= -x\sin\phi + y\cos\phi \qquad & y &= x'\sin\phi + y'\cos\phi \end{aligned} \tag{6.1}$$

For an infinitesimal angle we put in (6.1), $\cos\delta\phi = 1$, $\sin\delta\phi = \delta\phi$ and obtain

$$\begin{aligned} x' &= x + y\delta\phi \qquad & x &= x' - y'\delta\phi \\ y' &= y - x\delta\phi \qquad & y &= y' + x'\delta\phi \end{aligned} \tag{6.2}$$

Adopting the active interpretation, we consider the change in a *scalar* field $f(x,y,z)$ when it is rigidly rotated. The new field at the point whose coordinates are (x', y', z') is the original field $f(x(x'))$. To find its value we make use of the transformation (6.2) and expand in a Taylor series. As explained in the text following (5.26), we should use in this case the rotation in the opposite direction to that in (6.2). Changing the sign of $\delta\phi$ we thus obtain

$$\begin{aligned} f(\mathbf{x}(\mathbf{x}')) &= f(x' + y'\delta\phi, y' - x'\delta\phi, z') \\ &= f(x', y', z') + y'\delta\phi\frac{\partial f}{\partial x'} - x'\delta\phi\frac{\partial f}{\partial y'} + \cdots \\ &= f(x', y', z') - \delta\phi\left(x'\frac{\partial}{\partial y'} - y'\frac{\partial}{\partial x'}\right) f(x', y', z') + \cdots \end{aligned} \tag{6.3}$$

The first order change of the field $f(x', y', z')$ due to the infinitesimal rotation is expressed in (6.3) as the action of a linear differential operator on $f(x', y', z')$. That operator is proportional to the z-component of the orbital angular momentum vector. Thus, the change in the function f at any point whose coordinates are x, y, z is given by

$$\delta f(x, y, z) = -\delta\phi\left(x\frac{\partial}{\partial y} - y\frac{\partial}{\partial x}\right) f = -i\delta\phi l_z f \tag{6.4}$$

Infinitesimal rotations around other axes lead to analogous expressions. An infinitesimal rotation around an axis in the direction of a unit vector $\mathbf{n}$ is thus given by the component of $\boldsymbol{l}$ in that direction

$$\boxed{\delta f = -i\delta\phi(\mathbf{n}\cdot\boldsymbol{l})f} \tag{6.5}$$

This infinitesimal rotation is equivalent to three consecutive infinitesimal rotations around the x, y, z axes by amounts $n_x\delta\phi$, $n_y\delta\phi$ and $n_z\delta\phi$ respectively. Indeed, to first order we can write (6.5) in the form

$$f + \delta f = (1 - i\delta\phi(\mathbf{n}\cdot\boldsymbol{l}))f = (1 - i\delta\phi n_x l_x)(1 - i\delta\phi n_y l_y)(1 - i\delta\phi n_z l_z)f \tag{6.6}$$

The infinitesimal rotations around the three axes thus combine to form a rotation around $\mathbf{n}$ like the addition of three vectors. This is consistent with infinitesimal rotations being generated by components of the orbital angular momentum vector. Note that this does not hold for finite rotations. Each of the latter may be graphically represented by an arrow pointing in the direction of the axis and length proportional to the angle of rotation around it. Nevertheless, the rotation which is the combination of two finite rotations (around different axes) is not given by the vector sum of the corresponding arrows. Simple examples demonstrate that the combination of two finite rotations is non-commutative, very much unlike addition of vectors.

If we consider a scalar field which is a function of several variables $\mathbf{x}^{(1)}, \mathbf{x}^{(2)}, \ldots$ the change in it due to an infinitesimal rotation is a straightforward generalization of (6.5). Such a function could be the wave function of several spinless particles. We consider the change in $f(\mathbf{x}^{(1)}, \mathbf{x}^{(2)}, \ldots)$ due to the *same* infinitesimal rotation, i.e. around the same axis $\mathbf{n}$ by the same angle $\delta\phi$. Following in this case the steps (6.3) to (6.5) we obtain

$$\delta f(\mathbf{x}^{(1)}, \mathbf{x}^{(2)}, \ldots) = \left\{1 - i\delta\phi\left(\mathbf{n} \cdot \sum \boldsymbol{l}^{(i)}\right)\right\} f(\mathbf{x}^{(1)}, \mathbf{x}^{(2)}, \ldots) \quad (6.7)$$

The change in this case is generated by the *total* orbital angular momentum of the system

$$\mathbf{L} = \sum \boldsymbol{l}^{(i)} \quad (6.8)$$

In the following we shall use $\mathbf{L}$ for the orbital angular momentum and not specify whether it belongs to a single particle or many particles.

We can now see how spherical harmonics and more generally spatial wave functions with given l and m, transform under infinitesimal rotations. Due to (6.5), the results of such transformations can be taken from (3.16) where the action of components of $\boldsymbol{l}$ on Y_{lm} is given. We thus obtain

$$\boxed{L_z \phi_{lm} = m\phi_{lm} \qquad L^{\pm}\phi_{lm} = \sqrt{l(l+1) - m(m \pm 1)}\,\phi_{l,m\pm 1}} \quad (6.9)$$

The relations (6.9) show that any set ϕ_{lm} transforms *irreducibly* under rotations.

From the commutation relations (3.3) follows, as is well known, that the eigenvalues of $\mathbf{L}^2$ are equal to $l(l+1)$ where l is an integer or half integer. It then follows that the matrices of L_x, L_y and L_z which commute with $\mathbf{L}^2$, have vanishing matrix elements connecting states with different eigenvalues of $\mathbf{L}^2$. The eigenvalues of each of them are integers or half integers ranging from $-l$ to $+l$. In a basis in which L_z is diagonal, $L^+ = L_x + iL_y$ and $L^- = L_x - iL_y$ have non-vanishing matrix elements given by (6.9). The matrices of L_x, L_y, L_z cannot be further reduced. They cannot be brought, by unitary transformations on the basis states, to a form in which there are two, or more, submatrices along the diagonal and vanishing matrix elements between different submatrices. In any scheme of states, one of them, say L_z, may be fully diagonalized by independent unitary transformations of the bases of individual submatrices. The set of states, for integral values of l, in which L_z is diagonal is the ϕ_{lm}. In this scheme, $L_x + iL_y$ and L_x–iL_y have non-diagonal matrix elements as given by (6.9). Thus, there are non-vanishing non-diagonal matrix elements of L_x and L_y between *any* state ϕ_{lm} and some other $\phi_{lm'}(m \neq m')$. Hence, it is not possible to further reduce the matrices L_x, L_y and L_z. The $2l + 1$ *independent* eigenfunctions of $\mathbf{L}^2$ with the same eigenvalue $l(l + 1)$ transform into each other under action of components of $\mathbf{L}$ according to (6.9). We interpret now these transformations as the results of infinitesimal rotations. Hence, the $2l + 1$ functions ϕ_{lm} transform irreducibly under infinitesimal rotations and therefore under all rotations.

Let us now turn to two-component spinor fields. We consider first the infinitesimal rotation given by (6.2). Adopting the active interpretation we can determine the change induced by that rotation, in both components, according to (6.4) obtaining

$$\psi + \delta_1\psi = \begin{pmatrix} (1 - i\delta\phi L_z)\phi_1(\mathbf{x}) \\ (1 - i\delta\phi L_z)\phi_2(\mathbf{x}) \end{pmatrix} = (1 - i\delta\phi L_z)\begin{pmatrix} \phi_1(\mathbf{x}) \\ \phi_2(\mathbf{x}) \end{pmatrix} \tag{6.10}$$

The change in $\phi_1(\mathbf{x})$ and $\phi_2(\mathbf{x})$ given by $\delta_1\psi$, is not the only change in the field $\psi(\mathbf{x})$ when it is rigidly rotated. A rigid rotation changes also the components in addition to the change in their dependence on $\mathbf{x}$. Hence, ψ is also transformed by the appropriate U-matrix introduced in Section 5.

The components of ψ transform by the U-matrix under the same rigid rotation given by (6.2) with a change of sign of $\delta\phi$. Whereas in (6.3) the coordinates x_i are expressed in terms of x'_k, the transformation matrix to be used in (5.34) is the one expressing x'_i in terms of the x_k coordinates.

Hence, from (5.34) we obtain the relations

$$U_z(\sigma_x - \delta\phi\sigma_y)U_z^\dagger = \sigma_x \qquad U_z(\sigma_y + \delta\phi\sigma_x)U_z^\dagger = \sigma_y \qquad U_z\sigma_z U_z^\dagger = \sigma_z \tag{6.11}$$

Since U_z corresponds to an infinitesimal rotation we express it as

$$U_z = 1 + i\delta\phi\Omega_z \qquad U_z^\dagger = 1 - i\delta\phi\Omega_z \tag{6.12}$$

where Ω_z is a 2×2 hermitean matrix which should be determined from the equations (6.11). These equations can now be written as

$$\begin{aligned}\sigma_x - \delta\phi\sigma_y &= U_z^\dagger\sigma_x U_z = (1 - i\delta\phi\Omega_z)\sigma_x(1 + i\delta\phi\Omega_z)\\ &= \sigma_x + i\delta\phi[\sigma_x, \Omega_z] + \cdots\\ \sigma_y + \delta\phi\sigma_x &= U_z^\dagger\sigma_y U_z = (1 - i\delta\phi\Omega_z)\sigma_y(1 + i\delta\phi\Omega_z)\\ &= \sigma_y + i\delta\phi[\sigma_y, \Omega_z] + \cdots\\ \sigma_z &= U_z^\dagger\sigma_z U_z = (1 - i\delta\phi\Omega_z)\sigma_z(1 + i\delta\phi\Omega_z)\\ &= \sigma_z + i\delta\phi[\sigma_z, \Omega_z] + \cdots\end{aligned} \tag{6.13}$$

Equating the infinitesimal elements on both sides of the equations (6.13) we obtain the conditions on Ω_z as follows

$$[\sigma_y, \Omega_z] = -i\sigma_x \qquad [\Omega_z, \sigma_x] = -i\sigma_y \qquad [\Omega_z, \sigma_z] = 0 \tag{6.14}$$

The well known commutation relations of the σ-matrices (see (3.19)) determine the solution of (6.14) to be given by

$$\Omega_z = -\tfrac{1}{2}\sigma_z \tag{6.15}$$

Thus the change in ψ due to the transformation of the components induced by the rotation is given by

$$\psi + \delta_2\psi = \left(1 - \frac{i}{2}\delta\phi\sigma_z\right)\begin{pmatrix}\phi_1(\mathbf{x})\\ \phi_2(\mathbf{x})\end{pmatrix} \tag{6.16}$$

The total change in a ψ induced by the rotation (6.2) is given by this change as well as by the change (6.10) due to change of coordinates

$\delta_1\psi$. Since both changes are infinitesimal we obtain

$$\begin{aligned}\psi + \delta\psi &= (1+\delta_2)(1+\delta_1)\psi \\ &= \left(1 - \frac{i}{2}\delta\phi\sigma_z\right)(1 - i\delta\phi L_z)\begin{pmatrix}\phi_1(\mathbf{x})\\ \phi_2(\mathbf{x})\end{pmatrix} \\ &= (1 - i\delta\phi(L_z + \tfrac{1}{2}\sigma_z))\begin{pmatrix}\phi_1(\mathbf{x})\\ \phi_2(\mathbf{x})\end{pmatrix}\end{aligned} \tag{6.17}$$

The total change (6.17) in the spinor field ψ is determined by the z-component of the vector $\mathbf{L} + \frac{1}{2}\boldsymbol{\sigma} = \mathbf{L} + \mathbf{s}$. Rotations around other axes are similarly given by components of $\mathbf{L} + \mathbf{s}$ along those axes. The vector $\mathbf{L} + \mathbf{s}$ is the total angular momentum

$$\mathbf{J} = \mathbf{L} + \mathbf{s} \tag{6.18}$$

The components of $\mathbf{L}$ and $\mathbf{s}$ act on different degrees of freedom and commute with each other. The commutation relations of the matrices $\frac{1}{2}\sigma_x, \frac{1}{2}\sigma_y, \frac{1}{2}\sigma_z$ given in (3.19) are the same as those of L_x, L_y, L_z. Hence, the components of $\mathbf{J}$ satisfy the standard commutation relations of angular momentum components

$$[J_x, J_y] = iJ_z \qquad [J_z, J_x] = iJ_y \qquad [J_y, J_z] = iJ_x \tag{6.19}$$

A set of spinor wave functions which is irreducible under rotations is no longer determined by l. We saw in Section 3 that for any given $l (l > 0)$ there are two possible values of j, which are equal to $l + \frac{1}{2}$ and $l - \frac{1}{2}$. The $2j+1$ states ψ_{jm} constructed there form a set which transforms irreducibly under rotations. Their transformations under infinitesimal rotations are given by the relations

$$J_z\psi_{jm} = m\psi_{jm} \qquad J^{\pm}\psi_{jm} = \sqrt{j(j+1) - m(m\pm 1)}\,\psi_{j,m\pm 1} \tag{6.20}$$

These relations for the states (3.30) and (3.31) follow from the relation (3.16) and the structure of the spin matrices (3.17). We interpret them now as the changes due to infinitesimal rotations according to (6.17) and (6.18).

As is well known, the relations (6.20) are direct consequences of the commutation relations (6.19) for either integral or half-integral value of j. They hold for any set of eigenstates of $\mathbf{J}^2$ and $\mathbf{J}_z$ with eigenvalues $j(j+1)$ and $m(-j \leq m \leq j)$ respectively. From the discussion following (6.9) follows that any set of states ψ_{jm} transforms irreducibly under the operations of J_x, J_y and J_z.

Once $\boldsymbol{l}$ and $\mathbf{s}$ are combined to form a total J, the physical nature of the field does not appear explicitly. The transformation properties under rotations, as given e.g. in (6.20), determine most of the derivations in the following. Fields with more components like the electric field $\mathbf{E}$ can be treated in a similar way. Written as a column with $2s+1$ rows their components (in the passive interpretation) transform under infinitesimal rotations by the action of three independent $(2s+1)\times(2s+1)$ unitary matrices. These matrices are representations of J_x, J_y and J_z in the space of $2s+1$ dimensions. The matrices $\frac{1}{2}\sigma_x, \frac{1}{2}\sigma_y, \frac{1}{2}\sigma_z$ in (3.17) are the matrices of J_x, J_y, J_z for $j=\frac{1}{2}$. The total change in the field (in the active interpretation) is given by these matrices, of the intrinsic spin, and components of the operator $\mathbf{L}$ of orbital angular momentum. We shall not deal here with cases for which $s > \frac{1}{2}$. To calculate electromagnetic properties and transitions, it is necessary to consider three dimensional vector fields. Where considering such transitions, we shall merely quote results obtained elsewhere.

In Section 5 we considered the behavior under rotations of a general class of irreducible tensorial sets. The discussion started with scalar fields which were wave functions of the Schrödinger equation. It was then extended to other physical tensor fields. It was pointed out there that there are other kinds of irreducible tensorial sets. In particular, we saw that the spherical harmonics $Y_{k\kappa}$ may play the role of *irreducible tensor operators* as in (5.17). It was explained that the change induced by three dimensional rotations in ψ_{jm} is fully determined by the value of j and m and independent of their physical nature. The same is true also for infinitesimal rotations but the notation used so far, in particular (6.20), is appropriate only to wave functions. This should be now clarified and adapted also to infinitesimal rotations applied to *operators*.

Let us first look at operators like $Y_{k\kappa}$ which are functions of space coordinates. The change in $Y_{k\kappa}$ due to infinitesimal rotations is given by the application of the linear differential operators l^+, l^- and l^0 to

$Y_{k\kappa}$ as in (3.16). If, however, $Y_{k\kappa}$ is an operator it will be multiplied on the right by a wave function. If only the change in $Y_{k\kappa}$ is considered, the differential operators should not act on that wave function. Since $(d/dx)Y\psi = (dY/dx)\psi + Y(d/dx)\psi$, the change in the operator is $(d/dx)Y - Yd/dx$, i.e. the commutator of the linear differential operator and the operator considered. Thus, in that case we should write instead of (3.16) the change in $Y_{k\kappa}$ as

$$[L^{\pm}, Y_{k\kappa}] = \sqrt{k(k+1) - \kappa(\kappa \pm 1)}Y_{k,\kappa\pm1} \qquad [L^0, Y_{k\kappa}] = \kappa Y_{k\kappa} \tag{6.21}$$

In (6.21) the commutation relations appear only as a device to prevent the linear differential operators to act on any function to the right of $Y_{k\kappa}$. There are other cases, however, that commutation relations are more essential to express the change in tensor operators induced by an infinitesimal rotation.

The angular momenta **s**, $\boldsymbol{l}$ and **j** are examples of tensor operators with rank 1 or vector operators. Under rotations, the components of each transform among themselves. The transformations due to finite rotations are given by (5.23) with $k = 1$. Under infinitesimal transformations, they should also transform like $\psi_{k=1,\kappa}$. Such transformation law is obtained by the commutation relations between components of **J** and the components of any vector **v**. That vector may be, for instance, **s**, $\boldsymbol{l}$, any linear combination like $\boldsymbol{\mu} = g_s\mathbf{s} + g_l\boldsymbol{l}$ of (3.32) and also **J** itself. It may also be the vector **r** or **p** or vectors constructed from their components. Thus, the change under infinitesimal rotations is given, exactly as the r.h.s. of (6.20) by

$$[J^0, v_0] = 0 \qquad [J^0, v_{\pm1}] = \pm v_{\pm1} \qquad [J^{\pm}, v_0] = \sqrt{2}v_{\pm1}$$
$$[J^+, v_{+1}] = [J^-, v_{-1}] = 0 \qquad [J^+, v_{-1}] = [J^-, v_{+1}] = \sqrt{2}v_0 \tag{6.22}$$

The commutation relations (6.19) may be expressed in terms of J^+, J^- and $J^0 = J_2$ as follows

$$\boxed{[J^+, J^-] = 2J^0 \qquad [J^0, J^+] = J^+ \qquad [J^0, J^-] = -J^-} \tag{6.23}$$

Comparing (6.23) with (6.22) we verify that the spherical components of $\mathbf{J}$ are given by

$$J_{+1} = -\frac{J^+}{\sqrt{2}} = -\frac{J_x + iJ_y}{\sqrt{2}} \qquad J_{-1} = \frac{J^-}{\sqrt{2}} = \frac{J_x - iJ_y}{\sqrt{2}} \qquad J_0 = J^0 = J_z$$

In general, considering a rotation $\mathcal{O}$ applied to $T_{k\kappa}\psi_{jm}$ we obtain

$$\mathcal{O}T_{k\kappa}\psi_{jm} = \mathcal{O}T_{k\kappa}\mathcal{O}^{-1}\mathcal{O}\psi_{jm}$$

The transformed operator is thus $\mathcal{O}T_{k\kappa}\mathcal{O}^{-1}$ and putting $\mathcal{O} = 1 + i\epsilon(\mathbf{n}\cdot\mathbf{J})$ we obtain

$$\mathcal{O}T_{k\kappa}\mathcal{O}^{-1} = T_{k\kappa} + i\epsilon[(\mathbf{n}\cdot\mathbf{J}), T_{k\kappa}] + \cdots \tag{6.24}$$

Hence, we can express the change in the tensor operator $T_{k\kappa}$ due to infinitesimal rotations as the analog of (6.20) by

$$\boxed{[J^0, T_{k\kappa}] = \kappa T_{k\kappa} \qquad [J^{\pm}, T_{k\kappa}] = \sqrt{k(k+1) - \kappa(\kappa \pm 1)}\, T_{k\kappa\pm 1}} \tag{6.25}$$

The behavior of any set of $2k + 1$ components, functions or operators, according to (6.25) may be used for the *definition* of an irreducible tensor operator of rank k.

It may be worth while to point out that from (6.20) follows the well known relation

$$\mathbf{J}^2\psi_{jm} = j(j+1)\psi_{jm}$$

of which (3.10) is a special case. There is no corresponding relation for irreducible tensor *operators*. In view of

$$\mathbf{J}^2 = \tfrac{1}{2}J^+J^- + \tfrac{1}{2}J^-J^+ + (J^0)^2 = J^+J^- + (J^0)^2 - J^0$$

we obtain from (6.25) the result

$$\begin{aligned}\left[\mathbf{J}^2, T^{(k)}_{\kappa}\right] &= \sqrt{k(k+1) - \kappa(\kappa - 1)}\, J^+ T^{(k)}_{\kappa-1} \\ &\quad + \sqrt{k(k+1) - \kappa(\kappa + 1)}\, T^{(k)}_{\kappa+1} J^- \\ &\quad + \kappa J^0 T^{(k)}_{\kappa} + \kappa T^{(k)}_{\kappa} J^0 - \kappa T^{(k)}_{\kappa}\end{aligned}$$

The reason for this different behavior is that $\mathbf{J}^2$ is not an infinitesimal transformation. In the case of orbital angular momentum, $\mathbf{L}^2$ is not a *linear* differential operator and, hence, $\mathbf{L}^2 Y_{k\kappa}\psi_{lm}$ is *not* equal to $(\mathbf{L}^2 Y_{k\kappa})\psi_{lm} + Y_{k\kappa}\mathbf{L}^2\psi_{lm}$.

All three-dimensional rotations form a group of continuous transformations or a *Lie group* denoted by $O(3)$. The infinitesimal elements of the group form the *Lie algebra* and those which form an independent and complete set are called *generators*. The usual operations of an algebra are addition and multiplication by a number. In a Lie algebra a further condition is imposed—*the commutator of any two elements must be equal to a linear combination of the elements*. It is sufficient to impose this condition on the basis elements of the algebra from which any element can be obtained as a linear combination. The commutator condition for the $O(3)$ Lie algebra is the relation (6.19) obeyed by the basis generators J_x, J_y, J_z. As we saw above, infinitesimal rotation around any axis $\mathbf{n}$ can be expressed as a linear combination of them according to (6.5) by

$$-i\delta\phi(n_x J_x + n_y J_y + n_z J_z) \tag{6.26}$$

We will see in the following how any finite rotation can also be generated from these operators.

The origin of the commutator condition can be seen from the following argument. Up to first order, infinitesimal rotations or any infinitesimal transformations commute

$$(1+\epsilon' A)(1+\epsilon'' B) = 1 + \epsilon' A + \epsilon'' B + \cdots = (1+\epsilon'' B)(1+\epsilon' A) \tag{6.27}$$

In (6.27), ϵ' and ϵ'' are infinitesimals. The inverse infinitesimal transformation is obtained by changing the sign of ϵ since to first order we obtain

$$(1+\epsilon' A)(1-\epsilon' A) = 1 + \cdots \tag{6.28}$$

Yet *finite* rotations around different axes do not commute and we must turn to them to understand the reason for the commutator condition. If α and β are any two finite transformations we construct $\gamma = \alpha\beta\alpha^{-1}\beta^{-1}$ and consider the infinitesimal transformation corresponding to it, $\gamma = 1 + \epsilon C$. If we put for $\alpha, \alpha^{-1}, \beta, \beta^{-1}$ their expressions from (6.27) and (6.28) in the expression of γ, we see that the first order terms in ϵ' and ϵ'' vanish. For α and β non-commuting, γ

is not equal to the identity transformation. Hence, we realize that we must go to higher order and that ϵ is of higher order than ϵ' and ϵ''.

From $\gamma = \alpha\beta\alpha^{-1}\beta^{-1}$ follows $\alpha\beta = \gamma\beta\alpha$. Putting now the first order expressions for α, β and γ we obtain

$$(1 + \epsilon' A)(1 + \epsilon'' B) = (1 + \epsilon C)(1 + \epsilon'' B)(1 + \epsilon' A) \qquad (6.29)$$

Also here the first order terms $\epsilon' A + \epsilon'' B$ on both sides of (6.29) cancel and we obtain up to order ϵ or $\epsilon'\epsilon''$ the relation

$$\epsilon C = \epsilon'\epsilon''(AB - BA) = \epsilon'\epsilon''[A, B] \qquad (6.30)$$

Thus, the infinitesimal element C of γ is given by the commutator $[A, B]$. As an element of the Lie algebra it can be expressed as a linear combination of the generators. Hence, the commutator condition is a *necessary* condition for the infinitesimal elements of a Lie group.

It is more complicated to prove that the commutator condition is *sufficient* to guarantee that the product of any two finite transformations generated from infinitesimal elements is another such transformation. From any infinitesimal element it is possible to generate a finite transformation. An infinitesimal element is associated with some continuous parameter of the group like the angle of rotation around a given axis. A transformation in which such a parameter is changed from 0 to θ may be constructed by successive applications of many infinitesimal transformations. They all involve the same generator and hence commute. If we divide θ by N and combine N transformations of the form $(1 + (\theta/N)A)$ we obtain for the finite transformation

$$\lim_{N\to\infty} \left(1 + \frac{\theta}{N} A\right)^N = e^{\theta A} = 1 + \theta A + \frac{\theta^2}{2} A^2 + \cdots \qquad (6.31)$$

The exponential function of the generator θA is defined by the well-known series expansion. The transformations to be considered here are unitary, $\alpha^\dagger = \alpha^{-1}$. From (6.28) it follows then that $A^\dagger = -A$ and hence A is equal to a hermitean operator multiplied by i. The eigenvalues λ_i of A are discrete and the corresponding matrices can be diagonalized. Hence, the matrix of $e^{\theta A}$ is diagonal in that case and its eigenvalues are simply $e^{\theta\lambda_i}$.

A rotation by a finite angle θ around an axis whose direction is that of the unit vector $\mathbf{n}$, is given according to (6.31) by

$$e^{-i\theta(\mathbf{n}\cdot\mathbf{J})} \qquad (6.32)$$

Any rotation can thus be described by its fixed axis, which is the eigenvector with eigenvalue $+1$ of the matrix a_{ik} in (5.2), and the angle θ of the rotation around it. Thus, any finite rotation can be obtained from the generators J_x, J_y and J_z.

As an illustration to (6.32) let us consider three dimensional rotations given by (5.2). An infinitesimal rotation around the z-axis is given by (6.2). From (6.2) when the sign of $\delta\phi$ is changed it follows, in view of (6.4), that the matrix of $J_z = l_z$ in the basis defined by x, y and z, is equal to

$$l_z = \begin{pmatrix} 0 & -i & 0 \\ i & 0 & 0 \\ 0 & 0 & 0 \end{pmatrix}$$

The square of this matrix of l_z is the matrix

$$l_z^2 = \begin{pmatrix} 1 & 0 & 0 \\ 0 & 1 & 0 \\ 0 & 0 & 0 \end{pmatrix} = l_z^{2n}$$

Hence (6.32) assumes in that case the form

$$\begin{aligned} e^{-i\theta l_z} &= \begin{pmatrix} 1 & 0 & 0 \\ 0 & 1 & 0 \\ 0 & 0 & 1 \end{pmatrix} + \sum_{\nu=1}^{\infty} \frac{(-i)^\nu \theta^\nu}{\nu!} l_z^\nu \\ &= \begin{pmatrix} 1 & 0 & 0 \\ 0 & 1 & 0 \\ 0 & 0 & 1 \end{pmatrix} + \sum_{n=1}^{\infty} \frac{(-1)^n \theta^{2n}}{(2n)!} l_z^2 - \sum_{n=0}^{\infty} \frac{i(-1)^n \theta^{2n+1}}{(2n+1)!} l_z \\ &= \begin{pmatrix} \cos\theta & -\sin\theta & 0 \\ \sin\theta & \cos\theta & 0 \\ 0 & 0 & 1 \end{pmatrix} \end{aligned} \tag{6.33}$$

The matrix (6.33) represents a rotation of the *field* by angle θ around the z-axis. This is equivalent to a rotation of the frame of reference by an angle $\phi = -\theta$ as defined by (6.1).

If A and B do not commute, the product of the finite transformations $\exp(\theta' A)$ and $\exp(\theta'' B)$ is not equal to $\exp(\theta' A + \theta'' B)$ (it holds only for infinitesimal θ' and θ''). The product should be, however, a finite transformation which may be expressed in terms of its infinitesimal element as $\exp(\theta C)$. The product may be expressed due to a theorem of Hausdorff as

$$e^{\theta' A} \times e^{\theta'' B} = e^{\theta' A + \theta'' B + K} \tag{6.34}$$

where the operator K is equal to a series of multiple commutators including $\theta'\theta''[A,B]$ and its commutators with A, B etc. Hence, if the commutator of any two infinitesimal elements is a linear combination of them, then all finite transformations generated from the infinitesimal ones form a Lie group. In particular, the product of any two finite transformations can be generated from an infinitesimal element. The commutator condition is also called the *integrability condition.*

As expressed by (6.32), any finite rotation can be uniquely specified by an axis $\mathbf{n}$ and angle θ of rotation around it. It can thus be represented by a direction in space and magnitude θ. As mentioned above, although vectors are also represented this way, *finite* rotations do not combine like vectors. Only infinitesimal rotations combine like vectors according to

$$(1 + i\epsilon(\mathbf{n} \cdot \mathbf{J}))(1 + i\epsilon'(\mathbf{n}' \cdot \mathbf{J})) = 1 + i(\epsilon\mathbf{n} + \epsilon'\mathbf{n}') \cdot \mathbf{J} + \cdots$$

Finite rotations combine according to (6.34) in which the non-vanishing operator K appears. A simple example may illustrate this point. A rotation by π around the z-axis is given by (6.33) for $\theta = \pi$. If that rotation is followed by a rotation by angle π around the y-axis the combined rotation is given by the following product of matrices

$$\begin{pmatrix} -1 & 0 & 0 \\ 0 & 1 & 0 \\ 0 & 0 & -1 \end{pmatrix} \begin{pmatrix} -1 & 0 & 0 \\ 0 & -1 & 0 \\ 0 & 0 & 1 \end{pmatrix} = \begin{pmatrix} 1 & 0 & 0 \\ 0 & -1 & 0 \\ 0 & 0 & -1 \end{pmatrix}$$

The resultant rotation is thus a rotation by angle π around the x-axis. Had those rotations combined like vectors their combination would have been a rotation by angle $\sqrt{2}\pi$ around an axis in the y, z plane.

The matrices which carry out the transformation induced by a finite rotation R on an irreducible tensorial set ψ_{jm} are Wigner's $D^{(j)}_{mm'}(R)$ matrices. These matrices constitute an *irreducible representation* of the group of three dimensional rotations $O(3)$ in the space of $2j + 1$ basis states ψ_{jm}.

Given a Lie algebra (of a semi-simple group) it is always possible to construct from the generators, which form an independent and complete set of infinitesimal elements, an algebraic function which commutes with all of them. Such a function of the generators is called a Casimir operator of the group. The construction of the Casimir operators is given in many books and will not be described here. By its construction, a Casimir operator has vanishing matrix elements between states of different irreducible representations. According to Schur's lemma, the submatrix of the Casimir operator constructed from the basis states of an irreducible representation of the group is diagonal and all its diagonal elements are equal. Thus, irreducible representations may be characterized by the eigenvalues of the Casimir operator. In the case of the $O(3)$ group of three dimensional rotations, the quadratic Casimir operator is given by

$$\mathbf{J}^2 = J_x^2 + J_y^2 + J_z^2$$

which is simply the square of the angular momentum vector. Its eigenvalues within the irreducible representation whose basis states are ψ_{jm} are all equal to $j(j+1)$. We shall later see more complicated examples of Casimir operators.

In (6.20) the transformation of various tensor fields under infinitesimal rotations is completely specified. Let us now consider the transformation of those fields under some finite rotations given by (6.32). We consider rotations around an arbitrary axis by an angle equal to 2π. Let us first look at irreducible tensors with rank j which is an integer k. Sets $\psi_{k\kappa}$ transform like the spherical harmonics of order k or like the polynomials $r^2 Y_{k\kappa}(\theta,\phi)$. This is true irrespective of the physical nature of the field. It can be a physical scalar field with orbital momentum k or a physical field with integral intrinsic spin k (revealed by the passive interpretation). The $\psi_{k\kappa}$ could also represent a field with intrinsic integer spin s, with $2s+1$ components, in states with orbital angular momentum l coupled to s to yield a total spin k. After rotation by 2π the coordinates x, y, z of any point return to their original values. Hence, the polynomials $r^2 Y_{k\kappa}(\theta,\phi)$ remain invariant under such a rotation. Certainly the physical situation should not change by a 2π rotation but also the *phase* of the components $\psi_{k\kappa}$ does not change.

We can choose the direction of $\mathbf{n}$ in (6.32) to be the z-direction of the frame of reference. Putting $\theta = 2\pi$ we obtain for the change in any function which is a linear combination of $\psi_{k\kappa}$ the result

$$e^{2\pi iJ_z}\left(\sum c_\kappa \psi_{k\kappa}\right) = \sum c_\kappa e^{2\pi\kappa i}\psi_{k\kappa} = \sum c_\kappa \psi_{k\kappa} \tag{6.35}$$

In fact, the requirement made in Section 3 that m in $Y_{lm}(\theta,\phi)$ should be an integer followed from the condition that $r^2Y_{lm}(\theta,\phi)$ be a single valued function of x,y,z. From that condition follows $e^{2\pi mi} = +1$.

The situation is different for spinor fields where j is half integral. It is sufficient to consider the $s = \frac{1}{2}$ field in the passive interpretation. The effect of a rotation by angle θ around the z-axis is given by

$$e^{-i\theta(1/2)\sigma_z}\psi = \exp\left[i\theta\begin{pmatrix} -\frac{1}{2} & 0 \\ 0 & \frac{1}{2}\end{pmatrix}\right]\begin{pmatrix}\phi_1 \\ \phi_2\end{pmatrix} = \begin{pmatrix} e^{-i\theta/2}\phi_1 \\ e^{i\theta/2}\phi_2\end{pmatrix} \tag{6.36}$$

From (6.36) follows that under a rotation by $\theta = 2\pi$, even though the frame of reference returns to its original position in space, the spinor wave function is multiplied by $e^{i\pi} = e^{-i\pi} = -1$. The physical state does not change but the spinor field is multiplied by -1. Since $J_z = L_z + s_z$, this behavior under 2π rotations holds for spinor wave functions with any value of the orbital angular momentum l. The irreducible representations of the group of three-dimensional rotations, $D^{(j)}_{mm'}(R)$ with half integral values of j form a *double valued* representation of $O(3)$. For example, a 2π rotation is the same as the identity rotation and yet *two* $D^{(j)}$ matrices correspond to it. These are diagonal matrices with equal diagonal elements which are $+1$ in one matrix and -1 in the other.

Two component spinors are the basis of a *single valued* irreducible representation of another Lie group. This is $U(2)$, the group of unitary transformations in a two-dimensional space. Unitary 2×2 matrices are a simple realization of the group and form the simplest irreducible representation of it. A unitary matrix U is defined by the relation $UU^\dagger = U^\dagger U = 1$. This imposes certain conditions on the matrix elements. Let us consider the infinitesimal elements of the group, $U = 1 + i\epsilon A$. As mentioned above, the following relation, for infinitesimal ϵ,

$$UU^\dagger = (1+i\epsilon A)(1-i\epsilon A^\dagger) = 1 + i\epsilon(A - A^\dagger) + \cdots = 1 \tag{6.37}$$

implies that $A^\dagger = A$, i.e. the matrix A is *hermitean*. In (6.37) the equality holds to first order in ϵ but also for finite transformations generated by a hermitean A, the relation

$$\exp(i\epsilon A)\exp(i\epsilon A)^\dagger = \exp(i\epsilon A)\exp(-i\epsilon A) = 1$$

is exact.

The general hermitean matrix can be parametrized as

$$\begin{aligned} A &= \begin{pmatrix} \alpha & \beta \\ \beta & \gamma \end{pmatrix} + \begin{pmatrix} 0 & -i\delta \\ i\delta & 0 \end{pmatrix} \\ &= \frac{\alpha+\gamma}{2}\begin{pmatrix} 1 & 0 \\ 0 & 1 \end{pmatrix} + \beta\begin{pmatrix} 0 & 1 \\ 1 & 0 \end{pmatrix} + \delta\begin{pmatrix} 0 & -i \\ i & 0 \end{pmatrix} + \frac{\alpha-\gamma}{2}\begin{pmatrix} 1 & 0 \\ 0 & -1 \end{pmatrix} \end{aligned} \tag{6.38}$$

Thus, an independent and complete basis of $U(2)$ generators can be chosen as $1, \sigma_x, \sigma_y, \sigma_z$. Previously we saw that $\frac{1}{2}\sigma_x, \frac{1}{2}\sigma_y, \frac{1}{2}\sigma_z$ are the standard generators of the group of three-dimensional rotations $O(3)$ in the space of two component spinors. Those matrices are also the generators of a subgroup of $U(2)$. This is the group $SU(2)$ of all 2×2 matrices obeying the extra condition $\det U = 1$. Such matrices form a group since $\det U_1 U_2 = \det U_1 \det U_2$. This is a meaningful subgroup since a unitary matrix must obey only the condition $|\det U| = 1$.

The Lie algebra of a subgroup is a subalgebra of the Lie algebra of the group. We should check which finite transformations generated by the infinitesimal elements (6.38) are members of $SU(2)$. For that we observe that

$$\det e^A = e^{\mathrm{tr} A} \tag{6.39}$$

as can be verified by bringing A to be diagonal (or to a form in which there are only zeros below or above the diagonal). The traces of $\sigma_x, \sigma_y, \sigma_z$ vanish and hence only the unit matrix generates matrices with $\det U \neq 1$. Thus, $\sigma_x, \sigma_y, \sigma_z$ are generators of $SU(2)$. To obtain generators which obey the standard commutation relations (6.19) we choose as the standard generators of $SU(2)$ the matrices

$$s_x = \tfrac{1}{2}\sigma_x, \qquad s_y = \tfrac{1}{2}\sigma_y, \qquad s_z = \tfrac{1}{2}\sigma_z$$

The irreducible representations of $SU(2)$ generated by these infinitesimal elements are single valued. The infinitesimal element $(1 + i\epsilon\frac{1}{2}\sigma_z)$ generates a finite transformation $\exp((i\epsilon/2)\sigma_z)$ which is formally identical to (6.36). The difference is that ϵ here is an arbitrary parameter and not an angle of rotation defined by (6.1). Hence, putting $\epsilon = 2\pi$ does not yield a transformation which should be equivalent to the identity transformation.

As far as infinitesimal transformations are considered, the $SU(2)$ generators $\frac{1}{2}\sigma_x, \frac{1}{2}\sigma_y, \frac{1}{2}\sigma_z$ satisfy the same commutation relations as in (6.19). We can add to them components of the orbital momentum $\mathbf{L}$ to obtain $\mathbf{J} = \mathbf{L} + \frac{1}{2}\boldsymbol{\sigma}$ whose components satisfy

$$[J_x, J_y] = iJ_z \qquad [J_z, J_x] = iJ_y \qquad [J_y, J_z] = iJ_x \tag{6.40}$$

which are identical to those in (6.19). Thus, the Lie algebra of $SU(2)$, given by (6.40), is the same as that of $O(3)$. The states ψ_{jm} for half integral j form a basis of an irreducible representation of $SU(2)$. The only difference between $SU(2)$ and $O(3)$ is for finite transformations which will not be considered henceforth. In the following we shall refer to J_x, J_y, J_z as generators of $O(3)$ for integral values of j and as generators of $O(3)$ or $SU(2)$ for half integral j values. Our interest in those Lie algebras is because of the paramount role that angular momentum plays in nuclear physics. To make use of it the only information required is contained in (6.40).

Let us nevertheless consider a finite rotation of which we will make use in the following section. A rotation by an angle π around the y-axis transforms x into $-x$, z into $-z$ while y remains unchanged. We first look at the effect of this rotation on components ψ_{jm} with integral spin j. This is given by the transformation of spherical harmonics under that rotation. The Y_{lm} were defined in Section 3 in terms of homogeneous polynomials in x, y, z or in $-(x+iy) = -r\sin\theta e^{i\phi}$, $x - iy = r\sin\theta e^{-i\phi}$ and $z = r\cos\theta$. Hence, as will be explained in more detail in the next section, $r^l Y_{lm}(\theta, \phi)$ is a linear combination of terms

$$[-(x+iy)]^{m_1}(x-iy)^{m_2} z^{l-m_1-m_2}$$

where $m = m_1 - m_2$. Under the rotation considered here, each of these terms is transformed into

$$(x-iy)^{m_1}[-(x+iy)]^{m_2} z^{l-m_1-m_2}(-1)^{l-m_1-m_2}$$

which is a term in the linear combination of $r^l Y_{l,-m}(\theta, \phi)$ multiplied by

$$(-1)^{l-m_1-m_2} = (-1)^{l-m+2m_2} = (-1)^{l-m} \tag{6.41}$$

We now consider the effect of this rotation on states of a particle with spin $s = \frac{1}{2}$. According to (6.32) the operator which induces this rotation is

$$e^{-i\pi s_y} = e^{-i\pi(1/2)\sigma_y} = \exp\frac{\pi}{2}\begin{pmatrix} 0 & -1 \\ 1 & 0 \end{pmatrix}$$

The square of the matrix in the exponent is equal to

$$\begin{pmatrix} 0 & -1 \\ 1 & 0 \end{pmatrix}^2 = \begin{pmatrix} -1 & 0 \\ 0 & -1 \end{pmatrix} = -\begin{pmatrix} 1 & 0 \\ 0 & 1 \end{pmatrix}$$

which is the unit matrix multiplied by -1. Hence, expanding the exponential we obtain

$$\exp\frac{\pi}{2}\begin{pmatrix} 0 & -1 \\ 1 & 0 \end{pmatrix} = \begin{pmatrix} 1 & 0 \\ 0 & 1 \end{pmatrix}\cos\frac{\pi}{2} + \begin{pmatrix} 0 & -1 \\ 1 & 0 \end{pmatrix}\sin\frac{\pi}{2} = \begin{pmatrix} 0 & -1 \\ 1 & 0 \end{pmatrix} \tag{6.42}$$

Applying the matrix on the r.h.s. of (6.42) to the columns (3.18) we obtain

$$\begin{pmatrix} 0 & -1 \\ 1 & 0 \end{pmatrix}\begin{pmatrix} 1 \\ 0 \end{pmatrix} = \begin{pmatrix} 0 \\ 1 \end{pmatrix} \qquad \begin{pmatrix} 0 & -1 \\ 1 & 0 \end{pmatrix}\begin{pmatrix} 0 \\ 1 \end{pmatrix} = \begin{pmatrix} -1 \\ 0 \end{pmatrix} = -\begin{pmatrix} 1 \\ 0 \end{pmatrix} \tag{6.43}$$

Finally we consider the effect of the rotation considered on $\psi_{nlj=l+1/2,m}$ defined by (3.30). The components of $\boldsymbol{l}$ and s commute and hence we obtain, observing (6.41) and (6.43)

$$\begin{aligned} e^{-i\pi J_y}\phi_{nlm-1/2}\chi_{1/2} &= e^{-i\pi l_y}\phi_{nlm-1/2}e^{-i\pi s_y}\chi_{1/2} \\ &= (-1)^{l-(m-1/2)}\phi_{nl,-(m-1/2)}\chi_{-1/2} \\ &= (-1)^{j-m}\phi_{nl,-(m-1/2)}\chi_{-1/2} \\ e^{-i\pi J_y}\phi_{nlm+1/2}\chi_{-1/2} &= e^{-i\pi l_y}\phi_{nlm+1/2}e^{-i\pi s_y}\chi_{-1/2} \\ &= (-1)^{l-(m+1/2)}\phi_{nl,-(m+1/2)}(-\chi_{1/2}) \\ &= (-1)^{j-m}\phi_{nl,-(m+1/2)}\chi_{1/2} \end{aligned}$$

Hence, the state (3.30) is transformed into

$$(-1)^{j-m}\left[\sqrt{\frac{l+\frac{1}{2}-m}{2l+1}}\,\phi_{nl,-m-1/2}\chi_{1/2} + \sqrt{\frac{l+\frac{1}{2}+m}{2l+1}}\,\phi_{nl,-m+1/2}\chi_{-1/2}\right]$$

$$= (-1)^{j-m}\psi_{nlj=l+1/2,-m} \qquad (6.44)$$

As emphasized above, the transformation properties under rotations of the m-components of any irreducible tensorial set of rank j depend only on j and m. We showed above that under a rotation by π around the y-axis the transformation of ψ_{jm}, either for integral value of j (as in (6.41)) or half integral j (as in (6.44)), is given by

$$\boxed{R_y(\pi)\psi_{jm} = (-1)^{j-m}\psi_{j,-m}} \qquad (6.45)$$

Therefore, the relation (6.45) holds for any value of spin j whether it is the rank of a tensor operator or the value of a spin obtained by coupling l and $s = \frac{1}{2}$ or in any other way. The spin may be that of a single particle or a group of particles. It can be directly verified, by using (6.4) and (6.43), that (6.45) holds in particular for the state (3.31) with $j = l - \frac{1}{2}$.

7

Coupling of Angular Momenta. 3j-Symbols

In Section 3 we saw how to combine the intrinsic spin $\mathbf{s}$ and the orbital angular momentum $\mathbf{L}$ to form eigenstates of $\mathbf{J}^2$ where $\mathbf{J} = \mathbf{L} + \mathbf{s}$. The coefficients in (3.30) and (3.31) are special cases of a general class of transformation coefficients. The states on the r.h.s. of (3.30) and (3.31) are eigenstates of $\mathbf{s}^2$, s_z, $\mathbf{L}^2$, L_z. On the left hand side there are eigenstates of $\mathbf{s}^2$, $\mathbf{L}^2$, $\mathbf{J}^2$ and J_z. In general, we may start with *any* two sets $\psi_{j_1 m_1}, \psi_{j_2 m_2}$ of two different systems. The spins j_1 and j_2 may be either orbital angular momenta or total spins of two particles, two intrinsic spins or one orbital angular momentum and one intrinsic spin. The latter case was considered in Section 3. In other cases j_1 may be the rank of an irreducible tensor operator and j_2 the angular momentum of a wave function or both j_1 and j_2 may be the ranks of two irreducible tensor operators. To simplify the notation we will use, for the transformation of the sets $\psi_{j_1 m_1}, \psi_{j_2 m_2}$ under infinitesimal rotations, the expressions (6.20). This will not detract from the applicability of the following results to other tensorial sets. As pointed out several times above, the following mathematical derivations have general validity and are independent of the physical nature of $\psi_{j_1 m_1}$ and $\psi_{j_2 m_2}$. They are due only to the transformation properties of the two sets under rotations.

A complete set of states of the combined system may be constructed by taking all products $\psi_{j_1 m_1}\psi_{j_2 m_2}$. Such states transform linearly under three dimensional rotations but the transformation is reducible unless $j_1 = 0$ or $j_2 = 0$. To decompose the set of product functions into sets which transform irreducibly we should construct states which are eigenstates of $\mathbf{j}^2 = (\mathbf{j}_1 + \mathbf{j}_2)^2$. This is a transformation between the scheme in which j_{1z} and j_{2z} are diagonal and the scheme in which $\mathbf{j}^2$ and j_z are diagonal. In both schemes $\mathbf{j}_1^2$ and $\mathbf{j}_2^2$ are diagonal with eigenvalues $j_1(j_1+1)$ and $j_2(j_2+1)$.

The coefficients of the transformation between these two schemes are called Clebsch-Gordan coefficients or *vector addition coefficients.* They are defined by

$$\boxed{\psi_{jm} = \sum_{m_1 m_2} \psi_{j_1 m_1}\psi_{j_2 m_2}(j_1 m_1 j_2 m_2 \mid j_1 j_2 j m)} \tag{7.1}$$

for values of j_1, j_2 and j which satisfy the triangular conditions

$$|j_1 - j_2| \le j \le j_1 + j_2.$$

The normalization of the coefficients is fixed by the requirement that ψ_{jm} be normalized provided $\psi_{j_1 m_1}$ and $\psi_{j_2 m_2}$ are normalized and belong to independent systems. A common case is that $\psi_{j_1 m_1}$ and $\psi_{j_2 m_2}$ are states of *two* nucleons and then (7.1) is an eigenstate of a rotationally invariant Hamiltonian of the combined system. The overall phase of the Clebsch-Gordan coefficients in (7.1) will be discussed below. The expression (7.1) is also called the *tensor product* of the sets with j_1 and j_2 and denoted by $\psi_{jm} = [\psi_{j_1} \times \psi_{j_2}]^{(j)}_m$.

To determine the actual values of the coefficients we make use of the relations (6.20) which express the change in the system under infinitesimal rotations. Operating on (7.1) with $j_z = j_{1z} + j_{2z}$ we obtain

$$m\psi_{jm} = \sum_{m_1 m_2} (m_1 + m_2)(j_1 m_1 j_2 m_2 \mid j_1 j_2 j m)\psi_{j_1 m_1}\psi_{j_2 m_2} \tag{7.2}$$

Hence, the coefficients vanish unless $m = m_1 + m_2$. This is perhaps the occasion to mention that there exist various notations for Clebsch-Gordan coefficients and care should be paid when employing them. In some cases the $j_1 j_2$ are not repeated on the right hand side of the symbol. This is reasonable but in some cases the coefficients are written as $(j_1 j_2 m_1 m_2 \mid jm)$ which may lead to confusion. From (7.2) follows that the quantum number m need not be explicitly displayed since it is equal to $m_1 + m_2$. Several notations actually omit the m from the symbol. This is, however, a very bad practice. As we shall see later, there are important cases where products of Clebsch-Gordan coefficients have to be summed. Keeping track of all equalities between z-projections of angular momenta is extremely cumbersome and may lead to errors. The procedure which we will adopt is to define the Clebsch-Gordan coefficients, as in (7.1), for *any* values of m_1, m_2 and m and define their values to be zero unless $m = m_1 + m_2$. Keeping m as an independent variable greatly simplifies the following derivations.

Other relations between the coefficients in (7.1) are obtained when $j^{\pm} = j_1^{\pm} + j_2^{\pm}$ are applied to it. According to (6.20) we obtain

$$\begin{aligned}\sqrt{j(j+1) - m(m \pm 1)}\psi_{jm\pm 1} &= \sum_{m_1 m_2} (j_1 m_1 j_2 m_2 \mid j_1 j_2 j m) \\ &\times \left(\sqrt{j_1(j_1+1) - m_1(m_1 \pm 1)}\psi_{j_1 m_1 \pm 1}\psi_{j_2 m_2}\right. \\ &\left.+ \sqrt{j_2(j_2+1) - m_2(m_2 \pm 1)}\psi_{j_1 m_1}\psi_{j_2 m_2 \pm 1}\right) \qquad (7.3)\end{aligned}$$

Starting from $m_1 = j_1$ and $m_2 = j_2$ we find that the maximum value of j is $j = j_1 + j_2$. There is only one such state, $\psi_{j_1 j_1}\psi_{j_2 j_2}$ and the Clebsch-Gordan coefficient is thus found to be

$$\boxed{(j_1 j_1 j_2 j_2 \mid j_1 j_2, j = j_1 + j_2, m = j_1 + j_2) = +1} \qquad (7.4)$$

Notice the phase of (7.4) which was *defined* to be +1. This choice uniquely determines all phases of the coefficients for $j = j_1 + j_2$ for lower values of m as we will now see.

Operating on (7.4) with $j^- = j_1^- + j_2^-$ we obtain the state with $j = j_1 + j_2$ and $m = j_1 + j_2 - 1$. From (7.3) we obtain

$$\sqrt{2(j_1 + j_2)}\psi_{j_1+j_2, j_1+j_2-1} = \sqrt{2j_1}\psi_{j_1, j_1-1}\psi_{j_2 j_2} + \sqrt{2j_2}\psi_{j_1, j_1}\psi_{j_2, j_2-1} \qquad (7.5)$$

From which the values of the Clebsch-Gordan coefficients are obtained to be

$$(j_1, j_1 - 1, j_2 j_2 \mid j_1 j_2, j = j_1 + j_2, m = j_1 + j_2 - 1) = \sqrt{\frac{j_1}{j_1 + j_2}}$$

$$(j_1 j_1 j_2, j_2 - 1 \mid j_1, j_2, j = j_1 + j_2, m = j_1 + j_2 - 1) = \sqrt{\frac{j_2}{j_1 + j_2}}$$

This can be repeated and leads to uniquely determined coefficients with $j = j_1 + j_2$ for all values of m.

In addition to the state with $j = j_1 + j_2$, $m = j_1 + j_2 - 1$, there is another state with $m = j_1 + j_2 - 1$ which is a linear combination of $\psi_{j_1, j_1 - 1}\psi_{j_2 j_2}$ and $\psi_{j_1 j_1}\psi_{j_2, j_2 - 1}$ orthogonal to (7.5). The state

$$\pm\sqrt{\frac{j_2}{j_1 + j_2}}\psi_{j_1, j_1 - 1}\psi_{j_2 j_2} \mp \sqrt{\frac{j_1}{j_1 + j_2}}\psi_{j_1, j_1}\psi_{j_2, j_2 - 1}$$

is according to (7.3) annihilated by $j_+ = j_{1+} + j_{2+}$ and hence has $j = j_1 + j_2 - 1$. The *overall phase* of that state, as well as other states is determined by the following *convention*

$$\sum_{m_1 m_2} m_1 (j_1 m_1 j_2 m_2 \mid j_1 j_2 j m)(j_1 m_1 j_2 m_2 \mid j_1 j_2, j - 1, m) \geq 0 \quad (7.6)$$

This is, naturally, an arbitrary choice and does not introduce any asymmetry in the *physics* of coupling j_1 to j_2. It is, however, a *consistent* convention (this will not be shown here) and together with (7.4) leads to a unique specification of phases for all Clebsch-Gordan coefficients.

The phases of (3.30) and (3.31) were chosen according to the convention (7.6). It is very important to realize that the order of coupling in this case is for $j_1 = \frac{1}{2}, j_2 = l$ and the coefficients are $(\frac{1}{2} m_s l m_l \mid \frac{1}{2} l j m)$. If the order is reversed, due to (7.6) a phase $(-1)^{1/2 + l - j}$ will be introduced. Hence, the phase of states with $j = l - \frac{1}{2}$ will be reversed. The state (3.31) with $j = l - \frac{1}{2}$ was constructed according to the prescription described above. Unfortunately, also the other order of coupling is frequently used. When comparing results of calculations, it is important to find out which convention is being used.

The procedure described above can be repeated and leads to uniquely defined Clebsch-Gordan coefficients for $j = j_1 + j_2 - 2$,

$j_1+j_2-3,\ldots,|j_1-j_2|$. Expanding the state $\psi_{j,m-1}$ on the l.h.s. of (7.3) in terms of Clebsch-Gordan coefficients we obtain

$$\sqrt{j(j+1)-m(m-1)}\sum_{m_1m_2}(j_1m_1j_2m_2\mid j_1j_2j,m-1)\psi_{j_1m_1}\psi_{j_2m_2}$$

$$=\sum_{m_1'm_2'}(j_1m_1'j_2m_2'\mid j_1j_2jm)\Big(\sqrt{j_1(j_1+1)-m_1'(m_1'-1)}\psi_{j_1m_1'-1}\psi_{j_2m_2'}$$

$$+\sqrt{j_2(j_2+1)-m_2'(m_2'-1)}\psi_{j_1m_1'}\psi_{j_2m_2'-1}\Big)$$

Multiplying this equation by $\psi^*_{j_1m_1}\psi^*_{j_2m_2}$ and integrating we obtain the following recursion relation between Clebsch-Gordan coefficients

$$\begin{aligned}\sqrt{j(j+1)-m(m-1)}&(j_1m_1j_2m_2\mid j_1j_2j,m-1)\\ &=\sqrt{j_1(j_1+1)-m_1(m_1+1)}(j_1,m_1+1,j_2m_2\mid j_1j_2jm)\\ &\quad+\sqrt{j_2(j_2+1)-m_2(m_2+1)}(j_1m_1j_2,m_2+1\mid j_1j_2jm)\end{aligned} \tag{7.7}$$

The relation (7.7) can be used to determine all Clebsch-Gordan coefficients for given j_1 and j_2. A simple conclusion which can be drawn from this construction is that all these coefficients have real values.

The transformation between the scheme characterized by $j_1m_1j_2m_2$ and the one characterized by j_1j_2jm leads from an orthogonal and complete basis to another such basis. The number of independent states in both is, of course, equal to

$$\sum_{j=j_1-j_2}^{j_1+j_2}(2j+1)=\frac{2j_2+1}{2}(2(j_1-j_2)+2(j_1+j_2)+2)$$

$$=(2j_1+1)(2j_2+1)$$

This is the order of the matrix whose elements are the coefficients $(j_1m_1j_2m_2\mid j_1j_2jm)$. Hence, the transformation (7.1) is a unitary transformation. Since its coefficients are real, the transformation is a (real) orthogonal one. From this follow the orthogonality relations of

Clebsch-Gordan coefficients

$$\sum_{m_1 m_2} (j_1 m_1 j_2 m_2 \mid j_1 j_2 j m)(j_1 m_1 j_2 m_2 \mid j_1 j_2 j' m') = \delta_{jj'} \delta_{mm'} \tag{7.8}$$

as well as the orthogonality conditions for the inverse transformation

$$\sum_{jm} (j_1 m_1 j_2 m_2 \mid j_1 j_2 j m)(j_1 m_1' j_2 m_2' \mid j_1 j_2 j m) = \delta_{m_1 m_1'} \delta_{m_2 m_2'} \tag{7.9}$$

The relations (7.8) and (7.9) hold only if j_1 and j_2 can couple to j in which case the coefficients are defined. Due to the reality of the coefficients we need not distinguish between $(j_1 m_1 j_2 m_2 \mid j_1 j_2 j m)$ and $(j_1 j_2 j m \mid j_1 m_1 j_2 m_2)$ as already adopted in (7.8) and (7.9). Using the inverse transformation we can expand a given product function $\psi_{j_1 m_1} \psi_{j_2 m_2}$ in terms of ψ_{jm} by

$$\psi_{j_1 m_1} \psi_{j_2 m_2} = \sum_{jm} (j_1 m_1 j_2 m_2 \mid j_1 j_2 j m) \psi_{jm} \tag{7.10}$$

The relation (7.10) holds for ψ_{jm} defined by (7.1) and may be verified by using (7.9).

If we couple $\mathbf{j}_2$ to $\mathbf{j}_1$ instead of $\mathbf{j}_1$ to $\mathbf{j}_2$ to form $\mathbf{j}$ we obtain the *same* set of states which includes one state for each j satisfying $|j_1 - j_2| \leq j \leq j_1 + j_2$. Hence the states $|j_1 j_2 j m\rangle$ and $|j_2 j_1 j m\rangle$ should be the same apart from a phase factor. From this follows that the Clebsch-Gordan coefficients

$$(j_1 m_1 j_2 m_2 \mid j_1 j_2 j m) \qquad \text{and} \qquad (j_2 m_2 j_1 m_1 \mid j_2 j_1 j m)$$

have the same absolute values and may differ by a sign which is independent of m_1 and m_2. The phase convention (7.6) leads to the relation

$$(j_2 m_2 j_1 m_1 \mid j_2 j_1 j m) = (-1)^{j_1+j_2-j}(j_1 m_1 j_2 m_2 \mid j_1 j_2 j m) \tag{7.11}$$

For different values of j_1 and j_2 the phase relation (7.11) is between two different coupling schemes and we may choose the over-all phase of states in either of them at our will. The choice (7.11) is due to (7.6) and turns out to be consistent with the phase relation (7.11) for $j_1 = j_2$ which will now be considered.

For $j_1 = j_2$ the relation (7.11) is not between different schemes. It yields the *relative* sign of coefficients $(j_1 m_1 j_1 m_2 \mid j_1 j_1 j m)$ which is independent of any phase convention. The phase relation (7.11) must hold also for $j_1 = j_2$ once it has been proved that the phase convention (7.6) is consistent. Still, it is not pleasing to use an argument involving phase convention as a proof. Let us therefore consider specifically the case in which $j_1 = j_2$.

Coupling $\psi_{j_1 m_1}(\mathbf{r}_1)$ and $\psi_{jm}(\mathbf{r}_2)$ to a state $\psi_{jm}(\mathbf{r}_1,\mathbf{r}_2)$ yields only one such state, irrespective of the order of couplings. Hence, interchanging m_1 and m_2 in the Clebsch-Gordan coefficients can lead at most to a change in sign which is independent of m_1 and m_2. If we put in (7.5), $j_1 = j_2$ we see that the relative sign of the two coefficients is indeed $+1$ according to (7.11). All other Clebsch-Gordan coefficients for $j = 2j_1$ are obtained by successive operations on (7.1) with the symmetric operator $j^- = j_1^- + j_2^-$. Hence, they all have the same sign and remain unchanged when m_1 and m_2 are interchanged. On the other hand, the coefficients of the state $j = 2j_1 - 1$, $m = 2j_1 - 1$ change sign when $m_1 = j_1 - 1$ and $m_2 = j_1$ are interchanged. Due to the symmetry of $j^- = j_1^- + j_2^-$ the coefficients $(j_1 m_1 j_1 m_2 \mid j_1 j_1 j = 2j_1 - 1, m)$ for all values of m, change sign when m_1 and m_2 are interchanged. Also this change of sign agrees with (7.11).

Let us now show by induction that (7.11) is indeed applicable to all states with $0 \leq j \leq 2j_1$. Assume that this has been shown for j-values $2j_1, 2j_1 - 1, \ldots, 2j_1 - r$ where r is an even number. Let us take the coefficients of these states for $m = 2j_1 - r - 1$ and arrange them in a matrix. The rows of this matrix are characterized by the values of j and the columns are characterized by the value of m_1 in an ascending order. Hence, the number of columns which is equal to the number of coefficients is $r + 2$. We then add another row, orthogonal to all other rows, which includes the coefficients of the state with

$j = m = 2j_1 - r - 1$. The determinant of the resulting matrix does not vanish. Since its rows are orthogonal and normalized vectors, its determinant is equal to $+1$ or -1 (it is actually equal to -1 due to (7.11) but this is irrelevant for the following argument). In the matrix thus constructed, interchanging all values of m_1 and m_2 is equivalent to interchanging all pairs of columns which are located at equal distances from the middle of the rows. Due to these interchanges the determinant of the matrix will be multiplied by $(-1)^p$ where $p = (r+2)/2$ is the number of such column pairs. On the other hand, interchanging all m_1 and m_2 causes a change of sign of the coefficients of all states for which $(-1)^{2j_1-j} = -1$ and $j \geq 2j_1 - r$. The number of such states is equal to $r/2 = p - 1$. Thus, due to the change of sign of these coefficients the determinant is multiplied by $(-1)^{p-1}$ only. Hence, the coefficients of the state with $j = 2j_1 - r - 1$ must change signs when m_1 and m_2 are interchanged. This establishes (7.11) also for this value of j. In the next step, when going from the matrix for $m = 2j_1 - r - 1$ to the one for $m = 2j_1 - r - 2$, the number p of column pairs does not increase. (There are now $r + 3$ columns which is an odd number and hence there is a column in the middle of all rows.) Hence, the state with $j = 2j_1 - r - 2$ has coefficients which do not change sign under interchange of m_1 and m_2. This proves the validity of (7.11) for that value of j. Thus, we proved by induction, for integral or half integral values of j_1, the relation

$$\boxed{(j_1 m_1' j_1 m_1 \mid j_1 j_1 j m) = (-1)^{2j_1 - j}(j_1 m_1 j_1 m_1' \mid j_1 j_1 j m)} \tag{7.12}$$

The relation (7.12) which yields the relative phases of Clebsch-Gordan coefficients has very important consequences. This is the reason for the special attention paid to its derivation. For two particles in the *same* j-orbit it prescribes the symmetry properties of their wave functions. If j_1 is the total spin and is half integral, (7.12) implies that states with even values of j are *antisymmetric* whereas states with odd j values are symmetric under exchange of the variables of the two particles. If we consider only the orbital part, states of two particles in the *same* l-orbit are symmetric when the total L is even and if L is odd they are antisymmetric. In the following we will deal in detail with such states.

We can use the relation (7.12) to obtain from the orthogonality relation (7.9) some specific relations for the case $j_1 = j_2$. We write

(7.9) as

$$\sum_{jm}(j_1m_1j_1m_1' \mid j_1j_1jm)(j_1m_2j_1m_2' \mid j_1j_1jm) = \delta_{m_1m_2}\delta_{m_1'm_2'} \tag{7.13}$$

Similarly

$$\sum_{jm}(j_1m_1'j_1m_1 \mid j_1j_1jm)(j_1m_2j_1m_2' \mid j_1j_1jm) = \delta_{m_1'm_2}\delta_{m_1m_2'} \tag{7.14}$$

Changing the order of m_1 and m_1' on the l.h.s. of (7.14) we obtain from it according to (7.12)

$$\sum_{jm}(-1)^{2j_1-j}(j_1m_1j_1m_1' \mid j_1j_1jm)(j_1m_2j_1m_2' \mid j_1j_1jm) = \delta_{m_1'm_2}\delta_{m_1m_2'} \tag{7.15}$$

Combining (7.13) and (7.15) we obtain two orthogonality relations which may be written as

$$\sum_{jm}(1 \pm (-1)^{2j_1-j})(j_1m_1j_1m_1' \mid j_1j_1jm)(j_1m_2j_1m_2' \mid j_1j_1jm)$$
$$= \delta_{m_1m_2}\delta_{m_1'm_2'} \pm \delta_{m_1m_2'}\delta_{m_1'm_2} \tag{7.16}$$

By putting $m_2 = m_1$ $m_2' = m_1'$ we obtain for $m_1 \neq m_1'$

$$\sum_{j\,\text{even}}(j_1m_1j_1m_1' \mid j_1j_1jm)^2 = \tfrac{1}{2}$$
$$\sum_{j\,\text{odd}}(j_1m_1j_1m_1' \mid j_1j_1jm)^2 = \tfrac{1}{2} \tag{7.17}$$

Thus, these Clebsch-Gordan coefficients with odd j and even j are separately normalized. The sum of squares of the coefficients

$$(j_1m_1j_1m_1 \mid j_1j_1jm)$$

can be obtained directly from (7.14). Due to (7.12) these coefficients vanish in the case $(-1)^{2j_1-j} = -1$. Hence, the sum of squares over j satisfying $(-1)^{2j_1-j} = +1$ is, according to (7.14), equal to 1.

There is another symmetry property of the Clebsch-Gordan coefficients which is independent of any phase convention. It concerns the relation between the coefficients

$$(j_1 m_1 j_2 m_2 \mid j_1 j_2 j m) \qquad \text{and} \qquad (j_1, -m_1 j_2, -m_2 \mid j_1 j_2 j, -m)$$

which both belong to the *same* coupling scheme. To obtain that relation we apply to (7.1) a rotation by angle π around the y-axis. According to (6.45) we obtain

$$(-1)^{j-m}\psi_{j,-m} = \sum (-1)^{j_1-m_1+j_2-m_2} \times (j_1 m_1 j_2 m_2 \mid j_1 j_2 j m)\psi_{j_1,-m_1}\psi_{j_2,-m_2}$$

Since for the non-vanishing terms $m = m_1 + m_2$ we obtain

$$\psi_{j,-m} = \sum (-1)^{j_1+j_2-j}(j_1 m_1 j_2 m_2 \mid j_1 j_2 j m)\psi_{j_1,-m_1}\psi_{j_2,-m_2}$$

which may be expressed as

$$\psi_{jm} = \sum (-1)^{j_1+j_2-j}(j_1, -m_1 j_2, -m_2 \mid j_1 j_2 j, -m)\psi_{j_1 m_1}\psi_{j_2 m_2}$$

Comparing this expression with (7.1) we obtain the symmetry property

$$\boxed{(j_1, -m_1, j_2, -m_2 \mid j_1 j_2 j, -m) = (-1)^{j_1+j_2-j}(j_1 m_1 j_2 m_2 \mid j_1 j_2 j m)} \tag{7.18}$$

There are other symmetries of the Clebsch-Gordan coefficients for which we have to consider j_1, j_2 and j in a more symmetrical way.

Let us consider a special case of two equal angular momenta coupled to total spin zero. The values of the Clebsch-Gordan coefficients can be directly obtained from (7.7). Putting there $j_1 = j_2$, $j = 0$, $m = 0$ and observing that for $m_1 + m_2 + 1 = 0$ the l.h.s. of (7.7) vanishes, we obtain

$$(j_1, m_1 + 1, j_1, -m_1 - 1 \mid j_1 j_1 00) + (j_1, m_1, j_1, -m_1 \mid j_1 j_1 00) = 0$$

Hence, these coefficients have the same absolute value and their signs alternate. From this fact, the phase convention (7.6) and the normalization condition we obtain

$$(jmjm' \mid jj00) = \frac{(-1)^{j-m}}{\sqrt{2j+1}}\delta_{m',-m} \tag{7.19}$$

For two vectors **u** and **v**, irreducible tensorial sets with $j = 1$, the tensor product (7.1) to total spin 0, is obtained by using the coefficients (7.19) to be equal to

$$\frac{1}{\sqrt{3}}\left[\left(-\frac{u_x + iu_y}{\sqrt{2}}\right)\left(\frac{v_x - iv_y}{\sqrt{2}}\right) - u_z v_z + \left(\frac{u_x - iu_y}{\sqrt{2}}\right)\left(-\frac{v_x + iv_y}{\sqrt{2}}\right)\right] = \frac{-1}{\sqrt{3}}(\mathbf{u}.\mathbf{v})$$

The coupling of any two irreducible tensorial sets is givn by (7.1). For the combination of two tensor oerators, it is convenient to use the notation

$$[\mathbf{T}^{(k_1)}(1) \times \mathbf{T}^{(k)}(2)]_{\kappa}^{(k)} = \sum_{\kappa_1\kappa_2}(k_1\kappa_1 k_2\kappa_2 \mid k_1 k_2 k \kappa) T_{\kappa_1}^{(k_1)}(1) T_{\kappa_2}^{(k_2)}(2)$$

The tensor product with coefficients (7.19) is proportional to the ordinary scalar product of the two vectors. In agreement with this case, we define a *scalar product* of any two tensorial sets with integral rank k as

$$(\mathbf{T}_1^{(k)} \cdot \mathbf{T}_2^{(k)}) = \sum(-1)^m T_{1,m}^{(k)} T_{2,-m}^{(k)} = (-1)^k\sqrt{2k+1}[\mathbf{T}_1^{(k)} \times \mathbf{T}_2^{(k)}]_0^{(0)} \tag{7.20}$$

A more symmetrical coupling of three angular momenta can be achieved by coupling $\mathbf{j}_1$, $\mathbf{j}_2$ and $\mathbf{j}$ to a total spin zero. Coupling first $\mathbf{j}_1$

and $\mathbf{j}_2$ and then $\mathbf{j}_3$ (the change in notation is in accordance with the desired symmetry) we may obtain a $J = 0$ state by

$$\sum_{m_1 m_2 m_3 m'_3} (j_1 m_1 j_2 m_2 \mid j_1 j_2 j_3 m'_3)(j_3 m'_3 j_3 m_3 \mid j_3 j_3 00) \psi_{j_1 m_1} \psi_{j_2 m_2} \psi_{j_3 m_3} \tag{7.21}$$

The Clebsch-Gordan coefficients for coupling to $J = 0$ are given by (7.19). With these values of the coefficients the $J = 0$, $M = 0$ state (7.21) becomes

$$\sum_{m_1 m_2 m_3} (-1)^{j_3 + m_3} \frac{(j_1 m_1 j_2 m_2 \mid j_1 j_2 j_3, -m_3)}{\sqrt{2j_3 + 1}} \psi_{j_1 m_1} \psi_{j_2 m_2} \psi_{j_3 m_3} \tag{7.22}$$

There is only one $J = 0$ state obtained by coupling $\mathbf{j}_1$, $\mathbf{j}_2$ and $\mathbf{j}_3$ to zero. Any other order of coupling could yield at most a change in the overall phase of (7.22). The components $\psi_{j_1 m_1}$, $\psi_{j_2 m_2}$ and $\psi_{j_3 m_3}$ appear in (7.22) in a rather symmetrical form. This symmetry does not change if the coefficients in (7.22) are multiplied by a phase factor independent of m_1, m_2, m_3. It turns out that a convenient such factor is $(-1)^{j_1 - j_2 + j_3}$. Multiplying (7.22) by it and observing that $(-1)^{2j_3 + 2m_3} = 1$ we obtain the coefficients in the expansion of the $J = 0$ state to be given by

$$\boxed{\begin{pmatrix} j_1 & j_2 & j_3 \\ m_1 & m_2 & m_3 \end{pmatrix} = \frac{(-1)^{j_1 - j_2 - m_3}}{\sqrt{2j_3 + 1}} (j_1 m_1 j_2 m_2 \mid j_1 j_2 j_3, -m_3)} \tag{7.23}$$

The coefficient in (7.23) is called Wigner's $3j$-symbol (Wigner 1940). It has symmetry properties like those of Clebsch-Gordan coefficients but now also j_3 may be interchanged with j_1 and j_2. Using the definition (7.23), the following symmetries of the $3j$-symbol follow. It is invariant under *any* cyclic permutation of its columns and under any transpositions of columns it acquires the same phase, symmetric in j_1, j_2, j_3,

$$\begin{pmatrix} j_1 & j_2 & j_3 \\ m_1 & m_2 & m_3 \end{pmatrix} = \begin{pmatrix} j_2 & j_3 & j_1 \\ m_2 & m_3 & m_1 \end{pmatrix} = \begin{pmatrix} j_3 & j_1 & j_2 \\ m_3 & m_1 & m_2 \end{pmatrix}$$
$$= (-1)^{j_1+j_2+j_3} \begin{pmatrix} j_2 & j_1 & j_3 \\ m_1 & m_3 & m_2 \end{pmatrix}$$
$$= (-1)^{j_1+j_2+j_3} \begin{pmatrix} j_3 & j_2 & j_1 \\ m_3 & m_2 & m_1 \end{pmatrix}$$
$$= (-1)^{j_1+j_2+j_3} \begin{pmatrix} j_1 & j_3 & j_2 \\ m_1 & m_3 & m_2 \end{pmatrix} \tag{7.24}$$

Hence, a $3j$ symbol with two equal columns vanishes if $(-1)^{j_1+j_2+j_3} = -1$.

As in the case of Clebsch-Gordan coefficients, we keep m_1, m_2 and m_3 as independent variables with the convention that the $3j$-symbol vanishes unless $m_1 + m_2 + m_3 = 0$. The three j-values must satisfy the conditions that they may be coupled to $J = 0$. These are the *triangular conditions* which are $j_i + j_k \geq j_r \geq |j_i - j_k|$ for any order of indices $1, 2, 3$. It is only for such j_1, j_2, j_3 that the $3j$-symbol is written down.

The orthogonality relations of the Clebsch-Gordan coefficients (7.8) and (7.9) give rise to the following relations satisfied by $3j$-symbols

$$\sum_{m_1 m_2} \begin{pmatrix} j_1 & j_2 & j_3 \\ m_1 & m_2 & m_3 \end{pmatrix} \begin{pmatrix} j_1 & j_2 & j_3' \\ m_1 & m_2 & m_3' \end{pmatrix} = \frac{1}{2j_3 + 1} \delta_{j_3 j_3'} \delta_{m_3 m_3'} \tag{7.25}$$

$$\sum_{j_3 m_3} (2j_3 + 1) \begin{pmatrix} j_1 & j_2 & j_3 \\ m_1 & m_2 & m_3 \end{pmatrix} \begin{pmatrix} j_1 & j_2 & j_3 \\ m_1' & m_2' & m_3 \end{pmatrix} = \delta_{m_1 m_1'} \delta_{m_2 m_2'} \tag{7.26}$$

Since the l.h.s. of (7.25) is independent of m_3, summing over it multiplies the r.h.s. by $2j_3+1$ yielding the result

$$\sum_{m_1 m_2 m_3} \begin{pmatrix} j_1 & j_2 & j_3 \\ m_1 & m_2 & m_3 \end{pmatrix} \begin{pmatrix} j_1 & j_2 & j_3 \\ m_1 & m_2 & m_3 \end{pmatrix} = 1 \tag{7.27}$$

Another symmetry of the $3j$-symbols follows from (7.18). Changing the signs of m_1, m_2, m_3 in (7.23) and using (7.18), we obtain the relation

$$\begin{pmatrix} j_1 & j_2 & j_3 \\ -m_1 & -m_2 & -m_3 \end{pmatrix} = (-1)^{j_1+j_2+j_3} \begin{pmatrix} j_1 & j_2 & j_3 \\ m_1 & m_2 & m_3 \end{pmatrix} \tag{7.28}$$

Also here, as in (7.24), the phase is symmetric in the three spins j_1, j_2, j_3.

The $3j$-symbols may be calculated from corresponding Clebsch-Gordan coefficients. The latter may be calculated following the method shown above. The numerical values of many $3j$-symbols are tabulated in several books (e.g. Rotenberg et al. 1959). For the sake of completeness we give below an explicit expression for the coefficients. The following formula, due to Racah (1942), is the solution of the recursion relations described above

$$\begin{aligned} \begin{pmatrix} j_1 & j_2 & j_3 \\ m_1 & m_2 & m_3 \end{pmatrix} &= (-1)^{j_1-j_2-m_3} \delta(m_1+m_2+m_3, 0) \\ &\times \left[\frac{(j_1+j_2-j_3)!(j_1-j_2+j_3)!(-j_1+j_2+j_3)!}{(j_1+j_2+j_3+1)!} \right]^{1/2} \\ &\times [(j_1+m_1)!(j_1-m_1)!(j_2+m_2)!(j_2-m_2)! \\ &\quad \times (j_3+m_3)!(j_3-m_3)!]^{1/2} \\ &\times \sum_z (-1)^z [z!(j_1+j_2-j_3+z)!(j_1-m_1-z)!(j_2+m_2-z)! \\ &\quad \times (j_3-j_2+m_1+z)!(j_3-j_1-m_2+z)!]^{-1} \end{aligned} \tag{7.29}$$

The introduction of the $3j$-symbols, due to their broader symmetry properties, results in a great simplification of the algebra of angular momenta. This will be evident in the following sections. Here, let us make use of the $3j$-symbols for an important application.

It was pointed out above that if the normalized *wave functions* of two independent systems $\psi_{j_1m_1}$ and $\psi_{j_2m_2}$ are combined, the ψ_{jm} wave function in (7.1) is also normalized. This fact found its mathematical expression in the relation (7.8). If the two tensorial sets $\psi_{j_1m_1}$ and $\psi_{j_2m_2}$ are not independent, ψ_{jm} given by (7.1) still transforms irreducibly under rotations. If ψ_{jm} is the resulting wave function, its normalization is not guaranteed by the Clebsch-Gordan coefficients in (7.1). To normalize it some factor may have to be applied.

Let us take up the case of two spherical harmonics $Y_{l'm'}(\mathbf{r})$ and $Y_{l''m''}(\mathbf{r})$ of the same variables $\theta\phi$. Coupling them according to (7.1) to a total orbital angular momentum l and projection m, must yield a set which is an eigenstate of l^2 with eigenvalue $l(l+1)$. Hence, the result is proportional to the spherical harmonic $Y_{lm}(\mathbf{r})$ of the same variables. We thus write

$$\boxed{C^l_{l'l''} Y_{lm}(\mathbf{r}) = \sum_{m'm''} (l'm'l''m'' \mid l'l''lm) Y_{l'm'}(\mathbf{r}) Y_{l''m''}(\mathbf{r})} \tag{7.30}$$

We state without proof that the normalization coefficient in (7.30) is given by

$$\boxed{C^l_{l'l''} = (-1)^l \sqrt{\frac{(2l'+1)(2l''+1)}{4\pi}} \begin{pmatrix} l & l' & l'' \\ 0 & 0 & 0 \end{pmatrix}} \tag{7.31}$$

From the symmetry property of the $3j$-symbols (7.28) follows that (7.31) vanishes unless $(-1)^{l+l'+l''} = +1$. This result is due to the fact that the *parity* of Y_{lm} is $(-1)^l$. The parity of the l.h.s. of (7.30) is $(-1)^l$ and it must be equal to that of the r.h.s. which is $(-1)^{l'+l''}$. For l values which do not satisfy the parity condition, the r.h.s of (7.30) vanishes. An algebraic expression for the $3j$-symbol in (7.31) is given in the Appendix.

A special case of (7.30) is worth noticing. If we put $l'' = l'$, $l = 0$ in (7.30) we obtain by using (7.23) and the special value (7.19) in (7.31) the following result for any value of l

$$\sum_m (-1)^{l-m} Y_{lm}(\theta,\phi) Y_{l,-m}(\theta,\phi) = \frac{2l+1}{4\pi} \tag{7.32}$$

In the next section we will make use of the transformation properties of irreducible tensorial sets, both operators and wave functions. These transformation properties which depend only on rank or angular momentum greatly simplify the calculation of matrix elements.

8

The Wigner-Eckart Theorem

The use of irreducible tensor operators is very convenient for systems whose eigenstates have definite values of J and M. The transformation properties of operators and wave functions, both kinds being components of irreducible tensors, allow the efficient use of geometrical properties. This leads to great simplifications in the calculation of matrix elements as we shall see in the present section.

In Section 5 we saw how cartesian tensors may be constructed from components of vectors and remarked about decomposing them into irreducible tensors. This problem can now be completely solved by using vector addition coefficients. Vectors like $\mathbf{x}$ or $\boldsymbol{\nabla}$ of one particle, or their sums for systems of particles, are irreducible tensors of rank 1. The external product of the components of two of them can be expanded, according to (7.1), as a linear combination of irreducible tensors of ranks 0,1,2. When these are multiplied by components of another vector the result can be expressed in terms of tensors with ranks $0,1,2,3$ etc. If products of components of k vectors are considered, there is only one linear combination which is an irreducible tensor of rank k. The $\kappa = k$ component of that tensor is the product of the $\kappa = 1$ components of all k vectors. The irreducible tensor operators whose matrix elements will be calculated in the following are all constructed this way. According to the discussion in Section 5 we may consider also spin vectors of particles as vectors transforming according to (5.2). Such vectors may also be used in the construction of irreducible tensor operators considered here.

We consider, in general, an orthonormal set of states characterized by the total J and M and other quantum numbers $\alpha, \alpha' \ldots$ which may be needed to uniquely specify the states of the system considered. For single particle states these could be the principal quantum number n and orbital angular momentum l. For systems of several particles more quantum numbers may be necessary, some of which we will meet later on. We consider matrix elements of an operator $T_\kappa^{(k)}$

$$\langle \alpha JM | T_\kappa^{(k)} | \alpha' J' M' \rangle = \int \psi_{\alpha JM}^* T_\kappa^{(k)} \psi_{\alpha' J' M'} \tag{8.1}$$

We can make use of the transformation properties of $T_\kappa^{(k)}$ and $\psi_{\alpha' J' M'}$ and by using Clebsch-Gordan coefficients, obtain the following expansion

$$T_\kappa^{(k)} \psi_{\alpha' J' M'} = \sum_{J'' M''} (k\kappa J' M' \mid k J' J'' M'') \psi_{J'' M''}$$

Like in the derivation of (7.30), normalization of $\psi_{J''M''}$ is not guaranteed by (7.1) since the tensor $T_\kappa^{(k)}$ *operates* on $\psi_{\alpha' J' M'}$. We may expand $\psi_{J''M''}$ as a linear combination of the complete set of orthogonal and normalized wave functions $\psi_{\alpha'' J'' M''}$ and obtain

$$T_\kappa^{(k)} \psi_{\alpha' J' M'} = \sum_{\alpha'' J'' M''} C_{\alpha''}^{\alpha'} (k\kappa J' M' \mid k J' J'' M'') \psi_{\alpha'' J'' M''} \tag{8.2}$$

The coefficients $C_{\alpha''}^{\alpha'}$ are due to the fact that there may be several orthogonal and normalized states with the same values of J'' and M''. They all have the same transformation properties and are multiplied on the r.h.s. of (8.2) by the same vector addition coefficient.

Substituting (8.2) into the integral in (8.1) we see that the only term which contributes must have $\alpha'' = \alpha$, $J'' = J$ and $M'' = M$. The integration yields the result

$$\langle \alpha JM | T_\kappa^{(k)} | \alpha' J' M' \rangle = C_\alpha^{\alpha'} (k\kappa J' M' \mid k J' JM) \tag{8.3}$$

All matrix elements of a tensor operator between states with given quantum numbers αJ, $\alpha' J'$ are proportional to Clebsch-Gordan coefficients. There is only one *physical* quantity in this case which is $C_\alpha^{\alpha'}$. The dependence on κ, M and M' is entirely determined by the *geo-*

metrical Clebsch-Gordan coefficient. Thus, non-vanishing matrix elements between the $\alpha J, \alpha' J'$ states of all irreducible tensor operators of given rank k are proportional. The result (8.3) is the famous Wigner-Eckart theorem (Eckart 1930 and explicitly by Wigner 1931). It may be written in a more symmetrical form by using $3j$-symbols instead of vector addition coefficients. Also the dependence of $C_{\alpha}^{\alpha'}$ on J, J' and the nature of $T_{\kappa}^{(k)}$ should be made more explicit. We thus rewrite (8.3) as

$$\boxed{(\alpha JM|T_{\kappa}^{(k)}|\alpha' J'M') = (-1)^{J-M}\begin{pmatrix} J & k & J' \\ -M & \kappa & M' \end{pmatrix}(\alpha J||\mathbf{T}^{(k)}||\alpha' J')} \tag{8.4}$$

All the *physics* is contained in $(\alpha J||\mathbf{T}^{(k)}||\alpha' J')$ which is called the *reduced matrix element.* The matrix elements (8.4) vanish unless J, k and J' satisfy the triangular conditions. The number of such matrix elements for which $\kappa + M' = M$ is given by $(J + k + J')(J + k + J' + 3) + 1 - 2(J(J+1) + k(k+1) + J'(J'+1))$. The Wigner-Eckart theorem (8.4) clearly demonstrates the usefulness of using operators which are irreducible tensors.

The geometrical coefficient in (8.4) indicates the coupling of three angular momenta, J, J' and k to a total spin zero. The reason for that is that unless the integrand on the r.h.s. of (8.1) is a scalar (rank 0 tensor) the integral vanishes. To see it, let us look at the integral of $T_{\kappa}^{(k)}$, which is a function of space coordinates, over all space. The integral will not change when we perform any rotation. Using the transformation law (5.26) we obtain

$$F_{\kappa}^{(k)} = \int T_{\kappa}^{(k)}(\mathbf{r}^{(i)}) = \sum_{\kappa'} \int T_{\kappa'}^{(k)}(\mathbf{r}^{(i)'}) D_{\kappa'\kappa}^{(k)}(R) = \sum_{\kappa'} F_{\kappa'}^{(k)} D_{\kappa'\kappa}^{(k)} \tag{8.5}$$

The result (8.5) for $k \neq 0$ implies that there is a vector $\mathbf{F}^{(k)}$ in the $(2k+1)$-dimensional space that all D-matrices leave invariant. The existence of an invariant one-dimensional subspace in the space with $2k+1$ dimensions contradicts the irreducibility of the D-matrices. Hence, only a scalar with $k = \kappa = 0$ can have a non-vanishing value of the integral. If the $\psi_{\alpha JM}$ are wave functions of a spinor field with spin s, the operator $T_\kappa^{(k)}$ in (8.1) is a $(2s+1)\times(2s+1)$ matrix with elements which operate on the space coordinates. Hence, $\psi^*_{\alpha JM} T_\kappa^{(k)} \psi_{\alpha' J' M'}$ is a function of the space coordinates of the particles in the system. This function may be expanded in terms of irreducible tensors $T_M^{(L)}(\mathbf{r}^{(i)})$ and the non-vanishing contribution to its integral over space is due to the $L = 0$ term whose coefficient is proportional to

$$\begin{pmatrix} J & k & J' \\ -M & \kappa & M' \end{pmatrix}.$$

A closer look at the $3j$-symbol in (8.4) shows that it displays the fact that $T_\kappa^{(k)}$ and $\psi_{\alpha' J' M'}$ are the κ and M' components of irreducible tensors respectively. This is not the case for $\psi^*_{\alpha JM}$ as M appears there with a minus sign. The reason for that as well as for the phase factor in (8.4) should be explained. As shown above, only a scalar function may yield upon integration a non-vanishing result. Since the integral over $\psi^*_{JM}\psi_{JM}$ is normalized to unity it follows that ψ^*_{JM} must be proportional to the component with $-M$ of an irreducible tensor $\tilde{\psi}_{J,-M}$ which transforms under rotations in the same way as $T_\kappa^{(k)}$ in (5.26) and $\psi_{\alpha' J' M'}$. In fact, we obtain for the spatial function due to the properties of the spherical harmonics

$$\phi^*_{nlm_l} = (-1)^{m_l}\phi_{nl,-m_l} \tag{8.6}$$

For such functions, the complete wave functions of spinless particles, the phase in (8.4), apart from an overall phase $(-1)^l$, arises in a natural way.

The situation for two component spinors is somewhat more complicated. The eigenstates $\chi_{1/2}$ and $\chi_{-1/2}$ introduced in Section 3 satisfy the orthogonality relations $\chi^*_{m_s}\chi_{m'_s} = \delta_{m_s m'_s}$ which are satisfied by taking for χ_{m_s} a column with two rows and for $\chi^*_{m_s}$ the transposed column. The columns $\chi_{1/2}$ and $\chi_{-1/2}$ transform under infinitesimal rotations according to (6.17) by $(1-\delta\theta(i/2)(\mathbf{n}\cdot\boldsymbol{\sigma}))\chi_{m_s}$ where $\mathbf{n}$ is a unit vector in the direction of the axis of rotation. On the other hand, the rows representing $\chi^*_{m_s}$ transform according to $\chi^*_{m_s}(1+\delta\theta(i/2)(\mathbf{n}\cdot\boldsymbol{\sigma}))$. The change in $\chi^*_{m_s}$ is thus given by

$$\chi^*_{m_s}\left(-\tfrac{1}{2}\sigma_z\right) = -m_s\chi^*_{m_s}$$

$$\chi^*_{1/2}\tfrac{1}{2}(-\sigma_x - i\sigma_y) = -\chi^*_{-1/2} \qquad \chi^*_{-1/2}\tfrac{1}{2}(-\sigma_x + i\sigma_y) = -\chi^*_{1/2}$$

Comparing this change with the standard change in (6.20) we see that $\chi^*_{m_s}$ is equal to $(-1)^{(1/2)-m_s}\tilde{\chi}_{-m_s}$ where the components $\tilde{\chi}_{m_s}$ transform according to the standard transformation (6.20).

We can now examine the transformation of ψ^*_{nljm} under rotations. Using vector addition coefficients we can express it as

$$\begin{aligned}
\psi^*_{nljm} &= \left[\sum_{m_s m_l} (\tfrac{1}{2}m_s l m_l \mid \tfrac{1}{2}l j m)\phi_{nlm_l}\chi_{m_s}\right]^* \\
&= \sum_{m_s m_l} (-1)^{-m_l+(1/2)-m_s}(\tfrac{1}{2}m_s l m_l \mid \tfrac{1}{2}l j m)\phi_{nl,-m_l}\tilde{\chi}_{-m_s} \\
&= \sum_{m_s m_l} (-1)^{j-l-(1/2)-m_l+(1/2)-m_s} \\
&\quad \times (\tfrac{1}{2},-m_s,l,-m_l \mid \tfrac{1}{2}l j,-m)\phi_{nl,-m_l}\tilde{\chi}_{-m_s} \\
&= (-1)^{j-l-m}\tilde{\psi}_{j,-m}
\end{aligned} \tag{8.7}$$

The result in (8.7) was obtained by using the symmetry (7.18) of the vector addition coefficients. The $\tilde{\psi}_{jm}$ in (8.7) transforms according to (6.20). Apart from an overall phase $(-1)^l$, the phase on the r.h.s. of (8.7) is the same as in (8.4). When the system includes several angular momenta coupled by vector addition coefficients, the total M is the sum of the individual m values. Hence, the phase in (8.4) is obtained in a natural way. The choice of the overall sign, common to all M values, is due to considerations of time reversal which will not be discussed here.

When two irreducible tensors ψ_{JM} and ψ'_{JM} are coupled according to (7.1) into a scalar, the coefficients are given by (7.19). The resulting scalar is thus given by

$$\frac{1}{\sqrt{2J+1}}\sum_M (-1)^{J-M}\psi'_{JM}\psi_{J-M}$$

When the normalization integral of the wave functions ψ_{JM} is calculated we can write the integrand as $\sum_M \psi^*_{JM}\psi_{JM}$. According to the preceding discussion this is indeed a scalar since it has the transformation properties of $\sum_M (-1)^{J-M}\tilde{\psi}_{J,-M}\psi_{JM}$.

The definition of the reduced matrix element in (8.4) is the standard one. Unfortunately, some authors refer to the $C_{\alpha}^{\alpha'}$ in (8.3) as the reduced matrix elements. The use of the Clebsch-Gordan coefficient in (8.3) for that definition introduces an ugly asymmetry between J and J' in the reduced matrix element. It makes the calculations more difficult and adds confusion by using an established term for a different quantity.

The definition of the spherical components of irreducible tensors follows the standard form of spherical harmonics. Although the latter are constructed from hermitean operators, x, y, z, they have complex coefficients and behave under complex conjugation according to (8.6). This is the behavior of components of any irreducible tensor constructed from hermitean cartesian components of vectors. Accordingly, an irreducible tensor operator is called *hermitean* if its components satisfy the similar relation

$$\boxed{(T_{\kappa}^{(k)})^{\dagger} = (-1)^{\kappa} T_{-\kappa}^{(k)}} \tag{8.8}$$

From (8.8) follows for the complex conjugate of the matrix element

$$\langle \alpha' J' M' | T_{\kappa}^{(k)} | \alpha J M \rangle^* = (-1)^{\kappa} \langle \alpha J M | T_{-\kappa}^{(k)} | \alpha' J' M' \rangle$$

and using the Wigner-Eckart theorem we obtain

$$(-1)^{J'-M'} (\alpha' J' \| \mathbf{T}^{(k)} \| \alpha J)^* \begin{pmatrix} J' & k & J \\ -M' & \kappa & M \end{pmatrix}$$

$$= (-1)^{\kappa+J-M} (\alpha J \| \mathbf{T}^{(k)} \| \alpha' J') \begin{pmatrix} J & k & J' \\ -M & -\kappa & M' \end{pmatrix}$$

Using the symmetry properties of the $3j$-symbols (7.28) we deduce

$$\boxed{(\alpha' J' \| \mathbf{T}^{(k)} \| \alpha J)^* = (-1)^{J-J'} (\alpha J \| \mathbf{T}^{(k)} \| \alpha' J')} \tag{8.9}$$

From (8.9) follows that the reduced matrix element of an hermitean tensor between two identical states with given αJ is *real*. This may also be seen from the fact that the $\kappa = 0$ component $T_0^{(k)}$ of such a tensor is an hermitean operator. Hence, it has real diagonal elements given by

$$\langle \alpha JM | T_0^{(k)} | \alpha JM \rangle = (-1)^{J-M} (\alpha J \| \mathbf{T}^{(k)} \| \alpha J) \begin{pmatrix} J & k & J \\ -M & 0 & M \end{pmatrix} \tag{8.10}$$

A simple consequence of the geometrical factors in the Wigner-Eckart theorem (8.4) is that the only non-vanishing matrix elements of $T_\kappa^{(k)}$ are between states which satisfy $M = \kappa + M'$ and that J, J' and k can combine to zero (the triangular condition in the $3j$-symbol). In particular, any scalar operator ($k = 0$) is diagonal in J and M. Due to the values of the $3j$-symbols obtained from (7.19), the value of the diagonal matrix elements of a scalar operator are independent of M. A scalar operator may have non-vanishing matrix elements between states with different additional quantum numbers α and α' provided $J' = J$ and $M' = M$.

The orthogonality relations of the $3j$-symbols may be used to express the reduced matrix element in terms of all ordinary matrix elements. Multiplying (8.4) by

$$(-1)^{J-M} \begin{pmatrix} J & k & J' \\ -M & \kappa & M' \end{pmatrix}$$

and summing over M and M' we obtain in view of (7.24) and (7.25) the result

$$\boxed{\sum_{MM'} (-1)^{J-M} \begin{pmatrix} J & k & J' \\ -M & \kappa & M' \end{pmatrix} \langle \alpha JM | T_\kappa^{(k)} | \alpha' J' M' \rangle = \frac{1}{2k+1} (\alpha J \| T^{(k)} \| \alpha' J')} \tag{8.11}$$

The reduced matrix elements of simple operators can be obtained directly from their definition. The trivial operator 1 is a scalar with non-vanishing matrix elements equal to 1 only between states satisfying $\alpha' = \alpha$, $J' = J$, $M' = M$. Putting 1 for $T_\kappa^{(k)}$ in (8.4) we obtain by using (7.19) the result

$$\boxed{(\alpha J\|1\|\alpha' J') = \sqrt{2J+1}\,\delta_{\alpha\alpha'}\,\delta_{JJ'}} \tag{8.12}$$

Note that the reduced matrix element of 1 is *not* 1.

The operator $\mathbf{J}$ is a vector, $k = 1$, with the special property that it has non-vanishing matrix elements only for states with $\alpha' = \alpha$, $J' = J$. We can make use of (8.10) for the J_0 component yielding M on the l.h.s. of (8.10). Using the value of

$$\begin{pmatrix} J & 1 & J \\ -M & 0 & M \end{pmatrix}$$

given in the Appendix we obtain

$$\boxed{(\alpha J\|\mathbf{J}\|\alpha' J') = \sqrt{J(J+1)(2J+1)}\,\delta_{\alpha\alpha'}\,\delta_{JJ'}} \tag{8.13}$$

The reduced matrix elements of spherical harmonics can be expressed by using the value of $C^l_{l'l''}$ in (7.31). Multiplying (7.30) by $Y^*_{lm}(\mathbf{r})$ and integrating we obtain

$$\begin{aligned}
C^l_{l'l''} &= \sum_{m'm''} (l'm'l''m'' \mid l'l''lm)\langle lm|Y_{l'm'}|l''m''\rangle \\
&= \sum_{m'm''} (-1)^{l'-l''-m}\sqrt{2l+1}\begin{pmatrix} l' & l'' & l \\ m' & m'' & -m \end{pmatrix}\langle lm|Y_{l'm'}|l''m''\rangle \\
&= \sqrt{2l+1}\sum_{m'm''} (-1)^{l'-l''-m}\begin{pmatrix} l & l' & l'' \\ -m & m' & m'' \end{pmatrix}\langle lm|Y_{l'm'}|l''m''\rangle
\end{aligned} \tag{8.14}$$

Using the Wigner-Eckart theorem (8.4), we obtain from (8.14) the result

$$\sqrt{2l+1}\sum_{m'm''}(-1)^{l'-l''-l}\begin{pmatrix} l & l' & l'' \\ -m & m' & m'' \end{pmatrix}$$

$$\times\begin{pmatrix} l & l' & l'' \\ -m & m' & m'' \end{pmatrix}(l\|\mathbf{Y}_{l'}\|l'')$$

$$= \frac{1}{\sqrt{2l+1}}(-1)^{l'+l''+l}(l\|\mathbf{Y}_{l'}\|l'')$$

The last equality is due to the orthogonality relations (7.25) of $3j$-symbols. Taking the value of $C^{l}_{l'l''}$ from (7.31) we obtain the general result

$$(l\|\mathbf{Y}_k\|l') = (-1)^l\sqrt{\frac{(2l+1)(2k+1)(2l'+1)}{4\pi}}\begin{pmatrix} l & k & l' \\ 0 & 0 & 0 \end{pmatrix} \tag{8.15}$$

From (8.15), due to the symmetry property (7.28) of $3j$-symbols, follows that matrix elements of $Y_{k\kappa}$ between states with l and l' vanish for $(-1)^{l+k+l'} = -1$. This also follows directly from parity considerations. The parity of a product of three spherical harmonics is $(-1)^{l+k+l'}$. If that product is integrated over all angles the integral vanishes if the integrand changes sign when $\mathbf{r}$ is replaced by $-\mathbf{r}$.

The Wigner-Eckart theorem may be used to prove that any operator can be expanded in terms of irreducible tensor operators. More precisely, what we can show is the following: Matrix elements of any operator between states αJM and $\alpha' J' M'$ can be reproduced by matrix elements of a linear combination of irreducible tensor operators. The set of matrices of all irreducible tensor operators $U^{(k)}_{\kappa}$ with k ranging between $|J - J'|$ and $J + J'$ forms a complete set in the space of the corresponding $(2J+1)\times(2J'+1)$ matrices.

Let us first define in the space considered *unit tensor operators*, $U^{(k)}_{\kappa}$, to be those whose reduced matrix elements are given by

$$(\alpha J\|U^{(k)}\|\alpha' J') = 1 \tag{8.16}$$

Matrix elements of these operators are give according to (8.4) by

$$\langle \alpha JM | U_\kappa^{(k)} | \alpha' J' M' \rangle = (-1)^{J-M} \begin{pmatrix} J & k & J' \\ -M & \kappa & M' \end{pmatrix}$$

If we multiply this equation by

$$(-1)^{J-M_0}(2k+1) \begin{pmatrix} J & k & J' \\ -M_0 & \kappa & M_0' \end{pmatrix}$$

and sum over k and κ we obtain, due to (7.26), the result

$$\sum_{k\kappa} (-1)^{J-M_0}(2k+1) \begin{pmatrix} J & k & J' \\ -M_0 & \kappa & M_0' \end{pmatrix} \langle \alpha JM | U_\kappa^{(k)} | \alpha' J' M' \rangle$$

$$= \delta_{MM_0} \delta_{M'M_0'} \tag{8.17}$$

From (8.17) follows that matrix elements of any operator $\mathcal{O}$ can be expressed by

$$\langle \alpha JM_0 | \mathcal{O} | \alpha' J' M_0' \rangle$$

$$= \sum_{k\kappa MM'} \langle \alpha JM | \mathcal{O} | \alpha' J' M' \rangle (-1)^{J-M}(2k+1)$$

$$\times \begin{pmatrix} J & k & J' \\ -M & \kappa & M' \end{pmatrix} \langle \alpha JM_0 | U_\kappa^{(k)} | \alpha' J' M_0' \rangle \tag{8.18}$$

The expansion coefficients in (8.18) depend on κ, M_0 and M_0'. Naturally, an arbitrary operator does not transform simply under rotations. If the operator $\mathcal{O}$ itself is an irreducible tensor $T_\kappa^{(k)}$, the sum in (8.18) reduces to

$$\langle \alpha JM | T_\kappa^{(k)} | \alpha' J' M' \rangle = (\alpha J \| \mathbf{T}^{(k)} \| \alpha' J') \langle \alpha JM | U_\kappa^{(k)} | \alpha' J' M' \rangle \tag{8.19}$$

which can be verified by using the orthogonality relations (7.25). The relation (8.19) is just the Wigner-Eckart theorem (8.4). Operators of physical importance in systems with definite angular momenta are irreducible tensors.

It should be realized that the unit tensor operator defined by (8.16) is not necessarily hermitean. Using the symmetry properties of the $3j$-symbols we obtain the following relation between its matrix elements

$$
\begin{aligned}
\langle \alpha JM|U_\kappa^{(k)}|\alpha' J' M'\rangle &= (-1)^{J-M}\begin{pmatrix} J & k & J' \\ -M & \kappa & M' \end{pmatrix} \\
&= (-1)^{J-M}\begin{pmatrix} J' & k & J \\ -M' & -\kappa & M \end{pmatrix} \\
&= (-1)^{J'-M'}\begin{pmatrix} J' & k & J \\ -M' & -\kappa & M \end{pmatrix}(-1)^{J-J'+\kappa} \\
&= (-1)^{J-J'+\kappa}\langle \alpha' J' M'|U_{-\kappa}^{(k)}|\alpha JM\rangle
\end{aligned}
$$

Hence, $U_\kappa^{(k)}$ is hermitean if $(-1)^{J-J'}=1$, it is anti-hermitean if $(-1)^{J-J'}=-1$. If $\mathbf{T}^{(k)}$ is hermitean then $(\alpha J\|\mathbf{T}^{(k)}\|\alpha' J')U_\kappa^{(k)}$ is hermitean due to (8.9).

As mentioned above, non-vanishing matrix elements of components of any irreducible tensor operators, with the same rank k and κ, between states with αJM and $\alpha' J' M'$, are proportional. In particular, this holds for vector operators whose matrix elements between states αJM and $\alpha JM'$ are proportional to matrix elements of $\mathbf{J}$ (matrix elements of $\mathbf{J}$ between states with different values of J or α vanish). Let us make use of this property to calculate the magnetic moment of a system composed of two independent parts with spins $\mathbf{J}_1$ and $\mathbf{J}_2$ coupled to a state with definite value of $\mathbf{J}^2=(\mathbf{J}_1+\mathbf{J}_2)^2$. The magnetic moment operator is the vector

$$
\boxed{\boldsymbol{\mu} = g_1\mathbf{J}_1 + g_2\mathbf{J}_2} \tag{8.20}
$$

where g_1 and g_2 are the g-factors of the two systems. A special case of this kind was considered in Section 3. Eq. (3.32) is a special case of (8.20) with $\mathbf{J}_1=\mathbf{l}$ and $\mathbf{J}_2=\mathbf{s}$.

We consider states in which $\mathbf{J}_1^2$ and $\mathbf{J}_2^2$ have definite eigenvalues, $J_1(J_1+1)$ and $J_2(J_2+1)$ respectively. Due to the Wigner-Eckart theorem, matrix elements of $\boldsymbol{\mu}$ between states with the same α and J are proportional to those of $\mathbf{J}$. If we denote by g the proportionality factor we obtain

$$\langle \alpha JM|\boldsymbol{\mu}|\alpha JM'\rangle = g\langle \alpha JM|\mathbf{J}|\alpha JM'\rangle = g_1\langle \alpha JM|\mathbf{J}_1|\alpha JM'\rangle + g_2\langle \alpha JM|\mathbf{J}_2|\alpha JM'\rangle \tag{8.21}$$

We can take the scalar product of (8.21) with the vector

$$\langle \alpha JM'|\mathbf{J}|\alpha JM''\rangle$$

and sum over M'. Since $\mathbf{J}$ does not have other non-vanishing matrix elements, we obtain this way the matrix product between components of $\mathbf{J}$ with themselves and with components of $\mathbf{J}_1$ and $\mathbf{J}_2$. Hence,

$$g\langle \alpha JM|\mathbf{J}^2|\alpha JM''\rangle = g_1\langle \alpha JM|\mathbf{J}_1\cdot\mathbf{J}|\alpha JM''\rangle + g_2\langle \alpha JM|\mathbf{J}_2\cdot\mathbf{J}|\alpha JM''\rangle$$

The operator $\mathbf{J}^2$ is diagonal in M with eigenvalues $J(J+1)$ and also the operators $\mathbf{J}_1\cdot\mathbf{J}$ and $\mathbf{J}_2\cdot\mathbf{J}$ are scalars and thus diagonal in M and their eigenvalues are independent of M. Hence, we obtain that within states with the definite value of J

$$g\mathbf{J}^2 = g_1\mathbf{J}_1\cdot\mathbf{J} + g_2\mathbf{J}_2\cdot\mathbf{J} \tag{8.22}$$

The scalar products on the r.h.s. of (8.22) can be directly evaluated by using the operator identities which follow from $\mathbf{J} = \mathbf{J}_1 + \mathbf{J}_2$

$$\mathbf{J}_1^2 = (\mathbf{J}-\mathbf{J}_2)^2 = \mathbf{J}^2 + \mathbf{J}_2^2 - 2\mathbf{J}_2\cdot\mathbf{J} \qquad \mathbf{J}_2^2 = (\mathbf{J}-\mathbf{J}_1)^2 = \mathbf{J}^2 + \mathbf{J}_1^2 - 2\mathbf{J}_1\cdot\mathbf{J}$$

The scalar products are thus diagonal in the scheme in which $\mathbf{J}_1^2$, $\mathbf{J}_2^2$ and $\mathbf{J}^2$ are diagonal. We can express (8.22) as

$$g\mathbf{J}^2 = \tfrac{1}{2}[g_1\{\mathbf{J}^2 + \mathbf{J}_1^2 - \mathbf{J}_2^2\} + g_2\{\mathbf{J}^2 + \mathbf{J}_2^2 - \mathbf{J}_1^2)\}] \tag{8.23}$$

Substituting in (8.23) the eigenvalues of $\mathbf{J}^2$,$\mathbf{J}_1^2$ and $\mathbf{J}_2^2$ we obtain the result

$$\boxed{\begin{aligned} gJ = \frac{1}{2(J+1)}[&g_1\{J(J+1) + J_1(J_1+1) - J_2(J_2+1)\} \\ &+ g_2\{J(J+1) + J_2(J_2+1) - J_1(J_1+1)\}] \end{aligned}} \tag{8.24}$$

The magnetic moment μ is defined as the expectation value of the zero component of (8.20) in the state with $M = J$. From (8.21) follows $\mu = gJ$. The results obtained above for the magnetic moment of a single nucleon (3.34) and (3.37) are special cases of (8.24). They are obtained by substituting $\boldsymbol{l}$ for $\mathbf{J}_1$, g_l for g_1, $\mathbf{s}$ for $\mathbf{J}_2$ and g_s for g_2.

In some special cases, the g-factors expressed by (8.24) have a simple form. If $J_1 = J_2$, then

$$g = \frac{g_1 + g_2}{2} \qquad \text{for} \quad J_1 = J_2 \tag{8.25}$$

Obviously if $g_1 = g_2$ then $g_1\mathbf{J}_1 + g_1\mathbf{J}_2 = g_1\mathbf{J}$ and $g = g_1$. If $J = J_1 + J_2$, the result (8.24) simplifies into

$$\mu = gJ = g_1 J_1 + g_2 J_2 = \mu_1 + \mu_2 \tag{8.26}$$

Let us apply the Lande formula (8.24) to the $J = 4$ ground state of ^{40}K. As explained in Section 1, low lying levels of that nucleus can be calculated from those of ^{38}Cl by adopting for ^{40}K the $1d^3_{3/2}$ proton configuration and a single $1f_{7/2}$ neutron. Since the agreement between calculated and experimental energies is very good, it is interesting to see whether similar agreement may be obtained for magnetic moments.

Putting in (8.24) the values $J = 4$, $J_1 = \frac{3}{2}$ and $J_2 = \frac{7}{2}$ we obtain for this case

$$\mu = g4 = \tfrac{1}{10}(8g_1 + 32g_2) \tag{8.27}$$

The g-factor of states of three $1d_{3/2}$ protons is the same as that of a single $1d_{3/2}$ proton. We may use (3.37) to obtain that magnetic moment as

$$\mu_{d_{3/2}} = g_1 \tfrac{3}{2} = \frac{3/2}{5/2}(3g_l^\pi - \tfrac{1}{2}g_s^\pi) = \tfrac{3}{5}(3 - 2.79) = .106 \text{ n.m.}$$

The magnetic moment of the $1f_{7/2}$ neutron, given by (2.1) is

$$\mu_{f_{7/2}} = g_2 \tfrac{7}{2} = 3g_l^\nu + \tfrac{1}{2}g_s^\nu = -1.91 \text{ n.m.}$$

Substituting the values of g_1 and g_2 from these magnetic moments in (8.27) we obtain for the $J = 4$ state the value $\mu = -1.69$ n.m. which agrees only roughly with the experimental value -1.298 n.m.

We may do better, however, if instead of the Schmidt values of the magnetic moments we use some measured values. For the magnetic moment of a $1d_{3/2}$ proton we may take the experimental moment of ^{39}K which is .392 n.m. Similarly, the experimental magnetic moment of ^{41}Ca, -1.595 n.m. may be adopted as the moment of a single $1f_{7/2}$ neutron. With the values of g_1 and g_2 determined from these measured moments, substituted into (8.27), the calculated magnetic moment of the ^{40}K ground state is -1.249 n.m. This value is in much better agreement with the experimental moment. In fact, measured magnetic moments of some K isotopes were used to *predict* fairly accurately the magnetic moment of ^{41}Ca before it was measured (Talmi and Unna 1960a).

Matrix elements of certain single nucleon operators determine the rates of electromagnetic (and other) transition probabilities. Such transitions are due to an operator $\mathbf{T}^{(k)}$ which acts on a state with given spin J_i (and other quantum numbers α_i) and leads to a state with J_f (and α_f). If the initial state has the projection M_i, the rate of the transition to a state with given M_f is proportional to

$$\sum_{\kappa} |\langle \alpha_f J_f M_f | T_{\kappa}^{(k)} | \alpha_i J_i M_i \rangle|^2 \tag{8.28}$$

The summation over κ is indicated according to our convention even though, due to the Wigner-Eckart theorem, the only non-vanishing term in (8.28) has $\kappa = M_f - M_i$. To obtain the transition rate to all M_f states we sum (8.28) over M_f. Using the Wigner-Eckart theorem (8.4) we obtain for the combined rate the result

$$\boxed{\begin{aligned}&(\alpha_f J_f \| \mathbf{T}^{(k)} \| \alpha_i J_i)^2 \sum_{\kappa M_f} \begin{pmatrix} J_f & k & J_i \\ -M_f & \kappa & M_i \end{pmatrix}^2 \\ &\quad = \frac{1}{2J_i + 1} (\alpha_f J_f \| \mathbf{T}^{(k)} \| \alpha_i J_i)^2\end{aligned}} \tag{8.29}$$

The rate (8.29) is independent of M_i and is proportional to the square of the reduced matrix element.

9

Two Nucleon Wave Functions. 9j-Symbols

States of nucleons moving independently in a central potential well have usually large degeneracies. In the case of closed shells there is a single state with total angular momentum $J = 0$. States of one nucleon outside closed shells have also well defined spins equal to that of the single nucleon. The same is true for states in which one nucleon is *missing* from closed shells. These are single *hole* states which will be considered later in detail. Whenever there are several nucleons outside closed shells there are several states which are allowed by the Pauli principle. If the Hamiltonian includes only the kinetic energies of nucleons and their potential energy in the central well, all such states are degenerate. A look at any nuclear level scheme clearly demonstrates that energy levels are far from degenerate. This should not be surprising since the central potential cannot replace exactly the mutual interactions of the nucleons.

Incorporation of residual interactions between nucleons easily removes this degeneracy. The form of these residual interactions strongly depends on the model used. In a naive approach, these are depicted as simple functions of the space coordinates and spin variables of the nucleons. In more sophisticated theories residual interactions are highly non-local operators (g-matrix) of which only matrix elements in a given configuration may be calculated. The most reliable way of determining these matrix elements is directly from experiment. In this section we will see how wave functions of *two* nucleons (outside closed shells) are determined by a general two-body interaction. In the next section we will show how to calculate matrix elements of some simple interactions. States of several nucleons outside closed shells *(valence* nucleons) will be considered in subsequent sections.

States of two independent nucleons in a central potential well can be constructed as product wave functions $\psi_{jm}(1)\psi_{j'm'}(2)$. The indices 1 and 2 stand for the space coordinates and spin variables of the two nucleons. The state $\psi_{jm}(2)\psi_{j'm'}(1)$ has the same energy in the central field. As is well known, this is not an independent state if the two nucleons are identical. The Pauli principle dictates that the wave function should be in that case *antisymmetric* under exchange of space coordinates and spin variables of the nucleons. A normalized such state is given by

$$\frac{1}{\sqrt{2}}(\psi_{jm}(1)\psi_{j'm'}(2) - \psi_{jm}(2)\psi_{j'm'}(1)) \tag{9.1}$$

If one nucleon is a proton and the other a neutron there is no *a priori* need to construct the state (9.1). It turns out, however, that also in this case it is very useful to construct states with definite symmetry. This is due to proton and neutron masses being nearly equal and to the *charge independence* of nuclear interactions. Apart from the Coulomb repulsion between protons, the nuclear interaction between two protons, two neutrons and a proton and neutron in the *same state* is the same. This is one of the strongest symmetries in nuclear physics. Consider a proton and a neutron outside the *same* closed shells of protons and neutrons. Apart from the Coulomb energy (and the small proton-neutron mass difference) the two states

$$\psi_{jm}(1_\pi)\psi_{j'm'}(2_\nu), \qquad \psi_{jm}(2_\nu)\psi_{j'm'}(1_\pi) \tag{9.2}$$

have the same energy in the central potential well. The subscripts π and ν indicate that the nucleon is a proton or neutron respectively. If now charge independent residual interactions are considered, the situation is modified. The diagonal elements of the interaction in the two states will be equal. There will be a non-vanishing matrix element of the interaction between these two states. Diagonalization of the 2×2 interaction matrix yields eigenstates which are linear combinations of the states (9.2). There is one linear combination in which the proton and neutron are in the same state as in the state of two identical nucleons. This is the antisymmetric state

$$\frac{1}{\sqrt{2}}(\psi_{jm}(1_\pi)\psi_{j'm'}(2_\nu) - \psi_{jm}(2_\nu)\psi_{j'm'}(1_\pi)) \tag{9.3}$$

Matrix elements of charge independent interactions between states (9.3) are the *same* as those between states (9.1).

The state orthogonal to (9.3) is *symmetric*

$$\frac{1}{\sqrt{2}}(\psi_{jm}(1_\pi)\psi_{j'm'}(2_\nu) + \psi_{jm}(2_\nu)\psi_{j'm'}(1_\pi)) \tag{9.4}$$

Two identical nucleons cannot be in a state like (9.4). Hence, matrix elements between states (9.4) need not be equal to those between the states (9.1) or (9.3). Any charge independent interaction has vanishing matrix elements between the states (9.3) and (9.4).

Due to charge independence of the strong nuclear interaction and the small proton-neutron mass difference, it is convenient to consider the proton and neutron as two states of the same particle—the *nucleon*. The two states of the nucleon differ by the electric charge $+e$ of the proton and 0 of the neutron. The other differences, like in magnetic moments, do not affect the eigenstates of nucleons which are determined by the strong nuclear interaction. There is an elegant way to describe this situation which is based on the properties of the Pauli spin matrices (3.17). We introduce an abstract two-dimensional charge space in which proton and neutron states are described by the columns

$$\eta_{1/2} = \begin{pmatrix} 1 \\ 0 \end{pmatrix} \quad \text{for a proton} \qquad \eta_{-1/2} = \begin{pmatrix} 0 \\ 1 \end{pmatrix} \quad \text{for a neutron} \tag{9.5}$$

The states (9.5) are considered as two-component spinors and in analogy with ordinary spin we introduce the three matrices

$$\boxed{\tau_1 = \begin{pmatrix} 0 & 1 \\ 1 & 0 \end{pmatrix} \qquad \tau_2 = \begin{pmatrix} 0 & -i \\ i & 0 \end{pmatrix} \qquad \tau_3 = \begin{pmatrix} 1 & 0 \\ 0 & -1 \end{pmatrix}} \tag{9.6}$$

These matrices, divided by 2, obey the commutation relations (6.40) of angular momentum. They are called *isospin* matrices (short for isotopic spin or isobaric spin). The indices 1,2,3 emphasize the fact that the isospin vector $\mathbf{t} = \frac{1}{2}\boldsymbol{\tau}$ is not a vector in ordinary space. The proton state belongs to the eigenvalue $+1$ of τ_3 and the neutron state to -1. The charge of a nucleon is given accordingly by $\frac{1}{2}(1 + \tau_3)$. This choice of signs is used in elementary particle theory. It was adopted in the books by de-Shalit and Talmi (1962) and de-Shalit and Feshbach (1974). In many text books on nuclear physics the definition is the opposite one. This does not change, of course, the results obtained.

We should then consider the isospin variable as another variable of the nucleon and make use of the isospin formalism to construct states which are symmetric or antisymmetric. Due to (9.6), we can apply the formalism developed above for angular momentum to the isospin vectors and states.

As will be discussed in detail in the following sections, a charge independent Hamiltonian is a scalar in the total isospin $\mathbf{T}$ of the nucleons. Eigenstates of such Hamiltonians are characterized by the eigenvalues of $\mathbf{T}^2$ which are $T(T+1)$. Eigenstates with given value of T and different values of M_T have equal eigenvalues since they differ only by the relative numbers of protons and neutrons ($M_T = \frac{1}{2}(Z - N)$).

The isospins of two nucleons $\mathbf{t}^{(1)} = \frac{1}{2}\boldsymbol{\tau}^{(1)}$ and $\mathbf{t}^{(2)} = \frac{1}{2}\boldsymbol{\tau}^{(2)}$ can couple to a total isospin $\mathbf{T}$. The eigenvalues of $\mathbf{T}^2$ in this case are $T(T+1)$ with $T = 1$ or $T = 0$. The isospin states with $T = 1$ are symmetric according to (7.12) whereas the states with $T = 0$ are antisymmetric under exchange of the charges. Indeed, the isospin state of two protons is $\eta_{1/2}(1)\eta_{1/2}(2)$ which is the $M_T = 1$ component of the $T = 1$ state. The state with $T = 1$, $M_T = -1$ is the isospin state of two neutrons. The $T = 1$, $M_T = 0$ state is a state symmetric in the charge variables of a proton and a neutron. This state must correspond to the antisymmetric state (9.3) which is the same space and spin state as (9.1). The symmetric state (9.4) corresponds to the isospin state with $M_T = 0$ but it has $T = 0$. That latter state has no corresponding state of two protons or two neutrons.

The use of isospin variables thus introduces a generalized Pauli principle. The states of two nucleons must be fully antisymmetric in the space coordinates, spin variables and isospin (or charge) variables. The states (9.1) and (9.3) should now be expressed as

$$\frac{1}{\sqrt{2}}[\psi_{jm}(1)\psi_{j'm'}(2) - \psi_{jm}(1)\psi_{j'm'}(2)]\eta_{T=1,M_T} \qquad M_T = 1,0,-1 \tag{9.7}$$

and (9.4) as

$$\frac{1}{\sqrt{2}}[\psi_{jm}(1)\psi_{j'm'}(2)+\psi_{jm}(1)\psi_{j'm'}(2)]\eta_{T=0,M_T=0} \tag{9.8}$$

The isospin states in (9.7) and (9.8) are given in terms of the appropriate Clebsch-Gordan coefficients as

$$\eta_{T=1,M_T=1}=\eta_{1/2}(1)\eta_{1/2}(2) \qquad \eta_{T=1,M_T=-1}=\eta_{-1/2}(1)\eta_{-1/2}(2)$$
$$\eta_{T=1,M_T=0}=\frac{1}{\sqrt{2}}(\eta_{1/2}(1)\eta_{-1/2}(2)+\eta_{1/2}(2)\eta_{-1/2}(1)) \tag{9.9}$$

and

$$\eta_{T=0,M_T=0}=\frac{1}{\sqrt{2}}(\eta_{1/2}(1)\eta_{-1/2}(2)-\eta_{1/2}(2)\eta_{-1/2}(1)) \tag{9.10}$$

A simple demonstration of charge independence of the nuclear Hamiltonian and usefulness of isospin, is shown in Fig. 9.1. Corresponding states of two holes in the closed shells of ^{16}O with $T=1$ and $M_T=1,0,-1$ are shown in spectra of ${}^{14}_{8}O_6$, ${}^{14}_{7}N_7$ and ${}^{14}_{6}C_8$ nuclei. Some average Coulomb energies of the protons were subtracted from energies of these states. This way, the fairly equal spacings between $T=1$ levels in the three nuclei are displayed. Coulomb energies depend on the orbits occupied by the protons and hence, some levels are shifted more than others by the electrostatic repulsion.

The product wave functions as well as their symmetrized or antisymmetrized versions (9.8) and (9.7) are eigenstates of the Hamiltonian which includes only the central potential well. They are not eigenstates of the residual interaction. That interaction is invariant under rotations and so is the Hamiltonian which includes it in addition to kinetic energies and the central potential well. We take the residual interaction as a perturbation added to H_0 which is the Hamiltonian with only kinetic energies and a central potential well. In any given *configuration* the occupation numbers of proton and neutron orbits are given. Hence, all states of a configuration have the same eigenvalue of H_0. In such a subspace of the nuclear Hamiltonian we should choose a scheme of states in which the perturbation is diagonal. The first order corrections due to the perturbation are then its expectation values in that scheme. The scheme in which a rotationally invariant residual interaction is diagonal is a set of states which have definite

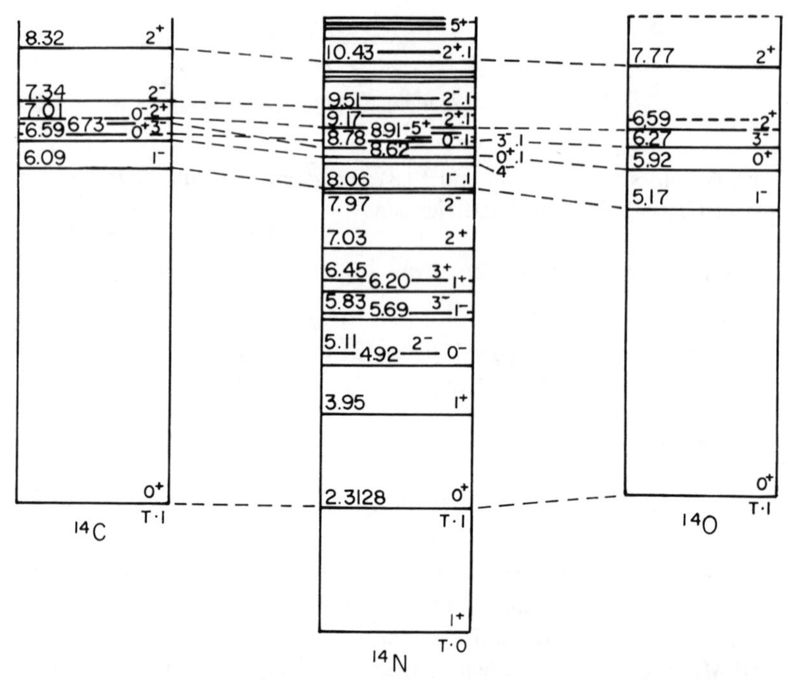

FIGURE 9.1. *Corresponding isospin $T = 1$ levels in $A = 14$ nuclei (in MeV).*

eigenvalues of $\mathbf{J}^2$. The eigenvalues $J(J+1)$ of the total angular momentum together with the isospin T, uniquely characterize the states of two nucleons. For several valence nucleons the situation is more complex and will be discussed in subsequent sections.

In the two nucleon jj' configuration, states with good angular momentum can be constructed from the states (9.7) and (9.8) as in (7.1)

$$\Psi_{jj'TM_TJM}(1,2) = \frac{1}{\sqrt{2}}\sum_{mm'}(jmj'm' \mid jj'JM) \times [\psi_{jm}(1)\psi_{j'm'}(2) + (-1)^T\psi_{jm}(2)\psi_{j'm'}(1)]\eta_{T,M_T} \tag{9.11}$$

The normalization factor $2^{-1/2}$ in (9.11) is due to the normalization of the Clebsch-Gordan coefficients and the fact that for $j' \neq j$ the two wave functions $\psi_{jm}(1)\psi_{j'm'}(2)$ and $\psi_{jm}(2)\psi_{j'm'}(1)$ are orthogonal. This is true also if $j' = j$ but $l' \neq l$ or even if only $n' \neq n$. The special case of nucleons in the *same* orbit will be discussed below. The set of states (9.11) diagonalise the residual interaction in the subspace of the nuclear Hamiltonian defined by the jj' configuration. There are *two* states (9.11) with given J ($|j - j'| \geq J \geq j + j'$) and M in that configuration, one with $T = 1$ and the other with $T = 0$. They may also be obtained from the product wave functions (9.2) by using vector addition coefficients. In that case, the 2×2 sub-matrix of the residual interaction, for given J and M, should be diagonalized. Using the isospin symmetry of the charge independent residual interaction, the resulting two eigenstates can be directly obtained, without any diagonalization, in terms of the isospin quantum number T as in (9.11).

Some care must be taken in the case of two nucleons in the *same* j-orbit. If $j = j'$ the vector addition coefficients have the symmetry properties (7.12). As long as $l \neq l'$ or even if $l = l'$ but $n \neq n'$, the expression (9.11) is valid for any value of J. The symmetric and antisymmetric combinations are then just those of the radial functions $(1/\sqrt{2})[R_{nl}(r_1)R_{n'l'}(r_2) \pm R_{nl}(r_2)R_{n'l'}(r_1)]$. If, however, $n' = n$, $l' = l$, $j' = j$ then for even values of J only the antisymmetric combination with ($T = 1$) does not vanish. For odd values of J only the symmetric combination (with $T = 0$) does not vanish. In both cases the normalization of the state is achieved by taking only the first term in the square brackets in (9.11) and omitting the $\sqrt{2}$ in the denominator.

The usefulness of isospin in the jj' configuration is due to the fact that states in which the proton is in the j-orbit and the neutron in the j'-orbit, or vice versa, are not good eigenstates of the proton neutron interaction. The latter admixes the states (9.2) to yield the states (9.3) and (9.4) or (9.11) in the isospin formalism. There are, however, cases in which it is meaningful to state that the proton is in the j-orbit and the neutron in the j'-orbit. This is the case, rather common among heavier stable nuclei, where the neutron j-orbit is completely filled. In such a case, isospin symmetry is not very useful and we may use for the j-proton j'-neutron configuration the states

$$\sum_{mm'} (jmj'm' \mid jj'JM)\psi_{jm}(1_\pi)\psi_{j'm'}(2_\nu) \tag{9.12}$$

The states (9.12) are fully characterized by J (and M). They correspond to many nucleon states with good isospin in which, in addition to the j-proton, all the $2j+1$ neutrons in the j-orbit must be considered. It is therefore more convenient to use the simpler prescription (9.12). The justification for (9.12) will be given in a subsequent section.

Isospin was introduced above, representing a symmetry of the nuclear Hamiltonian which is a consequence of charge independence. It is very useful but its use is not mandatory. As explained above, symmetric and antisymmetric states of two nucleons may be obtained by diagonalization of the matrix of a charge independent Hamiltonian. This is also the case in many nucleon configurations. There are, however, processes in nuclei for which the use of isospin is not just convenient but is necessary. These are due to weak interactions, like β-decay, in which a neutron is changed into a proton (with the emission of an electron and antineutrino) or a proton is changed into a neutron (with the emission of a positron and neutrino). Such transitions may be described by using the operator $t^+ = t_1 + it_2$ which acting on the isospin states (9.5) changes a neutron state into a proton state. The change of a proton state into a state of a neutron can be obtained by acting with the operator $t^- = t_1 - it_2$.

The theory of β-decay will not be reviewed here. A detailed description may be found in de-Shalit and Feshbach (1974). We limit the discussion to cases in which the energy of the emitted leptons is small and nucleons may be treated non-relativistically. The matrix element for the transition becomes then equal to the product of a matrix element of a simple operator between initial and final states of the nucleus and a certain function of the energy released in the decay. The simplest nuclear operators, giving rise to *allowed* transitions do not depend on the nucleon coordinates.

The simplest operator, in the case of the transition of a neutron to a proton is

$$\boxed{\sum_i t^+(i) = \frac{1}{2}\sum_i \tau^+(i) = T^+} \tag{9.13}$$

The operator (9.13) gives rise to *Fermi transitions*. It is a component of the isospin vector $\mathbf{T}$ and hence, does not change the state of the nucleus. The selection rules for such transitions are $\Delta J = 0$ and $\Delta T = 0$, only the value of $M_T = \frac{1}{2}(Z - N)$ increases by one. The transition probability is proportional to the square of the matrix element (9.13) which is equal to

$$T^-T^+ = \mathbf{T}^2 - (T^0)^2 - T^0 = T(T+1) - M_T^{(i)}(M_T^{(i)}+1) = T(T+1) - M_T^{(i)}M_T^{(f)} \tag{9.14}$$

The superscripts i and f refer to the initial and final states. The result (9.14) is independent of the detailed structure of the states and is determined by the value of the isospin T and its projection $T_3 = T^0$.

In other allowed transitions the spatial states of the nucleons are not changed but their intrinsic spins may participate. The operator for *Gamow-Teller transitions* is defined by

$$2\sum_i t^+(i)\mathbf{s}(i) = \frac{1}{2}\sum_i \tau^+(i)\boldsymbol{\sigma}(i) \tag{9.15}$$

The operator (9.15) is a vector in ordinary space and may change the angular momentum of the nucleus by one unit, $\Delta J = \pm 1, 0$. Due to the Wigner-Eckart theorem it has vanishing matrix elements between two states with $J_i = J_f = 0$. The operator (9.15) is no longer proportional to a component of the total isospin $\mathbf{T}$ and hence, may change the isospin, $\Delta T = \pm 1, 0$. Rates of Gamow-Teller transitions cannot be expressed in a simple general formula. In states of a single proton or single neutron outside closed shells the rate is proportional to

$$\langle\sigma\rangle^2 = \frac{1}{4}\sum_{\kappa m_f} |\langle T = \tfrac{1}{2}, M_{T_f} = \tfrac{1}{2}, j_f m_f|\tau^+\sigma_\kappa|T = \tfrac{1}{2}, M_{T_i} = -\tfrac{1}{2}, j_i m_i\rangle|^2 = \sum_{\kappa m_f} |\langle j_f m_f|\sigma_\kappa|j_i m_i\rangle|^2 = \frac{1}{2j_i+1}|(j_f\|\boldsymbol{\sigma}\|j_i)|^2 \tag{9.16}$$

The last equality is due to the Wigner-Eckart theorem according to (8.29). Both spins of the initial and final states j_i and j_f, must belong the same l-orbit and the j-values are then equal to $l \pm \frac{1}{2}$.

For states of many nucleons there is a simple sum rule for Gamow-Teller transitions (Gaarde et al. 1980). It involves matrix elements of (9.15) and of its hermitean conjugate between the initial nuclear state and *all* possible final states. Most of these transitions cannot take place by β-decay due to energy considerations but the corresponding final states may sometimes be reached by other processes. We consider the difference

$$\frac{1}{4}\sum_f \left|\langle f|\sum_k \tau^+(k)\boldsymbol{\sigma}(k)|i\rangle\right|^2 - \frac{1}{4}\sum_f \left|\langle f|\sum_k \tau^-(k)\boldsymbol{\sigma}(k)|i\rangle\right|^2 \tag{9.17}$$

The expression (9.17) may be written as

$$\begin{aligned}
&\frac{1}{4}\sum_f \langle i|\sum_k \tau^-(k)\boldsymbol{\sigma}(k)|f\rangle \cdot \langle f|\sum_j \tau^+(j)\boldsymbol{\sigma}(j)|i\rangle \\
&\qquad - \frac{1}{4}\sum_f \langle i|\sum_k \tau^+(k)\boldsymbol{\sigma}(k)|f\rangle \cdot \langle f|\sum_j \tau^-(j)\boldsymbol{\sigma}(j)|i\rangle \\
&\quad = \frac{1}{4}\langle i|\sum_{j,k} \tau^-(k)\tau^+(j)\boldsymbol{\sigma}(k)\cdot\boldsymbol{\sigma}(j)|i\rangle \\
&\qquad - \frac{1}{4}\langle i|\sum_{j,k} \tau^+(k)\tau^-(j)\boldsymbol{\sigma}(k)\cdot\boldsymbol{\sigma}(j)|i\rangle
\end{aligned}$$

The operators $\tau^+(j)$ and $\tau^-(k)$ for different nucleons commute and hence, the $j \neq k$ terms are cancelled in the difference. The remaining terms, with $j = k$, are equal to

$$\boxed{\begin{aligned}
&\frac{1}{4}\langle i|\sum_k (\tau^-(k)\tau^+(k) - \tau^+(k)\tau^-(k))\boldsymbol{\sigma}(k)\cdot\boldsymbol{\sigma}(k)|i\rangle \\
&\qquad = -3\langle i|\sum_h \tau_3(k)|i\rangle = -3\,\langle i|T_3|i\rangle = 3(N - Z)
\end{aligned}} \tag{9.18}$$

In obtaining (9.18) we used the equality $\boldsymbol{\sigma} \cdot \boldsymbol{\sigma} = \sum_i \sigma_i^2 = 3$.

Let us now return to two nucleon wave functions. The scheme (9.11) or (9.12) in which the total spins of the nucleons j and j' have definite values and are coupled to a resultant spin J is called *jj-coupling*. The wave functions ψ_{jm} and $\psi_{j'm'}$ were constructed according to (3.30) and (3.31) by coupling the intrinsic spins $\mathbf{s}$ and $\mathbf{s}'$ to the orbital angular momenta $\boldsymbol{l}$ and $\boldsymbol{l}'$ respectively. In this scheme, the spin-orbit interaction is diagonal. Sometimes, for actual calculations, it is more convenient to express the states (9.11) and (9.12) in another coupling scheme. In that scheme, the spins $\mathbf{s}$ and $\mathbf{s}'$ are first coupled to form a state with definite $\mathbf{S}^2 = (\mathbf{s} + \mathbf{s}')^2$ and the orbital angular momenta $\boldsymbol{l}$ and $\boldsymbol{l}'$ are coupled to form a state with definite $\mathbf{L}^2 = (\boldsymbol{l} + \boldsymbol{l}')^2$. The spin state with definite S is then coupled to the state of spatial coordinates with definite L to form a state with definite total J. This scheme is called *Russel-Saunders scheme* or *LS-coupling*. In the absence of a strong spin orbit interaction, or in the presence of certain strong mutual interactions, orbits are characterized by their l values and LS-coupling is a more convenient starting point for calculations. This is the case in most *atomic spectra*. It is somewhat confusing that a strong spin-orbit interaction is sometimes referred to as "LS-coupling" which breaks the LS-coupling scheme leading to jj-coupling.

The spin states of two nucleons are obtained by using Clebsch-Gordan coefficients

$$\chi_{SM_S} = \sum_{m_s m_s'} (\tfrac{1}{2} m_s \tfrac{1}{2} m_s' \mid \tfrac{1}{2} \tfrac{1}{2} S M_S) \chi_{m_s}(1) \chi_{m_s'}(2) \tag{9.19}$$

According to (7.12) such states are symmetric for $S = 1$ and antisymmetric if $S = 0$. The coupling of $\boldsymbol{l}$ and $\boldsymbol{l}'$ to $\mathbf{L}$ is obtained by the expression

$$\Phi_{ll'LM_L}(1,2) = \frac{1}{\sqrt{2}} \sum (l m_l l' m_l' \mid l l' L M_L) \times [\phi_{lm_l}(\mathbf{r}_1)\phi_{l'm_l'}(\mathbf{r}_2) \pm \phi_{lm_l}(\mathbf{r}_2)\phi_{l'm_l'}(\mathbf{r}_1)] \tag{9.20}$$

Both symmetric and antisymmetric combinations are given in (9.20) since they should be multiplied by functions of spin variables and isospin variables so that the total wave function will be antisymmetric. The case where both nucleons are in the *same* l-orbit is a special case. Due to (7.12) states with even values of L are symmetric and those with odd L values are antisymmetric. Hence, in both cases only one combination in (9.20) does not vanish and its normalization is achieved by taking only the first term in the square brackets omitting the factor $\sqrt{2}$ in the denominator. Finally, states with definite values of J (and M) are obtained by coupling $\mathbf{S}$ and $\mathbf{L}$ as follows

$$\Phi_{ll'SLJM} = \sum_{M_S M_L} (SM_S LM_L \mid SLJM)\chi_{SM_S}\Phi_{ll'LM_L} \tag{9.21}$$

As we saw above, the states (9.21) with $S = 1$ and plus sign in (9.20) should be multiplied by antisymmetric $T = 0$ isospin states to ensure full antisymmetry. States (9.21) with $S = 1$ and a minus sign in (9.20) should be multiplied by the $T = 1$ states. States with $S = 0$ and plus sign should be multiplied by the $T = 1$ states whereas states with $S = 0$ and minus sign should be multiplied by $T = 0$ isospin states. Remembering these rules, to simplify the notation we will not carry isospin states along. Whereas $\mathbf{s},\mathbf{s}',\boldsymbol{l},\boldsymbol{l}'$ may be coupled in different schemes to produce states with total $\mathbf{J}$, the isospin $\mathbf{T}$ is not an angular momentum in three-dimensional space and is not coupled to any other vector.

The set of states (9.21) is an orthogonal and complete set of symmetric and antisymmetric states in the space coordinates and spin variables of the two nucleons in the lj and $l'j'$ orbits. The same is true of the set of jj-coupling states in (9.11) namely,

$$\Psi_{ll'jj'JM}(1,2) = \frac{1}{\sqrt{2}}\sum_{mm'}(jmj'm' \mid jj'JM) \times [\psi_{jm}(1)\psi_{j'm'}(2) \pm \psi_{jm}(2)\psi_{j'm'}(1)] \tag{9.22}$$

Both sets are normalized and hence, there is a unitary transformation which carries one set into the other. The transformation carrying symmetric states in one scheme to symmetric ones in the other is the same as that between antisymmetric states and is given by

$$\boxed{\Psi_{ll'jj'JM} = \sum_{S,L} \langle \tfrac{1}{2}l(j)\tfrac{1}{2}l'(j')J \mid \tfrac{1}{2}\tfrac{1}{2}(S)ll'(L)J\rangle \Psi_{ll'SLJM}} \tag{9.23}$$

The coefficients of this transformation are independent of M since states on both sides of (9.23) transform in the same way under rotations. Operating with J^+ or J^- on both sides of (9.17) clearly demonstrates this point. The transformation (9.23) between LS-coupling and jj-coupling is between two schemes in which linear combinations of the product wave functions $\chi_{m_s}(1)\chi_{m_s'}(2)\phi_{lm_l}(1)\phi_{l'm_l'}(2)$ are taken with Clebsch-Gordan coefficients. Since those are real so are the coefficients of the transformation (9.23). In the following these coefficients will be studied in detail.

It is customary to denote LS-coupling wave functions by ${}^{2S+1}L_J$ ($2S+1$ is called the multiplicity). Since S can be equal to 1 or 0, L may be equal to $J, J-1$ or $J+1$. If the state (9.22) is antisymmetric then on the r.h.s. of (9.23), the state with $S=0$, $L=J$ has a spatial function (9.20) which is symmetric and in the states with $S=1$, $L=J-1, J, J+1$ the spatial function is antisymmetric. On the other hand, if the state (9.22) is symmetric then the $S=0$, $L=J$ state in (9.23) has antisymmetric spatial function whereas the $S=1$ states with $L=J-1, J, J+1$ have symmetric spatial functions. The cases in which the nucleons are in the *same* l-orbit are special. The states (9.20) with $l=l'$ and the *same* radial functions are, according to (7.12), symmetric for even values of L and antisymmetric for odd values. In that case, for antisymmetric states (9.22), with J even, states with $S=0$ on the r.h.s. of (9.23) must have even values of L, i.e. $L=J$. On the other hand, $S=1$ states can have only odd values of L, namely $L=J-1$, $J+1$. For symmetric states (9.22) with J odd, $S=1$ states must have even values of L which may be $L=J-1$, $J+1$. On the other hand, $S=0$ states must have odd values of L and $L=J$ is the only possibility.

The transformation coefficients in (9.23) are listed in Table 9.1. In the special cases of nucleons in the same orbit, $n'=n$, $l'=l$ the normalization of the coefficients is different and attention must be paid to it. If $j'=j=l\pm\frac{1}{2}$ both sides of (9.23) are normalized by the same factor. If, however, $j=l\pm\frac{1}{2}$ and $j'=l\mp\frac{1}{2}$, the coefficients of the transformation should be multiplied by $\sqrt{2}$. In that case, for even values of J, the ${}^3J$ state is fully symmetric whereas the ${}^3(J+1), {}^3(J-1)$ and ${}^1J$ are antisymmetric. For odd values of J the situation is reversed. In either case, the $j=l+\frac{1}{2}$, $j'=l-\frac{1}{2}$ states are given by the coefficients in *either* the first 3 columns *or* the fourth column of Table 9.1 according to their symmetry.

To illustrate the use of the transformation from LS-coupling to jj-coupling, we consider a few simple examples. Consider one nucleon in the $p_{3/2}$ orbit and another in the $d_{5/2}$ orbit. To find the expansion of

TABLE 9.1 *Transformation coefficients between jj-coupling and LS-coupling schemes.* $\Sigma = l + l' + 1$, $\Delta = l - l'$. *For* $l = l'$, $n = n'$, *the second row should be multiplied by* $\sqrt{2}$ *and the third row ignored.*

LS-coupling / jj-coupling	$^3(J+1)$ $S=1, L=J+1$	^1J $S=0, L=J$	$^3(J-1)$ $S=1, L=J-1$	^3J $S=1, L=J$
$j = l + \frac{1}{2}\ j' = l' + \frac{1}{2}$	$-\sqrt{\frac{((J+1)^2-\Delta^2)(\Sigma-J)(\Sigma-J-1)}{2(\Sigma^2-\Delta^2)(J+1)(2J+1)}}$	$\sqrt{\frac{(\Sigma-J)(\Sigma+J+1)}{2(\Sigma^2-\Delta^2)}}$	$\sqrt{\frac{(J^2-\Delta^2)(\Sigma+J)(\Sigma+J+1)}{2(\Sigma^2-\Delta^2)J(2J+1)}}$	$-\Delta\sqrt{\frac{(\Sigma-J)(\Sigma+J+1)}{2(\Sigma^2-\Delta^2)J(J+1)}}$
$j = l + \frac{1}{2}\ j' = l' - \frac{1}{2}$	$\sqrt{\frac{\left(\Sigma^2-(J+1)^2\right)(J-\Delta)(J-\Delta+1)}{2(\Sigma^2-\Delta^2)(J+1)(2J+1)}}$	$-\sqrt{\frac{(J-\Delta)(J+\Delta+1)}{2(\Sigma^2-\Delta^2)}}$	$\sqrt{\frac{\left(\Sigma^2-J^2\right)(J+\Delta)(J+\Delta+1)}{2(\Sigma^2-\Delta^2)J(2J+1)}}$	$\Sigma\sqrt{\frac{(J-\Delta)(J+\Delta+1)}{2(\Sigma^2-\Delta^2)J(J+1)}}$
$j = l - \frac{1}{2}\ j' = l' + \frac{1}{2}$	$-\sqrt{\frac{\left(\Sigma^2-(J+1)^2\right)(J+\Delta)(J+\Delta+1)}{2(\Sigma^2-\Delta^2)(J+1)(2J+1)}}$	$\sqrt{\frac{(J+\Delta)(J-\Delta+1)}{2(\Sigma^2-\Delta^2)}}$	$-\sqrt{\frac{\left(\Sigma^2-J^2\right)(J-\Delta)(J-\Delta+1)}{2(\Sigma^2-\Delta^2)J(2J+1)}}$	$\Sigma\sqrt{\frac{(J+\Delta)(J-\Delta+1)}{2(\Sigma^2-\Delta^2)J(J+1)}}$
$j = l - \frac{1}{2}\ j' = l' - \frac{1}{2}$	$\sqrt{\frac{((J+1)^2-\Delta^2)(\Sigma+J)(\Sigma+J+1)}{2(\Sigma^2-\Delta^2)(J+1)(2J+1)}}$	$\sqrt{\frac{(\Sigma+J)(\Sigma-J-1)}{2(\Sigma^2-\Delta^2)}}$	$-\sqrt{\frac{(J^2-\Delta^2)(\Sigma-J)(\Sigma-J-1)}{2(\Sigma^2-\Delta^2)J(2J+1)}}$	$\Delta\sqrt{\frac{(\Sigma+J)(\Sigma-J-1)}{2(\Sigma^2-\Delta^2)J(J+1)}}$

the state in which the two nucleons are coupled to a state with $J = 2$, we look at the top row of Table 9.1. In this case, $\Sigma = 1 + 2 + 1 = 4$, $\Delta = 1 - 2 = -1$, and the LS-coupling states which appear in the expansion have $S = 0$, $L = 2$ and $S = 1$, $L = 3,2,1$. It is customary to denote states with $L = 0,1,2,3,4,\ldots$ by $S,P,D,F,G,\ldots$. Using this notation, the transformation is given by

$$|p_{3/2}d_{5/2}J = 2\rangle = -\frac{2\sqrt{2}}{15}\,{}^3F_2 + \sqrt{\frac{7}{15}}\,{}^1D_2 + \frac{\sqrt{42}}{10}\,{}^3P_2 + \frac{1}{6}\sqrt{\frac{14}{5}}\,{}^3D_2$$

If the $J = 2$ state has isospin $T = 1$, it is antisymmetric in space and spin coordinates. Hence, in the ${}^3F$, ${}^3P$ and ${}^3D$ states, the spatial parts are antisymmetric whereas the ${}^1D$ state is space symmetric. If, however, the $J = 2$ state has $T = 0$, the ${}^3F$, ${}^3P$ and ${}^3D$ have spatially symmetric wave functions and the ${}^1D$ state is space antisymmetric.

If both nucleons are in a $p_{3/2}$ orbit and are coupled to a state with $J = 1$ we obtain from the top row of Table 9.1, by putting $\Sigma = 3$, $\Delta = 0$, the following expansion in LS-coupling wave functions

$$|p^2_{3/2}J = 1\rangle = -\frac{1}{3}\sqrt{\frac{2}{3}}\,{}^3D_1 + \frac{\sqrt{5}}{3}\,{}^1P_1 + \frac{1}{3}\sqrt{\frac{10}{3}}\,{}^3S_1$$

The coefficient of ${}^3P$ vanishes in this case since that state has an antisymmetric space and spin wave function whereas the $p^2_{3/2}J = 1$ state is *symmetric* (it has $T = 0$). Let us now consider one nucleon is in a $p_{3/2}$ orbit and the other in the $p_{1/2}$ orbit with the *same* principal quantum number. The expansion in LS-coupling states is found in the second row with the coefficients multiplied by $\sqrt{2}$. States with given value of J may be either symmetric with $T = 0$ or antisymmetric with $T = 1$. The $T = 0$ symmetric state with $J = 1$ is equal to the following linear combination of the symmetric states ${}^3D_1$, ${}^1P_1$ and ${}^3S_1$

$$|p_{3/2}p_{1/2}T = 0, J = 1\rangle = \frac{1}{3}\sqrt{\frac{5}{3}}\,{}^3D_1 - \frac{\sqrt{2}}{3}\,{}^1P_1 + \frac{4}{3\sqrt{3}}\,{}^3S_1$$

The antisymmetric state with $T = 1$ is equal to the only antisymmetric LS-coupling state with $J = 1$ in the p^2 configuration

$$|p_{3/2}p_{1/2}T = 1, J = 1\rangle = {}^3P_1.$$

The transformation (9.23) is a special case of a *change of coupling* transformation. Starting with four angular momenta $j_1 j_2 j_3 j_4$ of independent systems we may couple them to a total J proceeding in several ways. We may couple first j_1 and j_2 to J_{12}, j_3 and j_4 to J_{34} and then couple J_{12} and J_{34} to J. Another equivalent scheme which is equally orthogonal and complete is obtained by coupling j_1 and j_3 to J_{13}, j_2 and j_4 to J_{24} and then J_{13} and J_{24} to J. There is a unitary transformation between these two schemes, namely

$$\boxed{\begin{aligned}&\psi(j_1 j_3(J_{13}) j_2 j_4(J_{24}) JM)\\ &\quad = \sum_{J_{12}J_{34}} \langle j_1 j_3(J_{13}) j_2 j_4(J_{24}) J \mid j_1 j_2(J_{12}) j_3 j_4(J_{24}) J\rangle\\ &\quad\quad \times \psi(j_1 j_2(J_{12}) j_3 j_4(J_{34}) JM)\end{aligned}} \tag{9.24}$$

To obtain an expression for the transformation coefficients we expand the functions on both sides of (9.24) using vector addition coefficients. The functions on the r.h.s. are expressed as follows

$$\begin{aligned}&\psi(j_1 j_2(J_{12}) j_3 j_4(J_{34}) JM)\\ &\quad = \sum (j_1 m_1 j_2 m_2 \mid j_1 j_2 J_{12} M_{12})(j_3 m_3 j_4 m_4 \mid j_3 j_4 J_{34} M_{34})\\ &\quad\quad \times (J_{12} M_{12} J_{34} M_{34} \mid J_{12} J_{34} JM) \psi_{j_1 m_1}(1) \psi_{j_2 m_2}(2) \psi_{j_3 m_3}(3) \psi_{j_4 m_4}(4)\end{aligned} \tag{9.25}$$

and a similar expansion is used for the function on the l.h.s. of (9.24). Equating on both sides of (9.24) the coefficients of $\psi_{j_1 m_1}(1)\psi_{j_2 m_2}(2)$ $\psi_{j_3 m_3}(3)\psi_{j_4 m_4}(4)$ we obtain a relation in which products of three $3j$-symbols on the l.h.s. are summed over M_{13}, M_{24} and on the r.h.s. products of the transformation coefficients and three $3j$-symbols summed over M_{12}, M_{34}. We now multiply both sides of that equation by three $3j$-symbols

$$\begin{pmatrix} j_1 & j_2 & J'_{12} \\ m_1 & m_2 & -M'_{12} \end{pmatrix}, \begin{pmatrix} j_3 & j_4 & J'_{34} \\ m_3 & m_4 & -M'_{34} \end{pmatrix} \text{ and } \begin{pmatrix} J_{12} & J_{34} & J' \\ M_{12} & M_{34} & -M' \end{pmatrix}$$

and sum over m_1, m_2, m_3, m_4 obtaining due to (7.25) $J'_{12} = J_{12}$, $M'_{12} = M_{12}$, $J'_{34} = J_{34}$, $M'_{34} = M_{34}$. We then sum over $M_{12}M_{34}$ obtaining $J' = J$, $M' = M$. Thus, using the orthogonality relation (7.25) we obtain the coefficients as sums of products of six $3j$-symbols. Such sums occur in any change-of-coupling transformations and are called $9j$-symbols. Indeed there are 9 spins involved in (9.17) namely, $j_1 j_2 j_3 j_4 J_{12} J_{34} J_{13} J_{24}$ and J. The definition of the $9j$-symbol is

$$\begin{Bmatrix} J_{11} & J_{12} & J_{13} \\ J_{21} & J_{22} & J_{23} \\ J_{31} & J_{32} & J_{33} \end{Bmatrix} = \sum_{\text{all } M_{ij}} \begin{pmatrix} J_{11} & J_{12} & J_{13} \\ M_{11} & M_{12} & M_{13} \end{pmatrix} \begin{pmatrix} J_{21} & J_{22} & J_{23} \\ M_{21} & M_{22} & M_{23} \end{pmatrix} \times \begin{pmatrix} J_{31} & J_{32} & J_{33} \\ M_{31} & M_{32} & M_{33} \end{pmatrix} \begin{pmatrix} J_{11} & J_{21} & J_{31} \\ M_{11} & M_{21} & M_{31} \end{pmatrix} \times \begin{pmatrix} J_{12} & J_{22} & J_{32} \\ M_{12} & M_{22} & M_{32} \end{pmatrix} \begin{pmatrix} J_{13} & J_{23} & J_{33} \\ M_{13} & M_{23} & M_{33} \end{pmatrix} \tag{9.26}$$

In terms of $9j$-symbols, the transformation coefficients in (9.24) are given by

$$\langle J_1 J_3 (J_{13}) J_2 J_4 (J_{24}) J \mid J_1 J_2 (J_{12}) J_3 J_4 (J_{34}) J \rangle = [(2J_{13}+1)(2J_{24}+1)(2J_{12}+1)(2J_{34}+1)]^{1/2} \times \begin{Bmatrix} J_1 & J_2 & J_{12} \\ J_3 & J_4 & J_{34} \\ J_{13} & J_{24} & J \end{Bmatrix} \tag{9.27}$$

The transformation coefficients (9.23) between jj-coupling and LS-coupling wave functions are obtained from (9.27) by putting $J_1 = J_3 = \frac{1}{2}$, $J_2 = l$, $J_4 = l'$, $J_{13} = S$, $J_{24} = L$, $J_{12} = j$ and $J_{34} = j'$. The $9j$-symbol was introduced by Wigner (1940) and independently by Schwinger (1952) and Arima et al. (1954).

The properties of the $9j$-symbols follow from their definition (9.26). The first thing to notice is that the spins in each row and in each column must satisfy the triangular condition (they may be coupled to total spin zero). Interchanging rows and columns does not change the r.h.s. of (9.26) and the $9j$-symbol does not change its value. Interchanging two rows or two columns multiplies the $9j$-symbol by $(-1)^s$ where s is the sum of all 9 spins. Hence, *a $9j$-symbol with two equal rows or columns vanishes.* Certain relations satisfied by $9j$-symbols follow from the fact that they appear in the coefficients of an orthogonal transformation (they are all real as seen from (9.26)). Other relations follow from the fact that the product of two such change-of-coupling transformations (9.23) gives another such transformation.

The orthogonality of the transformation (9.27) yields the orthogonality relation

$$\sum_{J_{13},J_{24}} (2J_{13}+1)(2J_{24}+1) \begin{Bmatrix} J_1 & J_2 & J_{12} \\ J_3 & J_4 & J_{34} \\ J_{13} & J_{24} & J \end{Bmatrix} \begin{Bmatrix} J_1 & J_2 & J'_{12} \\ J_3 & J_4 & J'_{34} \\ J_{13} & J_{24} & J \end{Bmatrix} = \frac{\delta(J_{12},J'_{12})\delta(J_{34},J'_{34})}{(2J_{12}+1)(2J_{34}+1)} \tag{9.28}$$

The transformation from $|J_1J_2(J_{12})J_3J_4(J_{34})J\rangle$ to $|J_1J_4(J_{14})J_2J_3(J_{23})J\rangle$ may be obtained as the product of the transformation

$$\langle J_1J_2(J_{12})J_3J_4(J_{34})J \mid J_1J_3(J_{13})J_2J_4(J_{24})J\rangle$$

by the transformation $\langle J_1J_3(J_{13})J_2J_4(J_{24})J \mid J_1J_4(J_{14})J_2J_3(J_{23})J\rangle$. Summing over the intermediate angular momenta J_{13}, J_{24} and paying attention to the order of couplings we obtain the following relation between $9j$-symbols

$$\sum_{J_{13},J_{24}} (-1)^{J_2+J_4+J_{24}+J_2+J_3+J_{23}}(2J_{13}+1)(2J_{24}+1) \times \begin{Bmatrix} J_1 & J_3 & J_{13} \\ J_2 & J_4 & J_{24} \\ J_{12} & J_{34} & J \end{Bmatrix} \begin{Bmatrix} J_1 & J_4 & J_{14} \\ J_3 & J_2 & J_{23} \\ J_{13} & J_{24} & J \end{Bmatrix} = (-1)^{J_3+J_4+J_{34}} \begin{Bmatrix} J_1 & J_4 & J_{14} \\ J_2 & J_3 & J_{23} \\ J_{12} & J_{34} & J \end{Bmatrix} \tag{9.29}$$

A useful relation between $3j$-symbols and $9j$-symbols may be obtained directly from (9.24). After expansion in terms of $3j$-symbols and equating coefficients of $\psi_{j_1m_1}(1)\psi_{j_2m_2}(2)\psi_{j_3m_3}(3)$ and $\psi_{j_4m_4}(4)$ we may multiply both sides of that equation by *two* $3j$-symbols,

$$\begin{pmatrix} j_1 & j_2 & J'_{12} \\ m_1 & m_2 & -M'_{12} \end{pmatrix} \quad \text{and} \quad \begin{pmatrix} j_3 & j_4 & J'_{34} \\ m_3 & m_4 & -M'_{34} \end{pmatrix}$$

Summation over m_1, m_2, m_3 and m_4 yields the relations

$$\begin{pmatrix} J_{12} & J_{34} & J \\ M_{12} & M_{34} & M \end{pmatrix} \begin{Bmatrix} j_1 & j_2 & J_{12} \\ j_3 & j_4 & J_{34} \\ J_{13} & J_{24} & J \end{Bmatrix} = \sum_{\substack{m_1m_2m_3m_4 \\ M_{13}M_{24}}} \begin{pmatrix} j_1 & j_2 & J_{12} \\ m_1 & m_2 & M_{12} \end{pmatrix} \begin{pmatrix} j_3 & j_4 & J_{34} \\ m_3 & m_4 & M_{34} \end{pmatrix} \times \begin{pmatrix} j_1 & j_3 & J_{13} \\ m_1 & m_3 & M_{13} \end{pmatrix} \begin{pmatrix} j_2 & j_4 & J_{24} \\ m_2 & m_4 & M_{24} \end{pmatrix} \begin{pmatrix} J_{13} & J_{24} & J \\ M_{13} & M_{24} & M \end{pmatrix} \tag{9.30}$$

Multiplying (9.30) by

$$(2J+1)\begin{pmatrix} J_{13} & J_{24} & J \\ M'_{13} & M'_{24} & M \end{pmatrix}$$

and summing over J and M we obtain by using (7.26) on the r.h.s. of (9.30) $M_{13} = M'_{13}$, $M_{24} = M'_{24}$. We thus obtain another relation between $3j$-symbols and $9j$-symbols

$$\sum_{JM}(2J+1)\begin{pmatrix} J_{12} & J_{34} & J \\ M_{12} & M_{34} & M \end{pmatrix}\begin{pmatrix} J_{13} & J_{24} & J \\ M_{13} & M_{24} & M \end{pmatrix} \times \begin{Bmatrix} j_1 & j_2 & J_{12} \\ j_3 & j_4 & J_{34} \\ J_{13} & J_{24} & J \end{Bmatrix}$$
$$= \sum_{m_1 m_2 m_3 m_4} \begin{pmatrix} j_1 & j_2 & J_{12} \\ m_1 & m_2 & M_{12} \end{pmatrix}\begin{pmatrix} j_3 & j_4 & J_{34} \\ m_3 & m_4 & M_{34} \end{pmatrix} \times \begin{pmatrix} j_1 & j_3 & J_{13} \\ m_1 & m_3 & M_{13} \end{pmatrix}\begin{pmatrix} j_2 & j_4 & J_{24} \\ m_2 & m_4 & M_{24} \end{pmatrix} \tag{9.31}$$

The $9j$-symbols appear in many problems. They define geometrical aspects of angular momenta independent of the physical characteristics of the problem. They are universal functions of their arguments and their values are tabulated in several publications.

We will now make use of $9j$-symbols to calculate the reduced matrix elements of an irreducible tensor operator obtained by tensor multiplication of two independent irreducible tensors. Consider two irreducible tensor operators $\mathbf{T}^{(k_1)}(1)$ and $\mathbf{T}^{(k_2)}(2)$ and products of their components $T^{(k_1)}_{\kappa_1}(1)T^{(k_2)}_{\kappa_2}(2)$. We can construct an irreducible tensor of rank k from these products by using Clebsch-Gordan coefficients.

According to (7.1) we obtain in this case

$$T_\kappa^{(k)}(1,2) = [\mathbf{T}^{(k_1)}(1) \times \mathbf{T}^{(k_2)}(2)]_\kappa^{(k)} = \sum_{\kappa_1\kappa_2} (k_1\kappa_1 k_2\kappa_2 \mid k_1 k_2 k \kappa) T_{\kappa_1}^{(k_1)}(1) T_{\kappa_2}^{(k_2)}(2) \tag{9.32}$$

The independent systems 1 and 2 in (9.32) may, for instance, be tensors constructed from the variables of two nucleons or the spin of a nucleon and its orbital angular momentum. If components of the two tensor operators do not commute, it is important to keep the order in which they appear on the r.h.s. of (9.32).

Due to the Wigner-Eckart theorem the *physical* properties of $T_{\kappa_1}^{(k_1)}(1)$ and $T_{\kappa_2}^{(k_2)}(2)$ are contained in their reduced matrix elements. The same is true of $\mathbf{T}_\kappa^{(k)}$ in states of the combined system. Hence, there should be a geometrical relation between the reduced matrix elements $(\alpha_1 J_1 \| \mathbf{T}^{(k_1)} \| \alpha_1' J_1')$ and $(\alpha_2 J_2 \| \mathbf{T}^{(k)}) \| \alpha_2' J_2')$ and the reduced matrix elements $(\alpha_1 J_1 \alpha_2 J_2 J \| \mathbf{T}^{(k)} \| \alpha_1' J_1' \alpha_2' J_2' J')$. Note that in a state obtained by (7.1) from states of system 1 with $\alpha_1 J_1$ and system 2 with $\alpha_2 J_2$, there is only *one* state with given total J and no further quantum numbers are necessary for unique specification of the state.

In this scheme let us calculate matrix elements of $T_\kappa^{(k)}$ defined by (9.32). They are defined by

$$\langle \alpha_1 J_1 \alpha_2 J_2 J M | [\mathbf{T}^{(k_1)}(1) \times \mathbf{T}^{(k_2)}(2)]_\kappa^{(k)} | \alpha_1' J_1' \alpha_2' J_2' J' M' \rangle \tag{9.33}$$

According to the Wigner-Eckart theorem, taking a matrix element amounts to coupling of three angular momenta to zero. In (9.33), the following spins are thus coupled to zero. First, J_1, J_2, J and J_1', J_2', J' due to the coupling scheme. Then J, k, and J' are coupled to zero due to the Wigner-Eckart theorem. We would like to express (9.33) in terms of matrix elements of $\mathbf{T}^{(k_1)}$ and $\mathbf{T}^{(k_2)}$. This introduces couplings of J_1, k_1, J_1' and J_2, k_2, J_2' to zero and in this scheme due to (9.32) also the coupling of k_1, k_2, k to zero. Thus, the problem is equivalent to the change of coupling transformation (9.24).

Expanding the states in (9.33) and using the Wigner-Eckart theorem, we obtain

$$\sum (J_1M_1J_2M_2 \mid J_1J_2JM)(J_1'M_1'J_2'M_2' \mid J_1'J_2'J'M')(-1)^{J-M}$$

$$\times \begin{pmatrix} J & k & J' \\ -M & \kappa & M' \end{pmatrix} (J\|\mathbf{T}^{(k)}\|J')$$

$$= \sum (k_1\kappa_1k_2\kappa_2 \mid k_1k_2k\kappa)(-1)^{J_1-M_1}$$

$$\times \begin{pmatrix} J_1 & k_1 & J_1' \\ -M_1 & \kappa_1 & M_1' \end{pmatrix} (-1)^{J_2-M_2}$$

$$\times \begin{pmatrix} J_2 & k_2 & J_2' \\ -M_2 & \kappa_2 & M_2' \end{pmatrix} (\alpha_1J_1\|\mathbf{T}^{(k_1)}\|\alpha_1'J_1')(\alpha_2J_2\|\mathbf{T}^{(k_2)}\|\alpha_2'J_2') \tag{9.34}$$

Multiplying the r.h.s. of (9.34) by three $3j$-symbols and summing over appropriate m-values we obtain the result

$$(\alpha_1J_1\alpha_2J_2J\|[\mathbf{T}^{(k_1)}(1) \times \mathbf{T}^{(k_2)}(2)]^{(k)}\|\alpha_1'J_1'\alpha_2'J_2'J')$$

$$= [(2J+1)(2k+1)(2J'+1)]^{1/2} \begin{Bmatrix} J_1 & J_2 & J \\ J_1' & J_2' & J' \\ k_1 & k_2 & k \end{Bmatrix}$$

$$\times (\alpha_1J_1\|\mathbf{T}^{(k_1)}\|\alpha_1'J_1')(\alpha_2J_2\|\mathbf{T}^{(k_2)}\|\alpha_2'J_2') \tag{9.35}$$

The physical contents of the combined tensor operator (9.32) is, according to the Wigner-Eckart theorem, completely determined by its reduced matrix element. This in turn is given, according to (9.35) by the product of the two reduced matrix elements. The geometrical aspects describing the couplings of individual spins to J and J', as well as the coupling due to (8.4) are all included in the $9j$-symbol. The latter is a universal function of its arguments entirely independent of the physical nature of the tensor operators.

It is worthwhile to point out that if both $\mathbf{T}^{(k_1)}$ and $\mathbf{T}^{(k_2)}$ are hermitean tensor operators according to (8.8), their tensor product is not always hermitean. Using the relation (8.9) and the symmetry properties of the $9j$-symbol we conclude that the tensor product in (9.35) is hermitean if $(-1)^{k_1+k_2+k} = 1$. If, however, $(-1)^{k_1+k_2+k} = -1$ the tensor product is *antihermitean*. Actually, this follows directly from the definition (9.32) and the relation (8.9).

In (3.2) the orbital angular momentum $\mathbf{L}$ was defined as the vector product

$$\mathbf{r} \times \mathbf{p} = \hbar \mathbf{L}$$

Both $\mathbf{r}$ and $\mathbf{p}$ are hermitean tensor operators of rank 1 and so is $\mathbf{L}$. How are these facts reconciled with the statement made above? Using the spherical components of $\mathbf{r}$ and $\mathbf{p}$ we may calculate their tensor product with rank $k = 1$. Using the actual values of the Clebsch-Gordan coefficients we find

$$\boxed{[\mathbf{r}^{(1)} \times \mathbf{p}^{(1)}]^{(1)}_{\kappa} = i\frac{\hbar}{\sqrt{2}} L^{(1)}_{\kappa}} \tag{9.36}$$

Hence, (9.36) is consistent with $\mathbf{r}$, $\mathbf{p}$ and $\mathbf{L}$ being hermitean operators.

In the next section we will see how a special case of $9j$-symbols plays an important role in the calculation of matrix elements of two-body interactions.

10

Matrix Elements of Two Nucleon Interactions. 6j-Symbols

As emphasized in the introduction, the two-nucleon interactions which may be used with shell model wave functions are *effective interactions*. They bear little resemblance to the interactions between free nucleons. The effective interaction may be obtained in nuclear many body theory as the result of drastic renormalization of the interaction between free nucleons. Given a subspace of the shell model, including one or several configurations, the theory should yield matrix elements of the effective interaction in that space. To this date, however, there is no reliable way to obtain such matrix elements by using methods of many body theory. The only practical way to use the shell model for calculation of nuclear energies is to determine matrix elements from experiment. When this is carried out self-consistently, we obtain a set of matrix elements between states of the configurations considered. There is very little chance that those may be calculated from a simple potential interaction.

Nevertheless, we show in this section how matrix elements of simple interactions, between states constructed in Section 9, can be calculated. Some of these interactions have been used by various authors. The consideration of simple interactions is a good introduction to fundamental concepts of nuclear spectroscopy.

The simplest interaction between two nucleons is a potential interaction which depends only on the distance between them. A simple example of such an interaction is the Coulomb potential. Matrix elements of this interaction are evaluated between states which are products, or sums of products, of single nucleon wave functions. Hence, it is worthwhile to express such an interaction in terms of the coordinates of the two nucleons. We thus express the interaction by

$$V(|\mathbf{r}_1 - \mathbf{r}_2|) = V\left(\sqrt{r_1^2 + r_2^2 - 2r_1 r_2 \cos\omega_{12}}\right) = \sum_{k=0}^{\infty} v_k(r_1, r_2) P_k(\cos\omega_{12}) \tag{10.1}$$

This is an expansion in terms of the Legendre polynomials whose argument is the cosine of the angle between the vectors $\mathbf{r}_1$ and $\mathbf{r}_2$ so that $\cos\omega_{12} = (\mathbf{r}_1 \cdot \mathbf{r}_2)/r_1 r_2$. The functions $v_k(r_1, r_2)$ are symmetric in r_1 and r_2. In the case of the Coulomb potential $1/|\mathbf{r}_1 - \mathbf{r}_2|$, these functions are given by

$$v_k(r_1, r_2) = \frac{r_<^k}{r_>^{k+1}} \qquad V(|\mathbf{r}_1 - \mathbf{r}_2|) = \frac{1}{|\mathbf{r}_1 - \mathbf{r}_2|} \tag{10.2}$$

In (10.2) $r_<$ is the smaller of r_1 and r_2 and $r_>$ is the larger one. The corresponding functions for the Yukawa potential $\exp(-\lambda|\mathbf{r}_1 - \mathbf{r}_2|)/|\mathbf{r}_1 - \mathbf{r}_2|$ are given in the literature. Due to the orthogonality relations of the Legendre polynomials, the $v_k(r_1, r_2)$ may be obtained from $V(|\mathbf{r}_1 - \mathbf{r}_2|)$ by the integral

$$v_k(r_1, r_2) = \frac{2k+1}{2} \int_{-1}^{+1} V(|\mathbf{r}_1 - \mathbf{r}_2|) P_k(\cos\omega_{12}) d\cos\omega_{12} \tag{10.3}$$

The Legendre polynomials in (10.1) can now be expressed by the addition theorem of spherical harmonics

$$P_k(\cos\omega_{12}) = \frac{4\pi}{2k+1}\sum_{\kappa}(-1)^{\kappa}Y_{k\kappa}(\theta_1\phi_1)Y_{k,-\kappa}(\theta_2\phi_2) \tag{10.4}$$

This expansion offers an example of the scalar product (7.20) between the irreducible tensors $\mathbf{Y}_k(1)$ and $\mathbf{Y}_k(2)$. The spherical harmonics in (10.4) play here the role of operators rather than that of wave functions. To make (10.4) equal to a scalar product, without the coefficient $4\pi/(2k+1)$, sometimes the definition $C_{k\kappa} = [4\pi/(2k+1)]Y_{k\kappa}$ is introduced. There is, however, hardly a need to introduce another symbol.

Matrix elements of the interaction (10.1) in which (10.4) is substituted for $P_k(\cos\omega_{12})$, can now be taken between the two nucleon states (9.7) and (9.8). Since the interaction does not contain isospin operators we obtain the following result between states with the same value of T

$$\begin{aligned}
&\frac{1}{2}\Big[\int \psi^*_{jm}(1)\psi^*_{j'm'}(2)V\psi_{jm_1}(1)\psi_{j'm'_1}(2) \\
&\qquad \mp \int \psi^*_{jm}(1)\psi^*_{j'm'}(2)V\psi_{jm_1}(2)\psi_{j'm'_1}(1) \\
&\qquad \mp \int \psi^*_{jm}(2)\psi_{j'm'}(1)V\psi_{jm_1}(1)\psi_{j'm'_1}(2) \\
&\qquad + \int \psi^*_{jm}(2)\psi_{j'm'}(1)V\psi_{jm_1}(2)\psi_{j'm'_1}(1)\Big] \\
&\quad = \int \psi^*_{jm}(1)\psi^*_{j'm'}(2)V\psi_{jm_1}(1)\psi_{j'm'_1}(2) \\
&\qquad \mp \int \psi_{jm}(1)\psi_{j'm'}(2)V\psi_{jm_1}(2)\psi_{j'm'_1}(1) \\
&\quad = \sum_k \frac{4\pi}{2k+1}\Big[\int \psi^*_{jm}(1)\psi^*_{j'm'}(2)v_k(\mathbf{Y}_k(1)\cdot\mathbf{Y}_k(2))\psi_{jm_1}(1)\psi_{j'm'_1}(2) \\
&\qquad \mp \int \psi^*_{jm}(1)\psi^*_{j'm'}(2)v_k(\mathbf{Y}_k(1)\cdot\mathbf{Y}_k(2))\psi_{jm_1}(2)\psi_{j'_1m'_1}(1)\Big]
\end{aligned} \tag{10.5}$$

The equality in (10.5) was obtained by interchanging the indices of nucleons 1 and 2 and is due to the symmetry of $V(|\mathbf{r}_1 - \mathbf{r}_2|)$. The upper (−) sign in (10.5) is for $T = 1$ antisymmetric states and the lower (+) sign is for symmetric states with $T = 0$. The terms

$$\int \psi^*_{jm}(1)\psi^*_{j'm'}(2)V\psi_{jm_1}(1)\psi_{j'm'_1}(2)$$

are called *direct terms* whereas terms

$$\int \psi^*_{jm}(1)\psi^*_{j'm'}(2)V\psi_{jm_1}(2)\psi_{j'm'_1}(1)$$

are called *exchange terms*.

Matrix elements of the interaction (10.1) taken between symmetric and antisymmetric states always vanish since the interaction is symmetric in the coordinates of the two nucleons. Such a matrix element is equal to (10.5) in which the last two terms on the l.h.s. have their signs changed. In that case the first term is cancelled by the last term and the two middle terms cancel each other. This result can also be formally obtained from the orthogonality of $T = 0$ and $T = 1$ states and the fact that any charge independent interaction is a scalar in isospin.

Before proceeding, let us consider matrix elements of the interaction (10.1) between states (9.12). Those states describe a proton in the j-orbit and a neutron in the j'-orbit when the neutron j-orbit is completely filled. Instead of (10.5) we obtain the result

$$\sum_k \frac{4\pi}{2k+1} \int \psi^*_{jm}(1)\psi^*_{j'm'}(2)v_k(\mathbf{Y}_k(1) \cdot \mathbf{Y}_k(2))\psi_{jm_1}(1)\psi_{j'm'_1}(2) \tag{10.6}$$

In (10.6) only the direct terms appear. The matrix element (10.6) can be simply related to those given by (10.5). It is equal to

$$\langle j_\pi m_1 j'_\nu m_2 | V | j_\pi m'_1 j'_\nu m'_2 \rangle$$
$$= \tfrac{1}{2}[\langle jm_1 j' m_2 T = 1 | V | jm'_1 j' m'_2 T = 1\rangle$$
$$+ \langle jm_1 j' m_2 T = 0 | V | jm'_1 j' m'_2 T = 0\rangle] \tag{10.7}$$

It is thus the average between the (charge independent) interaction in the corresponding states with $T = 1$ and $T = 0$.

We shall now proceed to calculate matrix elements of (10.1) in the scheme of functions (9.11) or (9.16) in which j and j' are coupled to a state with a definite value of J. The integrals over $v_k(r_1, r_2)$ multiplied by the four radial functions in (10.5) are independent of m, m', m_1, m_1'. The integration over the various angles should next be carried out. The matrix elements to be evaluated first are those of the direct term, namely

$$\langle jj'JM|\mathbf{Y}_k(1)\cdot\mathbf{Y}_k(2)|jj'JM\rangle$$

These matrix elements are a special case of a more general expression which will now be described.

Consider the matrix element of a scalar product of two irreducible tensor operators

$$\begin{aligned}
&\langle \alpha_1 J_1 \alpha_2 J_2 JM|\mathbf{T}^{(k)}(1)\cdot\mathbf{T}^{(k)}(2)|\alpha_1' J_1' \alpha_2' J_2' JM\rangle \\
&\quad = (-1)^k\sqrt{2k+1}\langle \alpha_1 J_1 \alpha_2 J_2 JM|[\mathbf{T}^{(k)}(1)\times\mathbf{T}^{(k)}(2)]_0^{(0)}|\alpha_1' J_1' \alpha_2' J_2' JM\rangle \\
&\quad = (-1)^k\sqrt{2k+1}(-1)^{J-M}\begin{pmatrix} J & 0 & J \\ -M & 0 & M \end{pmatrix} \\
&\qquad \times (\alpha_1 J_1 \alpha_2 J_2 J\|[\mathbf{T}^{(k)}(1)\times\mathbf{T}^{(k)}(2)]^{(0)}\|\alpha_1' J_1' \alpha_2' J_2' J) \\
&\quad = (-1)^k\sqrt{2k+1}\frac{1}{\sqrt{2J+1}} \\
&\qquad \times (\alpha_1 J_1 \alpha_2 J_2 J\|[\mathbf{T}^{(k)}(1)\times\mathbf{T}^{(k)}(2)]^{(0)}\|\alpha_1' J_1' \alpha_2' J_2' J) \qquad (10.8)
\end{aligned}$$

The last equality was obtained by using the value of the $3j$-symbol related to the Clebsch-Gordan coefficient (7.19). The reduced matrix element in (10.8) is a special case of the general result (9.25) obtained by putting there $k_1 = k_2$, $J = J'$ and $k = 0$. This reduces the number of independent J-values in the 9-j symbol in (9.25) to six only. This is a very special $9j$-symbol and deserves a special notation. The $6j$-symbol (Wigner 1940) or *Racah coefficient* (Racah, 1942) is defined

by

$$\begin{Bmatrix} J_1 & J_2 & J \\ J_2' & J_1' & k \end{Bmatrix} = (-1)^{J_2+J+J_1'+k}[(2J+1)(2k+1)]^{1/2} \begin{Bmatrix} J_1 & J_2 & J \\ J_1' & J_2' & J \\ k & k & 0 \end{Bmatrix} \tag{10.9}$$

It appears in many applications, the first of them was in the calculation of matrix elements of scalar products as in (10.8) (Racah 1942).

The symmetry properties of the $6j$-symbol follow directly from its definition (10.9). There are four triangular conditions that its six arguments must satisfy. These include the j-values in the upper row and other three conditions. The latter must be satisfied by two J-values in the lower row and the one J-value in the upper row which is not written above them. Interchanging any two columns may change the sign of the $9j$-symbol in (10.9) but that change is compensated by the change in the phase factor in front of it. Thus, any permutation of columns of the $6j$-symbol leaves its value unchanged. Interchanging in the $9j$-symbol the upper two rows, yields a $6j$-symbol where J_1J_2 are interchanged with $J_2'J_1'$ written below them. Interchanging rows and columns in that $9j$-symbol does not change its value nor the phase factor. This is equivalent to replacing in the $6j$-symbol J_2J by $J_1'k$ written below them. Hence, the value of a $6j$-symbol does not change if two of its arguments in the upper row are interchanged with those below them. These operations complete the list of allowed symmetry transformations of the arguments in the $6j$-symbol. *None of these changes its sign.*

Like $9j$-symbols, the $6j$-symbols appear in coefficients of orthogonal transformations between different coupling schemes. If in (9.21) we put $J = 0$, $J_{12} = J_{34}$, $J_{13} = J_{24}$ we obtain, with a slight change of notation, the result

$$\begin{aligned} &\langle J_1J_3(J')J_2J_4(J')0 \mid J_1J_2(J)J_3J_4(J)0\rangle \\ &\quad = (-1)^{J_2+J_3+J+J'}\sqrt{(2J+1)(2J'+1)} \begin{Bmatrix} J_1 & J_2 & J \\ J_4 & J_3 & J' \end{Bmatrix} \end{aligned} \tag{10.10}$$

Instead of coupling J_1, J_2, J_3, J_4 to $J = 0$ we can couple three of them to obtain the fourth angular momentum. The transformation between two ways of such couplings is given by

$$\boxed{\begin{aligned}&\langle J_1J_3(J_{13})J_2J \mid J_1J_2(J_{12})J_3J\rangle \\ &\quad = (-1)^{J_2+J_3+J_{12}+J_{13}}\sqrt{(2J_{12}+1)(2J_{13}+1)}\begin{Bmatrix} J_1 & J_2 & J_{12} \\ J & J_3 & J_{13} \end{Bmatrix}\end{aligned}} \tag{10.11}$$

The coefficients of this transformation are the same as in (10.10). Another possible change of coupling transformation is obtained by keeping the order of couplings. That one is expressed by

$$\boxed{\begin{aligned}&\langle J_1J_2(J_{12})J_3J \mid J_1, J_2J_3(J_{23})J\rangle \\ &\quad = (-1)^{J_1+J_2+J_3+J}\sqrt{(2J_{12}+1)(2J_{23}+1)}\begin{Bmatrix} J_1 & J_2 & J_{12} \\ J_3 & J & J_{23} \end{Bmatrix}\end{aligned}} \tag{10.12}$$

In all these transformations the positions of the various J-values in the $6j$-symbol are fixed by the couplings. The numerical coefficients involve the intermediate angular momenta. The only factor to pay attention to is the phase factor. To compensate for this inconvenience, the $6j$-symbol does not change its sign in any allowed operation on its arguments.

From the orthogonality of the transformation (10.10) we obtain the following relation between $6j$-symbols

$$\boxed{\sum_J (2J+1)\begin{Bmatrix} J_1 & J_2 & J \\ J_1' & J_2' & J' \end{Bmatrix}\begin{Bmatrix} J_1 & J_2 & J \\ J_1' & J_2' & J'' \end{Bmatrix} = \frac{\delta_{J'J''}}{2J'+1}} \tag{10.13}$$

Another useful relation follows from the fact that successive applications of the transformation (10.10) yields a transformation to a third coupling scheme. Thus

$$\sum_J \langle J_1J_2(J')J_3J_4(J')0 \mid J_1J_3(J)J_2J_4(J)0\rangle$$

$$\times \langle J_1J_3(J)J_2J_4(J)0 \mid J_1J_4(J'')J_2J_3(J'')0\rangle$$

$$= \langle J_1J_2(J')J_3J_4(J')0 \mid J_1J_4(J'')J_2J_3(J'')0\rangle$$

Taking care of the changes of order of coupling necessary to use (10.10) we obtain due to (7.11) the following sum rule

$$\sum_J (-1)^{J+J'+J''}(2J+1)\begin{Bmatrix} J_1 & J_2 & J' \\ J_4 & J_3 & J \end{Bmatrix}\begin{Bmatrix} J_1 & J_4 & J'' \\ J_2 & J_3 & J \end{Bmatrix}$$

$$= \begin{Bmatrix} J_1 & J_2 & J' \\ J_3 & J_4 & J'' \end{Bmatrix} \tag{10.14}$$

Another useful relation between $6j$-symbols is the Biedenharn (1953)–Elliott (1953) sum rule. It can be obtained by considering two different sets of successive transformations leading from one specific scheme to another. In one scheme J_1 and J_2 are coupled to J_{12}, J_3 and J_4 to J_{34} and J_{12} and J_{34} are coupled to the total J. In the other scheme J_1 and J_3 are coupled to J_{13} which is coupled to J_2 to yield a definite J_{132} which is then coupled to J_4 to yield the total J. This transformation may be obtained as the product of two transformations as follows

$$\langle J_1J_3(J_{13})J_2(J_{132})J_4J \mid J_1J_2(J_{12})J_3J_4(J_{34})J\rangle$$

$$= \langle J_1J_3(J_{13})J_2(J_{132})J_4J \mid J_1J_2(J_{12})J_3(J_{132})J_4J\rangle$$

$$\times \langle J_1J_2(J_{12})J_3(J_{132})J_4J \mid J_1J_2(J_{12})J_3J_4(J_{34})J\rangle \tag{10.15}$$

Alternatively, the same transformation may be obtained as the product of *three* transformations, summed over an intermediate angular momentum J_{134} as given by

$$\sum_{J_{134}} \langle J_1J_3(J_{13})J_2(J_{132})J_4J \mid J_1J_3(J_{13})J_4(J_{134})J_2J\rangle$$
$$\times \langle J_1J_3(J_{13})J_4(J_{134})J_2J \mid J_3J_4(J_{34})J_1(J_{134})J_2J\rangle$$
$$\times \langle J_3J_4(J_{34})J_1(J_{134})J_2J \mid J_1J_2(J_{12})J_3J_4(J_{34})J_2J\rangle \quad (10.16)$$

Each of the transformations on the r.h.s. of (10.15) and in (10.16) is a change of coupling of three angular momenta given by either (10.11) or (10.12). Equating (10.15) to (10.16) and substituting the actual transformation coefficients, we obtain a product of two $6j$-symbols equal to a sum of products of three $6j$-symbols. In a simpler notation this relation is

$$\begin{Bmatrix} J_1 & J_2 & J_3 \\ J_1' & J_2' & J_3' \end{Bmatrix}\begin{Bmatrix} J_1 & J_2 & J_3 \\ J_1'' & J_2'' & J_3'' \end{Bmatrix}$$
$$= \sum_J (-1)^{J_1+J_2+J_3+J_1'+J_2'+J_3'+J_1''+J_2''+J_3''+J}(2J+1)$$
$$\times \begin{Bmatrix} J_1 & J_2' & J_3' \\ J & J_3'' & J_2'' \end{Bmatrix}\begin{Bmatrix} J_1' & J_2 & J_3' \\ J_3'' & J & J_1'' \end{Bmatrix}\begin{Bmatrix} J_1' & J_2' & J_3 \\ J_2'' & J_1'' & J \end{Bmatrix} \quad (10.17)$$

If we put in (9.20) $J_{33} = 0$, $J_{13} = J_{23}$ and $J_{31} = J_{32}$ we obtain the $6j$-symbol as a sum of products of $3j$-symbols. The two $3j$-symbols which include J_{33} are equal to those in (7.19). Thus, a $6j$-symbol is essentially equal to a sum of products of four $3j$-symbols. The actual expression is

$$\begin{Bmatrix} J_1 & J_2 & J_3 \\ J'_1 & J'_2 & J'_3 \end{Bmatrix} = \sum_{\text{all } M_i, M'_i} (-1)^{J'_1+J'_2+J'_3+M'_1+M'_2+M'_3} \times \begin{pmatrix} J_1 & J_2 & J_3 \\ M_1 & M_2 & M_3 \end{pmatrix} \begin{pmatrix} J_1 & J'_2 & J'_3 \\ M_1 & M'_2 & -M'_3 \end{pmatrix} \times \begin{pmatrix} J'_1 & J_2 & J'_3 \\ -M'_1 & M_2 & M'_3 \end{pmatrix} \begin{pmatrix} J'_1 & J'_2 & J_3 \\ M'_1 & -M'_2 & M_3 \end{pmatrix} \tag{10.18}$$

We can use (10.18) to obtain the numerical value of a special $6j$-symbol, one of whose arguments is zero. If we put in (10.18) $J'_3 = 0$, two $3j$-symbols are explicitly given by (7.19). We can sum the products of the other two $3j$-symbols and the sum is given by (7.25). We thus obtain the result

$$\begin{Bmatrix} J_1 & J_1 & 0 \\ J_2 & J_2 & J_3 \end{Bmatrix} = \frac{(-1)^{J_1+J_2+J_3}}{\sqrt{(2J_1+1)(2J_2+1)}} \tag{10.19}$$

From this result follow two special sum rules. If in (10.13) we put $J'' = 0$ we obtain

$$\sum_J (-1)^{J_1+J_2+J}(2J+1) \begin{Bmatrix} J_1 & J_1 & J' \\ J_2 & J_2 & J \end{Bmatrix} = \sqrt{(2J_1+1)(2J_2+1)}\,\delta_{J',0} \tag{10.20}$$

Putting $J'' = 0$ in the sum rule (10.14) we obtain due to (10.19) the special case

$$\boxed{\sum_J (2J+1)\begin{Bmatrix} J_1 & J_2 & J' \\ J_1 & J_2 & J \end{Bmatrix} = (-1)^{2(J_1+J_2)}} \tag{10.21}$$

In the special case of $J_1 = J_2$, the sum and difference of (10.20) and (10.21) yield simple results for separate summations of

$$(2J+1)\begin{Bmatrix} J_1 & J_1 & J \\ J_1 & J_1 & J' \end{Bmatrix}$$

over odd and even values of J.

To derive useful relations between $6j$- and $3j$-symbols we may put $J = 0$ in (9.24) and (9.25). To derive those relations directly we start from

$$\boxed{\begin{aligned}\psi(J_1J_3(J')J_2J_4(J')0) = \sum (-1)^{J_2+J_3+J+J'} \sqrt{(2J+1)(2J'+1)} \\ \times \begin{Bmatrix} J_1 & J_2 & J \\ J_4 & J_3 & J' \end{Bmatrix} \psi(J_1J_2(J)J_3J_4(J)0)\end{aligned}} \tag{10.22}$$

Expanding in terms of Clebsch-Gordan coefficients and equating coefficients of $\psi_{J_1M_1}\psi_{J_2M_2}\psi_{J_3M_3}\psi_{J_4M_4}$ on both sides we obtain

$$\begin{aligned}&\sum_{M'} \frac{(-1)^{J'-M'}}{\sqrt{2J'+1}} (J_1M_1J_3M_3 \mid J_1J_3J'M')(J_2M_2J_4M_4 \mid J_2J_4J', -M') \\ &\quad = \sum_M (-1)^{J_2+J_3+J+J'} \sqrt{(2J+1)(2J'+1)} \begin{Bmatrix} J_1 & J_2 & J \\ J_4 & J_3 & J' \end{Bmatrix} \\ &\quad \times \frac{(-1)^{J-M}}{\sqrt{2J+1}} (J_1M_1J_2M_2 \mid J_1J_2JM)(J_3M_3J_4M_4 \mid J_3J_4J, -M)\end{aligned}$$

Multiplying both sides by $(J_1M_1J_2M_2 \mid J_1J_2JM_0)$ and summing over M_1,M_2 we obtain, in view of the orthogonality relations of vector addition coefficients (7.8), the result is

$$(-1)^{J_2+J_3+J+J'}\sqrt{2J'+1}(-1)^{J-M_0}(J_3M_3J_4M_4 \mid J_3J_4J,-M_0)\begin{Bmatrix} J_3 & J_4 & J \\ J_2 & J_1 & J' \end{Bmatrix}$$

$$= \sum_{M_1M_2M'} \frac{(-1)^{J'-M'}}{\sqrt{2J'+1}}(J_1M_1J_3M_3 \mid J_1J_3J'M')$$

$$\times (J_2M_2J_4M_4 \mid J_2J_4J',-M')(J_3M_3J_4M_4 \mid J_3J_4JM_0)$$

Expressing the Clebsch-Gordan coefficients in terms of $3j$-symbols we obtain the useful relation

$$\begin{pmatrix} J_1 & J_2 & J_3 \\ M_1 & M_2 & M_3 \end{pmatrix}\begin{Bmatrix} J_1 & J_2 & J_3 \\ J_1' & J_2' & J_3' \end{Bmatrix}$$

$$= \sum_{M_1'M_2'M_3'} (-1)^{J_1'+J_2'+J_3'+M_1'+M_2'+M_3'}$$

$$\times \begin{pmatrix} J_1 & J_2' & J_3' \\ M_1 & M_2' & -M_3' \end{pmatrix}\begin{pmatrix} J_1' & J_2 & J_3' \\ -M_1' & M_2 & M_3' \end{pmatrix}$$

$$\times \begin{pmatrix} J_1' & J_2' & J_3 \\ M_1' & -M_2' & M_3 \end{pmatrix} \tag{10.23}$$

If we multiply both sides of (10.23) by

$$\begin{pmatrix} J_1 & J_2 & J_3 \\ M_1 & M_2 & M_3 \end{pmatrix}$$

and sum over $M_1M_2M_3$ we obtain, in view of (7.27), the expression (10.18). If, however, we multiply both sides of (10.23) by

$$(-1)^{J_3-M_3}(2J_3+1)\begin{pmatrix} J_1' & J_2' & J_3 \\ M_1'' & M_2'' & -M_3 \end{pmatrix}$$

and sum over J_3 and M_3 we obtain due to (7.26) another useful relationship which can be written as

$$\sum_{M_3}\begin{pmatrix} J_1 & J_2 & J_3 \\ M_1 & M_2 & M_3 \end{pmatrix}\begin{pmatrix} J_1' & J_2' & J_3 \\ M_1' & M_2' & -M_3 \end{pmatrix}$$
$$= \sum_{J_3' M_3'} (-1)^{J_3+J_3'+M_1+M_1'}(2J_3'+1)\begin{Bmatrix} J_1 & J_2 & J_3 \\ J_1' & J_2' & J_3' \end{Bmatrix}$$
$$\times \begin{pmatrix} J_1' & J_2 & J_3' \\ M_1' & M_2 & M_3' \end{pmatrix}\begin{pmatrix} J_1 & J_2' & J_3' \\ M_1 & M_2' & -M_3' \end{pmatrix} \tag{10.24}$$

The derivation of (10.23), (10.24) as well as (10.18) show the advantage of keeping *all* m-values as independent variables. In the various summations it would have been extremely difficult to keep track of which combinations of m-values are equal to others. Several interesting relations exist between $6j$-symbols and $9j$-symbols. Some of these are listed in the Appendix.

The $6j$-symbols appear in many calculations. They express geometrical relations which are independent of the physical characteristics of the given problem. Their tabulated values can be used in many applications. They were introduced by Racah (1942) for calculating matrix elements of two-body interactions like (10.8). The original definition of Racah was related to the $6j$-symbol, which was independently introduced by Wigner, by

$$W(j_1 j_2 j_3 j_4; j_5 j_6) = (-1)^{j_1+j_2+j_3+j_4}\begin{Bmatrix} j_1 & j_2 & j_5 \\ j_4 & j_3 & j_6 \end{Bmatrix} \tag{10.25}$$

It is clear that the W-coefficients have less symmetry in their arguments than the $6j$-symbols. This fact makes them less convenient in applications. Racah himself abandoned his original definition and used $6j$-symbols in the notation $W(j_1 j_2 j_3 / j_4 j_5 j_6)$ (Fano and Racah 1959). The fact that some authors still use W-coefficients must be ascribed to misguided conservatism.

The sum over the four $3j$-symbols in (10.18) can be carried out (Racah 1942) and yields an explicit formula for the $6j$-symbol in terms of its arguments. This formula is

$$\begin{Bmatrix} J_1 & J_2 & J_3 \\ j_1 & j_2 & j_3 \end{Bmatrix} = \Delta(J_1J_2J_3)\Delta(J_1j_2j_3)\Delta(j_1J_2j_3)\Delta(j_1j_2J_3)$$
$$\times \sum_t (-1)^t(t+1)![(t-J_1-J_2-J_3)!(t-J_1-j_2-j_3)!$$
$$\times (t-j_1-J_2-j_3)!(t-j_1-j_2-J_3)!(j_1+j_2+J_1+J_2-t)!$$
$$\times (j_2+j_3+J_2+J_3-t)!(j_1+j_3+J_1+J_3-t)!]^{-1} \qquad (10.26)$$

The summation in (10.26) is over all values of integers t for which the arguments of the factorial functions are non-negative. The functions $\Delta(j_1j_2j_3)$ are defined by

$$\Delta(j_1j_2j_3) = \left[\frac{(j_1+j_2-j_3)!(j_1-j_2+j_3)!(-j_1+j_2+j_3)!}{(j_1+j_2+j_3+1)!}\right]^{1/2}$$

Let us now return to the physical problem for which Racah (1942) introduced the $6j$-symbols. Using (10.8) and (9.35) for the special case given in (10.9) we find the general result

$$\boxed{\begin{aligned} &\langle \alpha_1J_1\alpha_2J_2JM|\mathbf{T}^{(k)}(1)\cdot\mathbf{U}^{(k)}(2)|\alpha_1'J_1'\alpha_2'J_2'JM\rangle \\ &= (-1)^{J_2+J+J_1'} \begin{Bmatrix} J_1 & J_2 & J \\ J_2' & J_1' & k \end{Bmatrix} \\ &\times (\alpha_1J_1\|\mathbf{T}^{(k)}\|\alpha_1'J_1')(\alpha_2J_2\|\mathbf{U}^{(k)}\|\alpha_2'J_2') \end{aligned}} \qquad (10.27)$$

In the formula (10.27) the physics is contained in the reduced matrix elements on the r.h.s. The geometrical factor of angular momentum couplings is given by the $6j$-symbol. The latter is a universal function which once computed may be used in various physical problems.

Before proceeding, let us consider another application of $6j$-symbols for the calculation of the reduced matrix-element of a tensor product of operators. Unlike the case considered in (9.32), we consider operators $T^{(k_1)}$ and $T^{(k_2)}$ both operating on the *same* system and the matrix element considered is

$$\langle \alpha JM|[\mathbf{T}^{(k_1)} \times \mathbf{T}^{(k_2)}]^{(k)}_\kappa|\alpha'J'M'\rangle$$

There are four $3j$-symbols involved in this calculation: one coupling k_1 and k_2 to k, another due to the Wigner-Eckart theorem coupling J and J' to k and two others. When a complete set of states with $\alpha''J''M''$ is introduced between $T^{(k_1)}_{\kappa_1}$ and $T^{(k_2)}_{\kappa_2}$ there are two $3j$-symbols due to the Wigner-Eckart theorem coupling J and J'' to k_1 and coupling J'' and J' to k_2. As a result, the reduced matrix element of the tensor product is given by

$$\boxed{\begin{aligned}&(\alpha J\|[\mathbf{T}^{(k_1)}\times\mathbf{T}^{(k_2)}]^{(k)}\|\alpha'J')\\&=(-1)^{J+J'+k}\sqrt{2k+1}\sum_{\alpha''J''}(\alpha J\|\mathbf{T}^{(k_1)}\|\alpha''J'')(\alpha''J''\|\mathbf{T}^{(k_2)}\|\alpha'J')\\&\times\begin{Bmatrix}k_1 & k_2 & k\\ J' & J & J''\end{Bmatrix}\end{aligned}} \tag{10.28}$$

In the special case of a scalar product of $\mathbf{T}^{(k)}$ and $\mathbf{U}^{(k)}$, the result (10.28) reduces to

$$\begin{aligned}(\alpha J\|[\mathbf{T}^{(k)}\times\mathbf{U}^{(k)}]^{(0)}\|\alpha'J)&=\frac{(-1)^k}{\sqrt{(2k+1)(2J+1)}}\\&\times\sum_{\alpha''J''}(-1)^{J''-J}(\alpha J\|\mathbf{T}^{(k)}\|\alpha''J'')(\alpha''J''\|\mathbf{U}^{(k)}\|\alpha'J)\end{aligned} \tag{10.29}$$

From (10.29) follows, in view of (8.4) and (7.20), that the matrix element of the scalar product is given by

$$\boxed{\begin{aligned}&\langle\alpha JM|(\mathbf{T}^{(k)}\cdot\mathbf{U}^{(k)})|\alpha'JM\rangle\\&=\frac{1}{2J+1}\sum_{\alpha''J''}(-1)^{J''-J}(\alpha J\|\mathbf{T}^{(k)}\|\alpha''J'')(\alpha''J''\|\mathbf{U}^{(k)}\|\alpha'J)\end{aligned}} \tag{10.30}$$

If $\mathbf{T}^{(k)}$ and $\mathbf{U}^{(k)}$ are hermitean tensors, the matrix element is given by

$$\langle \alpha JM|\mathbf{T}^{(k)} \cdot \mathbf{U}^{(k)}|\alpha' JM\rangle$$

$$= \frac{1}{2J+1} \sum_{\alpha'' J''} (\alpha J\|\mathbf{T}^{(k)}\|\alpha'' J'')(\alpha' J\|\mathbf{U}^{(k)}\|\alpha'' J'')^* \quad (10.31)$$

We can apply (10.27) to the matrix element appearing in the direct term of (10.5) and obtain

$$\langle jj'JM|\mathbf{Y}_k(1) \cdot \mathbf{Y}_k(2)|jj'JM\rangle$$

$$= (-1)^{j+j'+J} \begin{Bmatrix} j & j' & J \\ j' & j & k \end{Bmatrix} (j\|\mathbf{Y}_k\|j)(j'\|\mathbf{Y}_k\|j') \quad (10.32)$$

In (10.32) it is clearly displayed that the sum over k in (10.5) or (10.6) is rather limited. The highest value of k for which there may be a non-vanishing contribution to the matrix element is not higher than $\min(2j, 2j')$. The direct term in (10.5) is a sum of matrix elements (10.32) multiplied by radial integrals. We may use the unit tensor operators defined by (8.16) to express the direct term as

$$\sum_k F^k (-1)^{j+j'+J} \begin{Bmatrix} j & j' & J \\ j' & j & k \end{Bmatrix}$$

$$= \sum_k F_k \langle jj'JM|\mathbf{u}^{(k)}(1) \cdot \mathbf{u}^{(k)}(2)|jj'JM\rangle \quad (10.33)$$

where the coefficients F^k are defined by

$$F^k = \frac{4\pi}{2k+1} (j\|\mathbf{Y}_k\|j)(j'\|\mathbf{Y}_k\|j')$$

$$\times \int R^2_{nlj}(r_1) R^2_{n'l'j'}(r_2) \nu_k(r_1, r_2)\, dr_1\, dr_2 \quad (10.34)$$

It should be noted that there are other definitions of the F^k. A common choice, adopted in de-Shalit and Talmi (1963), is to denote by F^k just the radial integral $\int R^2_{nlj}(r_1) R^2_{n'l'j'}(r_2) \nu_k(r_1, r_2)\, dr_1\, dr_2$.

To calculate the exchange term with the help of (10.27) we must first change the order of coupling. The spin of nucleon 1 must be coupled to the spin of nucleon 2 in *that order* in both states defining the matrix element. This introduces a phase given by (7.11) and the matrix elements appearing in the exchange term have the form

$$(-1)^{j+j'-J}\langle jj'JM|\mathbf{Y}_k(1)\cdot\mathbf{Y}_k(2)|j'jJM\rangle$$

$$=(-1)^{j-j'}\begin{Bmatrix} j & j' & J \\ j & j' & k \end{Bmatrix}(j\|\mathbf{Y}_k\|j')(j'\|\mathbf{Y}_k\|j)$$

$$=\begin{Bmatrix} j & j' & J \\ j & j' & k \end{Bmatrix}(j\|\mathbf{Y}_k\|j')^2 \tag{10.35}$$

The last equality in (10.35) is due to the relation (8.9) of reduced matrix elements of hermitean operators applied to the case of *real* reduced matrix elements.

The exchange term in (10.5) can thus be expressed as

$$\sum_k G^k \begin{Bmatrix} j & j' & J \\ j & j' & k \end{Bmatrix} \tag{10.36}$$

where the coefficients in the expansion (10.36) are given by

$$\boxed{\begin{aligned} G^k &= \frac{4\pi}{2k+1}(j\|\mathbf{Y}_k\|j')^2 \\ &\quad\times\int R_{nlj}(r_1)R_{n'l'j'}(r_1)R_{nlj}(r_2)R_{n'l'j'}(r_2)\,dr_1\,dr_2 \end{aligned}} \tag{10.37}$$

The $6j$-symbols in (10.36) clearly show that the range of k for which there may be non-vanishing contributions to the sum is $|j-j'|\le k\le j+j'$.

The expansion (10.36) of the exchange term does not look like (10.33). The direct term there can be obtained from a linear combination of scalar products of unit tensors. We shall prove in a subsequent section that *any* two body interaction in a given configuration may be expressed as such a linear combination. Here we will show that this holds for (10.36) and obtain explicitly the desired expansion. We use the sum rule (10.14) to obtain

$$\begin{Bmatrix} j & j' & J \\ j & j' & k \end{Bmatrix} = \sum_r (-1)^{r+k+J}(2r+1)\begin{Bmatrix} j & j' & J \\ j' & j & r \end{Bmatrix}\begin{Bmatrix} j & j' & k \\ j' & j & r \end{Bmatrix}$$

$$= \sum_r (-1)^{j+j'+r+k}(2r+1)\begin{Bmatrix} j & j' & k \\ j' & j & r \end{Bmatrix}$$

$$\times \langle jj'JM|\mathbf{u}^{(r)}(1)\cdot\mathbf{u}^{(r)}(2)|jj'JM\rangle \tag{10.38}$$

The exchange term can thus be expressed as the direct term of *another* two body interaction

$$\sum F'^{k}\langle jj'JM|\mathbf{u}^{(k)}(1)\cdot\mathbf{u}^{(k)}(2)|jj'JM\rangle \tag{10.39}$$

where the coefficients F'^{k} are given by

$$F'^{k} = \sum_t (-1)^{j+j'+k+t}(2k+1)\begin{Bmatrix} j & j' & t \\ j' & j & k \end{Bmatrix} G^t \tag{10.40}$$

It is worthwhile to point out that the interaction (10.39) cannot be expressed in general, as a function of $|\mathbf{r}_1 - \mathbf{r}_2|$.

In the coefficients F^k in (10.33) and F'^{k} in (10.40), the reduced matrix elements of $\mathbf{Y}_k$ are between states of single nucleon states with given j, j'. Eq. (8.15) gives the expansion of reduced matrix elements of $\mathbf{Y}_k$ between states with l, l'. There must be a simple relation between these two sets of reduced matrix elements. To obtain this relation we refer again to (9.31) and apply it to the case that $J_1 = J_1' = \frac{1}{2}$ coupled to $J_2 = l$ and to $J_2' = l'$ to form $J = j$, $J' = j'$ and the tensor $\mathbf{T}^{(k_1)}$ is 1, i.e. $k_1 = 0$. In the 9j-symbol on the r.h.s. of (9.35) one argument is zero ($k_1 = 0$) and hence it reduces according to (10.9) to a 6j-symbol. The general result in such cases is the following formula, for any two independent systems 1 and 2

$$(\alpha_1 J_1 \alpha_1 J_2 J \| \mathbf{T}^{(k)}(2) \| \alpha_1' J_1' \alpha_2' J_2' J') = (-1)^{J_1+J_2'+J+k}\sqrt{(2J+1)(2J'+1)} \times (\alpha_2 J_2 \| \mathbf{T}^{(k)} \| \alpha_2' J_2') \begin{Bmatrix} J_2 & J & J_1 \\ J' & J_2' & k \end{Bmatrix} \delta_{\alpha_1 \alpha_1'} \delta_{J_1 J_1'} \tag{10.41}$$

A similar relation holds for the reduced matrix element in a combined system for a tensor operating only on system 1. The phase factor is different and the formula is given by

$$(\alpha_1 J_1 \alpha_2 J_2 J \| \mathbf{T}^{(k)}(1) \| \alpha_1' J_1' \alpha_2' J_2') = (-1)^{J_1+J_2+J'+k}\sqrt{(2J+1)(2J'+1)} \times (\alpha_1 J_1 \| \mathbf{T}^{(k)} \| \alpha_1' J_1') \begin{Bmatrix} J_1 & J & J_2 \\ J' & J_1' & k \end{Bmatrix} \delta_{\alpha_2 \alpha_2'} \delta_{J_2 J_2'} \tag{10.42}$$

Applying (10.41) to the reduced matrix element of $\mathbf{Y}_k$, we put $J_1 = J_1' = \frac{1}{2}$, $J_2 = l$, $J_2' = l'$, $J = j$, $J' = j'$ and obtain

$$(\tfrac{1}{2} l j \| \mathbf{Y}^{(k)} \| \tfrac{1}{2} l' j') = (-1)^{(1/2)+l'+j+k}\sqrt{(2j+1)(2j'+1)} \times \begin{Bmatrix} l & j & \frac{1}{2} \\ j' & l' & k \end{Bmatrix} (l \| \mathbf{Y}_k \| l') \tag{10.43}$$

Recalling (8.15) we can express the reduced matrix element as

$$(\tfrac{1}{2}lj\|\mathbf{Y}_k\|\tfrac{1}{2}l'j') = (-1)^{l+l'+(1/2)+j+k} \times \sqrt{\frac{(2j+1)(2j'+1)(2l+1)(2k+1)(2l'+1)}{4\pi}} \times \begin{pmatrix} l & k & l' \\ 0 & 0 & 0 \end{pmatrix} \begin{Bmatrix} l & j & \tfrac{1}{2} \\ j' & l' & k \end{Bmatrix} \tag{10.44}$$

From (10.44) follow further restrictions on the values of k which appear in the summations in (10.33) and (10.39). The reduced matrix element (10.44) vanishes if $(-1)^{l+k+l'} = -1$. Hence, in the direct term (10.33) the only k values of terms which may not vanish are those for which $(-1)^{2l+k} = 1$ and $(-1)^{2l'+k} = 1$. These values of k are even. On the other hand, in the exchange term (10.39) the only relevant k values are those for which $(-1)^{l+l'+k} = +1$.

The expression (10.44) may be cast in a somewhat simpler form. We can apply in it the relation (10.23) to the product of $3j$-symbol and $6j$-symbol in (10.44). Using the values of $3j$-symbols from the Appendix, we obtain

$$\begin{pmatrix} l & l' & k \\ 0 & 0 & 0 \end{pmatrix} \begin{Bmatrix} l & l' & k \\ j' & j & \tfrac{1}{2} \end{Bmatrix} = -\tfrac{1}{2}[1+(-1)^{l+l'+k}] \times [(2l+1)(2l'+1)]^{-1/2} \begin{pmatrix} j & j' & k \\ \tfrac{1}{2} & -\tfrac{1}{2} & 0 \end{pmatrix} \tag{10.45}$$

Substituting from (10.45) into (10.44) we obtain another expression for the reduced matrix element of $\mathbf{Y}_k$ between states with j and j'

$$
\boxed{(\tfrac{1}{2}lj\|\mathbf{Y}_k\|\tfrac{1}{2}l'j') = (-1)^{j-1/2}\sqrt{\frac{(2j+1)(2k+1)(2j'+1)}{4\pi}} \times \tfrac{1}{2}[1+(-1)^{l+l'+k}]\begin{pmatrix} j & j' & k \\ \frac{1}{2} & -\frac{1}{2} & 0 \end{pmatrix}}
\tag{10.46}
$$

The result (10.46) is essentially independent of l and l'. In the cases of interest, where $(-1)^{l+l'+k} = 1$, the reduced matrix element of $\mathbf{Y}_k$ is a function of j and j' and is independent of l, l'. We can apply (10.46) to the calculation of single nucleon quadrupole moments defined by the expectation value of the operator (5.17) with $\kappa = 0$ in the state with $m = j$. Using the Wigner-Eckart theorem and the actual values of the $3j$-symbols

$$
\begin{pmatrix} j & j & 2 \\ \frac{1}{2} & -\frac{1}{2} & 0 \end{pmatrix} \quad \text{and} \quad \begin{pmatrix} j & 2 & j \\ -j & 0 & j \end{pmatrix}
$$

we obtain

$$
\boxed{\langle \tfrac{1}{2}ljm = j|Q_0^{(2)}|\tfrac{1}{2}ljm = j\rangle = -\frac{2j-1}{2(j+1)}e\langle r^2\rangle}
\tag{10.47}
$$

In (10.47) the integral of r^2 multiplied by $R^2_{nlj}(r)$ is denoted by $\langle r^2 \rangle$.

Matrix elements between two different states of the quadrupole operator (5.17) determine the rates of *electric quadrupole transitions* between these states. Operators defining other, static or dynamic, multipole moments of the *charge* distribution determine rates of other *electric quadrupole transitions*. We shall not present here the theory of electromagnetic transitions. It is described in detail in several books (e.g. de-Shalit and Talmi (1963), Bohr and Mottelson (1969) and de-Shalit and Feshbach (1974)). We will only show how matrix elements of the various operators which determine transition rates are calculated.

The rate of an electric transition of multipole order L between a state with spin J_i and a state with spin J_f is proportional, in view of

(8.29), to

$$B(EL) = \frac{1}{2J_i+1}\left|\left(J_f\left\|\sum_i \frac{e_i}{e} r_i^L \mathbf{Y}_L(\theta_i\phi_i)\right\|J_i\right)\right|^2 \tag{10.48}$$

In (10.48) the e_i are the (effective) charges of the nucleons and e is the electric charge of the proton. In practically all observed transitions, the wave length of the emitted photon is large compared to nuclear dimensions, i.e.

$$\frac{\lambda}{2\pi} = \frac{1}{k} = \frac{\hbar c}{\Delta E} \gg R \tag{10.49}$$

In (10.49), ΔE is the energy difference between initial and final nuclear states (ignoring the energy of the recoiling nucleus) and R is the nuclear radius. In the *long wave approximation*, when (10.49) holds, the transition probability per unit time is given by

$$T(L) = 8\pi c \frac{e^2}{\hbar c} \frac{L+1}{L[(2L+1)!!]^2} k^{2L+1} B(L) \tag{10.50}$$

where the wave number k is defined by (10.49).

The *reduced transition rate* $B(L)$ in (10.50) is equal to (10.48) for electric transitions. Other electromagnetic transitions are due to static and dynamic magnetic moments of the nucleus. The reduced transition rates of such *magnetic* transitions are given by

$$B(ML) = \frac{1}{2J_i+1}\left(\frac{\hbar}{mc}\right)^2 L(2L+1) \times \left|\left(J_f\left\|\sum_i r_i^{L-1}\left[\mathbf{Y}_{L-1}(\theta_i\phi_i)\times\left(\frac{g_{l_i}}{L+1}\mathbf{l}_i + \frac{1}{2}g_{s_i}\mathbf{s}_i\right)\right]^{(L)}\right\|J_i\right)\right|^2 \tag{10.51}$$

where m is the nucleon mass, g_{l_i} the orbital g-factor of the i-th nucleon and g_{s_i} is its spin g-factor.

From (10.48) and (10.51) and the Wigner-Eckart theorem follow general selection rules on electromagnetic transitions. An electric or magnetic transition of multiple order L may take place only if the following inequalities are satisfied

$$|J_i - J_f| \leq L \leq J_i + J_f \tag{10.52}$$

Other selection rules concern the change of parity between the initial and final states. That change of parity for *electric transitions* should be equal to the parity of the spherical harmonics in (10.48), i.e. $(-1)^L$. Since acting by $\boldsymbol{l}_i$ or $\mathbf{s}_i$ on any state does not change its parity, it follows from (10.51) that the change of parity by a *magnetic transition* is equal to $(-1)^{L-1}$.

In the case of $M1$ transitions, $L = 1$, the operator in (10.51) is proportional to $\sum_i (g_{l_i}\boldsymbol{l}_i + g_{s_i}\mathbf{s}_i)$. In jj-coupling states of n identical nucleons in the j-orbit, matrix elements of the $M1$ operator are proportional, according to the Wigner-Eckart theorem, to $\sum_i g_j \mathbf{j}_i = g_j \sum_i \mathbf{j}_i = g_j\mathbf{J}$—where g_j is given by (3.34) or (3.37). Hence, in the long wave approximation, no $M1$ transitions can take place between any two eigenstates of the j^n configuration of identical nucleons. We note in passing, that the g-factors of all states of such configurations are equal to the g-factor of a single j-nucleon.

In subsequent sections it will be shown how to express matrix elements of operators like those in (10.48) or (10.51) in many nucleon systems, in terms of single nucleon matrix elements. Here we evaluate matrix elements of these operators between states of a single nucleon. To obtain the reduced rate of single nucleon electric transitions we use (10.46) and obtain

$$\boxed{B_{s.p}(EL) = \frac{1}{4\pi}(2L+1)(2j_f+1)\begin{pmatrix} j_i & j_f & L \\ \frac{1}{2} & -\frac{1}{2} & 0 \end{pmatrix}^2 \langle r^L\rangle^2} \tag{10.53}$$

The radial integral in (10.53) is defined by

$$\langle r^L\rangle = \int_0^\infty R_{n_i l_i j_i}(r) R_{n_f l_f j_f}(r) r^L dr \tag{10.54}$$

It should be kept in mind that the rate (10.53) vanishes unless, in addition to (10.52) with $J_i = j_i$, $J_f = j_f$, also the following conditions hold

$$|l_i - l_f| \leq L \leq l_i + l_f, \qquad (-1)^{l_i+l_f+L} = 1 \quad \text{for electric transitions} \tag{10.55}$$

To obtain the reduced rates of single nucleon magnetic transitions we first rewrite the tensor product in (10.51) as

$$\left[\mathbf{Y}_{L-1} \times \left(\frac{g_l}{L+1}\boldsymbol{l} + \frac{1}{2}g_s\mathbf{s}\right)\right]^{(L)}$$
$$= \frac{g_l}{L+1}[\mathbf{Y}_{L-1} \times \mathbf{j}]^{(L)} + \left(\frac{1}{2}g_s - \frac{g_l}{L+1}\right)[\mathbf{Y}_{L-1} \times \mathbf{s}]^{(L)} \tag{10.56}$$

The reduced matrix element of the first term on the r.h.s. of (10.56) may be obtained by using (10.28). We notice that the summation there is restricted to only one term with $J'' = J_i = j_i$ and then we use (10.46) and (8.13). In the second term we change the order of coupling and use (9.35) where $T^{(k_1)}$ is replaced by $\mathbf{s}$ and $\mathbf{T}^{(k_2)}$ by $\mathbf{Y}_{L-1}$. The reduced matrix elements of these operators are given by (10.46) and (8.13) respectively. Combining both reduced matrix elements we obtain the single nucleon reduced transition rate in the form

$$B_{s.p.}(ML) = \left(\frac{\hbar}{mc}\right)^2 \frac{L(2L+1)}{2j_i+1}\left[\frac{g_l}{L+1}(-1)^{j_i+j_f+L}\sqrt{2L+1}\right.$$
$$\times (j_f\|\mathbf{Y}_{L-1}\|j_i)\sqrt{j_i(j_i+1)(2j_i+1)}\begin{Bmatrix} L-1 & 1 & L \\ j_i & j_f & j_i \end{Bmatrix}$$
$$+\left(\frac{1}{2}g_s - \frac{g_l}{L+1}\right)\sqrt{(2j_i+1)(2L+1)(2j_f+1)}\,(l_f\|\mathbf{Y}_{L-1}\|l_i)$$
$$\left.\times\sqrt{\frac{3}{2}}\begin{Bmatrix} \frac{1}{2} & l_f & j_f \\ \frac{1}{2} & l_i & j_i \\ 1 & L-1 & L \end{Bmatrix}\right]^2 \langle r^{L-1}\rangle^2 \tag{10.57}$$

The reduced rate (10.51) vanishes unless, in addition to (10.52) with $J_i = j_i$, $J_f = j_f$, also the following relations hold

$$
\begin{aligned}
&|l_i - l_f| \leq L - 1 \leq l_i + l_f, \\
&(-1)^{l_i + l_f + L - 1} = 1 \quad \text{for magnetic transitions}
\end{aligned}
\tag{10.58}
$$

The result (10.57) given in de-Shalit and Talmi (1963) may be greatly simplified (Bohr and Mottelson (1969), de-Shalit and Feshbach (1974)). In the first term in the square bracket we use (10.46) and the explicit expression of the $6j$-symbol. In the second term we use (8.15) for the reduced matrix element and then the identity (9.30) to obtain

$$
\begin{aligned}
&\begin{pmatrix} l_f & l_i & L-1 \\ 0 & 0 & 0 \end{pmatrix} \begin{Bmatrix} \frac{1}{2} & l_f & j_f \\ \frac{1}{2} & l_i & j_i \\ 1 & L-1 & L \end{Bmatrix} \\
&= \begin{pmatrix} l_f & l_i & L-1 \\ 0 & 0 & 0 \end{pmatrix} \begin{Bmatrix} j_f & \frac{1}{2} & l_f \\ j_i & \frac{1}{2} & l_i \\ L & 1 & L-1 \end{Bmatrix} \\
&= \sum \begin{pmatrix} j_f & \frac{1}{2} & l_f \\ m_1 & m_2 & 0 \end{pmatrix} \begin{pmatrix} j_i & \frac{1}{2} & l_i \\ m_3 & m_4 & 0 \end{pmatrix} \begin{pmatrix} j_f & j_i & L \\ m_1 & m_3 & M \end{pmatrix} \\
&\times \begin{pmatrix} \frac{1}{2} & \frac{1}{2} & 1 \\ m_2 & m_4 & M' \end{pmatrix} \begin{pmatrix} L & 1 & L-1 \\ M & M' & 0 \end{pmatrix}
\end{aligned}
\tag{10.59}
$$

The non-vanishing $3j$-symbols in (10.59) must have $m_2 = \pm\frac{1}{2}$, $m_4 = \pm\frac{1}{2}$, $m_1 = -m_2$, $m_3 = -m_4$ and $M = -M'$. Hence, the only terms on the r.h.s. of (10.59) with non-vanishing contributions have $m_1 = m_3 = \frac{1}{2}$, $m_2 = m_4 = -\frac{1}{2}$ $(M = -1,\ M' = 1)$, $m_1 = m_3 = -\frac{1}{2}$, $m_2 = m_4 = \frac{1}{2}$ $(M = 1,\ M' = -1)$, $m_1 = \frac{1}{2}$, $m_3 = -\frac{1}{2}$, $m_2 = -\frac{1}{2}$, $m_4 = \frac{1}{2}$ $(M = M' = 0)$ and $m_1 = -\frac{1}{2}$, $m_3 = \frac{1}{2}$, $m_2 = \frac{1}{2}$, $m_4 = -\frac{1}{2}$ $(M = M' = 0)$.

We now use a relation between $3j$-symbols. In the recursion relation (7.7) we put $m_1 = m_2 = -\frac{1}{2}$ and express the result in terms of $3j$-symbols. After slight rearrangements of the arguments we obtain the following relation

$$\begin{pmatrix} j & j' & J \\ \frac{1}{2} & \frac{1}{2} & -1 \end{pmatrix} = -\frac{1}{2\sqrt{J(J+1)}}\{(2j+1)(-1)^{j+j'+J} + (2j'+1)\}$$
$$\times \begin{pmatrix} j & j' & J \\ \frac{1}{2} & -\frac{1}{2} & 0 \end{pmatrix} \tag{10.60}$$

We now use this relation, the symmetry properties of $3j$-symbols and the actual values of some of them. Keeping in mind the selection rules (10.58) we obtain for (10.57) the following expression

$$\boxed{\begin{aligned} B_{s.p.}(ML) = {} & \left(\frac{\hbar}{mc}\right)^2 \frac{L(2L+1)}{2j_i+1}\frac{1}{4}\frac{(2j_i+1)(2j_f+1)}{4\pi L} \\ & \times \left[\frac{g_l}{L+1}\begin{pmatrix} j_i & j_f & L-1 \\ \frac{1}{2} & -\frac{1}{2} & 0 \end{pmatrix}\right. \\ & \times ((j_i+j_f+L+1)(j_i+L-j_f) \\ & \times (j_f+L-j_i)(j_i+j_f-L+1))^{1/2} \\ & + \left(\frac{1}{2}g_s - \frac{g_l}{L+1}\right)\begin{pmatrix} j_i & j_f & L \\ \frac{1}{2} & -\frac{1}{2} & 0 \end{pmatrix} \\ & \times (L + \tfrac{1}{2}\{(-1)^{(1/2)+l_i-j_i}(2j_i+1) \\ & \left. + (-1)^{(1/2)+l_f-j_f}(2j_f+1)\})\right]^2 \\ & \times \langle r^{L-1}\rangle^2 \end{aligned}} \tag{10.61}$$

The interesting case of a single nucleon magnetic transition is when $L = |j_i - j_f|$. This is the transition whose rate is the fastest. If we denote the larger of j_i, j_f by $j_>$ and the smaller by $j_<$, we may write $j_> - j_< = L$. From all possible combinations of $j_> = l_> \pm \frac{1}{2}$ and $j_< =$

$l_< \pm \frac{1}{2}$ the only one consistent with the selection rules (10.58) is $j_> = l_> + \frac{1}{2}, j_< = l_< - \frac{1}{2}$ in which case $l_> - l_< = L - 1$. In that case, the first term in (10.61) vanishes due to vanishing of the $3j$-symbol in that term $(L - 1 < |j_i - j_f|)$. The reduced transition rate is then given by

$$B_{s.p.}(ML) = \left(\frac{\hbar}{2mc}\right)^2 \left(\frac{1}{2}g_s - \frac{g_l}{L+1}\right)^2 \frac{4L^2(2L+1)}{4\pi} \times (2j_f+1) \begin{pmatrix} j_i & j_f & L \\ \frac{1}{2} & -\frac{1}{2} & 0 \end{pmatrix}^2 \langle r^{L-1} \rangle^2 \tag{10.62}$$

For single proton transitions we put in (10.62) and in preceding expressions of $B(ML)$, just the pure numbers $\frac{1}{2}g_s = 2.79$, $g_l = 1$ whereas for single neutron transitions, $g_l = 0$, $\frac{1}{2}g_s = -1.91$. The factor e^2 in the expression of the square of a nuclear magneton appears explicitly in (10.50) and the factor $(\hbar/2mc)^2$ appears explicitly in (10.62) and in the other expressions of $B(ML)$.

After this digression, let us return to two-nucleon wave functions and matrix elements. The case of two nucleons in the same orbit, $n' = n$, $l' = l$, $j' = j$ is a special case. As noted in the preceding section, in that case, the wave function (9.11) with $T = 1$ vanishes if J is odd due to the symmetry properties (7.12) of the Clebsch-Gordan coefficients. The states with even values of J do not vanish but to be properly normalized, only the first term in the square brackets in (9.11) should be taken and the factor $\sqrt{2}$ omitted. For $T = 0$ states, the only non-vanishing wave functions (9.11) are those with *odd* values of J. Also there, the normalized wave functions are obtained by keeping only the first term in the square brackets in (9.11) and omitting the factor $\sqrt{2}$. In both cases, the matrix element of a two-body interaction between such states is equal to the *direct term* (10.33) *only*. The exchange term is equal in that case to the direct term but its contribution should not be included because of the different normalization of (9.11).

The simple interaction (10.1) depends only on the distance between the nucleons. More complicated interactions are functions of also the (intrinsic) spin vectors of the nucleons. Such interactions occur also in atomic spectroscopy where the main interaction is the Coulomb repulsion between nucleons. They are not due to magnetic spin-spin interaction between electrons. Such terms in the interaction arise from the antisymmetry of the wave functions as will be shown later in this section. In nuclear physics, along with ordinary interactions like (10.1) also *exchange forces* have been used. Among these, *spin exchange* interactions depend explicitly on the spin value S of two nucleons. In states with $S = 1$ the potential is $V_1(|\mathbf{r}_1 - \mathbf{r}_2|)$ whereas for $S = 0$ states it is $V_0(|\mathbf{r}_1 - \mathbf{r}_2|)$. Such an interaction can be expressed in terms of the *spin exchange operator* P_{12}^{σ} which has the eigenvalue $+1$ for symmetric states (with $S = 1$) and -1 for antisymmetric ones ($S = 0$). This expression is as follows

$$\begin{aligned} V(1,2) &= \tfrac{1}{2}(1 + P_{12}^{\sigma})V_1(|\mathbf{r}_1 - \mathbf{r}_2|) + \tfrac{1}{2}(1 - P_{12}^{\sigma})V_0(|\mathbf{r}_1 - \mathbf{r}_2|) \\ &= \tfrac{1}{2}[V_1(|\mathbf{r}_1 - \mathbf{r}_2|) + V_0(|\mathbf{r}_1 - \mathbf{r}_2|)] \\ &\quad + P_{12}^{\sigma}\tfrac{1}{2}[V_1(|\mathbf{r}_1 - \mathbf{r}_2|) - V_0(|\mathbf{r}_1 - \mathbf{r}_2|)] \end{aligned} \tag{10.63}$$

The spin exchange operator may be explicitly constructed from the spin vectors $\mathbf{s}_1$ and $\mathbf{s}_2$ of the two nucleons as

$$P_{12}^{\sigma} = \tfrac{1}{2}(1 + 4(\mathbf{s}_1 \cdot \mathbf{s}_2)) \tag{10.64}$$

To verify that the operator (10.64) has indeed the property of the spin exchange operator we make use of the operator identity

$$2(\mathbf{s}_1 \cdot \mathbf{s}_2) = (\mathbf{s}_1 + \mathbf{s}_2)^2 - \mathbf{s}_1^2 - \mathbf{s}_2^2 = S(S+1) - \tfrac{3}{2} \tag{10.65}$$

From (10.65) follows that $2(\mathbf{s}_1 \cdot \mathbf{s}_2)$ is diagonal if $\mathbf{S}^2 = (\mathbf{s}_1 + \mathbf{s}_2)^2$ is diagonal and its eigenvalues are $2 - \frac{3}{2} = \frac{1}{2}$ for $S = 1$ and $-\frac{3}{2}$ for $S = 0$. Hence, the eigenvalues of (10.64) are $+1$ and -1 respectively.

Introducing the expression (10.64) of P_{12}^{σ} into (10.63) we obtain in addition to an ordinary interaction $(3V_1 + V_0)/4$, also an interaction which has the form $(\mathbf{s}_1 \cdot \mathbf{s}_2)V(|\mathbf{r}_1 - \mathbf{r}_2|)$. Such interactions may also arise directly from various theories. If we expand the spatial part as in (10.1) we obtain for it the expansion in scalar products $\mathbf{Y}_k(1) \cdot \mathbf{Y}_k(2)$. Multiplied by the scalar product of the spins, we find in the k-th term the product

$$(\mathbf{s}_1 \cdot \mathbf{s}_2)(\mathbf{Y}_k(1) \cdot \mathbf{Y}_k(2)) \tag{10.66}$$

In LS-coupling, matrix elements of the product (10.66) can be readily evaluated as we shall see later on. In jj-coupling we must first transform it to a form to which we will be able to apply (10.27). This is achieved by a change of coupling from (10.66) to one in which $\mathbf{s}_1$ is coupled to $\mathbf{Y}_k(1)$ to form a tensor $\mathbf{T}^{(r)}(1)$, irreducible with respect to $\mathbf{j}_1$, and $\mathbf{s}_2$ is coupled to $\mathbf{Y}_k(2)$ to form $\mathbf{T}^{(r)}(2)$. This change-of-coupling transformation is, apart from numerical factors, given by (10.10). Hence we obtain

$$(\mathbf{s}_1 \cdot \mathbf{s}_2)(\mathbf{Y}_k(1) \cdot \mathbf{Y}_k(2)) = \sum_r ([\mathbf{s}_1 \times \mathbf{Y}_k(1)]^{(r)} \cdot [\mathbf{s}_2 \cdot \mathbf{Y}_k(2)]^{(r)})(-1)^{k+r+1} \tag{10.67}$$

In the summation in (10.67) the possible values of r are $r = k - 1, k, k + 1$. To evaluate matrix elements of the scalar products in (10.67) we use the general formula (10.27). The reduced matrix elements which should be evaluated are given by (9.35) as

$$(\tfrac{1}{2}lj\|[\mathbf{s} \times \mathbf{Y}_k]^{(r)}\|\tfrac{1}{2}l'j') = \sqrt{(2j+1)(2r+1)(2j'+1)} \times \begin{Bmatrix} \frac{1}{2} & l & j \\ \frac{1}{2} & l' & j' \\ 1 & k & r \end{Bmatrix} (\tfrac{1}{2}\|\mathbf{s}\|\tfrac{1}{2})(l\|\mathbf{Y}_k\|l') \tag{10.68}$$

The reduced matrix element of $\mathbf{s}$ is given by (8.13) as $\sqrt{3/2}$ and that of $\mathbf{Y}_k$ is given by (8.15).

A more complicated dependence on the spins of interacting nucleons appears in *non-central interactions*. Both the interactions (10.1) and (10.63) are scalars in the spin variables as well as in the spatial coordinates and are called *central interactions*. All admissible two-body interactions must be scalars with respect to the total angular momentum. Such scalars may be constructed as scalar products of a tensor constructed from the spin vectors and a tensor with the same rank in ordinary space. Two spin vectors $\mathbf{s}_1$ and $\mathbf{s}_2$ may combine to form irreducible tensors with ranks $k = 0,1,2$. The case $k = 0$ refers to central interactions considered above. An example of a non-central interaction with $k = 1$ is offered by the mutual spin orbit interaction

$$((\mathbf{s}_1 + \mathbf{s}_2) \cdot \mathbf{L}_{12}) V_{SO}(|\mathbf{r}_1 - \mathbf{r}_2|) \tag{10.69}$$

The L_{12} in (10.69) is the angular momentum associated with the relative coordinate $\mathbf{r}_1 - \mathbf{r}_2$, namely

$$\mathbf{L}_{12} = \frac{1}{2}(\mathbf{r}_1 - \mathbf{r}_2) \times (\mathbf{p}_1 - \mathbf{p}_2) \tag{10.70}$$

Also in this case we can start with the expansion of $V_{SO}(|\mathbf{r}_1 - \mathbf{r}_2|)$ and use change-of-coupling transformations to obtain (10.69) as a linear combination of scalar products of irreducible tensor operators with respect to $\mathbf{j}_1$ and $\mathbf{j}_2$. The situation in LS-coupling is much simpler as we shall soon see. It may be worthwhile to calculate matrix elements of (10.69) in that scheme and then transform to jj-coupling.

Still more complicated dependence on spins is exhibited by tensor forces where $k = 2$. Such interactions occur in electrodynamics where they express the interaction between two magnetic dipoles. These magnetic interactions between atomic electrons are negligible but in nuclei the tensor forces due to *nuclear* interactions play a very important role. Their effect has been first observed in the deuteron and since then they have always been under consideration. The usual form of tensor forces is given by

$$\left[\frac{(\mathbf{s}_1 \cdot (\mathbf{r}_1 - \mathbf{r}_2)) \cdot (\mathbf{s}_2 \cdot (\mathbf{r}_1 - \mathbf{r}_2))}{|\mathbf{r}_1 - \mathbf{r}_2|^2} - \frac{1}{3}(\mathbf{s}_1 \cdot \mathbf{s}_2)\right] V_T(|\mathbf{r}_1 - \mathbf{r}_2|) \tag{10.71}$$

To verify that this is indeed a scalar product of a $k = 2$ tensor in spins and a $k = 2$ tensor in space coordinates, we recouple the vectors in (10.71). Making use of (10.67) we obtain

$$\begin{aligned}
&(\mathbf{s}_1 \cdot (\mathbf{r}_1 - \mathbf{r}_2))(\mathbf{s}_2 \cdot (\mathbf{r}_1 - \mathbf{r}_2)) \\
&\quad = ([\mathbf{s}_1 \times \mathbf{s}_2]^{(2)} \cdot [(\mathbf{r}_1 - \mathbf{r}_2) \times (\mathbf{r}_1 - \mathbf{r}_2)]^{(2)}) \\
&\qquad - ([\mathbf{s}_1 \times \mathbf{s}_2]^{(1)} \cdot [(\mathbf{r}_1 - \mathbf{r}_2) \times (\mathbf{r}_1 - \mathbf{r}_2)]^{(1)}) \\
&\qquad + [\mathbf{s}_1 \times \mathbf{s}_2]^{(0)} [(\mathbf{r}_1 - \mathbf{r}_2) \times (\mathbf{r}_1 - \mathbf{r}_2)]^{(0)}
\end{aligned} \tag{10.72}$$

The second term on the r.h.s. of (10.72) vanishes since the vector product of a vector with itself vanishes if its components commute.

The last term, divided by $|\mathbf{r}_1 - \mathbf{r}_2|^2$ is simply cancelled by the scalar which is the second term in (10.71). Thus, (10.71) has indeed the transformation properties stated above. Also in this case it is possible to expand $V_T(|\mathbf{r}_1 - \mathbf{r}_2|)$ in a linear combination of scalar products of $\mathbf{Y}_k(1)$ and $\mathbf{Y}_k(2)$ as carried out above for $V(|\mathbf{r}_1 - \mathbf{r}_2|)$ in (10.1). It is then possible to expand $[(\mathbf{r}_1 - \mathbf{r}_2) \times (\mathbf{r}_1 - \mathbf{r}_2)]^{(2)}(\mathbf{Y}_k(1) \cdot \mathbf{Y}_k(2))$ as a linear combination of products of the form $\left[\mathbf{Y}_k(1) \times \mathbf{Y}_{k'}(2)\right]^{(2)}$. Then it is possible to use (10.10) to change the couplings in the scalar products $([\mathbf{s}_1 \times \mathbf{s}_2]^{(2)} \cdot [\mathbf{Y}_k(1) \times \mathbf{Y}_{k'}(2)]^{(2)})$ into a combination of scalar products of the form $([\mathbf{s}_1 \times \mathbf{Y}_k(1)]^{(r)} \cdot [\mathbf{s}_2 \times \mathbf{Y}_{k'}(2)]^{(r)})$. We shall return later to this expansion. Also in this case the calculation of matrix elements in LS-coupling is much simpler. We will now consider matrix elements of two-body interactions in this coupling scheme.

The interactions considered above, (10.1), (10.69) and (10.71) are scalar products of a tensor of rank k ($k = 0, 1, 2$) in the intrinsic spins of the nucleons, $\mathcal{T}^{(k)}$, and a tensor with the same rank in the space coordinates, $\mathcal{W}^{(k)}$. We apply the formula (10.27) to LS-coupling and obtain

$$\langle SLJM|(\mathcal{T}^{(k)} \cdot \mathcal{W}^{(k)})|S'L'JM\rangle = (-1)^{L+S'+J} \begin{Bmatrix} S & L & J \\ L' & S' & k \end{Bmatrix} \times (S\|\mathcal{T}^{(k)}\|S')(L\|\mathcal{W}^{(k)}\|L') \tag{10.73}$$

The reduced matrix elements of the spin operator are given by (10.27) for $k = 0$ and by (9.35) for $k = 2$. In the case, $k = 1$, of the spin operator in (10.69), the reduced matrix element is given directly by (8.13). Actually the case of a central interaction with $k = 0$ is even simpler. In (10.63) we can replace P_{12}^{σ} by $(-1)^{S+1}$. The matrix elements of the tensor $\mathcal{W}^{(k)}$ may be similarly calculated.

Let us consider the case of an ordinary, spin independent potential interaction. Since it contains no spin operators it must be a scalar under spatial rotations and thus commute with components of $\mathbf{L}$. Hence, and in agreement with the Wigner-Eckart theorem for $k = 0$, it may have non-vanishing matrix elements only between states with the same value of L. A more general central interaction like (10.63) or (10.66), is a scalar in ordinary space and a scalar with respect to spin. Hence, its non-vanishing matrix elements between LS-coupling wave functions, including spin states, are diagonal both in L and in S. The matrix elements of such an interaction are obtained from (10.73) by putting $k = 0$, $S' = S$, $L' = L$ to be equal to

$$\frac{1}{\sqrt{(2S+1)(2L+1)}}(S\|\mathcal{T}^{(0)}\|S)(L\|\mathcal{W}^{(0)}\|L) \tag{10.74}$$

and are *independent* of J. If there are, in addition, small non-central interactions, all states with the same S and L, with J satisfying $|S-L| \geq J \geq S+L$ will be nearly degenerate. Such states form a *multiplet*. Also in LS-coupling, matrix elements between symmetric states and antisymmetric states in space coordinates, vanish due to the symmetry of an ordinary (potential) interaction in the space coordinates.

From (10.74) it is clearly seen that any central interaction is diagonal in the LS-coupling scheme. Even in the case of non-central interaction, with $k = 1$ and $k = 2$ in (10.73), the diagonal as well as non-diagonal matrix elements assume a simple form. In the jj-coupling scheme, the single nucleon spin orbit interactions are diagonal. Nucleon-nucleon interactions have non-diagonal matrix elements in that scheme. If the effective mutual interaction of the nucleons is large compared with the spin-orbit splittings it tends to break the jj-coupling scheme and make wave functions move closer to the LS-coupling limit. From the experimental evidence it seems that jj-coupling is the prevalent coupling scheme in nuclei. Still, the mutual interaction may cause in certain cases important deviations from pure jj-coupling. Experimental evidence, some of which will be described in the following, indicates that the higher the isospin of the state, the closer it is to the jj-coupling limit.

Between states (9.14), matrix elements of that interaction are obtained by a procedure similar to that adopted above for jj-coupling. They are given by

$$\int \Phi^*_{ll'LM_L} V(|\mathbf{r}_1 - \mathbf{r}_2|)\Phi_{ll'LM}$$
$$= \langle l_1 l'_2 LM_L|V|l_1 l'_2 LM_L\rangle \pm \langle l_1 l'_2 LM_L|V|l_2 l'_1 LM_L\rangle \tag{10.75}$$

The indices 1 and 2 of l and l' indicate that the state with this angular momentum is of nucleon 1 or nucleon 2. Thus, these matrix elements are linear combinations of direct and exchange terms. The total wave function must be fully antisymmetric. Hence, a plus sign appears in (10.75) if the spin-isospin part of the wave function is antisymmetric. This occurs for symmetric spin function ($S = 1$) and antisymmetric isospin ($T = 0$) or vice versa, $S = 0$, $T = 1$. The minus sign appears in (10.74) if the spin-isospin part is symmetric, which occurs for $S = 1, T = 1$ or $S = 0, T = 0$. We may thus express the matrix element (10.75) in various spin and isospin states as

$$\begin{aligned}\langle l_1 l'_2 LM_L|V|l_1 l'_2 LM_L\rangle &- (-1)^{S+T}\langle l_1 l'_2 LM_L|V|l_2 l'_1 LM_L\rangle \\ &= \langle ll'LM_L|V|ll'LM_L\rangle - (-1)^{S+T+l+l'-L}\langle ll'LM_L|V|l'lLM_L\rangle\end{aligned} \tag{10.76}$$

In the last equality the order of coupling on the r.h.s. of the exchange term has been changed in accordance with (7.11). This will make it possible to apply to that term the formula (10.27).

Expanding $V(|\mathbf{r}_1 - \mathbf{r}_2|)$ according to (10.1) and (10.4) we obtain for the direct term the result

$$\begin{aligned}\langle ll'LM_L|V|ll'LM_L\rangle &= \sum_k \frac{4\pi}{2k+1}(l\|\mathbf{Y}_k\|l)(l'\|\mathbf{Y}_k\|l') \\ &\quad\times (-1)^{l+l'+L}\begin{Bmatrix} l & l' & L \\ l' & l & k \end{Bmatrix} \\ &\quad\times \int R^2_{nl}(r_1)R^2_{n'l'}(r_2)v_k(r_1,r_2)\,dr_1\,dr_2 \\ &= \sum_k F^k(-1)^{l+l'+L}\begin{Bmatrix} l & l' & L \\ l' & l & k \end{Bmatrix}\end{aligned} \tag{10.77}$$

The exchange term can be similarly brought into the form

$$\begin{aligned}&(-1)^{l+l'+L}\langle ll'LM_L|V|l'lLM_L\rangle \\ &\quad= \sum_k \frac{4\pi}{2k+1}(l\|\mathbf{Y}_k\|l')^2\begin{Bmatrix} l & l' & L \\ l & l' & k \end{Bmatrix} \\ &\quad\times \int R_{nl}(r_1)R_{n'l'}(r_1)R_{nl}(r_2)R_{n'l'}(r_2)v_k(r_1,r_2)\,dr_1\,dr_2 \\ &\quad= \sum_k G^k\begin{Bmatrix} l & l' & L \\ l & l' & k \end{Bmatrix}\end{aligned} \tag{10.78}$$

We shall not further discuss here these matrix elements in the general case. In the next section we will consider a very special interaction.

The case of two nucleons in the same orbit, $n' = n$, $l' = l$, is a special case. Due to the symmetry property (7.12) of the Clebsch-Gordan coefficients, states with *even* values of L are symmetric and have $S = 0$, $T = 1$ or $S = 1$, $T = 0$ whereas the states with *odd* values of L are antisymmetric and are multiplied by $S = 0$, $T = 0$ or $S = 1$, $T = 1$ spin-isospin states. In both cases, only the first term in the square brackets of (9.14) should be kept and the factor $\sqrt{2}$ omitted. Hence, in both cases the matrix element is equal to the direct term.

In (10.77) and (10.78), the notation F^k and G^k was used for the coefficients of the expansion. Comparing with (10.34) and (10.37) we see that they are not equal to the coefficients in jj-coupling for which the same notation has been used. With little care no confusion should arise and taking such care is more than compensated by not introducing more symbols.

11

Short Range Potentials—the δ-Interaction

As mentioned above, in the case of a Hamiltonian of nucleons moving independently in a central potential well, all states of several nucleons in the same j-orbit are degenerate. Such degeneracy is never observed in actual nuclei and is easily removed if the mutual interaction of the nucleons is added to the Hamiltonian. In order to obtain a rough idea about the order of levels in a two nucleon configuration, let us consider a specific case of mutual interaction. An ordinary interaction whose range is short in comparison with the spatial extension of the single nucleon wave functions may be approximated by a *zero range interaction*. This is an extreme limit and some care must be taken when using it. In the present context, however, for evaluating matrix elements within a given configuration, a $\delta(1,2) = \delta(|\mathbf{r}_1 - \mathbf{r}_2|)$ potential behaves very regularly. Such a potential has been used since the early days of the shell model. The results reproduce fairly well certain features of nuclear spectra as we shall see in the following.

To calculate matrix elements of the δ-potential, it is more convenient to use the LS-coupling scheme. States of two nucleons in any l, l' orbits are antisymmetric in space coordinates if the spin-isospin states are symmetric ($S = 1$, $T = 1$ or $S = 0$, $T = 0$). The contributions to the integral (10.74) over a δ-function arise only from the point

$\mathbf{r}_1 = \mathbf{r}_2$. Hence, matrix elements of the δ-potential vanish between *any* two such space antisymmetric states. Non-vanishing matrix elements of that potential can be obtained only between the *space symmetric* states with $S = 1$, $T = 0$ or $S = 1$, $T = 0$. Let us now consider such matrix elements for two *different* l, l' orbits

$$\int \Phi^{s*}_{ll'LM_L}(1,2)\delta(1,2)\Phi^{s}_{ll'LM_L}(1,2) \tag{11.1}$$

where $\Phi^{s}_{ll'LM_L}$ is the symmetric function as given in (9.14)

$$\Phi^{s}_{ll'LM_L}(1,2) = \frac{1}{\sqrt{2}}\sum_{mm'}(lml'm' \mid ll'LM_L) \times [\phi_{nlm}(\mathbf{r}_1)\phi_{n'l'm'}(\mathbf{r}_2) + \phi_{nlm}(\mathbf{r}_2)\phi_{n'l'm'}(\mathbf{r}_1)] \tag{11.2}$$

Since the contributions to the integral (11.1) are only from the point $\mathbf{r}_1 = \mathbf{r}_2$, we can replace (11.2) in the integrand by

$$\sqrt{2}\sum_{mm'}(lml'm \mid ll'LM_L)\phi_{nlm}(r,\theta,\phi)\phi_{n'l'm'}(r,\theta,\phi) = \frac{1}{r^2}\sqrt{2}R_{nl}(r)R_{n'l'}(r)\sum_{mm'}(lml'm' \mid ll'LM_L)Y_{lm}(\theta,\phi)Y_{l'm'}(\theta,\phi) \tag{11.3}$$

The last equality follows from the form (3.7) of single nucleon wave functions. The sum of products of spherical harmonics, according to (7.31), is given by

$$(-1)^L\sqrt{\frac{(2l+1)(2l'+1)}{4\pi}}\begin{pmatrix} l & L & l' \\ 0 & 0 & 0 \end{pmatrix} Y_{lm}(\theta,\phi) \tag{11.4}$$

The δ-potential can be expressed in spherical coordinates as

$$\frac{1}{r_1 r_2}\delta(r_1 - r_2)\delta(\cos\theta_1 - \cos\theta_2)\delta(\phi_1 - \phi_2) \tag{11.5}$$

Substituting the expression (11.4) into (11.3) and the latter as well as (11.5) into the integral (11.1), we obtain the value

$$\boxed{2\frac{(2l+1)(2l'+1)}{4\pi}\begin{pmatrix} l & l' & L \\ 0 & 0 & 0 \end{pmatrix}^2 \int R_{nl}^2(r)R_{n'l'}^2(r)\frac{1}{r^2}dr} \tag{11.6}$$

A direct consequence of the result (11.6) is that the integral (11.1) over a δ-potential vanishes for all states for which L satisfies $(-1)^{l+l'+L} = -1$. The only non-vanishing matrix elements of a δ-potential are those for which

$$(-1)^{l+l'+L} = +1 \tag{11.7}$$

The factor 2 in (11.6) arises from the fact that for a δ-potential the direct and exchange terms are equal. This fact would have been obvious if we have kept the form (11.2) of the wave functions with $\mathbf{r}_1 = \mathbf{r}_2$. To emphasize this point we can write (11.6) explicitly in terms of S and T as in (10.75). The energy shift of levels with given L, S and T from the sum of single nucleon energies in l and l' orbits, due to the δ-potential is (Pryce 1952, Talmi 1953)

$$\begin{aligned}\Delta E_{ll'TSL} &= (1-(-1)^{S+T})(2l+1)(2l'+1)\begin{pmatrix} l & l' & L \\ 0 & 0 & 0 \end{pmatrix}^2 \\ &\quad\times\frac{1}{4\pi}\int R_{nl}^2(r)R_{n'l'}^2(r)\frac{1}{r^2}dr \\ &= (1-(-1)^{S+T})C_0(2l+1)(2l'+1)\begin{pmatrix} l & l' & L \\ 0 & 0 & 0 \end{pmatrix}^2\end{aligned} \tag{11.8}$$

The radial integral C_0 depends on nl and $n'l'$ but this dependence will not be explicitly displayed.

The fact that the expectation value of the δ-potential vanishes for some states symmetric in space coordinates (those that do not satisfy (11.7)) may seem surprising. It should be realized, however, that whereas space antisymmetric states *must* vanish at $\mathbf{r}_1 = \mathbf{r}_2$, space symmetric states *may* also vanish at $\mathbf{r}_1 = \mathbf{r}_2$. There are symmetric states which are actually antisymmetric in r_1, r_2 and antisymmetric in $\theta_1\phi_1$,

$\theta_2\phi_2$. Such symmetric states vanish when either $r_1 = r_2$ or $\theta_1 = \theta_2$, $\phi_1 = \phi_2$ and *a fortiori* if $\mathbf{r}_1 = \mathbf{r}_2$. Coming back to (11.2) we realize from (7.11) that if $l' = l$ but $n' \neq n$ and $(-1)^{l+l-L} = -1$ then the wave function is antisymmetric with respect to interchange of the angles $\theta_1\phi_1$ with $\theta_2\phi_2$. Due to the overall symmetry of (11.2), its radial part is $[R_{nl}(r_1)R_{n'l}(r_2) - R_{nl}(r_2)R_{n'l}(r_1)]/\sqrt{2}$.

If the l-orbit is identical to the l'-orbit (i.e. $l = l'$, $n = n'$) we may still use the states (11.2) with the normalization factor $\frac{1}{2}$ rather than $1/\sqrt{2}$. In that case, states with odd values of L are antisymmetric in space coordinates. Only states with even values of L have non-vanishing matrix elements of the δ-potential. In these latter states, $(-1)^{T+S} = -1$ and we obtain the energy shift of such levels due to the δ-potential, by putting $l' = l$ in (11.8) and dividing by 2, as

$$\boxed{\Delta E_{l^2L} = \tfrac{1}{2}(1 + (-1)^L)C_0(2l+1)^2 \begin{pmatrix} l & l & L \\ 0 & 0 & 0 \end{pmatrix}^2} \tag{11.9}$$

The coefficient C_0 in (11.9) is equal in this case to

$$C_0 = \frac{1}{4\pi} \int R_{nl}^4(r) \frac{1}{r^2} dr \tag{11.10}$$

The special case of $L = 0$ is particularly simple. Substituting into (11.9) the value of the $3j$-symbol taken from (7.19) we obtain

$$\Delta E_{l^2L=0} = (2l+1)C_0 \tag{11.11}$$

It is now a simple matter to obtain matrix elements of the δ-potential in jj-coupling. If we use the expansion of these functions given by (9.17), we see that for $T = 1$, only one term in it, the one with $S = 0, L = J$, may contribute. Hence, we conclude that $T = 1$ matrix elements of the δ-potential vanish unless

$$\boxed{(-1)^{l+l'+J} = 1} \tag{11.12}$$

In even-even nuclei, states with odd parity could be due to the excitation of one nucleon to an orbit with different parity. For short-ranged interaction the lowest odd parity states in even-even nuclei are expected to have, according to (11.12), odd values of J (Talmi 1953). This feature is indeed consistent with the experimental evidence (Glaubman 1953).

The transformation coefficient between the

$$\psi_{ll'jj'JM} \qquad \text{and} \qquad \psi_{ll'S=0,L=J,JM} \qquad \text{for} \quad l \neq l'$$

is given according to (9.23), (9.27) and (10.9) by

$$\sqrt{(2j+1)(2J+1)(2j'+1)} \begin{Bmatrix} \frac{1}{2} & l & j \\ \frac{1}{2} & l' & j' \\ 0 & J & J \end{Bmatrix}$$

$$= (-1)^{j+(1/2)+l'+J} \sqrt{\frac{(2j+1)(2j'+1)}{2}} \begin{Bmatrix} l & j & \frac{1}{2} \\ j' & l' & J \end{Bmatrix} \tag{11.13}$$

From this coefficient and (11.8) we obtain the energy shift for $T = 1$ states with J satisfying (11.12), the result

$$C_0(2j+1)(2j'+1)(2l+1)(2l'+1) \left[\begin{pmatrix} l & l' & L \\ 0 & 0 & 0 \end{pmatrix} \begin{Bmatrix} l & j & \frac{1}{2} \\ j' & l' & J \end{Bmatrix} \right]^2 \tag{11.14}$$

This result can be further modified. The product of the $3j$-symbol and $6j$-symbol in (11.14) was evaluated in (10.45). Using it for $J = L$ values satisfying (11.12) we obtain (de-Shalit 1953)

$$\boxed{\Delta E_{jj'T=1,J} = C_0(2j+1)(2j'+1) \begin{pmatrix} j & j' & J \\ \frac{1}{2} & -\frac{1}{2} & 0 \end{pmatrix}^2} \tag{11.15}$$

If the nucleons are in the *same* l-orbit we should use (11.9) rather than (11.8). If in that case $j' \neq j$ (i.e. $j = l \pm \frac{1}{2}$, $j' = l \mp \frac{1}{2}$) the transformation coefficient (11.13) should be multiplied by $\sqrt{2}$ as explained in introducing Table 9.1. In such cases, (11.15) is still the correct expression. If, however, $j' = j$ (i.e. $j = l \pm \frac{1}{2}$, $j' = l \pm \frac{1}{2}$) then the correct result for the expectation value of the δ-potential is given by

$$\boxed{\Delta E_{j^2J} = \tfrac{1}{2}C_0(2j+1)^2 \begin{pmatrix} j & j & J \\ \frac{1}{2} & -\frac{1}{2} & 0 \end{pmatrix}^2} \tag{11.16}$$

The result (11.16) assumes a very simple form for the case $J = 0$. In that case we use the value of the $3j$-symbol from (7.19) and obtain

$$\Delta E_{jjT=1,J=0} = \tfrac{1}{2}(2j+1)C_0 \tag{11.17}$$

The method used above for $T = 1$ states is less convenient for $T = 0$ states. Such states with non-vanishing matrix elements of the δ-potential must be symmetric in space coordinates and thus must have $S = 1$. In the expansion (9.17) there are now several states with given J. Another approach is to use the tensor expansion of the δ-potential. Due to (11.5) this is given by

$$\boxed{\frac{1}{2\pi}\frac{\delta(r_1 - r_2)}{r_1 r_2}\delta(\cos\omega_{12} - 1) = \frac{\delta(r_1 - r_2)}{r_1 r_2}\sum_k \frac{2k+1}{4\pi}P_k(\cos\omega_{12})} \tag{11.18}$$

Comparing with (10.1) we find that in this case

$$v_k(r_1, r_2) = \frac{2k+1}{4\pi}\frac{\delta(r_1 - r_2)}{r_1 r_2} \tag{11.19}$$

With the $v_k(r_1, r_2)$ given by (11.19) we see that the radial integrals in (10.34) are equal to

$$\int R_{nlj}^2(r_1)R_{n'l'j'}^2(r_2)v_k(r_1, r_2)\,dr_1\,dr_2$$

$$= (2k+1)\frac{1}{4\pi}\int R_{nlj}^2(r)R_{n'l'j'}^2(r)\frac{1}{r^2}\,dr = (2k+1)C_0 \tag{11.20}$$

In the short range limit, the radial integrals are proportional to $(2k + 1)$. This dependence should be contrasted with the one in the long range limit. If the potential of the two body interaction may be approximated by a constant C in the region where the radial functions do not vanish, we obtain from (10.3)

$$v_k(r_1, r_2) = \frac{2k+1}{2} C \int_{-1}^{+1} P_k(\cos\omega_{12}) d\cos\omega_{12} = C\delta_{k0} \tag{11.21}$$

Thus, in the long range limit only the radial integral with $k = 0$ does not vanish.

Let us first use LS-coupling wave functions. The direct term is given according to (10.77) by using there the expression (11.19) for $v_k(r_1, r_2)$, as

$$\sum_k (-1)^{l+l'+L} \begin{Bmatrix} l & l' & L \\ l' & l & k \end{Bmatrix} (l\|\mathbf{Y}_k\|l)(l'\|\mathbf{Y}_k\|l')$$

$$\times \frac{4\pi}{2k+1} \int R_{nl}^2(r) R_{n'l'}^2(r) \frac{2k+1}{4\pi} \frac{dr}{r^2} \tag{11.22}$$

Substituting for the reduced matrix elements their values from (8.15) we obtain for the direct term (11.22) the form

$$\frac{1}{4\pi} \int R_{nl}^2(r) R_{n'l'}^2(r) \frac{dr}{r^2} (2l+1)(2l'+1) \sum_k (-1)^L (2k+1)$$

$$\times \begin{pmatrix} l & l & k \\ 0 & 0 & 0 \end{pmatrix} \begin{pmatrix} l' & l' & k \\ 0 & 0 & 0 \end{pmatrix} \begin{Bmatrix} l & l' & L \\ l' & l & k \end{Bmatrix}$$

$$= C_0 (2l+1)(2l'+1) \begin{pmatrix} l & l' & L \\ 0 & 0 & 0 \end{pmatrix} \begin{pmatrix} l' & l & L \\ 0 & 0 & 0 \end{pmatrix} \tag{11.23}$$

The last equality in (11.23) is due to the identity (10.24) and to the fact that in the summation on the l.h.s. of (11.23) unless k is even the $3j$-symbols vanish. The coefficient C_0 was defined in (11.8).

The exchange term as in (10.78) may be similarly expressed as

$$
\begin{aligned}
4\pi C_0(-1)^{l+l'+L}&\sum_k(-1)^L(l\|\mathbf{Y}_k\|l')(l'\|\mathbf{Y}_k\|l)\begin{Bmatrix} l & l' & L \\ l' & l & k \end{Bmatrix} \\
&= C_0(-1)^L(2l+1)(2l'+1)\sum_k(-1)^L(2k+1) \\
&\quad\times\begin{pmatrix} l & l' & k \\ 0 & 0 & 0 \end{pmatrix}\begin{pmatrix} l & l' & k \\ 0 & 0 & 0 \end{pmatrix}\begin{Bmatrix} l & l' & L \\ l & l' & k \end{Bmatrix} \\
&= (-1)^{l+l'+L}C_0(2l+1)(2l'+1)\begin{pmatrix} l & l' & L \\ 0 & 0 & 0 \end{pmatrix}\begin{pmatrix} l & l' & L \\ 0 & 0 & 0 \end{pmatrix}
\end{aligned}
\tag{11.24}
$$

The last equality in (11.24) is due to the identity (10.24) and to the fact that unless $(-1)^{l+l'+k} = +1$ the $3j$-symbols in the summation vanish. We see that the sum of the direct and exchange terms, for $(-1)^{l+l'+L} = 1$, is equal to the result (11.6) and to zero for $(-1)^{l+l'+L} = -1$. This can be summarized as in (10.76) for the case of *different* l and l'-orbits by

$$
\Delta E_{ll'TSL} = (1-(-1)^{S+T})C_0(2l+1)(2l'+1)\begin{pmatrix} l & l' & L \\ 0 & 0 & 0 \end{pmatrix}^2
\tag{11.25}
$$

which is identical with (11.8). The situation is more complicated in jj-coupling to which we now turn.

In the jj-coupling scheme we first consider different j, j' orbits. The direct term is given according to (10.33) and (10.34) by

$$
\begin{aligned}
\sum_k&(-1)^{j+j'+J}(lj\|\mathbf{Y}_k\|lj)(l'j'\|\mathbf{Y}_k\|l'j')\begin{Bmatrix} j & j' & J \\ j' & j & k \end{Bmatrix} \\
&\quad\times\frac{4\pi}{2k+1}\int R_{nl}^2(r)R_{n'l'}^2(r)\frac{2k+1}{4\pi}\frac{dr}{r^2} \\
&= 4\pi C_0\sum_k(-1)^{j+j'+J}(lj\|\mathbf{Y}_k\|lj)(l'j'\|\mathbf{Y}_k\|l'j')\begin{Bmatrix} j & j' & J \\ j' & j & k \end{Bmatrix}
\end{aligned}
\tag{11.26}
$$

With the values (10.46) for the reduced matrix elements we obtain for (11.26) the expression

$$
\begin{aligned}
&C_0(2j+1)(2j'+1)\sum_k \tfrac{1}{2}(1+(-1)^k)(-1)^{J+1}(2k+1) \\
&\quad\times \begin{pmatrix} j & j & k \\ \frac{1}{2} & -\frac{1}{2} & 0 \end{pmatrix} \begin{pmatrix} j' & j' & k \\ \frac{1}{2} & -\frac{1}{2} & 0 \end{pmatrix} \begin{Bmatrix} j & j' & J \\ j' & j & k \end{Bmatrix} \\
&= \tfrac{1}{2}C_0(2j+1)(2j'+1)\sum_k (-1)^{J+k}(2k+1) \\
&\quad\times \begin{pmatrix} j & j & k \\ \frac{1}{2} & -\frac{1}{2} & 0 \end{pmatrix} \begin{pmatrix} j' & j' & k \\ -\frac{1}{2} & \frac{1}{2} & 0 \end{pmatrix} \begin{Bmatrix} j & j' & J \\ j' & j & k \end{Bmatrix} \\
&\quad+ \tfrac{1}{2}C_0(2j+1)(2j'+1)\sum_k (-1)^{J+k+1}(2k+1) \\
&\quad\times \begin{pmatrix} j & j & k \\ \frac{1}{2} & -\frac{1}{2} & 0 \end{pmatrix} \begin{pmatrix} j' & j' & k \\ \frac{1}{2} & -\frac{1}{2} & 0 \end{pmatrix} \begin{Bmatrix} j & j' & J \\ j' & j & k \end{Bmatrix}
\end{aligned}
\tag{11.27}
$$

In the first summation on the r.h.s. of (11.27) the relation (7.28) was used for one of the $3j$-symbols. We now make use of the identity (10.24) to obtain for (11.27) the form

$$
\begin{aligned}
&\tfrac{1}{2}C_0(2j+1)(2j'+1)\begin{pmatrix} j & j' & J \\ \frac{1}{2} & \frac{1}{2} & -1 \end{pmatrix}\begin{pmatrix} j' & j & J \\ -\frac{1}{2} & -\frac{1}{2} & 1 \end{pmatrix} \\
&\quad+ \tfrac{1}{2}C_0(2j+1)(2j'+1)\begin{pmatrix} j & j' & J \\ \frac{1}{2} & -\frac{1}{2} & 0 \end{pmatrix}\begin{pmatrix} j' & j & J \\ \frac{1}{2} & -\frac{1}{2} & 0 \end{pmatrix}
\end{aligned}
\tag{11.28}
$$

The result (11.28) may be further simplified by using the relation (10.60) between two $3j$-symbols, i.e.

$$
\begin{aligned}
\begin{pmatrix} j & j' & J \\ \frac{1}{2} & \frac{1}{2} & -1 \end{pmatrix} &= -\frac{1}{2\sqrt{J(J+1)}}\{(2j+1)(-1)^{j+j'+J} + (2j'+1)\} \\
&\quad\times \begin{pmatrix} j & j' & J \\ \frac{1}{2} & -\frac{1}{2} & 0 \end{pmatrix}
\end{aligned}
\tag{11.29}
$$

Introducing this value of the $3j$-symbol into (11.28) we obtain for the direct term, after using the symmetry properties of the $3j$-symbols, the following result (de-Shalit 1953)

$$\tfrac{1}{2}C_0(2j+1)(2j'+1)\begin{pmatrix} j & j' & J \\ \tfrac{1}{2} & -\tfrac{1}{2} & 0 \end{pmatrix}^2 \times \left[1+\frac{\{(2j+1)+(-1)^{j+j'+J}(2j'+1)\}^2}{4J(J+1)}\right] \tag{11.30}$$

The exchange term can be similarly handled. According to (10.36) and (10.37) it is given for the $v_k(r_1,r_2)$ of (11.19) by

$$\begin{aligned} 4\pi C_0 \sum_k (lj\|\mathbf{Y}_k\|l'j')^2 \begin{Bmatrix} j & j' & J \\ j & j' & k \end{Bmatrix} &= C_0(2j+1)(2j'+1)\sum_k \tfrac{1}{2}(1+(-1)^{l+l'+k}) \\ &\quad\times \begin{pmatrix} j & j' & k \\ \tfrac{1}{2} & -\tfrac{1}{2} & 0 \end{pmatrix}^2 \begin{Bmatrix} j & j' & J \\ j & j' & k \end{Bmatrix} \\ &= \tfrac{1}{2}C_0(2j+1)(2j'+1) \\ &\quad\times \left[\begin{pmatrix} j & j' & J \\ \tfrac{1}{2} & \tfrac{1}{2} & -1 \end{pmatrix}^2 - (-1)^{l+l'+J}\begin{pmatrix} j & j' & J \\ \tfrac{1}{2} & -\tfrac{1}{2} & 0 \end{pmatrix}^2\right] \end{aligned} \tag{11.31}$$

Also in deriving (11.31) use has been made of the identity (10.24). To use it, various symmetry operations were carried out on the $3j$-symbols paying careful attention to the phases. The expression (11.31) may be further simplified by using (11.29) leading to the following result for the exchange term

$$-\tfrac{1}{2}C_0(2j+1)(2j'+1)\begin{pmatrix} j & j' & J \\ \tfrac{1}{2} & -\tfrac{1}{2} & 0 \end{pmatrix}^2 \times \left[(-1)^{l+l'+J}-\frac{\{(2j+1)+(-1)^{j+j'+J}(2j'+1)\}^2}{4J(J+1)}\right] \tag{11.32}$$

In the case of $T = 1$ states, the exchange term should be *subtracted* from the direct term. Looking at (11.30) and (11.32) we see that the result is zero if $(-1)^{l+l'+J} = -1$ and for $(-1)^{l+l'+J} = +1$ the result is identical with (11.15) obtained above.

In the case of $T = 0$ states, there are non-vanishing matrix elements for both signs of $(-1)^{l+l'+J}$. In this case the exchange term (11.32) should be *added* to the direct term (11.30). For $(-1)^{l+l'+J} = +1$ the result is

$$\Delta E_{jj'T=0,J} = C_0(2j+1)(2j'+1)\begin{pmatrix} j & j' & J \\ \frac{1}{2} & -\frac{1}{2} & 0 \end{pmatrix}^2 \times \frac{\{(2j+1)+(-1)^{j+j'+J}(2j'+1)\}^2}{4J(J+1)} \quad \text{for} \quad (-1)^{l+l'+J} = 1 \tag{11.33}$$

In the case of the opposite sign the exchange term becomes equal to the direct term and the result is

$$\Delta E_{jj'T=0,J} = C_0(2j+1)(2j'+1)\begin{pmatrix} j & j' & J \\ \frac{1}{2} & -\frac{1}{2} & 0 \end{pmatrix}^2 \times \left[1+\frac{\{(2j+1)+(-1)^{j+j'+J}(2j'+1)\}^2}{4J(J+1)}\right] \quad \text{for} \quad (-1)^{l+l'+J} = -1 \tag{11.34}$$

It is worth while to point out that the $3j$-symbol in (11.34) does not vanish in spite of the relation $(-1)^{l+l'+J} = -1$. The expression (10.45) was obtained by taking l and l' to be the orbital angular momenta of the j- and j'-orbits respectively. If in the present case, we put in it $k = J$, both sides vanish. The relation (10.45) holds, however, for *any* values of $l = l_1$, $l' = l_2$ which satisfy with $k = J$ the triangular condition *and* are related to j, j' by $j = l_1 \pm \frac{1}{2}$, $j' = l_2 \pm \frac{1}{2}$. For instance, if $j = l + \frac{1}{2}$, $j' = l' + \frac{1}{2}$, it is possible to use in (10.45) the values $l_1 = j + \frac{1}{2}$, $l_2 = j' - \frac{1}{2}$ and obtain from it the value of the $3j$-symbol in (11.34). As explained above, non-vanishing contributions to (11.34) are due to ${}^3L_J$ states with $L = J + 1$ and $L = J - 1$, for which the $3j$-symbols in (11.8) do not vanish.

In the case considered above, the interaction energy in the state (9.12) of a proton in the j-orbit and a neutron in the j'-orbit (and the neutron j-orbit completely filled), is equal to the direct term only. In the case of a δ-interaction, we obtain from (11.30) the result (de-Shalit 1953)

$$\boxed{\begin{aligned}\Delta E_{j_\pi j'_\nu J} &= \tfrac{1}{2}C_0(2j+1)(2j'+1)\begin{pmatrix} j & j' & J \\ \frac{1}{2} & -\frac{1}{2} & 0 \end{pmatrix}^2 \\ &\quad\times\left[1 + \frac{\{(2j+1) + (-1)^{j+j'+J}(2j'+1)\}^2}{4J(J+1)}\right]\end{aligned}} \tag{11.35}$$

It was shown in (10.7) that the interaction energy in this case is the *average* between the interaction energies in the corresponding states with $T = 1$ and $T = 0$. Indeed, (11.35) is equal to the average of (11.33) and (11.15) for $(-1)^{l+l'+J} = 1$ and to the average of (11.34) and zero (the value for $T = 1$) for $(-1)^{l+l'+J} = -1$.

We considered above two body interactions with explicit dependence on the spin operators, $(\mathbf{s}_1 \cdot \mathbf{s}_2)V(|\mathbf{r}_1 - \mathbf{r}_2|)$. If the potential $V(|\mathbf{r}_1 - \mathbf{r}_2|)$ is short-ranged it can be approximated by the δ-potential. In that case it is not necessary to resort to the expansion (10.67) for calculating matrix elements. We make use of the fact that the δ-potential has non-vanishing matrix elements only for states which are symmetric in the space coordinates. As noted above, such states must be antisymmetric in the spin-isospin part and hence must have $S = 0$, $T = 1$ or $S = 1$, $T = 0$. From (10.65) follows that the eigenvalues of $(\mathbf{s}_1 \cdot \mathbf{s}_2)$ are $-\frac{3}{4}$ for $S = 0$ and $\frac{1}{4}$ for $S = 1$. Hence we can replace the interaction $4(\mathbf{s}_1 \cdot \mathbf{s}_2)\delta(|\mathbf{r}_1 - \mathbf{r}_2|)$ by

$$-3\delta(|\mathbf{r}_1 - \mathbf{r}_2|) \qquad \text{for} \quad T = 1 \text{ states}$$

and by

$$\delta(|\mathbf{r}_1 - \mathbf{r}_2|) \qquad \text{for} \quad T = 0 \text{ states}$$

Thus, the eigenvalues of the interaction $4(\mathbf{s}_1 \cdot \mathbf{s}_2)\delta(|\mathbf{r}_1 - \mathbf{r}_2|)$ can be read directly from (11.15) for $T = 1$ states and from (11.33) and (11.34) for $T = 0$ states.

We can apply now these results to the case of a proton in a j-orbit and a neutron in the j'-orbit. The matrix elements of the interaction $4(\mathbf{s}_1 \cdot \mathbf{s}_2)\delta(|\mathbf{r}_1 - \mathbf{r}_2|)$ can be written down as the averages of the $T = 1$ and $T = 0$ matrix elements in the states with the same j, j' and J. We thus obtain (de-Shalit 1953)

$$\Delta E'_{j_\pi j'_\nu J} = \tfrac{1}{2}C_0(2j+1)(2j'+1)\begin{pmatrix} j & j' & J \\ \tfrac{1}{2} & -\tfrac{1}{2} & 0 \end{pmatrix}^2 \times \left[-3 + \frac{\{(2j+1) + (-1)^{j+j'+J}(2j'+1)\}^2}{4J(J+1)}\right] \quad \text{for} \quad (-1)^{l+l'+J} = 1 \tag{11.36}$$

For the opposite sign, the result is just one half of (11.34) namely (de-Shalit 1953)

$$\Delta E'_{j_\pi j_\nu J} = \tfrac{1}{2}C_0(2j+1)(2j'+1)\begin{pmatrix} j & j' & J \\ \tfrac{1}{2} & -\tfrac{1}{2} & 0 \end{pmatrix}^2 \times \left[1 + \frac{\{(2j+1) + (-1)^{j+j'+J}(2j'+1)\}^2}{4J(J+1)}\right] \quad \text{for} \quad (-1)^{l+l'+J} = -1 \tag{11.37}$$

The results obtained above are greatly simplified when we consider two nucleons in the same orbit $n' = n$, $l' = l$, $j' = j$. In that case all states with even values of J, for which $(-1)^{2l+J} = 1$, are antisymmetric and hence have $T = 1$. All states with odd values of J, for which $(-1)^{2l+J} = -1$, are symmetric with isospin $T = 0$. From this follows that the eigenvalues of the δ-potential for $T = 1$ states are given by (11.16) as

$$\frac{1}{2}C_0(2j+1)^2 \begin{pmatrix} j & j & J \\ \frac{1}{2} & -\frac{1}{2} & 0 \end{pmatrix}^2 \qquad T = 1,\ J \text{ even} \tag{11.38}$$

and for $T = 0$ states are given by (11.34) for $j' = j$ divided by 2 as

$$\frac{1}{2}C_0(2j+1)^2 \begin{pmatrix} j & j & J \\ \frac{1}{2} & -\frac{1}{2} & 0 \end{pmatrix}^2 \left[1 + \frac{(2j+1)^2}{J(J+1)}\right] \qquad T = 0,\ J \text{ odd} \tag{11.39}$$

These results may be used to obtain the eigenvalues of the interaction $4(\mathbf{s}_1 \cdot \mathbf{s}_2)\delta(|\mathbf{r}_1 - \mathbf{r}_2|)$ as explained above.

Let us now consider the features of some nuclear spectra arising from the use of an attractive δ-interaction. The energy eigenvalues given by the various expressions derived above should be multiplied by a negative constant. We start with two identical nucleons ($T = 1$) in the same j-orbit. All states allowed by the Pauli principle in this case have even values of J, $J = 0, 2, \ldots, 2j - 1$. As an example, consider two nucleons in the $j = \frac{9}{2}$ orbit. The energy level spacings of two identical nucleons in the $j = \frac{9}{2}$-orbit due to the δ-interaction are given by (11.38) with $j = \frac{9}{2}$ and even values of J. These spacings are shown in Fig. 11.1 in column *c*. The most important feature is that the lowest level has $J = 0$ as is indeed the case in *all* even-even nuclei. The other levels are higher in monotonic order. Also this feature is in qualitative agreement with experiment. The quantitative agreement with actual data is, however, not very good.

The experimental level spacings for two protons in the $1g_{9/2}$-orbit are shown in column *a* of Fig. 11.1. These level spacings were extracted from a detailed analysis of experimental level spacings in several Zr isotopes which will be discussed later in detail. In column *e*

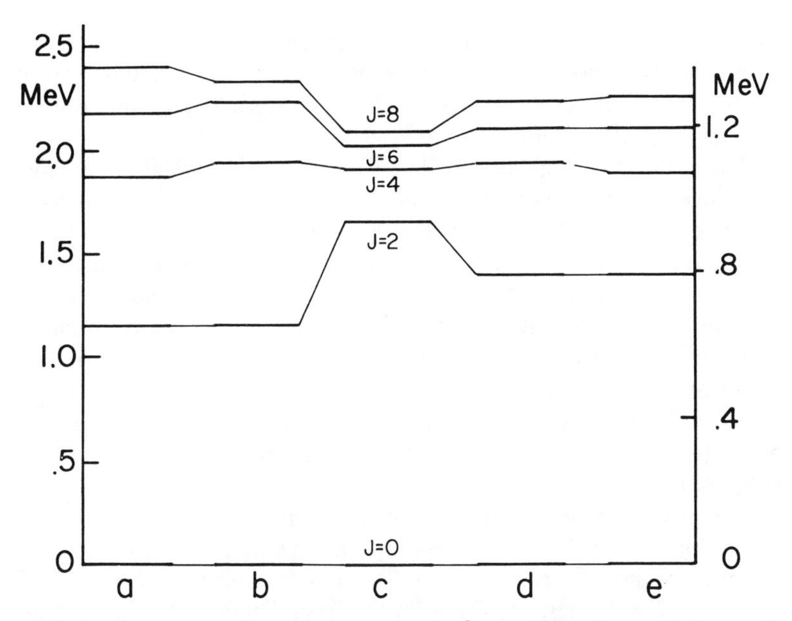

FIGURE 11.1. *Experimental $T = 1$ levels of $g^2_{9/2}$ configurations and those calculated from a δ-potential.*

of Fig. 11.1, experimental level spacings of ^{210}Pb are plotted. These levels should be due to two neutrons in the $2g_{9/2}$ orbit outside the closed shells of ^{208}Pb. The energy scales for the two sets of experimental energies were chosen so that the calculated levels in column *c* could be used for both sets. The scale for the calculated levels has been arbitrarily determined. Its actual value depends on the coefficient of the δ-potential which has not been specified so far, as well as the value of the radial integral in C_0. Since the real nuclear interaction is certainly not the δ-potential, it is not worthwhile to obtain a better estimate of these quantities. By comparison of the level diagrams we see that the levels due to the δ-interaction with $J = 2, 4, 6, 8$ are much more compressed than those observed in actual nuclei. This compression is due to the fact that the δ-interaction yields a spacing between the $J = 0$ and $J = 2$ levels which is much bigger than other spacings. If another interaction is added which will reduce only the 0–2 spacing, the coefficient of the δ-interaction may be increased and other level spacings will agree better with experiment. Such an additional interaction will be introduced later. Meanwhile let us observe that this indeed improves the situation. In column *b*, the 0–2 spacing was fixed in this fashion and the coefficient of the δ-potential was chosen to give the best fit to the experimental spacings, taken from Zr isotopes. The levels in column d were determined in this way to give the best agreement with the ^{210}Pb levels.

The order of levels as calculated from the δ-interaction remains unchanged if we use potentials which have ranges different from zero. As long as the potential has the minimum at $r = 0$, is monotonically increasing and has a finite range, the order of levels with even J-values remains as in Fig. 11.1. A typical such potential is an attractive Yukawa potential $\exp(-\lambda r)/r$ or even a Gaussian potential $\exp(-\lambda r^2)$. If we add to the δ-potential a certain amount of the interaction $4(\mathbf{s}_1 \cdot \mathbf{s}_2)\delta(|\mathbf{r}_1 - \mathbf{r}_2|)$ discussed above, the order and relative spacings do not change (if, however, the coefficient of the latter interaction exceeds 3 times that of the δ-potential the order of levels is reversed).

The next example is that of two nucleons in the $1f_{7/2}$ orbit in states with both $T = 1$ and $T = 0$. The $T = 1$ levels with $J = 0, 2, 4, 6$ may be taken from the spectrum of ^{42}Ca. The $T = 0$ levels with $J = 1, 3, 5, 7$, as well as $T = 1$ ones, are found in the level diagram of ^{42}Sc. In both cases we assume that these levels are due to the $(1f_{7/2})^2$ configuration. Later we will see arguments supporting this assignment. Experimental levels of ^{42}Sc are plotted in column b of Fig. 11.2. Among them, the $T = 1$ level spacings agree well with those observed in ^{42}Ca (in column c) as required by charge independence.

Now we consider $T = 0$ levels together with $T = 1$ ones. Adding the $4(\mathbf{s}_1 \cdot \mathbf{s}_2)\delta(|\mathbf{r}_1 - \mathbf{r}_2|)$ interaction affects these levels differently. According to the discussion presented above, the latter interaction is equivalent to $-3\delta(|\mathbf{r}_1 - \mathbf{r}_2|)$ for $T = 1$ levels but for $T = 0$ levels it is equivalent to $\delta(|\mathbf{r}_1 - \mathbf{r}_2|)$. If the coefficients of these two interactions are x and y, the energies of $T = 1$ levels are given by (11.15) multiplied by $(x - 3y)$ whereas those of $T = 0$ levels are multiplied by $(x + y)$. The values of x and y (which include also the common coefficient C_0) were determined so that the calculated energy spacings give the best agreement with the experimental ones. The results are plotted in column a of Fig. 11.2. Also here we see nice agreement with the observed *order* of levels but the actual calculated spacings are not in good agreement with the experimental ones.

Let us now look at energy levels of a proton in a j-orbit and a neutron in a j'-orbit. The eigenvalues of the δ-interaction are given in this case by (11.35) and those of the $4(\mathbf{s}_1 \cdot \mathbf{s}_2)\delta(|\mathbf{r}_1 - \mathbf{r}_2|)$ interaction are given by (11.36) or (11.37). Energy levels of a proton in the $1d_{3/2}$ or-

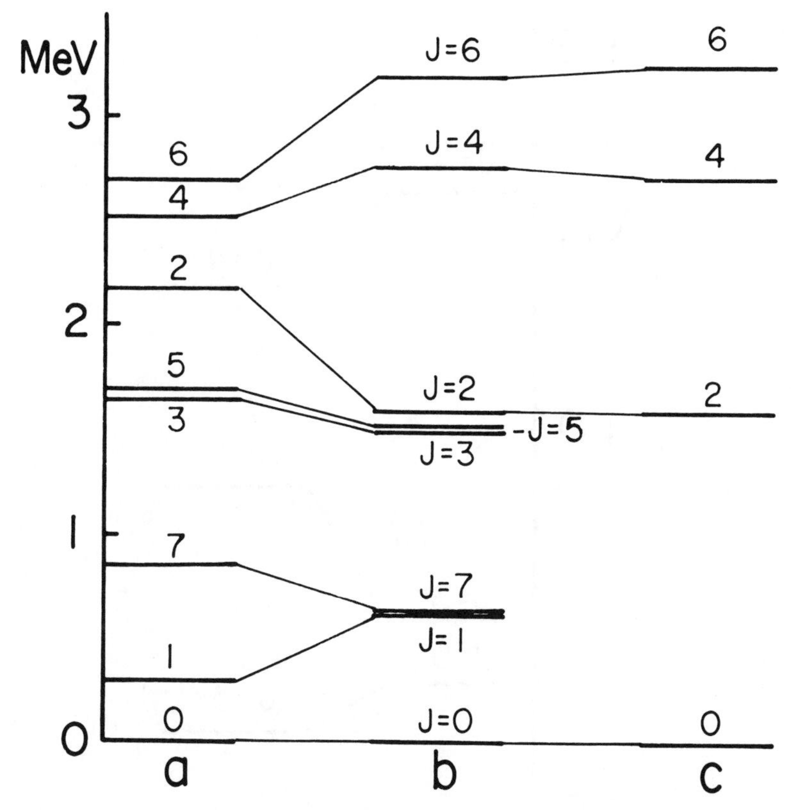

FIGURE 11.2. *Experimental* $(1f_{7/2})^2$ *levels, with* $T = 0$ *and* $T = 1$, *and those calculated from zero range interactions.*

bit and a neutron in the $1f_{7/2}$ orbit can be taken from ^{38}Cl where the neutron $1d_{3/2}$ orbit is completely filled. They are plotted in column *a* of Fig. 11.3. In column *b* are plotted level spacings calculated for a linear combination of the two interactions whose coefficients were determined to yield the best agreement with the experimental spacings. That linear combination yields the right order of levels but the spacings do not agree well with the experimental ones. This should be contrasted with the very good agreement presented in Section 1 between experimental energies of ^{38}Cl and those calculated from matrix elements determined from the ^{40}K spectrum.

In Fig. 11.4 similar results are plotted for a proton in the $1h_{9/2}$ orbit and a neutron in the $2g_{9/2}$ orbit outside the closed shells of ^{208}Pb. The

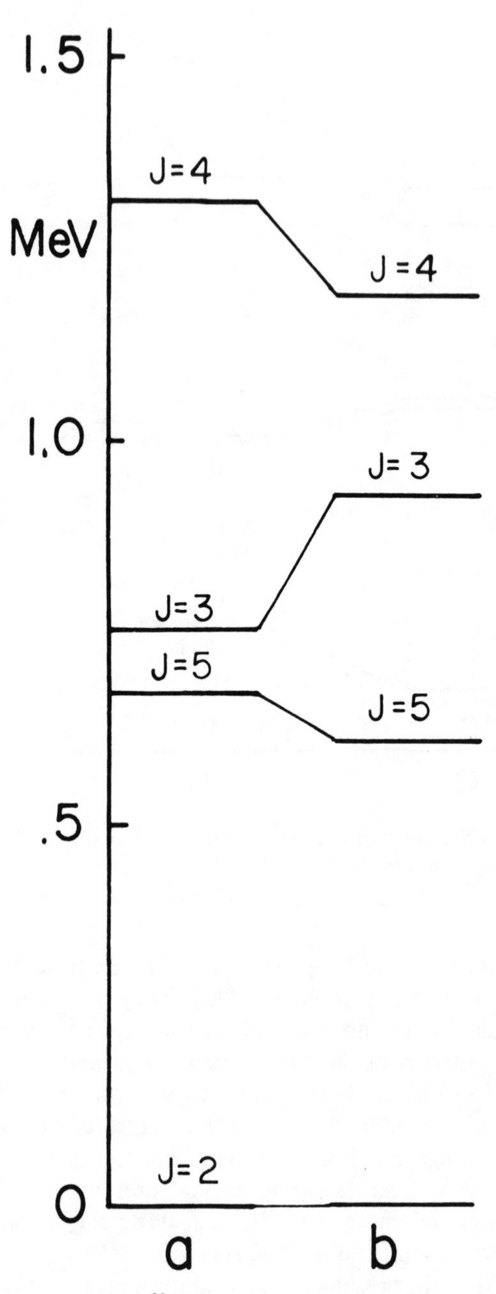

FIGURE 11.3. *Experimental* ^{38}Cl *levels and those calculated from zero range interactions.*

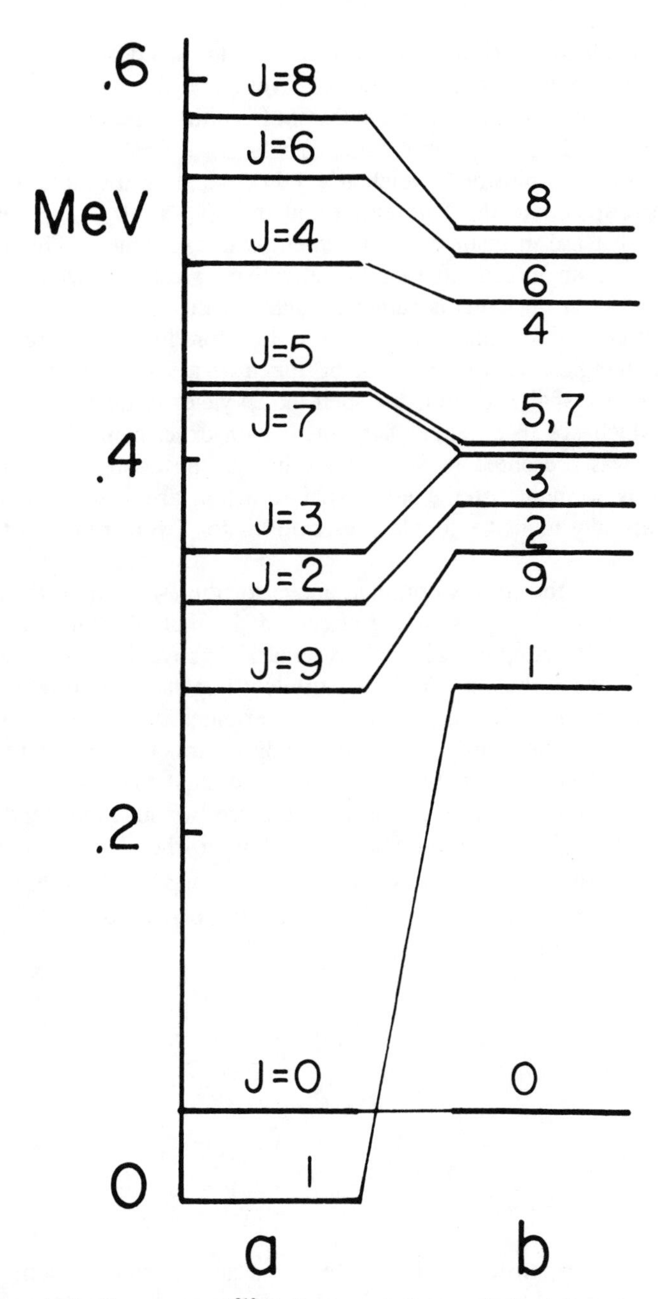

FIGURE 11.4. *Experimental* ^{210}Bi *levels and those calculated from zero range interactions.*

linear combination of the two interactions determined by the best fit to the data yields the correct order of levels. The calculated spacings (column *b*), however, are rather different from the experimental ones observed in ^{210}Bi (plotted in column a).

The examples shown above clearly indicate that even simple minded interactions, like the δ-potential and the $4(\mathbf{s}_1 \cdot \mathbf{s})\delta(|\mathbf{r}_1 - \mathbf{r}_2|)$, have features that are in qualitative agreement with experiment. This may be taken as an indication that the effective or renormalized interaction between nucleons is rather regular. At the same time we see that simple interactions are not able to reproduce quantitatively nuclear energies. This should not be taken as a failure of the shell model. As we shall see later, the shell model yields in many cases energies which are in excellent agreement with experiment. One such example was presented in Section 1. This can be done, however, if the matrix elements of the effective interaction are specified much more carefully than the simple prescriptions for a zero range potential.

We can use the results obtained above for the δ-potential, to look in some simple examples, at the effects of the mutual interaction of nucleons on the coupling scheme. As remarked above, if the spin-orbit splitting between the $l + \frac{1}{2}$ and $l - \frac{1}{2}$ orbits is large compared with the mutual interaction, the eigenstates are determined by the jj-coupling scheme. In the opposite limit, a two-body central interaction is diagonal in LS-coupling states. Let us consider first the $T = 1,\ J = 0$ states of two nucleons in the l-orbit. There are two such independent states. In the jj-coupling scheme they belong to the $(l + \frac{1}{2})^2$ configuration and the $(l - \frac{1}{2})^2$ configuration. In LS-coupling those are the $^1S_0(S = 0,\ L = 0)$ and $^3P_0$ $(S = 1,\ L = 1)$ states. According to Table 9.1, the jj-coupling states are equal to

$$|(l + \tfrac{1}{2})^2 T = 1,\ J = 0\rangle = \sqrt{\frac{l+1}{2l+1}}\,{}^1S_0 - \sqrt{\frac{l}{2l+1}}\,{}^3P_0$$

$$|(l - \tfrac{1}{2})^2 T + 1\ , J = 0\rangle = \sqrt{\frac{l}{2l+1}}\,{}^1S_0 + \sqrt{\frac{l+1}{2l+1}}\,{}^3P_0 \tag{11.40}$$

Using the expansions (11.40), we can calculate matrix elements of the δ-potential between these states. By using the results (11.9) and (11.11) the following matrix elements, all for $T = 1,\ J = 0$, are obtained

$$\begin{aligned}
\langle (l+\tfrac{1}{2})^2|\delta|(l+\tfrac{1}{2})^2\rangle &= \frac{l+1}{2l+1}\langle {}^1S_0|\delta|{}^1S_0\rangle + \frac{l}{2l+1}\langle {}^3P_0|\delta|{}^3P_0\rangle \\
&= \frac{l+1}{2l+1}(2l+1)C_0 = (l+1)C_0 \\
\langle (l+\tfrac{1}{2})^2|\delta|(l-\tfrac{1}{2})^2\rangle &= \frac{\sqrt{l(l+1)}}{2l+1}\langle {}^1S_0|\delta|{}^1S_0\rangle \\
&\quad - \frac{\sqrt{l(l+1)}}{2l+1}\langle {}^3P_0|\delta|{}^3P_0\rangle \\
&= \sqrt{l(l+1)}C_0 \\
\langle (l-\tfrac{1}{2})^2|\delta|(l-\tfrac{1}{2})^2\rangle &= \frac{l}{2l+1}\langle {}^1S_0|\delta|{}^1S_0\rangle + \frac{l+1}{2l+1}\langle {}^3P_0|\delta|{}^3P_0\rangle \\
&= lC_0
\end{aligned} \tag{11.41}$$

The diagonal elements in (11.41) are equal to (11.17) for $j = l + \frac{1}{2}$ and for $j = l - \frac{1}{2}$ respectively. We can now construct the sub-matrix of the Hamiltonian in the space of the two states (11.40). The single nucleon spin-orbit interaction is diagonal in the jj-coupling states (11.40). According to (3.27), it is equal to $2al$ in the $(l+\frac{1}{2})^2$ state and to $-2a(l+1)$ in the state of the $(l-\frac{1}{2})^2$ configuration. Hence, the matrix, whose rows and columns are defined by the states (11.40), is equal to

$$\begin{array}{c|cc|}
 & (l-\frac{1}{2})^2 & (l+\frac{1}{2})^2 \\
(l-\frac{1}{2})^2 & -2(l+1)a + lC_0 & \sqrt{l(l+1)}C_0 \\
(l+\frac{1}{2})^2 & \sqrt{l(l+1)}C_0 & 2la + (l+1)C_0
\end{array}$$

The eigenvalues of (11.42) are obtained as the solutions of the quadratic equations

$$X^2 - ((2l+1)C_0 - 2a)X - 2(2l+1)aC_0 - 4l(l+1)a^2 = 0 \tag{11.43}$$

In the limit in which $C_0 = 0$, the eigenvalues are $2la$ and $-2(l+1)a$ as in (3.27). In the other limit, when $a = 0$, they are equal to $(2l+1)C_0$, as in (11.11) for the ${}^3S$ state and 0 for the ${}^1P$ state.

Another example is offered by the $T = 0$, $J = 1$ states in the l^2 configuration. There are three such states which belong to the

$(l+\frac{1}{2})^2$, $(l+\frac{1}{2})(l-\frac{1}{2})$ and $(l-\frac{1}{2})^2$ configurations in jj-coupling. They can be expressed in terms of the LS-coupling states $^3S_1$, $^1P_1$ and $^3D_1$, by the coefficients in Table 9.1. These expansions are given by

$$|(l+\tfrac{1}{2})^2 T=0,\ J=1\rangle = \frac{1}{(2l+1)\sqrt{3}}\Big[\sqrt{(l+1)(2l+3)}\,{}^3S_1 + \sqrt{3l(2l+3)}\,{}^1P_1 - \sqrt{2l(2l-1)}\,{}^3D_1\Big]$$

$$|(l+\tfrac{1}{2})(l-\tfrac{1}{2})T=0,\ J=1\rangle = \frac{1}{(2l+1)\sqrt{3}}\Big[\sqrt{8l(l+1)}\,{}^3S_1 - \sqrt{6}\,{}^1P_1 + \sqrt{(2l-1)(2l+3)}\,{}^3D_1\Big]$$

$$|(l-\tfrac{1}{2})^2 T=0,\ J=1\rangle = \frac{1}{(2l+1)\sqrt{3}}\Big[-\sqrt{l(2l-1)}\,{}^3S_1 + \sqrt{3(l+1)(2l-1)}\,{}^1P_1 + \sqrt{2(l+1)(2l+3)}\,{}^3D_1\Big] \qquad (11.44)$$

We can now construct the general matrix of the Hamiltonian defined by the states (11.44), which is analogous to (11.42). We express it in terms of matrix elements of a general central interaction V, for which we use the notation

$$\langle {}^3S_1|V|{}^3S_1\rangle = V({}^3S) \qquad \langle {}^1P_1|V|{}^1P_1\rangle = V({}^1P)$$
$$\langle {}^3D_1|V|{}^3D_1\rangle = V({}^3D).$$

Matrix elements of the two body interaction between the $T=0$, $J=1$ states (11.44) can thus be expressed as follows

$$\langle (l+\tfrac{1}{2})^2|V|(l+\tfrac{1}{2})^2\rangle = \frac{1}{3(2l+1)^2}[(l+1)(2l+3)V({}^3S) + 3l(2l+3)V({}^1P) + 2l(2l-1)V({}^3D)]$$

$$\langle(l+\tfrac{1}{2})^2|V|(l+\tfrac{1}{2})(l-\tfrac{1}{2})\rangle$$
$$=\frac{1}{3(2l+1)^2}[2(l+1)\sqrt{2l(2l+3)}V(^3S)$$
$$-3\sqrt{2l(2l+3)}V(^1P)$$
$$-(2l-1)\sqrt{2l(2l+3)}V(^3D)]$$

$$\langle(l+\tfrac{1}{2})^2|V|(l-\tfrac{1}{2})^2\rangle$$
$$=\frac{1}{3(2l+1)^2}[-\sqrt{l(l+1)(2l-1)(2l+3)}V(^3S)$$
$$+3\sqrt{l(l+1)(2l-1)(2l+3)}V(^1P)$$
$$-2\sqrt{l(l+1)(2l-1)(2l+3)}V(^3D)]$$

$$\langle(l+\tfrac{1}{2})(l-\tfrac{1}{2})|V|(l+\tfrac{1}{2})(l-\tfrac{1}{2})\rangle$$
$$=\frac{1}{3(2l+1)^2}[8l(l+1)V(^3S)+6V(^1P)$$
$$+(2l-1)(2l+3)V(^3D)]$$

$$\langle(l+\tfrac{1}{2})(l-\tfrac{1}{2})|V|(l-\tfrac{1}{2})^2\rangle$$
$$=\frac{1}{3(2l+1)^2}[-2l\sqrt{2(l+1)(2l-1)}V(^3S)$$
$$-3\sqrt{2(l+1)(2l-1)}V(^1P)$$
$$+(2l+3)\sqrt{2(l+1)(2l-1)}V(^3D)]$$

$$\langle(l-\tfrac{1}{2})^2|V|(l-\tfrac{1}{2})^2\rangle$$
$$=\frac{1}{3(2l+1)^2}[l(2l-1)V(^3S)+3(l+1)(2l-1)V(^1P)$$
$$+2(l+1)(2l+3)V(^3D)]$$

The diagonal elements of the spin-orbit interaction are $-2(l+1)a$, $-a$ and $2la$ in the $(l-\frac{1}{2})^2$, $(l+\frac{1}{2})(l-\frac{1}{2})$ and $(l+\frac{1}{2})^2$ configurations respectively.

To write the matrix of the Hamiltonian in the case of the δ-potential, we use (11.11) and (11.9) with the values of the $3j$-symbols for $L=2$. In this case, $V(^3S)=(2l+1)C_0$, $V(^1P)=0$ and $V(^3D)$ $=l(l+1)(2l+1)C_0/(2l-1)(2l+3)$. The matrix, labelled by the $(l+\frac{1}{2})^2$, $(l+\frac{1}{2})(l-\frac{1}{2})$ and $(l-\frac{1}{2})^2$ configurations is equal to

	$(l-\frac{1}{2})^2$	$(l+\frac{1}{2})(l-\frac{1}{2})$	$(l+\frac{1}{2})^2$
$(l-\frac{1}{2}^2)$	$-2(l+1)a+\frac{l(2l^2+1)}{(2l+1)(2l-1)}C_0$	$-\frac{l(l-1)\sqrt{2(l+1)(2l-1)}}{(2l+1)(2l-1)}C_0$	$-\frac{(2l^2+2l-1)\sqrt{l(l+1)(2l-1)(2l+3)}}{(2l+1)(2l-1)(2l+3)}C_0$
$(l+\frac{1}{2})(l-\frac{1}{2})$	$-\frac{l(l-1)\sqrt{2(l+1)(2l-1)}}{(2l+1)(2l-1)}C_0$	$-a+\frac{3l(l+1)}{2l+1}C_0$	$\frac{(l+1)(l+2)\sqrt{2l(2l+3)}}{(2l+1)(2l+3)}C_0$
$(l+\frac{1}{2})^2$	$-\frac{(2l^2+2l-1)\sqrt{l(l+1)(2l-1)(2l+3)}}{(2l+1)(2l-1)(2l+3)}C_0$	$\frac{(l+1)(l+2)\sqrt{2l(2l+3)}}{(2l+1)(2l+3)}C_0$	$2la+\frac{(l+1)(2l^2+4l+3)}{(2l+1)(2l+3)}C_0$

In the limit $|C_0| \ll |a|$, and $a < 0$, the lowest state with $T = 0$, $J = 1$ belongs to the $(l + \frac{1}{2})^2$ configuration. In the opposite limit, the matrix with $a = 0$ may be diagonalized and, for $C_0 < 0$, the lowest eigenstate is the ${}^3S_1$ state.

It is interesting to look at situations in which $|C_0|$ is smaller than $|a|$ but not negligible. There are then deviations from the pure jj-coupling scheme. The diagonal element of the δ-interaction in the state of the $(l + \frac{1}{2})(l - \frac{1}{2})$ configuration is considerably bigger, in absolute value, than the diagonal elements of the other configurations. With $C_0 < 0$ which is not very small, the difference between diagonal elements of the $(l + \frac{1}{2})(l - \frac{1}{2})$ and $(l + \frac{1}{2})^2$ configurations may decrease significantly. The non-diagonal element between these two states is also large. The spin-orbit splitting between these two states is only $(2l + 1)a$ and hence, we expect considerable mixing between these two $T = 0$, $J = 1$ states due to the mutual interaction.

This situation should be contrasted with the one, described above, for the $T = 1$, $J = 0$ states. There, the gap between the states of the $(l - \frac{1}{2})^2$ and $(l + \frac{1}{2})^2$ configuration is twice as large, $2(2l + 1)a$, and it increases due to the δ-interaction. Hence, the admixtures in this case seem to be rather small and no large deviation from pure jj-coupling is expected.

The gap between the $(l - \frac{1}{2})^2$ $T = 0$, $J = 1$ state and the $(l + \frac{1}{2})^2$ $T = 0$, $J = 1$ state is equal to $2(2l + 1)a$ due to the spin-orbit interaction, and does not decrease by the δ-interaction. Hence, for values of $|C_0|$ which are not very large, the $(l - \frac{1}{2})^2$ state is not expected to be strongly admixed with the $T = 0$, $J = 1$ state of the $(l + \frac{1}{2})^2$ configuration.

The different response of jj-coupling states with $T = 1$ and $T = 0$ to the mutual interaction has important implications also for many nucleon configurations. These will be discussed in Section 38.

12

The Pairing Interaction and the Surface Delta Interaction

In Section 10 we derived general expressions for matrix elements of certain interactions between states of a given two nucleon configuration. In many cases of nuclear spectra, as we shall see later in detail, the actual eigenstates are admixtures of states in several shell model configurations. In such cases, it is necessary to evaluate matrix elements between different configurations of two nucleons. The states we consider are in the jj-coupling scheme. The wave functions of space and spin variables for different j_1, j_2 orbits are given by (9.11) as

$$\psi_{j_1j_2TJM} = \frac{1}{\sqrt{2}} \sum_{m_1m_2} (j_1m_1j_2m_2 \mid j_1j_2JM)$$
$$\times [\psi_{j_1m_1}(1)\psi_{j_2m_2}(2) + (-1)^T \psi_{j_1m_1}(2)\psi_{j_2m_2}(1)] \quad (12.1)$$

The wave function (12.1) is properly normalized if the j_1-orbit and j_2-orbit are different (i.e. at least members of one pair among nn', ll' and jj' are different). If the two nucleons are in the same orbit then states with even values of J have $T = 1$ and are antisymmetric and states with odd J values are symmetric ($T = 0$). In that case the normalized wave functions are given by

$$\psi_{j^2TJM} = \sum_{m_1m_2} (jm_1jm_2 \mid jjJM)\psi_{jm_1}(1)\psi_{jm_2}(2) \qquad (-1)^{T+J} = -1 \tag{12.2}$$

Let us first consider matrix elements between states (12.1) which are characterized by pairs of different j_1j_2 and $j_1'j_2'$ orbits. As in Section 10 we obtain a direct term and an exchange term. In the latter we change the order of coupling and obtain for the matrix element the form

$$\langle j_1j_2JM|V|j_1'j_2'JM\rangle + (-1)^T\langle j_1j_2JM|V|j_2'j_1'JM\rangle(-1)^{j_1'+j_2'-J} \tag{12.3}$$

Using the expansion given by (10.1) and (10.4) of the two body interaction and formula (10.27) we obtain for the matrix element (12.3) the expansion

$$\sum_k (-1)^{j_2+j_1'+J}\begin{Bmatrix} j_1 & j_2 & J \\ j_2' & j_1' & k \end{Bmatrix} F^k + (-1)^T \sum_k \begin{Bmatrix} j_1 & j_2 & J \\ j_1' & j_2' & k \end{Bmatrix} G^k \tag{12.4}$$

In (12.4) the coefficients F^k and G^k are given by

$$F^k = \frac{4\pi}{2k+1}(j_1\|\mathbf{Y}_k\|j_1')(j_2\|\mathbf{Y}_k\|j_2')$$
$$\times \int R_{n_1l_1j_1}(r_1)R_{n_1'l_1'j_1'}(r_1)R_{n_2l_2j_2}(r_2)R_{n_2'l_2'j_2'}(r_2)v_k(r_1,r_2)\,dr_1\,dr_2 \tag{12.5}$$

$$G^k = \frac{4\pi}{2k+1}(j_1\|\mathbf{Y}_k\|j_2')(j_1'\|\mathbf{Y}_k\|j_2)$$
$$\times \int R_{n_1l_1j_1}(r_1)R_{n_2'l_2'j_2'}(r_1)R_{n_1'l_1'j_1'}(r_2)R_{n_2l_2j_2}(r_2)v_k(r_1,r_2)\,dr_1\,dr_2 \tag{12.6}$$

The expression of G^k in (12.6) was obtained by using the relation (8.9).

Also in the present, more general case, we can use the identity (10.14) of $6j$-symbols to cast the exchange term in the form of the direct term. That identity may be written as

$$\begin{Bmatrix} j_1 & j_2 & J \\ j_1' & j_2' & k \end{Bmatrix} = \sum(-1)^{J+k+r}(2r+1)\begin{Bmatrix} j_1 & j_2 & J \\ j_2' & j_1' & r \end{Bmatrix}\begin{Bmatrix} j_1 & j_2' & k \\ j_2 & j_1' & r \end{Bmatrix}$$

Hence, if we define

$$F'^k = \sum_t (-1)^{j_2+j_1'+k+t}(2k+1)\begin{Bmatrix} j_1 & j_2' & t \\ j_2 & j_1' & k \end{Bmatrix} G^t \qquad (12.7)$$

We can express the matrix element (12.4) as

$$\sum_k (F^k + (-1)^T F'^k)(-1)^{j_2+j_1'+J}\begin{Bmatrix} j_1 & j_2 & J \\ j_2' & j_1' & k \end{Bmatrix}$$

$$= \sum_k (F^k + (-1)^T F'^k)\langle j_1 j_2 JM|\mathbf{u}^{(k)}(1)\cdot\mathbf{u}^{(k)}(2)|j_1' j_2' JM\rangle$$

$$(12.8)$$

The unit tensor operators in (12.8) are defined as in (8.16) by

$$(j\|\mathbf{u}^{(k)}(1)\|j') = \delta_{jj_1}\delta_{j'j_1'} \qquad (j\|\mathbf{u}^{(k)}(2)\|j') = \delta_{jj_2}\delta_{j'j_2'}$$

It should be noted that the states between which the matrix elements on the r.h.s. of (12.8) are evaluated are *not antisymmetrized.*

Let us now evaluate the matrix element between a state (12.1) with different j', j''-orbits and a state (12.2). We obtain for the matrix element the form

$$\frac{1}{\sqrt{2}}[\langle j_{12}^2 JM|V|j_1' j_2'' JM\rangle + (-1)^T\langle j_{12}^2 JM|V|j_2' j_1'' JM\rangle] \quad (12.9)$$

The subscripts 1, 2 and 12 in (12.9) indicate the indices of nucleons occupying the orbits j', j'' and j. On the r.h.s. of the exchange term nucleon 2 occupies the j'-orbit whereas nucleon 1 is in the j''-orbit. We can now interchange nucleons 1 and 2 in the state (12.2) on the l.h.s. of the exchange term. This leads according to the symmetry of the Clebsch-Gordan coefficients to a factor $(-1)^{2j+J}$. After this interchange the exchange matrix element $\langle j_{21}^2 JM|V|j_2' j_1'' JM\rangle$ becomes equal to the first term in (12.9) which is the direct term. They differ only by the indices assigned to the two nucleons. The phase factor in (12.9) becomes $(-1)^{T+J+1}$ which is equal to 1 according to (12.2). Hence, the matrix element becomes equal to

$$\sqrt{2}\langle j^2 JM|V|j'j''JM\rangle \tag{12.10}$$

where $|j'j''JM\rangle$ in the matrix element (12.10) is not antisymmetric. In (12.10) the order of nucleons is the same on both sides. Formula (10.27) may now be directly applied to yield for (12.10) the expansion

$$\sum_k \sqrt{2}F^k\langle j^2 JM|\mathbf{u}^{(k)}(1)\cdot\mathbf{u}^{(k)}(2)|j'j''JM\rangle \tag{12.11}$$

where F^k is obtained by putting in (12.5) $n_1 = n_1' = n$, $l_1 = l_1' = l$, $j_1 = j_1' = j$ and $j_2 = j'$, $j_2' = j''$.

Let us define the matrix element of any two body interaction between antisymmetric and normalized states in terms of the tensor expansion as

$$\boxed{\begin{aligned}&\langle j_1 j_2 TM_T JM|V|j_1' j_2' TM_T JM\rangle_a\\&\quad=\sum_k F_k\langle j_1 j_2 JM|\mathbf{u}^{(k)}(1)\cdot\mathbf{u}^{(k)}(2)|j_1' j_2' JM\rangle\\&\quad=\sum_k F_k(-1)^{j_2+j_1'+J}\begin{Bmatrix} j_1 & j_2 & J\\ j_2' & j_1' & k\end{Bmatrix}\end{aligned}} \tag{12.12}$$

The coefficients F_k are equal to $F^k + (-1)^T F'^k$ in (12.8) for states with either $T = 1$ or $T = 0$, to $\sqrt{2}F^k$ in (12.11) or to F^k when matrix elements between states of the j^2 configuration and j'^2 configuration are evaluated (as explained in the discussion following (10.62)). The states between which matrix elements on the r.h.s. of (12.12) are evaluated are not antisymmetric if the two nucleons are not in the same orbit.

The relation (12.12) is between two sets, $F_k = F_k(j_1 j_2 j_1' j_2')$ and $V_{TJ}(j_1 j_2 j_1' j_2') = \langle j_1 j_2 TM_T JM|V|j_1' j_2' TM_T JM\rangle_a$. It can be seen from the $6j$-symbol in (12.12) that both sets have equal numbers of elements. As discussed in more detail below, if $j_1 \neq j_2$, *and* $j_1' \neq j_2'$, there are equal numbers of matrix elements with *either* $T = 1$ or $T = 0$.

In other cases, where $(-1)^{T+J} = -1$, matrix elements between states with both $T = 1$ and $T = 0$ must be included. In fact, the $6j$-symbols carry out a real orthogonal transformation between the F_k and V_J in the given space. Given the set F_k, the V_J matrix elements are uniquely determined. On the other hand, if a complete set of matrix elements V_J is given, the F_k are uniquely determined. This follows simply by taking the inverse transformation to (12.12). Multiplying (12.12) by

$$(-1)^{j_2+j_1'+J}(2J+1)\begin{Bmatrix} j_1 & j_2 & J \\ j_2' & j_1' & k_0 \end{Bmatrix}$$

and summing over J we obtain, due to the orthogonality relation (10.13), the result

$$\begin{aligned}\sum_J (-1)^{j_2+j_1'+J}(2J+1)\begin{Bmatrix} j_1 & j_2 & J \\ j_2' & j_1' & k_0 \end{Bmatrix} V_J &= \sum_j (2J+1)\begin{Bmatrix} j_1 & j_2 & J \\ j_2' & j_1' & k_0 \end{Bmatrix}\begin{Bmatrix} j_1 & j_2 & J \\ j_2' & j_1' & k \end{Bmatrix} F_k \\ &= \frac{F_{k_0}}{2k_0+1}\end{aligned}$$

which can be written as

$$\boxed{\begin{aligned} F_k(j_1j_2j_1'j_2') = (2k+1)\sum_J (-1)^{j_2+j_1'+J}(2J+1) \\ \times \begin{Bmatrix} j_1 & j_2 & J \\ j_2' & j_1' & k \end{Bmatrix} V_J(j_1j_2j_1'j_2') \end{aligned}} \qquad (12.13)$$

The unique determination of the coefficients F_k of the tensor expansion (12.12) by the matrix elements of the two body interaction is based on including V_J with all possible J values in (12.13). If $j_1 \neq j_2$ *and* $j_1' \neq j_2'$, there is a complete set of V_J matrix elements, with all possible J-values, *both* for $T = 1$ and $T = 0$ states. In that case, the F_k depend on the isospin, as explicitly shown in (12.8), and hence, there are *different* expansions for interactions in $T = 1$ and $T = 0$ states. If we consider states (9.12) in which there is a proton in the j-orbit and a neutron in the j'-orbit, the matrix elements are given by the direct term only. Also in such cases, there is one complete set of matrix elements and the relation (12.13) may be safely used. The case of either j_1 and j_2 or j_1' and j_2' denoting identical orbits needs special attention. The wave functions (12.2) with even values of J are antisymmetric and must be multiplied by $T = 1$ isospin state. The symmetric wave functions with odd values of J are multiplied by $T = 0$ isospin states. Hence, to obtain in such a case the F_k, we have to include in the summation (12.13) both $T = 1$ *and* $T = 0$ states. The resulting coefficients are $F_k = \sqrt{2}F^k$ if only one pair $j_1 j_2$ or $j_1' j_2'$ are a pair of identical orbits. If the j_1-orbit and j_2-orbit are the same *and* j_1'-orbit and j_2'-orbit are the same then the resulting coefficient in (12.13) is $F_k = F^k$ (of the direct term only).

Let us restrict our attention to states *within* the j^2 configuration. As explained above, the k values which appear in the expansion (12.12) for non-vanishing terms, are limited by the relation $0 \leq k \leq 2j$. If the two body interaction is a potential like (10.1), it follows from (10.36) that $k \leq 2l$. More important, we see that $(-1)^{2l+k}$ must be equal to $+1$ for a non-vanishing reduced matrix element. Hence, only *even* values of k appear in that case in the expansion (12.12). In the case of other interactions, this is no longer the case.

We saw that in jj-coupling the expansion of the interaction $(\mathbf{s}_1 \cdot \mathbf{s}_2) \cdot V(|\mathbf{r}_1 - \mathbf{r}_2|)$ can be carried out by using (10.67). The F^k in this case, given by (10.34) or (12.5) contain products of reduced matrix elements given by (10.68). For $l = l'$, $j = j'$ this reduced matrix element is given by

$$
\left(\tfrac{1}{2}lj\|[\mathbf{s} \times \mathbf{Y}_k]^{(r)}\|\tfrac{1}{2}lj\right) = (2j+1)\sqrt{2r+1}\left(\tfrac{1}{2}\|\mathbf{s}\|\tfrac{1}{2}\right) \times (l\|\mathbf{Y}_k\|l) \begin{Bmatrix} \frac{1}{2} & l & j \\ \frac{1}{2} & l & j \\ 1 & k & r \end{Bmatrix} \tag{12.14}
$$

The $9j$-symbol in (12.14) has two equal rows and thus vanishes if $(-1)^{2+2l+2j+k+r} = -1$. The k values appearing in non-vanishing reduced matrix elements (8.15) are even and hence (12.14) vanishes unless r is *odd*.

This feature, of having only odd ranks of tensors in the expansion (12.12) is shared also by the tensor forces (10.71) considered above.

In the discussion following (10.72) it was explained that (10.71) may be expanded in terms of scalar products

$$([\mathbf{s}_1 \times \mathbf{s}_2]^{(2)} \cdot [\mathbf{Y}_k(1) \times \mathbf{Y}_{k'}(2)]^{(2)})$$

Applying a change of coupling transformation like (10.22) we obtain an expansion in terms of the following scalar products

$$([\mathbf{s}_1 \times \mathbf{Y}_k(1)]^{(r)} \cdot [\mathbf{s}_2 \times \mathbf{Y}_{k'}(2)]^{(r)})$$

The reduced matrix elements of the irreducible tensor operators $[\mathbf{s} \times \mathbf{Y}_k]^{(r)}$ are given by (12.14) and hence vanish unless r is odd. Interactions which can be expanded as linear combinations of scalar products of tensor operators with odd ranks are called *odd tensor interactions*. Such interactions have special properties which we will explore in subsequent sections.

In the preceding section we considered the δ-potential in detail. Like any ordinary interaction (10.1) its expansion, given by (11.26) contains only terms with *even* values of k. Yet, in the discussion following (11.35) it was shown that for $T = 1$ states the δ-potential is equivalent to the odd tensor interaction $-(\frac{4}{3})(\mathbf{s}_1 \cdot \mathbf{s}_2)\delta(|\mathbf{r}_1 - \mathbf{r}_2|)$. For $T = 0$ states, it is equivalent to $4(\mathbf{s}_1 \cdot \mathbf{s}_2)\delta(|\mathbf{r}_1 - \mathbf{r}_2|)$. As explained above, the expansion (12.2) in the case of j^2 configurations is unique only if both $T = 1$ and $T = 0$ matrix elements are included. Only then is it possible to invert (12.12) and obtain in a unique way the F_k coefficients given by (12.13). If we restrict our attention to only $T = 1$ states of the j^2 configuration, the expansion is not unique. There are only $j + \frac{1}{2}$ matrix elements with $T = 1$ and they cannot determine uniquely the $2j + 1$ coefficients F_k. This is clearly demonstrated by the fact that the δ-interaction may be expanded in the $T = 1$ subspace also in terms of odd tensors. As long as we limit the discussion to $T = 1$ states, the δ-interaction may be thus considered as an odd tensor interaction (Racah and Talmi 1952).

The result presented above for the δ-potential in $T = 1$ states can be generalized as follows. We can make use of the Majorana *space exchange* operator. The latter is defined by $P_{12}^x f(\mathbf{r}_1, \mathbf{r}_2) = f(\mathbf{r}_2, \mathbf{r}_1)$. Due to the total antisymmetry of the wave function we obtain

$$P_{12}^x = -P_{12}^\sigma = -\tfrac{1}{2}(1 + 4\mathbf{s}_1 \cdot \mathbf{s}_2) \qquad \text{for} \quad T = 1 \text{ states} \tag{12.15}$$

From (12.15) follows for $T = 1$ states the relation

$$4(\mathbf{s}_1 \cdot \mathbf{s}_2)V(|\mathbf{r}_1 - \mathbf{r}_2|) = -(1 + 2P_{12}^x)V(|\mathbf{r}_1 - \mathbf{r}_2|) \qquad (12.16)$$

Hence, the interaction $(1 + 2P_{12}^x)V(|\mathbf{r}_1 - \mathbf{r}_2)$ is equivalent, in $T = 1$ states, to an odd tensor interaction for *any* shape or range of the potential $V(|\mathbf{r}_1 - \mathbf{r}_2|)$. If, however, the potential $V(|\mathbf{r}_1 - \mathbf{r}_2|)$ in (12.16) is the δ-potential, it has non-vanishing matrix elements only between states symmetric in the space coordinates. When P_{12}^x acts on such states it leaves them invariant and hence follows the equivalence of

$$4(\mathbf{s}_1 \cdot \mathbf{s}_2)\delta(|\mathbf{r}_1 - \mathbf{r}_2|) \quad \text{and} \quad -3\delta(|\mathbf{r}_1 - \mathbf{r}_2|) \quad \text{for} \quad T = 1 \text{ states}$$

In the discussion of Fig. 11.1 it was pointed out that there exists an interaction which can change the 0–2 spacings without affecting the spacings between the $J = 2$ and $J = 4, 6, 8$ levels. Such an interaction can be defined in terms of its matrix elements

$$\boxed{\langle j^2 JM|P|j^2 JM\rangle = (2j + 1)\delta_{J0}} \qquad (12.17)$$

It is called the *pairing interaction* and is considered to be a crude approximation to the δ-interaction. We saw that the spacings between two nucleon levels with $J = 2, 4, 6, 8$ are much smaller than the 0–2 spacing. In the case of the pairing interaction, the levels with $J \neq 0$ are degenerate with eigenvalue zero. We shall later see that the pairing interaction, as well as the δ interaction, are rather poor approximations of the effective interaction in nuclei. Nevertheless, the operator defined by (12.17) plays an important role in nuclear spectroscopy. Therefore it is worthwhile to study it in some detail.

Let us consider the tensor expansion of the operator defined by (12.17). The coefficients F_k can be calculated from (12.13) in which the matrix elements (12.17) have been introduced. The result is

$$F_k = (2k + 1)\sum_J (-1)^{2j+J}(2J + 1) \times \begin{Bmatrix} j & j & J \\ j & j & k \end{Bmatrix} (2j + 1)\delta_{J0} = (2k + 1)(-1)^k$$

Hence, matrix elements of the operator in (12.17) are given, according to (12.12) by the expansion

$$\sum_k (-1)^{2j+J+k}(2k+1)\begin{Bmatrix} j & j & J \\ j & j & k \end{Bmatrix}$$
$$= \sum_k (-1)^k (2k+1)\langle j^2 JM|\mathbf{u}^{(k)}(1)\cdot\mathbf{u}^{(k)}(2)|j^2 JM\rangle \tag{12.18}$$

The l.h.s. of (12.18) is a special case of the sum rule (10.20) and hence it is indeed equal to $(2j+1)\delta_{J0}$. The expansion of the pairing interaction given by (12.18) holds for even values of J as well as for odd values. It includes all scalar products of tensors of rank k which satisfy $0 \leq k \leq 2j$ with coefficients $(2k+1)(-1)^k$.

If we limit our attention to $T = 1$ states only, the expansion is not unique. We recall another sum rule of $6j$-symbols (10.21) which assumes for the present case the form

$$1 = \sum_k (2k+1)\begin{Bmatrix} j & j & J \\ j & j & k \end{Bmatrix}$$

For $T = 1$ states with even values of J we can rewrite this identity as

$$-1 = \sum_k (-1)^{2j+J}(2k+1)\begin{Bmatrix} j & j & J \\ j & j & k \end{Bmatrix}$$
$$= \sum_k (2k+1)\langle j^2 JM|\mathbf{u}^{(k)}(1)\cdot\mathbf{u}^{(k)}(2)|j^2 JM\rangle \tag{12.19}$$

Subtracting (12.19) from (12.18) we obtain the expansion of the operator P in (12.17) in terms of odd tensors

$$\boxed{-1 - 2\sum_{k\ \mathrm{odd}} (2k+1)\langle j^2 JM|\mathbf{u}^{(k)}(1)\cdot\mathbf{u}^{(k)}(2)|j^2 JM\rangle} \tag{12.20}$$

Thus, apart from the constant term -1, the pairing interaction in $T = 1$ states is an odd tensor interaction. This fact will have significant importance in subsequent sections. Another expansion can be obtained by *adding* (12.19) to (12.18) yielding for the matrix elements (12.17) the expansion

$$1 + 2 \sum_{k \text{ even}} (2k+1) \langle j^2 JM | \mathbf{u}^{(k)}(1) \cdot \mathbf{u}^{(k)}(2) | j^2 JM \rangle \qquad (12.21)$$

in terms of *even* tensors only. The pairing interaction is an example of a *non-local* interaction which cannot be expressed in terms of the space coordinates of the two nucleons. It can be defined only by its matrix elements (12.17).

As emphasized above, matrix elements of an interaction between states with $T = 1$ do not determine uniquely the expansion (12.12). This situation is changed if we consider expansions in terms of even tensors only. We will prove that a rotationally invariant two-body interaction in the set of states with $T = 1$ only, can be uniquely expressed as a linear combination of scalar products of even tensors. Using the identity (10.13) we obtain in analogy with (12.18) the result

$$\delta_{JJ_0} = (2J_0+1) \sum_k (2k+1) \begin{Bmatrix} j & j & k \\ j & j & J_0 \end{Bmatrix} \begin{Bmatrix} j & j & k \\ j & j & J \end{Bmatrix}$$

$$= -(2J_0+1) \sum_k (2k+1) \begin{Bmatrix} j & j & k \\ j & j & J_0 \end{Bmatrix}$$

$$\times \langle j^2 JM | \mathbf{u}^{(k)}(1) \cdot \mathbf{u}^{(k)}(2) | j^2 JM \rangle \qquad (12.22)$$

The last equality was obtained due to $(-1)^{2j+J} = -1$. Another relation follows from the sum rule (10.14) for $6j$-symbols which assumes the form

$$\sum_k (-1)^{k+J_0+J} (2k+1) \begin{Bmatrix} j & j & J_0 \\ j & j & k \end{Bmatrix} \begin{Bmatrix} j & j & J \\ j & j & k \end{Bmatrix} = \begin{Bmatrix} j & j & J \\ j & j & J_0 \end{Bmatrix}$$

For J, J_0 even it can be rewritten as

$$(2J_0+1) \sum_k (-1)^k \left[(2k+1) \begin{Bmatrix} j & j & J_0 \\ j & j & k \end{Bmatrix} - \delta_{J_0 k} \right]$$

$$\times \langle j^2 JM | \mathbf{u}^{(k)}(1) \cdot \mathbf{u}^{(k)}(2) | j^2 JM \rangle = 0 \qquad (12.23)$$

Subtracting (12.23) from (12.22) we obtain the expansion

$$\boxed{\begin{aligned}\delta_{JJ_0} = \sum_{k\,\text{even}} (2J_0+1)\left[\delta_{kJ_0} - 2(2k+1)\begin{Bmatrix} j & j & J_0 \\ j & j & k \end{Bmatrix}\right] \\ \times \langle j^2JM|\mathbf{u}^{(k)}(1)\cdot\mathbf{u}^{(k)}(2)|j^2JM\rangle\end{aligned}} \tag{12.24}$$

Thus, any interaction which has in the j^2 configuration the matrix elements $V_{J_0} = \langle j^2T=1J_0M|V|j^2T=1J_0M\rangle$ can be expressed by multiplying (12.24) by V_{J_0} and summing over J_0 as

$$\sum_k\left(\sum_{J_0}(2J_0+1)V_{J_0}\left[\delta_{kJ_0} - 2(2k+1)\begin{Bmatrix} j & j & J_0 \\ j & j & k \end{Bmatrix}\right]\right)(\mathbf{u}^{(k)}(1)\cdot\mathbf{u}^{(k)}(2)) \tag{12.25}$$

The pairing interaction is a special case of (12.25) obtained by putting $V_{J_0} = (2j+1)\delta_{J_0 0}$.

We may consider the matrix elements $\langle j^2JM|\mathbf{u}^{(k)}(1)\cdot\mathbf{u}^{(k)}(2)|j^2JM\rangle$ for even values of k and J, as $j+\frac{1}{2}$ basis vectors characterized by k, in a space with $j+\frac{1}{2}$ dimensions labelled by the various J values. The matrix elements of arbitrary interactions are arbitrary vectors in that space. The fact that any interaction may be expanded in terms of even tensors demonstrates the fact that these basis vectors are linearly independent. From this follows the fact that the coefficients in the expansion (12.25) are uniquely determined by the set of V_J.

The pairing interaction (12.17) defined within the j^2 configuration can be simply generalized for several configurations. To do it we first calculate matrix elements of (12.17) in the $jmjm'$ scheme. The transformation coefficients from the j^2JM scheme are given, as explained in Section 7, by Clebsch-Gordan coefficients. We thus obtain

$$\begin{aligned}&\langle jmjm'|P|jm''jm'''\rangle \\ &\quad= \sum_{JM}(jmjm' \mid jjJM)\langle j^2JM|P|j^2JM\rangle(jm''jm''' \mid jjJM)\end{aligned}$$

Substituting the matrix elements from (12.17) we obtain

$$\langle jmjm'|P|jm''jm'''\rangle = (jmjm' \mid jj00)(2j+1)(jm''jm''' \mid jj00) = (-1)^{j-m+j-m''}\delta_{m',-m}\delta_{m''',-m''} \tag{12.26}$$

Apart from the phase $(-1)^{m+m''+1}$, all non-vanishing matrix elements are equal. From such a matrix, the result (12.17) follows directly. In such a matrix all rows (and columns) are proportional and thus its rank is 1. It has only *one* non-vanishing eigenvalue which is equal to the trace of the matrix which is $2j+1$. This eigenvalue belongs to the state with $J=0$ since all $M \neq 0$ matrix elements vanish according to (12.26).

If several j-orbits are considered together, a simple generalization of (12.26) is defined by

$$\langle jmj'm'|P|j''m''j'''m'''\rangle = (-1)^{j-m+j''-m''}\delta_{jj'}\delta_{j''j'''}\delta_{m',-m}\delta_{m''',-m''} \tag{12.27}$$

Matrix elements of P vanish unless j and j' (as well as j'' and j''') denote the *same orbit*. The matrix (12.27) is of order

$$\sum_j (2j+1) = 2\Omega \tag{12.28}$$

where the summation is over all orbits connected by elements (12.27). All rows (and columns) of that matrix are proportional to each other and its rank is 1. It has only *one* non-vanishing matrix element equal to the trace of the matrix. All elements in the diagonal are equal to 1 and the trace is equal to 2Ω defined by (12.28). The non-vanishing eigenvalue belongs to a state with $J = 0$ since all matrix elements (12.27) between states with $M \neq 0$ vanish. Let us now examine the situation in which the eigenvalue (12.28), multiplied by the strength of the interaction, may be added to the energy of two nucleons in the central potential well.

As explained above, states of one nucleon in the j-orbit and another in the j'-orbit are degenerate if the Hamiltonian contains only the central potential. All such states have the same single nucleon energies $\epsilon_j + \epsilon_{j'}$. When the residual interaction is introduced, the degeneracy is removed, in general, and the resulting eigenstates have definite values of J (and M). The corresponding eigenvalues are given by first order perturbation theory as the expectation value in those states of the two-body interaction. The residual interaction may well have, however, non-vanishing non-diagonal matrix elements between states, with given J (and M) in the jj' configuration and those in higher $j''j'''$ configurations. The difference between the unperturbed energies $\epsilon_{j''} + \epsilon_{j'''} - (\epsilon_j + \epsilon_{j'})$ may be large compared with the non-diagonal matrix-element

$$|\langle jj'JM|V|j''j'''JM\rangle| \ll \epsilon_{j''} + \epsilon_{j'''} - (\epsilon_j + \epsilon_{j'}) \qquad (12.29)$$

If (12.29) holds, the effect of the $j''j'''$-configuration on the lower one may be neglected or taken into account, explicitly or implicitly, as a second order perturbation. There may be cases, however, in which (12.29) is not satisfied and then the subspace of the shell model considered should be enlarged.

In such cases the submatrix of the Hamiltonian to be diagonalized includes several configurations. For given J (and M) the rows and columns of this submatrix are labeled by the various two nucleon configurations. The corresponding matrix elements are then given by

$$\langle jj'JM|V|j''j'''JM\rangle + (\epsilon_j + \epsilon_{j'})(\delta_{jj''}\delta_{j'j'''} + \delta_{jj'''}\delta_{j'j''}) \qquad (12.30)$$

As repeatedly emphasized above, we do not have detailed knowledge about the residual or effective two-body interaction in nuclei. In the case where mixing of configurations is necessary, the information required is even less available. Only in a few cases of actual nuclei such detailed information has been extracted from experimental data. For some simple interactions, such as those described above, it is possible to diagonalize *exactly* matrices like the one in (12.30). This is possible under the condition that the energy differences between single nucleon energies ϵ_j are small compared to non-diagonal matrix elements. This condition is the opposite of the condition (12.29).

If all j-orbits connected by matrix elements (12.27) have the same single nucleon energies, then the matrix (12.30) can be diagonalized exactly. The matrix (12.27) has *one* eigenvalue equal to (12.28) and all other eigenvalues vanish. If the pairing interaction is multiplied by a negative constant G_0, the gain in energy due to it is given by $2\Omega G_0$. All nucleon pairs in the various orbits contribute coherently to yield that eigenvalue.

In order to find the eigenstate which belongs to this eigenvalue we diagonalize the matrix (12.27) in two steps. It is sufficient to consider the submatrix of (12.27) defined by $M = 0 = m + m' = m'' + m''$. In that $M = 0$ matrix, the interaction within each j^2 configuration is first diagonalized. As a result, the only non-vanishing diagonal matrix element in the j^2-configuration is characterized by the $J = 0$ state. There are, however, non-vanishing non-diagonal matrix elements connecting it to $J = 0$ states in other j'^2 configurations. The elements of the sub-matrix of (12.27), characterized by $J = 0$ ($M = 0$) are obtained by using the inverse of the transformation used to derive (12.26) as

$$\begin{aligned}
&\langle j^2 J = 0 M = 0 | P | j'^2 J = 0 M = 0 \rangle \\
&= \sum_{mm'} \frac{(-1)^{j-m}}{\sqrt{2j+1}} (-1)^{j-m+j'-m'} \frac{(-1)^{j'-m'}}{\sqrt{2j'+1}} \\
&= \sqrt{(2j+1)(2j'+1)}
\end{aligned} \tag{12.31}$$

As seen from (12.31), the trace of this smaller matrix is equal to the trace of the matrix (12.27). The matrix (12.31) is also a *separable matrix*, with rank 1. Hence, the only non-vanishing eigenvalue is equal to the trace (12.28). The (unnormalized) eigenstate corresponding to it is given by

$$\sum_j \sqrt{2j+1}\,\psi_{j^2 J=0, M=0} \tag{12.32}$$

as can be easily verified. The eigenvalues of all states with $J \neq 0$ vanish. Also the eigenvalues of all $J = 0$ states which are orthogonal to the special $J = 0$ state (12.32) vanish. The spectrum of the two nucleon system considered here is very simple. The ground state has $J = 0$ which is lower than the sum of two single nucleon energies by $2G_0\Omega$. All other states are degenerate and have zero interaction energy.

It is worthwhile to point out an interesting aspect of the diagonalization of the matrix (12.26) as well as (12.27). The interaction in (12.26) is charge independent. It has the same matrix elements for states of two protons, two neutrons or a proton and neutron characterized by $jmjm'$ and $jm''jm'''$. This is also true for the interaction (12.27). We explained that this symmetry of the Hamiltonian leads to eigenstates characterized by the total isospin, $T = 1$ or $T = 0$ in the case of two nucleons. Yet in deriving the eigenvalue (12.28) we found it *simpler* not to make use of that symmetry. Had we used wave functions with definite isospin T, like those in (9.7) and (9.8), for calculating matrix elements of P we would obtain zero for $T = 0$ states and *twice* the matrix elements (12.26) or (12.27) for $T = 1$ states. Hence, the $T = 1$ submatrix of the pairing interaction would have been of order $j + \frac{1}{2}$ only, but the diagonal elements would have been equal to 2. Thus, the trace of the $T = 1$ matrix in the j^2 configuration would be still $2j + 1$ and the trace of the $T = 1$ matrix for several orbits would be equal to (12.28).

The pairing interaction (12.27), apart form the phases, is very similar to the electron-electron interaction introduced by Bardeen, Cooper and Schrieffer (1957) in their theory of superconductivity. In that theory the analogs of the states $|jmj, -m\rangle$ appearing in (12.27), are states of electrons with opposite linear momenta $\mathbf{k}$ and $-\mathbf{k}$.

Another interaction, more interesting than the pairing interaction, which can be easily diagonalized in the subspace of several orbits, is a generalization of the δ-interaction considered above. Let us first consider matrix elements of the δ-potential between LS-coupling wave functions characterized by $ll'LM_L$. For simplicity, we assume that different orbits are characterized by different l-values. Going back to Section 11, we recall the vanishing of matrix elements of the $\delta(|\mathbf{r}_1 - \mathbf{r}_2|)$ potential between states which are not symmetric in space coordinates. The symmetric states (11.2) for $\mathbf{r}_1 = \mathbf{r}_2$ and (11.3) were transformed into a product of radial functions and the angular function (11.4). Using such wave functions for the states $ll'LM_L$ and $l''l'''LM_L$ we find that the matrix element of the δ-potential between them is given by

$$\boxed{\begin{aligned}&[(1+\delta_{ll'})(1+\delta_{l''l'''})]^{-1/2}2(-1)^{l+l''}\\ &\quad\times\sqrt{(2l+1)(2l'+1)(2l''+1)(2l'''+1)}\\ &\quad\times\begin{pmatrix} l & l' & L \\ 0 & 0 & 0 \end{pmatrix}\begin{pmatrix} l'' & l''' & L \\ 0 & 0 & 0 \end{pmatrix}\\ &\quad\times\frac{1}{4\pi}\int R_{nl}(r)R_{n'l'}(r)R_{n''l''}(r)R_{n'''l'''}(r)\frac{dr}{r^2}\end{aligned}} \tag{12.33}$$

In non-vanishing matrix elements (12.33) the conditions

$$(-1)^{l+l'+L}=1$$

as well as

$$(-1)^{l''+l'''+L}=1$$

must be obeyed.

Apart from the radial integral, the matrix whose elements are given by (12.33) is a separable one. It is convenient to assume that all radial integrals for a group of single nucleon states are equal to some C_0. In that case the interaction is called a *surface delta interaction* (SDI) (Green and Moszkowski 1965). If the single nucleon energies of these orbits are degenerate, the diagonalization of the Hamiltonian is straightforward since only the matrix (12.33) should be diagonalized. For each given L, the rank of the matrix (12.33), with equal radial integrals, is 1. There is only *one* non-vanishing eigenvalue, equal to the trace of the matrix. The matrix elements (12.33) were calculated by using symmetric states of nucleons in l, l' orbits. Hence, the rows and columns of the matrix to be diagonalized, are labeled by l, l' where $l \leq l'$. The diagonal matrix elements in (12.33) are equal to

$$2C_0\frac{(2l+1)(2l'+1)}{1+\delta_{ll'}}\begin{pmatrix} l & l' & L \\ 0 & 0 & 0 \end{pmatrix}^2 \tag{12.34}$$

Hence, the trace of the matrix is equal to

$$\boxed{\begin{aligned} 2C_0 \sum_{l \leq l'} \frac{(2l+1)(2l'+1)}{1+\delta_{ll'}} \begin{pmatrix} l & l' & L \\ 0 & 0 & 0 \end{pmatrix}^2 &= C_0 \sum_{l} (2l+1)^2 \begin{pmatrix} l & l & L \\ 0 & 0 & 0 \end{pmatrix}^2 \\ &\quad + 2C_0 \sum_{l<l'} (2l+1)(2l'+1) \begin{pmatrix} l & l' & L \\ 0 & 0 & 0 \end{pmatrix}^2 \\ &= C_0 \sum_{l} (2l+1)^2 \begin{pmatrix} l & l & L \\ 0 & 0 & 0 \end{pmatrix}^2 \\ &\quad + C_0 \sum_{l \neq l'} (2l+1)(2l'+1) \begin{pmatrix} l & l' & L \\ 0 & 0 & 0 \end{pmatrix}^2 \\ &= C_0 \sum_{l,l'} (2l+1)(2l'+1) \begin{pmatrix} l & l' & L \\ 0 & 0 & 0 \end{pmatrix}^2 \end{aligned}} \tag{12.35}$$

The eigenvalue (12.35) belongs to the (unnormalized) eigenstate

$$\boxed{\sum_{l \leq l'} (-1)^l \sqrt{(2l+1)(2l'+1)} \begin{pmatrix} l & l' & L \\ 0 & 0 & 0 \end{pmatrix} \Phi^s_{ll'LM}} \tag{12.36}$$

as can be directly verified. The eigenvalues (12.35) which do not vanish are those for which $(-1)^{l+l'+L} = 1$.

The behavior of the surface delta interaction in jj-coupling is similar to the situation in LS-coupling. As explained in Section 11, matrix elements between states with isospin $T = 1$ are easier to calculate.

The symmetric functions (11.4) must then be multiplied by spin singlet states with $S = 0$. When transforming from the states defined by $ll'SM_SLM_L$ to jj-coupling states defined by $ll'jj'JM$, there is only one term which may contribute to a state with given J. This is the state with $S = 0$, $L = J$. The transformation coefficient is given by (11.13). Hence, we obtain for the matrix element between $T = 1$ states with $jj'JM$ and $j''j'''JM$, which satisfy the relation $(-1)^{l+l'+J} = (-1)^{l''+l'''+J} = 1$, the expression

$$\begin{aligned}
&(-1)^{j+j''+1+l'+l'''}C_0\left[(1+\delta_{jj'})(1+\delta_{j''j'''})\right]^{-1/2} \\
&\quad\times\sqrt{(2j+1)(2j'+1)(2j''+1)(2j'''+1)} \\
&\quad\times\sqrt{(2l+1)(2l'+1)(2l''+1)(2l'''+1)} \\
&\quad\times\begin{pmatrix} l & l' & L \\ 0 & 0 & 0 \end{pmatrix}\begin{Bmatrix} l & j & \frac{1}{2} \\ j' & l' & J \end{Bmatrix}\begin{pmatrix} l'' & l''' & L \\ 0 & 0 & 0 \end{pmatrix}\begin{Bmatrix} l'' & j'' & \frac{1}{2} \\ j''' & l''' & J \end{Bmatrix}
\end{aligned} \tag{12.37}$$

Using the identity (10.45) for the product of a $3j$-symbol by a $6j$-symbol in (12.37) we obtain a more concise result which is a generalization of (11.15)

$$\begin{aligned}
&C_0(-1)^{j+j''+1+l'+l'''}\sqrt{\frac{(2j+1)(2j'+1)}{1+\delta_{jj'}}\frac{(2j''+1)(2j'''+1)}{1+\delta_{j''j'''}}} \\
&\quad\times\begin{pmatrix} j & j' & J \\ \frac{1}{2} & -\frac{1}{2} & 0 \end{pmatrix}\begin{pmatrix} j'' & j''' & J \\ \frac{1}{2} & -\frac{1}{2} & 0 \end{pmatrix} \\
&\qquad\qquad\text{for}\quad (-1)^{l+l'+J} = (-1)^{l''+l'''+J} = +1
\end{aligned} \tag{12.38}$$

The result (12.38), as well as (12.37), is clearly an element of a separable matrix of rank 1. For each J, there is at most one non-vanishing eigenvalue given by the trace of the matrix (12.38) which may be expressed like (12.35) as

$$\frac{1}{2}C_0 \sum_{jj'} (2j+1)(2j'+1) \begin{pmatrix} j & j' & J \\ \frac{1}{2} & -\frac{1}{2} & 0 \end{pmatrix}^2 \qquad (-1)^{l+l'+J} = 1 \tag{12.39}$$

The $T = 1$ eigenstate (unnormalized) which belongs to the eigenvalue (12.39) is given by

$$\sum_{j \leq j'} (-1)^{j-(1/2)+l'} \sqrt{(2j+1)(2j'+1)} \begin{pmatrix} j & j' & J \\ \frac{1}{2} & -\frac{1}{2} & 0 \end{pmatrix} \Psi^a_{jj'JM} \qquad (-1)^{l+l'+J} = 1 \tag{12.40}$$

The two nucleon wave functions in (12.40) are antisymmetric and normalized. The state (12.40) is an eigenstate with given J of the Hamiltonian of the two nucleons if all orbits included in the summation are degenerate and have the same single nucleon energies.

A special case of (12.40) is worth noting. For $J = 0$, (and only one orbit for given value of j) we use the value of the $3j$-symbol (7.14) to obtain the $T = 1$, $J = 0$ unnormalized ground state as

$$\sum_j (-1)^l \sqrt{2j+1} \Psi_{j^2 J=0, M=0} \tag{12.41}$$

The eigenvalue which belongs to the state (12.41) is given by (12.39) for $j = j', J = 0$ as

$$\boxed{\tfrac{1}{2}C_0 \sum_j (2j+1)^2 \begin{pmatrix} j & j & 0 \\ \frac{1}{2} & -\frac{1}{2} & 0 \end{pmatrix} = \tfrac{1}{2}C_0 \sum_j (2j+1) = C_0\Omega} \tag{12.42}$$

Apart from the phase factor $(-1)^l$, the eigenstate (12.41) is the same as (12.32) obtained above for the pairing interaction. Such pair states with $J = 0$ have very special properties which we will explore in detail in subsequent sections. Here, let us only look at the case in which the summation in (12.41) includes both spin-orbit partners $j = l + \frac{1}{2}$ and $j' = l - \frac{1}{2}$. Their coefficients, $\sqrt{2l+2}$ and $\sqrt{2l}$ respectively, are proportional to the coefficients of transformation from the state $jj'J = 0$ to the state with $S = 0$, $L = 0$, $J = 0$. As mentioned in Section 9, it is customary to denote states with given S, L and J by the symbol $^{2S+1}L_J$ and instead of the value of L, to write for $L = 0, 1, 2, 3, 4, \ldots$ the letters $S, P, D, F, G, \ldots$. The expansion of the $^1S_0$ state ($S = 0$, $L = 0$, $J = 0$) in terms of the states

$$(l + \tfrac{1}{2})^2 \; J = 0 \qquad \text{and} \qquad (l - \tfrac{1}{2})^2 \; J = 0$$

is given in Table 9.1. The normalized coefficients are

$$\sqrt{(l+1)/(2l+1)} \qquad \text{and} \qquad \sqrt{l/(2l+1)}$$

respectively. Similar results hold also for other J-values. This is to be expected in the absence of spin-orbit interaction in the l^2 configuration. The eigenstates of any rotationally invariant potential interaction (10.1), are characterized by definite values of S and L.

To calculate matrix elements of the surface delta interaction between $T = 0$ states we calculate the direct and exchange terms

$$\begin{aligned} &\langle j_1 j_2' JM | \delta(|\mathbf{r}_1 - \mathbf{r}_2|) | j_1'' j_2''' JM \rangle \\ &\qquad + (-1)^T \langle j_1 j_2' JM | \delta(|\mathbf{r}_1 - \mathbf{r}_2|) | j_2'' j_1''' JM \rangle \end{aligned} \tag{12.43}$$

Using the expansion (11.18) of the δ-potential and (10.32) we obtain the direct term in the form

$$\frac{4\pi C_0(-1)^{j'+j''+J}}{\sqrt{(1+\delta_{jj'})(1+\delta_{j''j'''})}}\sum_k (j\|\mathbf{Y}_k\|j'')(j'\|\mathbf{Y}_k\|j''')\begin{Bmatrix} j & j' & J \\ j''' & j'' & k \end{Bmatrix}$$

$$= C_0(-1)^{j+j''+J}\sqrt{\frac{(2j+1)(2j'+1)(2j''+1)(2j'''+1)}{(1+\delta_{jj'})(1+\delta_{j''j'''})}}$$

$$\times \sum_k (2k+1)\begin{pmatrix} j & k & j'' \\ -\frac{1}{2} & 0 & \frac{1}{2} \end{pmatrix}\begin{pmatrix} j' & k & j''' \\ -\frac{1}{2} & 0 & \frac{1}{2} \end{pmatrix}\begin{Bmatrix} j & j' & J \\ j''' & j'' & k \end{Bmatrix}$$

$$\times \tfrac{1}{4}(1+(-1)^{l+l''+k})(1+(-1)^{l'+l'''+k}) \qquad (12.44)$$

The last equality was obtained by substituting the values of the reduced matrix elements (10.46). We can use now the identity (10.24) and obtain for the direct term the form

$$\tfrac{1}{2}C_0\sqrt{\frac{(2j+1)(2j'+1)(2j''+1)(2j'''+1)}{(1+\delta_{jj'})(1+\delta_{j''j'''})}}$$

$$\times\left[\begin{pmatrix} j & j' & J \\ -\frac{1}{2} & -\frac{1}{2} & 1 \end{pmatrix}\begin{pmatrix} j''' & j'' & J \\ \frac{1}{2} & \frac{1}{2} & -1 \end{pmatrix} - (-1)^{j+j''+l+l''}\right.$$

$$\left.\times\begin{pmatrix} j & j' & J \\ \frac{1}{2} & -\frac{1}{2} & 0 \end{pmatrix}\begin{pmatrix} j''' & j' & J \\ \frac{1}{2} & -\frac{1}{2} & 0 \end{pmatrix}\right]$$

$$= \tfrac{1}{2}C_0\sqrt{\frac{(2j+1)(2j'+1)(2j''+1)(2j'''+1)}{(1+\delta_{jj'})(1+\delta_{j''j'''})}}$$

$$\times\begin{pmatrix} j & j' & J \\ \frac{1}{2} & -\frac{1}{2} & 0 \end{pmatrix}\begin{pmatrix} j'' & j''' & J \\ \frac{1}{2} & -\frac{1}{2} & 0 \end{pmatrix}$$

$$\times\left((-1)^{j+j''+1+l+l''}\right.$$

$$\left. + \frac{\{(2j+1)+(2j'+1)(-1)^{j+j'+J}\}\times\{(2j''+1)+(2j'''+1)(-1)^{j''+j'''+J}\}}{4J(J+1)}\right)$$

$$(12.45)$$

The last equality in (12.45) is due to the relation (11.29). Eq. (12.45) reduces to (11.30) if we put $j'' = j$, $l'' = l$, $j''' = j'$, $l''' = l'$ and $j \neq j'$.

The exchange term in (12.43) can be similarly evaluated and we obtain for it the form

$$\frac{1}{2}C_0\sqrt{\frac{(2j+1)(2j'+1)(2j''+1)(2j'''+1)}{(1+\delta_{jj'})(1+\delta_{j''j'''})}}\begin{pmatrix} j & j' & J \\ \frac{1}{2} & -\frac{1}{2} & 0 \end{pmatrix}\begin{pmatrix} j'' & j''' & J \\ \frac{1}{2} & -\frac{1}{2} & 0 \end{pmatrix}$$
$$\times\left((-1)^{j+j''+l+l'''+J} + \frac{\begin{array}{c}\{(2j+1)+(2j'+1)(-1)^{j+j'+J}\} \\ \times\{(2j''+1)+(2j'''+1)(-1)^{j''+j'''+J}\}\end{array}}{4J(J+1)}\right) \tag{12.46}$$

which for $j'' \neq j$, $j''' \neq j'$, $l'' = l$, $l''' = l'$ and $j \neq j'$ reduces to (11.32). For $T = 1$ states we subtract the exchange term (12.46) from the direct term (12.45). The result is zero if $(-1)^{l''+l'''+J} = -1$ (and then also $(-1)^{l+l'+J} = -1$). If $(-1)^{l+l'+J} = (-1)^{l''+l'''+J} = 1$ the result is equal to (12.38) derived above. To obtain $T = 0$ matrix elements (12.46) and (12.45) should be *added*. The result is

$$\boxed{\begin{array}{l}\frac{1}{2}C_0\sqrt{\frac{(2j+1)(2j'+1)(2j''+1)(2j'''+1)}{(1+\delta_{jj'})(1+\delta_{j''j'''})}} \\ \quad\times\begin{pmatrix} j & j' & J \\ \frac{1}{2} & -\frac{1}{2} & 0 \end{pmatrix}\begin{pmatrix} j'' & j''' & J \\ \frac{1}{2} & -\frac{1}{2} & 0 \end{pmatrix} \\ \quad\times\Bigg[(-1)^{j+j''+l+l''+1}(1-(-1)^{l''+l'''+J}) \\ \qquad + 2\frac{\begin{array}{c}\{(2j+1)+(2j'+1)(-1)^{j+j'+J}\} \\ \times\{(2j''+1)+(2j'''+1)(-1)^{j''+j'''+J}\}\end{array}}{4J(J+1)}\Bigg]\end{array}} \tag{12.47}$$

In the case $(-1)^{l''+l'''+J} = (-1)^{l+l'+J} = 1$, (12.47) becomes an element of a separable matrix for given J with rank 1 whose rows and columns are labeled by j, j' and j'', j'''. This occurs as a result of the radial integrals being independent of j, j', j'', j'''. Only one eigenvalue of such a matrix element does not vanish and it is equal to the trace of the matrix which is

$$\boxed{\frac{1}{2}C_0 \sum_{jj'} (2j+1)(2j'+1) \begin{pmatrix} j & j' & J \\ \frac{1}{2} & -\frac{1}{2} & 0 \end{pmatrix}^2 \times \frac{\{(2j+1)+(2j'+1)(-1)^{j+j'+J}\}^2}{4J(J+1)}} \tag{12.48}$$

The (unnormalized) eigenstate, with given J, which belongs to the eigenvalue (12.48) is

$$\boxed{\sum_{j\leq j'} \sqrt{(2j+1)(2j'+1)} \begin{pmatrix} j & j' & J \\ \frac{1}{2} & -\frac{1}{2} & 0 \end{pmatrix} \times \{(2j+1)+(2j'+1)(-1)^{j+j'+J}\} \Psi^{s}_{jj'JM}} \tag{12.49}$$

The two nucleon wave functions in (12.49) are symmetric and normalized.

Various applications of the surface delta interaction to nuclear levels have been published. In a modified version of SDI, there are different coefficients for $T = 1$ and $T = 0$ states and constant interactions which depend on T are added. A detailed description of such applications may be found in a book by Brussaard and Glaudemans (1977).

13

Two Nucleons in a Harmonic Oscillator Potential Well

Looking at matrix elements of a two-body interaction we see that simple interactions are functions of $\mathbf{r}_1 - \mathbf{r}_2$ whereas the wave functions are sums of products of wave functions of $\mathbf{r}_1$ and of $\mathbf{r}_2$. In Section 10 it was found useful to expand the interaction in terms of functions of $\mathbf{r}_1$ and $\mathbf{r}_2$. It became then a straightforward matter to use methods of tensor algebra to carry out the integrations over $d^3\mathbf{r}_1$ and $d^3\mathbf{r}_2$. Another approach is to make use of the fact that the interaction is a function of the relative coordinate $\mathbf{r}_1 - \mathbf{r}_2$. This is certainly the case for central interactions $V(|\mathbf{r}_1 - \mathbf{r}_2|)$ with or without spin operators. It is also true for tensor forces (10.71) for which, as shown in (10.72), the potential $V_T(|\mathbf{r}_1 - \mathbf{r}_2|)$ is multiplied by $([\mathbf{s}_1 \times \mathbf{s}_2]^{(2)} \cdot [(\mathbf{r}_1 - \mathbf{r}_2) \times (\mathbf{r}_1 - \mathbf{r}_2)]^{(2)})$. The mutual spin-orbit interaction (10.69) is slightly more complicated but it is expressed in terms of the orbital angular momentum (10.70) which is associated with the relative coordinate. In an alternative approach, the interaction is kept intact and it is attempted to expand the shell model wave functions of two nucleons in terms of the relative coordinate $\mathbf{r} = \mathbf{r}_1 - \mathbf{r}_2$ and the center-of-mass coordinate $\mathbf{R} = (\mathbf{r}_1 + \mathbf{r}_2)/2$. Such an expansion is always possible but in the general case it is rather complicated. It turns out that if the single nucleon wave functions are determined by a harmonic oscillator potential well, as described in Section 4, such an expansion can be conveniently carried out. The wave functions discussed in Section 4 do not have the correct asymptotic behavior. Still, they may be good approximations for wave functions of low lying levels in the region where they are large and contribute most to the matrix elements.

Consider the Hamiltonian of two nucleons moving in an oscillator potential

$$\boxed{\frac{1}{2m}\mathbf{p}_1^2 + \frac{1}{2}m\omega^2\mathbf{r}_1^2 + \frac{1}{2m}\mathbf{p}_2^2 + \frac{1}{2}m\omega^2\mathbf{r}_2^2} \tag{13.1}$$

We introduce relative and center-of-mass coordinates and the cannonically conjugate momenta by

$$\boxed{\begin{array}{ll} \mathbf{r} = \mathbf{r}_2 - \mathbf{r}_1 & \mathbf{p} = \frac{1}{2}(\mathbf{p}_2 - \mathbf{p}_1) \\ \mathbf{R} = \frac{1}{2}(\mathbf{r}_1 + \mathbf{r}_2) & \mathbf{P} = \mathbf{p}_1 + \mathbf{p}_2 \end{array}} \tag{13.2}$$

This is a canonical transformation as can be verified by calculating the commutation relations between components of **p** and **r** as well as of **P** and **R** and verifying that components of **P** commute with those of **r** and components of **p** commute with **R** components. In these variables the Hamiltonian (13.1) may be expressed by

$$\boxed{\frac{1}{4m}\mathbf{P}^2 + m\omega^2\mathbf{R}^2 + \frac{1}{m}\mathbf{p}^2 + \frac{m}{4}\omega^2\mathbf{r}^2} \tag{13.3}$$

The transformation (13.2) is a special case (for $A = 2$) of the one introduced in Section 4. Eigenstates of these Hamiltonians may be thus expressed either as functions of $\mathbf{r}_1$ and $\mathbf{r}_2$ or of **r** and **R**. From (13.3) follows that the mass in the part of the Hamiltonian due to **P** and **R** is 2m which is the *total* mass of the two nucleons. The mass associated with the relative coordinate and momentum is $m/2$ which is the *reduced* mass of the two nucleons. The frequency ω associated with these two parts of the Hamiltonian is the *same*. For the single nucleons in (13.1) the square of that frequency is given by the product of the coefficient of $\mathbf{p}^2$ by that of $\mathbf{r}^2$ multiplied by 4. For the center of mass part of (13.3) we find $4(\frac{1}{4}m)m\omega^2 = \omega^2$ while for the relative part of that Hamiltonian we obtain the same value $4(1/m)m\omega^2/4 = \omega^2$.

The set of orthogonal eigenstates which belong to a given eigenvalue of (13.1) or (13.3) may be expressed by product wave functions of $\mathbf{r}_1$ and $\mathbf{r}_2$ or by products of wave functions of $\mathbf{r}$ and $\mathbf{R}$. There is a linear unitary transformation which carries one such orthogonal basis into the other. In the adopted choice of wave functions this transformation is real and orthogonal. Before considering these transformations let us prove that the nice property of the harmonic oscillator potential is not shared by any other potential. We start from a single nucleon potential $U(r_1) + U(r_2)$ and assume that it can be expressed as

$$U(r_1) + U(r_2) = V(r) + W(R) \tag{13.4}$$

Let us take the derivative of (13.4) with respect to the variable $\cos\omega_{12}$ which appears in

$$r = |\mathbf{r}_1 - \mathbf{r}_2| = [\mathbf{r}_1^2 + \mathbf{r}_2^2 - 2(\mathbf{r}_1 \cdot \mathbf{r}_2)]^{1/2} = [r_1^2 + r_2^2 - 2r_1 r_2 \cos\omega_{12}]^{1/2}$$

$$R = \frac{1}{2}|\mathbf{r}_1 + \mathbf{r}_2| = \left[\left(\frac{r_1}{2}\right)^2 + \left(\frac{r_2}{2}\right)^2 + 2\left(\frac{r_1}{2}\right)\left(\frac{r_2}{2}\right)\cos\omega_{12}\right]^{1/2}$$

The l.h.s of (13.4) is independent of $\cos\omega$ and the derivative vanishes. Thus we obtain

$$-\frac{r_1 r_2}{r}V'(r) + \frac{r_1 r_2}{4R}W'(R) = 0 \qquad \text{or} \qquad \frac{V'(r)}{r} = \frac{W'(R)}{4R} \tag{13.5}$$

Since $\mathbf{r}$ and $\mathbf{R}$ are independent variables, the last equality in (13.5) implies that both ratios are equal to the same constant $k/2$. Hence

$$V'(r) = \left(\frac{k}{2}\right) r \qquad W'(R) = 4\left(\frac{k}{2}\right) R$$

from which follows

$$V(r) = \tfrac{1}{4}kr^2 + \text{const.} \qquad W(R) = kR^2 + \text{const.} \tag{13.6}$$

From (13.6) follows by expressing $\mathbf{r}_1$ and $\mathbf{r}_2$ in terms of $\mathbf{r}$ and $\mathbf{R}$

$$U(r_1) + U(r_2) = \tfrac{1}{2}kr_1^2 + \tfrac{1}{2}kr_2^2 + \text{const.} \tag{13.7}$$

The constant in (13.7) which is the sum of constants in (13.6) may be taken to be zero. Thus, the only potential U which satisfies (13.4) is the harmonic oscillator potential (Talmi 1952).

The relation (13.4) for harmonic oscillator potentials has the same form as the transformation of the kinetic energy part of the Hamiltonian from (13.1) to (13.3). This is due to the symmetric appearance of the momentum $\mathbf{p}$ and the coordinate vector $\mathbf{r}$ in the oscillator Hamiltonian. Eigenstates of the kinetic energy of two free nucleons may thus be expressed as functions of $\mathbf{R}$ and $\mathbf{r}$. Such eigenstates describe free nucleons but they appear also in the system called *nuclear matter*. This infinite system is considered to be *translationally* invariant and hence eigenstates of two nucleons are simply

$$e^{i\mathbf{k}_1\cdot\mathbf{r}_1}e^{i\mathbf{k}_2\cdot\mathbf{r}_2} = e^{i(\mathbf{k}_1\cdot\mathbf{r}_1+\mathbf{k}_2\cdot\mathbf{r}_2)}$$

Such states can be expressed in terms of $\mathbf{R}$ and $\mathbf{r}$. Substituting $\mathbf{r}_1 = \frac{1}{2}(2\mathbf{R}-\mathbf{r})$, $\mathbf{r}_2 = \frac{1}{2}(2\mathbf{R}+\mathbf{r})$ we obtain the same eigenstate written as

$$e^{i(\mathbf{K}\cdot\mathbf{R}+\mathbf{k}\cdot\mathbf{r})}$$

where $\mathbf{K} = \mathbf{k}_1 + \mathbf{k}_2$ and $\mathbf{k} = \frac{1}{2}(\mathbf{k}_2 - \mathbf{k}_1)$. The concept of nuclear matter is difficult to apply to finite nuclei and we will not consider it further.

The eigenstates of (13.1) are characterized by $N_1l_1m_1$ and $N_2l_2m_2$ where N_1 and N_2 are the principal quantum numbers defined in Section 4 by $N = 2(n-1)+l$ (n is the number of nodes in the radial functions including the one at $r = 0$ but not the point at infinity). The corresponding eigenvalues are given by

$$\hbar\omega(N_1 + N_2 + 3) \tag{13.8}$$

The *same* eigenvalues may be expressed by the principal quantum numbers $\mathcal{N}$ and n of the radial functions of R and r by

$$\hbar\omega(\mathcal{N} + n + 3) \tag{13.9}$$

The finite set of eigenstates which belong to (13.8) can be expressed as linear combinations of the finite set of states with quantum numbers in (13.9) provided (13.8) and (13.9) are equal

$$\boxed{\mathcal{N} + n = N_1 + N_2} \tag{13.10}$$

In addition to (13.10) the transformation between the two sets should be between states with the same value of the total orbital momentum L and its projection along the z-axis, M_L. Moreover, the dependence of the transformation coefficients on $\mathcal{M}$ and m, which are the z-projections of $\mathcal{L}$ and l, should be given by Clebsch-Gordan coefficients. We can thus express any state of $\mathbf{r}_1$ and $\mathbf{r}_2$ as

$$\boxed{\begin{aligned}\Psi_{N_1l_1N_2l_2LM_L}(\mathbf{r}_1,\mathbf{r}_2) = \sum_{\substack{\mathcal{N}\mathcal{L}nl\\ \mathcal{M}m}} a^{N_1l_1N_2l_2}_{\mathcal{N}\mathcal{L}nlL}(\mathcal{L}\mathcal{M}lm \mid \mathcal{L}lLM_L)\\ \times\, \phi_{\mathcal{N}\mathcal{L}\mathcal{M}}(\mathbf{R})\phi_{nlm}(\mathbf{r})\end{aligned}} \tag{13.11}$$

There is a rather limited number of non-vanishing terms in the expansion (13.11). They are limited by (13.10) as well as by the requirement that $\mathcal{L}$ and l must couple to the given $\mathbf{L}$. Another limitation arises from symmetry considerations. Under exchange of $\mathbf{r}_1$ and $\mathbf{r}_2$, $\mathbf{R}$ remains unchanged, whereas $\mathbf{r}$ goes to $-\mathbf{r}$. Hence, if we use the expansion (13.11) for *symmetric* states, r.h.s. must contain only states with l values satisfying $(-1)^l = 1$. If *antisymmetric* states are expanded according to (13.11), only states with l values satisfying $(-1)^l = -1$ take part in the expansion. The radial parts of the functions on the r.h.s. of (13.11), are given by (4.6). In view of (13.3) the oscillator constant ν in (4.6) given by (4.4) for single nucleon wave functions $m\omega/2h$, should be replaced by $m\omega/\hbar$ for the radial function of R and by $m\omega/4\hbar$ for the radial function of r.

Using the transformed wave functions according to (13.11), the calculation of matrix elements of two-body interactions can be carried out in a straightforward manner. The mutual interaction depends on $\mathbf{r}$ and $\mathbf{p}$ and hence, the integration over $\mathbf{R}$ may be carried out right away. In the LS-coupling scheme, matrix elements are given by (10.73). The calculation of matrix elements of the spatial part in

(10.73) becomes, after the integration over **R**, equal to

$$\langle N_1 l_1 N_2 l_2 L M_L | \mathcal{W}_\kappa^{(k)} | N_1' l_1' N_2' l_2' L' M_L' \rangle = \sum_{\substack{\mathcal{N}\mathcal{L}\mathcal{M} \\ n l n' l' m m'}} a_{\mathcal{N}\mathcal{L} n l L}^{N_1 l_1 N_2 l_2} a_{\mathcal{N}\mathcal{L} n' l' L'}^{N_1' l_1' N_2' l_2'} (\mathcal{L}\mathcal{M} l m \mid \mathcal{L} l L M_L) \times (\mathcal{L}\mathcal{M} l' m' \mid \mathcal{L} l' L' M_L') \times \langle n l m | \mathcal{W}_\kappa^{(k)} | n' l' m' \rangle \tag{13.12}$$

As we saw in Section 10, the spatial part of the interaction is a function of only the relative coordinate (and relative momentum in the case $k = 1$, mutual spin orbit interaction). Thus, to evaluate the matrix element (13.12) only *one* integration over the relative coordinate $\mathbf{r}$ is needed.

In the special case of a central interaction, $k = 0$, the matrix element (13.12) is greatly simplified. For $V(|\mathbf{r}_1 - \mathbf{r}_2|) = V(r)$, l' must be equal to l on the r.h.s. of (13.12), m' must be equal to m and the matrix element is *independent* of m. The summation over $\mathcal{M}$ and m can be carried out as well as the integration over the angles of $\mathbf{r}$. Due to the orthogonality relations of the Clebsch-Gordan coefficients (7.8) we obtain

$$\langle N_1 l_1 N_2 l_2 L M_L | V(r) | N_1' l_1' N_2' l_2' L' M_L' \rangle = \delta_{LL'} \delta_{M_L M_L'} \sum_{\mathcal{N}\mathcal{L} n n' l} a_{\mathcal{N}\mathcal{L} n l L}^{N_1 l_1 N_2 l_2} a_{\mathcal{N}\mathcal{L} n' l L}^{N_1' l_1' N_2' l_2'} \times \int_0^\infty R_{nl}(r) R_{n'l}(r) V(r) \, dr \tag{13.13}$$

To calculate the matrix element of (13.13) only *one* integration over r is required.

The radial functions in (13.13) are given by (4.6) where the parameter $\nu = m\omega/2\hbar$ of (4.4) is replaced by $m\omega/4\hbar = \nu/2$. They are equal to a polynomial in r^2 multiplying $r^l e^{-(\nu/2)r^2}$. Therefore, it is always possible to express the radial integral in (13.13) by a linear combination of the simple radial integral

$$I_l = \int_0^\infty R^2_{n=1,l}(r)V(r)\,dr = \frac{2^{l+2}\nu^{l+\frac{3}{2}}}{\sqrt{\pi}(2l+1)!!}\int_0^\infty e^{-\nu r^2} r^{2l+2} V(r)\,dr \tag{13.14}$$

The last equality in (13.14) is due to (4.7) in which 2ν is replaced by ν. The coefficients of expansion of the matrix element (13.13) in terms of I_l are the *same* for *all* potential interactions. The physical properties follow from the values of the various I_l integrals. If the interaction has a very short range and can be replaced by the δ-potential (11.18) the integrand in (13.14) contains the factor r^{2l} and hence only I_0 does not vanish. For a more realistic but short ranged interaction, I_0 is large compared with I_1, I_1 is larger than I_2 etc. The other extreme limit is a potential which can be approximated by a constant in the region where the radial functions do not vanish. In that long range limit all I_l are equal to that constant.

The integrals I_l can be directly evaluated for the Coulomb interaction between two protons. They are given by

$$I_l = e^2\int_0^\infty e^{-\nu r^2} r^{2l+1}\,dr = \frac{2^{l+1}l!}{(2l+1)!!}e^2\sqrt{\frac{\nu}{\pi}} \tag{13.15}$$

In the harmonic oscillator potential well, the Coulomb energy between protons can be thus related to the size of the nucleus as determined by (4.11). This is to be expected since there is only one parameter of length in the harmonic oscillator potential which is given by $\nu^{-1/2}$.

As explained above, if harmonic oscillator wave functions are used, the expansion (10.1) is completely unnecessary. If, however, matrix elements of a certain interaction have been calculated in terms of the radial integrals F^k in (10.34) and G^k in (10.37), these radial integrals may be expressed in terms of the I_l. The functions $v_k(r_1, r_2)$ may be expressed in terms of $V(r)$ by (10.3) and $P_k(\cos\omega_{12})$ in it may be expressed by using

$$r_1 r_2 \cos\omega_{12} = \mathbf{r}_1\cdot\mathbf{r}_2 = \tfrac{1}{4}(4R^2 - r^2)$$

Finally, the products of radial functions in (10.34) and (10.37) may be expressed in terms of functions of **R** and **r**.

The transformation of harmonic oscillator wave functions of $\mathbf{r}_1$ and $\mathbf{r}_2$ into those of **R** and **r** was introduced and applied to nuclear spectroscopy, specially for non-central interactions, in 1952. Transformation coefficients for some simple configurations in the $l_1m_1l_2m_2$ scheme were tabulated (Talmi 1952). Since then the transformation has been taken up by several authors and many transformation coefficients have been calculated and tabulated (Moshinsky 1959, Balashov and Eltekov 1960, Lawson and Goeppert-Mayer 1960, Brody and Moshinsky 1960, Arima and Terasawa 1960). The transformation has been generalized to unequal masses (Smirnov 1961, 1962, Kumar 1966, Gal 1968) and even used in the kinetic theory of gases (Kumar 1967). It has been used also in the nuclear many-body theory (Dawson, Talmi and Walecka 1962, Barrett et al. 1971). The list of papers dealing with the transformation, its mathematical properties and various applications is too long to be included here.

14

Determinantal Many Nucleon Wave Functions

Let us now consider states of several nucleons in a central potential well. We shall first characterize these states in the m-scheme in which the quantum numbers of n nucleons are $j_1m_1, j_2m_2, \ldots j_nm_n$. Each of the j-orbits is characterized by its quantum numbers n and l but in the following we omit them for brevity. We should only remember that when we consider $j_i = j_k$ we mean identical orbits, namely also $n_i = n_k$, $l_i = l_k$. If we consider both protons and neutrons, the charge or isospin quantum numbers of the nucleons $m_{\tau_1}, m_{\tau_2}, \ldots, m_{\tau_n}$ should also be specified. We shall first deal with identical nucleons and the case of protons and neutrons will be taken up later on.

A normalized antisymmetric wave function of the system can be written down as

$$\boxed{\frac{1}{\sqrt{n!}}\Psi_n = \frac{1}{\sqrt{n!}}\sum_P (-1)^P P\psi_{j_1m_1}(1)\psi_{j_2m_2}(2)\ldots\psi_{j_nm_n}(n)} \tag{14.1}$$

In (14.1) the summation is over all permutations P of the nucleon coordinates 1 to n. The phase $(-1)^P$ is defined to be $+1$ for an even permutation and -1 for an odd one. The antisymmetric wave function (14.1) vanishes if two of the jm quantum numbers are equal. If they are all different, each of the $n!$ permutations yields a state orthogonal to all others from which follows the normalization of (14.1). The wave functions (14.1) with all possible allowed values of $m_1 m_2, \ldots, m_n$ form an *orthogonal* and *complete* set of states for n-nucleons in the orbits characterized by $j_1, j_2, \ldots, j_n$. The wave functions (14.1) may be conveniently expressed as a *Slater determinant*

$$\boxed{\frac{1}{\sqrt{n!}}\Psi_n = \frac{1}{\sqrt{n!}}\begin{vmatrix} \psi_{j_1 m_1}(1) & \psi_{j_2 m_2}(1) & \cdots & \psi_{j_n m_n}(1) \\ \psi_{j_1 m_1}(2) & \psi_{j_2 m_2}(2) & \cdots & \psi_{j_n m_n}(2) \\ \vdots & \vdots & & \vdots \\ \psi_{j_1 m_1}(n) & \psi_{j_2 m_2}(n) & \cdots & \psi_{j_n m_n}(n) \end{vmatrix}} \tag{14.2}$$

The wave function (14.1) or (14.2) may be used for the calculation of matrix elements of single nucleon operators and two-nucleon operators. Consider first the single nucleon operator

$$\mathbf{F} = \sum_{i=1}^{n} \mathbf{f}(i) \tag{14.3}$$

The matrix element between two states (14.2), Ψ and Ψ', is given by $(1/n!)\int \Psi^* \mathbf{F} \Psi'$. The integrand is a *symmetric* function of all nucleon coordinates. Hence, we can evaluate it by taking *one* of the $\mathbf{f}(i)$, say $\mathbf{f}(1)$, and multiply the result by n. The matrix elements of $\mathbf{F}$ between two states (14.2) are then given by

$$n\frac{1}{n!}\int \Psi_n^* \mathbf{f}(1) \Psi_n' \tag{14.4}$$

In (14.4) we now expand the determinants Ψ_n and Ψ_n' according to their *first* rows obtaining for (14.4) the result

$$\frac{1}{(n-1)!}\sum_{k,k'}(-1)^{k+k'}\int \psi_{j_k m_k}^*(1)\mathbf{f}(1)\psi_{j_{k'} m_{k'}}(1)\Psi_{n-1}^*(k)\Psi_{n-1}'(k') \tag{14.5}$$

In (14.5), $\Psi_{n-1}(k)$ and $\Psi'_{n-1}(k')$ are the subdeterminants obtained from (14.2) by removing the first row and the k-th and k'-th columns respectively. The integration over all nucleon coordinates $2, 3, \ldots, n$ in each term in (14.5) can be carried out. It yields $(n-1)!$ if and only if all quantum numbers which appear in the sub-determinants are the same. Hence, we conclude that the matrix element (14.5) vanishes if the quantum numbers in Ψ_n and Ψ'_n differ by more than *one* set of single nucleon quantum numbers jm. If there is one pair $j_k m_k, j'_{k'}, m'_{k'}$ which are different, the matrix element (14.5) becomes equal to

$$(-1)^{k+k'} \langle j_k m_k | \mathbf{f} | j'_{k'} m'_{k'} \rangle$$

Without loss of generality we can rearrange the columns of (14.2) to put the two different sets in the same column $k' = k$. The matrix element (14.4) becomes then equal to

$$\langle j_k m_k | \mathbf{f} | j'_k m'_k \rangle \tag{14.6}$$

If all sets of single nucleon quantum numbers in Ψ_n and Ψ'_n are the same, we obtain for the expectation value of $\mathbf{F}$ in the state (14.2) the result

$$\sum_k \langle j_k m_k | \mathbf{f} | j_k m_k \rangle \tag{14.7}$$

If we put $\mathbf{f} = 1$ in (14.6) we conclude that two wave functions (14.1) or (14.2) are orthogonal if they have different sets of single nucleon quantum numbers. If in (14.7) we put $\mathbf{f} = 1/n$, which is equivalent to $\mathbf{F} = 1$. we verify the normalization of (14.1). It is worthwhile to point out that the results (14.6) and (14.7) may be obtained from product wave functions $\prod_{k=1}^{n} \psi_{j_k m_k}(k)$ *without* antisymmetrization. This of course is true provided all sets $j_k m_k$ are different.

The observation made above, about the matrix elements (14.6) and (14.7) is true if all single nucleon orbits are orthogonal (and normalized). It is quite possible to use a non-orthogonal basis for constructing the states (14.1) or (14.2). In fact, if we start with an orthogonal basis of states $\psi_{n_i l_i j_i m_i}$ we may add to a given column in (14.2) any linear combination of the other columns. This will not change the value of the determinant (14.2). The results (14.6) and (14.7), however, will not be obtained from a non-orthogonal basis of states $\psi_{j_i m_i}$. Normally, it will not be practical to admix single nucleon states with different values of l, j and m. It may occasionally be convenient to take linear combinations of single nucleon states with different radial functions. Should this be done, it is important to carry out the antisymmetrization explicitly. It is also necessary to calculate explicitly the normalization of the resulting wave functions. The normalization of a determinantal wave function like (14.2) constructed with a non-orthogonal basis (with normalized states) is no longer given by $(n!)^{-1/2}$.

The calculation of matrix elements of a two-body operator in the scheme of states (14.2) can be carried out in a similar fashion. Consider the symmetric operator

$$\mathbf{G} = \sum_{i<k} \mathbf{g}(i,k) \tag{14.8}$$

Matrix elements of $\mathbf{G}$ may be calculated by picking one of the $n(n-1)/2$ different terms $\mathbf{g}(i,k)$, $\mathbf{g}(1,2)$ say, calculating its matrix element and multiplying the result by $n(n-1)/2$. We thus start with the matrix element

$$\frac{n(n-1)}{2}\frac{1}{n!}\int \Psi_n^* \mathbf{g}(1,2)\Psi_n' \tag{14.9}$$

In (14.9) we expand the determinants Ψ_n and Ψ_n' according to their first *two* rows obtaining

$$\frac{1}{2(n-2)!}\sum_{k<k',i<i'}(-1)^{k+k'+i+i'}\int \begin{vmatrix} \psi_{j_k m_k}(1) & \psi_{j_{k'} m_{k'}}(1) \\ \psi_{j_k m_k}(2) & \psi_{j_{k'} m_{k'}}(2) \end{vmatrix}^*$$
$$\times\, \mathbf{g}(1,2) \begin{vmatrix} \psi_{j_i m_i}(1) & \psi_{j_{i'} m_{i'}}(1) \\ \psi_{j_i m_i}(2) & \psi_{j_{i'} m_{i'}}(2) \end{vmatrix} \Psi_{n-2}^*(k,k')\Psi_{n-2}'(i,i') \tag{14.10}$$

In (14.10), $\Psi_{n-2}(k,k')$ is the subdeterminant obtained from (14.2) by removing the first two rows and the k-th and k'-th columns. The definition of $\Psi_{n-2}'(i,i')$ is similar. In each term of the summation in (14.10), intergration over nucleon coordinates $3,\ldots,n$ yields zero unless the sets of single nucleon quantum numbers in $\Psi_{n-2}(k,k')$ and $\Psi_{n-2}'(i,i')$ are identical. If they are the same, the integration yields $(n-2)!$. Hence, the matrix element (14.9) vanishes if more than *two* sets in Ψ_n and Ψ_n' are different.

If there are two sets, say $j_1m_1j_2m_2$ and $j'_1m'_1j'_2m'_2$, which are different we obtain for the matrix element (14.9) the result

$$\frac{1}{2}\int\begin{vmatrix}\psi_{j_1m_1}(1) & \psi_{j_2m_2}(1)\\ \psi_{j_1m_1}(2) & \psi_{j_2m_2}(2)\end{vmatrix}^* \mathbf{g}(1,2)\begin{vmatrix}\psi_{j'_1m'_1}(1) & \psi_{j'_2m'_2}(1)\\ \psi_{j'_1m'_1}(2) & \psi_{j'_2m'_2}(2)\end{vmatrix}$$
$$= \int \psi^*_{j_1m_1}(1)\psi^*_{j_2m_2}(2)\mathbf{g}(1,2)\psi_{j'_1m'_1}(1)\psi_{j'_2m'_2}(2)$$
$$- \int \psi^*_{j_1m_1}(1)\psi^*_{j_2m_2}(2)\mathbf{g}(1,2)\psi_{j'_1m'_1}(2)\psi_{j'_2m'_2}(1) \qquad (14.11)$$

The result (14.11) is the difference between the direct term and the exchange term. It is identical with the expressions obtained in Section 10 for two nucleon configurations.

If the sets of single nucleon quantum numbers in Ψ_n and Ψ'_n differ by only *one* set, j_1m_1 and $j'_1m'_1$ say, we obtain for the matrix element (14.9) the linear combination

$$\frac{1}{2}\sum_{k=2}^{n}\int\begin{vmatrix}\psi_{j_1m_1}(1) & \psi_{j_km_k}(1)\\ \psi_{j_1m_1}(2) & \psi_{j_km_k}(2)\end{vmatrix}^* \mathbf{g}(1,2)\begin{vmatrix}\psi_{j'_1m'_1}(1) & \psi_{j_km_k}(1)\\ \psi_{j'_1m'_1}(2) & \psi_{j_km_k}(2)\end{vmatrix}$$
$$= \sum_{k=2}^{n}\left[\int \psi^*_{j_1m_1}(1)\psi^*_{j_km_k}(2)\mathbf{g}(1,2)\psi_{j'_1m'_1}(1)\psi_{j_km_k}(2)\right.$$
$$\left. - \int \psi^*_{j_1m_1}(1)\psi^*_{j_km_k}(2)\mathbf{g}(1,2)\psi_{j'_1m'_1}(2)\psi_{j_km_k}(1)\right]$$
$$(14.12)$$

If all sets coincide, $\Psi_n = \Psi'_n$, the expectation value of G in the state (14.2) is given by

$$\sum_{i<k}^{n}\left[\int \psi^*_{j_im_i}(1)\psi^*_{j_km_k}(2)\mathbf{g}(1,2)\psi_{j_im_i}(1)\psi_{j_km_k}(2)\right.$$
$$\left. - \int \psi^*_{j_im_i}(1)\psi^*_{j_km_k}(2)\mathbf{g}(1,2)\psi_{j_im_i}(2)\psi_{j_km_k}(1)\right] \qquad (14.13)$$

Also the matrix elements of **G**, given by (14.11), (14.12) and (14.13), can be obtained by using product wave functions $\prod_{k=1}^{n}\psi_{j_km_k}(k)$ provided in the final step, the two nucleon matrix elements are calculated with antisymmetrized wave functions. If all sets j_km_k are different, the antisymmetrization may be carried out only in the last stage. The results (14.11), (14.12) and (14.13) demonstrate the fact, mentioned in Section 1, that matrix elements of a two-body interaction in a system with n nucleons are linear combinations of two-nucleon matrix elements.

The results about matrix elements of single nucleon and two-nucleon operators are simple, yet their use is rather limited. States like (14.1) or (14.2) in general, are not eigenstates of $\mathbf{J}^2$. Eigenstates of the shell model Hamiltonian which is invariant under rotations must have well-defined angular momenta and usually states with different values of J are far from degenerate. We must learn how to form linear combinations of states like (14.1) which have J as a good quantum number. This problem will be addressed later on. Still, there are cases where we can make use of the m-scheme wave functions (14.1) to obtain interesting results. Obviously, these are cases in which the wave function (14.1) has a definite value of J.

The simplest such system is where identical nucleons fill completely a given j-orbit. The wave function (14.1) is in this case given by

$$\frac{1}{(2j+1)!}\sum(-1)^P P\psi_{j,-j}(1)\psi_{j,-j+1}(2)\ldots\psi_{jj}(2j+1) \tag{14.14}$$

There is only one such possible state of the j^{2j+1} configuration. By construction it has $M = 0$ and hence also $J = 0$. To be fully convinced, operate on (14.14) with either $J^+ = \sum j_i^+$ (or J^-). This gives $2j+1$ terms in each of which the state of one nucleon is changed into an occupied state and therefore must vanish due to the antisymmetrization. Due to the Wigner-Eckart theorem (8.4), such a configuration may have non-vanishing expectation value only for scalar operators. The interaction energy in the state (14.14) is given according to (14.13) by

$$\sum_{m<m'}^{2j+1}\left[\int\psi^*_{jm}(1)\psi^*_{jm'}(2)V(1,2)\psi_{jm}(1)\psi_{jm'}(2)\right.$$
$$\left.-\int\psi^*_{jm}(1)\psi^*_{jm'}(2)V(1,2)\psi_{jm}(2)\psi_{jm'}(1)\right] \tag{14.15}$$

The sum over direct and exchange terms in (14.15) is the *trace* of the matrix constructed from $V(1,2)$ by using antisymmetric wave functions characterized by jm, jm'. The matrix of $V(1,2)$ can be constructed by using antisymmetric wave-functions in the j^2JM scheme. Since the trace is invariant under unitary transformations, the interaction energy of the closed (fully occupied) j-orbit (14.15) may be expressed by

$$\boxed{\sum_{J\,\text{even},M} \langle j^2JM|V|j^2JM\rangle = \sum_{J\,\text{even}} (2J+1)\langle j^2JM_0|V|j^2JM_0\rangle} \tag{14.16}$$

where M_0 in (14.16) is any one of the possible M values.

As an exercise, it is possible to derive (14.16) from (14.15) by using the actual values of the transformation coefficients. We first include in the summation in (14.15) the terms with $m' = m$ which vanish since for them the direct term is equal to the exchange term. We then take in the summation all m, m' pairs, rather than those with $m < m'$, and divide the result by 2. We then use Clebsch-Gordan coefficients and obtain from (14.15) the expression

$$\frac{1}{2}\sum_{mm'}\sum_{JM,J'M'} [(jmjm' \mid jjJM)(jmjm' \mid jjJ'M')\langle j^2JM|V|j^2J'M'\rangle$$

$$-(jmjm' \mid jjJM)(jm'jm \mid jjJ'M')\langle j^2JM|V|j^2J'M'\rangle]$$

$$=\frac{1}{2}\sum_{mm'}\sum_{JM}[(jmjm' \mid jjJM)^2\langle j^2JM|V|j^2JM\rangle - (-1)^{2j-J}$$

$$\times (jmjm' \mid jjJM)^2\langle j^2JM|V|j^2JM\rangle] \tag{14.17}$$

The last equality in (14.17) follows from the rotational invariance of $V(1,2)$ leading to $J' = J$, $M' = M$ and the symmetry property (7.12) of the Clebsch-Gordan coefficients. Summing over mm' on the r.h.s. of (14.17) yields

$$\frac{1}{2}\sum_{JM}[\langle j^2JM|V|j^2JM\rangle + (-1)^J\langle j^2JM|V|j^2JM\rangle]$$

which vanishes for odd values of J and for even J values is identical to (14.16).

A simple check on the result (14.16) is obtained by putting all matrix elements $\langle j^2JM|V|j^2JM\rangle$ equal to 1. We then obtain

$$\sum_{J\,\text{even}}(2J+1)=\left(j+\frac{1}{2}\right)\frac{1+2(2j-1)+1}{2}=\frac{1}{2}(2j+1)2j \tag{14.18}$$

The number $j(2j+1)$ is the total number of interactions equal to the number $\frac{1}{2}n(n-1)$ of different pairs among n nucleons for $n=2j+1$.

Recalling the tensor expansion (12.12) of any two-body interaction we can insert it in (14.16) and obtain for the latter the expression

$$\begin{aligned}\sum_{J\,\text{even}}&(2J+1)\sum_k(-1)^{J+2j}F_k\begin{Bmatrix} j & j & J\\ j & j & k\end{Bmatrix}\\ &=\frac{1}{2}\sum_k F_k\left(\sum_J(2J+1)(-1)^{2j+J}\begin{Bmatrix} j & j & J\\ j & j & k\end{Bmatrix}\right.\\ &\qquad\left.-\sum_J(2J+1)\begin{Bmatrix} j & j & J\\ j & j & k\end{Bmatrix}\right)\\ &=\frac{1}{2}(2j+1)F_0-\frac{1}{2}\sum_k F_k=\frac{1}{2}2jF_0-\frac{1}{2}\sum_{k>0}F_k\end{aligned} \tag{14.19}$$

The last equality in (14.19) follows from the sum rules (10.17) and (10.18) of $6j$-symbols.

It may be worth while to trace the origin of the various terms on the r.h.s. of (14.19) by using directly the tensor expansion (12.12) for $V(1,2)$. The total interaction can then be expressed as

$$\begin{aligned}\sum_{i<i'}^{n}V(i,i')&=\sum_k F_k\sum_{i<i'}^{n}(\mathbf{u}^{(k)}(i)\cdot\mathbf{u}^{(k)}(i'))\\ &=\sum_k F_k\left[\frac{1}{2}\left(\sum_i^n\mathbf{u}^{(k)}(i)\right)\cdot\left(\sum_i^n\mathbf{u}^{(k)}(i)\right)\right.\\ &\qquad\left.-\frac{1}{2}\sum_i^n\mathbf{u}^{(k)}(i)\cdot\mathbf{u}^{(k)}(i)\right]\end{aligned} \tag{14.20}$$

If we take the expectation value of (14.20) for the single $J=0$ state (14.14) of the j^{2j+1} configuration we conclude from the Wigner-Eckart theorem that the only non-vanishing matrix element of $\sum \mathbf{u}^{(k)}(i)$ is

$$\langle j^{2j+1}J = 0M = 0|\sum_i^n \mathbf{u}^{(k)}(i)|j^{2j+1}J = 0M = 0\rangle$$

$$= \delta_{k0}\delta_{\kappa 0}\langle j^{2j+1}J = 0M = 0|\sum_i^n \mathbf{u}_0^{(0)}|j^{2j+1}J = 0M = 0\rangle$$

The single nucleon scalar $\mathbf{u}^{(k)}(i)\cdot\mathbf{u}^{(k)}(i)$ contributes to the expectation value according to (14.7) and (10.31)

$$\sum_m \langle jm|\mathbf{u}^{(k)}(i)\cdot\mathbf{u}^{(k)}(i)|jm\rangle$$

$$= (2j+1)\langle jm_0|\mathbf{u}^{(k)}(i)\cdot\mathbf{u}^{(k)}(i)|jm_0\rangle = 1$$

The matrix element of $\sum \mathbf{u}^{(0)}(i)$ due to (14.7) and the Wigner-Eckart theorem is equal to

$$\langle j^{2j+1}J = 0M = 0|\sum_i^n \mathbf{u}^{(0)}(i)|j^{2j+1}J = 0M = 0\rangle$$

$$= (2j+1)\langle jm_0|\mathbf{u}^{(0)}|jm_0\rangle = \sqrt{2j+1}$$

If we collect all terms we obtain the matrix element of (14.20) directly in the form (14.19). Thus, the F_0 appears in (14.19) as the coefficient of the sole non-vanishing term among the scalar products $(\sum \mathbf{u}^{(k)}(i))\cdot(\sum \mathbf{u}^{(k)}(i))$. The other F_k (including also F_0) appear as coefficients of the self interaction terms on the r.h.s. of (14.20). This is relevant in particular to the coefficient F_0 which is the only one among the F_k that does not vanish in the limit of a long range potential. If all matrix elements $\langle j^2JM|V|j^2JM\rangle$ are equal to 1, then $F_0 = 2j+1$ and the r.h.s. of (14.19) is equal to the number of interactions $\frac{1}{2}(2j+1)2j$.

If we consider both protons and neutrons we may construct fully antisymmetric states as in (14.1) or (14.2) by multiplying each single nucleon state by an isospin state. We remarked earlier that the states (14.1) do not have, in general, a definite value of the total spin J. Similarly, the analogous states where isospin states are included, do not have, in general, a definite isospin T. There are cases, however, in which such states have definite isospins. The simplest of these is the state in which the j-orbit is completely filled with *both* protons and neutrons. The wave function may be written as

$$\boxed{\frac{1}{\sqrt{[2(2j+1)]!}}\sum_P(-1)^P P\prod_{i=1}^{2(2j+1)}\psi_{m_{t_i} jm_i}(i)} \tag{14.21}$$

where i is the nucleon number going from 1 to $2(2j+1)$. The state (14.21) is the *only* state, allowed by the Pauli principle, of the $j^{2(2j+1)}$ configuration. It has $M=0$ and $M_T=0$ and hence $J=0$ as well as $T=0$.

Following the steps taken to obtain (14.16), we use (14.21) to obtain the total charge independent interaction energy in the full j-orbit in the form

$$\begin{aligned}\sum_{TJ}&(2T+1)(2J+1)\langle j^2TM_{T_0}JM_0|V|j^2TM_{T_0}JM_0\rangle\\ &=3\sum_{J\text{ even}}(2J+1)\langle j^2JM_0|V|j^2JM_0\rangle\\ &\quad+\sum_{J\text{ odd}}(2J+1)\langle j^2JM_0|V|j^2JM_0\rangle\end{aligned} \tag{14.22}$$

The equality in (14.22) follows from the fact that $T=1$ states in the j^2 configuration have even values of J whereas $T=0$ states have odd J values. The factors 3 and 1 in (14.22) arise from $2T+1$ for $T=1$ and $T=0$ respectively. Comparing (14.22) to the result (14.16) we see that the total interaction can be expressed as a sum of two parts. One is the sum of the interaction energy of the protons and that of the neutrons *each* given by (14.16) (they are equal for charge independent interactions). The second part can be written as

$$\begin{aligned}\sum_{\text{all } J}&(2J+1)\langle j^2JM_0|V|j^2JM_0\rangle\\ &=\sum_k F_k\sum_J(2J+1)(-1)^{2j+J}\begin{Bmatrix} j & j & J\\ j & j & k\end{Bmatrix}\\ &=(2j+1)F_0\end{aligned} \tag{14.23}$$

The last equality in (14.23) follows from the sum rule (10.20). The term (14.23) is the value of the interaction energy between protons and neutrons. In this case, no self interactions appear in the tensor expansion unlike in the expansion (14.20). Hence, only the monopole term contributes to the proton neutron interaction in the closed j-orbit.

A check similar to (14.18), which was applied to (14.16), may be applied also here. Putting *all* two nucleon matrix elements in (14.23) equal to 1 we obtain

$$\sum_{J=0}^{2j}(2J+1)=(2j+1)\frac{1+2(2j)+1}{2}=(2j+1)^2 \qquad (14.24)$$

The number in (14.24) is the total number of interactions between $2j+1$ protons and $2j+1$ neutrons. If we add to that twice the number in (14.18) we obtain $(4j+2)(4j+1)/2$ which is the total number of two nucleon interactions $n(n-1)/2$ for $n=2(2j+1)$.

Other states in which wave functions in the m-scheme have definite values of J and T are those in which several j-orbits are completely filled. In order to calculate the interaction energy between closed j-orbits we consider first the state of *identical* nucleons completely filling the j-orbit and the j'-orbit. The normalized wave function is given by

$$\frac{1}{\sqrt{(2j+1+2j'+1)!}}\sum_P(-1)^P P\prod_{m=-j}^{j}\psi_{jm}(i_m)\prod_{m'=-j'}^{j'}\psi_{j'm'}(i'_{m'}) \qquad (14.25)$$

where i_m is the nucleon number going from $i_{-j}=1$ to $i_j=2j+1$ and $i'_{m'}$ similarly defined.

The expectation value of the interaction energy in the state (14.25) is given, according to (14.13), as a sum of three terms. In the first term $j_i=j_k=j$ and the interaction is within the j^{2j+1} configuration as given by (14.15) or (14.16). In the third term $j_i=j_k=j'$ and the interaction is between the $2j'+1$ nucleons in the j'-orbit. The second term is the one in which $j_i=j$ and $j_k=j'$ and is equal to

$$\sum_{mm'}\left[\int\psi^*_{jm}(1)\psi^*_{j'm'}(2)V(1,2)\psi_{jm}(1)\psi_{j'm'}(2)\right.$$
$$\left.-\int\psi^*_{jm}(1)\psi^*_{j'm'}(2)V(1,2)\psi_{jm}(2)\psi_{j'm'}(1)\right] \qquad (14.26)$$

The term (14.26) is the interaction between nucleons in the j-orbit with those in the j'-orbit. The sum (14.26) is the trace of the matrix of $V(1,2)$ constructed from $T=1$ antisymmetric states in the jj' configuration. The same trace is given in the $jj'T=1, JM$ scheme by

$$\sum_{|j-j'|}^{j+j'} (2J+1)\langle jj'T=1, JM|V|jj'T=1, JM\rangle \qquad (14.27)$$

Putting all matrix-elements in (14.27) equal to 1 we obtain the total number of interactions between j and j' nucleons to be

$$\begin{aligned}\sum_{|j-j'|}^{j+j'} (2J+1) &= \left[2\mathrm{Min}(j,j')+1\right] \\ &\quad\times \frac{2\left[\mathrm{Max}(j,j')-\mathrm{Min}(j,j')\right]+1+2j+2j'+1}{2} \\ &= (2j+1)(2j'+1)\end{aligned}$$

which is a check on the result (14.27). If we use the tensor expansion (12.12) for $V(1,2)$ we obtain in view of (10.20) for (14.27) the expression

$$\begin{aligned}&\sum_k F_k(jj'T=1)\sum_J(-1)^{j+j'+J}(2J+1)\begin{Bmatrix} j & j' & J \\ j' & j & k \end{Bmatrix} \\ &\quad= \sqrt{(2j+1)(2j'+1)}F_0(jj'T=1) \\ &\quad= \sqrt{(2j+1)(2j'+1)}(F^0-F'^0) \\ &\quad= \sqrt{(2j+1)(2j'+1)}F^0-\sum_k G^k \end{aligned} \qquad (14.28)$$

The last equalities in (14.28) are due to (12.7) and (12.8). On the r.h.s. of (14.28) the contributions of the direct term and exchange terms are explicitly displayed.

Finally, let us compute the interaction energy between the protons in the j-orbit which is completely filled by protons and neutrons and the neutrons in the j'-orbit. As explained in Section 10, the interaction between a j-proton and j'-neutron in that case is given by (10.7). To simplify matters we do not use the isospin formalism and obtain an expression equal to the direct term in (14.26). Hence, the expectation value of the interaction is given by (14.27) in which $T = 1$ matrix elements are replaced by the average of $T = 1$ and $T = 0$ matrix elements. We thus obtain

$$\sum_{|j-j'|}^{j+j'} (2J+1)\tfrac{1}{2}\big[\langle jj'T=1,JM|V|jj'T=1,JM\rangle + \langle jj'T=0,JM|V|jj'T=0,JM\rangle\big] \tag{14.29}$$

Also in this case, putting all matrix elements in (14.29) equal to 1 we obtain $(2j+1)(2j'+1)$ which is the total number of interactions.

A more interesting case in which the state (14.1) is an eigenvalue of $\mathbf{J}^2$ is that of a single nucleon outside closed shells. Let us consider first the case in which the j'-orbit is completely filled with identical nucleons and there is one j-nucleon identical to them. The wave function

$$\frac{1}{\sqrt{(2j'+2)!}}\sum_P (-1)^P P\psi_{jm}(0)\prod_{m'}\psi_{j'm'}(i_{m'}) \tag{14.30}$$

where nucleon coordinates range form 0 to $2j'+1$, has the quantum numbers $J = j$, $M = m$. This follows from the fact that each term in the summation in (14.30) is a product of ψ_{jm} by a wave function with $J = 0$, $M = 0$. This can be checked by operating on (14.30) with $J^+ = j_0^+ + j_1^+ + \cdots + j_{2j'+1}^+$ (or with J^-). The only single nucleon state that can be changed without making (14.30) vanish is ψ_{jm}.

Matrix elements of single nucleon operators between states (14.30), in which the j-nucleon occupies the m and m' states, are given according to (14.6) and (14.7) by

$$\langle jm|\mathbf{f}|jm'\rangle + \langle j'^{2j'+1}J=0,M=0|\mathbf{F}|j'^{2j'+1}J=0,m=0\rangle\delta_{mm'} \tag{14.31}$$

The result (14.31) is a sum of matrix elements of the single j-nucleon and of the closed j'-orbit. Unless $\mathbf{F}$ is a scalar operator, the second term vanishes according to the Wigner-Eckart theorem and (14.31) is entirely due to the single nucleon. This justifies ignoring the closed shells in calculating magnetic moments or quadrupole moments of nuclei. In such cases, the magnetic moment of the nucleus is equal to that of a single nucleon.

The expectation value of a two-body interaction in the state (14.30) is given by (14.13). It has two terms, one in which $j_i = j$, $m_i = m$, $j_k = j'$, $m_k = m'$, and the other in which $j_i = j_k = j'$. The first term is the interaction of the j-nucleon with the $2j' + 1$ nucleons in the j'-orbit whereas the second term is the mutual interactions of the latter. In view of (14.16) we thus obtain for the expectation value the expression

$$\boxed{\begin{aligned}&\sum_{m'} \langle jmj'm'T = 1|V|jmj'm'T = 1\rangle \\ &\qquad + \sum_{J\,\text{even}} (2J + 1)\langle j'^2 T = 1, JM|V|j'^2 T = 1, JM\rangle\end{aligned}} \tag{14.32}$$

The interaction is rotationally invariant – a scalar operator – and due to the Wigner-Eckart theorem its expectation value is independent of $M = m$. Therefore we can evaluate (14.32) by summing the first term in it over m and dividing the outcome by $2j + 1$. The second term in (14.32) is independent of m and does not change. The first term becomes, in view of (14.27), equal to

$$\begin{aligned}&\frac{1}{2j+1}\sum_{mm'} \langle jmjm'T = 1|V|jmjm'T = 1\rangle \\ &\quad = \frac{1}{2j+1}\sum_{|j-j'|}^{j+j'} (2J + 1)\langle jj'T = 1, JM|V|jj'T = 1, JM\rangle \\ &\quad = (2j' + 1)\bar{V}(jj'T = 1)\end{aligned} \tag{14.33}$$

where $\bar{V}(jj'T = 1)$ is the *average* interaction energy between nucleons in the j, j' orbits in $T = 1$ states. Due to the independence of m of the interaction of the j-nucleon and the closed j'-orbit, (14.33) is equal to the interaction between the closed j-orbit and the closed j'-orbit divided by $2j + 1$.

The interactions of a single j-nucleon with the $2(2j'+1)$ protons and neutrons in the fully closed j'-orbit can be similarly obtained. It is also independent of m and is given by $1/(2j+1)$ of the interaction between $2j+1$ identical nucleons in the j-orbit and the $2j'+1$ protons and $2j'+1$ neutrons in the j'-orbit. It is thus obtained by combining (14.27) and (14.29) and dividing by $2j+1$. We thus obtain for that interaction the value

$$\boxed{\begin{aligned}\frac{1}{2j+1}\sum_{|j-j'|}^{j+j'}(2J+1)[\tfrac{3}{2}\langle jj'T=1,JM|V|jj'T=1,JM\rangle \\ +\tfrac{1}{2}\langle jj'T=0,JM|V|jj'T=0,JM\rangle]\end{aligned}} \tag{14.34}$$

To check the expression (14.34) we put all matrix elements in it equal to 1 obtaining $2(2j+1)(2j'+1)/(2j+1)=2(2j'+1)$ which is the number of interactions between the j-nucleon and all nucleons in the j'-orbit. The interaction energy (14.34) is equal to the sum of two terms. The first is (14.33) which is the interaction of a j-nucleon and the j'-nucleons identical to it. The second term is equal to (14.29) divided by $2j+1$. This second term is the interaction of a single j-nucleon with the j'-nucleons of the other kind. That interaction energy in the case of a j-proton and j'-neutrons is thus equal to

$$\boxed{\frac{1}{2j+1}\sum_{|j-j'|}^{j+j'}(2J+1)\langle j_\pi j'_\nu JM|V|j_\pi j'_\nu JM\rangle=(2j'+1)\bar{V}(j_\pi j'_\nu)} \tag{14.35}$$

where the matrix elements are given by (10.7). In (14.35) $\bar{V}(j_\pi j'_\nu)$ is the average interaction energy between a proton in the j-orbit and a neutron in the j'-orbit.

There is an important consequence of the independence of the projection m on the interaction of a single j-nucleon with closed shells. If there are several nucleons in given orbits their interaction with the closed shells is independent of the state into which they combine. For example, a j-nucleon and j'-nucleon outside closed shells may form states with given J with $|j - j'| \leq J \leq j + j'$. Their mutual interaction may well depend on J but their interaction with the closed shells does not. The latter interaction is just the sum of (scalar) single nucleon energies. The closed shells thus supply the central potential well in which the *valence* nucleons move. If the effective or residual interaction is known sufficiently well, the Schrödinger equation of a single nucleon in that potential may be solved to yield wave functions and energies of single nucleons in any unoccupied j-orbit. Using effective interactions determined from experiment, single nucleon energies may be obtained by the difference between binding energies of the nucleus with one valence nucleon and the nucleus with closed shells only. Such single nucleon energies, ϵ_j, include the interaction of the valence nucleon with the closed shells as well as the expectation value of the kinetic energy in the valence orbit. If there are several valence nucleons the contribution of single nucleon energies to the total binding energy is just $\sum \epsilon_j$.

Another case which is as simple as that of one valence nucleon, is that of a *hole* state in which one nucleon is *missing* from closed shells. Due to the antisymmetry of fermion wave functions, dictated by the Pauli principle, such a state behaves very much like a single nucleon state. Consider the wave function

$$\frac{1}{\sqrt{2j!}} \sum_P (-1)^P P \psi_{j,-j}(1) \ldots \psi_{j,m-1}(j+m) \psi_{j,m+1}(j+m+1) \ldots \psi_{jj}(2j) \tag{14.36}$$

The state (14.36) has $M = -m$ and a definite value of $J = j$. This follows from the fact that by multiplying (14.36) by ψ_{jm} and antisymmetrizing the only non-vanishing state of the j^{2j+1} configuration has total spin zero. Operating on (14.36) by $J^+ = \sum_{i=1}^{2j} j_i^+$, the only single nucleon state that can change without making the resulting state vanish, is $\psi_{j,m-1}$ which becomes proportional to ψ_{jm}. The resulting hole state has thus $M = -m + 1$. Continuing this process we reach the state where the missing state from (14.36) is $\psi_{j,-j}$ and then $M = j$. The next step annihilates the state which confirms the assignment $J = j$ for (14.36).

Matrix elements of single nucleon operators in the state (14.36) can be expressed in terms of matrix elements between states of the missing nucleon. Let us consider a single nucleon operator which is a component of an irreducible tensor of rank k. In order to make use of (14.7) we consider expectation values of the operator $\sum_i f_\kappa^{(k)}(i)$ in the states (14.36) with $\mu = m$. The only non-vanishing such expectation values are for the $\kappa = 0$ components. The expectation value in the state (14.36), in which the single nucleon state with $-m$ is unoccupied, is equal, according to (14.7) to

$$\langle j^{2j+1} J = 0, M = 0 | \sum_{i=1}^{2j+1} f_0^{(k)}(i) | j^{2j+1} J = 0, M = 0 \rangle$$

$$- \langle j, -m | f_0^{(k)} | j, -m \rangle$$

Using the Wigner-Eckart theorem and the symmetries of $3j$-symbols we can express the second term as

$$- \langle j, -m | f_0^{(k)} | j, -m \rangle = -(-1)^{j+m} \begin{pmatrix} j & k & j \\ m & 0 & -m \end{pmatrix} (j \| \mathbf{f}^{(k)} \| j)$$

$$= -(-1)^k (-1)^{j-m} \begin{pmatrix} j & k & j \\ -m & 0 & m \end{pmatrix} (j \| \mathbf{f}^{(k)} \| j)$$

$$= (-1)^{k+1} \langle jm | f_0^{(k)} | jm \rangle$$

For $k > 0$, the expectation value in the state of the j^{2j+1} configuration, with $J = 0$, $M = 0$, vanishes and we obtain the relation

$$\langle j^{2j} J = jM = m | \sum_i f_0^{(k)}(i) | j^{2j} J = jM = m \rangle$$

$$= (-1)^{k+1} \langle jm | f_0^{(k)} | jm \rangle \qquad k > 0 \tag{14.37}$$

The relation (14.37) may be applied to magnetic moments and quadrupole moments. They are defined by the expectation values of the corresponding operators, with $k = 1$ and $k = 2$, respectively, in the state with $M = J$. We see that the magnetic moment of a single

j-hole state is equal to that of a single j-nucleon. The quadrupole moment of a single j-hole is equal in magnitude but has the opposite sign to that of a single j-nucleon. For $k = 0$, the relation is simply given by

$$\langle j^{2j} J = jM = m | \sum_i f_0^{(0)}(i) | j^{2j} J = jM = m \rangle = 2j \langle jm | f_0^{(0)} | jm \rangle$$

If there are several holes in the j-orbit, there are similar relations between expectation values of $\kappa = 0$ components of irreducible tensor operators. Using (14.7), we define in this case

$$\langle j^{2j+1-n} m_{n+1} m_{n+2} \ldots m_{2j+1} | \sum_i f_0^{(k)}(i) | j^{2j+1-n} m_{n+1} m_{n+2} \ldots m_{2j+1} \rangle$$

$$= \langle j^{2j+1} J = 0, M = 0 | \sum_i f_0^{(k)}(i) | j^{2j+1} J = 0, M = 0 \rangle$$

$$- \sum_{i=1}^{n} \langle j, -m_i | f_0^{(k)}(i) | j, -m_i \rangle$$

where the summation in the second term on the r.h.s. is over the n missing nucleons. Using the Wigner-Eckart theorem and the symmetries of $3j$-symbols, as in deriving (14.37), we obtain, for $k > 0$ where the expectation value in the $J = 0$, $M = 0$ vanishes, the result

$$\langle j^{2j+1} m_{n+1} \ldots m_{2j+1} | \sum_i f_0^{(k)}(i) | j^{2j+1} m_{n+1} \ldots m_{2j+1} \rangle$$

$$= (-1)^{k+1} \langle j^n m_1 \ldots m_n | \sum_i f_0^{(k)}(i) | j^n m_1 \ldots m_n \rangle \qquad k > 0 \qquad (14.38)$$

To obtain states with definite values of J and given M, we should take linear combinations of states in the m scheme (with $\sum_i m_i = M$). Each m-scheme state of the $2j + 1 - n$ nucleons is equally well defined by the corresponding m-scheme state of the n holes, with the same value of M. The linear combination of the nucleon states yielding a state with definite value of J, can be applied to the corresponding m-scheme hole states and yields a state with the same value of J.

Applying such linear transformations to the states in (14.38) we obtain

$$\langle j^{2j+1-n}JM|\sum_i f_0^{(k)}(i)|j^{2j+1-n}JM\rangle$$
$$= (-1)^{k+1}\langle j^n JM|\sum_i f_0^{(k)}(i)|j^n JM\rangle \qquad k>0$$

which is a generalization of (14.37). Due to the Wigner-Eckart theorem, we obtain the following relation between reduced matrix elements of the nucleon configuration and the hole configuration

$$\left(j^{2j+1-n}J\|\sum_i f_0^{(k)}(i)\|j^{2j+1-n}J\right)$$
$$= (-1)^{k+1}\left(j^n J\|\sum_i f_0^{(k)}(i)\|j^n J\right) \qquad k>0$$

(14.39)

The expectation values of a rotationally invariant two-body interaction in the state (14.36) is independent of $M = m$. Hence, it can be obtained by adding the interaction energies for all values of m and divide the result by $2j+1$. If we use (14.13) we realize that in the sum over m, interaction matrix elements between all m, m' pairs appear an equal number of times. Simple counting shows that this sum is equal to (14.15) multiplied by $2j-1$. From this follows that the interaction energy in the state (14.36) is equal to

$$\frac{2j-1}{2j+1}\sum_{J\,\text{even}}(2J+1)\langle j^2JM|V|j^2JM\rangle =$$
$$= \left(1-\frac{2}{2j+1}\right)\sum_{J\,\text{even}}(2J+1)\langle j^2JM|V|j^2JM\rangle$$

(14.40)

Hence, the single hole energy, measured from the interaction energy of the closed shell, is equal to

$$-\frac{2}{2j+1}\sum_{J \text{ even}}(2J+1)\langle j^2 JM|V|j^2 JM\rangle$$

Putting all matrix elements in (14.38) equal to 1 we obtain for it the value

$$(2j-1)j\frac{2j+1}{2j+1} = 2j\frac{2j-1}{2}$$

which is equal to the number of interactions of n nucleons for $n = 2j$.

The interaction energy in a configuration with several holes in the j-orbit is given by the summation in (14.15) from which the terms with either m or m', equal to any of the m_i, have been removed. We can express these summations by first replacing a summation $\sum_{m<m'}$ by $\frac{1}{2}\sum_{m,m'}$ (the terms with $m = m'$ have vanishing contributions) and then use the equality

$$\sum_{m \neq m_i, m' \neq m_i} = \sum_{m,m'} - \sum_{m=m_1,m'} - \sum_{m,m'=m_1}$$
$$- \sum_{m=m_2,m'} - \sum_{m,m'=m_2} - \cdots - \sum_{m=m_n,m'} - \sum_{m,m'=m_n} + \sum_{m_i,m'_i}$$

where the states with m_i, m'_i are *unoccupied*. This means that the interaction energy is equal to the interaction energy of the closed j-orbit minus n times the single hole energy (14.40) to which the interaction energy between the missing nucleons is *added*. Hence, the interaction energy of several holes is equal to the interaction energy of *nucleons* in the same m-states.

Before leaving the m-scheme and developing methods of constructing antisymmetric states with definite values of J (and T) we still look at another simple case. Consider a proton hole in the fully occupied j-orbit and a neutron in the j'-orbit. The interaction of the j' neutron with the $2j$ protons in the j-orbit is equal to its interaction with the closed j-orbit minus the interaction it would have with the missing

j-nucleon. In the m-scheme we thus obtain

$$\boxed{\begin{aligned}&\langle j^{2j}(j)mj'm'|\sum_{i=1}^{2j} V(0,i)|j^{2j}(j)mj'm'\rangle\\ &\quad=\frac{1}{2j'+1}\sum_{|j-j'|}^{j+j'}(2J+1)\langle jj'JM|V|jj'JM\rangle\\ &\quad-\langle j,-mj'm'|V|j,-mj'm'\rangle\end{aligned}} \tag{14.41}$$

In writing (14.41) we suppressed the indices π and ν and made use of (14.35). In the state $|j^{2j}(j)m\rangle$ the nucleon is missing from the state with $-m$. Using Clebsch-Gordan coefficients in the transformation (7.10) and the definition (7.23) we obtain for the second term in (14.41) the expression

$$\begin{aligned}&\langle j,-mj'm'|V|j,-mj'm'\rangle\\ &\quad=\sum_{J'M'}(2J'+1)\begin{pmatrix} j & j' & J'\\ -m & m' & -M'\end{pmatrix}^2\langle jj'J'M'|V|jj'J'M'\rangle\end{aligned} \tag{14.42}$$

In order to apply the standard formalism to (14.41) we must transform (14.42) into a form in which the projection of j will be m rather than $-m$. We recall the identity (10.24) and obtain

$$\begin{aligned}&\begin{pmatrix} j & j' & J'\\ -m & m' & -M'\end{pmatrix}^2\\ &\quad=(-1)^{j+j'+J'}\begin{pmatrix} j & j' & J'\\ m & -m' & M'\end{pmatrix}\begin{pmatrix} j & j' & J'\\ -m & m' & -M'\end{pmatrix}\\ &\quad=(-1)^{j+j'+J'}\sum_{JM}(-1)^{J+J'}(2J+1)\begin{Bmatrix} j & j' & J'\\ j & j' & J\end{Bmatrix}\\ &\qquad\times\begin{pmatrix} j & j' & J\\ -m & -m' & M\end{pmatrix}\begin{pmatrix} j & j' & J\\ m & m' & -M\end{pmatrix}\\ &\quad=\sum_{JM}(2J+1)\begin{Bmatrix} j & j' & J\\ j & j' & J'\end{Bmatrix}\begin{pmatrix} j & j' & J\\ m & m' & -M\end{pmatrix}^2\end{aligned} \tag{14.43}$$

Substituting (14.43) into (14.42) we obtain

$$\langle j,-mj'm'|V|j,-mj'm'\rangle = \sum_{JM}(2J+1)\begin{pmatrix} j & j' & J \\ m & m' & -M \end{pmatrix}^2$$

$$\times\left(\sum_{J'}(2J'+1)\begin{Bmatrix} j & j' & J \\ j & j' & J' \end{Bmatrix}\langle jj'J'M'|V|jj'J'M'\rangle\right) \tag{14.44}$$

Comparing (14.44) with (14.42), we see that the former can be viewed as matrix elements $\langle jmj'm'|V'|jmj'm'\rangle$ of an interaction V' whose matrix elements in the JM-scheme are given by

$$\langle jj'JM|V'|jj'JM\rangle = \sum_{J'}(2J'+1)\begin{Bmatrix} j & j' & J \\ j & j' & J' \end{Bmatrix}\langle jj'J'M'|V|jj'J'M'\rangle \tag{14.45}$$

Using this modified interaction in the r.h.s. of (14.41) we can construct states of the $j_\pi^{2j}j_\nu'$ configuration with definite J by the usual procedure of using vector addition coefficients. The interaction energies in these hole nucleon states are thus given by

$$\langle j^{2j}(j)j'JM|\sum_{i=1}^{2j}V(0,i)|j^{2j}(j)j'JM\rangle$$

$$= \frac{1}{2j'+1}\sum_{|j-j'|}^{j+j'}(2J+1)\langle jj'JM|V|jj'JM\rangle$$

$$-\sum_{J'}(2J'+1)\begin{Bmatrix} j & j' & J \\ j & j' & J' \end{Bmatrix}\langle jj'J'M'|V|jj'J'M'\rangle \tag{14.46}$$

This relation is due to Pandya (1956) who did not publish his complicated derivation. Simple different proofs were given by Talmi (1960), de Shalit and Talmi (1963) and Bertsch (1972).

The formula (14.46) for nucleon-hole interaction matrix elements is applicable also to cases in which the nucleon is a j-proton and the hole is due to a neutron missing from the *same* j-orbit. The simplest way to see it is by not using the isospin formalism even though the interaction is charge independent. The proton neutron states in that case with even values of J, correspond to $T = 1$ states whereas the ones with odd J correspond to $T = 0$ states. Still there is no need to specify the isospins of the nucleon-hole states since also in that case there is only one state with any given value of J. In Section 18 nucleon-hole interactions will be further discussed and formulae analogous to (14.46) will be derived.

The formula (14.46) demonstrates the fact that nucleon-hole levels do not bear a simple relation to the corresponding nucleon-nucleon levels. There are however special cases in which there is a direct simple relation between these two sets of levels. To find these cases, let us use the tensor expansion (12.12) of the matrix elements in the second term in (14.46). We thus obtain, by using the sum rule (10.14) the result

$$-\sum_k F_k \sum_{J'} (-1)^{j+j'+J'}(2J'+1)\begin{Bmatrix} j & j' & J' \\ j & j' & J \end{Bmatrix}\begin{Bmatrix} j & j' & J' \\ j' & j & k \end{Bmatrix}$$

$$= -\sum_k F_k (-1)^{j+j'+J+k}\begin{Bmatrix} j & j' & J \\ j' & j & k \end{Bmatrix} \tag{14.47}$$

Comparing (14.47) with (12.12) we see that the r.h.s. of the former is exactly equal to the matrix element (12.12) if $(-1)^k = -1$. It is equal to it apart from a minus sign if $(-1)^k = 1$. Since the first term in (14.46) is a constant, independent of J, we arrive at the following conclusion. If the expansion (12.12) contains only odd tensors ($(-1)^k = -1$) levels of nucleon-hole configurations have the *same order* and *spacings* as in the nucleon-nucleon configuration. If the expansion contains only even tensors ($(-1)^k = 1$) level spacings in the two configurations are equal but the *order* is *reversed*.

15

Nucleons in the Same Orbit. Coefficients of Fractional Parentage

In the preceding section it was explained that, in general, wave functions in the m-scheme are not eigenstates of $\mathbf{J}^2$. In the m-scheme, wave functions may be easily antisymmetrized and then the problem is to construct linear combinations of them to obtain states with definite J. Instead, by using vector addition coefficients we can easily construct wave functions with definite values of J and M and then proceed to antisymmetrize them. In the present section we use this approach and learn how to construct antisymmetric wave functions and how to make use of them.

Let us consider n identical nucleons in orbits characterized by $j_1, j_2, \ldots, j_n$. We can first couple $\psi_{j_1}(1)$ and $\psi_{j_2}(2)$ with Clebsch-Gordan coefficients to form a state with definite value of spin J_{12}, then couple J_{12} to j_3 to form J_{123} and so on. We obtain a state characterized by $J_{12}, J_{123}, \ldots$ and, of course, the total J. Any two such states are orthogonal if the value of J or the value of any intermediate spin in one state is different from the corresponding one in the other. This may no longer be the case if antisymmetrization is applied to such two states as will be evident later. If, however, the n orbits involved are all different, such two wave functions will be still orthogonal even after antisymmetrization. To see it we observe that to check orthogonality of two wave functions it is sufficient to antisymmetrize only one of them. The integration over a product of an antisymmetric function and any other function yields a non-vanishing contribution only from the component of the other function which is fully antisymmetric. Any other component contributes an integral that changes sign under exchange of at least one pair of nucleon coordinates. This amounts to only renaming variables of integration and the integral should remain unchanged. If its sign is changed it must vanish. Hence, we may calculate the integral over the product of the functions

$$\psi[j_1(1)j_2(2)J_{12}j_3(3)J_{123}\ldots JM] \tag{15.1}$$

with another such function which has been antisymmetrized. If all the n orbits are different, the only one among the $n!$ terms of the antisymmetrized function which may be non-orthogonal to (15.1) has nucleon 1 in orbit j_1, nucleon 2 in orbit j_2 and so on. In all other terms, one nucleon is in an orbit different from the orbit it occupies in (15.1). If any of the intermediate spins $J_{12\ldots k}$ or $J = J_{12\ldots n}$ are different in the two states considered, integration over nucleon coordinates 1 to k or 1 to n, yields zero due to the orthogonality of two states with different values of J.

A direct consequence of this discussion is that for different orbits $j_1, j_2, \ldots, j_n$, the intermediate spins $J_{12\ldots}$ (and J) specify uniquely the antisymmetrized states and may serve as quantum numbers (or labels) for them. This, however, is no longer the case for nucleons in the same j-orbit as will become evident in the following. Most of this section is occupied with methods to construct antisymmetric states with definite J of n nucleons in the same orbit. If there are n_1 nucleons in the j_1-orbit, n_2 nucleons in the j_2-orbit etc., we may need quantum numbers α_1 in addition to J_1 to specify antisymmetric states of the $j_1^{n_1}$ configuration, $\alpha_2 J_2$ for the $j_2^{n_2}$ configuration and so on. Having succeeded in doing that, we may obtain states with given J by coupling states of these various configurations as indicated by

$$\psi(j_1^{n_1}(\alpha_1 J_1)j_2^{n_2}(\alpha_2 J_2)J_{12}j_3^{n_3}(\alpha_3 J_3)J_{123}\ldots JM) \tag{15.2}$$

The discussion above shows that states (15.2) which differ in either one of the $\alpha_i J_i$ or the $J_{12\ldots k}$ (as well as J and M) are orthogonal to each other. This orthogonality holds even if these states are antisymmetrized. Hence, once the problem of constructing antisymmetric states of n nucleons in the same j-orbit is solved, the intermediate spins $J_{12\ldots k}$ in (15.2) furnish a set of quantum numbers. We turn now to the problem of nucleons in the same orbit.

Let us consider first the case of three identical nucleons in a j-orbit. In Section 9 we saw how antisymmetric states of *two* nucleons with definite values of J may be constructed. The normalized antisymmetric states of two j-nucleons are constructed as

$$\psi(j_{1,2}^2 JM) = \sum (jmjm' \mid jjJM)\psi_{jm}(1)\psi_{jm'}(2) \qquad (15.3)$$

with even values of J. Since we consider here only the j-orbit we put the nucleon number as a subscript. We proceed to construct a state with three nucleons like (15.1) as the unnormalized state

$$\psi(j_{1,2}^2(J_0)j_3JM) = \sum_{M_0 m} (J_0M_0jm \mid J_0jJM)\psi(j_{1,2}^2J_0M_0)\psi_{jm}(3) \qquad (15.4)$$

The state (15.4) for even values of J_0 is antisymmetric in nucleons 1 and 2 by the construction (15.3). To make it fully antisymmetric we need only to permute nucleons 1,3 and 2,3. The antisymmetrization operation is thus defined by

$$\begin{aligned} \mathcal{A}\psi(j_{1,2}^2(J_0)j_3JM) = \psi(j_{1,2}^2(J_0)j_3JM) - \psi(j_{1,3}^2(J_0)j_2JM) \\ - \psi(j_{3,2}^2(J_0)j_1JM) \end{aligned} \qquad (15.5)$$

The three wave functions on the r.h.s. of (15.5) are not orthogonal as we shall immediately see. To express (15.5) as a sum of orthogonal terms we restore the order of couplings by using the transformation (10.11) thus obtaining for (15.5) the expression

$$\begin{aligned} &\psi(j_{1,2}(J_0)j_3JM) - \sum_{J_1}(-1)^{2j+J_0+J_1}\sqrt{(2J_0+1)(2J_1+1)} \\ &\times \begin{Bmatrix} j & j & J_0 \\ J & j & J_1 \end{Bmatrix} \psi(j_{1,2}^2(J_1)j_3JM) \\ &+ \sum_{J_1}(-1)^{J_1+1}(-1)^{2j+J_0+J_1}\sqrt{(2J_0+1)(2J_1+1)} \\ &\times \begin{Bmatrix} j & j & J_0 \\ J & j & J_1 \end{Bmatrix} \psi(j_{1,2}^2(J_1)j_3JM) \end{aligned} \qquad (15.6)$$

In the last term of (15.6) the order of nucleons coupled to J_0 was changed yielding a minus sign and then the order of nucleons 1 and 2 coupled to J_1 was changed from which the phase $(-1)^{J_1+1}$ arose. The two sums in (15.6) cancel each other for odd values of J_1 and they add for even values of J_1 yielding the result

$$\mathcal{A}\psi(j_{1,2}^2(J_0)j_3JM) = \sum_{J_1 \text{ even}} \left[\delta_{J_1J_0} + 2\sqrt{(2J_0+1)(2J_1+1)}\begin{Bmatrix} j & j & J_0 \\ J & j & J_1 \end{Bmatrix}\right] \times \psi(j_{1,2}^2(J_1)j_3JM) \tag{15.7}$$

In the linear combination in (15.7) we see that the last two terms in (15.6) have non-vanishing amplitudes which are equal to the first term with $J_1 = J_0$. On the other hand, wave functions with different values of J_1 in (15.7) are orthogonal. Hence, if the given state is not allowed by the Pauli principle, all coefficients in (15.7) vanish.

The normalization of the state (15.7) may be directly calculated. The sum of squares of the coefficients is given by

$$1 + 4(2J_0+1)\begin{Bmatrix} j & j & J_0 \\ J & j & J_0 \end{Bmatrix} + 4(2J_0+1)\sum_{J_1 \text{ even}}(2J_1+1)\begin{Bmatrix} j & j & J_0 \\ J & j & J_1 \end{Bmatrix}^2$$

The summation over J_1, even, can be expressed as

$$4(2J_0+1)\frac{1}{2}\left[\sum_{J_1}(2J_1+1)\begin{Bmatrix} j & j & J_0 \\ J & j & J_1 \end{Bmatrix}^2 + \sum_{J_1}(-1)^{J_1}(2J_1+1)\begin{Bmatrix} j & j & J_0 \\ J & j & J_1 \end{Bmatrix}^2\right]$$
$$= 2(2J_0+1)\left[\frac{1}{2J_0+1} + \begin{Bmatrix} j & j & J_0 \\ J & j & J_0 \end{Bmatrix}\right]$$

The last equality follows from the identities (10.13) and (10.14) of $6j$-symbols. Collecting all terms we obtain the following result for the normalization factor of (15.7)

$$\mathcal{N}^{-2} = 3 + 6(2J_0+1)\begin{Bmatrix} j & j & J_0 \\ J & j & J_0 \end{Bmatrix} \tag{15.8}$$

The fully antisymmetric and normalized wave function with spin J obtained from (15.4) is thus given by

$$\boxed{\psi(j^3[J_0]JM) = \sum [j^2(J_1)jJ|\}j^3[J_0]J]\psi(j^2_{1,2}(J_1)j_3JM)} \tag{15.9}$$

The coefficients of the orthogonal terms in (15.9) are given by

$$\boxed{\begin{aligned}[j^2(J_1)jJ|\}j^3[J_0]J] = &\left[\delta_{J_1J_0} + 2\sqrt{(2J_0+1)(2J_1+1)}\begin{Bmatrix} j & j & J_0 \\ J & j & J_1 \end{Bmatrix}\right] \\ &\times\left[3 + 6(2J_0+1)\begin{Bmatrix} j & j & J_0 \\ J & j & J_0 \end{Bmatrix}\right]^{-1/2}\end{aligned}} \tag{15.10}$$

The coefficients (15.10) are called *coefficients of fractional parentage* (c.f.p.). The spin J_0 from which we started, uniquely specifies the state with given J and is kept as the label of the state. It is called the *principal parent*. In the present case, however, it is no longer true that states (15.10) with different principal parents are orthogonal.

The wave function (15.10) has very interesting features. It is fully antisymmetric and yet nucleon number 3 occupies a special position. This turns out to be very convenient in applications. The price to be paid for this is that the expansion includes several "fractional parents" with various values of J_1. Starting with other values of J_0 may lead to other states like (15.10) which are independent. There may be several independent states of the j^3 configurations with the same value of J. In such cases, other quantum numbers α should be found which characterize an orthogonal and normalized basis for these states. We shall return to this point in subsequent sections.

Coefficients of fractional parentage were first introduced by Bacher and Goudsmit (1934) for atomic spectroscopy. They were extensively used by Racah (1943) (for atomic spectroscopy) who developed methods for calculating their values. The explicit formula (15.10) was derived by Redmond (1954) and by Schwartz and de-Shalit (1954). Efficient ways to calculate c.f.p. were developed by Bayman and Lande (1966).

A special case of c.f.p. is particularly simple. If we take the principal parent to be the $J_0 = 0$ state of the j^2 configuration, then (15.10) defines a state with $J = j$ in the j^3 configuration. Inserting in (15.10) the values $J_0 = 0$, $J = j$, we use the special value (10.19) of the $6j$-symbol to obtain

$$[j^2(0)jJ = j|\}j^3[J_0 = 0]J = j] = \sqrt{\frac{2j-1}{3(2j+1)}}$$

$$[j^2(J_1)jJ = j|\}j^3[J_0 = 0]J = j] = -\frac{2\sqrt{2J_1+1}}{\sqrt{3(2j+1)(2j-1)}} \qquad J_1 > 0, \text{ even}$$

(15.11)

The state whose c.f.p. are given by (15.11) and its generalization to $n > 3$ play a special role in nuclear spectroscopy as will be explained in a subsequent section.

A simple consequence of (15.7) is that there is no antisymmetric state with $J = \frac{1}{2}$ in *any* j^3 configuration. There is only one possible value of J_0 or J_1 which is either $j + \frac{1}{2}$ if it is even or $j - \frac{1}{2}$ if it is even. Starting from the proper value of $J_0 = j + \frac{1}{2}$ or $J_0 = j - \frac{1}{2}$, we find by looking at the algebraic expression of

$$\begin{Bmatrix} j & j & j \pm \frac{1}{2} \\ \frac{1}{2} & j & j \pm \frac{1}{2} \end{Bmatrix}$$

that in either case

$$1 + 2(2(j \pm \tfrac{1}{2}) + 1) \begin{Bmatrix} j & j & j \pm \frac{1}{2} \\ \frac{1}{2} & j & j \pm \frac{1}{2} \end{Bmatrix} = 0$$

Another simple consequence of (15.7) is the vanishing of certain $6j$-symbols. The maximum J value of an antisymmetric state of the

j^3 configuration is given by $J_{\max} = M_{\max} = j + j - 1 + j - 2 = 3j - 3$. There is only *one* antisymmetric state with $M = M_{\max} - 1$ obtained by antisymmetrizing the state with $m_1 = j$, $m_2 = j - 1$ and $m_3 = j - 3$. Hence this is the state with $J = M_{\max}$, $M = M_{\max} - 1$ and there is no antisymmetric state with $J = 3j - 4$ in any j^3 configuration. There are two possible parents (for $j > \frac{3}{2}$) of such a state, had it existed, namely, $J_1 = 2j - 1$ and $J_1 = 2j - 3$. We can choose one of these to be the principal parent and then the c.f.p. of the other should vanish. According to (15.7) we obtain

$$\boxed{\begin{Bmatrix} j & j & 2j-1 \\ 3j-4 & j & 2j-3 \end{Bmatrix} = 0 \qquad \text{for any} \quad j > \tfrac{3}{2}}$$

If we consider states of particles in the l-orbit which are fully antisymmetric in their space coordinates we may use a similar argument. In that case, due to the symmetry properties of Clebsch-Gordan coefficients space antisymmetric states of two particles have odd values of L_0. The same considerations lead to vanishing of the $6j$-symbol written above also for values of j which are integers.

The generalization of coefficients of fractional parentage can be readily defined for $n > 3$. We start from a complete and orthogonal set of antisymmetric states in the j^{n-1} configuration. These are characterized by the angular momentum J_1 and if there are several orthogonal states with the same value of J_1, also by an additional label α_1. The set of states obtained by coupling the spin of the n-th nucleon to J_1 to form a state with total spin J is

$$\psi(j^{n-1}(\alpha_1 J_1) j_n JM) = \sum_{M_1 m} (J_1 M_1 jm \mid J_1 jJM) \times \psi(j^{n-1}\alpha_1 J_1 M_1)\psi_{jm}(n) \qquad (15.12)$$

The set (15.12) is a complete (and orthogonal) set of states of n nucleons which are antisymmetric in the coordinates of the first $n - 1$ nucleons. The fully antisymmetric states in the j^n configuration, for given J, form a *subspace* of the space spanned by the states (15.12). Hence, it is possible to *project* the space of fully antisymmetric states out of the space spanned by the set (15.12). In other words, any state with given J which is fully antisymmetric in all n nucleons can be expressed as a linear combination of the states (15.12)

$$\psi(j^n\alpha JM) = \sum_{\alpha_1 J_1} [j^{n-1}(\alpha_1 J_1)jJ|\}j^n\alpha J]\psi(j^{n-1}(\alpha_1 J_1)j_n JM) \tag{15.13}$$

The coefficients in (15.13) are the $n \to n-1$ coefficients of fractional parentage (c.f.p.). They are coefficients in a linear combination of tensorial sets with rank J and hence, are independent of M.The fact that in the fully antisymmetric wave function (15.13) the n-th nucleon occupies a special position, makes it possible to apply directly the powerful methods of tensor algebra.

The additional labels α on the l.h.s. of (15.13) distinguish between orthogonal states of the j^n configuration which may have the same value of the total J. Due to the orthogonality of the states (15.12) with different values of J_1 or α_1, the orthogonality of two states (15.13) with different labels α leads to the following relation between the c.f.p.

$$\sum_{\alpha_1 J_1} [j^{n-1}(\alpha_1 J_1)jJ|\}j^n\alpha J][j^{n-1}(\alpha_1 J_1)jJ|\}j^n\alpha' J] = \delta_{\alpha\alpha'} \tag{15.14}$$

States with different values of J are orthogonal irrespective of the values of the coefficients in (15.13). Hence, no orthogonality relations for the c.f.p. follow from this orthogonality. In writing the expression (15.14) we made use of the fact that the c.f.p. in (15.13) are real numbers. This fact is demonstrated for $n = 3$ in (15.11) and will be shown below also for higher values of n.

Let us now see how useful is the expansion (15.13). Consider the single nucleon tensor operator of rank k

$$\mathbf{F}^{(k)} = \sum_{i=1}^{n} \mathbf{f}^{(k)}(i) \tag{15.15}$$

When matrix elements of this operator are calculated, the following integral should be evaluated

$$\int \psi^*(j^n\alpha JM)F_\kappa^{(k)}\psi(j^n\alpha' J'M') \tag{15.16}$$

Since both wave functions are fully antisymmetric and the operator $\mathbf{F}^{(k)}$ is symmetric, the integrand in (15.16) is *symmetric* in all nucleon coordinates. Hence, the matrix element (15.16) is equal to the matrix element of any of the $\mathbf{f}^{(k)}(i)$ multiplied by n. Naturally, we choose $i = n$ and obtain, in view of (10.41), the result

$$\begin{aligned}(j^n\alpha J\|\mathbf{F}^{(k)}\|j^n\alpha' J') &= n\sum_{\alpha_1 J_1}[j^{n-1}(\alpha_1 J_1)jJ|\}j^n\alpha J][j^{n-1}(\alpha_1 J_1)jJ|\}j^n\alpha' J'] \\ &\times(-1)^{J_1+j+J+k}\sqrt{(2J+1)(2J'+1)}\begin{Bmatrix} j & J & J_1 \\ J' & j & k \end{Bmatrix}(j\|\mathbf{f}^{(k)}\|j)\end{aligned} \tag{15.17}$$

The result (15.17) assumes a particularly simple form for a scalar operator, $k = 0$. Putting $k = 0$, we obtain, due to (15.14), a simple relation between the reduced matrix elements of $\mathbf{F}^{(0)}$ and $\mathbf{f}^{(0)}$. The relation thus obtained is even simpler for ordinary matrix elements, namely

$$\langle j^n\alpha JM|\sum_{i=1}^{n}\mathbf{f}^{(0)}(i)|j^n\alpha' J'M'\rangle = n\langle jm|\mathbf{f}^{(0)}|jm\rangle\,\delta_{\alpha\alpha'}\,\delta_{JJ'}\,\delta_{MM'} \tag{15.18}$$

The geometrical meaning of this relation is simple. Matrix elements of a scalar operator are independent of m and hence, each of the $\mathbf{f}^{(0)}(i)$ contributes equally to the diagonal matrix element of $\mathbf{F}^{(0)}$.

The expansion (15.13) in terms of c.f.p. can be used to calculate also matrix elements of two-body operators. Matrix elements of the two-body operator

$$\mathbf{G} = \sum_{i<k}^{n}\mathbf{g}(i,k) \tag{15.19}$$

between fully antisymmetric states of the j^n configuration are equal to the matrix elements between those states of any $\mathbf{g}(i,k)$ multiplied by $n(n-1)/2$. This follows from the symmetry of the operator (15.19) and the resulting symmetry of the integrand in

$$\langle j^n \alpha JM|\mathbf{G}|j^n \alpha' J'M'\rangle = \int \psi^*(j^n \alpha JM) \sum \mathbf{g}(i,k) \psi(j^n \alpha' J'M') \tag{15.20}$$

Hence, we can calculate the contribution to (15.20) of the $(n-1)(n-2)/2$ terms in $\sum_{i<k}^{n-1} \mathbf{g}(i,k)$ and multiply the result by $n/(n-2)$ to obtain the full value of (15.20). For a scalar operator $\mathbf{G}$, we thus obtain

$$\begin{aligned}&\langle j^n \alpha JM|\mathbf{G}|j^n \alpha' JM\rangle \\ &\quad = \frac{n}{n-2} \sum_{\alpha_1 \alpha_1' J_1} [j^{n-1}(\alpha_1 J_1) jJ|\} j^n \alpha J][j^{n-1}(\alpha_1' J_1) jJ|\} j^n \alpha' J] \\ &\qquad \times \langle j^{n-1} \alpha_1 J_1 M_1| \sum_{i<k}^{n-1} \mathbf{g}(i,k) | j^{n-1} \alpha_1' J_1 M_1\rangle\end{aligned} \tag{15.21}$$

To obtain (15.21) it is possible to use (10.42) but it is simpler to derive it directly. Using the expansion (15.13) we obtain an expression like (15.21) in which the matrix element has the form

$$\begin{aligned}&\langle j^{n-1}(\alpha_1 J_1) j_n JM| \sum_{i<k}^{n-1} \mathbf{g}(i,k) | j^{n-1}(\alpha_1' J_1) j_n JM\rangle \\ &\quad = \sum_{\substack{M_1 m \\ M_1' m'}} (J_1 M_1 jm \mid J_1 jJM)(J_1 M_1' jm' \mid J_1 jJM) \\ &\qquad \times \langle j^{n-1} \alpha_1 J_1 M_1 j_n m| \sum_{i<k}^{n-1} \mathbf{g}(i,k) | j^{n-1} \alpha_1' J_1 M_1' j_n m'\rangle\end{aligned}$$

Since $\sum_{i<k}^{n-1} \mathbf{g}(i,k)$ is independent of the coordinates of the n-th nucleon and is a scalar operator, $m' = m$, $M_1' = M_1$ and its matrix elements are independent of M_1. The orthogonality relations of the Clebsch-Gordan coefficients yield in this case (15.21). The matrix elements in the j^{n-1} configuration on the r.h.s. of (15.21) can be expressed in terms of those in the j^{n-2} configuration. This procedure can be continued until the matrix element (15.21) is expressed in terms of the eigenvalues $\langle j^2JM|\mathbf{G}|j^2JM\rangle$ in the j^2 configuration.

It may be worthwhile to point out that the formula (15.21) may be applied to scalar operators which depend on the variables of more than two particles. If, for instance, $\mathbf{G}$ is a $(n-1)$-body operator then (15.21) gives its matrix elements in the j^n configuration in terms of its matrix elements in the j^{n-1} configuration. In that case, $\sum \mathbf{g}(i,k)$ in (15.21) should be replaced by $\mathbf{g}(1,\ldots,n-1)$. In general, to calculate matrix elements of a k-body operator the procedure may be continued until matrix elements in the j^k configuration are reached.

The relation (15.14) can be used as a simple check of c.f.p. Another useful check follows from (15.21) if we apply it to the operators $(\mathbf{j}_i \cdot \mathbf{j}_k)$. The matrix elements of $\mathbf{J}^2$ are given by $J(J+1)\delta_{\alpha\alpha'}$. These matrix elements may be calculated from $\mathbf{J}^2 = 2\sum_{i<k}(\mathbf{j}_i \cdot \mathbf{j}_k) + \sum_i \mathbf{j}_i^2 = 2\sum_{i<k}(\mathbf{j}_i \cdot \mathbf{j}_k) + nj(j+1)$. Applying (15.21) to the *two-body* operator $\mathbf{J}^2 - nj(j+1)$ we obtain (due to (15.14)), the relation

$$\boxed{\begin{aligned}&\left[J(J+1) + \frac{n}{n-2}j(j+1)\right]\delta_{\alpha\alpha'}\\ &\quad= \frac{n}{n-2}\sum_{\alpha_1 J_1}[j^{n-1}(\alpha_1 J_1)jJ|\}j^n\alpha J]\\ &\qquad\times[j^{n-1}(\alpha_1 J_1)jJ|\}j^n\alpha' J]J_1(J_1+1)\end{aligned}} \tag{15.22}$$

As mentioned above, the expansion of the fully antisymmetric states in the j^n configuration in terms of the states (15.12) is a projection operation. From this follow interesting properties of the c.f.p. In general, a projection is not a unitary transformation. The number of independent states in the projected space is usually smaller than in the original space. If the basis in the original space is constructed from a basis in the projected space augmented by the necessary number of orthogonal states, the projection operator is diagonal. It has the eigenvalues 1 corresponding to the basis states in the projected space and all other eigenvalues vanish. Hence, a projection operator, $\mathcal{P}$, in any basis, satisfies the relation

$$\mathcal{P}^2 = \mathcal{P} \tag{15.23}$$

We may start from one state (15.12) and antisymmetrize it. It is sufficient to exchange the coordinates of the n-th nucleon with those of each of the nucleons with $i \leq n-1$. This antisymmetrization is thus given by

$$\begin{aligned} &\mathcal{A}\psi(j^{n-1}(\alpha_0 J_0)j_n JM) \\ &\quad = \psi(j^{n-1}(\alpha_0 J_0)j_n JM) - \sum_{i=1}^{n-1} \psi(j^{n-1}(\alpha_0 J_0)j_i JM) \end{aligned} \tag{15.24}$$

The state $\psi(j^{n-1}\alpha_0 J_0)$ is the principal parent in the expansion of (15.24) in c.f.p. That expansion is obtained by restoring the order of coupling of the i-th and n-th nucleons by a change of coupling transformation. The labels of the resulting antisymmetric state, provided it does not vanish, may be chosen to be $\alpha_0 J_0$ of the principal parent. If the antisymmetrized $\mathcal{A}$ in (15.24) is applied to a fully antisymmetric state it yields the same state multiplied by n. Hence $\mathcal{A}/n$ is a projection operator $\mathcal{P}$ from which we can derive the following relations.

The normalization coefficient in (15.10) is simply related to the c.f.p. of the principal parent. This relation can now be generalized to higher values of n. Any c.f.p. may be obtained from the fully antisymmetric and normalized state by the overlap integral

$$[j^{n-1}(\alpha_1 J_1)jJ|\}j^n \alpha J] = \int \psi^*(j^n \alpha JM)\psi(j^{n-1}(\alpha_1 J_1)j_n JM) \tag{15.25}$$

The normalization coefficient $\mathcal{N}_{\alpha_0 J_0 J}$ makes the state (15.24) normalized. To obtain its relation to the c.f.p. of the principal parent, we put in (15.25) $\alpha_1 = \alpha_0$, $J_1 = J_0$ and use for α the labels $\alpha_0 J_0$. We then obtain

$$
\begin{aligned}
&[j^{n-1}(\alpha_0 J_0)jJ|\}j^n[\alpha_0 J_0]J] \\
&\quad = \int \left\{ \frac{\mathcal{A}}{n} \psi^*(j^n[\alpha_0 J_0]JM) \right\} \psi(j^{n-1}(\alpha_0 J_0)j_n JM) \\
&\quad = \frac{1}{n} \int \psi^*(j^n[\alpha_0 J_0]JM) \mathcal{A} \psi(j^{n-1}(\alpha_0 J_0)j_n JM) = \frac{1}{n} \frac{1}{\mathcal{N}_{\alpha_0 J_0 J}}
\end{aligned}
\tag{15.26}
$$

In deriving (15.26) we used the relation (15.23) for the projection operator $\mathcal{P} = \mathcal{A}/n$. In this way we obtain the relation

$$
\boxed{n\mathcal{N}_{\alpha_0 J_0 J}[j^{n-1}(\alpha_0 J_0)jJ|\}j^n[\alpha_0 J_0]J] = 1}
\tag{15.27}
$$

We obtained in (15.8) the special case of (15.27) for $n = 3$ by direct computation.

The relation (15.27) can be used to derive a recursion relation for the c.f.p. The derivation is given in detail in de-Shalit and Talmi (1963) and will be only outlined here. In the wave function (15.24) the states of the j^{n-1} configuration in the sum are expressed with the help of $n-1 \to n-2$ c.f.p. in terms of the j^{n-2} configuration. In those states, necessary permutations are carried out so that the nucleon coupled last is the *n-th* nucleon. The order of coupling of the n-th nucleon and the i-th nucleon is changed by the transformation (10.11) leading to the following expression for (15.24)

$$
\begin{aligned}
&\mathcal{A}\psi(j^{n-1}(\alpha_0 J_0)j_n JM) \\
&\quad = \psi(j^{n-1}(\alpha_0 J_0)j_n JM) - \sum_{\alpha_2 J_2} [j^{n-2}(\alpha_2 J_2)jJ_0|\}j^{n-1}\alpha_0 J_0] \\
&\qquad \times \sum_{J_1} \langle J_2 j(J_0)jJ \mid J_2 j(J_1)jJ \rangle \\
&\qquad \times \Big[\psi(j^{n-2}(\alpha_2 J_2)j_{n-1}(J_1)j_n JM) \\
&\qquad\qquad - \sum_i \psi(j^{n-2}(\alpha_2 J_2)j_i(J_1)j_n JM) \Big]
\end{aligned}
\tag{15.28}
$$

The expression in the square brackets in (15.28) is simply

$$\mathcal{A}_1\psi(j^{n-2}(\alpha_2 J_2)j_{n-1}(J_1)j_n JM)$$

where $\mathcal{A}_1$ is the antisymmetrizer for the first $n-1$ nucleons. Multiplying both sides of (15.28) by $\psi^*(j^{n-1}(\alpha_1 J_1)j_n JM)$ and integrating, we obtain in view of (15.27) and the actual values of the transformation coefficients (10.11) the following relation between $n \to n-1$ and $n-1 \to n-2$ c.f.p.

$$n[j^{n-1}(\alpha_0 J_0)jJ|\}j^n[\alpha_0 J_0]J][j^{n-1}(\alpha_1 J_1)jJ|\}j^n[\alpha_0 J_0]J]$$
$$= \delta_{\alpha_1\alpha_0}\delta_{J_1 J_0} + (n-1)\sum_{\alpha_2 J_2}(-1)^{J_0+J_1}\sqrt{(2J_0+1)(2J_1+1)}$$
$$\times \begin{Bmatrix} J_2 & j & J_1 \\ J & j & J_0 \end{Bmatrix} [j^{n-2}(\alpha_2 J_2)jJ_0|\}j^{n-1}\alpha_0 J_0]$$
$$\times [j^{n-2}(\alpha_2 J_2)jJ_1|\}j^{n-1}\alpha_1 J_1] \tag{15.29}$$

The relation (15.29) is a recursion relation for the c.f.p. It is due to Redmond (1954). If the c.f.p. in the j^{n-1} configuration are known, those in the j^n configuration may be calculated. By putting in (15.29), $\alpha_1 = \alpha_0$, $J_1 = J_0$, the c.f.p. of the principal parent is determined and once it is known, all other c.f.p. can be obtained. The c.f.p. in the j^{n-1} configuration are similarly determined from those in the j^{n-2} configuration. This way we finally reach the j^2 configuration for which

$$[jjJ|\}j^2 J] = 1 \qquad \text{for } J \text{ even} \tag{15.30}$$

A simple consequence of (15.29) is that all the c.f.p. are real numbers.

It is important to realize that the recursion relation (15.29) yields c.f.p. for states which are characterized by their principal parents. Having obtained all (antisymmetric) independent states for given J, it is possible to choose any orthogonal basis with a set of labels α.

These basis states may then be used to obtain the c.f.p. in the j^{n+1} configuration.

The notion of fractional parentage states and coefficients can be extended. Any fully antisymmetric state in the j^n configuration can be expressed as a linear combination of states in which $n-k$ nucleons are coupled to a fully antisymmetric state characterized by $\alpha_2 J_2$ and the other k nucleons are antisymmetrically coupled to states with $\alpha' J'$. The spins J_2 and J' are coupled to yield the total spin J. A useful expansion of this kind is for $k=2$, namely

$$\begin{aligned}\psi(j^n\alpha JM) = \sum_{\alpha_2 J_2 J'} [j^{n-2}(\alpha_2 J_2)j^2(J')J|\}j^n\alpha J] \\ \times \psi(j^{n-2}(\alpha_2 J_2)j^2_{n-1,n}(J')JM)\end{aligned} \tag{15.31}$$

The $n \to n-2$ c.f.p. defined by (15.31) may be conveniently used to calculate directly matrix elements of two-body operators $\mathbf{G}$. We simply calculate the matrix element of $\mathbf{g}(n-1,n)$ and multiply it by $n(n-1)/2$. For a scalar operator $\mathbf{G}$ we thus obtain

$$\begin{aligned}&\langle j^n\alpha JM|\sum_{i<k}^{n}\mathbf{g}(i,k)|j^n\alpha' JM\rangle \\ &= \frac{n(n-1)}{2}\sum_{\alpha_2 J_2 J'} [j^{n-2}(\alpha_2 J_2)j^2(J')J|\}j^n\alpha J] \\ &\times [j^{n-2}(\alpha_2 J_2)j^2(J')J|\}j^n\alpha' J] \\ &\times \langle j^2 J'M'|\mathbf{g}(n-1,n)|j^2 J'M'\rangle\end{aligned} \tag{15.32}$$

We shall not consider such c.f.p. further. It is only worthwhile to point out that the $n \to n-2$ c.f.p. in (15.31) can be expressed in terms of ordinary $n \rightarrow n-1$ c.f.p. (Schwartz and de-Shalit 1954) by

$$\begin{aligned}
&[j^{n-2}(\alpha_2 J_2)j^2(J')J|\}j^n\alpha J] \\
&\quad = \sum_{\alpha_1 J_1} [j^{n-2}(\alpha_2 J_2)jJ_1|\}j^{n-1}\alpha_1 J_1][j^{n-1}(\alpha_1 J_1)jJ|\}j^n\alpha J] \\
&\quad \times (-1)^{J_2+J+2j}\sqrt{(2J_1+1)(2J'+1)}\begin{Bmatrix} J_2 & j & J_1 \\ j & J & J' \end{Bmatrix}
\end{aligned} \tag{15.33}$$

When protons and neutrons occupy the same j-orbit, fully antisymmetric states with definite isospin can be constructed by using c.f.p. We can define in analogy with (15.13) the general expansion

$$\begin{aligned}
\psi(j^n\alpha TM_T JM) = \sum_{\alpha_1 T_1 J_1} &[j^{n-1}(\alpha_1 T_1 J_1)jTJ|\}j^n\alpha TJ] \\
&\times \psi(j^{n-1}(\alpha_1 T_1 J_1)j_n TM_T JM)
\end{aligned} \tag{15.34}$$

The c.f.p. in (15.34) obey orthonormality relations, like those in (15.14)

$$\begin{aligned}
\sum_{\alpha_1 T_1 J_1} &[j^{n-1}(\alpha_1 T_1 J_1)jTJ|\}j^n\alpha TJ] \\
&\times [j^{n-1}(\alpha_1 T_1 J_1)jTJ|\}j^n\alpha' TJ] = \delta_{\alpha\alpha'}
\end{aligned}$$

The expansion (15.34) may be used for evaluating matrix elements of single nucleon operators and two-body interactions in the same way as formulae (15.17) and (15.21) were obtained.

In practice, in analogy with (15.5), we can start with a state with given T_0, J_0 (and if necessary an additional label α_0) and couple to T_0 the isospin $\frac{1}{2}$ and to J_0 the spin j of the n-th nucleon. We then carry out antisymmetrization with respect to space, spin and isospin variables of the nucleons. At this stage we apply a change of coupling transformation which is a product of (10.11) for the spins as well as for the isospins. As a result we obtain a linear combination which is finally normalized to yield the c.f.p. in (15.34).

Let us apply this procedure to the case of $n = 3$ nucleons. Following the steps leading to (15.10) we obtain in the present case the result

$$[j^2(T_1J_1)jTJ|\}j^3[T_0J_0]TJ]$$
$$= \left[\delta_{T_1T_0}\delta_{J_1J_0} - 2\sqrt{(2T_1+1)(2T_0+1)(2J_1+1)(2J_0+1)}\right.$$
$$\left.\times \begin{Bmatrix} \frac{1}{2} & \frac{1}{2} & T_1 \\ T & \frac{1}{2} & T_0 \end{Bmatrix} \begin{Bmatrix} j & j & J_1 \\ J & j & J_0 \end{Bmatrix}\right]$$
$$\times \left[3 - 6(2T_0+1)(2J_0+1) \begin{Bmatrix} \frac{1}{2} & \frac{1}{2} & T_0 \\ T & \frac{1}{2} & T_0 \end{Bmatrix} \begin{Bmatrix} j & j & J_0 \\ J & j & J_0 \end{Bmatrix}\right]^{-1/2} \tag{15.35}$$

If we start with $T_0 = 1$ and $T = \frac{3}{2}$, the only possible value of T_1 is $T_1 = 1$. This case includes three identical j-nucleons for which we obtained the result (15.10) above. In fact, if we insert in (15.35) the value of the $6j$-symbol

$$\begin{Bmatrix} \frac{1}{2} & \frac{1}{2} & 1 \\ \frac{3}{2} & \frac{1}{2} & 1 \end{Bmatrix} = -\frac{1}{3}$$

we obtain for the c.f.p. $[j^2(T_1 = 1, J_1)jT = \frac{3}{2}J|\}j^3[T_0 = 1, J_0]T = \frac{3}{2}, J]$ the values (15.10).

Another simple result follows from (15.35) by starting with $T_0 = 1$, $J_0 = 0$ and obtaining a state with total isospin $T = \frac{1}{2}$ and $J = j$. Using in (15.35) the values

$$\begin{Bmatrix} \frac{1}{2} & \frac{1}{2} & 1 \\ \frac{1}{2} & \frac{1}{2} & 1 \end{Bmatrix} = \frac{1}{6} \quad \text{and} \quad \begin{Bmatrix} \frac{1}{2} & \frac{1}{2} & 0 \\ \frac{1}{2} & \frac{1}{2} & 1 \end{Bmatrix} = \frac{1}{2},$$

we obtain for this case the following c.f.p.

$$[j^2(T_1J_1)jT=\tfrac{1}{2}J=j|\}j^3[10]T=\tfrac{1}{2}J=j|]$$

$$=\begin{cases}\sqrt{\dfrac{2(j+1)}{3(2j+1)}} & \text{for}\quad T_1=1,\ J_1=0\\[2ex] \dfrac{\sqrt{2J_1+1}}{\sqrt{6(j+1)(2j+1)}} & \text{for}\quad T_1=1,\ J_1>0\,\text{even}\\[2ex] -\dfrac{\sqrt{2J_1+1}}{\sqrt{2(j+1)(2j+1)}} & \text{for}\quad T_1=0,\ J_1\,\text{odd}\end{cases}$$

(15.36)

Using wave functions (15.34) we can write down a simple relation which may be used as a check for c.f.p. We can use the analog of (15.21) to calculate matrix elements of the simple two-body interaction $(\mathbf{t}_i\cdot\mathbf{t}_k)$. We obtain in analogy with (15.22) the relation

$$\left[T(T+1)+\frac{n}{n-2}\times\frac{3}{4}\right]\delta_{\alpha\alpha'}$$

$$=\frac{n}{n-2}\sum_{\alpha_1T_1J_1}[j^{n-1}(\alpha_1T_1J_1)jTJ|\}j^n\alpha TJ]$$

$$\times[j^{n-1}(\alpha_1T_1J_1)jTJ|\}j^n\alpha'TJ]T_1(T_1+1) \qquad (15.37)$$

In the case of maximum isospin, $T=n/2$, $T_1=(n-1)/2$, (15.37) reduces to the normalization relation of the c.f.p.

Let us apply the c.f.p. (15.36) to the calculation of the magnetic moments of a nucleus in which there are a proton and two neutrons in the j-orbit outside closed shells. The magnetic moment of a single j-nucleon is, according to the Wigner-Eckart theorem, proportional to its angular momentum, $\boldsymbol{\mu}_\pi=g_\pi\mathbf{j}$ and $\boldsymbol{\mu}_\nu=g_\nu\mathbf{j}$. We can therefore express the magnetic moment operator of the nucleus by

$$
\begin{aligned}
\boldsymbol{\mu} &= g_\pi \sum_{\text{protons}} \mathbf{j}_i + g_\nu \sum_{\text{neutrons}} \mathbf{j}_i \\
&= g_\pi \sum_i^n \frac{1}{2}(1+\tau_3(i))\mathbf{j}_i + g_\nu \sum_i^n \frac{1}{2}(1-\tau_3(i))\mathbf{j}_i \\
&= \frac{g_\pi + g_\nu}{2} \sum \mathbf{j}_i + \frac{g_\pi - g_\nu}{2} \sum_i^n \tau_3(i)\mathbf{j}_i \\
&= \frac{g_\pi + g_\nu}{2} \mathbf{J} + \frac{g_\pi - g_\nu}{2} \sum_i^n \tau_3(i)\mathbf{j}_i
\end{aligned} \tag{15.38}
$$

To evaluate the second term on the r.h.s. of (15.38) we use c.f.p. according to a generalization of (15.17) for protons and neutrons. The use of (10.41) for space and spin variables as well as for isospins yields the result

$$
\begin{aligned}
&\left(j^n \alpha T J \left\| \sum_i^n \mathbf{h}^{(k')}(i)\mathbf{f}^{(k)}(i) \right\| j^n \alpha' T' J'\right) \\
&= n \sum_{\alpha_1 T_1 J_1} [j^{n-1}(\alpha_1 T_1 J_1) j T J |\} j^n \alpha T J] \\
&\quad \times [j^{n-1}(\alpha_1 T_1 J_1) j T' J' |\} j^n \alpha' T' J'] \\
&\quad \times (-1)^{T_1 + (1/2) + T + k' + J_1 + j + J + k} \\
&\quad \times \sqrt{(2T+1)(2T'+1)(2J+1)(2J'+1)} \\
&\quad \times \begin{Bmatrix} \frac{1}{2} & T & T_1 \\ T' & \frac{1}{2} & k' \end{Bmatrix} \begin{Bmatrix} j & J & J_1 \\ J' & j & k \end{Bmatrix} (\tfrac{1}{2} \| \mathbf{h}^{(k')} \| \tfrac{1}{2})(j \| \mathbf{f}^{(k)} \| j)
\end{aligned} \tag{15.39}
$$

The operator $\mathbf{h}^{(k')}$ is an irreducible tensor of rank k' in isospin space. The rank k' may be either 1 or 0. If $k' = 0$ then $T' = T$ and if $h_0^{(0)} = 1$ the product of isospin coefficients reduces to $\sqrt{2T+1} = (T\|1\|T)$.

To obtain the magnetic moment, we need the matrix element $\langle TM_T|\boldsymbol{\mu}|TM_T\rangle$ rather than the reduced matrix element with respect to isospin. Hence, when applying (15.39) to the operator $\sum \tau_3(i)\mathbf{j}_i$, we should multiply the reduced matrix element by

$$(-1)^{T-M_T}\begin{pmatrix} T & 1 & T \\ -M_T & 0 & M_T \end{pmatrix}.$$

Recalling the values of the reduced matrix elements

$$(\tfrac{1}{2}\|\boldsymbol{\tau}\|\tfrac{1}{2}) = \sqrt{6}, \qquad (J\|\mathbf{J}\|J) = \sqrt{J(J+1)(2J+1)},$$

we obtain from (15.38) the expression

$$\begin{aligned}
&\left(g - \frac{g_\pi + g_\nu}{2}\right)(J\|\mathbf{J}\|J) \\
&\quad = \frac{g_\pi - g_\nu}{2}\left(j^n\alpha TJ\left\|\sum_{i=1}^{n}\boldsymbol{\tau}(i)\mathbf{j}_i\right\|j^n\alpha TJ\right) \\
&\qquad \times (-1)^{T-M_T}\begin{pmatrix} T & 1 & T \\ -M_T & 0 & M_T \end{pmatrix} \\
&\quad = \frac{g_\pi - g_\nu}{2}(-1)^{T-M_T}\begin{pmatrix} T & 1 & T \\ -M_T & 0 & M_T \end{pmatrix} \\
&\qquad \times n\sqrt{6}\sqrt{j(j+1)(2j+1)(2T+1)(2J+1)} \\
&\qquad \times \sum_{\alpha_1 T_1 J_1}(-1)^{T_1+(1/2)+T+J_1+j+J}[j^{n-1}(\alpha_1 T_1 J_1)jTJ|\}j^n\alpha TJ]^2 \\
&\qquad \times \begin{Bmatrix} \frac{1}{2} & T & T_1 \\ T & \frac{1}{2} & 1 \end{Bmatrix}\begin{Bmatrix} j & J & J_1 \\ J & j & 1 \end{Bmatrix}
\end{aligned} \tag{15.40}$$

Substituting in (15.40) the actual value of the $3j$-symbol, we obtain from it, in view of $\mu = gJ$, the final general result

$$g = \frac{g_\pi + g_\nu}{2} + \frac{g_\pi - g_\nu}{2} M_T n\sqrt{6}\sqrt{j(j+1)(2j+1)}$$
$$\times \sqrt{\frac{2T+1}{T(T+1)}}\sqrt{\frac{2J+1}{J(J+1)}}$$
$$\times \sum_{\alpha_1 T_1 J_1} (-1)^{T_1+(1/2)+T+J_1+j+J}$$
$$\times [j^{n-1}(\alpha_1 T_1 J_1) jTJ|\}j^n \alpha TJ]^2$$
$$\times \begin{Bmatrix} \frac{1}{2} & T & T_1 \\ T & \frac{1}{2} & 1 \end{Bmatrix} \begin{Bmatrix} j & J & J_1 \\ J & j & 1 \end{Bmatrix} \tag{15.41}$$

We can now apply (15.41) to the case of $n = 3$, $T = \frac{1}{2}$ and $J = j$. Substituting the values of the c.f.p. (15.36), we may use the sum rules (10.20) and (10.21) for $6j$-symbols and obtain the result

$$g = \frac{g_\pi + g_\nu}{2} - M_T \frac{g_\pi - g_\nu}{3} \frac{j+4}{j+1} \tag{15.42}$$

In the case of one proton and two neutrons, $M_T = -\frac{1}{2}$ and the g-factor is equal to

$$g = \frac{(4j+7)g_\pi + (2j-1)g_\nu}{6(j+1)} \tag{15.43}$$

In the case of one neutron and two protons, $M_T = \frac{1}{2}$, we obtain the result (15.43) in which g_π and g_ν are interchanged. From (15.43) follows that the two neutrons are not coupled to $J = 0$ and thus, the g-factor is not that of the odd proton. This effect is due to charge independence and demonstrates that an extreme single nucleon picture of the nucleus cannot be valid. If $j = l + \frac{1}{2}$, g_π is positive whereas g_ν

is negative and the g-factor in (15.43) is *smaller* than the value for a single proton. Similarly, in the case of one neutron and two protons, the g-factor (15.43) is higher, less negative than g_ν of a single neutron. The changes in g-factors given by (15.43) are in the same direction as experimentally observed. For $j = l - \frac{1}{2}$, the effect is considerably smaller.

Let us use the expression (15.38) to derive a selection rule on M1 transitions. The M1 transition operator is proportional, according to (10.51), to $\boldsymbol{\mu}$. The first term on the r.h.s. of (15.38) has vanishing matrix elements between any two orthogonal states with definite values of J. The second term is a vector, rank $k' = 1$ tensor, in isospin. In self conjugate nuclei, $N = Z$, $M_T = 0$. Matrix elements of this isospin vector between states with the same isospin T are proportional, according to the Wigner-Eckart theorem to

$$\begin{pmatrix} T & 1 & T \\ 0 & 0 & 0 \end{pmatrix}$$

which vanishes for any value of (integral) isospin T. Hence, in the long wave approximation, no $M1$ transitions with $\Delta T = 0$ can take place between any two states of such nuclei.

A similar selection rule holds for E1 transitions in self conjugate nuclei. The transition operator is proportional to

$$\sum_{\text{protons}} \mathbf{r}_i = \sum_i \frac{1}{2}[1 + \tau_3(i)]\mathbf{r}_i = \frac{1}{2}\sum_i \mathbf{r}_i + \frac{1}{2}\sum_i \tau_3(i)\mathbf{r}_i$$

The first term on the r.h.s. is proportional to the center of mass coordinate which has vanishing matrix elements between any two orthogonal intrinsic states of the nucleus. The second term is an isospin vector and has vanishing matrix elements between any two states with the same value of T as in the case of M1 transitions.

Any two states with different values of the total isospin T are orthogonal. This orthogonality is due to the isospin parts of the wave functions and hence, it is not possible to derive directly from it orthogonality relations between c.f.p. Still, it may be verified that the sum of products of c.f.p. in (15.36) by corresponding c.f.p. in (15.11) vanishes. Let us look at the reason for this feature.

States with definite isospins may be eigenstates of a charge independent Hamiltonian. Consider two protons and one neutron in the j-orbit. As explained above, the use of isospin introduces great simplifications but it is not mandatory. If we do not use isospin, we may obtain all states of the system by coupling all states of the two protons to the states of the neutron. A complete basis of states is thus given by

$$\psi(j_\pi^2(J_1)j_\nu JM) \tag{15.44}$$

Eigenstates of any charge independent Hamiltonian can be expressed as

$$\sum_{J_1 \text{ even}} C^\alpha_{J_1J}\psi(j_\pi^2(J_1)j_\nu JM) \tag{15.45}$$

The labels α are certain quantum numbers specifying, together with J, the eigenstates. The orthogonality of different normalized eigenstates implies the relation

$$\sum_{J_1 \text{ even}} C^\alpha_{J_1J}C^{\alpha'}_{J_1J} = \delta_{\alpha\alpha'} \tag{15.46}$$

Some of the states (15.45) correspond to $T = \frac{1}{2}$ states while others correspond to $T = \frac{3}{2}$ states. We can multiply each term in (15.45) by the isospin function which, according to (9.5), is $\eta_{1/2}(1)\eta_{1/2}(2)\eta_{-1/2}(3)$, and antisymmetrize the wave function. In this way, a state with definite isospin T will be obtained which is an eigenstate of a charge independent Hamiltonian with the same eigenvalue as that of (15.45).

We can now consider a complete wave function (15.34), for $n = 3$, which is an eigenstate of a charge independent Hamiltonian H. As will be shown in the beginning of Section 26, such a Hamiltonian can be expressed in a form in which isospin operators do not appear. We start from the eigenvalue equation

$$[H - E_\alpha]\sum_{T_1J_1}[j^2(T_1J_1)jTJ|\}j^3\alpha TJ]\psi[j_{12}^2(T_1J_1)j_3TM_TJM] = 0 \tag{15.47}$$

We now multiply (15.47) from the left by the isospin state $\eta^*_{1/2}(1)$ $\eta^*_{1/2}(2)\eta^*_{-1/2}(3)$. In the expansion in (15.47) there may be non-vanishing c.f.p. with $T_1 = 0$. They are multiplied, however, by isospin functions $[\eta_{1/2}(1)\eta_{-1/2}(2) - \eta_{1/2}(2)\eta_{-1/2}(1)]$ which are orthogonal to

$\eta_{1/2}(1)\eta_{1/2}(2)$. Due to the fact that H does not contain isospin operators, we obtain from (15.47) the result

$$(H - E_\alpha) \sum_{J_1 \text{ even}} [j^2(T_1 = 1J_1)jTJ|\}j^3\alpha TJ]\, \psi[j_{12}^2(J_1)j_3JM] = 0 \tag{15.48}$$

where isospin states do not appear. We may safely assume that all eigenvalues E_α (for given J) are unequal. Comparing (15.48) with the expansion (15.45) of the eigenstate, we see that the c.f.p., for even values of J_1, are *proportional* to the coefficients $C^\alpha_{J_1J}$. If in (15.48) $T = \frac{3}{2}$, there are only c.f.p. with even values of J_1 in the expansion and hence, the c.f.p. are *equal* to the (normalized) coefficients $C^\alpha_{J_1J}$. In the case of $T = \frac{1}{2}$, there are also non-vanishing c.f.p. with $T_1 = 0$, J_1 odd. However, the orthogonality of c.f.p. for $T = \frac{3}{2}$ and $T = \frac{1}{2}$ states follows from (15.46) which involves only even values of J_1.

The relation between c.f.p. in the isospin formalism and expansion coefficients of eigenstates of a charge independent Hamiltonian in terms of proton-neutron states may be generalized. We shall not discuss here higher numbers of protons and neutrons.

16

Examples of Effective Interactions in Nuclei and Atoms

To give a simple illustration of the approach described above, we consider Zr isotopes (proton number $Z = 40$) with valence neutrons beyond the closed shells of $N = 50$. The ^{90}Zr nucleus does not have closed proton shells and yet it exhibits some properties of a doubly magic nucleus. Its first excited state has $J^\pi = 0^+$ and lies 1.76 MeV above the ground state. The next higher state, 2^+, lies at 2.19 MeV. We can thus tentatively consider states with neutron numbers beyond 50 as states of valence neutrons.

Let us first repeat that the binding energy of ^{90}Zr cannot be reliably calculated. As explained above, the interaction between free nucleons has not been well determined. Moreover, no reliable approximation procedures are available to calculate binding energies of finite nuclei from any given nucleon-nucleon interaction.

The ground state of ^{91}Zr is a $\frac{5}{2}^+$ ($J = \frac{5}{2}$ and positive parity) state due, according to the shell model, to a single $2d_{5/2}$ neutron outside the closed shells of $N = 50$. The *separation energy* of that nucleon given by B.E.(^{91}Zr) – B.E.(^{90}Zr) is 4.48 MeV. Although much smaller than the binding energy of ^{90}Zr (which in 784 MeV), it cannot be reliably calculated. We must then turn to an easier problem—calculation of energy levels due to the interaction of several valence neutrons.

The first excited state of ^{91}Zr is a $(\frac{1}{2})^+$ state at 1.21 MeV due to the valence neutron occupying the $3s_{1/2}$ orbit. The next state at 1.47 MeV is another $(\frac{5}{2})^+$ state which could be due to the $2d_{5/2}$ neutron coupled to the first 0^+ excited state in ^{90}Zr. The next higher level at 1.88 MeV has $(\frac{7}{2})^+$ interpreted as due to the valence neutron in the $1g_{7/2}$ orbit. Thus, the state of a neutron in the $2d_{5/2}$ orbit in ^{91}Zr may not be sufficiently separated from other states. Still, we can try to check whether low-lying states of Zr isotopes beyond ^{90}Zr can be described by $(2d_{5/2})^n$ configurations of valence neutrons (Talmi 1962a).

In ^{92}Zr we expect to find the $(2d_{5/2})^2$ neutron configuration whose states, allowed by the Pauli principle, have $J = 0, 2, 4$. Experimentally we find a 2^+ level at .93 MeV and a 4^+ level at 1.49 MeV above the 0^+ ground state. Even these energy spacings or the difference between the ^{92}Zr and ^{91}Zr binding energies cannot be calculated by the nuclear many body theory. We may assume that these energy spacings are indeed due to the interaction energy of the two $2d_{5/2}$ valence neutrons. If we adopt this interpretation we must make sure that it is a consistent one. If we consider only two body effective interactions, level spacings in all $(2d_{5/2})^n$ configurations can be calculated from the ^{92}Zr level spacings. The consistency check is thus the comparison between levels calculated in this way and the experimental levels. The same check should also be applied to binding energy differences but this will be postponed to a subsequent section where special consideration will be given to binding energies.

When calculating binding energies, the single $2d_{5/2}$ neutron energies should be considered. From the binding energy difference of ^{91}Zr and ^{90}Zr we obtain for it the value $\epsilon = 4.48$ MeV. In all states of a given $(2d_{5/2})^n$ configuration there is the same contribution $n\epsilon$ of single neutron energies. They do not contribute to spacings of energy levels considered here.

The simplest configuration whose level spacings can be obtained from those in the two nucleon $(2d_{5/2})^2$ configuration is $(2d_{5/2})^4$. Since there can be only 6 neutrons in a $d_{5/2}$ orbit, that configuration is just of two holes. As explained in Section 14, the level spacings of such a configuration are identical to those of the two nucleon $(2d_{5/2})^2$ configuration. This two hole configuration is expected in ^{94}Zr and we can compare the positions of its 2^+ and 4^+ levels with those of ^{92}Zr. Indeed, the experimentally observed 2^+ and 4^+ levels of ^{94}Zr are at .92 MeV and 1.47 MeV respectively above the 0^+ ground state.

The spectra of ^{96}Zr and ^{95}Zr are consistent with predictions of $(2d_{5/2})^n$ configurations. The ground state of the full orbit in ^{96}Zr is well separated from excited states and so is the single hole $(\frac{5}{2})^+$ ground state in ^{95}Zr. Their binding energies will be considered in a subsequent section. The only other configuration which should be now checked is $(2d_{5/2})^3$ expected in ^{93}Zr. Let us calculate its energy levels and then compare with experimental data.

There are three antisymmetric states of a $(\frac{5}{2})^3$ configuration with spins $J = \frac{3}{2}, \frac{5}{2}, \frac{9}{2}$. This fact can be verified by the vanishing of (15.7) when we put in it $j = \frac{5}{2}$, $J = \frac{1}{2}, \frac{7}{2}$ for any even value of J_0 and by the fact that for $J = \frac{3}{2}, \frac{5}{2}, \frac{9}{2}$ there is only one independent state (15.7) for any even value of J_0. A simpler way is to count the number of independent states in the m-scheme. The maximum value of M is for $m_1 = \frac{5}{2}$, $m_2 = \frac{3}{2}$, $m_3 = \frac{1}{2}$. Hence, the antisymmetrized state $(\frac{5}{2}, \frac{3}{2}, \frac{1}{2})$ must be the $J = \frac{9}{2}$, $M = \frac{9}{2}$ state. There is only one independent state with $M = \frac{7}{2}$, which is $(\frac{5}{2}, \frac{3}{2}, -\frac{1}{2})$, and, when antisymmetrized, it is the $J = \frac{9}{2}$, $M = \frac{7}{2}$ state. There are *two* independent states with $M = \frac{5}{2}$, $(\frac{5}{2}, \frac{3}{2}, -\frac{3}{2})$ and $(\frac{5}{2}, \frac{1}{2}, -\frac{1}{2})$. One linear combination of these states after antisymmetrization is the $M = \frac{5}{2}$ state with $J = \frac{9}{2}$ and the other orthogonal combination must be a $J = \frac{5}{2}$, $M = \frac{5}{2}$ state. The continuation of this procedure yields the allowed states with $J = \frac{9}{2}, \frac{5}{2}, \frac{3}{2}$.

The expansion of these states in terms of c.f.p. is given for $J = \frac{5}{2}$ by (15.11) and for $J = \frac{3}{2}, \frac{9}{2}$ by (15.10) with $J_0 = 2$ as follows:

$$\psi(5/2^3 J = 5/2, M) = \frac{\sqrt{2}}{3}\psi(5/2^2(0)5/2_3 J = 5/2, M)$$

$$- \frac{\sqrt{10}}{6}\psi(5/2^2(2)5/2_3 J = 5/2, M)$$

$$- \frac{\sqrt{2}}{2}\psi(5/2^2(4)5/2_3 J = 5/2, M)$$

$$\psi(5/2^3 J = 3/2, M) = \sqrt{\frac{5}{7}}\psi(5/2^2(2)5/2_3 J = 3/2, M)$$

$$- \sqrt{\frac{2}{7}}\psi(5/2^2(4)5/2_3 J = 3/2, M)$$

$$\psi(5/2^3 J = 9/2, M) = \sqrt{\frac{3}{14}}\psi(5/2^2(2)5/2_3 J = 9/2, M)$$
$$- \sqrt{\frac{11}{14}}\psi(5/2^2(4)5/2_3 J = 9/2, M)$$

The interaction energies in these states are expressed in terms of the two-body matrix elements by (15.21) as follows

$$E_{5/2} = \tfrac{2}{3}V_0 + \tfrac{5}{6}V_2 + \tfrac{3}{2}V_4$$
$$E_{3/2} = \tfrac{15}{7}V_2 + \tfrac{6}{7}V_4 \tag{16.1}$$
$$E_{9/2} = \tfrac{9}{14}V_2 + \tfrac{33}{14}V_4$$

where in (16.1) we used the notation

$$V_J = \langle 5/2^2 JM|V|5/2^2 JM\rangle.$$

From (16.1) follow the relations between level spacings

$$E_{3/2} - E_{5/2} = \tfrac{2}{3}(V_2 - V_0) - \tfrac{9}{14}(V_4 - V_2)$$
$$E_{9/2} - E_{5/2} = \tfrac{2}{3}(V_2 - V_0) + \tfrac{6}{7}(V_4 - V_2) \tag{16.2}$$

Using the values $V_2 - V_0 = .93$ MeV and $V_4 - V_2 = .56$ MeV from ^{92}Zr (or $V_2 - V_0 = .92$ and $V_4 - V_2 = .55$ from ^{94}Zr) we obtain that the ground state of ^{93}Zr is the $(\frac{5}{2})^+$ state and that the $(\frac{3}{2})^+$ state should lie at .26 MeV above it. The experimental spectrum of ^{93}Zr has indeed a $(\frac{5}{2})^+$ ground state and a $(\frac{3}{2})^+$ state .27 MeV above it. This excited state is definitely not a state in which a valence neutron occupies the $2d_{3/2}$ orbit. The spin-orbit splitting should be much stronger (the lowest $(\frac{3}{2})^+$ state in ^{91}Zr lies at 2.04 MeV above the $(\frac{5}{2})^+$ ground state).

The $(\frac{9}{2})^+$ state in ^{93}Zr is predicted to be at 1.09 MeV above the ground state. No such state has yet been found by experiment. The ^{93}Zr nucleus is unstable and various direct reactions on neighboring nuclei could lead into such $(\frac{9}{2})^+$ states only through a small admixture of a $1g_{9/2}$ neutron (the $1g_{9/2}$ neutron orbit is closed at $N = 50$). It is perhaps interesting to note that a $(\frac{9}{2})^+$ state was observed in ^{95}Mo at .95 MeV above the ground state. In that nucleus, proton excitations must be taken into consideration and are coupled to states of the valence neutrons. Still, in ^{95}Mo there is a $(\frac{3}{2})^+$ state which lies .20 MeV above the $(\frac{5}{2})^+$ ground state, also fairly close to the calculated position for ^{93}Zr.

Thus, energies of low-lying levels of Zr isotopes can be adequately described by simple shell model $(2d_{5/2})^n$ configurations provided an effective interaction is used. The effective interaction can be determined in this case from the experimental data in a consistent way.

We can use this simple case to demonstrate some properties of the effective interaction. From the analysis described above we determined only two differences of matrix elements from which we cannot make detailed deductions about the effective interaction. Still we can rule out certain simple-minded interactions. To do so we choose a different parametrization of the effective interaction. The following method of calculation was introduced by Racah (1952).

We note that *any* two-body interaction which has the same matrix elements in the two nucleon j^2 configuration may be used to calculate matrix elements in all j^n configurations. We take the following two-body interaction which has three free parameters

$$V_{12} = a + 2b(\mathbf{j}_1 \cdot \mathbf{j}_2) + cP_{12} \tag{16.3}$$

where P_{12} is the pairing interaction defined by (12.17). The parameter a is the coefficient of a scalar interaction which is independent of the value of the total J. It can be visualized as the long range limit of a potential like $\exp(-\lambda r_{12}^2)$ with a small enough λ so that the potential is practically constant in the region where the nuclear wave functions are important. The interaction $(\mathbf{j}_1 \cdot \mathbf{j}_2)$ may be considered as the long range limit of the spin-spin interaction $(\mathbf{s}_1 \cdot \mathbf{s}_2)V(r_{12})$ (see (12.16)). The potential may be then replaced by a constant. The spin operators may be replaced, due to the Wigner-Eckart theorem within the j^n configuration, by the operators $\mathbf{j}_1$ and $\mathbf{j}_2$.

It is important to realize that (16.3) is not a specific interaction like those mentioned here. As will be shown below (in (16.6)), matrix elements of *any* two body interaction within $(\frac{5}{2})^n$ configurations of identical nucleons are equal to those of (16.3). It was pointed out in Section 12 that the tensor expansion of an interaction is not unique if limited to j^2 states with $T = 1$. This is clearly demonstrated in the special case considered here. Any two body interaction is equivalent, within $(\frac{5}{2})^n$ configurations with maximum T to (16.3). Different interactions will have different coefficients a, b and c. The interaction (16.3) has in its expansion a scalar term (the term with coefficient a and a term in P_{12}) and in addition only odd tensor terms (in P_{12} and the $(\mathbf{j}_1 \cdot \mathbf{j}_2)$ term).

In the j^n configuration the sum of the interactions (16.3) is given by the operator

$$\begin{aligned}
&\sum_{i<k}^{n} a + b\sum_{i<k}^{n} 2(\mathbf{j}_i \cdot \mathbf{j}_k) + c\sum_{i<k}^{n} P_{ik} \\
&= \frac{n(n-1)}{2}a + b\left[\left(\sum_{i=1}^{n}\mathbf{j}_i\right)^2 - \sum_{i=1}^{n}\mathbf{j}_i^2\right] + c\sum_{i<k}^{n} P_{ik} \\
&= \frac{n(n-1)}{2}a + b[J(J+1) - nj(j+1)] + c\sum_{i<k}^{n} P_{ik}
\end{aligned} \tag{16.4}$$

Let us first make sure that the interaction (16.3) can reproduce in the $(\frac{5}{2})^2$ configuration *any* values of V_0, V_2 and V_4. Putting $n = 2$ in (16.4) we obtain, in view of (12.17)

$$\begin{aligned}
V_0 &= a - \tfrac{35}{2}b + 6c \\
V_2 &= a - \tfrac{23}{2}b \\
V_4 &= a + \tfrac{5}{2}b
\end{aligned} \tag{16.5}$$

Equations (16.5) may be solved for a, b and c and the results are

$$\begin{aligned}
a &= \tfrac{1}{28}(5V_2 + 23V_4) \\
b &= \tfrac{1}{14}(V_4 - V_2) \\
c &= \tfrac{1}{6}(V_0 - \tfrac{10}{7}V_2 + \tfrac{3}{7}V_4)
\end{aligned} \tag{16.6}$$

Hence, the interaction (16.3) is the most general two-body interaction within the $(\frac{5}{2})^n$ configuration.

To apply (16.4) to the case $n = 3$ we need to know the eigenvalues of the pairing interaction. These are given by a simple general expression which will be derived in a subsequent section. For the case at hand, it is sufficient to see from the c.f.p. expansion of the states given above, that due to (15.21) or (16.1), the eigenvalues of cP for the $J = \frac{3}{2}$ and $J = \frac{9}{2}$ states vanish. For the state with $J = \frac{5}{2}$, we obtain from (15.21) or (16.1) the eigenvalue $(\frac{2}{3})6c = 4c$. We can then use (16.4) to obtain for the eigenvalues in the $(\frac{5}{2})^3$ configuration the expressions

$$
\begin{aligned}
E_{5/2} &= 3a - \tfrac{35}{2}b + 4c \\
E_{3/2} &= 3a - \tfrac{45}{2}b \qquad (16.7) \\
E_{9/2} &= 3a - \tfrac{3}{2}b
\end{aligned}
$$

We can now consider two limits of the interaction (16.3). In one we put $b = 0$ (i.e. $V_4 = V_2$) and in the other $c = 0$ (energy spacings are independent of a in any limit). We can interpolate between these two limits by varying b while keeping $V_2 - V_0 = 6(b - c)$ constant and equal to .93 MeV. This way we obtain a graphic illustration of the calculation of $(\frac{5}{2})^3$ levels as seen in Fig. 16.1. The main point of Fig. 16.1, as well as of the direct calculation presented above, is that the pairing interaction and also the short range δ-potential are poor approximations of the effective nuclear interaction. We saw earlier that in j^2 configurations such interactions predict a 0-2 splitting much bigger than the spacings of $J = 2, 4, \ldots, 2j - 1$ levels. Here we see that such interactions in j^3 configurations give rise also to level spacings between J states with $J \neq j$ which are much smaller than the gap between these states and the $J = j$ ground state. The pairing interaction is quoted as the reason for an "energy gap" between the $J = 0$ ground state and states with $J > 0$. It is rarely mentioned that according to the pairing interaction a similar gap should exist between the $J = j$ ground state and all other degenerate states of the j^3 configuration. According to (15.21) the gap between the $J = j$ state given by (15.11) and $J \neq j$ states should be equal to the gap in the j^2 configuration multiplied by $(2j - 1)/(2j + 1)$ (equal to $\frac{2}{3}$ for the case of $j = \frac{5}{2}$ considered above). This is far from being the case. The occurrence of low-lying states with $J = j - 1$ in j^3 configurations is not unique to

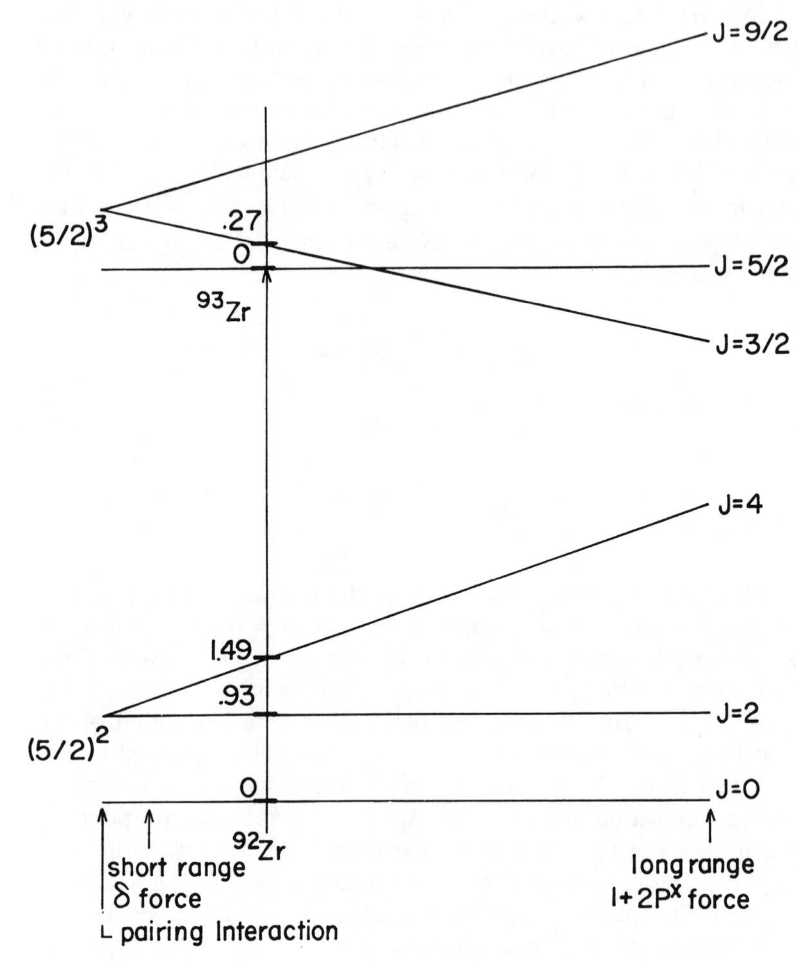

FIGURE 16.1. *Levels of* $(\frac{5}{2})^2$ *and* $(\frac{5}{2})^3$ *configurations.*

^{93}Zr. It is seen in many nuclei where j^3 configurations of identical nucleons have been identified.

Schematic interactions like the pairing interaction or the δ-potential encounter difficulties with the "energy gap" both in even-even and odd mass nuclei. In the former, the 2–4 spacings may be rather large whereas in the latter, low lying levels with $J = j - 1$ occur. Different mechanisms had to be invoked to explain these two features. On the other hand, if we try to learn from experiment the nature of the effective interaction, we find that these two features are related. As in the case of Zr isotopes, an effective interaction obtained from experimental level spacings in even nuclei yields low lying levels with $J = j - 1$ in odd nuclei.

The shell model of the nucleus was introduced by analogy with the atomic shell model. Shells of electrons in atoms are determined by the strong attractive electric potential of the heavy nucleus and by their mutual Coulomb repulsion. The periodic table of the elements is a clear demonstration of shell structure in atoms. It is therefore instructive to examine how well simple shell model wave functions describe real states of electrons in atoms. What we shall actually see is whether atomic energy levels can be calculated by using *simple* shell model wave functions with the Coulomb repulsion as the residual interaction between electrons.

The prevalent coupling scheme in atoms is LS-coupling described in Section 9. In order to obtain fully antisymmetric many particle wave functions, we can use coefficients of fractional parentage like those defined in Section 15. This procedure is described in detail by de-Shalit and Talmi (1963). We shall only point out that here we couple the n-th particle to antisymmetric states of the l^{n-1} configuration characterized by S_0, M_{S_0} and L_0, M_{L_0} (and possibly additional labels α_0). In doing so, S_0 should be coupled to the spin $\frac{1}{2}$ of the particle to yield total S, L_0 should be coupled to l to yield total L and only then S and L may be coupled to the total J. Hence, when changing the order of coupling, the transformation coefficients are products of the coefficients of the transformation of the orbital angular momenta and that of the intrinsic spins. Thus, antisymmetric states of the l^3 configuration are given in analogy to (15.9) and (15.10) by

$$\boxed{\begin{aligned}&\psi(l^3[S_0L_0]SLJM)\\ &\qquad = \sum_{S_1L_1}[l^2(S_1L_1)lSL|\}l^3[S_0L_0]SL]\psi(l^2(S_1L_1)l_3SLJM)\end{aligned}} \tag{16.8}$$

where the c.f.p. are given by

$$[l^2(S_1L_1)lSL|\}l^3[S_0L_0]SL]$$

$$= \left[\delta_{S_1S_0}\delta_{L_1L_0} + 2\sqrt{(2S_0+1)(2S_1+1)(2L_0+1)(2L_1+1)} \times \begin{Bmatrix} \frac{1}{2} & \frac{1}{2} & S_1 \\ S & \frac{1}{2} & S_0 \end{Bmatrix} \begin{Bmatrix} l & l & L_1 \\ L & l & L_0 \end{Bmatrix}\right] \times \left[3 + 6(2S_0+1)(2L_0+1) \times \begin{Bmatrix} \frac{1}{2} & \frac{1}{2} & S_0 \\ S & \frac{1}{2} & S_0 \end{Bmatrix} \begin{Bmatrix} l & l & L_0 \\ L & l & L_0 \end{Bmatrix}\right]^{-1/2} \tag{16.9}$$

We shall consider here a simple case in which we will not need this formula.

The simple example we consider is the l^n configuration with $l = 1$. There are three antisymmetric states in the p^2 (l^2 with $l = 1$) configuration of electrons (or any identical spin $\frac{1}{2}$ fermions). According to the discussion in Section 9, states with $S = 0$ must have even values of L, i.e. $L = 0$ or $L = 2$ and $S = 1$ implies odd L, i.e. $L = 1$. With central interactions only (no spin-orbit or tensor interactions) the energies of these states, as well as those of any state in the l^n configuration, do not depend on J. Hence, any two-body interaction which reproduces the energies of the p^2 states can be used to calculate energies of all p^n configurations. A simple such interaction can be conveniently chosen as (Racah 1952)

$$V_{12} = a + 2b(\mathbf{s}_1 \cdot \mathbf{s}_2) + 2c(\boldsymbol{l}_1 \cdot \boldsymbol{l}_2) \tag{16.10}$$

In the p^n configuration, $\sum V_{ik}$ has eigenstates with given S and L whose eigenvalues are given by

$$\frac{n(n-1)}{2}a + b[S(S+1) - \tfrac{3}{4}n] + c[L(L+1) - 2n] \tag{16.11}$$

As will be shown by (16.13), the interaction (16.10) is equivalent to any two body interaction between p-electrons. As in the case of jj-coupling discussed in Section 12, the tensor expansion is not unique if limited to l^n configurations of identical particles. The Coulomb interaction, for instance, may be expanded in a very different form. Still, within the electronic p^n configuration it can have the expansion (16.10) which includes a scalar and two odd tensor terms. The magnitudes of the parameters a, b and c depend on the particular interaction and the radial functions of the electrons. Thus, the term in (16.10), whose coefficient is b, does not represent a real spin-spin interaction. The explicit dependence of (16.11) on S is usually due to the *symmetry* of the various states. As we shall see in a subsequent section, the higher the value of S, the higher the spin symmetry and the lower the spatial symmetry. Hence, in states with high values of S, the interaction is expected to be weaker. This leads in the case of the repulsion between electrons, to Hund's rule in atomic spectroscopy.

The parameters a, b and c can be determined to reproduce the energies of the p^2 states. Putting in (16.11) $n = 2$ we obtain

$$\begin{aligned} &S = 0,\ L = 0 \qquad V(^1S) = a - \tfrac{3}{2}b - 4c \\ &S = 0,\ L = 2 \qquad V(^1D) = a - \tfrac{3}{2}b + 2c \\ &S = 1,\ L = 1 \qquad V(^3P) = a + \tfrac{1}{2}b - 2c \end{aligned} \tag{16.12}$$

In (16.12) we use for the various states the customary notation $^{2S+1}L_J$ where $L = 0,1,2$ is replaced by the letters S,P,D. We omit the value of J since the energies are taken to be independent of it, as is the case for the examples given below. The equations (16.12) can be solved for any values of $V(^1S)$, $V(^1D)$ and $V(^3P)$ to yield

$$\begin{aligned} a &= \tfrac{3}{4}V(^3P) + \tfrac{5}{12}V(^1D) - \tfrac{1}{6}V(^1S) \\ b &= \tfrac{1}{2}(V^3P) - \tfrac{1}{6}V(^1D) - \tfrac{1}{3}V(^1S) \\ c &= \tfrac{1}{6}V(^1D) - \tfrac{1}{6}V(^1S) \end{aligned} \tag{16.13}$$

The parameter a contributes equally to all states and is eliminated when we consider energy spacings.

In the p^3 configuration there are three antisymmetric states with $S = \frac{1}{2}$, $L = 1(^2P)$, $S = \frac{1}{2}$, $L = 2(^2D)$ and $S = \frac{3}{2}$, $L = 0(^4S)$. In terms of a, b, c their energies are given by (16.11) with $n = 3$ as

$$E(^2P) = 3a - \tfrac{3}{2}b - 4c$$
$$E(^2D) = 3a - \tfrac{3}{2}b \qquad (16.14)$$
$$E(^4S) = 3a + \tfrac{3}{2}b - 6c$$

Starting from the experimentally observed spacings in the p^2 configuration we can calculate the spacings of the p^3 levels. In the calculation of level spacings there is no contribution from the single p electron energies and they have been omitted from (16.14). Let us first draw some consequences of the assumption that the residual interaction is given by the Coulomb potential.

The Coulomb repulsion between electrons, $e^2/|\mathbf{r}_1 - \mathbf{r}_2|$, can be expanded according to (10.1) and (10.4). Due to (8.15) the only tensors whose scalar products appear in the expansion have even ranks k. The interaction energies in the p^2 states are then given by (10.77). If we introduce the notation $F^0 = 3F_0$ and $F^2 = 30F_2$ we obtain

$$V(^1S) = F_0 + 10F_2 \qquad V(^1D) = F_0 + F_2 \qquad V(^3P) = F_0 - 5F_2 \qquad (16.15)$$

The state with the higher value of S has indeed the lowest energy according to Hund's rule. The values of the parameters a, b and c in this case are obtained by substituting from (16.15) into (16.13) and are given by

$$a = F_0 - 5F_2 \qquad b = -6F_2 \qquad c = -\tfrac{3}{2}F_2 \qquad (16.16)$$

In order to calculate spacings of energy levels the radial integral in F_2 should be computed. For this, the radial parts of the p electron wave functions must be known. This involves a very complicated and approximate calculation which has not been possible to carry out with sufficient accuracy. We can, however, deduce from (16.11) and (16.16) the predicted *ratios* between level spacings. For the Coulomb interaction *all* level spacings depend linearly on F_2 and it is eliminated from their ratios. For the p^2 configuration the predicted ratios are

$$\frac{V(^1S) - V(^1D)}{V(^1D) - V(^3P)} = \frac{9F_2}{6F_2} = 1.5 \qquad (16.17)$$

TABLE 16.1 *Experimental ratios between level energies in p^2 and p^3 configurations*

Atom	Configuration	Ratio (16.17)	Atom	Configuration	Ratio (16.18)
CI	$1p^2$	1.13			
NII	$1p^2$	1.14	NI	$1p^3$	.500
OIII	$1p^2$	1.14	OII	$1p^3$	.509
SiI	$2p^2$	1.48	SII	$2p^3$	.651

In the p^3 configuration the predicted ratio of level spacings is obtained from (16.14) and (16.16) to be

$$\frac{E(^2P)-E(^2D)}{E(^2D)-E(^4S)} = \frac{6F_2}{9F_2} = \frac{2}{3} \tag{16.18}$$

We can now check whether these ratios agree with the experimental data.

The comparison between these predictions and experiment is given in the classical book of Condon and Shortley (1935). They list the ratios of experimental level-spacings in cases where LS-coupling is a good approximation. Their results are quoted in Table 16.1. The roman numerals I, II, III following the name of the element, indicate the degree of ionization of the given atom. The neutral atom is denoted by I, once ionized by II and twice ionized by III.

The clear deviations from the predicted values of 1.5 and .667 are attributed by Condon and Shortley to perturbations from other configurations. What is, however, remarkable, as was shown by Racah (1964), is that these perturbations simply *renormalize* the two-body interaction between the electrons.

Once the Coulomb interaction is renormalized, we should remove the restriction to even tensors in its expansion. In the general case there is no fixed ratio between level spacings in the p^2 configuration. If, however, the two-body interactions are taken from the *observed* spacings in the p^2 configuration, the ratios of level spacings in the p^3 configuration can be calculated. In fact, if we calculate the ratio (16.18) from (16.14) with a, b, c given by (16.13) we obtain the following relation between the ratios (16.17) and (16.18)

$$\frac{E(^2P)-E(^2D)}{E(^2D)-E(^4S)} = \frac{4}{9}\frac{V(^1S)-V(^1D)}{V(^1D)-V(^3P)} \tag{16.19}$$

Using the ratios given in Table 16.1 for p^2 configurations we find

$$1.14 \times \tfrac{4}{9} = .507 \qquad \text{and} \qquad 1.48 \times \tfrac{4}{9} = .658$$

in very good agreement with the experimental ratios in p^3 configurations.

This case as well as other examples quoted by Racah (1964), clearly demonstrate the need for renormalization of the effective interaction. This is evident even in atomic spectra where the shell model is considered to be a very good approximation.

17

Second Quantization. Single Nucleon and Two-Body Operators

In Section 14 we considered many nucleon states as the antisymmetrized products of single nucleon wave functions. Due to the antisymmetrization, the wave function is completely determined by the single nucleon states which are occupied. This suggests that a direct way to deal with many nucleon states is by using a representation in which occupied single nucleon states, rather than nucleon coordinates, are specified. To do this we have to define a fixed set of single nucleon states. The results (14.6), (14.7) indicate that we should provide matrix elements of relevant single nucleon operators, calculated in the basis of these single nucleon wave functions. The results (14.11), (14.12) and (14.13) show that we should also calculate matrix elements of two-body operators in the basis of single nucleon wave functions. These calculations should be performed in the way shown in preceding sections. They are carried out in *configuration space* by using wave functions of space coordinates as well as spin (and isospin) variables. Equipped with these matrix elements we can introduce the representation in terms of occupied states by using the formalism of second quantization. This formalism will now be briefly described without entering into serious discussions and rigorous proofs.

We start with the vacuum state $|0\rangle$ in which there are only closed shells. The state of a single valence nucleon in the j-orbit with given m is obtained by acting on the vacuum state with the *creation operator* a^{+}_{jm}. We should actually specify the j-orbit also by its quantum number n and l but we omit them for the sake of conciseness. In the following, if we write $j' = j$ it means also $n' = n$, $l' = l$ whereas if $j' \neq j$, it means that either one or more of n', l', j' are different from n, l, j. If both protons and neutrons are considered we should also specify the isospin state given by m_t (either $+\frac{1}{2}$ or $-\frac{1}{2}$). For the sake of simplified notation we may or may not include it explicitly.

States of several valence nucleons are given by applying to the vacuum state several creation operators

$$a^{+}_{j_1m_1}a^{+}_{j_2m_2}\cdots a^{+}_{j_nm_n}|0\rangle \tag{17.1}$$

The state (17.1) corresponds to the fully antisymmetric wave function (14.2) in configuration space. The antisymmetric wave function (14.1) changes sign under interchange of the coordinates of any two nucleons. Such an interchange is equivalent to the interchange of two *rows* in the Slater determinant (14.2). The antisymmetry may equally well be demonstrated by interchanging any two labels of single nucleon states amounting to the interchange of two columns in (14.2). Hence, the correspondence between (17.1) and (14.2) requires that the former changes sign under interchange of any two creation operators in (17.1).

Thus, the fermion nature of the nucleons described by (17.1) is guaranteed if $a^{+}_{jm}a^{+}_{j'm'} = -a^{+}_{j'm'}a^{+}_{jm}$, i.e. by the *anticommutation relation*

$$\boxed{a^{+}_{jm}a^{+}_{j'm'} + a^{+}_{j'm'}a^{+}_{jm} = \{a^{+}_{jm}, a^{+}_{j'm'}\} = 0} \tag{17.2}$$

In particular, the antisymmetry thus implies $(a^{+}_{jm})^2 = 0$ which expresses the Pauli principle that no two identical fermions may occupy the same jm state. The creation operator a^{+}_{jm} transforms the state in which the single nucleon jm state is empty to the state in which it is occupied. The inverse transformation is carried out by the hermitean conjugate $(a^{+}_{jm})^{\dagger} = a_{jm}$. Taking the hermitean conjugate of (17.2) we obtain

$$\boxed{a_{jm}a_{j'm'} + a_{j'm'}a_{jm} = \{a_{jm}, a_{j'm'}\} = 0} \tag{17.3}$$

Acting on the vacuum state by any *annihilation operation* a_{jm} we obtain

$$a_{jm}|0\rangle = 0$$

The operator which transforms the empty jm state into an occupied one *and* the occupied jm state into an empty one is thus $a^{+}_{jm} + a_{jm}$. This transformation is unitary, from which follows

$$(a^{+}_{jm} + a_{jm})^{\dagger}(a^{+}_{jm} + a_{jm}) = (a_{jm} + a^{+}_{jm})(a^{+}_{jm} + a_{jm}) = 1$$

From (17.2) and (17.3) follows $(a^{+}_{jm})^2 = a^2_{jm} = 0$ and hence we obtain the anticommutation relation

$$a^{+}_{jm}a_{jm} + a_{jm}a^{+}_{jm} = \{a^{+}_{jm}, a_{jm}\} = 1$$

When $\{a^{+}_{jm}, a_{jm}\}$ acts on any state, if the jm single nucleon state is empty then $a^{+}_{jm}a_{jm}$ has the eigenvalue 0 and $a_{jm}a^{+}_{jm}$ has eigenvalue 1. If the jm single nucleon state is occupied then $a^{+}_{jm}a_{jm}$ has eigenvalue 1 and $a_{jm}a^{+}_{jm}$ eigenvalue 0.

The anticommutation relations of any a^{+}_{jm} and $a_{j'm'}$ are given by

$$\boxed{a^{+}_{jm}a_{j'm'} + a_{j'm'}a^{+}_{jm} = \{a^{+}_{jm}, a_{j'm'}\} = \delta_{jj'}\delta_{mm'}} \qquad (17.4)$$

The anticommutation relation (17.4) for $j' \neq j$ or $m' \neq m$ follows from a simple consideration. When $a^{+}_{jm}a_{j'm'} + a_{j'm'}a^{+}_{jm}$ for $j' \neq j$ or $m' \neq m$, is applied to any state it vanishes unless the $j'm'$ single nucleon state is occupied. Hence, in order to check the validity of (17.4) we should apply it to a state in which the $j'm'$ state is occupied. Such a state is created by applying $a^{+}_{j'm'}$ to a state in which the $j'm'$ state is empty. Hence, we evaluate the operator

$$(a^{+}_{jm}a_{j'm'} + a_{j'm'}a^{+}_{jm})a^{+}_{j'm'} = a^{+}_{jm}a_{j'm'}a^{+}_{j'm'} - a_{j'm'}a^{+}_{j'm'}a^{+}_{jm} \qquad (17.5)$$

applied to some state in which the $j'm'$ single nucleon state is empty. The equality in (17.5) follows from the anticommutation relations

(17.2). The action of $a_{j'm'}a^{+}_{j'm'}$ on any such state simply multiplies it by +1. Thus, (17.5) vanishes in accordance with (17.4).

The anticommutation relation (17.5) together with (17.4) determine the normalization of the state (17.1). Consider first a single nucleon state $a^{+}_{jm}|0\rangle$ for which we obtain

$$\langle 0|(a^{+}_{jm})^{\dagger}a^{+}_{jm}|0\rangle = \langle 0|a_{jm}a^{+}_{jm}|0\rangle = \langle 0|1-a^{+}_{jm}a_{jm}|0\rangle = \langle 0 \mid 0\rangle = 1 \tag{17.6}$$

The state $a^{+}_{jm}|0\rangle$ thus corresponds to the normalized single nucleon wave function ψ_{jm} in configuration space. The same procedure may be used to evaluate the norm of (17.1) given by

$$\begin{aligned} &\langle 0|a_{j_nm_n}\ldots a_{j_2m_2}a_{j_1m_1}a^{+}_{j_1m_1}a^{+}_{a_{j_2m_2}}\cdots a^{+}_{j_nm_n}|0\rangle \\ &\quad = \langle 0|a_{j_nm_n}\ldots a_{j_2m_2}(1-a^{+}_{j_1m_1}a_{j_1m_1})a^{+}_{j_2m_2}\cdots a^{+}_{j_nm_n}|0\rangle \end{aligned} \tag{17.7}$$

The first term on the r.h.s. of (17.7) is the norm of a state with $n-1$ nucleons. The second term vanishes since $a_{j_1m_1}$ may be anticommuted with all other creation operators (all labels jm must be different; otherwise the state (17.1) vanishes due to (17.2)) until it annihilates the vacuum state. This procedure can be repeated until we arrive at (17.6). Hence, the norm of the state (17.1) is unity and it corresponds to the normalized many nucleon wave function (14.2).

Single nucleon operators have a simple form in the formalism of second quantization. They should have non-vanishing matrix elements between states of a single nucleon. Hence, they are given by linear combinations of products of a creation operator and annihilation operator in the normal order $a^{+}_{jm}a_{j'm'}$ (the other order may have non-vanishing vacuum expectation values). The coefficients of these linear combinations are matrix elements of the corresponding operator in configuration space. We thus define

$$\boxed{\mathbf{F} = \sum_{jm,j'm'} \langle jm|\mathbf{F}|j'm'\rangle a^{+}_{jm}a_{j'm'}} \tag{17.8}$$

Matrix elements of this operator between single nucleon states are indeed given by

$$\begin{aligned}\langle 0|a_{j_0m_0}\mathbf{F}a^{+}_{j_0'm_0'}|0\rangle &= \sum_{jmj'm'} \langle jm|\mathbf{F}|j'm'\rangle\langle 0|a_{j_0m_0}a^{+}_{jm}a_{j'm'}a^{+}_{j_0'm_0'}|0\rangle \\ &= \sum_{jmj'm'} \langle jm|\mathbf{F}|j'm'\rangle(\delta_{jj_0}\delta_{mm_0} - a^{+}_{jm}a_{j_0m_0}) \\ &\quad\times(\delta_{j'j_0'}\delta_{m'm_0'} - a^{+}_{j_0'm_0'}a_{j'm'})|0\rangle \\ &= \langle j_0m_0|\mathbf{F}|j_0'm_0'\rangle \end{aligned} \tag{17.9}$$

All other terms on the r.h.s. of (17.9) vanish due to $a_{jm}|0\rangle = 0$ and its hermitean conjugate $\langle 0|a^{+}_{jm} = 0$. Strictly, we should have introduced another symbol for the operator (17.8) which corresponds to the operator $\mathbf{F}$ in configuration space. We use the same symbol to simplify the notation since no actual confusion may be caused by this practice.

Matrix elements of the operator $\mathbf{F}$ given by (17.8) may be calculated between two states (17.1). By using the anticommutation relations the results (14.6) and (14.7) may be derived.

Let us now consider single nucleon operators which play an important role here, namely irreducible tensor operators. According to (17.8) we define the κ component of a rank k tensor by

$$\boxed{T^{(k)}_{\kappa} = \sum_{jmj'm'} \langle jm|T^{(k)}_{\kappa}|j'm'\rangle a^{+}_{jm}a_{j'm'}}$$

In this expression the matrix elements are calculated in configuration space. The operator $T^{(k)}_{\kappa}$ is expressed here in the form of second quantization. Using the Wigner-Eckart theorem (8.4) we obtain

$$T^{(k)}_{\kappa} = \sum_{jmj'm'} (-1)^{j-m}\begin{pmatrix} j & k & j' \\ -m & \kappa & m' \end{pmatrix}(j\|T^{(k)}\|j')a^{+}_{jm}a_{j'm'} \tag{17.10}$$

The hermitean conjugate of (17.10) is equal to

$$(T_\kappa^{(k)})^\dagger = \sum_{jmj'm'} (-1)^{j-m}(j\|T^{(k)}\|j')\begin{pmatrix} j & k & j' \\ -m & \kappa & m' \end{pmatrix} a^\dagger_{j'm'}a_{jm}$$

$$= \sum_{jmj'm'} (j\|T^{(k)}\|j')(-1)^{j-m}\begin{pmatrix} j' & k & j \\ -m' & -\kappa & m \end{pmatrix} a^\dagger_{j'm'}a_{jm}$$

The last equality is due to the symmetry properties (7.24) and (7.28). We recall the relation (8.9) for reduced matrix elements of hermitean tensor operators. If the tensor whose matrix elements appear in (17.10) is hermitean we may use (8.9) and then see that the tensor (17.10) satisfies

$$(T_\kappa^{(k)})^\dagger = (-1)^{m-m'}T_{-\kappa}^{(k)} = (-1)^\kappa T_{-\kappa}^{(k)}$$

which is identical with (8.8) and thus is a component of a hermitean tensor operator.

As in Section 8, we can interpret the expression (17.10) as a certain coupling of $a^\dagger_{jm}$ and $a_{j'm'}$. To make it explicit we replace in (17.10) the $3j$-symbol by the corresponding Clebsch-Gordan coefficient. This yields

$$T_\kappa^{(k)} = \frac{1}{\sqrt{2k+1}}\sum_{jj'}(j\|T^{(k)}\|j') \times \sum_{mm'}(-1)^{j'-m'}(jmj',-m' \mid jj'k\kappa)a^\dagger_{jm}a_{j'm'} \tag{17.11}$$

Each side of (17.11) is the κ component of an irreducible tensor of rank k. Looking at the r.h.s. of (17.11) we see that $a_{j'm'}$ cannot transform under rotations like the m' component of the standard irreducible tensor $a^\dagger_{j'm'}$. This is not surprising since $a^\dagger_{j'm'}|0\rangle$ is a state which transforms like the m' component of an irreducible tensor of rank j'. On the other hand, $a_{j'm'}$ annihilates a nucleon in the state $j'm'$ and hence, adds the value $-m'$ to the total M. Indeed, if we define

$$\boxed{\tilde{a}_{jm} = (-1)^{j+m} a_{j-m}} \tag{17.12}$$

we can rewrite (17.11) as

$$\begin{aligned} T_\kappa^{(k)} &= \frac{1}{\sqrt{2k+1}} \sum_{jj'} (j \| T^{(k)} \| j') \sum_{nm'} (jmj'm' \mid jj'k\kappa) a_{jm}^{+} \tilde{a}_{j'm'} \\ &= \frac{1}{\sqrt{2k+1}} \sum_{jj'} (j \| T^{(k)} \| j') [a_j^{+} \times \tilde{a}_{j'}]_\kappa^{(k)} \end{aligned} \tag{17.13}$$

Thus, (17.12) is the m component of an irreducible tensor of rank j. The phase in (17.12) was chosen to make the overall sign of (17.13) positive.

From (17.13) follows that the unit tensor operators introduced in (8.16) are given in the present formalism by

$$u_\kappa^{(k)} = \frac{1}{\sqrt{2k+1}} ([a_j^{+} \times \tilde{a}_{j'}]_\kappa^{(k)} + (-1)^{j-j'} [a_{j'}^{+} \times \tilde{a}_j]_\kappa^{(k)}) \tag{17.14}$$

We notice that, due to the symmetry properties of the Clebsch-Gordan coefficients, it follows that

$$[(a_j^{+} \times \tilde{a}_{j'})_\kappa^{(k)}]^{\dagger} = (-1)^{j-j'+\kappa} (a_{j'}^{+} \times \tilde{a}_j)_{-\kappa}^{(k)}$$

Hence, the operator (17.14) is hermitean if $(-1)^{j-j'} = 1$ and is antihermitean if $(-1)^{j-j'} = -1$. Only when (17.14) is multiplied by the reduced matrix element of a hermitean tensor $(j \| T^{(k)} \| j')$ it becomes a component of a hermitean tensor. The simplest single nucleon operator is a scalar, $k = 0$. Recalling that

$$(j \| 1 \| j') = \sqrt{2j+1} \delta_{jj'}$$

we obtain by inserting this value in (17.13) the operator

$$\sum_{jm}(-1)^{j-m}a^{+}_{jm}\tilde{a}_{j-m} = \sum_{jm} a^{+}_{jm}a_{jm} = \sum_{j}\hat{n}_j = \hat{n} \tag{17.15}$$

The operator thus obtained is the *number operator*. The eigenvalue of each $a^{+}_{jm}a_{jm}$ is 0 if the jm state is empty and it is equal to 1 if the jm state is occupied. Thus, the eigenvalues of $\hat{n}$ are equal to the total number of nucleons.

Let us calculate the commutation relations of $\hat{n}$ with a^{+}_{jm} and a_{jm}. We first realize that the product $a^{+}_{jm}a_{jm}$ commutes with $a^{+}_{j'm'}$ or $a_{j'm'}$ if $j' \neq j$ or $m' \neq m$. This is due to the anticommutation relations (17.2) and (17.5). We obtain then

$$[\hat{n}, a^{+}_{jm}] = [a^{+}_{jm}a_{jm}, a^{+}_{jm}] \qquad [\hat{n}, a_{jm}] = [a^{+}_{jm}a_{jm}, a_{jm}] \tag{17.16}$$

To evaluate the commutators in (17.16) it is convenient to use the formulae

$$\begin{aligned} [ab,c] &= abc - cab = abc + acb - acb - cab = a\{b,c\} - \{a,c\}b \\ [a,bc] &= abc - bca = abc + bac - bac - bca = \{a,b\}c - b\{a,c\} \end{aligned} \tag{17.17}$$

Using (17.17) we obtain the following relations

$$[\hat{n}, a^{+}_{jm}] = a^{+}_{jm} \qquad [\hat{n}, a_{jm}] = -a_{jm} \tag{17.18}$$

which express the fact that a^{+}_{jm} is a creation operator and a_{jm} is an annihilation operator.

A single nucleon operator of great importance is the angular momentum vector **J**. It is a tensor operator of rank 1 and has non-vanishing reduced matrix elements between states with the same value of J, given by (8.13). According to (17.13) we obtain for J the expression

$$\boxed{\mathbf{J} = \frac{1}{\sqrt{3}} \sum_j \sqrt{j(j+1)(2j+1)}[a_j^\dagger \times \tilde{a}_j]^{(1)}} \tag{17.19}$$

The operators $\sqrt{j(j+1)(2j+1)/3}[a_j^\dagger \times \tilde{a}_j]^{(1)}$ are representations of $\mathbf{J}$ in the $2j+1$ dimensional space of a single j-nucleon. It is a simple excercise to calculate the commutation relations between components of (17.19) and $a_{jm}^\dagger$, $\tilde{a}_{jm}$. Using the formulae (17.17) we obtain

$$[J^0, a_{jm}^\dagger] = m a_{jm}^\dagger \qquad [J^\pm, a_{jm}^\dagger] = \sqrt{j(j+1) - m(m\pm 1)}\, a_{jm\pm 1}^\dagger$$

$$[J^0, \tilde{a}_{jm}] = m \tilde{a}_{jm} \qquad [J^\pm, \tilde{a}_{jm}] = \sqrt{j(j+1) - m(m\pm 1)}\, \tilde{a}_{jm\pm 1} \tag{17.20}$$

The commutation relations (17.20) are the analogs of (6.25). They clearly demonstrate that $a_{jm}^\dagger$ and $\tilde{a}_{jm}$ transform irreducibly under rotations. They are irreducible tensor operators of rank j. In Section 6 the tensor operators we considered had only integral ranks k. Those operators have non-vanishing matrix elements only between two states with the same number of nucleons. The latter may have either integral or half integral values of J. The operators $a_{jm}^\dagger$ and $\tilde{a}_{jm}$ change the number of nucleons by one and hence have non-vanishing matrix elements between two states, one of which has an integral value of J and the other half integral J value. In view of (17.20), properties of tensor operators which follow from their transformation properties, like Wigner-Eckart theorem, are shared also by these tensor operators.

As mentioned above, the calculation of matrix elements of single nucleon operators between states (17.1) is straightforward. We should remember, however, that the states (17.1) like the states (14.2) do not usually have a definite value of angular momentum. In order to obtain eigenstates of the nuclear Hamiltonian we must take, in general, linear combinations of the states (17.1). We repeat here briefly the considerations presented in the beginning of Section 15. Using Clebsch-Gordan coefficients we can construct states with definite angular momentum by successive couplings of creation operators

$$[\cdots[[a_{j_1}^\dagger \times a_{j_2}^\dagger]^{(J_{12})} \times a_{j_3}^\dagger]^{(J_{123})} \times \cdots]_M^{(J)}|0\rangle \tag{17.21}$$

As long as all the j-orbits in (17.21) are different, this set of states has very nice properties. The states (17.21) are antisymmetric due to the anticommutation relations of the a^{+}_{jm}. All states obtained by combining the various j_i to all possible intermediate states do not vanish. They are normalized to 1 due to the properties of the vector addition coefficients. Any two such states which differ by any of the intermediate angular momenta $J_{12}, J_{123}, \ldots$ or J and M are orthogonal.

These nice properties of the states (17.21) do not persist if some of the j-orbits are identical. Some of the states may vanish and the intermediate angular momenta no longer uniquely determine orthogonal states. If we consider two identical nucleons in the same j-orbit there is no difficulty. The states

$$\boxed{A^{+}(j^2JM)|0\rangle = \frac{1}{\sqrt{2}}[a^{+}_{j} \times a^{+}_{j}]^{(J)}_{M}|0\rangle} \tag{17.22}$$

for even values of J ($J = 0, 2, 4, \ldots, 2j-1$) exist and are normalized, whereas states with odd values of J vanish. The normalization factor $2^{-\frac{1}{2}}$ is due to the fact that both nucleons are in the same j-orbit. In other cases, $j_1 \neq j_2$ the following state is normalized to 1

$$\boxed{\begin{aligned} A^{+}(j_1j_2JM)|0\rangle &= (a^{+}_{j_1} \times a^{+}_{j_2})^{(J)}_{M}|0\rangle \\ &= \sum (j_1m_1j_2m_2 \mid j_1j_2JM)a^{+}_{j_1m_1}a^{+}_{j_2m_2}|0\rangle \end{aligned}} \tag{17.23}$$

To show how the normalization facors of the operators in (17.22) and (17.23) were obtained we evaluate the commutator

$$\begin{aligned} &[(A^{+}(j_3j_4J'M'))^{\dagger}, A^{+}(j_1j_2JM)] \\ &\quad = \sum (j_3m_3j_4m_4 \mid j_3j_4J'M')(j_1m_1j_2m_2 \mid j_1j_2JM) \\ &\quad \times [a_{j_4m_4}a_{j_3m_3}, a^{+}_{j_1m_1}a^{+}_{j_2m_2}] \end{aligned}$$

The commutator on the r.h.s. may be evaluated by using (17.17) as follows

$$
\begin{aligned}
&[a_{j_4m_4}a_{j_3m_3}, a^+_{j_1m_1}a^+_{j_2m_2}] \\
&\quad = a_{j_4m_4}[a_{j_3m_3}, a^+_{j_1m_1}a^+_{j_2m_2}] + [a_{j_4m_4}, a^+_{j_1m_1}a^+_{j_2m_2}]a_{j_3m_3} \\
&\quad = a_{j_4m_4}\delta_{j_1j_3}\delta_{m_1m_3}a^+_{j_2m_2} - a_{j_4m_4}a^+_{j_1m_1}\delta_{j_2j_3}\delta_{m_2m_3} \\
&\qquad + \delta_{j_1j_4}\delta_{m_1m_4}a^+_{j_2m_2}a_{j_3m_3} - a^+_{j_1m_1}\delta_{j_2j_4}\delta_{m_2m_4}a_{j_3m_3} \\
&\quad = \delta_{j_1j_3}\delta_{m_1m_3}\delta_{j_2j_4}\delta_{m_2m_4} - \delta_{j_1j_3}\delta_{m_1m_3}a^+_{j_2m_2}a_{j_4m_4} \\
&\qquad - \delta_{j_2j_3}\delta_{m_2m_3}\delta_{j_1j_4}\delta_{m_1m_4} + \delta_{j_2j_3}\delta_{m_2m_3}a^+_{j_1m_1}a_{j_4m_4} \\
&\qquad + \delta_{j_1j_4}\delta_{m_1m_4}a^+_{j_2m_2}a_{j_3m_3} - \delta_{j_2j_4}\delta_{m_2m_4}a^+_{j_1m_1}a_{j_3m_3}
\end{aligned}
$$

Substituting these expressions and using the orthogonality of Clebsch-Gordan coefficients we obtain

$$
\begin{aligned}
&[(A^+(j_3j_4J'M'))^\dagger, A^+(j_1j_2JM)] \\
&\quad = (\delta_{j_1j_3}\delta_{j_2j_4} - (-1)^{j_1+j_2-J'}\delta_{j_1j_4}\delta_{j_2j_3})\delta_{JJ'}\delta_{MM'} \\
&\qquad - \delta_{j_1j_3}\sum(j_1m_1j_4m_4 \mid j_1j_4J'M')(j_1m_1j_2m_2 \mid j_1j_2JM)a^+_{j_2m_2}a_{j_4m_4} \\
&\qquad + \delta j_2j_3\sum(j_2m_2j_4m_4 \mid j_2j_4J'M')(j_1m_1j_2m_2 \mid j_1j_2JM)a^+_{j_1m_1}a_{j_4m_4} \\
&\qquad + \delta_{j_1j_4}\sum(j_3m_3j_1m_1 \mid j_3j_1J'M')(j_1m_1j_2m_2 \mid j_1j_1JM)a^+_{j_2m_2}a_{j_3m_3} \\
&\qquad - \delta_{j_2j_4}\sum(j_3m_3j_2m_2 \mid j_3j_2J'M')(j_1m_1j_2m_2 \mid j_1j_2JM)a^+_{j_1m_1}a_{j_3m_3}
\end{aligned}
\tag{17.24}
$$

The norm of the state (17.23) is obtained by putting $j_3 = j_1$, $j_4 = j_2$ and taking the vacuum expectation value of (17.24). Hence, for different orbits j_1 and j_2 the normalization factor is 1. If, however, both nucleons are in the same orbit, the expectation value of (17.24) becomes equal to 2 for even values of J leading to the factor $2^{-1/2}$ in (17.22). For odd values of J, (17.24) vanishes if $j_1 = j_2$. The equality in (17.24) follows from the orthogonality and normalization properties of the vector addition coefficients (7.9) and their symmetry property (7.12).

If we couple a third nucleon we may obtain fully antisymmetric states with total J by

$$\left(\sum_{M_0 m}(J_0M_0jm \mid J_0jJM)A^+(j^2JM)a^+_{jm}\right)|0\rangle \tag{17.25}$$

The state (17.25) may vanish if an antisymmetric state with given J does not exist in the j^3 configuration (like $J=\frac{1}{2},\frac{7}{2}$ for $j=\frac{5}{2}$). If the state (17.25) does not vanish, its normalization must be calculated. The properties of Clebsch-Gordan coefficients no longer guarantee that the norm is equal to 1. The calculation of matrix elements of single nucleon operators (and two-body interactions) between states (17.25) with $n>3$ is usually not very convenient. There are special cases which are very important where this problem can be easily solved. In general, however, such calculations become complex and difficult.

In Section 15 we saw that the use of c.f.p. greatly facilitates the calculation of matrix elements of single nucleon operators and two-body interactions in j^n configurations. Let us make the connection between c.f.p. and the formalism of second quantization. It will turn out that this connection is important for many applications.

Starting with the definition (15.25) of the general c.f.p. we can use the projection operator $\mathcal{A}/n$ to obtain, in the same way as (15.26) was obtained, the relation

$$\begin{aligned}[j^{n-1}(\alpha_1J_1)jJ|\}j^n\alpha J] &= \int \psi^*(j^n\alpha JM)\psi(j^{n-1}(\alpha_1J_1)j_nJM)\\ &= \int\left\{\frac{\mathcal{A}}{n}\psi^*(j^n\alpha JM)\right\}\psi(j^{n-1}(\alpha_1J_1)j_nJM)\\ &= \frac{1}{n}\int \psi^*(j^n\alpha JM)\mathcal{A}\psi(j^{n-1}(\alpha_1J_1)j_nJM)\end{aligned} \tag{17.26}$$

The integral on the r.h.s. of (17.26) should be related to the matrix element

$$\langle 0|A(n,\alpha JM)(A^+(n-1,\alpha_1J_1)\times a^+_j)^{(J)}_M|0\rangle \tag{17.27}$$

where $A^+(n,\alpha JM)$ creates a fully antisymmetric state of n nucleons and $(A^+(n-1,\alpha_1 J_1)\times a_j^+)_M^{(J)}$ corresponds to the antisymmetrized state $\mathcal{A}\psi(j^{n-1}(\alpha_1 J_1)j_n JM)$.

Before equating (17.26) and (17.27), however, we must take into account the normalization of states in the two formalisms. To do this it is sufficient to compare states corresponding to products of single nucleon states. All states can be constructed by taking linear combinations of such products. The antisymmetrization of a product wave function by

$$\frac{1}{n}\mathcal{A}\psi_{\alpha_1}(1)\psi_{\alpha_2}(2)\ldots\psi_{\alpha_n}(n) \tag{17.28}$$

leads to a linear combination of n orthogonal functions and the normalization coefficient of (17.28) is given by $(n/n^2)^{-1/2}=\sqrt{n}$. On the other hand, the normalization of the corresponding state $a^+_{\alpha_1}a^+_{\alpha_2}\ldots a^+_{\alpha_n}|0\rangle$ is simply 1. Hence, the c.f.p. in (17.26) is equal to the matrix element (17.27) divided by $\sqrt{n}$. We thus obtain

$$\begin{aligned}
&[j^{n-1}(\alpha_1 J_1)jJ|\}j^n\alpha J]\\
&\quad=\frac{1}{\sqrt{n}}\langle 0|A(n,\alpha JM)(A^+(n-1,\alpha_1 J_1)\times a_j^+))_M^{(J)}|0\rangle\\
&\quad=\frac{1}{\sqrt{n}}\sum_{M_1 m}(J_1M_1jm\mid J_1jJM)\\
&\qquad\times\langle 0|A(n,\alpha JM)A^+(n-1,\alpha_1J_1M_1)a^+_{jm}|0\rangle\\
&\quad=\frac{(-1)^{n-1}}{\sqrt{n}}\sum_{M_1 m}(J_1M_1jm\mid J_1jJM)\\
&\qquad\times\langle 0|A(n,\alpha JM)a^+_{jm}A^+(n-1,\alpha_1J_1M_1)|0\rangle\\
&\quad=\frac{(-1)^{n-1}}{\sqrt{n}}\sum_{M_1 m}(J_1M_1jm\mid J_1jJM)\langle j^n\alpha JM|a^+_{jm}|j^{n-1}\alpha_1J_1M_1\rangle
\end{aligned} \tag{17.29}$$

The phase $(-1)^{n-1}$ is due to anticommuting a^+_{jm} through the $n-1$ creation operators in $A^+(n-1,\alpha_1 J_1 M_1)$.

The matrix elements on the r.h.s. of (17.29) are matrix elements of the irreducible tensor operators a^+_{jm} of rank j. The sum over M_1

and m yields the reduced matrix element of $a_j^\dagger$. Summing the r.h.s. of (17.29) over M and dividing by $2J+1$ we can use (8.11). Taking into account the summation over m, yielding a factor $2j+1$, we can express (17.29) in the form

$$[j^{n-1}(\alpha_1 J_1)jJ|\}j^n\alpha J] = \frac{(-1)^n}{\sqrt{n}}\frac{1}{\sqrt{2J+1}}(j^n\alpha J\|a_j^\dagger\|j^{n-1}\alpha_1 J_1) \tag{17.30}$$

Two-body interactions can also be conveniently expressed in terms of creation and annihilation operators. A true two-body operator should have vanishing vacuum expectation values and vanishing matrix elements between single nucleon states. It must then be a linear combination of products of two creation operators and two annihilation operators in the normal order. We consider rotationally invariant interactions and it is therefore convenient to express the two-body operators by using the operators (17.22) and (17.23) and their hermitean conjugates. We thus use

$$\begin{aligned} A^+(j_1j_2JM) &= \left(1+\delta_{j_1j_2}\right)^{-1/2}(a_{j_1}^\dagger \times a_{j_2}^\dagger)_M^{(J)} \\ &= (1+\delta_{j_1j_2})^{-1/2}\sum(j_1m_1j_2m_2 \mid j_1j_2JM)a_{j_1m_1}^\dagger a_{j_2m_2}^\dagger \end{aligned} \tag{17.31}$$

and its hermitean conjugate

$$\begin{aligned} [A^+(j_1j_2JM)]^\dagger &= (1+\delta_{j_1j_2})^{-1/2}\sum(j_1m_1j_2m_2 \mid j_1j_2JM)a_{j_2m_2}a_{j_1m_1} \\ &= A(j_1j_2JM) \end{aligned} \tag{17.32}$$

The pair annihilation operator $A(j_1j_2JM)$ is defined by (17.32) and hence, the order of the annihilation operators in it is not the one in which their quantum numbers appear in the vector addition coefficients. Due to (17.24), the operators (17.31) acting on the vacuum create a state which is normalized to unity. It is, of course, antisymmetric and has angular momentum J and z-projection M.

A two-body operator which has non-vanishing two nucleon matrix elements between states with definite J and M may be expressed as

$$\boxed{\mathbf{G} = \sum_{j_1 \leq j_2, j'_1 \leq j'_2 JM} V_J(j_1 j_2 j'_1 j'_2) A^+(j_1 j_2 JM) A(j'_1 j'_2 JM)} \tag{17.33}$$

The matrix elements of **G** between such states are independent of M and are given by $V_J(j_1 j_2 j'_1 j'_2)$. These are matrix elements between two antisymmetric and normalized states of identical nucleons and include both the direct term and exchange term. Indeed, calculating matrix elements of (17.33) between the states

$$A^+(jj'J_0M_0)|0\rangle \qquad \text{and} \qquad A^+(j_0 j'_0 J_0 M_0|0\rangle$$

we obtain due to (17.24) the result $V_{J_0}(j_0 j'_0 j j')$. The restriction $j_1 \leq j_2$, $j'_1 \leq j'_2$ is due to the fact that

$$A^+(j_2 j_1 JM) = (-1)^{j_1+j_2-J+1} A^+(j_1 j_2 JM)$$

Both operators create the same state and only one of them should be considered. Due to the restriction $j_1 \leq j_2$, $j'_1 \leq j'_2$ there is only one contribution of $V_J(j_1 j_2 j'_1 j'_2)$ to the matrix element between given two states. In writing $j_1 \leq j_2$, $j'_1 \leq j'_2$ we assume that no two different orbits have the same value of j. This is usually the case since we calculate the matrix of **G** in some shell model subspace with only a few configurations. If, however, there are two different orbits with the same value of j, we introduce some order among them (e.g. with respect to n and l). Looking at (17.33) we verify that **G** is hermitean provided $V_J(j_1 j_2 j'_1 j'_2) = V_J(j'_1 j'_2 j_1 j_2)$.

If we remove the restrictions $j_1 \leq j_2$, $j'_1 \leq j'_2$, matrix elements of G between any two states with $j_1 \neq j_2$, $j'_1 \neq j'_2$ will be multiplied by 4. For $j_1 = j_2$, $j'_1 \neq j'_2$ or $j_1 \neq j_2$, $j'_1 = j'_2$, there will be only two contributions to the matrix elements, e.g. $V_J(j_1 j_2 j^2)$ and $V_J(j_2 j_1 j^2)$. For $j_1 = j_2$ and $j'_1 = j'_2$ there will be only one contribution, $V_J(j_1^2, j_1'^2)$. Hence, we can express the two-body operator (17.33) by

$$\boxed{\mathbf{G}=\frac{1}{4}\sum_{j_1j_2j_1'j_2'JM}(1+\delta_{j_1j_2})(1+\delta_{j_1'j_2'})V_J(j_1j_2j_1'j_2') \times A^+(j_1j_2JM)A(j_1'j_2'JM)} \tag{17.34}$$

The matrix elements $V_J(j_1j_2j_1'j_2')$ between antisymmetric and normalized states were described in Sections 10 and 12. Here we only relate them to matrix elements of **G** between antisymmetric (and normalized) states in the m-scheme. In configuration space a state with $j_1 \neq j_2$ is given by

$$\sum_{m_1m_2}(j_1m_1j_2m_2 \mid j_1j_2JM)\frac{1}{\sqrt{2}}[\psi_{j_1m_1}(1)\psi_{j_2m_2}(2)-\psi_{j_2m_2}(1)\psi_{j_1m_1}(2)] \tag{17.35}$$

If, however, $j_1 = j_2$ (and also $n_1 = n_2$, $l_1 = l_2$) the normalized state

$$\sum_{m_1m_2}(jm_1jm_2 \mid jjJM)\psi_{jm_1}(1)\psi_{jm_2}(2) \tag{17.36}$$

is antisymmetric if J is even. In this case it can be written as

$$\frac{1}{\sqrt{2}}\sum_{m_1m_2}(jm_1jm_2 \mid jjJM)\frac{1}{\sqrt{2}}[\psi_{jm_1}(1)\psi_{jm_2}(2)-\psi_{jm_2}(1)\psi_{jm_1}(2)] \qquad J \text{ even} \tag{17.37}$$

Hence, if we calculate matrix elements of **G** between states (17.35) we obtain

$$V_J(j_1j_2j_1'j_2') = \sum_{m_1m_2m_1'm_2'}(j_1m_1j_2m_2 \mid j_1j_2JM)(j_1'm_1'j_2'm_2' \mid j_1'j_2'JM) \times \langle j_1m_1j_2m_2|\mathbf{G}|j_1'm_1'j_2'm_2'\rangle_a \tag{17.38}$$

The m-scheme matrix elements of **G** in (17.38) are written with a subscript a since they are taken between antisymmetric and normalized states. Similarly, matrix elements of **G** between a state (17.35) and a state (17.37) for even values of J are given by

$$V_J(j_1j_2j^2) = \frac{1}{\sqrt{2}} \sum_{m_1m_2m_1'm_2'} (j_1m_1j_2m_2 \mid j_1j_2JM)(jm_1'jm_2' \mid jjJM) \times \langle j_1m_1j_2m_2|\mathbf{G}|jm_1'jm_2'\rangle_a \tag{17.39}$$

Finally, matrix elements between two states (17.37) are equal to

$$V_J(j^2j^2) = \frac{1}{2} \sum_{m_1m_2m_1'm_2'} (jm_1jm_2 \mid jjJM)(jm_1'jm_2' \mid jjJM) \times \langle jm_1jm_2|\mathbf{G}|jm_1'jm_2'\rangle_a \tag{17.40}$$

We can now express (17.33) in the m-scheme. Substituting in (17.33) the definitions (17.31), (17.32) and the results (17.38), (17.39) and (17.40), we obtain by using the orthogonality relations of Clebsch-Gordan coefficients the expression

$$\begin{aligned}\mathbf{G} = \sum_{m_1m_2m_1'm_2'} \Bigg[& \sum_{j_1<j_2, j_1'<j_2'} \langle j_1m_1j_2m_2|\mathbf{G}|j_1'm_1'j_2'm_2'\rangle_a \\ & \times a^+_{j_1m_1}a^+_{j_2m_2}a_{j_2'm_2'}a'_{j_1m_1} \\ & + \frac{1}{2} \sum_{j_1<j_2 j} \langle j_1m_1j_2m_2|\mathbf{G}|jm_1'jm_2'\rangle_a a^+_{j_1m_1}a^+_{j_2m_2}a_{jm_2'}a_{jm_1'} \\ & + \frac{1}{2} \sum_{j, j_1<j_2} \langle jm_1jm_2|\mathbf{G}|j_1m_1'j_2m_2'\rangle_a a^+_{jm_1}a^+_{jm_2}a_{j_2m_2'}a_{j_1m_1'} \\ & + \frac{1}{4} \sum_j \langle jm_1jm_2|\mathbf{G}|jm_1'jm_2'\rangle_a a^+_{jm_1}a^+_{jm_2}a_{jm_2'}a_{jm_1'} \Bigg]\end{aligned} \tag{17.41}$$

The expression (17.41) may be simplified by removing the restriction $j_1 \leq j_2$, $j_1' \leq j_2'$. The first term in (17.41) is then multiplied by $2 \times 2 = 4$ and the second and third terms are multiplied by 2. We thus obtain

$$\mathbf{G} = \frac{1}{4} \sum_{j_1 j_2 j_1' j_2'} \sum_{m_1 m_2 m_1' m_2'} \langle j_1 m_1 j_2 m_2 | \mathbf{G} | j_1' m_1' j_2' m_2' \rangle_a \times a_{j_1 m_1}^+ a_{j_2 m_2}^+ a_{j_2' m_2'} a_{j_1' m_1'} \tag{17.42}$$

In (17.42) always $m_1 \neq m_2$ and $m_1' \neq m_2'$ and hence, there is no distinction between the cases where $j_1 = j_2$ or $j_1' = j_2'$ and the other cases.

The expression (17.33) for identical nucleons can be directly generalized to the case of isospin formalism for protons and neutrons. We first introduce creation and annihilation operators specified not only by j (which includes n and l) and m but also by the isospin quantum number μ. As introduced in Section 9, μ is the eigenvalue of t_0. It is equal to $+\frac{1}{2}$ for a state of a proton and $-\frac{1}{2}$ for a state of a neutron.

Due to the generalized Pauli principle, these operators satisfy anticommutation relations similar to those of (17.2), (17.3) and (17.4). These are given by

$$\{a_{jm\mu}^+, a_{j'm'\mu'}\} = \delta_{jj'} \delta_{mm'} \delta_{\mu\mu'}$$
$$\{a_{jm,\mu}^+, a_{j'm'\mu'}^+\} = \{a_{jm\mu}, a_{j'm'\mu'}\} = 0 \tag{17.43}$$

The components of the isospin vector in the present formalism are given by the operators

$$T^0 = \sum_{jm\mu} \mu a_{jm\mu}^+ a_{jm\mu} = \frac{1}{2} \sum_{jm} (a_{jm1/2}^+ a_{jm1/2} - a_{jm,-1/2}^+ a_{jm,-1/2})$$
$$T^+ = \sum_{jm} a_{jm1/2}^+ a_{jm,-1/2} \qquad T^- = \sum_{jm} a_{jm,-1/2}^+ a_{jm1/2} \tag{17.44}$$

This form of $\mathbf{T}$ can be verified by calculating the commutators of the operators in (17.44) and their action on a proton state and a neutron state (defined by $a_{jm1/2}^+|0\rangle$ and $a_{jm,-1/2}^+|0\rangle$ respectively).

In analogy with (17.31) we define the operator creating a two nucleon state with given J, M, T and M_T as

$$A^+(jj'TM_TJM) = (1+\delta_{jj'})^{-1/2} \sum_{mm'\mu\mu'} (jmj'm' \mid jj'JM) \times (\tfrac{1}{2}\mu\tfrac{1}{2}\mu' \mid \tfrac{1}{2}\tfrac{1}{2}TM_T) a^+_{jm\mu} a^+_{j'm'\mu'} \tag{17.45}$$

Using the operator (17.45) and its hermitean conjugate

$$A(jj'TM_TJM),$$

we can express any charge independent two nucleon interaction by the analog of (17.33) as

$$\mathbf{G} = \sum_{j_1 \leq j_2, j'_1 \leq j'_2 TM_TJM} V_{TJ}(j_1j_2j'_1j'_2) A^+(j_1j_2TM_TJM) \times A(j'_1j'_2TM_TJM) \tag{17.46}$$

To verify that the operator (17.46) has the matrix elements

$$V_{TJ}(j_1j_2j'_1j'_2)$$

between two states with given T, M_T and J, M we calculate the following expression

$$\begin{aligned} A(j'_1j'_2T'M'_TJ'M')A^+(jj'TM_TJM)|0\rangle &= [(1+\delta_{j'_1j'_2})(1+\delta_{jj'})]^{-1/2} \\ &\times \sum_{\substack{m'_1m'_2mm' \\ \mu'_1\mu'_2\mu\mu'}} (j'_1m'_1j'_2m'_2 \mid j'_1j'_2J'M)(jmj'm' \mid jj'JM) \\ &\times (\tfrac{1}{2}\mu'_1\tfrac{1}{2}\mu'_2 \mid \tfrac{1}{2}\tfrac{1}{2}T'M'_T)(\tfrac{1}{2}\mu\tfrac{1}{2}\mu' \mid \tfrac{1}{2}\tfrac{1}{2}TM_T) \\ &\times a_{j'_2m'_2\mu'_2} a_{j'_1m'_1\mu'_1} a^+_{jm\mu} a^+_{j'm'\mu'} |0\rangle \end{aligned} \tag{17.47}$$

Anticommuting the annihilation operators to the right, or using (17.17) we obtain for the product of annihilation and creation operators operating on the vacuum the result

$$(\delta_{jj_1'}\delta_{mm_1'}\delta_{\mu\mu_1'}\delta_{j'j_2'}\delta_{m'm_2'}\delta_{\mu'\mu_2'} - \delta_{jj_2'}\delta_{mm_2'}\delta_{\mu\mu_2'}\delta_{j'j_1'}\delta_{m'm_1'}\delta_{\mu'\mu_1'})|0\rangle \tag{17.48}$$

Recalling the restriction in (17.46), $j_1' \leq j_2'$, and using the convention $j \leq j'$, we see that if $j_1' = j_2' = j = j'$ both terms in (17.48) do not vanish. In any other case, only the first term in (17.48) does not vanish. Substituting (17.48) into (17.47) and carrying out the summations we obtain, in the case of non-identical orbits, due to the orthogonality relations of the vector addition coefficients, the result

$$\begin{aligned} &A(j_1'j_2'T'M_T'J'M')A^+(jj'TM_TJM)|0\rangle \\ &\quad = \delta_{jj_1'}\delta_{j'j_2'}\delta_{T'T}\delta_{M_T'M_T}\delta_{J'J}\delta_{M'M}|0\rangle \end{aligned} \tag{17.49}$$

If $j_1' = j_2' = j' = j$ the r.h.s. of (17.49) is equal to

$$\tfrac{1}{2}(1-(-1)^{T+J})\delta_{T'T}\delta_{M_T'M_T}\delta_{J'J}\delta_{M'M}|0\rangle$$

Hence, using the result (17.49) and its hermitean conjugate relation, we see that **G** given by (17.46) yields the correct two nucleon matrix elements.

The state $A^+(jj'TM_TJM)|0\rangle$ describes two nucleons in the j and j' orbits outside closed shells with isospin T. If both nucleons are protons or neutrons, the state has $T = 1$ and $M_T = 1$ or $M_T = -1$ respectively. If there are a proton and a neutron outside closed shells, $M_T = 0$, and T may be either 1 or 0. We may think of a state in which the proton is in the j-orbit and the neutron in the j'-orbit. If the Hamiltonian is charge independent, the state with a neutron in the j-orbit and a proton in the j'-orbit has the same expectation value. A charge independent two nucleon interaction will mix these two states and yield two orthogonal eigenstates which have $T = 0$ and $T = 1$. This is the situation in light nuclei, for instance in ${}^{42}_{21}\mathrm{Sc}_{21}$. The states in which one valence nucleon is in the $1f_{7/2}$ orbit and the other in the higher $2p_{3/2}$ orbit are such $T = 0$ and $T = 1$ states.

There are however nuclear states in which it is known that the proton and the neutron are in definite orbits. For example, consider the ground state of ${}^{50}_{21}\mathrm{Sc}_{29}$. The valence proton is in its lowest state which is in the $1f_{7/2}$ orbit and the valence neutron cannot enter the full $1f_{7/2}$ neutron orbit and must occupy a higher orbit like the $2p_{3/2}$ one. This situation is not in contradiction with charge independence. The system is no longer that of two nucleons but in the correct isospin formalism we must consider 10 nucleons. Yet in Section 9 we wrote a simple two nucleon wave function for such a situation and in Section 10 we showed how to calculate energy levels of such a configuration. We will now justify formally those results. First we show that such states and similar ones have indeed definite isospins.

Consider low lying levels of a nucleus in which valence protons occupy the $j_1, j_1', \ldots$ orbits whereas these orbits are fully occupied by neutrons and valence neutrons are in higher orbits $j_2, j_2', \ldots$ in which there are no protons. All these states have the *same* value of the total isospin

$$T = |M_T| = \tfrac{1}{2}(N - Z) \tag{17.50}$$

To prove (17.50) we first recall that $T \geq |M_T| = (N - Z)/2$. We can apply to all the states thus characterized, the operator T^- which changes one of the protons into a neutron. Since all protons are in orbits which are fully occupied by neutrons, due to the Pauli principle, the operation with T^- annihilates the states. Hence, the lowest possible M_T value of these states is $M_T = -(N - Z)/2$. Thus T cannot be larger than $(N - Z)/2$ from which follows (17.50). Another way to see that $T \leq (N - Z)/2$ is by coupling the isospins T_j of the various j-orbits. In any j-orbit, fully occupied by neutrons, the isospin must have the value $T_j = (2j + 1 - Z_j)/2$ where Z_j is the number of j-protons. In each neutron valence orbit the isospin is equal to one half of N_j— the number of j-neutrons. The total isospin cannot exceed the sum of all these T_j isospins. Hence,

$$T \leq \sum T_j = \sum_{j_1} \frac{1}{2}(2j_1 + 1 - Z_j) + \sum_{j_2} \frac{1}{2} N_{j_2} = \frac{1}{2}(N - Z)$$

which also yields (17.50). Since all these states have the same total isospin, charge independence does not lead to a finer classification of states. Thus, there is no advantage in using the isospin formalism in these cases.

Let us now use (17.46) to calculate the expectation values of the interaction between a j-proton and a j'-neutron outside the fully occupied neutron j-orbit. For simplicity, we take j' to be smaller than j. If the opposite is true, we can impose in (17.46) the restriction

$j_1 \geq j_2$, $j_1' \geq j_2'$ without changing any physical consequence. The relevant part of (17.46) which contributes to the configuration considered, is given by $j_1 = j_1' = j'$, $j_2 = j_2' = j$. The interactions between the j proton and the full j-neutron orbit is a constant term whose value was calculated in Section 14. The operator acting on the core which has $T' = (2j+1)/2$ and $M_{T'} = -(2j+1)/2$ to create the j-proton j'-neutron state with given J, M, is given by

$$(a^+_{j',-1/2} \times a^+_{j1/2})^{(J)}_M = \sum_{mm'} (j'm'jm \mid j'jJM) a^+_{j'm',-1/2} a^+_{jm1/2} \tag{17.51}$$

The state thus constructed is the only state with given J, M of the $j^{2j+2}j'$ configuration with $T = -M_T = (2j+1)/2$. It is therefore an eigenstate of the submatrix of the Hamiltonian defined by that configuration. To calculate the eigenvalue of the two nucleon interaction we apply to that state the relevant part of (17.46) obtaining

$$\sum_{TM_TJ_1M_1} V_{TJ_1}(j'jj'j) A^+(j'jTM_TJ_1M_1) A(j'jTM_TJ_1M_1) \times (a^+_{j',-1/2} \times a^+_{j1/2})^{(J)}_M |c\rangle \tag{17.52}$$

where $|c\rangle$ denotes the core state. Let us first evaluate the expression

$$\begin{aligned} A(j'jTM_TJ_1M_1)&(a^+_{j',-1/2} \times a^+_{j1/2})^{(J)}_M |c\rangle \\ &= \sum_{\substack{m_1'm_1m'm \\ \mu'\mu}} (j'm_1'jm_1 \mid j'jJ_1M_1)(j'm'jm \mid j'jJM)(\tfrac{1}{2}\mu'\tfrac{1}{2}\mu \mid \tfrac{1}{2}\tfrac{1}{2}TM_T) \\ &\quad \times a_{jm_1\mu} a_{j'm_1'\mu'} a^+_{j'm',-1/2} a^+_{jm1/2} |c\rangle \\ &= \sum (j'm_1'jm_1 \mid j'jJ_1M_1)(j'm'jm \mid j'jJM)(\tfrac{1}{2}\mu'\tfrac{1}{2}\mu \mid \tfrac{1}{2}\tfrac{1}{2}TM_T) \\ &\quad \times (a_{jm_1\mu} a^+_{jm1/2} \delta_{m_1'm'} \delta_{-1/2,\mu'} + a_{jm_1\mu} a^+_{j'm',-1/2} a^+_{jm1/2} a_{j'm_1'\mu'}) |c\rangle \end{aligned} \tag{17.53}$$

The second term on the r.h.s. of (17.53) yields zero since there are no protons or neutrons in the j'-orbit of the core. Carrying out the summation over m_1' and μ' we substitute (17.53) into (17.52) and obtain

$$\sum V_{TJ_1}(j'jj'j)(j'm_0'jm_0 \mid j'jJ_1M_1)(j'm'jm_1 \mid j'jJ_1M_1) \times (j'm'jm \mid j'jJM)(\tfrac{1}{2}\mu_0'\tfrac{1}{2}\mu_0 \mid \tfrac{1}{2}\tfrac{1}{2}TM_T)(\tfrac{1}{2},-\tfrac{1}{2}\tfrac{1}{2}\mu \mid \tfrac{1}{2}\tfrac{1}{2}TM_T) \times a^{+}_{j'm_0'\mu_0'}a^{+}_{jm_0\mu_0}a_{jm_1\mu}a^{+}_{jm1/2}|c\rangle \tag{17.54}$$

Let us now see for which values of μ_0' the expression (17.54) does not vanish. For $\mu_0' = \frac{1}{2}$ we obtain $\mu_0 = M_T - \frac{1}{2}$ and $\mu = M_T + \frac{1}{2}$. Since $|\mu_0| = |\mu| = \frac{1}{2}$, it follows that $M_T = 0$, $\mu = \frac{1}{2}$ and $\mu_0 = -\frac{1}{2}$. In that case however

$$a^{+}_{j'm_0'1/2}a^{+}_{jm_0,-1/2}a_{jm_11/2}a^{+}_{jm1/2}|c\rangle = a^{+}_{j'm_0'1/2}a_{jm_11/2}a^{+}_{jm1/2}a^{+}_{jm_0,-1/2}|c\rangle = 0$$

since the neutron j-orbit is completely filled in the core state $|c\rangle$. Hence, the only contribution to (17.54) is due to the term with $\mu_0' = -1/2$ in which case $\mu_0 = \mu$. To evaluate (17.54) we use the anticommutation relations (17.43) to obtain

$$\begin{aligned} a^{+}_{jm_0\mu}a_{jm_1\mu}a^{+}_{jm1/2}|c\rangle &= \delta_{\mu 1/2}a^{+}_{jm_01/2}(\delta_{mm_1} - a^{+}_{jm1/2}a_{jm_11/2})|c\rangle \\ &\quad + \delta_{\mu,-1/2}(\delta_{m_0m_1} - a_{jm_1,-1/2}a^{+}_{jm_0,-1/2})a^{+}_{jm1/2}|0\rangle \\ &= (\delta_{\mu 1/2}\delta_{mm_1}a^{+}_{jm_01/2} + \delta_{\mu,-1/2}\delta_{m_0m_1}a^{+}_{jm1/2})|c\rangle \end{aligned} \tag{17.55}$$

The last equality in (17.55) is due to the fact that in $|c\rangle$ there are no j-protons and the neutron j-orbit is full. Substituting (17.55) in (17.54) we obtain

$$\begin{aligned} &\sum V_{TJ_1}(j'jj'j)(j'm_0'jm_0 \mid j'jJ_1M_1)(j'm'jm \mid j'jJ_1M_1) \\ &\quad \times (j'm'jm \mid j'jJM) \\ &\quad \times (\tfrac{1}{2},-\tfrac{1}{2}\tfrac{1}{2}\tfrac{1}{2} \mid \tfrac{1}{2}\tfrac{1}{2}TM_T)^2 a^{+}_{j'm_0',-1/2}a^{+}_{jm_01/2}|c\rangle \\ &\quad + \sum V_{TJ_1}(j'jj'j)(j'm_0'jm_1 \mid j'jJ_1M_1) \\ &\quad \times (j'm'jm_1 \mid j'jJ_1M_1)(j'm'jm \mid j'jJM) \\ &\quad \times (\tfrac{1}{2},-\tfrac{1}{2}\tfrac{1}{2},-\tfrac{1}{2} \mid \tfrac{1}{2}\tfrac{1}{2}TM_T)^2 a^{+}_{j'm_0',-1/2}a^{+}_{jm1/2}|c\rangle \end{aligned} \tag{17.56}$$

In the first term in (17.56) the sum over m and m' yields, due to (7.8), $\delta_{J_1J}\delta_{M_1M}$. In that term $M_T = 0$ and hence

$$(\tfrac{1}{2}, -\tfrac{1}{2}, \tfrac{1}{2}\tfrac{1}{2} \mid \tfrac{1}{2}\tfrac{1}{2}T0)^2 = \tfrac{1}{2}$$

for both $T = 0$ and $T = 1$. In the second term, $M_T = -1$ and $T = 1$ so that $(\tfrac{1}{2}, -\tfrac{1}{2}\tfrac{1}{2}, -\tfrac{1}{2} \mid \tfrac{1}{2}\tfrac{1}{2}1, -1)^2 = 1$. Summing over m_1 and M_1 in the second term we obtain

$$\sum V_{T=1,J_1}(j'jj'j)(2J_1+1)\begin{pmatrix} j' & j & J_1 \\ m'_0 & m_1 & -M_1 \end{pmatrix}\begin{pmatrix} j' & j & J_1 \\ m' & m_1 & -M_1 \end{pmatrix}$$
$$\times (j'm'jm \mid j'jJM)a^{+}_{j'm'_0-1/2}a^{+}_{jm1/2}|c\rangle$$
$$= \sum V_{T=1,J_1}(j'jj'j)\frac{2J_1+1}{2j'+1}(j'm'jm \mid j'jJM)a^{+}_{j'm',-1/2}a^{+}_{jm1/2}|c\rangle \tag{17.57}$$

The last quality in (17.57) is due to (7.25). Thus we obtain for (17.52) the result

$$\frac{1}{2}\sum_{T} V_{TJ}(j'jj'j)(a^{+}_{j',-\frac{1}{2}} \times a^{+}_{j\frac{1}{2}})^{(J)}_{M}|c\rangle$$
$$+ \frac{1}{2j'+1}\sum_{J_1}(2J_1+1)V_{T=1,J_1}(j'jj'j)(a^{+}_{j',-1/2} \times a^{+}_{j1/2})^{(J)}_{M}|c\rangle \tag{17.58}$$

Hence, the state $(a^{+}_{j',-\frac{1}{2}} \times a^{+}_{j\frac{1}{2}})^{(J)}_{M}|c\rangle$ is an eigenstate and its eigenvalue is given by

$$\boxed{\tfrac{1}{2}[V_{T=1,J}(j'jj'j) + V_{T=0,J}(j'jj'j)] + \frac{1}{2j'+1}\sum_{J_1} V_{T=1,J_1}(j'jj'j)} \tag{17.59}$$

The second term in (17.59) is, according to Section 14, the interaction energy between a j'-neutron and the closed j-neutron orbit. The first term in (17.59) is the interaction energy between the j-proton and the j'-neutron. It is indeed identical with the result (10.7) obtained above from physical considerations. Here we showed that it follows from a detailed description of the many nucleon configuration.

18

Two Nucleon Operators. Particle Hole Interactions

In (17.33) or (17.46) the two nucleon interaction is expressed in terms of its matrix elements. In Sections 10 and 12 we saw how such an interaction can be expressed in terms of the coefficients F_k of an expansion in scalar products of irreducible tensors. We will now obtain the analogous expansion in the formalism of second quantization. Before doing that, we transform (17.33) into a form that clearly displays the fact that it is a scalar (rotationally invariant) operator. Recalling the definition (17.12) we can express $A(jj'JM)$ by:

$$\begin{aligned} A(jj'JM) &= -(1+\delta_{jj'})^{-1/2}\sum_{mm'}(j,-mj,'-m' \mid jj'JM)a_{j,-m}a_{j',-m'} \\ &= -\sum_{mm'}(1+\delta_{jj'})^{-1/2}(-1)^{j+j'-J+j+m+j'+m'} \\ &\quad \times (jmj'm' \mid jj'J,-M)\tilde{a}_{jm}\tilde{a}_{j'm'} \\ &= -(1+\delta_{jj'})^{-1/2}(-1)^{J+M}(\tilde{a}_j \times \tilde{a}_{j'})^{(J)}_{-M} \qquad (18.1) \end{aligned}$$

The minus sign on the r.h.s. of (18.1) is due to interchanging of $a_{j'm'}$ and a_{jm}.

Making use of (18.1) and recalling (17.31), we can express **G** in (17.33) by the expression

$$G = -\sum_{j_1 \leq j_2, j_1' \leq j_2', J} V_J(j_1 j_2 j_1' j_2')[(1+\delta_{j_1 j_2})(1+\delta_{j_1' j_2'})]^{-1/2} \times (-1)^J (a_{j_1}^+ \times a_{j_2}^+)^{(J)} \cdot (\tilde{a}_{j_1'} \times \tilde{a}_{j_2'})^{(J)} \tag{18.2}$$

Our aim is to recouple the operators in the scalar products in (18.2) in order to obtain scalar products of $(a_{j_1}^+ \times \tilde{a}_{j_1'})^{(k)}$ and $(a_{j_2}^+ \times \tilde{a}_{j_2'})^{(k)}$. This change of coupling transformation is given by (10.10) but, due to the anticommutation relations (17.4), some care must be taken. We therefore carry out the transformation in detail. The scalar products in (18.2) are written explicitly as:

$$\begin{aligned}
&(-1)^J (a_{j_1}^+ \times a_{j_2}^+)^{(J)} \cdot (\tilde{a}_{j_1'} \times \tilde{a}_{j_2'})^{(J)} \\
&\quad = \sqrt{2J+1} \sum_{\substack{M \\ m_1 m_2 m_1' m_2'}} \frac{(-1)^{J-M}}{\sqrt{2J+1}} (j_1 m_1 j_2 m_2 \mid j_1 j_2 J M) \\
&\qquad \times (j_1' m_1' j_2' m_2' \mid j_1' j_2' J, -M) a_{j_1 m_1}^+ a_{j_2 m_2}^+ \tilde{a}_{j_1' m_1'} \tilde{a}_{j_2' m_2'}
\end{aligned} \tag{18.3}$$

We can now apply to (18.3) the transformation (10.10) and obtain the detailed expression

$$\begin{aligned}
&\sqrt{2J+1} \sum_{k\kappa} (-1)^{j_2 + j_1' + J + k} \sqrt{(2J+1)(2k+1)} \begin{Bmatrix} j_1 & j_2 & J \\ j_2' & j_1' & k \end{Bmatrix} \\
&\quad \times \sum_{m_1 m_2 m_1' m_2'} \frac{(-1)^{k-\kappa}}{\sqrt{2k+1}} (j_1 m_1 j_1' m_1' \mid j_1 j_1' k \kappa) \\
&\quad \times (j_2 m_2 j_2' m_2' \mid j_2 j_2' k, -\kappa) a_{j_1 m_1}^+ a_{j_2 m_2}^+ \tilde{a}_{j_1' m_1'} \tilde{a}_{j_2', m_2'}
\end{aligned} \tag{18.4}$$

We now interchange $a_{j_2 m_2}^+$ and $\tilde{a}_{j_1' m_1'} = (-1)^{j_1' + m_1'} a_{j_1', -m_1'}$ obtaining due to (17.4)

$$a_{j_2 m_2}^+ \tilde{a}_{j_1' m_1'} = -\tilde{a}_{j_1' m_1'} a_{j_2 m_2}^+ + (-1)^{j_1' + m_1'} \delta_{j_2 j_1'} \delta_{m_2, -m_1'} \tag{18.5}$$

Substituting (18.5) into (18.4) we obtain a sum of two terms. The first term is equal to

$$-\sum_k (-1)^{j_2+j_1'+J}(2J+1)\begin{Bmatrix} j_1 & j_2 & J \\ j_2' & j_1' & k \end{Bmatrix}$$

$$\times \sum_{m_1 m_2 m_1' m_2' \kappa} (-1)^{\kappa}(j_1 m_1 j_1' m_1' \mid j_1 j_1' k \kappa)(j_2 m_2 j_2' m_2' \mid j_2 j_2' k, -\kappa)$$

$$\times a^{+}_{j_1 m_1} \bar{a}_{j_1' m_1'} a^{+}_{j_2 m_2} \bar{a}_{j_2' m_2'}$$

$$= -\sum_k (-1)^{j_2+j_1'+J}(2J+1)\begin{Bmatrix} j_1 & j_2 & J \\ j_2' & j_1' & k \end{Bmatrix}$$

$$\times (a^{+}_{j_1} \times \bar{a}_{j_1'})^{(k)} \cdot (a^{+}_{j_2} \times \bar{a}_{j_2'})^{(k)} \tag{18.6}$$

The second term is equal to

$$\sum_k (-1)^{j_2+j_1'+J}(2J+1)\begin{Bmatrix} j_1 & j_2 & J \\ j_2' & j_1' & k \end{Bmatrix} \delta_{j_2 j_1'}$$

$$\times \sum_{m_1 m_2 m_2' \kappa} (-1)^{j_2-m_2+j_2'+m_2'+\kappa}(j_1 m_1 j_2, -m_2 \mid j_1 j_2 k \kappa)$$

$$\times (j_2 m_2 j_2' m_2' \mid j_2 j_2' k, -\kappa) a^{+}_{j_1 m_1} a_{j_2', -m_2'} \tag{18.7}$$

The summation over m_2 and κ in (18.7) can be evaluated by expressing the Clebsch-Gordan coefficients in terms of $3j$-symbols. Due to the orthogonality relation (7.25) we obtain for that sum the result

$$(-1)^{j_2+j_1+1}\frac{2k+1}{2j_1+1}\delta_{j_2' j_1}\delta_{m_2', -m_1}$$

and (18.7) becomes equal to

$$\delta_{j_1 j_2'}\delta_{j_2 j_1'}\sum_k (-1)^{j_1+j_2+J}(2J+1)\begin{Bmatrix} j_1 & j_2 & J \\ j_1 & j_2 & k \end{Bmatrix}\frac{2k+1}{2j_1+1}\sum_{m_1} a^{+}_{j_1 m_1} a_{j_1 m_1} \tag{18.8}$$

Since according to (18.2), $j_1 \leq j_2$ and $j_1' = j_2 \leq j_2' = j_1$, the only non-vanishing contributions to (18.8) come from terms where $j_1 = j_2 = j_1' = j_2'$. The summation over m_1 is equal to the number operator $\hat{n}_{j_1}$ defined in (17.15).

We now collect all terms and express the two body interaction **G** as

$$\mathbf{G} = \sum_{j_1 \le j_2, j_1' \le j_2', k} [(1+\delta_{j_1 j_2})(1+\delta_{j_1' j_2'})]^{-1/2}$$
$$\times \left(\sum_J (-1)^{j_2+j_1'+J}(2J+1) \begin{Bmatrix} j_1 & j_2 & J \\ j_2' & j_1' & k \end{Bmatrix} V_J(j_1 j_2 j_1' j_2') \right)$$
$$\times (a_{j_1}^{+} \times \tilde{a}_{j_1'})^{(k)} \cdot (a_{j_2}^{+} \times \tilde{a}_{j_2'})^{(k)}$$
$$-\frac{1}{2}\sum_{j,k} \left(\sum_J (-1)^{J+1}(2J+1) \begin{Bmatrix} j & j & J \\ j & j & k \end{Bmatrix} V_J(jjjj) \right)$$
$$\times \frac{2k+1}{2j+1} \sum_m a_{jm}^{+} a_{jm} \tag{18.9}$$

The coefficients in (18.9) are simply related to the F_k defined by (12.13) and are given by

$$\sum_J (-1)^{j_2+j_1'+J}(2J+1) \begin{Bmatrix} j_1 & j_2 & J \\ j_2' & j_1' & k \end{Bmatrix} V_J(j_1 j_2 j_1' j_2') = \frac{1}{2k+1} F_k(j_1 j_2 j_1' j_2') \tag{18.10}$$

Hence, the expansion (18.9) may be expressed as

$$\mathbf{G} = \sum_{j_1 \le j_2, j_1' \le j_2' k} [(1+\delta_{j_1 j_2})(1+\delta_{j_1' j_2'})]^{-1/2}$$
$$\times \frac{1}{2k+1} F_k(j_1 j_2 j_1' j_2')(a_{j_1}^{+} \times \tilde{a}_{j_1})^{(k)} \cdot (a_{j_2}^{+} \times \tilde{a}_{j_2'})^{(k)}$$
$$-\frac{1}{2}\sum_{j,k} \frac{1}{2j+1} F_k(jjjj)\hat{n}_j \tag{18.11}$$

The expansion (18.11) is a demonstration of the fact that *any* two body scalar interaction can be expressed as a linear combination of scalar products of single nucleon irreducible tensor operators and an additional single nucleon scalar operator. The single nucleon terms in (18.11) are due to the fact that the scalar products of single nucleon operators are no longer pure two body interactions. They include also spurious interactions of nucleons with themselves. These self-interaction terms are exactly cancelled by the second term in (18.11) which includes the number operators $\hat{n}_j$. In configuration space, the equivalent formulation is given in (14.20) where the self-interaction terms are clearly displayed. The expression (18.11) is equivalent to the expansion (14.20) or (12.12) in configuration space. The coefficient $2k+1$ in the denominator in the first term in (18.11) is due to the tensor operators used here whose reduced matrix elements are given by (17.14) as $\sqrt{2k+1}$ whereas the operators $\mathbf{u}^{(k)}$ in (12.12) have reduced matrix elements equal to 1.

In the shell model subspaces where $j_1 \neq j_2$, $j_1' \neq j_2'$ the expansion (18.11) of a two nucleon interaction is unique. The two nucleon matrix elements are given in terms of the F_k by (12.12). That relation may be inverted yielding (12.13) in which the F_k are uniquely determined by the $V_J(j_1 j_2 j_1' j_2')$. It was mentioned above that if the j_1-orbit and j_2-orbit are identical (or if j_1' and j_2' denote identical orbits), this is no longer the case if we consider only the $T=1$ states with even values of J. In that case, the F_k are uniquely determined if we include in the summation (12.13), in addition to $T=1$ states, also $T=0$ states with odd J values.

Starting from $\mathbf{G}$ defined by (17.33) or (18.2) we obtain the definition of $F_k(jjjj)$ from (18.10) as

$$F_k(jjjj) = (2k+1)\sum_{J\,\text{even}}(-1)^{J+1}(2J+1) \times \begin{Bmatrix} j & j & J \\ j & j & k \end{Bmatrix} V_J(jjjj) \tag{18.12}$$

The coefficients defined by (18.12) are thus uniquely defined by a two-nucleon interaction which for even values of J has the matrix elements which appear in (18.2) and $V_J(jjjj)=0$ for odd values of J. Indeed, if we calculate two nucleon interactions from (18.12) by using (12.12) we obtain, in view of the orthogonality relation of $6j$-symbols (10.13), the result

$$\sum_k (-1)^{J+1} \begin{Bmatrix} j & j & J \\ j & j & k \end{Bmatrix} F_k(jjjj)$$

$$= \sum_{J' \text{ even},k} (2J+1) V_{J'}(jjjj)(2k+1)$$

$$\times \begin{Bmatrix} j & j & J' \\ j & j & k \end{Bmatrix} \begin{Bmatrix} j & j & J \\ j & j & k \end{Bmatrix} = \sum_{J' \text{ even}} V_{J'}(jjjj)\delta_{JJ'} \tag{18.13}$$

For even values of J, the r.h.s. of (18.13) is equal to $V_J(jjjj)$ but it vanishes for odd J values.

The validity of the tensor expansion (18.11) of a two nucleon interaction does not depend on this special choice of F_k for $j_1 = j_2$ or $j'_1 = j'_2$. In fact, we can start with the expansion (18.11) as the definition of **G** with arbitrary values of the coefficients F_k and then use the transformation (10.10) to bring **G** into the form (18.2). The two body matrix elements will be given in terms of the F_k by (12.12). In the subspace where all $j_1 j_2 j'_1 j'_2$ orbits are identical, this relation is

$$V_J(jjjj) = \sum_k (-1)^{J+1} \begin{Bmatrix} j & j & J \\ j & j & k \end{Bmatrix} F_k(jjjj) \tag{18.14}$$

The expansion (18.14) will, in general, no longer vanish for odd values of J. The coefficients $F_k(jjjj)$ in this case will be given by (18.12) only if the summation will include also odd values of J. In the expression (18.2) of **G**, however, odd values of J will not appear if $j_1 = j_2$ or $j'_1 = j'_2$. Their matrix elements will be multiplied by either $A^+(j^2JM)$ or $A(j^2JM)$ and both operators vanish for odd values of J due to the Pauli principle.

A simple application of the expansion (18.11) is the evaluation of the interaction energy of a closed j-orbit of identical nucleons. The only state of that j^{2j+1} configuration is a $J = 0$ state. When we apply to it the operator $(a_j^\dagger \times \tilde{a}_j)^{(k)}_\kappa$ it must lead, according to the Wigner-Eckart theorem to a state with $J = k (M = \kappa)$. The only such state is the $J = 0$ state itself and hence, the only non-vanishing contribution comes from the $k = 0$ term in (18.11). That contribution is thus

equal to

$$\frac{1}{2}F_0(jjjj)(a_j^\dagger \times \tilde{a}_j)^{(k=0)}(a_j^\dagger \times \tilde{a}_j)^{(k=0)} - \frac{1}{2}\frac{1}{2j+1}\sum_k F_k(jjjj))\hat{n}_j$$

$$= \frac{1}{2(2j+1)}F_0(jjjj)\hat{n}_j^2 - \frac{1}{2(2j+1)}\sum_k F_k(jjjj))\hat{n}_j$$

Recalling that the eigenvalues of $\hat{n}_j = \sum a_{jm}^\dagger a_{jm}$ are the nucleon numbers in the j-orbit we obtain by putting $n = 2j+1$ the following result

$$\frac{2j+1}{2}F_0(jjjj) - \frac{1}{2}\sum_k F_k(jjjj) \tag{18.15}$$

Using (18.12) we obtain $F_0(jjjj) = (\sum(2J+1)V_J(jjjj))/(2j+1)$ and

$$\sum_k F_k(jjjj) = \sum_J (-1)^{J+1}(2J+1)V_J(jjjj)\sum_k(2k+1)\begin{Bmatrix} j & j & J \\ j & j & k \end{Bmatrix}$$

$$= -\sum_J (-1)^J(2J+1)V_J(jjjj) \tag{18.16}$$

The last equality in (18.16) is due to the sum rule (10.18). We thus obtain from (18.15) that the interaction energy of the j^{2j+1} configuration of identical nucleons is equal to

$$\sum_{J\,\text{even}} (2J+1)V_J(jjjj) \tag{18.17}$$

The results (18.15) and (18.17) are identical with (14.16), or (14.19), obtained above.

A more interesting application of (18.11) is for calculating the interaction between a nucleon in the j'-orbit and a *hole* in the closed j-orbit. This problem has been discussed in Section 14 but it is worthwhile to derive it in the formalism of second quantization. A state of a j'-nucleon coupled to a j-hole state to form a state with given value of J can be expressed as

$$(a_{j'}^\dagger \times \tilde{a}_j)_M^{(J)}|j^{2j+1}\rangle \tag{18.18}$$

To make efficient use of (18.11) it is convenient to consider the terms with $j_2' = j'$ and $j_2 = j$. This dictates the choice $j_1' = j$ and $j_1 = j'$.

Since $j' \neq j$, this is in contradiction with our restriction to $j_1 \le j_2$, $j'_1 \le j'_2$ and we must make a slight modification of (18.11).

We can start with **G** expressed by the same form as (17.33) but with the restriction $j_1 \le j_2$ and $j'_1 \ge j'_2$. This choice is useful for $j' < j$, if however $j' > j$ we can introduce the restriction $j_1 \ge j_2$, $j'_1 \le j'_2$. All these restrictions lead to the same results. We then proceed as before arriving at the expression (18.6). For the second term of (18.6), however, we obtain a non-negative contribution to (18.18) also from terms with $j_1 = j$ and $j_2 = j'$. Adding such terms to (18.6) we obtain for the part of **G** which contributes to the j'-nucleon j-hole configuration considered here, the expression

$$\sum_k \frac{1}{2k+1} F_k(j'jjj')(a^+_{j'} \times \tilde{a}_j)^{(k)} \cdot (a^+_j \times \tilde{a}^{(k)}_{j'}) - \frac{(-1)^{j'-j}}{2j+1}\left(\sum_k F_k(j'jjj')\right)\hat{n}_{j'} \tag{18.19}$$

This expression is the interaction between the j'-nucleon and the $2j$ nucleons in the j-orbit. Other terms of **G** have matrix elements leading from the given nucleon-hole configuration to other configurations. These are ignored here since we calculate matrix elements of the interaction *within* the given configuration.

We now apply (18.19) to the state (18.18). To do this we calculate the following commutation relation

$$\begin{aligned}
&[(a^+_j \times \tilde{a}_{j'})^{(k)}_\kappa, (a^+_{j'} \times \tilde{a}_j)^{(J)}_M] \\
&\quad = \sum (-1)^{j'+m'+j+m_1}(jmj'm' \mid jj'k\kappa)(j'm'_1jm_1 \mid j'jJM) \\
&\quad \times [a^+_{jm}a_{j',-m'}, a^+_{j'm'_1}a_{j,-m_1}]
\end{aligned} \tag{18.20}$$

Using the formulae (17.17) we obtain for the commutator on the r.h.s. of (18.20) the result

$$\begin{aligned}
&a^+_{jm}a_{j,-m_1}\delta_{-m',m'_1} - a^+_{j'_1m'_1}a_{j',-m'}\delta_{m,-m_1} \\
&\quad = \delta_{m,-m_1}\delta_{-m',m'_1} - a_{j,-m_1}a^+_{jm}\delta_{-m',m'_1} - a^+_{j'_1m'_1}a_{j',-m'}\delta_{m,-m_1}
\end{aligned} \tag{18.21}$$

When the r.h.s. of (18.21) is applied to the state $|j^{2j+1}\rangle$, the last two terms vanish. This is due to the fact that the state has all m-states of the j-orbit occupied and it does not contain any j'-nucleons. What is left is just $\delta_{m,-m_1}\delta_{-m',m'_1}$ which yields, upon substitution in (18.20), the result

$$\sum_{mm'}(-1)^{j'+m'+j-m}(jmj'm' \mid jj'k\kappa)(j',-m'j,-m \mid j'jJM)$$

$$=\sum_{mm'}(-1)^{j'-j+\kappa}(jmj'm' \mid jj'k\kappa)(jmj'm' \mid jj'J,-M)$$

$$=(-1)^{j'-j+\kappa}\delta_{kJ}\delta_{\kappa,-M} \tag{18.22}$$

In obtaining (18.22) we made use of the symmetry properties (7.11) and (7.18) of Clebsch-Gordan coefficients and their orthogonality relation (7.8). Hence, the first term of (18.19) when applied to the state (18.18) becomes equal to

$$(-1)^{j'-j}\sum_{k\kappa}\frac{1}{2k+1}F_k(j'jjj')(a^\dagger_{j'}\times\tilde{a}_j)^{(k)}_{-\kappa}\delta_{kJ}\delta_{\kappa,-M}|j^{2j+1}\rangle$$

$$=(-1)^{j'-j}\frac{1}{2J+1}F_J(j'jjj')(a^\dagger_{j'}\times\tilde{a}_j)^{(J)}_M|j^{2j+1}\rangle \tag{18.23}$$

We see that $F_J(j'jjj')/(2J+1)$ plays the role of the interaction between a nucleon and a hole and it is often referred to as the particle-hole interaction. When the second term of (18.19) is applied to (18.18) the number operator $\hat{n}_{j'}$ yields 1 and hence, the eigenvalue of (18.19) in the state (18.18) is given by

$$(-1)^{j'-j}\frac{1}{2J+1}F_J(j'jjj')-\frac{(-1)^{j'-j}}{2j'+1}\sum_k F_k(j'jjj') \tag{18.24}$$

We can express the result (18.24) in terms of two-nucleon interaction energies in the $j'j$ configuration. We use the definition (18.10) to obtain

$$F_k(j'jjj')=(2k+1)\sum_{J'}(-1)^{j+j+J'}(2J'+1)\begin{Bmatrix} j' & j & J' \\ j' & j & k \end{Bmatrix}V_{J'}(j'jjj') \tag{18.25}$$

In (18.25) the two nucleon matrix elements are defined by

$$\begin{aligned} V_{J'}(j'jjj') &= \langle j'jJ'M'|\mathbf{G}|jj'J'M'\rangle_a \\ &= (-1)^{j+j'+J'+1}\langle j'jJ'M'|\mathbf{G}|j'jJ'M'\rangle_a \end{aligned} \tag{18.26}$$

The last equality in (18.26) follows from the symmetry properties of the Clebsch-Gordan coefficients and the antisymmetry of the m-scheme wave functions. We can thus rewrite (18.25) as

$$F_k(j'jjj') = (-1)^{j'-j+1}(2k+1)\sum_{J'}(2J'+1)\begin{Bmatrix} j' & j & J' \\ j' & j & k \end{Bmatrix} V_{J'}(j'jj'j) \tag{18.27}$$

Introducing the expression (18.27) for the F_k in (18.24) we obtain the final expression

$$\begin{aligned} &\frac{1}{2j'+1}\sum_{J'}(2J'+1)V_{J'}(j'jj'j)\sum_{k}(2k+1)\begin{Bmatrix} j' & j & J' \\ j' & j & k \end{Bmatrix} \\ &\qquad -\sum_{J'}(2J'+1)\begin{Bmatrix} j' & j & J' \\ j' & j & J \end{Bmatrix} V_{J'}(j'jj'j) \\ &\qquad = \frac{1}{2j'+1}\sum_{J'}(2J'+1)V_{J'}(j'jj'j) - \sum_{J'}(2J'+1) \\ &\qquad \times \begin{Bmatrix} j' & j & J' \\ j' & j & J \end{Bmatrix} V_{J'}(j'jj'j) \end{aligned} \tag{18.28}$$

The equality in (18.28) is due to the sum rule (10.18) of $6j$-symbols. The relation (18.28) is the Pandya relation (14.46) obtained above. Some of its implications were discussed in Section 14. For deriving (14.46) we considered a situation in which a neutron was coupled with a proton hole. The matrix elements $V_{J'}(j'jj'j)$ appearing in it are then given by (10.7). If, however, the hole and particle are of the same kind, (14.46) should hold if the matrix elements $V_{J'}(j'jj'j)$ are calculated with antisymmetric wave functions. This is the case in the derivation of (18.28).

The relation (18.28) expresses the interaction between a proton hole and a proton or a neutron hole and a neutron. Such configurations arise when the protons are in closed shells and there are no protons in the j-orbit whereas the neutron j-orbit is full. We then consider particle hole states in which a j-neutron is raised to a higher j'-orbit (in which there are no protons). An example for such a situation is ${}^{48}_{20}Ca_{28}$ with a full $1f_{7/2}$ neutron orbit and we consider excited states in which a neutron is raised from that orbit into the $1f_{5/2}$-orbit or $2p_{3/2}$-orbit. In such cases, the protons are in closed shells and the states are obtained by the various couplings of neutrons. The isospin formalism may be used but in such cases it does not introduce any simplification as explained in detail in Section 17.

There are, however, cases in which the symmetry associated with isospin is very useful as we saw in the case of two nucleon configurations. Consider a nucleus like ${}^{40}_{20}Ca_{20}$ in whose ground state protons and neutrons fill completely the same orbits. Consider now excitations in which a nucleon is excited into a higher orbit, $1f_{7/2}$ say. Due to charge independence, a state in which a proton is excited is not an eigenstate of the nuclear Hamiltonian, nor is a state in which a neutron is excited an eigenstate. One correct eigenstate is a linear combination of these two states with total isospin $T = 0$ and the other eigenstate is the orthogonal linear combination with isospin $T = 1$. To describe correctly such excitations we need a generalization of (18.28) incorporating the isospin degree of freedom.

We start with the two-nucleon interaction (17.46) with the restriction $j_1 \le j_2$, $j'_1 \ge j'_2$. In order to transform $\mathbf{G}$ into the form which displays the particle hole interaction, we first introduce, in analogy with (17.12) the operators

$$\tilde{a}_{jm\mu} = (-1)^{j+m+(1/2)+\mu} a_{j,-m,-\mu} \tag{18.29}$$

Like in (18.1) we now bring the conjugate of the operator (17.45) into the form

$$\begin{aligned} A(j_1 j_2 T M_T J M) &= \frac{(-1)^{J+M+T+M_T+1}}{\sqrt{1+\delta_{j_1 j_2}}} \sum (j_1 m_1 j_2 m_2 \mid j_1 j_2 J, -M) \\ &\quad \times (\tfrac{1}{2}\mu_1 \tfrac{1}{2}\mu_2 \mid \tfrac{1}{2}\tfrac{1}{2} T, -M_T) \tilde{a}_{j_1 m_1 \mu_1} \tilde{a}_{j_2 m_2 \mu_2} \\ &= -(-1)^{J+M+T+M_T} (\tilde{a}_{j_1} \times \tilde{a}_{j_2})^{(T,J)}_{-M_T,-M} \end{aligned} \tag{18.30}$$

The operator **G** in (17.46) may now be expressed as

$$\mathbf{G} = -\sum_{\substack{j_1 \le j_2, j_1' \ge j_2' \\ JM}} \frac{(-1)^{J+T}}{\sqrt{(1+\delta_{j_1 j_2})(1+\delta_{j_1' j_2'})}} V_{TJ}(j_1 j_2 j_1' j_2') \times (a_{j_1}^{+} \times a_{j_2}^{+})^{(T,J)} \cdot (\tilde{a}_{j_1'} \times \tilde{a}_{j_2'})^{(T,J)} \tag{18.31}$$

where the scalar product simply means

$$(a_{j_1}^{+} \times a_{j_2}^{+})^{(T,J)} \cdot (\tilde{a}_{j_1'} \times \tilde{a}_{j_2'})^{(T,J)} = \sum (-1)^{M_T+M} (a_{j_1}^{+} \times a_{j_2}^{+})^{(T,J)}_{M_T,M} (\tilde{a}_{j_1'} \times \tilde{a}_{j_2'})^{(T,J)}_{-M_T,-M}$$

The j'-nucleon j-hole state is created by the operator $(a_{j'}^{+} \times \tilde{a}_j)^{(T,J)}$ acting on the ground state of a nucleus with completely filled orbits to be specified later on. The relevant part of **G** in (18.31) is thus specified by $j_1 = j'$, $j_2 = j$ and $j_1' = j$, $j_2' = j'$ if $j' < j$. If, however, $j' > j$, we restrict the summation in (18.31) to $j_1 \ge j_2$, $j_1' \le j_2'$.

Starting from (18.31) with $j_1 = j_2' = j'$, $j_2 = j_1' = j$, we proceed as in the derivation of (18.6) and (18.19) from (18.2). We carry out the transformation (10.10) on both the angular momenta as well as the isospins, obtaining for the relevant part of **G** (which contributes to the particle hole configuration considered) the expression

$$\begin{aligned}
&\sum_{T_1 k, TJ} (-1)^{J+T}(2J+1)(2T+1) \begin{Bmatrix} j' & j & J \\ j' & j & k \end{Bmatrix} \begin{Bmatrix} \frac{1}{2} & \frac{1}{2} & T \\ \frac{1}{2} & \frac{1}{2} & T_1 \end{Bmatrix} \\
&\quad \times V_{TJ}(j'jjj')(a_{j'}^{+} \times \tilde{a}_j)^{(T_1 k)} \cdot (a_j^{+} \times \tilde{a}_{j'})^{(T_1 k)} \\
&- \sum_{T_1 k, TJ} (-1)^{J+T}(2J+1)(2T+1) \begin{Bmatrix} j' & j & J \\ j' & j & k \end{Bmatrix} \begin{Bmatrix} \frac{1}{2} & \frac{1}{2} & T \\ \frac{1}{2} & \frac{1}{2} & T_1 \end{Bmatrix} \\
&\quad \times V_{TJ}(j'jjj') \frac{2k+1}{2j'+1} \frac{2T_1+1}{2} \hat{n}_{j'}
\end{aligned} \tag{18.32}$$

To apply (18.32) to a state, created by acting with $(a^{+}_{j'} \times \tilde{a}_j)^{(T,J)}_{M_T,M}$ on the ground state of a nucleus, we calculate the commutation relation

$$[(a^{+}_{j} \times \tilde{a}_{j'})^{(T_1,k)}_{M_{T_1},\kappa}, (a^{+}_{j'} \times \tilde{a}_j)^{(T,J)}_{M_T,M}]$$
$$= \sum (-1)^{j'+m'+j+m_1+(1/2)+\mu'+(1/2)+\mu_1}$$
$$\times (jmj'm' \mid jj'k\kappa)(j'm'_1 jm_1 \mid j'jJM)$$
$$\times (\tfrac{1}{2}\mu\tfrac{1}{2}\mu' \mid \tfrac{1}{2}\tfrac{1}{2}T_1M_{T_1})(\tfrac{1}{2}\mu'_1\tfrac{1}{2}\mu_1 \mid \tfrac{1}{2}\tfrac{1}{2}TM_T)$$
$$\times [a^{+}_{jm\mu}a_{j',-m',-\mu'}, a^{+}_{j'm'_1\mu'_1}a_{j,-m_1,-\mu_1}] \tag{18.33}$$

Using the anticommutation relations (17.43) and the formula (17.17) we obtain for the commutator on the r.h.s. of (18.33) the result

$$a^{+}_{jm\mu}a_{j,-m_1,-\mu_1}\delta_{-m'm'_1}\delta_{-\mu'\mu'_1} - a^{+}_{j'm'_1\mu'_1}a_{j',-m',-\mu'}\delta_{-mm_1}\delta_{-\mu\mu_1}$$
$$= \delta_{-m_1m}\delta_{-m'm'_1}\delta_{-\mu_1\mu}\delta_{-\mu'\mu'_1} - a_{j,-m_1,-\mu_1}a^{+}_{jm\mu}\delta_{-m'm'_1}\delta_{-\mu'\mu'_1}$$
$$- a^{+}_{j'm'_1\mu'_1}a_{j',-m',-\mu'}\delta_{-mm_1}\delta_{-\mu\mu_1} \tag{18.34}$$

In order to obtain a particle hole state we apply the operator $(a^{+}_{j'} \times \tilde{a}_j)^{(T,J)}$ to a given core. By definition, the core does not have any nucleons in the j'-orbit. If in addition, the j-orbit in the core is completely filled by protons and neutrons, the last two terms on the r.h.s. of (18.34) vanish when applied to the core state. We will also consider later the case in which there are no protons in the j-orbit whereas the j-neutron orbit is completely filled. This is the case for which we obtained the expression (18.28). In that case the second term on the r.h.s. of (18.34) will vanish only if $\mu = -\frac{1}{2}$.

With this in mind we keep only the first term on the r.h.s. of (18.34). Substituting it in (18.33) we obtain, in the same way that (18.22) was obtained, the following result for the commutator (18.33)

$$\sum (-1)^{j'+m'+j-m+(1/2)+\mu'+(1/2)-\mu}(jmj'm' \mid jj'k\kappa)$$
$$\times (j',-m'j,-m \mid j'jJM)$$
$$\times (\tfrac{1}{2}\mu\tfrac{1}{2}\mu' \mid \tfrac{1}{2}\tfrac{1}{2}T_1M_{T_1})(\tfrac{1}{2},-\mu'\tfrac{1}{2},-\mu \mid \tfrac{1}{2}\tfrac{1}{2}TM_T)$$
$$= (-1)^{j'-j+M+M_T}\delta_{kJ}\delta_{\kappa,-M}\,\delta_{T_1T}\,\delta_{M_{T_1},-M_T} \tag{18.35}$$

We can insert the result (18.35) into (18.32) but before that let us express $V_{TJ}(j'jjj')$ by a simpler matrix element, making use of (7.11)

$$\begin{aligned} V_{TJ}(j'jjj') &= \langle j'jTM_TJM|\mathbf{G}|jj'TM_TJM\rangle_a \\ &= (-1)^{j+j'-J+(1/2)+(1/2)-T+1}\langle j'jTM_TJM|\mathbf{G}|j'jTM_TJM\rangle_a \\ &= (-1)^{j+j'+J+T}V_{TJ}(j'jj'j) \end{aligned} \tag{18.36}$$

Incorporating (18.35) and (18.36) into (18.32) we finally obtain for the particle-hole interaction the value

$$\begin{aligned} &-\sum_{T'J'}(2J'+1)(2T'+1)\begin{Bmatrix} j' & j & J' \\ j' & j & J \end{Bmatrix}\begin{Bmatrix} \frac{1}{2} & \frac{1}{2} & T' \\ \frac{1}{2} & \frac{1}{2} & T \end{Bmatrix}V_{T'J'}(j'jj'j) \\ &\quad+\sum_{T'J'}(2J'+1)(2T'+1)V_{T'J'}(j'jj'j)\sum_{T_1k}\frac{2k+1}{2j'+1}\frac{2T_1+1}{2} \\ &\quad\times\begin{Bmatrix} j' & j & J' \\ j' & j & k \end{Bmatrix}\begin{Bmatrix} \frac{1}{2} & \frac{1}{2} & T' \\ \frac{1}{2} & \frac{1}{2} & T_1 \end{Bmatrix} \\ &=\frac{1}{2(2j'+1)}\sum_{T'J'}(2J'+1)(2T'+1)V_{T'J'}(j'jj'j) \\ &\quad-\sum_{T'J'}(2J'+1)(2T'+1)\begin{Bmatrix} j' & j & J' \\ j' & j & J \end{Bmatrix} \\ &\quad\times\begin{Bmatrix} \frac{1}{2} & \frac{1}{2} & T' \\ \frac{1}{2} & \frac{1}{2} & T \end{Bmatrix}V_{T'J'}(j'jj'j) \end{aligned} \tag{18.37}$$

The last equality in (18.37) is due to the sum rule (10.21) of the $6j$-symbols. In obtaining (18.37) we made use of the fact that the eigenvalue of $\hat{n}_{j'}$ in the second term of (18.32) is equal to 1 in the state with one j' nucleon.

Let us now examine the result (18.37). Its value depends on the isospin T of the particle-hole pair. If in the core the j-orbit is filled by protons and neutrons, its isospin is zero. There are two sets of states of the $j^{-1}j'$ configuration, those with $T = 1$ and the others, which usually lie lower, with $T = 0$. The first term on the r.h.s. of (18.37) is independent of T. It is equal to

$$\frac{3}{2(2j'+1)}\sum_{J'}(2J'+1)V_{T'=1,J'}(j'jj'j) + \frac{1}{2(2j'+1)}\sum_{J'}(2J'+1)V_{T'=0,J'}(j'jj'j) \qquad (18.38)$$

which is equal, according to (14.34), to the interaction of one j'-nucleon with the full j-orbit. It can be written as

$$\frac{1}{2j'+1}\sum_{J'}(2J'+1)V_{T=1}(j'jj'j) + \frac{1}{2j'+1}\sum_{J'}(2J'+1)\tfrac{1}{2}[V_{T=1}(j'jj'j)+V_{T=0}(j'jj'j)] \qquad (18.39)$$

where the first term is the interaction of a j' nucleon with the j-nucleons of the same kind and the other is its interaction with the other kind.

We now look at the values of the second term on the r.h.s. of (18.37). If we consider the states with $T = 0$ we obtain for it the expression

$$\boxed{-\frac{3}{2}\sum_{J'}(2J'+1)\begin{Bmatrix} j' & j & J' \\ j' & j & J \end{Bmatrix} V_{T'=1,J'}(j'jj'j) + \frac{1}{2}\sum_{J'}(2J'+1)\begin{Bmatrix} j' & j & J' \\ j' & j & J \end{Bmatrix} V_{T'=0,J'}(j'jj'j)} \qquad (18.40)$$

For the $T = 1$ states that term is equal to

$$
\boxed{
\begin{aligned}
&-\frac{1}{2}\sum_{J'}(2J'+1)\begin{Bmatrix} j' & j & J' \\ j' & j & J \end{Bmatrix} V_{T'=1,J'}(j'jj'j) \\
&\qquad -\frac{1}{2}\sum_{J'}(2J'+1)\begin{Bmatrix} j' & j & J' \\ j' & j & J \end{Bmatrix} V_{T'=0,J'}(j'jj'j)
\end{aligned}
}
\tag{18.41}
$$

To understand the physical meaning of (18.40) and (18.41) we look again at ${}^{40}_{20}\mathrm{Ca}_{20}$ and excitations from its ground state, in which a $1d_{3/2}$ nucleon is raised into the $1f_{7/2}$ orbit. If the particle-hole operator $(a^{+}_{j'} \times \tilde{a}_j)^{(T,J)}_{M_T,M}$ has $T = 0$, the state created by it is an excited state of ${}^{40}\mathrm{Ca}$. This is also the case for $T = 1$ and the component with $M_T = 0$. If, however, the $M_T = -1$ component is applied to the ground state of ${}^{40}\mathrm{Ca}$, we obtain states in ${}^{40}_{19}\mathrm{K}_{21}$ in which there is a $1d_{3/2}$ proton hole and a $1f_{7/2}$ neutron. The particle-hole interaction energy in this case is expressed in terms of the interaction energies between a j'-neutron and j-protons which is given by (10.7) or (17.59). The result (18.41) for $T = 1$ states which is independent of M_T, is given indeed by the Pandya relation (14.46) where the particle-particle matrix elements are the averages of those with $T = 0$ and $T = 1$. From the derivation of (14.46) follows that it holds, with matrix elements $V_{J'}(j'jj'j)$ given by (10.7), for states of a j-proton hole and j'-neutron also if there are other orbits completely filled by neutrons. Use of isospin is not necessary in such cases and it actually complicates the derivation. In the following there are two examples in which isospin is used to derive results which were obtained above. These should be considered as exercises.

The particle-hole interaction for the case of identical nucleons was obtained above and expressed by (18.28). That result should also be obtainable from the isospin formalism. For that case we consider a core which has no protons in the j-orbit whereas that orbit is completely filled by neutrons. The isospin of the core is thus given by $T' = \frac{1}{2}(2j+1)$. States in which one neutron is raised to a higher j'-orbit have the same isospin T' and are obtained by action on the core state $|c\rangle$ by the particle hole creation operator

$$(a^{+}_{j',-1/2} \times \tilde{a}_{j1/2})^{(J)}_{M}|c\rangle$$
$$= \left(\sum_{mm'}(j'm'jm \mid j'jJM)a^{+}_{j'm',-1/2}\tilde{a}_{jm1/2}\right)|c\rangle \quad (18.42)$$

The operator $\tilde{a}_{jm,1/2} = (-1)^{j+m+(1/2)+(1/2)}a_{j,-m,-1/2}$ annihilates a j-neutron in the core. In contrast with the particle hole creation operator considered above, the operator (18.42) does not have a definite rank in isospin space.

We now apply the part of (18.32) which is the two-body interaction between j and j' nucleons to the particle hole state (18.42). We first consider the expression

$$(a^{+}_{j} \times \tilde{a}_{j'})^{(T_1k)}_{M_{T_1}\kappa}(a^{+}_{j',-1/2} \times \tilde{a}_{j1/2})^{(J)}_{M}|c\rangle$$
$$= \sum(jm_1j'm'_1 \mid jj'k\kappa)(j'm'jm \mid j'jJM)(\tfrac{1}{2}\mu_1\tfrac{1}{2}\mu'_1 \mid \tfrac{1}{2}\tfrac{1}{2}T_1M_{T_1})$$
$$\times (-1)^{j'+m'_1+j+m+(1/2)+\mu'_1+1}$$
$$\times a^{+}_{jm_1\mu_1}a_{j',-m'_1,-\mu'_1}a^{+}_{j'm',-1/2}a_{j,-m,-1/2}|c\rangle \quad (18.43)$$

By interchanging creation and annihilation operators we first obtain

$$a^{+}_{jm_1\mu_1}a_{j',-m'_1,-\mu'_1}a^{+}_{j'm',-1/2}a_{j,-m,-1/2}|c\rangle$$
$$= (a^{+}_{jm_1\mu_1}a_{j,-m,-1/2}\delta_{-m'_1m'}\delta_{\mu'_11/2}$$
$$+ a^{+}_{jm_1\mu_1}a^{+}_{j'm',-1/2}a_{j,-m,-1/2}a_{j',-m'_1,-\mu'_1})|c\rangle \quad (18.44)$$

The second term on the r.h.s. of (18.44) vanishes since there are no j'-nucleons in the core state. Substituting (18.44) into (18.43) we multiply it from the left by $(-1)^{M_{T_1}+\kappa}(a^{+}_{j'} \times \tilde{a}_{j})^{(T_1k)}_{-M_{T_1},-\kappa}$ and sum over M_{T_1},κ, thereby obtaining

$$\sum(-1)^{j'-m'+j+m+M_{T_1}+\kappa+j+m_0+(1/2)+\mu_0}(j'm'_0jm_0 \mid j'jk,-\kappa)$$
$$\times (jm_1j',-m' \mid jj'k\kappa)(j'm'jm \mid j'jJM)$$
$$\times (\tfrac{1}{2}\mu'_0\tfrac{1}{2}\mu_0 \mid \tfrac{1}{2}\tfrac{1}{2}T_1,-M_{T_1})(\tfrac{1}{2}\mu_1\tfrac{1}{2}\tfrac{1}{2} \mid \tfrac{1}{2}\tfrac{1}{2}T_1M_{T_1})$$
$$\times a^{+}_{j'm'_0\mu'_0}a_{j,-m_0,-\mu_0}a^{+}_{jm_1\mu_1}a_{j,-m,-1/2}|c\rangle \quad (18.45)$$

We first notice that if $\mu_1 = \frac{1}{2}$ then $M_{T_1} = 1$ (and $T_1 = 1$) which implies $\mu'_0 = \mu_0 = -\frac{1}{2}$. The isospin Clebsch-Gordan coefficients in (18.45) are then equal to 1. Making now use of the anticommutation relations (17.43) we obtain

$$
\begin{aligned}
& a_{j,-m_0,-\mu_0} a^{+}_{jm_1\mu_1} a_{j,-m,-1/2}|c\rangle \\
&\quad = \delta_{\mu_1 1/2}(\delta_{-m_0 m_1} a_{j,-m,-1/2} + a^{+}_{jm_1 1/2} a_{j,-m,-1/2} a_{j,-m_0 1/2})|c\rangle \\
&\qquad + \delta_{\mu_1,-1/2}(a_{j,-m_0,-\mu_0}\delta_{-mm_1} - a_{j,-m_0,-\mu_0} a_{j,-m,-1/2} a^{+}_{jm_1,-1/2})|c\rangle \\
&\quad = (\delta_{\mu_1 1/2}\delta_{-m_0 m_1} a_{j,-m,-1/2} + \delta_{\mu_1,-1/2}\delta_{-mm_1} a_{j,-m_0,-\mu_0})|c\rangle
\end{aligned}
\tag{18.46}
$$

The last equality in (18.46) follows from the fact that in the core state there are no j-protons while the neutron j-orbit is completely filled. Due to this fact, in the second term on the r.h.s. of (18.46) there is a non-vanishing contribution only for $\mu_0 = \frac{1}{2}$. In that term, due to $\mu_1 = -\frac{1}{2}$, $M_{T_1} = 0$, we obtain $\mu'_0 = -\frac{1}{2}$. The product of isospin Clebsch-Gordan coefficients in (18.45) is then equal to $\frac{1}{2}$. Substituting the r.h.s. of (18.46) into (18.45) we obtain for it the expression

$$
\begin{aligned}
& \Big[\sum (-1)^{j'+m+1} \delta_{T_1 1} (j'm'_0 j,-m_1 \mid j'jk,-\kappa)(jm_1 j',-m' \mid jj'k\kappa) \\
&\quad \times (j'm'jm \mid j'jJM) a^{+}_{j'm'_0,-1/2} a_{j,-m,-1/2} \\
&\quad + \sum (-1)^{j'+m_0+1} \frac{1}{2} (j'm'_0 jm_0 \mid j'jk,-\kappa)(j,-mj',-m' \mid jj'k\kappa) \\
&\quad \times (j'm'jm \mid j'jJM) a^{+}_{j'm'_0,-1/2} a_{j,-m_0,-1/2} \Big] |c\rangle
\end{aligned}
\tag{18.47}
$$

In the second term of (18.47), the summation over m, m' yields, due to the orthogonality of Clebsch-Gordan coefficients, $\delta_{kJ}\,\delta_{\kappa,-M}$. In the first summation in (18.47) we express two Clebsch-Gordan coefficients, by two $3j$-symbols and summing their product over m_1 and

κ yields $\delta_{m_0' m'}/(2j'+1)$. Hence, we obtain for (18.47) the expression

$$\Bigg[\delta_{T_1 1}(-1)^{j-j'}\frac{2k+1}{2j'+1}\sum (j'm'jm \mid j'jJM)a^{+}_{j'm',-1/2}$$
$$\times(-1)^{j+m+1}a_{j,-m,-1/2}+\frac{1}{2}\delta_{kJ}(-1)^{j-j'}$$
$$\times\frac{1}{2}\sum (j'm_0'jm_0 \mid j'jJM)a^{+}_{j'm_0',-1/2}$$
$$\times(-1)^{j+m_0+1}a_{j,-m_0,-1/2}\Bigg]|c\rangle$$
$$=(-1)^{j-j'}\left[\delta_{T_1 1}\frac{2k+1}{2j'+1}+\frac{1}{2}\delta_{kJ}\right](a^{+}_{j',-1/2}\times \tilde{a}_{j1/2})^{(J)}_M|c\rangle \tag{18.48}$$

We now substitute (18.48) into the first part of (18.32) and obtain, due to (18.36) for the eigenvalue of the state (18.42) the result

$$-\sum_{kT'J'}(2J'+1)(2T'+1)\begin{Bmatrix} j' & j & J' \\ j' & j & k \end{Bmatrix}\frac{2k+1}{2j'+1}$$
$$\times\begin{Bmatrix} \frac{1}{2} & \frac{1}{2} & T' \\ \frac{1}{2} & \frac{1}{2} & 1 \end{Bmatrix}V_{T'J'}(j'jj'j)$$
$$-\frac{1}{2}\sum_{T_1T'J'}(2J'+1)(2T'+1)\begin{Bmatrix} j' & j & J' \\ j' & j & J \end{Bmatrix}$$
$$\times\begin{Bmatrix} \frac{1}{2} & \frac{1}{2} & T' \\ \frac{1}{2} & \frac{1}{2} & T_1 \end{Bmatrix}V_{T'J'}(j'jj'j) \tag{18.49}$$

The first term in (18.49) is independent of J and we consider it later. The second term can be simplified by carrying out explicitly the summation over T_1. For $T'=0$ that summation yields zero due to

$$\begin{Bmatrix} \frac{1}{2} & \frac{1}{2} & 0 \\ \frac{1}{2} & \frac{1}{2} & 0 \end{Bmatrix}+\begin{Bmatrix} \frac{1}{2} & \frac{1}{2} & 0 \\ \frac{1}{2} & \frac{1}{2} & 1 \end{Bmatrix}=0$$

For $T' = 1$ the summation yields

$$\begin{Bmatrix} \frac{1}{2} & \frac{1}{2} & 1 \\ \frac{1}{2} & \frac{1}{2} & 0 \end{Bmatrix} + \begin{Bmatrix} \frac{1}{2} & \frac{1}{2} & 1 \\ \frac{1}{2} & \frac{1}{2} & 1 \end{Bmatrix} = \tfrac{1}{2} + \tfrac{1}{6} = \tfrac{2}{3}$$

Hence the second term is equal to

$$-\sum_{J'}(2J'+1)\begin{Bmatrix} j' & j & J' \\ j' & j & J \end{Bmatrix} V_{T'=1,J'}(j'jj'j) \tag{18.50}$$

which is identical to the result in (18.28) obtained above.

In the first term in (18.49) the summation over k may be carried out yielding for it, by use of the sum rule (10.21), the result

$$-\frac{1}{2j'+1}\sum_{T'J'}(2T'+1)\begin{Bmatrix} \frac{1}{2} & \frac{1}{2} & T' \\ \frac{1}{2} & \frac{1}{2} & 1 \end{Bmatrix}(2J'+1)V_{T'J'}(j'jj'j)$$

$$= \frac{-1}{2(2j'+1)}\sum_{J'}(2J'+1)[V_{T'=1J'}(j'jj'j) + V_{T'=0J'}(j'jj'j)] \tag{18.51}$$

We now combine (18.51) with the second term of (18.32) which is equal, for one j' nucleon, to (18.38). Their sum is equal to

$$\frac{1}{2j'+1}\sum_{J'}(2J'+1)V_{T=1}(j'jj'j) \tag{18.52}$$

which is the interaction of a j' neutron with the closed j-neutron orbit as given in Section 14 as well as by the first term on the r.h.s. of (18.28).

In Section 14 we remarked that the Pandya relation (14.46) may be applied also to the case where the particle is a j-proton and the hole is one neutron missing from the *same* j-orbit. It was pointed out that this relation follows directly if we do not use isospin. Yet, also this case may be considered within the framework of the isospin formalism. A simple example is offered by the nucleus ${}^{48}_{21}Sc_{27}$ whose low lying levels are due to a proton and a neutron hole in the $1f_{7/2}$ orbit. The core state to which the particle hole operator is applied, is the ground state of ${}^{48}_{20}Ca_{28}$.

The j-proton j-neutron hole states which we consider are defined by

$$(a^{+}_{j1/2} \times \tilde{a}_{j1/2})^{(J)}_{M}|c\rangle = \sum (jmjm' \mid jjJM) a^{+}_{jm1/2} \tilde{a}_{jm'1/2}|c\rangle \tag{18.53}$$

where the core state $|c\rangle$ satisfies the relations

$$a^{+}_{jm,-1/2}|c\rangle = 0 \qquad a_{jm1/2}|c\rangle = 0 \tag{18.54}$$

which mean that there are no j-protons in $|c\rangle$ and the neutron j-orbit is completely filled. In the case treated above, leading to (18.49) and (18.52), all states (18.42) have the same isospin as the core state $T_0 = (2j+1)/2$. The core state has $M_{T_0} = -(2j+1)/2$ and the particle-hole operator in (18.42) with $M_T = 0$, does not change it. The states of $2j+1$ nucleons with $M_{T_0} = -(2j+1)/2$ must have $T_0 = (2j+1)/2$ since no lower value of M_{T_0} may be reached. In the present case, the operator in (18.53) has $M_T = 1$ and the isospin of the resulting particle hole states may be equal to T_0 or to $T_0 - 1$. It turns out that we do not have to specify the isospins of the states (18.53) in order to calculate their interaction energies. The reason is that there is only *one* state (18.53) with given J (and M) and we shall later see which is the isospin of each of these states.

To calculate the interaction energy in the state (18.53) we can use the part of the operator (17.46) within the j^n configuration. That part is given by

$$\sum_{\substack{T'J' \\ M'_T M'}} V_{T'J'}(jjjj) A^{+}(j^2 T' M'_T J' M') A(j^2 T' M'_T J' M') \tag{18.55}$$

Applying (18.55) to the state (18.53) we obtain the expression

$$\begin{aligned} &\frac{1}{2} \sum V_{T'J'}(jjjj)(jm_0 jm'_0 \mid jjJ'M')(jm_1 jm'_1 \mid jjJ'M') \\ &\times (-1)^{j+m'+1}(jmjm' \mid jjJM) \\ &\times (\tfrac{1}{2}\mu_0 \tfrac{1}{2}\mu'_0 \mid \tfrac{1}{2}\tfrac{1}{2}T'M'_T)(\tfrac{1}{2}\mu_1 \tfrac{1}{2}\mu'_1 \mid \tfrac{1}{2}\tfrac{1}{2}T'M'_T) \\ &\times a^{+}_{jm_0\mu_0} a^{+}_{jm'_0\mu'_0} a_{jm'_1\mu'_1} a_{jm_1\mu_1} a^{+}_{jm1/2} a_{j,-m',-1/2}|c\rangle \end{aligned} \tag{18.56}$$

By interchanging the order of creation and annihilation operators it is possible to express the product of these operators in (18.56) as the sum of three terms

$$a^{+}_{jm_0\mu_0}a^{+}_{jm'_0\mu'_0}a_{jm'_1\mu'_1}\delta_{mm_1}\delta_{\mu_1 1/2}a_{j,-m',-1/2}|c\rangle \tag{18.57}$$

$$-a^{+}_{jm_0\mu_0}a^{+}_{jm'_0\mu'_0}a_{jm_1\mu_1}\delta_{mm'_1}\delta_{\mu'_1 1/2}a_{j,-m',-1/2}|c\rangle \tag{18.58}$$

$$a^{+}_{jm1/2}a^{+}_{jm_0\mu_0}a^{+}_{jm'_0\mu'_0}a_{jm'_1\mu'_1}a_{jm_1\mu_1}a_{j,-m',-1/2}|c\rangle \tag{18.59}$$

From (18.54) we conclude that in (18.59) the only non-vanishing contribution comes from terms with $\mu_1 = \mu'_1 = -\frac{1}{2}$. This implies $M'_T = -1$, $T' = 1$ and hence $\mu'_0 = \mu_0 = -\frac{1}{2}$. Similarly, the only contributing terms in (18.57) are those with $\mu'_1 = -\frac{1}{2}$ and since $\mu_1 = \frac{1}{2}$, $M'_T = 0$ and hence $\mu_0 = \pm\frac{1}{2}$, $\mu'_0 = \mp\frac{1}{2}$. In (18.58), we must have $\mu_1 = -\frac{1}{2}$, $\mu'_1 = \frac{1}{2}$ and also here $M'_T = 0$ and $\mu_0 = \pm\frac{1}{2}$, $\mu'_0 = \mp\frac{1}{2}$.

We may now simplify (18.57), (18.58) and (18.59) by interchanging the creation operators with $\mu = -\frac{1}{2}$ bringing them to the right so that (18.54) may be used. Using the symmetry and orthogonality properties of $3j$-symbols as well as the value

$$\begin{aligned}(\tfrac{1}{2}\mu\tfrac{1}{2},-\mu \mid \tfrac{1}{2}\tfrac{1}{2}T'M'_T = 0) &= (-1)^{(1/2)-\mu}/\sqrt{2}\\ &= (-1)^{T'+1}(\tfrac{1}{2},-\mu\tfrac{1}{2}\mu \mid \tfrac{1}{2}\tfrac{1}{2}T'M'_T = 0),\end{aligned}$$

we obtain the contribution of (18.59), when it is substituted in (18.56), as equal to

$$\left[\frac{-2}{2j+1}\sum_{J'}(2J'+1)V_{T'=1,J'}(jjjj) + \sum_{J'}(2J'+1)V_{T'=1,J'}(jjjj)\right]$$
$$\times (a^{+}_{j1/2} \times \tilde{a}_{j1/2})^{(J)}_{M}|c\rangle \tag{18.60}$$

The other two terms (18.57) and (18.58), when substituted into (18.56), give the result

$$\left[\frac{1}{2j+1}\sum_{T'J'}(2J'+1)V_{T'J'}(jjjj)+\sum(-1)^{T'+1+j+m'+1+M}\right.$$

$$\times V_{T'J'}(jjjj)(2J'+1)\sqrt{2J+1}\times\begin{pmatrix} j & j & J \\ m & m' & -M \end{pmatrix}$$

$$\left.\times\begin{pmatrix} j & j & J' \\ m_0 & -m' & -M' \end{pmatrix}\begin{pmatrix} j & j & J' \\ -m & -m_1 & M' \end{pmatrix}\right]$$

$$\times a^{+}_{jm_0 1/2}a_{jm_1,-1/2}|c\rangle \tag{18.61}$$

Due to the relation (10.23) the last term in (18.61) can be expressed as

$$\left(\sum_{T'J'}(-1)^{2j+J'+T'+1}(2J'+1)\begin{Bmatrix} j & j & J' \\ j & j & J \end{Bmatrix}V_{T'J'}(jjjj)\right)$$

$$\times(a^{+}_{j1/2}\times\tilde{a}_{j1/2})^{(J)}_{M}|c\rangle \tag{18.62}$$

Recalling the symmetry property of the wave functions in the j^2 configuration, discussed in Section 9, we obtain $(-1)^{J'+T'}=-1$. We can now add (18.60) and (18.61) to obtain the eigenvalue of (18.55) for the state (18.53) as

$$\frac{2j-1}{2j+1}\sum_{J'}(2J'+1)V_{T'=1,J'}(jjjj)+\frac{1}{2j+1}\sum_{T'J'}(2J'+1)V_{T'J'}(jjjj)$$

$$-\sum_{T'J'}(2J'+1)\begin{Bmatrix} j & j & J' \\ j & j & J \end{Bmatrix}V_{T'J'}(jjjj) \tag{18.63}$$

The first term of (18.63) is the interaction energy of the $2j$ neutrons as given by (14.37). The other two terms of (18.63) are equal to the r.h.s. of (14.46) which gives the particle-hole interaction energy. The summations in (18.63) are over T' and J' but they amount to summing over J' only. States of the j^2 configuration with even values of J' have $T'=1$ and those with odd J' have $T'=0$.

Let us conclude by assigning the isospin values T to the various states (18.53). Let us first consider the state with $J = 0$ given by

$$\left(\sum_m \frac{(-1)^{j-m}}{\sqrt{2j+1}} a^+_{jm1/2}\tilde{a}_{j,-m1/2}\right)|c\rangle$$
$$= \frac{1}{\sqrt{2j+1}}\left(\sum_m a^+_{jm1/2}a_{jm,-1/2}\right)|c\rangle \qquad (18.64)$$

The operator acting on $|c\rangle$ in (18.64) is equal, apart from the factor $(2j+1)^{-1/2}$, to the operator T^+ in (17.44) acting on $|c\rangle$. Hence, the state (18.64) has the same isospin as $|c\rangle$, $T_0 = (2j+1)/2$ and $M_{T_0} = -(2j+1)/2 + 1$. Its interaction energy, as given by (18.63) with $J = 0$, is indeed equal to

$$\frac{2j-1}{2j+1}\sum_{J'}(2J'+1)V_{T'=1,J}(jjjj) + \frac{1}{2j+1}\sum_{T'J'}(2J'+1)V_{T'J'}(jjjj)$$
$$-\sum_{T'J'}(2J'+1)\frac{(-1)^{2j+J'}}{2j+1}V_{T'J'}(jjjj)$$
$$= \frac{2j-1}{2j+1}\sum_{J'}(2J'+1)V_{T'=1,J'}(jjjj)$$
$$+\frac{2}{2j+1}\sum_{J'}(2J'+1)V_{T'=1,J'}(jjjj) \qquad (18.65)$$

The last equality in (18.65) is due to the relation $(-1)^{T'+J'} = -1$. The r.h.s. of (18.65) includes only interactions in $T' = 1$ states and is equal to the interaction energy of the j^{2j+1} configuration in $|c\rangle$ as given by (14.16).

Let us now determine the isospins of other states (18.53) with $J > 0$. There is only one state with $T_0 = (2j+1)/2$ in the j^{2j+1} configuration as can be seen by going to the state with $M_{T_0} = -(2j+1)/2$ which is the full j-neutron orbit, $|c\rangle$. Hence, all other states (18.53) with $J > 0$ must have $T = T_0 - 1 = (2j-1)/2$. As mentioned above, the isospin T of the state (18.53) does not enter explicitly into the Pandya relation and the interaction energies of all states are givevn by the same formula (18.65).

In the expressions of **G** as a linear combination of scalar products of irreducible tensor operators we introduced the restrictions $j_1 \leq j_2$,

$j_1' \leq j_2'$ or $j_1 \geq j_2$, $j_1' \leq j_2'$ to avoid double counting of the same matrix elements. As a result, some irreducible tensors appear in the expansion but their hermitean conjugates do not, even though the operator **G** is hermitean. Along with $(a_{j_1}^{+} \times \tilde{a}_{j_1'})^{(k)}$ the term with $(a_{j_1'}^{+} \times \tilde{a}_{j_1})^{(k)}$ will appear if $j_1' \leq j_2$ and $j_1 \leq j_2'$ but not in other cases. This can be fixed if we start from (17.34) instead of (17.33) and then obtain, instead of (18.11), the expression

$$\mathbf{G} = \sum_{j_1 j_2 j_1' j_2' k} \left[(1+\delta_{j_1 j_2})(1+\delta_{j_1' j_2'})\right]^{1/2} F_k(j_1 j_2 j_1' j_2') \times (a_{j_1}^{+} \times \tilde{a}_{j_1'})^{(k)} \cdot (a_{j_2}^{+} \times \tilde{a}_{j_2'})^{(k)} - \frac{1}{2} \sum_{jk} \frac{1}{2j+1} F_k(jjjj) \hat{n}_j \tag{18.66}$$

19

Nucleon Pairing and Seniority

The calculation of fractional parentage coefficients described in Section 15 starts from a certain principal parent. It was explained that antisymmetric states arising from different principal parents need not be independent. For example, in the $(\frac{5}{2})^3$ configuration of identical nucleons there is only one antisymmetric state with $J = \frac{5}{2}$ (and given M). The same state can be obtained from $(\frac{5}{2})^2$ principal parents with either $J_0 = 0$, $J_0 = 2$ or $J_0 = 4$. In other cases, however, there may be *several* independent states with the same value of J (and M) in the j^n configuration. For example, in the $(\frac{9}{2})^3$ configuration of identical nucleons there are two antisymmetric states with $J = \frac{9}{2}$ (and M). In such a case it is necessary to find an orthogonal basis for these two $J = \frac{9}{2}$ states. Another example is the $(\frac{7}{2})^4$ configuration of identical nucleons which has two states with $J = 2$ and two states with $J = 4$. It is convenient to find basis states in such cases which can be characterized (or labelled) by the eigenvalues of a simple operator. Such quantum numbers then uniquely characterize the various states. The Hamiltonian matrix within the j^n configuration of identical nucleons has then its rows and columns labelled by these quantum numbers in addition to J (and M). The submatrix for a given J (and M) which should be diagonalized, has rows and columns labelled by these additional quantum numbers. We shall now describe a scheme that can distinguish in simple cases of practical importance, between states with the same value of J. It will also facilitate the calculation of matrix elements of various operators. It will turn out that this scheme and its generalizations give a very good description of certain nuclear states and energies.

The *seniority scheme* was introduced by Racah (1943) for the classification of states in atomic spectra. His aim was to find an additional quantum number in order to distinguish between states of electron l^n configurations which have the same values of S, L and J (and M). This problem arises more frequently in LS-coupling, which is the prevalent coupling scheme of atomic electrons, than in jj-coupling. As mentioned above, this problem arises in j^n configurations of identical nucleons only for $j \geq \frac{7}{2}$. The seniority scheme for l^n configurations was introduced by Racah (1943) and for j^n configurations it was introduced by Racah (1952) and independently by Flowers (1952). In this section we shall consider j^n configurations of identical nucleons using the formalism introduced by Kerman (1961). The seniority scheme for protons and neutrons in the same orbit will be described in a subsequent section.

The scheme introduced by Racah is based on the idea of *pairing* of particles into $J = 0$ pairs. Loosely speaking, the seniority quantum number v (Vetek in Hebrew means seniority), which will be defined below, is equal to the number of unpaired particles in the j^n configuration. The state with $J = j$ of one particle has obviously no pairs and is assigned $v = 1$. In the j^2 configuration there is complete pairing in the $J = 0$ state and its seniority is $v = 0$. In all other j^2 states, with $J = 2, 4, \ldots, 2j-1$, there are no pairs coupled to $J = 0$ and their seniority is $v = 2$.

To make the notion of pairing and seniority more precise we introduce the pair creation operator

$$\boxed{\begin{aligned} S_j^+ &= \sqrt{\frac{2j+1}{2}} A^+(j^2 J = 0, M = 0) \\ &= \frac{1}{2}\sum (-1)^{j-m} a_{jm}^+ a_{j,-m}^+ \end{aligned}} \tag{19.1}$$

Acting on the vacuum state, (19.1) creates a state with $J = 0$ which is not normalized to 1. The reason for the normalization of (19.1) will become clear in the following. The hermitean conjugate of (19.1), $S_j^- = (S_j^+)^\dagger$, is the annihilation operator of a $J = 0$ pair. We can thus define states with seniority v in the j^v configuration by

$$\boxed{S_j^- |j^v, v, J, M\rangle = 0} \tag{19.2}$$

If to such a state we add pairs of particles coupled to $J = 0$ by

$$(S_j^+)^{(n-v)/2} |j^v vJM\rangle \tag{19.3}$$

we obtain a state with the same seniority v in the j^n configuration. By this definition, in states of j^n configurations with n even (odd) the seniorities v are even (odd). In the example mentioned above, there are two independent states with $J = \frac{9}{2}$ in the $(\frac{9}{2})^3$ configuration. We can form a linear combination of these two states which is annihilated by $S_{9/2}^-$. This state has seniority $v = 3$. The orthogonal $J = \frac{9}{2}$ state, as will be shown below, may be obtained by operating with $S_{9/2}^+$ on the state of a $j = \frac{9}{2}$ single particle state and thus has seniority $v = 1$.

In order to prove the consistency of the scheme, Racah (1943) introduced a special hermitean operator. The seniority scheme is defined as the scheme of eigenstates of the operator

$$\boxed{\begin{aligned} P = 2S_j^+ S_j^- &= (2j+1)A^+(j^2 J = 0, M = 0) \\ &\times A(j^2 J = 0, M = 0) \end{aligned}} \tag{19.4}$$

Comparing this expression with (17.33) we see that the eigenvalues of (19.4) in the j^2 states are given by

$$(2j+1)\delta_{J0}\delta_{M0} \tag{19.5}$$

Thus, the operator P introduced by Racah is the pairing interaction (12.17). There is an elegant way to calculate the eigenvalues of P in the states (19.3) and prove that all eigenstates of P have this form. The commutation relation between S_j^+ and S_j^- is directly calculated to be

$$[S_j^+, S_j^-] = \frac{1}{4}\sum_{mm'}(-1)^{j-m+j-m'}[a_{jm}^+ a_{j,-m}^+, a_{j,-m'} a_{jm'}]$$
$$= \sum_m a_{jm}^+ a_{jm} - \frac{2j+1}{2} \tag{19.6}$$

Apart from a constant term, the commutator is equal to the number operator $\hat{n}_j$ defined by (17.15). Its commutation relations with S_j^+ and S_j^- follow directly from (17.18). Hence, if we define the r.h.s. of (19.6) to be equal to $2S_j^0$, we obtain the following set of commutation relations

$$[S_j^+, S_j^-] = 2S_j^0 \qquad [S_j^0, S_j^+] = S_j^+ \qquad [S_j^0, S_j^-] = -S_j^- \tag{19.7}$$

Thus, the three operators (19.7) are generators of the Lie algebra of $SU(2)$ or $O(3)$, as defined by (6.19). The properties of the angular momentum algebra are well known. We can thus make use of these *quasi spin operators*, S_j^+, S_j^- and S_j^0, to calculate the eigenvalues of (19.4) and to derive many properties of the seniority scheme.

A straightforward calculation yields

$$S_j^+ S_j^- = \mathbf{S}_j^2 - (S_j^0)^2 + S_j^0 = \mathbf{S}^2 - S_j^0(S_j^0 - 1)$$
$$= \mathbf{S}_j^2 - \frac{1}{4}\left(\frac{2j+1}{2} - n\right)\left(\frac{2j+5}{2} - n\right) \tag{19.8}$$

The eigenvalues of the square of the quasi-spin vector $\mathbf{S}_j$ are given by $s(s+1)$ where s is an integer or half-integer. The eigenvalues of P are thus determined by s and n to be

$$s(s+1) - \frac{1}{4}\left(\frac{2j+1}{2} - n\right)\left(\frac{2j+5}{2} - n\right) \tag{19.9}$$

To express these eigenvalues by the seniority quantum number v, we apply (19.8) to a state $|j^v vJM\rangle$ which according to the definition (19.2) belongs to the eigenvalue 0 of P. We thus obtain

$$0 = s(s+1) - \frac{1}{4}\left(\frac{2j+1}{2} - v\right)\left(\frac{2j+5}{2} - v\right)$$

This is a quadratic equation which determines s in terms of v. Since $s \geq 0$, the negative solution is ruled out and hence

$$\boxed{s = \frac{1}{2}\left(\frac{2j+1}{2} - v\right)} \tag{19.10}$$

From this follows that v cannot exceed $(2j+1)/2$ which is the number of identical nucleons in the middle of the j-orbit. The eigenvalues of P are thus functions of v and n (or s and n) given by

$$\begin{aligned} P(n,v) &= 2s(s+1) - \frac{1}{2}\left(\frac{2j+1}{2} - n\right)\left(\frac{2j+5}{2} - n\right) \\ &= \frac{n-v}{2}(2j+3-n-v) \end{aligned} \tag{19.11}$$

The states (19.3) are eigenstates of v (with eigenvalues (19.11)) since operating on the state (19.2) by S_j^+ changes only the eigenvalue of S_j^0 (i.e. n) but not the value of s (or v). This holds for any state with definite seniority in the j^n configuration. Operating on it by S_j^+ or S_j^- (unless $n = v$) does not change the seniority.

The operator P in (19.4) measures the amount of $J = 0$ pairing in a given state. This amount, however, is not proportional to $n - v$. It is only as long as both n and v are small compared to $2j + 1$ that the eigenvalues (19.11) are given approximately by $(n-v)(2j+1)/2$. There is in (19.11) a term quadratic in n which is due to the Pauli principle. Due to the limited number of m-states available, $J = 0$ pairs cannot be independent of each other and the amount of pairing, for given v, decreases if n exceeds $(2j+3)/2$. In fact, starting from the vacuum state which has seniority $v = 0$ according to (19.2) we can operate on it by $(S_j^+)^{(2j+1)/2}$ to obtain the state of a full j-orbit. The eigenvalue of P for that state is given by (19.11) with $v = 0$ and $n = 2j + 1$ to be equal to $2j + 1$. This is equal to the eigenvalue of P for *one* $J = 0$ pair as given by (19.5) (or by (19.11) with $n = 2$, $v = 0$). Indeed, if all m-states are occupied, any state with given $J = 0, 2, \ldots, 2j - 1$ occurs with the weight $2J + 1$. Only one pair may be coupled to $J = 0$.

In nuclei, pairing and seniority play a much more important role than in atoms. The most interesting states in semi-magic nuclei are determined by maximum pairing and thus have the lowest seniorities. In atoms, where the mutual interaction of electrons is repulsive, states with lowest seniorities lie highest and may not even be observed. In even semi-magic nuclei the $J = 0$ ground state may be assigned seniority $v = 0$. It is obtained from the vacuum state by

$$(S_j^+)^{n/2}|0\rangle \tag{19.12}$$

The vacuum state $|0\rangle$ in (19.12) is any state with closed shells and no j-nucleons. Similarly, for even values of n, $2j - 1 \geq n \geq 2$ there should be states with $J = 2, 4, \ldots, 2j - 1$ and seniority $v = 2$ defined by

$$(S_j^+)^{(n-2)/2} A^+(j^2 JM)|0\rangle \tag{19.13}$$

For any odd value of n there is a state with $J = j$ and seniority $v = 1$ defined by

$$(S_j^+)^{(n-1)/2} a_{jm}^+|0\rangle \tag{19.14}$$

In the following, we shall present arguments for identifying these states with actual states observed in semi-magic nuclei, where levels are due to configurations of identical nucleons. This means that states with definite seniorities are eigenstates of the $T = 1$ part of the effective nuclear Hamiltonian. In the next section we will derive properties of shell model Hamiltonians which are diagonal in the seniority scheme.

We can now consider many particle states in the seniority scheme. We saw that states with different seniorities in the j^n configuration are orthogonal since they belong to different eigenvalues of the hermitean operator (19.4). Starting with a state with v and given J (and M) in the j^v configuration we obtain by operating on it by S_j^+ a state with the same seniority v, as well as J (and M), in the j^{v+2} configuration. If there is in that configuration another state with the same value of J (and M) which is annihilated by S_j^-, it has the eigenvalue zero of

(19.4) and hence seniority $v+2$. This process may be repeated and thus all j^n configurations can be reached.

It may happen that in a given j^v configuration there are two or more independent states with seniority v which have the same value of J (and M). For example there are two $v=4$ states with $J=2$ and two $v=4$ states with $J=4$ in the $(\frac{9}{2})^4$ configuration. Another example is offered by the $(\frac{11}{2})^4$ configuration with two $J=12$ states with seniority $v=4$. In that case, an orthogonal basis for such states must be chosen and the basis states are characterized by some additional label or quantum number α. It is a remarkable property of the seniority scheme that states obtained from any two such orthogonal j^v states, are still orthogonal after the S_j^+ operator has been applied to them $(n-v)/2$ times. Thus, the labels introduced in the j^v configuration may serve for the states in any j^n configuration (with even $n-v$ values).

To prove this statement, let us start from two such orthogonal and normalized states $|j^v v\alpha JM\rangle$ and $|j^v v\alpha' JM\rangle$ and calculate the overlap

$$\langle j^v v\alpha JM|(S_j^-)^r(S_j^+)^r|j^v v\alpha' JM\rangle \tag{19.15}$$

For $\alpha' \equiv \alpha$, the overlap (19.15) gives the normalization factor for the states $(S_j^+)^{(n-v)/2}|j^v v\alpha JM\rangle$ if we put $r=(n-v)/2$. By commuting one S_j^- operator with one S_j^+ operator we obtain for (19.15) the result

$$\langle j^v v\alpha JM|(S_j^-)^{r-1}([S_j^-,S_j^+]+S_j^+S_j^-)(S_j^+)^{r-1}|j^v v\alpha' JM\rangle \tag{19.16}$$

The commutator $[S_j^-,S_j^+]$ is equal to $-2S_j^0$ and the eigenvalues of $S_j^+S_j^-$ are given by (19.11). We thus obtain a recursion relation

$$\begin{aligned}\langle j^v v\alpha JM|&(S_j^-)^r(S_j^+)^r|j^v v\alpha' JM\rangle\\ &=\tfrac{1}{2}[r(2j+3-2v-2r)]\langle j^v v\alpha JM|(S_j^-)^{r-1}(S_j^+)^{r-1}|j^v\alpha' JM\rangle\end{aligned} \tag{19.17}$$

This relation may be repeated from $r=(n-v)/2$ down to $r=1$ leading to the final expression for (19.15)

$$\boxed{\begin{aligned}\langle j^v v\alpha JM|&(S_j^-)^{(n-v)/2}(S_j^+)^{(n-v)/2}|j^v v\alpha' JM\rangle\\ &=\mathcal{N}_{n,v}^{-2}\langle j^v v\alpha JM\mid j^v v\alpha' JM\rangle=\mathcal{N}_{n,v}^{-2}\delta_{\alpha\alpha'}\end{aligned}} \tag{19.18}$$

The relation (19.18) proves our assertion about orthogonality of the states $(S_j^+)^{(n-v)/2}|j^v v\alpha JM\rangle$ and $(S_j^+)^{(n-v)/2}|j^v v\alpha' JM\rangle$. If we take $\alpha \equiv \alpha'$ we obtain the normalization factor which is independent of α, J (and M) given by

$$\mathcal{N}_{n,v}^{-2} = \prod_{r=1}^{(n-v)/2} r\left(\frac{2j+3}{2} - v - r\right) = \left(\frac{n-v}{2}\right)! \frac{\left(\frac{2j+1}{2} - v\right)!}{\left(\frac{2j+1-n-v}{2}\right)!} \tag{19.19}$$

Thus, the state $\mathcal{N}_{n,v}(S_j^+)^{(n-v)/2}|j^v v\alpha JM\rangle$ is normalized to unity.

The quasi-spin operators in (19.7) are very useful in obtaining properties of states and observables in the seniority scheme. In the case of angular momentum we considered the transformation properties of states and operators under rotations. The components of the angular momentum are generators of infinitesimal rotations and the behaviour of operators under the latter is determined by their commutation relations with those components. We introduce here irreducible tensor operators whose components transform among themselves under infinitesimal rotations defined by the quasi-spin operators. Thus, the commutation relations of the σ component of an irreducible tensor operator of rank s with the quasi-spin operators should be, in analogy with (6.25), as follows:

$$\begin{aligned}
&[S_j^0, T_\sigma^{(s)}] = \sigma T_\sigma^{(s)} \\
&[S_j^+, T_\sigma^{(s)}] = \sqrt{s(s+1) - \sigma(\sigma+1)}\, T_{\sigma+1}^{(s)} \\
&[S_j^-, T_\sigma^{(s)}] = \sqrt{s(s+1) - \sigma(\sigma-1)}\, T_{\sigma-1}^{(s)}
\end{aligned} \tag{19.20}$$

We begin with the simplest operators—creation and annihilation operators. Their commutation relations with the quasi-spin operators determine the behavior of single nucleon states with respect to these rotations. We calculate directly

$$[S_j^0, a_{jm}^+] = \tfrac{1}{2} a_{jm}^+ \qquad [S_j^+, a_{jm}^+] = 0$$
$$[S_j^-, a_{jm}^+] = (-1)^{j-m} a_{j,-m} = -\tilde{a}_{jm} \tag{19.21}$$

Similarly, we obtain

$$[S_j^0, (-1)^{j-m} a_{j,-m}] = -\tfrac{1}{2}(-1)^{j-m} a_{j,-m}$$
$$[S_{j,}^- (-1)^{j-m} a_{j,-m}] = 0$$
$$[S_j^+, (-1)^{j-m} a_{j,-m}] = a_{jm}^+ \tag{19.22}$$

We see that a_{jm}^+ and $(-1)^{j-m} a_{j,-m} = -\tilde{a}_{jm}$ are the $\sigma = \frac{1}{2}$ and $\sigma = -\frac{1}{2}$ components of an irreducible quasi-spin tensor of rank $s = \frac{1}{2}$.

We can use the Wigner-Eckart theorem to relate matrix elements of quasi-spin tensors between states with given s and s' with different values of S_j^0. Starting with the creation operators a_{jm}^+ we may relate in this way c.f.p. of states in different j^n configurations. We start with the relation between c.f.p. and reduced matrix elements of a_j^+. In the seniority scheme, (17.30) may be written explicitly as

$$[j^{n-1}(v_1\alpha_1 J_1) jJ |\} j^n v\alpha J]$$
$$= \frac{(-1)^n}{\sqrt{n}} \frac{1}{\sqrt{2J+1}} (j^n v\alpha J \| a_j^+ \| j^{n-1} v_1 \alpha_1 J_1) \tag{19.23}$$

The matrix element on the r.h.s. of (19.23) is reduced with respect to rotations in the three dimensional space, namely with respect to M.

In quasi-spin space a^+_{jm} is the $\sigma = \frac{1}{2}$ component of rank $s = \frac{1}{2}$ tensor. It connects a state with

$$s = \frac{1}{2}\left(\frac{2j+1}{2} - v\right) \quad \text{and} \quad S^0_j = \frac{1}{2}\left(n - \frac{2j+1}{2}\right)$$

to a state with

$$s_1 = \frac{1}{2}\left(\frac{2j+1}{2} - v_1\right) \quad \text{and} \quad S^0_j = \frac{1}{2}\left(n - 1 - \frac{2j+1}{2}\right).$$

We refer in the following to the dependence on S^0_j, i.e. on n, for given values of α, J and $\alpha_1 J_1$ and will not explicitly indicate them.

Using the Wigner-Eckart theorem in quasi-spin space for the matrix element in (19.23), we obtain

$$\begin{aligned}
&(j^n v\alpha J \| a^+_j \| j^{n-1}\alpha_1 v_1 J_1) \\
&= \langle s, S^0_j | (a^+_j)^{s=1/2}_{\sigma=1/2} | s_1, S^0_j - \tfrac{1}{2}\rangle \\
&= (-1)^{s-S^0_j} \begin{pmatrix} s & \frac{1}{2} & s_1 \\ -S^0_j & \frac{1}{2} & S^0_j - \frac{1}{2} \end{pmatrix} (s \| a^+_j \| s_1) \qquad (19.24)
\end{aligned}$$

There are, in general, two possible values of s_1 which differ from s by $\frac{1}{2}$. These are $s_1 = s - \frac{1}{2}$ and $s_1 = s + \frac{1}{2}$ which correspond to seniorities $v_1 = v + 1$ and $v_1 = v - 1$ respectively. We can now use the algebraic formulae for $3j$-symbols to obtain for $v_1 = v - 1$ the result

$$\begin{aligned}
&\langle s, S^0_j | (a^+_j)^{s=1/2}_{\sigma=1/2} | s + \tfrac{1}{2}, S^0_j - \tfrac{1}{2}\rangle \\
&= \sqrt{\frac{s - S^0_j + 1}{(2s+1)(2s+2)}} (s \| a^+_j \| s + \tfrac{1}{2}) \\
&= \sqrt{\frac{(2j + 3 - n - v)}{2(2s+1)(2s+2)}} (s \| a^+_j \| s + \tfrac{1}{2}) \qquad (19.25)
\end{aligned}$$

We can apply (19.25) to the case $n = v$ ($S_j^0 = -s$) and express the reduced matrix element in (19.24) in terms of

$$\langle s, -s|(a_j^+)^{s=1/2}_{\sigma=1/2}|s+\tfrac{1}{2}, -s-\tfrac{1}{2}\rangle = (j^v v\alpha J\|a_j^+\|j^{v-1}v-1, \alpha_1 J_1)$$

This gives, by using (19.23) the following relation between the c.f.p. in the j^n configuration in terms of those in the j^v configuration

$$\boxed{\begin{aligned}&[j^{n-1}(v-1, \alpha_1 J_1)jJ|\}j^n v\alpha J]\\ &\quad = \sqrt{\frac{v(2j+3-n-v)}{n(2j+3-2v)}}[j^{v-1}(v-1, \alpha_1 J_1)jJ|\}j^v v\alpha J]\end{aligned}}$$

(19.26)

From (19.26) we obtain for the special case $v = 1$, $J = j$ an explicit formula for the c.f.p.

$$[j^{n-1}(v_1 = 0, J_1 = 0)jJ = j|\}j^n v = 1, J = j] = \sqrt{\frac{2j+2-n}{n(2j+1)}}$$

The other possible value of s_1 in (19.24) is $s_1 = s - \frac{1}{2}$ ($v_1 = v+1$). The Wigner-Eckart theorem in this case is

$$\begin{aligned}&\langle s, S_j^0|(a_j^+)^{s=1/2}_{\sigma=1/2}|s-\tfrac{1}{2}, S_j^0-\tfrac{1}{2}\rangle\\ &\quad = -\sqrt{\frac{s+S_j^0}{2s(2s+1)}}(s\|a_j^+\|s-\tfrac{1}{2})\\ &\quad = -\sqrt{\frac{n-v}{2s(2s+1)}}(s\|a_j^+\|s-\tfrac{1}{2}) \qquad (19.27)\end{aligned}$$

The lowest configuration with seniority $v+1$ states is j^{v+1}. Therefore, we cannot reduce the c.f.p. in the j^n configuration to those in the j^v one. We can only reduce them into c.f.p. in the j^{v+2} configuration. Dividing the r.h.s. of (19.27) by the same expression for $n = v+2$, i.e.

$S_j^0 = -s + 1$, we obtain, in view of (19.23) the relation

$$[j^{n-1}(v+1, \alpha_1 J_1) jJ|\} j^n v\alpha J] = \sqrt{\frac{(v+2)(n-v)}{2n}} [j^{v+1}(v+1, \alpha_1 J_1) jJ|\} j^{v+2} v\alpha J] \tag{19.28}$$

From (19.28) we obtain the value of the special c.f.p. with $v = 0$, $J = 0$

$$[j^{n-1}(v_1 = 1, J_1 = j) jJ = 0|\} j^n v = 0, J = 0] = [jjJ = 0|\} j^2 J = 0] = 1.$$

From the normalization of the c.f.p. it follows that the state with $v_1 = 1$, $J_1 = j$ is the only (fractional) parent of the $v = 0$, $J = 0$ state. In other words, the antisymmetric wave function of the $v = 0$, $J = 0$ state is given by

$$\psi(j^n v = 0, J = 0, M = 0) = \psi(j^{n-1}(v_1 = 1, J_1 = j) j_n J = 0, M = 0)$$

The properties of the pairing operator P defined by (19.4) may be used to obtain an explicit formula for a special $n \to n-2$ c.f.p. If we use (15.32) to calculate the eigenvalues $P(n, v)$ we obtain

$$P(n, v) = \frac{n(n-1)}{2} \sum_{v_1 \alpha_1 J_1 J'} [j^{n-2}(v_1 \alpha_1 J_1) j^2(J') J|\} j^n v J]^2 \times \langle j^2 J' M' | P | j^2 J' M' \rangle$$

Since the two particle matrix element of P vanishes for $J' > 0$ and is equal to $(2j+1)$ for $J' = 0$, the only non-vanishing c.f.p. are those with $v_1 = v, \alpha_1 \equiv \alpha, J_1 = J$ and in view of (19.11) we obtain

$$[j^{n-2}(v\alpha J) j^2(0) J|\} j^n v\alpha J]^2 = \frac{(n-v)(2j+3-n-v)}{n(n-1)(2j+1)} \tag{19.29}$$

This c.f.p. depends on n and v but it is *independent* of J and α. This feature has important consequences for the discussion in Section 21.

The c.f.p. on the r.h.s. of (19.28) applies to the j^{v+2} configuration which is not the lowest one in which states with seniority v appear. We can further reduce this c.f.p. and express it in terms of a c.f.p. in the j^{v+1} configuration. According to (19.23) this c.f.p. can be expressed as

$$[j^{v+1}(v+1,\alpha_1 J_1)jJ|\}j^{v+2}v\alpha J] = \frac{(-1)^{v+2}}{\sqrt{v+2}}\frac{1}{\sqrt{2J+1}}$$
$$\times \sum_{M_1 m M} (-1)^{J-M} \begin{pmatrix} J & j & J_1 \\ -M & m & M_1 \end{pmatrix}$$
$$\times \mathcal{N}_{v+2,v} \langle j^v v\alpha JM | S_j^- a_{jm}^\dagger | j^{v+1} v+1, \alpha_1 J_1 M_1 \rangle \tag{19.30}$$

In (19.30), the normalization coefficient is defined by (19.22) and the reduced matrix element in (19.23) has been expressed by (8.11).

We now commute S_j^- and $a_{jm}^\dagger$ in the matrix element on the r.h.s. of (19.29). Using (19.21) we obtain

$$S_j^- a_{jm}^\dagger = [S_j^-, a_{jm}^\dagger] + a_{jm}^\dagger S_j^- = (-1)^{j-m} a_{j,-m} + a_{jm}^\dagger S_j^-$$

The operator S_j^- annihilates the state with seniority $v+1$ in the j^{v+1} configuration. Thus, we obtain

$$\langle j^v v\alpha JM | S_j^- a_{jm}^\dagger | j^{v+1} v+1, \alpha_1 J_1 M_1 \rangle$$
$$= (-1)^{j-m} \langle j^v v\alpha JM | a_{j,-m} | j^{v+1} v+1, \alpha_1 J_1 M_1 \rangle$$

These matrix elements are real and hence obey the relation

$$\langle j^v v\alpha JM | a_{j,-m} | j^{v+1} v+1, \alpha_1 J_1 M_1 \rangle$$
$$= \langle j^{v+1} v+1, \alpha_1 J_1 M_1 | a_{j,-m}^\dagger | j^v v\alpha JM \rangle$$

Substituting this expression into (19.30) we obtain the expression

$$\frac{(-1)^{v+2}}{\sqrt{v+2}}\frac{1}{\sqrt{2J+1}}\mathcal{N}_{v+2,v}\sum_{M_1 m M}(-1)^{J-M}\begin{pmatrix} J & j & J_1 \\ -M & m & M_1 \end{pmatrix}$$

$$\times(-1)^{j-m}\langle j^{v+1}v+1,\alpha_1 J_1 M_1|a^{\dagger}_{j,-m}|j^v v\alpha JM\rangle$$

$$=\frac{(-1)^{v+2}}{\sqrt{v+2}}\frac{1}{\sqrt{2J+1}}\mathcal{N}_{v+2,v}(-1)^{J-j-J_1}\sum_{M_1 m M}(-1)^{J_1-M_1}$$

$$\times\begin{pmatrix} J_1 & j & J \\ -M_1 & -m & M \end{pmatrix}$$

$$\times\langle j^{v+1},v+1,\alpha_1 J_1 M_1|a^{\dagger}_{j,-m}|j^v v\alpha JM\rangle$$

$$=\frac{(-1)^{v+2}}{\sqrt{v+2}}\frac{1}{\sqrt{2J+1}}\mathcal{N}_{v+2,v}(-1)^{J-j-J_1}(j^{v+1}v+1,\alpha_1 J_1\|a^{\dagger}_j\|j^v v\alpha J)$$

Substituting in this expression the value $\mathcal{N}_{v+2,v}=[(2j+1-2v)/2]^{-1/2}$ from (19.19) we obtain in view of (19.23) for $n=v+1$ the relation

$$\begin{aligned}&[j^{v+1}(v+1,\alpha_1 J_1)jJ|\}j^{v+2}v\alpha J]\\&\quad=(-1)^{J+j-J_1}\sqrt{\frac{2(2J_1+1)(v+1)}{(2J+1)(v+2)(2j+1-2v)}}\\&\quad\times[j^v(v\alpha J)jJ_1|\}j^{v+1}v+1,\alpha_1 J_1]\end{aligned} \tag{19.31}$$

The relation (19.31) is rather different from (19.26) and (19.28). In both sides of those relations are coefficients of expansion of states with even (odd) number of particles in terms of states with odd (even) particle number. The relation (19.31), however, has on the l.h.s. the expansion coefficients of states of even (odd) configurations in terms of states of odd (even) ones, whereas on the r.h.s. the c.f.p. relate states in odd (even) configurations to states in even (odd) ones. Relations similar to (19.31) for any value of n may be obtained by using on both sides of (19.31) the relations (19.26) and (19.28). For $n=3$,

(19.31) reduces to the c.f.p. in (15.11). Combining (19.28) with (19.31) we obtain for the case $v = 1$, $J = j$ the explicit expression

$$[j^{n-1}(v_1 = 2, J_1)jJ = j|\}j^n v = 1, J = j] = -\sqrt{\frac{2(n-1)(2J_1+1)}{n(2j+1)(2j-1)}}$$

$$J_1 > 0 \text{ even}$$

We may use (19.31) to derive a different type of orthogonality relations between c.f.p. in the seniority scheme. These relations are derived from the summation

$$\sum_{\alpha J}(2J+1)[j^{n-1}(v_1\alpha_1 J_1)jJ|\}j^n v\alpha J][j^{n-1}(v_1', \alpha_1' J_1)jJ|\}j^n v\alpha J]$$

by choosing $v_1 = v_1' = v+1, v_1 = v_1' = v-1$ and $v_1 = v-1, v_1' = v+1$. In the first case we reduce the c.f.p. by using (19.28) and then applying to them (19.31) to obtain

$$\frac{(n-v)(v+2)}{2n}\sum_{\alpha J}(2J+1)[j^{v+1}(v+1, \alpha_1 J_1)jJ|\}j^{v+2}v\alpha J]$$
$$\times [j^{v+1}(v+1, \alpha_1' J_1)jJ|\}^{v+2}v\alpha J]$$
$$= \frac{(n-v)(v+1)(2J_1+1)}{n(2j+1-2v)}\sum_{\alpha J}[j^v(v\alpha J)jJ_1|\}j^{v+1}v+1, \alpha_1 J_1]$$
$$\times [j^v(v\alpha J)jJ_1|\}j^{v+1}v+1, \alpha_1' J_1]$$

The sum over αJ of products of c.f.p. in the j^{v+1} configuration is, due to the ordinary orthogonality relations (15.14), equal to $\delta_{\alpha_1\alpha_1'}$. We thus obtain

$$\boxed{\begin{aligned}&\sum_{\alpha J}(2J+1)[j^{n-1}(v+1, \alpha_1 J_1)jJ|\}j^n v\alpha J]\\ &\qquad \times [j^{n-1}(v+1, \alpha_1' J_1)jJ|\}j^n v\alpha J]\\ &\quad = \frac{(n-v)(v+1)}{n(2j+1-2v)}(2J_1+1)\delta_{\alpha_1\alpha_1'}\end{aligned}} \quad (19.32)$$

In the other case, where $v_1 = v_1' = v - 1$, we first reduce the c.f.p. by using (19.26). We then use (19.31) for $v-1$ replacing v and obtain

$$\frac{v(2j+3-n-v)}{n(2j+3-2v)}\sum_{\alpha J}(2J+1)[j^{v-1}(v-1,\alpha_1 J_1)jJ|\}j^v v\alpha J]$$
$$\times [j^{v-1}(v-1,\alpha_1' J_1)jJ|\}j^v v\alpha J]$$
$$=\frac{(v+1)(2j+3-n-v)(2J_1+1)}{2n}$$
$$\times\sum_{\alpha J}[j^v(v\alpha J)jJ_1|\}j^{v+1}v-1,\alpha_1 J_1]$$
$$\times [j^v(v\alpha J)jJ_1|\}j^{v+1}v-1,\alpha_1' J_1]$$

The sum over αJ of products of c.f.p. in the j^{v+1} configuration may be complemented by products of c.f.p. of parent states with seniority $v-2$ to yield, due to (15.14), $\delta_{\alpha_1\alpha_1'}$. We thus obtain for that sum

$$\frac{(v+1)(2j+3-n-v)(2J_1+1)}{2n}$$
$$\times\Big\{\delta_{\alpha_1\alpha_1'}-\sum_{\alpha J}[j^v(v-2,\alpha J)jJ_1|\}j^{v+1}v-1,\alpha_1 J_1]$$
$$\times [j^v(v-2,\alpha J)jJ_1|\}^{v+1}v-1,\alpha_1' J_1]\Big\}$$
$$=\frac{(v+1)(2j+3-n-v)(2J_1+1)}{2n}$$
$$\times\Big\{\delta_{\alpha_1\alpha_1'}-\frac{(v-1)(2j+3-2v)}{(v+1)(2j+5-2v)}$$
$$\times\sum_{\alpha J}[j^{v-2}(v-2,\alpha J)jJ_1|\}j^{v-1}v-1,\alpha_1 J_1]$$
$$\times [j^{v-2}(v-2,\alpha J)jJ_1|\}j^{v-1}v-1,\alpha_1' J_1]\Big\}$$

The last equality was obtained by using (19.26) for $n = v+1$ and seniority $v-1$. The resulting sum over αJ of products of c.f.p. in the j^{v-1} configuration is equal, due to (15.14), to $\delta_{\alpha_1\alpha_1'}$. As a result we obtain the orthogonality relation

$$\boxed{\begin{aligned}&\sum_{\alpha J}(2J+1)[j^{n-1}(v-1,\alpha_1 J_1)jJ|\}j^n v\alpha J]\\ &\qquad\times[j^{n-1}(v-1,\alpha_1' J_1)jJ|\}j^n v\alpha J]\\ &\qquad=\frac{(2j+3-n-v)(2j+4-v)(2J_1+1)}{n(2j+5-2v)}\delta_{\alpha_1\alpha_1'}\end{aligned}} \tag{19.33}$$

In the last case, where $v_1 = v-1, v_1' = v+1$, we first use (19.26) and (19.28) to reduce the c.f.p. to those in the j^v and j^{v-2} configurations respectively. We then use (19.31) for seniorities $v-1$ and v respectively to obtain a sum of bilinear products of c.f.p. in the j^{v+1} configuration of states with seniorities $v-1$ and $v+1$ respectively. Due to the orthogonality of states with different seniorities we obtain the orthogonality relation

$$\boxed{\begin{aligned}&\sum_{\alpha J}(2J+1)[j^{n-1}(v-1,\alpha_1 J_1)jJ|\}j^n v\alpha J]\\ &\qquad\times[j^{n-1}(v+1,\alpha_1' J_1)jJ|\}j^n v\alpha J]=0\end{aligned}} \tag{19.34}$$

We saw above that creation and annihilation operators are the $\pm\frac{1}{2}$ components of quasi-spin tensors of rank $s=\frac{1}{2}$. Hence, higher rank quasi-spin tensors may be formed by taking linear combinations of products of creation and annihilation operators. The products appearing in the σ component contain a certain number α of annihilation operators and $\alpha+2\sigma$ creation operators. Hence, only $\sigma=0$ components have non-vanishing matrix elements between states with the same number of particles.

As we saw, S_j^+ is proportional (with factor $-2^{-1/2}$) to the $\sigma=1$ component of a special tensor with rank $k=1$ which is the quasi-spin vector $\mathbf{S}_j$. We can consider also operators which create a pair of particles with $J>0$. Recalling the definition (17.22) for even values of J, we obtain

$$[S_j^+, A^+(j^2JM)] = 0 \qquad [S_j^0, A^+(j^2JM)] = A^+(j^2JM)$$

Hence, $A^+(j^2JM)$ is the $\sigma = 1$ component of an irreducible quasi-spin tensor operator with rank $s = 1$. To find the $\sigma = 0$ component of that tensor we calculate

$$[S_j^-, A^+(j^2JM)] = \frac{1}{\sqrt{2}} \sum_{mm'} (jmjm' \mid jjJM)[S_j^-, a_{jm}^+ a_{jm'}^+]$$

$$= \sqrt{2} \left[-(a_j^+ \times \tilde{a}_j)_M^{(J)} + \frac{\sqrt{2j+1}}{2} \delta_{J0} \delta_{M0} \right]$$

In view of (19.20), the $\sigma = 0$ component is thus equal to

$$-(a_j^+ \times \tilde{a}_j)_M^{(J)} + \frac{\sqrt{2j+1}}{2} \delta_{J0} \delta_{M0} \tag{19.35}$$

The operator (19.35) is a single particle operator which is the M-th component of an ordinary tensor operator with rank J. Thus, single particle tensor operators with *even* ranks $k > 0$ are the $\sigma = 0$ components of quasi-spin tensors of rank $s = 1$. The case of $k = 0$ is special since in this case a constant term must be added. For $J = 0$, (19.35) is equal to $-2S_j^0$ divided by $\sqrt{2j+1}$. The $\sigma = -1$ component of the quasi-spin tensor whose $\sigma = 1,0$ components are given by $A^+(j^2JM)$ and (19.35) respectively, is equal to the commutator of S_j^- and (19.35) divided by $\sqrt{2}$. It is equal to $(-1)^{M+1}A(j^2J, -M) = (1/\sqrt{2})(\tilde{a}_j \times \tilde{a}_j)_M^{(J)}$.

Tensor operators with *odd* ranks k have also $\sigma = 0$ as any product of one creation and one annihilation operators. They cannot, however, have quasi-spin rank $s = 1$ since $A^+(j^2JM)$ vanishes for odd values of J. Hence, such odd tensor operators must have quasi-spin ranks $s = 0$. We can check this statement by taking the commutator of S_j^+ with (19.35). We obtain

$$-[S_j^+, (a_j^+ \times \tilde{a}_j)_\kappa^{(k)}] = \sqrt{2} A^+(j^2 k \kappa) \tag{19.36}$$

which vanishes for odd values of k. Thus, odd rank tensor operators are *quasi-spin scalars*.

Single particle operators determine electromagnetic moments and transition probabilities. They also appear in tensor expansions of two body interactions as described in Section 18. The fact that odd tensor operators are quasi-spin scalars has important consequences for the seniority scheme in j^n configurations. According to the Wigner-Eckart theorem applied to the quasi-spin algebra, matrix elements of odd tensor operators vanish between states with different values of s (or seniority v). Moreover, their non-vanishing matrix elements are independent of the values of S_j^0, i.e. of n. We summarize these properties by

$$\boxed{\langle j^n v\alpha JM|T_\kappa^{(k)}|j^n v'\alpha' J'M'\rangle = \langle j^v v\alpha JM|T_\kappa^{(k)}|j^v v\alpha' J'M'\rangle\delta_{vv'} \qquad k \text{ odd}}$$

(19.37)

The α, α' are additional labels which may be necessary to distinguish between states with the same values of v, J (and M).

Even tensor operators are quasi-spin tensors with rank $s = 1$ and therefore, according to the Wigner-Eckart theorem, may have non-vanishing matrix elements between states with s and s' differing at most by 1. Thus, matrix elements of even tensor operators between two states of the j^n configuration vanish if the seniorities v, v' of the two states differ by more than 2. Moreover, the Wigner-Eckart theorem determines the dependence on n of non-vanishing matrix elements according to

$$\langle s, S_j^0|T_{\sigma=0}^{s=1}|s', S_j^0\rangle = (-1)^{s-S_j^0}\begin{pmatrix} s & 1 & s' \\ -S_j^0 & 0 & S_j^0 \end{pmatrix}(s\|T^{s=1}\|s') \tag{19.38}$$

The dependence of (19.38) on n is due to the $3j$-symbol and is different for the cases $s' = s$ and $s' = s \pm 1$. Substituting the actual values of the $3j$-symbols we obtain, for $s' = s$, the result

$$\begin{aligned} \langle s, S_j^0|T_{\sigma=0}^{s=1}|s, S_j^0\rangle &= \frac{2S_j^0}{\sqrt{2s(2s+1)(2s+2)}}(s\|T^{s=1}\|s) \\ &= \left(n - \frac{2j+1}{2}\right)\frac{(s\|T^{s=1}\|s)}{\sqrt{2s(2s+1)(2s+2)}} \end{aligned} \tag{19.39}$$

From (19.39) we obtain the dependence on n of matrix elements of even tensors with $k > 0$

$$\langle j^n v\alpha JM|T_\kappa^{(k)}|j^n v\alpha' J'M'\rangle = \frac{2j+1-2n}{2j+1-2v}\langle j^v v\alpha JM|T_\kappa^{(k)}|j^v v\alpha' J'M'\rangle \qquad k > 0 \text{ even} \tag{19.40}$$

From (19.40) follows that matrix elements of even tensor operators, $k > 0$, between states with the same seniority v, are equal in absolute value but have opposite signs in the j^n and j^{2j+1-n} configurations. This is in agreement with the relation (14.39) obtained above. As a function of n, the matrix element (19.40) changes its sign at the middle of the orbit. In particular, all even rank tensors, with $k > 0$, have vanishing matrix elements in the $j^{(2j+1)/2}$ configuration. This follows from (19.40) only for states with $v < (2j+1)/2$, but it is true also in states with $v = (2j+1)/2$. In such states, $s = 0$ due to (19.10) and matrix elements between $s = 0$ states of a quasi-spin tensor of rank $s = 1$ must vanish.

The case of $k = 0$ is a special one since, according to (19.35), the rank $k = 0$ tensor is equal to the $\sigma = 0$ component of a quasi-spin tensor of rank $s = 1$ to which a constant term is added. The simplest way to calculate matrix elements of $T_0^{(0)}$ is to recall (17.15)

$$T_0^{(0)} = \sum_{mm'}(jmjm' \mid jj00)a_{jm}^{+}\tilde{a}_{jm'} = \frac{1}{\sqrt{2j+1}}\sum_m a_{jm}^{+}a_{jm} \tag{19.41}$$

Thus, a single particle operator with rank $k = 0$ is simply proportional to the number operator. It has only diagonal elements in *any* scheme, which are all equal and proportional to n. As remarked above, the reason for the constant term in (19.35) is simple. We can express (19.35) explicitly by

$$-\frac{1}{\sqrt{2j+1}}\left(\sum_m a_{jm}^{+}a_{jm} - \frac{2j+1}{2}\right) = -\frac{2}{\sqrt{2j+1}}S_j^0 \tag{19.42}$$

Thus, apart from a constant factor, the $k = 0$ tensor operator is proportional to the $\sigma = 0$ component of the quasi-spin vector $\mathbf{S}_j$.

The other case to be considered for even tensors with $k > 0$ is $s' = s + 1$, i.e. $v' = v - 2$. Inserting the actual values of the $3j$-symbols into (19.38), we obtain

$$\langle s, S_j^0 | T_{\sigma=0}^{s=1} | s+1, S_j^0 \rangle$$
$$= -\sqrt{\frac{4(s+S_j^0+1)(s-S_j^0+1)}{2(2s+1)(2s+2)(2s+3)}}(s\|T^{s=1}\|s+1)$$
$$= -\sqrt{\frac{(n-v+2)(2j+3-n-v)}{2(2s+1)(2s+2)(2s+3)}}(s\|T^{s=1}\|s+1) \tag{19.43}$$

The last equality in (19.43) is due to the relation (19.10) between s and v. From (19.43) follows for any even tensor

$$\boxed{\begin{aligned}&\langle j^n v\alpha JM|T_\kappa^{(k)}|j^n v-2, \alpha' J' M'\rangle\\ &= \sqrt{\frac{(n-v+2)(2j+3-n-v)}{2(2j+3-2v)}}\\ &\quad\times \langle j^v v\alpha JM|T_\kappa^{(k)}|j^v v-2, \alpha' J' M'\rangle \qquad k \text{ even}\end{aligned}} \tag{19.44}$$

The coefficient of the matrix element on the r.h.s. of (19.44) is symmetric between n and $2j+1-n$. The expression (19.44) is the same for states of n nucleons and states of n nucleon holes. In (19.40) there is a change of sign between states of the j^n configuration and states of the complementary j^{2j+1-n} configuration. That is in agreement with the result (14.39) which refers to expectation values. The matrix elements in (19.44), however, are all non-diagonal. Thus their signs depend on the phase relation between nucleon and hole states for the two states of the j^n configuration.

The results (19.39), (19.40) and (19.44) demonstrate the power and elegance of the seniority scheme. Once matrix elements have been calculated in the lowest j^v configuration, they are completely determined in any j^n configuration by explicit and simple expressions involving only n and v. It is therefore very convenient to use the formalism of the seniority scheme even if there is no need for additional quantum numbers like in j^n configurations with $j \leq \frac{5}{2}$.

Let us now consider two-body operators in the seniority scheme. The most important among those is the two nucleon interaction Hamiltonian. Within the j^n configurations the two-nucleon interaction (17.33) may be expressed as

$$V = \sum_{JM} V_J A^+(j^2JM)A(j^2JM) \tag{19.45}$$

In (19.45), to simplify the notation, V_J stands for $V_J(jjjj)$. Since $A^+(j^2JM)$ is the $\sigma = 1$ component and $A(j^2JM)$ the $\sigma = -1$ component of quasi-spin tensors of rank $s = 1$, the Hamiltonian (19.45) has $\sigma = 0$ and may be a linear combination of quasi-spin tensors with ranks $s = 0$, $s = 1$, or $s = 2$. Hence, any two-body interaction can have non-vanishing matrix elements between states whose seniorities differ at most by 4.

In order to obtain the dependence of matrix elements of the two-body interaction on n, it is more convenient to start with the tensor expansion (18.11). Within j^n configurations, this expansion of any two-body interaction is given by

$$V = \frac{1}{2}\sum_k \frac{F_k}{2k+1}(a_j^+ \times \tilde{a}_j)^{(k)} \cdot (a_j^+ \times \tilde{a}_j)^{(k)} - \frac{1}{2(2j+1)}\sum_k F_k \hat{n}_j \tag{19.46}$$

where F_k stands for $F_k(jjjj)$. The last term in (19.46) is proportional to n and thus, apart from a constant, proportional to S_j^0. The coefficient of S_j^0 in (19.46) is given, according to (18.16) by

$$\frac{1}{(2j+1)}\sum_{J'}(-1)^{J'}(2J'+1)V_{J'} \tag{19.47}$$

This is, however, not the only term in (19.46) which has quasispin rank $s = 1$. The first term is a product of two $\sigma = 0$ components of tensors of rank $s = 0$ in quasi-spin if k is odd and rank $s = 1$ if k is even and $k > 0$. Hence, the scalar product of any two odd tensors has rank $s = 0$. The product of two components of any even tensors with $k > 0$ may have rank $s = 0$, $s = 1$ or $s = 2$ which are proportional to

$$\begin{pmatrix} 1 & 1 & s \\ 0 & 0 & 0 \end{pmatrix}.$$

Since

$$\begin{pmatrix} 1 & 1 & 1 \\ 0 & 0 & 0 \end{pmatrix} = 0$$

due to the symmetry property (7.24) of $3j$-symbols, the scalar products in (19.46), with $k > 0$, may contain only quasi-spin tensors with ranks $s = 0$ and $s = 2$.

We must still look at the product of rank $k = 0$ tensors in the first part of (19.46). Due to (19.42) we may express the $k = 0$ tensors by

$$(a_j^+ \times \tilde{a}_j)_0^{k=0} = \frac{1}{\sqrt{2j+1}}\hat{n}_j = \frac{2S_j^0}{\sqrt{2j+1}} + \frac{\sqrt{2j+1}}{2}$$

The term with $k = 0$ in the first part of (19.46) is then equal to

$$\tfrac{1}{2}F_0(a_j^+ \times \tilde{a}_j)_0^{(0)}(a_j^+ \times \tilde{a}_j)_0^{(0)}$$

$$= \frac{F_0}{2(2j+1)}\left[4(S_j^0)^2 + 2(2j+1)S_j^0 + \frac{(2j+1)^2}{4}\right] \tag{19.48}$$

The first term on the r.h.s. of (19.48) is indeed a linear combination of quasi-spin tensors with ranks $s = 2$ and $s = 0$. The second term, however, is the $\sigma = 0$ component of a special quasi-spin tensor of rank $s = 1$, namely the quasi-spin vector $\mathbf{S}_j$. Hence, the $s = 1$ part of V is proportional to S_j^0 and the proportionality factor is the sum of (19.47) and the coefficient of S_j^0 in (19.48). We thus obtain for this sum in view of (18.16) the value

$$F_0 - \frac{1}{2j+1}\sum_k F_k$$

$$= \frac{1}{2j+1}\left(\sum_{J'}(2J'+1)V_{J'} + \sum_{J'}(-1)^{J'}(2J'+1)V_{J'}\right)$$

$$= \frac{2}{2j+1}\sum_{J'\,\text{even}}(2J'+1)V_{J'} = E_0$$

(19.49)

Let us just make sure that the contribution of the $k = 0$ term is the same for all states and is proportional to $n(n-1)$. Putting $F_k = 0$ for all $k \neq 0$ in (19.46) we obtain indeed

$$\frac{F_0}{2}\frac{n^2}{2j+1} - \frac{F_0}{2(2j+1)}n = \frac{F_0}{(2j+1)}\frac{n(n-1)}{2}.$$

Let us now consider matrix elements of V between states with seniorities v and $v-4$, i.e. s and $s+2$. Only the $s = 2$ part of V may contribute to such matrix elements. Making use of the Wigner-Eckart theorem for the $SU(2)$ quasi-spin group we obtain

$$\langle s, S_j^0 | V_{\sigma=0}^{s=2} | s+2, S_j^0\rangle = (-1)^{s-S_j^0}\begin{pmatrix} s & 2 & s+2 \\ -S_j^0 & 0 & S_j^0 \end{pmatrix}(s\|V^{s=2}\|s+2)$$

$$= \sqrt{\frac{6(s+S_j^0+2)(s+S_j^0+1)(s-S_j^0+2)(s-S_j^0+1)}{(2s+5)(2s+4)(2s+3)(2s+2)(2s+1)}}$$

$$\times (s\|V^{s=2}\|s+2)$$

$$= \sqrt{\tfrac{3}{8}(n-v+4)(n-v+2)(2j+3-n-v)(2j+5-n-v)}$$

$$\times \frac{(s\|V^{s=2}\|s+2)}{\sqrt{(2s+1)(2s+2)(2s+3)(2s+4)(2s+5)}} \tag{19.50}$$

From (19.50) we deduce the recursion relation for the matrix elements of V between states with v and $v-4$

$$\langle j^n v\alpha JM|V|j^n v-4,\alpha' JM\rangle = \sqrt{\frac{(n-v+2)(n-v+4)(2j+3-n-v)(2j+5-n-v)}{8(2j+3-2v)(2j+5-2v)}} \times \langle j^v v\alpha JM|V|j^v v-4,\alpha' JM\rangle \tag{19.51}$$

Next we consider matrix elements of V between states with seniorities v and $v-2$. Both $V^{s=2}$ and $V^{s=1}$ parts of V may contribute to these matrix elements. We saw above, however, that $V^{s=1}$ is proportional to $\hat{n}_j$ and has no non-vanishing non-diagonal elements in any scheme. Hence, we obtain from the Wigner-Eckart theorem

$$\langle s, S_j^0|V^{s=2}|s+1, S_j^0\rangle = (-1)^{s-S_j^0}\begin{pmatrix} s & 2 & s+1 \\ -S_j^0 & 0 & S_j^0 \end{pmatrix}(s\|V^{s=2}\|s+1)$$

$$= 2S_j^0\sqrt{\frac{6(s+S_j^0+1)(s-S_j^0+1)}{(2s+4)(2s+3)(2s+2)(2s+1)2s}}(s\|V^{s=2}\|s+1)$$

$$= \left(n-\frac{(2j+1)}{2}\right)\sqrt{\frac{3}{2}(n-v+2)(2j+3-n-v)} \times \frac{(s\|V^{s=2}\|s+1)}{\sqrt{2s(2s+1)(2s+2)(2s+3)(2s+4)}} \tag{19.52}$$

From (19.52) follows the recursion relation

$$\boxed{\begin{aligned}&\langle j^n v\alpha JM|V|j^n v-2,\alpha' JM\rangle\\ &\qquad=\frac{2j+1-2n}{2j+1-2v}\sqrt{\frac{(n-v+2)(2j+3-n-v)}{2(2j+3-2v)}}\\ &\qquad\times\langle j^v v\alpha JM|V|j^v v-2,\alpha' JM\rangle\end{aligned}} \tag{19.53}$$

From (19.53) follows an interesting consequence for j^n configurations with $j=\frac{7}{2}$. We remarked above that in the $(\frac{7}{2})^4$ configuration there are two states with $J=2$ and two states with $J=4$. If we construct states with $J=2$ (or $J=4$) with seniorities $v=2$ and $v=4$, it follows from (19.53) that the matrix element of V between these states vanishes. This is the case in general in the middle of the orbit where $n=(2j+1)/2$. These $J=2$ and $J=4$ states are the only cases where states with the same value of J appear in any $(\frac{7}{2})^n$ configuration. Hence, we may use the quantum numbers of the seniority scheme to uniquely characterize all eigenstates of *any* two-body interaction in $(\frac{7}{2})^n$ configurations.

Finally we consider matrix elements of V between any two states with the same seniority v. To these matrix elements, parts of V with all possible values of quasi-spin rank, $s=0,1,2$, may contribute. We express V, according to the preceding discussion by

$$V=V^{s=0}+E_0 S_j^0+V^{s=2}+C \tag{19.54}$$

where C is a constant arising from the constant term in S_j^0, like in (19.48). Matrix elements of the quasi-spin scalar $V^{s=0}$ between states with the same seniority v are independent of S_j^0, i.e. of n. They are equal to those in the j^v configuration in which $S_j^0=-s$. Using the Wigner-Eckart theorem we obtain for the matrix elements between states with the same seniority the result

$$\begin{aligned}\langle s,S_j^0|V|s,S_j^0\rangle &= \langle s,-s|V^{s=0}|s,-s\rangle+E_0S_j^0\\ &\quad+(-1)^{s-S_j^0}\begin{pmatrix} s & 2 & s\\ -S_j^0 & 0 & S_j^0\end{pmatrix}(s\|V^{s=2}\|s)+C\end{aligned}$$

$$= \langle s,-s|V^{s=0}|s,-s\rangle + E_0 S_j^0 + \frac{2[3(S_j^0)^2 - s(s+1)]}{\sqrt{(2s+3)(2s+2)(2s+1)(2s)(2s-1)}} \times (s\|V^{s=2}\|s) + C \qquad (19.55)$$

The matrix elements in the j^n configuration defined by (19.50) and (19.52) have been reduced to matrix elements in *one* configuration, chosen to be the j^v one. In the matrix element (19.55), however, there are two terms with $s = 0$ and $s = 2$ with very different dependence on n. We may reduce the matrix element by using *two* configurations which could be conveniently chosen as the j^v and j^{v+2} configurations. We thus write (19.55) for these two configurations as

$$\langle s,-s|V|s,-s\rangle = \langle s,-s|V^{s=0}|s,-s\rangle - E_0 s + \frac{2(3s^2 - s(s+1))(s\|V^{s=2}\|s)}{\sqrt{(2s-1)(2s)(2s+1)(2s+2)(2s+3)}} + C \qquad (19.56)$$

$$\langle s,-s+1|V|s,-s+1\rangle = \langle s,-s|V^{s=0}|s,-s\rangle - E_0 s + E_0 + \frac{2(3(s-1)^2 - s(s+1))(s\|V^{s=2}\|s)}{\sqrt{(2s-1)(2s)(2s+1)(2s+2)(2s+3)}} + C \qquad (19.57)$$

Subtracting one equation from the other we obtain

$$\frac{(s\|V^{s=2}\|s)}{\sqrt{(2s-1)(2s)(2s+1)(2s+2)(2s+3)}} = -\frac{1}{6(2s-1)}[\langle s,-s+1|V|s,-s+1\rangle - \langle s,-s|V|s,-s\rangle - E_0] \qquad (19.58)$$

From (19.58) and (19.56) we obtain also

$$\langle s,-s|V^{s=0}|s,-s\rangle = \langle s,-s|V|s,-s\rangle + E_0 s + \frac{s}{3}[\langle s,-s+1|V|s,-s+1\rangle - \langle s,-s|V|s,-s\rangle - E_0] - C \qquad (19.59)$$

Substituting (19.58) and (19.59) into (19.55) we obtain

$$\langle s, S_j^0|V|s, S_j^0\rangle = \langle s,-s|V|s,-s\rangle + E_0(S_j^0+s)$$
$$-\frac{(S_j^0)^2-s^2}{(2s-1)}[\langle s,-s+1|V|s,-s+1\rangle - \langle s,-s|V|s,-s\rangle - E_0] \tag{19.60}$$

The reduction formula (19.60) may now be expressed in terms of n and v and the quantum numbers $\alpha, \alpha' J, M$ which were omitted for brevity. We thus obtain

$$\boxed{\begin{aligned}&\langle j^n v\alpha JM|V|j^n v\alpha' JM\rangle = \langle j^v v\alpha JM|V|j^v v\alpha' JM\rangle \\ &\quad + E_0\frac{n-v}{2}\delta_{\alpha\alpha'} + \frac{(n-v)(2j+1-n-v)}{2(2j-1-2v)} \\ &\quad \times [\langle j^{v+2} v\alpha JM|V|j^{v+2} v\alpha' JM\rangle \\ &\qquad - \langle j^v v\alpha JM|V|j^v v\alpha' JM\rangle - E_0\delta_{\alpha\alpha'}]\end{aligned}} \tag{19.61}$$

In the next section we will consider Hamiltonians which are diagonal in the seniority scheme. In such cases, (19.61) will be greatly simplified.

20

Hamiltonians Which are Diagonal in the Seniority Scheme

Odd tensor operators play an important role in the seniority scheme. In the previous section we showed that components of irreducible tensor operators with odd ranks are quasi-spin scalars, $s = 0$. This follows from the fact that if k is odd, $(a_j^\dagger \times \tilde{a}_j)_\kappa^{(k)}$ commutes with S_j^+ (or S_j^-). In a subsequent section we will see how odd tensor operators may be used to define the seniority scheme and to derive additional properties. Here we just point out that two-body interactions including in their expansion only odd tensors have special properties in the seniority scheme. Looking at (19.46) we see that such *odd tensor interactions* V^{odd} are a sum of a quasi-spin scalar and a term proportional to n. In order to determine the proportionality factor of the latter term, we apply V^{odd} to the state with seniority $v = 0$ (and $J = 0$) in the j^2 configuration. The eigenvalue of $V^{s=0}$ for this state is equal to that of the state with $v = 0$ in any j^n configuration with even n and in particular, to that of the vacuum state $|0\rangle$. Hence, the eigenvalue of $V^{s=0}$ is zero and the term in V^{odd} proportional to n must have the eigenvalue $V_0 = V_{J=0}(jjjj)$. Hence, for an odd tensor interaction we deduce from (19.46) the relation

$$-\frac{1}{2j+1}\sum_{k\,\text{odd}} F_k = V_0 \tag{20.1}$$

and the coefficient of n is $\frac{1}{2}V_0$. The relation (20.1) may be obtained directly by using the properties of $6j$-symbols.

Thus, we can express any V^{odd} by

$$V^{\text{odd}} = V^{s=0} + \tfrac{1}{2}V_0\hat{n}_j \tag{20.2}$$

From (20.2) follows the *pairing property* of any odd tensor interaction

$$\boxed{\begin{aligned}&(j^n v\alpha JM|V^{\text{odd}}|j^n v'\alpha' JM)\\ &\quad = \left[(j^v v\alpha JM|V^{\text{odd}}|j^v v\alpha' JM) + \frac{n-v}{2}V_0\,\delta_{\alpha\alpha'}\right]\delta_{vv'}\end{aligned}} \tag{20.3}$$

In (20.2), V^{odd} is a two-body interaction. Hence, the quasi-spin scalar $V^{s=0}$ contains also a term proportional to $\hat{n}_j$ which is cancelled by the second term on the r.h.s. of (20.2). The term $(\frac{1}{2})V_0\hat{n}_j$ is equal to the sum of the $\sigma = 0$ component of a quasi-spin vector, $V_0S_j^0$, and a constant term $(\frac{1}{4})V_0(2j+1)$.

The interaction (20.2) is not the most general interaction which is diagonal in the seniority scheme. In the previous section we saw that the term with $(a_j^+ \times \tilde{a}_j)^{(0)} \cdot (a_j^+ \times \tilde{a}_j)^{(0)}$ in (19.46) contributes $F_0n^2/2(2j+1)$. As indicated in (19.48), the term with $s = 2$, $\sigma = 0$ (as well as the $s = 1$, $\sigma = 0$ term) due to the $k = 0$ term, is the quasi-spin tensor (and vector) constructed from the quasi-spin vector $\mathbf{S}_j$ itself. Such terms are diagonal in the seniority scheme, as well as in any other scheme with fixed number of particles. The $k = 0$ term thus contributes to any matrix element the same contribution, independent of J equal to

$$\frac{F_0}{2j+1}\,\frac{n(n-1)}{2} \tag{20.4}$$

For $n = 2$ this contribution is equal to $F_0/(2j+1)$ and it is the same for all matrix elements $V_J = \langle j^2 JM|V|j^2 JM\rangle$.

If we add the contribution of the $k = 0$ term to an odd tensor interaction, we obtain a more general interaction which is diagonal in the seniority scheme. We shall now study the properties of this interaction. It is sufficient to consider such interactions since we will prove below that:

Any two body interaction which is diagonal in the seniority seniority scheme can be expressed as the sum of an odd tensor interaction and a $k = 0$ interaction.

We saw above that such interactions are linear combinations of a quasi-spin scalar $V^{s=0}$ and terms linear and quadratic in n. The expression (20.2) no longer holds in this more general case and we will now evaluate the coefficients of n and n^2. Putting $J = 0$ in (18.14) we obtain now, instead of (20.1), the relation

$$V_0 = -\frac{1}{2j+1}\sum_{k\,\text{odd}} F_k + \frac{1}{2j+1}F_0 \tag{20.5}$$

Since $F_k = 0$ for $k > 0$, even, we may write this result as

$$V_0 = -\frac{1}{2j+1}\sum_{k} F_k + \frac{2}{2j+1}F_0$$

Recalling (18.16) we transform this expression into

$$V_0 = \frac{1}{2j+1}\sum_{J\,\text{even}}(2J+1)V_J - \frac{1}{2j+1}\sum_{J\,\text{odd}}(2J+1)V_J + \frac{2}{2j+1}F_0$$

$$= \frac{2}{2j+1}\sum_{J\,\text{even}}(2J+1)V_J - F_0 + \frac{2}{2j+1}F_0 \tag{20.6}$$

The last equality is due to (18.12) for $k = 0$. Using (20.6), we can express F_0 in terms of V_0 and the other V_J with J even. Since V_0 occupies a special position in (20.6), as is the general case in the seniority scheme, and all other V_J appear in the same combination, we introduce the notation

$$\bar{V}_2 = \sum_{J>0\,\text{even}} (2J+1)V_J \Big/ \sum_{J>0\,\text{even}} (2J+1)$$
$$= \frac{1}{(j+1)(2j-1)} \sum_{J>0\,\text{even}} (2J+1)V_J \tag{20.7}$$

We now obtain for the coefficient of the quadratic term $n(n-1)/2$ in (20.4) or (19.46) the value

$$\frac{1}{2j+1}F_0 = \frac{2(j+1)\bar{V}_2 - V_0}{2j+1} = \alpha \tag{20.8}$$

The coefficient of the linear term $n/2$ in (19.46) is given by using (20.5) as

$$-\frac{1}{2j+1}\sum_{k\,\text{odd}} F_k = V_0 - \frac{1}{2j+1}F_0 = \frac{2(j+1)}{2j+1}(V_0 - \bar{V}_2) = \beta \tag{20.9}$$

The last equality is obtained by substitution of the value of F_0 from (20.8). For an odd tensor interaction, according to (20.3), the quadratic term vanishes. Hence, the $k = 0$ part contributes only to α but not to β. Since it contributes equally to all V_J its contribution disappears from $V_0 - \bar{V}_2$.

The most general two-body interaction which is diagonal in the seniority scheme can thus be expressed by

$$V = V^{s=0} + \frac{n(n-1)}{2}\alpha + \frac{n}{2}\beta \tag{20.10}$$

where the coefficients α and β are defined in terms of the V_J by (20.8) and (20.9). An example of such an interaction is the pairing interaction (19.4). In (19.8) it is expressed in terms of the quasi-spin scalar $\mathbf{S}_j^2$ and the $\sigma = 0$ components of a vector and $s = 2$ tensor, constructed from components of $\mathbf{S}_j$. We also recall the expansion of the pairing interaction P in terms of odd tensors and $k = 0$ tensor given by (12.20). There, $F_0/(2j+1) = -1$ and this is the coefficient of the quadratic term (putting $V_0 = (2j+1)$, $\bar{V}_2 = 0$ in (20.8) yields $\alpha = -1$). The coefficient of the linear term is $\beta = 2j+2$ and hence the eigenvalues of P are equal to

$$P(n,v) = \frac{n-v}{2}(2j+2) - \frac{n(n-1)}{2} + \frac{v(v+1)}{2} \qquad (20.11)$$

This result is identical to (19.11) but it shows the separate contributions to P of the odd tensor interaction and of the $k = 0$ term.

An example of an odd tensor interaction is the δ-potential which can be expressed, according to (12.16) as an odd tensor interaction. We also saw in Section 12 that any interaction of the form $(\sigma_1 \cdot \sigma_2) \cdot V(r_{12})$ is an odd tensor interaction. Finally, the tensor forces introduced in (10.47) were shown there to be an odd tensor interaction within the j^n configuration. As demonstrated by (20.3), odd tensor interactions have the pairing property and energies in j^n configurations depend on n only through the pairing term. The coefficient α of the quadratic term (defined by 20.8) vanishes for such interactions. It was explained in Section 19 that the repulsive quadratic term in the case of the pairing interaction is due to the Pauli principle. In the general case of Hamiltonians which are diagonal in the seniority scheme, the coefficient α may be repulsive, may vanish and may even be attractive even though the Pauli principle is strictly obeyed.

Due to the Pauli principle, as more identical nucleons occupy the j-orbit, the amount of couplings into $J = 0$ pairs cannot increase linearly with n. The total number of interactions is $n(n-1)/2$ but they include an increasing number of matrix elements in states with $J > 0$. Hence, the sign and size of the quadratic term depend on the relative strength of the interaction in states with $J > 0$ compared to V_0. As shown above, the coefficient α of the quadratic term is equal to the linear combination (20.8) of two nucleon matrix elements. In the case of the pairing interaction $\bar{V}_2 = 0$ and $\alpha > 0$ whereas for the δ-potential, or any odd tensor interaction, $2(j+1)\bar{V}_2 = V_0$ and $\alpha = 0$. For more attractive values of $\bar{V}_2$ relative to V_0 the coefficient α becomes negative. In the limit of a long range attractive interaction where all V_J are equal, $V_J = a$, the coefficient of the quadratic term, $\alpha = a$, is attractive.

It should be realized that interactions which are diagonal in the seniority scheme form a large and rich class. There are not very many restrictions on the two nucleon matrix elements of such interactions. These will be discussed in the following. The various V_J may be very far apart with arbitrarily large spacings between them. This should be contrasted with the pairing interaction (19.4) which has all energies V_J for $J = 2,\ldots,2j-1$ degenerate. This is far from the experimental situation which makes the pairing interaction a rather poor substitute for the effective interaction in nuclei (we saw some of its weaknesses in Section 16). On the other hand, we will demonstrate that the actual effective interaction as determined consistently from experimental data, is diagonal in the seniority scheme to a very good approximation.

The $s = 0$ part of the two body interaction (20.10), $V^{s=0}$, is diagonal in the seniority scheme and its matrix elements are independent of n. If there are several states with the same seniority v and J (and M) in the j^v configuration, we choose a label α to distinguish between such states which form an orthogonal basis. Without loss of generality, we may assume that $V^{s=0}$ is diagonal in the basis chosen in the j^v configuration. In that case, the states $|j^n v\alpha JM\rangle$ are eigenstates of (20.10) and the eigenvalues are given by

$$\boxed{\begin{aligned} \langle j^n v\alpha JM|V|j^n v\alpha JM\rangle &= \langle j^v v\alpha JM|V^{s=0}|j^v v\alpha JM\rangle \\ &\quad + \frac{n(n-1)}{2}\alpha + \frac{n}{2}\beta \\ &= \langle j^v v\alpha JM|V|j^v v\alpha JM\rangle + \frac{n(n-1)}{2}\alpha \\ &\quad + \frac{n}{2}\beta - \frac{v(v-1)}{2}\alpha - \frac{v}{2}\beta \end{aligned}} \tag{20.12}$$

From (20.12) follow important properties of Hamiltonians which are diagonal in the seniority scheme. The first property is as follows.

Energy differences between levels of the j^n configuration with any seniorities $v' \leq v$ are equal to those in the j^v configuration.

These energy differences are determined only by the $V^{s=0}$ part of V. The terms $\alpha n(n-1)/2$, $\beta n/2$ as well as $\alpha v(v-1)/2$ and $\beta v/2$ contribute equally to all levels of the j^n configuration and j^v configuration respectively. Hence, they do not contribute to energy differences. In other words, spacings between energy levels are independent of n in all j^n configurations in which they appear.

Of particular interest are the levels in semi-magic nuclei with $v = 2$ and $J = 2, 4, \ldots, 2j-1$. They appear in all j^n configurations with n even and $2j+1 > n > 0$. If the two-body interaction is diagonal in the seniority scheme, they should have the same spacings (also between them and the $J = 0$, $v = 0$ ground state) in all nuclei in which a given j-orbit is being filled. There may be other states with these spins but they should have higher seniorities $v \geq 4$. Such higher seniority states cannot decay into the ground state by EL electromagnetic transitions. The operator which determines such transitions is a single particle operator and cannot connect states with seniorities differing by more than 2. For example, only the $v = 2$, $J = 2$ state can decay into the $v = 0$, $J = 0$ ground state by an electric quadrupole E2 transition.

Another important consequence of (20.12) follows for interaction energies of the lowest states. These are the $v = 0$, $J = 0$ states of even j^n configurations and the $v = 1$, $J = j$ states of odd n ones. Matrix elements of the two body interaction V vanish in the j^v configuration for either $v = 0$ or $v = 1$. For $v = 0$ this is the vacuum state with no j-nucleons and for $v = 1$ it is the state of a *single* j-nucleon. We thus obtain from (20.12) the result

$$\boxed{V(j^n g.s.) = \frac{n(n-1)}{2}\alpha + \frac{n-v}{2}\beta = \frac{n(n-1)}{2}\alpha + \left[\frac{n}{2}\right]\beta} \tag{20.13}$$

In (20.13), $[n/2]$ is the largest integer not exceeding $n/2$. It is given, for integral n, by $(n - \frac{1}{2}(1-(-1)^n))/2$. The very simple expression (20.13) includes a quadratic term in n and a *pairing term*. It demonstrates the elegance and power of the seniority scheme. It holds for *any* two body interaction which is diagonal in the seniority scheme. The interaction could be central or non-central, local or non-local. Energies of all levels in j^n configurations are functions of the two nucleon matrix elements V_J. For ground states, with lowest seniorities, only two independent combinations of the V_J, given by (20.8) and (20.9) appear in the simple formula (20.13).

We may add to (20.13) the single nucleon energies which have the form

$$\epsilon_j \sum_m a^+_{jm} a_{jm}$$

This term contributes $n\epsilon_j$ to all states of the j^n configuration. The expression of the binding energy of ground states due to n identical nucleons in the j-orbit is thus

$$\boxed{\text{B.E.}(j^n g.s.) = \text{B.E.}(n=0) + n\epsilon_j + \frac{n(n-1)}{2}\alpha + \left[\frac{n}{2}\right]\beta} \tag{20.14}$$

The term B.E.$(n = 0)$ is the total energy (binding energy) of the vacuum state $|0\rangle$. It is the ground state of the closed shells nucleus from which all j-nucleons have been removed. In a subsequent section we prove that if ground state energies are given by (20.14) for even values of n, the Hamiltonian is diagonal in the seniority scheme.

It is worthwhile to point out that once the general expression (20.13) has been established it is a simple matter to calculate the values of α and β. We apply (20.13) to the $v = 0$, $J = 0$ state of the j^2 configuration and to the $v = 0$, $J = 0$ state of the closed orbit j^{2j+1} configuration. We obtain accordingly

$$\alpha + \beta = V_0$$

$$\frac{(2j+1)2j}{2}\alpha + \frac{2j+1}{2}\beta = \sum_{J\,\text{even}} (2J+1)V_J \tag{20.15}$$

$$= V_0 + \sum_{J>0\,\text{even}} (2J+1)V_J$$

From these two equations the values of α and β can be directly calculated and the results (20.8) and (20.9) are reproduced.

We can use now (20.14) to discuss energies of ground states of $(1d_{5/2})^n$ neutron configurations in Zr isotopes introduced in Section 16. As mentioned above, there are no two independent states with the same value of J (and M) in such configurations. Hence, all these states may be assigned definite seniorities and any interaction is diagonal in the seniority scheme within $(\frac{5}{2})^n$ configurations. To check whether (20.14) does indeed agree with the shell model configuration adopted for the nuclei considered in Section 16, we adopt a simple procedure. We first consider single nucleon *separation energies* defined by B.E.$(j^n g.s.)$ – B.E.$(j^{n-1} g.s.)$. Separation energies calculated from (20.14) are given by

$$\text{B.E.}(j^n_{g.s.}) - \text{B.E.}(j^{n-1}_{g.s.}) = \epsilon_j + (n-1)\alpha + \frac{1+(-1)^n}{2}\beta \tag{20.16}$$

Hence, according to (20.14), separation energies (20.16) plotted as a function of n should lie on two straight and parallel lines. The slope of the line is given by α whereas the spacing of the lines for any n is equal to β. The experimentally observed separation energies of Zr isotopes ^{87}Zr to ^{98}Zr are plotted in Fig. 20.1. The straight lines drawn are those which give the best fit to the experimental points. We see the sharp drops of experimental separation energies beyond $N = 50$ displaying the closed shells at that magic number. A similar drop, considerably smaller, occurs beyond $N = 56$ where the $2d_{5/2}$ neutron orbit is taken to be completely filled. The agreement between experimental data and the fit by (20.14) is excellent.

In Fig. 20.1 the absolute values of binding energies were plotted. With this convention, the α coefficient is negative, i.e. repulsive. It is rather small in comparison with the value of β which is positive, i.e. attractive. This is the place to explain that this feature, which occurs in all semi-magic nuclei, is due to the seniority scheme. A strong and attractive pairing term, giving the well known odd-even variation of binding energies is not due to the pairing interaction or to a short range interaction. Looking at the expression (20.9) for β we see that the size of this term depends on the spacing between the j^2 $J = 0$ state and the center of mass of the other j^2 states. If $\bar{V}_2$ is much higher than V_0, then β will be large and attractive. The spacings of the $v = 2$ levels, with $J = 2, 4, \ldots, 2j-1$ may be large or small without changing the value of β. In fact, we should compare the value of β derived from binding energies to the value of (20.9) in which the value of $V_0 - \bar{V}_2$ taken from ^{92}Zr or ^{94}Zr is substituted. From binding energies, the best fit is obtained with the value of $\beta = 1.638$ MeV which is

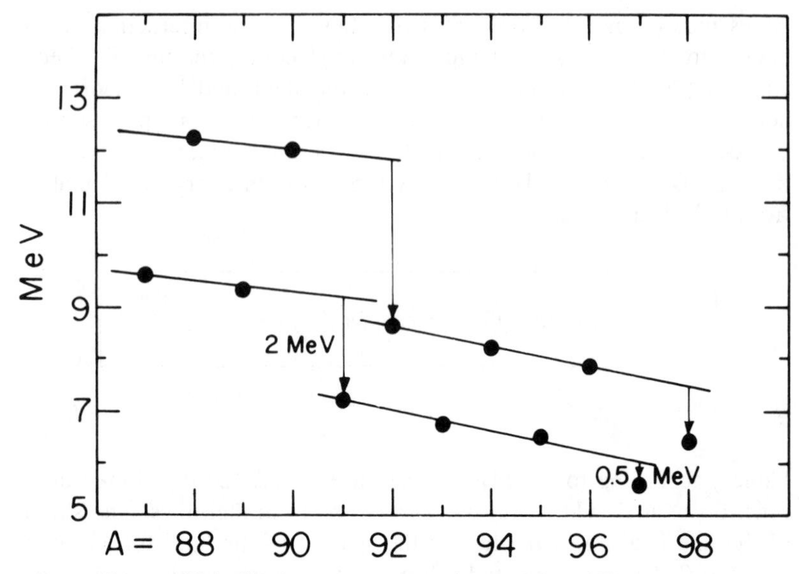

FIGURE 20.1. *Single neutron separation energies of* Zr *isotopes.*

in fair agreement with the value $\beta = 1.50$ MeV obtained from ^{92}Zr and ^{94}Zr levels by using (20.9). Actually, by using $\beta = 1.50$ MeV in (20.16) we also obtain fairly good agreement with the experimental binding energies. The value of α determined from binding energies turns out to be $\alpha = -.198$ MeV (and $\epsilon_{1d_{5/2}} = 7.194$ MeV). As evident from (20.8), the value of α cannot be computed from level spacings. It must be determined from binding energies.

The core term B.E.($n = 0$) in (20.14) is taken to be the binding energy of the nucleus with closed shells and no j-nucleons. If the core is *polarized* by the valence j-nucleons, its binding energy may well depend on their number n. A linear change in n is simply absorbed by ϵ_j and forms part of its renormalization. As explained in preceding sections, instead of considering the complicated wave function of the polarized core, we use simple shell model wave functions of a core with closed shells and valence j-nucleons. Similarly, a quadratic change in B.E.($n = 0$) may appear implicitly as a renormalization of the coefficient α in (20.14). Also changes which are linear in n in the single nucleon energy ϵ_j contribute to α. More complicated dependence on n may lead to renormalization of β in the pairing term. Some kinds of core polarizations will be considered in Section 32.

The remarkably simple formula (20.14) holds for eigenvalues of any shell model Hamiltonian which is diagonal in the seniority scheme. It has, however, more general validity. The expression (20.13) is equal to the *expectation value* of any two body interaction between states with lowest seniorities $v = 0$ or $v = 1$. We have calculated in the preceding section such expectation values as given by (19.61). If we put there $v = 0$ we obtain

$$\langle j^n v = 0, J = 0, M = 0|V|j^n v = 0, J = 0, M = 0\rangle$$

$$= \frac{n}{2}E_0 + \frac{n(2j+1-n)}{2(2j-1)}(V_0 - E_0)$$

$$= \frac{n}{2}\frac{2jV_0 - E_0}{2j-1} + \frac{n(n-1)}{2}\frac{E_0 - V_0}{2j-1} \tag{20.17}$$

According to (19.49), E_0 may be expressed in terms of α and β by

$$E_0 = 2j\alpha + \beta$$

According to (20.15), $V_0 = \alpha + \beta$ and we obtain for the r.h.s. of (20.17) the simple expression

$$\frac{n(n-1)}{2}\alpha + \frac{n}{2}\beta \qquad n \text{ even}$$

which is identical with (20.13) for even values of n.

The case of the states $v = 1$, $J = j$ in odd j^n configurations is slightly more complicated. Putting $v = 1$ in (19.61) we obtain, due to $\langle jv = 1, J = j, M|V|jv = 1, J = j, M\rangle = 0$,

$$\langle j^n v = 1, J = j, M|V|j^n v = 1, J = j, M\rangle$$

$$= \frac{n-1}{2}E_0 + \frac{(n-1)(2j-n)}{2(2j-3)}$$

$$\times [\langle j^3 v = 1, J = j, M|V|j^3 v = 1, J = j, M\rangle - E_0] \tag{20.18}$$

To evaluate (20.18) we must calculate the expectation value of V in the $v = 1, J = j$ state of the j^3 configuration. We recall our results in Section 15 for the c.f.p. of the $j^3, J = j$ state with principal parent

$J_0 = 0$. According to (19.3) this is the state with $v = 1$, $J = j$ in the j^3 configuration and its c.f.p. are given by (15.11). Using these c.f.p. we obtain from (15.11) the result

$$\langle j^3 v = 1, J = jM | V | j^3 v = 1, J = jM \rangle$$

$$= 3\frac{2j-1}{3(2j+1)}V_0 + 3\frac{4}{3(2j+1)(2j-1)} \sum_{J_1 > 0,\,\text{even}} (2J_1+1)V_{J_1}$$

$$= \frac{2}{2j-1}E_0 + \frac{2j-3}{2j-1}V_0 \tag{20.19}$$

When (20.19) substituted into (20.18) we obtain, by expressing E_0 and V_0 in terms of α and β, the result

$$\frac{n-1}{2}\frac{(2jV_0 - E_0)}{2j-1} + \frac{n(n-1)}{2}\frac{E_0 - V_0}{2j-1}$$

$$= \frac{n-1}{2}\beta + \frac{n(n-1)}{2}\alpha \qquad n \text{ odd}$$

which is identical with (20.13) for odd values of n. We thus proved that (20.13) is equal to the expectation value in states with lowest seniorities of *any* two-body interaction.

The fact proved above in detail, follows directly from the Wigner–Eckart theorem for the $SU(2)$ quasi-spin group. Matrix elements of any rotationally invariant $s = 2$, $\sigma = 0$ quasi-spin tensor between states with the same seniority, $s' = s$, are proportional to those of $[\mathbf{S}_j \times \mathbf{S}_j]^{(2)}_0$ or $3(S_j^0)^2 - \mathbf{S}^2$. The latter operator includes also $s = 1$, $\sigma = 0$ and $s = 0$ components (a term linear in n and a constant term). Apart from these, its matrix elements are given by (20.13).

We should now prove that the most general two body interaction which is diagonal in the seniority scheme can be expressed by (20.10). In fact, we will prove that any interaction diagonal in the seniority scheme may be expressed as an odd tensor interaction and a $k = 0$ interaction. We start with the expression (19.45) for the general two-body interaction

$$\sum_{J\,\text{even},M} V_J A^+(j^2JM)A(j^2JM) \tag{20.20}$$

Each term in (20.20) is a product of a $\sigma = 1$ and $\sigma = -1$ components of a quasi-spin tensor of rank $s = 1$. Such a product is equal to a linear combination of $\sigma = 0$ components of quasi-spin tensors with ranks $s = 0$, $s = 1$ and $s = 2$. A *sufficient* condition for (20.20) to be diagonal in the seniority scheme is the vanishing of the $s = 2$ part. In the preceding section we saw that a $k = 0$ term contributes according to (19.48) a special $s = 2$ tensor constructed from the components of $\mathbf{S}_j$. Such a term is clearly diagonal in the seniority scheme (and, in fact, in any scheme). Hence, we can add to the interaction a $k = 0$ term with arbitrary coefficient and still obtain an interaction which is diagonal in the seniority scheme.

To obtain the $s = 2$ part of (20.20) we first express it according to (18.2) for $j_1 = j_2 = j_1' = j_2' = j$, as

$$-\frac{1}{2}\sum_{J\text{ even}} V_J(-1)^M(a_j^\dagger \times a_j^\dagger)^{(J)}_M(\tilde{a}_j \times \tilde{a}_j)^{(J)}_{-M}$$

We recall that $(1/\sqrt{2})(a_j^\dagger \times a_j^\dagger)^{(J)}_M$ for J even, is the $\sigma = 1$ component of a quasi-spin tensor of rank $s = 1$, $T^{s=1}_{\sigma=1}(JM)$ and $-(1/\sqrt{2})(-1)^M\cdot(\tilde{a}_j \times \tilde{a}_j)^{(J)}_{-M}$ is the corresponding $T^{s=1}_{\sigma=-1}(J,-M)$ component (see the discussion of (19.35)). We can then expand (20.20) as a linear combination of $\sigma = 0$ components of quasi-spin tensors with ranks $s = 0, 1, 2$. Using actual values of Clebsch-Gordan coefficients we obtain

$$\begin{aligned}\sum_{J\text{ even},M} & V_J A^+(j^2JM)A(j^2JM)\\ &= -\sum_{JM} V_J(-1)^M T^{s=1}_{\sigma=1}(JM)T^{s=1}_{\sigma=-1}(J,-M)\\ &= \frac{1}{\sqrt{3}}V^{s=0}_{\sigma=0} + \frac{1}{\sqrt{2}}V^{s=1}_{\sigma=0} + \frac{1}{\sqrt{6}}V^{s=2}_{\sigma=0}\end{aligned}$$

Similarly, by taking the products of $T^{s=1}_{\sigma=-1}(JM)$ and $T^{s=1}_{\sigma=1}(J,-M)$ we obtain

$$\begin{aligned}\sum_{J\text{ even},M} & V_J A(j^2JM)A^+(j^2JM)\\ &= -\sum_{JM} V_J(-1)^M T^{s=1}_{\sigma=-1}(JM)T^{s=1}_{\sigma=1}(J,-M)\\ &= \frac{1}{\sqrt{3}}V^{s=0}_{\sigma=0} - \frac{1}{\sqrt{2}}V^{s=1}_{\sigma=0} + \frac{1}{\sqrt{6}}V^{s=2}_{\sigma=0}\end{aligned}$$

We now take the products of $T^{s=1}_{\sigma=0}(JM)$ and $T^{s=1}_{\sigma=0}(J,-M)$ as given in (19.35) and obtain

$$\sum_{J\,\text{even}} V_J\left((a_j^+\times\tilde{a}_j)^{(J)}-\frac{\sqrt{2j+1}}{2}\delta_{J0}\right)$$

$$\cdot\left((a_j^+\times\tilde{a}_j)^{(J)}-\frac{\sqrt{2j+1}}{2}\delta_{J0}\right)$$

$$=\sum_{JM} V_J(-1)^M T^{s=1}_{\sigma=0}(JM)T^{s=1}_{\sigma=0}(J,-M)$$

$$=\frac{1}{\sqrt{3}}V^{s=0}_{\sigma=0}-\sqrt{\frac{2}{3}}V^{s=2}_{\sigma=0}$$

From these three equations we can eliminate $V^{s=0}_{\sigma=0}$ and $V^{s=1}_{\sigma=1}$ and obtain the following expression for $V^{s=2}_{\sigma=0}$.

$$\sum_{J\,\text{even}} V_J\left\{\sum_M A^+(j^2JM)A(j^2JM)+\sum_M A(j^2JM)A^+(j^2JM)\right.$$

$$-2\left((a_j^+\times\tilde{a}_j)^{(J)}-\frac{\sqrt{2j+1}}{2}\delta_{J0}\right)$$

$$\left.\cdot\left((a_j^+\times\tilde{a}_j)^{(J)}-\frac{\sqrt{2j+1}}{2}\delta_{J0}\right)\right\}=\sqrt{6}V^{s=2}_{\sigma=0} \qquad (20.21)$$

The condition for vanishing of the $s=2$ part of (20.20) is thus the vanishing of (20.21). This leads to certain conditions on the two particle interaction energies V_J. In order to obtain them we rearrange the terms in (20.21), making use of the following commutation relation obtained from (17.24) for the case $j_1=j_2=j_3=j_4=j$,

$$[A(j^2JM),A^+(j^2J'M')]$$

$$=\delta_{JJ'}\delta_{MM'}-2\sqrt{(2J+1)(2J'+1)}\sum_{\mu mm'}(-1)^{M+M'}$$

$$\times\begin{pmatrix} j & j & J\\ \mu & m' & -M\end{pmatrix}\begin{pmatrix} j & j & J'\\ \mu & m & -M'\end{pmatrix}a^+_{jm}a_{jm'} \qquad (20.22)$$

In the case considered here, $J = J'$ and $M = M'$ and hence, using the orthogonality relations of the $3j$ symbols, the sum over M and μ yields the following expression for (20.22)

$$\sum_M [A(j^2JM), A^+(j^2JM)]$$

$$= 2J + 1 - 2(2J+1)\frac{1}{2j+1}\sum a^+_{jm}a_{jm'}\delta_{mm'}$$

$$= (2J+1)\left(1 - \frac{2}{2j+1}\sum a^+_{jm}a_{jm}\right)$$

$$= -4\frac{(2J+1)}{2j+1}S^0_j \tag{20.23}$$

Interchanging in (20.21) $A(j^2JM)$ and $A^+(j^2JM)$ we obtain by using (20.23) the following form for the condition that $V^{s=2}_{\sigma=0}$ vanishes

$$\sum V_J \left[\sum_M A^+(j^2JM)A(j^2JM) - (a^+_j \times \tilde{a}_j)^{(J)} \cdot (a^+_j \times \tilde{a}_j)^{(J)}\right]$$
$$- \frac{2}{2j+1}\left(\sum (2J+1)V_J\right)S^0_j$$
$$+ V_0\sqrt{2j+1}(a^+_j \times \tilde{a}_j)^{(0)} - V_0\frac{(2j+1)}{4} = 0 \tag{20.24}$$

By applying (20.24) to the vacuum we see that the constant term in (20.24) must vanish, which yields the condition

$$V_0 = \frac{2}{2j+1}\sum (2J+1)V_J = E_0 \tag{20.25}$$

This condition, due to (20.17), is equivalent to putting $\alpha = 0$ in (20.8). The condition (20.25) holds for V_J (including V_0) which are obtained from an interaction V for which $V^{s=2}$ vanishes. To such an interaction we can always add a term $V'_0P/(2j+1)$ with arbitrary value of V'_0. The resulting interaction will still be diagonal in the seniority scheme. Keeping this in mind, we now substitute the value of V_0 given by (20.25) in (20.24) obtaining the result

$$\sum V_J A^+(j^2JM)A(j^2JM)$$

$$= \sum_{J\,\text{even}} V_J(a_j^\dagger \times \tilde{a}_j)^{(J)} \cdot (a_j^\dagger \times \tilde{a}_j)^{(J)}$$

$$- \frac{1}{2j+1}\left(\sum_{J\,\text{even}} (2J+1)V_J\right)\sum a_{jm}^\dagger a_{jm} \qquad (20.26)$$

The l.h.s. of (20.26) is the interaction(20.20) from which we started. The r.h.s. is therefore equal to the expansion (19.46) of that V in terms of scalar products of irreducible tensors with even ranks k whose coefficients are given by

$$F_k = 2(2k+1)V_k \qquad k \text{ even} \qquad (20.27)$$

We can now use (20.27) to obtain further conditions on the V_J. The relation (18.14) can be thus expressed as

$$V_J + 2\sum_{k\,\text{even}} (2k+1)\begin{Bmatrix} j & j & J \\ j & j & k \end{Bmatrix} V_k = 0 \qquad J \text{ even} \qquad (20.28)$$

For $J = 0$, (20.28) reduces to the condition (20.25) obtained above. The even tensor interaction (20.26) whose eigenvalues V_J satisfy the conditions (20.27) and (20.28) can be expressed as an odd tensor interaction. To prove it we first notice that the interaction whose coefficients are given by (20.27) has non-vanishing eigenvalues also for j^2 states with odd values of J. They are given, according to (18.14) by

$$V_J = \sum_{k\,\text{even}} 2(2k+1)\begin{Bmatrix} j & j & J \\ j & j & k \end{Bmatrix} V_k \qquad J \text{ odd} \qquad (20.29)$$

The expansion coefficients $F_k = 2(2k+1)V_k$, k even, are given by (18.12) as a sum over *both* even and odd values of J. Thus,

$$F_k = 2(2k+1)V_k = -\sum_{J\,\text{even}} (2k+1)(2J+1)\begin{Bmatrix} j & j & J \\ j & j & k \end{Bmatrix} V_J$$

$$+ \sum_{J\,\text{odd}} (2k+1)(2J+1)\begin{Bmatrix} j & j & J \\ j & j & k \end{Bmatrix} V_J \qquad (20.30)$$

As pointed out above, the expansion in scalar products of irreducible tensors is not unique if we limit our attention to j^2 states with $T=1$ and even values of J. Thus, the expression of the F_k in terms of the V_J given by (20.30) is not unique. In fact, an equivalent expansion is given by (20.28) (where J and k are interchanged) as

$$2(2k+1)V_k = F_k = -4\sum_{J\,\text{even}}(2k+1)(2J+1)\begin{Bmatrix} j & j & J \\ j & j & k \end{Bmatrix} V_J \qquad k \text{ even} \tag{20.31}$$

This is an expansion of the interaction in $T=1$ states in terms of scalar products of even tensor operators. From (20.30) and (20.31) we can eliminate the summation over even values of J. We thus obtain, for even values of J the expansion

$$V_J = \sum_{k\,\text{odd}} \begin{Bmatrix} j & j & J \\ j & j & k \end{Bmatrix} \tfrac{2}{3}(2k+1)V_k \qquad J \text{ even} \tag{20.32}$$

where V_k for odd values of k, is given by the r.h.s. of (20.29) in terms of V_J, J even, as

$$V_k = \sum_{J\,\text{even}} 2(2J+1)\begin{Bmatrix} j & j & k \\ j & j & J \end{Bmatrix} V_J \qquad k \text{ odd}$$

Thus, $T=1$ matrix elements of the interaction (20.20), with vanishing $V^{s=2}$ part, can be obtained according to (20.32) from an expansion in scalar products of *odd* tensor operators. The coefficients of this equivalent expansion are given by comparing (20.32) with (18.14), as

$$\begin{aligned} F'_k &= -\tfrac{2}{3}(2k+1)V_k \\ &= -\tfrac{4}{3}(2k+1)\sum_{J\,\text{even}}(2J+1)\begin{Bmatrix} j & j & k \\ j & j & J \end{Bmatrix} V_J \qquad k \text{ odd} \end{aligned} \tag{20.33}$$

The last equality in (20.33) is due to the relation (20.29).

The conditions (20.28) obtained above, impose conditions on the eigenvalues V_J of a two body interaction whose $V^{s=2}$ part vansihes. These conditions are not useful criteria to determine whether a given interaction is diagonal in the seniority scheme. As remarked above we can add to the given interaction the pairing interaction with arbitrary coefficient without spoiling its being diagonal in the seniority scheme. This addition will change V_0 but not the values of the other V_J. As shown above, if the conditions (20.28) hold, the interaction is equivalent to an odd tensor interaction. We showed in (12.20) that the pairing interaction is the sum of an odd tensor interaction and a $k = 0$ term. A $k = 0$ term adds to the interaction a special $\sigma = 0$ component of a tensor with rank $s = 2$ constructed from components of the quasi-spin vector $\mathbf{S}_j$. This latter term, if present, invalidates the conditions (20.28). Adding a $k = 0$ term to the interaction leads to a non vanishing α coefficient which is in contrast to (20.25). Adding a constant term α to all V_J, however, keeps the interaction diagonal in the seniority scheme. This shows that the criteria for an interaction to be diagonal in the seniority scheme cannot depend on the eigenvalue V_0. They should depend only on energy spacings between the j^2 states with $J > 0$, even, i.e. seniority $v = 2$ states.

We can use the conditions (20.28) which hold for an odd tensor interaction to derive conditions on the V_J which do not change when a $k = 0$ term (or the pairing interaction) is added to the interaction considered. Let us denote the eigenvalues of any odd tensor interaction by V'_J. These V'_J satisfy the conditions (20.28) and in particular, for $J = 0$

$$
\begin{aligned}
V'_0 = E'_0 &= \frac{2}{2j+1} \sum_{J\,\text{even}} (2J+1)V'_J \\
&= \frac{2}{2j+1} V'_0 + \frac{2}{2j+1} \sum_{J>0\,\text{even}} (2J+1)V'_J \\
&= \frac{2}{2j+1} V'_0 + \frac{2(j+1)(2j-1)}{2j+1} \bar{V}'_2
\end{aligned}
\tag{20.34}
$$

From this follows the condition

$$
V'_0 = 2(j+1)\bar{V}'_2 \tag{20.35}
$$

which, according to (20.8), is equivalent to putting $\alpha = 0$. Substituting this value of V_0' into the conditions (20.28) for $J > 0$, even, we obtain

$$V_J' + 2 \sum_{k>0\,\text{even}} (2k+1) \begin{Bmatrix} j & j & J \\ j & j & k \end{Bmatrix} V_k' - \frac{4(j+1)}{2j+1} \bar{V}_2'$$

$$= V_J' - \bar{V}_2' + 2 \sum_{k>0\,\text{even}} (2k+1) \begin{Bmatrix} j & j & J \\ j & j & k \end{Bmatrix} (V_k' - \bar{V}_2') = 0 \tag{20.36}$$

The last equality in (20.36) follows from combining the sum rules (10.17) and (10.18). The conditions (20.36) which hold for any odd tensor interaction depend only on spacings of j^2 levels with $J > 0$ ($v = 2$). They hold also if we add to the odd tensor interaction a $k = 0$ term with arbitrary coefficient. Hence, the conditions

$$\boxed{V_J - \bar{V}_2 + 2 \sum_{k>0\,\text{even}} (2k+1) \begin{Bmatrix} j & j & J \\ j & j & k \end{Bmatrix} (V_k - \bar{V}_2) = 0 \qquad J > 0 \text{ even}} \tag{20.37}$$

are *sufficient* conditions for any two body interaction to be diagonal in the seniority scheme.

The conditions (20.28) are *sufficient* to guarantee that a two body interaction is diagonal in the seniorty scheme. Moreover, such an interaction can be expressed as an odd tensor interaction. By adding a $k = 0$ term to such an interaction, the conditions (20.37) are still obeyed and the resulting interaction is still diagonal in the seniority scheme. These conditions are also *necessary* conditions. Any interaction that is diagonal in the seniority scheme must satisfy them. To see it we show that the only $s = 2$ term in V which is diagonal in the seniority scheme must be due to a $k = 0$ interaction.

Matrix elements of any quasi-spin tensor with rank $s = 2$ between states with the same value of s are proportional, according to the Wigner-Eckart theorem, to those of $(\mathbf{S}_j \times \mathbf{S}_j)^{(2)}$. If that tensor $V^{s=2}$ is diagonal in the seniority scheme, its matrix elements between states with different values of s vanish. Thus, the $\sigma = 0$ component of such a tensor can be replaced by a multiple of $3(S_j^0)^2 - \mathbf{S}_j^2$ which is equal to $\sqrt{6}(\mathbf{S}_j \times \mathbf{S}_j)_0^{(2)}$. Any $V^{s=2}$ part of the interaction which is diagonal in the seniority scheme can thus be replaced by

$$3(S_j^0)^2 - \mathbf{S}_j^2 = 2(S_j^0)^2 + S_j^0 - S_j^+ S_j^- = 2(S_j^0)^2 + S_j^0 - \tfrac{1}{2}P \tag{20.38}$$

multiplied by a constant coefficient. The pairing interaction P is the sum of an odd tensor interaction and a $k = 0$ term. The other terms on the r.h.s. of (20.38) depend only on n and are also due to a $k = 0$ interaction. The two-body part of $2(S_j^0)^2 + S_j^0$ is equal to $n(n-1)/2$ which is due to a $k = 0$ interaction as seen e.g. in (20.4). The single nucleon part and the constant term are cancelled by similar terms in the $s = 0$ and $s = 1$ parts of the two-body interaction (20.20). Hence, any two-body interaction which is diagonal in the seniority scheme can be expressed as the sum of an odd tensor interaction and a $k = 0$ (monopole) interaction. We will next show also in another way that (20.37) are *necessary* conditions. This will offer a look at these conditions from a different point of view.

The simplest j^n configuration in which states with the same value of J and different seniorities may appear is the j^3 configuration. This configuration has always a state with $J = j$ and $v = 1$ and for $j \geq \frac{9}{2}$ there are also other orthogonal states with $J = j$ which must have $v = 3$. Any two body interaction which is diagonal in the seniority scheme must have vanishing matrix elements between j^3 $J = j$ states with seniorities $v = 3$ and the j^3 $J = j$ state with $v = 1$. Let us calculate such matrix elements which are non-diagonal in the seniority. The j^3 state with $J = j$, $v = 1$ has the principal parent $J_0 = 0$ and its c.f.p. are given by (15.11). In order to construct a state with $v = 3$, we must start with some j^2 J_0 state with $J_0 > 0$. Such states whose c.f.p. are given by (15.10) with $J = j$ are, however, not orthogonal to the $v = 1$ state. In order to obtain a state with $v = 3$, which belongs to the eigenvalue 0 of P, we have to construct a state for which the c.f.p. for $J_1 = 0$ vanishes. To achieve this goal, the simplest way is to start with any $J_0 > 0$ and construct the antisymmetric state which is not yet normalized as given by (15.7) for $J = j$. The c.f.p. for $J_1 = 0$ in that state is equal to $-2\sqrt{2J_0 + 1}/(2j + 1)$ whereas the c.f.p. for $J_1 > 0$ are given by

$$\delta_{J_1J_0} + 2\sqrt{(2J_0+1)(2J_1+1)}\begin{Bmatrix} j & j & J_1 \\ j & j & J_0 \end{Bmatrix}.$$

If we add to it the $v = 1$ state, whose c.f.p. are given by (15.11), multiplied by $2\sqrt{3(2J_0+1)}/\sqrt{(2j+1)(2j-1)}$ we obtain the unnormalized state with vanishing c.f.p. for $J_1 = 0$, whose c.f.p. for $J_1 > 0$ are proportional to

$$\delta_{J_1J_0} + 2\sqrt{(2J_0+1)(2J_1+1)}\begin{Bmatrix} j & j & J_1 \\ j & j & J_0 \end{Bmatrix} - \frac{4\sqrt{(2J_0+1)(2J_1+1)}}{(2j+1)(2j-1)}$$

$$J_0, J_1 > 0, \text{ even} \qquad (20.39)$$

If the coefficients (20.39) do not all vanish they define a state with $J = j, v = 3$ in the j^3 configuration. The normalization of the state with c.f.p. proportional to (20.39) can be directly calculated using the relations (10.13), (10.14) as well as (10.17), (10.18) of $6j$-symbols. It may also be obtained directly from the value of (20.39) for $J_1 = J_0$ according to (15.26). The c.f.p. of the normalized state with $J = j$, $v = 3$ obtained from J_0 is thus given by

$$[j^2(J_1)jJ = j|\}j^3[J_0]v = 3J = j]$$

$$= \left[\delta_{J_1J_0} + 2\sqrt{(2J_0+1)(2J_1+1)}\begin{Bmatrix} j & j & J_1 \\ j & j & J_0 \end{Bmatrix} - \frac{4\sqrt{(2J_0+1)(2J_1+1)}}{(2j+1)(2j-1)}\right]$$

$$\times\left[3 + 6(2J_0+1)\begin{Bmatrix} j & j & J_0 \\ j & j & J_0 \end{Bmatrix} - \frac{12(2J_0+1)}{(2j+1)(2j-1)}\right]^{-1/2} \qquad (20.40)$$

The matrix element of any interaction V between the state defined by the c.f.p. (20.40) and the $J = j$, $v = 1$ state can be now calculated by using (15.21) for $n = 3$. Substituting into it the c.f.p. (20.40) and the c.f.p. (15.11) we obtain

$$\frac{-6\sqrt{2J_0+1}}{\sqrt{3(2j+1)(2j-1)}}\left[V_{J_0}+2\sum_{J_1>0\,\text{even}}(2J_1+1)\right.$$
$$\left.\times\begin{Bmatrix} j & j & J_1 \\ j & j & J_0 \end{Bmatrix}V_{J_1}-\frac{4(j+1)}{2j+1}\bar{V}_2\right]$$
$$\times\left[3+6(2J_0+1)\begin{Bmatrix} j & j & J_0 \\ j & j & J_0 \end{Bmatrix}-\frac{12(2J_0+1)}{(2j+1)(2j-1)}\right]^{-1/2}$$
$$=-2\left[V_{J_0}-\bar{V}_2+2\sum_{J_1>0\,\text{even}}(2J_1+1)\begin{Bmatrix} j & j & J_1 \\ j & j & J_0 \end{Bmatrix}(V_{J_1}-\bar{V}_2)\right]$$
$$\times\sqrt{\frac{(2J_0+1)}{(2j+1)(2j-1)}}$$
$$\times\left[1+2(2J_0+1)\begin{Bmatrix} j & j & J_0 \\ j & j & J_0 \end{Bmatrix}-\frac{4(2J_0+1)}{(2j+1)(2j-1)}\right]^{-1/2} \tag{20.41}$$

The last equality in (20.41) was obtained by using the sum rules (10.17) and (10.18) of $6j$-symbols.

The first conclusion we draw from (20.41) is that when it is equal to zero, i.e. the interaction is diagonal in the seniority scheme, the conditions (20.37) must hold. Hence, these conditions are *necessary* as well as *sufficient* for the interaction to be diagonal with respect to seniority. We also see that not all the conditions (20.37) are independent. Due to (20.40), we see that the number of independent conditions is equal to the number of independent states with $J=j$, $v=3$ in the j^3 configuration. For $j\leq\frac{7}{2}$ there is no antisymmetric j^3 state with $J=j$, $v=3$. Hence, any two body interaction is diagonal in the seniority scheme for j^n configurations if $j\leq\frac{7}{2}$. For $j=\frac{9}{2}$ there is one j^3 state with $J=\frac{9}{2}$, $v=3$ and hence two body energies V_J obtained from an interaction which is diagonal in the seniority scheme must obey one condition.

The number of conditions on the j^2 energies V_J may be obtained by counting states in the m-scheme in the j^3 configuration. The number of states with $M=j+1$ is equal to the number of states with $J\geq j+1$. When this number is subtracted from the number of states with $M = j$ (equal to the number of $J = j$ states), we obtain the number of j^3 states with $J = j$. One of these is the $J = j$ state with seniority $v = 1$ and the number of $J = j$, $v = 3$ states is equal to $[(2j-3)/6]$—the largest integer not exceeding $(2j-3)/6$ (Ginocchio 1993). Hence, the number of conditions on the V_J of an interaction which is diagonal in the seniority scheme is less than one third of their total number, $(2j+1)/2$, for any value of j.

We saw above that (20.37) are sufficient conditions that a two body interaction be diagonal in the seniority scheme. Their appearance as factors in (20.41) leads to a simple consequence. We can use (15.21) to express matrix elements in the j^n configuration in terms of those in the j^{n-1} configuration. This may be continued until the j^3 configuration is reached. Hence, matrix elements between states with different seniorities in any j^n configuration may be expressed as linear combinations of the matrix elements (20.41) in the j^3 configuration. Vanishing of (20.41) thus guarantees vanishing of matrix elements between states with different seniorities in *all* j^n configurations. The expressions of j^n matrix elements between states with different seniorities may not include j^3 matrix elements between states with the same seniority. Had that been the case, conditions (20.37) could not guarantee that those j^n matrix elements vanish. As we saw above, the conditions (20.37) are sufficient conditions and no extra conditions may be imposed on the V_J for the interaction to be diagonal with respect to seniority.

A simple example is offered by $J = 0$ states in the j^4 configuration. By adding a particle the seniority must change by one (corresponding to a change of s by $\frac{1}{2}$). Hence, the state with $J = 0$, $v = 0$ in any even j^n configuration may have as its only (fractional) parent the state with $J = j$, $v = 1$ in the j^{n-1} configuration. Thus, the state

$$\psi(j^n v = 0, J = 0, M = 0) = \psi(j^{n-1}(v = 1, J = j)j_n J = 0, M = 0) \tag{20.42}$$

is fully antisymmetric as obtained above from (19.28). There is no $J = 0$ with $v = 2$ in the j^4 configuration and hence, other j^4 states with $J = 0$ must have $v = 4$. The (fractional) parents of those must be j^3 states with $J = j$, $v = 3$. In fact, to each state with $J = j$, $v = 3$ in the j^3 configuration corresponds an independent state with $J = 0$, $v = 4$ in the j^4 configuration. To see it, we use the recursion formula (15.28) for c.f.p. for $n = 4$, $J = 0$, $J_1 = J_0 = j$. We can then express it as

$$4[j^3(\alpha_0)jJ = 0|\}j^4[\alpha_0]J = 0][j^3(\alpha_1)jJ = 0|\}j^4[\alpha_0]J = 0]$$
$$= \delta_{\alpha_0\alpha_1} + 3\sum_{\alpha_2 J_2}[j^2(\alpha_2 J_2)jJ_0 = j|\}j^3\alpha_0 J_0 = j]$$
$$\times [j^2(\alpha_2 J_2)jJ_1 = j|\}j^3\alpha_1 J_1 = j] \tag{20.43}$$

Due to the orthogonality of states in the j^3 configuration we see from (20.43) that the c.f.p. for the principal parent $\alpha_1 = \alpha_0$ is equal to 1 and all other c.f.p. vanish if $\alpha_1 \neq \alpha_0$. Hence all states

$$\psi(j^4(v = 3, \alpha, J = j)j_4 J = 0, M = 0)$$

are fully antisymmetric and have seniority $v = 4$. According to (15.14) they are orthogonal to each other. The matrix elements of a two body interaction between any of these $v = 4$ states and the state (20.42) with $v = 0$ are given according to (15.21) by twice the corresponding matrix elements (20.41).

As explained above, the effective interaction in nuclei can be reliably determined only from experiment. Its matrix elements cannot be obtained from some simple interaction. As a result, it cannot be expected that even if the seniority scheme gives a good description of nuclear states, the conditions (20.37) will be satisfied *exactly*. In practice, we need a quantitative measure for determining how good is the seniority quantum number. We should calculate matrix elements of the effective interaction which are non-diagonal in seniority. We should then check how small they are in relation to the energy difference between the corresponding diagonal elements. Such non-diagonal matrix elements are given by (20.41). Nondiagonal matrix elements in higher j^n configurations may be expresed as linear combinations of them by using (15.21). In the following section we shall see examples of j^n configurations of identical nucleons in actual nuclei.

21

Examples of j^n Configurations in Nuclei

Seniority was introduced by Racah (1943) as an additional quantum number, to distinguish between states of a given configuration which have the same values of angular momentum. In configurations of identical nucleons the simplest such situation arises in the $(\frac{7}{2})^4$ configuration for $J = 2$ and $J = 4$ states. The seniority scheme, however, is very useful also for states of the nuclear shell model which do not need an additional quantum number for their characterization. This is particularly true in semi-magic nuclei where, as mentioned above, ground states have lowest seniorities $v = 0$ in even nuclei and $v = 1$ in odd ones.

Let us go back to $(2d_{5/2})^n$ configurations considered in Section 16. For any value of n there is only one state with given value of J allowed by the Pauli principle. Hence, all states of $(\frac{5}{2})^n$ configurations have definite values of the seniority quantum number. In particular, for n even, states with $J = 0$ have seniority $v = 0$ and states with $J = \frac{5}{2}$ occurring in configurations with n odd, have seniority $v = 1$. Any Hamiltonian is trivially diagonal in the seniority scheme within $(\frac{5}{2})^n$ configurations. Hence, if the shell model Hamiltonian includes single nucleon energies and two-body interactions, its eigenvalues in $v = 0$ and $v = 1$ states are given by (20.14). Single nucleon separation energies are then given by (20.16) and should lie on two parallel straight lines. In Fig. 20.1 there is a plot of neutron separation energies in Zr nuclei and the agreement with this prediction is rather good.

The next simple case is that of $(\frac{7}{2})^n$ configurations of identical nucleons (Lawson and Uretsky 1957a, Talmi 1957). In the $(\frac{7}{2})^4$ configuration there are *two* states with $J = 2$ and two states with $J = 4$. Still, as explained in Section 19 following (19.53) and in Section 20 following (20.41), any two-body interaction in $(\frac{7}{2})^n$ configurations is diagonal in the seniority scheme. Binding energies of semi-magic nuclei with valence identical nucleons are then given by putting $j = \frac{7}{2}$ in (20.14) which is

$$\boxed{\text{B.E.}(j^n g.s.) = \text{B.E.}(n=0) + n\epsilon_j + \frac{n(n-1)}{2}\alpha + \left[\frac{n}{2}\right]\beta} \tag{21.1}$$

The coefficients α and β are defined by (20.8) and (20.9) respectively. Single nucleon separation energies were calculated from (20.14) to be given by (20.16), namely

$$\boxed{\text{B.E.}(j^n g.s.) - \text{B.E.}(j^{n-1} g.s.) = \epsilon_j + (n-1)\alpha + \frac{1+(-1)^n}{2}\beta} \tag{21.2}$$

The $1f^n_{7/2}$ configurations of identical nucleons in actual nuclei should occur in calcium isotopes ($Z = 20$) or in nuclei with neutron number $N = 28$. The ground states of odd Ca nuclei have indeed $J = \frac{7}{2}$ and odd parity. This is also the case for nuclei with $N = 28$ neutrons and an odd number of protons between 20 and 28. The measured neutron separation energies are plotted in Fig. 21.1 where they seem to lie fairly well on two straight and parallel lines. The lines connect the values of (21.2) for fixed values of the coefficients ϵ, α and β. As emphasized on several occasions above, there is no reliable way to calculate values of these coefficients from the interaction between free nucleons. The method suggested here is to determine matrix elements of the effective interaction from experiment in a consistent

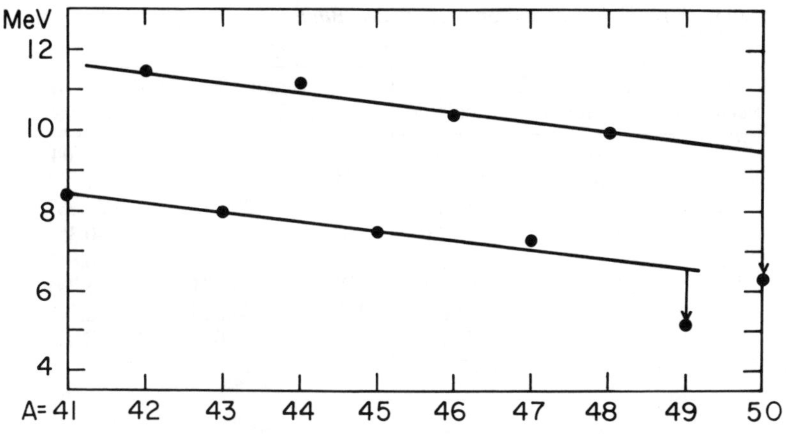

FIGURE 21.1. *Single neutron separation energies of* Ca *isotopes.*

way. One consintency check is to see whether good agreement with experiment is obtained by calculating binding energies from (21.1) with values of ϵ, α and β which give the best fit to experimental energies. The parameters in Fig. 21.1 are determined by the best fit to the binding energies of Ca isotopes from which the binding energy of ^{40}Ca was subtracted. The values which yield the best fit are

$$\epsilon_{f_{7/2}} = 8.423 \text{ MeV} \qquad \alpha = -.227 \text{ MeV} \qquad \beta = 3.233 \text{ MeV} \tag{21.3}$$

The agreement between binding energies and those calculated from (21.1) using the values (21.3) is very good as seen in Table 21.1. The root mean square (r.m.s.) deviation defined by

$$\left(\sum_{i=1}^{N} (E_i^{\text{exp}} - E_i^{\text{cal}})^2 / (N-k) \right)^{1/2}$$

where $k = 3$ is the number of free parameters in the fit, is .093 MeV. As in the case of $(2d_{5/2})^n$ configurations, the pairing term β is strong and attractive and the coefficient of the quadratic term α is small and repulsive. We take binding energies to be *positive* quantities which leads to the signs of α and β as they appear in (21.3). The linear combination of two-body matrix elements which is defined in (20.8) as α can be determined only if the actual values of the matrix elements

TABLE 21.1 *Experimental and Calculated Binding Energies in the* $1f_{7/2}$ *Shell (in* MeV)

Binding Energy-B.E.(^{40}Ca)			Binding Energy-B.E.(^{48}Ca)		
Nucleus	Experimental	Calculated	Nucleus	Experimental	Calculated
^{41}Ca	8.363	8.423	^{49}Sc	9.628	9.704
^{42}Ca	19.844	19.852	^{50}Ti	21.789	21.761
^{43}Ca	27.777	27.822	^{51}V	29.852	29.883
^{44}Ca	38.909	38.798	^{52}Cr	40.357	40.358
^{45}Ca	46.324	46.314	^{53}Mn	46.918	46.898
^{46}Ca	56.720	56.837	^{54}Fe	55.771	55.791
^{47}Ca	63.996	63.901	^{55}Co	60.835	60.748
^{48}Ca	73.990	73.991	^{56}Ni	68.000	68.059

are known. Hence, it can be determined only if binding energies are considered. On the other hand, the coefficient β of the pairing term is given by (20.9) and is thus determined by *differences* between matrix elements. Therefore, its value given in (21.3) may be compared with the expression (20.9) where measured spacings of energy levels are substituted. Spectra of nuclei in which $(1f_{7/2})^n$ configurations appear will be considered below. To make the comparison between (20.9) and (21.3) we may use energy levels of the two $1f_{7/2}$ neutron configuration in ^{42}Ca. These have the spins $J = 0$ (ground state), $J = 2$ at 1.525 MeV, $J = 4$ at 2.752 MeV and $J = 6$ at 3.189 MeV above the ground state. Hence, the center of mass of the levels with seniority $v = 2$, defined by (20.7) as

$$\bar{V}_2 = \frac{1}{(j+1)(2j-1)} \sum_{J>0,\,\mathrm{even}} (2J+1)V_J \tag{21.4}$$

lies 2.735 MeV above the $J = 0$, $v = 0$ ground state. The coefficient β is defined by (20.9) as

$$\beta = \frac{2(j+1)}{2j+1}(V_0 - \bar{V}_2) \tag{21.5}$$

Putting the value $V_0 - \bar{V}_2 = -2.735$ into (21.5) and reversing the sign (due to our taking binding energies as positive quantities) we obtain

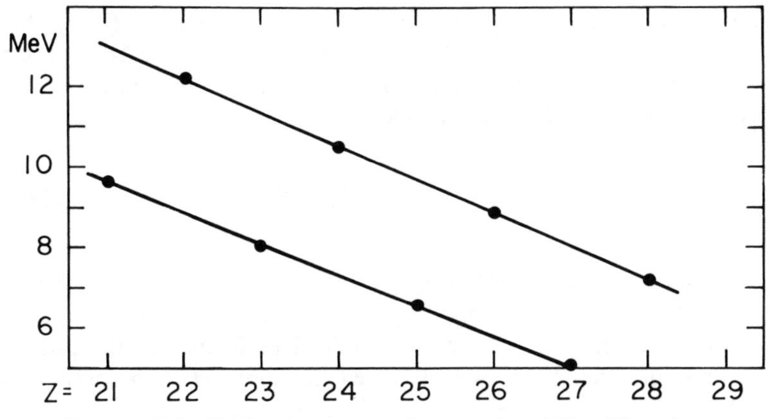

FIGURE 21.2. *Single proton separation energies of $N = 28$ nuclei.*

the value $9 \times 2.735/8 = 3.077$ MeV. This is in fair agreement with the value of β in (21.3). As we shall see later, the states in Ca isotopes cannot belong to *pure* $(1f_{7/2})^n$ configurations. In view of this fact the agreement between the values of β from (21.3) and (21.5) is satisfactory.

Let us now consider ground state energies of nuclei with closed neutron shells, $N = 28$. Single proton separation energies are plotted in Fig. 21.2 and lie rather well on two straight and parallel lines in agreement with (21.2). The parameters ϵ, α and β can be determined by using (21.1) to obtain the best fit to differences between binding energies of nuclei with proton number $28 \geq Z \geq 21$ and the binding energy of ^{48}Ca. These coefficients turn out to be

$$\epsilon_{f_{7/2}} = 9.704 \text{ MeV} \qquad \alpha = -.791 \text{ MeV} \qquad \beta = 3.144 \text{ MeV} \tag{21.6}$$

The agreement between experimental binding energies and those calculated by using the values (21.6) in (21.1) is very good as can be seen in Table 21.1. The r.m.s. deviation is only .064 MeV.

The fact that $\epsilon_{f_{7/2}}$ of a proton (for $N = 28$) is not very different from $\epsilon_{f_{7/2}}$ of a neutron (for $Z = 20$) is due to the electrostatic repulsion of the protons. The value of $\epsilon_{f_{7/2}}$ in (21.6) contains also the interaction of one $1f_{7/2}$ proton with the closed $1f_{7/2}$ neutron orbit. It is not much higher (in absolute value) than the neutron $\epsilon_{f_{7/2}}$ only due to the Coulomb repulsion between the $1f_{7/2}$ proton and the closed proton shells. To appreciate the size of this effect we can compare the proton separation energy in ^{41}Sc, 1.086 MeV, to $\epsilon_{f_{7/2}}$ in (21.6). The difference of 8.618 MeV is due to the interaction of one $1f_{7/2}$ proton with the 8 neutrons completely filling the $1f_{7/2}$ orbit. As we saw in Section 14, this interaction is equal to

$$\sum_{J=0}^{7}(2J+1)V(j_\pi j_\nu J) = 8\bar{V}(j_\pi j_\nu) \tag{21.7}$$

where j_π denotes the proton $1f_{7/2}$ orbit and j_ν denotes the $1f_{7/2}$ neutron orbit. Thus, the average $1f_{7/2}$ proton—$1f_{7/2}$ neutron interaction is *attractive* and amounts to 1.08 MeV. This feature of the proton-neutron interaction will be discussed in more detail in subsequent sections.

The electrostatic repulsion between protons in j^n configurations is considered as a perturbation. Its contribution to energies are given by its expectation values in states with definite seniorities. As explained in Section 20, the expectation values of *any* two-body interaction in states with lowest seniorities have the form (20.13). Hence, the contribution of the Coulomb interaction is absorbed into the coefficients C, α and β. In the special case of $1f_{7/2}$ protons, the electrostatic repulsion, like any two-body interaction, is diagonal in the seniority scheme.

The electrostatic repulsion between protons leads to a significant difference between the value of α for neutrons in (21.3) and that of the protons in (21.6). The latter includes a large contribution from the two proton matrix elements of the Coulomb repulsion. There may also be other effects which contribute to the difference. The effective interaction may not be the same since the cores are different. In particular, the polarization of the different cores due to valence nucleons may also not be the same. Still, the main bulk of the difference should be due to the Coulomb repulsion. We may learn about the size of that repulsion by comparing energies of ^{42}Ca, ^{42}Sc and ^{42}Ti. The ground states of all three nuclei have $J = 0$ and $T = 1$. Hence, the binding energy difference B.E.(^{42}Sc)–B.E.(^{42}Ti) should be larger than the difference B.E.(^{42}Ca)–B.E.(^{42}Sc) due to the Coulomb repulsion of the two $1f_{7/2}$ protons in the $J = 0$ state. Experimentally the difference of those binding energy differences amounts to $7.784 - 7.206 = .578$ MeV, rather close to .564 MeV, the difference between the values of α in (21.3) and (21.6).

The value of β, the coefficient of the pairing term in (21.6), may be compared to the expression (21.5) for which experimental information about level spacings may be used. In ^{50}Ti the lowest levels have the spins $J = 0$ (ground state), $J = 2$ at 1.554 MeV, $J = 4$ at 2.675 MeV and $J = 6$ at 3.198 MeV above the ground state. The center of mass of these levels with seniority $v = 2$ lies at 2.719 MeV above the ground state. Thus, the value $V_0 - \bar{V}_2 = -2.719$ MeV can be substituted into (21.5) with $j = \frac{7}{2}$. Reversing the sign, due to binding energies taken as positive numbers, we obtain for β the calculated value of $9 \times 2.719/8 = 3.059$ MeV. This is in fair agreement with the value of β in (21.6). The difference between the values of β in (21.3) and in (21.6) is only .089 MeV. In the case of protons there is a contribution to β from the electrostatic repulsion.

The repulsive nature of the coefficient α is consistent with general features of nuclear binding energies. Saturation implies that the latter are roughly proportional to $N + Z$ for $N = Z$ nuclei. As nucleons of one kind are added to, or removed from a nucleus with $N = Z$, the absolute value of the binding energy decreases by a term roughly proportional to $(N - Z)^2$. Such a term leads to a quadratic repulsive term in the binding energy formula (21.1).

We now turn our attention to energies of excited states in $1f^n_{7/2}$ configurations. In $(1f_{7/2})^2$ and $(1f_{7/2})^6 \equiv (1f_{7/2})^{-2}$ configurations only excited levels with seniority $v = 2$ should be observed. In $(1f_{7/2})^4$ configurations, there are in addition to the $J = 2, 4, 6$ states with seniority $v = 2$, also $v = 4$ states with $J = 2, 4, 5, 8$. If the mutual effective interaction has only two-body terms then, as explained in Section 20, the eigenstates of the Hamiltonian have definite seniorities. In that case, the spacings between states with seniorities $v = 2$ and $v = 0$ of $(f_{7/2})^n$ configurations should be independent of n. The experimental levels in even-even nuclei are shown in Fig. 21.3 for both proton and neutron $(1f_{7/2})^n$ configurations. It shows that the $J = 2$ and $J = 6$ levels have only roughly constant spacings to ground states. There are *two* low lying $J = 4$ levels in $(1f_{7/2})^4$ configurations which roughly agree with the calculation described below.

In the middle of Fig. 21.3 a set of energies for the $J = 2, 4, 6$ levels with seniority $v = 2$ was adopted to approximately represent the experimental data. Using these energies as the $(1f_{7/2})^2$ matrix elements, levels with seniority $v = 4$ were calculated. It turned out that the $J = 4$, $v = 4$ level should be somewhat *lower* than the $J = 4$, $v = 2$ level. The other $v = 4$ levels are calculated to lie very much higher, the $J = 2$, $v = 4$ level at 3.031 MeV, the $J = 5$, $v = 4$ level at 3.565 MeV and the $J = 8$, $v = 4$ level at 5.164 MeV above the ground state. Thus, the two low lying $J = 4$ levels observed in ^{44}Ca and ^{52}Cr may be assigned the $(1f_{7/2})^4$ configuration. The spacings between the two $J = 4$

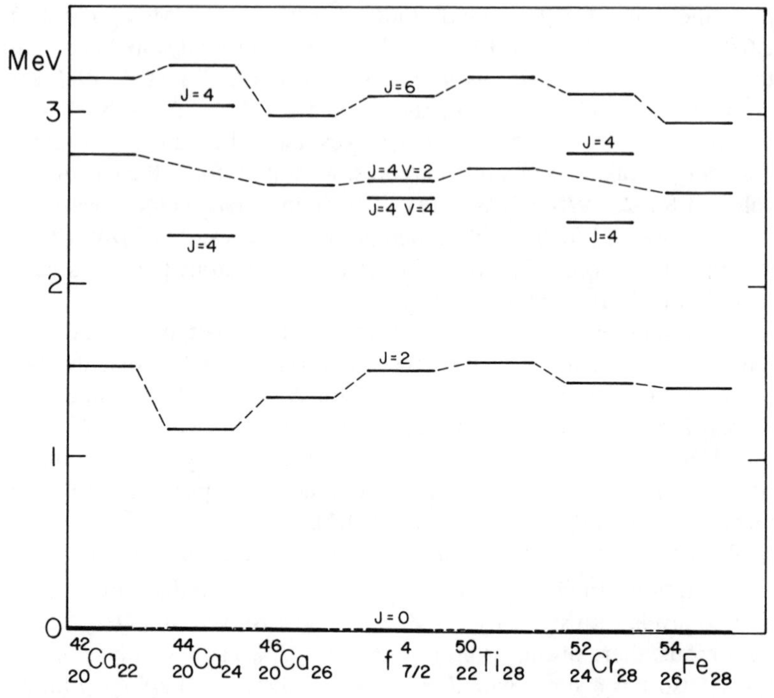

FIGURE 21.3. *Levels of even $(1f_{7/2})^n$ configurations of protons and of neutrons.*

observed levels in both nuclei are considerably larger than the calculated one. Since the calculated positions of the $J = 4$ levels with $v = 2$ and $v = 4$ are rather close, they may be pushed apart if there is a non-vanishing matrix element of the effective interaction connecting them. Such a matrix element could be due to three-body interactions or simply due to interaction with *other configurations*. There are some indications from E2 transition probabilities in ^{52}Cr that there is indeed some mixing between the $J = 4$ levels with seniorities $v = 2$ and $v = 4$ (Talmi 1962b). The fact that the description in terms of pure $1f^n_{7/2}$ configurations for nuclei in this region is not very good is clearly evident in odd-even nuclei which we will now consider.

The matrix elements of the $(1f_{7/2})^2$ configuration adopted in Fig. 21.3 may be used to calculate energy levels of $(1f_{7/2})^3$ and $(1f_{7/2})^5 \equiv (1f_{7/2})^{-3}$ configurations. The calculated positions of the $v = 3$ levels with $J = \frac{3}{2}, \frac{9}{2}, \frac{11}{2}, \frac{15}{2}$ above the $J = \frac{7}{2}$, $v = 1$ ground state are plotted in Fig. 21.4 where also the experimental levels are shown. The level

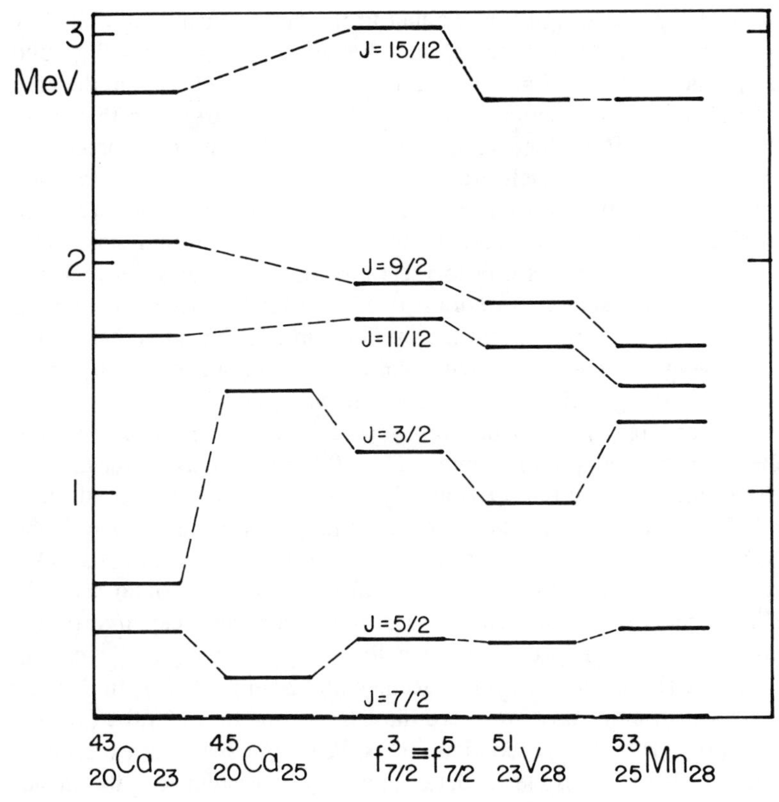

FIGURE 21.4. *Levels of odd* $(1/f_{7/2})^n$ *configurations of protons and of neutrons.*

spacings should have been the same for $n = 3$ and $n = 5$ but in the experimental data only the same order of levels is observed. In particular, the $J = \frac{3}{2}$ level exhibits large variations which may be attributed to the proximity of the $2p_{3/2}$ and $1f_{7/2}$ orbits. It is clear that there are strong perturbations from other configurations and thus there is only rough agreement between calculated and experimental level spacings.

One important feature of Fig. 21.4 is the very low lying $J = \frac{5}{2}$ level which is observed experimentally and is in good agreement with the shell model calculation. The calculated position is obtained from using the *experimental information* on two nucleon matrix elements. If instead, the pairing interaction (or the δ-interaction) is adopted, all seniority $v = 3$ levels must be degenerate (or nearly so) at the position which, due to (19.11), is equal to the "energy gap" multiplied by $\frac{3}{4}$. That "energy gap" is the spacing between the $v = 0$, $J = 0$ ground state and the $v = 2$, $J = 2, 4, \ldots$ levels which are degenerate in the case of the pairing interaction. From Fig. 21.3 it is clearly seen that those levels are far from degenerate in $(1f_{7/2})^n$ configurations in actual nuclei. Still, all $v = 2$ levels are appreciably higher than the $v = 0$, $J = 0$ ground state. In odd-even nuclei, however, the $J = j - 1$ level with seniority $v = 3$ is very close to the $J = j$, $v = 1$ ground state. This small spacing which is due to the rather large spacings between the $v = 2$ levels, displays more dramatically the inadequacy of the pairing interaction for representing the effective interaction in actual nuclei. Low lying $J = j - 1$ states with seniority $v = 3$ appear also in proton $(1g_{9/2})^n$ configurations which we will now consider.

Proton $(1g_{9/2})^n$ configurations appear in nuclei with neutron number $N = 50$ and proton number $Z < 50$. The nucleus $^{88}_{38}\mathrm{Sr}_{50}$ may be considered to have the $2p_{3/2}$ and $1f_{5/2}$ orbits outside the closed shells of $Z = 28$, completely filled. In $^{89}_{39}\mathrm{Y}_{50}$ the ground state has $J = \frac{1}{2}$ and negative parity as due to a single proton in the $2p_{1/2}$ orbit. A $J = 9/2$ state with positive parity lies .909 MeV above the ground state. In $^{90}_{40}\mathrm{Zr}_{50}$ the first excited state has $J = 0$ which has been interpreted as due to strong mixing between the $(2p_{1/2})^2$ and $(1g_{9/2})^2$ configurations (Ford 1955, Bayman, Reiner and Sheline 1959). In $N = 50$ nuclei with more protons, there must be mixing between the $(1g_{9/2})^n$ and $(1g_{9/2})^{n-2}(2p_{1/2})^2$ configurations. From the available experimental data it was possible to determine in a consistent way the diagonal as well as the non-diagonal matrix elements and obtain a good description of energy levels and electromagnetic transitions (Talmi and Unna (1960b), Cohen, Lawson, Macfarlane and Soga (1964), Auerbach and Talmi (1965), Gloeckner and Serduke (1974)). The results of these calculations will not be presented here. Instead, we will use an important feature of the mixing of these configurations to extract directly some matrix elements in the pure $(1g_{9/2})^n$ proton configurations.

The $(2p_{1/2})$ configuration has only one antisymmetric state with $J = 0$. Hence, in the case of two valence nucleons outside $^{88}\mathrm{Sr}$, the states with $J = 2, 4, 6, 8$ are $v = 2$ states of the pure $(1g_{9/2})^n$ configuration. The only state of that configuration affected by the proximity of the $2p_{1/2}$ orbit is the $J = 0$ state. Thus, we may take the spacings between the $J = 2, 4, 6, 8$ levels in $^{90}\mathrm{Zr}$ to be the unperturbed levels of the $(1g_{9/2})^2$ configuration. As we shall now see, if the two body effective interaction is diagonal in the seniority scheme in $(1g_{9/2})^n$

configurations, these spacings should be equal to those in other nuclei with more valence protons.

Interactions between different configurations will be presented in detail in Section 32. Here we shall only consider the special case where there is mixing between the state $|j^n\alpha JM\rangle$ and the state $|j^{n-2}(\alpha_1 J)j'^2(J=0)JM\rangle$. The non-diagonal matrix element is

$$\mathcal{N}\int\psi^*(j^n\alpha JM)(\sum V_{ik})\mathcal{A}\psi(j^{n-2}(\alpha_1 J)j'^2_{n-1,n}(0)JM) \quad (21.8)$$

The antisymmetrizer $\mathcal{A}$ in (21.8) is defined by

$$\mathcal{A}\psi(j^{n-2}(\alpha_1 J)j'^2_{n-1,n}(0)JM) = \sum(-1)^P\psi(j^{n-2}(\alpha_1 J)j'^2_{ik}(0)JM) \quad (21.9)$$

where i,k are any pair of indices from 1 to n and the permutation P replaces $n-1,n$ by i,k. Any two terms on the r.h.s. of (21.9) are orthogonal since at least one nucleon is in a different orbit in the two terms. There are $n(n-1)/2$ such terms and thus the normalization factor $\mathcal{N}$ is $(n(n-1)/2)^{-1/2}$. We now remove in (21.8) the projection operator $\mathcal{A}/(n(n-1)/2)$ from ψ and let it operate on ψ^*. This way we obtain the matrix element (21.8) in the form

$$\sqrt{\frac{n(n-1)}{2}}\int\psi^*(j^n\alpha JM)(\sum V_{ik})\psi(j^{n-2}(\alpha_1 J)j'^2_{n-1,n}(0)JM) \quad (21.10)$$

The only term which may contribute to the integral in (21.10) is $V_{n-1,n}$. All other terms vanish when the integration is carried out on the variables of nucleon $n-1$ or n. Expanding $\psi^*(j^n\alpha JM)$ in $n\to n-2$ coefficients of fractional parentage we obtain for the matrix element (21.8) the result

$$\sqrt{\frac{n(n-1)}{2}}[j^{n-2}(\alpha_1 J)j^2(0)J|\}j^n\alpha J]\langle j^2J=0|V|j'^2J=0\rangle \quad (21.11)$$

In the case of the seniority scheme, we may use v_1 and v instead of α_1 and α. The c.f.p. in (21.11) vanishes unless $v_1=v$ and for

$v_1 = v$ it is given by (19.29). It depends on n and v but is *independent* of J.

The diagonal matrix elements of the two nucleon interaction in the states $|j^{n-2}(vJ)j'^2(0)JM\rangle$, according to (14.33), are equal to

$$E'_J = E(j^{n-2}vJ) + E(j'^2J = 0) + 2(n-2)\bar{V}(jj') \qquad (21.12)$$

where the average jj' interaction $\bar{V}(jj')$ is defined by (14.27) divided by $(2j+1)(2j'+1)$. In the seniority scheme, the energy E_J of the state $|j^n vJM\rangle$, according to (20.14) is equal to

$$E_J = E'_J + 2\epsilon_j + (2n-3)\alpha + \beta - E(j'^2J = 0) - 2(n-2)\bar{V}(jj') \qquad (21.13)$$

where E'_J is the energy of the state in the $j^{n-2}j'^2(0)$ configuration. Thus, the difference between E_J and E'_J is also *independent* of J. The 2×2 matrix for each value of v and J, whose diagonal matrix elements are given by (21.12) and (21.13) and the non-diagonal element is (21.11), should now be diagonalized. This matrix is equal to the unit matrix multiplied by E_J, or E'_J, plus a matrix whose elements, for a given v, are independent of J. The result of the diagonalization is that the downward shifts of the lower eigenvalues are also *independent* of J.

It follows that if the effective two-nucleon interaction is diagonal in the seniority scheme, spacings between perturbed levels which have the same seniority should be independent of n. In particular, they should be equal to those in the *unperturbed* j^n configuration. This is true for mixing of states of the j^n configuration with states $|j^{n-2}(vJ)j'^2(0)JM\rangle$. There is only one antisymmetric state, with $J = 0$, of the $(2p_{1/2})^2$ configuration so that the result stated above holds in the case considered here, provided the Hamiltonian is diagonal in the seniority scheme within $(1g_{9/2})^n$ configurations.

As mentioned in Section 20, the lowest j-value for which a two-body interaction may not be diagonal in the seniority scheme is $j = \frac{9}{2}$. In order to find out whether seniority is a good quantum number in $(1g_{9/2})^n$ configurations, we should determine values of matrix elements which are non-diagonal in the security scheme. The matrix element between the $J = \frac{9}{2}$, $v = 3$ state and the $J = \frac{9}{2}$, $v = 1$ state is given by (20.41) in terms of spacings between $v = 2$ levels. Since there is only *one* $J = \frac{9}{2}$ state with seniority $v = 3$, we may choose $J_0 = 2$ in

(20.41). Substituting the actual values of the $6j$-symbols we obtain for that matrix element the expression

$$-\frac{1}{20\sqrt{429}}[230V_2 - 18V_4 + 832V_6 + 408V_8 - 1452\bar{V}_2]$$
$$= -\frac{1}{20\sqrt{429}}[65V_2 - 315V_4 + 403V_6 - 153V_8] \qquad (21.14)$$

The matrix element (21.14) may now be calculated by using the $J = 2, 4, 6, 8$ energy levels of ^{90}Zr. The expression (21.14) as well as (20.41) is determined only by the *spacings* between those levels. Using these spacings, (21.14) becomes equal to .032 MeV. This value should be compared with the difference between the diagonal matrix elements of the interaction in the $J = \frac{9}{2}$ states with seniorities $v = 1$ and $v = 3$. That difference depends also on the spacing between the $J = 0$ and $v = 2$ levels in the $(1g_{9/2})^2$ configuration. Using for $V_0 - V_2$ the value -1.1 MeV obtained from the detailed analysis (Gloeckner and Serduke 1974) we obtain the difference between those diagonal matrix elements as equal to 1.67 MeV. Reducing the absolute value of $V_0 - V_2$ by .25 MeV reduces the difference between the diagonal matrix elements to 1.47 MeV. Thus, the non-diagonal matrix element is a tiny fraction, .019 (or .022) of the difference between diagonal elements. Hence, seniority is a very good quantum number indeed. The weight of the $v = 3$ component in the lowest $J = \frac{9}{2}$ eigenstate is only .04 to .05 *percent*.

As a result, we expect to find in nuclei with $N = 50$ and $40 < Z < 50$ the same level spacings between $J = 2, 4, 6, 8$ states as those between $v = 2$ levels in ^{90}Zr. Levels of these nuclei are plotted in Fig. 21.5 from which it is seen that those level spacings are fairly independent of Z. This provides a *consistency check* on the configuration assignment and on the assumption of effective two-body interactions. Any of those spacings may be used to determine the matrix element between the $v = 1$ and $v = 3$ states with $J = \frac{9}{2}$. The values obtained from ^{92}Mo, ^{94}Rh and ^{96}Pd are $-.023$, $-.015$ and $-.05$ MeV respectively. Still, even in the case of ^{96}Pd the non-diagonal matrix element is only .033 of the difference between diagonal elements. Thus, in the $(1g_{9/2})^3$ configuration the lowest $J = \frac{9}{2}$ eigenstate is made of 99.9% of the pure seniority $v = 1$ state.

The observed states in nuclei with odd values of $Z > 39$ are linear combinations of the $(1g_{9/2})^{n-2}(2p_{1/2})^2$ configurations. According to the discussion above, the shifts of the lower eigenvalues depend on n

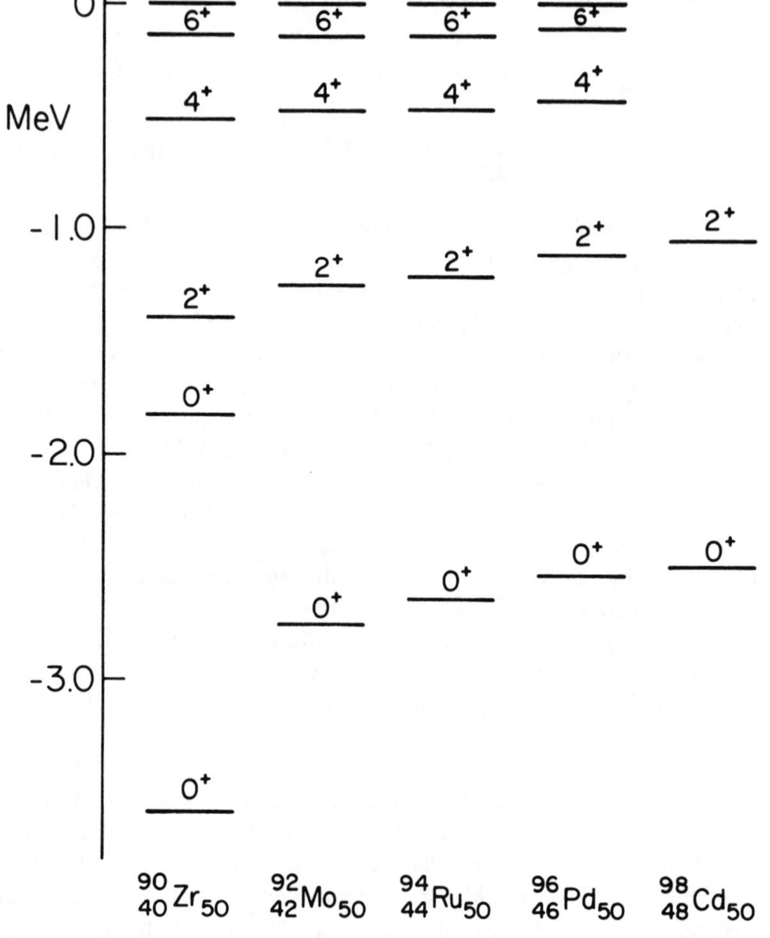

FIGURE 21.5. *Levels of even* $(1g_{9/2})^n$ *proton configurations in* $N = 50$ *nuclei.*

and v but are *independent* of J. In the case $n = 3$ the only state of the $(1g_{9/2})(2p_{1/2})^2$ configuration interacts with the $v = 1$, $J = \frac{9}{2}$ state of the higher $(1g_{9/2})^3$ configuration. The positions of the $v = 3$ levels of the latter configuration are not affected at all. In nuclei with $9 > n > 3$, positions of all levels are affected by configuration mixing but corresponding level spacings are the same as those of $v = 3$ levels in a *pure* $(1g_{9/2})^3$ configuration. In Fig. 21.6 levels of ^{91}Nb to ^{97}Ag are

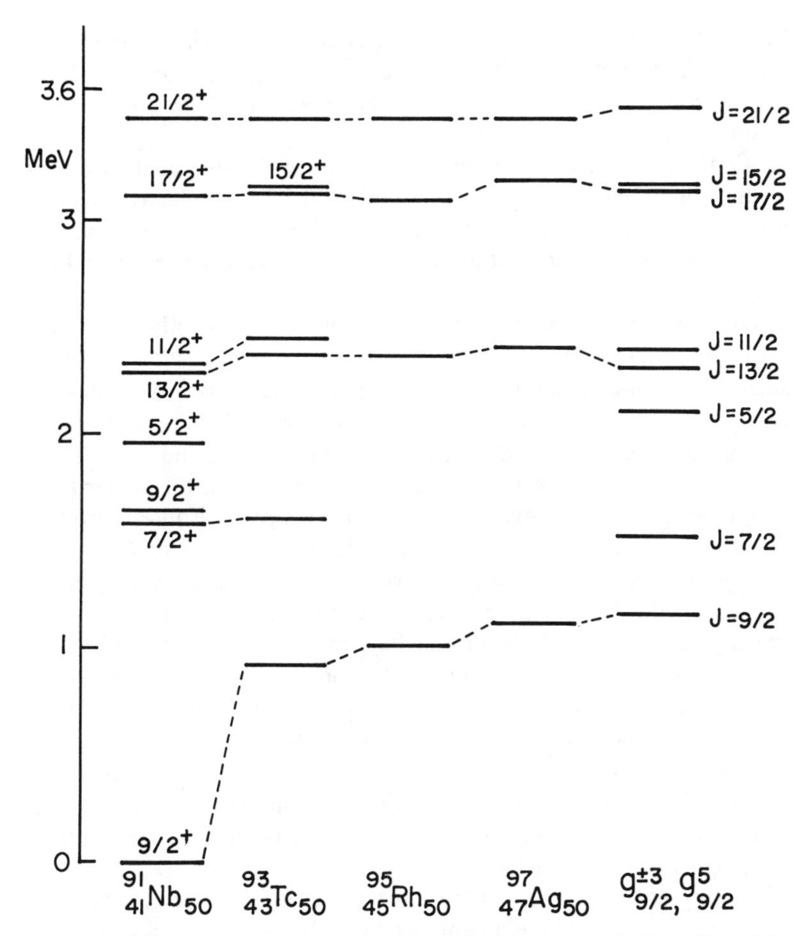

FIGURE 21.6. *Levels of odd* $(1g_{9/2})^n$ *proton configurations in* $N = 50$ *nuclei.*

plotted and we see that level spacings are fairly equal. The position of the lowest $J = \frac{9}{2}$ state must be calculated by taking into account configuration mixing which will not be done here. Spacings of other levels may be calculated from level spacings in Fig. 21.5. The results of such a calculation are presented in Fig. 21.6. The agreement between experimental and calculated levels is rather good providing another consistency check on the simple shell model approach described here. The $v = 1$, $J = \frac{9}{2}$ state of the $(1g_{9/2})^3$ configuration in ^{91}Nb is pushed upwards due to interaction with the lower $1g_{9/2}(2p_{1/2})^2$ state.

In ^{43}Tc the lowest $J = \frac{9}{2}$ state, as well as other levels shown, belong to a linear combination of $(1g_{9/2})^3(2p_{9/2})^2$ and $(1g_{9/2})^5$ configurations. As a result, the $J = \frac{9}{2}$ ground state lies *lower* than the unperturbed position relative to other states.

A rather sensitive test of the amplitudes of the two configurations in the states described above is provided by rates of electromagnetic transitions. Also for these, the agreement between experimental and calculated rates is good. Some of these transitions will be described in Section 31.

Another example of j^n configurations with $j = \frac{9}{2}$ is offered by protons occupying the $1h_{9/2}$ orbit outside the closed proton and neutron shells of $^{208}_{82}$Pb$_{126}$. The ground state of $^{209}_{83}$Bi$_{126}$ has $J = \frac{9}{2}$ and negative parity due to a single proton in the $1h_{9/2}$ orbit. A state with $J = \frac{7}{2}$ and negative parity lies .897 MeV above the ^{209}Bi ground state. Thus, the $2f_{7/2}$ orbit may not be sufficiently far away. Still, the spectra observed in nuclei with several valence protons may be interpreted as due to $(1h_{9/2})^n$ configurations. Looking for the expected states in even configurations, we find low lying levels with $J = 0, 2, 4, 6, 8$ in $^{210}_{84}$Po$_{126}$, $^{212}_{86}$Rn$_{126}$ and $^{214}_{88}$Ra$_{126}$ rather well separated from higher levels. The spacings of these levels in the nuclei considered, shown in Fig. 21.7 are not the same but vary appreciably from one nucleus to the other. If we would like to assign to them $(1h_{9/2})^n$ configurations we must realize that matrix elements of the effective interaction in this case are not the same for these nuclei. Since the change is fairly regular, we may check the consistency of the configuration assignment by taking matrix elements from an even nucleus for calculating energy levels of the neighboring odd nucleus.

Spacing of $v = 2$ levels in the three nuclei in Fig. 21.7 may be used to calculate the matrix element between the $v = 1$ and $v = 3$ states with $J = \frac{9}{2}$ in the $(1h_{9/2})^3$ configuration. We find for ^{210}Po, ^{212}Rn and ^{214}Ra the values $-.041$, $-.036$ and $-.050$ MeV respectively. These values are very small compared to the differences of diagonal matrix elements which are 1.125, 1.228 and 1.336 MeV respectively. Thus, also in this case, the effective interaction is diagonal to a high degree of accuracy in the seniority scheme.

Calculated levels of $(1h_{9/2})^n$ configurations for odd values of n are presented in Fig. 21.8. The three hole configuration $(1h_{9/2})^5$ has the same level spacings as in the $(1h_{9/2})^3$ configuration. To calculate level spacings in $^{211}_{85}$At$_{126}$ we used matrix elements taken from the ^{210}Po levels, for $^{213}_{87}$Fr$_{126}$ we used the ^{212}Rn levels and for ^{215}Ac, the ^{214}Ra levels. The agreement between calculated positions of lev-

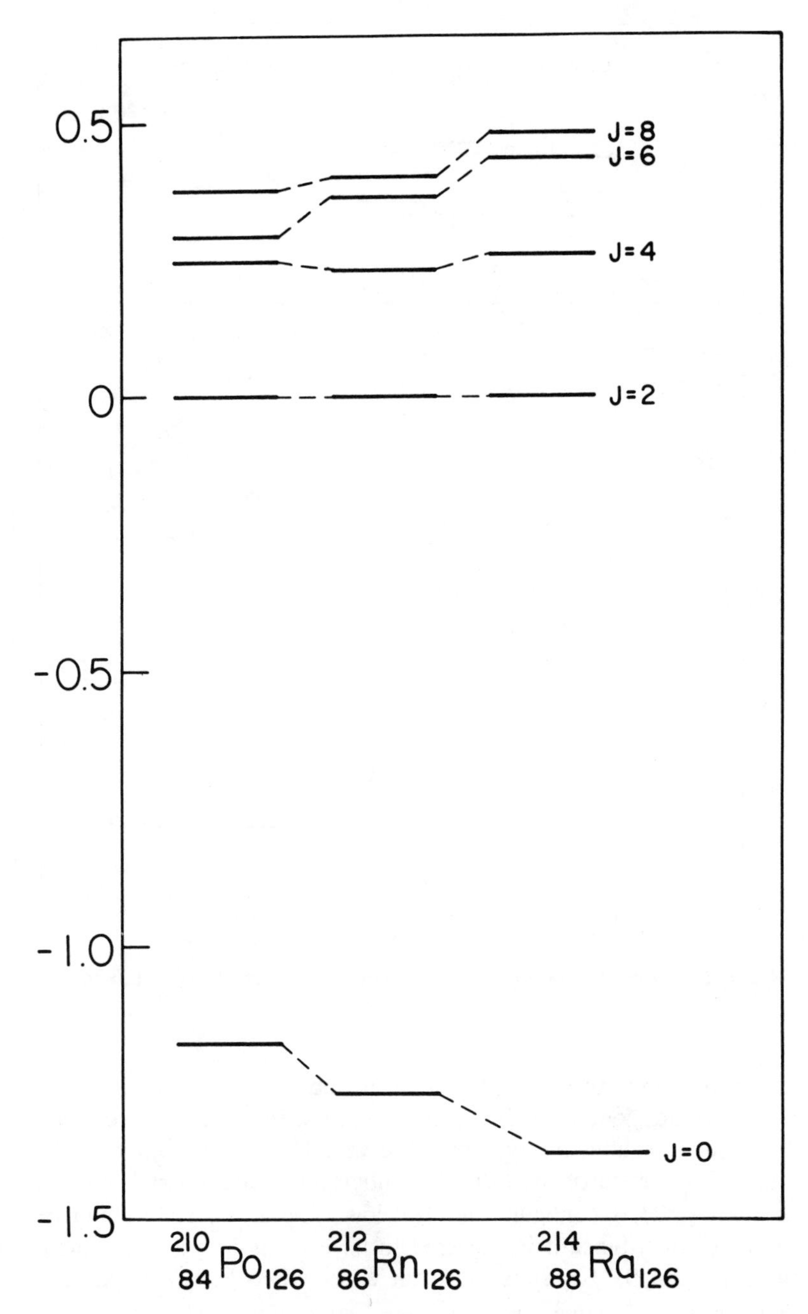

FIGURE 21.7. *Levels of even* $(1h_{9/2})^n$ *proton configurations in* $N = 126$ *nuclei.*

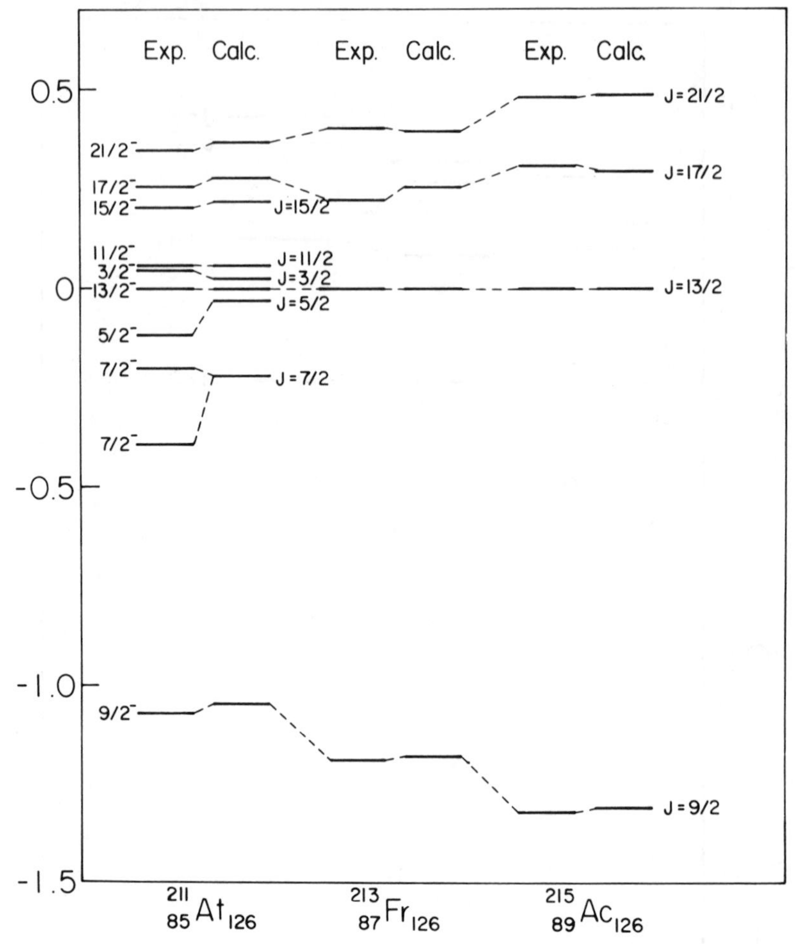

FIGURE 21.8. *Levels of odd* $(1h_{9/2})^n$ *proton configurations in* $N = 126$ *nuclei.*

els, in relation to the $v = 3$, $J = \frac{13}{2}$ level, and those experimental positions is rather good apart from one exception. In ^{211}At *two* levels with $J = \frac{7}{2}$ (and negative parity) were observed. Theoretically, there is the $v = 3$, $J = \frac{7}{2}$ state of the $(1h_{9/2})^3$ configuration and a $J = \frac{7}{2}$ state of the $(1h_{9/2})^2 2f_{7/2}$ configuration. The low lying $J = \frac{7}{2}$ state could be the latter one. Since it is expected to have as its largest component the $1h^2_{9/2}(J = 0)2f_{7/2}$ state, it is not expected to have a large non-diagonal matrix element with the $v = 3$ state of the $(1h_{9/2})^3$ configu-

TABLE 21.2 *Magnetic Moments of* $(1h_{9/2})^n$ *Proton Configurations*

Nucleus	Spin of state	Magnetic moment	g-factor
$^{209}_{83}Bi_{126}$	$\frac{9}{2}$	4.1106(2)	.913
$^{210}_{84}Po_{126}$	6	5.48(5)	.913(8)
	8	7.35(5)	.919(6)
$^{211}_{85}At_{126}$	$\frac{21}{2}$	9.54(9)	.909(9)
$^{212}_{86}Rn_{126}$	4	4.04(24)	1.01(6)
	6	5.454(48)	.909(8)
	8	7.152(16)	.894(2)
$^{213}_{87}Fr_{126}$	$\frac{9}{2}$	4.02(8)	.893(18)
$^{214}_{88}Ra_{126}$	8	7.080(32)	.885(4)
$^{215}_{89}Ac_{126}$	$\frac{17}{2}$	7.82(16)	.92(2)
	$\frac{21}{2}$	9.66(20)	.92(2)

ration which has no $1h_{9/2}$ pairs coupled to $J = 0$. This follows from (21.11) if we put $n = 3$, $j \equiv 1h_{9/2}$, $j' \equiv 2f_{7/2}$ which leads to vanishing c.f.p. for $\alpha = v = 3$.

The magnetic moment of the $J = \frac{9}{2}$ ground state of $^{209}_{83}Bi_{216}$ is 4.1106 n.m. The calculated value for a proton in a $h_{9/2}$ orbit is given by (3.37) for $l = 5$ as 2.63 n.m. Possible explanations for this discrepancy are discussed in Section 31. Here we would like to check whether by adopting the experimental value as the *effective* magnetic moment of a $1h_{9/2}$ proton, moments of other states could be calculated. As explained in Section 8, within j^n configurations of identical nucleons the magnetic moment operator is given by

$$\mu = \sum_{i=1}^{n} g\mathbf{j}_i = g \sum_{i=1}^{n} \mathbf{j}_i = g\mathbf{J} \tag{21.15}$$

Hence, the g-factors of all states of the j^n configuration of identical nucleons should be equal to the g-factor of a single j-nucleon. Thus, in the present case we expect the g-factors of all states of the $(1h_{9/2})^n$ configuration to be equal to $4.1106/(\frac{9}{2}) = .913$ n.m. Available experimental data with the quoted errors are presented in Table 21.2. It can be seen that all measured g-factors are equal within a few percent. There is possibly a larger discrepancy for the $J = 4$ state in ^{212}Rn but the experimental error is rather large in that case.

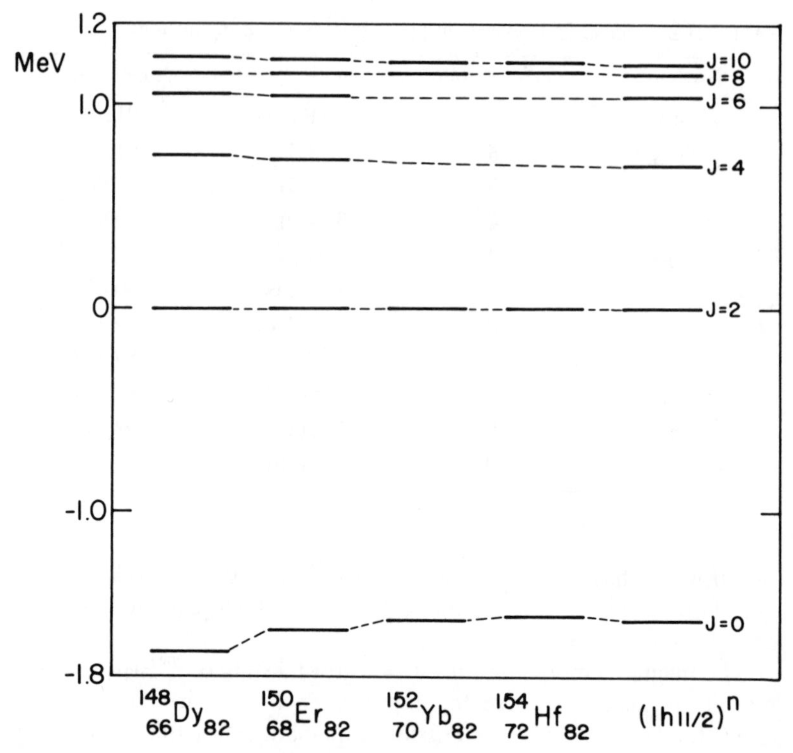

FIGURE 21.9. *Levels of even* $(1h_{11/2})^n$ *proton configurations in* $N = 82$ *nuclei.*

Another example of pure j^n configurations is offered by nuclei with proton number $Z > 64$ and neutron number $N = 82$. Proton number $Z = 64$ is in the middle of the major shell between $Z = 50$ and $Z = 82$. Yet, it was found experimentally in the pioneering work of Kleinheinz *et al.* (1978) that the ground state of ${}^{146}_{64}\mathrm{Gd}_{82}$ shares several features of a doubly magic nucleus. It seems that for $N = 82$ the $1g_{7/2}$ and $2d_{5/2}$ orbits are completely filled at proton number $Z = 64$ and the other orbits in the shell, $1h_{11/2}$, $2d_{3/2}$ and $3s_{1/2}$ are empty. If protons are added to ${}^{146}_{64}\mathrm{Gd}_{82}$, they seem to go into the $1h_{11/2}$ orbit. In ${}^{148}_{66}\mathrm{Dy}_{82}$ states with $J = 0, 2, 4, 6, 8, 10$ expected in the $(1h_{11/2})^2$ configuration, have been observed, which are well separated from higher levels. These levels were observed in ${}^{150}_{68}\mathrm{Er}_{82}$ and heavier elements. The low lying experimental levels of these even-even nuclei are presented in Fig. 21.9. Detailed theoretical calculations were carried out by Blomqvist (1984).

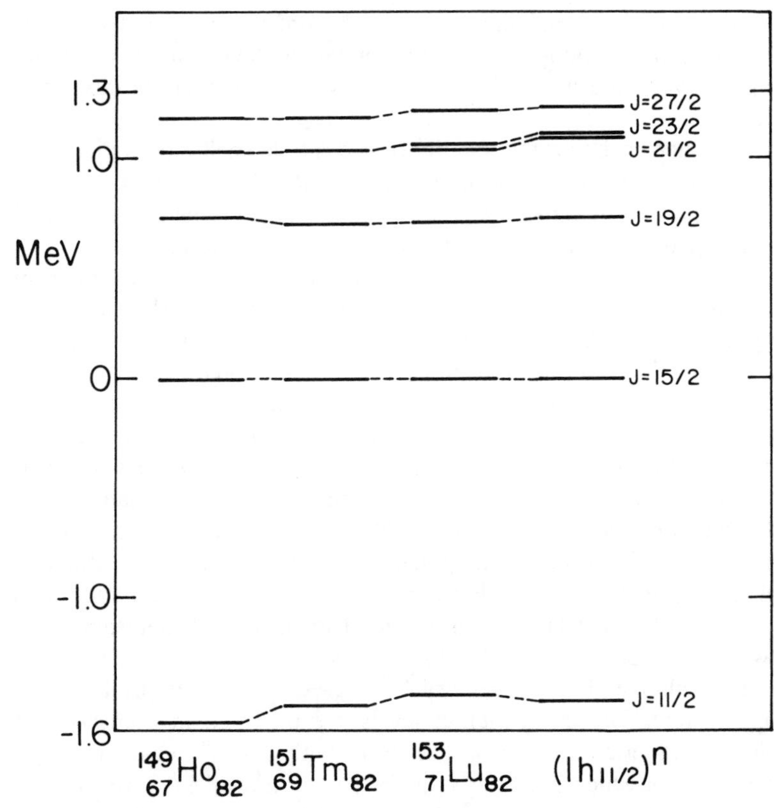

FIGURE 21.10. *Levels of odd* $(1h_{11/2})^n$ *proton configurations in* $N = 82$ *nuclei.*

Level spacings, specially between $J = 2, 4, 6, 8, 10$ states are fairly independent of proton number. This is an indication that the effective interaction is diagonal in the seniority scheme. We may use the spacings between $v = 2$ levels in ^{148}Dy (or in any other nucleus shown in Fig. 21.9) to calculate the matrix element between the $v = 1$, $J = \frac{11}{2}$ state and the other $J = \frac{11}{2}$ state with $v = 3$ in the $(1h_{11/2})^3$ configuration. Using (20.41) for $j = \frac{11}{2}$ we obtain the value of that matrix to be only a few keV as compared to about 2 MeV difference between diagonal matrix elements (Lawson 1981). Thus, the effective interaction is practically diagonal in the seniority scheme. As a result, the calculated positions of the $v = 2$ levels above the $v = 0$ ground state are exactly equal to those in the $(1h_{11/2})^2$ configuration. The fairly constant spacings seen in Fig. 21.9 provide a *consistency check* on the assumption of simple $(1h_{11/2})^n$ configurations and effective two-body interactions. The somewhat larger deviations in the spacings between the $v = 0$, $J = 0$ and the $v = 2$ states will be discussed below.

Another check on the validity of the simple shell model picture is provided by nuclei in which there are *odd* $(1h_{11/2})^n$ configurations. Since the interaction is diagonal in the seniority scheme, calculated spacings of $v = 1$ and $v = 3$ levels in such configurations are equal to those in the $(1h_{11/2})^3$ configuration. Some levels calculated from adopted level spacings in the $(1h_{11/2})^2$ configuration, shown on the right of Fig. 21.9, are presented in Fig. 21.10 along with experimental levels. There is fair agreement between calculated levels and those observed experimentally. In these configurations there are *two* states with $J = \frac{15}{2}$ and seniority $v = 3$. Two independent states with $J = \frac{15}{2}$ may be obtained from two different principal parents (like $J_0 = 2$ and $J_0 = 8$) and then orthogonalized. The eigenstates of the Hamiltonian are then obtained by diagonalizing the 2×2 matrix of the interaction constructed from these orthogonal states. In Fig. 21.10 only the *lowest* $J = \frac{15}{2}$ level is shown. For $J = \frac{15}{2}$, besides seniority, an additional quantum number (label) α is required to distinguish between these two states. There is no fundamental way to define such a quantum number and the necessary label used may be simply lower or upper state. The experimental levels are observed in a cascade of electromagnetic transitions. The level with $J = \frac{19}{2}$ decays into the $J = \frac{15}{2}$ level which indicates that the $J = \frac{17}{2}$ level lies above the $J = \frac{19}{2}$ level. Also the $J = \frac{15}{2}$ level does not decay into the $J = \frac{13}{2}$ level which must lie above it. These features are nicely reproduced by the calculation.

The change with the proton number in the position of the $v = 0$, $J = 0$ state relative to the $v = 2$ levels, although rather small, may be due to configuration mixing. This effect may be similar to the very pronounced effect in $(1g_{9/2})^n$ proton configurations. It may arise if the states which interact with those of the $(1h_{11/2})^n$ configuration have the form $|1h_{11/2}^{n-2}(v\alpha J)j'^2(0)JM\rangle$. The higher orbits in the same major shell are $2d_{3/2}$ and $3s_{1/2}$. Two protons in the $3s_{1/2}$ orbit have only one antisymmetric state with $J = 0$ and thus may play the role of j'. The interaction with states $|1h_{11/2}^{n-2}(v\alpha J)2d_{3/2}^2(0)JM\rangle$ may also contribute to variations in ground state energies as seen in Fig. 21.9. The shifts due to interaction and mixing with other configurations are slightly different in the various nuclei presented in Fig. 21.9. When we go to odd configurations, such mixings affect the spacings between $J = \frac{11}{2}$

ground states and states corresponding to $v = 3$ states. We see indeed in Fig. 21.10 a variation of these spacings as a function of proton number which is very similar to the one in Fig. 21.9. The admixture of configurations based on orbits whose energies are not very high may be appreciable. It may, however, be completely masked if its effect on energy levels may be reproduced by renormalization of the matrix elements of the two body effective interaction.

A more sensitive test of the wave functions is offered by transition possibilities. Expectation values of the Hamiltonian are *stationary* at the correct eigenstates. This property is usually not shared by single particle operators whose matrix elements determine rates of various transitions. Small admixtures in the wave functions may change appreciably rates of such transitions. Thus, acceptable wave functions must first yield correctly the energy levels of the system. A more stringent test of wave functions is their use in successful calculations of transition rates.

There are several electric quadrupole transitions (E2) between states of $(1h_{11/2})^n$ configurations considered here whose rates have been measured (McNeill et al. 1989). Let us consider such transitions between a state with J_i and another state with spin J_f, both with the same seniority v. According to (8.29) the reduced transition rate is given by

$$B(\mathrm{E2}) = \frac{1}{2J_i+1}\left|(J_f\|\sum_i r_i^2\mathbf{Y}_2(\theta_i\phi_i)\|J_i)\right|^2 \tag{21.16}$$

According to (19.40) we can express the reduced matrix element in (21.16) by

$$\left(j^n vJ_f\left\|\sum_i r_i^2\mathbf{Y}_2(\theta_i\phi_i)\right\|j^n vJ_i\right) = \frac{2j+1-2n}{2j+1-2v}\left(j^v vJ_f\left\|\sum_i r_i^2\mathbf{Y}_2(\theta_i\phi_i)\right\|j^v vJ_i\right) \tag{21.17}$$

The reduced matrix elements in the j^v configuration may be calculated by using c.f.p. according to (15.17) to obtain

$$\begin{aligned}\left(j^\nu vJ_f\left\|\sum_i r_i^2\mathbf{Y}_2(\theta_i\phi_i)\right\|j^\nu vJ_i\right)\\&=\nu\sum_{v_1J_1}[j^{\nu-1}(v_1J_1)jJ_i|\}j^\nu vJ_i][j^{\nu-1}(v_1J_1)jJ_f|\}j^\nu vJ_f]\\&\quad\times(-1)^{J_1+j+J_f+2}\sqrt{(2J_i+1)(2J_f+1)}\\&\quad\times\begin{Bmatrix}j & J_i & J_1\\ J_f & j & 2\end{Bmatrix}(j\|r^2\mathbf{Y}_2\|j)\end{aligned}\tag{21.18}$$

The reduced matrix element of the single j-nucleon is given by (10.38) as

$$(j\|r^2\mathbf{Y}_2\|j)=(-1)^{j-1/2}(2j+1)\sqrt{\frac{5}{4\pi}}\begin{pmatrix}j & 2 & j\\ -\frac{1}{2} & 0 & \frac{1}{2}\end{pmatrix}\langle r^2\rangle\tag{21.19}$$

In (21.19), $\langle r^2\rangle$ is the integral of r^2 multiplied by the square of the radial part of the j-nucleon wave function.

In nuclei with even $(1h_{11/2})^n$ configurations, E2 transitions have been measured between $v=2$ states with $J_i=10$ and $J_f=8$. The c.f.p. in (21.18), for $v=2$, $v_1=1$, $J_1=\frac{11}{2}$ are equal to 1. Using the actual values of the $6j$-symbol

$$\begin{Bmatrix}\frac{11}{2} & 10 & \frac{11}{2}\\ 8 & \frac{11}{2} & 2\end{Bmatrix}$$

and the $3j$-symbol

$$\begin{pmatrix}\frac{11}{2} & 2 & \frac{11}{2}\\ -\frac{1}{2} & 0 & \frac{1}{2}\end{pmatrix}$$

we obtain the value $.05733\langle r^2\rangle^2/\pi$ for, the reduced transition rate (21.16) for the $(1h_{11/2})^2$ proton configuration in ${}^{148}_{66}\mathrm{Dy}_{82}$. The experimental B(E2) value was found to be 43 ± 3 fm^4. If the effective charge of the $1h_{11/2}$ proton is equal to the charge of a free proton we obtain from it $\langle r^2\rangle^{1/2}=6.96$ fm. This value is larger than the expected radius of the $1h_{11/2}$ orbit and indicates an effective charge of the $1h_{11/2}$ protons which is larger than the free proton charge.

In the case of odd nuclei, E2 transition rates have been measured between the $J_i = \frac{27}{2}$ and $J_f = \frac{23}{2}$ states. The non-vanishing c.f.p. in (21.18) for $v = 3$ have $v_1 = 2$ and are given by (15.10) if we adopt the value $J_0 = 8$ or $J_0 = 10$. In this case there is no parent with $J_1 = 0$ (and $v_1 = 0$) and hence the resulting states have seniority $v = 3$. With these values of the c.f.p. and the actual values of the $6j$-symbols (and $3j$-symbol in (21.19)), the reduced transition rate (21.16) in ${}^{149}_{67}\text{Ho}_{82}$, with the $(1h_{11/2})^3$ configuration, is equal to $.12188\langle r^2\rangle^2/\pi$. The experimental rate is 88 ± 6 fm^4 from which we obtain $\langle r^2\rangle^2 = 2270 \pm 160$ fm^4 which compares nicely with the value $\langle r^2\rangle^2 = 2350 \pm 160$ fm^4 obtained from ${}^{148}_{66}\text{Dy}_{82}$. In ${}^{148}_{66}\text{Dy}_{82}$ the states with $J = 8$ and $J = 10$ should be *pure* $v = 2$ states of the $(1h_{11/2})^2$ configuration if the $Z = 64$ protons are in closed shells. Configurations with two protons in the $3s_{1/2}$ and $2d_{3/2}$ orbits do not have such high spin states. Similarly, the states with $J = \frac{27}{2}$ and $J = \frac{23}{2}$ (and negative parity) must also be states with $v = 3$ of the *pure* $(1h_{11/2})^3$ configuration. Hence, the agreement between the calculated and experimental ratios of transition rates in these nuclei is a further check on the consistency of the simple shell model picture. A more interesting feature of the transition rates is their dependence on n which we now consider.

For $(1h_{11/2})^n$ configurations with even n we obtain, by using (21.17) for the $B(\text{E2})$ values of the transitions between the $v = 2$ levels $J_i = 10$, $J_f = 8$, the result

$$\left(\frac{12-2n}{8}\right)^2 \times .05733\frac{\langle r^2\rangle^2}{\pi} = \left(\frac{6-n}{4}\right)^2 43 \text{ fm}^4 \qquad (21.20)$$

For $n = 4$ occurring in ${}^{150}_{68}\text{Er}_{82}$ the calculated value is 10.8 which is rather close to the experimental value of $11.3 \pm .4$ fm^4. In the middle of the orbit, for $n = 6$, the $B(\text{E2})$ should vanish. Experimentally, the measured $B(\text{E2})$ value for ${}^{152}_{70}\text{Yb}_{82}$ is small $.9 \pm .1$ fm^4. The occurrence of such a slow E2 transition is rather exceptional and supports the simple picture presented here. Beyond the middle of the orbit, for $n = 8$ the calculated $B(\text{E2})$ value is the same as for $n = 4$, namely 10.8 fm^4. The experimental $B(\text{E2})$ value in ${}^{154}_{72}\text{Hf}_{82}$ is not known very accurately. It is quoted as 2.9 ± 1.4 fm^4 and thus considerably lower than the calculated value. This may indicate appreciable admixtures of other configurations but there are not enough experimental data to determine them in a consistent way.

A similar picture is found in odd $(1h_{11/2})^n$ configurations. The $B(\text{E2})$ values of the transitions between the $v = 3$ levels with $J_i = \frac{27}{2}$

and $J_f = \frac{23}{2}$ are given according to (21.17) by

$$\left(\frac{12-2n}{6}\right)^2 \times .12188\frac{\langle r^2\rangle^2}{\pi} = \left(\frac{6-n}{3}\right)^2 91 \text{ fm}^4 \tag{21.21}$$

In ${}^{151}_{69}\text{Tm}_{82}$, n is equal to 5 and the value of the $B(\text{E2})$ calculated from (21.21) is 10.1 which fairly agrees with the measured value 11.7 fm^4. Beyond the middle of the orbit for $n = 7$ the calculated $B(\text{E2})$ value is equal to that for $n = 5$, i.e. 10.1 fm^4. The experimental value is, however, much smaller, only $.45 \pm .09$ fm^4. If this experimental value is accurate, it indicates appreciable admixtures of other configurations. There are not enough experimental data to determine such admixtures in a consistent way.

The simple picture of $(1h_{11/2})^n$ proton configurations may be applied to the analysis of certain allowed Gamow-Teller β-decays (Nolte et al. 1982). These take place between $J = 0$ ground states of even-even nuclei with $N = 82$ to $J = 1$ states with positive parity, in odd-odd nuclei with $N = 83$. They may be interpreted as transitions between the $(1h_{11/2})^n$ proton configuration to states of the $(1h_{11/2})^{n-1}$ proton configuration coupled to the state of a single $1h_{9/2}$ neutron. Such decays were considered in Section 9 and were shown to be due to the matrix element of the operator (9.15). The inverse of the reduced transition rate is denoted by ft.

The matrix elements for the Gamow-Teller β-decays of $J = 0$ states of proton $1h^n_{11/2}$ configurations to $J = 1$ states of $n-1$ protons and a $1h_{9/2}$ neutron are related to single nucleon matrix elements by (31.32). In applying that expression, c.f.p. with isospin must be used, taking into account the full neutron $1h_{11/2}$-orbit. The nucleon number n appearing in (31.32) is in the present case equal to $2j+1+n = 12+n$. The isospins are equal to $T = (2j+1-n)/2$, $T' = (2j+2-(n-1)/2 = T+1$ and $T_1 = (2j+1-(n-1)/2 = T+\frac{1}{2}$. The relevant c.f.p. are then given by (27.4). Substituting the values of these c.f.p. as well as the values of the various coefficients in (31.32) we obtain that the square of the matrix element (reduced with respect to spin but not with respect to isospin) is equal to $n(\frac{20}{11})$. For $n = 2$, the transition rate should be proportional to $\frac{40}{11}$. We can compare it to the value 6 calculated for the ${}^1\text{S}_0$ to ${}^3\text{S}_1$ transition expected in the ${}^6\text{He}$ decay, where $\log ft = 2.9$. We expect accordingly the log ft of the decay of the $J = 0$ ground state of ${}^{148}_{66}\text{Dy}_{82}$ to the $J = 1$ state in ${}^{148}_{65}\text{Tb}_{83}$ to be 3.1. The experimental rate was found to be log ft = 3.9.

There may be several reasons why this value is larger by about 6 than the expected value. The simplest of which is poor overlap between the radial parts of the $1h_{11/2}$ proton and the $1h_{9/2}$ neutron wave functions. The single nucleon transition may be strongly renormalized by many-body effects. It is an experimental fact that measured Gamow-Teller matrix elements are usually smaller than expected. The simple approach we can take is to adopt the ft-value from the ^{148}Dy β-decay as due to renormalization of the single nucleon operator. We can then look at the consequences of this assumption for Gamow-Teller transitions in other nuclei.

The linear dependence on n of the β-transition rates, quoted above, may be obtained from a simple argument. The matrix element for the transition is a product of matrix elements for annihilating a $1h_{11/2}$-proton and for creating a neutron in the *empty* $1h_{9/2}$ orbit. The dependence on n is determined by the matrix elements (31.33) for removing a proton from the $(1h_{11/2})^n$ configuration. In the case considered here, with $v = 0$, $J = 0$, this matrix element is given by the square root of (31.34). Thus, the matrix elements for the transitions considered here are equal to $\sqrt{n}$ multiplied by a constant factor. The transition rates are then proportional to n and the ft-values proportional to $1/n$. The measured rate of the β-decay from $^{150}_{68}\mathrm{Eu}_{82}$ to $^{150}_{67}\mathrm{Ho}_{83}$ yields $\log \mathrm{ft} = 3.6$. The decay of $^{152}_{70}\mathrm{Yb}_{82}$ to $^{152}_{69}\mathrm{Tm}_{83}$ has not been definitely determined but the quoted experimental result is $\log \mathrm{ft} = 3.4$. Thus, for $n = 2, 4, 6$ the experimental data are $\mathrm{ft} = 7945, 3980, 2510$. These values are fairly proportional to $1/n$ as can be seen by comparing them to $15900/n$. For $n = 2, 4, 6$ the latter values are 7950, 3975, 2650.

Before concluding this section let us return to binding energies. In the early days of the nuclear shell model physicists had in their minds the analogy to atomic shell structure. A dramatic display of electron shells in atoms is offered by separation energies of electrons (*ionization* energies). As seen in Fig. 21.11, there is a steep rise in separation energy as closed shells are approached followed by a sharp drop beyond Z of a noble gas. This picture was carried over into the nuclear domain. In many books, statements are made about the extra large binding energies of nuclei with closed shells. The experimental data contradict this naive analogy with atoms. We saw above that the coefficient α in the binding energy formula is *repulsive*. Thus, adding more nucleons of the same kind *reduces* the separation energy when a doubly magic nucleus is approached. This is opposite to the behavior in the atomic case.

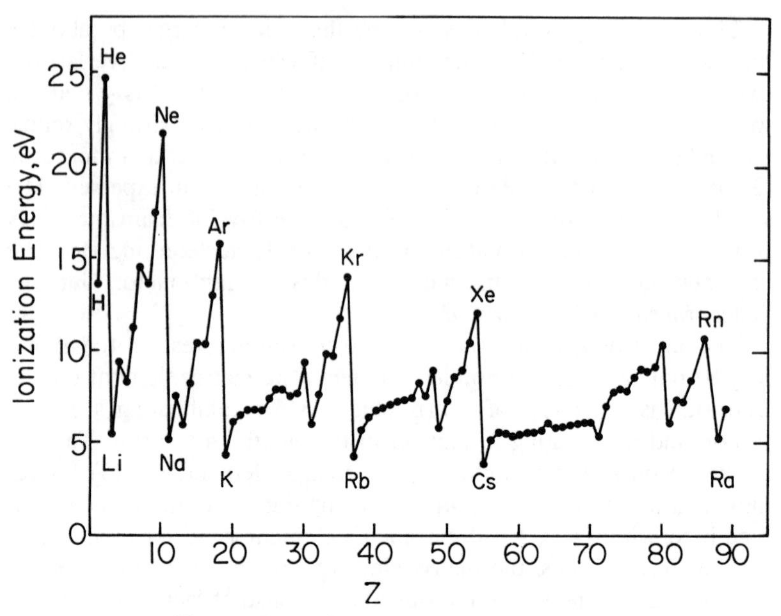

FIGURE 21.11. *Ionization energies of atoms.*

The shell structure in nuclei is revealed by looking at separation energies just beyond a magic number. There is a large drop as can be seen in Fig. 20.1. Such drops are characteristic of closed shells in nuclei as well as in atoms. They are evident also in separation energies of pairs of identical nucleons, as in Fig. 23.1, where the large odd even variation of single nucleon separation energies is eliminated. These drops are responsible for the extra stability of magic nuclei. As will be pointed out in the following sections, apart from the strong and attractive pairing energy the average interaction between identical nucleons is repulsive. For identical nucleons in the same j-orbit, this repulsion finds its expression in the repulsive nature of the coefficient α. The interaction between identical nucleons in different orbits is repulsive on the average as will be discussed in detail in Section 31. The central potential well in which nucleons move is due to the attractive interaction between protons and neutrons which will be discussed in Section 31 and in Sections 37 and 38.

Equipped with this information about nuclear binding energies we may return to atoms. The mutual interaction of electrons is totally *repulsive*. Where, then, is the origin of large binding energies at closed shells? The answer is simple, in Fig. 21.11 ionization energies of *neu-*

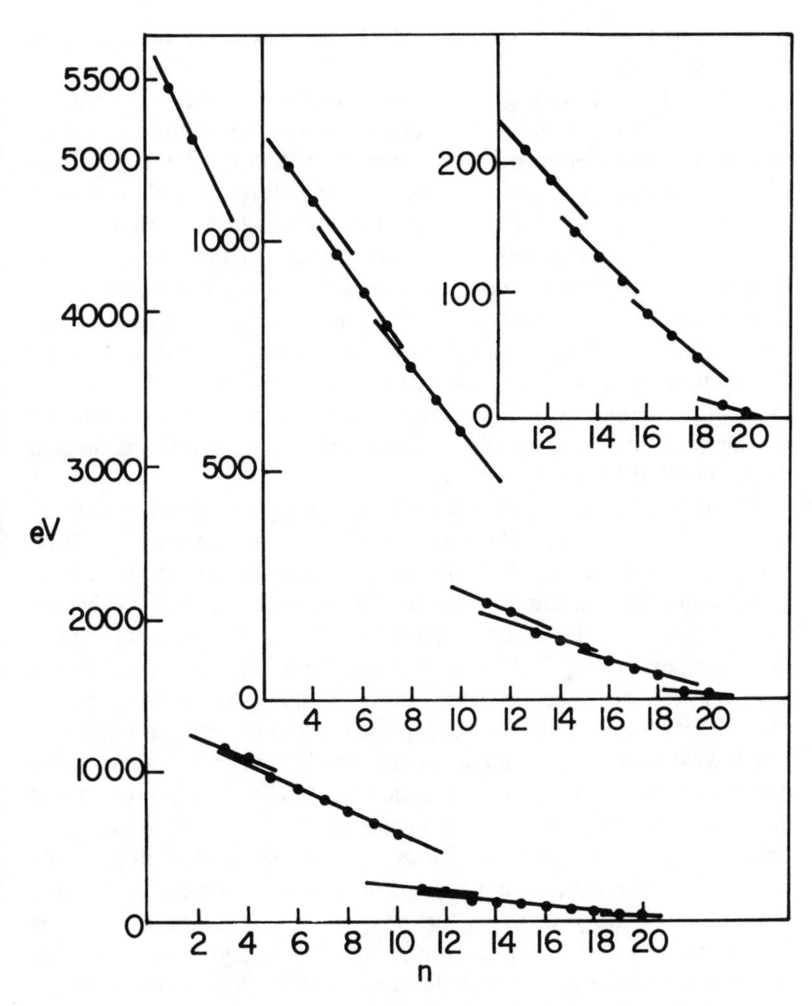

FIGURE 21.12. *Successive ionization energies of the* Ca *atom.*

tral atoms are plotted. When going from one atom to the next, approaching shell closure, the central charge of the nucleus is increased by one unit. The additional attractive charge in the center is more effective than the repulsion of the added electron (screening of the central charge by electrons in the *same* orbit is not very effective). When an electron shell is completely filled, the next electron must occupy an orbit with higher kinetic energy and the central charge is well screened by electrons in closed shells. Hence, there is a large drop in ionization energies.

It should be obvious that the large increase in ionization energies when approaching a closed shell could not be due to the mutual interaction of the electrons which is *repulsive*. The Coulomb repulsion itself should lead to ionization energies *decreasing* as a shell closure is approached. Beyond a closed shell there should be a sharp drop. We would thus see a behavior rather similar to that of separation energies of identical nucleons in nuclei. To display this behavior we should plot ionization energies of electrons moving in a potential well due to a *fixed* central charge Z. Successive ionization energies of the Ca atom, with $Z = 20$, are plotted in Fig. 21.12. We see there that ionization energies go indeed down as a function of electron number n. The various closed shells are characterized by large drops beyond them (Talmi 1969, 1985).

There is no need to plot separately ionization energies for odd and even electron numbers. There is no odd even variation of separation energies. Due to the repulsive nature of the mutual interaction of electrons, states with maximum pairing (lowest seniority) are the *highest* states. There are, however, other effects in the behavior of ionization energies. In Fig. 21.12 the small gap between s and p orbits are clearly seen. There are also small breaks in the middle of the p-orbits. These breaks are due to the Pauli principle and the mutual repulsion. The lowest states are as antisymmetric as possible in the space coordinates. In the p^2 and p^3 configuration the lowest states, as described in Section 16, are ${}^3P(S = 1, L = 1)$ and ${}^4S(S = \frac{3}{2}, L = 0)$ respectively. The wave functions of these states are fully symmetric in the spin variables and hence fully antisymmetric in space coordinates. Beyond the middle of the p orbit, in the p^4, p^5 and p^6 configurations, it is impossible to construct states which are fully antisymmetric in space coordinates (there are only 3 different spatial single particle states, $m_l = \pm 1, 0$). Thus, the electron configurations in the second half of the p shell are less bound than those in the first half.

22

The Quasi-Spin Scheme

In preceding sections we considered the seniority scheme for j^n configurations of identical nucleons. We saw examples where nuclear states may be fairly well approximated by such configurations. We also saw how well predictions of the seniority scheme agree with actual data. There are, however, semi-magic nuclei whose states cannot be described by pure j^n configurations. Yet, as we shall see later on, many properties of the seniority scheme are observed also in some cases where there is appreciable configurations mixing. The concept of seniority should then be generalized to include multi-configuration situations.

A straightforward generalization of the seniority scheme, due to Kerman (1961), can be obtained as follows. Instead of the pair creation operator S_j^+ consider the sum

$$\boxed{S^+ = \sum_j S_j^+} \tag{22.1}$$

where the summation extends over all j-orbits which take part in the configurations considered. These orbits may be, for instance, all orbits in a major shell. The commutation relation of the operator (22.1) with its hermitean conjugate $(\mathcal{S}^+)^\dagger = \mathcal{S}^-$ is given by

$$[\mathcal{S}^+, \mathcal{S}^-] = \sum_{jm} a^+_{jm} a_{jm} - \frac{1}{2}\sum_j (2j+1) = \sum_j \hat{n}_j - \Omega = 2\mathcal{S}^0 \tag{22.2}$$

The operator $\mathcal{S}^0$ defined by (22.2) satisfies with $\mathcal{S}^+$ and $\mathcal{S}^-$ the commutation relations

$$[\mathcal{S}^0, \mathcal{S}^+] = \mathcal{S}^+ \qquad [\mathcal{S}^0, \mathcal{S}^-] = -\mathcal{S}^- \tag{22.3}$$

The commutation relations (22.2) and (22.3) are identical with those in (19.7). Hence, $\mathcal{S}^+$, $\mathcal{S}^-$ and $\mathcal{S}^0$ are generators of the $SU(2)$ Lie algebra. They are simply components of the total quasi-spin of the system $\boldsymbol{\mathcal{S}} = \mathbf{S}_{j_1} + \mathbf{S}_{j_2} + \cdots$.

From this follows that the system of eigenstates of $\boldsymbol{\mathcal{S}}^2$, or those of $2\mathcal{S}^+\mathcal{S}^-$, defines a seniority scheme in the space of states of identical nucleons in the various j-orbits. This scheme has all features of the seniority scheme in a single j-orbit which follow from the $SU(2)$ properties. We may call it the *quasi-spin scheme* as it was introduced by Kerman (1961) in terms of the quasi-spin operators $\mathcal{S}^+$, $\mathcal{S}^-$ and $\mathcal{S}^0$. The eigenvalues of the pairing interaction

$$2\mathcal{S}^+\mathcal{S}^- = 2(\boldsymbol{\mathcal{S}}^2 - \mathcal{S}^0(\mathcal{S}^0 - 1)) \tag{22.4}$$

are given by

$$2s(s+1) - \tfrac{1}{2}(\Omega - n)(\Omega + 2 - n) = \frac{n - v}{2}(2\Omega + 2 - n - v) \tag{22.5}$$

This is the analog of (19.11) where $n = n_j$ is replaced by $n = \sum n_j$ and $2j+1$ is replaced by $2\Omega = \sum(2j+1)$. The pairing interaction (22.4) is the one described in Section 12 for the two nucleon configuration.

In addition to the pairing interaction in the case of degenerate single nucleon energies, which is diagonal in the quasi-spin scheme, there are other interactions that share this property. Hamiltonians which are diagonal in the quasi-spin scheme have eigenstates with very special configuration mixing. In the $v = 0$ eigenstate, $(\sum S_j^+)^N|0\rangle$, the states of various configurations which are mixed, have all the form $(S_j^+)^{N_j}|0\rangle$. Hence, each state $(S_j^+)^{N_j}|0\rangle$ is an eigenstate of the submatrices of the Hamiltonian defined by the j^{2N_j} configuration. As stated in Section 20 and proved in Section 23, from this fact follows that the submatrices of the Hamiltonian within the various j^n configurations are diagonal in the seniority scheme. In Section 19 we saw that odd tensors are quasi-spin scalars and thus, odd tensor interactions are diagonal in the seniority scheme. We shall now see the generalization of these properties to the case of the quasi-spin scheme with several j-orbits.

Let us first consider a single nucleon tensor operator defined by

$$T_\kappa^{(k)} = \sum_{j,j'} \frac{(j\|\mathbf{T}^{(k)}\|j')}{\sqrt{2k+1}}(a_j^+ \times \tilde{a}_{j'})_\kappa^{(k)} \tag{22.6}$$

and calculate its commutation relation with $\mathcal{S}^+$. Due to (22.1) and the commutation relations (19.22) we obtain

$$[\mathcal{S}^+, T_\kappa^{(k)}] = \sum_{j'} [S_{j'}^+, T_\kappa^{(k)}]$$

$$= \frac{1}{\sqrt{2k+1}} \sum_{\substack{j,j' \\ mm'}} (j\|\mathbf{T}^{(k)}\|j')(jmj'm'|jj'k\kappa)a_{jm}^+(-1)^{j'+m'}$$

$$\times [S_{j'}^+, a_{j'-m'}]$$

$$= \frac{1}{\sqrt{2k+1}} \sum_{jj'} (j\|\mathbf{T}^{(k)}\|j')(a_j^+ \times a_{j'}^+)_\kappa^{(k)} \tag{22.7}$$

If $T_\kappa^{(k)}$ is a scalar in quasi-spin, the r.h.s. of (22.7) must vanish. The terms with $j = j'$ vanish if k is odd as we saw in Section 19. The other terms, for which $j' \neq j$, can be expressed as

$$\frac{1}{\sqrt{2k+1}}\sum_{j<j'}[(j\|\mathbf{T}^{(k)}\|j')(a_j^+ \times a_{j'}^+)_\kappa^{(k)} + (j'\|\mathbf{T}^{(k)}\|j)(a_{j'}^+ \times a_j^+)_\kappa^{(k)}]$$

$$= \frac{1}{\sqrt{2k+1}}\sum_{j<j'}[(j\|\mathbf{T}^{(k)}\|j') + (-1)^{j+j'-k+1}(j'\|\mathbf{T}^{(k)}\|j)]$$

$$\times (a_j^+ \times a_{j'}^+)_\kappa^{(k)} \tag{22.8}$$

We now recall the relation (8.9) between reduced matrix elements of hermitean tensor operators. If $\mathbf{T}^{(k)}$ is hermitean we obtain for the r.h.s. of (22.8) the result

$$\frac{1}{\sqrt{2k+1}}\sum_{j<j'}[(1+(-1)^k](j\|\mathbf{T}^{(k)}\|j')(a_j^+ \times a_{j'}^+)_\kappa^{(k)} \tag{22.9}$$

and conclude that odd hermitiean tensors are quasi-spin scalars

$$[\mathcal{S}^+, T_\kappa^{(k)}] = 0 \qquad k \text{ odd for hermitean } \mathbf{T}^{(k)} \tag{22.10}$$

This is a simple generalization of the result obtained in Section 19 for $j = j'$. In the case with several j-orbits, there is another class of single nucleon tensors which are quasi-spin scalars. If $\mathbf{T}^{(k)}$ in (22.8) is anti-hermitean we obtain instead of (22.9) the result

$$\frac{1}{\sqrt{2k+1}}\sum_{j<j'}[1+(-1)^{k+1}](j\|\mathbf{T}^{(k)}\|j')(a_j^+ \times a_{j'}^+)_\kappa^{(k)}$$

and reach the following conclusion

$$[\mathcal{S}^+, T_\kappa^{(k)}] = 0 \qquad k \text{ even for antihermitean } \mathbf{T}^{(k)} \tag{22.11}$$

The commutation relations of $\mathcal{S}^+$, $\mathcal{S}^-$ and $\mathcal{S}^0$ with any of the $a_{jm}^+, \tilde{a}_{jm}$ are identical to those given by (19.21) and (19.22). Hence,

a^+_{jm} is the $+\frac{1}{2}$ component of a quasi-spin tensor with rank $\frac{1}{2}$ and $-\tilde{a}_{jm}$ is the $-\frac{1}{2}$ component. Thus, the tensor with ordinary rank k, defined by $\sum_{jj'}(a^+_j \times \tilde{a}_{j'})^{(k)}$, is a linear combination of a quasi-spin scalar and a quasi-spin vector (rank 1 tensor). If $\mathbf{T}^{(k)}$ is a hermitean tensor, the linear combination (22.6) is a quasi-spin scalar if k is odd. If, however, k is even the linear combination (22.6) is the $\sigma = 0$ component of a quasi-spin vector. On the other hand, if $\mathbf{T}^{(k)}$ is antihermitean then (22.6) is a quasi-spin scalar if k is even and if k is odd it is the $\sigma = 0$ component of a vector with respect to quasi-spin.

When we considered the seniority scheme in a single j-orbit, it turned out to be convenient to use the tensor expansion of two body interactions. This led to the expression of the interaction as the sum of scalar products of single nucleon operators from which self-interaction terms are subtracted. A similar expression was used in (14.20) for the direct derivation of the total interaction of closed shells given by (14.19). When considering several orbits, however, the situation is more complicated. The coefficients $F_k(j_1j_2j'_1j'_2)$ are in general, not independent of the orbits involved nor can they be expressed as a product of a term depending on j_1, j'_1 and a term which depends on j_2, j'_2.

Still, there may be interesting cases where it is possible to obtain for certain two body interactions expressions similar to those obtained for a single j-orbit. For discussing such cases it is not convenient to use the expansion (12.12). There, the coefficients $F_k(j_1j_2j'_1j'_2)$ defined by (12.13) are related to matrix elements of the interaction between fully antisymmetric (and normalized) states. It is simpler to start from the original tensor expansion like

$$\mathbf{G}_{12} = \sum_k \frac{4\pi}{2k+1} v_k(r_1, r_2)(\mathbf{T}^{(k)}(1) \cdot \mathbf{T}^{(k)}(2)) \qquad (22.12)$$

where $\mathbf{T}^{(k)}$ is equal to $\mathbf{Y}_k$ for a potential interaction like (10.4). For a spin dependent interaction like (10.43), $\mathbf{T}^{(k)}$ should be replaced by $\mathbf{T}^{(1l)k} = [\mathbf{s} \times \mathbf{Y}^l]^{(k)}$, $v_k(r_1, r_2)$ by $v_l(r_1, r_2)$ and each term in (22.12) be multiplied by $(-1)^{l+k+1}$ (the summation is then over l and k).

Matrix elements of (22.12) taken between jj' and $j_0j'_0$ states have the form of

$$\langle jj'JM|\mathbf{T}^{(k)}(1) \cdot \mathbf{T}^{(k)}(2)|j_0j'_0JM\rangle \qquad (22.13)$$

multiplied by radial integrals over $4\pi v_k/(2k+1)$. Only if $v_k(r_1, r_2)$ is a product $v_k(r_1)v_k(r_2)$ are the radial integrals products of two terms, one which depends on j, j_0 and the other on j', j'_0. There are such cases and one of them will be considered in a subsequent section.

Another case in which it is possible to express the two body interaction as a sum of scalar products of single nucleon operators (from which self interaction terms are subtracted) is where *all* radial integrals are taken to be equal. Such an assumption is made in the definition of the surface delta interaction (SDI) described in Section 12. Let us now consider the case in which all direct and exchange radial integrals are equal. For simplicity we assume that among the orbits considered there are no two with the same values of l and j (and different n values). Once the radial integrations have been carried out, matrix elements of the operator

$$\mathbf{G}_{12} = \sum_k g_k(\mathbf{T}^{(k)}(1) \cdot \mathbf{T}^k(2)) \tag{22.14}$$

should be evaluated between antisymmetric wave functions of angles and spins.

We can use the expression (22.14) and obtain from it the expression for configurations with several nucleons. In analogy to (14.20) and (18.11) we obtain

$$\begin{aligned}\sum_{i<j} \mathbf{G}_{ij} &= \sum_k g_k \sum_{i<j} (\mathbf{T}^{(k)}(i) \cdot \mathbf{T}^{(k)}(j)) \\ &= \sum_k g_k \frac{1}{2}\left[\left(\sum_i \mathbf{T}^{(k)}(i)\right) \cdot \left(\sum_j \mathbf{T}^{(k)}(j)\right) - \sum_i (\mathbf{T}^{(k)}(i) \cdot \mathbf{T}^{(k)}(i))\right]\end{aligned} \tag{22.15}$$

We can now express (22.15) in the formalism of second quantization by

$$\begin{aligned}&\frac{1}{2}\sum_k g_k \left(\sum_{j_1, j'_1} \frac{(j_1\|\mathbf{T}^{(k)}\|j'_1)}{\sqrt{2k+1}}(a^+_{j_1} \times \tilde{a}_{j'_1})^{(k)}\right) \\ &\cdot \left(\sum_{j_2, j'_2} \frac{(j_2\|\mathbf{T}^{(k)}\|j'_2)}{\sqrt{2k+1}}(a^+_{j_2} \times \tilde{a}_{j'_2})^{(k)}\right) \\ &- \frac{1}{2}\sum_k g_k \sum_j \langle jm|\mathbf{T}^{(k)} \cdot \mathbf{T}^{(k)}|jm\rangle \sum_m a^+_{jm} a_{jm}\end{aligned} \tag{22.16}$$

The expression (22.16) was obtained from (22.15) by using the prescriptions (17.8) and (17.10) for expressing single nucleon operators in terms of creation and annihilation operators.

The interaction we consider is to be diagonalized in a certain shell model subspace. Matrix elements of $\mathbf{T}^{(k)}(1)$ and $\mathbf{T}^{(k)}(2)$ in (22.13) are evaluated only between j-orbits included in this sub-space. Hence, when considering states of several nucleons, this should also be the case for matrix elements of the operators in (22.15). The scalar products $(\sum \mathbf{T}^{(k)}(i)) \cdot (\sum \mathbf{T}^{(k)}(i))$ and $\mathbf{T}^{(k)}(i) \cdot \mathbf{T}^{(k)}(i)$ should be evaluated as products of the submatrices of these operators defined by the shell model sub-space. This is the sub-space defined by orbits appearing in the definition (22.1). They are not products of the full operators. From this follows that the summations in (22.16) are over j and j' which belong to the shell model sub-space considered.

From the expression (22.16) of the two-body interaction we can derive sufficient conditions for it to be diagonal in the quasi-spin scheme. If the single nucleon operators in (22.16) have odd ranks k, if they are hermitean, and even ranks k, if they are antihermitean, they are, according to (22.10) and (22.11), quasi-spin scalars. The first term in (22.16) is then also a quasi-spin scalar and hence diagonal in the quasi-spin scheme. The second term is proportional to a linear combination of the terms $\sum a^{\dagger}_{jm} a_{jm} = 2\mathcal{S}^0_j + (2j+1)/2$ for the various values of j. Apart from a constant, it is the $\sigma = 0$ component of a quasi-spin vector. It is diagonal in the quasi-spin scheme only if it is proportional to $2\mathcal{S}^0 + \Omega = \sum_j [2\mathcal{S}^0_j + (2j+1)/2]$.

In fact, if the first term in (22.16) is a quasi-spin scalar, the second term may be always made proportional to $2\mathcal{S}^0 + \Omega$ by adding to the two-body interaction (22.16) properly chosen single nucleon energies. The latter are given by $\sum_j \epsilon_j \sum_m a^{\dagger}_{jm} a_{jm}$ and may be adjusted to make all single nucleon terms in the Hamiltonian equal.

The coefficients g_k in (22.16) are the radial integrals which are taken to be independent of the various orbits. In the general case of an interaction which is diagonal in the quasi-spin scheme, these coefficients may have different values for the various values of k.

It was explained above that the surface delta interaction (SDI) can be expressed in the form (22.16). We can now determine the quasi-spin properties of the SDI (Arvieu and Moszkowski 1966). Looking at (22.8) we see that the commutation relations with the single nucleon operators appearing in (22.16) are determined by the relation between $(j\|\mathbf{T}^{(k)}\|j')$ and $(j'\|\mathbf{T}^{(k)}\|j)$. Let us examine that relation in the case of SDI. In the discussion following (12.16) it was shown that for $T = 1$

states

$$\delta(|\mathbf{r}_1-\mathbf{r}_2|) \text{ is equivalent to } -\tfrac{4}{3}(\mathbf{s}_1\cdot\mathbf{s}_2)\delta(|\mathbf{r}_1-\mathbf{r}_2|) \quad \text{for} \quad T=1 \text{ states} \tag{22.17}$$

It was then shown that due to the reduced matrix element (12.14), $(\mathbf{s}_1\cdot\mathbf{s}_2)V(|\mathbf{r}_1-\mathbf{r}_2|)$ has in its expansion within $(lj)^2$ states scalar products of only odd tensors.

In the present case we start with the expansion

$$\begin{aligned}(\mathbf{s}_1\cdot\mathbf{s}_2)V(\mathbf{r}_1-\mathbf{r}_2) &= (\mathbf{s}_1\cdot\mathbf{s}_2)\sum_k \frac{4\pi}{2k+1}v_k(r_1,r_2)(\mathbf{Y}_k(1)\cdot\mathbf{Y}_k(2)) \\ &= \sum_{k,r}(-1)^{k+r+1}\frac{4\pi}{2k+1}v_k(r_1,r_2) \\ &\quad \times([\mathbf{s}_1\times\mathbf{Y}_k(1)]^{(r)}\cdot[\mathbf{s}_2\times\mathbf{Y}_k(2)]^{(r)})\end{aligned} \tag{22.18}$$

The last equality in (22.18) is due to the change of coupling transformation (10.10). The reduced matrix elements of the tensor operators in (22.18) are given by (10.68) as

$$\boxed{\begin{aligned}(\tfrac{1}{2}lj\|[\mathbf{s}\times\mathbf{Y}_k]^{(r)}\|\tfrac{1}{2}l'j') &= \sqrt{(2j+1)(2r+1)(2j'+1)} \\ &\quad\times\begin{Bmatrix}\frac{1}{2} & l & j \\ \frac{1}{2} & l' & j' \\ 1 & k & r\end{Bmatrix}(\tfrac{1}{2}\|\mathbf{s}\|\tfrac{1}{2})(l\|\mathbf{Y}_k\|l')\end{aligned}} \tag{22.19}$$

Using the expression (8.15) for the reduced matrix element of $\mathbf{Y}_k$ and the symmetry properties of the $9j$-symbol we obtain

$$\begin{aligned}(\tfrac{1}{2}l'j'\|[\mathbf{s}\times\mathbf{Y}_k]^{(r)}\|\tfrac{1}{2}lj) &= (-1)^{l'-l}(-1)^{l+l'+k+j+j'+r} \\ &\quad\times(\tfrac{1}{2}lj\|[\mathbf{s}\times\mathbf{Y}_k]^{(r)}\|\tfrac{1}{2}l'j')\end{aligned} \tag{22.20}$$

The phase factor on the r.h.s. of (22.20) is equal to

$$(-1)^{j+j'+k+r} = (-1)^{j-j'}(-1)^{k+r+1} \tag{22.21}$$

This is consistent with the phase (8.9) for a real hermitean tensor provided $(-1)^{k+r+1} = +1$. If, however, $(-1)^{k+r+1} = -1$ the tensor $[\mathbf{s} \times \mathbf{Y}_k]^{(r)}$ is antihermitean and the phase (22.21) is opposite to the one in (8.9). This shows that we cannot apply directly the results (22.10) and (22.11) to the present case. It is only if $(-1)^{l+l'} = +1$ that k is even (due to (8.15)) and $[\mathbf{s} \times \mathbf{Y}_k]^{(r)}$ is hermitean for odd r and antihermitean for even r. If $(-1)^{l+l'} = -1$, however, the opposite is true. We shall now see that a slight modification of the operator S^+ is sufficient to make SDI diagonal in a quasi-spin scheme.

Let us return to the phase in (22.20). Recalling (8.15) we see that the matrix element (22.19) vanishes unless $(-1)^{l+l'+k} = 1$. Hence, the phase in (22.20) is equal to $(-1)^{l'-l+j+j'+r}$. When multiplied by the phase $(-1)^{j+j'-r+1}$ which appears in (22.8) it becomes equal to $(-1)^{l'-l+1}$. Thus, for vanishing of (22.8) the phase $(-1)^{l'-l}$ must be eliminated. This can be done by using instead of the operator S^+ in (22.1) a modified pair creation operator defined by

$$\boxed{S_1^+ = \sum_j (-1)^{l_j} S_j^+} \tag{22.22}$$

In (22.22) l_j is the orbital angular momentum of the j-orbit. In fact, we recall that the two nucleon $T = 1$, $J = 0$ ground state of SDI, given by (12.40), is created by S_1^+ as defined by (22.22).

This modification leads to the same $SU(2)$ commutation relations between S_1^+, $S_1^- = (S_1^+)^\dagger$ and $S_1^0 = \frac{1}{2}[S_1^+, S_1^-] = S^0$ as can be easily verified. It amounts to a change of axes in the definition of quasi-spin for orbits with $(-1)^l = -1$. This change is a rotation around the 3-axis by the angle π. It transforms S_j^+ into $-S_j^+$, S_j^- into $-S_j^-$ while S_j^0 remains unchanged. Since the quasi-spin operators of the various j-orbits are completely independent, the phases of the various S_j^+ operators in (22.1) may be chosen arbitrarily. The actual form of the resulting wave functions $S^+|0\rangle$ may well depend on these phases. The phases in (22.22) follow from an attractive short-ranged interaction. Hence, they lead to a pair ground state $S_1^+|0\rangle$ with stronger spatial correlations. The dependence of such correlations on the phases may be demonstrated by a simple example.

Consider the two orbits $j = l + \frac{1}{2}$ and $j' = l - \frac{1}{2}$ which are the spin-orbit partners of the same l-orbit. This case was considered at the end of Section 12. The pair creation operator (22.1) has in this case the form

$$\mathcal{S}^+ = \mathcal{S}_1^+ = S_{l+1/2}^+ + S_{l-1/2}^+ \tag{22.23}$$

The resulting (unnormalized) state is

$$\sqrt{2(l+\tfrac{1}{2})+1}\,|\,(l+\tfrac{1}{2})^2 J=0\rangle + \sqrt{2(l-\tfrac{1}{2})+1}\,|\,(l-\tfrac{1}{2})^2 J=0\rangle \tag{22.24}$$

Recalling the expansion of jj-coupling states in terms of LS-coupling ones, we find that (22.24) is the $|l^{2\,1}S_0\rangle$ state. In this example the quasi-spin scheme is the seniority scheme in l^n configurations which will be described in more detail in a subsequent section. If, however, the signs of the two states in (22.24) are opposite, the state becomes a linear combination of $|l^{2\,1}S_0\rangle$ and $|l^{2\,3}P_0\rangle$. The amplitude of the $|l^{2\,1}S_0\rangle$ is only $1/(2l+1)$ and hence that state is mostly a $|l^{2\,3}P_0\rangle$ state. The lowest state of the l^2 configuration for a two body potential interaction, which is attractive and has short range, is $|l^{2\,1}S_0\rangle$ in which state the two nucleons are closer. In the $|l^{2\,3}P_0\rangle$ the angular part of the wave function is antisymmetric and the nucleons are rather apart.

We have shown that the single nucleon operator

$$\sum_{jj'}(\tfrac{1}{2}lj\|[\mathbf{s}\times\mathbf{Y}_k]^{(r)}\|\tfrac{1}{2}l'j')(a_j^\dagger\times\tilde{a}_{j'})^{(r)} \tag{22.25}$$

is a scalar with respect to the quasi-spin of SDI. Its components commute with $\mathcal{S}_1^+$ defined by (22.22), for any value of k and r. This has direct consequences for magnetic multipole transitions, considered in Section 10. Rates of these transitions are determined by matrix elements of the operator (10.56). The second term in (10.56) is equal, up to a phase factor, to (22.25). Since the operator (22.25) commutes with $\mathcal{S}_1^+$, also the second term in (10.56), which is of physical interest, commutes with $\mathcal{S}_1^+$. Matrix elements of scalars in quasi-spin, like (10.56), are independent of the eigenvalues of $\mathcal{S}^0$. We thus obtain the following result.

Rates of magnetic multipole transitions, between corresponding states, are independent of nucleon number n.

The dependence on n of *electric* multipole transitions is rather different. Rates of those transitions are given by (10.48) where matrix elements of the hermitean operator $\mathbf{Y}_L$ appear. The commutator of an hermitean operator $\mathbf{T}^{(k)}$ and $\mathcal{S}^+$ is given by (22.9). If we calculate the commutator of $\mathbf{T}^{(L)}$ and $\mathcal{S}_1^+$, we obtain instead of $1+(-1)^L$ in (22.9), the factor $1+(-1)^{l+l'+L}$. Matrix elements of Y_{LM} between states with l,l' vanish unless $(-1)^{l+l'+L} = +1$. Hence, the commutator of $\mathcal{S}_1^+$ and Y_{LM} does not vanish for any value of L. Thus, the operators which determine rates of electric multiple transitions are not scalars in the quasispin defined by $\mathcal{S}_1^+$. They are the $\sigma = 0$ components of quasi-spin vectors and the dependence on n of their matrix elements is determined accordingly. We shall not consider further these operators.

Let us now return to the SDI Hamiltonian. The single nucleon operator

$$\sum_{kr}\sum_{jj'}(\tfrac{1}{2}lj\|[\mathbf{s}\times\mathbf{Y}_k]^{(r)}\|\tfrac{1}{2}l'j')(a_j^+\times\tilde{a}_{j'})^{(r)} \tag{22.26}$$

which appears in the expansion (22.16) of SDI is a scalar with respect to the quasi-spin defined by (22.22). Each term in (22.26) with given k and r commutes with $\mathcal{S}_1^+$. Hence, the first term in (22.16) is also a quasi-spin scalar. To prove that SDI is indeed diagonal in this quasi-spin scheme, we have to prove that the second term in (22.16) is proportional to $2\mathcal{S}_1^0+\Omega = 2\mathcal{S}^0+\Omega$. In other words, we should prove that the coefficients of the various $\hat{n}_j = \sum a_{jm}^+a_{jm}$ terms are all equal.

For given k and r, the coefficient of $\sum a_{jm}^+a_{jm}$ can be expressed with the help of (10.30) as

$$\begin{aligned}(-1)^{k+r+1}&\langle jm_0|([\mathbf{s}\times\mathbf{Y}_k]^{(r)}\cdot[\mathbf{s}\times\mathbf{Y}_k]^{(r)})\mid jm_0\rangle\\ &=\frac{1}{2j+1}\sum_{j'}(-1)^{j-j'+k+r+1}\\ &\quad\times(\tfrac{1}{2}lj\|[\mathbf{s}\times\mathbf{Y}_k]^{(r)}\|\tfrac{1}{2}l'j')(\tfrac{1}{2}l'j'\|[\mathbf{s}\times\mathbf{Y}_k]^{(r)}\|\tfrac{1}{2}lj)\end{aligned} \tag{22.27}$$

Using the relation (22.20) we can express the r.h.s. of (22.27) as

$$\frac{1}{2j+1}\sum_{j'}(\tfrac{1}{2}lj\|[\mathbf{s}\times\mathbf{Y}_k]^{(r)}\|\tfrac{1}{2}lj')^2 \tag{22.28}$$

In the case of the SDI all g_k are equal. This may be seen from the special case of a single j-orbit. Since in the case of SDI the radial integrals are taken to be equal for all orbits, they are equal to those in which wave functions of only one j-orbit appear. In that case we see from (11.18) that all the g_k are equal and may be replaced by one coefficient g. The second term in (22.16) thus becomes equal to

$$-\frac{1}{2}\sum_j g\frac{1}{2j+1}\left[\sum_{j'}(\tfrac{1}{2}lj\|[\mathbf{s}\times\mathbf{Y}_k]^{(r)}\|\tfrac{1}{2}l'j')^2\right]\sum a^+_{jm}a_{jm}$$
$$=\sum_j c_j\left(S^0_j+\frac{2j+1}{4}\right) \tag{22.29}$$

The values of the c_j, which turn out to be all equal, may be calculated from the actual values (22.19) by using identities involving $9j$-symbols (Arvieu and Moszkowski 1966). We can demonstrate this fact in a much simpler way.

The SDI interaction Hamiltonian may be expressed as

$$\boxed{H_{\mathrm{SDI}}=H^{(s=0)}+\sum_j c_j\left(S^0_j+\frac{2j+1}{4}\right)} \tag{22.30}$$

where $H^{(s=0)}$ is a quasi-spin scalar. We saw in Section 12 that $\mathcal{S}^+_1|0\rangle$ is an eigenstate of the SDI Hamiltonian in the two nucleon configuration with the eigenvalue $C_0\sum_j(2j+1)=C_0\Omega$. From this fact, namely

$$\boxed{H_{\mathrm{SDI}}\mathcal{S}^+_1|0\rangle=C_0\Omega\mathcal{S}^+_1|0\rangle} \tag{22.31}$$

we obtain

$$\begin{aligned}H_{\mathrm{SDI}}\mathcal{S}^+_1|0\rangle&=[H_{\mathrm{SDI}},\mathcal{S}^+_1]|0\rangle\\&=[H^{(s=0)},\mathcal{S}^+_1]|0\rangle+\sum_j c_j[S^0_j,\mathcal{S}^+_1]|0\rangle\\&=\sum c_j(-1)^{l_j}[S^0_j,S^+_j]|0\rangle=\sum c_j(-1)^{l_j}S^+_j|0\rangle\\&=C_0\Omega\sum_j(-1)^{l_j}S^+_j|0\rangle\end{aligned} \tag{22.32}$$

From the last equality in (22.32) follows, due to the orthogonality of the various $S^+_j|0\rangle$ states, that all the c_j are indeed equal and are given

by

$$c_j = C_0\Omega$$

From the equality of the c_j coefficients we obtain the eigenvalues of the SDI Hamiltonian in all states with seniority $v = 0$. The quasi-spin scalar part in (22.30) vanishes when applied to the state $(S_1^+)^N|0\rangle$. The eigenvalues of the second part which is equal to

$$2C_0\Omega\sum_j\left(S_j^0 + \frac{2j+1}{4}\right) = \tfrac{1}{2}C_0\Omega\sum_j \hat{n}_j \tag{22.33}$$

are equal to $C_0\Omega N$. The coefficient α of the quadratic term in the interaction energy of ground states vanishes in the case of SDI, as was the case for the δ-potential in a single j-orbit. According to the pairing property which holds also in the quasi-spin scheme for SDI, the eigenvalues for states with seniority $v = 1$ are equal to $\frac{1}{2}C_0\Omega(n-1)$ if n is the odd number of nucleons. This fact may be directly verified keeping in mind that H_{SDI} contains only two-body interactions and vanishes when applied to the state $a_{jm}^+|0\rangle$. We obtain accordingly

$$\boxed{\begin{aligned} H_{\mathrm{SDI}}(S_1^+)^N a_{jm}^+|0\rangle &= [H_{\mathrm{SDI}},(S_1^+)^N]a_{jm}^+|0\rangle + (S_1^+)^N H_{\mathrm{SDI}}a_{jm}^+|0\rangle \\ &= C_0\Omega N(S_1^+)^N a_{jm}^+|0\rangle \end{aligned}}$$

The last equality is due to (22.32) and (22.33). Thus, the eigenvalue of H_{SDI} in $v = 1$ states are equal to $C_0\Omega N = \frac{1}{2}C_0\Omega(n-1)$ where $n = 2N+1$. As in the seniority scheme, level spacings are independent of the number of particles. In particular, the $v = 2$ levels are at the same positions relative to the $v = 0$ ground state for all even particle numbers. These are the $J = 2, 4, \ldots$ levels with non-vanishing eigenvalues (12.39) as well as the other states with vanishing eigenvalues in the two particle configuration.

The pairing interaction and the surface delta interaction were shown above to be diagonal in the quasi-spin scheme. This feature holds provided all single nucleon energies of the orbits included in the calculation are equal. If non degenerate single nucleon energies are added to such Hamiltonians they are no longer diagonal in the quasi-spin scheme. Still, it may be worthwhile to point out that in such cases it is possible to introduce a modification of the two-body interaction that will restore the diagonality of the Hamiltonian.

Consider a Hamiltonian H_0 which is diagonal in the quasi-spin scheme. If we add to it non-degenerate single nucleon energies ϵ_j we obtain the Hamiltonian

$$H_0 + \sum_j \epsilon_j \hat{n}_j = H_0 + \sum_j \epsilon_j \sum_m a^+_{jm} a_{jm} \tag{22.34}$$

We can now add to (22.34) a two-body interaction to make it equal to H_0 to which a quasi-spin scalar has been added. We recall that for each j-orbit (see (19.8))

$$\begin{aligned} 2\mathbf{S}_j^2 &= 2S_j^+ S_j^- + 2(S_j^0)^2 - 2S_j^0 \\ &= 2S_j^+ S_j^- + \frac{1}{2}\left(\hat{n}_j - \frac{2j+1}{2}\right)^2 - \left(\hat{n}_j - \frac{2j+1}{2}\right) \\ &= 2S_j^+ S_j^- + \tfrac{1}{2}\hat{n}_j(\hat{n}_j - 1) - (j+1)\hat{n}_j \\ &\quad + \frac{(2j+1)^2}{8} + \frac{2j+1}{2} \end{aligned}$$

We can now obtain a Hamiltonian which is diagonal in the seniority scheme given by

$$\begin{aligned} H_0 - \sum_j \frac{\epsilon_j}{j+1}[2S_j^+ S_j^- + \tfrac{1}{2}\hat{n}_j(\hat{n}_j - 1)] + \sum_j \epsilon_j \hat{n}_j \\ = H_0 - \sum_j \frac{\epsilon_j}{j+1}(2\mathbf{S}_j^2 - \tfrac{1}{8}(2j+1)(2j+5)) \end{aligned} \tag{22.35}$$

The Hamiltonian (23.35) is manifestly diagonal in the quasi-spin scheme since $[S^+, \mathbf{S}_j^2] = [S_j^+, \mathbf{S}_j^2] = 0$. It is equal to (23.34) to which two body interactions, those in the square brackets on the l.h.s. of (23.35) were added. Those two body interactions act only within j-orbits and have no matrix elements between different orbits. It can be directly verified that in the two nucleon configurations, these two body interactions exactly cancel the contribution of the single nucleon energies to the diagonal elements. Hence $S^+|0\rangle$ is the eigenstate of (23.35) in the $J = 0$ ground state of the two nucleon configuration. Still, the Hamiltonian H defined by (23.35) is not identical to H_0. The eigenvalues of the single nucleon states $a^+_{jm}|0\rangle$ are obtained by using

$[\hat{n}_j, a^+_{jm}] = a^+_{jm}$. They are given by

$$\begin{aligned}
Ha^+_{jm}|0\rangle &= [H, a^+_{jm}]|0\rangle \\
&= [H_0, a^+_{jm}]\,|\,0\rangle - \frac{\epsilon_j}{j+1} \\
&\quad \times (2S^+_j(-1)^{j-m}a_{j,-m} + a^+_{jm}\hat{n}_j)|0\rangle + \epsilon_j a^+_{jm}|0\rangle \\
&= [H_0, a^+_{jm}]|0\rangle + \epsilon_j a^+_{jm}|0\rangle
\end{aligned} \tag{22.36}$$

The two body interaction in (22.35) does not contribute to the eigenvalue of the single nucleon state $a^+_{jm}|0\rangle$. In the case of the pairing interaction (22.4), or the SDI, $H_0 a^+_{jm}|0\rangle = 0$ and the single nucleon energies are the ϵ_j.

In a special case SDI has some interesting features. The case of two orbits, $j = l + \frac{1}{2}$, $j' = l - \frac{1}{2}$ which arise form the *same* l-orbit was considered above. The pair state $S^+_1|0\rangle$ in that case, given by (22.24), is the ^1S state ($S = 0$, $L = 0$) of the l^2 configuration. In actual nuclei, due to the strong spin-orbit interaction, such two orbits have widely spaced single nucleon energies. There are, however, other pairs of orbits whose single nucleon energies are rather close. In some of those pairs, j and j' differ by one unit, $|j - j'| = 1$. A nice example is offered by the $2d_{5/2}$ and $1g_{7/2}$ proton orbits in nuclei beyond $Z = 50$ and closed neutron shells with $N = 82$. The mixing of configurations, based on these orbits, by the SDI has a particularly simple form.

To calculate matrix elements we first expand the δ-interaction in terms of scalar products of spherical harmonics. All radial integrals are replaced in this case by a constant and hence, only the reduced matrix elements of the spherical harmonics should be considered. They are given by (10.38) where it was noticed that they depend only on j, j' and not on the values of l, l' provided $(-1)^{l+l'+k} = 1$. Hence, the SDI Hamiltonian matrices defined by $(g_{7/2})^{n_1}(d_{5/2})^{n_2}$ configurations are *exactly* equal (apart from an overall multiplicative constant) to those of $(f_{7/2})^{n_1}(f_{5/2})^{n_2}$ configurations. We may thus introduce a *pseudo* orbital angular momentum $\tilde{l} = 3$ which is coupled to a *pseudo-spin* $\tilde{s} = \frac{1}{2}$ to represent the $g_{7/2}$ and $d_{5/2}$ orbits (Hecht and Adler 1969, Arima et al. 1969).

The δ-potential is independent of the spins of the particles. In fact, as explained in Section 10, any central interaction may be expressed in a spin-independent form. Hence, due to the near equality of single nucleon energies, the scheme of eigenstates is very close to LS-coupling where $\tilde{S}$ and $\tilde{L}$ are good quantum numbers. This is true also for more general interactions like those with unequal g_k coefficients in (22.16). The problem thus reduces to calculating eigenstates and eigenvalues of the SDI in $\tilde{l}^n$ configurations with $\tilde{l} = 3$.

The $S_1^+|0\rangle$ pair state with $j = \tilde{l} + \frac{1}{2}$ and $j' = \tilde{l} - \frac{1}{2}$ is given by (22.24) which is the $\tilde{L} = 0, \tilde{S} = 0$ state of the $\tilde{l}^2$ configuration. The $L = 0$, $S = 0$ pair state plays in l^n configurations the same role as the $J = 0$ pair state in j^n configurations. As explained in detail in Section 28, it is the basis of the seniority scheme in l^n configurations. Thus, the quasi-spin scheme in the $\tilde{l} \pm \frac{1}{2}$ case is equivalent to the seniority scheme in l^n configurations.

The quasi-spin scheme is an elegant generalization of the seniority scheme in a single j-orbit. It is, however, not sufficiently general to describe the experimental data. If the pairing interaction or the somewhat more realistic SDI are adopted, it is difficult to adopt, even approximately, degenerate single nucleon orbits. As we saw, this particular difficulty may be cured by the prescription (22.33). Still, it is rather difficult to imagine that in the pair creation operator the amplitudes of the various orbits are all equal. The strongest objection against the use of the quasi-spin scheme in nuclei arises from the experimentally observed level spacings. Had the shell model Hamiltonian been diagonal in the quasi-spin scheme, level spacings would have been independent of the number n of identical nucleons. This would hold not only for even values of n but also for odd-even nuclei. This is in sharp contradiction with experiment. Observed level spacings in odd mass semi-magic nuclei change drastically with n (cf. Figs. 24.6, 24.7, 24.8 and 24.9).

It is necessary to consider a generalization of the seniority scheme which will be able to account for the experimental data. Such a generalization will be introduced in the next section.

23

Generalized Seniority. Ground States of Semi-Magic Nuclei

At the end of last section it was pointed out that the quasi-spin scheme is not sufficiently general to describe correctly nuclear spectra. In ground states, $(\mathcal{S}^+)^N|0\rangle$, with seniority $v = 0$, the operator $\mathcal{S}^+$ in (22.1) gives rise to mixing of configurations where nucleons are in several j-orbits. Yet this mixing is restricted by the equal amplitudes of the various S_j^+ in (22.1). It turns out, as will be explained in the following, that this gives an over simplified description of eigenstates of the shell model Hamiltonian. The quasi-spin scheme is indeed a generalization of seniority in a single j-orbit to the case of several orbits. It seems, however, that also a generalization of the quasi-spin scheme is needed.

A pair creation operator which is a generalization of (22.1) may be written as

$$\boxed{S^+ = \sum_j \alpha_j S_j^+} \tag{23.1}$$

The summation is over all j-orbits which contribute to the configurations considered. The operator $\mathcal{S}^+$ in (22.1) is a special case of (23.1) with equal α_j coefficients. We consider now pair creation operators with real coefficients α_j which need not be equal.

It is important to realize that S^+ with unequal coefficients (apart from a change of sign) is no longer a component of quasi-spin. The commutation relation of S^+ with its hermitean conjugate $S = (S^+)^\dagger = \sum \alpha_j S_j^-$ is equal to $\sum \alpha_j^2 S_j^0$. The commutator of that operator, in which not all α_j^2 values are equal, with either S^+ or S yields operators whose coefficients are higher powers of the α_j. Thus, S^+, S and $\frac{1}{2}[S^+,S]$ are not generators of the $SU(2)$ Lie algebra. We can no longer use the elegant and powerful methods of group theory. This means bad news to those who are interested only in applications of group theory or to those who believe that states of actual nuclei should fit some group theoretical scheme. Those with such opinions are warned not to follow the path starting here. Those who are interested primarily in nuclear structure and would like to understand the origin of certain regularities observed in nuclei, are invited to see how the operator (23.1) can be used for that purpose.

We consider here states of $n = 2N$ identical nucleons outside closed shells whereas nucleons of the other kind are in closed shells. Such configurations occur in semi-magic nuclei. Let us first consider ground states of such nuclei. In analogy to ground states with $J = 0$, and seniority $v = 0$ (19.12), in a pure j^n configuration, we construct the states

$$(S^+)^N|0\rangle \tag{23.2}$$

The states (23.2) may be described as states with *generalized seniority* zero. Such states were shown (Mottelson 1958) to be the components with $2N$ particles of the ground state wave functions of the Bardeen, Cooper and Schrieffer (1957) theory of super-conductivity. They were later used as variational wave functions for certain Hamiltonians. The coefficients α_j were taken to be variational parameters (Gambhir, Rimini and Weber 1969, Lorazo 1970). The approach described here (Talmi 1971) is rather different.

We try to see whether the states (23.2) can be *exact* eigenstates of the shell model Hamiltonian of identical nucleons. Thus, let us see what are the conditions on the shell model Hamiltonian H which guarantee that the states (23.2) are its eigenstates. We shall soon see that these conditions lead to interesting results. As before, the vacuum state $|0\rangle$ is the ground state of the nucleus with only closed shells. It is taken to be an eigenstate of H and the corresponding eigenvalue E_0 is the binding energy of that doubly magic nucleus. This will be taken as a constant and may be conveniently set to zero (by replacing H by $H - E_0$), i.e.

$$H|0\rangle = 0 \tag{23.3}$$

The conditions that the states (23.2) are eigenstates of H are

$$\boxed{H(S^+)^N|0\rangle = E_N(S^+)^N|0\rangle} \tag{23.4}$$

Let us first look at the condition (23.4) for $N = 1$, namely

$$\boxed{HS^+|0\rangle = [H,S^+]|0\rangle = V_0S^+|0\rangle} \tag{23.5}$$

In (23.5) we made use of the normalization (23.3). The first condition, (23.5), is that the pair state created by S^+ should be an eigenstate of H (whose eigenvalue is denoted by V_0). For the applications we take the state in (23.5) to be the lowest state of the two nucleon configuration. This is, however, irrelevant for the following derivations.

The next condition is derived by putting $N = 2$ in (23.4). We obtain

$$\begin{aligned} H(S^+)^2|0\rangle &= [H,S^+]S^+|0\rangle + S^+HS^+|0\rangle \\ &= [[H,S^+],S^+]|0\rangle + S^+[H,S^+]|0\rangle + S^+HS^+|0\rangle \\ &= [[H,S^+],S^+]|0\rangle + 2V_0(S^+)^2|0\rangle \end{aligned}$$

The last equality follows from the condition (23.5). We see that the state with $N = 2$ is an eigenstate of H if

$$[[H,S^+],S^+]|0\rangle = W(S^+)^2|0\rangle \tag{23.6}$$

where W is a numerical coefficient. If the shell model Hamiltonian H contains single nucleon terms and two-body interactions, its double commutator with S^+ contains only products of four creation operators. Hence, if (23.6) is satisfied, it will be satisfied if we replace in it the vacuum state with any other state. We can thus write the condition (23.6) for $N = 2$ as the operator condition

$$\boxed{[[H,S^+],S^+] = W(S^+)^2} \tag{23.7}$$

Having derived the conditions (23.4) for $N = 1$ and $N = 2$, we may ask whether the substitution in (23.4) of $N = 3, 4, \ldots$ will yield further conditions on H. Fortunately this is not so. First we prove the following lemma by induction

$$[H,(S^+)^N] = \sum_{\nu=1}^{N} \binom{N}{\nu} (S^+)^{N-\nu}[\cdots[H,S^+],\ldots,S^+] \tag{23.8}$$

where the ν-th term contains the multiple commutator of H with ν operators S^+. The relation (23.8) holds trivially for $N = 1$. For $N = 2$ it can be directly verified

$$[H,(S^+)^2] = [H,S^+]S^+ + S^+[H,S^+] = [[H,S^+],S^+] + 2S^+[H,S^+]$$

If (23.8) holds for $N - 1 \geq 2$ we show that it holds also for N. We obtain

$$\begin{aligned}
[H,(S^+)^N] &= [H,(S^+)^{N-1}]S^+ + (S^+)^{N-1}[H,S^+] \\
&= \sum_{\nu=1}^{N-1} \binom{N-1}{\nu} (S^+)^{N-1-\nu}[\cdots[H,S^+],\ldots,S^+]S^+ \\
&\quad + (S^+)^{N-1}[H,S^+] \\
&= \sum_{\nu=1}^{N-1} \binom{N-1}{\nu} (S^+)^{N-1-\nu}[[\cdots[H,S^+],\ldots,S^+],S^+] \\
&\quad + \sum_{\nu=1}^{N-1} \binom{N-1}{\nu} (S^+)^{N-\nu}[\cdots[H,S^+],\ldots,S^+] \\
&\quad + (S^+)^{N-1}[H,S^+]
\end{aligned}$$

Combining terms with the same powers of $(S^+)^{N-\nu}$ we obtain

$$(N-1)(S^+)^{N-1}[H,S^+] + \sum_{\nu=2}^{N-1}\left\{\binom{N-1}{\nu-1}+\binom{N-1}{\nu}\right\}$$

$$\times (S^+)^{N-\nu}[\cdots[H,S^+],\ldots,S^+] + (S^+)^{N-1}[H,S^+]$$

$$+ [[\cdots[H,S^+],\ldots,S^+],S^+]$$

$$= N(S^+)^{N-1}[H,S^+] + \sum_{\nu=2}^{N}\binom{N}{\nu}(S^+)^{N-\nu}$$

$$\times [[\cdots[H,S^+],\ldots,S^+],S^+]$$

$$= \sum_{\nu=1}^{N}\binom{N}{\nu}(S^+)^{N-\nu}[\cdots[H,S^+],\ldots,S^+]$$

Hence, we proved by induction that (23.8) holds for *any* value of N.

If H contains only one nucleon terms and two-body interactions the commutators in (23.8) with $\nu = 3,\ldots,N$ vanish. Hence, if the conditions (23.5) and (23.7) hold, we obtain by applying (23.8) to the vacuum state, for any N, the result

$$\boxed{\begin{aligned} H(S^+)^N|0\rangle &= [H,(S^+)^N]|0\rangle \\ &= N(S^+)^{N-1}[H,S^+]|0\rangle \\ &\quad + \tfrac{1}{2}N(N-1)(S^+)^{N-2}[[H,S^+],S^+]|0\rangle \\ &= (NV_0 + \tfrac{1}{2}N(N-1)W)(S^+)^N|0\rangle \end{aligned}} \tag{23.9}$$

Thus, conditions (23.5) and (23.7), or simply the requirements that $S^+|0\rangle$ and $(S^+)^2|0\rangle$ are eigenstates of H, are sufficient to guarantee that all states (23.2) will be eigenstates. This is like the situation in seniority for a single j-orbit. There it was enough to check the j^2 and j^4 (or j^3) configurations to find out whether H is diagonal in the seniority scheme in all j^n configurations.

The expression (23.9) for the eigenvalues is an exact linear and quadratic function of the number of pairs N. It has the same form as the expression (20.14) for the eigenvalues of a Hamiltonian which is diagonal in the seniority scheme in j^n configurations. If we put the following into (20.14)

$$N = n/2, \qquad 2\epsilon_j + \alpha + \beta = V_0, \qquad 4\alpha = W \tag{23.10}$$

we obtain (23.9). The same expression holds for ground state energies of even nuclei in the quasi-spin scheme. In those schemes, the simple dependence on particle number is due to properties of the $SU(2)$ Lie algebra. It is indeed remarkable that the simple formula for the interaction in ground states of even configurations survives the generalization (23.1). It holds for any shell model Hamiltonian that satisfies conditions (23.5) and (23.7). The exact linear and quadratic dependence on N of the eigenvalues holds also for binding energies and nuclear masses. It is worthwhile to emphasize that this dependence on N holds only if the coefficients α_j in (23.1) are *constant* throughout the major shell being filled and do not change with N.

The conditions (23.5) and (23.7) have a simple meaning in the quasi-spin scheme. If in (23.1) all α_j are equal (apart from a change of sign) we obtain the pair creation operator $\mathcal{S}^+$ of (22.1). Any shell model Hamiltonian may then be expressed as a sum of a quasi-spin scalar and the $\sigma = 0$ components of a quasi-spin vector and a quasi-spin tensor with rank $s = 2$.

$$H^{(s=0)} + H^{(s=1)}_{\sigma=0} + H^{(s=2)}_{\sigma=0}$$

In Section 20 we showed that any Hamiltonian which is diagonal in the seniority scheme can be expressed as a linear combination of a quasi-spin scalar and special $\sigma = 0$ components of a quasi-spin vector and rank $s = 2$ tensor. In the quasi-spin scheme we may consider such Hamiltonians expressed as

$$H = H^{(s=0)} + x\mathcal{S}^0 + y(\mathcal{S}^0)^2 \tag{23.11}$$

Using the commutation relations (22.3) we can verify that the Hamiltonian (23.11) satisfies with $\mathcal{S}^+$ the relation

$$\begin{aligned}[H,\mathcal{S}^+]|0\rangle &= (x\mathcal{S}^+ + y\mathcal{S}^+\mathcal{S}^0 + y\mathcal{S}^0\mathcal{S}^+)|0\rangle \\ &= (x\mathcal{S}^+ + y\mathcal{S}^+ + 2y\mathcal{S}^+\mathcal{S}^0)|0\rangle = (x + y - y\Omega)\mathcal{S}^+|0\rangle\end{aligned} \tag{23.12}$$

The double commutator of (23.11) and $\mathcal{S}^+$ is equal to

$$[[H,\mathcal{S}^+],\mathcal{S}^+] = 2y(\mathcal{S}^+)^2 \tag{23.13}$$

From (23.12) and (23.13) follows, by comparison with the conditions (23.5) and (23.7), that H in (23.11) satisfies those conditions, it has eigenvalues given by (23.9) and x and y are given by

$$y = \tfrac{1}{2}W \qquad x = V_0 + \frac{W}{2}(\Omega - 1) \tag{23.14}$$

The same result may be obtained by substituting in (23.11) for $\mathcal{S}^0$ its expression $(\hat{n} - \Omega)/2$. The expression (23.9) for the eigenvalues follows in this case from the fact that $H^{(s=0)}|0\rangle = 0$.

The inverse is also true. If a shell model Hamiltonian satisfies the conditions (23.5) and (23.7) with the special pair operator $\mathcal{S}^+$, then it is diagonal in the quasi-spin scheme. This holds particularly in the case of seniority in j^n configurations in which case only one α_j coefficient does not vanish. In (23.7) or (23.13) the $\sigma = 2$ component is obtained from $H^{(s=2)}_{\sigma=0}$ by taking twice the commutator with $\mathcal{S}^+$. If (23.13) is satisfied, $H^{(s=2)}_{\sigma=2}$ is equal to $W(\mathcal{S}^+)^2/2\sqrt{6}$ and hence, by operating twice with $\mathcal{S}^-$, the original component $H^{(s=2)}_{\sigma=0}$ is obtained to be equal to

$$\frac{W}{2}[\tfrac{2}{3}(\mathcal{S}^0)^2 - \tfrac{1}{6}\mathcal{S}^+\mathcal{S}^- - \tfrac{1}{6}\mathcal{S}^-\mathcal{S}^+] = \frac{W}{2}[(\mathcal{S}^0)^2 - \tfrac{1}{3}\mathcal{S}^2]$$

From condition (23.5) we now obtain (by using $[\mathcal{S}^2,\mathcal{S}^+] = 0$)

$$\begin{aligned}
&[H^{(s=1)}_{\sigma=0},\mathcal{S}^+]|0\rangle + \frac{W}{2}[(\mathcal{S}^0)^2,\mathcal{S}^+]|0\rangle \\
&\quad = [H^{s=1}_{\sigma=0},\mathcal{S}^+]|0\rangle + \frac{W}{2}(2\mathcal{S}^+\mathcal{S}^0 + \mathcal{S}^+)|0\rangle \\
&\quad = [H^{(s=1)}_{\sigma=0},\mathcal{S}^+]|0\rangle - \frac{W}{2}(\Omega - 1)\mathcal{S}^+|0\rangle = V_0\mathcal{S}^+|0\rangle
\end{aligned} \tag{23.15}$$

If $H^{(s=1)}_{\sigma=0}$ is due only to single nucleon terms (we actually proved it in the discussion following (19.46)), its commutator with $\mathcal{S}^+$ contains products of two creation operators. In that case we obtain from

(23.15) the operator relation

$$[H^{(s=1)}_{\sigma=0}, \mathcal{S}^+] = \left(V_0 - \frac{W}{2}(\Omega - 1)\right)\mathcal{S}^+ \qquad (23.16)$$

The l.h.s. of (23.16) is equal to $-\sqrt{2}H^{(s=1)}_{\sigma=1}$ and since it is proportional to $\mathcal{S}^+$ we conclude, by taking its commutator with $\mathcal{S}^-$, that $H^{(s=1)}_{\sigma=0}$ is proportional to $\mathcal{S}^0$. This proves the assertion stated above.

Before leaving the quasi-spin scheme we repeat a statement made in Section 22. From the form (23.11) follows that spacings between levels are determined only by $H^{(s=0)}$. The eigenvalues of a quasi-spin scalar depend only on the value s of the total quasi-spin of the corresponding eigenstates but are independent of the values of $\mathcal{S}^0$ or nucleon number. Hence, if the shell model Hamiltonian is diagonal in the quasi-spin scheme, level spacings in even-even as well as in odd-even semi-magic nuclei should be independent of nucleon number. This is in contradiction with experimental level spacings as shown in Figs. 24.6, 24.7, 24.8 and 24.9.

There is, however, a price to be paid for leaving the quasi-spin scheme. If the α_j are not equal, acting by the creation operators S^+ does not yield an orthogonal set of states. In the seniority scheme, as well as in the quasi-spin scheme, consider two orthogonal states with seniorities s and s'

$$\langle n, \alpha, sJM \mid n, \alpha', s'JM\rangle = 0$$

The states $\mathcal{S}^+|n,\alpha,s,J,M\rangle$ and $\mathcal{S}^+|n,\alpha',s',JM\rangle$ are also orthogonal. This is trivial if $s' \neq s$ since the resulting states have different seniorities s and s'. If $s = s'$ this was demonstrated in the derivation of (19.18). In the case of generalized seniority defined by (23.1) with unequal α_j coefficients, no general scheme may be defined. Two orthogonal states with given n, J, M need not remain orthogonal when S^+ is applied to them. This may be demonstrated by a simple example.

Consider $J = 0$ states which are orthogonal to the state $S^+|0\rangle$. In the quasi-spin scheme they all have seniorities $v = 2$. If there are k orbits with non-vanishing α_j coefficients, there are exactly $k - 1$ such linearly independent states. Each such state may be expressed as $A^{(i)+}_{J=0} = \sum_j \alpha^{(i)}_j S^+_j|0\rangle$ $(i = 1,\dots,k-1)$. The orthogonality relations imply

$$\sum_j \alpha_j\alpha^{(i)}_j\langle 0|S^-_j S^+_j|0\rangle = \frac{1}{2}\sum(2j+1)\alpha_j\alpha^{(i)}_j = 0 \qquad (23.17)$$

These equations mean that the vector with k components α_j is orthogonal to the $k-1$ vectors with components $(2j+1)\alpha_j^{(i)}$. We now calculate the overlap of $(S^+)^2|0\rangle$ and $S^+A_{J=0}^{(i)+}|0\rangle$ given by

$$\begin{aligned}\langle 0|S^2S^+A_{J=0}^{(i)+}|0\rangle &= \langle 0|S[S,S^+]A_{J=0}^{(i)+}|0\rangle + \langle 0|[S,S^+]SA_{J=0}^{(i)+}|0\rangle \\ &= \langle 0|[S,[S,S^+]]A_{J=0}^{(i)+}|0\rangle + 2\langle 0|[S,S^+]SA_{J=0}^{(i)+}|0\rangle\end{aligned} \tag{23.18}$$

Recalling that $[S,S^+] = -2\sum_j \alpha_j^2 S_j^0$ from which follows $[S,[S,S^+]] = -2\sum \alpha_j^3 S_j^-$, we obtain for (23.18) the expression

$$\begin{aligned}&-2\sum \alpha_j^3\alpha_j^{(i)}\langle 0|S_j^- S_j^+|0\rangle + \sum_j \alpha_j^2(2j+1)\langle 0|SA_{J=0}^{(i)+}|0\rangle \\ &= -\sum \alpha_j^3\alpha_j^{(i)}(2j+1)\end{aligned} \tag{23.19}$$

If the states are orthogonal, the r.h.s. of (23.19) vanishes. In that case we find that the vector whose components are α_j^3 is orthogonal to the $k-1$ vectors with components $(2j+1)\alpha_j^{(i)}$. Hence, that vector is proportional to the vector with components α_j which implies that all α_j^2 must be equal. Thus, apart from possible change of sign, all α_j coefficients must be equal. If they are not equal, at least one of the states $S^+A_{J=0}^{(i)+}|0\rangle$ is not orthogonal to $(S^+)^2|0\rangle$. Thus, we cannot assign the states $(S^+)^N A_{J=0}^{(i)+}|0\rangle$ generalized seniorities $v=2$ if v is to distinguish between orthogonal states. There is only one set of states with $J=0$, defined by (23.2), which constitutes a meaningful generalization of seniority $v=0$ states.

Let us turn back to the general case where the coefficients α_j in (23.1) are not equal. As mentioned above, we mean by that unequal apart from a change of sign or unequal squares α_j^2 of the coefficients in (23.1). From now on when we refer to unequal α_j coefficients it should be understood in this sense and the qualification "apart from change of sign" will be omitted. If the shell model Hamiltonian satisfies conditions (23.5) and (23.7) it has eigenstates and eigenvalues given by (23.9). As in Section 20, we can use this expression to obtain a mass formula or a formula for binding energies. There we had to add $n\epsilon_j$ to the eigenvalues of the two-body interaction given by (20.13). In the present case, however, the expression (23.9), as well as V_0 for $N=1$, are the eigenvalues of the shell model Hamiltonian which includes the single nucleon energies. Although the single nucleon energies are usually not degenerate, the linear and quadratic dependence of (23.9) is a direct consequence of (23.5) and (23.7). Binding energies of even semi-magic nuclei due to a shell model Hamiltonian satisfying these conditions can thus be expressed by

$$\boxed{\text{B.E.}(n = 2N) = \text{B.E.}(N = 0) + NV_0 + \tfrac{1}{2}N(N-1)W} \tag{23.20}$$

We shall see in the following how well (23.20) fits experimental binding energies.

A striking feature of (23.20) is the total absence of sub-shell effects. Even though some orbits have lower single nucleon energies than others, the occupation numbers of orbits change smoothly with N and no sharp closures of j-orbits occur. Looking closely at the structure of the eigenstates (23.2) it is clear that orbits with larger α_j coefficients have larger occupation numbers as the major shell is being filled. This follows from the expansion

$$(S^+)^N|0\rangle = \left(\sum \alpha_j S_j^+\right)^N |0\rangle = \sum_{\nu_1,\nu_2,\ldots,\nu_k} \frac{N!}{\nu_1!\nu_2!\cdots\nu_k!} \times \alpha_{j_1}^{\nu_1}\alpha_{j_2}^{\nu_2}\cdots\alpha_{j_k}^{\nu_k}(S_{j_1}^+)^{\nu_1}(S_{j_2}^+)^{\nu_2}\cdots(S_{j_k}^+)^{\nu_k}|0\rangle \tag{23.21}$$

As $N = \sum \nu_i$ increases, the Pauli principle makes terms in (23.21) vanish if $\nu_1 > 2j_1 + 1$, $\nu_2 > 2j_2 + 1, \ldots, \nu_k > 2j_k + 1$. This makes the occupation numbers of other orbits, even with small α_j, grow too.

The absence of sub-shell closures in (23.20) can be made more explicit by looking at separation energies of pairs of identical nucleons. From (23.20) follows that those separation energies are given by

$$\boxed{\text{B.E.}(N) - \text{B.E.}(N-1) = V_0 + (N-1)W} \tag{23.22}$$

If the conditions (23.5) and (23.7) are satisfied, pair separation energies should lie on a straight line. No breaks in the curve of (23.22) must show up. We shall later see how well (23.22) fits experimental pair separation energies.

It should be realized that the absence of effects due to sub-shell closures is a characteristic feature of the very special configuration mixing prescribed by (23.2). It is due to the constancy of the α_j coefficients and their independence of N. If the coefficients α_j are considered as variational parameters for some Hamiltonian, they may turn out to depend on N. In that case the smooth behavior of binding energies need not follow. Discontinuities in the binding energy curve may then show up. In fact, if variational wave functions like (23.2) are used across the closure of a major shell, they may reproduce the rather large discontinuity. The parameters α_j will then change when going from one nucleus to another.

A simple example may help to clarify this point. Consider two orbits j and j' whose single nucleon energies differ by a large energy gap ϵ. The Hamiltonian of the system H_0 is the sum of Hamiltonians which act within j^n configurations and within j'^n configurations. It is taken to be diagonal in the seniority scheme in each of these configurations. If mixing of such configurations is due to an interaction acting only in $J = 0$ states, the Hamiltonian can be expressed as

$$H = H_0 + G(S_j^+ S_{j'}^- + S_{j'}^+ S_j^-) \tag{23.23}$$

We now use as variational function $(S_j^+ + \alpha S_{j'}^+)^N|0\rangle$ and determine α which yields the lowest value of the following expectation value

$$\frac{\langle 0|(S_j^- + \alpha S_{j'}^-)^N[H_0 + G(S_j^+ S_{j'}^- + S_{j'}^+ S_j^-)](S_j^+ + \alpha S_{j'}^+)^N|0\rangle}{\langle 0|(S_j^- + \alpha S_{j'}^-)^N(S_j^+ + \alpha S_{j'}^+)^N|0\rangle} \tag{23.24}$$

We take G to be very small compared to 2ϵ and keep in (23.24) only the terms up to second order in α. Naturally, we obtain in this way the results of second order perturbation theory. We thus obtain from (23.24), for $N \leq (2j+1)/2$, the expression

$$\begin{aligned}
&[\langle 0|(S_j^-)^N H_0(S_j^+)^N|0\rangle + 2\alpha N G\langle 0|(S_j^-)^{N-1} S_{j'}^- S_{j'}^+ S_j^-(S_j^+)^N|0\rangle + \\
&+ \alpha^2 N^2\langle 0|(S_j^-)^{N-1} S_{j'}^- H_0(S_j^+)^{N-1} S_{j'}^+|0\rangle] \\
&\times [\langle 0|(S_j^-)^N(S_j^+)^N|0\rangle + \alpha^2 N^2\langle 0|(S_j^-)^{N-1} S_{j'}^-(S_j^+)^{N-1} S_{j'}^+|0\rangle]^{-1}
\end{aligned} \tag{23.25}$$

Recalling the result

$$\langle 0|(S_j^-)^N(S_j^+)^N|0\rangle = N!\left(\frac{2j+1}{2}\right)!\Big/\left(\frac{2j+1}{2}-N\right)!$$
$$= N!\Omega_j!/(\Omega_j - N)!$$

given by (19.19) for $v = 0$, we obtain due to the fact that $S_j^{\pm}$ and $S_{j'}^{\pm}$ commute, that (23.25) is equal to

$$\left[E_N\frac{N!\Omega!}{(\Omega_j - N)!} + 2\alpha NG\frac{N!\Omega_j!}{(\Omega_j - N)!}\Omega_{j'}\right.$$
$$\left. + \alpha^2 N^2\frac{(N-1)!\Omega_j!}{(\Omega_j - N + 1)!}\Omega_{j'}(E_{N-1} + E_1')\right]$$
$$\times\left[\frac{N!\Omega_j!}{(\Omega_j - N)!} + \alpha^2 N^2\frac{(N-1)!\Omega_j!}{(\Omega_j - N + 1)!}\Omega_{j'}\right]^{-1}$$
$$= \left[E_N + 2\alpha NG\Omega_{j'} + \alpha^2 N^2\frac{\Omega_{j'}}{N(\Omega_j - N + 1)}(E_{N-1} + E_1')\right]$$
$$\times\left[1 + \alpha^2 N^2\frac{\Omega_{j'}}{N(\Omega_j - N + 1)}\right]^{-1} \tag{23.26}$$

In (23.26) E_N is the energy of the $v = 0$, $J = 0$ state of the j^{2N} configuration and E_1' is the energy of the $j'^2 J = 0$ state. The latter includes in addition to the interaction energy $V_{j'}$ in the $J = 0$ state also 2ϵ.

The terms up to α^2 in (23.26) may be expressed as

$$E_N + 2\alpha NG\Omega_{j'} + \alpha^2\frac{N\Omega_{j'}}{(\Omega_j - N + 1)}(E_{N-1} + E_1' - E_N) \tag{23.27}$$

The value of α which minimizes (23.27) is thus given by

$$\alpha = \frac{(\Omega_j - N + 1)G}{E_N - (E_{N-1} + E_1')} = \frac{(\Omega_j - N + 1)G}{V_j + (N-1)W_j - V_{j'} - 2\epsilon} \tag{23.28}$$

The interaction energy E_N was expressed according to (20.13) or (23.9) with $V_j = V_0$ and $W_j = W$ for the j^n configuration. For a large gap ϵ between single nucleon energies of the j- and j'-orbits, α in (23.28) is indeed small if G is small in comparison with ϵ.

A direct conclusion which may be drawn from (23.28) is that the configuration mixing in (23.4) with constant coefficients α_j cannot be considered as a small perturbation in any given configuration. The coefficient (23.28), derived in perturbation theory, has a definite dependence on N. Even if we ignore the N dependence in the denominator (e.g. if for the given H_0 the W_j term vanishes), still the numerator depends on N. Only for values of N, small compared to Ω_j, is the coefficient (23.28) approximately constant. It is then approximately equal to the value of α obtained by imposing the conditions $[H,S^+]|0\rangle = V_0S^+|0\rangle$ on the Hamiltonian (23.23). For small values of α the value obtained in that case is $\alpha = G\Omega_j/(V_j - V_{j'} - 2\epsilon)$. For such values of N, the main configuration is indeed the j^{2N} configuration and the amount of mixing of the $j^{2N-2}j'^2$ configuration is small. As N increases however, the change in (23.28) becomes appreciable. Thus, if perturbation theory may be applied, the coefficients are not constant and the simple expression (23.9) does not hold.

The energies due to the variational calculation for small α do not follow the simple result (23.9). Putting the value of α from (23.28) into (23.27) we obtain the result

$$E_N + \frac{NG^2(\Omega_j - N + 1)\Omega_{j'}}{E_N - (E_{N-1} + E'_1)} = E_N + \frac{NG^2(\Omega_j - N + 1)\Omega_{j'}}{V_j + (N-1)W_j - V_{j'} - 2\epsilon} \tag{23.29}$$

The dependence on N of (23.29) is complicated but if we put here $W_j = 0$ it simplifies into a linear and quadratic function of N. This is a simple demonstration of a general feature discussed in Section 32. The effects of small admixtures of configurations obtained by the excitation of two particles may appear as simple renormalization of the two body effective interaction. This smooth dependence, however, will not continue beyond $N = \Omega_j$. To see it we calculate the expectation value (23.24) for $N = \Omega_j + 1$.

In the binomial expansion of $(S_j^+ + \alpha' S_{j'}^+)^{\Omega_j+1}$ the first term $(S_j^+)^{\Omega_j+1}$ vanishes due to the Pauli principle. Dividing the state $(S^+)^{\Omega_j+1}|0\rangle$ by $\alpha'(\Omega_j + 1)$ we obtain for the expectation value (23.24) with no higher than α'^2 terms, the expression

$$\frac{\langle 0|(S_j^{\Omega_j}S_{j'} + \frac{1}{2}\alpha'\Omega_j S_j^{\Omega_j-1}S_{j'}^2)H((S_j^+)^{\Omega_j}S_{j'}^+ + \frac{1}{2}\alpha'\Omega_j(S_j^+)^{\Omega_j-1}(S_{j'}^+)^2)|0\rangle}{\langle 0|S_j^{\Omega_j}S_{j'}(S_j^+)^{\Omega_j}S_{j'}^+|0\rangle + \frac{1}{4}\alpha'^2\Omega_j^2\langle 0|S_j^{\Omega_j-1}S_{j'}^2(S_j^+)^{\Omega_j-1}(S_{j'}^+)^2|0\rangle} \tag{23.30}$$

The evaluation of (23.30) is very similar to the derivation of (23.27) from (23.25). This way we obtain for the expectation value (23.30) up to α'^2 terms, the expression

$$E_{\Omega_j} + E_1' + 2\alpha' G\Omega_j(\Omega_{j'} - 1) + \tfrac{1}{2}\alpha'^2\Omega_j(\Omega_{j'} - 1)(E_{\Omega_j - 1} + E_2' - E_{\Omega_j} - E_1') \tag{23.31}$$

The value of α' which minimizes (23.31) is

$$\alpha' = \frac{2G}{E_{\Omega_j} - E_{\Omega_j - 1} + E_1' - E_2'} = \frac{2G}{V_j + (\Omega_j - 1)W_j - V_{j'} - W_{j'} - 2\epsilon} \tag{23.32}$$

The value of α' is very similar to (23.28) for $N = \Omega_j - 1$. It is identical to it if we put $W_j = W_{j'}$. The energy of the state considered, obtained by substituting α' from (23.32) into (23.31), is (for $W_j = W_{j'} = 0$) equal to

$$E_{\Omega_j} + E_1' + \frac{2G^2\Omega_j(\Omega_{j'} - 1)}{V_j - V_{j'} - 2\epsilon} = E_{\Omega_j} + V_{j'} + 2\epsilon + \frac{2G^2\Omega_j(\Omega_{j'} - 1)}{V_j - V_{j'} - 2\epsilon} \tag{23.33}$$

We can now look at the behavior of separation energies as a function of N. Putting $W_j = 0$ we obtain from (23.29) the pair separation energies given by

$$V_j + \frac{G^2\Omega_j\Omega_{j'}}{V_j - V_{j'} - 2\epsilon} - \frac{2G^2\Omega_{j'}}{V_j - V_{j'} - 2\epsilon}(N - 1) \tag{23.34}$$

The pair separation energy for $N = \Omega_j + 1$ is the difference between (23.33) and (23.29) for $N = \Omega_j$. Putting also $W_{j'} = 0$ we obtain for that difference the value

$$V_{j'} + 2\epsilon + \frac{G^2\Omega_j\Omega_{j'}}{V_j - V_{j'} - 2\epsilon} - \frac{2G^2\Omega_j}{V_j - V_{j'} - 2\epsilon} \tag{23.35}$$

The main difference between (23.35) and (23.34) for $N = \Omega_j$ is 2ϵ. This difference in a pair of single nucleon energies is much larger not only in comparison to G but also to $|V_j - V_{j'}|$. This is necessary for keeping the denominators in (23.28) and (23.32) sufficiently large. Hence, when going from $N = \Omega_j$ to $N = \Omega_j + 1$ there is a large break in the curve of pair separation energies. Such breaks never occur if the coefficients α_j are constant throughout the major shell.

Let us now return to the general case where the coefficients α_j in (23.1) are constant and not equal. In the cases of j^n configurations described in Section 21 it was possible to extract from the experimental data matrix elements of the effective interaction. It was then possible to determine whether that interaction is diagonal in the seniority scheme by comparing non-diagonal matrix elements in the j^n configuration to differences between diagonal elements. In the situation considered here it is necessary to find the values of single nucleon energies and of both diagonal and non-diagonal matrix elements of the two-body interaction. We can then check whether conditions (23.5) and (23.7) are satisfied. We first solve the two nucleon problem obtaining the values of α_j. Then it is possible to check whether the condition (23.7) is satisfied. This is equivalent to checking whether the state $(S^+)^2|0\rangle$ is indeed an eigenstate of the shell model Hamiltonian.

The possibility just described is, in practice, very difficult to follow. As emphasized again and again, the only reliable way of determining matrix elements of the effective interaction is from experimental data. As the number of orbits in the configurations considered is increased, the number of matrix elements rapidly rises. In addition to single nucleon energies and diagonal matrix elements of the two-body effective interactions, also non-diagonal elements must be determined. Only in very simple cases is it possible to determine all matrix elements which are needed from experiment. It is easier to check whether the predictions from conditions (23.5) and (23.7), as well as other conditions to be introduced later, agree with experimental data in even-even semi-magic nuclei. As in Section 21, we look at separation energies. In the present case, however, we consider pair separation energies which are given by (23.22).

The first example we consider is actually not an example of generalized seniority. It is the case of the $1f_{7/2}$ neutron orbit in Ca isotopes. The evidence for $1f^n_{7/2}$ neutron configurations was presented in Section 21. Here we examine ground states of even Ca isotopes from a somewhat different point of view. We already saw the behavior of binding energies *within* the $1f_{7/2}$ shell. Let us now examine whether any admixtures of other orbits assume the form (23.2). In other words,

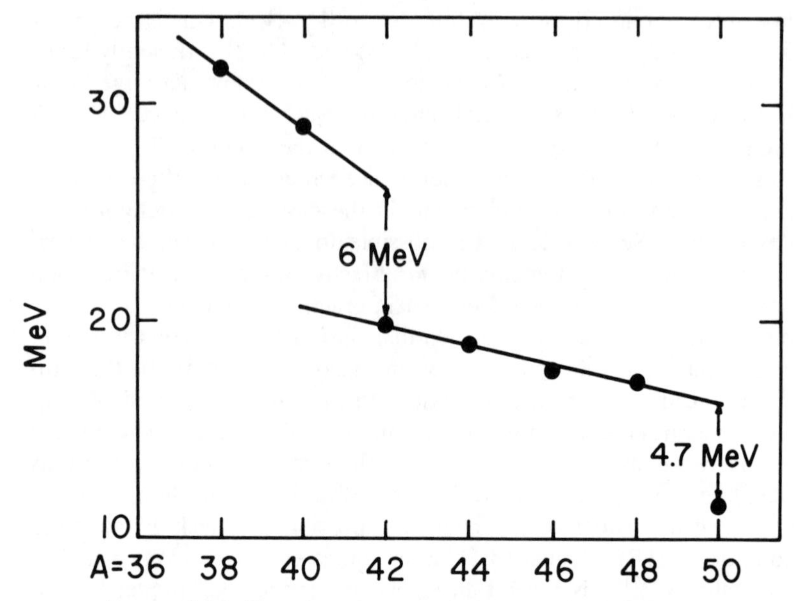

FIGURE 23.1. *Neutron pair separation energies of* Ca *isotopes.*

we check whether the curve of neutron pair separation energies continues smoothly beyond neutron number 28. The experimental separation energies are shown in Fig. 23.1. The large drops beyond neutron numbers 20 and 28 clearly display the validity of the description in terms of $(1f_{7/2})^n$ configurations. This does not imply that the $1f_{7/2}$ shell is pure. The effective matrix elements, V_0 and W, may include some configuration mixings but those cannot assume the strong form (23.2).

A possible argument against determination of matrix elements of the effective interaction from experiment is associated with *pseudonium*. This is a fictitious nucleus in which strong configuration mixing yields binding energies and level spacings which look like those of pure j^n configurations (Cohen et al. 1966). In an example, configurations based on $1d_{3/2}$ and $1f_{7/2}$ orbits of identical nucleons were mixed in this fashion. The resulting levels of even nuclei and odd parity levels in odd nuclei between nucleon numbers 20 to 28 looked like those of $(f_{7/2})^n$ configurations. Actually, the simplest case of such a situation is in eigenstates of the quasi-spin scheme discussed in Section 22. In that case, binding energies follow the formula (20.14) and level spacings are independent of n. This behavior, however, should persist

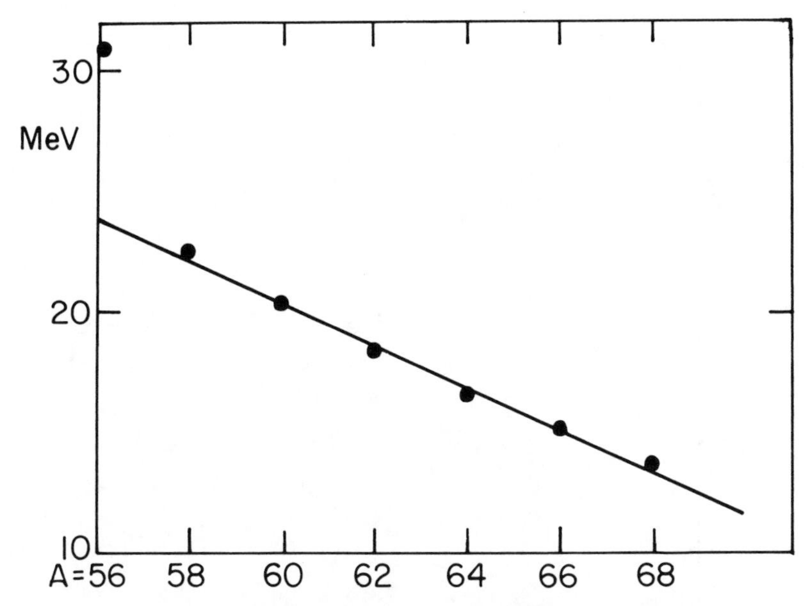

FIGURE 23.2. *Neutron pair separation energies of* Ni *isotopes.*

throughout the region where valence nucleons occupy all orbits with non-vanishing coefficients in the S^+ operator. This yields a simple test for pseudonium like situations. In the example quoted above, binding energies given by (20.14) and level spacings independent of n should be evident not only between nucleon numbers 20 to 28 but from 16 to 28. The separation energies in Fig. 23.1 clearly rule out such strong admixtures into ground states of $(1f_{7/2})^n$ configurations.

Let us now consider Ni isotopes beyond neutron number 28. The lowest orbits available for the valence neutrons are $2p_{3/2}$, $1f_{5/2}$ and $2p_{1/2}$. Let us look at Fig. 23.2 where neutron pair separation energies are plotted. The experimental points lie on a smooth and fairly straight line. No breaks corresponding to subshell closures can be detected. Actually, shell model calculations have been carried out for Ni isotopes (Auerbach 1966, Cohen et al. 1967). Matrix elements of the effective interaction, both diagonal and non-diagonal, which yield the experimental energy levels were determined. It turned out that ground states of even Ni isotopes may be very well approximated by $(S^+)^N|0\rangle$ where the pair creation operator is given by

$$S^+ = 2.75S^+_{p_{3/2}} + 1.57S^+_{f_{5/2}} + 1.58S^+_{p_{1/2}} \qquad (23.36)$$

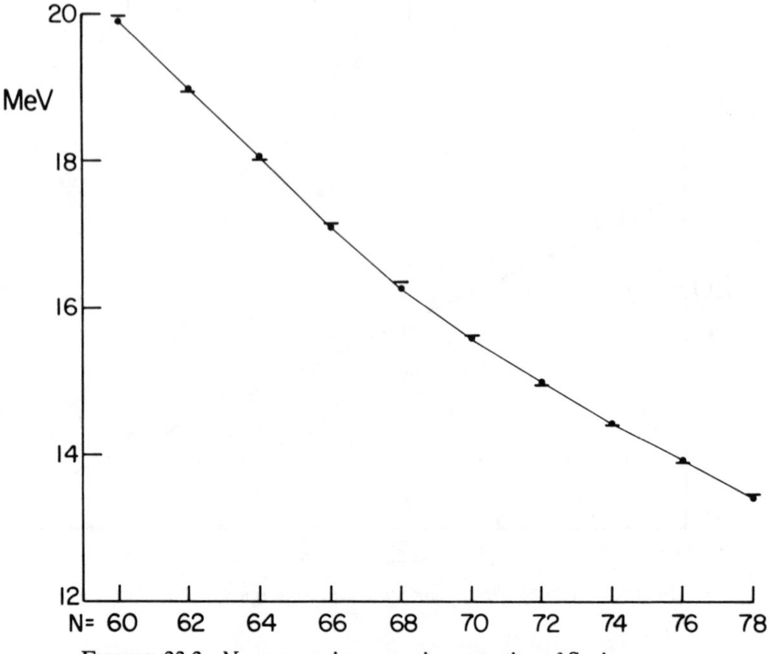

FIGURE 23.3. *Neutron pair separation energies of* Sn *isotopes.*

The experimental data available could not yield the values of all matrix elements required for spectra of Ni isotopes. For ground states, however, the situation is better and the necessary matrix elements were determined with a higher degree of confidence. As in the cases described in Section 21, the coefficient W of the quadratic term in the binding energy formula is repulsive.

In larger shells no complete shell model calculations could have been carried out. In Sn isotopes, for instance, the neutron valence orbits are in the major shell between magic numbers 50 and 82. These are the $1g_{7/2}$, $2d_{5/2}$, $3s_{1/2}$, $2d_{3/2}$ and $1h_{11/2}$ orbits. Binding energies of all tin isotopes from ^{100}Sn to ^{132}Sn have been measured. In Fig. 23.3 the experimental neutron pair separation energies are plotted as a function of N. There are no breaks in the smooth curve of separation energies. There are, however, deviations from linearity and the continuous line is not just the linear function of N as in (23.22). A term proportional to N^2 with a rather small coefficient has been added. There are several possibilities for the origin of such a term in (23.22) or a cubic N^3 term in (23.20). As noted already in Section 20, linear and quadratic terms in the binding energy formula arise from single nucleon terms and two-body interactions in the seniority scheme (or in generalized seniority). As explained there, the core, the doubly magic nucleus with $N = 0$, may be *polarized* by the valence nucleons and its contribution to binding energies may be a function of N. Such polarization effects may contribute to the linear and quadratic terms in (23.20) and also give rise to a cubic term. This polarization may also modify the single nucleon energies. Along with changes in the linear term this may give rise to quadratic and possibly to cubic terms in N.

Polarization effects as described above give rise to small three-body terms in the effective interaction which are independent of the state considered. It is worthwhile to mention that even if there are more general three-body effective interactions their contribution may still be proportional to N^3 (in addition to linear and quadratic terms). If (23.2) are eigenstates of such a shell model Hamiltonian, we derive in addition to (23.5) instead of (23.7) the condition

$$[[H,S^+],S^+]|0\rangle = W(S^+)|0\rangle \tag{23.37}$$

From the case with $N = 3$ we derive

$$[[[H,S^+],S^+],S^+] = U(S^+)^3 \tag{23.38}$$

while higher commutators vanish. In view of the lemma (23.8) we obtain the eigenvalues of such Hamiltonians to be given not by (23.9) but by

$$NV_0 + \tfrac{1}{2}N(N-1)W + \tfrac{1}{6}N(N-1)(N-2)U \tag{23.39}$$

In the case of Fig. 23.3 the best fit to experimental energies is obtained for the following values

$$V_0 = 24.68 \qquad W = -.295 \qquad U = .0721 \text{ MeV}$$

In practice, we find that three-body effective interactions are small and they will be ignored in the following. The values of V_0, W and U listed above were determined from binding energies which have been taken as *positive* quantities. Actually, the coefficient W is repulsive whereas V_0 which includes contributions of single nucleon energies is attractive.

As we saw, shell model Hamiltonians satisfying the conditions (23.5) and (23.7) have special properties. The states (23.2) are eigenstates of such Hamiltonians and the eigenvalues (23.9) are exact linear and quadratic functions of N. The states (23.2) thus share some of the properties of the ground states with seniority $v = 0$ in even j^n configurations or in the quasi-spin scheme. The conditions (23.5) and (23.7) impose definite relations between matrix elements of the shell model Hamiltonian and we will outline here some of them.

We recall the general form (17.33) of a two-body interaction. In order to evaluate its commutator with S^+ we recall the commutation relations (19.22) which, in the case considered, lead to

$$\begin{aligned}[A(j_1 j_2 JM), S_j^+] &= \sqrt{\frac{2j+1}{2}}\delta_{j_1 j}\delta_{j_2 j}\delta_{J0} \\ &- \frac{\delta_{j_1 j}}{\sqrt{1+\delta_{jj_2}}}\sum_{m_2 m}(-1)^{j-m}(jmj_2m_2 \mid jj_2JM)a^{\dagger}_{j,-m}a_{j_2m_2} \\ &+ \frac{\delta_{j_2 j}}{\sqrt{1+\delta_{jj_1}}}\sum_{m_1 m}(-1)^{j-m}(j_1m_1jm \mid j_1jJM)a^{\dagger}_{j,-m}a_{j_1m_1}\end{aligned} \tag{23.40}$$

We now consider the condition (23.7) and focus our attention on the term $W\alpha_j^2(S_j^+)^2$ on the r.h.s. Taking (23.40) into account we see that it must be equal to the double commutator of H with the $\alpha_j S_j^+$ term in each of the S^+ operators. Thus we obtain

$$\boxed{\alpha_j^2[[H, S_j^+], S_j^+] = W\alpha_j^2(S_j^+)^2} \tag{23.41}$$

for each j whose coefficient α_j does not vanish. As discussed above, see (23.15) and (23.16), we deduce from (23.41) that, within the j^n configuration, the $s = 2, \sigma = 0$ part of H is proportional to $3(S_j^0)^2 - \mathbf{S}_j^2$. Apart from a constant, the $s = 1, \sigma = 0$ part is, as we saw in Section 20, proportional to S_j^0. Hence, the Hamiltonian satisfying the conditions (23.5) and (23.7) is diagonal in the seniority scheme within each j^n configuration for which $\alpha_j \neq 0$. Moreover, we conclude from (23.41) that the W_j, the coefficients of the quadratic terms in the interaction energies of the $J = 0$, $v = 0$ states of j^n configurations, must all be equal, $W_j = W$.

This result has an important consequence. It can be used to show how the conditions (23.5) and (23.7) lead in cases of physical interest to a *repulsive* W coefficient. Consider a shell model Hamiltonian H whose submatrices within each j^n configuration are diagonal in the seniority scheme and the coefficients of the quadratic terms are all equal, $W_j = W$. Such Hamiltonians have non-diagonal matrix elements between different configurations and, unlike (23.23), include interactions between particles in different orbits. The eigenvalues for $v = 0$ states in each j^n configuration are given by (20.14) or (23.9) as

$$V_j N_j + \tfrac{1}{2} N_j (N_j - 1) W$$

where V_j includes, unlike in (23.23), also single nucleon energies. In the presence of configuration mixing it is not possible to calculate directly the lowest eigenvalues of H but in the special case of a closed major shell the calculation is straightforward.

The state in which all j-orbits are completely filled is $\prod_j (S_j^+)^{\Omega_j}|0\rangle$. The eigenvalue of the Hamiltonian considered here for this state is

$$\sum_j V_j \Omega_j + \frac{W}{2} \sum_j \Omega_j(\Omega_j - 1) + \sum_{j<j'} 4\bar{V}_{jj'} \Omega_j \Omega_{j'} \qquad (23.42)$$

The last term is due to interactions between particles in different orbits. Its simple form is given in Section 14 where interactions of identical particles in closed orbits were calculated. The average interaction $\bar{V}_{jj'}$ is accordingly defined by (14.35) as

$$\bar{V}_{jj'} = \frac{1}{(2j+1)(2j'+1)} \sum_{|j-j'|}^{j+j'} (2J+1) V(jj'J)$$

Only diagonal matrix elements appear in (23.42) since there is only *one* state of the fully closed major shell. If the Hamiltonian satisfies conditions (23.5) and (23.7), the same eigenvalue is equal, according to (23.9) to

$$V_0 \sum_j \Omega_j + \frac{W}{2} \left(\sum_j \Omega_j \right) \left(\sum_j \Omega_j - 1 \right) \qquad (23.43)$$

The effects of non-diagonal elements show up in this expression through V_0 whereas they are absent from (23.42). By putting (23.43) equal to (23.42) we obtain the following equation

$$\boxed{\sum_j \Omega_j(V_j - V_0) = \sum_{j<j'} \Omega_j \Omega_{j'}(W - 4\bar{V}_{jj'})} \tag{23.44}$$

In any non-trivial case, V_0 is lower than any of the V_j due to mixing of two-particle configurations. Thus, the l.h.s. of (23.44) is positive which implies that W should be more repulsive than some average of the $\bar{V}_{jj'}$ interactions multiplied by 4. In the case of SDI it was shown in Section 22 that $W = 0$. This is in agreement with the fact that for SDI, the average interactions $\bar{V}_{jj'}$ are attractive. This feature of SDI is in contradiction with experimental data. We find, as discussed in Section 31, that the average interaction $\bar{V}_{jj'}$ between identical nucleons in different orbits is repulsive. Hence, their average contribution to the r.h.s. of (23.44) is negative. This contribution should then be more than compensated by a larger repulsive contribution of the W coefficient. It is worth while to point out that adding a constant term W_0 to SDI does not change the situation. Such a term has equal contributions to (23.42) and (23.43), $\frac{1}{2}(\sum \Omega_j)(\sum \Omega_j - 1)W_0$, and hence does not change the condition (23.44).

It was pointed out in Section 20 that, within j^n configurations, the coefficient W_j of the quadratic term may have an arbitrary value. It can be repulsive, attractive or zero. In the case of generalized seniority, the situation is rather different. The states (23.2), $(S^+)^N|0\rangle$ cannot be the *lowest* eigenstates unless $\sum_{jj'} \Omega_j \Omega_{j'}(W - 4\bar{V}_{jj'}) > 0$. In actual nuclei, as mentioned above, and as discussed in Section 31, the average interaction $\bar{V}_{jj'}$ between identical nucleons in different orbits is repulsive. Hence, the agreement between predictions of generalized seniority and experiment determines the repulsive nature of W.

Other conditions that follow from (23.5) and (23.7) relate matrix elements between different configurations. Their derivation (Talmi 1971, and Shlomo and Talmi 1972) will not be given here but we list them, for the sake of completeness, as follows

$$2(\alpha_j^2 + \alpha_{j'}^2)\frac{V(j^2 j'^2 J = 0)}{\sqrt{(2j+1)(2j'+1)}} - \alpha_j \alpha_{j'} \frac{4}{(2j+1)(2j'+1)}$$
$$\times \sum_{J'} (2J'+1) V(jj'jj'J') + \alpha_j \alpha_{j'} W = 0 \tag{23.45}$$

$$(\alpha_j^2 + \alpha_{j'}^2)V(j^2 j'^2 J) - 2\alpha_j\alpha_{j'} \times \sum_{J'} (2J'+1)(-1)^{j+j'-J'} \begin{Bmatrix} j & j' & J' \\ j' & j & J \end{Bmatrix} V(jj'jj'J) = 0 \qquad J > 0 \text{ even} \tag{23.46}$$

Another condition, independent of the actual values of the α_j coefficients, is

$$\alpha_j(\alpha_j + \alpha_{j'}) \left[V(j^2 jj'J) + 2 \sum_{J' \text{ even}} (2J'+1) \begin{Bmatrix} j & j & J' \\ j & j' & J \end{Bmatrix} V(j^2 jj'J') \right] = 0 \qquad J \text{ even} \tag{23.47}$$

Other, more complicated conditions are

$$(\alpha_{j_1}^2 + \alpha_{j_2}\alpha_{j_3})V(j_1 j_1 j_2 j_3 J) + \sqrt{2}\alpha_{j_1}(\alpha_{j_2} + \alpha_{j_3}) \times \sum_{J'} (2J'+1) \begin{Bmatrix} j_1 & j_1 & J \\ j_2 & j_3 & J' \end{Bmatrix} V(j_1 j_3 j_2 j_1 J') = 0 \tag{23.48}$$

$$\begin{aligned}(\alpha_{j_1}\alpha_{j_2} + \alpha_{j_3}\alpha_{j_4})V(j_1 j_2 j_3 j_4 J) &+ (\alpha_{j_1}\alpha_{j_4} + \alpha_{j_2}\alpha_{j_3}) \times \sum_{J'} (2J'+1) \begin{Bmatrix} j_1 & j_2 & J \\ j_3 & j_4 & J' \end{Bmatrix} V(j_1 j_4 j_3 j_2 J') \\ &- (-1)^{j_3+j_4-J}(\alpha_{j_1}\alpha_{j_3} + \alpha_{j_2}\alpha_{j_4}) \sum_{J'} (2J'+1) \\ &\times \begin{Bmatrix} j_1 & j_2 & J \\ j_4 & j_3 & J' \end{Bmatrix} V(j_1 j_3 j_4 j_2 J') = 0\end{aligned} \tag{23.49}$$

If the parities of all j-orbits are not equal, some of the matrix elements in these conditions must vanish. For instance, if the j- and j'-orbits have different parities in (23.47) that condition is meaningless. If we make all coefficients α_j equal we obtain sufficient conditions for the Hamiltonian to be diagonal in the quasi-spin scheme.

The condition (23.45) may be expressed as

$$\frac{2(\alpha_j^2 + \alpha_{j'}^2)}{\sqrt{(2j+1)(2j'+1)}} \alpha_j\alpha_{j'} V(j^2 j'^2 J = 0) = \alpha_j^2 \alpha_{j'}^2 (4\bar{V}_{jj'} - W)$$

If only α_j and $\alpha_{j'}$ in the operator S^+ do not vanish, the sign of $\alpha_j\alpha_{j'}V(j^2j'^2J=0)$ is always negative. In that case, (23.45) expresses the requirement that W should be more repulsive than $4\bar{V}_{jj'}$. In the case of the pairing interaction and degenerate single nucleon energies, $\alpha_j^2=\alpha_{j'}^2$, $\bar{V}_{jj'}=0$, $V(j^2j'^2J=0)\alpha_j\alpha_{j'}=-G\sqrt{(2j+1)(2j'+1)}$ and we obtain from (23.45) the value $W=-4G$.

Looking at equations (23.45) to (23.49) we see that the simple looking conditions (23.5) and (23.7) impose many complicated relations between matrix elements of the effective interaction. These relations involve not only matrix elements of the two body interaction. They depend explicitly on the α_j coefficients which are determined by the solution of the two particle problem. That solution depends not only on the two-body matrix elements but also on single nucleon energies. The question now arises what is the physical reason that shell model Hamiltonians seem to obey these rather complex conditions. One necessary condition guarantees that the Hamiltonian sub matrices defined by the various j^n configurations are diagonal in the seniority scheme. As we saw in Section 21, this seems to be a general property of the effective two body interaction. The other conditions in which also single nucleon energies appear implicitly (through the α_j coefficients) cannot have a general characterization. Although single nucleon energies are also determined by the two-body nuclear interaction, there is no simple, or even clear way, to derive them from the interaction between free nucleons.

The conditions (23.5) and (23.7) lead to a set of eigenvalues of the Hamiltonian which are exact linear and quadratic functions of N. Is the inverse statement correct? If the lowest eigenvalues of the nuclear Hamiltonian are given by (23.9) does it follow that the corresponding eigenstates must have the form (23.2)? We shall now prove that if the eigenstate of the two nucleon system with eigenvalue V_0 is $\mathcal{S}^+|0\rangle$ then this is indeed the case. This applies also to the special case where only *one* of the α_j coefficients is different from zero. We prove this statement by using as variational trial functions the states $(\mathcal{S}^+)^N|0\rangle$. Using the lemma (23.8) we obtain

$$\begin{aligned}H(\mathcal{S}^+)^N|0\rangle &= N(\mathcal{S}^+)^{N-1}[H,\mathcal{S}^+]|0\rangle \\ &\quad + \tfrac{1}{2}N(N-1)(\mathcal{S}^+)^{N-2}[[H,\mathcal{S}^+],\mathcal{S}^+]|0\rangle \\ &= NV_0(\mathcal{S}^+)^N|0\rangle + \tfrac{1}{2}N(N-1)(\mathcal{S}^+)^{N-2}[[H,\mathcal{S}^+],\mathcal{S}^+]|0\rangle\end{aligned} \tag{23.50}$$

We can express the double commutator by

$$[[H,\mathcal{S}^+],\mathcal{S}^+] = \lambda(\mathcal{S}^+)^2 + B^+ \tag{23.51}$$

where B^+ creates a four particle state orthogonal to $(\mathcal{S}^+)^2|0\rangle$

$$\langle 0|\mathcal{S}^2 B^+|0\rangle = 0 \tag{23.52}$$

Substituting from (23.51) into (23.50) we obtain

$$H(\mathcal{S}^+)^N|0\rangle = NV_0(\mathcal{S}^+)^N|0\rangle + \tfrac{1}{2}N(N-1)\lambda(\mathcal{S}^+)^N|0\rangle \\ + \tfrac{1}{2}N(N-1)(\mathcal{S}^+)^{N-2}B^+|0\rangle \tag{23.53}$$

If we take the overlap of (23.51) with $(\mathcal{S}^+)^N|0\rangle$ we obtain

$$\langle 0|\mathcal{S}^N H(\mathcal{S}^+)^N|0\rangle = (NV_0 + \tfrac{1}{2}N(N-1)\lambda)\langle 0|\mathcal{S}^N(\mathcal{S}^+)^N|0\rangle \tag{23.54}$$

The vanishing of $\langle 0|\mathcal{S}^N(\mathcal{S}^+)^{N-2}B^+|0\rangle$ follows from (23.52) due to the properties of $\mathcal{S}^+$ as a quasi-spin component. We now look at the only state of the completely closed major shell. This state is given by $(\mathcal{S}^+)^{\Omega}|0\rangle$ where $\Omega = \sum_j \Omega_j$. In fact, the only term in the expansion (23.21) for $N = \Omega$ is the one with $\nu_1 = \Omega_{j_1}, \nu_2 = \Omega_{j_2}, \ldots$. Thus, according to (23.9)

$$H(\mathcal{S}^+)^{\Omega}|0\rangle = (\Omega V_0 + \tfrac{1}{2}\Omega(\Omega-1)W)(\mathcal{S}^+)^{\Omega}|0\rangle \tag{23.55}$$

and the overlap of (23.55) with $(\mathcal{S}^+)^{\Omega}$ is given by

$$\langle 0|\mathcal{S}^{\Omega}H(\mathcal{S}^+)^{\Omega}|0\rangle = (\Omega V_0 + \tfrac{1}{2}\Omega(\Omega-1)W)\langle 0|\mathcal{S}^{\Omega}(\mathcal{S}^+)^{\Omega}|0\rangle \tag{23.56}$$

Comparing (23.56) with (23.54) for $N = \Omega$ we conclude that $\lambda = W$. Hence, for any value of N, the expectation value of H in the state $(\mathcal{S}^+)^N|0\rangle$ is equal to the lowest eigenvalue. These states are thus the *exact* eigenstates of H and conditions (23.5) and (23.7) are satisfied ($B^+ = 0$).

If the α_j coefficients are not equal, the argument does not apply. The vanishing of $\langle 0|\mathcal{S}^N(\mathcal{S}^+)^{N-2}B^+|0\rangle$ cannot be deduced from the vanishing of (23.52). Indeed, it is not true that if the lowest eigenvalues are linear and quadratic functions of N the form (23.2) follows for the eigenstates. Counter examples have been constructed (Talmi

1971). Still, if the lowest eigenvalues are those in (23.9), very many conditions are imposed on the effective nuclear Hamiltonian. In the case of generalized seniority they can all be neatly expressed by (23.5) and (23.7). If the latter conditions do not hold, another set of conditions must be fulfilled. The virtue of (23.5) and (23.7) is that they lead to a simple structure of ground states expressed by (23.2). That structure is based on nucleon pairing and is a direct generalization of the seniority scheme. If a phenomenological interaction is sought which gives rise to the observed behavior of binding energies, it may well fulfill the conditions (23.5) and (23.7) and have eigenstates (23.2). In fact, this was the case for the effective interaction fitted to Ni isotopes. The Hamiltonian which was deduced from experiment has indeed, as was found much later, ground states of the form (23.2).

It was remarked above that there is only one $J = 0$ state which may be assigned generalized seniority zero. It is not possible, in general, to define a complete scheme, like the quasi-spin scheme, for other $J = 0$ states. In the next section we shall see how states with $J > 0$ may be constructed which correspond to $v = 2$ states in the quasi-spin scheme.

24

Generalized Seniority. Other States of Semi-Magic Nuclei

In Section 23 it was pointed out that if the coefficients in the pair creation operator S^+ are unequal it is not possible to define a complete scheme of generalized seniority. Unlike the case of the quasi-spin scheme, there is no set of $J = 0$ states with generalized seniority $v = 2$ (or $v = 4, \ldots$). Among all $J = 0$ states of the system of identical valence nucleons, it is possible to pick out *one* state which shares important features with the seniority $v = 0$ state in the quasi-spin scheme. That state, with generalized seniority $v = 0$, has been the subject of the preceding section. Here, we shall consider states of such two nucleon systems with $J = 2, 4, \ldots$. It turns out that it is possible to define *one* state for each spin which is the analog of a $v = 2$ state in the quasi-spin scheme. These states may be assigned generalized seniority $v = 2$. In the following, we focus our attention on $J = 2$ states which are better known and are of greater interest. Similar results may be obtained also for $J = 4$ and higher even values of J.

We start by considering the operator which creates a pair of identical nucleons with $J = 2$ in the jj' configuration. A normalized $J = 2$

state is created by

$$\boxed{\begin{aligned} D^+_M(jj') &= A^+(jj'J = 2, M) \\ &= (1 + \delta_{jj'})^{-1/2} \sum_{mm'} (jmj'm' \mid jj'2M) a^+_{jm} a^+_{j'm'} \end{aligned}} \tag{24.1}$$

The lowest $J = 2$ state of two identical nucleons occupying a major shell is a coherent admixture of such states (24.1). The lowest eigenstate of a shell model Hamiltonian H may thus be expressed as

$$D^+_M = \sum_{j \le j'} \beta_{jj'} D^+_M(jj') \tag{24.2}$$

The amplitudes $\beta_{jj'}$ in (24.2) are determined by the eigenvalue equation

$$HD^+_M|0\rangle = V_2 D^+_M|0\rangle \tag{24.3}$$

In the quasi-spin scheme, states with seniority $v = 2$, are obtained from any two nucleon state with $v = 2$ by applying to it the operator $(\mathcal{S}^+)^{N-1}$. If we replace $\mathcal{S}^+$ by the general operator S^+ this is no longer the case. We can still start from the eigenstate (24.3) and apply to it $(S^+)^{N-1}$. Such states will be useful and may deserve to be assigned generalized seniority $v = 2$ if they turn out to be eigenstates of the shell model Hamiltonian. Let us therefore examine the conditions that this indeed is the case. We shall then learn the properties of such states and their eigenvalues.

We consider shell model Hamiltonians which satisfy, in addition to (24.3), also

$$\boxed{H(S^+)^{N-1} D^+_M|0\rangle = E^{(2)}_N (S^+)^{N-1} D^+_M|0\rangle} \tag{24.4}$$

The shell model Hamiltonians which we consider here satisfy the conditions (23.5) and (23.7). Therefore, we use the lemma (23.8) and express the l.h.s. of (24.4) by

$$H(S^+)^{N-1}D^+_M|0\rangle = (S^+)^{N-1}HD^+_M|0\rangle + (N-1)(S^+)^{N-2}[H,S^+]D^+_M|0\rangle + \tfrac{1}{2}(N-1)(N-2)(S^+)^{N-3}[[H,S^+],S^+]D^+_M|0\rangle \tag{24.5}$$

Using (23.5) and (23.7) as well as (24.3) we obtain for (24.5) the expression

$$V_2(S^+)^{N-1}D^+_M|0\rangle + (N-1)V_0(S^+)^{N-1}D^+_M|0\rangle + (N-1)(S^+)^{N-2}[[H,S^+],D^+_M]|0\rangle + \tfrac{1}{2}(N-1)(N-2)W(S^+)^{N-1}D^+_M|0\rangle \tag{24.6}$$

Comparing (24.4) with (24.6) we see that the states $(S^+)^{N-1}D^+_M|0\rangle$ are eigenstates provided the following condition is satisfied

$$[[H,S^+],D^+_M] = \lambda S^+D^+_M \tag{24.7}$$

From (24.6) follows that (24.7) should hold when both sides act on the vacuum state. If, however, the shell model Hamiltonian contains only single nucleon terms and two body interactions, the double commutator in (24.7) contains only products of four creation operators. Hence, if (24.7) holds when acting on the vacuum state, it holds when applied to any state and may be written as an operator equation (like the condition (23.7)).

The actual value of λ is determined by the condition (23.7) to be $\lambda = W$. To see it, we recall that if (23.7) is satisfied, the submatrix of the Hamiltonian within any j^n configuration is diagonal in the seniority scheme. We also saw that the coefficients of the quadratic terms in the various j-orbits are all equal, $W_j = W$. The eigenvalue of that submatrix in the state $S^+_j D^+_M(j^2)|0\rangle$ is thus equal, according to (20.14) and (23.10), to $4\epsilon_j + V(j^2J=0) + V(j^2J=2) + W$. Note that in the case of a single j-orbit, $S^+ = S^+_j, V_0$ includes also two single nucleon energies in addition to the two nucleon interaction, $V_0 = 2\epsilon_j + V(j^2J=0)$. We look now at the term $\lambda\alpha_j\beta_{jj}S^+_jD^+_M(j^2)$ on the r.h.s. of (24.7). The only way it can be obtained on the l.h.s. is by commuting H with $\alpha_j S_j^+$ and $\beta_{jj} D_M^+(j^2)$. We can thus write

$$\alpha_j \beta_{jj} [[H, S_j^+], D_M^+(j^2)] = \alpha_j \beta_{jj} \lambda S_j^+ D_M^+(j^2) \tag{24.8}$$

Using the relation (24.8) in (24.6) with $N = 2$, we obtain the eigenvalue of the submatrix which belongs to the state $S_j^+ D_M^+(j^2)|0\rangle$ equal to $4\epsilon_j + V(j^2 J = 0) + V(j^2 J = 2) + \lambda$. Comparing this eigenvalue to the one obtained above for that state, we conclude that $\lambda = W$ and write condition (24.7) as

$$\boxed{[[H, S^+], D_M^+] = W S^+ D_M^+} \tag{24.9}$$

In conclusion, we find that if the shell model Hamiltonian satisfies along with (23.5) and (23.7) also (24.3) and (24.9), the states in (24.4) are eigenstates. The corresponding eigenvalues are given according to (24.6) and (24.9) by the expression

$$\boxed{E_N^{(2)} = (N-1)V_0 + V_2 + \tfrac{1}{2}N(N-1)W = E_N + V_2 - V_0} \tag{24.10}$$

where E_N, the eigenvalue of the ground state, is given by (23.9). The result (24.10) means that

the spacing between $J = 0$ ground states and
first excited $J = 2$ states is independent of N

This remarkable result is the same as in the quasi-spin scheme. There, as already mentioned in the case of single j-orbit, all level spacings are independent of the particle number. Here this is no longer the case but this important feature is still shared by a special set of $J = 2$ states. These $J = 2$ states may then be assigned generalized seniority $v = 2$ (Talmi 1971).

In Section 21 we saw how experimental level spacings in j^n configurations in semi-magic nuceli are fairly independent of nucleon number. This feature was a clear signature of the seniority scheme. In that case, as well as in the quasi-spin scheme, this is due to the quasi-spin

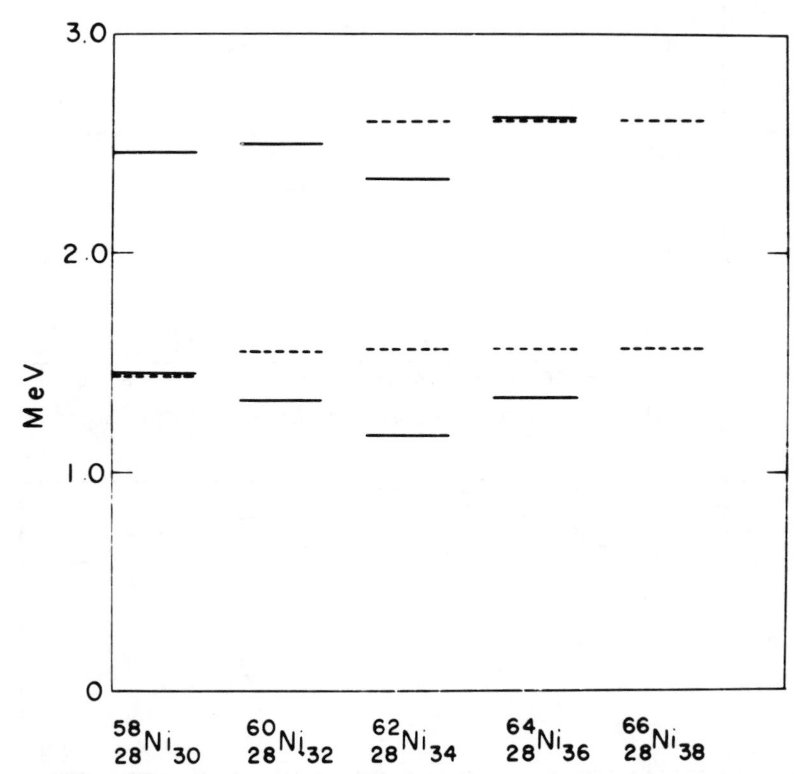

FIGURE 24.1. *Experimental and calculated $J = 2$ and $J = 4$ levels of even* Ni *isotopes.*

scalar term in the Hamiltonian. The other terms are proportional to n and n^2 and do not affect level spacings. Such group theoretical notions are no longer applicable to the case of generalized seniority. Still, this feature seems to survive the generalization for a special set of $J = 2$ states. We shall now see whether the lowest $J = 2$ states in semi-magic nuclei are adequately described by (24.4). As mentioned above, there are no reliable shell model calculations in most interesting cases. What we can do, however, is to compare the prediction (24.10) with experiment.

In Fig. 24.1 the experimental positions of $J = 2^+$ and $J = 4^+$ levels in Ni isotopes (solid lines) are plotted as a function of neutron number. They are at fairly constant height above the $J = 0$ ground states. This is also the case for the level positions (dashed lines) calculated

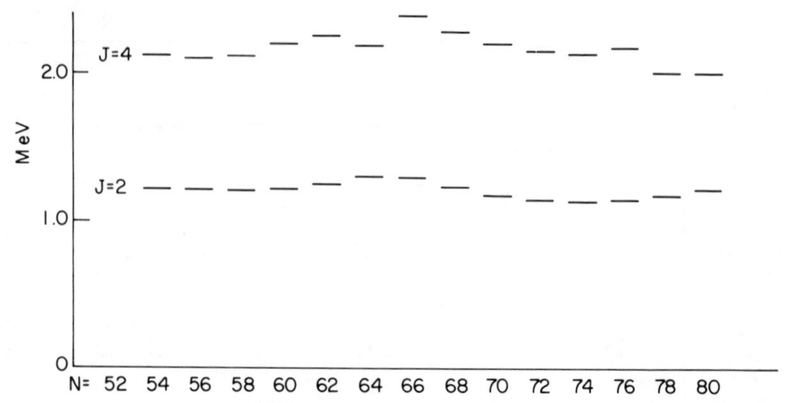

FIGURE 24.2. *Experimental $J = 2$ and $J = 4$ levels of even* Sn *isotopes.*

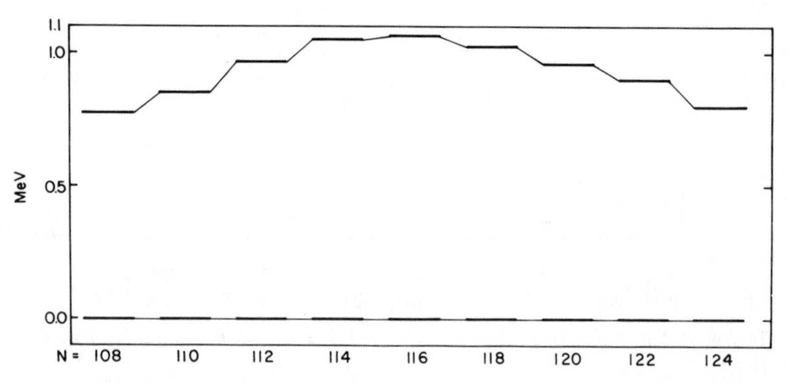

FIGURE 24.3. *Experimental $J = 2$ levels of even* Pb *isotopes.*

from a shell model Hamiltonian fitted to yield the experimental levels (Auerbach 1966, Cohen et al. 1967). A more impressive group of nuclei are the Sn isotopes. No comprehensive shell model calculations have been carried out for them. The experimentally observed 0–2 spacings in even Sn isotopes, plotted in Fig. 24.2 are fairly constant throughout the complete shell from neutron number 52 to 80. Like the case of binding energies discussed in Section 23, no breaks in these 0–2 spacings corresponding to subshell closures may be detected. For other major shells no complete information is available.

In Fig. 24.3, positions of $J = 2^+$ levels in Pb nuclei are plotted. The neutron major shell in those nuclei is between 82 and 126. It comprises the $1h_{9/2}$, $2f_{7/2}$, $2f_{5/2}$, $3p_{3/2}$, $3p_{1/2}$ and $1i_{13/2}$ orbits. Also these 0–2 spacings are fairly constant as a function of neutron number. An-

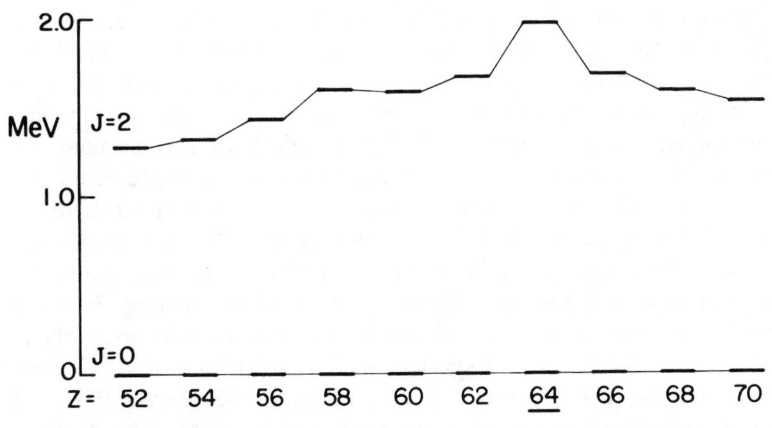

FIGURE 24.4. *Experimental* $J = 2$ *levels of even* $N = 82$ *nuclei.*

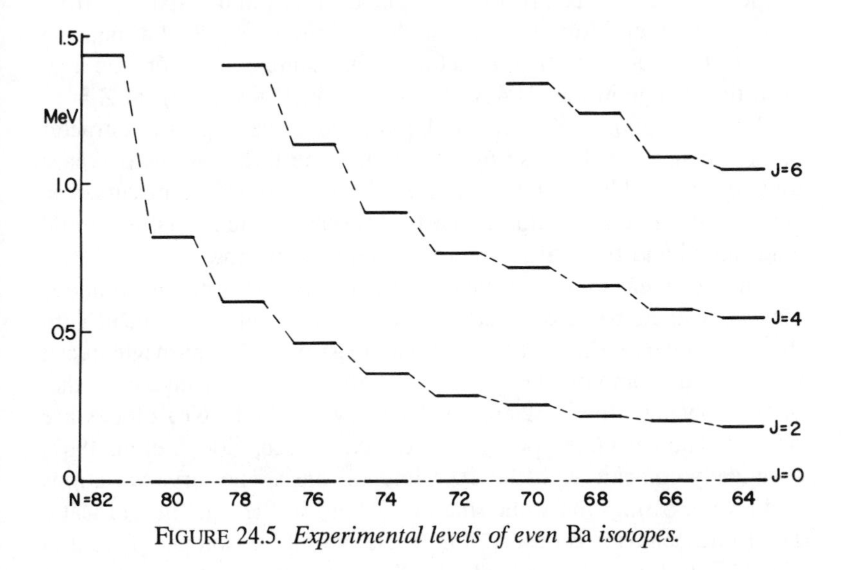

FIGURE 24.5. *Experimental levels of even* Ba *isotopes.*

other example is offered by nuclei (isotones) with neutron magic number 82. There is a definite trend for the 0–2 spacings plotted in Fig. 24.4 to increase as a function of proton number. There is also a clear, though not very large, discontinuity at $Z = 64$. This should be associated with the quasi-magic nature of $^{146}_{64}Gd_{82}$ discussed in Section 21.

The fair constancy of 0–2 spacings is presented here as evidence for the structure (24.4) of the $J = 2$ wave functions. Clearly the 0–2 spacings in actual semi-magic nuclei are not exactly equal. As emphasized on several occasions, nuclei are very complex physical systems. This is true not only about their composition but also about the mutual strong interactions between their building blocks. Any simple model cannot be expected to agree exactly with experiment. We should find, however, criteria to decide whether any predicted feature is indeed satisfactorily observed. In the case considered here, we would like to decide whether 0–2 spacings in semi-magic nuclei may be described as constant. We can compare the behavior of these spacings in semi-magic nuclei with the behavior of 0–2 spacings in nuclei with both valence protons and valence neutrons outside closed shells. We can compare the 0–2 spacings in Sn isotopes ($Z = 50$) to those with $Z = 56$. The experimental positions of $J = 2^+$ levels above the ground states of Ba isotopes are plotted in Fig. 24.5. The dramatic decrease of the 0–2 spacing as the neutron number changes from the magic number 82 is clearly evident. The decrease in 0–2 spacings from neutron number 82 to about the middle of the major shell at neutron number 64 is by more than a factor 8. This is much more drastic than even the change in Fig. 24.4 where 0–2 spacings increase from $Z = 52$ to $Z = 62$ by about 35%. Also to be noticed is the opposite character of the change. In the case of Ba isotopes the 0–2 spacings decrease towards the middle of the major shell. This trend will be discussed in detail in subsequent sections. Apart from semi-magic nuclei, this trend is universal and there are numerous examples to show it.

Another measure of constancy of the 0–2 spacings discussed above may be gained by looking at level spacings of other of nuclei with the same valence shell. Very good data exist in odd semi-magic nuclei and we can compare the behavior of their level spacings with that of 0–2 spacings. In Fig. 24.6, low-lying levels of odd Ni isotopes are plotted. The calculated positions (Auerbach 1966, Cohen et al. 1967) practically coincide with the experimental ones. The large variations and level crossings make the small variations in Fig. 24.1 insignificant. The same behavior is observed in levels of odd Sn isotopes plotted in Fig. 24.7. In Fig. 24.8, levels of odd Pb isotopes are plotted and in Fig. 24.9, those of odd nuclei with neutron number 82. In all these plots the large variations in spacings and order of levels are evident. In view of this it is justified to consider the 0–2 spacings in even semi-magic nuclei as fairly constant.

An important conclusion can be drawn from the behavior of level spacings in odd semi-magic nuclei. The quasi-spin scheme is a very poor approximation to nuclear states. Had all α_j coefficients been

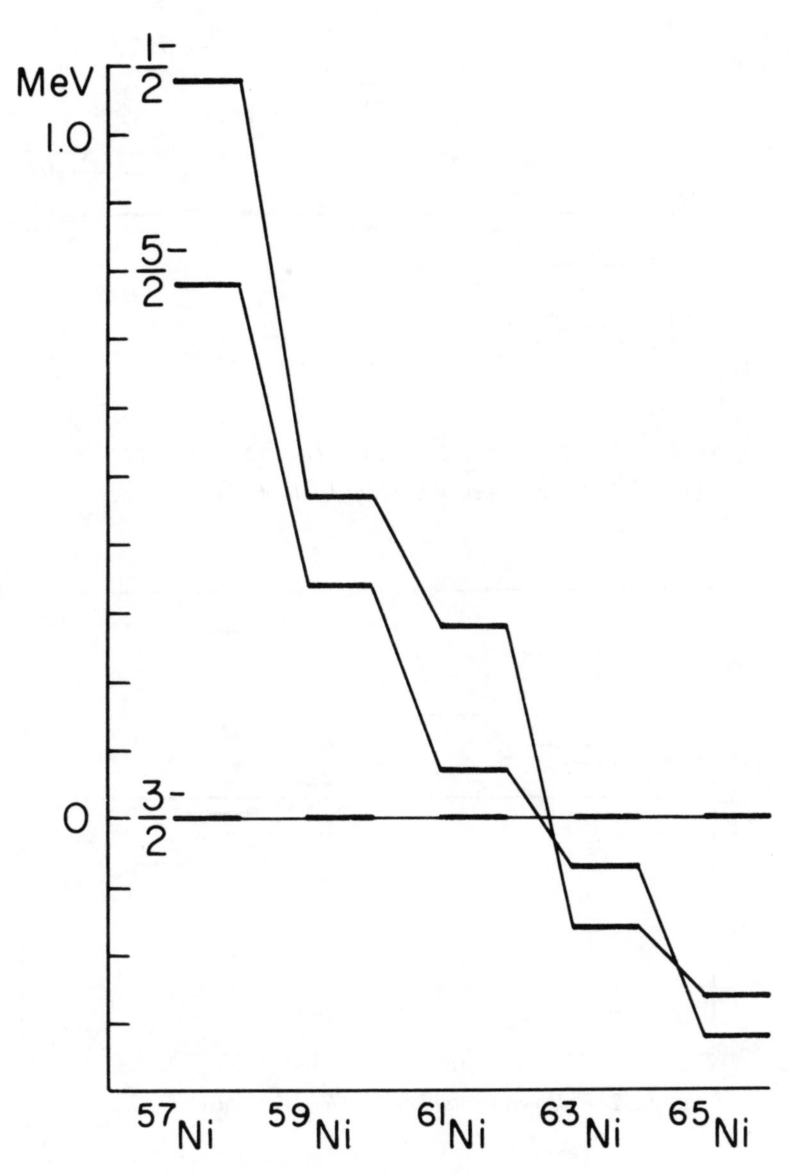

FIGURE 24.6. *Low lying levels of odd* Ni *isotopes.*

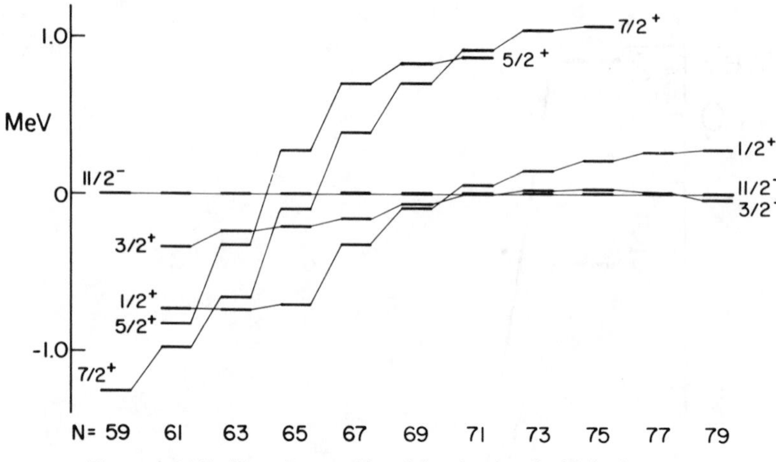

FIGURE 24.7. *Experimental low lying levels of odd* Sn *isotopes.*

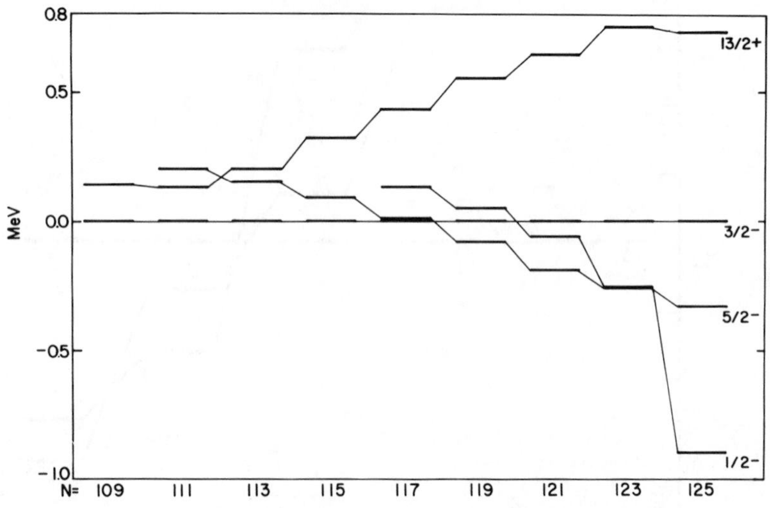

FIGURE 24.8. *Experimental low lying levels of odd* Pb *isotopes.*

equal, all level spacings, in even as well as in odd semi-magic nuclei, would have been constant, independent of nucleon number. The large variations in level spacings in odd Ni isotopes are consistent with the α_j in (23.36) being unequal. They are in good agreement with detailed shell model calculations where those are available. In fact, in

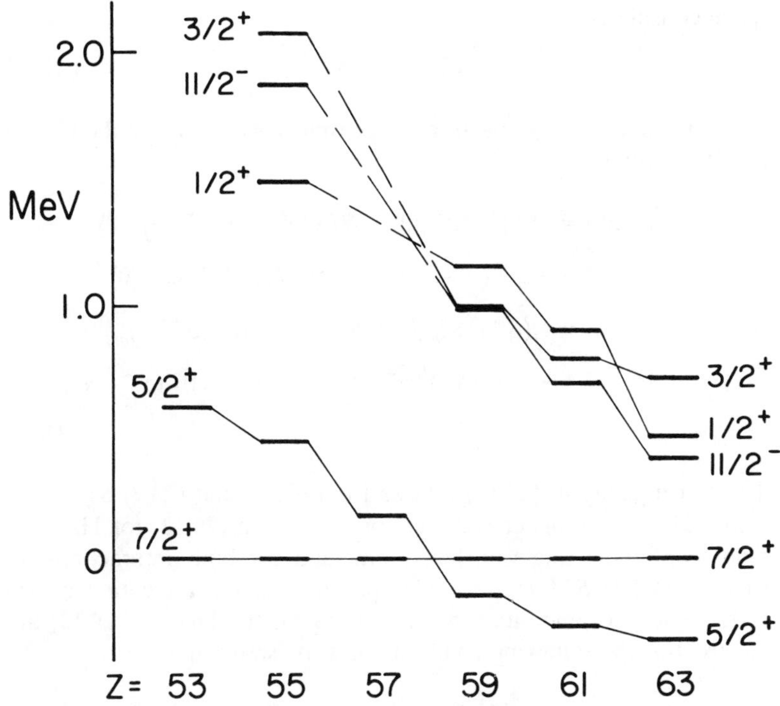

FIGURE 24.9. *Experimental low lying level of odd nuclei with $N = 82$.*

Fig. 24.6, no distinction was made between experimental level spacings and those calculated from a shell model Hamiltonian (Auerbach 1966, Cohen et al. 1967). The agreement between calculated and experimental values is very good indeed. The variation of level spacings in odd nuclei is thus a strong argument against equal α_j coefficients.

In the quasi-spin scheme, some states in odd nuclei, where the spin is equal to that of a single nucleon, have seniority $v = 1$. As we saw in Section 22, their energies need not be degenerate. The spacings of such levels are constant if the Hamiltonian is diagonal in the quasi-spin scheme. When we go over to pair creation operators S^+ with unequal α_j coefficients, it seems that such states cannot be assigned generalized seniority $v = 1$. This is indeed the case. We will prove now that if the states $(S^+)^N a^+_{jm}|0\rangle$ are eigenstates of the shell model Hamiltonian, all α_j coefficients must be equal.

Let us consider the set of single nucleon states $a^+_{jm}|0\rangle$. They are all eigenstates of the submatrix of the Hamiltonian in the shell model space considered

$$Ha^{+}_{jm}|0\rangle = \epsilon_j a^{+}_{jm}|0\rangle \tag{24.11}$$

If also the states $(S^+)^N a^{+}_{jm}|0\rangle$ are eigenstates, we use again the lemma (23.8) to obtain

$$\begin{aligned}H(S^+)^N a^{+}_{jm}|0\rangle &= (S^+)^N Ha^{+}_{jm}|0\rangle + N(S^+)^{N-1}[H,S^+]a^{+}_{jm}|0\rangle \\ &\quad + \tfrac{1}{2}N(N-1)(S^+)^{N-2}[[H,S^+],S^+]a^{+}_{jm}|0\rangle \\ &= \epsilon_j(S^+)^N a^{+}_{jm}|0\rangle + N(S^+)^{N-1}[[H,S^+],a^{+}_{jm}]|0\rangle \\ &\quad + NV_0(S^+)^N a^{+}_{jm}|0\rangle + \tfrac{1}{2}N(N-1)W(S^+)^N a^{+}_{jm}|0\rangle\end{aligned} \tag{24.12}$$

The last equality in (24.12) follows from (23.5) and (23.7) as well as from (24.11). No higher commutators appear in (24.12) for Hamiltonians containing single nucleon terms and two-body interactions. In such cases $[[H,S^+],a^{+}_{jm}]$ contains products of three creation operators. Hence, to guarantee that the states on the l.h.s. of (24.12) are eigenstates, the following condition must be satisfied

$$[[H,S^+],a^{+}_{jm}] = \lambda S^+ a^{+}_{jm} \tag{24.13}$$

The term $\lambda\alpha_j S^{+}_{j} a^{+}_{jm}$ on the r.h.s. of (24.13) can be obtained on the l.h.s. only by commuting H with S^{+}_{j} and a^{+}_{jm}. Due to (23.7) the Hamiltonian matrix within j^n configurations is diagonal in the seniority scheme. Its eigenvalue in the $v=1$ state $S^{+}_{j}a^{+}_{jm}|0\rangle$ is given by (20.14) as $3\epsilon_j + 3\alpha + \beta$ which, according to (23.10), is equal to $V(j^2J=0)+$ $3\epsilon_j + \frac{1}{2}W$. Note that in the case of a single j-orbit, $S^+ = S^{+}_{j}, V_0$ includes also two single nucleon energies in addition to the two body interaction, $V_0 = 2\epsilon_j + V(j^2J=0)$. If we use (24.12) and (24.13) to calculate the eigenvalue of the state $S^+a^{+}_{jm}|0\rangle$ we obtain $V(j^2J=0)+3\epsilon_j+\lambda$. We conclude that λ in (24.13) is equal to $W/2$ and the condition (24.13) may be written as

$$[[H,S^+],a^{+}_{jm}] = \tfrac{1}{2}WS^+a^{+}_{jm} \tag{24.14}$$

The condition (24.14) if satisfied, has important consequences. Any eigenstate of H with n nucleons may be constructed as a linear combination of products of n operators a^{+}_{jm}. Let us consider any of these states $B_n^+|0\rangle$

$$HB_n^+|0\rangle = EB_n^+|0\rangle \tag{24.15}$$

If (24.14) is satisfied, also $(S^+)^N B_n^+|0\rangle$ is an eigenstate of H. To see it we apply the lemma (23.8) to that state and obtain as in (24.12)

$$\begin{aligned} H(S^+)^N B_n^+|0\rangle &= (S^+)^N HB_n^+|0\rangle + N(S^+)^{N-1}[H,S^+]B_n^+|0\rangle \\ &\quad + \tfrac{1}{2}N(N-1)(S^+)^{N-2}[[H,S^+],S^+]B_n^+|0\rangle \\ &= E(S^+)^N B_n^+|0\rangle + N(S^+)^{N-1}[[H,S^+],B_n^+]|0\rangle \\ &\quad + NV_0(S^+)^N B_n^+|0\rangle + \tfrac{1}{2}N(N-1)W(S^+)^N B_n^+|0\rangle \end{aligned} \tag{24.16}$$

The evaluation of the double commutator $[[H,S^+],B_n^+]$ can be carried out with the help of (24.14). The commutator of $[H,S^+]$ with every product of n operators a_{jm}^+ can be carried out successively

$$[[H,S^+],a_{jm}^+a_{j'm'}^+\cdots] = [[H,S^+],a_{jm}^+]a_{j'm'}^+\cdots + a_{jm}^+[[H,S^+],a_{j'm'}^+\cdots] \tag{24.17}$$

If the conditions (24.14) are satisfied, then (24.17) becomes proportional to $S^+a_{jm}^+a_{j'm'}^+\ldots$ with the proportionality factor $nW/2$. Hence from (24.14) follows the result

$$[[H,S^+],B_n^+] = \frac{n}{2}WS^+B_n^+ \tag{24.18}$$

Thus, the state on the l.h.s. of (24.16) is an eigenstate and the corresponding eigenvalue is equal to

$$E + NV_0 + \tfrac{1}{2}N(N-1)W + \tfrac{1}{2}nNW \tag{24.19}$$

The eigenvalues (24.19) for the various eigenstates $(S^+)^N B_n^+|0\rangle$ differ only by the various E values. Therefore, their spacings are independent of N. This is precisely the situation in the quasi-spin scheme. We may suspect that the conditions (24.14) for all j-orbits in a major shell may be satisfied only if all α_j coefficients in the S^+ operator are equal. To show that this is indeed the case, we may choose a small subset of the operators B_n^+ with the values $n = 2$ $(N = 1)$. We look at the states $A^{(i)+}_{J=0}|0\rangle$ considered in Section 23 (eq. (23.17)). We proved there that if $S^+A^{(i)+}_{J=0}|0\rangle$ are eigenstates of H, for $i = 1,\ldots,k-1$ then the squares of all α_j coefficients must be equal. Hence, from (24.14) it follows that, apart from a possible change of sign, all α_j coefficients must be equal and the Hamiltonian is diagonal in the quasi-spin scheme.

It was explained above that the coefficients α_j used for the description of actual semi-magic nuclei are not equal. Hence, eigenstates of odd semi-magic nuclei cannot be represented by the states $(S^+)^N a^+_{jm} \mid 0\rangle$. The simple binding energy formula (20.14), which holds also in the quasi-spin scheme, cannot be reliably used for calculating energies of these states. Due to the properties of the interaction between identical nucleons, described in Sections 20 and 31, qualitative features of (20.14) hold also for actual semi-magic nuclei. There is a large odd-even variation in binding energies of these nuclei. This is true if we consider in odd nuclei the lowest states with J equal to one of the single nucleon spins j. To compute separation energies in this case, the excitation energy of the $J = j$ state should be subtracted from the binding energy. This feature holds also if we consider ground states of odd nuclei, irrespective of their spins, the lines connecting the plotted experimental values in each set, however, are not straight and parallel as in Fig. 20.1 and Figs. 21.1 and 21.2. Such lines are not smooth as the line in Fig. 23.3 where its curvature is due to a small cubic term in the binding energy formula (23.39).

The observation made above raises an important question. Are there shell model Hamiltonians that satisfy the conditions (23.5) and (23.7) with S^+ operators whose α_j coefficients are not all equal? Could the stringent conditions (23.44) to (23.48) be satisfied without enforcing equality of all α_j? The answer to this question has been established. There are shell model Hamiltonians that obey exactly the conditions (23.5) and (23.7) (Talmi (1971 and 1975)). The effective shell model Hamiltonian constructed for Ni isotopes by Auerbach (1967) satisfies those conditions approximately. Slight changes in its parameters would make it satisfy conditions (23.5) and (23.7) exactly. A simple example is offered by the Hamiltonian (23.23).

If we define the pair creation operator by $S^+ = S^+_j + \alpha S^+_{j'}$, we obtain the following commutation relations of (23.23) and S^+

$$[H,S^+] = [H_0,S^+_j] + \alpha[H_0,S^+_{j'}] - 2GS^+_{j'}S^0_j - 2\alpha GS^+_j S^0_{j'} \tag{24.20}$$

$$[[H,S^+],S^+] = [[H_0,S_j^+],S_j^+] + \alpha^2[[H_0,S_{j'}^+],S_{j'}^+]$$
$$-2GS_{j'}^+S_j^+ - 2\alpha^2GS_j^+S_{j'}^+ \tag{24.21}$$

When (24.20) is applied to the vacuum state we obtain

$$\begin{aligned}[H,S^+]|0\rangle &= [H_0,S_j^+]|0\rangle + \alpha[H_0,S_{j'}^+]|0\rangle \\ &\quad + GS_{j'}^+\Omega_j|0\rangle + \alpha GS_j^+\Omega_{j'}|0\rangle \\ &= V_jS_j^+|0\rangle + \alpha V_{j'}S_{j'}^+|0\rangle + G\Omega_jS_{j'}^+|0\rangle + \alpha G\Omega_{j'}S_j^+|0\rangle\end{aligned} \tag{24.22}$$

The last equality in (24.22) follows from the fact that H_0 does not admix the j- and j'-orbits. The eigenvalues V_j and $V_{j'}$ include here, unlike in Section 23, also the single nucleon energies. The coefficient α can now be determined from the condition (23.5) which states that $S^+|0\rangle$ is an eigenstate of H with eigenvalue V_0. Comparing the r.h.s. of (24.22) with that of (23.5) we obtain two equations for the coefficients of the orthogonal states $S_j^+|0\rangle$ and $S_{j'}^+|0\rangle$ respectively. These are

$$V_j + \alpha G\Omega_{j'} = V_0 \qquad \alpha V_{j'} + G\Omega_j = \alpha V_0 \tag{24.23}$$

From these equations we can eliminate V_0 obtaining an equation for α in terms of the parameters of the Hamiltonian

$$\alpha^2G\Omega_{j'} + \alpha(V_j - V_{j'}) - G\Omega_j = 0 \tag{24.24}$$

There are two solutions of the quadratic equation for α. The value of α corresponding to the lower eigenvalue V_0 is given by

$$\alpha = \frac{-(V_j - V_{j'}) - \sqrt{(V_j - V_{j'})^2 + 4G^2\Omega_j\Omega_{j'}}}{2G\Omega_{j'}} \tag{24.25}$$

For small values of G, if $4G^2\Omega_j\Omega_{j'} \ll (V_j - V_{j'})^2$ and $V_j \leq V_{j'}$, the value of α may be approximated by $G\Omega_j/(V_j - V_{j'})$. Unlike the expression of α in (23.28) which was obtained by perturbation theory, (24.25) does not depend on N.

If we compare the r.h.s. of (24.21) to that of (23.7) we obtain

$$W_j(S_j^+)^2 + \alpha^2W_{j'}(S_{j'}^+)^2 - 2GS_{j'}^+S_j^+ - 2\alpha^2GS_j^+S_{j'}^+$$
$$= W((S_j^+)^2 + \alpha^2(S_{j'}^+)^2 + 2\alpha S_j^+S_{j'}^+) \tag{24.26}$$

In the l.h.s. of (24.26) we made use of the fact that H_0 is diagonal in the seniority scheme within the j^n and j'^n configurations. We conclude from (24.26) that $W_j = W_{j'} = W$ as stated above in the general case. The other equation following from (24.26) yields

$$G = -\frac{\alpha}{1+\alpha^2}W \tag{24.27}$$

This equation is a special case of (23.44).

We can construct examples of Hamiltonians satisfying conditions (24.23) and (24.26) by starting from H_0 which is diagonal in the seniority scheme within j^n and j'^n configurations. We further arrange the equality $W_{j'} = W_j = W$. As explained in Section 23, W must be repulsive, $W > 0$. We now determine G which satisfies the condition (24.27). For small values of G, the coefficient α, determined from (24.25), is small. As G increases, α increases too and its extremal value is $|\alpha| = 1$. Hence, for sufficiently large W, there is always a value of G for which (24.27) is satisfied. Since the coefficient α is determined by (24.25), conditions (24.23) are also satisfied.

It is worthwhile to point out that for *any* value of G, there is mixing of the j^n configuration and configurations in which 2, 4 or more nucleons are moved into the j'-orbit. If G is small, its effect may be treated in perturbation theory as shown in Section 23. Once it satisfies condition (24.27) its effect can no longer be considered as a small perturbation. Still, the transition from the $v = 0$, $J = 0$ state of the pure j^n configuration to the state $(S_j^+ + \alpha S_{j'}^+)^N|0\rangle$ is rather smooth as the parameter G changes from 0 to the value for which (24.27) is satisfied.

Another way of constructing Hamiltonians which satisfy conditions (24.23) and (24.26) and hence (23.5) and (23.7) is by starting from H_0 with some values of W_j and $W_{j'}$. For a given value of G we calculate α from (24.25) and then determine W from (24.27). To satisfy the condition (24.26) we add to H_0 certain single particle energies and two-body interactions. These additional terms contribute to W_j and $W_{j'}$ but do not change V_j and $V_{j'}$ and hence do not change the value of α. These terms are as follows

$$H' = \tfrac{1}{8}(\hat{n}_j(\hat{n}_j-1)-\hat{n}_j)W'_j + \tfrac{1}{8}(\hat{n}_{j'}(\hat{n}_{j'}-1)-\hat{n}_{j'})W'_{j'} \tag{24.28}$$

It can be directly verified that $[H', S^+] = 0$ and $[[H', S^+], S^+] = W'_j(S_j^+)^2 + W'_{j'}(S_{j'}^+)^2$. Hence, the double commutator condition is satisfied if we choose $W'_j = W - W_j$ and $W'_{j'} = W - W_{j'}$ where W is determined by (24.27).

We can now ask whether the extra conditions for the states in (24.4) to be eigenstates, namely (24.3) and (24.9), are not too stringent. They are certainly satisfied in the case of the quasi-spin scheme. The question is whether they can be satisfied, along with (23.5) and (23.7), by shell model Hamiltonians and S^+ operators with unequal α_j coefficients. Some simple examples of such Hamiltonians were given by Shlomo and Talmi (1972). In that paper, detailed conditions involving the α_j and $\beta_{jj'}$ coefficients and two-body matrix elements were derived. They will not be listed here. It is clear that the more j-orbits are admixed, the easier it is to satisfy these conditions. In the example given by Talmi (1975), the Hamiltonian satisfies conditions (23.5) and (23.7) but it cannot satisfy (24.3) and (24.9).

We saw that if the α_j coefficients are all equal, conditions (23.5) and (23.7) guarantee that the Hamiltonian is diagonal in the quasi-spin scheme. In that case, any state $(S^+)^N B_n^+|0\rangle$ is an eigenstate provided $B_n^+|0\rangle$ is an eigenstate of H. No more independent conditions are necessary. If not all α_j coefficients in (23.5) are equal, the conditions (24.3) and (24.9) introduce further limitations on the shell model Hamiltonian. If we try to construct states analogous to (24.2) with $J = 4$ or higher spins, more conditions are imposed on the shell model Hamiltonian. From the experimental point of view, there is clear information only about the first excited states with $J = 2$ in semi-magic nuclei. They are the lowest excited states and their fairly constant height above the ground states follows directly from the structure (24.4).

Let us come back to the question how well does (24.4) describe the structure of first-excited $J = 2^+$ states in semi-magic nuclei. This question could have been simply answered if the shell model Hamiltonian had been known. As emphasized throughout this book, apart from some simple cases, we do not have the information about matrix elements of the effective interaction. Hence, there is no direct way to check whether conditions (24.3) and (24.9) and even (23.5) and (23.7), are satisfied by the shell model Hamiltonian. Certainly there is no sense in using for such a check some simple interactions whose main virtue is the convenience in calculating matrix elements. There is no sense in checking whether the states (24.4) are eigenstates of such interactions with constant coefficients $\beta_{jj'}$. Also the question whether the states (23.2) are ground states of such interactions with constant coefficients α_j is irrelevant.

This should be pointed out specially against using for such checks simple interactions which are diagonal in the quasi-spin scheme for *degenerate* single nucleon orbits. This includes the pairing interaction and the surface delta interaction. Once this degeneracy is removed, as dictated by experiment, such interactions cannot have the states (23.2), with constant α_j coefficients, as eigenstates. To prove this statement we consider a Hamiltonian H_0 which is diagonal in the quasi-spin scheme satisfying

$$H_0 S^+|0\rangle = V_0' S^+|0\rangle \qquad [[H_0, S^+], S^+] = W(S^+)^2 \tag{24.29}$$

If we add to H_0 a set of non-degenerate single nucleon energies H' we first find

$$(H_0 + H')S^+|0\rangle = V_0 S^+|0\rangle \tag{24.30}$$

The α_j coefficients in the S^+ operator (24.30) may not be all equal. We now try to see whether $H_0 + H'$ can satisfy (23.7) with the S^+ pair operator. Since H' is a linear combination of products of a creation and annihilation operators, its double commutator with S^+ vanishes. We thus obtain from (23.7) the condition

$$[[H_0 + H', S^+], S^+] = [[H_0, S^+], S^+] = W(S^+)^2 \tag{24.31}$$

The coefficient W in (24.31) must be the same as in (24.29) since in either case it is equal to the W_j of the individual orbits which must be all equal.

From (24.31) follows the condition (23.44) for matrix elements of H_0. That condition may be simply written as

$$(\alpha_j^2 + \alpha_{j'}^2)X + 2\alpha_j\alpha_{j'}Y = 0 \tag{24.32}$$

Similarly, the double commutation relation in (24.29) may be written as (24.32) with $\alpha_j = \alpha_{j'} = 1$

$$2X + 2Y = 0 \tag{24.33}$$

with the same values of X and Y. Multiplying (24.33) by $\alpha_j\alpha_{j'}$ and subtracting it from (24.32) we obtain

$$(\alpha_j - \alpha_{j'})^2 X = 0 \tag{24.34}$$

Since $S^+|0\rangle$ is an eigenstate of $H_0 + H'$, there must be a sufficient number of non-vanishing matrix elements $V(j^2 j'^2 J = 0)$ to connect any j-orbit, directly or indirectly, with all the others. We thus conclude from (24.34) that if (24.31) is satisfied, all α_j coefficients must be equal. If there is a sign difference, $\alpha_j = -\alpha_{j'} = 1$ we obtain instead of (24.33) the equation $2X - 2Y = 0$ and the condition $(\alpha_j + \alpha_{j'})^2 X = 0$. Hence, if H_0 satisfies the conditions (23.5) and (23.7) with equal α_j coefficients, the Hamiltonian obtained by adding to H_0 non-degenerate single nucleon energies *cannot* satisfy those conditions.

There is another property of $J = 2$ states with seniority $v = 2$ in j^n configurations which is shared also by the states (24.4) with generalized seniority $v = 2$. Among all $J = 2$ states in the j^n configuration with n even, there is only *one* state connected to the $J = 0$, $v = 0$ ground state by a non-vanishing matrix element of a single nucleon operator. Such an operator has $k = 2$ and thus is a quasi-spin vector. It can change a state with seniority $v = 0$ only into one with seniority $v = 2$. There is only one $J = 2$ state in any j^n configuration, n even, with seniority $v = 2$. All other $J = 2$ states have seniorities $v = 4, 6, \ldots$ and vanishing matrix elements between them and the ground state of any single nucleon operator. A consequence of this fact is that only the $J = 2$ state with seniority $v = 2$ may decay directly to the ground state by emitting $E2$ electromagnetic radiation. The operator whose matrix elements determine this transition is usually taken to be a single nucleon operator. The latter is proportional, within the j^n configuration to $(a_j^+ \times \tilde{a}_j)^{(2)}$.

In the quasi-spin scheme for several j-orbits there are several states with $J = 2$ and seniority $v = 2$. They may all be connected to the $J = 0$, $v = 0$ ground state by non-vanishing matrix elements of a single nucleon operator. In the generalization of seniority considered here, there is *one* very special $J = 2$ state constructed according to (24.4). We shall now look at a single nucleon operator connecting it to the ground state (23.2).

Consider a single nucleon quadrupole operator $\mathbf{Q}$ and its action on the state (23.2)

$$Q_M (S^+)^N |0\rangle = N (S^+)^{N-1} [Q_M, S^+] |0\rangle \qquad (24.35)$$

In deriving (24.35) we made use of the lemma (23.8) and the fact that the commutator $[Q_M, S^+]$ contains products of two creation operators. All higher commutators in (23.8) vanish and also $Q_M|0\rangle$ is equal to zero. We see that (24.35) is proportional to (24.4) if the commutator of Q_M and S^+ is proportional to D_M. We thus take $\mathbf{Q}$ to be the special quadrupole operator defined by

$$\boxed{[Q_M, S^+] = D_M^+} \qquad (24.36)$$

The operator $\mathbf{Q}$ acting on the state (23.2) transforms it into the state (24.4) with $J = 2$ and generalized seniority $v = 2$. Hence, no other $J = 2$ state has non-vanishing matrix elements of $\mathbf{Q}$ connecting it to the ground state. If the operator generating electric quadrupole transitions is proportional to $\mathbf{Q}$, the only $J = 2$ state decaying directly and strongly to the ground state is the state (24.4). This behavior is actually seen in Ni and Sn isotopes and to a lesser extent in semi-magic nuclei with $N = 82$ neutrons.

The shell model Hamiltonian of identical nucleons need not include any quadrupole interactions. As explained in Section 19, within each j^n configuration the interaction may be expanded in terms of scalar products of *odd* tensors in addition to a $k = 0$ term. Unlike certain model Hamiltonians, H is not constructed by using the operator $\mathbf{Q}$. Yet, if $\mathbf{Q}$ is defined by (24.36) and H satisfies the conditions (23.5) and (23.7) as well as (24.3) and (24.9), a simple result is obtained if $[H, Q_M]$ is applied to the state (23.2). We then obtain

$$\boxed{\begin{aligned}[H, Q_M](S^+)^N|0\rangle &= HQ_M(S^+)^N|0\rangle - Q_M H(S^+)^N|0\rangle \\ &= (E_N + V_2 - V_0)Q_M(S^+)^N|0\rangle \\ &\quad - E_N Q_M(S^+)^N|0\rangle \\ &= (V_2 - V_0)Q_M(S^+)^N|0\rangle\end{aligned}} \tag{24.37}$$

It seems that acting on the state (23.2), the operator $\mathbf{Q}$ obeys a *linear equation of motion*. In some approximate calculations it has been attempted to establish an equation of the form

$$[H, Q_M] = \hbar\omega Q_M \tag{24.38}$$

If (24.38) holds, then Q_M generates a harmonic vibration with frequency ω. The expression (24.37) differs from (24.38) by being an *exact* result which is, however, not an operator equation. It holds only when applied to the state (23.2) with generalized seniority $v = 0$. Hence, the state $Q_M(S^+)^N|0\rangle$ should not be considered as an harmonic vibrational state. The expression (24.37) reiterates the constance of the 0–2 spacing which is independent of N.

In the case of equal coefficients the operator **Q** can be obtained as $Q_M = \frac{1}{2}[D_M^+, S^-]$. This is due to the $SU(2)$ commutation relations (22.2), (22.3) and the Jacobi identity as follows

$$\frac{1}{2}[[D_M^+, S^-], S^+] = -\frac{1}{2}[[S^+, D_M^+], S^-] - \frac{1}{2}[[S^-, S^+], D_M^+]$$
$$= [S^0, D_M^+] = D_M^+ \qquad (24.39)$$

In the general case, we express the quadrupole operator **Q** by

$$Q_M = \sum_{jj'} \gamma_{jj'} (a_j^+ \times \tilde{a}_{j'})_M^{(2)} \qquad (24.40)$$

To determine the coefficients $\gamma_{jj'}$ in terms of α_j and $\beta_{jj'}$ we make use of the definition (24.36). The commutator of Q_M and S^+ is evaluated by using (23.1) and (19.22) and is equal to

$$[Q_M, S^+] = \sum_j \gamma_{jj} \alpha_j (a_j^+ \times a_j^+)_M^{(2)}$$
$$+ \sum_{j<j'} (\gamma_{jj'} \alpha_{j'} + (-1)^{j+j'+1} \gamma_{j'j} \alpha_j)(a_j^+ \times a_{j'}^+)_M^{(2)} \qquad (24.41)$$

Comparing the r.h.s. of (24.41) to (24.2) we obtain the following relations between the various coefficients

$$\gamma_{jj} \alpha_j = \frac{1}{\sqrt{2}} \beta_{jj}$$
$$(\gamma_{jj'} \alpha_{j'} + (-1)^{j+j'+1} \gamma_{j'j} \alpha_j) = \beta_{jj'} \qquad j < j' \qquad (24.42)$$

The coefficients $\gamma_{jj'} (j \neq j')$ cannot be determined from (24.42) but if we demand that Q_M be hermitean, we obtain, according to (17.13) and (8.9)

$$\gamma_{j'j} = (-1)^{j-j'} \gamma_{jj'} \qquad (24.43)$$

Substituting (24.43) in (24.42) the values of $\gamma_{jj'}$ are uniquely determined by the α_j and $\beta_{jj'}$ to be

$$\boxed{\begin{aligned} \gamma_{jj}\alpha_j &= \frac{1}{\sqrt{2}}\beta_{jj} \\ \gamma_{jj'}(\alpha_{j'}+\alpha_j) &= \beta_{jj'} \end{aligned}} \tag{24.44}$$

Any state of $n = 2N$ identical fermions distributed on a given set of j-orbits may be equivalently expressed as a state of $2\Omega - 2N$ *holes* in the closed shell containing these j-orbits. In the cases of generalized seniority 0 and 2, this description assumes a simple form which we will now derive. The closed shell may be created by acting with $(S^+)^\Omega$ on the vacuum state. Let us look for an operator S' creating a state of two holes which is equivalent to $(S^+)^{\Omega-1}|0\rangle$. A simple form for S' is a linear combination of the pair annihilation operators S_j^-

$$S' = \sum \alpha'_j S_j^- \tag{24.45}$$

The commutation relations of the operator S' with the S^+ operators follow from (19.7) and are given by

$$\begin{aligned} [S', S^+] &= \sum_{jj'} \alpha'_j \alpha_{j'} [S_j^-, S_{j'}^+] = -2\sum_j \alpha'_j \alpha_j S_j^0 \\ [[S', S^+], S^+] &= -2\sum_{jj'} \alpha'_j \alpha_j \alpha_{j'} [S_j^0, S_{j'}^+] = -2\sum \alpha'_j \alpha_j^2 S_j^+ \end{aligned} \tag{24.46}$$

All higher commutators vanish. We can now use the lemma (23.8) to evaluate the result of applying S' to the closed shell

$$\begin{aligned} S'(S^+)^\Omega|0\rangle &= -2\Omega(S^+)^{\Omega-1}\left(\sum_j \alpha'_j \alpha_j S_j^0\right)|0\rangle \\ &\quad - \Omega(\Omega-1)(S^+)^{\Omega-2}\left(\sum_j \alpha'_j \alpha_j^2 S_j^+\right)|0\rangle \end{aligned} \tag{24.47}$$

The r.h.s. of (24.47) is proportional to $(S^+)^{\Omega-1}|0\rangle$ if for every non-vanishing α_j coefficient we set $\alpha'_j = 1/\alpha_j$ and hence (Talmi 1982)

$$\boxed{S' = \sum_j \frac{1}{\alpha_j} S_j^-} \tag{24.48}$$

It was explained in Section 23 that the coefficients α_j are determined by diagonalization of the shell model Hamiltonian in the two nucleon configuration. Similarly, the coefficients $\alpha'_j = 1/\alpha_j$ may be determined by diagonalizing the shell model Hamiltonian in the configuration of two *holes*. The hole-hole interactions are the same as the particle-particle interactions. The single hole energies, however, are not simply related to the single particle energies. Single hole energies are equal to the energy of the closed shell from which the interaction of the appropriate particle with all others, as well as the single nucleon energy, should be subtracted. From (24.48) we see that unless all α_j are equal, single hole energies are not equal to single particle energies.

To obtain the operator creating a hole pair with $J = 2$ we make use of the operator Q_M defined by (24.36) and define a $\tilde{D}'_M$ operator by

$$\boxed{\tilde{D}'_M = [Q_M, S']} \tag{24.49}$$

The commutation relations of the operator $\tilde{D}'_M$ with the S^+ operators are obtained from (24.36) and the Jacobi identity to be

$$[\tilde{D}'_M, S^+] = [[Q_M, S'], S^+] = -[[S', S^+], Q_M] - [[S^+, Q_M], S']$$

$$= 2\Big[\sum_j S_j^0, Q_M\Big] + [D_M^+, S'] = [D_M^+, S']$$

$$[[\tilde{D}'_M, S^+], S^+] = [[D_M^+, S'], S^+] = -[[S^+, D_M^+], S'] - [[S', S^+], D_M^+]$$

$$= 2\Big[\sum S_j^0, D_M^+\Big] = 2D_M^+$$

Using these commutators, we obtain, due to the lemma (23.8),

$$\begin{aligned}\tilde{D}'_M(S^+)^N|0\rangle &= N(S^+)^{N-1}[\tilde{D}'_M, S^+]|0\rangle \\ &\quad + \tfrac{1}{2}N(N-1)(S^+)^{N-2}[[\tilde{D}'_M, S^+], S^+]|0\rangle \\ &= N(N-1)(S^+)^{N-2}D^+_M|0\rangle \end{aligned} \tag{24.50}$$

The term linear in N vanishes due to (17.24).

From (24.47), where N may be substituted for Ω, it follows that the states $(S^+)^N|0\rangle$ may be obtained by operating on $(S^+)^\Omega|0\rangle$ by $(S')^{\Omega-N}$. Due to (24.50) the states with generalized seniority 2, $(S^+)^{N-1}D^+_M|0\rangle$, may be similarly obtained by operating with $\tilde{D}'_M(S')^{\Omega-N-1}$ on $(S^+)^\Omega|0\rangle$. In particular, the state of two holes coupled to $J = 2$ is equal to $\tilde{D}'_M(S^+)^\Omega|0\rangle$. Since that state is an eigenstate of the shell model Hamiltonian, the coefficients of $\tilde{D}'_M$ may be determined by diagonalizing the submatrix of the two hole configuration. These coefficients, however, are determined by (24.49) in terms of the α_j and $\gamma_{jj'}$ (or $\beta_{jj'}$ coefficients). Let us define the coefficients of $\tilde{D}'_M$ by

$$\boxed{\tilde{D}'_M = \sum_{j\leq j'} \beta'_{jj'} \frac{1}{\sqrt{1+\delta_{jj'}}}(\tilde{a}_j \times \tilde{a}_{j'})^{(2)}_M} \tag{24.51}$$

Substituting (24.51) into (24.49) we obtain, by using (19.16) and comparing the various terms, the following values of the $\beta'_{jj'}$ coefficients

$$\boxed{\begin{aligned} \beta'_{jj} &= \frac{\sqrt{2}\gamma_{jj}}{\alpha_j} = \frac{\beta_{jj}}{\alpha_j^2} \\ \beta'_{jj'} &= \gamma_{jj'}\left(\frac{1}{\alpha_j} + \frac{1}{\alpha_{j'}}\right) = \frac{\gamma_{jj'}(\alpha_j + \alpha_{j'})}{\alpha_j \alpha_{j'}} = \frac{\beta_{jj'}}{\alpha_j \alpha_{j'}} \end{aligned}} \tag{24.52}$$

The last equalities are due to the relations (24.44).

States with generalized seniorities 0 and 2 were introduced above and in Section 23 as generalizations of states in the seniority scheme. It is important to realize that by choosing shell model Hamiltonians which have those states as eigenstates, a tremendous simplification is obtained. As long as eigenstates belong to a definite j^n configuration, the seniority scheme supplies quantum numbers (labels) which distinguish states which have the same value of J. It also supplies a convenient scheme in which construction of states and calculation of matrix elements is greatly simplified. Still, in all j^n configurations, which may be observed in actual nuclei, there are only several states with the same value of J. Thus, there is no real problem in diagonalizing the submatrix of the shell model Hamiltonian within a given j^n configuration which is characterized by a given J (and M). The number of two-body matrix elements needed to construct any such sub-matrix is only $(2j+1)/2$. They may be determined from experimental data as shown in Section 21.

The situation is rather different if eigenstates are linear combinations of states in several configurations. The number of two-body matrix elements increases rapidly since in addition to diagonal matrix elements, there are non-vanishing non-diagonal elements between different configurations. The construction of the Hamiltonian matrices becomes rather involved and if there are several valence nucleons then the size of matrices with given J becomes very very large. As an example, let us consider Sn isotopes for which the description in terms of generalized seniority seems to be useful. In ^{112}Sn there are 12 valence neutrons which may occupy any of the orbits in the major shell between magic numbers 50 and 82. These orbits are

$$2d_{5/2}, 1g_{7/2}, 3s_{1/2}, 2d_{3/2} \text{ and } 1h_{11/2} \tag{24.53}$$

There are 5 states with $J = 0$ for two valence neutrons. Four valence neutrons can couple to yield 14 states with $J = 0$. The number of $J = 0$ states grows rapidly with the number of valence neutrons. For 12 valence neutrons there are

56,907 states with $J = 0$ and positive parity,

267,720 states with $J = 2$ and positive parity,

426,558 states with $J = 4$ and positive parity,

To construct the Hamiltonian submatrices, the values of 160 two body matrix elements of the effective interaction (of which 56 are diagonal) and 5 single neutron energies are required. These are difficult to determine reliably form measured spectra of Sn isotopes. The matrices must be calculated and then diagonalized. It is not trivial to diagonalize a $10^5 \times 10^5$ matrix and not easy to determine properties of a wave function with 10^5 components.

The simple dependence of binding energies on the number of valence nucleons given by (23.20) or the constant 0–2 spacings expressed by (24.10) may emerge from exact diagonalization of those giant matrices. The reason for these simple features, however, would be hardly evident from the list of quarter of a million $J = 2$ components. The simple prescriptions of generalized seniority (23.2) and (24.4) give rise to these features in a straightforward and clear manner. The state (23.2) with $N = 6$ is a $J = 0$ eigenstate of the shell model Hamiltonian of ^{112}Sn and thus it is exactly decoupled from the other 56,906 states with $J = 0$. Similarly, the eigenstate (24.4) with $N = 6$ is decoupled from the other 267,719 states with $J = 2$ in ^{112}Sn. We may consider generalized seniority as a *truncation scheme* in which the problem of finding the $J = 0$ ground states and lowest $J = 2$ levels is reduced to finding the values of the few α_j and $\beta_{jj'}$ coefficients. Hamiltonians which have lowest eigenstates described by generalized seniority are certainly only approximations to the exact shell model Hamiltonians with effective interactions. The agreement with experiment shown above and in Section 23, indicates that these Hamiltonians are good approximations to the exact ones. The exact eigenstates (23.9) and (24.4) of these (approximate) Hamiltonians exhibit the simple features observed experimentally.

The smooth and regular behavior of binding energies like in Fig. 23.3 and of 0–2 spacings exhibited in Fig. 24.2 is a strong argument against a truncation scheme based on reducing the number of orbits. There are no breaks in separation energies nor in the 0–2 spacings which would indicate that only some of the orbits should be considered for certain nuclei. In the prescriptions (23.2) and (24.4) the occupation numbers in different orbits do change indeed as a function of N but the contributions of all orbits are combined in yielding (23.20) and (24.10).

As mentioned above, the 0-2 spacings in even-even nuclei decrease significantly when valence neutrons (or protons) are added to a nucleus which has several valence protons (or neutrons). An argument has been made that this decrease is caused by non-diagonal matrix elements between the various $J = 2$ states obtained by coupling proton excitations (with various values of J_π) and neutron excitations (with various J_ν). It was argued that the gain in energy of the lowest $J = 2$ state is higher than that of the $J = 0$ ground state because of the larger number of $J = 2$ states. The examples of generalized seniority shows that an argument based merely on numbers of states is not necessarily correct. In the example presented above, the number of $J = 2$ states is almost 5 times bigger than the number of $J = 0$ states. Still, the gain in energy due to mixing of many of these states is *exactly* the same for the lowest $J = 0$ and $J = 2$ states.

The situation becomes much more complicated if there are both protons and neutrons outside closed shells. Consider for example $^{154}_{62}Sm_{92}$ in which there are 12 valence protons in the orbits listed in (24.53) and 10 valence neutrons outside the closed shells of 82 which occupy the orbits

$$1h_{9/2}, 2f_{7/2}, 2f_{5/2}, 3p_{3/2}, 3p_{1/2} \text{ and } 1i_{13/2} \tag{24.54}$$

The number of positive parity states of these valence nucleons is

$$\begin{array}{rl} 41,654,193,517,797 & \text{with} \quad J = 0 \\ 346,132,052,934,889 & \text{with} \quad J = 2 \\ 530,897,397,260,575 & \text{with} \quad J = 4 \end{array}$$

Clearly, it is impossible to handle matrices of this order. A drastic truncation scheme must be found to enable any possible shell model approach to this system. We shall come back to this problem in a subsequent section. Before that we return to the seniority scheme in j^n configurations and consider it from a different point of view. This will enable us to introduce the seniority scheme for both protons and neutrons in the j-orbit as well as in l^n configurations.

25

Seniority and the $Sp(2j+1)$ Group

In this section we come back to consider the seniority scheme in j^n configurations from another point of view. The remarkable relation (20.14) is the simple expression for certain eigenvalues of Hamiltonians which are diagonal in the seniority scheme. All matrix elements in j^n configurations can be expressed as linear combinations of two body matrix elements $V_J = \langle j^2JM|V|j^2JM\rangle$. Yet, the interaction in states with lowest seniorities, $v=0$ and $v=1$, is a linear combination of only two parameters. These are V_0 and the linear combination $\bar{V}_2$ defined by (20.7). This property is a special case of a more general feature of the seniority scheme, namely

$$\text{The average interaction energy of a set of states with given seniority } v \text{ in the } j^n \text{ configuration is a linear combination of only } V_0 \text{ and } \bar{V}_2 \tag{25.1}$$

If the interaction is diagonal in the seniority scheme, the averages are averages of eigenvalues of the interaction. In other cases, (25.1) holds for the averages of the *expectation values* of the interaction in the states with definite seniorities. The coefficients of the linear combination in (25.1) are functions of n and v but they are the same for all two body interactions.

We shall discuss the property (25.1) in some detail but first let us draw a direct consequence from it. Any two body operator which has the same values of V_0 and $\bar{V}_2$ as those of a given interaction, has the same average interaction in any group of states with seniority v in any j^n configuration. In particular, the linear combination of the pairing interaction and a constant interaction may be taken to be such an operator. If the coefficients of these interactions are a and b then the eigenvalues are given by (19.11) as

$$a\frac{n(n-1)}{2} + b\frac{n-v}{2}(2j+3-n-v) \tag{25.2}$$

The coefficients a and b may be determined from V_0 and $\bar{V}_2$. Putting $n = 2$ in (25.2) we obtain for $v = 0$ and $v = 2$ the eigenvalues

$$\begin{aligned} a + (2j+1)b &= V_0 \\ a &= \bar{V}_2 \end{aligned} \tag{25.3}$$

From these equations a and b can be expressed in terms of V_0 and $\bar{V}_2$. The general formula for average interaction energies in sets of j^n states with seniority v becomes then equal to

$$\boxed{\bar{V}(j^n, v) = \frac{n(n-1)}{2}\bar{V}_2 + \frac{n-v}{2}(2j+3-n-v)\frac{V_0 - \bar{V}_2}{2j+1}} \tag{25.4}$$

There is only *one* state with $v = 0$ or $v = 1$ in any j^n configuration and the average energy is thus equal to the energy of that state. In fact, putting $v = 0$ or $v = 1$ in (25.4) reproduces the result (20.13) previously obtained.

The statement (25.1) concerns the whole set of states with given seniority in the *same* j^n configuration. In Section 19 we showed that each state with given v and J (and additional quantum number α if necessary) is a member of a set forming the basis of an irreducible representation of the $SU(2)$ Lie algebra (19.7). The other members of that basis are states with the *same* quantum numbers v and J (and α) in *other* j^n configurations. These states transform among themselves when the generators S_j^0, S_j^+ and S_j^- of the $SU(2)$ algebra act on them.

If we consider *all* states with given v in *one* j^n configuration we must look for another group whose generators transform these states among themselves.

In Section 19 we considered operators that just do that. We saw that irreducible tensor operators with odd ranks k are quasi-spin scalars. They have nonvanishing matrix elements only between states with the same seniority v (and n). Thus, they transform states of the j^n configuration with the same v and *all* values of J and M among themselves. As we shall discuss in more detail, such transformations are irreducible. The components of all odd tensors constitute the infinitesimal generators of a Lie group. To characterize this group more generally, we can define it as the group under whose finite elements the antisymmetric $J=0$ state of two j-nucleons is invariant. In fact, we saw in Section 19 that

$$\boxed{T_\kappa^{(k)}|j^2J=0\rangle=0 \qquad k \text{ odd}} \tag{25.5}$$

From (25.5) follows for the finite element generated by $T_\kappa^{(k)}$

$$\boxed{e^{\alpha T_\kappa^{(k)}}|j^2J=0\rangle=|j^2J=0\rangle \qquad k \text{ odd}} \tag{25.6}$$

To complete the discussion, let us go back one step and look at all irreducible tensor operators $(a_j^\dagger\times\tilde{a}_j)_\kappa^{(k)}$ with any rank k. The following considerations, as well as similar ones in the following sections, are based on the approach of Racah presented in his lectures on Group Theory and Spectroscopy. These lectures were given at the Institute for Advanced Study in Princeton in the spring of 1951. The mimeographed lecture notes had a wide circulation and were eventually published (Racah 1965).

The operators $(a_j^\dagger\times\tilde{a}_j)_\kappa^{(k)}$ transform any state of a single j-nucleon into another state. When acting on a state of the j^n configuration they transform it to a linear combination of such states. They are actually the infinitesimal elements of a certain group or the generators of its Lie algebra. In fact, calculating the commutation relation of two such operators we obtain, by using the identity (10.24), the following result

$$[(a_j^+ \times \tilde{a}_j)^{(k')}_{\kappa'}, (a_j^+ \times \tilde{a}_j)^{(k'')}_{\kappa''}] = \sqrt{(2k'+1)(2k''+1)} \sum_{k\kappa} (-1)^{k+1} \times (1-(-1)^{k'+k''+k}) \begin{Bmatrix} k' & k'' & k \\ j & j & j \end{Bmatrix} \times (k'\kappa' k''\kappa'' \mid k'k''k\kappa)(a_j^+ \times \tilde{a}_j)^{(k)}_{\kappa} \tag{25.7}$$

The commutator of any two such components is a linear combination of these components. If both k' and k'' are odd, then the r.h.s. of (25.7) contains only tensors with odd ranks k. Thus, the components of odd tensors constitute a Lie sub-algebra as is already evident from the property (25.5) of odd tensors.

The special case of (25.7) for $k' = 1$ should yield the commutation relations of the total spin $\mathbf{J}$ with any tensor operator. The r.h.s. of (25.7) vanishes in this case unless $k = k''$. By using the actual values of the $6j$- and $3j$-symbols and (17.9), we obtain from (25.7) the commutation relations (6.25).

Before proceeding with the Lie algebra (25.7) and the sub-algebra of odd tensors, one point must be clarified. Due to the Wigner-Eckart theorem, the commutation relations (25.7) hold, within j^n configurations, for all single nucleon irreducible tensor operators $T^{(k)}_{\kappa}$. Since the irreducible tensors are single nucleon operators, it is enough to check (25.7) between any two single j-nucleon states. If, however, the products of matrix elements on the l.h.s. of (25.7) are summed over *all* intermediate j (and m) values, the results may be rather different. For instance, two spherical harmonics $Y_{k'\kappa'}(\mathbf{r})$ and $Y_{k''\kappa''}(\mathbf{r})$ surely commute. Thus, the following matrix element between the single nucleon jm and $j'm'$ state vanishes

$$\sum_{j''m''} [\langle jm|Y_{k'\kappa'}|j''m''\rangle\langle j''m''|Y_{k''\kappa''}|j'm'\rangle - \langle jm|Y_{k''\kappa''}|j''m''\rangle\langle j''m''|Y_{k'\kappa'}|j'm'\rangle] \tag{25.8}$$

for any j, j' and m, m'. If, however, for $j' = j$, the summation on (25.8) is restricted to $j'' = j$ then the result corresponding to (25.7) is obtained.

Which is the group whose Lie algebra is expressed by (25.7)? Actually, in Section 6 we were introduced to a special case of such a group for $j = \frac{1}{2}$. There we considered the unit matrix along with the three matrices $\frac{1}{2}\sigma_1, \frac{1}{2}\sigma_2, \frac{1}{2}\sigma_3$ which are the matrices of J_x, J_y, J_z in the $(2\frac{1}{2} + 1 = 2)$-dimensional space. Thus, these four matrices are proportional to the matrices of the $k = 0$ and $k = 1$ (i.e. $\mathbf{J}$) irreducible tensor operators in that space. For $j = \frac{1}{2}$, these are the only possible irreducible tensor operators and the group they generate has been shown in Section 6 to be the $U(2)$ Lie algebra. Also in the general case, for any j, the irreducible tensor operators $(a_j^+ \times \tilde{a}_j)_\kappa^{(k)}$ are the elements of the Lie algebra of the group $U(2j+1)$ of unitary transformations in the $(2j+1)$-dimensional space. To show this, let us recall (6.37) from which it follows that any unitary transformation may be generated from the infinitesimal transformation $1 + i\epsilon A$ where the matrix A is hermitean.

The most general hermitean matrix is the sum of a real symmetric matrix and a real skew-symmetric matrix multiplied by i. The number of independent parameters in a real symmetric matrix is $f(f+1)/2$ where $f = 2j + 1$ is the order of the matrix (only the diagonal elements and those below the diagonal should be counted). The corresponding number for a real skew-symmetric matrix is $f(f-1)/2$ (the number of elements below the vanishing diagonal). The sum is simply f^2 which is the total number of independent elements of a real matrix. In fact, given a real matrix B, a symmetric matrix can be defined by $B + \tilde{B}$ ($\tilde{B}$ is the transposed matrix) and a skew-symmetric matrix is obtained as $B - \tilde{B}$. Let us now see that any real matrix of order $2j + 1$ can be expressed as a linear combination with real coefficients of the matrices of all irreducible tensors operators. The total number of their components is indeed $\sum_{k=0}^{2j}(2k+1) = ((2j+1)/2)(4j+2) = (2j+1)^2$.

All irreducible tensor operators are linear combinations of $a_{jm}^+ a_{jm'}$. When such an operator acts on a state of a single j-nucleon in the m'-state, it transforms it to a state with m. Let the jm be represented by a column with $2j + 1$ rows whose elements are zeroes with the exception of 1 in the m-th row. The matrix of that operator has elements which are zeroes with the exception of the element in the m-th row and m'-th column which is equal to 1. The most general real matrix in the $(2j+1)$-dimensional space of single nucleon states can be expressed as a linear combination, with real coefficients, of the matrices of all $a^{+}_{jm}a_{jm'}$ as follows

$$\sum_{mm'} \langle jm|B|jm'\rangle a^{+}_{jm}a_{jm'} \tag{25.9}$$

The real matrix whose element in the m_0-th row and m'_0-th column is 1 and all other elements vanish may be expressed in terms of irreducible tensor operators. Starting from (25.9) we can expand $a^{+}_{jm}a_{jm'}$ in terms of the tensors $(a^{+}_{j} \times \tilde{a}_j)^{(k)}_{\kappa}$. Starting instead with the expansion

$$\sum_{k\kappa} f_{k\kappa}\langle jm|(a^{+}_{j} \times \tilde{a}_j)^{(k)}_{\kappa}|jm'\rangle$$
$$= \sum_{k\kappa} f_{k\kappa}\sqrt{2k+1}(-1)^{j-m}\begin{pmatrix} j & k & j \\ -m & \kappa & m' \end{pmatrix} \tag{25.10}$$

we take the coefficients in (25.10) to be given by

$$f_{k\kappa} = (-1)^{j-m_0}\sqrt{2k+1}\begin{pmatrix} j & k & j \\ -m_0 & \kappa & m'_0 \end{pmatrix}$$

We then obtain by using the orthogonality relations of $3j$-symbols for the matrix element (25.10) the result

$$\sum_{k\kappa}(-1)^{j-m_0+j-m}(2k+1)\begin{pmatrix} j & k & j \\ -m_0 & \kappa & m'_0 \end{pmatrix}\begin{pmatrix} j & k & j \\ -m & \kappa & m' \end{pmatrix}$$
$$= \delta_{mm_0}\delta_{m'm'_0} \tag{25.11}$$

Thus, any real matrix B may be expressed by the linear combination

$$B = \sum_{k\kappa}\sum_{m_0m'_0} \langle jm_0|B|jm'_0\rangle(-1)^{j-m_0}$$
$$\times \begin{pmatrix} j & k & j \\ -m_0 & \kappa & m'_0 \end{pmatrix}(a^{+}_{j} \times \tilde{a}_j)^{(k)}_{\kappa} \tag{25.12}$$

The tensor $(a^{+}_{j} \times \tilde{a}_j)^{(k)}_{\kappa}$ is hermitean since, as can be directly verified,

$$[(a^{+}_{j} \times \tilde{a}_j)^{(k)}_{\kappa}]^{\dagger} = (-1)^{\kappa}(a^{+}_{j} \times \tilde{a}_j)^{(k)}_{-\kappa} \tag{25.13}$$

The matrix of $(a_j^+ \times \tilde{a}_j)^{(k)}_\kappa$ is real and hence the relation (25.13) holds if the transposed matrix is written on the l.h.s. Thus, any real symmetric matrix can be expressed as a linear combination of

$$(a_j^+ \times \tilde{a}_j)^{(k)}_\kappa + (-1)^\kappa (a_j^+ \times \tilde{a}_j)^{(k)}_{-\kappa} \tag{25.14}$$

The general real skew-symmetric matrix can be expressed as the linear combination

$$(a_j^+ \times \tilde{a}_j)^{(k)}_\kappa - (-1)^\kappa (a_j^+ \times \tilde{a}_j)^{(k)}_{-\kappa} \tag{25.15}$$

This way, all operators $(a_j^+ \times \tilde{a}_j)^{(k)}_\kappa$ may be taken as the generators of the $U(2j+1)$—the unitary group in $2j+1$ dimensions.

The group $U(2j+1)$ is not a *semi-simple* Lie group since it has an invariant abelian subgroup. Among all generators $(a_j^+ \times \tilde{a}_j)^{(k)}_\kappa$ there is one which commutes with all the others. This is the scalar operator $(a_j^+ \times \tilde{a}_j)^{(0)}_0 = \hat{n}_j/\sqrt{(2j+1)}$ which generates that subgroup. The number operator commutes with all single nucleon operators since they are products of a creation and an annihilation operator. This fact is consistent with the commutation relation (25.7). If $k'=0$ then $1-(-1)^{k''+k} = 1-(-1)^{2k} = 0$. We may remove the $k=0$ operator as was done for $U(2)$ in Section 6. Also here, the generators with $k>0$ belong to the group $SU(2j+1)$. This is the group of unitary transformations whose matrices have determinants equal to 1. We recall (6.39) stating that $\det\exp A = \exp\operatorname{tr} A$ and calculate the following trace

$$\begin{aligned}\sum_m \langle jm|(a_j^+ \times \tilde{a}_j)^{(k)}_\kappa|jm\rangle &= \sum_{m\mu\mu'} (j\mu j\mu' \mid jjk\kappa)\langle jm|a^+_{j\mu}\tilde{a}_{j\mu'}|jm\rangle \\ &= \sum_m (jmj,-m \mid jjk\kappa)(-1)^{j-m} \\ &= \sqrt{2j+1}\delta_{k0}\,\delta_{\kappa 0}\end{aligned} \tag{25.16}$$

The last equality in (25.16) follows from the orthogonality relation (7.8) of the Clebsch-Gordan coefficients with k and with 0. Hence, the only generator whose trace does not vanish is the one with $k=0$. All other generators, with $k>0$, have zero trace and the determinants of exponentials of their matrices are equal to 1.

We saw in Section 6 an example of the quadratic Casimir operator constructed from the generators of the Lie algebra of a semi-simple group. We shall not describe here the general prescription for constructing such operators but write down in the present case the operator

$$\boxed{\sum_k (a_j^\dagger \times \tilde{a}_j)^{(k)} \cdot (a_j^\dagger \times \tilde{a}_j)^{(k)}} \tag{25.17}$$

and show that it commutes with all $U(2j+1)$ generators. Strictly speaking, the $k = 0$ term in (25.17) should be removed to obtain the Casimir operator of the semi-simple group $SU(2j+1)$. It is equal to $\hat{n}_j^2/(2j+1)$ and commutes with all generators, hence it could be added to the $k > 0$ terms in (25.17) with any coefficient. The definition (25.17) will turn out to yield simple eigenvalues for that operator.

Let us calculate the commutator of (25.17) with any generator. We obtain by using (25.7) the following expression

$$\begin{aligned}
&\left[\sum_{k'} (a_j^\dagger \times \tilde{a}_j)^{(k')} \cdot (a_j^\dagger \times \tilde{a}_j)^{(k')}, (a_j^\dagger \times \tilde{a}_j)^{(k'')}_{\kappa''}\right] \\
&\quad = \sum_{k'\kappa'} (-1)^{\kappa'} (a_j^\dagger \times \tilde{a}_j)^{(k')}_{\kappa'} [(a_j^\dagger \times \tilde{a}_j)^{(k')}_{-\kappa'}, (a_j^\dagger \times \tilde{a}_j)^{(k'')}_{\kappa''}] \\
&\qquad + \sum_{k'\kappa'} (-1)^{\kappa'} [(a_j^\dagger \times \tilde{a}_j)^{(k')}_{\kappa'}, (a_j^\dagger \times \tilde{a}_j)^{(k'')}_{\kappa''}] (a_j^\dagger \times \tilde{a}_j)^{(k')}_{-\kappa'} \\
&\quad = \sqrt{(2k'+1)(2k''+1)} \sum_{k'\kappa' k\kappa} (-1)^{\kappa'+k+1} (1 - (-1)^{k'+k''+k}) \\
&\qquad \times \begin{Bmatrix} k' & k'' & k \\ j & j & j \end{Bmatrix} (k', -\kappa' k'' \kappa'' \mid k' k'' k \kappa) (a_j^\dagger \times \tilde{a}_j)^{(k')}_{\kappa'} (a_j^\dagger \times \tilde{a}_j)^{(k)}_{\kappa} \\
&\qquad + \sqrt{(2k'+1)(2k''+1)} \sum_{k'\kappa' k\kappa} (-1)^{\kappa'+k+1} (1 - (-1)^{k'+k''+k}) \\
&\qquad \times \begin{Bmatrix} k' & k'' & k \\ j & j & j \end{Bmatrix} (k' \kappa' k'' \kappa'' \mid k' k'' k \kappa) (a_j^\dagger \times \tilde{a}_j)^{(k)}_{\kappa} (a_j^\dagger \times \tilde{a}_j)^{(k')}_{-\kappa'}
\end{aligned} \tag{25.18}$$

In order to evaluate (25.18) we change in the last term on the r.h.s. κ' into $-\kappa'$. We then notice that k', κ', k, κ are all dummy indices so we interchange in the last term on the r.h.s. of (25.18) k' and k, κ' and

κ. Then we obtain for (25.18) the expression

$$\sqrt{(2k''+1)} \sum_{k'\kappa' k\kappa} (1-(-1)^{k'+k''+k}) \begin{Bmatrix} k' & k'' & k \\ j & j & j \end{Bmatrix}$$
$$\times (a_j^\dagger \times \tilde{a}_j)^{(k')}_{\kappa'} (a_j^\dagger \times \tilde{a}_j)^{(k)}_{\kappa}$$
$$\times [\sqrt{2k'+1}(-1)^{\kappa'+k+1}(k',-\kappa' k''\kappa'' \mid k'k''k\kappa)$$
$$+ \sqrt{2k+1}(-1)^{\kappa+k'+1}(k,-\kappa k''\kappa'' \mid kk''k'\kappa')] \tag{25.19}$$

The coefficient in the square brackets in (25.19) can be now calculated by expressing the Clebsch-Gordan coefficients in terms of $3j$-symbols. It becomes equal to

$$\sqrt{(2k+1)(2k'+1)} \Bigg[(-1)^{\kappa'+k+1+k'-k''+\kappa} \begin{pmatrix} k' & k'' & k \\ -\kappa' & \kappa'' & -\kappa \end{pmatrix}$$
$$+ (-1)^{\kappa+k'+1+k-k''+\kappa'} \begin{pmatrix} k & k'' & k' \\ -\kappa & \kappa'' & -\kappa' \end{pmatrix} \Bigg]$$
$$= \sqrt{(2k+1)(2k'+1)} \begin{pmatrix} k' & k'' & k \\ -\kappa' & \kappa'' & -\kappa \end{pmatrix}$$
$$\times [(-1)^{\kappa'+k+1+k'-k''+\kappa} + (-1)^{\kappa+k'+1+k-k''+\kappa'+k+k'+k''}]$$
$$= \sqrt{(2k+1)(2k'+1)} \begin{pmatrix} k' & k'' & k \\ -\kappa' & \kappa'' & -\kappa \end{pmatrix}$$
$$\times (-1)^{\kappa+\kappa'+1+k+k'-k''} (1+(-1)^{k+k'+k''})$$

Substituting this expression into (25.19) we see that each term in the summation contains the product $(1-(-1)^{k+k'+k''})(1+(-1)^{k+k'+k''}) = 0$. Hence, (25.19) as well as (25.18) vanishes and the operator (25.17) commutes with all the generators $(a_j^\dagger \times \tilde{a}_j)^{(k)}_{\kappa}$.

The fact that the Casimir operator (25.17) commutes with all $U(2j+1)$ generators follows from the commutation relations (25.7). It is true for any set of operators $T^{(k)}_{\kappa}$ satisfying those commutation relations. In the special case considered here, of generators defined by fermion creation and annihilation operators, this fact may be demonstrated in a straightforward manner. This procedure will also yield the eigenvalues of the operator (25.17).

We begin with the operator (25.17) and express it as

$$\sum_{k\kappa}(-1)^{\kappa}(a_j^+ \times \tilde{a}_j)_{\kappa}^{(k)}(a_j^+ \times \tilde{a}_j)_{-\kappa}^{(k)}$$
$$= \sum_{k\kappa\mu\mu'\nu\nu'} (-1)^{\kappa+j+\mu'+j+\nu'}(j\mu j\mu' \mid jjk\kappa)$$
$$\times (j\nu j\nu' \mid jjk,-\kappa) a_{j\mu}^+ a_{j,-\mu'} a_{j\nu}^+ a_{j,-\nu'}$$
$$= \sum_{k\kappa\mu\mu'\nu\nu'} (-1)^{\mu+\mu'+\mu'+\nu'+1}(j\mu j\mu' \mid jjk\kappa)$$
$$\times (j,-\nu' j,-\nu \mid jjk\kappa) a_{j\mu}^+ a_{j,-\mu'} a_{j\nu}^+ a_{j,-\nu'} \qquad (25.20)$$

The last equality in (25.20) follows from the symmetry properties (7.11) and (7.18) of the Clebsch-Gordan coefficients. Expressing the latter in terms of $3j$-symbols and summing over k and κ we obtain, due to (7.26), for (20.25) the form

$$\sum_{\mu\mu'\nu\nu'} (-1)^{\mu+\nu'}\delta_{\mu,-\nu'}\delta_{\mu',-\nu} a_{j\mu}^+ a_{j,-\mu'} a_{j\nu}^+ a_{j,-\nu'}$$
$$= \sum_{\mu\mu'} a_{j\mu}^+ a_{j,-\mu'} a_{j,-\mu'}^+ a_{j\mu}$$
$$= (2j+1)\sum_{\mu} a_{j\mu}^+ a_{j\mu} - \sum_{\mu\mu'} a_{j\mu}^+ a_{j,-\mu'}^+ a_{j,-\mu'} a_{j\mu}$$
$$= (2j+1)\hat{n} + \sum_{\mu} a_{j\mu}^+ a_{j\mu} - \sum_{\mu\mu'} a_{j\mu}^+ a_{j\mu} a_{j,-\mu'}^+ a_{j,-\mu'}$$
$$= (2j+1)\hat{n} - \hat{n}(\hat{n}-1) \qquad (25.21)$$

The operator (25.17) can thus be expressed in terms of the number operator $\hat{n}$ and its square which commute with all generators $(a_j^+ \times \tilde{a}_j)_{\kappa}^{(k)}$. The eigenvalues of (25.17) depend only on the number of j-nucleons and are equal to $(2j+1)n - n(n-1)$. This result, as well as the expression (25.21), hold *only* for fully antisymmetric states of identical nucleons.

According to Schur's lemma, a matrix which commutes with all matrices which form an irreducible representation of a group is proportional to the unit matrix. Hence, the sub-matrix of the Casimir operator defined by an irreducible representation is diagonal and all its elements are equal. The eigenvalues of the Casimir operator thus supply labels for these irreducible representations. If the Hamiltonian is constructed from the generators of $U(2j+1)$ it has no non-vanishing non-diagonal elements connecting states which belong to two different irreducible representations. Hence, each eigenstate of such a Hamiltonian must belong to a group of states which transform according to a definite irreducible representation of $U(2j+1)$. The labels of that irreducible representation may serve as additional quantum numbers for the eigenstate.

We do not gain any additional classification of states in the present case. This should not be surprising since all allowed states of j^n configurations of identical nucleons belong to *one* irreducible representation of $U(2j+1)$. Those representations can be characterized, as we shall see in the next section, by their symmetry properties. Allowed states are fully antisymmetric and they transform among themselves according to *one* irreducible representation of $U(2j+1)$. We shall meet more complicated cases in subsequent sections but here we just notice that the operators $T_\kappa^{(k)}$ are *symmetric* in all nucleons and hence, acting on an antisymmetric state yield an antisymmetric state. A more interesting situation arises when we consider the subgroup of $U(2j+1)$ whose generators are the components of odd tensors only.

The generators $(a_j^\dagger \times \tilde{a}_j)_\kappa^{(k)}$, with odd ranks k, form a Lie algebra of a group of transformations which according to (25.6) leave invariant the $J=0$ state of the j^2 configuration. The state $|j^2 J=0\rangle$ is antisymmetric under exchange of the two nucleons. A group whose transformations leave invariant a bilinear antisymmetric form is called a *symplectic group*. The group generated by the operators $(a_j^\dagger \times \tilde{a}_j)_\kappa^{(k)}$, k odd, is the symplectic group in $2j+1$ dimensions—$Sp(2j+1)$. This group is a sub-group of the unitary group $U(2j+1)$. As we saw in Section 19, all generators $(a_j^\dagger \times \tilde{a}_j)_\kappa^{(k)}$, k odd, have vanishing matrix elements between states with different seniorities. We conclude that within j^n configurations of identical nucleons the irreducible representations of $Sp(2j+1)$ are characterized by the seniority quantum number v. We will show this fact in a different way which will provide another approach to the seniority scheme.

The quadratic Casimir operator of $Sp(2j+1)$ can be directly written down by limiting the summation in (25.17) to odd tensors only

$$\boxed{C_{Sp(2j+1)} = 2\sum_{k\,\mathrm{odd}} (a_j^\dagger \times \tilde{a}_j)^{(k)} \cdot (a_j^\dagger \times \tilde{a}_j)^{(k)}} \tag{25.22}$$

To show that it commutes with all generators $(a_j^\dagger \times \tilde{a}_j)^{(k)}_\kappa$, k odd, we use again the derivation (25.18). Since in the present case both k' and k'' are odd, the only k values appearing in the summations on the r.h.s. of (25.18) are also odd. Hence, we can interchange k and k', κ and κ' in the last term on the r.h.s. of (25.18) and obtain for the commutator the result (25.19) which is equal to zero for k, k' and k'' odd.

The eigenvalues of (25.22) can be obtained from the expansion (12.20) of the pairing interaction. If we introduce it to the formula (12.20) we obtain

$$C_{Sp(2j+1)} = -P - \frac{n(n-1)}{2} + n\frac{1}{2j+1}\sum_{k\text{ odd}}(2k+1) \qquad (25.23)$$

Hence, states of j^n configurations of identical nucleons with given seniority v are eigenstates of (25.23). The eigenvalues of the Casimir operator in such states are given by using (19.11)

$$\boxed{\begin{aligned} &-\frac{n-v}{2}(2j+3-n-v) - \frac{n(n-1)}{2} + n(j+1) \\ &\quad = \frac{v}{2}(2j+3-v) = C_{Sp(2j+1)}(v) \end{aligned}} \qquad (25.24)$$

The eigenvalues (25.24) depend only on v and are the same for all states with the same seniority. This is in accordance with the property stated above that the Casimir operator is proportional to the unit matrix in the space spanned by an irreducible representation of the group considered and all its equal eigenvalues depend only on the irreducible representation.

Going from the group $U(2j+1)$ to its subgroup $Sp(2j+1)$ introduces a finer classification of states. Sets of states which transform irreducibly under the operations of the $U(2j+1)$ group need not do so under operations of a subgroup. Thus, the fully antisymmetric irreducible representations of $U(2j+1)$ which are simply characterized by n, split into sets of irreducible representations of $Sp(2j+1)$ which are characterized by the seniority v. If the Hamiltonian is constructed from generators of $Sp(2j+1)$, its eigenstates have definite seniorities. If this is not the case, the scheme introduced by this procedure may be used as a basis for constructing the matrix of the Hamiltonian. The matrix constructed in the seniority scheme may then be diagonalized.

We can find now a new interpretation to the results (19.37) obtained above for odd tensor operators. Since they are quasi-spin scalars, their non-vanishing matrix elements are only between states with the same seniority v and these are independent of n. Those properties follow directly from the fact that odd tensor operators are generators of $Sp(2j+1)$. By their definition, generators of a group transform among themselves states which belong to a given irreducible representation of the group. These transformations depend only on the irreducible representation and not on the particular realization of the group and irreducible representation considered. Matrix elements of a generator between a given state and other orthogonal states are coefficients of the expansion of the state obtained by the generator acting on the given state. Hence, these matrix elements vanish if the other states do not belong to the same irreducible representation as the given state. The non-vanishing matrix elements depend only on the labels of states which form a basis of the irreducible representation to which the given state belongs. The Casimir operator which is constructed from the generators shares also this property. It has only non-vanishing diagonal matrix elements, which are all equal, in states of a given irreducible representation. These eigenvalues depend only on the quantum numbers which characterize the irreducible representation. In the present case, matrix elements of odd tensor operators depend only on the seniority v, which specifies the irreducible representation, and the labels α, J within it. They are independent of the particular j^n configuration in which the states with seniority v appear. Hence, matrix elements between states with given v in the j^n configuration are equal to the corresponding ones in the j^v configuration.

An important feature of the $Sp(2j+1)$ group is that it contains as a subgroup the group $O(3)$ (or $SU(2)$) of three-dimensional rotations. All generators $(a_j^\dagger \times \tilde{a}_j)^{(k)}_\kappa$, k odd, annihilate the state $|j^2 J=0\rangle$ as in (25.5). But only three of them, those with $k=1$, generate infinitesimal transformations in the $(2j+1)$-dimensional space which are due to infinitesimal rotations in the three-dimensional space. The generators of $SU(2)$ (or $O(3)$) are the components of $\mathbf{J}$. The spherical components of $\mathbf{J}$ are given according to (17.19) by

$$\boxed{\frac{1}{\sqrt{3}}\sqrt{j(j+1)(2j+1)}(a_j^\dagger \times \tilde{a}_j)^{(1)}_\kappa} \qquad (25.25)$$

If we go over from the group $Sp(2j+1), j > \frac{1}{2}$ to its $SU(2)$ sub-group, the sets of states with given v in the j^n configuration split into sets of states characterized by J, each set having $2J+1$ states with various M values. In the case $j = \frac{1}{2}$, the generators (25.25) of $SU(2)$ are the same as those of $Sp(2)$ and hence $SU(2) \equiv Sp(2)$. For higher values of j there may be several independent sets of states with given value of J included in the same set of states with seniority v. In such cases, additional quantum numbers are necessary to uniquely specify the states. These may be conveniently introduced in the j^v configuration and were denoted by α in Section 19.

The last step of the classification scheme obtained by the *chain* of groups

$$U(2j+1) \supset Sp(2j+1) \supset SU(2) \tag{25.26}$$

points to an important distinction between these groups. Any shell model Hamiltonian is invariant under three-dimensional rotations. Hence, according to Schur's lemma, its eigenvalues for the $2J+1$ states, with given J and different M values, must be equal. Shell model Hamiltonians within j^n configurations do not, in general, commute with generators of the $U(2j+1)$ or $Sp(2j+1)$ groups. This is due to the fact that the operations of $(a_j^+ \times \tilde{a}_j)^{(k)}_\kappa$ on a state with given J and M yield a linear combination of states with different values of J and M. In general, such states do not have the same eigenvalues of the shell model Hamiltonian. Hence, such shell model Hamiltonian cannot commute with the generators $(a_j^+ \times \tilde{a}_j)^{(k)}_\kappa$. Had the Hamiltonian commuted with the generators of a group, its eigenvalues for all states in the same irreducible representation would have been degenerate due to Schur's lemma. If the Hamiltonian is constructed from the generators of such a group, it just has no non-vanishing non-diagonal elements between states which belong to different irreducible representations. If the Hamiltonian is constructed from generators of a subgroup its symmetry is *increased*. There are more restrictions on non-vanishing matrix elements. This way more quantum numbers may be introduced. We shall later see other examples of this procedure.

There is an alternative way to interpret a group chain like (25.26). We can consider Hamiltonians whose matrices, within j^n configurations of identical nucleons, commute with all generators of $U(2j+1)$. Such Hamiltonians which contain single nucleon terms and two-body interactions must then be a linear combination of the quadratic Casimir operator and the number operator. Their eigenvalues are then given by the sum of linear and quadratic terms in n. When going to $Sp(2j+1)$ it is required that the Hamiltonian commutes with its generators. Since the latter commute with the Casimir operator of $U(2j+1)$, we may add to it $C_{Sp(2j+1)}$ with an arbitrary coefficient. This is sometimes referred to as *breaking* of the $U(2j+1)$ symmetry. Finally, the Casimir operator of $SU(2)$ may be added so that a term proportional to $J(J+1)$ is added to the Hamiltonian. The eigenvalues of such Hamiltonians are thus given by the rather schematic expression

$$\alpha\frac{n(n-1)}{2}+\frac{\beta}{2}n+\gamma\frac{v}{2}(2j+3-v)+\delta J(J+1) \qquad (25.27)$$

with arbitrary coefficients. It is clear that only a very restricted class of Hamiltonians have eigenvalues which may be expressed by (25.27). For $j\leq\frac{5}{2}$ it is possible to express any Hamiltonian with single nucleon energies and two-body interactions in the form (25.27). In fact, for $j=\frac{5}{2}$, the most general two-body interaction (16.4) is equivalent to (25.27) with $\alpha=a-c/2$, $\beta=2bj(j+1)+2(j+1)c$, $\gamma=-c$ and $\delta=b$. For $j\geq\frac{7}{2}$, however, there are at least 4 two-body matrix elements (in states with $J=0,2,4,6,\ldots$) and one single nucleon energy. These cannot be reproduced, in general, by the 4 coefficients in (25.27).

Let us examine more closely the relation (25.23) between the Casimir operator of $Sp(2j+1)$ and the pairing interaction. The pairing interaction is invariant under any operation that leaves the state $|j^2J=0\rangle$ invariant. Hence, it commutes with all generators of $Sp(2j+1)$ and is therefore proportional to the unit matrix in the space of all states with given v in the j^n configuration (Schur's lemma). This feature is characteristic of the (quadratic) Casimir operator. Hence, there should be a simple connection between the pairing interaction and the quadratic Casimir operator of $Sp(2j+1)$. One difference is that the eigenvalues of the Casimir operator depend only on the irreducible representation of $Sp(2j+1)$ and hence on v. The eigenvalues of the pairing interaction may depend also on the irreducible representation of $U(2j+1)$. That dependence may be expressed by linear and quadratic terms in n. In fact, we know that if $v=n$ the eigenvalues of the pairing interaction P vanish. Since both these operators are quadratic functions of the generators, we can express the pairing interaction up to a multiplicative factor by

$$F(n)-C_{Sp(2j+1)}(v) \qquad (25.28)$$

The requirement that for $v = n$ the expression (25.28) should vanish yields for the pairing interaction the result

$$\boxed{P(n,v) = C_{Sp(2j+1)}(n) - C_{Sp(2j+1)}(v)} \tag{25.29}$$

The multiplicative constant may be determined from the case $n = 2$, $v = 0$ and is equal to 1. As is directly verified, the result (25.29) is the same as (19.11) obtained above. The result (25.29) is actually a direct consequence of (25.24). The relation between Casimir operators and pairing interactions is, however, rather general and we shall see more cases of it in subsequent sections.

Having presented the seniority scheme as the scheme of states characterized by the irreducible representations of the $Sp(2j+1)$ group, we return to the property (25.1) of averages. It asserts that the average interaction energy of *all* states with given seniority v in the j^n configuration of identical nucleons is a linear combination of V_0 and $\bar{V}_2$. Those parameters are the averages of interaction energies in $T = 1$ states of the j^2 configuration with seniorities $v = 0$ and $v = 2$ respectively. This is proved in detail by direct methods in de-Shalit and Talmi (1963). Here we present only the group theoretical meaning of this feature.

The irreducible tensor operators $T_\kappa^{(k)}$ are defined by their transformation properties under three-dimensional rotations. Such operators are grouped into sets of $2k+1$ components which transform linearly and irreducibly under $O(3)$ (or $SU(2)$) transformations. In Section 19 we considered the $SU(2)$ operations carried out by applying S_j^+, S_j^- and S_j^0 to the various states. We classify accordingly operators as quasi-spin scalars or tensors with various ranks. We can consider transformation properties of operators within the j^n configurations under transformations due to $U(2j+1)$ or $Sp(2j+1)$ generators. States of j^n configurations can be grouped into sets which transform irreducibly under the group operations. Similarly, we can consider sets of operators which transform irreducibly under the operations of the group. These sets are components of irreducible tensors with respect to the group considered. The number of components in each set as well as the characterization of the sets depend on the group considered. There are, however, sets which have only *one* component and are invariant under the operations of the group. They are *scalars* with respect to the group and commute with its generators.

In general, shell model Hamiltonians do not commute with the $(a_j^+ \times \tilde{a}_j)^{(k)}_\kappa$ generators, they are not just scalars of $U(2j+1)$ or $Sp(2j+1)$. The only two-body interactions which commute with all $(a_j^+ \times \tilde{a}_j)^{(k)}_\kappa$ generators are the two body parts of the Casimir operator (25.17) and of $(a_j^+ \times \tilde{a}_j)^{(0)}_0 (a_j^+ \times \tilde{a}_j)^{(0)}_0$. Both have eigenvalues which are proportional to $n(n-1)/2$. The only two body interaction which commutes with all $Sp(2j+1)$ generators $(a_j^+ \times \tilde{a}_j)^{(k)}_\kappa$, k odd, in addition to those, is the two body part of $C_{Sp(2j+1)}$ in (25.22). According to (25.23), the expression

$$C_{Sp(2j+1)} - n\frac{1}{2j+1}\sum_{k\,\text{odd}}(2k+1) = C_{Sp(2j+1)} - n(j+1) \tag{25.30}$$

is equal to the two-body interaction $-(P + n(n-1)/2)$.

Two irreducible tensor operators of $O(3)$ (or $SU(2)$) obey the orthogonality relation

$$\begin{aligned}\sum_{MM'} &\langle JM|T^{(k)}_\kappa|JM'\rangle\langle JM|T^{(k')}_{\kappa'}|JM'\rangle \\ &= (J\|\mathbf{T}^{(k)}\|J)(J\|\mathbf{T}^{(k')}\|J)\sum_{MM'}\begin{pmatrix} J & k & J \\ -M & \kappa & M'\end{pmatrix}\begin{pmatrix} J & k' & J \\ -M & \kappa' & M'\end{pmatrix} \\ &= \frac{1}{2k+1}(J\|\mathbf{T}^{(k)}\|J)(J\|\mathbf{T}^{(k')}\|J)\delta_{kk'}\,\delta_{\kappa\kappa'}\end{aligned} \tag{25.31}$$

The orthogonality follows from the Wigner-Eckart theorem (8.4) for the $O(3)$ group and the orthogonality of the vector addition coefficients. Those coefficients decompose the product of any two components of irreducible $O(3)$ representations into a linear combination of components of irreducible representations. These concepts can be generalized to other Lie groups and lead to analogous orthogonality relations between diagonal matrix elements of operators which belong to different irreducible representations of the group considered.

A special case of (25.31) is obtained when $k'=0$, $\kappa'=0$ and $T^{(k')}_{\kappa'}$ is a scalar, invariant under $O(3)$ rotations. We obtain in that case

$$\sum_M \langle JM|T_\kappa^{(k)}|JM\rangle = (J\|\mathbf{T}^{(k)}\|J)\sum_M (-1)^{J-M}\begin{pmatrix} J & k & J \\ -M & \kappa & M \end{pmatrix}$$

$$= \sqrt{2J+1}(J\|\mathbf{T}^{(k)}\|J)\delta_{k0}\delta_{\kappa 0} \qquad (25.32)$$

The sum (or equally well, the average) of the diagonal elements in a given $O(3)$ irreducible representation of any *irreducible* tensor operator which is not invariant is zero. This feature holds also for other groups. Also in those cases, a scalar commuting with all generators is proportional to the unit matrix within the space spanned by an irreducible representation. Hence, the matrix within an irreducible representation of any operator which is not a scalar has a vanishing *trace*.

Considering irreducible representations of $U(2j+1)$, we deduce that the matrix of any operator which is not a scalar in the space of all (antisymmetric) states of the j^n configuration must have zero trace. Hence, the trace of any two-body interaction in the j^n configuration of identical nucleons is equal to the trace of the $U(2j+1)$ scalars contained in that interaction. As we saw above, matrix elements of those scalars are proportional to $n(n-1)/2$. The average interaction energy is equal to the trace divided by the number of states and therefore equal to the diagonal elements. We thus deduce that the average interaction energy of *all* states of the j^n configuration is given by

$$\boxed{\overline{\langle j^n JM|\sum_{i<k} V_{ik}|j^n JM\rangle} = \alpha\frac{n(n-1)}{2}} \qquad (25.33)$$

The coefficient α, which is independent of n, may be determined by putting in (25.33) $n=2$. It is thus equal to

$$\boxed{\begin{aligned}\alpha &= \sum_{J\text{ even},M} \langle j^2 JM|V|j^2 JM\rangle \Big/ \sum_{J\text{ even}} (2J+1) \\ &= \frac{1}{j(2j+1)}\sum_{J\text{ even}} (2J+1)V_J\end{aligned}} \qquad (25.34)$$

If we consider averages of states within irreducible representations of $Sp(2j+1)$ we must consider only *its* scalars. The only contribution to the trace of the matrix defined by such representations of any two-body interaction is due to $U(2j+1)$ scalars and the $Sp(2j+1)$ scalar (25.24). The averages of interaction energies are thus given by (25.33) to which the eigenvalues of (25.30) are added. These two terms may be combined to give for the averages the expression

$$a\frac{n(n-1)}{2}+b\frac{n-v}{2}(2j+3-n-v) \qquad (25.35)$$

In (25.35) the coefficient a is equal to the sum of α in (25.34) and b which is the coefficient of the two-body scalar interaction (25.30). The expression (25.35) is identical with (25.2) which was discussed in detail above.

In Section 19, we saw how the $SU(2)$ generators S_j^+, S_j^- and S_j^0 transform states of various j^n configurations with *given* seniority v, spin J (and M) and additional quantum numbers, if necessary, among themselves. In the present section we saw how the generators of $Sp(2j+1)$ transform irreducibly among themselves states in a *given* j^n configuration and given seniority v with various values of J (and M) and additional quantum numbers. As we saw in Section 19, S_j^+, S_j^- and S_j^0 commute with the generators $(a_j^\dagger \times \tilde{a}_j)_\kappa^{(k)}$, k odd. The group whose members transform among themselves states with seniority v and various J (and M) values in all j^n configurations is the direct product $SU(2)\otimes Sp(2j+1)$.

26

Seniority With Isospin. The $U(2(2j+1))$ *Group and Subgroups*

The j^n configuration with both protons and neutrons offers a more interesting application of group theory. This approach will be used here to introduce the seniority scheme in such configurations. We consider Hamiltonians with charge independent interactions between nucleons without the Coulomb repulsion between protons which must be considered separately. Any charge independent shell model Hamiltonian may be constructed without using explicitly the isospin operators. The nucleon interactions may include as a factor the scalar product $\boldsymbol{\tau}_1 \cdot \boldsymbol{\tau}_2$. This, however, could be expressed in terms of the *charge exchange operator*, analogous to the spin exchange operator (10.64),

$$P^{\tau} = \tfrac{1}{2}(1 + \boldsymbol{\tau}_1 \cdot \boldsymbol{\tau}_2) = \tfrac{1}{2}(1 + 4\mathbf{t}_1 \cdot \mathbf{t}_2) \tag{26.1}$$

Due to the generalized Pauli principle introduced in Section 9, the charge exchange operator may be replaced for allowed states by $-P^{\sigma}P^{x}$.

Hamiltonians which do not include isospin operators are invariant under permutations of the space and spin variables of protons and neutrons. If we want to exploit this symmetry, we must consider the spin and space variables of nucleons separately from their isospin variables. It is thus not convenient to use the formalism of second quantization and therefore, we consider in this section states in configuration space. If the Hamiltonian is invariant under a permutation P then if ϕ is one of its eigenstates, $P\phi$ is also an eigenstate with the same eigenvalue. This has been shown for rotations in Section 5. It is a general feature as follows from operating on $H\phi = E\phi$ by P obtaining

$$PH\phi = PHP^{-1}P\phi = HP\phi = EP\phi \tag{26.2}$$

In deriving (26.2) we used the invariance of H under permutations, $PHP^{-1} = H$. In (26.2), ϕ is a function of the space and spin variables only.

The state $P\phi$ obtained from ϕ by a permutation of nucleon space and spin variables cannot be *another* physical eigenstate of the Hamiltonian. Nucleons are indistinguishable and the numbers attached to their variables have therefore no importance. As in the case of two nucleons, discussed in Section 9, the way out of this unphysical degeneracy is by imposing a generalized Pauli principle. The total wave function, including isospin variables, must be antisymmetric with respect to interchange of nucleon coordinates. This is accomplished by constructing a linear combination of a set of independent functions among $\phi, P\phi, P'\phi, \ldots$ multiplied by appropriate isospin functions. In Section 9 we considered two nucleon wave functions. For $n = 2$, the wave function ϕ may be either symmetric or antisymmetric. If it is antisymmetric it should be multiplied by the symmetric $T = 1$ isospin function. A symmetric ϕ should be multiplied by the antisymmetric $T = 0$ isospin function. For any number n of nucleons, if ϕ is fully antisymmetric, $P\phi$ is the same function apart from possible change of sign. Such a ϕ should be multiplied by a fully symmetric isospin function (with $T = n/2$).

For $n \geq 3$, however, there are states which have a more complicated symmetry type. In such cases, $P\phi$ for some permutation P may be a function ϕ' which is independent of $\phi, P\phi'$ may be linearly independent of both ϕ and ϕ', etc. For regular Hamiltonians there is only a finite number of independent wave functions which belong to a given eigenvalue E. These functions transform among themselves under permutations of nucleon (space and spin) variables. In the absence of accidental degeneracies, a situation which may be achieved by slightly changing the Hamiltonian if necessary, these transformations are irreducible. These wave functions, however, do not describe different physical states. They must be multiplied by appropriate isospin

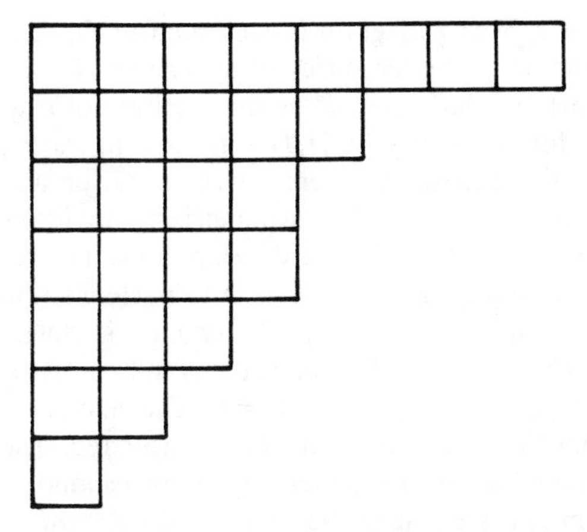

FIGURE 26.1. *A Young diagram.*

functions and then added to yield a fully antisymmetric wave function of *all* nucleon variables. In the following we will describe the way this is accomplished.

Wave functions of space and spin variables which have different symmetry types belong to different eigenvalues of some charge independent Hamiltonian. In the following, such Hamiltonians will be explicitly constructed. Therefore, wave functions with different symmetry types are orthogonal. This orthogonality holds even without the multiplication of such functions by isospin functions. In Section 15, we saw an example of this kind of orthogonality. The eigenvalues of charge independent Hamiltonians are determined by the space and spin eigenstates. Hence, they are independent of the M_T values of the isospin states which are used to obtain fully antisymmetric states.

The eigenfunctions ϕ of H can thus be classified according to the irreducible representations of the group of permutations under which they transform among themselves. These irreducible representations determine some symmetry type which may be either symmetric, antisymmetric or of some *mixed symmetry*. The various symmetries may be characterized by *Young diagrams* or *tableaux* (sometimes called shapes or patterns). These are composed of k rows of squares of lengths f_i with the following restrictions

$$f_1 \geq f_2 \geq \cdots \geq f_k \qquad \sum_{i=1}^{k} f_i = n \tag{26.3}$$

An example of a Young diagram is presented in Fig. 26.1. Different Young diagrams characterize different symmetry types. The various Young diagrams are characterized by the partition of n to the f_i. The notation used for them may be $[f_1 f_2 \cdots f_k]$ and in short $[f]$. In each square, one of the nucleon numbers is written. Given any function ϕ, the given symmetry is obtained by symmetrizing it with respect to the nucleons whose numbers appear in the same row. The resulting function is *then* antisymmetrized with respect to nucleons whose numbers appear in the same column. Thus, fully symmetric states are characterized by $f_1 = n$, $f_2 = \cdots = f_k = 0$, whereas fully antisymmetric ones by $f_1 = f_2 = \cdots f_n = 1$, $f_{n+1} = \cdots = f_k = 0$. The number of rows k is the total number of single nucleon states considered. The number of independent functions of a symmetry type corresponding to a given Young diagram $[f]$ is equal to the number of *standard* arrangements in it, $n_{[f]}$. In a standard arrangement the numbers in each row increase from left to right and in each column they increase from top to bottom. It follows immediately that there is only one standard arrangement for the fully symmetric or fully antisymmetric diagrams. For $n = 3$, the possible Young tableaux and standard arrangements are given below

$$\begin{array}{|c|c|c|}\hline 1&2&3\\\hline\end{array}\;,\quad \begin{array}{|c|c|}\hline 1&2\\\hline 3\\\cline{1-1}\end{array}\quad \begin{array}{|c|c|}\hline 1&3\\\hline 2\\\cline{1-1}\end{array}\;,\quad \begin{array}{|c|}\hline 1\\\hline 2\\\hline 3\\\hline\end{array}$$

There is only one independent symmetric function (denoted by [3]), one antisymmetric function (whose diagram is [1 1 1]) whereas there are two independent functions with mixed symmetry (denoted by [2 1]).

If in ϕ all nucleons occupy different states, *each* standard arrangement of a diagram $[f]$ gives rise to an independent irreducible representation of the group of permutations. Having applied $[f]$ with a given standard arrangement to such a ϕ we may apply to it all possible $n!$ permutations. Among those, there are $n_{[f]}$ linearly independent functions which may be taken to be the basis functions for the irreducible representation characterized by $[f]$. All other functions $P\phi$ can be expressed as linear combinations of these basis functions. Starting from another standard arrangement we obtain in this way an independent set of functions which transform among themselves under permutations according to the *same* irreducible representation.

There are such $n_{[f]}$ independent sets each of which has a basis of $n_{[f]}$ functions. The operation of $[f]$ on ϕ is a linear combination of various $P\phi$ functions. Hence, the total number of basis functions for all Young diagrams $[f]$ and all standard arrangements is equal to the total number of linearly independent functions $P\phi$. If in ϕ all nucleons occupy different states, any two functions $P\phi$ and $P'\phi$ $(P \neq P')$ are orthogonal since at least two nucleons occupy different states in the two functions. Hence, the total number of linearly independent functions $P\phi$ is the number of permutations $n!$. We thus obtain the relation

$$\boxed{\sum_{[f]}(n_{[f]})^2 = n!} \tag{26.4}$$

In the example shown above for $n = 3$ we obtain according to (26.4) the relation $1 + 2^2 + 1 = 6 = 3!$

If not all single nucleon states in ϕ are different, the number of basis functions in any irreducible representation characterized by $[f]$ is still $n_{[f]}$. The number of independent sets, however, is less than $n_{[f]}$. Not all standard arrangements yield non-vanishing independent sets. The number of standard arrangements of a given Young diagram is given by the general formula

$$\boxed{n_{[f_1 f_2 \cdots f_k]} = n! \frac{\prod_{i<j}^{k}(f_i - f_j + j - i)}{\prod_{i=1}^{k}(f_i + k - i)!}} \tag{26.5}$$

The number of standard arrangements depends on the given Young tableau. We can formally add to it rows with zero squares. This formal procedure should not change the number of standard arrangements but it increases the value of k in (26.5). The number of standard arrangements given by (26.5) is, however, independent of it. For instance, if $f_3 = \cdots = f_k = 0$, $n = f_1 + f_2$, (26.5) reduces to

$$\binom{f_1 + f_2}{f_1} \frac{f_1 - f_2 + 1}{f_1 + 1}.$$

The wave functions ϕ in (26.2) which transform according to a given irreducible representation of the group of permutations can be expressed as linear combinations of a set of basis functions ϕ_i. In order to obtain an antisymmetric state, these ϕ_i should be multiplied by appropriate isospin functions η_i and then added. The prescription is that the set η_i are basis states of the irreducible representation characterized by the *dual* Young diagram. To obtain the latter, rows and columns of the original diagram should be interchanged. Intuitively this means that the isospin function is symmetric where the space and spin function is antisymmetric and vice versa. Any standard arrangement of a Young diagram yields one standard arrangement of the dual diagram when rows and columns are interchanged. Thus, the number of basis functions $n_{[f]}$ is the same for $[f]$ and its dual diagram. The symmetry in space and spin variables is thus uniquely determined by the symmetry of the isospin functions. Since isospin functions are the same for all problems it may be convenient to use *their* symmetry properties to characterize the symmetry in the space and spin variables. The requirement of total antisymmetry introduces limitations on the allowed irreducible representations. Since there are only two possible isospin states, $m_t = +\frac{1}{2}$, and $m_t = -\frac{1}{2}$ for each nucleon, not more than two nucleon isospin variables can be found in antisymmetric states. Hence, the Young tableaux of isospin functions must not have more than *two rows*. This imposes on the space and spin functions the restriction that they cannot belong to irreducible representations whose Young diagrams have more than *two columns*.

Let us now consider the isospin functions which are obtained from a Young diagram with two rows of lengths f_1 and f_2 ($f_1 \geq f_2$). Among all these isospin functions, those with the highest M_T value are obtained from a function η_0 in which all f_1 nucleons whose numbers appear in the top row are in $m_t = \frac{1}{2}$ states, whereas the other f_2 nucleons are in $m_t = -\frac{1}{2}$ states. No more nucleons can be in $m_t = +\frac{1}{2}$ states since then at least two $m_t = \frac{1}{2}$ nucleons will be in the same column and the function will vanish upon antisymmetrization. Operating with T^+ on η_0 will thus annihilate the resulting wave functions. Since the function η_0 and those obtained from it by applying to it the operations of $[f_1 f_2]$ have $M_T = \frac{1}{2}(f_1 - f_2)$, they all have $T = \frac{1}{2}(f_1 - f_2)$. The Young diagram which is uniquely characterized by f_1 and f_2 may be equally well specified by n and T which are given by

$$\boxed{f_1 + f_2 = n \qquad \tfrac{1}{2}(f_1 - f_2) = T} \tag{26.6}$$

Given any standard arrangement of the $[f_1 f_2]$ diagram, we can always find a function η_0 with the property described above. The set of independent isospin functions η_i, obtained from η_0 by applying the operations prescribed by $[f_1 f_2]$ and all possible permutations, has $n_{[f_1 f_2]}$ members in it. It can be shown that starting from any other standard arrangement the *same* set of η_i is obtained. Hence, isospin functions with T given by (26.6) and $M_T = T$ are uniquely determined by n and T. This special feature is due to the fact that only two states are available for the isospin of the nucleon.

Other isospin functions with the symmetry $[f_1 f_2]$ with lower values of M_T are also eigenstates of $\mathbf{T}^2$ with the same value (26.6) of T. To see it, we operate on the functions η_i, with $M_T = T$, by the operator $T^- = \sum_{i=1}^n t_i^-$. Since T^- is a symmetric operator, we may operate with it on η_0 and afterwards apply to $T^-\eta_0$ the operations of $[f_1 f_2]$. The function $T^-\eta_0$ is a linear combination of functions in which there are $f_1 - 1$ nucleons with $m_t = \frac{1}{2}$ and $f_2 + 1$ nucleons in $m_t = -\frac{1}{2}$ states. Among these functions, only those in which one of the last $f_1 - f_2$ nucleons in the top row has $m_t = -\frac{1}{2}$ will survive the antisymmetrization prescribed by $[f_1 f_2]$. The set of independent isospin functions has in this case $M_T = \frac{1}{2}((f_1 - 1) - (f_2 + 1)) = \frac{1}{2}((f_1 - f_2) - 1)$ and T given by (26.6). Also in this case, of $M_T = T - 1$, it can be shown that there is only one set of $n_{[f_1 f_2]}$ independent isospin functions. There are no isospin functions with the symmetry of $[f_1 f_2]$ which have isospin $T - 1$. This procedure may be repeated until in $(T^-)^{(f_1 - f_2)}\eta_0$, all nucleons whose numbers are in the last $f_1 - f_2$ squares of the top row are in $m_t = -\frac{1}{2}$ states. In that case we obtain $M_T = \frac{1}{2}[(f_1 - (f_1 - f_2)) - (f_2 + (f_1 - f_2))] = -\frac{1}{2}(f_1 - f_2) = -T$. Thus, for each value of M_T which satisfies $-\frac{1}{2}(f_1 - f_2) \leq M_T \leq \frac{1}{2}(f_1 - f_2)$ there is one set of linearly independent $n_{[f_1 f_2]}$ isospin functions which transform irreducibly among themselves under permutations.

The relation (26.6) shows how the symmetry of the isospin functions, which in turn determines the symmetry properties of the space and spin functions, is uniquely determined by n and T. The most symmetric isospin function is characterized by $f_1 = n$, $f_2 = 0$ and $T = n/2$. In that case, the space and spin function is fully antisymmetric. Isospin functions with lower values of T, have lower symmetry and accordingly, the space and spin functions are more symmetric. The lower the isospin T, the higher the symmetry of the space and spin functions. Effective nuclear interactions, which are stronger in space symmetric states of nucleons, yield the well known observation about isospins. Apart from special cases in light nuclei, states with higher isospins T lie higher in energy. This point will be discussed in more detail below.

In order to illustrate the procedure described above, we consider a state of three nucleons occupying three different states. To simplify the expressions, we denote a state in which nucleons 1,2,3 are in the first, second and third states respectively by (1 2 3). A permutation of nucleon numbers will change the order within the brackets. The symmetric and antisymmetric combinations are straightforward. We shall only describe the functions with mixed symmetry corresponding to $[f_1 f_2] = [21]$. Starting with (1 2 3) we apply to it the operations prescribed by one standard arrangement. This may be expressed by

$$\begin{array}{|c|c|}\hline 1 & 2 \\ \hline 3 & \multicolumn{1}{c}{} \\ \cline{1-1} \end{array}(1\ 2\ 3) = (1\ 2\ 3) + (2\ 1\ 3) - (3\ 2\ 1) - (2\ 3\ 1) = \phi_1$$

Acting on ϕ_1 by the permutation P_{12} in which 1 and 2 are transposed we obtain

$$P_{12}\phi_1 = (2\ 1\ 3) + (1\ 2\ 3) - (3\ 1\ 2) - (1\ 3\ 2) = \phi_2$$

The function ϕ_2 is linearly independent of ϕ_1 but not orthogonal to it. Orthogonalization may be carried out in a straightforward manner but we will not bother with it. The effect of other transpositions is as follows

$$P_{12}\phi_2 = \phi_1 \qquad P_{13}\phi_1 = -\phi_1 \qquad P_{13}\phi_2 = -\phi_1 + \phi_2$$

$$P_{23}\phi_1 = \phi_1 - \phi_2 \qquad P_{23}\phi_2 = -\phi_2$$

All permutations may be obtained as products of transpositions and thus, the irreducible representation is defined. Note that P_{12}^2, P_{13}^2 and P_{23}^2 are equivalent to 1 on both ϕ_1 and ϕ_2 as they should be.

In the basis of these ϕ_1 and ϕ_2 the matrices of the three transpositions in the irreducible representation [21] are thus given by

$$P_{12} = \begin{pmatrix} 0 & 1 \\ 1 & 0 \end{pmatrix} \qquad P_{13} = \begin{pmatrix} -1 & -1 \\ 0 & 1 \end{pmatrix} \qquad P_{23} = \begin{pmatrix} 1 & 0 \\ -1 & -1 \end{pmatrix}$$

From these, the matrices of the other permutations can be obtained by matrix multiplication. Had we constructed from ϕ_1, ϕ_2 an orthogonal and normalized basis, any permutation (which is just renaming of nucleon coordinates) operating on such basis functions would yield orthogonal functions. Hence, in such a basis, permutations are represented by real orthogonal matrices. Matrices representing transpositions, whose squares are equal to 1, are represented by symmetric matrices (with determinants equal to -1).

Let us now obtain the independent irreducible representation due to the dual diagram. In the particular case considered here, it has the same shape but a different standard arrangement. Acting on the same function (1 2 3) it yields

$$\begin{array}{|c|c|}\hline 1 & 3 \\ \hline 2 & \multicolumn{1}{c}{} \\ \cline{1-1} \end{array}(1\ 2\ 3) = (1\ 2\ 3) + (3\ 2\ 1) - (2\ 1\ 3) - (3\ 1\ 2) = \phi_2' - \phi_1'$$

Acting on this function by P_{13} we obtain

$$P_{13}(\phi_2' - \phi_1') = (1\ 2\ 3) + (3\ 2\ 1) - (2\ 3\ 1) - (1\ 3\ 2) = \phi_2'$$

Hence, we obtain

$$\phi_1' = (2\ 1\ 3) + (3\ 1\ 2) - (2\ 3\ 1) - (1\ 3\ 2)$$

We see that ϕ_1' and ϕ_2' are linearly independent of ϕ_1 and ϕ_2 but not orthogonal to them. The choice of basis functions is arbitrary. The choice of ϕ_1' and ϕ_2' made here is intended to demonstrate the equivalence of the irreducible representations whose bases are ϕ_1, ϕ_2 and ϕ_1', ϕ_2'. In fact, we obtain by applying the three transpositions the following transformations of ϕ_1', ϕ_2'

$$P_{12}\phi_2' = \phi_1' \quad P_{13}\phi_1' = -\phi_1' \quad P_{13}\phi_2' = -\phi_1' + \phi_2'$$
$$P_{23}\phi_1' = \phi_1' - \phi_2' \quad P_{23}\phi_2' = -\phi_2'$$

These relations are identical to those obtained above for the ϕ_1, ϕ_2. Hence, the equivalence of the irreducible representations of the group of permutations for which ϕ_1, ϕ_2 and ϕ_1', ϕ_2' are bases.

The choice of bases for the irreducible representations of the permutation group is arbitrary. Also the use of other standard arrangements to obtain ϕ_1' and ϕ_2' is not mandatory. For example, the function ϕ_1' may be obtained by

$$\begin{array}{|c|c|}\hline 1 & 2 \\ \hline 3 & \multicolumn{1}{c}{} \\ \cline{1-1} \end{array}(1\ 3\ 2) = (1\ 3\ 2) + (2\ 3\ 1) - (3\ 1\ 2) - (2\ 1\ 3) = -\phi_1'$$

Thus, the standard arrangement yielding ϕ_1 when applied to (1 2 3), yields when applied to the function (1 3 2), the function ϕ_1' which belongs to an equivalent but independent irreducible representation.

We turn now to isospin functions η. There are only two possible states for the nucleon isospin so that two of the states in the function (1 2 3) must be identical. Let us take the first two states to be the same for which we may use the notation $(12,3) \equiv (21,3)$. The isospin states corresponding to ϕ_1, ϕ_2 are η_1 and η_2 given by

$$(12,3) + (21,3) - (32,1) - (23,1) = 2(12,3) - 2(23,1) = 2\eta_1$$

$$(12,3) + (21,3) - (31,2) - (13,2) = 2(12,3) - 2(13,2) = 2\eta_2$$

The states η_1', η_2' corresponding to ϕ_1', ϕ_2' are no longer independent states since we find

$$\eta_1' = (21,3) + (31,2) - (23,1) - (13,2) = (12,3) - (23,1) = \eta_1$$

$$\eta_2' = (12,3) + (32,1) - (23,1) - (13,2) = (12,3) - (13,2) = \eta_2$$

We can now combine the isospin functions η_1, η_2 with space and spin functions to form wave functions which are fully antisymmetric. There is a general detailed prescription which we will not describe here. We only write down the combination of products of space and spin functions derived from one standard arrangement by isospin functions derived from the dual tableau. We verify that the combination

$$\eta_1'\phi_2 - \eta_2'\phi_1 = \eta_1\phi_2 - \eta_2\phi_1$$

is antisymmetric by applying to it the various permutations (the three transpositions are sufficient). For instance

$$P_{13}(\eta_1\phi_2 - \eta_2\phi_1) = (P_{13}\eta_1)(P_{13}\phi_2) - (P_{13}\eta_2)(P_{13}\phi_1)$$

$$= -\eta_1(\phi_2 - \phi_1) + (\eta_2 - \eta_1)\phi_1 = -\eta_1\phi_2 + \eta_2\phi_1$$

The other independent antisymmetric state is

$$\eta_1\phi_2' - \eta_2\phi_1'$$

If the first two states in the function (1 2 3) of the space and spin coordinates are identical, then ϕ_1' becomes equal to $\frac{1}{2}\phi_1, \phi_2'$ to $\frac{1}{2}\phi_2$ and the two antisymmetric states become identical.

Let us make these considerations more definite by considering $M_T = \frac{1}{2}$ states of two protons and one neutron in a $j = \frac{5}{2}$ orbit. The two protons can couple to states with $J = 0$, $J = 2$ and $J = 4$. The

$j = \frac{5}{2}$ neutron can couple to the proton states to yield states with $J = \frac{13}{2}, \frac{11}{2}, \frac{9}{2}, \frac{9}{2}, \ldots$. We saw in Section 16 that according to the Pauli principle three identical $j = \frac{5}{2}$ nucleons, may form antisymmetric states with $J = \frac{9}{2}, \frac{5}{2}, \frac{3}{2}$. Hence, one of the $J = \frac{9}{2}$ states in the system considered here is the $M_T = \frac{1}{2}$ state with $T = \frac{3}{2}$ whereas the $J = \frac{9}{2}$ state orthogonal to it should have $T = \frac{1}{2}$ like the $J = \frac{13}{2}$ and $J = \frac{11}{2}$ states. Such states may be also considered in the m-scheme. The only state with $M = \frac{13}{2}$ is the one in which the protons are in $m = \frac{5}{2}$ and $m' = \frac{3}{2}$ states and the neutron in the $m = \frac{5}{2}$ state. Hence, this state, $(\frac{5}{2}\ \frac{3}{2})\frac{5}{2}$, is the $M = \frac{13}{2}$ state with $J = \frac{13}{2}$. Similarly, one linear combination of $(\frac{5}{2}\ \frac{1}{2})\frac{5}{2}$ and $(\frac{5}{2}\ \frac{3}{2})\frac{3}{2}$ states is the $M = \frac{11}{2}$ component of the $J = \frac{13}{2}$ state and the orthogonal state is the one with $J = \frac{11}{2}$. There are four states with $M = \frac{9}{2}$, namely $(\frac{5}{2}\ \frac{3}{2})\frac{1}{2}$, $(\frac{5}{2}\ \frac{1}{2})\frac{3}{2}$, $(\frac{5}{2}, -\frac{1}{2})\frac{5}{2}$ and $(\frac{3}{2}\ \frac{1}{2})\frac{5}{2}$. Orthogonal linear combinations of those are the $M = \frac{9}{2}$ components of the $J = \frac{13}{2}, \frac{11}{2}, \frac{9}{2}$ states with $T = \frac{1}{2}$ and the $T = \frac{3}{2}$, $J = \frac{9}{2}$ state. To construct states which have the correct isospins we make use of the formalism developed above.

We consider the space and spin functions of the nucleons *without* the isospin labels. In this approach the ϕ_i of the $M = \frac{13}{2}$ state are denoted by

$$\phi_1 = (\tfrac{5}{2}(1)\tfrac{5}{2}(2)\tfrac{3}{2}(3)) - (\tfrac{5}{2}(2)\tfrac{5}{2}(3)\tfrac{3}{2}(1))$$

$$\phi_2 = (\tfrac{5}{2}(1)\tfrac{5}{2}(2)\tfrac{3}{2}(3)) - (\tfrac{5}{2}(1)\tfrac{5}{2}(3)\tfrac{3}{2}(2))$$

The isospin functions are similarly constructed as

$$\eta_1 = (1_+2_+3_-) - (2_+3_+1_-) \qquad \eta_2 = (1_+2_+3_-) - (1_+3_+2_-)$$

The antisymmetric state with $T = \frac{1}{2}$, $J = \frac{13}{2}$, $M = \frac{13}{2}$ is thus given by

$$\eta_1\phi_2 - \eta_2\phi_1$$

Similar expressions can be written down for the $M = \frac{11}{2}$ states. In the $M = \frac{9}{2}$ case the possible m-values are $\frac{5}{2}, \frac{3}{2}, \frac{1}{2}$ and $\frac{5}{2}, \frac{5}{2}, -\frac{1}{2}$. The antisymmetric state in the case $\frac{5}{2}, \frac{5}{2}, -\frac{1}{2}$ may be constructed just like the $\frac{5}{2}, \frac{5}{2}, \frac{3}{2}$ case. The other set of m-values $\frac{5}{2}, \frac{3}{2}, \frac{1}{2}$ yields *three* independent states. One of them can be constructed by applying to it the antisymmetrization procedure prescribed by the [1 1 1] diagram. The resulting state should then be multiplied by the fully symmetric isospin function

$(1_+2_+3_+)$ to obtain the $T = \frac{3}{2}$ state (with $J = \frac{9}{2}$). One $T = \frac{1}{2}$ state is obtained from

$$\phi_1 = (\tfrac{5}{2}(1)\tfrac{3}{2}(2)\tfrac{1}{2}(3)) + (\tfrac{5}{2}(2)\tfrac{3}{2}(1)\tfrac{1}{2}(3))$$

$$- (\tfrac{5}{2}(3)\tfrac{3}{2}(2)\tfrac{1}{2}(1)) - (\tfrac{5}{2}(2)\tfrac{3}{2}(3)\tfrac{1}{2}(1))$$

$$\phi_2 = (\tfrac{5}{2}(1)\tfrac{3}{2}(2)\tfrac{1}{2}(3)) + (\tfrac{5}{2}(2)\tfrac{3}{2}(1)\tfrac{1}{2}(3))$$

$$- (\tfrac{5}{2}(3)\tfrac{3}{2}(1)\tfrac{1}{2}(2)) - (\tfrac{5}{2}(1)\tfrac{3}{2}(3)\tfrac{1}{2}(2))$$

as the antisymmetric combination

$$\eta_1\phi_2 - \eta_2\phi_1$$

For the other independent state with $T = \frac{1}{2}$, we construct the functions

$$\phi_1' = (\tfrac{5}{2}(2)\tfrac{3}{2}(1)\tfrac{1}{2}(3)) + (\tfrac{5}{2}(3)\tfrac{3}{2}(1)\tfrac{1}{2}(2))$$

$$- (\tfrac{5}{2}(2)\tfrac{3}{2}(3)\tfrac{1}{2}(1)) - (\tfrac{5}{2}(1)\tfrac{3}{2}(3)\tfrac{1}{2}(2))$$

$$\phi_2' = (\tfrac{5}{2}(1)\tfrac{3}{2}(2)\tfrac{1}{2}(3)) + (\tfrac{5}{2}(3)\tfrac{3}{2}(2)\tfrac{1}{2}(1))$$

$$- (\tfrac{5}{2}(1)\tfrac{3}{2}(3)\tfrac{1}{2}(2)) - (\tfrac{5}{2}(2)\tfrac{3}{2}(3)\tfrac{1}{2}(1))$$

The antisymmetric combination is then given by

$$\eta_1\phi_2' - \eta_2\phi_1'$$

The states thus obtained are fully antisymmetric and have definite values of T. For simplicity we used the m-scheme and hence the states have definite M values but are linear combinations of states with definite values of J. The three $T = \frac{1}{2}$, $M_T = \frac{1}{2}$, $M = \frac{9}{2}$ states with definite values of J are the $J = \frac{13}{2}$, $J = \frac{11}{2}$ and $J = \frac{9}{2}$ states.

Up until now this discussion has not been limited to the j^n configuration. The eigenstates of the shell model Hamiltonian may contain any admixtures of configurations. The space and spin functions still possess a definite symmetry type. Due to the requirement of full antisymmetry, that type may be described by the symmetry type of the isospin functions. The latter are uniquely characterized by n and T. The eigenvalues of charge independent Hamiltonians are determined by the space and spin eigenfunctions and hence, are independent of the M_T value of the isospin state. If no further information on the wave functions is available, this is the only general result. If, however, the wave functions belong to the j^n configuration, further classification of eigenstates may be achieved. We shall now turn our attention to this problem.

The aim in this section is to introduce the seniority scheme for states of j-protons and j-neutrons. Yet, so far, only the effect of permutations has been discussed in great detail. The reason will become clear very soon. Also in the present case, as in the preceding section, we start by considering the unitary group in $2j+1$ dimensions $U(2j+1)$. We consider unitary transformations in the space spanned by the $2j+1$ space and spin states of a single nucleon. Such transformations induce linear transformations among products of single nucleon wave functions which form states of the j^n configuration. For identical nucleons, or for maximum isospin $T = n/2$, eigenstates of the Hamiltonian are fully antisymmetric in space and spin variables of the nucleons. They belong to the fully antisymmetric irreducible representation of $U(2j+1)$. In the present case, however, other types of symmetry play an important role. The question then arises as to which irreducible representations of $U(2j+1)$ do eigenstates of the Hamiltonian belong. The answer is straightforward. The irreducible representations of $U(2j+1)$ are uniquely characterized by the Young tableaux which determine their symmetry properties under permutations. The $U(2j+1)$ transformations act in the same way on all nucleon space and spin variables. Hence, acting on a state with given symmetry type they do not change that symmetry.

There is a big difference between the action of permutations on eigenstates of the Hamiltonian and the action of unitary transformations on them. Charge independent Hamiltonians are invariant under permutations of space and spin variables. Hence, the action of any permutation on an eigenstate of such a Hamiltonian yields an eigenstate with the same eigenvalue. Such Hamiltonians are, in general, not invariant under the $U(2j+1)$ transformations. As we saw in Section 25, the generators of $U(2j+1)$ are all irreducible tensor operators $U_\kappa^{(k)}$ acting in the $(2j+1)$-dimensional space of single nucleon states. Any charge independent Hamiltonian may be constructed as a linear combination of scalar products of these generators. Hence, such Hamiltonians have no non-vanishing elements between states which belong to two different irreducible representations of $U(2j+1)$. Yet, in general, states within an irreducible representation are far from degenerate. The action of any $U_\kappa^{(k)}$ on a space and spin function with given symmetry usually yields a linear combination of states with different values of J and M. For instance, *all* antisymmetric states transform irreducibly among themselves under operations of $U(2j+1)$. Shell model Hamiltonians which commute with all generators $U_\kappa^{(k)}$ have very special properties. According to Schur's lemma they must have the *same* eigenvalues in all states which belong to an irreducible representation of $U(2j+1)$. This implies that all eigenvalues of antisymmetric states in the j^2 configuration, with J even, must be equal. Similarly, all eigenvalues of the symmetric j^2 states, with J odd, must also be equal to each other.

In Section 25 the only irreducible representation that appeared was the fully antisymmetric one. The irreducible representations of $U(2j+1)$ in the present case are more general. The symmetry properties of allowed space and spin functions are determined, as we saw above, by Young diagrams with two *columns* of length f_1 and f_2. Hence, Also the irreducible representations of $U(2j+1)$ are uniquely characterized by Young diagrams with *two columns* of *lengths* f_1 and f_2, $f_1 \geq f_2$, $f_1 + f_2 = n$. The length f_1 cannot exceed the number of single nucleon space and spin states, namely $2j+1$. Under $U(2j+1)$ transformations, states of a j^n configuration with given symmetry transform among themselves. These transformations, however, are not necessarily irreducible. They may be reduced to several *equivalent* irreducible representations, the number of which is equal to $n_{[f_1f_2]}$ given by (26.5). The number $n_{[f_1f_2]}$ of standard arrangements of the Young diagram with two *rows* is equal to that number for the dual diagram with two *columns* of lengths f_1 and f_2. Thus the *number* of equivalent irreducible representations of $U(2j+1)$ is equal to the *dimension* of the irreducible representation $[f_1f_2]$ of the group of permutations given by (26.5). The dimension of the irreducible representation of $U(2j+1)$ denoted by $N_{[f]}$ depends not only on f_1 and f_2 but also on the dimension k of the space considered, i.e. $k = 2j+1$. The general formula for $N_{[f]}$ is

$$\boxed{N_{[f]} = \prod_{1 \leq i < h \leq k} \left(\frac{f_i - f_h + h - i}{h - i} \right)} \tag{26.7}$$

The total number of independent basis functions is obtained by multiplying $N_{[f]}$ by the number $n_{[f]}$ of such irreducible $U(2j+1)$ representations and summing over all partitions $[f]$. The number of all independent states constructed from products of n wave functions, each of which has k independent states is k^n. Thus, the following equation holds

$$\sum_{[f]} N_{[f]} n_{[f]} = k^n \tag{26.8}$$

In the case of $k = 4(j = \frac{3}{2})$, $n = 3$ we obtain from (26.5), or from a simple consideration, that $n_{[3]} = 1$ $n_{[111]} = 1$, i.e. the symmetric or antisymmetric representation of the group of permutations is one-dimensional. For mixed symmetry states we obtain, as shown above, $n_{[21]} = 2$. For the dimensions of the corresponding $U(2j+1)$ irreducible representations we obtain from (26.7) $N_{[3]} = 20$, $N_{[111]} = 4$, $N_{[21]} = 20$. In this case, (26.8) reads $20 + 2 \times 20 + 4 = 64 = 4^3$.

The basis functions of each irreducible representation of $U(2j+1)$ may range, in general, over states with several different J and M values. The $n_{[f]}$ representations of $U(2j+1)$ due to the Young diagram $[f]$ are all equivalent. We state without proof that in each of them the basis functions $\phi_i^{(1)}, \phi_i^{(2)} \ldots, \phi_i^{(N_{[f]})}$ $(i = 1, \ldots, n_{[f]})$ may be arranged in the following order. The functions $\phi_i^{(1)}$ transform irreducibly among themselves under permutations, the $\phi_i^{(2)}$ transform in the same manner and so do all sets of $\phi_i^{(h)}$ for every $1 \le h \le N_{[f]}$. The various functions $\phi_i^{(h)}$ with given i and various values of h transform irreducibly among themselves under $U(2j+1)$ transformations. The $U(2j+1)$ irreducible representations are equivalent and all $\phi_i^{(h)}$ with various indices i undergo the same transformations. If P is any permutation, acting by it on a given $\phi_i^{(h)}$ yields

$$P\phi_i^{(h)} = \sum P_{ij}\phi_j^{(h)}$$

for any h. The P_{ij} are elements of the matrices which form irreducible representation of the group of permutations. If U is a unitary transformation we obtain

$$UP\phi_i^{(h)} = U\sum_j P_{ij}\phi_j^{(h)} = \sum_j P_{ij}U\phi_j^{(h)} = \sum_{jh'} P_{ij}U_{hh'}\phi_j^{(h')}$$

This is equal also to

$$PU\phi_i^{(h)} = P\sum_{h'} U_{hh'}\phi_i^{(h')} = \sum_{jh'} U_{hh'}P_{ij}\phi_j^{(h')}$$

Thus, the functions $U\phi_i^{(h)}$ for any h (and U) transform under all permutations like $\phi_i^{(h)}$.

To obtain states which are fully antisymmetric, the functions $\phi_i^{(h)}$ for a given value of h should be multiplied by isospin functions η_i and summed over i. The resulting states are characterized by n and T as explained above. In the special case of fully antisymmetric space and spin functions, $f_1 = f_2 = \cdots = f_n = 1$, $f_{n+1} = \cdots = f_r = 0$, they are simply multiplied by fully symmetric functions whose Young diagram is given by $f_1 = n, f_2 = 0$. Eventually, eigenstates of the Hamiltonian with definite values of J and M will be constructed but this will be carried out only at the last stage.

The generators of $U(2j+1)$ are given in configuration space by $U_\kappa^{(k)}$. We can express the commutation relations between generators of $U(2j+1)$ by replacing in (25.7) the tensor operators $(a_j^\dagger \times \tilde{a}_j)_\kappa^{(k)}$ by the unit tensor operators $U_\kappa^{(k)}$ defined by $U_\kappa^{(k)} = \sum U_\kappa^{(k)}(i)$. We thus obtain

$$\boxed{\begin{aligned}[U_{\kappa'}^{(k')}, U_{\kappa''}^{(k'')}] &= \sum_{k\kappa}(-1)^{k+1}\sqrt{2k+1}(1-(-1)^{k+k'+k''}) \\ &\quad \times \begin{Bmatrix} k & k' & k'' \\ j & j & j \end{Bmatrix} (k'\kappa'k''\kappa'' \mid k'k''k\kappa)U_\kappa^{(k)}\end{aligned}} \qquad (26.9)$$

The commutation relation (25.7) was derived by using the anti-commutation relations of the $a_{jm}^\dagger$ and a_{jm}. It is, however, due to the Wigner-Eckart theorem and using it yields a direct derivation of (26.9). The tensors $U_\kappa^{(k)}$ are, in configuration space, equal to the sums $\sum U_\kappa^{(k)}(i)$. Since $U_\kappa^{(k)}(i)$ and $U_\kappa^{(k)}(i')$ commute for different nucleon numbers i and i', it is sufficient to prove that (26.9) holds between single nucleon states. We obtain, by using the Wigner-Eckart theorem for the matrix elements of the unit tensor operators $U_\kappa^{(k)}$, the following expression

$$\langle jm|[u_{\kappa'}^{(k')},u_{\kappa''}^{(k'')}]|jm'\rangle = \sum_{m''}\langle jm|u_{\kappa'}^{(k')}|jm''\rangle\langle jm''|u_{\kappa''}^{(k'')}|jm'\rangle$$

$$-\sum_{m''}\langle jm|u_{\kappa''}^{(k'')}|jm''\rangle\langle jm''|u_{\kappa'}^{(k')}|jm'\rangle$$

$$=\sum_{m''}(-1)^{j-m+j-m''}\begin{pmatrix} j & k' & j \\ -m & \kappa' & m'' \end{pmatrix}\begin{pmatrix} j & k'' & j \\ -m'' & \kappa'' & m' \end{pmatrix}$$

$$-\sum_{m''}(-1)^{j-m+j-m''}\begin{pmatrix} j & k'' & j \\ -m & \kappa'' & m'' \end{pmatrix}\begin{pmatrix} j & k' & j \\ -m'' & \kappa' & m' \end{pmatrix}$$

Using the identity (10.24), the symmetry properties of $3j$-symbols and their relation to Clebsch-Gordan coefficients we obtain (26.9).

The operator, analogous to (25.17), which commutes with all generators of $U(2j+1)$ can now be expressed as

$$\boxed{\sum_k (2k+1)(\mathbf{U}^{(k)}\cdot\mathbf{U}^{(k)})} \tag{26.10}$$

This can be seen directly from the derivations in Section 25 when $(a_j^\dagger \times \tilde{a}_j)_\kappa^{(k)}$ are replaced by $U_\kappa^{(k)}$. The eigenvalues of (26.10) may be found by using its matrix elements in the j^2 configuration. We obtain by using (10.27) and (10.28) the result

$$\langle j^2JM|\sum_k (2k+1)(\mathbf{U}^{(k)}\cdot\mathbf{U}^{(k)})|j^2JM\rangle$$

$$= 2(-1)^{J+1}\sum_k (2k+1)\begin{Bmatrix} j & j & J \\ j & j & k \end{Bmatrix}$$

$$+\frac{2}{2j+1}\sum_k (2k+1) = 2(-1)^{J+1} + 2(2j+1) \tag{26.11}$$

The last equality is due to the sum rule (10.21) of $6j$-symbols. As discussed in Section 9, the j^2 states with even values of J have $T=1$ and those with odd J have $T=0$. Hence, for allowed states, $(-1)^{J+1} = (-1)^T$. The phase $(-1)^T$ may be expressed in terms of the charge exchange operator (26.1). The eigenvalues of P^τ in j^2 states with $T=1$

are $+1$ and those in $T = 0$ states are -1. Thus, $(-1)^T$ may be replaced by $-P^\tau$ and the eigenvalues of (26.10) in j^n states with isospin T are given by

$$-2\sum_{i<k} P^\tau_{ik} + (2j+1)n$$
$$= -\sum_{i<k}(1 + 4\mathbf{t}_i \cdot \mathbf{t}_k) + (2j+1)n$$
$$= -\frac{n(n-1)}{2} - 2(T(T+1) - \tfrac{3}{4}n) + (2j+1)n \qquad (26.12)$$

For maximum isospin, $T = n/2$, (26.12) reduces to the expression obtained from (25.21). It is important to emphasize that the eigenvalues (26.12) of the Casimir operator (26.10) are special cases for irreducible representations of $U(2j+1)$ characterized by Young diagrams with only two columns of lengths f_1 and f_2 given by (26.6).

Let us consider all states of a single nucleon in the j-orbit. There are $2(2j+1)$ such states of space, spin and charge (or isospin). We may then consider unitary transformations in the $(2(2j+1))$-dimensional space which transform states of a single j-nucleon with $m_t = \frac{1}{2}$ and $m_t = -\frac{1}{2}$. They induce linear transformations also among the states of the j^n configuration. Due to the generalized Pauli principle only states which are fully antisymmetric in space, spin and isospin variables are allowed. All these states thus transform irreducibly under transformations of the group $U(2(2j+1))$. If we consider charge independent Hamiltonians, their eigenstates belong to irreducible representations of $U(2j+1)$ multiplied by isospin functions with the dual symmetry type. The latter belong to irreducible representations of another unitary group acting in the two-dimensional space of the isospin variables of a single nucleon. This is the $U(2)$ group which is isomorphic to the $U(2)$ group considered in Section 5. The discussion presented above of $U(2j+1)$ irreducible representations holds for the isospin functions if we put $j = \frac{1}{2}$ and will not be repeated here. Hence, the eigenstates of a charge independent Hamiltonian belong to irreducible representations of a group which is the direct product of $U(2)$ and $U(2j+1)$. This is a subgroup of $U(2(2j+1))$,

$$U(2(2j+1)) \supset U(2) \otimes U(2j+1) \qquad (26.13)$$

The fully antisymmetric irreducible representation of $U(2(2j+1))$ splits into several irreducible representations of the subgroup $U(2) \otimes$

$U(2j+1)$. Due to the requirement of full antisymmetry, the irreducible representations of $U(2)$ and $U(2j+1)$ are obtained from dual Young diagrams. We may thus characterize the symmetry type in space and spin variables by the irreducible representations of $U(2)$. The latter are uniquely specified by n and T. Both numbers are needed here since the group considered is $U(2)$. If only $SU(2)$ is considered, T completely characterizes the irreducible representations as we saw in Section 5 for the case of instrinsic spin S.

The seniority scheme can now be introduced into j^n configurations of protons and neutrons. In Section 25 this was carried out by going from $U(2j+1)$ to its subgroup $Sp(2j+1)$. Here we go from the group $U(2) \otimes U(2j+1)$ to its subgroup $U(2) \otimes Sp(2j+1)$ continuing the chain of groups (26.13) into

$$U(2(2j+1)) \supset U(2) \otimes U(2j+1) \supset U(2) \otimes Sp(2j+1) \tag{26.14}$$

The group $Sp(2j+1)$ is defined as the group that leaves invariant the bilinear antisymmetric form $|j^2 J = 0, M = 0\rangle$. The infinitesimal elements of $Sp(2j+1)$ must annihilate that state. Among all $U(2j+1)$ generators $U_\kappa^{(k)}$, those with odd ranks k do indeed satisfy

$$\boxed{U_\kappa^{(k)} |j^2 J = 0, M = 0\rangle = 0 \qquad k \text{ odd}} \tag{26.15}$$

This relation holds also in the present case of protons and neutrons. The operator $U_\kappa^{(k)}$ acting on $|j^2 J = 0, M = 0\rangle$ must yield, according to the Wigner-Eckart theorem, a state with $J = k$. That state, however, is symmetric in the space and spin variables of the nucleons. Hence, it cannot arise from action of the symmetric operator $U_\kappa^{(k)}$ on the antisymmetric state $|j^2 J = 0, M = 0\rangle$. Odd rank tensor operators $U_\kappa^{(k)}$ are generators of the Lie algebra of $Sp(2j+1)$. Operators satisfying (26.15) form a subalgebra of the $U(2j+1)$ Lie algebra as can also be seen from the commutation relations (26.9).

The seniority scheme is thus defined as the scheme of bases of irreducible representations of $Sp(2j+1)$. Since the $U_\kappa^{(k)}$, k odd, are generators of $Sp(2j+1)$ it follows that they are diagonal in the seniority scheme also in the present case of both protons and neutrons. Let us first look at the irreducible representations of $Sp(2j+1)$ for identical nucleons, $T = n/2$. Fully antisymmetric states with seniority v in the

j^n configuration transform under operations of $Sp(2j+1)$ like antisymmetric states with seniority v in the j^v configuration. Indeed, we obtain for odd values of k

$$\boxed{\begin{aligned} U_\kappa^{(k)}|j^n v\alpha JM\rangle &= \sum_{\alpha' J' M'} \langle j^n v\alpha' J'M'|U_\kappa^{(k)}|j^n v\alpha JM\rangle |j^n v\alpha' J'M'\rangle \\ &= \sum_{\alpha' J' M'} \langle j^v v\alpha' J'M'|U_\kappa^{(k)}|j^v v\alpha' J'M'\rangle |j^n v\alpha' J'M'\rangle \\ & \qquad\qquad\qquad\qquad k \text{ odd} \end{aligned}} \tag{26.16}$$

Going from the j^v to the j^n configuration simply adds $(n-v)/2$ pairs with $J=0$ which are invariant under the $Sp(2j+1)$ transformations. The irreducible representation of $U(2j+1)$ characterized by the Young diagram of one column of length n splits into representations of $Sp(2j+1)$. Each of the latter is characterized by the seniority v which in turn characterizes antisymmetric states of the j^v configuration. These antisymmetric states may be characterized by a Young diagram with the one column of length v, $v \leq n$ and $n-v$ is even. We recall that due to the symmetry between particles and holes, the seniority v cannot exceed the value $(2j+1)/2$. Also in the present case of both protons and neutrons, the irreducible representations of $Sp(2j+1)$ may be characterized by Young diagrams as we shall soon see. The state with seniority $v=0$ in the j^n configuration, constructed with $J=0$ pairs, is itself the basis of an irreducible representation of $Sp(2j+1)$. It transforms into itself like the vacuum state. The corresponding Young diagram has no squares in it and is usually denoted by (0).

In the present case we start with irreducible representations of $U(2j+1)$ characterized by Young diagrams with two columns of lengths f_1 and f_2, $f_1 \geq f_2$, $f_1 + f_2 = n$. When going from $U(2j+1)$ to the sub-group $Sp(2j+1)$ each of these may split into several irreducible representations of the sub-group. The irreducible representations of $Sp(2j+1)$ are determined by symmetry properties of states in a j^v configuration with $v \leq n$ and $n-v$ even. In the present case, however, the states in the j^v configuration need not be fully antisymmetric. The Young diagrams which characterize the irreducible representations of $Sp(2j+1)$ have two columns of lengths v_1 and v_2, $v_1 \geq v_2$ which also satisfy $v_1 \leq f_1$, $v_2 \leq f_2$. The total number of nucleons in the j^v configuration is thus given by

$$\boxed{v_1 + v_2 = v} \tag{26.17}$$

As in the case of identical nucleons, v_1 cannot exceed the value of $(2j+1)/2$. If we consider states with maximum isospin, $T = n/2$, then $f_1 = n$, $f_2 = 0$ and accordingly $v_1 = v$, $v_2 = 0$. The cases in which $v_2 \neq 0$ appear whenever there are both protons and neutrons and we shall now look at such cases.

When going from the j^v configuration to the j^n configuration we add $(n-v)/2$ nucleon pairs. These pair states have $J = 0$ and hence are invariant under the transformations induced by three-dimensional rotations in the $(2j+1)$-dimensional space of j-nucleon states. Therefore, the J values of states do not change by addition of pairs. The $J = 0$ pair states are also invariant under $Sp(2j+1)$ transformations and hence do not change the irreducible representation to which the states belong. These pairs, however, have $T = 1$ and hence, the isospin may change as such pairs are added. In the case of maximum isospin, the total T increases by 1 when each pair is added. This is no longer the case here. Starting from a state with given isospin in the j^v configuration, the j^n state obtained from it by adding pairs may have a different value T of the isospin. Looking at the original state in the j^v configuration we can determine its isospin from its symmetry properties. Applying (26.4) to the j^v configuration we conclude that $(v_1 - v_2)/2$ is the value of its isospin. That isospin is called the *reduced isospin* and is denoted by t (Flowers 1952). We thus complement (26.17) to expressing both v and t by

$$\boxed{v_1 + v_2 = v \qquad \tfrac{1}{2}(v_1 - v_2) = t} \tag{26.18}$$

We see that either v_1 and v_2, or v and t, uniquely specify the irreducible representations of $Sp(2j+1)$ in the case of non-identical nucleons. There may be, however, several ways of coupling of the $(n-v)/2$ pairs to obtain the same value of total isospin T from the initial isospin t. In other words, when going from $U(2j+1)$ to $Sp(2j+1)$, a given irreducible representation of the former is split into several irreducible representations of the latter, some of which may be equivalent. We shall not deal here with this possible complication which cannot arise in the case of maximum isospin $T = n/2$.

The j^2 state with $J = 0$ has $T = 1$. It has, however, seniority $v = 0$ and hence also $t = 0$ (from (26.18) follows $t \leq \frac{1}{2}v$). Hence, all lowest seniority $J = 0$ states in j^n configurations have $v = 0, t = 0$ irrespective of their isospin values. The j^2 states with $J = 2, 4, \ldots, 2j - 1$ also have $T = 1$. Since they have seniority $v = 2$, their reduced isospins are given by $t = 1$. The j^2 states with odd values of J ($J = 1, 3, \ldots, 2j$) have also seniority $v = 2$. Their reduced isospin is thus equal to their total isospin $t = T = 0$. Such states in j^n configurations may have various values of isospin T but they all have $v = 2, t = 0$. Similarly, all $J = j$ states in odd j^n configurations with seniority $v = 1$ have $t = \frac{1}{2}$ like the state of a single j-nucleon.

We can now calculate the eigenvalues of the Casimir operator of the $Sp(2j + 1)$ group in the irreducible representations defined by v and t. The quadratic Casimir operator may be defined according to (25.22) by

$$\boxed{C_{Sp(2j+1)} = 2\sum_{k \text{ odd}}(2k + 1)(\mathbf{U}^{(k)} \cdot \mathbf{U}^{(k)})} \tag{26.19}$$

We may obtain this operator by combining (26.10) with the tensor expansion of the pairing interaction given in (12.18). From (26.11) and (26.12) we obtain for (26.10) the expression

$$\sum_k (2k + 1)(\mathbf{U}^{(k)} \cdot \mathbf{U}^{(k)}) = -\sum_{i<h}(1 + 4(\mathbf{t}_i \cdot \mathbf{t}_h)) + (2j + 1)n \tag{26.20}$$

From the expansion (19.18) we obtain

$$\sum_k (-1)^k (2k + 1)(\mathbf{U}^{(k)} \cdot \mathbf{U}^{(k)}) = 2P - n \tag{26.21}$$

where P is the pairing interaction. From (26.20) and (26.21) we obtain for the Casimir operator the expression

$$C_{Sp(2j+1)} = -2P - \sum_{i<h}(1 + 4(\mathbf{t}_i \cdot \mathbf{t}_h)) + 2(j + 1)n \tag{26.22}$$

The Casimir operator (26.19) is a linear combination of scalar products of odd irreducible tensor operators. Hence, its eigenvalues in states with given v and t in the j^n configuration, are the same as in those states in the j^v configuration. We now evaluate (26.22) in both configurations and equate the eigenvalues. Recalling that in states with seniority v in the j^v configuration there are no pairs with $J = 0$, we see that the eigenvalues of P in such states vanish. We thus obtain

$$-2P(n,T,v,t) - \frac{n(n-1)}{2} - 2(T(T+1) - \tfrac{3}{4}n) + 2(j+1)n$$
$$= -\frac{v(v-1)}{2} - 2(t(t+1) - \tfrac{3}{4}n) + 2(j+1)v \qquad (26.23)$$

From (26.23) we obtain the eigenvalues of the pairing interaction to be

$$\boxed{\begin{aligned} P(n,T,v,t) &= -\frac{n(n-1)}{4} - (T(T+1) - \tfrac{3}{4}n) + n(j+1) \\ &\quad + \frac{v(v-1)}{4} + (t(t+1) - \tfrac{3}{4}v) - v(j+1) \\ &= \frac{n}{4}(4j+8-n) - T(T+1) - \frac{v}{4}(4j+8-v) + t(t+1) \\ &= \frac{n-v}{4}(4j+8-n-v) - T(T+1) + t(t+1) \end{aligned}}$$

(26.24)

The eigenvalues of $C_{Sp(2j+1)}$ are then obtained by putting (26.24) into (26.22) for the j^n configuration which yields

$$\boxed{\begin{aligned} C_{Sp(2j+1)}(v,t) &= -\frac{v(v-1)}{2} - 2(t(t+1) - \tfrac{3}{4}v) + 2v(j+1) = \\ &= \frac{v}{2}(4j+8-v) - 2t(t+1) \end{aligned}}$$

(26.25)

The pairing interaction measures the amount of couplings into $J = 0$ pairs. Its eigenvalues do not depend on J of the state. They are the same for all states which belong to the same irreducible representation of $Sp(2j+1)$. This is also the basic property of the Casimir operator but unlike it, the eigenvalues of P may depend also on the irreducible representation of $U(2j+1)$ to which the states belong. Both operators are quadratic functions of the generators which indicates a linear relation between them. Up to a multiplicative constant we may express P by the analog of (25.28) as

$$F(f_1, f_2) - C_{Sp(2j+1)}(\nu_1, \nu_2) \tag{26.26}$$

For $\nu_1 = f_1, \nu_2 = f_2$, n is equal to ν and the eigenvalues of P are zero. From the vanishing of (26.26) we obtain $F(f_1, f_2) = C_{Sp(2j+1)}(f_1, f_2)$. The multiplicative constant may be determined from the case $n = 2$. The eigenvalues of the pairing interaction are then given by the analog of (25.29) as

$$\boxed{\begin{aligned} P(n, T, \nu, t) &= \tfrac{1}{2}[C_{Sp(2j+1)}(f_1, f_2) - C_{Sp(2j+1)}(\nu_1, \nu_2)] \\ &= \tfrac{1}{2}[C_{Sp(2j+1)}(n, T) - C_{Sp(2j+1)}(\nu, t)] \end{aligned}} \tag{26.27}$$

This result is identical with the expression (26.24) obtained above.

The $SU(2)$ (or $O(3)$) group with the generators (25.25) is a subgroup of $Sp(2j+1)$. The odd tensor operators, with $k = 1$ (see (25.25)), induce in the $(2j+1)$-dimensional space of a single nucleon, the transformation due to a three-dimensional rotation. By going from $U(2j+1)$ first to $Sp(2j+1)$, additional quantum numbers, those of the seniority scheme, have been derived. The last step in the classification scheme is obtained by going from $Sp(2j+1)$ to its $SU(2)$ subgroup. The irreducible representations of $Sp(2j+1)$ may split into irreducible representation of $SU(2)$ which are characterized by J. There may be several states with the same value of J which are contained in the same irreducible representation of $Sp(2j+1)$. Additional quantum numbers (labels) may then be introduced to distinguish between them. This may be done in the j^ν configuration and will hold for all j^n configurations. This last step is a continuation of the group chain

(26.14) into

$$U(2(2j+1)) \supset U_T(2) \otimes U(2j+1)$$
$$\supset U_T(2) \otimes Sp(2j+1) \supset U_T(2) \otimes SU(2) \qquad (26.28)$$

where the subscript T was added to $U(2)$ of isospin to distinguish it from $SU(2)$ of spatial rotations.

Let us make use of the fact that odd tensors are generators of $Sp(2j+1)$ to derive a reduction relation between their matrix elements. We will generalize (26.16) to the case with $T < n$. The expression (26.16) follows from (19.37) or from the discussion of matrix elements of generators following (25.24). Here we write down explicitly the analogous expression which follows from the transformation properties under $Sp(2j+1)$ transformations. States of the j^n configuration which belong to the irreducible representation of $Sp(2j+1)$ defined by v, t transform among themselves when acted on by $U_\kappa^{(k)}, k$ *odd*. The transformation coefficients are determined by the irreducible representation and are independent of n, T or M_T. Hence, they are the same as those in the j^v configuration with $T = t$. Matrix elements of space and spin operators are independent of the charge state M_T. We thus obtain

$$U_\kappa^{(k)}|j^n T M_T v t \alpha J M\rangle$$
$$= \sum_{\alpha'' J'' M''} \langle j^n T M_T v t \alpha'' J'' M''|U_\kappa^{(k)}|j^n T M_T v t \alpha J M\rangle$$
$$\times |j^n T M_T v t \alpha'' J'' M''\rangle$$
$$= \sum_{\alpha'' J'' M''} \langle j^v T = t, M_t v t \alpha'' J'' M''|U_\kappa^{(k)}|j^v T = t, M_t v t \alpha J M\rangle$$
$$\times |j^n T M_T v t \alpha'' J'' M''\rangle \qquad k \text{ odd} \qquad (26.29)$$

Taking the scalar product of (26.29) with $\langle j^n T M_T v t \alpha' J' M'|$ we obtain

$$\langle j^n T M_T v t \alpha' J' M'|U_\kappa^{(k)}|j^n T M_T v t \alpha J M\rangle$$
$$= \langle j^v T = t, M_t v t \alpha' J' M'|U_\kappa^{(k)}|j^v T = t, M_t v t \alpha J M\rangle \qquad (26.30)$$

Thus, odd tensor operators are diagonal in the seniority scheme and their matrix elements are independent of n and T. Odd tensor interactions constructed from them are diagonal in the seniority scheme and have the pairing property, analogous to (20.3).

In addition to odd tensor interactions, any $k = 0$ tensor interaction is diagonal in the seniority scheme (and in any scheme of states with definite values of J). There is a linear combination of scalar products of *even* tensor operators which is diagonal in the seniority scheme. Adding (26.20) to (26.21), we obtain

$$2 \sum_{k \text{ even}} (2k + 1)(\mathbf{U}^{(k)} \cdot \mathbf{U}^{(k)}) = 2P + 2jn - \sum_{i<h} (1 + 4(\mathbf{t}_i \cdot \mathbf{t}_h)) \tag{26.31}$$

The r.h.s. of (26.31), for states with definite isospins T, is diagonal in the seniority scheme. It was shown (de-Shalit and Talmi 1963) that the tensor expansion of *any* two-body interaction which is diagonal in the seniority scheme, may include terms with even tensors only if they are proportional to the linear combination on the l.h.s. of (26.31). Due to (26.31) this linear combination of even tensor interactions may be replaced by the sum of an odd tensor interaction, a $k = 0$ tensor interaction and isospin dependent interaction $(\tau_1 \cdot \tau_2)$ multiplied by a constant factor.

In the next section we will apply the seniority scheme developed above to the study of some properties of actual nuclei.

27

Applications of Seniority With Isospin. Symmetry Energy

For states with $T < n/2$ in nuclear j^n configurations, seniority does not play the important role which it does for states of identical nucleons ($T = n/2$). Nevertheless, certain features of the seniority scheme for valence protons and neutrons may give some insight into properties of actual nuclei. In this section we discuss some applications of seniority, mostly to binding energies of nuclei. These are energies of states with lowest seniority, namely $J = 0$, $v = 0$, $t = 0$ states of even nuclei and states with $J = j$, $v = 1$, $t = \frac{1}{2}$ in odd nuclei. We begin by a brief description of these states and a calculation of the magnetic moments of odd-even nuclei.

We first consider states with $v = 0$, $t = 0$, $J = 0$ in the j^n configuration. The total isospin T of such states may have only certain values. The state with $T = n/2$ is allowed since it is the state with $v = 0$, $J = 0$ of identical j-nucleons. To find the other allowed values we proceed as follows. We start from a wave function in which nucleons 1 and 2 are coupled to a state with $T = 1$, $J = 0$, nucleons 3,4 are coupled to such a state and so on. The isospins of the pairs are coupled to the total isospin T. We now perform the antisymmetrization in two steps. We first permute the space, spin and isospin variables of both nucleons in any $T = 1$, $J = 0$ pair with any two others which are also in a $T = 1$, $J = 0$ state. Such permutations are even and the permuted functions should be all *added*. In the second step we consider the other permutations necessary for obtaining a fully antisymmetric function. Looking at the result of the first step, we see a function which is symmetric in the various $T = 1$, $J = 0$ pairs. This yields a restriction on the allowed values of the total isospin T. The allowed states are those for which $n/2$ pairs each with $T = 1$ can couple symmetrically.

To determine such allowed states, we may use the M_T-scheme. The state with maximum value of M_T, which is equal to $n/2$, is the state in which all M_T of the pairs are equal to 1. Since this is the highest value of M_T, the state has $T = M_T = n/2$. There is only one symmetric state with $M_T = n/2 - 1$ obtained by $n/2 - 1$ M_T projections equal to 1 and the M_T of one pair equal to 0. Hence, that state is the $M_T = n/2 - 1$ state with $T = n/2$. In the case of $M_T = n/2 - 2$ there are *two* symmetric states. In one state, $n/2 - 1$ projections are equal to 1 and the M_T of one pair is equal to -1. In the other state, $n/2 - 2$ projections are equal to 1 and the M_T of two pairs are equal to 0. Hence, one linear combination of these $M_T = n/2 - 2$ states has $T = n/2$ and the one orthogonal to it has $T = n/2 - 2$.

Counting symmetric states can be continued. Consider states with $M_T = n/2 - 2k$, where $k = 1, \ldots, n/4$ or $(n-2)/4$. There is a state with $2k$ pairs whose M_T projections are equal to 0, whereas the other pairs have $M_T = 1$. There is an independent state with $2k - 2$ pairs with $M_T = 0$ (and two pairs with $M_T = 0$ replaced by a pair with $M_T = 1$ and another pair with $M_T = -1$), an independent state with $2k - 4$ pairs with $M_T = 0$ and so on. the number of such states is thus equal to $k + 1$. If we consider symmetric states with $M_T = n/2 - 2k - 1$, the number of pairs with $M_T = 0$ is odd in that case and all states must have at least one pair with $M_T = 0$. Therefore, their number is still equal to $k + 1$. Hence, we obtain the allowed values of T of states with $v = 0$, $t = 0$ in the j^n configuration to be

$$T = \frac{n}{2}, \frac{n}{2} - 2, \frac{n}{2} - 4, \ldots, 1 \text{ or } 0 \tag{27.1}$$

Any one of these allowed states may be expressed in terms of c.f.p. as follows

$$\psi(j^n v=0,t=0,T,J=0)$$

$$= [j^{n-1}(v_1=1,t_1=\tfrac{1}{2},T_1=T+\tfrac{1}{2},J=j)jT,J=0|\}j^n v=0,t=0,T,J=0]$$

$$\times\, \psi(j^{n-1}(v_1=1,t_1=\tfrac{1}{2},T_1=T+\tfrac{1}{2},J=j)j_nTJ=0)$$

$$+ [j^{n-1}(v_1=1,t_1=\tfrac{1}{2},T_1=T-\tfrac{1}{2},J_1=j)jT,J=0|\}j^n v=0,t=0,T,J=0]$$

$$\times\, \psi(j^{n-1}(v_1=1,t_1=\tfrac{1}{2},T_1=T-\tfrac{1}{2},J=j)j_nTJ=0) \qquad (27.2)$$

The squares of these c.f.p. may be directly calculated by using the orthogonality of the c.f.p. and the relation (15.37). Denoting the squares of the c.f.p. in (27.2) by $X(T+\frac{1}{2})$ and $X(T-\frac{1}{2})$ respectively, we obtain

$$X(T+\tfrac{1}{2})+X(T-\tfrac{1}{2})=1$$

$$\frac{n-2}{n}\left(T(T+1)+\frac{n}{n-2}\times\frac{3}{4}\right) \qquad (27.3)$$

$$=(T^2+2T+\tfrac{3}{4})X(T+\tfrac{1}{2})+(T^2-\tfrac{1}{4})X(T-\tfrac{1}{2})$$

The solution of these equations yields the following values of the c.f.p.

$$[j^{n-1}(v_1=1,t_1=\tfrac{1}{2},T_1=T+\tfrac{1}{2},J=j)jT,J=0|\}j^n v=0,t=0,T,J=0]^2 = \frac{(n-2T)(T+1)}{n(2T+1)}$$

(27.4)

$$[j^{n-1}(v_1=1,t_1=\tfrac{1}{2},T_1=T-\tfrac{1}{2},J=j)jT,J=0|\}j^n v=0,t=0,T,J=0]^2 = \frac{(n+2T+2)T}{n(2T+1)}$$

(27.5)

In the special case of $T = n/2$, the c.f.p. (27.4) vanishes whereas the c.f.p. (27.5) becomes equal to 1 as in (29.42). In the case $T = 0$, if possible, (27.5) vanishes whereas (27.4) is equal to 1.

Next we consider states in odd j^n configurations with lowest seniorities, $v = 1$, $t = \frac{1}{2}$ (and $J = j$). Possible (fractional) parents of such states may have seniority quantum numbers $v_1 = 0$, $t_1 = 0$ ($J_1 = 0$), $v_1 = 2$, $t_1 = 1$ ($J_1 > 0$, even) and $v_1 = 2$, $t_1 = 0$ (J_1 odd). A state with given T may have, in general, fractional parents with $T_1 = T + \frac{1}{2}$ and $T_1 = T - \frac{1}{2}$. There are, however, restrictions on the values of T_1 in the parents with $t = 0$. The parent with $v_1 = 0$, $t_1 = 0$ must have, according to (27.1), $T_1 = (n-1)/2, (n-5)/2, \dots$. Therefore, if in the state considered

$$\boxed{T = n/2, (n-4)/2, \dots \qquad \text{Case I}}$$

then the only possible value of T_1 is $T_1 = T - \frac{1}{2}$. On the other hand, if

$$\boxed{T = (n-2)/2, (n-6)/2, \dots \qquad \text{Case II}}$$

then the only possible isospins of the $v = 0$, $t = 0$ states are $T_1 = T + \frac{1}{2}$. The dependence of T_1 of parents with $v = 2$, $t = 0$ is opposite to that. In those parent states, two nucleons are coupled to $T = t = 0$ and only $n - 3$ nucleons carry the isospin T. The same considerations leading to (27.1), yield the possible values of T_1 in such cases to be $T_1 = (n-3)/2, (n-7)/2 \dots$. As a result, in Case I, the possible values of T_1 of parents with $v = 2$, $t = 0$ are $T_1 = T + \frac{1}{2}$, whereas in Case II they are $T_1 = T - \frac{1}{2}$.

We next consider the dependence of the c.f.p. of the $v = 1$, $t = \frac{1}{2}$ state on J_1. For $n = 3$ this dependence is given in (15.36) for $J_1 > 0$, even and J_1 odd, by $\sqrt{2J_1 + 1}$. For larger values of n, there are in the seniority scheme reduction formulae for the c.f.p. As in Section 19, the factors yielding the reduction of $v = 1$, $t = \frac{1}{2}$ c.f.p. from the j^n configuration to the $j^{v+2} = j^3$ configuration depend on the seniority quantum numbers v_1, t_1, as well as on n and T, but *not* on the value of J_1. Hence, the dependence on J_1 of the c.f.p. considered here, is still given by $\sqrt{2J_1 + 1}$. These properties of the c.f.p. allow a simple calculation of the c.f.p. (or rather the squares of c.f.p.) of the $v=1$, $t=\frac{1}{2}$ state.

The state with $v=1$, $t=\frac{1}{2}, J=j$ in the j^n configuration can be obtained from a principal parent with $v_0=0$, $t_0=0$, $J_0=0$ in the j^{n-1} configuration. Since the $n-1 \rightarrow n-2$ c.f.p. of this principal parent are given by (27.4) and (27.5), we may use a recursion formula for c.f.p. to determine the $n \rightarrow n-1$ c.f.p. of that principal parent. The recursion formula is a generalization of (15.29) to the case with isospin. In the same way that (15.29) was derived, we obtain here the general recursion relation

$$\boxed{\begin{aligned} & n[j^{n-1}(\alpha_0 T_0 J_0) jTJ|\} j^n[\alpha_0 T_0 J_0] jTJ] \\ & \quad \times [j^{n-1}(\alpha_1 T_1 J_1) jTJ|\} j^n[\alpha_0 T_0 J_0] TJ] \\ & \quad = \delta_{\alpha_1 \alpha_0} \delta_{T_1 T_0} \delta_{J_1 J_0} - (n-1) \sum (-1)^{T_0+T_1+J_0+J_1} \\ & \quad \times \sqrt{(2T_0+1)(2T_1+1)(2J_0+1)(2J_1+1)} \\ & \quad \times \begin{Bmatrix} T_2 & \frac{1}{2} & T_1 \\ T & \frac{1}{2} & T_0 \end{Bmatrix} \begin{Bmatrix} J_2 & j & J_1 \\ J & j & J_0 \end{Bmatrix} \\ & \quad \times [j^{n-2}(\alpha_2 T_2 J_2) jT_0 J_0|\} j^{n-1} \alpha_0 T_0 J_0] \\ & \quad \times [j^{n-2}(\alpha_2 T_2 J_2) jT_1 J_1|\} j^{n-1} \alpha_1 T_1 J_1] \end{aligned}} \tag{27.6}$$

In the case considered here,

$$\alpha_0 \equiv (v_0 = 0,\ t_0 = 0), \qquad J_0 = 0,$$

and

$$\alpha_2 \equiv (v_2 = 1,\ t_2 = \tfrac{1}{2}), \qquad J_2 = j$$

If we put in (27.6) these values, as well as $\alpha_1 \equiv \alpha_0,\ T_1 = T_0,\ J_1 = J_0$ we obtain from (27.6) the expression

$$n[j^{n-1}(v_0 = 0, t_0 = 0, T_0, J_0 = 0) jTJ = j|\} j^n v = 1, t = \tfrac{1}{2}, T, J = j]^2$$

$$= 1 + \frac{(n-1)}{2j+1}(2T_0+1)\sum_{T_2}\begin{Bmatrix} T_2 & \frac{1}{2} & T_0 \\ T & \frac{1}{2} & T_0 \end{Bmatrix}$$

$$\times [j^{n-2}(v_2 = 1, t_2 = \tfrac{1}{2}, T_2, J_2 = j) jT_0J_0 = 0|\} j^{n-1} v_0 = 0, t_0 = 0, T_0J_0 = 0]^2 \tag{27.7}$$

As explained above, the allowed values of T_0 in Case I is $T_0 = T - \frac{1}{2}$, whereas in Case II, $T_0 = T + \frac{1}{2}$. The $n-2 \to n-1$ c.f.p. on the r.h.s. of (27.7) are given by (27.4) for $T_2 = T_0 + \frac{1}{2}$ and by (27.5) for the case $T_2 = T_0 - \frac{1}{2}$. With these values of c.f.p. and the actual values of the $6j$ symbols, we obtain for the squares of the c.f.p. on the l.h.s. of (27.7) the results

$$\begin{aligned} &\frac{4j+4-n-2T}{2n(2j+1)} \qquad \text{Case I} \\ &\frac{4j+6-n+2T}{2n(2j+1)} \qquad \text{Case II} \end{aligned} \tag{27.8}$$

The squares of the other c.f.p. may now be obtained by using various relations between them. One relation is the normalization of the c.f.p. The other relation is (15.37) and the third relation is (15.22). If we use (15.21) for the pairing interaction, whose eigenvalues are given by (26.24) we do not obtain another independent relation. The summations over J_1 in these relations can be carried out directly. Thus, these three relations are sufficient to determine, apart from sign, the coefficients of $\sqrt{2J_1+1}$ in the c.f.p. with $J_1 > 0$, even, with $T_1 = T + \frac{1}{2}$ and $T_1 = T - \frac{1}{2}$ as well as the coefficient of $\sqrt{2J_1+1}$ in the c.f.p. with odd values of J_1. There is only one coefficient for J_1 odd, since in that case $T_1 = T + \frac{1}{2}$ in Case I and $T_1 = T - \frac{1}{2}$ in Case II.

We can thus express the square of the c.f.p. of the state with $v = 1$, $t = \frac{1}{2}, T$ and $J = j$ in the j^n configuration as follows.

Case I $\quad T = \frac{n}{2}, \frac{n-4}{2}, \ldots$

$$[j^{n-1}(v_1 = 0, t_1 = 0, T_1 = T - \tfrac{1}{2}, J_1 = 0)jTJ = j|\}j^n v = 1, t = \tfrac{1}{2}, T, J = j]^2$$

$$= \frac{4j + 4 - n - 2T}{2n(2j + 1)}$$

$$[j^{n-1}(v_1 = 2, t_1 = 0, T_1 = T + \tfrac{1}{2}, J_1)jTJ = j|\}j^n v = 1, t = \tfrac{1}{2}, T, J = j]^2$$

$$= \frac{(n - 2T)(2J_1 + 1)}{4n(j + 1)(2j + 1)} \qquad J_1 \text{ odd}$$

$$[j^{n-1}(v_1 = 2, t_1 = 1, T_1 = T + \tfrac{1}{2}, J_1)jTJ = j|\}j^n v = 1, t = \tfrac{1}{2}, T, J = j]^2$$

$$= \frac{(n - 2T)(2T + 3)(2J_1 + 1)}{4n(j + 1)(2j - 1)(2T + 1)} \qquad J_1 > 0, \text{ even}$$

$$[j^{n-1}(v_1 = 2, t_1 = 1, T_1 = T - \tfrac{1}{2}, J_1)jTJ = j|\}j^n v = 1, t = \tfrac{1}{2}, T, J = j]^2$$

$$= \frac{\{4(j + 1)(2T^2 + nT - 1) + n - 2T\}(2J_1 + 1)}{2n(j + 1)(2j - 1)(2j + 1)(2T + 1)} \qquad J_1 > 0, \text{ even}$$

(27.9)

For states with maximum isospin, $T = n/2$, only the first and fourth c.f.p. do not vanish. The first c.f.p. is then equal to (19.26) in which we put $v = 1$, $J = j$. The last c.f.p. is equal, for $T = n/2$, to the result of (19.28) with $v = 1$, $J = j$, combined with (15.11).

In the other case we obtain

Case II $\quad T = \frac{n-2}{2}, \frac{n-6}{2}, \ldots$

$$[j^{n-1}(v_1 = 0, t_1 = 0, T_1 = T + \tfrac{1}{2}, J_1 = 0)jTJ = j|\}j^n v = 1, t = \tfrac{1}{2}, T, J = j]^2$$

$$= \frac{4j + 6 - n + 2T}{2n(2j+1)}$$

$$[j^{n-1}(v_1 = 2, t_1 = 0, T_1 = T - \tfrac{1}{2}, J_1)jTJ = j|\}j^n v = 1, t = \tfrac{1}{2}, T, J = j]^2$$

$$= \frac{(n + 2T + 2)(2J_1 + 1)}{4n(j+1)(2j+1)} \qquad J_1 \text{ odd}$$

$$[j^{n-1}(v_1 = 2, t_1 = 1, T_1 = T + \tfrac{1}{2}, J_1)jTJ = j|\}j^n v = 1, t = \tfrac{1}{2}, T, J = j]^2$$

$$= \frac{\{4(j+1)[(T+1)(n - 2T - 2) + 1] - (n + 2T + 2)\}(2J_1 + 1)}{2n(j+1)(2j-1)(2j+1)(2T+1)}$$

$$J_1 > 0, \text{ even}$$

$$[j^{n-1}(v_1 = 2, t_1 = 1, T_1 = T - \tfrac{1}{2}, J_1)jTJ|\}j^n v = 1, t = \tfrac{1}{2}, T, J = j]^2$$

$$= \frac{(n + 2T + 2)(2T - 1)(2J_1 + 1)}{4n(j+1)(2j-1)(2T+1)} \qquad J_1 > 0, \text{ even}$$

(27.10)

In the special case of $n = 3$, $T = \frac{1}{2}$, the last c.f.p. in (27.10) vanishes and the first three c.f.p. reduce to (15.36) obtained above.

In (27.9) and (27.10) only the squares of c.f.p. are given. Still, there are applications for which this information is sufficient. We can apply these squares of c.f.p. for the calculation of magnetic moments of states with $v = 1$, $t = \frac{1}{2}$ and $J = j$. We make use of the general formula (15.41) and put in it $J = j$. There are two possible values of $T_1, T_1 = T + \frac{1}{2}$ and $T_1 = T - \frac{1}{2}$ and we substitute the actual values of the $6j$-symbols. We then substitute the values of c.f.p. and obtain terms with $J_1 = 0$, J_1 odd and J_1 even. The summations over J_1 may be carried out by using the sum rules (10.20) and (10.21). We then obtain for the g-factors the following formulae

$$\boxed{\begin{gathered} g = \frac{g_\pi + g_\nu}{2} + M_T \frac{g_\pi - g_\nu}{2} \frac{4(j+1)(T+1) - n + 2T}{4(j+1)T(T+1)} \\ \text{Case I} \qquad T = \frac{n}{2}, \frac{n-4}{2}, \ldots \end{gathered}} \tag{27.11}$$

$$\boxed{\begin{gathered} g = \frac{g_\pi + g_\nu}{2} - M_T \frac{g_\pi - g_\nu}{2} \frac{4(j+1)T + n + 2T + 2}{4(j+1)T(T+1)} \\ \text{Case II} \qquad T = \frac{n-2}{2}, \frac{n-6}{2}, \ldots \end{gathered}} \tag{27.12}$$

In the case of maximum isospin, $T = n/2$, the result (27.11) reduces to

$$g = \frac{g_\pi + g_\nu}{2} + \frac{2M_T}{n} \frac{g_\pi - g_\nu}{2} \tag{27.13}$$

In the case of only j-protons, $M_T = n/2$, (27.13) reduces to $g = g_\pi$, whereas for j-neutrons only, $M_T = -n/2$, it reduces to $g = g_\nu$. In the case of $n = 3$, $T = \frac{1}{2}$, (27.12) reduces to (15.42) obtained above. The formulae (27.11) and (27.12) are due to Teitelbaum (1954).

In order to apply (27.11) and (27.12) to ground state of nuclei, we consider the case with Z' protons and N' neutrons in the j-orbit. The isospin T in those expressions, in the lowest states, is equal to

$T = |Z' - N'|/2$. We take Z' to be odd and N' to be even. The following results hold for the other case, N' odd and Z' even, if we interchange Z' and N' as well as g_π and g_ν. Since in the states considered here, with $v = 1$, $t = \frac{1}{2}$, the total spin J is equal to j, (27.11) and (27.12) hold also for magnetic moments if g, g_π, g_ν are replaced by μ, μ_π, μ_ν respectively. According to Case I and Case II, we distinguish between two cases. I. $Z' > N'$. In this case $T = M_T = (Z' - N')/2 = (n - 2N')/2$, and since N' is even, this situation corresponds to Case I. The expression (27.11) assumes then the form

$$\boxed{\mu = \mu_\pi \left(1 - \frac{N'}{4(j+1)(T+1)}\right) + \mu_\nu \frac{N'}{4(j+1)(T+1)}} \tag{27.14}$$

II. $Z' < N'$. In this case $T = -M_T = (N' - Z')/2 = (n - 2Z')/2$, and due to Z' being odd, this corresponds to Case II. The expression (27.12) reduces in this case to

$$\boxed{\mu = \mu_\pi \left(1 - \frac{2j+1-N'}{4(j+1)(T+1)}\right) + \mu_\nu \frac{2j+1-N'}{4(j+1)(T+1)}} \tag{27.15}$$

Looking at (27.14) and (27.15), we see that the neutrons are not in a $J = 0$ state and they contribute to the magnetic moment. If $N' = 0$ in (27.14) we obtain, $\mu = \mu_\pi$. Also if $N' = 2j + 1$, the neutron j-orbit completely filled, (27.15) reduces to $\mu = \mu_\pi$. In other cases, there are deviations of μ from the single proton value which are in qualitative agreement with experimental data. For given value of Z' if $Z' > N'$, the larger the N', the larger the deviation and if $Z' < N'$, the smaller the N', the larger the deviation. An example of such behavior is seen in Sc isotopes. The (effective) magnetic moment of a single $1f_{7/2}$ proton can be determined from μ of the $^{41}_{21}\mathrm{Sc}$ ground state, $\mu = 5.54$ n.m. The magnetic moments of $^{43}\mathrm{Sc}$, $^{45}\mathrm{Sc}$ and $^{47}\mathrm{Sc}$ are 4.62, 4.76 and 5.34 n.m. respectively. In most stable nuclei $N' > Z'$ and in them we find an example of the dependence on Z' in (27.15) where Z' replaces N' and μ_π and μ_ν are interchanged. The magnetic moments of nuclei with 27 neutrons in $^{47}_{20}\mathrm{Ca}_{27}$, $^{49}_{22}\mathrm{Ti}_{27}$ and $^{51}_{24}\mathrm{Cr}_{27}$ are -1.38, $-.110$ and $-.93$ n.m. respectively.

The largest deviation from single nucleon moments are found in *mirror nuclei* where $N' = Z' - 1$ and $N' = Z' + 1$. In the examples presented above, the nucleus ^{43}Sc is a member of a mirror pair. The other member is $^{43}_{22}\text{Ti}_{21}$ whose magnetic moment has not been measured. According to (27.14) and (27.15) the *sum* of magnetic moments of members of mirror pairs should be equal to $\mu_\pi + \mu_\nu$. The sum should show no deviation from the sum of calculated, or from effective, single nucleon moments. Measured magnetic moments of several mirror pairs of light nuclei have been measured. Deviations of sums of magnetic moments of mirror pairs are indeed smaller than those of the individual members.

In the preceding section we saw that any interaction which is a linear combination of scalar products of odd tensor operators is diagonal in the seniority scheme. This remains true if a constant interaction (a $k = 0$ term) is added with an arbitrary coefficient, leading to a term proportional to $n(n-1)/2$. In the present case, with both protons and neutrons in the j-orbit, another interaction may be added without destroying this property. We may add the Casimir operator of $U(2j+1)$ defined in (26.10) or simply the two-body interaction $\sum \mathbf{t}_i \cdot \mathbf{t}_h$ with an arbitrary coefficient. This leads to a $T(T+1)$ term in the eigenvalues of the shell model Hamiltonian. If we restrict our choice of interactions to those which do not contain explicitly the isospin operators, we may replace the latter interaction by another one. Instead of using the expression (26.12), we may add the equivalent expression (26.10) with an arbitrary coefficient. Since the odd tensor terms and the $k = 0$ term may be included with arbitrary coefficients, we may simply add to the interaction, with arbitrary coefficient, the special linear combination of even tensor interactions

$$\sum_{k>0,\,\text{even}} (2k+1)(\mathbf{u}_1^{(k)} \cdot \mathbf{u}_2^{(k)}) \tag{27.16}$$

It was shown by de-Shalit and Talmi (1963) that *any* interaction which is diagonal in the seniority scheme can be expressed as a sum of an odd tensor interaction, a $k = 0$ term and a term proportional to (27.16).

As we saw in preceding sections, the seniority scheme is very useful in the treatment of j^n configurations of identical nucleons in actual nuclei. In cases with $j \geq \frac{9}{2}$ two body interactions may have non-vanishing matrix elements which are non-diagonal in the seniority scheme. We saw that the experimentally determined values of such non-diagonal matrix elements are very small compared to the difference between the corresponding diagonal elements. As a result, seniority is a good quantum number in those cases. The natural question which arises here is how good is the seniority scheme for states with $T < n/2$. Given the two body matrix elements with both $T = 1$ and $T = 0$, the size of matrix elements which are non-diagonal in the seniority scheme may be calculated. We will address this question in the following but before that, we consider a special case. For $j \leq \frac{3}{2}$ there are no states with the same values of T and J in any j^n configuration. Hence, any two body interaction is trivially diagonal in the seniority scheme. Let us make use of this fact for j^n configurations with $j = \frac{3}{2}$ (the $j = \frac{1}{2}$ case is of little interest).

Another way to see that any two body interaction is indeed diagonal in $(\frac{3}{2})^n$ configurations is to construct a simple expression for any such interaction. There are four states in $(\frac{3}{2})^n$ configurations—those with $J = 0, 2$ with $T = 1$ and $J = 1, 3$ with $T = 0$. Any two-body scalar operator which has in these states the same eigenvalues as the given interaction may be used for calculating energies in all $(\frac{3}{2})^n$ configurations. Such as operator can be defined by the linear combination (Racah 1952)

$$a + b2(\mathbf{t}_1 \cdot \mathbf{t}_2) + c2(\mathbf{j}_1 \cdot \mathbf{j}_2) + dP_{12} \tag{27.17}$$

where P_{12} is the pairing interaction. The operator (27.17) is equal to the operator (16.3) to which the isospin term has been added. The coefficients in (27.17) may be determined to reproduce the eigenvalues $V_{TJ} = \langle j^2TJM|V|j^2TJM\rangle$. To do that, we first sum (27.17) over all nucleon pairs obtaining

$$\frac{n(n-1)}{2}a + b[T(T+1) - \tfrac{3}{4}n] + c[J(J+1) - \tfrac{15}{4}n] \\ + d\left[\frac{n-v}{4}(14 - n - v) - T(T+1) + t(t+1)\right] \tag{27.18}$$

In view of (26.12) and (26.25), the eigenvalues (27.18) are eigenvalues of a linear combination of the Casimir operators of $U(2(2j+1))$ and its chain of subgroups $U(2j+1)$, $Sp(2j+1)$ and $SU(2)$. In (27.18), only two-body interactions are included which implies that the linear Casimir operator of $U(2(2j+1))$, proportional to n, appears in the linear combination multiplied by an appropriate coefficient. It cancels the single nucleon terms of the quadratic Casimir operators. To include single particle energies, the term $\epsilon_j n$ should be added to (27.18). For $j > \frac{3}{2}$, (27.18) is no longer equivalent to the most general, charge independent, two-body interaction.

Putting in (27.18) $n = 2$ and the appropriate values of T and J we obtain the equations

$$V_{10} = a + \tfrac{1}{2}b - \tfrac{15}{2}c + 4d \qquad V_{12} = a + \tfrac{1}{2}b - \tfrac{3}{2}c$$
$$V_{01} = a - \tfrac{3}{2}b - \tfrac{11}{2}c \qquad V_{03} = a - \tfrac{3}{2}b + \tfrac{9}{2}c \tag{27.19}$$

These equations can be solved for any values of the given two-body matrix elements. The parameters of (27.18) are then given by

$$a = \tfrac{1}{4}(3V_{12} + V_{03}) \qquad b = \tfrac{1}{10}(5V_{12} - 2V_{03} - 3V_{01})$$
$$c = \tfrac{1}{10}(V_{03} - V_{01}) \qquad d = \tfrac{1}{20}(5V_{10} - 5V_{12} + 3V_{03} - 3V_{01}) \tag{27.20}$$

As in the case of identical nucleons, the energy eigenvalues (27.18) assume a simple form when applied to states with lowest seniorities. In states with $J = 0$, $v = 0$, $t = 0$, we obtain for (27.18) the form

$$\frac{n(n-1)}{2}a + b[T(T+1) - \tfrac{3}{4}n] - \tfrac{15}{4}cn - d[T(T+1) - \tfrac{3}{4}n] + d\frac{n}{4}(11 - n)$$

In $J = j$ states with $v = 1$, $t = \frac{1}{2}$, (27.18) assumes the form

$$\frac{n(n-1)}{2}a + b[T(T+1) - \tfrac{3}{4}n] - \tfrac{15}{4}c(n-1) - d[T(T+1) - \tfrac{3}{4}n] + d\frac{n-1}{4}(10 - n)$$

Both expressions may be combined in the single expression

$$\frac{n(n-1)}{2}(a - \tfrac{1}{2}d) + (b - d)[T(T+1) - \tfrac{3}{4}n] + (5d - \tfrac{15}{2}c)\left[\frac{n}{2}\right] \tag{27.21}$$

where $[n/2]$ is the largest integer not exceeding $n/2$. If the states with lowest seniorities are the ground states, binding energies in $(\frac{3}{2})^n$ configurations from which that of the closed shells was subtracted are then given by

$$\boxed{nC + \frac{n(n-1)}{2}\alpha + \beta[T(T+1) - \tfrac{3}{4}n] + \gamma\left[\frac{n}{2}\right] + \text{Coulomb Energy}} \tag{27.22}$$

The coefficient C in (27.22) is the single nucleon energy. The coefficients α, β, and γ in (27.22) are the following combinations of a, b, c and d

$$\alpha = a - \tfrac{1}{2}d \qquad \beta = b - d \qquad \gamma = 5d - \tfrac{15}{2}c \tag{27.23}$$

The dependence of energies only on α, β and γ, is not limited to states with lowest seniorities. For instance, from (27.19) follows the relation

$$V_{10} - V_{12} = -6c + 4d = \tfrac{4}{5}\gamma \tag{27.24}$$

It is relatively easy to obtain the expressions (27.18) and (27.22). It is much more difficult to find nuclear $(\frac{3}{2})^n$ configurations of protons and neutrons to apply to them those formulae. Protons and neutrons occupy the same j-orbit only in light nuclei. In heavier nuclei, due to the Coulomb energy, there is a neutron excess in stable nuclei and valence protons and neutrons occupy different orbits. The lowest orbit with $j = \frac{3}{2}$ is the $1p_{3/2}$ orbit. In the lightest nuclei, where this orbit is being filled, the spin-orbit interaction is not sufficiently strong. Those light nuclei do not exhibit the characteristic features of jj-coupling. The next $j = \frac{3}{2}$ orbit to be considered is the $1d_{3/2}$ orbit. In nuclei with proton or neutron numbers between 16 and 20, where the $1d_{3/2}$ should be the valence orbit, the $2s_{1/2}$ and $1d_{5/2}$ orbits are not very far in energy. It is not expected that the description of those nuclei in terms of pure $(1d_{3/2})^n$ configurations will be a good approximation. Still, we examine these nuclei and see how well their energies may be described by (27.18) and (27.22). We shall begin by looking at binding energies since ground states may be less affected by mixing with other configurations involving $2s_{1/2}$ and $1d_{5/2}$ orbits. A special case is the ground state of ^{36}Cl which has $J = 2$ and thus must be assigned $v = 2$, $t = 1$. Therefore, its binding energy is not given by (27.22). Inserting $v = 2$, $t = 1$ into (27.18) for $n = 4$, $T = 1$, we obtain for the interaction energy the expression

$$6a - b - 9c + 4d = 6(a - \tfrac{1}{2}d) - (b - d) + (6d - 9c) = 6\alpha - \beta + \tfrac{6}{5}\gamma \tag{27.25}$$

Thus, also the binding energy of ^{36}Cl may be expressed in terms of C, α, β and γ by (27.25). We shall first determine them from binding energies of other nuclei and then check how well the ^{36}Cl binding energy is reproduced.

We take binding energies of the nuclei considered and subtract from them the binding energy of the nucleus $^{32}_{16}\mathrm{S}_{16}$ where the $1d_{5/2}$ and $2s_{1/2}$ orbits are assumed to be closed. We then subtract from these binding energy differences the Coulomb energy of the $2d_{3/2}$ protons. This is taken to be the same for all isotopes and may be obtained from differences between binding energies of mirror odd-even nuclei or analog states of even-even nuclei. We then determine the parameters C, α, β and γ by a least squares fit of (27.22) (less the Coulomb energies) to the experimental binding energy differences. The values of the parameters which best reproduce the data are

$$C = 8.679 \qquad \alpha = .105 \qquad \beta = -1.851 \qquad \gamma = 3.538 \text{ MeV} \tag{27.26}$$

It should be kept in mind that the binding energies have been taken as *positive* numbers.

The binding energy differences calculated by (27.22) with the parameters (27.26), including the Coulomb energies, are given along with the experimental values in Table 27.1. As can be seen, the agreement between calculated and experimental values is rather good. The root mean square (r.m.s.) deviation is defined by

$$\left[\sum_{i=1}^{N}(\Delta E_i)^2/(N-k)\right]^{1/2} \tag{27.27}$$

where ΔE_i are the differences between calculated and experimental values, N is the number of data and k is the number of free parameters. The r.m.s. deviation turns out to be rather small, its value being .149 MeV. We may use the values (27.26) to calculate the binding energy of ^{36}Cl. Putting the values (27.26) of the parameters into (27.25)

TABLE 27.1 *Binding energies of nuclei in the* $1d_{3/2}$ *shell. The binding energy of* ^{32}S *was subtracted from the experimental values. Energies are given in* MeV.

Nucleus	Binding Energy Experimental	Binding Energy Calculated	Nucleus	Binding Energy Experimental	Binding Energy Calculated
^{33}S	8.641	8.679	^{35}Ar	19.682	19.493
^{34}S	20.058	20.076	^{36}Ar	34.936	34.801
^{35}S	27.044	27.113	^{37}Ar	43.723	43.900
			^{38}Ar	55.562	55.715
^{36}S	36.933	36.869	^{37}K	36.793	36.943
^{33}Cl	2.276	2.396	^{38}K $J=0$	48.736	48.759
^{34}Cl $J=0$	13.784	13.793	^{39}K	61.943	61.768
^{35}Cl	26.429	26.383	^{39}Ca	54.630	54.586
^{37}Cl	45.320	45.237	^{40}Ca	70.272	70.314

we obtain 35.16 MeV as compared with the experimental value of 35.01 MeV.

In spite of the rather good agreement between calculated and experimental binding energies in Table 27.1 it cannot be concluded that $(1d_{3/2})^n$ configurations give adequate description of nuclei in that region. Had that been the case, according to (27.18) $J = 2$ states with $v = 2$, $t = 1$ should lie above the $J = 0$, $v = 0$, $t = 0$ states in the same j^n configuration and isospin T, at the position given by (27.19). The value of γ in (27.26) yields the expected energy spacing to be equal to $0.8\gamma = 2.83$ MeV. This is considerably higher than the measured values of 2.09 – 2.17 MeV in ^{34}Ar, ^{34}S, ^{34}Cl and ^{38}Ar or 2.27 MeV in ^{38}K. In ^{36}Ar we expect also the same spacing but experimentally the value of 1.97 is found. This is not surprising since $(\frac{1}{2})^+$ levels and $(\frac{5}{2})^+$ levels do not lie very high above $(\frac{3}{2})^+$ levels in odd nuclei. Admixtures due to $2s_{1/2}$ and $1d_{5/2}$ orbits are expected. In ^{33}S the $(\frac{1}{2})^+$ level is at 0.84 MeV above the $(\frac{3}{2})^+$ ground state. A $(\frac{5}{2})^+$ level lies at 1.97 MeV. The situation is certainly better in ^{37}Cl where those levels are at 1.73 and 3.09 MeV respectively. Only in ^{39}K those spacing are rather large, 2.52 and higher than 4 MeV respectively. Excited states are closer in energy to states of higher configurations and may be more affected by mixing with them. Still, the features displayed by the mass formula (27.22) with parameters like (27.26) represent the general trends of nuclear binding energies as will be discussed in more detail below. For the moment let us look at some excited states in the nuclei considered here and obtain some semi-quantitative information on the parameters in (27.18). This will clarify some features observed in nuclei with $(1d_{3/2})^n$ configurations.

In the $(1d_{3/2})^2$ configuration in ^{34}Cl the $J = 3$ level lies 0.146 MeV above the $J = 0$ ground state. In the corresponding hole configuration in ^{38}K the $J = 3$ state is the ground state whereas the $J = 0$ level lies 0.130 MeV above it. If we make these two levels have the same energy, $V_{10} = V_{03}$, we obtain from (27.19) a linear relation between b, c, and d. We may then use the expression (27.23) of β and γ together with their experimental values (27.11) to determine the values of the parameters a, b, c and d. These values turn out to be

$$a = .549 \qquad b = -.872 \qquad c = .181 \qquad d = .979 \text{ MeV} \tag{27.28}$$

Using these values in the expression (27.18) we may calculate all levels in $(1d_{3/2})^n$ configurations.

An immediate conclusion follows directly from (27.19). The spacing between the $J = 3$ and $J = 1$ levels in $(1d_{3/2})^n$ configurations is equal to $10c = 1.81$ MeV. Since binding energies were taken to be positive, this means that the $J = 3$ has *more* binding energy. The $J = 1$ state is thus predicted to lie 1.81 MeV *above* the $J = 3$ state. This prediction is only qualitatively correct. In the two hole configuration in ^{38}K there is a $J = 1^+$ level .46 MeV above the $J = 3^+$ ground state. Another $J = 1^+$ state lies 1.7 MeV above the ground state. These two $J = 1$ states are probably strongly admixed. In ^{34}Cl there are two $J = 1^+$ levels at .32 and .52 MeV above the $J = 3$ level. From (27.18) we see that the $J = 3^+ - J = 1^+$ spacing should be the same in *all* $(1d_{3/2})^n$ configurations. In ^{36}Cl there are two 1^+ states, one lies .38 and the other .81 MeV above the 3^+ level.

Another feature of ^{36}Cl is worth examining. Its ground state, like the ground state of the mirror nucleus ^{36}K, has $J = 2$. This $J = 2$ state, like the $J = 3,1$ states has $T = 1$. The $J = 0$ state, however, has $T = 2$ and should therefore lie, according to (27.18), $6c - 4b = 4.21$ MeV above the $J = 2$ ground state. This is, in fact, the analog of the ^{36}S ground state and their energies differ only by the Coulomb energy. Both $J = 2$ and $J = 3$ states have $T = 1$ and seniority $v = 2$ but differ in their reduced isospins which are $t = 1$ for the $J = 2$ state and $t = 0$ for the $J = 3$ one. Hence, the $J = 3$ state should lie $2d - 6c = 0.871$ MeV *above* the $J = 2$ state. Indeed, the ground state of ^{36}Cl has $J = 2$ and a $J = 3^+$ level lies 0.778 MeV above it.

Three nucleon $(1d_{3/2})^3$ configurations with $T = \frac{1}{2}$ are expected in ^{35}Cl and in ^{37}Ar (the hole configuration $1d_{3/2}^{-3}$). The allowed states of two neutrons and a proton in the $j = \frac{3}{2}$ orbit are obtained by coupling the allowed neutron spins $J_\nu = 0, 2$ with $J_\pi = \frac{3}{2}$. There are two $J = \frac{3}{2}$ states, both with seniority quantum numbers $v = 1$, $t = \frac{1}{2}$. One of them is a $T = \frac{3}{2}$ state, the analog of the ^{35}S ground state, and the other has $T = \frac{1}{2}$. The states with $J = \frac{1}{2}, \frac{3}{2}, \frac{5}{2}, \frac{7}{2}$ have all $T = \frac{1}{2}$, $v = 3$ and hence $t = \frac{1}{2}$. Using the formula (27.18) with the values (27.28) of the parameters we find that the $J = \frac{3}{2}$ state with $v = 1$, $t = \frac{1}{2}$ is the ground state and the $J = \frac{7}{2}, J = \frac{5}{2}, J = \frac{1}{2}$ states are expected at 2.72, 3.90, 5.44 MeV above it. The $J = \frac{7}{2}$ state is observed in ^{35}Cl at 2.645 MeV above the ground state and in ^{37}Ar at 2.217 MeV above its ground state. The low lying $(\frac{1}{2})^+$ and $(\frac{5}{2})^+$ levels in both nuclei are most probably due to single nucleon excitations from the $2s_{1/2}$ and $1d_{5/2}$ orbits.

In ^{36}Ar we expect to see the $(1d_{3/2})^4$ configuration. The allowed states of two protons and two neutrons are obtained by coupling the allowed states of the protons with $J_\pi = 0, 2$ with those of the neutrons, $J_\nu = 0, 2$. We thus find two $J = 0$ states, three $J = 2$ states and single states with $J = 1, 3, 4$. One $J = 0$ state has $T = 2$ and is the analog of the ^{36}S ground state. The states with $J = 1, 3$ and one $J = 2$ state have $T = 1$ and are the analogs of the ^{36}Cl levels. These states lie high in energy. The remaining states with $T = 0$ are the $J = 0$ state with $v = 0$, $t = 0$, one $J = 2$ state with $v = 2$, $t = 1$ and states with $J = 2, 4$ which have $v = 4$, $t = 0$. Using (27.18) with the values (27.28) we find that the ground state is the $J = 0$ state and the $J = 2$, $v = 2$, $t = 1$ state is calculated to lie $0.8\gamma = 2.83$ MeV above it. Experimentally it is at 1.97 MeV above the $J = 0$ ground state. The calculated position of the state with $J = 4$, $v = 4$, $t = 0$ is at 6.17 MeV above the ground state, and the one with $J = 2$, $v = 4$, $t = 0$ at 8.7 Mev. The observed $J = 4$ level in ^{36}Ar is at a much lower energy, 4.41 MeV and there is a $J = 2$ level at 4.44 MeV above the ground state.

From the comparison made above, it is clear that the description of the nuclei considered here in terms of pure $(1d_{3/2})^n$ configurations is not a very good one. Detailed shell model calculations (Wildenthal 1984) in which a very good description was obtained for nuclei between ^{16}O and ^{40}Ca have demonstrated strong configuration mixing involving the $1d_{5/2}$, $2s_{1/2}$ and $1d_{3/2}$ orbits. Still, several features of energy levels follow from the simple, very approximate description. This is particularly true for binding energies which we will now consider for more complicated j^n configurations.

In j^n configurations with $j \geq \frac{5}{2}$ a two body interaction need not be diagonal in the seniority scheme. We must refer to the experimental data to find out the nature and properties of the effective interaction between protons and neutrons in $T = 0$ states. An important property of Hamiltonians which are diagonal in the seniority scheme holds also in states with $T < n/2$:

Spacings between j^n levels with the same
T, v and t are independent of n

This property holds in case of the interactions considered in the beginning of this section. According to (26.30) matrix elements of odd tensor interactions are independent of n and T. Other interactions which may be added, contribute only linear and quadratic terms in n and a $T(T+1)$ term. Neither of these interactions affect *spacings* between energy levels with the same value of T. For example, for Hamiltonians diagonal in the seniority scheme, levels with $T = t = 1$, $v = 2$, $J > 0$, even, in j^2 configurations should be equal to those in the proton-particle neutron-hole configuration. All these levels in the particle-hole configuration have the same value of isospin $T = (2j-1)/2$ (the $J = 0$, $v = 0$, $t = 0$ level has the higher isospin value $T = (2j+1)/2$ and lies much higher). We can check whether this feature is indeed observed in spectra of actual nuclei.

We saw in Section 21 that level spacings in $(1f_{7/2})^n$ configurations with $T = n/2$ are fairly independent of n. That result should hold in that case for any two body interaction. This is no longer the case if both protons and neutrons occupy the $1f_{7/2}$ orbits. We can compare the spacings between the $J = 2,4,6$ levels with $T = 1$ in ^{42}Ca to those levels with $T = 3$ in ^{48}Sc. This comparison is made in Fig. 27.1 and it is evident that seniority must be badly broken by the $T = 0$ effective interactions in $(1f_{7/2})^n$ configurations.

The situation in the $1g_{9/2}$ orbit is very similar. In Section 21 it was demonstrated that seniority is a good quantum number in $(1g_{9/2})^n$ configurations with $T = n/2$. To test the situation in $T < n/2$ states, we can compare the spacings between $J = 2,4,6,8$ levels with $T = 1$, $v = 2$, $t = 1$ and those levels with $T = 4$ in the particle-hole configuration in $^{90}_{41}$Nb$_{49}$. Looking at Fig. 27.2 we see that seniority is badly broken also in this case. Not only are the spacings not equal, they are almost reversed. The lowest j value for which seniority (with isospin) may be broken is $j = \frac{5}{2}$. The evidence in nuclei where valence nucleons are expected to occupy the $1d_{5/2}$ orbit, also indicates that the $T = 0$ part of the interaction is *not* diagonal in the seniority scheme. Because of configuration mixings the evidence is not as striking as the one presented here.

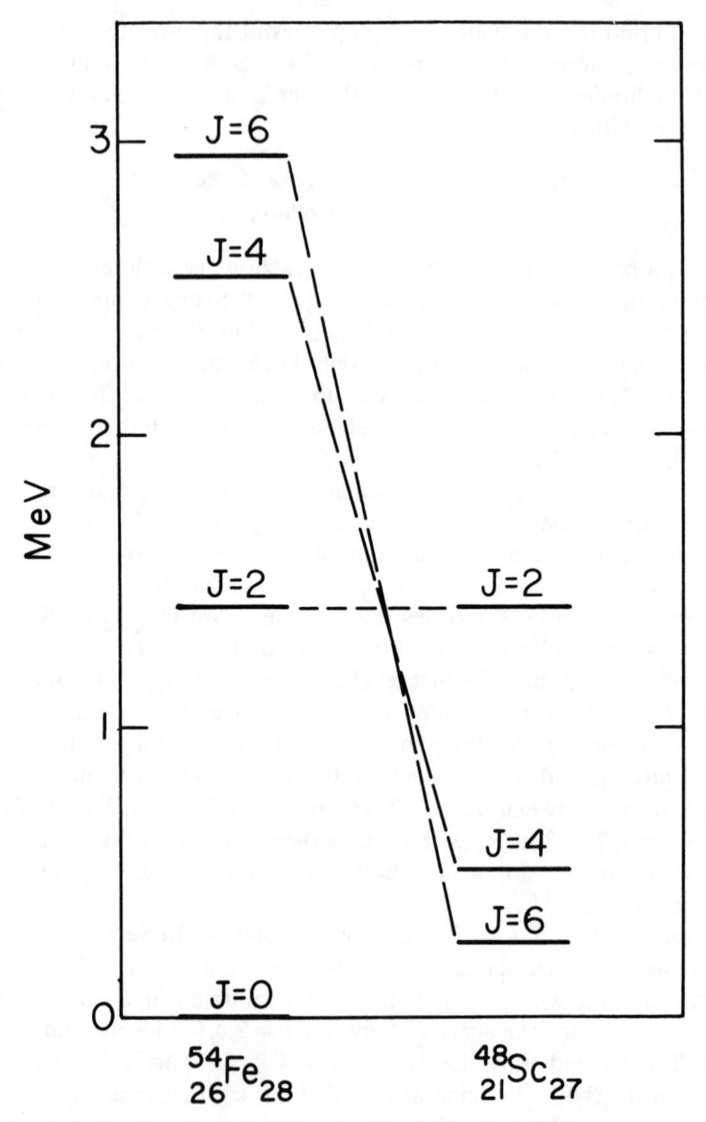

FIGURE 27.1. *Experimental two nucleon and nucleon-hole levels in the* $1f_{7/2}$ *shell.*

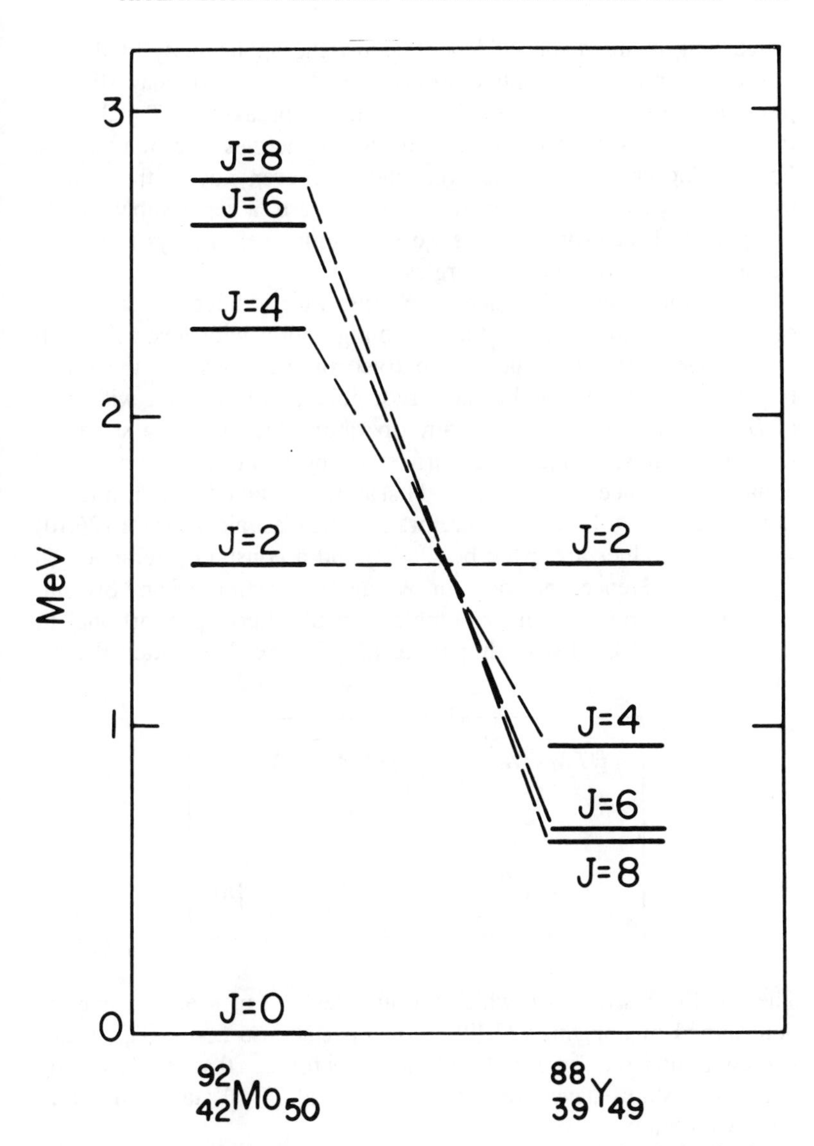

FIGURE 27.2. *Experimental two nucleon and nucleon-hole levels in the* $1g_{9/2}$ *orbit.*

Keeping in mind the evidence against the applicability of the seniority scheme to j^n configurations with $T < n/2$, we may still explore its consequences. The effect of seniority breaking of $T = 0$ effective interactions may not be very strong on energies of ground states. For binding energies of odd-even and even-even nuclei, the seniority scheme yields here, as in the $T = n/2$ case, a remarkably simple expression. That expression, as we shall soon see, displays some important features of binding energies.

As in the case of identical nucleons, we consider first averages of energies of all states in the j^n configuration with given T. Such an average is proportional to the trace of the matrix of the effective interaction defined by the irreducible representation with n,T of $U(2j+1)$. Such a trace of any operator which is not a scalar of $U(2j+1)$ vanishes. Hence the trace of any two-body interaction is equal to the trace of the $U(2j+1)$ scalars contained in that interaction. The $U(2j+1)$ scalar operators are the Casimir operator (26.10) whose eigenvalues are given by (26.12) and a constant interaction (a $k = 0$ term). Hence, the trace, or average interaction, of any two nucleon interaction is a linear combination of a term proportional to $n(n-1)/2$ and the two-body part of (26.12). We thus obtain the result

$$\boxed{\begin{aligned}&\overline{\langle j^n T M_T J M | \sum_{i<h} V_{ih} | j^n T M_T J M \rangle}\\ &\qquad = \alpha \frac{n(n-1)}{2} + \beta[T(T+1) - \tfrac{3}{4}n]\end{aligned}} \tag{27.29}$$

The coefficients α and β which are independent of n and T, may be determined by applying (27.29) to the special case $n = 2$. This yields two equations, one for $T = 1$ and the other for $T = 0$ states. Recalling that $T = 1$ states have even values of J and $T = 0$ states have odd ones, we obtain

$$\alpha + \tfrac{1}{2}\beta = \sum_{J\,\text{even},M} \langle j^2 JM | V | j^2 JM \rangle \Big/ \sum_{J\,\text{even}} (2J+1)$$

$$= \frac{1}{j(2j+1)} \sum_{J\,\text{even}} (2J+1) V_J$$

$$\alpha - \tfrac{3}{2}\beta = \sum_{J\,\text{odd},M} \langle j^2JM|V|j^2JM\rangle \Big/ \sum_{J\,\text{odd}}(2J+1)$$

$$= \frac{1}{(j+1)(2j+1)} \sum_{J\,\text{odd}}(2J+1)V_J$$

From these equations we obtain for the coefficients in (27.14) the expressions

$$\boxed{\begin{aligned}\alpha &= \frac{3}{4j(2j+1)}\sum_{J\,\text{even}}(2J+1)V_J \\ &\quad + \frac{1}{4(j+1)(2j+1)}\sum_{J\,\text{odd}}(2J+1)V_J \\ \beta &= \frac{1}{2j(2j+1)}\sum_{J\,\text{even}}(2J+1)V_J \\ &\quad - \frac{1}{2(j+1)(2j+1)}\sum_{J\,\text{odd}}(2J+1)V_J\end{aligned}} \qquad (27.30)$$

We now consider in j^n configurations average interaction energies of states of with given T and v,t. Such an average is proportional to the trace of the matrix of the two body interaction in the space of a given irreducible representation of $Sp(2j+1)$. Hence, only the $Sp(2j+1)$ scalars contained in that interaction contribute to the trace. The average of any two body interaction is thus a linear combination of the eigenvalues of these scalars in the given irreducible representation. The $Sp(2j+1)$ scalars are the Casimir operator of $U(2j+1)$ of which $Sp(2j+1)$ is a subgroup, the $k=0$ term and $C_{Sp(2j+1)}$ defined by (26.19) whose eigenvalues are given by (26.25). Taking the two body part of these three terms, after slight rearrangements of the coefficients, we may express the average energies of any two body interaction by

$$\overline{\langle j^n vtTM_TJM|\sum_{i<h} V_{ih}|j^n vtTM_TJM\rangle}$$

$$= \alpha' \frac{n(n-1)}{2} + \beta'[T(T+1) - \tfrac{3}{4}n]$$

$$+ \gamma' \left[\frac{n-v}{2}(2j+2) - \frac{n(n-1)}{4} - T(T+1) \right.$$

$$\left. + \frac{3}{4}n + \frac{v(v-1)}{4} + t(t+1) - \frac{3}{4}v \right]$$

$$= \alpha \frac{n(n-1)}{2} + \beta[T(T+1) - \tfrac{3}{4}n]$$

$$+ \gamma \frac{n-v}{2} + \frac{\gamma}{2j+2} \left[\frac{v(v-1)}{4} + t(t+1) - \frac{3}{4}v \right] \quad (27.31)$$

The coefficients α, β, and γ in (27.31) are independent of n, T, v, t. They may be determined by applying (27.31) to the case of $n = 2$ where we obtain three equations for them. For $J = 0$ we put in (27.31) $T = 1$, $v = 0$, $t = 0$, for the average of $T = 1$ states with even values of J we put $v = 2$, $t = 1$ and for the average of those with odd J values, $v = 2$, $T = t = 0$. Solving the equations

$$\alpha + \tfrac{1}{2}\beta + \gamma = V_0$$

$$\alpha + \frac{1}{2}\beta + \frac{\gamma}{2j+2}$$

$$= \frac{1}{(j+1)(2j-1)} \sum_{J>0 \text{ even}} (2J+1)V_J = \bar{V}_2$$

$$\alpha - \frac{3}{2}\beta - \frac{\gamma}{2j+2}$$

$$= \frac{1}{(j+1)(2j+1)} \sum_{J \text{ odd}} (2J+1)V_J = \bar{V}_1 \quad (27.32)$$

we obtain the following values of the coefficients

$$\alpha = \frac{1}{4(2j+1)}[(6j+5)\bar{V}_2 + (2j+1)\bar{V}_1 - 2V_0]$$

$$\beta = \frac{1}{2(2j+1)}[(2j+3)\bar{V}_2 - (2j+1)\bar{V}_1 - 2V_0]$$

$$\gamma = \frac{2(j+1)}{2j+1}(V_0 - \bar{V}_2) \tag{27.33}$$

Thus, the average energies of any two body interaction in a group of states in the j^n configuration with given T and seniority quantum numbers v,t are equal to the linear combination of V_0, $\bar{V}_1$ and $\bar{V}_2$ given by (27.31).

The relation (27.31) holds for *any* two body interaction. If it is diagonal in the seniority scheme, the averages in (27.31) are averages of its eigenvalues. If it is not diagonal in the seniority scheme, it will have non-vanishing matrix elements connecting states with given v,t to others with different seniority quantum numbers. In that case, the relation (27.31) holds for the *expectation values* of the interaction defined by the states with given v and t. The trace is invariant under unitary transformations on the matrix considered. Hence, it is not even important to bring to diagonal form the Hamiltonian sub matrix defined by the states with given v and t.

The expression (27.31) assumes a particularly simple form for states with lowest seniorities in even-even and odd-even nuclei. The last term on the r.h.s. of (27.31) vanishes for states with $v = 0$, $t = 0$ and states with $v = 1$, $t = \frac{1}{2}$. There is only one state, with $J = 0$ with $v = 0$, $t = 0$ with given T in j^n configurations with n even. There is also only one state, with $J = j$, with $v = 1$, $t = \frac{1}{2}$ with given T in odd j^n configurations. Hence, in those cases, the r.h.s. of (27.31) is equal to the eigenvalue in such a state of the two-body interaction which is diagonal in the seniority scheme. If the interaction is not, (27.31) gives the expectation value in such a state. Hence, by adding single nucleon energies and the Coulomb energy, the seniority scheme leads to the binding energy formula (Talmi and Thieberger 1956).

$$\begin{aligned} \text{B.E.}(j^n g.s.) = \text{B.E.}(n=0) + nC + \frac{n(n-1)}{2}\alpha \\ + \beta[T(T+1) - \tfrac{3}{4}n] + \gamma\left[\frac{n}{2}\right] \\ + \text{Coulomb energy} \end{aligned} \tag{27.34}$$

The term whose coefficient is γ is the $\gamma(n-v)/2$ term in (27.31).

The binding energy formula (27.34) is the extension of (20.14) for nuclei with $T < n/2$. For $T = n/2$ it reduces to (20.14) with $\alpha + \beta/2$ becoming the coefficient of the quadratic term in n. Like (20.14), (27.34) contains a quadratic term and a pairing term. In addition, it contains the *symmetry energy* term proportional to $T(T+1)$. The formula (27.22) which was derived for the $j = \frac{3}{2}$ orbit is identical with (27.34). The values (27.26) of the coefficients are typical for the behavior of binding energies. The coefficient of the quadratic term is small, the pairing term is large and attractive and the coefficient β is large and repulsive. This implies that the higher the T, the higher the binding energy which is in agreement with experiment. Apart from the small quadratic term, (27.34) gives rise to *exact saturation* for binding energies of nuclei with valence j-nucleons.

The value of α, the coefficient of the quadratic term is small but it is attractive. We saw in Section 21 that for identical nucleons, i.e. for $T = n/2$, the quadratic term is repulsive. This feature is due to the existence of the large symmetry energy term in (27.34). For $T = n/2$, the quadratic dependence has the coefficient $\alpha + \beta/2$. Since β is rather large and repulsive, the quadratic term in nuclei with $T = n/2$ is repulsive. This is consistent with the observed facts about nuclear saturation.

We may try to apply the binding energy formula to actual nuclei. We can try to fit (27.34) to binding energies in some j^n configurations with parameters C, α, β, γ which best reproduce the data (Talmi and Thieberger 1956). We can consider for instance the $1d_{5/2}$-orbit and repeat the analysis we did for the $1d_{3/2}$ orbit. Using binding energies of nuclei between ^{16}O and ^{28}Si we obtain the best fit with the following values of the parameters

$$\alpha = .325 \qquad \beta = -2.728 \qquad \gamma = 5.697 \qquad C = 4.799 \text{ MeV}$$

These values show the same qualitative features as those in (27.26). The agreement between calculated and experimental binding energies is only fair. The r.m.s. deviation, defined by (27.27), is in this case equal to 1.22 MeV, more than 7 times larger than in the $1d_{3/2}$ orbit. This agreement may still be impressive since the binding energies considered here range between 4 MeV for ^{17}O to 136 MeV for ^{28}Si (from the actual binding energies the binding energy of ^{16}O was subtracted). Similar results were obtained for the $1f_{7/2}$ shell. This may suggest that ground states are less affected by configuration mixings than excited states. Taking the expectation values of the two body interaction in a state with definite seniority may not be too bad an approximation. Breakdown of seniority leads to drastic changes in level spacings but ground state energies may not be affected much.

For $j > \frac{3}{2}$ the information about ground states included in (27.34) cannot be used to calculate energies of excited states. The formula (27.31) may be used to calculate average energies of groups of states with the same v, t (and T). It may then be applied also to odd odd nuclei. In the case of $(1d_{3/2})^n$ configurations we concluded that the ground state of $^{36}_{17}\mathrm{Cl}_{19}$ should have $J = 2$ which was lower than the $J = 3$, $J = 1$ states. The reason was due to the $t(t+1)$ term in the pairing energy which is equal to $2\gamma(2j+2)$ for $J = 2$ and vanishes for $J = 1, 3$. In j^n configurations with $j > \frac{3}{2}$ in odd-odd nuclei, with $N \neq Z$, we can only deduce that the center of mass of the levels with $J = 2, 4, \ldots, 2j-1$ which have $v = 2$, $t = 1$ should be lower than that of the $J = 1, 3, \ldots, 2j$ levels with $v = 2$, $t = 0$. Both these groups have the same value of isospin $T = \frac{1}{2}|N - Z|$ and hence have the same symmetry energy. The difference in the position of their centers of mass arises only from the pairing term. This prediction agrees well with the observed spectra of $(1d_{5/2})^n$ configurations in ^{20}F and ^{24}Na and of $(1f_{7/2})^n$ configurations in ^{44}Sc, ^{46}Sc, ^{48}Sc and ^{48}V. In nuclei with $N = Z$ the situation is different. There, states with odd values of J have isospins $T = 0$ and hence, are lower, due to the symmetry energy than states with even values of $J > 0$ which have isospins $T = 1$.

The binding energy formula (27.34) with its basic interaction terms may be a good approximation even in the presence of configuration mixings. Using it and its generalizations to protons and neutrons in different shells, the Garvey-Kelson relations (1966) between nuclear binding energies as well as other relations were derived (Garvey et al., 1969). We shall not further discuss the mass formula (27.34). Still, as a simple exercise, we tried to fit (27.34) to nuclei between ^{28}Si and ^{36}Ar. Just by looking at the data and due to detailed calculations (Wildenthal 1984) it is clear that in these nuclei there is much configuration mixing. Nevertheless, we took all binding energies, irrespective of the ground state spin being $\frac{3}{2}$ or $\frac{1}{2}$, and tried to reproduce them with (27.34). The values of the parameters which gave the best fit are

$$C = 8.466 \qquad \alpha = .072 \qquad \beta = -2.038 \qquad \gamma = 3.359 \text{ MeV}$$

These values are very similar to those in (27.26). The agreement between calculated and experimental values is not at all bad. The r.m.s. deviation turns out to be 0.23 MeV, only 50% larger than the one in $1d_{3/2}$ orbit. This demonstrates that (27.34) may be useful even where there is no a priori reason for it to work. It also serves as a warning against attaching too much significance to the good agreement obtained for binding energies of $(1d_{3/2})^n$ configurations.

28

Seniority in l^n Configurations With $T = n/2$. The $O(2l+1)$ Group

The classification of states according to the irreducible representations of the unitary group and its sub-groups may be applied to the case of LS-coupling. For this case, in atomic spectroscopy, it was introduced by Racah (1943). Let us first consider the simpler case of identical nucleons or maximum isospin, $T = n/2$. Some Hamiltonians may be expressed in that case in terms of the space coordinates of the particles and their momenta. Some explicit dependence on spin operators which they may have, can be eliminated in some cases. If such dependence occurs through the scalar product $\mathbf{s}_1 \cdot \mathbf{s}_2$, the interactions can be expressed in terms of the spin exchange operator (10.64). Due to the Pauli principle, the action of P^{σ}_{12} on any allowed wave function may be obtained by applying to it $-P^{x}_{12}$, the Majorana operator which exchanges the *space* coordinates of particles 1 and 2. If the dependence on the spins is more complicated, as in the case of spin orbit interaction or tensor forces, this is no longer the case. The Hamiltonians we consider in this section are those which contain central interactions only. They may be expressed by operators acting only on space coordinates and thus are invariant under permutations of the space coordinates of the particles. It follows that any eigenstate, which is a function of space coordinates, is transformed by a permutation to another eigenstate with the same eigenvalue. Among all eigenstates which belong to the same eigenvalue we can define a set of functions ϕ_i which form a basis. All other functions are linear combinations of these basis functions. Apart from possible accidental degeneracies, these functions ϕ_i form a basis of an irreducible representation of the group of permutations. In the beginning of Section 26 we saw examples of such considerations. In the following, we shall use a similar approach. The steps we take are very similar to those taken in Section 26.

As in Section 26, the linearly independent functions ϕ_i do not represent different *physical* eigenstates. They must be multiplied by appropriate spin functions to obtain a fully antisymmetric state. The functions ϕ_i have a certain symmetry type which can be expressed by a Young diagram. All functions ϕ_i with a given symmetry type must be multiplied by spin functions which belong to the *dual* Young diagram. This guarantees that the spin functions are symmetric where the space functions are antisymmetric and vice versa.

There are only two possible states for the spin of each particle. As a result, spin functions may be antisymmetrized only with respect to two particles. The Young tableaux must therefore contain at most *two* rows, of lengths f_1 and f_2 ($f_1 \geq f_2$). The sum of f_1 and f_2 is equal to n, the total number of particles. Among all spin functions with the $[f_1 f_2]$ symmetry, the one with maximum value of M_S may have f_1 particles in $m_s = +\frac{1}{2}$ states and f_2 particles in $m_s = -\frac{1}{2}$ states. No more particles may be in states with $m_s = \frac{1}{2}$ since otherwise the function will vanish when the antisymmetrization prescribed by $[f_1 f_2]$ will be applied. Hence, applying to any of these spin functions the operator S^+ gives zero. We conclude that such functions have $S = M_s = \frac{1}{2}(f_1 - f_2)$ so that

$$\boxed{f_1 + f_2 = n \qquad \tfrac{1}{2}(f_1 - f_2) = S} \tag{28.1}$$

We start with a spin function χ in which the first f_1 nucleons are in $m_s = \frac{1}{2}$ states and the others in $m_s = -\frac{1}{2}$ states and apply to it the symmetry operations prescribed by $[f_1 f_2]$ followed by any permutation. It can be shown that there is only one set of linearly independent spin functions χ_i whose number is $n_{[f_1 f_2]}$—the number of standard arrangements in the given Young diagram.

Operating on that spin function χ with S^- and repeating the procedure we obtain $M_s = \frac{1}{2}(f_1 - f_2) - 1$ for which there is also only one set of $n_{[f_1f_2]}$ linearly independent spin functions. Hence, these functions are the $M_s = S - 1$ spin functions where S is given by (28.1). This may be continued until $f_1 - f_2$ particles, in addition to the initial f_2 particles are in the $m_s = -\frac{1}{2}$ state. In this case, $M_s = \frac{1}{2}[f_2 - (f_1 - f_2) - f_2] = -\frac{1}{2}(f_1 - f_2)$ which is equal to $-S$. No more particles may be moved from $m_s = \frac{1}{2}$ to $m_s = -\frac{1}{2}$ states, otherwise the functions vanish upon the antisymmetrization prescribed by $[f_1f_2]$.

The space functions that may be combined with these spin functions to form antisymmetric states must have only *two columns* of length f_1 and f_2. They are uniquely characterized by n and S as given in (28.1). In this way, the spin determines uniquely the symmetry of the eigenstates in the space coordinates. The higher the spin, the more symmetric the spin functions and the less symmetric the space functions. The eigenvalues of the interaction energy depend on the spatial symmetry. Hence, for identical particles there may be strong dependence of the interaction energy on the spin value of the state. That dependence is totally due to the symmetry properties and not to an explicit magnetic spin-spin interaction or any other dynamical effect. The mutual interaction of atomic electrons is the repulsive Coulomb potential. The repulsion is stronger if the electrons are closer on the average as they are in certain symmetric functions of their space coordinates. The lesser the spatial symmetry, the lesser the repulsion. This is the basis of Hund's rule in atomic spectra: *states with higher spins lie lower*.

The considerations described above hold for any Hamiltonian which is a function of the space coordinates of the particles and their conjugate momenta. The important feature of such Hamiltonians is the fact that they are invariant under permutations of space coordinates of the particles. The Hamiltonians considered here are shell model Hamiltonians but these considerations are not limited to them. Even when we limit our attention to shell model Hamiltonians, their eigenstates may be arbitrary admixtures of various configurations. All such Hamiltonians are rotationally invariant and hence the total orbital angular momentum L of any eigenstate is a good quantum number. Thus, the only quantum numbers are S specifying the spatial symmetry and L. The energies of all states obtained by coupling L and S to a total J are equal. There are $2S + 1$ states (if $S \geq L$) in each *multiplet* of states. If eigenstates of the shell model Hamiltonian belong to a given l^n configuration, we can derive additional quantum numbers to characterize the states. These additional labels are obtained by introducing the seniority scheme which we now proceed to do.

We start by considering unitary transformations in the $(2l+1)$-dimensional space spanned by space functions of a single nucleon. These transformations are elements of the $U(2l+1)$ Lie group. If the wave functions of each nucleon undergo a unitary transformation, wave functions of the l^n configuration also transform linearly among themselves. These transformations are also unitary but, in general, they are not irreducible. The irreducible representations of $U(2l+1)$ may be classified according to their symmetry type. The $U(2l+1)$ transformations are *symmetric* in all nucleon space coordinates and thus *commute* with permutations of nucleon coordinates. The generators of $U(2l+1)$ are the irreducible tensor operators $U_\kappa^{(k)}$. They transform among themselves states with different values of L and M but do not change the symmetry type of states.

Following the considerations of Section 26, we start from a given function ϕ and apply to it the permutations prescribed by a given standard arrangement of a Young diagram. We symmetrize ϕ with respect to nucleon numbers written in rows of the tableau and *then* antisymmetrize with respect to nucleons whose numbers are written in the columns. We apply to the resulting function all permutations (actually it is enough to consider all transpositions) obtaining a set of functions. We choose among those, basis functions ϕ_i, $i = 1, \ldots, n_{[f]}$ so that any other function of the set is equal to a linear combination of the ϕ_i. If in ϕ all quantum numbers of the occupied single particle states are different, the various standard arrangements of the same shape $[f]$ yield in this way $n_{[f]}$ equivalent but independent sets. If the single particle states are not all different, there are less than $n_{[f]}$ independent sets. The number of basis functions in each set is always equal to $n_{[f]}$ which is the dimension of the irreducible representation of the permutation group defined by $[f]$.

The $U(2l+1)$ elements transform states with given symmetry $[f]$ of the l^n configuration into states with the same symmetry. Thus, the partition $[f] \equiv [f_1 f_2 \cdots f_{2l+1}]$ uniquely specifies the $U(2l+1)$ irreducible representation. The transformations induced by $U(2l+1)$ elements on states with symmetry type $[f]$ of the l^n configuration are *not* irreducible. They contain exactly $n_{[f]}$ equivalent irreducible representations of $U(2j+1)$. The number of independent functions in each of these, $N_{[f]}$, is given by (26.7). To obtain the bases of these irreducible representations we construct all independent bases ϕ_i corresponding to $[f]$ for all states of the l^n configuration. This set of bases includes states with all values of L and M consistent with the symmetry of $[f]$. We enumerate the various bases by a superscript

$h = 1, \ldots, N_{[f]}$ so each basis function is specified as $\phi_i^{(h)}$. Permutations of particle coordinates transform irreducibly the functions $\phi_i^{(h)}$ *with the same* h among themselves. All such transformations are equivalent and hence, it can be arranged that all sets $\phi_i^{(h)}$ with the various h values, transform in the same way. With this arrangement, $U(2l+1)$ elements transform irreducibly the various $\phi_i^{(h)}$ with fixed i among themselves. The transformation law, for a given $U(2l+1)$ element, is the same for the $\phi_i^{(h)}$ with all i values. This arrangement is possible due to the fact that $U(2l+1)$ transformations commute with permutations of particle numbers.

To obtain fully antisymmetric functions, the spatial functions $\phi_i^{(h)}$ for any h must be multiplied by the spin functions χ_i obtained from the dual Young tableau and summed over i. This imposes severe limitations on the acceptable irreducible representations of $U(2l+1)$. As described above, the symmetry of spin functions may be obtained from Young diagrams with at most two rows with lengths f_1 and f_2. The dual diagrams which prescribe the symmetry of the spatial functions may have at most two columns whose lengths are f_1 and f_2. The corresponding diagram has f_2 rows with two squares and $f_1 - f_2$ rows with one square. The total length of the first column may not exceed $2l+1$ which is the number of single nucleon spatial states. The fully antisymmetric spatial function is thus characterized by the partition $1 + 1 + \cdots + 1 = n$. Instead of specifying the irreducible representations by f_1, f_2 (either lengths of rows or columns) we may use n and S as given by (28.1). As we saw in Section 26 for isospin functions, the spin functions with given symmetry transform among themselves under unitary transformations induced by unitary transformations in the two-dimensional space of single particle spin functions. The group $U(2)$ and its generators have been considered in Section 6.

All space and spin states of a single particle span a space with $2(2l+1)$ dimensions. Unitary transformation in that space, which are elements of $U(2(2l+1))$, induce unitary transformation in the space of l^n wave functions which are products of n single particle states. There is a simple way to characterize the transformation properties of states in l^n configurations of identical particles. According to the Pauli principle allowed states should be fully antisymmetric. The number of standard arrangements of a Young diagram with a single column is just 1. Hence, all antisymmetric states transform irreducibly under action of $U(2(2l+1))$. No new classification arises from considering the $U(2(2l+1))$ group of transformations. The discussion above, leading to definite symmetry types of space wave functions is due to the restriction to a subgroup of $U(2(2l+1))$, namely

$$U(2(2l+1)) \supset U_S(2) \otimes U(2l+1) \tag{28.2}$$

States may be classified by their spatial symmetry which is closely related to the value of the intrinsic spin S. The chain of groups (28.2) is useful if the Hamiltonian does not contain explicitly the spin operators.

Let us consider a concrete example to demonstrate the procedures described above. In section 16 we considered three electrons or identical nucleons in the p-orbit ($l = 1$). We showed how to construct antisymmetric states by using coefficients of fractional parentage. Now we consider separately space functions and spin functions. Wave functions in the m-scheme will be denoted by

$$(ijk) = \phi_{l=1,m=1}(i)\phi_{l=1,m=0}(j)\phi_{l=1,m=-1}(k)$$

where i, j and k stand for space coordinates of particles i, j and k. Young diagrams will be applied to such product states which will be symmetrized and then antisymmetrized according to the usual prescription. Not more than two particles may occupy the same m state since otherwise the corresponding spin functions vanish upon antisymmetrization.

The maximum value of $M_L = m_1 + m_2 + m_3$ can thus be obtained for a state with $m_1 = m_2 = 1$, $m_3 = 0$ denoted by $(12, 3, \ )$. That state cannot be antisymmetrized since two m-states are equal. Hence, no state corresponding to

$$\begin{array}{|c|}\hline 1 \\ \hline 2 \\ \hline 3 \\ \hline \end{array} \quad \text{exists with} \quad S = \tfrac{3}{2} \quad \text{and} \quad M = 2$$

States with $S = \frac{1}{2}$ may be obtained by

$$\begin{array}{|c|c|}\hline 1 & 2 \\ \hline 3 & \multicolumn{1}{c}{} \\ \cline{1-1} \end{array}(12, 3, \ ) = (12, 3, \ ) - (23, 1, \ ) = \phi_1^{(2)}$$

where the superscript 2 indicates the M_L value. Another independent function of the same irreducible representation is obtained as

$$\phi_2^{(2)} = P_{23}\phi_1^{(2)} = (13,2,\ ) - (23,1,\ )$$

Since in these functions two m-states are identical there are no other irreducible representations of the group of permutations with the [21] symmetry. We find the following transformations under permutations

$$P_{12}\phi_1^{(2)} = \phi_1^{(2)} - \phi_2^{(2)} \qquad P_{13}\phi_1^{(2)} = -\phi_1^{(2)} \qquad P_{23}\phi_1^{(2)} = \phi_2^{(2)}$$

$$P_{12}\phi_2^{(2)} = -\phi_2^{(2)} \qquad P_{13}\phi_2^{(2)} = \phi_2^{(2)} - \phi_1^{(2)} \qquad P_{23}\phi_2^{(2)} = \phi_1^{(2)}$$

Both $\phi_1^{(2)}$ and $\phi_2^{(2)}$ belong to the state the $L = M_L = 2$. The fully antisymmetric state is obtained by multiplying $\phi_1^{(2)}$ and $\phi_2^{(2)}$ by spin functions which are the analogues of isospin functions η_1 and η_2 from Section 26. The spin functions are thus given by

$$\chi_1 = (12,3) - (23,2) \qquad \chi_2 = (13,2) - (23,1)$$

where (12,3) indicates $m_s = \frac{1}{2}$ for particles 1, 2 and $m_s = -\frac{1}{2}$ for particle 3. The antisymmetric state with $S = \frac{1}{2}$, $M_S = \frac{1}{2}$, $L = 2$, $M_L = 2(^2D)$ is

$$\chi_1\phi_2^{(2)} - \chi_2\phi_1^{(2)}$$

as can be verified by applying to it the various permutations.

The $M_L = 1$ functions of the $L = 2$ states are obtained by applying to $\phi_1^{(2)}$ and $\phi_2^{(2)}$ the operator $L^- = l_1^- + l_2^- + l_3^-$. Application of L^- to $\phi_1^{(2)}$ and $\phi_2^{(2)}$ yields, apart from a constant factor of $\sqrt{2}$, the two functions.

$$\phi_1^{(1)} = (1,23,\ ) + (2,13,\ ) - (2,13,\ ) - (3,12,\ ) + (12,\ ,3) - (23,\ ,1)$$

$$= (1,23,\ ) - (3,12,\ ) + (12,\ ,3) - (23,\ ,1)$$

$$\phi_2^{(1)} = (1,23,\ ) - (2,13,\ ) + (13,\ ,2) - (23,\ ,1)$$

These functions transform under permutations like $\phi_1^{(2)}$, $\phi_2^{(2)}$ and form a basis for an irreducible representation of the group of permutations. The fully antisymmetric state with $S = \frac{1}{2}$, $L = 2$, $M_L = 1$ is given by

$$\chi_1\phi_2^{(1)} - \chi_2\phi_1^{(1)}$$

In both $\phi_1^{(1)}$ and $\phi_2^{(1)}$ there are equal m states of two particles so it could be expected that no other independent irreducible representation may be constructed. The functions $\phi_1^{(1)}$ and $\phi_2^{(1)}$, however, are linear combinations of *two* independent wave functions, one with one particle in the $m = 1$ state and two particles in the $m = 0$ state and the other with two particles in the $m = 1$ state and one in the $m = -1$ state. Hence, two orthogonal functions to $\phi_1^{(1)}$ and $\phi_2^{(2)}$ may be constructed as follows

$$\phi_1^{(1)'} = (1,23,\ ) - (3,12,\ ) - (12,\ ,3) + (23,\ ,1)$$

$$\phi_2^{(1)'} = (1,23,\ ) - (2,13,\ ) - (13,\ ,2) + (23,\ ,1)$$

These functions transform under permutations like $\phi_1^{(1)}$, $\phi_2^{(1)}$ and form a basis for an equivalent but independent irreducible representation. They belong to a state with $M_L = 1$ and $L = 1$ as can be verified by their vanishing when L^+ is applied to them. The fully antisymmetric state with $S = \frac{1}{2}$, $M_S = \frac{1}{2}$, $L = 1$, $M_L = 1(^2P)$ is given by

$$\chi_1\phi_2^{(1)'} - \chi_2\phi_1^{(1)'}$$

The $M_L = 0$ components of the $L = 2$ and $L = 1$ states are obtained by application of L^- to the various functions. From $\phi_1^{(1)}$ and $\phi_2^{(1)}$ we obtain

$$\phi_1^{(0)} = 2(1,2,3) - 2(3,2,1) + (1,3,2) + (2,1,3) - (3,1,2) - (2,3,1)$$

$$\phi_2^{(0)} = 2(1,3,2) - 2(2,3,1) + (1,2,3) + (3,1,2) - (2,1,3) - (3,2,1)$$

These functions transform under permutations as $\phi_1^{(1)}, \phi_2^{(1)}$ or $\phi_1^{(2)}, \phi_2^{(2)}$. The antisymmetric state with $S = \frac{1}{2}$, $M_S = \frac{1}{2}$, $L = 2$, $M_L = 0$ is thus given by

$$\chi_1\phi_2^{(0)} - \chi_2\phi_1^{(0)}$$

The functions with $M_L = 0$ of the $L = 1$ state are obtained by applying L^- to $\phi_1^{(1)'}$ and $\phi_2^{(1)'}$ and are given by

$$\phi_1^{(0)'} = (1,3,2) + (2,3,1) - (3,1,2) - (2,1,3)$$

$$\phi_2^{(0)'} = (1,2,3) + (3,2,1) - (3,1,2) - (2,1,3)$$

Also these functions transform under permutations in the same manner as $\phi_1^{(1)'}, \phi_2^{(1)'}$ or $\phi_1^{(0)}, \phi_2^{(0)}$. The antisymmetric state with $S = \frac{1}{2}$, $M_S = \frac{1}{2}$, $L = 1$, $M_L = 0$ is thus given by

$$\chi_1 \phi_2^{(0)'} - \chi_2 \phi_1^{(0)'}$$

This procedure can be continued to states with $M_L = -1$ and $M_L = -2$ yielding all components of the ${}^2D$ and ${}^2P$ states.

The L^- operator is proportional, within the p^n configurations, to the component $U_{-1}^{(1)}$ of the $k = 1$ tensor operator $\mathbf{U}^{(1)}$. The $U_\kappa^{(1)}$ operators are among the generators of $U(2 \times 1 + 1) = U(3)$ Lie group. As we saw, L^- transforms states with given symmetry properties into others with the same properties. Other generators $U_\kappa^{(2)}$ may transform states with $L = 2$ into a linear combination of $L = 2$ and $L = 1$ states. Still, they may transform $\phi_1^{(2)}$ into linear combinations of $\phi_1^{(M_L)}$ and $\phi_1^{(M_L)'}$ but those will not contain $\phi_2^{(M_L)}$ nor $\phi_2^{(M_L)'}$. The total number of independent functions in the irreducible representations of $U(3)$ with $[f] = [21]$ is given by putting $f_1 = 2$, $f_2 = 1$, $k = 3$ in (26.7). The result is 8 which is equal to the number of components of the states with $L = 2$ and $L = 1$, $5 + 3 = 8$.

To obtain states with $M_S = -\frac{1}{2}$ of all L, M_L states described above, the *same* functions of the space coordinates should be multiplied by appropriate spin functions obtained from χ_1 and χ_2 by the operation of S^-. The eigenvalues of the Hamiltonians considered depend only on the spatial functions and, for the value of S, are independent of M_S.

In the various functions for the case $M_L = 0$ the three particles are in different m-states. Hence, it is possible to construct a state which is fully antisymmetric in the space coordinates of the particles. It can be written down as a Slater determinant in the single nucleon wave functions ϕ_{lm}

$$\begin{vmatrix} \phi_{11}(1) & \phi_{10}(1) & \phi_{1,-1}(1) \\ \phi_{11}(2) & \phi_{10}(2) & \phi_{1,-1}(2) \\ \phi_{11}(3) & \phi_{10}(3) & \phi_{1,-1}(3) \end{vmatrix}$$

$$= (1,2,3) - (2,1,3) - (1,3,2) - (3,2,1) + (3,1,2) + (2,3,1)$$

This $M_L = 0$ function belongs to a $L = 0$ state and is annihilated by L^+. To form a totally antisymmetric state, this state should be multiplied by a fully symmetric spin state with $f_1 = 3$, $f_2 = 0$. Thus, this $L = 0$ state of the p^3 configuration has $S = \frac{3}{2}$ and is denoted by ${}^4S$.

Let us again emphasize the important distinction between the action of permutations of particle space coordinates and the action of $U(2l + 1)$ transformations. Any Hamiltonian which is a function of space coordinates and momenta commutes with these permutations. Hence, such permutations transform an eigenstate with given symmetry type, L, M and any additional quantum numbers into an eigenstate with the *same* quantum numbers. On the other hand, $U(2l + 1)$ transformations admix states with same symmetry type but generally with different values of L and M. This is due to the fact that shell model Hamiltonians do not generally commute with $U(2l + 1)$ transformations. Any Hamiltonian with single particle terms and two-body interactions, which does not contain spin operators, may be constructed, within l^n configurations, from the $U(2l + 1)$ generators $U_\kappa^{(k)}$. The eigenstates of such a Hamiltonian belong to a definite irreducible representation of $U(2l + 1)$. We saw that those representations are uniquely specified by n and S. A Hamiltonian which commutes with all $U(2l + 1)$ generators must have the same eigenvalues for all states of an irreducible representation. This is a very special case which does not apply to actual physical systems. The usefulness of $U(2l + 1)$ lies in the fact that it has a subgroup, to be considered below, which may supply additional quantum numbers. That subgroup will define the seniority scheme in l^n configurations.

The generators of $U(2l + 1)$ are, like those of $U(2j + 1)$, the irreducible tensor operators $U_\kappa^{(k)}$. In analogy with (26.9) their commutation relations are given by

$$[U_{\kappa'}^{(k')}, U_{\kappa''}^{(k'')}] = \sum_{k\kappa} (-1)^k (1 - (-1)^{k'+k''+k}) \times \sqrt{2k+1} \begin{Bmatrix} k' & k'' & k \\ l & l & l \end{Bmatrix} (k'\kappa' k''\kappa'' \mid k'k''k\kappa) U_\kappa^{(k)} \tag{28.3}$$

This relation holds only within l^n configurations. The single particle unit tensor operators should have reduced matrix elements given by

$$(l\|u_\kappa^{(k)}\|l') = \delta_{ll'}$$

for given l and all l'. The relation (28.3) can be proved exactly like (26.9). There is an overall change of sign on the r.h.s. of (28.3) which is due to $(-1)^{2j} = -1 = -(-1)^{2l}$.

The Casimir operator of $U(2l+1)$ can be written in the same form as (26.10). The operator

$$\boxed{\sum_k (2k+1)(\mathbf{U}^{(k)} \cdot \mathbf{U}^{(k)})} \tag{28.4}$$

commutes with all the $U_\kappa^{(k)}$ as can be verified by using the commutation relations (28.3). The matrix elements of (28.4) in the l^2 configuration are given by using (10.27) and (10.30) as

$$\begin{aligned}
&\langle l^2LM|\sum_k (2k+1)(\mathbf{U}^{(k)} \cdot \mathbf{U}^{(k)})|l^2LM\rangle \\
&\quad = 2(-1)^L \sum_k (2k+1) \\
&\qquad \times \begin{Bmatrix} l & l & L \\ l & l & k \end{Bmatrix} + \frac{2}{2l+1}\sum_k (2k+1) \\
&\quad = 2(-1)^L + 2(2l+1)
\end{aligned} \tag{28.5}$$

The last equality is due to the sum rule (10.21). The requirement of antisymmetry implies $(-1)^{L+S} = 1$ and hence, $(-1)^L$ may be replaced by $(-1)^S$. We may then use the spin exchange operator (10.64) to replace $(-1)^S$ by $-P^\sigma$. We thus obtain for the eigenvalues of (28.4) the result

$$
\begin{aligned}
&\langle l^n S M_S L M_L | \sum_k (2k+1) \mathbf{U}^{(k)} \cdot \mathbf{U}^{(k)} | l^n S M_S L M_L \rangle \\
&\quad = \langle l^n S M_S L M_L | 2 \sum_{i<j,k} (2k+1) \mathbf{u}^{(k)}(i) \cdot \mathbf{u}^{(k)}(j) | l^n S M_S L M_L \rangle \\
&\quad + \langle l^n S M_S L M_L | \sum_{k,i} (2k+1) \mathbf{u}^{(k)}(i) \cdot \mathbf{u}^{(k)}(i) | l^n S M_S L M_L \rangle \\
&\quad = \frac{n}{2l+1} \sum_k (2k+1) \\
&\quad + 2 \langle l^n S M_S L M_L | \sum_{i<j} (-\tfrac{1}{2} - 2\mathbf{s}_i \cdot \mathbf{s}_j) | l^n S M_S L M_L \rangle \\
&\quad = (2l+1)n - \frac{n(n-1)}{2} - 2[S(S+1) - \tfrac{3}{4}n]
\end{aligned}
\tag{28.6}
$$

Strictly speaking, the operator (28.4) is not a Casimir operator since $U(2l+1)$ is not a semi-simple group. It contains the Abelian invariant subgroup generated by $U_0^{(0)} = \hat{n}/\sqrt{(2l+1)}$ which commutes with all generators. Removing it, the operators $U_\kappa^{(k)}$, $k > 0$, generate a subgroup of $U(2l+1)$ denoted by $SU(2l+1)$ whose matrices have their determinants equal to 1. As mentioned in Section 6, if A is a matrix $\det(\exp A) = \exp(\operatorname{tr} A)$. The trace of any generator is given by

$$
\sum_m \langle lm | U_\kappa^{(k)} | lm \rangle = \sum (-1)^{l-m} \begin{pmatrix} l & k & l \\ -m & \kappa & m \end{pmatrix} = \frac{1}{2l+1} \delta_{k0} \delta_{\kappa 0}
\tag{28.7}
$$

The Casimir operator of $SU(2l+1)$ is thus equal to (28.4) with the $k = 0$ term not included in the sum. The $k = 0$ term $U_0^{(0)} U_0^{(0)} =$

$\hat{n}^2/(2l+1)$ may be added with an arbitrary coefficient to $C_{SU(2l+1)}$. The choice of coefficient used in (28.4) leads to simple eigenvalues.

The $U(2l+1)$ generators $U_\kappa^{(1)}$ are proportional within l^n configurations, to the components of the angular momentum

$$\boxed{\mathbf{L} = \sqrt{l(l+1)(2l+1)}\mathbf{U}^{(1)}} \tag{28.8}$$

Hence, the three generators $U_\kappa^{(1)}$ form a sub-algebra of the $U(2l+1)$ Lie algebra. They generate in the space of l^n states, the linear transformations due to rotations of the particle three-dimensional coordinates. The matrices of $U_\kappa^{(1)}$ form irreducible representations of the group $O(3)$ of three-dimensional rotations characterized by L. Each such representation has $(2L+1)$ basis states. Going from $U(2l+1)$ to the $O(3)$ subgroup the irreducible representations of $U(2l+1)$ split into irreducible $O(3)$ representations. There are usually several of these with the same value of L and additional quantum numbers are necessary to uniquely specify the states. The introduction of seniority by Racah (1943) was motivated by the need for additional quantum numbers in atomic spectroscopy. One way of achieving this goal is by going from $U(2l+1)$ into $O(3)$ in two steps via an intermediate subgroup.

A subgroup of $U(2l+1)$ whose existence follows directly from the commutation relations (28.3) is generated by odd tensor operators. If k' and k'' in (28.3) are odd so must be k on the right hand side. Thus, odd tensor operators are the elements of a sub-algebra. A direct way to see this fact is by observing that such operators annihilate the $|l^2L=0\rangle$ state

$$\boxed{U_\kappa^{(k)}|l^2L=0\rangle = 0 \qquad k \text{ odd}} \tag{28.9}$$

Due to the Wigner-Eckart theorem, the l.h.s. of (28.9) must have $J=k$. If k is odd that state is *antisymmetric* in the space coordinates of the two particles and cannot be obtained from the *symmetric* $|l^2L=0\rangle$ state by acting on it with the *symmetric* operator $U_\kappa^{(k)}$. From (28.9) follows that the Lie group generated by the $U_\kappa^{(k)}, k$ odd, leaves the $|l^2L=0\rangle$ state invariant.

$$e^{i\alpha U_\kappa^{(k)}}|l^2L=0\rangle = \left(1+\sum_{\nu=1}^{\infty}\frac{(i\alpha)^\nu}{\nu!}(U_\kappa^{(k)})^\nu\right)|l^2L=0\rangle = |l^2L=0\rangle \tag{28.10}$$

Since the tensor (28.8) is odd, this group has $O(3)$ as a subgroup.

To see which group is generated by odd tensor operators we write explicitly the $|l^2L=0\rangle$ state. It is equal to

$$\frac{1}{\sqrt{2l+1}}\sum(-1)^{l-\lambda}\phi_{l\lambda}(1)\phi_{l,-\lambda}(2) \tag{28.11}$$

The $\phi_{l\lambda}$, like the spherical harmonics, are complex functions. We can express (28.11) in terms of the following real functions

$$\begin{aligned} \phi'_{l0} &= \phi_{l0} \\ \phi'_{l\lambda} &= \frac{1}{\sqrt{2}}(\phi_{l\lambda}+(-1)^\lambda\phi_{l,-\lambda}) \qquad \lambda>0 \\ \phi'_{l\bar{\lambda}} &= \frac{-i}{\sqrt{2}}(\phi_{l\lambda}-(-1)^\lambda\phi_{l,-\lambda}) \qquad \lambda>0 \end{aligned} \tag{28.12}$$

The state (28.11) may be expressed as

$$\frac{(-1)^l}{\sqrt{2l+1}}\left[\phi'_{l0}(1)\phi'_{l0}(2)+\sum_{\lambda>0}\phi'_{l\lambda}(1)\phi'_{l\lambda}(2)+\sum_{\lambda>0}\phi'_{l\bar{\lambda}}(1)\phi'_{l\bar{\lambda}}(2)\right] \tag{28.13}$$

We see that the transformations of the group considered, leave invariant a real and symmetric bilinear form. The transformations generated by $U_\kappa^{(k)}, k$ odd, are thus real orthogonal transformations, i.e. rotations, in the $(2l+1)$-dimensional space spanned by the single particle functions (28.12). This Lie group is denoted by $O(2l+1)$ and the chain of groups is $U(2l+1) \supset O(2l+1) \supset O(3)$.

This situation should be contrasted with that of $U(2j+1)$ subgroups. There the subgroup of $U(2j+1)$ which contains $O(3)$ (or $SU(2)$) as a subgroup is the symplectic group $Sp(2j+1)$ and not the rotation group $O(2j+1)$. To see the difference between the two cases we recall that any unitary transformation can be expressed as $\exp(i\alpha A)$ where $A = A^\dagger$ is a hermitean matrix. Real orthogonal transformations in a real basis may be obtained from a matrix A which is a skew-symmetric matrix multiplied by i. The number of independent elements in a skew-symmetric matrix of order r is $r(r-1)2$. For $r = 2l+1$ this is the number of components of $\mathbf{U}^{(k)}$ tensors, with k odd, $\sum_{k=1}^{2l-1}(2k+1) = 2l(2l+1)/2$. For even values of $r = 2j+1$, the number of components of odd $\mathbf{U}^{(k)}$ tensors (which include the $k = 1$ generators of $O(3)$) is $\sum_{k=1}^{2j}(2k+1) = (j+1)(2j+1)$. That number is equal to $r(r+1)/2$ which is the number of generators of $Sp(2j+1)$ and not of the different group $O(2j+1)$.

The irreducible representations of $O(2l+1)$ according to which a given state transforms serve as additional quantum numbers. If the Hamiltonian is constructed from the generators of $O(2l+1)$, each of its eigenstates belongs to an irreducible representation of that group. The quantum numbers thereby obtained are thus good quantum numbers. If this is not the case, the seniority scheme of the $O(2l+1)$ irreducible representations provides an orthogonal and complete set of states. For given S and L, this set defines a sub-matrix of the Hamiltonian which may be later diagonalized.

Let us emphasize again that only very special Hamiltonians commute with all $O(2l+1)$ generators. Rotationally invariant Hamiltonians which are functions of only space coordinates and conjugate momenta commute with components of $\mathbf{L}$ (generators of $O(3)$). They are also invariant under permutations of particle coordinates. Some of these Hamiltonians, constructed from $O(2l+1)$ generators, have vanishing matrix elements between states which belong to different irreducible representations. Hence, eigenstates of such Hamiltonians belong to definite $O(2l+1)$ irreducible representations. These Hamiltonians do not commute, in general, with the $O(2l+1)$ generators. Those that do commute have, due to Schur's lemma, the same eigenvalues in all states which belong to a given irreducible representation.

Let us now consider the irreducible representations of $O(2l+1)$ which belong to an irreducible representation of $U(2l+1)$ whose Young diagram has two columns of lengths f_1 and f_2. The $O(2l+1)$ transformations are symmetric in all particles as are all $U(2l+1)$ transformations. Hence, the irreducible representations of $O(2l+1)$ may also be characterized by their symmetry properties. The $O(2l+1)$ transformations leave the symmetric state $|l^2 L = 0\rangle$ invariant. Hence, such pair states may be removed from any state without changing its transformation properties under $O(2l+1)$ transformations. Removal of a pair of nucleons in a symmetric state amounts to removing a row with two squares from the Young diagram with two columns of lengths f_1 and f_2. Hence, the Young diagrams which characterize the irreducible representations of $O(2l+1)$ have two columns of lengths v_1 and v_2 such that $f_1 - v_1 = f_2 - v_2$. The transformation properties of some l^n wave functions under $O(2l+1)$ transformations are the same as some of those of the l^v configuration where the seniority v is given by $v = v_1 + v_2$. The requirement that $f_1 - f_2 = v_1 - v_2$ has a very simple physical meaning. The $|l^2 L = 0\rangle$ pairs have, according to the Pauli principle $S = 0$. Hence, the states of the l^v configuration which correspond to those in the l^n configuration must have the same value of the total intrinsic spin S. Thus, we can write

$$\boxed{v_1 + v_2 = v \qquad \tfrac{1}{2}(v_1 - v_2) = S} \tag{28.14}$$

There is a slight complication concerning the choice of Young diagrams for the irreducible representations of $O(2l+1)$. From $v_1 + v_2 = v \leq 2l+1$ follows $v_2 \leq 2l+1-v_1$. Hence, given any Young diagram with column lengths $v_1, v_2 (v_1 \geq v_2)$ there is another possible diagram with column lengths $v_1' = 2l+1-v_1, v_2$. These two diagrams which are always different are called *associate diagrams*. The irreducible representations which are defined by associate diagrams are equivalent for $O(2l+1)$ transformations whose determinants are equal to $+1$. These transformations are rotations in the $2l+1$ dimensional space and exclude the reflection through the origin. From (28.7) follows that all transformations generated by $U_\kappa^{(k)}$, k odd, have determinants equal to $+1$. The group should be called $SO(2l+1)$ but we omit this detail for brevity. Thus, only one diagram of the two is needed for characterization of the irreducible $O(2l+1)$ representation. It is customary to choose the one for which the first column is *shorter*. Hence, if $v_1 \leq l$ the column lengths denoted by $a+b$ and a are given by v_1 and v_2 in (28.14). If, however, $v_1 > l$ the first column has length $a+b = 2l+1-v_1$ and the second column has length $a = v_2$. The lengths of the representative Young diagram $a+b$, a in both cases are given by a a and b determined by

$$\boxed{a = \tfrac{1}{2}v - S \qquad b = \mathrm{Min}(2S, 2l+1-v)} \tag{28.15}$$

The number a is the number of rows with two squares in the representative Young diagram and b is the number of one square rows. We may denote the Young diagrams by $(w_1 w_2 \cdots w_l)$ where $w_1 = \cdots = w_a = 2$, $w_{a+1} = \cdots = w_{a+b} = 1$, $w_{a+b+1} = \cdots = w_l = 0$. Usually, the zeroes are omitted.

The state of a single particle with $S = \frac{1}{2}$, $L = l$ has seniority $v = 1$ so that $a = 0$, $b = 1$ (for $l > 0$) and the irreducible representation is $(100\cdots0) = (1)$. All states in odd l^n configurations with these quantum numbers also belong to that representation. The $|l^2 L = 0\rangle$ state is invariant under $O(2l+1)$ transformation and hence transforms like the state with no particles. Thus, $L = 0$ states of l^n configurations with seniority $v = 0$ belong to the Young diagram given by $a = 0$, $b = 0$ denoted by $(0,\ldots,0) = (0)$. For $l > 0$ there are states with seniority $v = 2$ with even values of L and $S = 0$. These quantum numbers yield $a = 1$, $b = 0$ and the irreducible representation is $(20\cdots0) = (2)$. The other states in the l^2 configuration and other l^n, n even, configurations, have odd L values and $S = 1$. They belong to the irreducible representation with $a = 0$, $b = 2$ (for $l > 1$), i.e. $(110\cdots0) = (11)$. In the case $l = 1$, for the $S = 1$, $L = 1$ (^3P) state with seniority $v = 2$ we obtain from (28.15) $a = 0$, $b = 1$. The irreducible representation of $O(2l+1)$ to which this state belongs is thus defined by $(10\cdots0) = (1)$ just like a single p-particle. Using cartesian components this state is the vector product of two sets of single p-states. It transforms under rotations like a state of a p-particle but changes sign under the reflection $\phi_{l=1,m} \to -\phi_{l=1,m}$.

Let us now look at the p^3 configuration considered above. The $S = \frac{1}{2}$ states ^2D and ^2P belong to the same irreducible representation of $U(2l+1)$. The generators of $O(2l+1) = O(3)$ in this case are simply $U_\kappa^{(1)}$ which are proportional to the components of $\mathbf{L}$. Hence, under $O(2l+1)$ transformations the 5 components of the $L = 2$ state transform among themselves and the 3 components of the $L = 1$ state transform among themselves. The latter transform exactly like any $L = 1$ state, for instance like the states of a single p-particle and the seniority of the ^2P state is thus $v = 1$. the ^2D state has seniority $v = 3$ and belongs to the irreducible representation with $a = 1$, $b = 0$, i.e. $(200) = (2)$. The ^4S state with $S = \frac{3}{2}$, $L = 0$ has according to (28.15), $a = 0$, $b = 0$ and belongs to the (0) representation. Obviously, its space function is invariant under $O(3)$ rotations.

A more interesting case is the d^3 configuration where $l = 2$. The following states are allowed by the Pauli principle. States with $S = \frac{3}{2}$, $L = 3$, $1(^4F, ^4P)$, one state of each value of $L = 5,4,3,1$ with

$S = \frac{1}{2}(^2H, ^2G, ^2F, ^2P)$ and *two* $S = \frac{1}{2}$ states with $L = 2$. Under $U(2l + 1) = U(5)$ transformations the spatial functions of ^4F and ^4P states transform among themselves and so do the functions of the $^2H, ^2G, ^2F, ^2D, ^2D$ and ^2P states. When going over to $O(2l + 1)$ the transformation of the $S = \frac{1}{2}$ states is no longer irreducible. The functions of one ^2D state, with seniority $v = 1$ transform irreducibly among themselves just like one d-particle, i.e. according to the irreducible representation (1). The other $S = \frac{1}{2}$ states have $v = 3$ and thus $a = 1$, $b = 1$ and the irreducible representation is (21). On the other hand, the $S = \frac{3}{2}$ states, also with $v = 3$, have, according to (25.8), $a = 0$, $b = 2$ and thus belong to the (11) irreducible representation. For $l > 2$ the $S = \frac{3}{2}$ states of the l^3 configuration have $v = 3$ and consequently $a = 0$, $b = 3$ and hence belong to the to (111) representation.

The seniority scheme in l^n configurations is defined by the irreducible representations of $O(2l + 1)$. The generators of $O(2l + 1)$ are the odd tensor operators $U_\kappa^{(k)}$, k odd. They are diagonal in the seniority scheme and their matrix elements are independent of n. Any Hamiltonian constructed from these generators has also this property. The final step in this classification scheme is going from $O(2l + 1)$ to its subgroup $O(3)$ which leads to states with definite values of L. The group chain (28.2) is thus continued into

$$U(2(2l + 1)) \supset U_S(2) \otimes U(2l + 1)$$

$$\supset U_S(2) \otimes O(2l + 1) \supset U_S(2) \otimes O(3) \qquad (28.16)$$

There is a Hamiltonian which commutes with all generators of $O(2l + 1)$. This is the Casimir operator of that group which, in analogy with (26.19), is defined by

$$C_{O(2l+1)} = 2 \sum_{k \text{ odd}} (2k + 1)(\mathbf{U}^{(k)} \cdot \mathbf{U}^{(k)}) \qquad (28.17)$$

The fact that (28.17) commutes with all generators $U_\kappa^{(k)}$, k odd, can be verified by using the commutation relations (28.3). It actually follows from the similar calculations for the operator (28.4) since the commutator of $U_\kappa^{(k)}$ with odd k with $U_{\kappa'}^{(k')}$, k' odd, yields odd tensors whereas its commutator with $U_{\kappa'}^{(k')}$, k' even, yields only even tensors.

The matrix elements of (28.17) in the l^2 configuration are obtained by using (10.27) and (10.30) as

$$\langle l^2LM|2\sum_{k \text{ odd}}(2k+1)\mathbf{U}^{(k)}\cdot\mathbf{U}^{(k)}|l^2LM\rangle$$

$$= \langle l^2LM|4\sum_{k \text{ odd}}(2k+1)(\mathbf{u}^{(k)}(1)\cdot\mathbf{u}^{(k)}(2)|l^2LM\rangle$$

$$+ 4\langle l^2LM|2\sum_{k \text{ odd}}(2k+1)\mathbf{u}^{(k)}(1)\cdot\mathbf{u}^{(k)}(1)|l^2LM\rangle$$

$$= (-1)^L 4\sum_{k \text{ odd}}(2k+1)\begin{Bmatrix} l & l & L \\ l & l & k \end{Bmatrix} + \frac{4}{2l+1}\sum_{k \text{ odd}}(2k+1)$$

$$= 4(-1)^L\left[\frac{1}{2}\sum_k(2k+1)\begin{Bmatrix} l & l & L \\ l & l & k \end{Bmatrix}\right.$$

$$\left. - \frac{1}{2}\sum_k(-1)^k(2k+1)\begin{Bmatrix} l & l & L \\ l & l & k \end{Bmatrix}\right] + 4l$$

$$= (-1)^L(2 - 2(2l+1)\delta_{L0}) + 4l \tag{28.18}$$

The last equality is due to the sum rules (10.20) and (10.21) of $6j$-symbols. We can replace again $(-1)^L$ by $(-1)^S = -(1 + 4(\mathbf{s}_1\cdot\mathbf{s}_2))/2$ and obtain for (28.18) the expression

$$\langle l^2LM|2\sum_{k \text{ odd}}(2k+1)(\mathbf{U}^{(k)}\cdot\mathbf{U}^{(k)})|l^2lM\rangle$$

$$= \langle l^2SLM| - 1 - 4(\mathbf{s}_1\cdot\mathbf{s}_2) - 2P|l^2SLM\rangle + 4l \tag{28.19}$$

In (28.19) the operator P is defined as the two body operator whose eigenvalues are given by

$$\boxed{\langle l^2LM|P|l^2LM\rangle = (2l+1)\delta_{L0}\delta_{M0}} \tag{28.20}$$

This is the LS-coupling analog of the pairing interaction in j^n configurations.

We can now use the properties of odd tensor interactions to obtain the eigenvalues of the pairing interaction in l^n states. According to (28.19) the two-body pairing interaction can be expressed as

$$P(1,2) = -\tfrac{1}{2} - 2(\mathbf{s}_1 \cdot \mathbf{s}_2) - 2\sum_{k\ \mathrm{odd}} (2k+1)(\mathbf{u}^{(k)}(1) \cdot \mathbf{u}^{(k)}(2)) \tag{28.21}$$

In the case of l^n configurations we obtain

$$\sum_{i<j}^{n} P(i,j) = -\tfrac{1}{4}n(n-1) - (\mathbf{S}^2 - \tfrac{3}{4}n) - \sum_{k\ \mathrm{odd}} (2k+1)(\mathbf{U}^{(k)} \cdot \mathbf{U}^{(k)}) + nl \tag{28.22}$$

The odd tensor summation in (28.22) is diagonal in the seniority scheme and its eigenvalues in l^n configurations are equal to the corresponding ones in the l^v configuration. In the latter, there are no pairs coupled to $L = 0$ and the pairing term vanishes. We subtract from (28.22) the same equation applied to the l^v configuration, obtaining for the eigenvalues of (28.22) the result

$$\boxed{\begin{aligned} &-\tfrac{1}{4}n(n-1) + \tfrac{3}{4}n + nl + \tfrac{1}{4}v(v-1) - \tfrac{3}{4}v - vl \\ &\qquad = \frac{n}{4}(4l+4-n) - \frac{v}{4}(4l+4-v) = \frac{n-v}{4}(4l+4-n-v) \end{aligned}} \tag{28.23}$$

This result is the analog of (19.11) in the case of j^n configurations.

Using the eigenvalues (28.23) we may obtain from (28.22) the eigenvalues of $C_{O(2l+1)}$ defined by (28.17). These eigenvalues are thus given by

$$\boxed{C_{O(2l+1)}(v,S) = \frac{v}{2}(4l+4-v) - 2S(S+1)} \tag{28.24}$$

The eigenvalues of the Casimir operator depend only on v and S or, according to (28.14) on the v_1, v_2 which determine the irreducible representation of $O(2l+1)$. The expression (28.24) is equal to the following formula for the eigenvalues

$$C_{O(2l+1)}(w_1,\dots,w_l) = \sum_{i=1}^{l} w_i(w_i + 2l + 1 - 2i) \tag{28.25}$$

Putting $w_1 = \cdots = w_a = 2$, $w_{a+1} = \cdots = w_{a+b} = 1$, $w_{a+b+1} = \cdots = w_l = 0$ we obtain for (28.25) the value

$$\sum_{i=1}^{a} 2(2l+3-2i) + \sum_{i=a+1}^{a+b} (2l+2-2i)$$
$$= a(4l+4-2a) + b(2l+1-2a-b)$$

which is equal to (28.24) for a and b given by (28.15).

We again verify that the eigenvalues of the pairing interaction are given in terms of those of the Casimir operator $C_{O(2l+1)}$ according to the analog of (26.27) by the expressions

$$\begin{aligned} P(n,S,v) &= \tfrac{1}{2}[C_{O(2l+1)}(f_1,f_2) - C_{O(2l+1)}(v_1,v_2)] \\ &= \tfrac{1}{2}[C_{O(2l+1)}(n,S) - C_{O(2l+1)}(v,S)] \end{aligned} \tag{28.26}$$

Odd tensor operators, being generators of $O(2l+1)$, play an important role in defining the seniority scheme in l^n configurations. Scalar products of odd tensor operators may be used to expand, within l^n configurations, Hamiltonians which are diagonal in the seniority scheme. An example is the pairing interaction (28.20) whose expansion is given by (28.21). In contrast to the non-local pairing interaction, potential interactions $V(|\mathbf{r}_1 - \mathbf{r}_2|)$ can be directly expanded as linear combinations of scalar products of *even* tensor operators. The reason is due to the expression (8.15) for the reduced matrix elements of spherical harmonics. They vanish for ranks k satisfying $(-1)^{2l+k} = -1$. In the case of j^n configurations the combination of even ranks k of spherical harmonics coupled to the spin vector $\mathbf{s}$ led to odd rank tensors according to (12.14). This construction was used in proving that the δ-potential interaction is diagonal in the seniority scheme.

Let us examine the situation in the case of interactions that may be expanded as a linear combination of scalar products $(\mathbf{s}_1 \cdot \mathbf{s}_2)(\mathbf{Y}^{(k)}(1) \cdot \mathbf{Y}^{(k)}(2))$ where k is even. We can define the *double tensors* $T^{(1,k)}_{\sigma\kappa} = \sum_i s_\sigma(i) Y^{(k)}_\kappa(i)$ where the first index 1 is the rank of the spin vector and k is the rank of the tensor in space coordinates. We can then define scalar products by

$$(\mathbf{T}^{(1,k)}(1) \cdot \mathbf{T}^{(1,k)}(2)) = (\mathbf{s}_1 \cdot \mathbf{s}_2)(\mathbf{Y}^{(k)}(1) \cdot \mathbf{Y}^{(k)}(2)) \tag{28.27}$$

Instead of (28.9) we may consider the action of $\mathbf{T}^{(1,k)}$ on the fully antisymmetric l^2 state with $S = 0$, $L = 0$. We thus obtain

$$\boxed{(\mathbf{T}^{(1,k)}(1) + \mathbf{T}^{(1,k)}(2))|l^{2\,1}S\rangle = 0 \qquad k \text{ even}} \tag{28.28}$$

The validity of (28.28) follows from the Wigner-Eckart theorem which makes the l.h.s. of (28.28) have $L = k$, $S = 1$. For k even, such a state, however, is *symmetric* in spin and space variables and cannot arise from the *antisymmetric* ${}^1S$ state by the action of the *symmetric* double tensor. From (28.28) follows that interactions which are linear combinations of various (28.27) terms are diagonal in the seniority scheme if $(-1)^{1+k} = -1$, i.e. k is even. As mentioned in the beginning of this section, the scalar products $(\mathbf{s}_1 \cdot \mathbf{s}_2)$ in (28.27) may be replaced by the spin exchange operator P^σ_{12} and hence by $-P^x_{12}$. The resulting interaction may then be expanded in scalar products of tensor operators $\mathbf{U}^{(k)}$ with odd ranks k.

The relation (28.28) indicates that it is possible to define the seniority scheme in l^n configuration by a group chain which is different from (28.16). The generators of the unitary group $U(2(2l+1))$ whose transformations operate on the spin and space states of nucleons are *all* double tensors $U^{(1,k)}_{\sigma\kappa} = \sum s_\sigma(i) U^{(k)}_\kappa(i)$ and all tensors $U^{(k)}_\kappa = \sum U^{(k)}_\kappa(i)$. Among those, the generators $U^{(k)}_\kappa$ with k odd and $U^{(1,k)}_{\sigma\kappa}$ with k even form a Lie subalgebra whose elements satisfy (28.9) and (28.28). The elements of the Lie group, which is a subgroup of $U(2(2l+1))$, generated by this subalgebra leave invariant the bilinear *antisymmetric* form $|l^{2\,1}S\rangle$. Hence, this is the symplectic group in $2(2l+1)$ dimensions – $Sp(2(2l+1))$. The antisymmetric irreducible representation of $U(2(2l+1))$ splits into irreducible representations of $Sp(2(2l+1))$ whose Young diagrams have only one column of length v. The seniority is thus introduced first and the spin S only at the next step by going from $Sp(2(2l+1))$ to $U_S(2) \otimes O(2l+1)$. At that stage, the states with given v are further characterized by the irreducible representations of $O(2l+1)$ with Young diagrams with two columns whose lengths are v_1 and v_2 ($v_1 + v_2 = v$) as described before. The group chain in this case is

$$U(2(2l+1)) \supset Sp(2(2l+1)) \supset U_S(2) \otimes O(2l+1) \supset U_S(2) \otimes O(3) \tag{28.29}$$

The final step in constructing eigenstates of the shell model Hamiltonian is to couple the **S** and orbital angular momentum **L** to a total spin **J**. The Hamiltonians considered in this section have degenerate eigenvalues in all states with given S and L and hence, they are independent of J. In practice, spin dependent interactions split the multiplet into $2S+1$ states (if $S \leq L$) with good total spin J. The last group in either chain (28.16) or (28.28) is $U_S(2) \otimes O(3)$ whose generators are $1, S_x, S_y, S_z, L_x, L_y, L_z$. Coupling **S** and L to **J** amounts to going from $U_S(2) \otimes O(3)$ to its subgroup whose generators are $J_x = S_x + L_x$, $J_y = S_y + L_y$, $J_z = S_z + L_z$. This is the $SU(2)$ algebra which was introduced and discussed in Section 6.

Generators of $Sp(2(2l+1))$, $U_\kappa^{(k)}$ with k odd and $U_{\sigma\kappa}^{(1,k)}$ with k even, have simple properties which are analogous to those of odd tensor operators in j^n configurations. They have non-vanishing matrix elements only between states with the same seniority v. Their non-vanishing matrix elements in the l^n configuration are equal to the corresponding ones in the l^v configuration. Hence, scalar products of these tensor operators are also diagonal in the seniority scheme and have the pairing property, analogous to (20.3). We will not express these features explicitly.

29

LS-Coupling of Protons and Neutrons. The $SU(4)$ Scheme

Let us now consider the case of both valence protons and valence neutrons outside closed shells within the LS-coupling scheme. We consider charge independent Hamiltonians which do not depend explicitly on the spin *and* isospin variables of the nucleons. Such Hamiltonians are invariant under permutations of the space coordinates of nucleons. As before, eigenfunctions of these Hamiltonians, being functions of space coordinates which are transformed into each other under permutations, have the same eigenvalue. Such eigenfunctions belong to irreducible representations of the group of permutations. Wave functions which are fully antisymmetric, according to the generalized Pauli principle, are linear combinations of these spatial eigenfunctions multiplied by spin-isospin functions which belong to the dual Young diagram. The theory of such Hamiltonians, presented in the following, is due to Wigner (1937) and Hund (1937).

In Section 26 we saw that charge independent Hamiltonians have isospin symmetry. Their eigenvalues are the same for all states which belong to an irreducible representation of $SU_T(2)$ characterized by the isospin T. They are simply independent of M_T. In Section 28 we considered Hamiltonians which do not contain explicitly the spin operators. Their eigenvalues are equal in all states, with various values of M_S, which belong to the same irreducible representation of $SU_S(2)$ characterized by S. In the present section we consider Hamiltonians which do not contain explicitly either spin or isospin operators. The eigenvalues of such Hamiltonians depend only on the functions of space coordinates and are independent of all possible states of spins and isospins. Spin and isospin wave functions which have a definite symmetry type should be multiplied by spatial wave functions to form states which are fully antisymmetric. In this way, the symmetry type of the spin-isospin functions determines the symmetry type of the space functions. We shall next consider the various symmetry types of spin-isospin states.

Spin and isospin states of a single nucleon may be specified by (m_t, m_s) and hence, there are four possible states for each nucleon. The Young diagrams which characterize the symmetry of spin-isospin functions under permutations cannot have columns which contain more than four squares. Other diagrams yield zero when spin-isospin functions are antisymmetrized. Hence, the allowed Young diagrams have four rows whose lengths satisfy the relations

$$\boxed{\begin{aligned} &f_1 \geq f_2 \geq f_3 \geq f_4 \\ &f_1 + f_2 + f_3 + f_4 = n \end{aligned}} \tag{29.1}$$

where n is the number of valence protons and neutrons.

In Section 26 we saw that isospin functions with given symmetry transform irreducibly among themselves under $U_T(2)$ transformations. In Section 28 we saw that spin functions transform irreducibly under $U_S(2)$ transformations. Here, spin-isospin functions with given symmetry (29.1) form bases which transform irreducibly under unitary transformations which are induced by unitary transformations in the four dimensional space of (m_t, m_s). The group of transformations is $U(4)$ and its irreducible representations are given by Young diagrams according to (29.1). The number of generators of $U(2j+1)$ was shown in Section 25 to be equal to $(2j+1)^2$. The $U(4)$ transformations are generated by infinitesimal operators whose number should be given for $j = \frac{3}{2}$ by $4^2 = 16$. These generators may be constructed by taking, for each nucleon, the (external) product of the $U_S(2)$ generators $1, \sigma_x, \sigma_y, \sigma_z$ and the $U_T(2)$ generators $1, \tau_1, \tau_2, \tau_3$. In the basis (m_s, m_t)

these $U(4)$ generators are defined by

$$\begin{aligned}\langle m_t m_s|\sigma_i|m_t' m_s'\rangle &= \langle m_s|\sigma_i|m_s'\rangle\delta_{m_t m_t'}\\ \langle m_t m_s|\tau_i|m_t' m_s'\rangle &= \langle m_t|\tau_i|m_t'\rangle\delta_{m_s m_s'}\\ \langle m_t m_s|\tau_i\sigma_j|m_t' m_s'\rangle &= \langle m_t|\tau_i|m_t'\rangle\langle m_s|\sigma_j|m_s'\rangle\end{aligned} \tag{29.2}$$

It is seen that in the customary choice for the $\boldsymbol{\sigma}$ and $\boldsymbol{\tau}$ matrices, σ_z, τ_3 and also $\sigma_z\tau_3$ are diagonal.

The $U(4)$ symmetry introduced here goes beyond ordinary LS-coupling. The $U_S(2)$ symmetry discussed in Section 28 yields the spin S as a good quantum number. The $U_T(2)$ symmetry, due to charge independence, discussed in Section 26 yields the isospin quantum number T. For the special Hamiltonians considered here, which act only on space coordinates, there is a larger symmetry than the one yielding S and T. This symmetry is expressed by the $U(4)$ group and its consequences are described in the following pages.

The spin-isospin states of a single nucleon have the trivial symmetry $f_1 = n = 1$, $f_2 = f_3 = f_4 = 0$. Let us see the structure of spin-isospin states for $n > 1$. Let us start from a state in which the nucleons are in the following single nucleon states. All f_1 nucleons whose numbers appear in the top row of the diagram defined by (29.1) have $m_t = \frac{1}{2}$, $m_s = \frac{1}{2}$, those in the second row are in the $m_t = \frac{1}{2}$, $m_s = -\frac{1}{2}$ state, in the third row in the $m_t = -\frac{1}{2}$, $m_s = \frac{1}{2}$ state and in the fourth they are in the $m_t = -\frac{1}{2}$, $m_s = -\frac{1}{2}$ state. To this state we apply the symmetrizations followed by antisymmetrizations prescribed by the diagram (29.1). The resulting state, as well as all states obtained from it by permutations, has definite eigenvalues of the operators $\frac{1}{2}\sum\tau_3(i)$ and $\frac{1}{2}\sum\sigma_z(i)$, namely

$$\begin{aligned}M_T &= \tfrac{1}{2}(f_1 + f_2 - f_3 - f_4)\\ M_S &= \tfrac{1}{2}(f_1 - f_2 + f_3 - f_4)\end{aligned} \tag{29.3}$$

Also a third operator has a definite eigenvalue since, due to (29.2), it is diagonal in the scheme in which $\frac{1}{2}\sum\tau_3(i)$ and $\frac{1}{2}\sum\sigma_z(i)$ are diagonal. The operator

$$\boxed{\mathbf{Y} = \frac{1}{2}\sum_i \tau_3(i)\sigma_z(i)} \tag{29.4}$$

is diagonal and its eigenvalues are given by

$$Y = \tfrac{1}{2}(f_1 - f_2 - f_3 + f_4) \tag{29.5}$$

The eigenvalues of T_3 and S_z in (29.3) and **Y** in (29.5) together with $\sum_{i=1}^{n} 1 = n = f_1 + f_2 + f_3 + f_4$, uniquely specify the irreducible representations of $U(4)$. By their construction (29.3), M_T and M_S are non-negative whereas Y in (29.5) may become negative.

The value of M_T in (29.3) is the maximum eigenvalue of T_3 in the functions with the symmetry (29.1). Hence, among all spin-isospin functions with this symmetry, the highest possible value of T is given by M_T in (29.3). Of all states with the symmetry (29.1) and this value of T, the value (29.3) of M_S is the highest. Hence, given the maximum value of T, the highest value of S in this set of states is equal to M_S in (29.3). There are usually several states with various isospins T and spins S among spin-isospin states with the symmetry type (29.1). The Hamiltonians considered here have the same energy eigenvalue in all states obtained by multiplying these spin-isospin states by a definite set of spatial eigenfunctions and summing. All states with given $U(4)$ symmetry and definite quantum numbers, including L, which specify the various spatial eigenfunctions, form a *super-multiplet*. This is analogous to spin multiplets where states of identical particles with given S and L form degenerate (or near degenerate) states with various values of J. If the maximum value of T is higher than zero, states of the given super-multiplet appear in several nuclei.

From (29.1) follows that M_T given by (29.3) is positive and hence $Z > N$. In most nuclei which can be studied experimentally there is, however, a neutron excess. To consider such nuclei we should put in the two top rows of the Young diagram numbers of nucleons in $m_t = -\frac{1}{2}$ states. This will make M_T in (29.3) negative and will change the sign of Y as given by (29.5).

The spatial functions which should be multiplied by spin-isospin functions with the symmetry (29.1) are obtained from the dual Young diagram. The diagram dual to (29.1) has four *columns* whose lengths are f_1, f_2, f_3, f_4. The length of the first column, f_1 cannot exceed the total number of single nucleon space functions. That number is usually denoted by 2Ω. If the spatial part is due to several l-orbits, 2Ω is equal to $\sum(2l + 1)$ where the summation is extended over all valence orbits. Linearly independent spatial functions which have a definite symmetry type form a basis of an irreducible representation of the unitary group whose transformations are induced by unitary transformations on the space of single nucleon space functions. That group is

$U(2\Omega)$ which in the special case of l^n configurations is $U(2l+1)$. If we consider unitary transformations among spin-isospin and space coordinates of the nucleons, the complete wave functions of the configuration with n nucleons belong to the fully antisymmetric irreducible representation of $U(4\times 2\Omega)$. In the case of Hamiltonians considered here, this irreducible representation splits into irreducible representations of $U(4)\otimes U(2\Omega)$.

The spin operators and isospin operators play the same role in $U(4)$ and its irreducible representations. They could be interchanged without making any change in the outcome of the discussion. Moreover, instead of defining $Y(i)=\frac{1}{2}\tau_3(i)\sigma_z(i)$ in terms of $\tau_3(i)$ and $\sigma_z(i)$ we could have started with $\sigma_z(i)$ and $Y(i)$ say, and define $\tau_3(i)$ by $2Y(i)\sigma_z(i)$. Hence, T_3, S_z and Y play equivalent roles and may be interchanged in considering a given supermultiplet. Accordingly, instead of (29.3), (29.5), we define three quantum numbers P, P' and P'' which depend on the Young diagram (29.1) but not on the special choice of single nucleon states made above. We thus introduce

$$\boxed{\begin{aligned} P &= \tfrac{1}{2}(f_1+f_2-f_3-f_4) \\ P' &= \tfrac{1}{2}(f_1-f_2+f_3-f_4) \\ P'' &= \tfrac{1}{2}(f_1-f_2-f_3+f_4) \end{aligned}} \tag{29.6}$$

From (29.3) follows the inequality $P\geq P'\geq P''$. Together with n, these quantum numbers determine uniquely the irreducible representations of $U(4)$.

The group $SU(4)$ is defined as the subgroup of $U(4)$ whose matrices have determinants equal to 1. As we saw above, such matrices are obtained from infinitesimal elements whose traces vanish. All generators (29.2) have this property and the only $U(4)$ generator with non-vanishing trace is $\hat{n}$. Hence if we remove $\sum_{i=1}^{n}1=\hat{n}$ from the generators, the other 15 generators form the Lie algebra of $SU(4)$. The irreducible representations of $SU(4)$ are thus determined by the *three* numbers P,P',P''. Young diagrams with the same P,P',P'' numbers yield equivalent irreducible representations of $SU(4)$. From the definition (29.6) follows that these diagrams may differ by the lengths of the rows. Subtracting from all f_i any number $f\leq f_4$, the values of P, P' and P'' are not changed and the $SU(4)$ irreducible representations are equivalent. Subtracting f_4 from all f_i leads to a Young diagram with *three* rows whose lengths are $f_1 - f_4$, $f_2 - f_4$ and $f_3 - f_4$.

The physical meaning of the quantum numbers (29.6) is straightforward. The highest value of T, or S, in any state which belongs to the given supermultiplet is P. The highest value of S (or T) in any state with $T = P$ (or $S = P$) is P'. In these states, the eigenvalue of Y is then equal to P''. In cases where $N > Z$, however, the eigenvalue of $\mathbf{Y}$ as defined by (29.4) is equal to $-P''$ rather than to (29.5).

The spin-isospin state of a single l-nucleon has the trivial symmetry $f_1 = 1$, $f_2 = f_3 = f_4 = 0$ and hence $P = P' = P'' = \frac{1}{2}$ or $(\frac{1}{2}\frac{1}{2}\frac{1}{2})$. This is consistent with that state having $T = \frac{1}{2}$, $S = \frac{1}{2}$. There are such states with $(PP'P'') = (\frac{1}{2}\frac{1}{2}\frac{1}{2})$ in any configuration with $n = 4k + 1$ and $f_1 = k + 1$, $f_2 = f_3 = f_4 = k$. Two nucleons may be in states which are symmetric in space coordinates and antisymmetric in spin-isospin variables or space antisymmetric and spin-isospin symmetric ones. The $U(4)$ irreducible representations of the former are given by [11] and those of the latter by [2]. The [11] irreducible representation is specified by $P = 1$, $P' = P'' = 0$ (100). Such states have $T = 1$, $S = 0$ or $T = 0$, $S = 1$ and in the l^2 configuration have even values of L. Using (26.7) we conclude that there should be 6 functions in the [11] irreducible representation which correspond to the 3 functions with $T = 1$, $S = 0$ and the 3 ones with $T = 0, S = 1$. The [2] irreducible representation, with $(PP'P'') = (111)$ contains states with the maximum values $T = 1$, $S = 1$. To check whether these state exhaust the supermultiplet, we use again (26.7) which yields 10 out of which only 9 are accounted for by $T = 1$, $S = 1$ states. Indeed, spin-isopsin symmetric states may have also $T = 0$, $S = 0$ and this adds another state. These spin-isospin symmetric states are combined with antisymmetric space functions which in the l^2 configuration have odd values of L. These (100) states as well as the (111) states are found in any configuration with nucleon number $n = 4k + 2$ with $f_1 = f_2 = k + 1$, $f_3 = f_4 = k$ or $f_1 = k + 2$, $f_2 = f_3 = f_4 = k$ respectively. Of the $n = 3$ states we mention only those which belong to the Young diagram with $f_1 = f_2 = f_3 = 1$, $f_4 = 0$ or $(PP'P'') = (\frac{1}{2}\frac{1}{2}, -\frac{1}{2})$. These states have $T = \frac{1}{2}$, $S = \frac{1}{2}$ and are found in all configurations with $n = 4k + 3$ with $f_1 = f_2 = f_3 = k + 1$, $f_4 = k$.

It was mentioned before that all states of a given supermultiplet are degenerate for the Hamiltonians considered above. This choice of Hamiltonians is referred to as the Wigner approximation since he introduced it and derived the super-multiplet (or $SU(4)$) scheme (Wigner 1937). The spin and isospin composition of a given super-multiplet are of interest for two reasons. The isospin T of any state determines in which nuclei that state may be observed, namely nuclei for which $T \geq \frac{1}{2}|N-Z|$. The other reason concerns levels in the same nucleus. It is not expected that in actual nuclei there will be strict degeneracy of all states in a given super-multiplet. Deviations from the Wigner approximation split the various states of a super-multiplet in a given nucleus with given T into several multiplets characterized by their S values. Assigning spins and isospins to the various states amounts to finding the decomposition of an irreducible representation of $SU(4)$ into those of its subgroup $SU_T(2) \otimes SU_S(2)$. The transformations of this subgroup are obtained by applying to the spin-isospin states (m_t, m_s) the generators $\tau_1, \tau_2, \tau_3, \sigma_x, \sigma_y, \sigma_z$ as given in (29.2).

As in the case of identical particles, if the interactions which depend only on space coordinates are attractive, the higher the symmetry of the spatial functions the lower the energy. The amount of symmetric couplings may be measured by the eigenvalues of the Majorana exchange operator P^x. It is defined by

$$P_{12}^x f(\mathbf{r}_1, \mathbf{r}_2) = f(\mathbf{r}_2, \mathbf{r}_1) \tag{29.7}$$

and has the eigenvalue 1 in a space symmetric state of two nucleons and -1 in an antisymmetric state. In the l^2 configuration we obtain the relation

$$\boxed{P_{12}^x |l^2 LM\rangle = (-1)^L |l^2 LM\rangle} \tag{29.8}$$

In the preceding section we saw from (28.5) that this operator is related to $C_{U(2l+1)}$. We could replace there $(-1)^L$ by $-P^\sigma$, the spin exchange operator. In the present case we proceed in a similar way. Due to the generalized Pauli principle, we can express P_{12}^x by

$$P_{12}^x = -P_{12}^\tau P_{12}^\sigma = -\tfrac{1}{4}(1 + \boldsymbol{\tau}(1) \cdot \boldsymbol{\tau}(2))(1 + \boldsymbol{\sigma}(1) \cdot \boldsymbol{\sigma}(2)) \tag{29.9}$$

We can then calculate the eigenvalues of $\sum P_{ij}^x$ from spin and isospin functions. These $U(4)$ functions are the same for any composition of spatial functions. Hence, the eigenvalues of P^x will be determined only by the symmetry properties of the states. Once the eigenvalues of P^x are calculated they may be applied to pure l^n configurations or any admixture of configurations.

From (29.9) we obtain the following expression for the Majorana operator

$$
\begin{aligned}
P^x = \sum_{i<j} P^x_{ij} &= -\frac{1}{4}\sum_{i<j}(1+\boldsymbol{\tau}(i)\cdot\boldsymbol{\tau}(j)+\boldsymbol{\sigma}(i)\cdot\boldsymbol{\sigma}(j) \\
&\qquad + \boldsymbol{\tau}(i)\cdot\boldsymbol{\tau}(j)\boldsymbol{\sigma}(i)\cdot\boldsymbol{\sigma}(j)) \\
&= -\frac{1}{8}n(n-1) - \frac{1}{8}\left[\left(\sum\boldsymbol{\tau}(i)\right)\left(\sum\boldsymbol{\tau}(j)\right) - 3n\right] \\
&\quad - \frac{1}{8}\left[\left(\sum\boldsymbol{\sigma}(i)\right)\cdot\left(\sum\boldsymbol{\sigma}(j)\right) - 3n\right] \\
&\quad - \frac{1}{8}\sum_{i\neq j}(\boldsymbol{\tau}(i)\cdot\boldsymbol{\tau}(j))(\boldsymbol{\sigma}(i)\cdot\boldsymbol{\sigma}(j)) \\
&= \frac{n(16-n)}{8} - \frac{1}{2}T(T+1) - \frac{1}{2}S(S+1) \\
&\quad - \frac{1}{8}\sum_{i,j}(\boldsymbol{\tau}(i)\cdot\boldsymbol{\tau}(j))(\boldsymbol{\sigma}(i)\cdot\boldsymbol{\sigma}(j))
\end{aligned} \tag{29.10}
$$

The sum of terms with $i = j$ on the r.h.s. of (29.10) is equal to $-9n/8$ which has been subtracted from the first term. In (29.10), S and T appear in a symmetric way. Among all spin isospin states of the supermultiplet considered we choose the one introduced above for obtaining (29.3). From (29.10) it can be seen that apart from linear and quadratic terms in n, P^x is equal to the Casimir operator of $U(4)$ expressed in terms of the generators (29.2). Hence, P^x is diagonal in the quantum numbers P, P', P'' and its eigenvalues do not depend on the particular choice of the state within the super-multiplet. Thus, we can express (29.10) as

$$
\begin{aligned}
P^x &= \frac{n(16-n)}{8} - \frac{1}{2}(P(P+1) + P'(P'+1)) \\
&\quad - \frac{1}{8}\sum_{i,j}(\boldsymbol{\tau}(i)\cdot\boldsymbol{\tau}(j))(\boldsymbol{\sigma}(i)\cdot\boldsymbol{\sigma}(j))
\end{aligned} \tag{29.11}
$$

The last term in (29.11) should now be evaluated for a state with given P, P', and P''. There are general formulae for expressing the eigenvalues of (29.11) in terms of P, P', P'' and n but the eigenvalues may be also evaluated in a straightforward manner.

The last term in (29.11) can be written as

$$-2\sum_{i,j}(t_3(i)t_3(j)+\tfrac{1}{2}t^+(i)t^-(j)+\tfrac{1}{2}t^-(i)t^+(j))$$
$$\times(s_z(i)s_z(j)+\tfrac{1}{2}s^+(i)s^-(j)+\tfrac{1}{2}s^-(i)s^+(j)) \quad (29.12)$$

The summation is over nine products and we deal with every one of them. The first product is equal to

$$-2\sum_{i,j}t_3(i)t_3(j)s_z(i)s_z(j)=-2\left(\sum_i t_3(i)s_z(i)\right)^2=-\tfrac{1}{2}\mathbf{Y}^2 \quad (29.13)$$

where $\mathbf{Y}$ is defined by (29.4). In the states considered, since T,S assume the values P,P', the eigenvalue of $\mathbf{Y}$ is P''. The state we chose has the maximum value of $M_T=T$. Acting on it with any $t^+(i)$ annihilates it, as can be seen by the fact that nucleons with $m_t=-\frac{1}{2}$ are in antisymmetric states with $m_t=\frac{1}{2}$ nucleons. From this follows that the product $t^+(i)t^-(j)$ gives a non-vanishing result only for $i=j$. Let us then consider the action of the product

$$\frac{1}{2}\sum_i t^+(i)t^-(i)s_z(i)s_z(i)=\frac{1}{8}\sum_i t^+(i)t^-(i) \quad (29.14)$$

which does not change the m_s states of nucleons. Of the f_1 nucleons with $m_t=\frac{1}{2}$, $m_s=\frac{1}{2}$ only f_1-f_3 may change their states to $m_t=-\frac{1}{2}$, $m_s=\frac{1}{2}$. The first f_3 nucleons whose numbers appear in the top row are in antisymmetric states with nucleons in the third row. Similarly, only the first f_2-f_4 nucleons in the second row may give non-vanishing contributions to (29.14). Hence, when (29.14) acts on the state considered, it yields the same state multiplied by

$$\tfrac{1}{8}(f_1-f_3)+\tfrac{1}{8}(f_2-f_4)=\tfrac{1}{4}P \quad (29.15)$$

Similar considerations hold for the spin operators in (29.12) and lead to the contribution of

$$\frac{1}{2}\sum t_3(i)t_3(i)s_z^+(i)s_z^-(i)=\frac{1}{8}\sum s_z^+(i)s_z^-(i)$$

being equal to

$$\tfrac{1}{8}(f_1-f_2)+\tfrac{1}{8}(f_3-f_4)=\tfrac{1}{4}P' \quad (29.16)$$

Another operator to be considered is

$$\frac{1}{4}\sum t^+(i)t^-(i)s^+(i)s^-(i) \tag{29.17}$$

Each term in (29.17) changes the $m_t = \frac{1}{2}$, $m_s = \frac{1}{2}$ state of a single nucleon into the state with $m_t = -\frac{1}{2}$, $m_s = -\frac{1}{2}$ and back to the original state. Due to the antisymmetry properties of the state considered, only $f_1 - f_4$ nucleons in the top row contribute non-vanishing eigenvalues whose sum is $\frac{1}{4}(f_1 - f_4)$. The last term with a non-negative contribution is

$$\frac{1}{4}\sum t^+(i)t^-(i)s^-(i)s^+(i) \tag{29.18}$$

It changes nucleons with $m_t = \frac{1}{2}, m_s = -\frac{1}{2}$ into nucleons with $m_t = -\frac{1}{2}$, $m_s = \frac{1}{2}$ and back to their original states. Hence, its non-vanishing contribution is equal to $\frac{1}{4}(f_2 - f_3)$. The sum of contributions of the operators (29.17) and (29.18) is given by

$$\tfrac{1}{4}(f_1 - f_4) + \tfrac{1}{4}(f_2 - f_3) = \tfrac{1}{2}P \tag{29.19}$$

Adding (29.13) to the sum of (29.15), (29.16) and (29.19) multiplied by -2 we obtain the contribution of the last term in (29.11) as

$$-\tfrac{1}{2}P''^2 - \tfrac{3}{2}P - \tfrac{1}{2}P'$$

Substituting this contribution into (29.11) we obtain for the eigenvalue of P^x the expression

$$-\frac{n(n-16)}{8} - \frac{1}{2}[P(P+4) + P'(P'+2) + P''^2] \tag{29.20}$$

In terms of f_1, f_2, f_3, f_4 these eigenvalues may be rewritten as the following expression

$$\begin{aligned} -\tfrac{1}{2}[f_1(f_1-1) + f_2(f_2-3) + f_3(f_3-5) + f_4(f_4-7)] \\ = f_2 + 2f_3 + 3f_4 - \tfrac{1}{2}(f_1(f_1-1) + f_2(f_2-1) \\ + f_3(f_3-1) + f_4(f_4-1)) \end{aligned} \tag{29.21}$$

The numbers $f_2 + 2f_3 + 3f_4$ may be interpreted as the number of antisymmetric couplings and $\sum \frac{1}{2}f_i(f_i - 1)$ as the number of symmetric couplings in the spin-isospin functions with the symmetry type

$[f_1 f_2 f_3 f_4]$. In the corresponding spatial functions the numbers of symmetric and antisymmetric couplings are reversed.

Looking at (29.20) we see that the smaller the quantum numbers P, P' and P'', the higher the symmetry of the spatial functions. The operator $-V_0 P^x$ may be considered as the long range limit of an attractive Majorana interaction. Its eigenvalues are lower for lower values of P, P', P''. We will discuss below energy formulae based on the $SU(4)$ scheme. Before continuing let us make use of (29.20) to deduce the eigenvalues of the Casimir operator of the $SU(4)$ group. We may use the definition

$$C_{SU(4)} = \left(\sum \boldsymbol{\tau}(i)\right)\cdot\left(\sum \boldsymbol{\tau}(i)\right) + \left(\sum \boldsymbol{\sigma}(i)\right)\cdot\left(\sum \boldsymbol{\sigma}(i)\right) + \sum_{i,j} (\boldsymbol{\tau}(i)\cdot\boldsymbol{\tau}(j))(\boldsymbol{\sigma}(i)\cdot\boldsymbol{\sigma}(j)) \tag{29.22}$$

From (29.10) we obtain the following relation

$$P^x = \frac{n(16-n)}{8} - \frac{1}{8}C_{SU(4)} \tag{29.23}$$

and the following expression for the eigenvalues of (29.22)

$$\boxed{4[P(P+4) + P'(P'+2) + P''^2]} \tag{29.24}$$

These eigenvalues depend indeed only on the irreducible representation of $SU(4)$ which is uniquely specified by the values of P, P', P''.

The eigenvalues (29.20) of the Majorana operator may be simply related in a certain approximation to the interaction energy of nuclear configurations. The two-body interactions considered in this section are independent of spin and isospin operators. Two simple interactions of this kind are a potential interaction $V_W(|\mathbf{r}_1 - \mathbf{r}_2|)$ which is sometimes called Wigner force and a space exchange interaction $P^x_{12} V_M(|\mathbf{r}_1 - \mathbf{r}_2|)$ (Majorana force). If the range of potentials V_W and V_M is large compared to the spatial extension of the nucleon wave functions and they are rather flat in that region, they may be replaced by constants. Interaction energies of states in given configurations (or given mixed configurations) could in that case be roughly expressed

by

$$a\frac{n(n-1)}{2}+bP^x=\left(a-\frac{b}{4}\right)\frac{n(n-1)}{2}+\frac{15}{8}bn$$
$$-\tfrac{1}{2}b[P(P+4)+P'(P'+2)+P''^2] \tag{29.25}$$

The interaction in a space symmetric state of two nucleons is $a+b$ whereas in a space antisymmetric state it is $a-b$. Hence, from nuclear systematics follows that b is an attractive coefficient. From (29.25) follows that the smaller the last term on the r.h.s. of (29.25) the lower the energy. In any given nucleus the supermultiplet with the lowest possible values of P, P', P'' is expected to be the lowest. For general potentials of the Wigner and Majorana forces, (29.25) holds for the *average* interaction energies of supermultiplets with given n and quantum numbers P, P', P''. This point will be further discussed in the special case of l^n configurations.

The lowest value of P is equal to the lowest possible value of T or S. Since the lowest value of T is limited by $T \geq \frac{1}{2}|N-Z|$, it is better to choose $P=T=\frac{1}{2}|N-Z|$ for the state with lowest interaction energy. The term $-(1/2)bT(T+4)$ is thus the symmetry energy which arises naturally in the supermultiplet scheme. For even-even nuclei we consider the super-multiplet with $f_2=f_1, f_4=f_3$, for which $P'=P''=0$ according to (29.6). In that case $f_1-f_3=f_2-f_4=\frac{1}{2}|N-Z|=P$ and the supermultiplet is $(PP'P'')=(\frac{1}{2}|N-Z|, 0, 0)$.

An odd-even nucleus with $P=T=\frac{1}{2}|N-Z|$ may have the lowest value of S equal to $\frac{1}{2}$. If we add a nucleon to an even-even nucleus with $(PP'P'')=(\frac{1}{2}|N-Z|, 0, 0)$, there are two possibilities. We may add a neutron if there is a neutron excess (a proton if $Z \geq N$) or add a proton if $N>Z$ (a neutron if $Z>N$). In the first case $f_1= f_2+1, f_3=f_4$ and T is increased by $\frac{1}{2}$ and in the other case $f_1=f_2$ but $f_3=f_4+1$ and T is decreased by $\frac{1}{2}$. In the first case we obtain from (29.6) $P'=\frac{1}{2}$, $P''=\frac{1}{2}$ and the supermultiplet is $(\frac{1}{2}|N-Z|, \frac{1}{2}, \frac{1}{2})$. In the other case $P'=\frac{1}{2}$, $P''=-\frac{1}{2}$ and the supermultiplet is $(\frac{1}{2}|N-Z|, \frac{1}{2}, -\frac{1}{2})$. Comparing the binding energy of the odd-even nucleus to that of the even-even one, apart from the smooth behavior with n and the symmetry energy in both cases, we see the same extra term

$$-\tfrac{1}{2}b(P'(P'+2)+P''^2)=-\tfrac{1}{2}b(\tfrac{5}{4}+\tfrac{1}{4})=-\tfrac{3}{4}b \tag{29.26}$$

This term gives rise to the well known odd-even variation in binding energies. It has been related in the previous sections to the pairing term in the seniority scheme. It is interesting to see that it also arises in the $SU(4)$ scheme as a result of the spatial symmetry of the nucleon wave functions.

The binding energy formula (29.25) for an attractive (long range) Majorana interaction has some important features observed in experiment. It contains a symmetry energy term which, by putting $P = T$, is proportional to $T(T+4)$. For $b < 0$, the lower the total isospin T, the lower the energy. It also exhibits variation between odd-even and even-even nuclei with the correct sign for $b < 0$. It is therefore not surprising that mass formulae based on $SU(4)$ symmetry may seem to be in reasonable agreement with experiment. This is specially true if the coefficient of the symmetry energy term and the odd-even variation are taken to be independent, as in (29.46) below. The $SU(4)$ scheme, however, is incompatible with observed spin dependent interactions and in particular with the strong spin-orbit interaction that gives rise to the magic numbers.

Another interesting physical observable may be directly derived in the super-multiplet scheme. The magnetic moment operator is given by

$$\boxed{\boldsymbol{\mu} = g_S \mathbf{S} + g_L \mathbf{L}} \tag{29.27}$$

To obtain the magnetic moment in a given state, we should evaluate independently g_S and g_L and then use (8.24) with $g_1 = g_S$, $J_1 = S$, $g_2 = g_L$, $J_2 = L$, to obtain the value of μ.

The magnetic moment operator associated with the spin $\mathbf{S}$ can be written as

$$\begin{aligned} g_s^{\pi} \sum_{\text{protons}} \mathbf{s}_i + g_s^{\nu} \sum_{\text{neutrons}} \mathbf{s}_j \\ &= \tfrac{1}{2} g_s^{\pi} \sum (1 + \tau_3^{(i)}) \mathbf{s}_i + \tfrac{1}{2} g_s^{\nu} \sum (1 - \tau_3^{(j)}) \mathbf{s}_j \\ &= \tfrac{1}{2}(g_s^{\pi} + g_s^{\nu}) \sum \mathbf{s}_i + \tfrac{1}{2}(g_s^{\pi} - g_s^{\nu}) \sum \tau_3^{(i)} \mathbf{s}_i \end{aligned} \tag{29.28}$$

To obtain the magnetic moment associated with $\mathbf{S}$, we take the z-component of (29.28) in the state with $S_z = S$. The contribution of the first term on the r.h.s. of (29.28) is proportional to S whereas the z-component of the second term is proportional to the operator $\mathbf{Y}$

defined by (29.4). The latter is diagonal in the super-multiplet scheme and hence in states with $P = \frac{1}{2}|N - Z|$, the magnetic moment due to (29.28) for $N > Z$ (in which case $Y = -P''$) is given by

$$\tfrac{1}{2}(g_s^\pi + g_s^\nu)S + \tfrac{1}{2}(g_s^\pi - g_s^\nu)Y = \tfrac{1}{2}(g_s^\pi + g_s^\nu)P' - \tfrac{1}{2}(g_s^\pi - g_s^\nu)P'' \tag{29.29}$$

Let us apply (29.29) to ground states of odd-even nuclei with $N > Z$. As we saw above, for odd proton nuclei $P' = \frac{1}{2}$, $P'' = -\frac{1}{2}$ whereas for odd neutron nuclei $P' = \frac{1}{2}$, $P'' = \frac{1}{2}$. Substituting these values in (29.29) we obtain

$$\tfrac{1}{2}(g_s^\pi + g_s^\nu)\tfrac{1}{2} + \tfrac{1}{2}(g_s^\pi - g_s^\nu)\tfrac{1}{2} = \tfrac{1}{2}g_s^\pi \quad \text{odd proton nuclei}$$

$$\tfrac{1}{2}(g_s^\pi + g_s^\nu)\tfrac{1}{2} - \tfrac{1}{2}(g_s^\pi - g_s^\nu)\tfrac{1}{2} = \tfrac{1}{2}g_s^\nu \quad \text{odd neutron nuclei}$$

Recalling that in such nuclei $S = \frac{1}{2}$, we find that g_S is equal to g_s^π for ground states of odd proton nuclei and to g_s^ν for odd neutron nuclei. This is the same result as for a single nucleon outside closed shells.

The value of g_L in the supermultiplet theory may be rather different from that of a single nucleon. Margenau and Wigner (1940) who carried out this calculation assumed that protons and neutrons contribute equally to g_L which assumption yields the value

$$g_L = \tfrac{1}{2}(g_L^\pi + g_L^\nu) = \tfrac{1}{2}$$

If $Z = N$ then this value is the exact one but in other cases it depends very much on the model space. Adopting this value, the magnetic moment due to (29.27) is given by expressions analogous to (3.34) and (3.37) for $J = L + \frac{1}{2}$ and $J = L - \frac{1}{2}$ respectively. Nuclear magnetic moments are thus given by

$$\boxed{\mu = \tfrac{1}{2}L + \tfrac{1}{2}g_s \qquad \text{for} \quad J = L + \tfrac{1}{2}}$$

and

$$\boxed{\mu = \frac{J}{J+1}\left(\tfrac{1}{2}(L+1) - \tfrac{1}{2}g_s\right) \qquad \text{for} \quad J = L - \tfrac{1}{2}}$$

where $g_s = g_s^{\pi}$ for odd proton nuclei and $g_s = g_s^{\nu}$ for those with an odd neutron. These formulae yield magnetic moments similar to those in Fig. 2.1 but the agreement with experiment is not as good as the formulae of Lande (1934) and Schmidt (1937).

In order to obtain detailed binding energy formulae we should leave the most general $SU(4)$ scheme and limit our attention to definite orbital configurations of the nucleons. In the next section we will look at a special case of configuration mixing but here we consider l^n configurations. According to the discussion above and in the preceding section, we have in this case the following group chain

$$U(4(2l+1)) \supset U(4) \otimes U(2l+1) \tag{29.30}$$

The generators of $U(2l+1)$ are, as before, all irreducible tensor operators $U_{\kappa}^{(k)}$ and its Casimir operator is given by (28.4) as

$$\boxed{\sum_k (2k+1)(\mathbf{U}^{(k)} \cdot \mathbf{U}^{(k)})} \tag{29.31}$$

The eigenvalues of (29.31) in the various irreducible representations of $U(2l+1)$ which are allowed by the Pauli principle may be evaluated by using the irreducible representation of $U(4)$ with the dual Young diagrams. The maximum length of a row in the latter cannot exceed $2l+1$ which is the number of independent orbital states of a single nucleon. Recalling (28.5) as well as (29.8), we can evaluate (29.31) in analogy with (28.6) as follows

$$\sum_k (2k+1)(\mathbf{U}^{(k)} \cdot \mathbf{U}^{(k)}) = 2\sum_{i<j,k} (2k+1)(\mathbf{u}^{(k)}(i) \cdot \mathbf{u}^{(k)}(j)) + (2l+1)n$$

$$= 2\sum_{i<j} P_{ij}^x + (2l+1)n \tag{29.32}$$

The eigenvalues of $P^x = \sum_{i<j} P_{ij}^x$ are given by (29.20) from which the eigenvalues of (29.32) are determined to be

$$\boxed{(2l+1)n - \frac{n(n-16)}{4} - [P(P+4) + P'(P'+2) + P''^2]} \tag{29.33}$$

In the case of identical nucleons, $T = n/2, f_3 = f_4 = 0$ in which case $P' = P'' = S$ and hence, (29.33) reduces to (28.6).

The Casimir operator of the semi-simple group $SU(2l + 1)$ is obtained from (29.31) by removing the term with $\mathbf{U}^{(0)}$ which is the generator of an Abelian invariant subgroup of $U(2l + 1)$. That term is equal to $\hat{n}^2/(2l + 1)$ and the resulting eigenvalues of $C_{SU(2l+1)}$ are thus given by

$$\boxed{\begin{array}{l}\dfrac{4l(l+1)}{2l+1}n - \dfrac{n(n-1)}{2l+1} - \dfrac{n(n-16)}{4} \\ \qquad\qquad - [P(P+4) + P'(P'+2) + P''^2]\end{array}} \tag{29.34}$$

The eigenvalues of $C_{SU(2l+1)}$ are given by (29.34) only for states whose symmetry in the space coordinates is defined by a Young diagram which is dual to that of the $SU(4)$ irreducible representation defined by P, P', P''.

We can now give (29.25) a more precise interpretation. We consider average energies of states which belong to the same irreducible representation of $U(2l + 1)$. The only terms in the Hamiltonian which contribute to these averages are operators which are scalars with respect to $U(2l + 1)$ transformations. They commute with all generators and hence have the same eigenvalues in all states within a given irreducible representation. The two-body interaction P_{12}^x as well as a constant interaction are two such scalars. Their coefficients in the case of a given two-body interaction may be determined from the two nucleon l^2 configuration. Let us denote by $\bar{V}_s$ the average interaction energy in all space symmetric l^2 states with even values of L. They have either $T = 1, S = 0$ or $T = 0, S = 1$ spin-isospin quantum numbers. Similarly, $\bar{V}_a$ will be the average of space antisymmetric l^2 states with odd values of L. The spins and isospins of those states are either $T = 1, S = 1$ or $T = 0, S = 0$. We can then determine the coefficients of 1 and P_{12}^x from the equations

$$\bar{V}_s = a + b \qquad \bar{V}_a = a - b \tag{29.35}$$

The averages in any l^n configuration are then given in terms of these a and b by

$$\bar{V}(l^n, PP'P'') = a\frac{n(n-1)}{2} + bP^x$$

$$= \left(a - \frac{b}{4}\right)\frac{n(n-1)}{2} + \frac{15}{8}bn$$

$$- \tfrac{1}{2}b[P(P+4) + P'(P'+2) + P''^2] \qquad (29.36)$$

To obtain a formula for binding energies we add to (29.36) the single l-nucleon terms which are equal to $n\epsilon_l$.

A more precise binding energy formula may be obtained by making a distinction between the $L = 0$ state and states with $L > 0$, even, in the l^2 configuration. As in the preceding section we introduce the seniority scheme by the irreducible representations of $O(2l+1)$ which is a subgroup of $U(2l+1)$. The generators of $O(2l+1)$ are irreducible tensor operators $U_\kappa^{(k)}$ with odd ranks k. They satisfy the relation (28.9), namely

$$U_\kappa^{(k)}|l^2 L = 0\rangle = 0 \qquad k \text{ odd} \qquad (29.37)$$

As explained there, this relation is based on the spatial symmetry of both the $l^2 L = 0$ state and the operator $U_\kappa^{(k)}$. Hence, it holds also in the present case with both protons and neutrons in the l-orbit.

The Casimir operator of $O(2l+1)$ is defined, as in (28.17), by

$$C_{O(2l+1)} = 2\sum_{k \text{ odd}}(2k+1)(\mathbf{U}^{(k)} \cdot \mathbf{U}^{(k)}) \qquad (29.38)$$

Its eigenvalues in the present case depend on the irreducible representations which are more complicated than those considered in Section 28. Following (28.18) we replace $(-1)^L$ in the l^2 configuration by $-(-1)^{S+T} = -(1/4)(1 + 4(\mathbf{s}_1 \cdot \mathbf{s}_2))(1 + 4(\mathbf{t}_1 \cdot \mathbf{t}_2))$. The analog of (28.19) becomes then

$$\langle l^2LM|2\sum_{k\ \mathrm{odd}}(2k+1)(\mathbf{U}^{(k)}\cdot\mathbf{U}^{(k)})|l^2LM\rangle$$
$$=\langle l^2TSLM|-\tfrac{1}{2}(1+4(\mathbf{s}_1\cdot\mathbf{s}_2))$$
$$\times(1+4(\mathbf{t}_1\cdot\mathbf{t}_2))-2P(1,2)|l^2TSLM\rangle+4l \qquad (29.39)$$

where $P(1,2)$ is the pairing interaction (28.20). From (29.39) follows that in the l^n configuration we obtain, in view of (29.10), the relation

$$\langle l^nPP'P''LM|2\sum_{k\ \mathrm{odd}}(2k+1)(\mathbf{U}^{(k)}\cdot\mathbf{U}^{(k)})|l^nPP'P''LM\rangle$$
$$=-\frac{n(n-16)}{4}-[P(P+4)+P'(P'+2)+P''^2]$$
$$-2\langle l^nPP'P''LM|\sum_{i<j}P(ij)|l^nPP'P''LM\rangle+2nl$$
$$(29.40)$$

Recalling the general relation between pairing interactions and Casimir operators, we conclude from (29.40) that the eigenvalues of $C_{O(2l+1)}$ in the irreducible representation whose Young diagram has four columns with lengths

$$\nu_1\geq\nu_2\geq\nu_3\geq\nu_4 \qquad \nu_1+\nu_2+\nu_3+\nu_4=\nu \qquad \nu_1\leq l$$
$$(29.41)$$

are given by

$$2l\nu-\frac{\nu(\nu-16)}{4}-p(p+4)-p'(p'+2)-p''^2 \qquad (29.42)$$

In (29.42) the quantum numbers p, p', p'' are defined, in analogy with (29.6) by

$$p=\tfrac{1}{2}(\nu_1+\nu_2-\nu_3-\nu_4)$$
$$p'=\tfrac{1}{2}(\nu_1-\nu_2+\nu_3-\nu_4) \qquad (29.43)$$
$$p''=\tfrac{1}{2}(\nu_1-\nu_2-\nu_3+\nu_4)$$

The result (29.42) is equal to the general formula (28.25) for the eigenvalues of $C_{O(2l+1)}$, namely

$$C_{O(2l+1)}(w_1, w_2, \ldots, w_l) = \sum_{i=1}^{l} w_i(w_i + 2l + 1 - 2i) \tag{29.44}$$

Putting in (29.44) the values

$$w_1 = \cdots = w_{v_4} = 4, \quad w_{v_4+1} = \cdots = w_{v_3} = 3, \quad w_{v_3+1} = \cdots = w_{v_2} = 2,$$
$$w_{v_2+1} = \cdots = w_{v_1} = 1, \quad w_{v_1+1} = \cdots = w_l = 0$$

we obtain the result (29.42). There is no simple way to derive the values of $(v_1 v_2 v_3 v_4)$ which characterize irreducible representations of $O(2l+1)$ which are obtained from a given $[f_1 f_2 f_3 f_4]$ irreducible representation of $U(2l+1)$. In the following we will obtain these values for the lowest states of even-even and odd-even nuclei.

We can now write down the expression for the average interaction energies of states of the l^n configuration with given P, P', P'' *and* a given irreducible representation of $O(2l+1)$ denoted by $W = (w_1, \ldots, w_l)$. Adding single l-nucleon terms, $n\epsilon_l$, we can express these averages in terms of the scalars appearing in (29.36) and the two body part of the Casimir operator of $O(2l+1)$. We thus obtain

$$\bar{V}(l^n PP'P''W) = a\frac{n(n-1)}{2} + b'P^x + \frac{c}{2}[C_{O(2l+1)} - 2nl] + n\epsilon_l \tag{29.45}$$

The coefficients a and b are no longer given by (29.35). They and the coefficient c can be determined from the eigenvalues of the Hamiltonian in the l^2 configuration. We denote by $\bar{V}_a$ the average interaction of the space antisymmetric states with $L = 1, 3, \ldots, 2l-1$ which belong to the supermultiplet $(PP'P'') = (111)$ or to $[f] = [2,0]$ of $SU(4)$. The seniority of these states is $v = 2$ and hence they have $v_1 = 2$, $v_2 = v_3 = v_4 = 0$ or $p = p' = p'' = 1$. The $l^2 L = 0$ state, with interaction energy V_0, belongs to $[f] = [11]$ and thus to the $(PP'P'') = (100)$ supermultiplet. It has seniority $v = 0$ and hence $v_1 = v_2 = v_3 = v_4 = 0$

or $p = p' = p'' = 0$. The other states in this super-multiplet have $L = 2, 4, \ldots, 2l$ and seniorities $v = 2$. Therefore, these states have $v_1 = v_2 = 1$, $v_3 = v_4 = 0$, and $p = 1$, $p' = p'' = 0$. If the average interaction of these states is denoted by $\bar{V}_s'$ we obtain from (29.45) for $n = 2$ the relations

$$\begin{aligned}
\bar{V}_a &= a - b - c \\
\bar{V}_s' &= a + b + c \\
V_0 &= a + b - 2lc
\end{aligned} \tag{29.46}$$

We can now apply the binding energy formula to ground states of even-even and odd-even nuclei. We consider interactions which are stronger in space symmetric states which makes the coefficient b in (29.46) positive (binding energies are taken here to be positive quantities). If the coefficient of $C_{O(2l+1)}$ in (29.45) is negative, the lowest l^2 state will have $L = 0$. In any l^n configuration with even n, there is a state with $L = 0$ (and $S = 0$) with seniority $v = 0$, which has the lowest eigenvalue of (29.45). In such states $T = \frac{1}{2}|N - Z|$ and $(PP'P'') = (\frac{1}{2}|N - Z|, 0, 0,)$ as discussed above. There is only one such state with $v = 0$ and hence, the average value (29.45) is equal to the interaction energy in that state. If the Hamiltonian is not diagonal in the seniority scheme, (29.43) is equal to its expectation value in that $v = 0$ state. When the values $P = \frac{1}{2}|N - Z|$, $P' = P'' = 0$, $p = p' = p'' = 0$ are substituted in (29.45) we obtain

$$n\epsilon_l + \frac{n(n-1)}{2}a + b\left[\frac{n(16-n)}{8} - \frac{1}{2}T(T+4)\right] - nlc$$

In the case of odd-even nuclei, there is a state with $S = \frac{1}{2}$, $L = l$ which has seniority $v = 1$. As discussed above, it belongs to the super-multiplet $(PP'P'') = (\frac{1}{2}|N - Z|, \frac{1}{2}, \pm\frac{1}{2})$ and has the lowest energy of that l^n configuration. Putting in (29.45) $P = \frac{1}{2}|N - Z| = T$, $P' = 1/2$, $P'' = \pm\frac{1}{2}$, as well as $v_1 = 1, v_2 = v_3 = v_4 = 0$ or $p = p' = p'' = \frac{1}{2}$ we obtain the result

$$\boxed{n\epsilon_l + \frac{n(n-1)}{2}a + b\left[\frac{n(16-n)}{8} - \frac{1}{2}T(T+4) - \frac{3}{4}\right] - nlc + lc} \tag{29.47}$$

Apart from the symmetry energy and linear and quadratic terms in n, there is an odd-even variation in binding energies equal to $-3b/4 + lc$.

The most interesting feature of the super-multiplet scheme is in its application to allowed beta decays. As explained in Section 9, allowed Fermi transitions can take place between states which differ only by the value of $T_z = M_T$. All other quantum numbers, including the isospin T must be the same. The operator which gives rise to allowed Gamow-Teller beta transitions is according to (9.15), proportional to $\sum \tau^+(i)\mathbf{s}(i)$ or $\sum \tau^-(i)\mathbf{s}(i)$ and hence, its components are among the generators (29.2) of $SU(4)$. From this follows an important result, namely,

allowed Gamow-Teller beta transitions may take place only between states which are members of the same super-multiplet.

The Gamow-Teller operator does not affect space coordinates and hence, cannot change the spatial functions. Alternatively, since its components are generators of $SU(4)$, they have non-vanishing matrix elements only between spin-isospin functions which belong to the same irreducible representation of $SU(4)$.

This $SU(4)$ selection rule accounts very well for the observed difference between *favored* (or super-allowed) and *unfavored* Gamow-Teller beta decays. Specially in light nuclei, the ft-values, which are inversely proportional to squares of the matrix elements, of the unfavored decays are bigger by a factor 100 than the favored ones. The nice distinction between these two kinds of transitions does not exist in pure jj-coupling. It seems, however, that some deviations from jj-coupling, caused by nucleon mutual interactions, may restore the distinction between the favored and unfavored beta decays (Talmi 1962c). Similar deviations, due to tensor interactions, may give rise, in certain cases, to complete cancellation of Gamow-Teller matrix elements like in the ^{14}C decay (Talmi 1983). The very big attenuation of that allowed transition cannot be reconciled with the super-multiplet scheme. In the $SU(4)$ the matrix element for that transition should have been large and could not change by small deviations from that scheme.

30

Special Proton Neutron Mixed Configurations. The $SU(3)$ Scheme

In preceding sections it was mentioned that additional quantum numbers are needed to distinguish between states with the same symmetry type of the spatial functions which have the same value of L. In l^n configurations some of these quantum numbers are supplied by the seniority scheme. We saw that the seniority scheme is based on the group $O(2l+1)$ which is intermediate between $U(2l+1)$ and $O(3)$, namely $U(2l+1) \supset O(2l+1) \supset O(3)$. When more than one l-orbit is active, no such intermediate groups are generally known. There is, however a special case in which a simple scheme was found. We shall briefly describe this case in which there is theoretical interest. The theory of this case is due to Elliott (1958). Many papers have been written on the Elliott scheme and it would be difficult to list them here.

The harmonic oscillator potential well leads to various symmetries of the shell model Hamiltonian. We made use in Section 13 of a certain symmetry which allowed the transformation to the center of mass and relative coordinates of two nucleons. The Hamiltonian expressed in terms of **R** and **r**, as well as their conjugate momenta, turned out to be essentially the oscillator Hamiltonian. The symmetry between

$\mathbf{r}^2$ and $\mathbf{p}^2$ can be utilized also in another way. The harmonic oscillator Hamiltonian (4.1) is invariant under certain transformations which admix the components of $\mathbf{r}$ and $\mathbf{p}$.

We may use the nucleon mass m as the unit mass and $1/\omega$ as the unit of time. In terms of these units, the single nucleon Hamiltonian (4.1) can be expressed as

$$\boxed{H_0 = \tfrac{1}{2}(\mathbf{p}^2 + \mathbf{r}^2) = \tfrac{1}{2}(\mathbf{p} + i\mathbf{r}) \cdot (\mathbf{p} - i\mathbf{r}) + \tfrac{3}{2}\hbar = \hbar(\mathbf{A}^\dagger \cdot \mathbf{A} + \tfrac{3}{2})} \tag{30.1}$$

The vector operators $\mathbf{A}^\dagger$ and $\mathbf{A}$ are defined by

$$\boxed{\mathbf{A}^\dagger = \frac{1}{\sqrt{2\hbar}}(\mathbf{p} + i\mathbf{r}) \qquad \mathbf{A} = \frac{1}{\sqrt{2\hbar}}(\mathbf{p} - i\mathbf{r})} \tag{30.2}$$

Their components satisfy the Bose commutation relations

$$\boxed{[A_i, A^\dagger{}_k] = \delta_{ik} \qquad [A_i, A_k] = [A^\dagger{}_i, A^\dagger{}_k] = 0} \tag{30.3}$$

The Hamiltonian (30.1) is invariant under a large group of transformations in the six dimensional space spanned by $\mathbf{p}$ and $\mathbf{r}$ which leave invariant the scalar product $\mathbf{p}^2 + \mathbf{r}^2$. Here we consider the $U(3)$ subgroup—the group of unitary transformations on the (complex) vector $\mathbf{p} - i\mathbf{r}$. Such unitary transformations leave invariant the scalar product $\mathbf{A}^\dagger \cdot \mathbf{A} = (1/2\hbar)(\mathbf{p}^2 + \mathbf{r}^2) - \frac{3}{2}$.

The infinitesimal elements of $U(3)$ may be taken as all 3×3 hermitean matrices as explained in Sections 25–27. We saw that these generators of $U(2l + 1)$ may also be taken as all components of irreducible tensor operators. In the present case, these are the $1 + 3 + 5 = 9$ components of the irreducible tensors of ranks $k = 0$, $k = 1$ and $k = 2$. These operators may be constructed from the nine independent components of the second rank cartesian tensor $A^\dagger{}_i A_k$ as was shown in Section 5. We first consider the three antisymmetric products

$$A^{\dagger}{}_x A_y - A^{\dagger}{}_y A_x = \frac{i}{\hbar}(xp_y - yp_x) = il_z$$

$$A^{\dagger}{}_y A_z - A^{\dagger}{}_z A_y = il_x \qquad A^{\dagger}{}_z A_x - A^{\dagger}{}_x A_z = il_y \tag{30.4}$$

which are proportional to components of the orbital angular momentum $\mathbf{l}$ (generators of the subgroup $O(3)$ of $U(3)$). The six symmetric products are

$$A^{\dagger}{}_i A_k + A^{\dagger}{}_k A_i = \frac{1}{\hbar}(p_i p_k + r_i r_k) \tag{30.5}$$

The trace of the tensor (30.5) is $\mathbf{p}^2 + \mathbf{r}^2$ (divided by $\hbar$) and it commutes with all $U(3)$ generators. It is a scalar equal to twice the Hamiltonian H_0 in (30.1) (divided by $\hbar$).

Apart from H_0 the generators of $U(3)$ may be taken as the three components of $\mathbf{l}$ and the five components of the traceless symmetric tensor. The latter can be written as the 5 components of the irreducible tensor with $k = 2$ which is

$$q^{(2)}_{\mu} = \frac{\sqrt{6}}{2\hbar}(\mathbf{r} \times \mathbf{r})^{(2)}_{\mu} + \frac{\sqrt{6}}{2\hbar}(\mathbf{p} \times \mathbf{p})^{(2)}_{\mu} \tag{30.6}$$

The factor $\sqrt{6}$ was introduced into the definition (30.6) to conform with standard definitions. The 8 operators (30.4) and (30.6) are generators of $SU(3)$. As explained in Section 6 and then in Sections 25–27 the only generator with non-vanishing trace is the scalar, $k = 0$, operator. It is the only one which generates a matrix whose determinant is different from 1. In the present case the scalar is proportional to H_0 whose eigenvalues are $\hbar(N + \frac{3}{2})$ and is thus essentially the number operator of the oscillator quanta.

Since H_0 commutes with all $U(3)$ generators, they transform single nucleon wave functions with given N into themselves. In single nucleon states with given N there are N oscillator quanta. These quanta are bosons (with $l = 1$) since their creation and annihilation operators satisfy the commutation relations (30.3). Hence, states of a single nucleon in the N-th oscillator shell are fully symmetric functions of N oscillator quanta. All symmetric states of N quanta transform irreducibly among themselves under $SU(3)$ transformations. This is a special case of $U(2l+1)$ transformations with $l = 1$ which is applicable to these $l = 1$ quanta or p-bosons. Since H_0 commutes with all $U(3)$ generators, its eigenvalues in all states which belong to an irreducible representation of $U(3)$ are equal. The degeneracy of these states whose eigenvalues are determined by N alone, is thus related to the invariance of H_0 under $U(3)$ transformations.

For $N = 0$ there are no p-bosons and the single nucleon state has $l = 0$. The state with $N = 1$ p-boson has $l = 1$. For higher values of N there are several single nucleon l-orbits. For $N = 2$ there is a d-orbit ($l = 2$) and s-orbit ($l = 0$), for $N = 3$ there are f- and p-orbits etc. The orbit with the highest l value for given N has $l = N$. Other l-values are $N-2, N-4, \ldots, 1$ or 0. These allowed values of l are well known from the simple theory of the harmonic oscillator. These are all fully symmetric states of N p-bosons and they all have the same eigenvalues of H_0 namely $\hbar(N+\frac{3}{2})$. We met a similar situation in Section 27 in the determination of allowed isospin T values of states with $v = 0$, $t = 0$ ($J = 0$).

All irreducible representations of $U(3)$ are characterized by Young tableaux with three rows whose lengths n_1, n_2, n_3 satisfy the relations

$$\boxed{n_1 \geq n_2 \geq n_3 \qquad n_1 + n_2 + n_3 = N} \tag{30.7}$$

In (30.7), N is the total number of $l = 1$ particles. The symmetric representations of N particles which constitute single nucleon states are characterized by $n_1 = N$, $n_2 = n_3 = 0$. When going over to $SU(3)$, N is no longer an eigenvalue of a generator. The irreducible representations of $SU(3)$ are characterized by *two* rows whose lengths f_1 and f_2 are defined by

$$\boxed{f_1 = n_1 - n_3 \qquad f_2 = n_2 - n_3} \tag{30.8}$$

Elliott (1958) who introduced the application of $SU(3)$ described in this section, defined the notation

$$\boxed{f_1 = \lambda + \mu \qquad f_2 = \mu} \tag{30.9}$$

The single nucleon states described above have $f_1 = n_1 = N$, $n_2 = n_3 = 0$ and their quantum numbers (λ, μ) are thus given by $(N, 0)$. States with several nucleons may belong to more complicated irreducible representations of $SU(3)$ with $\mu \neq 0$. We will now consider such states.

If the Hamiltonian of each nucleon is given by (30.1), states of n nucleons in the oscillator shell with given N are all degenerate. All these states have the same eigenvalue $n(N + \frac{3}{2})\hbar$. As mentioned above, the Hamiltonian is invariant under $U(3)$ transformations carried out on every nucleon. Such $U(3)$ transformations may be carried out *independently* on each nucleon. We will consider, however, for the following applications, $U(3)$ transformations which are carried out simultaneously on *all* nucleons. The $U(3)$ transformations give rise to linear transformations among all wave functions which are products of n single nucleon states. These transformations are, in general, not irreducible and split into several irreducible ones characterized by certain values of (λ, μ).

The quadratic Casimir operator of $SU(3)$ is constructed as the linear combination of the squares of the generators. For n nucleons the $SU(3)$ generators are

$$\boxed{\mathbf{L} = \sum_{i=1}^{n} \mathbf{l}(i) \qquad Q_\mu^{(2)} = \sum_{i=1}^{n} q_\mu^{(2)}(i)} \tag{30.10}$$

where the quadrupole operator of each nucleon is defined by (30.6). The quadrupole and orbital angular momentum operators of different nucleons commute. Creation operators of oscillator quanta in states of one nucleon commute with annihilation operators of quanta in states of another nucleon. Accordingly, the p-particles which are the oscillator quanta are no longer bosons of the same kind and may couple also to antisymmetric states in addition to symmetric ones. The Casimir operator is defined by

$$\boxed{C_{SU(3)} = \tfrac{1}{4}(\mathbf{Q}^{(2)} \cdot \mathbf{Q}^{(2)}) + \tfrac{3}{4}(\mathbf{L} \cdot \mathbf{L})} \tag{30.11}$$

It is a scalar operator and hence commutes with components of **L**. Due to the commutation relations (30.3) the operator (30.11) commutes also with all $Q_\mu^{(2)}$ components. The term $(\mathbf{L} \cdot \mathbf{L}) = \mathbf{L}^2$ is diagonal in any state with definite value of L. The overall normalization of (30.1) was chosen to yield simple expressions for the eigenvalues.

To calculate the eigenvalues of the Casimir operator (30.11) we recall the considerations for $U(4)$ in Section 29. The expression (29.21) multiplied by –1 is equal to the eigenvalues of the operator exchanging the variables of nucleons in states which belong to the irreducible representation of $U(4)$ defined by $[f_1 f_2 f_3 f_4]$. It is equal to the number of symmetric couplings from which the number of antisymmetric couplings is subtracted. This operator is equal, according to (29.23) to

$$\frac{n(n-16)}{8} + \frac{1}{8} C_{SU(4)}$$

The eigenvalues of this exchange operator in the case of $U(3)$ considered here are, in analogy with (29.21) given by

$$\tfrac{1}{2} n_1(n_1 - 1) + \tfrac{1}{2} n_2(n_2 - 1) + \tfrac{1}{2} n_3(n_3 - 1) - n_2 - 2n_3 \tag{30.12}$$

Using (30.7) and the definition (30.8) we can express n_1, n_2, n_3 in terms of N and f_1, f_2 as

$$n_3 = \tfrac{1}{3}(N - f_1 - f_2) \qquad n_1 = \tfrac{1}{3}(N + 2f_1 - f_2) \qquad n_2 = \tfrac{1}{3}(N - f_1 + 2f_2)$$

Thus the eigenvalues (30.12) are equal to

$$\tfrac{1}{6} N(N-9) + \tfrac{1}{3}(f_1^2 - f_1 f_2 + f_2^2 + 3f_1)$$

and in terms of λ, μ defined in (30.9) these eigenvalues are equal to

$$\tfrac{1}{6} N(N-9) + \tfrac{1}{3}(\lambda^2 + \lambda\mu + \mu^2 + 3(\lambda + \mu)) \tag{30.13}$$

Due to the overall normalization of (30.11), the expression (30.13) is for the eigenvalues of the operator

$$\frac{N(N-9)}{6}+\frac{1}{3}C_{SU(3)}$$

which is the analog of (29.23). Removing the terms which depend on N we obtain the eigenvalues of the Casimir operator (30.11) in states of the $SU(3)$ irreducible representation (λ,μ), given by the formula

$$\boxed{C(\lambda,\mu)=\lambda^2+\lambda\mu+\mu^2+3(\lambda+\mu)} \tag{30.14}$$

The expression (30.14) is symmetric in λ and μ.

In the case of a single nucleon, the Casimir operator (30.11) may be transformed into a simpler form. Using the change-of-coupling transformation (10.10) or (10.67), the scalar products of $[\mathbf{p}\times\mathbf{p}]^{(2)}$, $[\mathbf{r}\times\mathbf{r}]^{(2)}$, with themselves can be expressed in terms of products of $\mathbf{p}^2$, $\mathbf{r}^2$ and l^2. Paying attention to the commutation relations between components of $\mathbf{p}$ and $\mathbf{r}$, we obtain for (30.11) the expression $H_0^2/\hbar^2-\frac{9}{4}$. Formula (30.14) may be verified by putting $\lambda=N$, $\mu=0$ obtaining N^2+3N which is equal to $(N+\frac{3}{2})^2-\frac{9}{4}$.

As long as the Hamiltonian is given by (30.1) the $SU(3)$ classification of states of n nucleons in the N-th oscillator shell is not very useful. The power of the $SU(3)$ scheme is demonstrated when certain two-body interactions between nucleons are introduced. Also in the present section we consider Hamiltonians which do not include spin and isospin operators. As described in Section 29, eigenstates of such Hamiltonians belong to bases of irreducible representations of $U(4)\otimes U(2\Omega)$. Here 2Ω is the total number of orbital m-states in the N-th oscillator shell, i.e.

$$2\Omega=\sum_{l=N,N-2,\ldots,1\text{ or }0}(2l+1)=(N+1)(N+2)/2 \tag{30.15}$$

The spatial parts of the nucleon wave functions form bases of irreducible representations of $U(2\Omega)$ characterized by their symmetry type. Each such basis may contain several states with the same value of L (and M). A more detailed characterization of such states could be provided by finding a subgroup of $U(2\Omega)$ which contains $O(3)$ as a subgroup. Two states with the same value of L contained in the same basis of an irreducible representation of $U(2\Omega)$, may belong to bases of two different irreducible representations of the intermediate group. The group $U(3)$, or $SU(3)$, is such an intermediate group. The $O(3)$ Lie algebra whose generators are listed in (30.4) is a subalgebra of $SU(3)$. On the other hand, the unitary transformations of $SU(3)$, generated by the action of (30.4) and (30.6) in the 2Ω-dimensional space of single nucleons states of the N-th oscillator shell, form a subset of the $U(2\Omega)$ transformations.

It is possible to construct Hamiltonians more interesting than just (30.1), whose eigenstates belong to definite irreducible representations of $SU(3)$. The (λ, μ) labels of those representations can then serve as exact additional quantum numbers for the various states. The prescription is rather simple—construct the Hamiltonian from generators of the $SU(3)$ group. We can thus introduce two-body interactions

$$\boxed{V_{ik} = a(\mathbf{q}^{(2)}(i) \cdot \mathbf{q}^{(2)}(k)) + b(\boldsymbol{l}_i \cdot \boldsymbol{l}_k)} \tag{30.16}$$

This interaction is the sum of a quadrupole-quadrupole interaction and a dipole-dipole interaction. It is, however, non-local since it includes (linear) momentum operators. It may be viewed as an effective interaction as those considered in preceding sections. The quadrupole term may also be shown to be equivalent to a simple local interaction. Let us first look at the eigenvalues of shell model Hamiltonians containing the interaction (30.16).

We first express the interaction (30.16) in the n nucleon configuration by

$$\begin{aligned}\sum_{i<k}^{n} V_{ik} &= \frac{1}{2}a(\mathbf{Q}^{(2)} \cdot \mathbf{Q}^{(2)}) - \frac{1}{2}a\sum_{i=1}^{a}\left(\mathbf{q}^{(2)}(i) \cdot \mathbf{q}^{(2)}(i)\right) \\ &\quad + \frac{1}{2}b(\mathbf{L} \cdot \mathbf{L}) - \frac{1}{2}b\sum_{i=1}^{n} \boldsymbol{l}_i^{(2)}\end{aligned} \tag{30.17}$$

In (30.17) single nucleon terms appear which may be directly calculated from the definition (30.6). We may, however, use the result (30.14) obtained above for the Casimir operator and apply it to the single nucleon states with $\lambda = N$, $\mu = 0$. This way we obtain the relation

$$\frac{1}{2}a\sum_{i=1}^{n}(q^{(2)}(i)\cdot q^{(2)}(i))+\frac{1}{2}b\sum_{i=1}^{n}l_i^2$$

$$=2a\sum_{i=1}^{n}C_{SU(3)}(i)-\frac{3}{2}a\sum_{i=1}^{n}l_i^2+\frac{1}{2}b\sum_{i=1}^{b}l_i^2$$

$$=2a(N^2+3N)n+\left(\frac{1}{2}b-\frac{3}{2}a\right)\sum_{i=1}^{n}l_i(l_i+1) \qquad (30.18)$$

The first term on the r.h.s. of (30.18) is due to n equal contributions. The second term is a sum of single nucleon terms which are, for $N>1$, not equal. To obtain a Hamiltonian constructed from the generators we first add to H_0 single nucleon energies which cancel the single nucleon terms in (30.17). Adding the two body interaction (30.17) to that Hamiltonian we obtain H_0 to which scalar products of $SU(3)$ generators are added.

We thus consider the Hamiltonian

$$\boxed{H=H_0+\tfrac{1}{2}a(\mathbf{Q}^{(2)}\cdot\mathbf{Q}^{(2)})+\tfrac{1}{2}b(\mathbf{L}\cdot\mathbf{L})} \qquad (30.19)$$

If $b=3a$, (30.19) is equal to $H_0+2aC_{SU(3)}$ and H commutes with all $SU(3)$ generators. Hence, the eigenstates of the Hamiltonian H belong to irreducible representations of $SU(3)$ and all states in the same irreducible representation are degenerate. In fact, the eigenvalues of states included in the (λ,μ) irreducible representation have the eigenvalues of H given by

$$n(N+\tfrac{3}{2})\hbar+2a(\lambda^2+\lambda\mu+\mu^2+3(\lambda+\mu)) \qquad (30.20)$$

If b is not equal to $3a$ the Hamiltonian in (30.19) does not commute with all $SU(3)$ generators. It can be expressed as

$$H=H_0+2aC_{SU(3)}+\tfrac{1}{2}(b-3a)\mathbf{L}^2 \qquad (30.21)$$

The operator $C_{SU(3)}$, commuting with all $SU(3)$ generators, commutes with $\mathbf{L}^2$ (which is the Casimir operator of $O(3)$). They are both diagonal in the scheme of irreducible representations of $SU(3)$ where the states have definite L values. The eigenstates in a given irreducible representation of $SU(3)$ are no longer degenerate, their eigenvalues are given by

$$n(N + \tfrac{3}{2})\hbar + 2a(\lambda^2 + \lambda\mu + \mu^2 + 3(\lambda + \mu)) + \tfrac{1}{2}(b - 3a)L(L + 1) \tag{30.22}$$

In the present case, there is a special feature which was not shared by the Hamiltonians with symmetry properties described in preceding sections. The expression (30.19) is the most general (scalar) Hamiltonian which can be constructed from $SU(3)$ generators. As explained above, this implies that the eigenstates of (30.19) belong to bases of irreducible representations of $SU(3)$. Yet, there is here a more explicit result. The Hamiltonian (30.19) may be expressed as the linear combination (30.21) of the Casimir operators of $SU(3)$ and its $O(3)$ subgroup. By their definition, these generators commute and eigenvalues of (30.21) are equal to the linear combination of the eigenvalues. This is how the explicit formula (30.22) was obtained. Hamiltonians with $SU(3)$ symmetry have thus this special property which is referred to as *dynamical symmetry*.

Finally, we can add to H in (30.19) another term which is also diagonal in the $SU(3)$ scheme. We can add to it the Majorana space exchange operator (29.10). According to (29.29) it is, apart from linear and quadratic terms in n, proportional to the Casimir operator of $SU(2\Omega)$. It thus commutes with all $SU(2\Omega)$ generators including those of the $SU(3)$ subgroup. The Casimir operators of any chain of groups commute. The chain of groups considered here is

$$U(2\Omega) \supset SU(3) \supset O(3) \tag{30.23}$$

With a coefficient of the proper sign, the Majorana operator lowers more the energies of states with higher spatial symmetry. As long as we confine our attention to states with given spatial symmetry, they have the same eigenvalues of the Majorana operator and hence we can consider only the terms included in (30.22).

Let us come back to the quadrupole interaction in (30.16) which includes besides nucleon coordinates also (linear) momenta. The quadrupole operator $(\mathbf{r} \times \mathbf{r})^{(2)}$ has non-vanishing matrix elements between different harmonic oscillator shells. The same is true for the operator $(\mathbf{p} \times \mathbf{p})^{(2)}$. It is only their sum whose matrix elements which are non-diagonal in N vanish, as follows directly from (30.6). Within the space of states with given N, matrix elements of $(\mathbf{r} \times \mathbf{r})^{(2)}$ and $(\mathbf{p} \times \mathbf{p})^{(2)}$ are proportional. This follows from the complete symmetry between **r** and **p** in the oscillator Hamiltonian (30.1). Between states in different N-shells their matrix elements are equal in magnitude but have opposite signs. Hence, within a given N shell, we can replace the quadrupole interaction in (30.16) by the equivalent *local* interaction

$$\boxed{\begin{aligned}&\frac{6a}{\hbar^2}((\mathbf{r}_i \times \mathbf{r}_i)^{(2)} \cdot (\mathbf{r}_k \times \mathbf{r}_k)^{(2)}) \\ &\qquad = \frac{a}{\hbar^2}\frac{16\pi}{5} r_i^2 r_k^2 (\mathbf{Y}^{(2)}(\theta_i, \phi_i) \cdot \mathbf{Y}^{(2)}(\theta_k, \phi_k))\end{aligned}} \tag{30.24}$$

The quadrupole interaction (30.24) has a simple and separable dependence on r_i and r_k. This is an example of a situation mentioned in Section 22. The interaction (30.24) grows indefinitely for large r_i or r_k but its matrix elements may have reasonable values due to the finite extension of single nucleon wave functions. In any case, it may be viewed, along with the dipole interaction in (30.16) as a possible approximation for the effective interaction. It should then be judged according to the consequences for nuclear energies. We shall discuss this point below but first point out that there is still a simple interpretation of (30.24) as a local interaction.

Consider a potential interaction $V(|\mathbf{r}_1 - \mathbf{r}_2|^2)$ which is regular at the point $\mathbf{r}_1 = \mathbf{r}_2$ and may be expanded as a power series in $(\mathbf{r}_1 - \mathbf{r}_2)^2$. The first few terms in the expansion are

$$V(0) + V'(0)(\mathbf{r}_i - \mathbf{r}_k)^2 + V''(0)(\mathbf{r}_i - \mathbf{r}_k)^4 \tag{30.25}$$

The first term is a constant interaction contributing $V(0)n(n-1)/2$ to the energy. The second term summed over all nucleons can be expressed as

$$V'(0)\sum_{i<k}^{A}(\mathbf{r}_i - \mathbf{r}_k)^2 = V'(0)A\sum_{i=1}^{A}\mathbf{r}_i^2 - V'(0)A^2\mathbf{R}^2 \tag{30.26}$$

where A is the total number of nucleons and $\mathbf{R}$ is the center of mass coordinate defined in Section 4. Thus, the second term in (30.25) leads to a modification of the single nucleon oscillator potential and to an extra oscillator potential acting on the center of mass coordinate. The latter term affects only the spurious center of mass motion considered in Section 4. The remaining term in (30.25) is equal to

$$\begin{aligned} V''(0)(\mathbf{r}_i - \mathbf{r}_k)^4 &= V''(0)(r_i^2 + r_k^2 - 2\mathbf{r}_i \cdot \mathbf{r}_k)^2 \\ &= V''(0)[(r_i^2 + r_k^2)^2 + 4(\mathbf{r}_i \cdot \mathbf{r}_k)^2 - 4(r_i^2 + r_k^2)(\mathbf{r}_i \cdot \mathbf{r}_k)] \end{aligned} \tag{30.27}$$

Matrix elements of the last term on the r.h.s. of (30.27) vanish between single nucleon states in an oscillator shell with given N since they have the same parity. The second term, due to (10.10) and vanishing of $(\mathbf{r}_i \times \mathbf{r}_i)^{(1)}$, is equal to

$$\begin{aligned} &V''(0)(\mathbf{r}_i \cdot \mathbf{r}_k)(\mathbf{r}_i \cdot \mathbf{r}_k) \\ &\quad = V''(0) r_i^2 r_k^2 + V''(0)((\mathbf{r}_i \times \mathbf{r}_i)^{(2)} \cdot (\mathbf{r}_k \times \mathbf{r}_k)^{(2)}) \end{aligned} \tag{30.28}$$

which contains, besides a $k = 0$ term, the quadrupole interaction (30.24).

Let us now look at the consequences of the form (30.22) of energy eigenvalues. All states in a given (λ, μ) irreducible representation of $SU(3)$ have the same eigenvalues of $C_{SU(3)}$ and hence their energies are proportional to $L(L+1)$. If the quadrupole interaction is attractive, $a < 0$, and b may be ignored, the higher the L, the higher the energy. These energies behave like energy levels of a rigid symmetric rotor whose moment of inertia $\mathcal{I}$ is given by $-3a/2 = \hbar^2/2\mathcal{I}$. To achieve a better picture of such rotational bands we should find out which L values belong to states within a given irreducible representation. We know already the values of L of states in irreducible representations $(\lambda, 0)$. The single nucleon states in the N-th oscillator shell, which belong to the $(N, 0)$ irreducible representation, have $L = N, N-2, \ldots, 1$ or 0. Thus,

$$\boxed{L = \lambda, \lambda - 2, \ldots, 1 \textit{ or } 0 \textit{ for states in the } (\lambda, 0) \textit{ irreducible representation}} \tag{30.29}$$

Although the result (30.29) was obtained above for fully symmetric irreducible representations $(\lambda, 0)$ of λ oscillator quanta of a single nucleon (p-bosons), it has general validity. No matter what the realization of the irreducible representation is, the L-values in (30.29) are obtained from the universal mathematical problem of the decomposition of $(\lambda,0)$ irreducible representations of $SU(3)$ into those of $O(3)$ characterized by L. The Young diagram corresponding to $(\lambda,0)$ has *one* row with λ squares and the irreducible representation is fully symmetric. In the case of states in (λ,μ) representations with $\mu \neq 0$ the situation is more complicated.

States characterized by (λ,μ) symmetry type can be constructed by coupling $(\lambda+2\mu)$ particles with $l=1$ (p-particles) to form states with that symmetry. The irreducible representations $(\lambda,0)$ are obtained by coupling λ p-particles to a fully symmetric state. The diagrams corresponding to the $(0,\mu)$ symmetry have μ columns with two squares. Particles in each column are coupled antisymmetrically and hence they are in a state with $L=1$. We recall that columns with three squares in the Young diagrams of $U(3)$ are omitted when going over to $SU(3)$ irreducible representations. This is due to the fact that three p-particles in a fully antisymmetric state have $L=0$ and constitute a full p-orbit. Hence states in the $(0,1)$ irreducible representation of $SU(3)$ are states of a *p-hole* and states of $(0,\mu)$ are those of μ p-holes. The latter can be constructed by coupling symmetrically μ p-holes (with $l=1$). The symmetry between λ and μ in (30.14) is thus the symmetry between particles and holes. Hence, the L-values of states in the $(0,\mu)$ irreducible representation are given, like in (30.29), by $L=\mu,\mu-2,\ldots,1$ or 0.

It should be emphasized that hole states in the present context are not due to the Pauli principle which does not hold for the oscillator p-particles. They arise from the antisymmetry of states of p-particles whose numbers are written, in the same column of Young diagrams. This is coupled with the fact that for those particles there are only 3 independent orbital states. Hence, states of three p-particles in one column are invariant under $SU(3)$ transformations.

To construct states with (λ,μ) symmetry we should first couple λ p-particles symmetrically, μ p-holes symmetrically and then combine the two groups paying attention to the given symmetry type. The (λ,μ) irreducible representation will be found when the product of $(\lambda,0)$ and $(0,\mu)$ irreducible representations is decomposed into irreducible representations. Let us first briefly describe how to carry out such multiplication.

Consider states of a set of particles characterized by a Young diagram of the Lie group $U(k)$ with rows $f_1,f_2,\ldots,f_k$ ($f_1 \geq f_2 \geq \cdots \geq f_k$) and states of other particles characterized by $[f'] = [f'_1 f'_2 \ldots f'_k]$. States of the combined system are obtained by multiplying states of one set by states of the other set. These product states transform linearly under $U(k)$ transformations but, in general, not irreducibly. The states of the combined system may be expressed as linear combinations of states which transform irreducibly under $U(k)$ transformations. The rule of decomposition of a product of two irreducible representations into a linear combination of irreducible ones is as follows. Write in the first row of the diagram of one system the same symbol a, in the second row b, in the third c, etc. Now add the f_1' squares of this diagram to the diagram of the other system observing two conditions. The first is that the resulting lengths of the k rows should remain monotonously non-increasing. The second condition is that no two squares, in which a is written, be added to the same column. In the next step, the f_2' squares (with b) are added observing the same conditions. In addition, there is another restriction on the addition of b-squares. Counting numbers of added symbols starting from right to left in the top row, then from right to left in the second row etc., at no point should the number of b-squares be larger than the number of a-squares. For instance, no b-squares may be added to the top row. In the next step, the f_3' squares with c are added following the same restrictions, also with respect to the number of squares with a and b symbols. If the same diagram appears in this process a certain number of times, the corresponding irreducible representation appears the same number of times in the decomposition of the direct product.

An example which illustrates this procedure is offered by taking the product of $[f] = [2,1]$ and $[f'] = [2,1]$. Following the rules above we obtain

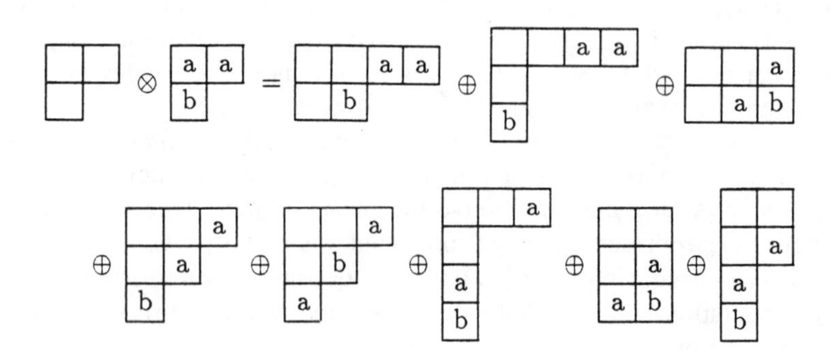

The results of this procedure are thus summarized by

$$[2,1] \otimes [2,1] = [4,2] \oplus [4,1,1] \oplus [3,3] \oplus 2[3,2,1] \oplus [3,1,1,1] \oplus [2,2,2] \oplus [2,2,1,1] \quad (30.30)$$

The irreducible representation $[3,2,1]$ appears twice in this example. If for a given group $U(k)$, the number of rows in one of the resulting diagrams exceeds k, the corresponding irreducible representation does not appear in the decomposition of the product.

Let us now come back to the product of the irreducible representations $(\lambda,0)$ and $(0,\mu)$. In the following we take $\lambda \geq \mu$, if however $\mu > \lambda$ we simply interchange the roles of λ and μ. Starting with the diagram $(0,\mu)$ we can add to its first row the λ squares obtaining the diagram of (λ,μ). Another possibility is to add only $\lambda-1$ squares to the top row. The last square can be added only to the first column. For $SU(3)$ this diagram is equivalent to a diagram from which the first column is removed, i.e. to the diagram of $(\lambda-1,\mu-1)$. The next possibility is to add to the top row $\lambda-2$ squares leading to the diagram of $(\lambda-2,\mu-2)$ irreducible representation, etc. The last possible arrangement is obtained by adding $\lambda-\mu$ squares to the top row and reaching the $(\lambda-\mu,0)$ irreducible representation. We thus obtain

$$(\lambda,0)\otimes(0,\mu)=(\lambda,\mu)\oplus(\lambda-1,\mu-1)\oplus(\lambda-2,\mu-2)\oplus\cdots\oplus(\lambda-\mu,0) \tag{30.31}$$

On the r.h.s. of (30.31) we see in addition to (λ,μ), the product of $(\lambda-1,0)$ and $(0,\mu-1)$. We thus obtain the recursion relation

$$(\lambda,0)\otimes(0,\mu)=(\lambda,\mu)\oplus[(\lambda-1,0)\otimes(0,\mu-1)] \tag{30.32}$$

The relation (30.32) may be used to determine the L-values of states included in the (λ,μ) irreducible representation. All states in $(\lambda,0)$ with $L_1=\lambda,\lambda-2,\ldots$ should be coupled with all states in $(0,\mu)$ with $L_2=\mu,\mu-2,\ldots$ to yield all states of the combined system with $\mathbf{L}=\mathbf{L}_1+\mathbf{L}_2$. The same should be done for the product $(\lambda-1,0)\otimes$ $(0,\mu-1)$. According to (30.32) the number of independent states with given L in the (λ,μ) irreducible representation is the number of L-states in the product $(\lambda,0)\otimes(0,\mu)$ from which the number of L-states in $(\lambda-1,0)\otimes(0,\mu-1)$ is subtracted.

For example, consider the product of representations $(\lambda,0)\otimes(0,1)$. The $(\lambda,0)$ states with $L_1=\lambda,\lambda-2,\ldots,1$ or 0, should be coupled with the only state $L_2=1$ in $(0,1)$. The resulting states have the L-values

$$\lambda+1,\lambda,\lambda-1,\lambda-1,\lambda-2,\lambda-3,\lambda-3,\lambda-4,\ldots,1,0 \text{ or } 1,1$$

The states of $(\lambda-1,0)\otimes(0,0)\equiv(\lambda-1,0)$ have the L-values $L=\lambda-1,\lambda-3,\ldots,0$ or 1. Hence, according to (30.32) the states included in $(\lambda,1)$ have the L-values

$$\lambda+1,\lambda,\lambda-1,\lambda-2,\ldots,1 \tag{30.33}$$

Similar considerations lead to the general result to be stated below.

The result for the L-content of the (λ,μ) irreducible representation, for $\lambda\geq\mu$, is as follows. Define an integer K by

$$\boxed{K=\mu,\mu-2,\ldots,1 \text{ or } 0} \tag{30.34}$$

the L values of the various states are given by

$$\boxed{L=K,K+1,K+2,\ldots,K+\lambda \qquad K\neq 0} \tag{30.35}$$

for $K\neq 0$, whereas for $K=0$ the states have L-values given by

$$\boxed{L=\lambda,\lambda-2,\lambda-4,\ldots,1 \text{ or } 0 \qquad K=0} \tag{30.36}$$

The result (30.29) is a special case of (30.34), and (30.36) for $\mu=0$ and (30.33) is a special case for $\mu=1$. If $\mu>\lambda$ the roles of λ and μ in (30.34), (30.35) and (30.36) are reversed. An immediate consequence of (30.34) and (30.35) is that the highest possible L value in the (λ,μ) irreducible representation is $\lambda+\mu$. If there are n nucleons in the N-th oscillator shell, their l-values satisfy $l\leq N$ and the total L cannot exceed nN. Hence we obtain the inequality $\lambda+\mu\leq nN$. Actually there is a stronger inequality for the values of λ and μ. Since each nucleon is described by a diagram with N squares, the total number of squares, $\lambda+2\mu$, cannot exceed nN. We thus obtain

$$\lambda+2\mu\leq nN \tag{30.37}$$

From (30.34) and (30.35) it is seen that unless $\mu=0$ or $\mu=1$ (for $\lambda\geq\mu$) there are several states with the same value of L in the irreducible representation of $SU(3)$ defined by (λ,μ). In other words, when going from $SU(3)$ to its subgroup $O(3)$, as in (30.23), a given irreducible representation of $SU(3)$ splits into several irreducible representation of $O(3)$, some of which may be equivalent. In such cases an additional quantum number is needed to distinguish between states with the same value of L. The integer K in (30.34) can serve as such a label. We shall not describe here how to construct an orthogonal set of states which belong to different values of K.

The formulae (30.34) and (30.35) can be proved by induction with respect to μ. A convenient way is to use (30.32) to write

$$(\lambda,\mu) = (\lambda,0)\otimes(0,\mu)\ominus(\lambda-1,0)\otimes(0,\mu-1) \qquad (30.38)$$

$$(\lambda,\mu-2) = (\lambda,0)\otimes(0,\mu-2)\ominus(\lambda-1,0)\otimes(0,\mu-3) \qquad (30.39)$$

Subtraction of (30.39) from (30.38) yields on the r.h.s. states with L-values obtained by coupling $L_1 = \lambda, \lambda-2, \ldots, 1$ or 0 with $L_2 = \mu$ from which states with L-values obtained by coupling $L_1 = \lambda-1$, $\lambda-3, \ldots, 0$ or 1 with $L_2 = \mu-1$ are removed. The remaining states have L-values given by

$$\lambda+\mu, \lambda+\mu-1, \lambda+\mu-2, \ldots, \mu \qquad (30.40)$$

If the states in the $(\lambda,\mu-2)$ irreducible representation have L-values given by (30.34) and (30.35) with $K = \mu-2, \mu-4, \ldots, 1$ or 0, the set of states (30.40) adds precisely the states with $K = \mu$ and hence (30.35) holds also for the (λ,μ) irreducible representation. Since the results (30.34),(30.35) hold for $\mu = 0$ according to (30.29) and for $\mu = 1$ as shown in (30.33) above, the results (30.34), (30.35) and (30.36) hold for any value of $\mu \leq \lambda$.

We can now return to the eigenvalues of the $SU(3)$ Hamiltonian (30.21). The spectrum is composed of a series of "rotational bands". For an attractive quadrupole interaction, $a < 0$, states of n nucleons in the N-th oscillator shell are lowest for the maximum value of $C_{SU(3)}$ in (30.22). Due to the inequality (30.37) the highest eigenvalue of $C_{SU(3)}$ is obtained for the maximum value of λ consistent with the Pauli principle. In the irreducible representation $(\lambda,0)$ the band has states with $L = \lambda, \lambda-2, \ldots, 1$ or 0 whose energies are proportional to $L(L+1)$ according to (30.22). In other irreducible representations, with $\mu \neq 0$, there are several bands each of which starts with one of the K-values in (30.34) and contains states with L-values given by (30.35) or (30.36). The energies within each band are proportional to $L(L+1)$ with the same proportionality factor (equal "moments of inertia"). States of the schematic Hamiltonian (30.21) with given L and within the same (λ,μ) representation are degenerate, their energies are independent of K. Due to this degeneracy, assigning states to the various bands within an irreducible representation is arbitrary to some extent. When an orthogonal basis is defined for (λ,μ) states with $\mu > 1$ (Vergados 1968), the state with $L = K_{\min}$ (0 or 1), is assigned a quantum number $\chi = K_{\min}$. Higher members of the band are obtained by applying to it components of $\mathbf{Q}^{(2)}$. From the state with $L = K_{\min} + \lambda$ which appears in (30.35) or (30.36), states with higher L values may be obtained in this way up to the state with $L = \mu + \lambda$. The states with $K_{\min} + 2$ are then orthogonalized to the $\chi = K_{\min}$ states and assigned $\chi = K_{\min} + 2$. That band continues up to the state with $L = \mu - 2 + \lambda$ which is orthogonal to the state with the same value of L and $\chi = K_{\min}$. The process is then continued for the remaining states.

The results obtained above about the L-content of the (λ,μ) irreducible representation have general validity as explained following (30.29). They follow only from the properties of the irreducible representations of the $SU(3)$ group and thus hold for nucleons in any N-th shell of the harmonic oscillator potential. They hold also in the case of *boson* models which will be introduced in subsequent sections. The allowed irreducible representations of $SU(3)$ for any given system, however, strongly depend on the nature of the system considered.

There is a simple physical meaning of the integer K which will be considered in more detail in Section 33 where the collective model of the nucleus will be briefly described. From (30.34), (30.35) and (30.36) it is seen that the value of K in any state is smaller than or equal to L of that state. Thus, it may be interpreted as the projection of L on an axis whose orientation is a function of the dynamical variables of the system. Here we only mention that if the state of the system is invariant under rotations around that axis, the result of applying an infinitesimal element is zero and hence $K = 0$. This is the case for a system with an axis of symmetry. In other cases, K may assume other values.

Let us first consider the applications of the $SU(3)$ scheme to the $1p$-shell, with $N = 1$, for protons and neutrons. There are only three levels in the p^2 configuration, the symmetric states with $L = 0$ and $L = 2$ which belong to the $(\lambda,\mu) = (2,0)$ representation and the antisymmetric $L = 1$ state of the $(\lambda,\mu) = (0,1)$ representation. Any two body interaction independent of spin and isospin operators is completely determined in p^n configurations by the eigenvalues V_0, V_2 and V_1 of these states. Any two body interaction which reproduces these eigenvalues may be used to calculate energies of all p^n states. A simple such interaction can be defined by

$$\alpha + \beta \mathbf{q}^{(2)}(1) \cdot \mathbf{q}^{(2)}(2) + \gamma \boldsymbol{l}_1 \cdot \boldsymbol{l}_2 \tag{30.41}$$

The eigenvalues of (30.41) in states of the p^n configuration are given by

$$\frac{n(n-1)}{2}\alpha + 2\beta C_{SU(3)} + \tfrac{1}{2}(\gamma - 3\beta)\mathbf{L}^2 - n[5\beta + \gamma] \tag{30.42}$$

Using (30.42) for the $n = 2$ case we obtain α, β, γ from the three equations

$$\begin{aligned} V_0 &= \alpha + 10\beta - 2\gamma \\ V_2 &= \alpha + \beta + \gamma \\ V_1 &= \alpha - 5\beta - \gamma \end{aligned} \tag{30.43}$$

We conclude from (30.42) that all states of p^n configurations have (λ, μ) and L as good quantum numbers for *any* two body interaction which is spin and isospin independent. The energy eigenvalues (30.42) depend only on λ, μ and L. This property of the p-orbit should not be surprising. In the case of the Hamiltonians considered here, states can be characterized by the irreducible representations of $U(2l+1)$ which is in the present case, $l = 1$, the group $U(3)$. In fact, we could have used in (30.42) instead of $C_{SU(3)}$ the Casimir operator of $SU(4)$ or the Majorana operator (29.10) as was originally done by Racah (1952).

To find an example of a band with more than two states we turn to the p^4 configuration. If the quadrupole interaction in (30.41) is attractive, $\beta < 0$, the lowest states should have the highest values of λ. All states fully symmetric in space coordinates belong to the irreducible representation $f_1 = 4$ of $U(2l+1) \equiv U(3)$ or to the $(\lambda, \mu) = (4, 0)$ irreducible representation of $SU(3)$. These are the states with $L = 0$, $L = 2$ and $L = 4$, with $T = 0$ and $S = 0$. Looking at the experimental spectra of ${}^8_4\mathrm{Be}_4$ we find such levels with energies roughly proportional to $L(L+1)$. Also in ${}^{12}_{6}\mathrm{C}_6$ with the possible $p^8 \equiv p^{-4}$ configuration, the lowest $J = 2$ level is at 4.44 MeV above the $J = 0$ ground state and the lowest $J = 4$ level is at 14.08 MeV. The ratio between these spacings, 3.18, is rather close to 3.33 which is due to $L(L+1)$ dependence. Other irreducible representations allowed for p^4 states are $(2,1)$, $(0,2)$ and $(1,0)$.

A more interesting case is offered by nucleons in the next oscillator shell with $N = 2$. This is the $2s, 1d$ shell and here the $SU(3)$ scheme introduces quantum numbers in addition to those of $U(6)$.

As we saw above, the latter are determined by the $SU(4)$ group of spin and isospin. A single $N = 2$ nucleon belongs to the irreducible representation $(2,0)$ of $SU(3)$ which includes $l = 0$ and $l = 2$ states. States of two nucleons belong to irreducible representations included in the decomposition of the product $(2,0) \otimes (2,0)$. The L values are obtained by coupling two s and d nucleons. Coupling d and d states yields symmetric states with $L = 0,2,4$ and antisymmetric ones with $L = 1,3$. Coupling s with d as well as d with s yields a symmetric $L = 2$ state and an antisymmetric $L = 2$ state. Coupling s and s yields a symmetric $L = 0$ state. Since

$$(2,0) \otimes (2,0) = (4,0) \oplus (2,1) \oplus (0,2)$$

we see that the symmetric states belong to $(4,0)$ (with $L = 0,2,4$) and $(0,2)$ (with $L = 0,2$) and the antisymmetric ones belong to $(2,1)$ with $L = 1,2,3$. The quantum numbers (λ, μ) thus distinguish between the two symmetric states with $L = 0$ and the two with $L = 2$.

A new feature can be seen at this stage. States in the $(4,0)$ irreducible representation are fully symmetric in the four $l = 1$ oscillator quanta from which the *two* $N = 2$ nucleons states are constructed. States in the $(0,2)$ irreducible representation, however, cannot be symmetric in these four p-particles. Nevertheless, these states are symmetric under exchange of *pairs* of p-particles coupled symmetrically to $l = 0$ and $l = 2$ nucleons. Similarly, the four p-particles in states of the $(2,1)$ irreducible representation have mixed symmetry but these states are antisymmetric under exchange of $l = 0$ and $l = 2$ *pairs* of p-particles. States which are symmetric in the s,d nucleons belong to a special class of irreducible representations. These may be obtained from the fully symmetric representation $(2n,0)$ by moving *two* squares from the top row into the second row. This yields the irreducible representations $(2n-4,2), (2n-8,4), \ldots$ down to $(0,n)$ for n-even and to $(2,n-1)$ for n odd. Two squares may now be moved into the third row of these $(2n-4k,2k)$ diagrams. This yields diagrams in which two columns are erased and the resulting diagrams are given by $(2n-6,0), (2n-10,2), \ldots$. This prescription holds, of course, only in the $N = 2$ shell. In the case where the s and d particles are fermions, which is considered here, the maximum number of nucleons which may be in a fully symmetric state is just $n = 4$. This limitation is due to the Pauli principle, dictating which $SU(4)$ spin-isospin functions should multiply the space functions. In the case of bosons there is no such limitation and these considerations hold for any number of s and d bosons, which will be considered in subsequent sections. For shells with higher values of N similar considerations may be carried out.

The fully symmetric states of *three* nucleons belong to the irreducible representation of $U(6)$ with $f_1 = 3$, $f_2 = f_3 = f_4 = f_5 = f_6 = 0$ (or $f_1 = f_2 = f_3 = 1$, $f_4 = 0$ in the dual irreducible representation of $SU(4)$ of spin and isospin). The irreducible representations of $SU(3)$ included in this $U(6)$ representation are characterized by $(6,0)$, $(2,2)$ and $(0,0)$. Thus, there are three fully symmetric $L = 0$ states in the $(2s,1d)^3$ configuration which maybe distinguished by their (λ,μ) quantum numbers. The symmetric $L = 2$ state in the $(6,0)$ irreducible representation is similarly distinguished from the two symmetric $L = 2$ states included in $(2,2)$. The latter $L = 2$ states, however, cannot be distinguished by $SU(3)$ quantum numbers. Hence, an additional quantum number is needed to distinguish between states with the same value of L in the same (λ,μ) irreducible representation. As mentioned above, it follows from (30.34), (30.35) and (30.36) that only *one* such number is needed, namely K.

All states of three $N = 2$ nucleons are obtained by decomposing the product of three $(2,0)$ irreducible representations. We thus obtain

$$[(4,0) \oplus (2,1) \oplus (0,2)] \otimes (2,0)$$
$$= (6,0) \oplus 2(4,1) \oplus 3(2,2) \oplus (0,0) \oplus 2(1,1) \oplus (3,0) \oplus (0,3)$$

Also this relation may be checked by coupling all L-states of $(s,d)^2$ to the $l = 0,2$ states of the third nucleon. The irreducible representations $(6,0)$, $(2,2)$ and $(0,0)$ are fully symmetric in the s and d nucleons and belong to the $[3]$ irreducible representation of $U(6)$. The other $U(6)$ irreducible representations are obtained by decomposing the product of three irreducible representations of $U(6)$ with $f_1 = 1, f_2 = \cdots = f_6 = 0$. These are in addition to the symmetric $[3]$ and antisymmetric $[111]$, also *two* independent irreducible representations with mixed symmetry $[21]$. It can be shown that the $SU(3)$ irreducible representations contained in $[111]$ are $(3,0)$ and $(0,3)$. Thus, each of the $[21]$ representations of $U(6)$ contains the $SU(3)$ irreducible representations $(4,1)$, $(2,2)$ and $(1,1)$.

It is interesting to note that the symmetry with respect to p-particles of the $(2,2)$ irreducible representation does not determine the symmetry of the states with respect to permutations of s and d nucleons. The same $(2,2)$ representation is included both in the $[3]$ symmetric representation of $U(6)$ and in the one with mixed symmetry $[21]$. It is thus possible to construct from states of 6 p-particles with

(2,2) symmetry states of three s or d nucleons which are either fully symmetric or states with the mixed symmetry [21]. Both sets of states transform in the same way under $SU(3)$ transformations but differently under *all* $U(6)$ transformations. The states in both cases have the same eigenvalues of the Casimir operator (30.11) but different eigenvalues of the Casimir operator of $SU(6)$ or the Majorana operator (29.10).

In general, there is no simple prescription for finding the (λ, μ) representations included in the set of states with given symmetry of n nucleons in the N-th oscillator shell. That symmetry is determined by the irreducible representation of $U(2\Omega)$ with $2\Omega = (N+1)(N+2)/2$ as calculated in (30.15). There are procedures which give the splitting of $U(2\Omega)$ irreducible representations into those of $SU(3)$ but they will not be described here. The interested reader is referred to available literature.

Looking at the eigenvalues (30.20) or (30.22), we see that as in the case of p^n configurations, if the quadrupole interaction in (30.16) is attractive, $a < 0$, the lowest band is obtained for the maximum value of λ. For $n = 2$, $\lambda_{max} = 2N$, for $n = 3$, $\lambda_{max} = 3N$ and for $n = 4$, $\lambda_{max} = 4N$ whereas $\mu = 0$. For larger numbers n, however, λ_{max} cannot increase. We recall that the longest row in the diagrams of $U(2\Omega)$ irreducible representations cannot exceed 4. This limit is set by the dual irreducible representation of $U(4)$ of spin and isospin which should be multiplied by basis functions of the $U(2\Omega)$ representation to obtain a totally antisymmetric state. In the case with $N = 2$, the lowest symmetric irreducible representation in the $(2s, 1d)^4$ configuration is (8,0). The "ground state band" has accordingly states with $L = 0, 2, 4, 6, 8$. The symmetric representations with smaller eigenvalues of $C_{SU(3)}$ are (4,2), (0,4) and (2,0).

We can refer to experimental data to check whether predictions of $SU(3)$ Hamiltonians agree with experiment. As we saw above, in the $1p$-shell there is no real test of the $SU(3)$ scheme. The region of nuclei to which $SU(3)$ has been mainly applied is the $2s, 1d$-shell. The fully symmetric representations of $U(6)$ occur for states with isospin $T = 0$. In $^{20}_{10}Ne_{10}$ the $(2s, 1d)^4$ configuration is expected and levels corresponding to the states in the (8,0) representation have been observed. These are the $J = 2$ state at 1.63, $J = 4$ at 4.25, $J = 6$ at 8.78 and $J = 8$ at 11.95 MeV above the $J = 0$ ground state. The energies of these states are fairly proportional to $J(J+1)$ and correspond to the ground state band of the $SU(3)$ scheme. The next even-even nucleus with $T = 0$ is $^{24}_{12}Mg_{12}$ where the most symmetric $U(6)$ irreducible representation is [44]. The irreducible representation of $SU(3)$ included in [44] which has the largest eigenvalue of $C_{SU(3)}$ is (8,4). The states expected to be the lowest ones are thus in three bands with $K = 0$, $L = 0,2,4,6,8$, with $K = 2$, $L = 2,3,4,5,6,7,8,9,10$ and with $K = 4$, $L = 4,5,\ldots,12$. Experimentally, the lowest levels corresponding to the $K = 0$ band are found with $J = 2$ at 1.37, $J = 4$ at 4.12 and $J = 6$ at 8.11 MeV above the $J = 0$ ground state. Levels corresponding to the $K = 2$ band may be identified with the $J = 2$ level at 4.24, $J = 3$ at 5.24 and $J = 4$ at 6.01 MeV above the ground state. The $K = 2$, $J = 2$ and $J = 4$ levels are far from being degenerate with the lowest $J = 2,4$ levels. Still, there is some correspondence between predicted bands and experimental levels.

The next even-even nucleus with $T = 0$ states is ${}^{28}_{14}\mathrm{Si}_{14}$ where the most symmetric states belong to the [44] representation of $U(6)$. The $SU(3)$ irreducible representations included in it which have the highest eigenvalues of $C_{SU(3)}$ are (12,0) and (0,12). Energies of states in these two representations should be degenerate according to the symmetry of (30.22) in λ and μ. The nucleus ${}^{28}\mathrm{Si}$ is assigned the configuration $(2s,1d)^{12}$ which is identical with the hole configuration $(2s,1d)^{-12}$. It is needless to point out that nothing of this degeneracy predicted in the $SU(3)$ scheme is actually observed. The experimental spectrum of ${}^{28}\mathrm{Si}$ does not essentially differ from that of ${}^{24}\mathrm{Mg}$.

The inadequacy of the $SU(3)$ scheme to give a good description of nuclei in the $2s,1d$ region is mostly revealed in states with higher isospins. The isospin of low lying levels in ${}^{20}_{8}\mathrm{O}_{12}$ is $T = 2$ and the most symmetric irreducible representation of $U(6)$ is thus [22] (the $(P,P',P'') = (2,0,0)$ supermultiplet of $SU(4)$). The irreducible representation of $SU(3)$ included in [22] which has the maximum eigenvalue of $C_{SU(3)}$ is $(\lambda,\mu) = (4,2)$. It contains states with $K = 0$, $L = 0,2,4$ and with $K = 2$, $L = 2,3,4,5,6$. Experimentally, there are levels with $J = 2$ at 1.67 and $J = 4$ at 3.57 MeV above the $J = 0$ ground state. Other $J = 2$ and $J = 4$ levels are far from being degenerate with the former ones. They are at 4.07 and 4.85 MeV respectively and no $J = 3$ level lies between them. An attempt to describe several $(2s,1d)$ nuclei in the $SU(3)$ scheme may be found in Harvey (1968).

The disagreement between experimental data and the rather schematic $SU(3)$ description should not be surprising. Spin dependent interactions, and in particular the large spin-orbit interaction, cannot be reconciled with the assumptions which form the basis of the $SU(3)$ scheme. In ${}^{17}_{8}\mathrm{O}_{9}$ the first excited $J^{\pi} = 3/2^{+}$ level lies 5.09 MeV above the $J^{\pi} = 5/2^+$ ground state. In $A = 39$ nuclei the $J^{\pi} = 5/2^+$ hole state lies at 5.26 MeV above the $J^{\pi} = 3/2^+$ ground state. Thus, the splitting between the single nucleon $1d_{5/2}$ and $1d_{3/2}$ orbits is at least 5 MeV. This is a rather large energy difference and it leads to strong deviations from the $SU(3)$ scheme, especially in odd-even nuclei. In even-even nuclei with high values of T, interactions between nucleons which are coupled in $T = 1$ states become important. As we will discuss in detail later on, the properties of such interactions are rather different from those of $T = 0$ interactions. As a result, the $SU(3)$ scheme gives a better qualitative description of nuclei with lower values of T.

The main value of the $SU(3)$ scheme is by supplying within the spherical shell model a prescription for configuration mixing which leads to features associated with collective motion. Rotational bands arise naturally in which energies behave like $J(J + 1)$ and there are enhanced intraband E2 transition probabilities. If the electromagnetic quadrupole operator is proportional to the generator $\mathbf{Q}^{(2)}$ in (30.10), it has non-vanishing matrix elements only between states in the same $SU(3)$ irreducible representation. Such transitions exhaust all the strength of the E2 transitions. These attractive features of the $SU(3)$ scheme are incorporated in the $SU(3)$ limit of the *interacting boson model* which will be described in a subsequent section.

31

Valence Protons and Neutrons in Different Orbits

So far, most of the discussion was concerned with states and matrix elements of nucleons in the same orbit. As emphasized in Section 15, it is in such configurations that the antisymmetrization may cause complications which were discussed in detail. Coupling of states in which nucleons occupy different orbits does not add any essential complication. Any antisymmetric state of the $j_1^{n_1}$ configuration with given J_1 (and additional quantum number α_1, if necessary) may be coupled to any allowed state with J_2 (and α_2) of the $j_2^{n_2}$ configuration with Clebsch-Gordan coefficients to yield states with all possible values of the total J. Any such state may be subsequently antisymmetrized with respect to the n_1 nucleons in the j_1-orbit and the n_2 nucleons in the j_2-orbit. This antisymmetrization yields always an allowed state with the given value of J. The expectation value of the Hamiltonian in such a state is made of three parts. The first is the single nucleon energies and two-body interactions of the j_1-nucleons in the state with $\alpha_1 J_1$. The second includes these terms in the state with $\alpha_2 J_2$ of the $j_2^{n_2}$ configuration. These two parts have the same values in all states with the various J values, $|J_1 - J_2| \leq J \leq J_1 + J_2$. Their evaluation has been extensively discussed in preceding sections. The third part is the interaction between the n_1 nucleons in the j_1-orbit and the n_2 ones in the j_2-orbit. We shall presently consider the calculation of this part in the expression

$$
\begin{aligned}
&\langle j_1^{n_1}(\alpha_1 J_1) j_2^{n_2}(\alpha_2 J_2) JM | \sum_{i<h}^{n_1+n_2} V_{ih} | j_1^{n_1}(\alpha_1 J_1) j_2^{n_2}(\alpha_2 J_2) JM \rangle \\
&\quad = \langle j_1^{n_1} \alpha_1 J_1 M_1 | \sum_{i<h}^{n_1} V_{ih} | j_1^{n_1} \alpha_1 J_1 M_1 \rangle \\
&\quad + \langle j_2^{n_2} \alpha_2 J_2 M_2 | \sum_{i>h=n_1+1}^{n_1+n_2} V_{ih} | j_2^{n_2} \alpha_2 J_2 M_2 \rangle \\
&\quad + \langle j_1^{n_1}(\alpha_1 J_1) j_2^{n_2}(\alpha_2 J_2) JM | \sum_{i=1}^{n_1} \sum_{h=n_1+1}^{n_1+n_2} V_{ih} | j_1^{n_1}(\alpha_1 J_1) j_2^{n_2}(\alpha_2 J_2) JM \rangle
\end{aligned}
\tag{31.1}
$$

From the considerations of Section 14 follows that the interaction energy between j_1 nucleons and j_2 nucleons may be calculated *before* antisymmetrization in terms of matrix elements in the $j_1 j_2$ configuration. Antisymmetrization should then be applied for calculating the two body matrix elements. If those are taken between $T = 1$ states, matrix elements should be evaluated between antisymmetric space and spin functions of the $j_1 j_2$ configuration. In matrix elements between states with $T = 0$, wave functions symmetric in space coordinates and spin variables should be used. Finally, there may be cases in which the j_1-nucleons are protons, the j_1 neutron orbit is full and the j_2 nucleons are neutrons. In that case, as explained in detail in Section 17, neither symmetrization nor antisymmetrization should be applied for calculating $j_1 j_2$ matrix elements. They turn out to be equal to the averages of the corresponding $T = 1$ and $T = 0$ matrix elements.

Let us start with configurations in which it is not necessary to use the isospin quantum number. These are either states of identical nucleons (with maximum isospin) or states in which valence protons and neutrons occupy different orbits and the isospin is given by $T = \frac{1}{2}|N - Z|$ as explained in Section 17. The states we now consider have the form

$$\psi(j_1^{n_1}(\alpha_1 J_1)j_2^{n_2}(\alpha_2 J_2)JM)$$
$$= \sum_{M_1 M_2} (J_1 M_1 J_2 M_2 \mid J_1 J_2 JM)\psi(j_1^{n_1}\alpha_1 J_1 M_1)\psi(j_2^{n_2}\alpha_2 J_2 M_2) \quad (31.2)$$

The interaction energy between nucleons in the two orbits is given by the expectation value

$$\langle j_1^{n_1}(\alpha_1 J_1)j_2^{n_2}(\alpha_2 J_2)JM|\sum_{i=1}^{n_1}\sum_{h=n_1+1}^{n_1+n_2} V_{ih}|j_1^{n_1}(\alpha_1 J_1)j_2^{n_2}(\alpha_2 J_2)JM\rangle$$
$$= n_1 n_2 \langle j_1^{n_1}(\alpha_1 J_1)j_2^{n_2}(\alpha_2 J_2)JM|V_{n_1,n_1+n_2}|j_1^{n_1}(\alpha_1 J_1)j_2^{n_2}(\alpha_2 J_2)JM\rangle \quad (31.3)$$

The last equality is due to the fact that (31.3) is symmetric in the first n_1 nucleons and in the last n_2 nucleons. We now expand the $j_1^{n_1}$ and $j_2^{n_2}$ states by using c.f.p. and obtain for (31.3) the expression

$$n_1 n_2 \sum_{\alpha_{11}J_{11}\alpha_{22}J_{22}} [j_1^{n_1-1}(\alpha_{11}J_{11})j_1 J_1|\}j_1^{n_1}\alpha_1 J_1]^2 [j_2^{n_2-1}(\alpha_{22}J_{22})j_2 J_2|\}j_2^{n_2}\alpha_2 J_2]^2$$
$$\times \left\{ \int \psi^*(j_1^{n_1-1}(\alpha_{11}J_{11})j_{1n_1}(J_1)j_2^{n_2-1}(\alpha_{22}J_{22})j_{2,n_1+n_2}(J_2)JM)V_{n_1,n_1+n_2} \right.$$
$$\left. \times \psi(j_1^{n_1-1}(\alpha_{11}J_{11})j_{1n_1}(J_1)j_2^{n_2-1}(\alpha_{22}J_{22})j_{2,n_1+n_2}(J_2)JM) \right\} \quad (31.4)$$

Since coordinates of nucleons 1 to $n_1 - 1$ and $n_1 + 1$ to $n_1 + n_2 - 1$ do not appear in V_{n_1,n_1+n_2}, the integration over them may be simply carried out. Hence, the states of the $j_1^{n_1-1}$ configuration in both ψ^* and ψ must be identical due to orthogonality of states with different $\alpha_{11}J_{11}$ values. The same is true for the $j_2^{n_2-1}$ states. Before carrying out the integration we should change the order of couplings in ψ^* and ψ so that nucleons n_1 and $n_1 + n_2$ are coupled to a state with definite J'. Due to the rotational invariance of V_{n_1,n_1+n_2}, the J' values in both ψ^* and ψ must be equal for a non-vanishing contribution. Making use of the transformation (9.24) we obtain for the integral in (31.4) the expression

$$\sum_{J_0 J'} \int \psi^* (j_1^{n_1-1}(\alpha_{11}J_{11}) j_2^{n_2-1}(\alpha_{22}J_{22})(J_0) j_{1n_1} j_{2,n_1+n_2}(J')JM)$$
$$\times V_{n_1,n_1+n_2} \psi (j_1^{n_1-1}(\alpha_{11}J_{11}) j_2^{n_2-1}(\alpha_{22}J_{22})(J_0) j_{1n_1} j_{2,n_1+n_2}(J')JM)$$
$$\times (2J_1+1)(2J_2+1)(2J_0+1)(2J'+1) \begin{Bmatrix} J_{11} & j_1 & J_1 \\ J_{22} & j_2 & J_2 \\ J_0 & J' & J \end{Bmatrix}^2 \tag{31.5}$$

In obtaining (31.5) we made use of the fact that the value of J_0 must be the same in ψ^* and ψ. Otherwise the matrix element vanishes upon integration. We can now integrate over the coordinates of the nucleons 1 to $n_1 - 1$ and $n_1 + 1$ to $n_1 + n_2 - 1$ obtaining for (31.5) the result

$$(2J_1+1)(2J_2+1) \sum_{J'=|j_1-j_2|}^{j_1+j_2} (2J'+1)\langle j_1 j_2 J'M'|V|j_1 j_2 J'M'\rangle$$
$$\times \sum_{J_0} (2J_0+1) \begin{Bmatrix} J_{11} & j_1 & J_1 \\ J_{22} & j_2 & J_2 \\ J_0 & J' & J \end{Bmatrix}^2 \tag{31.6}$$

The expectation value (31.3) may thus be expressed as

$$\boxed{\begin{aligned} & n_1 n_2 \sum_{\alpha_{11}J_{11}\alpha_{22}J_{22}} [j_1^{n_1-1}(\alpha_{11}J_{11}) j_1 J_1 | \} j_1^{n_1} \alpha_1 J_1]^2 \\ & \times [j_2^{n_2-1}(\alpha_{22}J_{22}) j_2 J_2 | \} j_2^{n_2} \alpha_2 J_2]^2 \\ & \times (2J_1+1)(2J_2+1) \sum_{J'=|j_1-j_2|}^{j_1+j_2} (2J'+1) V_{J'}(j_1 j_2) \\ & \times \sum_{J_0} (2J_0+1) \begin{Bmatrix} J_{11} & j_1 & J_1 \\ J_{22} & j_2 & J_2 \\ J_0 & J' & J \end{Bmatrix}^2 \end{aligned}} \tag{31.7}$$

The interaction between j_1 and j_2 nucleons could be weak compared to the interaction within $j_1^{n_1}$ and $j_2^{n_2}$ configurations. The states with various J-values, $|J_1 - J_2| \leq J \leq J_1 + J_2$, then form a multiplet whose level spacings are small compared to energy differences between it and multiplets based on other J_1 and J_2 states. The averages of the interaction energies (31.7) in each multiplet have a very simple property. If we multiply (31.7) by $2J + 1$ and sum over J and J_0 we can use the orthogonality relation of $9j$-symbols (9.28) to obtain

$$\sum_{J J_0}(2J+1)(2J_0+1)\begin{Bmatrix} J_{11} & j_1 & J_1 \\ J_{22} & j_2 & J_2 \\ J_0 & J' & J \end{Bmatrix}^2 = \frac{1}{(2j_1+1)(2j_2+1)}$$

We can now carry out the summations over $\alpha_{11}J_{11}$ and $\alpha_{22}J_{22}$ and obtain by using the orthogonality relations of the c.f.p. the result

$$n_1 n_2 (2J_1+1)(2J_2+1)\frac{1}{(2j_1+1)(2j_2+1)}\sum_{J'=|j_1-j_2|}^{j_1+j_2}(2J'+1)V_{J'}(j_1 j_2) \tag{31.8}$$

To obtain the *average* interaction energy in the multiplet we should divide (31.8) by $\sum(2J+1) = (2J_1+1)(2J_2+1)$. We thus obtain the final expression

$$\begin{aligned} &\frac{1}{(2J_1+1)(2J_2+1)}\sum_{J=|J_1-J_2|}^{J_1+J_2}(2J+1) \\ &\langle j_1^{n_1}(\alpha_1 J_1) j_2^{n_2}(\alpha_2 J_2) JM | \sum_{i=1}^{n_1}\sum_{h=n_1+1}^{n_1+n_2} V_{ih} | j_1^{n_1}(\alpha_1 J_1) j_2^{n_2}(\alpha_2 J_2) JM\rangle \\ &= n_1 n_2 \frac{1}{(2j_1+1)(2j_2+1)}\sum_{J'=|j_1-j_2|}^{j_1+j_2}(2J'+1)V_{J'}(j_1 j_2) \\ &= n_1 n_2 \bar{V}(j_1 j_2) \end{aligned} \tag{31.9}$$

where the average interaction energy $\bar{V}(j_1 j_2)$ was defined by (14.33). Thus, the average interaction energies in all multiplets are equal, independent of J_1 and J_2. They are given by $n_1 n_2$ multiplying the average interaction energy in the $j_1 j_2$ configuration. Energy differences between *centers of mass* of multiplets are thus determined by energies of the $\alpha_1 J_1$ and $\alpha_2 J_2$ states. The latter are determined by the interactions within the $j_1^{n_1}$ and $j_2^{n_2}$ configurations.

The result (31.9) is called the *center of mass theorem* (Lawson and Uretsky 1957b). Its origin is easy to trace. When taking averages, only the scalar term in the tensor expansion of the interaction gives a non-vanishing contribution. In fact, recalling (12.13) we obtain the average interaction energy (31.9) in the form

$$\boxed{n_1 n_2 F_0(j_1 j_2 j_1 j_2)/\sqrt{(2j_1+1)(2j_2+1)}} \tag{31.10}$$

The factor $n_1 n_2$ in (31.9) and (31.10) is the number of interactions between j_1 and j_2 nucleons in the given configuration.

Another case where the only non-vanishing contribution to (31.7) is due to the $k = 0$ term in the tensor expansion of the interaction, is obtained for $J_1 = 0$. In this case there is only *one* state in the multiplet, with $J = J_2$, and hence its interaction energy is equal to (31.9) or (31.10). It may be also directly obtained from (31.7) by putting there $J_1 = 0$, $J_{11} = j_1$, $J_2 = J$. Due to (10.9) we obtain

$$\begin{aligned}
&\langle j_1^{n_1}(\alpha_1 J_1 = 0) j_2^{n_2}(\alpha_2 J_2 = J) JM | \sum_{i=1}^{n_1} \sum_{n=n_1+1}^{n_1+n_2} V_{ih} \\
&\qquad | j_1^{n_1}(\alpha_1 J_1 = 0) j_2^{n_2}(\alpha_2 J_2 = J) JM \rangle \\
&\quad = n_1 n_2 \sum_{\alpha_{11}\alpha_{22}J_{22}} [j_1^{n_1-1}(\alpha_{11} J_{11} = j_1) j_1 J_1 | \} j_1^{n_1} \alpha_1 J_1 = 0]^2 \\
&\qquad \times [j_2^{n_2-1}(\alpha_{22} J_{22}) j_2 J_2 | \} j_2^{n_2} \alpha_2 J_2 = J]^2 \\
&\qquad \times (2J+1) \sum_{J'=|j_1-j_2|}^{j_1+j_2} (2J'+1) V_{J'}(j_1 j_2) \frac{1}{(2J+1)(2j_1+1)} \\
&\qquad \times \sum_{J_0} (2J_0+1) \begin{Bmatrix} J_{22} & j_2 & J \\ J' & J_0 & j_1 \end{Bmatrix}^2
\end{aligned} \tag{31.11}$$

The orthogonality relation of the $6j$-symbols (10.13) yields $(2j_2+1)^{-1}$ for the summation over J_0 on the r.h.s. of (31.11). Now the summation over α_{11} and over $\alpha_{22}J_{22}$ in the c.f.p. may be carried out yielding 1 due to the normalization of the c.f.p. The matrix element (31.11) then becomes equal to

$$\boxed{\begin{aligned} & n_1n_2\frac{1}{(2j_1+1)(2j_2+1)}\sum_{J'=|j_1-j_2|}^{j_1+j_2}(2J'+1)V_{J'}(j_1j_2) \\ &\quad = n_1n_2\bar{V}(j_1j_2) \\ &\quad = n_1n_2F_0(j_1j_2j_1j_2)/\sqrt{(2j_1+1)(2j_2+1)} \end{aligned}} \tag{31.12}$$

which is identical to (31.10). The last equality is due to (12.13). The result (31.12) is thus independent of J_2. This is to be expected. The state $j_1^{n_1}(\alpha_1J_1=0)$ has spherical symmetry and hence the interactions of the j_1 nucleons with those in the j_2-orbit are independent of the orientation of the latter. The interaction is thus the same in all $j_2^{n_2}(\alpha_2J_2)$ states.

The usefulness of the center of mass theorem is rather limited. It was obtained by calculating the expectation value (31.3) but the state (31.2) is, in general, not an eigenstate of the matrix of the Hamiltonian defined by the $j_1^{n_1}j_2^{n_2}$ configuration. The interaction (31.1) may have non-vanishing non-diagonal matrix elements between the state (31.2) and states with other values of α_1J_1 and α_2J_2. These non-diagonal elements are not due to configuration mixing, they connect states which are within the same $j_1^{n_1}j_2^{n_2}$ configuration. If these non-diagonal matrix elements are small compared with spacings of the corresponding diagonal elements the situation is referred to as the *weak coupling* case. Ignoring the non-diagonal elements may occasionally be a good approximation but is certainly not consistent with the shell model. We will now evaluate those matrix elements.

We consider the matrix element

$$\begin{aligned} &\langle j_1^{n_1}(\alpha_1J_1)j_2^{n_2}(\alpha_2J_2)JM|\sum_{i=1}^{n_1}\sum_{h=n_1+1}^{n_1+n_2}V_{ih}|j_1^{n_1}(\alpha_1'J_1')j_2^{n_2}(\alpha_2'J_2')JM\rangle \\ &\quad = n_1n_2\langle j_1^{n_1}(\alpha_1J_1)j_2^{n_2}(\alpha_2J_2)JM|V_{n_1,n_1+1}|j_1^{n_1}(\alpha_1'J_1')j_2^{n_2}(\alpha_2'J_2')JM\rangle \end{aligned} \tag{31.13}$$

We follow the steps taken to derive (31.7) using c.f.p. and recoupling j_{1,n_1} and j_{2,n_1+n_2} to a state with spin J'. The change of coupling transformations (9.21) in ψ^* and in ψ are now different. Instead of (31.7) we obtain for the matrix element (31.13) the result

$$\begin{aligned}
& n_1 n_2 \sum_{\alpha_{11}J_{11}\alpha_{22}J_{22}} [j_1^{n_1-1}(\alpha_{11}J_{11})j_1J_1|\}j_1^{n_1}\alpha_1J_1] \\
& \times [j_1^{n_1-1}(\alpha_{11}J_{11})j_1J_1'|\}j_1^{n_1}\alpha_1'J_1'] \\
& \times [j_2^{n_2-1}(\alpha_{22}J_{22})j_2J_2|\}j_2^{n_2}\alpha_2J_2] \\
& \times [j_2^{n_2-1}(\alpha_{22}J_{22})j_2J_2'|\}j_2^{n_2}\alpha_2'J_2'] \\
& \times \sqrt{(2J_1+1)(2J_2+1)(2J_1'+1)(2J_2'+1)} \\
& \times \sum_{J'=|j_1-j_2|}^{j_1+j_2} (2J'+1)V_{J'}(j_1j_2) \\
& \times \sum_{J_0}(2J_0+1)\begin{Bmatrix} J_{11} & j_1 & J_1 \\ J_{22} & j_2 & J_2 \\ J_0 & J' & J \end{Bmatrix}\begin{Bmatrix} J_{11} & j_1 & J_1' \\ J_{22} & j_2 & J_2' \\ J_0 & J' & J \end{Bmatrix}
\end{aligned} \tag{31.14}$$

In the special case $\alpha_1 \equiv \alpha_1'$, $J_1 = J_1'$, $\alpha_2 \equiv \alpha_2'$, $J_2 = J_2'$, the expression (31.14) reduces to (31.7). Putting $V_J(j_1j_2) = 1$ for all values of J', summing over J' and J_0 yields, according to (9.28), that (31.14) vanishes unless $J_1' = J_1$, $J_2' = J_2$. Due to the orthogonality of c.f.p., the non-diagonal matrix elements vanish whereas the diagonal elements are equal to n_1n_2.

The matrix of the shell model Hamiltonian within the $j_1^{n_1}j_2^{n_2}$ configuration is reduced to submatrices defined by J (and M). The rows and columns of each submatrix are labelled by $\alpha_1J_1\alpha_2J_2J$ (and M). Its non-diagonal elements are given by (31.14) whereas its diagonal elements are equal to the sum of (31.14) (or (31.7)) and the energy eigenvalues of the state with α_1J_1 in the $j_1^{n_1}$ configuration and of the state with α_2J_2 in the $j_2^{n_2}$ configuration. The latter eigenvalues are due to single nucleon energies and two body interactions within the two configurations.

Let us look at a special case in which there is only one j_1 nucleon, $n_1 = 1$, $J_1 = j_1$. In that case there is only one trivial c.f.p. equal to 1 for $J_{11} = 0$. Thus, we obtain $J_0 = J_{22}$ and the $9j$-symbols reduce, according to (10.9), to $6j$-symbols. With a slight change of notation, (31.14) assumes then the simpler form

$$\begin{aligned}
&\langle j'j^n(\alpha_1 J_1)JM|\sum_{h=2}^{n+1} V_{1h}|j'j^n(\alpha_1' J_1')JM\rangle \\
&= n\sum_{\alpha_{11}J_{11}} [j^{n-1}(\alpha_{11}J_{11})jJ_1|\}j^n\alpha_1 J_1] \\
&\quad\times [j^{n-1}(\alpha_{11}J_{11})jJ_1'|\}j^n\alpha_1' J_1'] \\
&\quad\times \sqrt{(2J_1+1)(2J_1'+1)} \sum_{J'=|j-j'|}^{j+j'} (2J'+1)V_{J'}(j'j) \\
&\quad\times \begin{Bmatrix} J_{11} & j & J_1 \\ j' & J & J' \end{Bmatrix} \begin{Bmatrix} J_{11} & j & J_1' \\ j' & J & J' \end{Bmatrix}
\end{aligned} \tag{31.15}$$

Putting all $V_{J'}(j'j) = 1$ in (31.15) yields, due to (10.13) and the orthogonality of the c.f.p., vanishing non-diagonal elements and diagonal elements equal to n, the number of $j'j$ interactions.

It is worthwhile to mention that there are cases where the center of mass theorem holds exactly and the states (31.2) are exact eigenstates of the shell model Hamiltonian. One such case, where $J_1 = 0$, was mentioned above. Another case of interest is where $j_2 = \frac{1}{2}$ and thus $n_2 = 1$ is the only case of interest (Talmi 1962a). In that case, α_1 and J_1 are exact quantum numbers and together with J they uniquely specify the eigenstates. To show these facts it is not necessary to use (31.15). We take into consideration the fact that for maximum isospin or protons and neutrons in different shells, there are only two states of the two nucleon configuration jj' if $j' = \frac{1}{2}$. Hence, any two body interaction that reproduces the interaction energies in these $J = j + \frac{1}{2}$ and $J = j - \frac{1}{2}$ states may be used to calculate eigenvalues and eigenstates of all $j^n j'$ configurations.

We introduce for this case the simple effective interaction

$$V = a + b2(\mathbf{j} \cdot \mathbf{j}') \tag{31.16}$$

In states of the $j^n j'$ configuration this interaction is equal to

$$na + b\sum_{i=1}^{n} 2(\mathbf{j}_i \cdot \mathbf{j}') = na + b(\mathbf{J}^2 - \mathbf{J}_1^2 - \mathbf{j}'^2) \tag{31.17}$$

The operator (31.17) is diagonal if $\mathbf{J}^2$ *and* $\mathbf{J}_1^2$ are diagonal and its eigenvalues are given by

$$\boxed{na + b(J(J+1) - J_0(J_0+1) - \tfrac{3}{4})} \tag{31.18}$$

where $J_0(J_0+1)$ is the eigenvalue of $\mathbf{J}_1^2$ in the eigenstate of the j^n configuration. The parameters a and b can be determined to yield, for $n = 1$, the eigenvalues $V_{j+1/2}(j\tfrac{1}{2})$ and $V_{j-1/2}(j\tfrac{1}{2})$. From (31.18) we obtain for $n = 1$, $J_0 = j$

$$V_{j+1/2}(j\tfrac{1}{2}) = a + bj, \qquad V_{j-1/2}(j\tfrac{1}{2}) = a - b(j+1)$$

which yield

$$\boxed{\begin{aligned} a &= \frac{1}{2j+1}[(j+1)V_{j+1/2}(j\tfrac{1}{2}) + jV_{j-1/2}(j\tfrac{1}{2})] \\ b &= \frac{1}{2j+1}[V_{j+1/2}(j\tfrac{1}{2}) - V_{j-1/2}(j\tfrac{1}{2})] \end{aligned}} \tag{31.19}$$

From (31.18) follows that any state $|j^n(\alpha_0 J_0)j' = \frac{1}{2}JM\rangle$ is an eigenstate of the jj' interaction and its eigenvalue is given explicitly by (31.18). In particular, the energy difference between the state with $J = J_0 + \frac{1}{2}$ and $J = J_0 - \frac{1}{2}$ is equal to $b(2J_0+1)$. If $b < 0$ the state with

TABLE 31.1 *Experimental and calculated levels of* ^{42}K *(in* MeV*)*

J of state		2	3	4	3	5	1	0
Excitation energy	calculated	—	.14	.26	.60	.77	1.12	1.19
	experimental	—	.108	.258	.639	.699	.844	1.113

the higher value of J is lower and if $b > 0$ the lower state has the lower value of the total spin J. The energy difference depends only on the value of J_0. It is independent of n or any other quantum number of the state in the j^n configuration. For instance, it is the same in the case of one j-nucleon ($n = 1$) and one j-nucleon hole ($n = 2j$). The center of mass of the two states, with $J = J_0 + \frac{1}{2}$ and $J = J_0 - \frac{1}{2}$, is obtained from (31.18) to be equal to na. In view of (31.19), na is equal to (31.9) if we put there $j_1 = j$, $n_1 = n$, $j_2 = \frac{1}{2}$, $n_2 = 1$.

Simple examples may demonstrate the procedure of diagonalization described above. We can apply (31.15) to nuclei near ^{40}Ca in which there are valence $1d_{3/2}$ protons and $1f_{7/2}$ neutrons. Consider for instance $^{42}_{19}\mathrm{K}_{23}$ which has one hole in the $1d_{3/2}$ orbit and three $1f_{7/2}$ neutrons. The levels of the $1f^3_{7/2}$ neutron configuration have been discussed in Section 21. The interaction between one $1d_{3/2}$ proton hole and $1f_{7/2}$ neutron may be obtained from levels of $^{40}_{19}\mathrm{K}_{21}$ discussed in Section 1. If only level *spacings* are calculated it is sufficient to take differences of the ^{40}K energy levels. Adding a constant term to all $V_J(1d^{-1}_{3/2}, 1f_{7/2})$ yields zero for non-diagonal matrix elements as mentioned above, and a constant term for the diagonal ones. The matrix due to (31.15) to whose diagonal elements are added energy differences in the $1f^3_{7/2}$ neutron configuration, can be constructed and diagonalized. The calculated spectrum is very different from that of ^{40}K. The ground state has $J = 2$ whereas in ^{40}K the $J = 2$ state is rather high. This demonstrates the fact that nuclear levels are states of all valence nucleons and the $f^3_{7/2}$ configuration cannot be replaced by a single $1f_{7/2}$ neutron. The agreement between the low lying calculated levels and those found by experiment is rather good as shown in Table 31.1 (Vichniac 1972).

Looking at the components of these states we recognize the effect of non-diagonal matrix elements. The states are linear combinations of wave functions where the $1f^3_{7/2}$ neutrons are in states with various values of J_1. Due to the proximity of the $1f^3_{7/2}$ states with $J_1 = \frac{7}{2}$ and $J_1 = \frac{5}{2}$ they have comparable amplitudes in the various low lying

TABLE 31.2 *Calculated amplitudes of* $|1d_{3/2}^{-1}1f_{7/2}^{3}(J_1)\rangle$ *states in eigenstates of* ^{42}K

J of state	Excitation energy	$J_1=\frac{3}{2}$	$J_1=\frac{5}{2}$	$J_1=\frac{7}{2}$	$J_1=\frac{9}{2}$	$J_1=\frac{11}{2}$
2	—	.135	.667	.730	—	—
3	.14	.100	.619	.777	−.065	—
4	.26	—	.442	.874	.076	−.189
3	.60	−.294	.748	−.540	.250	—
5	.77	—	—	.920	.310	−.370
1	1.12	.677	.736	—	—	—

states of ^{42}K. The amplitudes of some states are listed in Table 31.2. In other K nuclei the effect of excited states of the $1f_{7/2}^{n}$ neutron configuration is not so pronounced. In the case of $^{41}_{19}K_{22}$ for instance, there is a large separation between the $J_1=0$ and $J_1=2$ levels of the $1f_{7/2}^{2}$ configuration. The calculated ground state has $J=\frac{3}{2}$, as experimentally observed and the amplitude of the $J_1=0$ state is calculated to be .918. The $J_1=2$ amplitude is only −.396, whereas the amplitudes of the $J_1=4$ and $J_1=6$ are much smaller.

Another example is offered by levels of ^{93}Mo where, to a certain approximation, the configuration is that of two $1g_{9/2}$ protons and one $2d_{5/2}$ neutron (Auerbach and Talmi 1964a). We shall later mention a better approximation for ^{93}Mo and other nuclei based on the discussion in Section 21. We now assume that the $2p_{1/2}$ proton orbit is closed at $Z=40$ and take the observed levels in ^{92}Mo as due to the $1g_{9/2}^{2}$ configuration. These levels are the $J=0$ (g.s.), $J=2$ (1.51 MeV), $J=4$ (2.28 MeV), $J=6$ (2.61 MeV) and $J=8$ (2.76 MeV). The interaction between a $1g_{9/2}$ proton and $2d_{5/2}$ neutron can accordingly be determined from the energy levels of $^{92}_{41}Nb_{51}$. These levels have spins $J=7$ (g.s.), $J=2$ (.135 MeV), $J=3$ (.285 MeV), $J=5$ (.357 MeV), $J=4$ (.480 MeV) and $J=6$ (.501 MeV).

We focus our attention on excitation energies of the various levels in ^{93}Mo. For this purpose it is sufficient to use energy spacings between levels of the relevant two nucleon configurations. This amounts to adding a constant term to the two-body interaction. Such a constant term added to the $1g_{9/2}-2d_{5/2}$ interaction yields zero non-diagonal elements and a constant term for the diagonal elements. Adding a constant term to the $1g_{9/2}^{2}$ energies simply adds a constant term to all diagonal matrix elements.

In the present case the c.f.p. in (31.15) are all equal to 1 and the matrix elements of the proton-neutron interaction are equal to

$$\langle 2d_{5/2}1g_{9/2}^2(J_1)JM|\sum_{i=2}^{3} V_{1i}|2d_{5/2}1g_{9/2}^2(J_1')JM\rangle$$

$$= 2\sqrt{(2J_1+1)(2J_1'+1)}\sum_{J'=0}^{9}(2J'+1)V_{J'}(2d_{5/2}1g_{9/2})$$

$$\times\begin{Bmatrix} \frac{9}{2} & \frac{9}{2} & J_1 \\ \frac{5}{2} & J & J' \end{Bmatrix}\begin{Bmatrix} \frac{9}{2} & \frac{9}{2} & J_1' \\ \frac{5}{2} & J & J' \end{Bmatrix} \qquad (31.20)$$

The matrix for each value of J has at most 5 rows and columns labelled by the values of J_1. To the diagonal elements of (31.20) the energy of the $2g_{9/2}^2$ proton configuration in the state with $J_1 = J_1'$ should be added and the resultant matrix diagonalized.

There is only one state with maximum value of J, namely $J = \frac{21}{2}$. It can be obtained only by taking $J_1 = J_1' = 8$. Hence, no matrix should be diagonalized and the interaction energy is given directly by (31.20). The values of J' consistent with $J = \frac{21}{2}$ and $J_1 = 8$ are only $J' = 7$ and $J' = 6$. Substituting in (31.20) the values of the $6j$-symbols we find the following interaction energy in the state with $J = \frac{21}{2}$

$$.357\,V_6(2d_{5/2}1g_{9/2}) + 1.643\,V_7(2d_{5/2}1g_{9/2}) \qquad (31.21)$$

Also the $J = \frac{19}{2}$ state can be constructed from $J_1 = J_1' = 8$ only. Its interaction energy is then given by (31.20) to be

$$.539V_5(2d_{5/2}1g_{9/2}) + .857V_6(2d_{5/2}1g_{9/2}) + .604V_7(2d_{5/2}1g_{9/2}) \qquad (31.22)$$

The interaction in the state with $J' = 7$ of the $2d_{5/2}1g_{9/2}$ configuration is much stronger than in the state with $J' = 6$ and also in the $J' = 5$ state. Hence, the $J = \frac{21}{2}$ level lies considerably lower than the one with $J = \frac{19}{2}$. When we add to (31.21) and (31.22) the excitation energy of the $1g_{9/2}^2$ state with $J_1 = 8$ we obtain for the state with $J = \frac{21}{2}$ the energy 2.939 MeV whereas the energy of the $J = \frac{19}{2}$ state is 3.382 MeV.

States with $J = \frac{17}{2}$ may be constructed from $1g_{9/2}^2$ states with $J_1 = 8$ and $J_1 = 6$. Although the $J_1 = 6$ is lower than the $J_1 = 8$ state, the interaction energy between the $2d_{5/2}$ neutron and $1g_{9/2}$ protons is weaker than in the $J = \frac{21}{2}$. In the latter all spins are as parallel as allowed by the Pauli principle and this maximizes the interaction in $2d_{5/2} - 1g_{9/2}$ states with $J' = 7$ which is strong. When the 2×2 matrix for the $J = \frac{17}{2}$ is constructed, with rows and columns labelled by $J_1 = 8$ and $J_1 = 6$, and diagonalized, we find the eigenvalues 2.956 MeV and 3.699 MeV. The situation is similar in the case of the $J = \frac{15}{2}$ state in which the spins $J_1 = 6$ and the $j' = \frac{5}{2}$ are less parallel. The eigenvalues in this case are 3.074 MeV and 3.501 MeV.

The next lower state has spin $J = \frac{13}{2}$. It can be obtained also from the $1g^2_{9/2}$ state with $J_1 = 4$ which is considerably lower than the $J_1 = 6$ and $J_1 = 8$ states. The eigenvalues of the 3×3 matrix are 2.724, 3.394 and 3.538 MeV. Thus, there is a "spin gap" in the *Yrast Levels* (lowest states for given value of total spin) from $J = \frac{21}{2}$ state to $J = \frac{13}{2}$. The fastest way for the $J = \frac{21}{2}$ of ^{93}Mo to decay to lower levels is by E4 (electric hexadecapole) transition. The nucleus is in that case in a *Yrast Trap* and the lifetime of that level measured experimentally is 6.9 hours.

In order to find the positions of the levels mentioned above, relative to the ^{93}Mo ground state, we have to diagonalize the matrices for all possible J values. The lowest eigenvalue turns out to belong to a state with $J = \frac{5}{2}$. This is not surprising since it can be constructed from the $1g^2_{9/2}$ state with $J_1 = 0$ which lies much lower than other states in ^{92}Mo. The lowest eigenvalue is .502 MeV and this should be subtracted from all eigenvalues to find the corresponding excitation energies. We may then compare the calculated positions with those determined by experiment. The actual situation is more complicated due to two reasons. One is the existence of low lying levels due to excitation of the single neutron into the $3s_{1/2}$, $1g_{7/2}$ and $2d_{3/2}$ orbits. The experimentally observed $J = \frac{1}{2}, \frac{7}{2}, \frac{3}{2}$ levels may be due, to a large or small extent, to such excitations. The other reason is that ^{92}Mo levels belong to admixtures of $(2p_{1/2})^2(1g_{9/2})^2$ and $(1g_{9/2})^4$ configurations. Taking this effect, discussed in Section 21, into account turns out to yield results which are very close to those described here (Auerbach and Talmi 1965, Ball et al. 1972, Gloeckner and Serduke 1974, Gross and Frenkel 1976). Keeping these facts in mind, we present in Table 31.3 the comparison between some calculated level positions and those observed by experiment. It is seen that the simple calculation yields good agreement with experiment and reproduces the observed "spin gap".

The results in Table 31.3 were obtained, as described above, by diagonalization of interaction matrices. The eigenstates with $J = \frac{5}{2}, \frac{9}{2}, \frac{13}{2}$,

TABLE 31.3 *Experimental and calculated levels of* ^{93}Mo *(in* MeV*)*

J of state		$\frac{5}{2}$	$\frac{9}{2}$	$\frac{5}{2}$	$\frac{11}{2}$	$\frac{13}{2}$	$\frac{15}{2}$	$\frac{5}{2}$	$\frac{17}{2}$	$\frac{19}{2}$	$\frac{21}{2}$
Excitation energy	calculated	—	1.538	1.545	2.171	2.221	2.571	2.444	2.454	2.880	2.437
	experimental	—	1.477	1.695	2.247	2.162		2.356	2.430		2.425

$\frac{17}{2}$ are admixtures of *several* of the states

$$|2d_{5/2}1g_{9/2}^2(J_1)JM\rangle$$

The actual situation is far from the weak coupling case. Still, it is worth while to notice that the calculated spacings between levels with $J = J_1 + \frac{5}{2}$ are rather close to spacings between levels with $J_1 = 0, 2, 4$, $6, 8$ in ^{92}Mo. This phenomenon is rather common in heavier nuclei where levels exhibit rotational character. Here, it arises from a rather simple shell model calculation.

The features of the proton-neutron interaction seen in the case described above are rather general. They give rise to spin gaps and yrast traps also in other nuclei. Let us look at another example in which the nuclei are close to the ^{208}Pb doubly magic nucleus. In ^{211}Po a long lived (25 sec) isomeric state lies 1.462 MeV above the $\frac{9}{2}^+$ ground state. The single valence neutron should occupy the $2g_{9/2}$ orbit whereas the two protons should be in the $1h_{9/2}$ orbit. Energies of the $J_1 = 0, 2, 4, 6, 8$ states of the $(1h_{9/2})^2$ configuration may be taken from ^{210}Po to be g.s., 1.181, 1.427, 1.473 and 1.557 MeV respectively. The interaction between a $2g_{9/2}$ neutron and $1h_{9/2}$ proton may be obtained from the levels of $^{210}_{83}\text{Bi}_{127}$. Apart from a constant term, the interaction energies in the various states are equal in MeV to $0(J' = 1)$, $.047(J' = 0)$, $.271(J' = 9)$, $.319(J' = 2)$, $.348(J' = 3)$, $.433(J' = 7)$, $.439(J' = 5)$, $.502(J' = 4)$, $.549(J' = 6)$, $.582(J' = 8)$. Also in this case the interaction energy in the parallel state of spins, $J'_{\text{max}} = 9$, is rather strong whereas that in the state with $J'_{\text{max}} - 1 = 8$ is the least one. In the original calculation (Auerbach and Talmi 1964b) it was assumed that the state with $J_1 = 8$ in ^{210}Po lies at 1.53 MeV above the ground state. The experimental position was not yet known and it was assumed that the 6–8 spacing ^{210}Po is not larger than the 4–6 spacing. The measured 6–8 spacing as given above is, however, larger than the 4–6 spacing. This feature is rather unusual and is not observed in other nuclei like ^{212}Rn or ^{214}Ra (Section 21). This may indicate that the assumed $(1h_{9/2})^2$ configuration for the protons is not sufficiently pure. In the absence of more detailed calculations we keep the value of the $J_1 = 8$ level at 1.557 MeV but for the following calculation we take the $J_1 = 6$ level of the $(1h_{9/2})^2$ configuration to lie 1.525 MeV above the $J_1 = 0$ state instead of the experimental value 1.473 MeV in ^{210}Po.

We use (31.15) (in which all non-vanishing c.f.p. are equal to 1) with the values of the proton-neutron interaction from ^{210}Bi and the values listed above for the $(1h_{9/2})^2$ proton configuration, to construct the sub-matrix of the Hamiltonian within the $(1h_{9/2})^2 2g_{9/2}$ configuration. The maximum value of the total spin is $J = \frac{25}{2}$ which can be obtained from only one state with $J_1 = 8$. In this state the proton and neutron spins are as parallel as allowed by the Pauli principle. Therefore, the energy of the $J = \frac{25}{2}$ state is lower by .06 MeV than that of the $J = \frac{23}{2}$ state which is also due only to $J_1 = 8$. Energies of the states with $J = \frac{21}{2}$ are obtained by diagonalization of a 2×2 matrix whose rows and columns are labelled by $J_1 = 8$ and $J_1 = 6$. The lowest eigenvalue of that matrix is higher by .001 MeV than the energy of the $J = \frac{25}{2}$ state. This result is due to the strong proton-neutron interaction in $J' = 9$ state compared to that in $J' = 8$, $J' = 6$ and also in $J' = 7$ states. This interaction compensates the fact that the $J_1 = 6$ of the $(1h_{9/2})^2$ configuration lies lower than the $J_1 = 8$ state. The lowest eigenvalue of the $J = \frac{19}{2}$ state also lies .006 MeV above the energy of the $J = \frac{25}{2}$ state. Only the $J = \frac{17}{2}$ state is calculated to lie lower than the $J = \frac{25}{2}$ state by .067 MeV. Thus, a large spin gap is obtained between the $J = \frac{25}{2}$ and the $J = \frac{17}{2}$ states. The state with $J = \frac{25}{2}$ is in an yrast trap. Experimentally its lifetime is determined by α-decay.

To calculate the spin of the ^{211}Po ground state and the position of the $J = \frac{25}{2}$ state above it, we diagonalize the energy matrices for all spins between $J = \frac{1}{2}$ and $J = \frac{25}{2}$. We find that the lowest state has $J = \frac{9}{2}$ which is not surprising. That state has a large amplitude of the $(1h_{9/2})^2$ state with $J_1 = 0$ which is considerably lower than all other states with $J_1 = 2, 4, 6, 8$. Subtracting the lowest eigenvalue from the calculated energy of the $J = \frac{25}{2}$ we find the value 1.502 MeV for the position of the $J = \frac{25}{2}$ state above the ^{211}Po ground state. This is in good agreement with the experimental value of 1.462 MeV.

A similar calculation yields an yrast trap at the $J = 16$ state in $^{212}_{84}Po_{128}$. It is attributed to the state $J_1 = 8$ of the $(1h_{9/2})^2$ proton configuration and the $J_2 = 8$ state of the $(2g_{9/2})^2$ neutron configuration. There is a state in ^{212}Po at 2.905 MeV above the ground state which has a 45 sec lifetime determined by α-decay. To carry out the calculation we use the matrix elements of the proton neutron interaction given by (31.14) in which the relevant c.f.p. with $J_{11} = \frac{9}{2}$ and $J_{22} = \frac{9}{2}$ are equal to 1. The proton neutron interaction is taken from ^{210}Bi as in the case of ^{211}Po. The energies of the $(1h_{9/2})^2$ proton configuration are those taken for ^{210}Po. For the diagonal elements we need also the energies of the $(2g_{9/2})^2$ neutron configuration. These are not well determined. In ^{210}Pb the low lying levels are at the following positions (in MeV) $J_1 = 2(0.800)$, $J_1 = 4(1.098)$, $J_1 = 6(1.195)$ and $J_1 = 8(1.278)$. The 6-8 spacing here is only slightly smaller than the 4-6 spacing. In ^{212}Pb the situation is similar to other cases and these energies are 0, 0.806, 1.117, 1.277 and 1.335 MeV. To reproduce a spin gap in ^{212}Po the 6-8 spacing must be smaller than in ^{210}Pb. Since we have no detailed knowledge of the structure of these levels in ^{210}Pb, we adopt the observed level spacings with the exception of the $J_1 = 6$ state of the $(2g_{9/2})^2$ neutron configuration which we take to be at 1.23 MeV above the $J_1 = 0$ state.

The state with maximum total spin has $J_{max} = 16$ and the spins of the two protons and two neutrons are as parallel as allowed by the Pauli principle. The energy of this state is lower by .25 MeV than that of the $J = 15$ state (which has also $J_1 = 8$, $J_2 = 8$). The calculated energy of the states with $J = 14$ is obtained by diagonalization of a 3×3 matrix (labelled by $J_1 = 8$ $J_2 = 8$, $J_1 = 8$ $J_2 = 6$ and $J_1 = 6$ $J_2 = 8$). The lowest eigenvalue is higher than the energy of the $J = 16$ state by .0005 MeV. The calculated position of the $J = 13$ state is considerably higher.

Only the lowest eigenvalue of the matrix with $J = 12$ is lower than the energy of the $J = 16$ state. Thus, there is a spin gap between the $J = 16$ state and the $J = 12$ state which can explain the long lifetime of the former. To find the excitation energy of the $J = 16$ state, the matrices for all J values should be diagonalized. The lowest eigenvalue is the one with $J = 0$ which is not surprising in view of the possibility to obtain such a state by coupling $J_1 = 0$ and $J_2 = 0$. Due to the interaction of the $J_1 = J_2 = 0$ state with other states for which $J_1 = J_2 \neq 0$, the eigenvalue is lower than the diagonal element by .168 MeV. The calculated position of the $J = 16$ state turns out to be 2.902 MeV above the $J = 0$ ground state which is in good agreement with the experimental position of the isomeric state at 2.905 MeV. It is still not known experimentally whether the spin of the isomeric state is $J = 16$ or $J = 18$.

It is interesting to note that the calculated position of the second state with $J = 0$ is 1.788 MeV. Experimentally a $J = 0$ state has been identified at 1.801 MeV above the ground state of ^{212}Po. The position of the first excited $J = 2$ level was calculated by Auerbach and Talmi (1964b) to be .68 MeV above the ground state. This is also in good agreement with the experimental value of .728 MeV.

Yrast traps seem to be due to the properties of the proton neutron interaction combined with those of the interaction between two identical nucleons. They should be found near magic nuclei where simple and rather pure shell model configurations are expected. As more nucleons are added, the level structure becomes complicated and the chances of spin gaps and yrast traps diminish. The simple calculations described above demonstrate the source and mechanism which give rise to yrast traps. They have a simple description in the shell model and occur where simple shell model configurations account for the observed spectra.

The existence of yrast traps follows from the *differences* in interaction energies of different proton neutron states. To obtain properties of ground states, it is necessary to know the values of those interaction energies. For the extraction of these values it is necessary to consider binding energies of nuclei. As an example, let us consider interaction energies between a $1g_{9/2}$ proton and a $2d_{5/2}$ neutron.

The interaction energies in the $1g_{9/2}$ proton $-2d_{5/2}$ neutron configuration may be obtained from ${}^{90}_{39}Y_{51}$. In that nucleus, positive parity levels which may belong to that configuration are (in MeV)

$$J' = 7(.682)\ , \qquad J' = 2(.777)\ , \qquad J' = 3(.954),$$

$$J' = 5(1.047), \qquad J' = 4(1.190), \qquad J' = 6(1.298).$$

The spacings between these levels are rather close to those in ^{92}Nb which were used above. This is in agreement with detailed calculations which incorporate mixing of configurations with $1g_{9/2}$ and $2p_{1/2}$ proton pairs. The total energy of each of these states (binding energy of ^{90}Y minus the excitation energy) is made of three parts. The first is the binding energy of the nucleus with closed shells ^{88}Sr. The second part is the sum of single $1g_{9/2}$ proton energy and single $2d_{5/2}$ neutron energy. The ground state of ^{89}Sr has $J = \frac{5}{2}$ and positive parity. Hence, the single neutron energy is simply equal to B.E.(^{89}Sr) – B.E.(^{88}Sr) = 6.365 MeV. The $1g_{9/2}$ single nucleon level lies .909 MeV above the ground state of ^{89}Y. Hence, the single $1g_{9/2}$ proton energy is equal to B.E.(^{89}Y) – B.E.(^{88}Sr) – .909 = 6.166 MeV. The third part is the interaction energy between $1g_{9/2}$ proton and $2d_{5/2}$ neutron.

TABLE 31.4 *Interaction energies between a $1g_{9/2}$ proton and a $2d_{5/2}$ neutron (in* MeV*)*

J' of state	2	3	4	5	6	7
$V_{J'}(1g_{9/2}, 2d_{5/2})$	−.624	−.447	−.211	−.354	−.103	−.719

To obtain, for instance, $V_7(1g_{9/2}2d_{5/2})$ we use the equation

$$\text{B.E.}(^{90}\text{Y}) - .682 = \text{B.E.}(^{88}\text{Sr}) + 6.365 + 6.166 + V_7(1g_{9/2}2d_{5/2}) \tag{31.23}$$

from which we obtain the result $V_7 = .719$ MeV. In (31.23) binding energies were taken as *positive* quantities. Hence, $V_7(1g_{9/2}2d_{5/2})$ is actually *attractive* and to make it clear we write

$$V_7(1g_{9/2}2d_{5/2}) = -.719 \text{ MeV}$$

The interaction energies in other states may be similarly obtained. The results are summarized in Table 31.4. From Table 31.4 we see that the interaction between a $1g_{9/2}$ proton and a $2d_{5/2}$ neutron is attractive in all states.

The attractive nature of the proton neutron interaction is a very general feature. We may see another example taken from the same nuclei. The ground state of ^{89}Y has $J = \frac{1}{2}$ and negative parity. It may be attributed to a single proton in the $2p_{1/2}$ orbit. The two lowest states of ^{90}Y have $J = 2$ (g.s) and $J = 3$ (.202 MeV) and negative parity, as expected from coupling of a $2p_{1/2}$ proton and a $2d_{5/2}$ neutron. In this case, the single $2p_{1/2}$ proton energy is equal to B.E.(^{89}Y) − B.E.(^{88}Sr) = 7.075 MeV. The procedure described above yields in the present case the results

$$V_2(2p_{1/2}, 2d_{5/2}) = -.492 \text{ MeV} \qquad V_3(2p_{1/2}, 2d_{5/2}) = -.290 \text{ MeV}$$

There are many other examples of the interaction between protons and neutrons in different orbits and they are practically all attractive.

The strength of attraction between protons and neutrons in different orbits depends on the orbits which they occupy. In the examples above, we see that the average attraction of $2d_{5/2}$ neutrons to $1g_{9/2}$ protons is stronger than that to $2p_{1/2}$ protons. An amusing example where such differences lead to drastic change in order of levels is offered by some light nuclei. In $^{13}_{6}\text{C}_7$ the ground state has $J^\pi = \frac{1}{2}^-$ which may be taken as due to a single neutron in the $1p_{1/2}$ orbit. The

$\frac{1}{2}^+$ level at 3.089 MeV may be due to a single $2s_{1/2}$ neutron outside the closed $1s_{1/2}$ and $1p_{3/2}$ orbits in ^{12}C. If one proton is removed, the $^{12}_{5}B_7$ nucleus is obtained in which the low lying levels are 1^+(g.s.), 2^+(.953 MeV), 2^-(1.674 MeV) and 1^-(2.621 MeV). The lower two levels may be due to the $J = \frac{3}{2}$ state of the $(2p_{3/2})^3$ proton configuration (a $1p_{3/2}$ proton hole) coupled to the $1p_{1/2}$ neutron. The higher levels may similarly be assigned the $1p_{3/2}^{-1}2s_{1/2}$ configuration. To obtain the interaction between a $1p_{3/2}$ proton and $1p_{1/2}$ or $2s_{1/2}$ neutron we may use the procedure described above or the results obtained earlier for a $j = \frac{1}{2}$ orbit. Let us, however, consider first the nucleus ^{11}Be obtained by removing another $1p_{3/2}$ proton.

The interaction of a $j' = \frac{1}{2}$ neutron with two $1p_{3/2}$ protons coupled to $J_0 = 0$ is given by (31.12) with $n_1 = 2$, $n_2 = 1$, $j_1 \equiv 1p_{3/2}$, $j_2 = j'$. In this case, as explained above, J_0 is a good quantum number. This can be seen in the present case directly from the fact that two $J = \frac{3}{2}$ protons may be coupled only to $J_0 = 0$ and $J_0 = 2$. Adding a $j' = \frac{1}{2}$ nucleon yields different total spins J when coupled to either of these values of J_0. The interaction energy is thus given by

$$\tfrac{1}{4}[3V_1(1p_{3/2}, j') + 5V_2(1p_{3/2}, j')] = 2a(1p_{3/2}, j') \qquad (31.24)$$

where $a(1p_{3/2}, j')$ is defined by (31.19). In ^{13}C where there are 4 protons in the $1p_{3/2}$-orbit the interaction energy is twice that value, i.e. $4a(1p_{3/2}, j')$. The center of mass of the two levels, for given j', with $J = 1$ and $J = 2$ in ^{12}B is equal to $3a(1p_{3/2}, j')$ as explained above. Thus, there is a *linear* relation between the interaction energies in ^{13}C, ^{12}B and ^{11}Be. Hence, there is a linear relation also between the *differences* of interaction energies for $j' \equiv 1p_{1/2}$ and $j' \equiv 2s_{1/2}$ in those nuclei. This relation includes the difference between the $\frac{1}{2}^+$ and $\frac{1}{2}^-$ states in ^{13}C, the difference between the centers of mass of the $1^+, 2^+$ levels and the $1^-, 2^-$ levels in ^{12}B and the $\frac{1}{2}^+ - \frac{1}{2}^-$ spacing in ^{11}Be. This linear relation was used by Talmi and Unna (1960c) to predict the level order in ^{11}Be as shown in Fig. 31.1. Rather contrary to naive expectations, the $2s_{1/2}$ orbit is calculated to lie *below* the $1p_{1/2}$ orbit in ^{11}Be. This was later verified by experiment. The ground state of ^{11}Be has indeed spin $\frac{1}{2}$ and positive parity. The $\frac{1}{2}^-$ state lies at 0.320 MeV above it in good agreement with the predicted value of 0.275 MeV.

The reason for this inversion of order of levels in different major shells is due to the $1p_{3/2}$ proton–$1p_{1/2}$ neutron interaction being stronger than the $1p_{3/2} - 2s_{1/2}$ interaction. Removing two $1p_{3/2}$ protons strongly modifies the attractive potential of the neutrons. It is

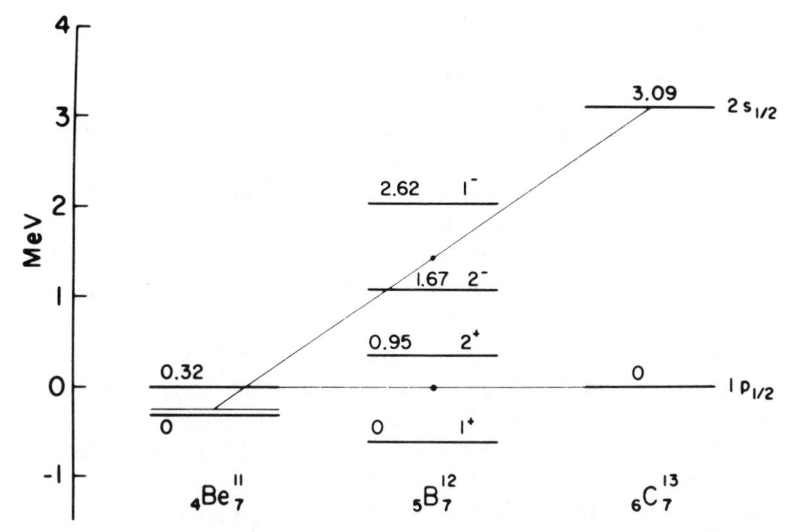

FIGURE 31.1. *Calculation of single neutron $1p_{1/2}$ and $2s_{1/2}$ levels of* ^{11}Be.

also possible to obtain the full values of matrix element of the proton neutron interaction in this case and calculate the binding energy of ^{11}Be. The linear relation may be used between the separation energies of the j' neutron in ^{13}C and ^{11}Be and the neutron separation energy in ^{12}B from which the center of mass positions have been subtracted. For $j' \equiv 2s_{1/2}$ a linear extrapolation from B.E.(^{13}C) – B.E.(^{12}C) – 3.089 = 1.857 MeV and B.E.(^{12}B) – B.E.(^{11}B) – $\frac{1}{8}(3E_{1^-} + 5E_{2^-}) = 1.367$ MeV yields 0.877 MeV for the $2s_{1/2}$ neutron separation energy in ^{11}Be. For $j' \equiv 1p_{1/2}$ we extrapolate from B.E.(^{13}C) – B.E.(^{12}C) = 4.946 MeV and B.E.(^{12}B) – B.E.(^{11}B) – $\frac{1}{8}(3E_{1^+} + 5E_{2^+}) = 2.774$ and obtain .602 MeV, which is smaller by 0.275 MeV than the $2s_{1/2}$ separation energy. This difference is, of course, identical to that obtained above. Positions of 1^-, 2^-, 1^+ and 2^+ levels of ^{12}B are denoted above by E_{1^-}, E_{2^-}, E_{1^+} and E_{2^+} respectively. The calculated value 0.877 MeV for B.E.(^{11}Be) – B.E.(^{10}Be) is in fair agreement with the experimental value of 0.504 MeV.

From the separation energy differences of ^{13}C and ^{11}Be we obtain $a(1p_{3/2}, 1p_{1/2}) = 2.381$ MeV and $a(1p_{3/2}, 2s_{1/2}) = 0.677$ MeV. Both these interactions are attractive. If we determine $b(1p_{3/2}, j')$, as defined in (31.19), from level spacings in ^{12}B we obtain the following values (with a minus sign for an attractive interaction) in MeV

$$V_1(1p_{3/2}, 1p_{1/2}) = -2.977 \qquad V_2(1p_{3/2}, 1p_{1/2}) = -2.024$$
$$V_1(1p_{3/2}, 2s_{1/2}) = -0.085 \qquad V_2(1p_{3/2}, 2s_{1/2}) = -1.032$$

The attractive proton neutron interactions considered above are equal to the average of the $T = 1$ and $T = 0$ interaction between nucleons in the orbits considered. More experimental information is required to obtain separately the $T = 0$ and $T = 1$ matrix elements of the effective interaction between nucleons in different orbits. Let us return to nuclei in the Zr region where we may be able to determine some $T = 1$ matrix elements. In $^{89}_{39}Y_{50}$ the ground state is due, as discussed above, to a proton in the $2p_{1/2}$ orbit. The $\frac{9}{2}^+$ state, .909 MeV above it is due to the single proton occupying the $1g_{9/2}$ orbit. In the two proton configuration in ^{90}Zr there are two negative parity states with $J = 5$ (at 2.319 MeV) and $J = 4$ (at 2.739 MeV). These are interpreted as states of the $2p_{1/2}1g_{9/2}$ proton configuration and their energies may yield information about the $T = 1$ part of two nucleon interactions.

Taking $^{88}_{38}Sr_{50}$ as the nucleus with closed shells, the separation energy of a $2p_{1/2}$ proton is 7.075 MeV and that of the $1g_{9/2}$ proton $7.075 - .909 = 6.166$ MeV as already mentioned above. The difference between the total energy of the 5^- and 4^- states in ^{90}Zr and the binding energy of ^{88}Sr should be equal to the sum of the separation energies and the proton-proton ($T = 1$) interaction energy. We thus obtain

$$\text{B.E.}(^{90}\text{Zr}) - \text{B.E.}(^{88}\text{Sr}) - E_{J'} = 7.075 + 6.166 + V_{J'}(2p_{1/2}, 1g_{9/2})$$

where $E_{J'}$ is the excitation energy of the state with J' above the ground state of ^{90}Zr. Using the appropriate energies, we find that both V_{5^-} and V_{4^-} are negative. Since binding energies were taken as positive quantities, the $T = 1$ interaction here is *repulsive*. The actual values, with the plus sign for repulsive terms, are equal to

$$V_5(2p_{1/2}1g_{9/2}) = 0.129 \text{ MeV} \qquad V_4(2p_{1/2}1g_{9/2}) = 0.549 \text{ MeV}.$$

We find that the average interaction energy is $a(2p_{1/2}1g_{9/2}) = +0.318$ MeV.

The average repulsion between two nucleons in different orbits in $T = 1$ states has been observed in all cases where it could have been isolated. It is consistent with our knowledge of saturation of nuclear energies. As we saw in Section 21, apart from the pairing term, nucleons in the same j-orbit repel each other. The quadratic term in the seniority binding energy formula is repulsive. This is the case in all nuclei with the exception of He isotopes. The decrease in average binding energy as more identical nucleons are added, should not be reversed or stopped at the beginning of a new orbit. As in the case of j-nucleons, the repulsion between protons includes also contributions of the repulsive Coulomb interaction. In the case of neutron configurations the repulsion is indeed weaker but its sign persists.

We have now determined which effective interactions within the framework of the shell model give rise to the central potential well which lies at its foundation. The interaction of a single j-nucleon with the $2j'+1$ nucleons in closed orbits was calculated in Section 14. In (31.9) derived above, in the case $n_1 = 1$, $n_2 = 2j_2 + 1$, that interaction is given by $(2j'+1)\bar{V}(jj')$. If the j'-nucleons are of the same kind as the single j-nucleon the average $\bar{V}(jj')$ is over all $T = 1$ states, $\bar{V}_{T=1}(jj')$. If it is of the other kind, $\bar{V}(jj')$ is equal to one half the sum of the average in all $T = 1$ states and the average in all $T = 0$ states, $\frac{1}{2}(\bar{V}_{T=1}(jj') + \bar{V}_{T=0}(jj'))$. The single j-nucleon energy in the shell model is equal to the expectation value of its kinetic energy in the j-orbit plus its effective interaction with all nucleons in closed shells. The latter contribution is the equivalent of the expectation value in the j-orbit of the potential energy of the central potential well. The effective interaction of a single nucleon with nucleons of the same kind is repulsive and is equivalent to a repulsive potential well. It is only the attractive proton-neutron interaction which gives rise to an equivalent attractive potential well which, for stable nuclei, more than compensates the $T = 1$ repulsion (Talmi 1962c).

Even if nucleons are put into a *fixed* (static) potential well which is charge symmetric, the lowest states for given $A = N + Z$ would be for $N = Z$. Such a picture is, however, incomplete. The equivalent central potential well is due to the interaction between protons and neutrons and hence, depends on their occupation numbers. In the ^{11}Be example considered above, we see the dynamic effects of the proton-neutron interaction on the depth and *shape* of the equivalent potential well. The existence of both protons and neutrons in nuclei has important consequences not only because of the Pauli principle. The different features of the $T = 1$ and $T = 0$ effective interactions have a profound effect on nuclear properties. These effects will be discussed in Section 38.

In deriving (31.10) from (31.9) we used the tensor expansion of the interaction between nucleons in the j_1 and j_2 orbits. We may use (12.13) in (31.14) to express also the matrix element (31.13) in terms of the $F_k(j_1j_2j_1j_2)$. It is simpler, however, to start from (31.13) and introduce the tensor expansion there, obtaining for it the expression

$$
\begin{aligned}
&\langle j_1^{n_1}(\alpha_1 J_1) j_2^{n_2}(\alpha_2 J_2) JM | \sum_k F_k(j_1j_2j_1j_2) \\
&\qquad \times \left(\sum_{i=1}^{n_1} \mathbf{u}^{(k)}(i) \right) \cdot \left(\sum_{h=n_1+1}^{n_1+n_2} \mathbf{u}^{(k)}(h) \right) | j_1^{n_1}(\alpha_1' J_1') j_2^{n_2}(\alpha_2' J_2') JM \rangle \\
&\quad = \sum_k F_k(j_1j_2j_1j_2)(-1)^{J_2+J+J_1'} \begin{Bmatrix} J_1 & J_2 & J \\ J_2' & J_1' & k \end{Bmatrix} \\
&\qquad \times \left(j_1^{n_1}\alpha_1 J_1 \middle\| \sum \mathbf{u}^{(k)}(i) \middle\| j_1^{n_1}\alpha_1' J_1' \right) \\
&\qquad \times \left(j_2^{n_2}\alpha_2 J_2 \middle\| \sum \mathbf{u}^{(k)}(h) \middle\| j_2^{n_2}\alpha_2' J_2' \right)
\end{aligned}
\tag{31.25}
$$

The last equality was obtained by using (10.27). We can now use (15.17) to express the reduced matrix elements on the r.h.s. of (31.25) in terms of c.f.p. in the $j_1^{n_1}$ configuration and the $j_2^{n_2}$ configuration. Recalling the definitions $(j_1\|u^{(k)}\|j_1) = 1$, $(j_2\|u^{(k)}\|j_2) = 1$, we obtain for the matrix element (31.25) the expression

$$
\boxed{
\begin{aligned}
&n_1 n_2 \sum_{\alpha_{11}J_{11}\alpha_{22}J_{22}} [j_1^{n_1-1}(\alpha_{11}J_{11})j_1J_1|\}j_1^{n_1}\alpha_1J_1] \\
&\quad \times [j_1^{n_1-1}(\alpha_{11}J_{11})j_1J_1'|\}j_1^{n_1}\alpha_1'J_1'] \\
&\quad \times [j_2^{n_2-1}(\alpha_{22}J_{22})j_2J_2|\}j_2^{n_2}\alpha_2J_2] \\
&\quad \times [j_2^{n_2-1}(\alpha_{22}J_{22})j_2J_2'|\}j_2^{n_2}\alpha_2'J_2'] \\
&\quad \times (-1)^{j_1+j_2+J_{11}+J_{22}+J_1+2J_2+J_1'+J} \\
&\quad \times [(2J_1+1)(2J_1'+1)(2J_2+1)(2J_2'+1)]^{1/2} \\
&\quad \times \sum_k F_k(j_1j_2j_1j_2) \begin{Bmatrix} J_1 & J_2 & J \\ J_2' & J_1' & k \end{Bmatrix} \\
&\quad \times \begin{Bmatrix} j_1 & J_1 & J_{11} \\ J_1' & j_1 & k \end{Bmatrix} \begin{Bmatrix} j_2 & J_2 & J_{22} \\ J_2' & j_2 & k \end{Bmatrix}
\end{aligned}
}
\tag{31.26}
$$

The products of two $9j$-symbols in (35.14) are replaced here by sums of products of three $6j$-symbols.

Finally, we can generalize (31.14) to the case of two groups of nucleons with definite (not necessarily maximum) isospins in different orbits. We can repeat the steps taken in the derivation of (31.14) with the changes introduced by isospin. When changing the order of coupling, the transformation is a product of two transformations, one for changing the coupling of spins and one for changing the coupling of isospins. We thus obtain the result

$$
\begin{aligned}
&\langle j_1^{n_1}(\alpha_1T_1J_1)j_2^{n_2}(\alpha_2T_2J_2)TM_TJM|\\
&\qquad\sum_{i=1}^{n_1}\sum_{h=n_1+1}^{n_1+n_2}V_{ih}|j_1^{n_1}(\alpha'_1T'_1J'_1)j_2^{n_2}(\alpha'_2T'_2J'_2)TM_TJM\rangle\\
&\quad=n_1n_2\sum_{\substack{\alpha_{11}T_{11}J_{11}\\ \alpha_{22}T_{22}J_{22}}}[j_1^{n_1-1}(\alpha_{11}T_{11}J_{11})j_1T_1J_1|\}j_1^{n_1}\alpha_1T_1J_1]\\
&\quad\times[j^{n_1-1}(\alpha_{11}T_{11}J_{11})j_1T'_1J'_1|\}j_1^{n_1}\alpha'_1T'_1J'_1]\\
&\quad\times[j_2^{n_2-1}(\alpha_{22}T_{22}J_{22})j_2T_2J_2|\}j_2^{n_2}\alpha_2T_2J_2]\\
&\quad\times[j_2^{n_2-1}(\alpha_{22}T_{22}J_{22})j_2T'_2J'_2|\}j_2^{n_2}\alpha'_2T'_2J'_2]\\
&\quad\times[(2T_1+1)(2T_2+1)(2T'_1+1)(2T'_2+1)\\
&\qquad\times(2J_1+1)(2J_2+1)(2J'_1+1)(2J'_2+1)]^{1/2}\\
&\quad\times\sum_{T'J'}(2T'+1)(2J'+1)V_{T'J'}(j_1j_2)\sum_{T_0}(2T_0+1)\\
&\quad\times\begin{Bmatrix}T_{11}&\frac{1}{2}&T_1\\ T_{22}&\frac{1}{2}&T_2\\ T_0&T'&T\end{Bmatrix}\begin{Bmatrix}T_{11}&\frac{1}{2}&T'_1\\ T_{22}&\frac{1}{2}&T'_2\\ T_0&T'&T\end{Bmatrix}\\
&\quad\times\sum_{J_0}(2J_0+1)\begin{Bmatrix}J_{11}&j_1&J_1\\ J_{22}&j_2&J_2\\ J_0&J'&J\end{Bmatrix}\begin{Bmatrix}J_{11}&j_1&J'_1\\ J_{22}&j_2&J'_2\\ J_0&J'&J\end{Bmatrix}
\end{aligned}
\tag{31.27}
$$

In the case of maximum isospins,

$$T_1 = n_1/2, \qquad T_2 = n_2/2, \qquad T = (n_1 + n_2)/2,$$

$$T_{11} = (n_1 - 1)/2, \qquad T_{22} = (n_2 - 1)/2$$

and hence

$$T_0 = (n_1 + n_2)/2 - 1 \qquad \text{and} \qquad T' = 1.$$

The summation over T', T_0, which is reduced to a single term, may then be carried out and due to the orthogonality relations of the $9j$-symbols, (9.28), all isospin factors are cancelled and (31.27) reduces to (31.14). Such a summation may be carried out also if $V_{T'J'}(j_1 j_2)$ is independent of T'. Due to (9.28), (31.37) vanishes unnless $T_1' = T_1$, $T_2' = T_2$ and the only dependence on isospin is in the c.f.p. An equivalent expression which is a generalization of (31.26) may be written down but we will not present it here.

Let us now consider matrix elements of single nucleon operators. These are necessary for calculating transition probabilities between states, like those of electromagnetic radiation. If the initial state and the final state are in the same configuration of n_1 protons in the j_1-orbit and n_2 neutrons in the j_2-orbit, such reduced matrix elements are given by

$$\begin{aligned}
&(j_1^{n_1}(\alpha_1 J_1) j_2^{n_2}(\alpha_2 J_2)J \| \mathbf{F}^{(k)}(1) + \mathbf{F}^{(k)}(2) \| j_1^{n_1}(\alpha_1' J_1') j_2^{n_2}(\alpha_2' J_2')J') \\
&= (-1)^{J_1+J_2+J'+k}\sqrt{(2J+1)(2J'+1)} \begin{Bmatrix} J_1 & J & J_2 \\ J' & J_1' & k \end{Bmatrix} \\
&\times (j_1^{n_1}\alpha_1 J_1 \| \mathbf{F}^{(k)}(1) \| j_1^{n_1}\alpha_1' J_1')\delta_{\alpha_2\alpha_2'}\delta_{J_2 J_2'} \\
&+ (-1)^{J_1+J_2'+J+k}\sqrt{(2J+1)(2J'+1)} \begin{Bmatrix} J_2 & J & J_1 \\ J' & J_2' & k \end{Bmatrix} \\
&\times (j_2^{n_2}\alpha_2 J_2 \| \mathbf{F}^{(k)}(2) \| j_2^{n_2}\alpha_2' J_2')\,\delta_{\alpha_1\alpha_1'}\delta_{J_1 J_1'}
\end{aligned} \qquad (31.28)$$

In (31.28) $\mathbf{F}^{(k)}(1)$ is the single nucleon operator acting in the $j_1^{n_1}$ configuration and $\mathbf{F}^{(k)}(2)$ is the one in the $j_2^{n_2}$ configuration. The equality in (31.28) is due to the relations (10.41) and (10.42).

A more interesting case is where the orbit of a nucleon is changed in the transition. Let us first consider the matrix elements between states of the j^n configuration and the $j^{n-1}j'$ configuration. If we include in the calculation also β-decay in which the charge state of a nucleon is changed, we have to use states with definite isospins. We thus consider the matrix element

$$\langle [j^{n-1}(\alpha_1 T_1 J_1) j' T' M_T' J' M']_a | \sum_{i=1}^{n} h(i) f_\kappa^{(k)}(i) | j^n \alpha T M_T J M \rangle$$

$$= \int \frac{1}{\sqrt{n}} \mathcal{A} \psi^*(j^{n-1}(\alpha_1 T_1 J_1) j_n' T' M_T' J' M')$$

$$\times \left(\sum_i h(i) f_\kappa^{(k)}(i) \right) \psi(j^n \alpha T M_T J M)$$

$$= n \int \frac{1}{\sqrt{n}} \mathcal{A} \psi^*(j^{n-1}(\alpha_1 T_1 J_1) j_n' T' M_T' J' M')$$

$$\times h(n) f_\kappa^{(k)}(n) \psi(j^n \alpha T M_T J M) \tag{31.29}$$

In (31.29) $h(i)$ is an isospin operator of a single nucleon which could be either a constant or proportional to a component of $\boldsymbol{\tau}$. The tensor of rank k, $\mathbf{f}^{(k)}$ operates on the space and spin coordinates. The antisymmetrizer $\mathcal{A}$ exchanges the label n with the other $n-1$ labels and is defined by

$$\mathcal{A} \psi(j^{n-1}(\alpha_1 T_1 J_1) j_n' T' M_T' J' M') = \psi(j^{n-1}(\alpha_1 T_1 J_1) j_n' T' M_T' J' M')$$

$$- \sum_{i=1}^{n-1} \psi(j^{n-1}(\alpha_1 T_1 J_1) j_i' T' M_T' J' M') \tag{31.30}$$

All the terms on the r.h.s. of (31.30) are orthogonal and hence the normalization of (31.30) is $1/\sqrt{n}$ which appears in (31.29). The wave function $\psi(j^n \alpha T M_T J M)$ is also antisymmetric and the single nucleon operator fully symmetric. Hence, each of its terms contributes the same amount from which follows the last equality in (30.29).

Of all terms on the r.h.s. of (31.30) only the first has a non-vanishing contribution to the integral on the r.h.s. of (31.29). In the other terms, the state of the i-th nucleon, $i \leq n-1$, cannot change from the j-orbit to the j'-orbit. We may expand in terms of c.f.p. and obtain for the matrix element (31.29) the expression

$$\sqrt{n}\int \psi^*(j^{n-1}(\alpha_1T_1J_1)j'_nT'M'_TJ'M')h(n)f^{(k)}_\kappa(n)\psi(j^n\alpha TM_TJM)$$
$$= \sqrt{n}[j^{n-1}(\alpha_1T_1J_1)jTJ|\}j^n\alpha TJ]$$
$$\times \int \psi^*(j^{n-1}(\alpha_1T_1J_1)j'_nT'M'_TJ'M')h(n)f^{(k)}_\kappa(n)$$
$$\times \psi(j^{n-1}(\alpha_1T_1J_1)j_nTM_TJM) \quad (31.31)$$

The terms with principal parents other than the one with $\alpha_1T_1J_1$ vanish upon integration over the coordinates of the first $n-1$ nucleons. To carry out this integration it is simpler to look at the matrix element (31.29) reduced with respect to space and spin variables but not with respect to isospin. That reduced matrix element is given according to (10.41) by

$$\sqrt{n}[j^{n-1}(\alpha_1T_1J_1)jTJ|\}j^n\alpha TJ]\langle T_1\tfrac{1}{2}T'M'_T|h(n)|T_1\tfrac{1}{2}TM_T\rangle$$
$$\times(-1)^{J_1+j+J'+k}\sqrt{(2J+1)(2J'+1)}$$
$$\times\begin{Bmatrix} j' & J' & J_1 \\ J & j & k \end{Bmatrix}(j'\|\mathbf{f}^{(k)}\|j) \quad (31.32)$$

If $h(n)$ is a tensor operator of rank 0 or 1, its matrix element may be expressed by using (10.41) and the Wigner-Eckart theorem. The expression (31.32) may be applied to the case of identical nucleons or maximum isospin in which case $h(n)$ is replaced by 1 and all isospins do not appear.

In (31.32) there are geometrical factors, like the $6j$-symbol, which depend on the relative orientation of the spins j, j' and J_1. There are also factors which depend on the structure of the j^n wave function, i.e. $\sqrt{n}$ and the c.f.p. The latter occur also in pick up and stripping reactions in which a nucleon is removed or added to the nucleus. Rates of those reactions depend on the square of the matrix element of a creation or annihilation operator. Let us consider the reduced matrix element of the creation operator. Due to (17.30) it may be expressed as

$$(n,\alpha J\|a_j^\dagger\|n-1,\alpha_1 J_1) = (-1)^n\sqrt{n}\sqrt{2J+1}[j^{n-1}(\alpha_1 J_1)jJ|\}j^n\alpha J] \tag{31.33}$$

In the case of states with lowest seniorities we obtain the dependence of the reaction rate on n by using (19.28) with $v = 0$ for even values of n and for odd n, we use (19.26) with $v = 1$. The squares of the reduced matrix elements are obtained in these cases to be

$$\boxed{(n, v = 0, J = 0\|a_j^\dagger\|n-1, v_1 = 1, J_1 = j)^2 = n \qquad n \text{ even}} \tag{31.34}$$

$$\boxed{(n, v = 1 J = j\|a_j^\dagger\|n-1, v_1 = 0, J_1 = 0)^2 = 2j+2-n \qquad n \text{ odd}} \tag{31.35}$$

The difference in the n-dependence of (31.34) and (31.35) can be attributed to the structure of states with seniority $v = 0$ and $v = 1$. In nucleon pick-up reactions, any nucleon may be removed from the $v = 0$ state to yield a $v = 1$, $J = j$ state. There are $n/2$ pairs in the $v = 0$ state and that rate is increasing proportionally to n. If, however, the nucleon is removed from a $v = 1$, $J = j$ state to leave the nucleus in a $v = 0$, $J = j$ state, only the "odd nucleon" may be removed. As there are more pairs in the $v = 0$ state the odd nucleon has less states available and the rate *decreases* with n. If we consider a stripping reaction, a nucleon is added to the state with seniority v_1. If $v_1 = 0$, $J_1 = 0$ the added nucleon must be inserted in a state not occupied by those in the $(n-1)/2$ pairs. Hence, the rate decreases with n as given by (31.35). If, however, $v_1 = 1$, $J_1 = j$ then adding a nucleon to yield $v = 0$, $J = 0$ state increases the number of pairs. The proportionality of the rate to n as in (31.34) indicates that the $J = 0$ pairs behave in some sense like bosons.

The more general matrix element of a single nucleon operator is between the $j_1^{n_1-1}j_2^{n_2+1}$ configuration and the $j_1^{n_1}j_2^{n_2}$ configuration. To obtain it, we proceed as in the derivation of (31.32). States of these configurations are obtained by starting from antisymmetric states in the $j_1^{n_1}$ and $j_2^{n_2}$ configurations, coupling them to states with total T and J (and M) like (31.2) and antisymmetrizing. The necessary antisymmetrization is carried out by the antisymmetrizer $\mathcal{A}$ defined by

$$\mathcal{A}_{nn'}\psi(j^n_{12\ldots n}(\alpha_1T_1J_1)j'^{n'}_{n+1,\ldots,n+n'}(\alpha_2T_2J_2)TM_TJM)$$
$$=\sum_P(-1)^P\psi(j^n_{i_1i_2\ldots i_n}(\alpha_1T_1J_1)j'^{n'}_{i_{n+1}\ldots i_{n+n'}}(\alpha_2T_2J_2)TM_TJM) \tag{31.36}$$

In (31.36) the permutations P replace the coordinates of nucleons 1 to n by the coordinates $i_1i_2\ldots i_n$. The sum is over all permutations for which the sets $i_1i_2\ldots i_n$ are *different*. Thus, all terms on the r.h.s. of (31.36) are orthogonal (and normalized). Their number is $(n+n')!/n!\,n'!$ and consequently the normalization factor for the wave function (31.36) is $[(n+n')!/n!\,n'!]^{-1/2}$.

We now proceed to calculate the matrix element

$$\langle[j_1^{n_1}(\alpha_1T_1J_1)j_2^{n_2}(\alpha_2T_2J_2)TM_TJM]_a|$$
$$\sum_{i=1}^{n_1+n_2}h(i)f_\kappa^{(k)}(i)|[j_1^{n_1-1}(\alpha'_1T'_1J'_1)j_2^{n_2+1}(\alpha'_2T'_2J'_2)T'M'_TJ'M']_a\rangle$$
$$=\binom{n_1+n_2}{n_1}^{-1/2}\binom{n_1+n_2}{n_1-1}^{-1/2}$$
$$\times\int\mathcal{A}_{n_1n_2}\psi^*(j_1^{n_1}(\alpha_1T_1J_1)j_2^{n_2}(\alpha_2T_2J_2)TM_TJM)\left(\sum_{i=1}^{n_1+n_2}h(i)f_\kappa^{(k)}(i)\right)$$
$$\times\mathcal{A}_{n_1-1,n_2+1}\psi(j_1^{n_1-1}(\alpha'_1T'_1J'_1)j_2^{n_2+1}(\alpha'_2T'_2J'_2)T'M'_TJ'M') \tag{31.37}$$

We can remove the antisymmetrizer $\mathcal{A}_{n_1n_2}$ from ψ^* and let it operate on the rest of the integrand which is fully antisymmetric. In analogy with the antisymmetrizer (15.23), the projection operator in this case is $\mathcal{A}_{n_1n_2}/((n_1+n_2)!/n_1!n_2!)$ and (31.37) is equal to

$$\binom{n_1+n_2}{n_1}^{1/2}\binom{n_1+n_2}{n_1-1}^{-1/2}\int \psi^*(j_1^{n_1}(\alpha_1T_1J_1)j_2^{n_2}(\alpha_2T_2J_2)TM_TJM)$$

$$\times\left(\sum_{i=1}^{n_1+n_2} h(i)f_\kappa^{(k)}(i)\right)$$

$$\times\,\mathcal{A}_{n_1-1,n_2+1}\psi(j_1^{n_1-1}(\alpha_1'T_1'J_1')j_2^{n_2+1}(\alpha_2'T_2'J_2')T'M_T'J'M') \qquad (31.38)$$

Of all terms in the summation over i in (31.38), only the terms with $i=1,2,\ldots,n_1$ make a non-vanishing contribution to the integral. Terms with higher values of i vanish upon the integration over the variables of the first n_1 nucleons since one of them must occupy different orbits in ψ^* and ψ. Since ψ^* is antisymmetric in the first n_1 nucleons and so is $\mathcal{A}_{n_1-1,n_2+1}\psi$, all terms with $i=1,2,\ldots,n_1$ contribute equally to the integral. We may keep only one of them, with $i=n_1$ for instance, and multiply the result by n_1. Of all $(n_1+n_2)!/(n_1-1)!(n_2+1)!$ wave functions included in $\mathcal{A}_{n_1-1,n_2+1}\psi$ only one of them makes a non-vanishing contribution to the matrix element of $h(n_1)f_\kappa^{(k)}(n_1)$. This is the wave function ψ in which the first n_1-1 nucleons occupy the j_1-orbit. All other functions yield zero when the integration over the first n_1-1 nucleons is performed. The matrix element (31.37) assumes thus the form

$$n_1\binom{n_1+n_2}{n_1}^{1/2}\binom{n_1+n_2}{n_1-1}^{-1/2}$$

$$\times\int \psi^*(j_{1;1,2,\ldots,n_1}^{n_1}(\alpha_1T_1J_1)j_{2;n_1+1,\ldots,n_1+n_2}^{n_2}(\alpha_2T_2J_2)TM_TJM)$$

$$\times\, h(n_1)f_\kappa^{(k)}(n_1)\psi(j_{1;1,2,\ldots,n_1-1}^{n_1-1}(\alpha_1'T_1'J_1')$$

$$\times\, j_{2;n_1,n_1+1,\ldots,n_1+n_2}^{n_2+1}(\alpha_2'T_2'J_2')T'M_T'J'M') \qquad (31.39)$$

The coefficient of the integral in (31.39) is equal to $\sqrt{n_1(n_2+1)}$.

To obtain the matrix element of $h(n_1)f_\kappa^{(k)}(n_1)$ we should decouple the n_1-th nucleon from the others. We expand the $j_1^{n_1}(\alpha_1T_1J_1)$ state and the $j_2^{n_2+1}(\alpha_2'T_2'J_2')$ state in terms of c.f.p. In the latter state we permute successively n_1 with n_1+1, n_1+2 up to n_1+n_2 obtaining the phase $(-1)^{n_2}$. The only terms in the expansion which give non-vanishing contributions to (31.39) are

$$
\begin{aligned}
&(-1)^{n_2}\sqrt{n_1(n_2+1)}[j_1^{n_1-1}(\alpha_1'T_1'J_1')j_1TJ|\}j_1^{n_1}\alpha TJ] \\
&\times [j_2^{n_2}(\alpha_2T_2J_2)j_2T_2'J_2'|\}j_2^{n_2+1}\alpha_2'T_2'J_2'] \\
&\times \int \psi^*([j_1^{n_1-1}(\alpha_1'T_1'J_1')j_{1;n_1}T_1J_1]j_2^{n_2}(\alpha_2T_2J_2)TM_TJM) \\
&\times h(n_1)f_\kappa^{(k)}(n_1) \\
&\times \psi(j_1^{n_1-1}(\alpha_1'T_1'J_1')[j_2^{n_2}(\alpha_2T_2J_2)j_{2;n_1}T_2'J_2']T'M_T'J'M')
\end{aligned}
\tag{31.40}
$$

We now carry out on ψ^* and ψ in (31.40) change of coupling transformations to put the single nucleon functions of the n_1-th nucleon in a convenient position. We thus obtain using (10.11) the expansion

$$
\begin{aligned}
&\psi^*([j_1^{n_1-1}(\alpha_1'T_1'J_1')j_{1;n_1}T_1J_1]j_2^{n_2}(\alpha_2T_2J_2)TM_TJM) \\
&\quad = \sum_{T_0J_0}(-1)^{\frac{1}{2}+T_2+T_1+T_0+j_1+J_2+J_1+J_0} \\
&\quad \times \sqrt{(2T_1+1)(2T_0+1)(2J_1+1)(2J_0+1)} \\
&\quad \times \begin{Bmatrix} T_1' & \frac{1}{2} & T_1 \\ T & T_2 & T_0 \end{Bmatrix} \times \begin{Bmatrix} J_1' & j_1 & J_1 \\ J & J_2 & J_0 \end{Bmatrix} \\
&\quad \times \psi^*(j_1^{n_1-1}(\alpha_1'T_1'J_1')j_2^{n_2}(\alpha_2T_2J_2)T_0J_0j_{1;n_1}TM_TJM)
\end{aligned}
\tag{31.41}
$$

Similarly, we make use of (10.12) to obtain the analogous expansion for ψ. When those expansions are put into (31.40), the integration over nucleon variables 1 to n_1-1 and n_1+1 to n_1+n_2 may be carried out. Hence, non-vanishing contributions may arise only from terms in which T_0J_0 are the same in the expansions of ψ^* and of ψ.

To obtain the expression of the matrix element (31.37) in terms of that of a single nucleon it is convenient to calculate the reduced matrix element. For that we take $h(i)$ to be a component of a tensor operator in isospin with rank $k'=0$ or $k'=1$. We may then use (10.41) to obtain a sum over J_0 of products of $6j$-symbols for spins and a similar sum for isospins. The sum of products of three $6j$-symbols is equal according to the Appendix to a $9j$-symbol. The final result for the reduced matrix element (31.37), with respect to spin and isospin is given by

$$
\begin{aligned}
&(-1)^{n_2}\sqrt{n_1(n_2+1)}[j_1^{n_1-1}(\alpha_1' T_1' J_1')j_1 T_1 J_1|\}j_1^{n_1}\alpha_1 T_1 J_1] \\
&\quad\times [j_2^{n_2}(\alpha_2 T_2 J_2)j_2 T_2' J_2'|\}j_2^{n_2+1}\alpha_2' T_2' J_2'] \\
&\quad\times (-1)^{T_2-T_2'+(1/2)+J_2-J_2'+j_2} \\
&\quad\times [(2T_1+1)(2T_2'+1)(2T+1)(2T'+1)(2J_1+1) \\
&\qquad\times (2J_2'+1)(2J+1)(2J'+1)]^{1/2} \\
&\quad\times \begin{Bmatrix} T_1 & T_2 & T \\ T_1' & T_2' & T' \\ \frac{1}{2} & \frac{1}{2} & k' \end{Bmatrix} \begin{Bmatrix} J_1 & J_2 & J \\ J_1' & J_2' & J' \\ j_1 & j_2 & k \end{Bmatrix} (\tfrac{1}{2}\|\mathbf{h}^{(k')}\|\tfrac{1}{2})(j_1\|\mathbf{f}^{(k)}\|j_2)
\end{aligned}
\tag{31.42}
$$

In the case where $n_2 = 0$, $J_2 = 0$ the matrix element (31.42) reduces to the result (31.32) obtained above.

Let us consider a special case of (31.42) in which the isospin formalism is not necessary. We look at valence identical nucleons (or states with the maximum isospins) and at states with lowest seniorities in all j^n configurations involved. If n_1 is odd and n_2 even, we put $J_1 = j_1$, $J_1' = 0$, $J_2 = 0$, $J_2' = j_2$. With these values and in view of (10.9) and (10.19), the matrix element (31.42) reduces to

$$
\begin{aligned}
&\sqrt{n_1(n_2+1)}[j_1^{n_1-1}(v_1'=0, J_1'=0)j_1 J_1 = j_1|\}j_1^{n_1} v_1 = 1, J_1 = j_1] \\
&\qquad\times [j_2^{n_2}(v_2=0, J_2=0)j_2 J_2' = j_2|\}j_2^{n_2+1} v_2' = 1, J_2' = j_2](j_1\|\mathbf{f}^{(k)}\|j_2) \\
&\quad= \sqrt{n_1(n_2+1)}\sqrt{\frac{2j_1+2-n_1}{n_1(2j_1+1)}}\sqrt{\frac{2j_2+1-n_2}{(n_2+1)(2j_2+1)}}(j_1\|\mathbf{f}^{(k)}\|j_2) \\
&\quad= \sqrt{\frac{(2j_1+2-n_1)(2j_2+1-n_2)}{(2j_1+1)(2j_2+1)}}(j_1\|\mathbf{f}^{(k)}\|j_2)
\end{aligned}
\tag{31.43}
$$

The values of c.f.p. in (31.43) were taken from (19.26) for $v = 1$ and $n = n_1$, $n = n_2 + 1$ respectively. If n_1 is even and n_2 is odd, $J_1 = 0$, $J_1' = j_1$, $J_2 = j_2$, $J_2' = 0$, we obtain from (31.42) the expression

$$\sqrt{n_1(n_2+1)}[j_1^{n_1-1}(v_1' = 1, J_1' = j_1)j_1 J_1 = 0|\}j_1^{n_1} v_1 = 0, J_1 = 0]$$

$$\times [j_2^{n_1}(v_2 = 1, J_2 = j_2)j_2 J_2' = 0|\}j_2^{n_2+1} v_2' = 0, J_2' = 0]$$

$$\times \frac{(-1)^{j_1+j_2+k}}{\sqrt{(2j_1+1)(2j_2+1)}}(j_1\|\mathbf{f}^{(k)}\|j_2)$$

$$= (-1)^{k+1}\frac{\sqrt{n_1(n_2+1)}}{\sqrt{(2j_1+1)(2j_2+1)}}(j_2\|\mathbf{f}^{(k)}\|j_1) \qquad (31.44)$$

The last equality in (31.44) was obtained by using (19.28) with $v = 0$ and (8.9). The square of the coefficient of the reduced matrix element on the r.h.s. of (31.43) is given in (37.38) of de-Shalit and Talmi (1963). The square of that coefficient in (31.44), multiplied by the geometrical factor $(2j_1+1)/(2j_2+1)$ should have been given by (37.39) of that reference. Due to a mistake, the denominator $(2j_1+1)(2j_2+1)$ is missing from (37.39) of that book.

Let us look at an example for a transition given by (31.43) and (31.44). We consider a group of identical nucleons where the isospin part is unnecessary. In some nuclei with neutron number 50, considered in Section 21, the proton $2p_{1/2}$ and $1g_{9/2}$ orbits may be occupied. Electromagnetic $M4$ transitions may take place between $\frac{1}{2}^-$ and $\frac{9}{2}^+$ states. As mentioned in Section 21, $\frac{1}{2}^-$ states are described by

$$|\tfrac{1}{2}^-\rangle = |g_{9/2}^n(v_0 = 0, J_0 = 0)p_{1/2} J = \tfrac{1}{2}, M\rangle \qquad (31.45)$$

The $\frac{9}{2}^+$ states are described by the following linear combination

$$|\tfrac{9}{2}^+\rangle = \alpha_n |g_{9/2}^{n+1} v = 1, J' = \tfrac{9}{2}, M'\rangle$$

$$+ \beta_n |g_{9/2}^{n-1}(v_1 = 1, J_1 = \tfrac{9}{2})p_{\frac{1}{2}}^2(J_2 = 0)J' = \tfrac{9}{2}, M'\rangle \qquad (31.46)$$

where $\alpha_n^2 + \beta_n^2 = 1$. The reduced matrix element for the transition is thus determined by

$$(\tfrac{1}{2}^{-}\|T(M4)\|\tfrac{9}{2}^{+})$$

$$= \alpha_n(g_{9/2}^n(v_0 = 0, J_0 = 0)p_{1/2}, J = \tfrac{1}{2}\|T(M4)\|g_{9/2}^{n+1} v = 1, J' = \tfrac{9}{2})$$

$$+ \beta_n(g_{9/2}^n(v_0 = 0, J_0 = 0)p_{1/2}J = \tfrac{1}{2}\|T(M4)\|$$

$$g_{9/2}^{n-1}(v_1 = 1, J_1 = \tfrac{9}{2})p_{1/2}^2(J_2 = 0)J' = \tfrac{9}{2}) \qquad (31.47)$$

The first term on the r.h.s. of (31.47) is given by (31.43) with $j_1 = \frac{1}{2}$, $n_1 = 1$, $j_2 = \frac{9}{2}$, $n_2 = n$. The second term is given by (31.44) with $j_1 = \frac{9}{2}$, $n_1 = n$, $j_2 = \frac{1}{2}$, $n_2 = 1$. We thus obtain for the reduced matrix element (31.47) the result

$$\alpha_n\sqrt{\frac{10-n}{10}}(p_{1/2}\|T(M4)\|g_{9/2})$$

$$+ \beta_n(-1)^{4+1}\sqrt{\frac{n}{10}}(p_{1/2}\|T(M4)\|g_{9/2})$$

$$= (p_{1/2}\|T(M4)\|g_{9/2})\left[\alpha_n\sqrt{1-\frac{n}{10}} - \beta_n\sqrt{\frac{n}{10}}\right] \qquad (31.48)$$

The values of α_n and β_n calculated (Gloeckner and Serduke 1974) from nuclear energies yield fair agreement with ratios of measured $B(M4)$ values in $N = 50$ nuclei and of the single proton transition in ^{89}Y. From nuclear energies only the absolute value of $\langle 2p_{1/2}^2 J = 0|V|1g_{9/2}^2 J = 0\rangle$ could be determined. The choice of its positive sign was made to give α_n and β_n opposite signs which fit the rates of $M4$ transitions. The calculated coefficients (31.48) for several $N = 50$ nuclei are fairly constant. In fact, they are constant, independent of n, if the states (31.46) have definite quasi-spin as described in Section 22. It was shown there that tensor operators whose matrix elements determine rates of magnetic transitions commute with S_1^+ and hence are scalars with respect to the quasi-spin scheme defined by (22.22). In the present case the quasi-spin pair creation operator is

$$S_1^+ = S_{9/2}^+ - S_{1/2}^+ \qquad (31.49)$$

The state (31.46) is obtained by

$$a^{+}_{9/2m}(S^{+}_{1})^{n/2}|0\rangle = a^{+}_{9/2m}(S^{+}_{9/2})^{n/2}|0\rangle - \frac{n}{2}a^{+}_{9/2,m}(S^{+}_{9/2})^{n/2-1}S^{+}_{1/2}|0\rangle \tag{31.50}$$

The amplitudes α_n and β_n in (31.46) may be calculated from (31.50) by using the normalization factor (19.19). The normalized amplitudes are thus given by

$$\alpha_n = \sqrt{\frac{10-n}{10}} \qquad \beta_n = -\sqrt{\frac{n}{10}}$$

and the reduced matrix element becomes equal, for any n, to

$$(p_{1/2}\|T(M4)\|g_{9/2}).$$

32

Configuration Mixing. Effective Operators

In preceding sections, examples of rather pure configurations were given. It was explained that pure shell model configurations may be used for the calculation of energies only if the effect of other configurations may be implicitly included as renormalization of the effective interaction. Configuration mixings were shown to be explicitly important in some cases but they were treated rather simply, in the formalism of generalized seniority. In many cases, however, there are clear effects of configuration mixing which cannot be replaced by simple renormalization and more detailed calculations must be carried out. Mixing of configurations has been considered in the shell model following two different directions. One way to deal with configuration mixing is to construct the shell model Hamiltonian in a subspace defined by all configurations involved. That submatrix of the Hamiltonian is then diagonalized. It is possible to carry out this procedure only in cases where the order of the matrices is manageable. This procedure has been adopted for nucleons occupying a rather small number of j-orbits whose single nucleon energies are close. The most complicated yet successful program has been carried out by Wildenthal et al. (1984) in the $1d_{5/2}, 2s_{1/2}, 1d_{3/2}$ shell.

Another direction considers mixing with configurations which lie at rather high energies. Perturbation theory may then be used and contributions due to many orders are summed. This is the approach used by nuclear many body theory to obtain from the strong and singular interaction between free nucleons a renormalized interaction which may be used in the shell model (g-matrix). These methods are described in detail in several books, e.g. Ring and Schuck (1980). In the following we will derive formulae for calculating matrix elements of two nucleon interactions between states in different configurations. These matrix elements may be used in any of the approaches described above. We will also use perturbation theory up to second order to demonstrate how single nucleon and two-body operators are affected. Some important features of renormalization may be exhibited by such simple calculations without entering the intricacies of the nuclear many body theory.

Non-vanishing matrix elements of a two-nucleon interaction occur only between states of configurations in which at most two nucleons occupy different orbits. In some cases, total spin and parity in one state may be equal to those in a state of another configuration in which only one nucleon occupies a different orbit. Such states may be connected by a non-vanishing matrix element of a two-nucleon interaction. We shall first consider matrix elements between configurations differing by orbits of two nucleons and later deal with matrix elements of the other kind.

Let us begin with the simple j^n configuration and consider matrix elements of a two-body interaction connecting it to the $j^{n-2}j'j''$ configuration. The matrix elements to be calculated have the form

$$\langle [j^{n-2}(\alpha_1 T_1 J_1)j'j''(T'J')TM_T JM]_a | \sum_{i<h}^{n} V_{ih} | j^n \alpha T M_T JM \rangle \quad (32.1)$$

The two body interaction is a scalar operator both with respect to $\mathbf{J}$ and the isospin. Hence, its matrix elements are diagonal in M and M_T and are independent of them. To save space we omit them in the following expressions. The antisymmetrized state of the $j^{n-2}j'j''$ configuration which appears in (32.1), is obtained by using the following antisymmetrizer

$$\begin{aligned} &\mathcal{A}\psi(j_{1,\ldots,n-2}^{n-2}(\alpha_1 T_1 J_1)j'_{n-1}j''_n(T'J')TJ) \\ &\qquad = \sum_P (-1)^P \psi(j_{i_1,\ldots,i_{n-2}}^{n-2}(\alpha_1 T_1 J_1)j'_{i_{n-1}}j''_{i_n}(T'J')TJ) \quad (32.2) \end{aligned}$$

The summation in (32.2) is over all permutations P which replace nucleon coordinates $1,\ldots,n-2$ by different sets $i_1,\ldots,i_{n-2}$ taken from the nucleon numbers 1 to n. If $j' \neq j''$ the permutations also antisymmetrize the wave function of the two nucleons in these orbits. Their number is therefore $n(n-1)/(1+\delta_{j'j''})$ and all terms on the r.h.s. of (32.2) are orthogonal. The normalization factor for the wave function (32.2) is thus $[n(n-1)/(1+\delta_{j'j''})]^{-1/2}$.

We start from

$$\left[\frac{n(n-1)}{1+\delta_{j'j''}}\right]^{-1/2}\int \mathcal{A}\psi^*(j^{n-2}_{1,\ldots,n-2}(\alpha_1 T_1 J_1)j'_{n-1}j''_n(T'J')TJ)$$
$$\times\left(\sum_{i<h}^{n} V_{ih}\right)\psi(j^n\alpha TJ) \tag{32.3}$$

and transform the projection operator $\mathcal{A}/[n(n-1)/(1+\delta_{j'j''})]$ from ψ^* to the rest of the integrand obtaining

$$\left[\frac{n(n-1)}{1+\delta_{j'j''}}\right]^{1/2}\int \psi^*(j^{n-2}_{1,\ldots,n-2}(\alpha_1 T_1 J_1)j'_{n-1}j''_n(T'J')TJ)$$
$$\times\left(\sum_{i<h}^{n} V_{ih}\right)\psi(j^n\alpha TJ) \tag{32.4}$$

In (32.4) nucleons $n-1$ and n are in different (orthogonal) orbits in ψ^* and ψ. Hence, the only non-vanishing contribution to the integral is due to the term $V_{n-1,n}$. Expanding $\psi(j^n\alpha TJ)$ in $n \to n-2$ c.f.p. and carrying out the integration, we obtain for the matrix element (32.1) the result

$$\boxed{\begin{aligned}&\sqrt{n(n-1)/2}[j^{n-2}(\alpha_1 T_1 J_1)j^2(T'J')TJ|\}j^n\alpha TJ]\\ &\times\langle [j'j''T'J']_a|V|j^2T'J'\rangle\end{aligned}} \tag{32.5}$$

The matrix element in (32.5) is calculated with the state of the two nucleons in the $j'j''$ orbits which is antisymmetric and *normalized*. Hence, if either $j' \equiv j''$ or $j' \neq j''$ the factor on the left of (32.5) is equal to $[n(n-1)/2]^{1/2}$. The result (32.5) is rather simple since the j', j'' orbits are occupied by one or two (if $j' \equiv j''$) nucleons. In other cases, the non-diagonal matrix elements are more complicated as we shall presently see.

We consider next matrix elements between states of the $j_1^{n_1} j_2^{n_2}$ configuration and states in the $j_1^{n_1-2} j_2^{n_2+2}$ configuration. We can express such matrix elements as

$$\langle [j_1^{n_1-2}(\alpha_1 T_1 J_1) j_2^{n_2+2}(\alpha_2 T_2 J_2) TJ]_a | \sum_{i<h}^{n_1+n_2} V_{ih} | [j_1^{n_1}(\alpha_1' T_1' J_1') j_2^{n_2}(\alpha_2' T_2' J_2') TJ]_a \rangle$$
$$= \binom{n_1+n_2}{n_1}^{-1/2} \binom{n_1+n_2}{n_1-2}^{-1/2}$$
$$\times \int \mathcal{A}_{n_1-2,n_2+2} \psi^*(j_{1;1,\ldots,n_1-2}^{n_1-2}(\alpha_1 T_1 J_1) j_{2;n_1+1,\ldots,n_1+n_2,n_1-1,n_1}^{n_2+2}(\alpha_2 T_2 J_2) TJ)$$
$$\times \left(\sum_{i<h}^{n_1+n_2} V_{ih} \right) \mathcal{A}_{n_1 n_2} \psi(j_{1;1,\ldots,n_1}^{n_1}(\alpha_1' T_1' J_1') j_{2;n_1+1,\ldots,n_1+n_2}^{n_2}(\alpha_2' T_2' J_2') TJ) \tag{32.6}$$

The antisymmetrizers $\mathcal{A}_{n_1-2,n_2+2}$ and $\mathcal{A}_{n_1 n_2}$ are defined in (31.36) where also the normalization factors are given. We now remove the projection operator $\mathcal{A}_{n_1-2,n_2+2}/[(n_1+n_2)!/(n_1-2)!(n_2+2)!]$ from $\psi*$ and let it act on the rest of the integrand obtaining

$$\binom{n_1+n_2}{n_1-2}^{1/2} \binom{n_1+n_2}{n_1}^{-1/2}$$
$$\times \int \psi^*(j_{1;1,\ldots,n_1-2}^{n_1-2}(\alpha_1 T_1 J_1) j_{2;n_1+1,\ldots,n_1+n_2,n_1-1,n_1}^{n_2+2}(\alpha_2 T_2 J_2) TJ)$$
$$\times \left(\sum_{i<h}^{n_1+n_2} V_{ih} \right) \mathcal{A}_{n_1 n_2} \psi(j_{1;1,\ldots,n_1}^{n_1}(\alpha_1' T_1' J_1') j_{2;n_1+1,\ldots,n_1+n_2}^{n_2}(\alpha_2' T_2' J_2') TJ) \tag{32.7}$$

Of all terms in the linear combination $\mathcal{A}_{n_1n_2}\psi$, the only non-vanishing contributions arise from terms in which nucleons 1 to n_1-2 are in the j_1-orbit. The only two nucleons which can change from the j_1-orbit in $\mathcal{A}_{n_1n_2}\psi$ to the j_2-orbit in ψ^* are those whose numbers range from n_1-1 to n_1+n_2. Since $\psi(j_2^{n_2+2}\alpha_2T_2J_2)$ is fully antisymmetric in these nucleons and so is the wave function $\mathcal{A}_{n_1n_2}\psi$, each of the $(n_2+2)!/2!\,n_2!$ pairs of nucleons makes the same contribution to the integral in (32.7). We can thus take one pair, e.g. n_1-1, n_1, calculate the integral and multiply the result by $(n_2+2)!/2!\,n_2!$. Among all the V_{ih}, the only non-vanishing contribution is due to V_{n_1-1,n_1}. We thus obtain for the matrix element (32.6) the expression

$$\binom{n_2+2}{2}\binom{n_1+n_2}{n_1}^{-1/2}\binom{n_1+n_2}{n_1-2}^{1/2}$$

$$\times\int\psi^*(j_{1;1,\ldots,n_1-2}^{n_1-2}(\alpha_1T_1J_1)j_{2;n_1+1,\ldots,n_1+n_2,n_1-1,n_1}^{n_2+2}(\alpha_2T_2J_2)TJ)$$

$$\times V_{n_1-1,n_1}\psi(j_{1;1,\ldots,n_1}^{n_1}(\alpha_1'T_1'J_1')j_{2;n_1+1,\ldots,n_1+n_2}^{n_2}(\alpha_2'T_2'J_2')TJ) \quad (32.8)$$

We now expand the $j_1^{n_1}$ and $j_2^{n_2+2}$ states in $n\rightarrow n-2$ c.f.p. and the only terms in (32.8) which do not vanish upon integration yield the result

$$\sqrt{\frac{n_1(n_1-1)}{2}\frac{(n_2+2)(n_2+1)}{2}}$$

$$\times\sum_{T'J'}[j_1^{n_1-2}(\alpha_1T_1J_1)j_1^2(T'J')T_1'J_1'|\}j_1^{n_1}\alpha_1'T_1'J_1']$$

$$\times[j_2^{n_2}(\alpha_2'T_2'J_2')j_2^2(T'J')T_2J_2|\}j_2^{n_2+2}\alpha_2T_2J_2]$$

$$\times\int\psi^*(j_1^{n_1-2}(\alpha_1T_1J_1)j_2^{n_2}(\alpha_2'T_2'J_2')j_{2;n_1-1,n_1}^2(T'J')T_2J_2)V_{n_1-1,n_1}$$

$$\times\psi((j_1^{n_1-2}(\alpha_1T_1J_1)j_{1;n_1-1,n_1}^2(T'J')T_1'J_1')j_2^{n_2}(\alpha_2'T_2'J_2')TJ) \quad (32.9)$$

Since V_{n_1-1,n_1} is a scalar charge independent operator, T' and J' must have the same values in ψ^* and ψ. To carry out the integrations in (32.9) we change the order of couplings. We first change in ψ^* the order of coupling of T_2', T' and J_2', J' obtaining the phase factor $(-1)^{T_2'+T'-T_2+J_2'+J'-J_2}$. We then apply to ψ the transformation (10.12) for spins and isospins. Now the integration may be carried out and we obtain for the matrix element (32.6) the result

$$\boxed{\begin{aligned}&\sqrt{\frac{n_1(n_1-1)}{2}}\sqrt{\frac{(n_2+1)(n_2+2)}{2}}\\ &\quad\times\sum_{T'J'}[j_1^{n_1-2}(\alpha_1T_1J_1)j_1^2(T'J')T_1'J_1'|\}j_1^{n_1}\alpha_1'T_1'J_1']\\ &\quad\times[j_2^{n_2}(\alpha_2'T_2'J_2')j_2^2(T'J')T_2J_2|\}j_2^{n_2+2}\alpha_2T_2J_2]\\ &\quad\times(-1)^{2J_2'+J_1-J_2+J+2T_2'+T_1-T_2+T}\\ &\quad\times\sqrt{(2T_1'+1)(2T_2+1)(2J_1'+1)(2J_2+1)}\\ &\quad\times\begin{Bmatrix}T_1 & T' & T_1'\\ T_2' & T & T_2\end{Bmatrix}\begin{Bmatrix}J_1 & J' & J_1'\\ J_2' & J & J_2\end{Bmatrix}\langle j_2^2T'J'|V|j_1^2T'J'\rangle\end{aligned}} \tag{32.10}$$

For the special case $n_2=0$, (32.10) reduces to (32.5).

Somewhat more complicated are matrix elements between states of the $j_1^{n_1}j_2^{n_2}j_3^{n_3}$ configuration and the $j_1^{n_1-2}j_2^{n_2+1}j_3^{n_3+1}$ configuration. By similar steps taken in deriving (32.10) also these may be calculated. The result is expressed in terms of $n\to n-2$ c.f.p. of the $j_1^{n_1}$ configuration and ordinary $n\to n-1$ c.f.p. in the $j_2^{n_2+1}$ and $j_3^{n_3+1}$ configurations. Another change of coupling transformation is necessary here which was not needed in deriving (32.10). This transformation leads to a state in which the j_2j_3 nucleons are coupled to a state with T'

and J'. The matrix elements are expressed as

$$
\begin{aligned}
&\langle [j_1^{n_1-2}(\alpha_1 T_1 J_1) j_2^{n_2+1}(\alpha_2 T_2 J_2) j_3^{n_3+1}(\alpha_3 T_3 J_3)(T_{23}J_{23})TJ]_a | \\
&\quad \sum_{i<h}^{n_1+n_2+n_3} V_{ih} | [j_1^{n_1}(\alpha_1' T_1' J_1') j_2^{n_2}(\alpha_2' T_2' J_2') j_3^{n_3}(\alpha_3' T_3' J_3')(T_{23}' J_{23}')TJ]_a \rangle \\
&= \sqrt{n_1(n_1-1)(n_2+1)(n_3+1)/2} \\
&\quad \times \sum_{T'J'} [j_1^{n_1-2}(\alpha_1 T_1 J_1) j_1^2(T'J') T_1' J_1' | \} j_1^{n_1} \alpha_1' T_1' J_1'] \\
&\quad \times [j_2^{n_2}(\alpha_2' T_2' J_2') j_2 T_2 J_2 | \} j_2^{n_2+1} \alpha_2 T_2 J_2] \\
&\quad \times [j_3^{n_3}(\alpha_3' T_3' J_3') j_3 T_3 J_3 | \} j_3^{n_3+1} \alpha_3 T_3 J_3] \\
&\quad \times (-1)^{2T_{23}' - T_{23} + T_1 + T + 2J_{23}' - J_{23} + J_1 + J} \\
&\quad \times [(2T_1'+1)(2T_2+1)(2T_3+1)(2T_{23}+1) \\
&\qquad \times (2T_{23}'+1)(2T'+1)]^{1/2} \\
&\quad \times [(2J_1'+1)(2J_2+1)(2J_3+1)(2J_{23}+1) \\
&\qquad \times (2J_{23}'+1)(2J'+1)]^{1/2} \\
&\quad \times \begin{Bmatrix} T_2' & \frac{1}{2} & T_2 \\ T_3' & \frac{1}{2} & T_3 \\ T_{23}' & T' & T_{23} \end{Bmatrix} \begin{Bmatrix} T_1 & T' & T_1' \\ T_{23}' & T & T_{23} \end{Bmatrix} \\
&\quad \times \begin{Bmatrix} J_2' & j_2 & J_2 \\ J_3' & j_3 & J_3 \\ J_{23}' & J' & J_{23} \end{Bmatrix} \begin{Bmatrix} J_1 & J' & J_1' \\ J_{23}' & J & J_{23} \end{Bmatrix} \langle [j_2 j_3 T'J']_a | V | j_1^2 T'J' \rangle
\end{aligned}
\tag{32.11}
$$

In the special case $n_2 = n_3 = 0$, $T'_2 = T'_3 = 0$, $J'_2 = J'_3 = 0$, $T'_{23} = 0$, $J'_{23} = 0$, (32.11) reduces to the result (32.5) obtained above.

Until now, to reach the other configuration, the two nucleons whose orbits were changed had to come from the same orbit. Another set of matrix elements between states in different configurations appears when the two nucleons are excited from different orbits. The simplest one is the matrix element

$$\langle[j_1^{n_1-1}(\alpha_1 T_1 J_1)j_2^{n_2-1}(\alpha_2 T_2 J_2)(T_{12}J_{12})j'_1 j'_2(T'J')TJ]_a|$$

$$\sum_{i<h}^{n_1+n_2} V_{ih}|[j_1^{n_1}(\alpha'_1 T'_1 J'_1)j_2^{n_2}(\alpha'_2 T'_2 J'_2)TJ]_a\rangle$$

$$= \left[\frac{(n_1+n_2)!}{(n_1-1)!(n_2-1)!(1+\delta_{j'_1 j'_2})}\right]^{-1/2} \binom{n_1+n_2}{n_1}^{-1/2}$$

$$\times \int \mathcal{A}'\psi^*(j_{1;1,\ldots,n_1-1}^{n_1-1}(\alpha_1 T_1 J_1)$$

$$j_{2;n_1+1,\ldots,n_1+n_2-1}^{n_2-1}(\alpha_2 T_2 J_2)(T_{12}J_{12})j'_1 j'_2(T'J')TJ)$$

$$\times \left(\sum_{i<h}^{n_1+n_2} V_{ih}\right) \mathcal{A}_{n_1 n_2}\psi(j_{1;1,\ldots,n_1}^{n_1}(\alpha'_1 T'_1 J'_1)j_{2;n_1+1,\ldots,n_1+n_2}^{n_2}(\alpha'_2 T'_2 J'_2)TJ)$$

$$(32.12)$$

In (32.12) the antisymmetrizer $\mathcal{A}_{n_1 n_2}$ and the corresponding normalization coefficient are defined by (31.36). The antisymmetrizer $\mathcal{A}'$ is a linear combination of permutations P with coefficients $(-1)^P$ where P replaces nucleon numbers $1,\ldots,n_1-1$ by different sets $i_1,\ldots,i_{n_1-1}$, and the numbers $n_1+1,\ldots,n_1+n_2-1$ by different sets $i_{n_1+1},\ldots,i_{n_1+n_2-1}$. If $j'_1 \neq j'_2$, $\mathcal{A}'$ also antisymmetrizes the wave function $\psi(j'_1 j'_2 T'J')$. All terms in $\mathcal{A}'\psi^*$ are orthogonal from which follows the normalization factor appearing in (32.12).

To evaluate (32.12) we remove the projection operator

$$\mathcal{A}'/[(n_1+n_2)!/(n_1-1)!(n_2-1)!(1+\delta_{j'_1 j'_2})]$$

from ψ^* and let it operate on the rest of the integrand which is fully antisymmetric obtaining

$$\left[\frac{(n_1+n_2)!}{(n_1-1)!(n_2-1)!(1+\delta_{j_1'j_2'})}\right]^{1/2}\binom{n_1+n_2}{n_1}^{-1/2}$$

$$\times\int\psi^*(j_{1;1,\ldots,n_1-1}^{n_1-1}(\alpha_1T_1J_1)$$

$$j_{2;n_1+1,\ldots,n_1+n_2-1}^{n_2-1}(\alpha_2T_2J_2)(T_{12}J_{12})j'_{1;n_1}j'_{2;n_1+n_2}(T'J')TJ)$$

$$\times\left(\sum_{i<h}^{n_1+n_2}V_{ih}\right)\mathcal{A}_{n_1n_2}\psi(j_{1;1,\ldots,n_1}^{n_1}(\alpha_1'T_1'J_1')j_{2;n_1+1,\ldots,n_1+n_2}^{n_2}(\alpha_2'T_2'J_2')TJ) \tag{32.13}$$

The only term in the two body interaction with non-vanishing contribution to the integral in (32.13) is V_{n_1,n_1+n_2} since nucleons n_1 and n_1+n_2 are in different orbits in ψ^* and ψ. From all terms in $\mathcal{A}_{n_1n_2}\psi$ the only ones that contribute to the integral in (32.13) are those in which nucleons 1 to n_1-1 are in the j_1-orbit and nucleons n_1+1 to n_1+n_2-1 are in the j_2-orbit. There are two such terms. To carry out the integration we expand $\psi(j_1^{n_1}\alpha_1'T_1'J_1')$ and $\psi(j_2^{n_2}\alpha_2'T_2'J_2')$ in c.f.p. and carry out a change of coupling transformation. This transformation is a product of the transformations (9.24) for spins and isospins. The final result for the matrix element (32.12) is

$$\boxed{\begin{aligned}&\sqrt{n_1n_2}[j_1^{n_1-1}(\alpha_1T_1J_1)j_1T_1'J_1'|\}j_1^{n_1}\alpha_1'T_1'J_1']\\&\times[j_2^{n_2-1}(\alpha_2T_2J_2)j_2T_2'J_2'|\}j_2^{n_2}\alpha_2'T_2'J_2']\\&\times[(2T_1'+1)(2T_2'+1)(2T_{12}+1)(2T'+1)\\&\qquad\times(2J_1'+1)(2J_2'+1)(2J_{12}+1)(2J'+1)]^{1/2}\\&\times\begin{Bmatrix}T_1&\frac{1}{2}&T_1'\\T_2&\frac{1}{2}&T_2'\\T_{12}&T'&T\end{Bmatrix}\begin{Bmatrix}J_1&j_1&J_1'\\J_2&j_2&J_2'\\J_{12}&J'&J\end{Bmatrix}\\&\times\langle[j_1'j_2'T'J']_a|V|[j_1j_2T'J']_a\rangle\end{aligned}} \tag{32.14}$$

The wave functions from which the two nucleon matrix element in (32.14) is calculated are antisymmetric and normalized.

Evaluation of matrix elements becomes more complicated if the nucleons from the j_1 and j_2 orbits are raised into partially occupied orbits. Let us consider matrix elements between states of the $j_1^{n_1} j_2^{n_2} j_3^{n_3}$ configuration and the $j_1^{n_1-1} j_2^{n_2-1} j_3^{n_3+2}$ configuration. Using methods similar to those explained above in this case we obtain

$$
\begin{aligned}
&\langle [j_1^{n_1-1}(\alpha_1 T_1 J_1) j_2^{n_2-1}(\alpha_2 T_2 J_2)(T_{12} J_{12}) j_3^{n_3+2}(\alpha_3 T_3 J_3) TJ]_a | \sum_{i<h}^{n_1+n_2+n_3} V_{ih} \\
&\qquad | [j_1^{n_1}(\alpha_1' T_1' J_1') j_2^{n_2}(\alpha_2' T_2' J_2')(T_{12}' J_{12}') j_3^{n_3}(\alpha_3' T_3' J_3') TJ]_a \rangle \\
&\qquad = \sqrt{n_1 n_2 (n_3+1)(n_3+2)/2} \\
&\qquad \times \sum_{T'J'} [j_1^{n_1-1}(\alpha_1 T_1 J_1) j_1 T_1' J_1' | \} j_1^{n_1} \alpha_1' T_1' J_1'] \\
&\qquad \times [j_2^{n_2-1}(\alpha_2 T_2 J_2) j_2 T_2' J_2' | \} j_2^{n_2} \alpha_2' T_2' J_2'] \\
&\qquad \times [j_3^{n_3}(\alpha_3' T_3' J_3') j_3^2 (T'J') T_3 J_3 | \} j_3^{n_3+2} \alpha_3 T_3 J_3] \\
&\qquad \times (-1)^{T_{12}+T_3'+T'+T+J_{12}+J_3'+J'+J} \\
&\qquad \times [(2T_1'+1)(2T_2'+1)(2T_{12}+1)(2T_{12}'+1) \\
&\qquad\qquad \times (2T_3+1)(2T'+1)(2J_1'+1)(2J_2'+1) \\
&\qquad\qquad \times (2J_{12}+1)(2J_{12}'+1)(2J_3+1)(2J'+1)]^{1/2} \\
&\qquad \times \begin{Bmatrix} T_1 & \frac{1}{2} & T_1' \\ T_2 & \frac{1}{2} & T_2' \\ T_{12} & T' & T \end{Bmatrix} \begin{Bmatrix} T_{12} & T' & T_{12}' \\ T_3' & T & T_3 \end{Bmatrix} \begin{Bmatrix} J_1 & j_1 & J_1' \\ J_2 & j_2 & J_2' \\ J_{12} & J' & J \end{Bmatrix} \\
&\qquad \times \begin{Bmatrix} J_{12} & J' & J_{12}' \\ J_3' & J & J_3 \end{Bmatrix} \langle j_3^2 T'J' | V | [j_1 j_2 T'J']_a \rangle
\end{aligned}
\tag{32.15}
$$

In the special case $n_3 = 0$, $T_3' = 0$, $J_3' = 0$, (32.15) reduces to the simpler expression (32.14).

Finally, we consider matrix elements between states of the $j_1^{n_1} j_2^{n_2} j_3^{n_3} j_4^{n_4}$ configuration and states of the $j_1^{n_1-1} j_2^{n_2-1} j_3^{n_3+1} j_4^{n_4+1}$ configuration. For these rather complicated matrix elements the methods used above yield

$$
\begin{aligned}
&\langle [j_1^{n_1-1}(\alpha_1 T_1 J_1) j_2^{n_2-1}(\alpha_2 T_2 J_2)(T_{12} J_{12}) \\
&\qquad j_3^{n_3+1}(\alpha_3 T_3 J_3) j_4^{n_4+1}(\alpha_4 T_4 J_4)(T_{34} J_{34}) TJ]_a | \\
&\qquad \sum_{i<h}^{n_1+n_2+n_3+n_4} V_{ih} | [j_1^{n_1}(\alpha_1' T_1' J_1') j_2^{n_2}(\alpha_2' T_2' J_2')(T_{12}' J_{12}') \\
&\qquad\qquad j_3^{n_3}(\alpha_3' T_3' J_3') j_4^{n_4}(\alpha_4' T_4' J_4')(T_{34}' J_{34}') TJ]_a \rangle \\
&= \sqrt{n_1 n_2 (n_3+1)(n_4+1)} \sum_{T'J'} [j_1^{n_1-1}(\alpha_1 T_1 J_1) j_1 T_1' J_1' | \} j_1^{n_1} \alpha_1' T_1' J_1'] \\
&\quad \times [j_2^{n_2-1}(\alpha_2 T_2 J_2) j_2 T_2' J_2' | \} j_2^{n_2} \alpha_2' T_2' J_2'] \\
&\quad \times [j_3^{n_3}(\alpha_3' T_3' J_3') j_3 T_3 J_3 | \} j_3^{n_3+1} \alpha_3 T_3 J_3] \\
&\quad \times [j_4^{n_4}(\alpha_4' T_4' J_4') j_4 T_4 J_4 | \} j_4^{n_4+1} \alpha_4 T_4 J_4] (-1)^{T_{12}+T_{34}+T+J_{12}+J_{34}+J} \\
&\quad \times [(2T_1'+1)(2T_2'+1)(2T_{12}+1)(2T_{12}'+1)(2T_3+1) \\
&\qquad \times (2T_4+1)(2T_{34}+1)(2T_{34}'+1)]^{1/2}(2T'+1) \\
&\quad \times [(2J_1'+1)(2J_2'+1)(2J_{12}+1)(2J_{12}'+1)(2J_3+1) \\
&\qquad \times (2J_4+1)(2J_{34}+1)(2J_{34}'+1)]^{1/2}(2J'+1) \\
&\quad \times \begin{Bmatrix} T_1 & \frac{1}{2} & T_1' \\ T_2 & \frac{1}{2} & T_2' \\ T_{12} & T' & T_{12}' \end{Bmatrix} \begin{Bmatrix} T_3' & \frac{1}{2} & T_3 \\ T_4' & \frac{1}{2} & T_4 \\ T_{34}' & T' & T_{34} \end{Bmatrix} \begin{Bmatrix} T_{12} & T' & T_{12}' \\ T_{34}' & T & T_{34} \end{Bmatrix} \\
&\quad \times \begin{Bmatrix} J_1 & j_1 & J_1' \\ J_2 & j_2 & J_2' \\ J_{12} & J' & J_{12}' \end{Bmatrix} \begin{Bmatrix} J_3' & j_3 & J_3 \\ J_4' & j_4 & J_4 \\ J_{34}' & J' & J_{34} \end{Bmatrix} \begin{Bmatrix} J_{12} & J' & J_{12}' \\ J_{34}' & J & J_{34} \end{Bmatrix} \\
&\quad \times \langle [j_3 j_4 T' J']_a | V | [j_1 j_2 T' J']_a \rangle
\end{aligned}
\tag{32.16}
$$

In the special case $n_3 = 0$, $n_4 = 0$, $T_3' = 0$, $J_3' = 0$, $T_4' = 0$, $J_4' = 0$, the result (32.16) reduces to the expression (32.14) obtained above.

This concludes the calculation of matrix elements between states in shell model configurations differing by the orbits of two nucleons. Before considering matrix elements of the other kind, in which the difference is in one orbit of a single nucleon, let us look at some applications of the matrix elements calculated so far. As mentioned above, perturbation theory may be used to calculate corrections to energies in a given configuration provided the non-diagonal matrix elements are small compared to differences in corresponding diagonal elements. This situation may arise only if the two-body interaction is sufficiently regular like the Coulomb repulsion between atomic electrons considered in Section 16. This is not the case for the strong and singular interaction between free nucleons. Perturbation theory may be applicable to the effective (renormalized) interaction which should be used in the shell model. The second order corrections to the energy, to be considered below, have interesting features which may be shared by the results of the sophisticated many body theory applied to the interaction between free nucleons.

Consider the simple j^n configuration and effects of second order of perturbation theory due to $j^{n-2}j'j''$ configurations. The matrix element (32.1) is expressed by (32.5). The second order corrections due to mixings with all states $\psi([j^{n-2}(\alpha_1T_1J_1)j'j''(T'J')TJ]_a)$ with all possible values of $\alpha_1T_1J_1T'J'$ are thus equal to

$$\frac{n(n-1)}{2}$$

$$\times \sum_{\alpha_1T_1J_1T'J'} \frac{[j^{n-2}(\alpha_1T_1J_1)j^2(T'J')TJ|\}j^n\alpha TJ]^2\langle[j'j''T'J']_a|V|j^2T'J'\rangle^2}{E(j^n\alpha TJ) - E([j^{n-2}(\alpha_1T_1J_1)j'j''(T'J')TJ]_a)} \tag{32.17}$$

If we start from two orthogonal states with the same values of TJ (and M_T, M) in the j^n configuration characterized by additional quantum numbers α and α' (like seniority) and include first order corrections to them due to admixtures of $j^{n-2}j'j''$ states, the resulting states need not be orthogonal. The second order correction to the matrix element between the state resulting from $\psi(j^n\alpha TJ)$ and the one resulting from $\psi(j^n\alpha'TJ)$ and orthogonalized to the former is given by the expression

$$\frac{n(n-1)}{2} \sum_{\alpha_1 T_1 J_1 T' J'} [j^{n-2}(\alpha_1 T_1 J_1) j^2(T'J')TJ|\}\alpha TJ]$$

$$\times [j^{n-2}(\alpha_1 T_1 J_1) j^2(T'J')TJ|\} j^n \alpha' TJ]$$

$$\times \frac{\langle [j' j'' T'J']_a | V | j^2 T'J' \rangle^2}{E(j^n \alpha TJ) - E([j^{n-2}(\alpha_1 T_1 J_1) j' j''(T'J')TJ]_a)}$$

(32.18)

Comparing (32.17) and (32.18) to (15.32) we see that the second order corrections to matrix elements between many nucleon states are expressed as corrections to the unperturbed two nucleon matrix elements in states with $T'J'$. If we collect contributions from several $j'j''$ configurations, the addition to the two nucleon matrix elements becomes equal to

$$\boxed{\delta V_{T'J'}(j^2) = \sum_{j'j''} \frac{\langle [j' j'' T'J']_a | V | j^2 T'J' \rangle^2}{E(j^n \alpha TJ) - E([j^{n-2}(\alpha_1 T_1 J_1) j' j''(T'J')TJ]_a)}}$$

(32.19)

We thus see how the two-nucleon interaction is renormalized by the interaction with other configurations. If energy differences between the j^n configuration and the excited $j^{n-2}j'j''$ configurations are large in comparison to energy spacings within these configurations, the energy denominators in (32.19) may be taken to be independent of $\alpha_1 T_1 J_1$, and TJ. In that case, the second order corrections $\delta V_{T'J'}(j^2)$ are the same in all states of the j^n configuration. This energy difference may be also fairly independent of n as in cases where it is determined mostly by differences in single nucleon energies. In that case, the resulting effective interaction is the same in all j^n configurations. In nuclear many body theory it is shown how certain higher order corrections can also be considered as modifications or renormalizations of the nucleon-nucleon interaction.

In the special case of the pairing interaction, the two nucleon matrix elements vanish unless $T' = 1$, $J' = 0$ (and $j' = j''$). The seniority is then a good quantum number in the j^n configurations and the additional quantum numbers α include v and the reduced isospin t. The second order corrections to matrix elements are also proportional to the pairing interaction. In the case of maximum isospin the c.f.p. in (32.18) are given by (19.29). The second order correction is thus equal to

$$\frac{n(n-1)}{2}\sum_{j'}[j^{n-2}(v\alpha J)j^2(J'=0)J|\}j^n v\alpha J]^2$$

$$\times \frac{(2j+1)(2j'+1)G}{E(j^n v\alpha J)-E([j^{n-2}(v\alpha J)j'^2(J'=0)J]_a)}$$

$$=\frac{n-v}{2}(2j+3-n-v)\sum_{j'}$$

$$\times \frac{(2j'+1)G}{E(j^n v\alpha J)-E([j^{n-2}(v\alpha J)j'^2(J'=0)J]_a)}$$

which, for fairly constant energy denominators, is clearly proportional to the eigenvalues $P(n, v)$ in (19.11) of the pairing interaction.

A case of more general interest is where the interaction is not restricted to the pairing interaction but the only non-vanishing two-body matrix elements in (32.5) have $T' = 1$, $J' = 0$ (and $j' = j''$). In the j^n configuration of identical nucleons, or maximum isospin, seniority is a good quantum number as demonstrated in Section 21. The c.f.p. in (32.5) with $J' = 0$ are then given by (19.29) and are the same for all states with given n and seniority v. They are independent of J and any additional quantum number α. In the Hamiltonian sub-matrices (characterized by v, J and α), which include in addition to the j^n configuration some other $j^{n-2}j'^2$ configurations, the non-diagonal elements thus depend only on v (and n). According to (31.12) the contributions of the jj' interaction to diagonal elements in different sub-matrices are equal. Hence, when these sub-matrices are exactly diagonalized, energy differences between lowest eigenvalues and diagonal matrix elements $\langle j^n v\alpha J|V|j^n v\alpha J\rangle$ are equal for all αJ states with the same seniority v. As a result, the spacings between levels obtained by diagonalization of submatrices with the same seniority are equal to corresponding level spacings in the pure j^n configuration. This fact has been mentioned in Section 21.

The situation is more complicated if the two nucleons are excited from the j_1-orbit to a partially occupied j_2 orbit. The matrix elements between states of the $j_1^{n_1}j_2^{n_2}$ configuration and states in the $j^{n_1-2}j^{n_2+2}$ configuration are expressed by (32.10). They are equal, in general, to linear combinations of terms with different $T'J'$ values and their squares do not have the form (15.31). Hence, their contribution up to second order in perturbation theory cannot be considered as a two-body operator. There are, however, special cases, more general than the $n_2 = 0$ case, in which the second order corrections assume the form of a nucleon-nucleon interaction. Consider the case, for even value of n_1, in which $J_1' = 0$. This occurs if the j_1-orbit is completely filled but also in other cases. Let us simplify the discussion by restricting it to identical nucleons (or maximum isospin) and omit the isospin labels. In that case, there is no summation over J', it being equal to J_1. Since $J_1' = 0$ implies $J_2' = J$ we obtain the matrix element (32.10) in the form

$$\sqrt{\frac{n_1(n_1-1)}{2}}[j_1^{n_1-2}(\alpha_1 J')j_1^2(J')J_1' = 0|\}j_1^{n_1}\alpha_1' J_1' = 0]$$
$$\times\sqrt{\frac{(n_2+1)(n_2+2)}{2}}[j_2^{n_2}(\alpha_2' J)j_2^2(J')J_2|\}j_2^{n_2+2}\alpha_2 J_2]$$
$$\times\sqrt{\frac{2J_2+1}{(2J+1)(2J'+1)}}\langle j_2^2 J'|V|j_1^2 J'\rangle \tag{32.20}$$

The second order correction due to the matrix elements (32.20) to the various states with possible values of J_2, J' is equal to

$$\sum_{J'}\frac{n_1(n_1-1)}{2}[j_1^{n_1-2}(\alpha_1 J')j_1^2(J')J_1' = 0|\}j_1^{n_1}\alpha_1' J_1' = 0]^2$$
$$\times\frac{(n_2+1)(n_2+2)}{2(2J+1)(2J'+1)}$$
$$\times\sum_{J_2}\frac{(2J_2+1)[j_2^{n_2}(\alpha_2' J)j_2^2(J')J_2|\}j_2^{n_2+2}\alpha_2 J_2]^2\langle j_2^2 J'|V|j_1^2 J'\rangle^2}{E([j_1^{n_1}(\alpha_1' J_1' = 0)j_2^{n_2}(\alpha_2' J)J]_a) - E([j_1^{n_1-2}(\alpha_1 J')j_1^{n_2+2}(\alpha_2 J_2)J]_a)} \tag{32.21}$$

This expression is similar to (15.32) but the two body operator is defined for the state with $J_1' = 0$ in the $j_1^{n_1}$ configuration. If that is the only state of that configuration, as is the case for a closed j_1 orbit, $n_1 = 2j_1 + 1$, we see here a modification to the two-body interaction in that state. In the special case in which $n_2 = 0$, $J = 0$, $J_2 = J'$ the c.f.p. in (32.34) is equal to 1 and the result reduces to (32.18) obtained above. If there are j_2 nucleons present the results are modified. This is an effect of the valence nucleons on the energy of the closed shells. In some cases, however, this effect may be expressed as an effective interaction in the $j_2^{n_2}$ configuration.

Instead of a systematic study, we consider a special case in which the only non-vanishing two-body matrix element in (32.21) is the one with $J' = 0$. This is the case for the pairing interaction considered in Section 12. The c.f.p. in the seniority scheme for $n_1 = 2j_1 + 1$ and $n_2 + 2$ are given in this case by (19.29). If we take the energy difference in the denominator of (32.21) to be a fixed quantity ΔE, we obtain for the second order correction in this case the value

$$\boxed{\frac{\langle j_2^2 J' = 0|V|j_1^2 J' = 0\rangle^2}{\Delta E}\left\{1 - n_2\frac{2}{2j_2+1} + \frac{1}{2j_2+1}P(n_2,v)\right\}} \tag{32.22}$$

In (32.22), $P(n_2, v)$ is the eigenvalue (19.11) of the pairing interaction in states with seniority v in the $j_2^{n_2}$ configuration. We identify in (32.22) a term independent of n_2 which is present when $n_2 = 0$, a term linear in n_2 which is a modification of the single j_2 nucleon energy and the two-body pairing interaction in the $j_2^{n_2}$ configuration.

If the only non-vanishing two-body matrix elements in (32.10) are those with $J' = 0$, the non-diagonal elements are greatly simplified. For identical nucleons, or maximum isospins, the c.f.p. are given by (19.29). Hence, the matrix elements (32.10) in that case depend only on v_1 and v_2, the seniorities of the states in the $j_1^{n_1}$ and $j_2^{n_2}$ configurations. Let us consider the Hamiltonian sub-matrices which include the configurations $j_1^{n_1}, j_1^{n_1-2}j_2^2, j_1^{n_1-4}j_2^4, \ldots$. The seniority of the $J' = 0$ state of the j_2^2 configuration is $v_2 = 0$. Hence, the non-vanishing non-diagonal elements between the $j_1^{n_1-2}j_2^2$ and $j^{n_1-4}j_2^4$ configurations have also $J_2' = J_2 = 0$ and $v_2 = v_2' = 0$. Thus, the seniority v_1 of the state in the $j_1^{n_1}$ configuration uniquely specifies all states in any sub-matrix. All non-diagonal elements of the various sub-matrices characterized by $v_1\alpha_1J_1$ are the same for all α_1J_1 states with the same seniority v_1. The differences between corresponding diagonal elements of these sub-matrices are due only to differences of the interaction energies in the various α_1J_1 states of the $j_1^{n_1}$ configuration. Hence, when these sub-matrices are exactly diagonalized, spacings between levels with the same seniority v_1 are equal to those between corresponding levels of the pure $j_1^{n_1}$ configuration. A special case of this behavior was considered above and mentioned in Section 21.

Second order corrections due to other configurations like those expressed in (32.11) are also complicated. They may become simplified in special cases but we will not consider them here. Our aim is to show that in some cases second order corrections may be incorporated into the effective two-body interaction between nucleons. In other cases the situation is more complicated. These considerations show the complexity of second order (and certainly of higher order) corrections in the case of actual nuclei. Most perturbation expansions of the nuclear many body theory have been developed for nuclear matter. That nuclear species is unfortunately not available in terrestrial laboratories. Before proceeding to the other kind of configuration mixing, let us look at another simple case where second order corrections may be expressed as a modification of two-body interactions.

Let us consider the perturbation due to the matrix element (32.12). We would like to obtain the second order correction to the matrix element between the state

$$\psi([j_1^{n_1}(\alpha_1'T_1'J_1')j_2^{n_2}(\alpha_2'T_2'J_2')TJ]_a)$$

and the state

$$\psi([j_1^{n_1}(\alpha_1''T_1''J_1'')j_2^{n_2}(\alpha_2''T_2''J_2'')TJ]_a)$$

The wave functions of these two states, if not the same, may become non-orthogonal when perturbed. In that case we orthogonalize the second state to the first one. The second order correction to the matrix element $\langle 1|V|2\rangle$ due to admixtures of the state $|i\rangle$ to the states $|1\rangle$ and $|2\rangle$ is then given by $\langle 1|V|i\rangle\langle i|V|2\rangle/(E_1 - E_i)$. By multiplying the expressions (32.14) for the two states, dividing by the appropriate energy difference and summing over all states with various

$\alpha_1 T_1 J_1 \alpha_2 T_2 J_2 T_{12} J_{12} T' J'$ of the perturbing configuration we obtain for the second order correction the expression

$$
\begin{aligned}
& n_1 n_2 \sum_{\substack{\alpha_1 T_1 J_1 \\ \alpha_2 T_2 J_1}} [j_1^{n_1-1}(\alpha_1 T_1 J_1) j_1 T_1' J_1' |\} j_1^{n_1} \alpha_1' T_1' J_1'] \\
& \times [j_1^{n_1-1}(\alpha_1 T_1 J_1) j_1 T_1'' J_1'' |\} j_1^{n_1} \alpha_1'' T_1'' J_1''] \\
& \times [j^{n_2-1}(\alpha_2 T_2 J_2) j_2 T_2' J_2' |\} j_2^{n_2} \alpha_2' T_2' J_2'] \\
& \times [j_2^{n_2-1}(\alpha_2 T_2 J_2) j_2 T_2'' J_2'' |\} j_2^{n_2} \alpha_2'' T_2'' J_2''] \\
& \times [(2T_1'+1)(2T_1''+1)(2T_2'+1)(2T_2''+1) \\
& \qquad \times (2J_1'+1)(2J_1''+1)(2J_2'+1)(2J_2''+1)]^{1/2} \\
& \times \sum_{T'J'} (2T'+1)(2J'+1) \frac{\langle [j_1' j_2' T' J']_a | V | [j_1 j_2 T' J']_a \rangle^2}{\Delta E} \\
& \times \sum_{T_{12} J_{12}} (2T_{12}+1) \\
& \times (2J_{12}+1) \begin{Bmatrix} T_1 & \frac{1}{2} & T_1' \\ T_2 & \frac{1}{2} & T_2' \\ T_{12} & T' & T \end{Bmatrix} \begin{Bmatrix} T_1 & \frac{1}{2} & T_1'' \\ T_2 & \frac{1}{2} & T_2'' \\ T_{12} & T' & T \end{Bmatrix} \\
& \times \begin{Bmatrix} J_1 & j_1 & J_1' \\ J_2 & j_2 & J_2' \\ J_{12} & J' & J \end{Bmatrix} \begin{Bmatrix} J_1 & j_1 & J_1'' \\ J_2 & j_2 & J_2'' \\ J_{12} & J' & J \end{Bmatrix}
\end{aligned}
\tag{32.23}
$$

where ΔE is defined as

$$
\begin{aligned}
\Delta E = {} & E([j_1^{n_1}(\alpha_1' T_1' J_1') j_2^{n_2}(\alpha_2' T_2' J_2') T J]_a) \\
& - E([j_1^{n_1-1}(\alpha_1 T_1 J_1) j_2^{n_2-1}(\alpha_2 T_2 J_2)(T_{12} J_{12}) j_1' j_2' (T' J') T J]_a)
\end{aligned}
\tag{32.24}
$$

Comparing (32.23) with (31.14) we see that the second order corrections appear in this case as a modification of the interaction between j_1 nucleons and j_2 nucleons. If the energy difference in (32.24) is large compared to energy differences between states of each configuration, we may replace it by a constant $\Delta E_{j'_1 j'_2}$. The modification is then the same for all states of the $j_1^{n_1} j_2^{n_2}$ configuration. The sum of corrections due to all excitations to various $j'_1 j'_2$ orbits is then equal to

$$\boxed{\delta V_{T'J'}(j_1 j_2) = \sum_{j'_1 j'_2} \frac{\langle [j'_1 j'_2 T'J']_a | V | [j_1 j_2 T'J']_a \rangle^2}{\Delta E_{j'_1 j'_2}}} \qquad (32.25)$$

If the various $\Delta E_{j'_1 j'_2}$ are fairly independent of n_1 and n_2 then the effective two body interaction (32.25) can be taken to be the same in all $j_1^{n_1} j_2^{n_2}$ configurations. This is the case if $\Delta E_{j'_1 j'_2}$ is essentially determined by the differences of single nucleon energies between the $j_1 j_2$ orbits and the $j'_1 j'_2$ orbits. We shall not discuss the more complicated cases where the matrix elements are given by (32.15) and (32.16) for the reasons explained above.

Let us now turn to interactions between configurations which differ by the orbit of *one* nucleon. We first consider the simple matrix element

$$\langle [j^{n-1}(\alpha_1 T_1 J_1) j' TJ]_a | \sum_{i<h}^{n} V_{ih} | j^n \alpha TJ \rangle \qquad (32.26)$$

The antisymmetric state of the $j^{n-1} j'$ configuration is obtained from the wave function

$$\mathcal{A}\psi(j^{n-1}(\alpha_1 T_1 J_1) j'_n TJ)$$
$$= \psi(j^{n-1}(\alpha_1 T_1 J_1) j'_n TJ) - \sum_{i=1}^{n-1} \psi(j^{n-1}(\alpha_1 T_1 J_1) j'_i TJ) \qquad (32.27)$$

The antisymmetrizer $\mathcal{A}$ interchanges the n-th nucleon with the first $n-1$ nucleons. All wave functions on the r.h.s. of (32.27) are orthogonal (and normalized) and the normalization factor of (32.27) is thus equal to $n^{-1/2}$. Hence, the matrix element (32.26) can be expressed as

$$n^{-1/2}\int \mathcal{A}\psi^*(j^{n_1}(\alpha_1T_1J_1)j'_nTJ)\left(\sum_{i<h}^{n}V_{ih}\right)\psi(j^n\alpha TJ)$$

$$= n^{1/2}\int \psi^*(j^{n-1}(\alpha_1T_1J_1)j'_nTJ)\left(\sum_{i<h}^{n}V_{ih}\right)\psi(j^n\alpha TJ) \tag{32.28}$$

The equality in (32.28) was obtained by transforming the projection operator $\mathcal{A}/n$ from ψ^* to the rest of the integrand.

The only terms which contribute to the integral on the r.h.s. of (32.28) are those with $h = n$. For other V_{ih} terms ($i < h$), the integral vanishes upon integration over the variables of the n-th nucleon. Both ψ^* and ψ in (32.28) are fully antisymmetric in the first $n-1$ nucleons and hence all V_{in} terms contribution equally. We can calculate the contribution due to one of them, like $V_{n-1,n}$, and multiply the result by $n-1$.

To evaluate the integral we use $n-1 \to n-2$ c.f.p. for states of the j^{n-1} configuration and $n \to n-2$ c.f.p. in the j^n configuration thus obtaining for the matrix element the expression

$$\begin{aligned}(n-1)\sqrt{n}\sum_{\alpha_2T_2J_2T'J'}&[j^{n-2}(\alpha_2T_2J_2)jT_1J_1|\}j^{n-1}\alpha_1T_1J_1]\\ &\times[j^{n-2}(\alpha_2T_2J_2)j^2(T'J')TJ|\}j^n\alpha TJ]\\ &\times\int\psi^*(j^{n-2}(\alpha_2T_2J_2)j_{n-1}(T_1J_1)j'_nTJ)\\ &\times V_{n-1,n}\psi(j^{n-2}(\alpha_2T_2J_2)j^2_{n-1,n}(T'J')TJ)\end{aligned} \tag{32.29}$$

When the integration over the coordinates of nucleons 1 to $n-2$ is performed, the only non-vanishing contributions come from terms in which $\alpha_2T_2J_2$ have the same values in ψ^* and ψ. To carry out that integration we first make a change of coupling transformation (10.12) on the spins and isospins in ψ^*. We then obtain for the matrix element

(32.26) the final form

$$
\begin{aligned}
&(n-1)\sqrt{n} \sum_{\alpha_2 T_2 J_2 T' J'} [j^{n-2}(\alpha_2 T_2 J_2) j T_1 J_1 | \} j^{n-1} \alpha_1 T_1 J_1] \\
&\times [j^{n-2}(\alpha_2 T_2 J_2) j^2 (T' J') T J | \} j^n \alpha T J] \\
&\times (-1)^{T_2 + (1/2) + (1/2) + T + J_2 + j + j' + J} \\
&\times [(2T_1+1)(2T'+1)(2J_1+1)(2J'+1)]^{1/2} \\
&\times \begin{Bmatrix} T_2 & \frac{1}{2} & T_1 \\ \frac{1}{2} & T & T' \end{Bmatrix} \begin{Bmatrix} J_2 & j & J_1 \\ j' & J & J' \end{Bmatrix} \frac{1}{\sqrt{2}} \langle [j j' T' J']_a | V | j^2 T' J' \rangle
\end{aligned}
\tag{32.30}
$$

The factor $1/\sqrt{2}$ in (32.30) is due to taking the two nucleon matrix element between antisymmetric *and normalized* wave functions.

A simpler expression is obtained for the matrix element (32.26) if we substitute in it the tensor expansion of the two body interaction. Let us consider *one* term, with given k, in that expansion and insert it in (32.28). We expand then the j^n states in c.f.p. and obtain for the matrix element the expression

$$
\begin{aligned}
&\sqrt{n} \sum_{\alpha_1' J_1'} [j^{n-1}(\alpha_1' T_1 J_1') j T J | \} j^n \alpha T J] F_k(j j j j') \\
&\quad \times \int \psi^*(j^{n-1}(\alpha_1 T_1 J_1) j_n' T J) \\
&\quad \times \left(\left(\sum_{i=1}^{n-1} \mathbf{u}^{(k)}(i) \right) \cdot \mathbf{u}^{(k)}(n) \right) \psi(j^{n-1}(\alpha_1' T_1 J_1') j_n T J)
\end{aligned}
\tag{32.31}
$$

The coefficient $F_k(jjjj')$ in (32.31) contains the radial integrals so that the operator $\mathbf{u}^{(k)}$ acts only on the angular and spin variables. There is no summation over T_1' in (32.31) since the single nucleon tensor operators $\mathbf{u}^{(k)}$ are taken to be scalars in isospin. As explained in

Section 20, the tensor expansion of matrix elements (32.30), between states of the j^2 configuration and any two-nucleon configuration, is the same one for $T' = 1$ and $T' = 0$ states. The expansion in such cases is in tensors which operate only on space and spin variables.

There are certain similarities between the expression (32.31) and the matrix element of a single nucleon operator between states of the j^n and $j^{n-1}j'$ configurations. That matrix element (31.31) is of an *external* operator or field of rank k (like a component of the electromagnetic field). In (32.31), the n-th nucleon is moved from the j-orbit into the j'-orbit by the field created by the other $n-1$ nucleons. These $n-1$ nucleons are in the same j-orbit in both ψ^* and ψ but they may be coupled to different states.

Let us consider a special case of (32.31) in which only the $k = 0$ term is considered. In this case the value of j' must be equal to that of j. For a parity conserving nuclear interaction also the values of l' and l must be equal. The j'-orbit may be different from the j-orbit only by the radial quantum number n. The matrix element (32.31) assumes then the form

$$\sqrt{n}[j^{n-1}(\alpha_1 T_1 J_1) jTJ|\}j^n \alpha TJ] F_0(jjjj')$$
$$\times \langle j^{n-1}\alpha_1 T_1 J_1 | \sum_{i=1}^{n-1} u_0^{(0)}(i) | j^{n-1}\alpha_1 T_1 J_1\rangle \langle n'lj|u_0^{(0)}|nlj\rangle \tag{32.32}$$

In (32.32) the values of M_{T_1}, M_1 and m were omitted for the sake of conciseness. The values of α_1' and J_1' in (32.31) must be equal to those of α_1 and J_1 respectively, since a scalar operator cannot change the coupling of angular momenta. The matrix elements of $\sum_{i=1}^{n-1} T_0^{(0)}(i)$ are independent of $\alpha_1 T_1 J_1$ as well as of M_1. Thus, we define the coefficient

$$C_{n-1} = (n-1)F_0\langle jm|u_0^{(0)}|jm\rangle\langle n'lj|T_0^{(0)}|nlj\rangle \tag{32.33}$$

The expression (32.32) is indeed similar to (31.32) for $k = 0$. It can be viewed as the non-diagonal matrix element of a central potential (since $k = 0$) between the j' and j states of a nucleon due to the action of the other $n-1$ nucleons. This potential cannot change the spin of a nucleon but may well have non-vanishing matrix elements between states with different radial quantum numbers.

If the admixtures of the states $n^{-1/2}\mathcal{A}\psi(j^{n-1}(\alpha_1T_1J_1)j'_nTJ)$ into the state $\psi(j^n\alpha TJ)$ are small and may be taken as a first order perturbation, their amplitudes are given by

$$\sqrt{n}[j^{n-1}(\alpha_1T_1J_1)jTJ|\}j^n\alpha TJ]\frac{C_{n-1}}{\Delta E} \tag{32.34}$$

The energy denominator in (32.34) is equal to $\Delta E = E(j^n\alpha TJ) - E([j^{n-1}(\alpha_1T_1J_1)j'TJ]_a)$. If ΔE is large compared to energy differences of the states with TJ in the $j^{n-1}j'$ configuration it may be approximated by a constant. In that case, the first order change in $\psi(j^n\alpha TJ)$ due to states with the various $\alpha_1T_1J_1$ quantum numbers is given by

$$\sqrt{n}C_{n-1}\frac{1}{\Delta E}n^{-1/2} \times \mathcal{A}\sum_{\alpha_1T_1J_1}[j^{n-1}(\alpha_1T_1J_1)jTJ|\}j^n\alpha TJ]\psi(j^{n-1}(\alpha_1T_1J_1)j'_nTJ) \tag{32.35}$$

The antisymmetric wave function in (32.35) is simply related to $\psi(j^n\alpha TJ)$. Since $l' = l$, $j' = j$, the angular and spin parts of these wave functions are the same. The only difference must be limited to the radial parts.

To see it more clearly, let us introduce the following notation for single nucleon wave functions of the j, j' orbits

$$\psi_{nljmm_t} = \frac{1}{r}R_{nlj}\bar{\psi}_{ljmm_t} = R_n(r)\bar{\psi}_{ljmm_t} \tag{32.36}$$

The radial function $R_n(r)$ depends on l, but since l is the same in the n and n' orbits we omit it for brevity. With this notation we may write

$$\psi(j^n\alpha TJ) = R_n(1)\ldots R_n(n)\bar{\psi}(j^n\alpha TJ) \tag{32.37}$$

where $\bar{\psi}$ is constructed from the angular, spin and isospin parts of the wave functions (32.36). The sum over $\alpha_1T_1J_1$ in (32.35) can then be expressed as

$$\sum_{\alpha_1T_1J_1}[j^{n-1}(\alpha_1T_1J_1)jTJ|\}j^n\alpha TJ]\psi(j^{n-1}(\alpha_1T_1J_1)j'_nTJ) = R_n(1)\ldots R_{n'}(n)\bar{\psi}(j^n\alpha TJ) \tag{32.38}$$

The function $\tilde{\psi}(j^n\alpha TJ)$ is fully antisymmetric in the angular coordinates, spin and isospin variables of the n nucleons. Hence, when the antisymmetrizer $\mathcal{A}$ is applied to (32.38) it must yield a function which is *symmetric* in the radial coordinates of the nucleons, namely

$$\left(\sum_P R_n(i_1)\cdots R_n(i_{n-1})R_{n'}(i_n)\right)\tilde{\psi}(j^n\alpha TJ) \qquad (32.39)$$

The summation in (32.39) is over all n permutations P, each of which substitutes for i_n a different number from 1 to n.

The first order change in $\psi(j^n\alpha TJ)$ given by (32.35) is thus equal to

$$\frac{C_{n-1}}{\Delta E}\left(\sum_P R_n(i_1)\cdots R_n(i_{n-1})R_{n'}(i_n)\right)\tilde{\psi}(j^n\alpha TJ) \qquad (32.40)$$

To first order in perturbation theory this change is equivalent to a modification of the radial parts of the single nucleon wave functions in the j-orbits given by

$$R'_n(r) = R_n(r) + \frac{C_{n-1}}{\Delta E}R_{n'}(r) \qquad (32.41)$$

Substituting the modified radial function (32.41) in the original function (32.37) we obtain

$$R'_N(1)\ldots R'_N(n)\tilde{\psi}(j^n\alpha TJ) = R_n(1)\cdots R_n(n)\tilde{\psi}(j^n\alpha TJ)$$
$$+\frac{C_{n-1}}{\Delta E}\left(\sum_P R_n(i_1)\cdots R_n(i_{n-1})R_{n'}(i_n)\right)\tilde{\psi}(j^n\alpha TJ)+\cdots \qquad (32.42)$$

The first order term in $C_{n-1}/\Delta E$ in (32.42) is indeed equal to the result (32.40) obtained above. If ΔE can be taken to be the same for all states with various αTJ then the modification (32.41) holds for all states of the j^n configuration.

Effects of configuration mixing of this kind may be eliminated if the central potential is modified to make (32.41) the correct radial part of the nucleons in the j-orbit. If the central field in which nucleons move is determined self-consistently, as in the Hartree-Fock approximation, first order corrections to $R_n(r)$ vanish. The self-consistent potential must depend on n as demonstrated by the dependence of C_{n-1} on n given by (32.33).

There is a more general case of (32.31) in which j' is equal to j and also $l' = l$. This is the case in which $J = 0$, k need not vanish and $J_1' = J_1 = j$. Using (10.27) we can express (32.31) in the form

$$\sqrt{n}\sum_{\alpha_1'J_1'}[j^{n-1}(\alpha_1'T_1J_1')jTJ|\}j^n\alpha TJ]F_k(jjjj')(-1)^{j'+J+J_1'}\begin{Bmatrix} J_1 & j' & J \\ j & J_1' & k \end{Bmatrix}$$

$$\times\left(j^{n-1}\alpha_1T_1J_1\left\|\sum_{i=1}^{n-1}\mathbf{u}^{(k)}(i)\right\|j^{n-1}\alpha_1'T_1J_1'\right)(j'\|\mathbf{u}^{(k)}(n)\|j) \quad (32.43)$$

In the case $J = 0$, (32.43) simplifies into

$$\sqrt{n}\sum_{\alpha_1'}[j^{n-1}(\alpha_1'T_1J_1' = j)jTJ = 0|\}j^n\alpha TJ = 0]$$

$$\times\frac{1}{2j+1}(-1)^kF_k(jjjj')$$

$$\times\left(j^{n-1}\alpha_1T_1J_1 = j\left\|\sum_{i=1}^{n-1}\mathbf{u}^{(k)}(i)\right\|j^{n-1}\alpha_1'T_1J_1' = j\right)$$

$$\times(n'lj\|\mathbf{u}^{(k)}(n)\|nlj) \quad (32.44)$$

This expression is very similar to (32.32) and under the conditions of perturbation theory leads to the same results. Thus, for $J = 0$, the first order perturbation theory leads to a modification of the radial parts of single nucleon wave functions. This holds in particular for closed shells.

In the general case, for admixtures obtained from terms with $k \neq 0$ in (32.31) it is no longer necessary for j' to be equal to j. Excitations of a single nucleon from the j-orbit to the j'-orbit, with $j' \neq j$ cannot be attributed to a change of the *central* potential well. If we try to build many nucleon wave functions modified by the interaction, from single nucleon wave functions in a given orbit, the latter involve admixtures of states with different values of j. Such single nucleon wave functions may be obtained by the solution of the Schrödinger equation in a *deformed* potential well. Many nucleon states, constructed from such single nucleon states which do not have a definite value of j, will in general not have a definite value of the total spin J. States with well defined angular momenta must be projected from such many nucleon states since only these may be exact eigenstates of the rotationally invariant shell model Hamiltonian. Starting with the state $\psi(j^n\alpha TJ)$ and constructing a self consistent potential to avoid single nucleon excitations, the resulting potential must be deformed. If a deformed potential is established, the lowest state of the many nucleon configuration is an *intrinsic* state which does not have a definite angular momentum. If states with definite values of J projected from it are approximately equal to eigenstates of the shell model Hamiltonian, the intrinsic state has a simple physical meaning. The projected states may be considered as due to adiabatic rotations of a deformed shape. The internal motion of nucleons is then decoupled to a large extent from the relatively slow rotations. The dynamical variables associated with the latter may be taken to be the Euler angles which define the orientation of the shape in space (Kurath and Picman 1959, Redlich 1958).

Matrix elements which lead to single nucleon excitations are thus seen to lead to the breakdown of the spherical shell model. We may consider, in particular, excitations from the $j = l + \frac{1}{2}$ orbit to the higher $j' = l - \frac{1}{2}$ orbit. The matrix elements of the two nucleon configurations which appear in (32.30) are linear combinations of some diagonal matrix elements of the central interaction in LS-coupling, $\langle l^2TSL|V|l^2TSL\rangle$. Therefore, the interaction between the j^n configuration and the $j^{n-1}j'$ configuration may be rather large in this case. Configurations in which *two* nucleons are excited from the $j = l + \frac{1}{2}$ orbit into the $j' = l - \frac{1}{2}$ orbit are expected to lie considerably higher due to the strong spin-orbit interaction. Their admixtures are expected to be less important and may be incorporated into the effective two body interaction. There is a rather strong dependence on the isospin T of these effects which was mentioned at the end of Section 11.

A more complicated matrix-element is between states of the $j_1^{n_1}j_2^{n_2}$ configuration and the $j_1^{n_1-1}j_2^{n_2+1}$ configuration. The special case with $n_2 = 0$ is (32.26). Use of antisymmetrizers like those employed in deriving (32.10), yields the following result for the k-th term in the tensor expansion of the interaction

$$F_k(jjjj')\langle[j_1^{n_1}(\alpha_1T_1J_1)j_2^{n_2}(\alpha_2T_2J_2)TJ]_a|$$

$$\sum_{i<h}^{n_1+n_2}(\mathbf{u}^{(k)}(i)\cdot\mathbf{u}^{(k)}(h))|[j_1^{n_1-1}(\alpha_1'T_1'J_1')j_2^{n_2+1}(\alpha_2'T_2'J_2')TJ]_a\rangle$$

$$= \sqrt{n_1(n_2+1)}(-1)^{n_2}$$

$$\times \sum_{\alpha_1''T_1''J_1''\alpha_2''T_2''J_2''} [j_1^{n_1-1}(\alpha_1''T_1''J_1'')j_1T_1J_1|\}j_1^{n_1}\alpha_1T_1J_1]$$

$$\times [j_2^{n_2}(\alpha_2T_2J_2)j_2T_2'J_2'|\}j_2^{n_2+1}\alpha_2'T_2'J_2']F_k(j_1j_2j_2j_2)$$

$$\times \int \psi^*(j_1^{n_1-1}(\alpha_1''T_1''J_1'')j_{1;n_1+n_2}(T_1J_1)j_2^{n_2}(\alpha_2T_2J_2)TJ)$$

$$\times \sum_{i=1}^{n_1+n_2-1}(\mathbf{u}^{(k)}(i)\cdot\mathbf{u}^{(k)}(n_1+n_2))$$

$$\times \psi(j_1^{n_1-1}(\alpha_1'T_1'J_1')j_2^{n_2}(\alpha_2''T_2''J_2'')j_{2;n_1+n_2}(T_2'J_2')TJ) \qquad (32.45)$$

To evaluate the matrix element given by the integral on the r.h.s. of (32.45) we use the change of coupling transformations (10.11) and (10.12). In the result, the various $6j$-symbols are multiplied by

$$\left(j_1^{n_1-1}(\alpha_1''T_1''J_1'')j_2^{n_2}(\alpha_2T_2J_2)T_0J_0\right.$$

$$\left.\left\|\sum_{i=1}^{n_1+n_2-1}\mathbf{u}^{(k)}(i)\right\| j_1^{n_1-1}(\alpha_1'T_1'J_1')j_2^{n_2}(\alpha_2''T_2''J_2'')T_0'J_0'\right)$$

$$\times (j_1\|\mathbf{u}^{(k)}\|j_2) \qquad (32.46)$$

From (32.46) we see that the change of a nucleon from the j_1-orbit to the j_2-orbit is due to its interaction with either the other n_1-1 nucleons in the j_1-orbit or with the n_2 j_2-nucleons. The reduced matrix element in (32.46) of the $j_1^{n_1-1}j_2^{n_2}$ configuration has been evaluated in (31.28). As explained in the discussion of (32.31), the non-vanishing matrix elements in (32.46) are those with $T_1''=T_1'$, $T_2''=T_2$ and $T_0'=T_0$.

The change of one nucleon from the j_1 orbit into the j_2-orbit may be due to the interaction of that nucleon with nucleons in another j_3-

orbit. The matrix elements to be considered are between states of the $j_1^{n_1} j_2^{n_2} j_3^{n_3}$ configuration and the $j_1^{n_1-1} j_2^{n_2+1} j_3^{n_3}$ configuration. The $j_3^{n_3}$ configuration is the same on both sides of the matrix element but its state may be changed by the interaction. If only the j_1 and j_2 nucleons take part in the interaction, the matrix elements are given by (32.45). We will write down the expression for that part of the matrix elements in which the j_3-nucleons participate. As above, we consider a term with given k in the tensor expansion of the interaction. We calculate its matrix element between the state

$$|[j_1^{n_1}(\alpha_1 T_1 J_1) j_2^{n_2}(\alpha_2 T_2 J_2) T_{12} J_{12} j_3^{n_3}(\alpha_3 T_3 J_3) TJ]_a\rangle$$

and the state

$$|[j_1^{n_1-1}(\alpha_1' T_1' J_1') j_2^{n_2+1}(\alpha_2' T_2' J_2') T_{12}' J_{12}' j_3^{n_3}(\alpha_3' T_3' J_3') TJ]_a\rangle$$

Using the methods explained above we obtain for the relevant part of the matrix element the expression

$$\begin{aligned}
&(-1)^{n_2}\sqrt{n_1(n_2+1)}\sum [j^{n_1-1}(\alpha_1' T_1' J_1') j_1 T_1 J_1 |\} j_1^{n_1} \alpha_1 T_1 J_1] \\
&\quad \times [j_2^{n_2}(\alpha_2 T_2 J_2) j_2 T_2' J_2' |\} j_2^{n_2+1} \alpha_2' T_2' J_2'] \\
&\quad \times \int \psi^*(j_1^{n_1-1}(\alpha_1' T_1' J_1') j_{1;n_1+n_2}(T_1 J_1) j_2^{n_2}(\alpha_2 T_2 J_2) T_{12} J_{12} \\
&\qquad j_3^{n_3}(\alpha_3 T_3 J_3) TJ) \left(\sum_{i=n_1+n_2+1}^{n_1+n_2+n_3} V_{n_1+n_2,i} \right) \\
&\quad \times \psi(j_1^{n_1-1}(\alpha_1' T_1' J_1') j_2^{n_2}(\alpha_2 T_2 J_2) j_{2;n_1+n_2}(T_2' J_2') T_{12}' J_{12}' j_3^{n_3}(\alpha_3' T_3' J_3') TJ)
\end{aligned} \tag{32.47}$$

The sum over i in (32.47) is only over nucleons in the j_3-orbit.

To further evaluate (32.47) we perform on ψ^* and ψ change of coupling transformations. We use (10.11) to recouple in ψ^*, J_1' and J_2 to some J_{12}'' and couple it to j_1 to obtain J_{12}. In ψ we use (10.12) to recouple J_1' and J_2 to the same J_{12}'' and couple it to j_2 to yield J_{12}'. Similar transformations are carried out on the isospins. We then pick from the tensor expansion of the interaction the term with given k value. The two-nucleon matrix elements which appear in (32.47) are $\langle j_1 j_3 T' J' | V | j_2 j_3 T' J' \rangle$. As explained in Section 12, the tensor expansion of such matrix elements is different for $T' = 1$ and $T' = 0$ states. Unlike the situation in (32.31) and (32.45), the expansion can no longer include only space and spin tensor operators. As explained in Section 26, a charge independent interaction may be expressed without isospin operators. Still, for the tensor expansion, isospin operators must be explicitly used. The coefficients $F_k(j_1j_3j_2j_3)$ for $T' = 1$ and $T' = 0$ interactions may be denoted by $F_{T'k}(j_1j_3j_2j_3)$ where T' may be 1 or 0. We may express the two-nucleon interaction by

$$\tfrac{1}{2}(1 + P_{12}^{\tau})\sum_k F_{1k}(j_1j_3j_2j_3)(\mathbf{u}^{(k)}(1)\cdot\mathbf{u}^{(k)}(2)) + \tfrac{1}{2}(1 - P_{12}^{\tau})\sum_k F_{0k}(j_1j_3j_2j_3)(\mathbf{u}^{(k)}(1)\cdot\mathbf{u}^{(k)}(2))$$

Using the expression of P^{τ} from (26.1) we obtain the expression

$$\sum_k \tfrac{1}{4}(3F_{1k} + F_{0k})(\mathbf{u}^{(k)}(1)\cdot\mathbf{u}^{(k)}(2)) + \sum_k \tfrac{1}{4}(F_{1k} - F_{0k})(\boldsymbol{\tau}_1\cdot\boldsymbol{\tau}_2)(\mathbf{u}^{(k)}(1)\cdot\mathbf{u}^{(k)}(2))$$

We now define the scalar products of double tensors

$$(\mathbf{T}^{(k'k)}(1)\cdot\mathbf{T}^{(k'k)}(2)) \quad \text{by} \quad (\mathbf{u}^{(k)}(1)\cdot\mathbf{u}^{(k)}(2)) \quad \text{for} \quad k' = 0$$

and by

$$(\boldsymbol{\tau}_1\cdot\boldsymbol{\tau}_2)(\mathbf{u}^{(k)}(1)\cdot\mathbf{u}^{(k)}(2)) \qquad \text{for} \quad k' = 1$$

and their coefficients $F_k^{(k')}$ by $F_k^{(0)} = \frac{1}{4}(3F_{1k} + F_{0k})$ and $F_k^{(1)} = \frac{1}{4}(F_{1k} - F_{0k})$. We then obtain the integral in (32.47) as the linear combination

$$\langle j_1^{n_1-1}(\alpha_1'T_1'J_1')j_2^{n_2}(\alpha_2T_2J_2)(T_{12}''J_{12}'')j_{1;n_1+n_2}T_{12}J_{12}j_3^{n_3}(\alpha_3T_3J_3)TJ| \sum_{k'k} F_k^{(k')}(j_1j_3j_2j_3)\mathbf{T}^{(k'k)}(n_1+n_2)\cdot\left(\sum_{i=n_1+n_2+1}^{n_1+n_2+n_3}\mathbf{T}^{(k'k)}(i)\right) |j_1^{n_1-1}(\alpha_1'T_1'J_1')j_2^{n_2}(\alpha_2T_2J_2)(T_{12}''J_{12}'')j_{2;n_1+n_2}T_{12}'J_{12}'j_3^{n_3}(\alpha_3'T_3'J_3')TJ\rangle \tag{32.48}$$

The matrix elements (32.48) may now be evaluated by using (10.12) and (10.27). We will not carry out this calculation but will consider an important special case.

To simplify the situation let us consider n identical nucleons in the j-orbit outside closed shells. We will calculate the effect of their interactions with the core nucleons to first order in perturbation theory. We will not use the isospin formalism and treat separately protons and neutrons. The matrix elements to be evaluated are expressed by (32.47) in a very special case. There are $n_3 = n$ nucleons in the $j_3 = j$-orbit, none in any unoccupied j_2-orbit, $n_2 = 0$, and the j_1-orbit is completely filled, $n_1 = 2j_1 + 1$. In this case the spin of the core J_c in the perturbed state is $J_c = J_{12} = 0$ and in the perturbing excited state is $J_c = J'_{12} = k$. Also $J_1 = 0$, $J_3 = J$ and $J_2 = 0$ and no recoupling transformations are necessary. Let us denote J'_3 by J_1 and α'_3 by α_1. The non-vanishing c.f.p. in (32.47) are all equal to 1 and in view of (32.48) we obtain using (10.27) the matrix element in this case to be

$$F_k(j_1 j j_2 j)(-1)^{J+J+k} \begin{Bmatrix} 0 & J & J \\ J_1 & k & k \end{Bmatrix}$$

$$\times (j_1^{2j_1+1} 0 \| \sum_i \mathbf{u}^{(k)}(i) \| j_1^{2j_1}(j_1) j_2 k)$$

$$\times (j^n \alpha J \| \sum_i \mathbf{u}^{(k)}(i) \| j^n \alpha_1 J_1)$$

$$= (-1)^{J-J_1} \frac{1}{\sqrt{(2J+1)(2k+1)}} F_k(j_1 j j_2 j)$$

$$\times (J_c = 0 \| \sum_i \mathbf{u}^{(k)}(i) \| j_1^{-1} j_2 J_c = k)$$

$$\times (j^n \alpha J \| \sum_i \mathbf{u}^{(k)}(i) \| j^n \alpha_1 J_1) \qquad (32.49)$$

The first order corrections to the state $|J_c = 0, \alpha J J M\rangle$, due to the various perturbations given by (32.49) which are due to all closed j_1-

orbits, are thus equal to

$$\sum_{j_1,j_2 \neq j\alpha_1 J_1} (-1)^{J-J_1} \frac{1}{\sqrt{(2J+1)(2k+1)}} F_k(j_1 j j_2 j)$$

$$\times (J_c = 0 \| \sum_i \mathbf{u}^{(k)}(i) \| j_1^{-1} j_2 J_c = k)$$

$$\times \frac{(j^n \alpha J \| \sum_i \mathbf{u}^{(k)}(i) \| j^n \alpha_1 J_1) | j_1^{-1} j_2 J_c = k, \alpha_1 J_1 J M\rangle}{\Delta E}$$

$$= \sum_{j_1 j_2 \neq j\alpha_1 J_1} \beta_k(j_1^{-1} j_2 \alpha J \alpha_1 J_1 J) | j_1^{-1} j_2 J_c = k, \alpha_1 J_1 J M\rangle \tag{32.50}$$

Each energy denominator ΔE in (32.50) is equal to the difference between the perturbed state and the perturbing state. It depends on the particular $j_1^{-1} j_2$ configuration of the core but also on the state with $\alpha_1 J_1$ in the j^n configuration. As on previous occasions, we assume that ΔE is large compared to energy differences in the j^n configuration both in the perturbed and the perturbing state. We shall therefore take it to be a constant independent of $\alpha_1 J_1$ as well as of αJ. The corrections (32.50) are due to *polarization* of the core by the valence j-nucleons.

We shall now see that the admixtures (32.50) may have an important effect on matrix elements of single nucleon operators. Let us calculate the reduced matrix element of a tensor operator $\sum \mathbf{F}^{(k)}(i)$ between two states which include the first order corrections (32.50). These states are

$$|\alpha J M\rangle = |J_c = 0, \alpha J J M\rangle$$

$$+ \sum_{j_1 j_2 \neq j\alpha_1 J_1} \beta_k(j_1^{-1} j_2 \alpha J \alpha_1 J_1 J) | j_1^{-1} j_2 J_c = k, \alpha_1 J_1 J M\rangle \tag{32.51}$$

$$|\alpha' J' M'\rangle = |J_c = 0, \alpha' J' J' M'\rangle + \sum_{j_1 j_2 \neq j \alpha_1' J_1'} \beta_k(j_1^{-1} j_2 \alpha' J' \alpha_1' J_1' J') |j_1^{-1} j_2 J_c = k, \alpha_1' J_1' J' M'\rangle \tag{32.52}$$

The reduced matrix element of $\sum \mathbf{F}^{(k)}(i)$ between the states (32.51) and (32.52) up to first order is given by

$$\begin{aligned}
&(\alpha J \| \sum_i \mathbf{F}^{(k)}(i) \| \alpha' J') \\
&= (J_c = 0 j^n \alpha J J \| \sum_i \mathbf{F}^{(k)}(i) \| J_c = 0 j^n \alpha' J' J') \\
&+ \sum_{j_1 j_2 \neq j \alpha_1' J_1'} \beta_k(j_1^{-1} j_2 \alpha' J' \alpha_1' J_1' J') \\
&\times (J_c = 0 j^n \alpha J J \| \sum_i \mathbf{F}^{(k)}(i) \| j_1^{-1} j_2 J_c = k j^n \alpha_1' J_1' J') \\
&+ \sum_{j_1 j_2 \neq j \alpha_1 J_1} \beta_k(j_1^{-1} j_2 \alpha J \alpha_1 J_1 J) \\
&\times (j_1^{-1} j_2 J_c = k j^n \alpha_1 J_1 J \| \sum_i \mathbf{F}^{(k)}(i) \| J_c = 0 j^n \alpha' J' J')
\end{aligned} \tag{32.53}$$

The summations over i in (32.53) are over all nucleons, both in the core and the valence j-nucleons but not all of them contribute to the various terms. In the first term on the r.h.s. of (32.53), the core nucleons cannot contribute (for $k \neq 0$) whereas in the other terms, only the core nucleons make non-vanishing contributions. Due to this fact, in the summation over $\alpha_1' J_1'$ only the terms with $\alpha_1' \equiv \alpha$, $J_1' = J$ may contribute and in the other sum only the terms with $\alpha_1 \equiv \alpha'$, $J_1 = J'$ may have non-vanishing contributions. Making use of (10.42) we now obtain for (32.53) the form

$$(\alpha J\|\sum \mathbf{F}^{(k)}(i)\|\alpha' J') = (j^n\alpha J\|\sum \mathbf{F}^{(k)}(i)\|j^n\alpha' J')$$

$$+\sum_{j_1j_2\neq j} \beta_k(j_1^{-1}j_2\alpha' J'\alpha J J')(-1)^{J+J'+k}\sqrt{(2J+1)(2J'+1)}$$

$$\times (J_c=0\|\sum \mathbf{F}^{(k)}(i)\|j_1^{-1}j_2J_c=k)\begin{Bmatrix} 0 & J & J \\ J' & k & k \end{Bmatrix}$$

$$+\sum_{j_1j_2\neq j} \beta_k(j_1^{-1}j_2\alpha J\alpha' J' J)(-1)^{k+J'+J'+k}\sqrt{(2J+1)(2J'+1)}$$

$$\times (j_1^{-1}j_2J_c=k\|\mathbf{F}^{(k)}(i)\|J_c=0)\begin{Bmatrix} k & J & J' \\ J' & 0 & k \end{Bmatrix} \quad (32.54)$$

Substituting in (32.54) the values of the β-coefficients from (32.50) and the values (10.19) of the $6j$-symbols we obtain for it the expression

$$(j^n\alpha J\|\mathbf{F}^{(k)}(i)\|j^n\alpha' J')$$

$$+\sum_{j_1j_2\neq j}\frac{1}{2k+1}(J_c=0\|\sum \mathbf{u}^{(k)}(i)\|j_1^{-1}j_2J_c=k)$$

$$\times (J_c=0\|\sum \mathbf{F}^{(k)}(i)\|j_1^{-1}j_2J_c=k)$$

$$\times \frac{1}{\Delta E}(-1)^{J'-J}(j^n\alpha' J'\|\sum \mathbf{u}^{(k)}(i)\|j^n\alpha J)F_k(j_1jj_2j)$$

$$+\sum_{j_1j_2\neq j}\frac{1}{2k+1}(J_c=0\|\sum \mathbf{u}^{(k)}(i)\|j_1^{-1}j_2J_c=k)$$

$$\times (j_1^{-1}j_2J_c=k\|\mathbf{F}^{(k)}(i)\|J_c=0)$$

$$\times \frac{1}{\Delta E}(-1)^{k}(j^n\alpha J\|\sum \mathbf{u}^{(k)}(i)\|j^n\alpha' J')F_k(j_1jj_2j) \quad (32.55)$$

The expression (32.55) can be further evaluated. We recall that due to the Wigner-Eckart theorem (8.4), the matrix elements, within the j^n configuration, of $\sum \mathbf{u}^{(k)}(i)$ and $\sum \mathbf{F}^{(k)}(i)$ are proportional. If the single nucleon matrix elements do not vanish, we can write

$$
\begin{aligned}
&(j^n \alpha J \| \sum \mathbf{u}^{(k)}(i) \| j^n \alpha' J') \\
&\quad = \frac{(j \| \mathbf{u}^{(k)} \| j)}{(j \| \mathbf{F}^{(k)} \| j)} (j^n \alpha J \| \sum \mathbf{F}^{(k)}(i) \| j^n \alpha' J') \\
&\quad = f_j^{(k)} (j^n \alpha J \| \sum \mathbf{F}^{(k)}(i) \| j^n \alpha' J')
\end{aligned}
\tag{32.56}
$$

Similarly, we can write

$$
\begin{aligned}
(J_c = 0 \| \sum \mathbf{u}^{(k)}(i) \| j_1^{-1} j_2 J_c = k) &= \frac{(j_1 \| \mathbf{u}^{(k)} \| j_2)}{(j_1 \| \mathbf{F}^{(k)} \| j_2)} \\
&\quad \times (J_c = 0 \| \sum \mathbf{F}^{(k)}(i) \| j_1^{-1} j_2 J_c = k) \\
&= f_{j_1 j_2}^{(k)} (J_c = 0 \| \sum \mathbf{F}^{(k)} \| j_1^{-1} j_2 J_c = k)
\end{aligned}
\tag{32.57}
$$

Using this notation, we can express (32.55) as

$$
\begin{aligned}
&(j^n \alpha J \| \sum \mathbf{F}^{(k)}(i) \| j^n \alpha' J') \\
&\quad \times \left[1 + 2 \frac{F_k(j_1 j j_2 j)}{2k+1} \sum_{j_1 j_2 \neq j} \frac{f_j^{(k)}}{f_{j_1 j_2}^{(k)}} \frac{1}{\Delta E} \right. \\
&\qquad \left. \times (J_c = 0 \| \mathbf{u}^{(k)}(i) \| j_1^{-1} j_2 J_c = k)^2 \right]
\end{aligned}
\tag{32.58}
$$

Thus, up to first order, the corrections due to polarization of the core to matrix elements of single nucleon operators are *proportional* to the matrix elements in the j^n configuration. The total polarization of the core due to the n valence nucleons is the sum of polarizations of the individual nucleons. If $\mathbf{F}^{(k)}$ is the operator of electromagnetic transition Ek, the proportionality in (32.58) may be expressed as a renormalization of the charge into an *effective charge*. It should be kept in mind that the effective charge may well be different for different multipole operators. It may depend on k as well as on the valence j-orbit. To the extent that ΔE is independent of n, the normalization factor is also independent of n. We shall discuss below the corrections (32.58) in various cases but before that, let us look at the polarization of the core by a single nucleon and its effect on single nucleon transitions.

The state of a single j-nucleon outside closed shells is modified by the core polarizations considered above. There are, however, first order corrections to that state (and also to states of several valence nucleons) which are due to mixing with configurations in which *two* nucleons are excited. A core nucleon in the j_1-orbit may move to an unoccupied j_2 orbit and the j-nucleon may move into another valence j'' orbit. Such matrix elements were defined in (32.12) but the order of couplings there is different from the one needed here. Instead, we may start from (32.48) where we replace the state $j_3^{n_3}(\alpha_3T_3J_3)$ by the state of a single j-nucleon and the $j_3^{n_3}(\alpha_3'T_3'J_3')$ state by the state of a nucleon in the j''-orbit. We then follow the procedure leading to (32.49) and (32.50) and obtain the first order corrections to the state $|J_c = 0jm\rangle$ to be given by

$$\sum_{j_1,j_2\neq j,j''} \frac{(-1)^{j-j''}}{\sqrt{(2j+1)(2k+1)}}(J_c = 0\|\sum_i \mathbf{u}^{(k)}(i)\|j_1^{-1}j_2J_c = k)$$
$$\times (j\|\mathbf{u}^{(k)}\|j'')F_k(j_1jj_2j'')$$
$$\times \frac{1}{\Delta E}|j_1^{-1}j_2J_c = k, j''J = j, M = m\rangle$$
$$= \sum_{j_1j_2\neq jj''} \beta_k(j_1^{-1}j_2jj''J = j)|j_1^{-1}j_2J_c = k, j''J = j, M = m\rangle \qquad (32.59)$$

Let us now calculate the reduced matrix element of a tensor operator $\sum \mathbf{F}^{(k)}(i)$ between two "single nucleon states" (like (32.59)). Each state is that of a single valence nucleon including the first order corrections discussed above. The two states are

$$|J_c = 0\,jm\rangle + \sum_{j_1j_2 \neq jj''} \beta_k(j_1^{-1}j_2jj''J = j) \times |j_1^{-1}j_2J_c = k, j''J = j, M = m\rangle \tag{32.60}$$

$$|J_c = 0\,j'm'\rangle + \sum_{j_1j_2 \neq j'j'''} \beta_k(j_1^{-1}j_2j'j'''J = j') \times |j_1^{-1}j_2J_c = k, j'''J = j', M = m'\rangle \tag{32.61}$$

The reduced matrix element of $\sum \mathbf{F}^{(k)}(i)$ between these states, up to first order in perturbation theory is given by

$$\begin{aligned}&(J_c = 0j\|\sum \mathbf{F}^{(k)}(i)\|J_c = 0j') + \sum_{j_1j_2 \neq j'j'''} \beta_k(j_1^{-1}j_2j'j'''J = j')\\&\times (J_c = 0j\|\sum \mathbf{F}^{(k)}(i)\|j_1^{-1}j_2J_c = kj'''J = j')\\&+ \sum_{jj_2 \neq j} \beta_k(j_1^{-1}j_2jj''J = j)\\&\times (j_1^{-1}j_2J_c = k, j''J = j\|\sum \mathbf{F}^{(k)}(i)\|J_c = 0\,j')\end{aligned} \tag{32.62}$$

As in (32.53) above, for $k \neq 0$, only the valence nucleon contributes to the first term in (32.62). In the other terms, only the core nucleons make a non-vanishing contribution. Hence, the only terms which should be kept in (32.62) are those in which $j''' = j$ and $j'' = j'$. Making use of (10.42) we obtain for (32.62) the form

$$\begin{aligned}&(j\|\mathbf{F}^{(k)}\|j') + \sum_{j_1j_2 \neq j'} \beta_k(j_1^{-1}j_2j'jJ = j')(-1)^{j+j'+k}\\&\times \sqrt{(2j+1)(2j'+1)}(J_c = 0\|\sum \mathbf{F}^{(k)}(i)\|j_1^{-1}j_2J_c = k)\\&\times \begin{Bmatrix} 0 & j & j \\ j' & k & k \end{Bmatrix} + \sum_{j_1j_2 \neq j} \beta_k(j_1^{-1}j_2jj'J = j)(-1)^{k+j'+j'+k}\\&\times \sqrt{(2j+1)(2j'+1)}(j_1^{-1}j_2J_c = k\|\sum \mathbf{F}^{(k)}(i)\|J_c = 0) \begin{Bmatrix} k & j & j' \\ j' & 0 & k \end{Bmatrix}\end{aligned} \tag{32.63}$$

From (32.63) we derive in the same way that (32.58) was derived from (32.54), using the definition of $f^{(k)}_{j_1 j_2}$ given in (32.57), the following expression for the reduced matrix element (32.62)

$$(j\|\mathbf{F}^{(k)}\|j')\left[1+F_k(j_1 j j_2 j')\frac{2f^{(k)}_{jj'}}{2k+1}\sum_{j_1 j_2\neq j,j'}\frac{1}{f^{(k)}_{j_1 j_2}}\frac{1}{\Delta E}\right. \\ \left.\times (J_c=0\|\sum \mathbf{u}^{(k)}(i)\|j_1^{-1}j_2 J_c=k)^2\right] \tag{32.64}$$

Also in this case, the single nucleon matrix element is renormalized into an effective one. The renormalization factor added to 1 in (32.64) is proportional to $f^{(k)}_{jj'}$ whereas in (32.58) it is proportional to $f^{(k)}_j$.

The results (32.58) and (32.64) are special cases of a more general feature. The matrix element of a single nucleon operator between any two states of valence configurations is renormalized in the same way by the first order polarization of the core. If the two configurations differ by the occupation numbers of the j and j' orbit, the renormalization factor is given by (32.64). If the matrix element is between two states of the j^n configuration, the renormalization factor is given by (32.58). Finally, the diagonal matrix element may be calculated in terms of $(j\|\mathbf{F}^{(k)}\|j)$ for all valence j-orbits and each of these is multiplied by the corresponding renormalization factor in (32.58). To show this fact, we may simply follow the steps taken to derive (32.58) from (32.53) or (32.64) from (32.62). This involves minute changes in the notation and we shall not repeat the procedure for the general case.

It is important to emphasize that the renormalization for different matrix elements may be different. This is true even if we take the various energy denominators ΔE to be large and replace them by a constant independent of the particular state of the valence nucleons. There are still differences in normalization factors due to the different values of $f^{(k)}_{jj'}$ and $f^{(k)}_j$. Every operator $\mathbf{F}^{(k)}$ should be replaced by a corresponding effective operator. There is no overall coefficient that may be expressed in terms of an effective charge for all electromagnetic multipoles.

The enhancement in certain moments and transitions may be appreciable even if we consider only the first order corrections. A good example is offered by electric quadrupole moments and transition probabilities which are appreciably enhanced in many nuclei. Such strong enhancements are expected if all contributions in (32.58) or (32.64) have the same sign. In that case all admixtures act *coherently* to enhance the effect. This happens, in particular, if the excitations of the core itself add coherently to yield an eigenstate with $J_c = k$ such that the matrix element of $\sum \mathbf{F}^{(k)}(i)$ between it and the $J_c = 0$ state is large. In other words, if the state created by acting with $\sum \mathbf{F}^{(k)}(i)$ on the $J_c = 0$ ground state is an (approximate) eigenstate of the nuclear Hamiltonian. Examples of such excitations are the giant dipole resonance (GDR) and giant quadrupole resonance (GQR) which will not be discussed here.

The renormalization considered here is due to first order perturbation theory. Higher orders may also contribute corrections which may be expressed as renormalization of single nucleon operators. We shall not discuss here such corrections. Like the difficulties encountered in the calculation of effective two body interactions between nucleons, there are difficulties in calculating various effective single nucleon operators. Attempts may be made to extract these effective operators from experiment. This is, however, more difficult than for effective interactions. The energy has the stationary property and is relatively less sensitive to small changes in the wave functions. Transition probabilities and diagonal matrix elements of other operators may be strongly affected by rather small admixtures in the nuclear wave functions. For these reasons not much work has been done in the determination of effective single nucleon operators from experimental data. Where some work has been done, it was assumed that the corrections are the same for all orbits considered, both for diagonal and non-diagonal matrix elements. Thus, only effective charges have been considered. In certain cases it was assumed that the corrections are the same for protons and neutrons although better agreement with experiment was obtained by making these corrections different.

Most of the attempts to determine effective charges were directed at the electric quadrupole moments and transitions. The values of these are enhanced over the values calculated with simple shell model wave functions. Also transitions due to a single neutron are rather strong and comparable to single proton transitions. In the case of the electric quadrupole operator the direct reduced matrix element which is the first term in (32.54) or (32.63) vanishes in the case of a neutron and is proportional to e for a proton. We can thus write

$$e_{\pi}^{eff} = e + \delta e_{\pi} \qquad e_{\nu}^{eff} = \delta e_{\nu} \tag{32.65}$$

Let us look in more detail at the corrections δe_{π} and δe_{ν}. The first remark to be made is that the renormalization of an operator $\sum \mathbf{F}^{(k)}(i)$ is expected to be stronger if the k-th term in the tensor expansion of the interaction is large. This is a rather general statement which cannot be made more precise without detailed knowledge of matrix elements of the nuclear interaction. The fact that electric quadrupole moments and transitions are enhanced may indicate that the quadrupole part of the interaction is rather large. This statement, however, should be made more precise.

The energy denominators in (32.58) and (32.64) are all *negative*. The ΔE appearing there are differences between energies of low lying states of the valence nucleons and states with higher energies in which the closed shells of the core are excited. The sign of the correction terms is expected to be *positive* for an *attractive* quadrupole part of the interaction. Of course, the sign actually depends on the relative signs of $f_j^{(k)}$ and $f_{j_1j_2}^{(k)}$ or $f_{jj'}^{(k)}$ and $f_{j_1j_2}^{(k)}$. For lack of more detailed knowledge about these coefficients, it is usually assumed that they have the same sign. This is consistent with the empirical observation of enhanced quadrupole moments and transitions.

To first order in perturbation theory, non-vanishing contributions to the electric quadrupole operator arise only from proton j_1-orbits in the summations in (32.58) and (32.64). As shown in Section 31, diagonal matrix elements of the proton-neutron interaction are strong and attractive. We may expect that also the corrections in δe_{ν} will be considerable and of the same sign as the proton charge. On the other hand, the diagonal matrix elements of the interaction between protons in different orbits are rather weak and repulsive. Thus, we may expect that δe_{π} will be smaller than δe_{ν} and perhaps have the opposite sign. The effective charges extracted empirically from experimental data of light and medium mass nuclei ($A = 60$) yield δe_{ν} and δe_{π} having the same sign as the proton charge $\delta e_{\nu} \sim 2\delta e_{\pi}$ to $3\delta e_{\pi}$. Typical values of δe_{ν} range between $0.5e$ and $2e$. As mentioned above, there may be other contributions in addition to the first order core polarization given by (32.58) and (32.64).

The feature that $\delta e_{\pi} \neq \delta e_{\nu}$ does not contradict isospin symmetry based on charge independence of the nuclear interaction. This is true even if in the ground state of a $T = \frac{1}{2}$ nucleus there is only a single nucleon outside a core with ground state isospin $T = 0$ (which must have equal numbers of protons and neutrons). To see it we go back to (32.48) and consider the case in which the j_1-orbit is fully occupied by protons and neutrons, the j_2-orbit is empty and one, or several, nucleons occupy the valence j-orbit. We then obtain expressions which are the analogs of (32.58) and (32.64) in the isospin formalism. Some states of the excited core may have $T = 0$. Such states, $|j_1^{-1} j_2 T_c = 0,\ J_c = k\rangle$, make the same contribution to the summations in the analogs of (32.58) and (32.64) for a valence proton and for a valence neutron. There are excited states, however, with $T = \frac{1}{2}$ where the isospin of the core is $T_c = 1$. The contribution of such states is different for a valence proton and for a valence neutron. The terms in (32.48) with $k' = 0$ have non-vanishing matrix elements between the perturbed state

$$|T_c = 0, J_c = 0, jT = \tfrac{1}{2}, M_T J = jM = m\rangle$$

and only states

$$|j_1^{-1} j_2 T_c J_c = k\, jT = \tfrac{1}{2}, M_T J = jM = m\rangle$$

for which $T_c = 0$. The terms with $k' = 1$, however, admix to the perturbed state excited states in which the isospin of the excited core must have $T_c = 1$.

In many nuclei with several valence protons and neutrons electric quadrupole transition probabilities are enhanced by factors of about 100 or more. Such enhancements cannot be due to large effective charges of individual nucleons. They must be due to the coherent or collective effects in the states of many nucleons. The shell model description of such states is very complex as remarked in Section 24. We will briefly describe in subsequent sections a certain model for such nuclei which may be related to the shell model.

Polarization of the core by valence nucleons may lead also to changes in the charge distribution. The charge distribution in the ground state may be modified by admixtures of core excited states whose charge distributions are different. Let us consider the charge radius of a nucleus which has valence neutrons in the j-orbit outside closed shells. Its dependence on the number of j-neutrons turns out to have interesting features.

We consider a state like (32.51) where the state of the j^n neutron configuration in the unperturbed state is characterized by the seniority

v, spin J_v and additional quantum numbers α, if necessary. The operator considered is $\sum \mathbf{r}_i^2$ where the summation is over protons only. Its expectation value in the state (32.51) is equal to

$$\langle v\alpha JM| \sum_{\text{protons}} \mathbf{r}_i^2 |v\alpha JM\rangle$$

$$= \langle J_c = 0, v\alpha J_v JM| \sum_{\text{protons}} \mathbf{r}_i^2 |J_c = 0, v\alpha J_v JM\rangle$$

$$+ 2 \sum_{j_1 j_2 \pm j v' \alpha' J_v'} \langle J_c = 0, v\alpha J_v JM| \sum_{\text{protons}} \mathbf{r}_i^2 |j_1^{-1} j_2 J_c = kv'\alpha' J_v' JM\rangle$$

$$\times \beta_k(j_1^{-1} j_2 v\alpha J_v v'\alpha' J_v' J)$$

$$+ \sum_{\substack{j_1 j_2 \neq j\, j_1' j_2' \neq j \\ v'\alpha' J_v' v''\alpha'' J_v''}} \langle j_1^{-1} j_2 J_c = kv'\alpha' J_v' JM|$$

$$\sum_{\text{protons}} \mathbf{r}_i^2 |j_1'^{-1} j_2' J_c = kv''\alpha'' J_v'' JM\rangle$$

$$\times \beta_k(j_1^{-1} j_2 J_c = kv\alpha J_v v'\alpha' J_v' J) \beta_k(j_1'^{-1} j_2' J_c = kv\alpha J_v v''\alpha'' J_v'' J) \tag{32.66}$$

To obtain the effect of perturbation theory up to the second order, the normalization of the state (32.51) should be taken into account. This will be carried out below. There are other first order corrections to the state (32.51) which do not contribute to the change in the single nucleon operator $\sum \mathbf{r}_i^2$. They may change the normalization of the state (32.51) but they will be ignored in the present discussion.

In the first term on the r.h.s. of (32.66), the protons are in closed shells and hence, it is equal to the square of the r.m.s. charge radius of the core (the nucleus with $n = 0$). The operator $\sum_{\text{protons}} \mathbf{r}_i^2$ is a scalar which leads to

$$\langle J_c = 0 v\alpha J_v JM| \sum_{\text{protons}} \mathbf{r}_i^2 |j_1^{-1} j_2 J_c = kv'\alpha' J_v' JM\rangle$$

$$= \langle J_c = 0| \sum_{\text{protons}} \mathbf{r}_i^2 |j_1^{-1} j_2 J_c = k\rangle \delta_{k0} \delta_{v'v} \delta_{\alpha'\alpha} \delta_{J_v' J_v} \tag{32.67}$$

The non-vanishing contributions to the second term on the r.h.s. of (32.66) are due only to states with $J_c = k = 0$ (and positive parity).

The value of the β-coefficient in this case with $j_2 = j_1$, $l_2 = l_1$ and $n_2 \neq n_1$ is given by (32.50) where we put $v_1 = v' = v$, $\alpha_1 \equiv \alpha' \equiv \alpha$ and $J_1 = J'_v = J_v$. As before, we take the energy differences between perturbed and perturbing states to be large compared to energy differences between states of the j^n configuration. The ΔE in (32.50) may then be replaced by a constant $\Delta E_{j_1^{-1}j_2}$. The second term on the r.h.s. of (32.66) can be expressed as

$$\delta_{k0}\frac{(j^n v\alpha J_v\|\sum_i \mathbf{u}^{(0)}(i)\|j^n v\alpha J_v)}{\sqrt{2J_v+1}}\sum_{j_1j_2\neq j}\frac{1}{\Delta E_{j_1^{-1}j_2}}F_0(j_1jj_2j)$$
$$\times (J_c=0\|\sum_i \mathbf{u}^{(0)}(i)\|j_1^{-1}j_2J_c=0)$$
$$\times (J_c=0\|\sum_{\text{protons}} \mathbf{r}_i^2\|j_1^{-1}j_2J_c=0)$$
$$=\delta_{k0}n\frac{1}{\sqrt{2j+1}}\sum_{j_1j_2\neq j}\frac{1}{\Delta E_{j_1^{-1}j_2}}F_0(jj_1jj_2)$$
$$\times (J_c=0\|\sum_i \mathbf{u}^{(0)}(i)\|j_1^{-1}j_2J_c=0)$$
$$\times (J_c=0\|\sum_{\text{protons}} \mathbf{r}_i^2\|j_1^{-1}j_2J_c=0) \tag{32.68}$$

The first order correction due to monopole polarization by the n valence j-neutrons to the square of the r.m.s. charge radius is proportional to n.

In the third term on the r.h.s. of (32.66), matrix elements of the scalar operator $\sum \mathbf{r}_i^2$ are equal to

$$\langle j_1^{-1}j_2J_c = kv'\alpha'J'_vJM|\sum_{\text{protons}} \mathbf{r}_i^2|j_1'^{-1}j_2'J_c = k'v''\alpha''J''_vJM\rangle$$
$$=\langle j_1^{-1}j_2J_c=k|\sum_{\text{protons}} \mathbf{r}_i^2|j_1'^{-1}j_2'J_c=k'\rangle\delta_{kk'}\delta_{v'v''}\delta_{\alpha'\alpha''}\delta_{J'_vJ''_v} \tag{32.69}$$

Due to (32.69) and (32.50), the third term on the r.h.s. of (32.66) can be expressed as

$$\sum_{\substack{j_1 j_2 \neq j\, j_1' j_2' \neq j \\ v'\alpha' J_v'}} \frac{\left(j^n v \alpha J_v \| \sum_i \mathbf{u}^{(k)}(i) \| j^n v' \alpha' J_v'\right)^2}{(2k+1)(2J_v+1)}$$

$$\times \frac{1}{\Delta E_{j^{-1} j_2} \Delta E_{j'^{-1} j_2'}} F_k(j_1 j j_2 j) F_k(j_1' j j_2' j)$$

$$\times (J_c = 0 \| \sum_i \mathbf{u}^{(k)}(i) \| j_1^{-1} j_2 J_c = k)$$

$$\times (J_c = 0 \| \sum_i \mathbf{u}^{(k)}(i) \| j_1'^{-1} j_2' J_c = k)$$

$$\times \langle j_1^{-1} j_2 J_c = k | \sum_{\text{protons}} \mathbf{r}_i^2 | j_1'^{-1} j_2' J_c = k \rangle \qquad (32.70)$$

The sum over $v'\alpha' J_v'$ may be carried out. Due to (10.31) we obtain

$$\frac{1}{2J_v+1} \sum_{v'\alpha' J_v'} (j^n v \alpha J_v \| \sum_i \mathbf{u}^{(k)}(i) \| j^n v' \alpha' J_v')^2$$

$$= \langle j^n v \alpha J_v M_v | \left(\sum_i \mathbf{u}^{(k)}(i)\right) \cdot \left(\sum_i \mathbf{u}^{(k)}(i)\right) | j^n v \alpha J_v M_v) \rangle \qquad (32.71)$$

The expression (32.71) appears in (32.66) multiplied by a sum, independent of v, α and J_v, equal to

$$\sum_{j_1 j_2 \neq j\, j_1' j_2' \neq j} \frac{F_k(j_1 j j_2 j) F_k(j_1' j j_2' j)}{(2k+1) \Delta E_{j_1^{-1} j_2} \Delta E_{j_1' - j_2'}} (J_c = 0 \| \sum_i \mathbf{u}^{(k)}(i) \| j_1^{-1} j_2 J_c = k)$$

$$\times (J_c = 0 \| \sum_i \mathbf{u}^{(k)}(i) \| j_1'^{-1} j_2' J_c = k)$$

$$\times \langle j_1^{-1} j_2 J_c = k | \sum_{\text{protons}} \mathbf{r}_i^2 | j_1'^{-1} j_2' J_c = k \rangle \qquad (32.72)$$

We now introduce the normalization of the state (32.51). The square of the norm of the state (32.51) is equal to

$$1 + \sum_{j_1 j_2 \neq j v' \alpha' J'_\nu} \beta_k (j_1^{-1} j_2 \nu \alpha J_\nu \nu' \alpha' J'_\nu J)^2$$

Therefore, to impose the correct normalization, up to second order in perturbation theory, on the matrix element (32.66), it should be multiplied by

$$1 - \sum_{j_1 j_2 \neq j v' \alpha' J'_\nu} \beta_k (j_1^{-1} j_2 \nu \alpha J_\nu \nu' \alpha' J'_\nu J)^2$$

The zero order term and the first order term in (32.66) remain unchanged. It is only the second order term which is modified. With the values of β-coefficients taken from (32.50) we obtain the correct second order correction as

$$\begin{aligned}
&\langle j^n \nu \alpha J_\nu M_\nu | \left(\sum_i \mathbf{u}^{(k)}(i) \right) \cdot \left(\sum_i \mathbf{u}^{(k)}(i) \right) | j^n \nu \alpha J_\nu M_\nu \rangle \\
&\times \sum_{j_1 j_2 \neq j \, j'_1 j'_2 \neq j} \frac{F_k(j_1 j j_2 j) F_k(j'_1 j j'_2 j)}{(2k+1) \Delta E_{j_1^{-1} j_2} \Delta E_{j_1'^{-1} j'_2}} \\
&\times (J_c = 0 \| \sum_i \mathbf{u}^{(k)} \| j_1^{-1} j_2 J_c = k) \\
&\times (J_c = 0 \| \sum_i \mathbf{u}^{(k)}(i) \| j_1'^{-1} j'_2 J_c = k) \\
&\times \left[\langle j_1^{-1} j_2 J_c = k | \sum_{\text{proton}} \mathbf{r}_i^2 | j_1'^{-1} j'_2 J_c = k \rangle \right. \\
&\qquad \left. - \delta_{j_1 j'_1} \delta_{j_2 j'_2} \langle J_c = 0 | \sum_{\text{proton}} \mathbf{r}_i^2 | J_c = 0 \rangle \right] \\
&= \langle j^n \nu \alpha J_\nu M_\nu | \left(\sum_i \mathbf{u}^{(k)}(i) \right) \cdot \left(\sum_i \mathbf{u}^{(k)}(i) \right) | j^n \nu \alpha J_\nu M_\nu \rangle C_k
\end{aligned} \tag{32.73}$$

The second order corrections are proportional to the matrix element on the r.h.s. of (32.71). It can be evaluated by using the results of Section 20. There is, however, a simpler procedure for state with lowest seniorities, $v = 0$, $J = 0$ and $v = 1$, $J = j$ in j^n configurations. This procedure clearly displays the dependence of (32.71) on n. As explained in Section 20, in the discussion of (20.17), (20.18) and (20.19), the expectation values in these states of any scalar two-body operator can be expressed as a linear combination of a term proportional to $n(n-1)$ and a pairing term proportional to $(n-v)/2 = [n/2]$. The scalar product on the r.h.s. of (32.71) contains also single nucleon terms and hence, to express its expectation values, a linear term in n should be added. The second order change in the square of the r.m.s. charge radius can be expressed as (Zamick 1971, Talmi 1984)

$$\boxed{a_1 n + a_2 \frac{n(n-1)}{2} + b\left[\frac{n}{2}\right]} \tag{32.74}$$

Squares of charge radii of isotopes show, according to (32.74), an odd-even variation on top of a linear and quadratic change in the number of valence neutrons (according to (32.68), the first order corrections are linear in n). This is in agreement with the odd-even effect which has been experimentally observed in isotopes of semi-magic nuclei Ca, Sn and Pb.

We can now see which multipoles in the tensor expansion of the proton-neutron interaction give rise to the odd-even variation. We can calculate the expectation values (32.71) in $v = 0$ and $v = 1$ states for $k = 0$, odd values of k and $k > 0$, even. For $k = 0$, (32.71) is equal, according to (19.41), to

$$n^2 \frac{(j\|\mathbf{u}^{(k)}\|j)^2}{2j+1} = n^2 \frac{1}{2j+1} \tag{32.75}$$

If k is odd, as we saw in Section 19, the tensor $\sum_i \mathbf{u}^{(k)}(i)$ is diagonal in the seniority scheme and its matrix elements are independent of n. According to (19.37), it has vanishing matrix elements between $v = 0$, $J = 0$ states and any state. Hence, for any even value of n we obtain from (32.74)

$$a_1 n + a_2 \frac{n(n-1)}{2} + b\frac{n}{2} = 0 \qquad k \text{ odd, } n \text{ even} \tag{32.76}$$

Equation (32.76), for $j > \frac{1}{2}$, can hold only if $a_2 = 0$ and $a_1 = -b/2$. For odd numbers n and $v = 1$, $J = j$, (32.71) is equal to $\langle jm|\mathbf{u}^{(k)} \cdot \mathbf{u}^{(k)}|jm\rangle = (j\|\mathbf{u}^{(k)}\|j)^2/(2j+1) = 1/(2j+1)$. Hence,

$$a_1 n - 2a_1\frac{n-1}{2} = \frac{1}{2j+1}C_k \qquad k \text{ odd, } n \text{ odd}$$

From which follows $b = -2a_1 = -2C_k/(2j+1)$. Thus, for odd ranks k there is an odd even variation, on top of a linear change, which is given by $b = -2C_k/(2j+1)$.

As will be shown in Section 38, large coefficients in the tensor expansion of proton-neutron interactions multiply scalar products of even rank tensors. Particularly large are the coefficients of the monopole term $k = 0$, and the quadrupole term, $k = 2$. To evaluate (32.71) for $k > 0$, even, we use the reduction formula (19.44) for even numbers n and $v = 2$, $J' = 0$. We then obtain for (32.71) the result

$$\begin{aligned}(j^n v = 0, J_v = 0\|\sum_i \mathbf{u}^{(k)}(i)\|j^n v = 2, J_v = k)^2 &= \frac{n(2j+1-n)}{2(2j-1)} \\ &\times (j^2 v = 0, J_v = 0\|\sum_i \mathbf{u}^{(k)}(i)\|j^2 v = 2, J_v = k)^2 \\ &= \frac{n(2j+1-n)}{2(2j-1)}\frac{4(j\|\mathbf{u}^{(k)}\|j)^2}{2j+1} = \frac{2n(2j+1-n)}{(2j+1)(2j-1)}\end{aligned} \tag{32.77}$$

The last equality is obtained by using the relations (10.41) and (10.42). From (32.77) and (32.74) we obtain

$$C_k\frac{2n(2j+1-n)}{(2j+1)(2j-1)} = a_1 n + a_2\frac{n(n-1)}{2} + b\frac{n}{2} \qquad n \text{ even} \tag{32.78}$$

Another simple relation is obtained in the case $n = 1$, namely

$$C_k\frac{(j\|\mathbf{u}^{(k)}\|j)^2}{2j+1} = \frac{C_k}{2j+1} = a_1$$

From (32.78) we obtain, for $j > \frac{1}{2}$, $a_2 = -4C_k/(2j+1)(2j-1)$ and $a_1 + b/2 = 4jC_k/(2j+1)(2j-1)$, and hence, $b = 2C_k/(2j-1)$. The second order corrections due to $k > 0$, even, terms are thus equal to

$$\boxed{\begin{array}{ll} \dfrac{2C_k}{(2j+1)(2j-1)} n(2j+1-n) & n \text{ even}, \\[2ex] \dfrac{2C_k}{(2j+1)(2j-1)} \left[n(2j+1-n) - \dfrac{2j+1}{2}\right] & n \text{ odd} \end{array}} \tag{32.79}$$

The odd-even variation arises from the fact that an even j^n configuration polarizes the core more effectively than an odd one.

Lack of knowledge of the effective interaction prevents us from making definite predictions about the size and even the sign of corrections due to core polarization. The observed changes in squares of r.m.s. charge radii of semi-magic nuclei with valence neutrons, can be nicely reproduced by the expression (32.74). The relative signs of the fitted parameters a_2 and b agree with those in (32.79), due to quadrupole core polarization, in the case of Ca and Pb isotopes (Talmi 1984). The absolute value of the fitted parameter a_2 is smaller than that of b but the ratio of the two does not agree with (32.79).

Another example where core polarization may lead to modified single nucleon operators is offered by magnetic moments of certain nuclei. Consider, for the sake of simplicity, a nucleus with one neutron in the j-orbit. Among all excitations from the $j_1 = l' + \frac{1}{2}$ orbit into higher j_2 orbits, the only ones that lead to first order corrections to magnetic moments are excitations into the $j_2 = l' - \frac{1}{2}$ orbit with the same radial quantum numbers. This is due to the fact that in $\boldsymbol{\mu} = \sum \boldsymbol{\mu}_i = \sum (g_l^{(i)} \boldsymbol{l}_i + g_s^{(i)} \mathbf{s}_i)$ neither $\boldsymbol{l}$ nor $\mathbf{s}$ can change the l-value of the j_1-orbit. We shall consider proton excitations and take for simplicity the proton $j_1 = l' + \frac{1}{2}$ orbit to be completely filled and the proton $j_2 = l' - \frac{1}{2}$ orbit to be empty. Due to the strong proton-neutron interaction such excitations are expected to be the most important ones and we consider them here.

The term in the tensor expansion of the proton-neutron interaction which has non vanishing matrix elements between the core states $|J_c = 0\rangle$ and $|(l' + \frac{1}{2})^{-1} l' - \frac{1}{2}, J_c = 1\rangle$ has $k = 1$. It arises from an interaction like $(\mathbf{s}_1 \cdot \mathbf{s}_2) V(|\mathbf{r}_1 - \mathbf{r}_2|)$ and in the terms like (10.67) of the expansion, the value $r = 1$ may be due to $k = 0$ as well as to $k = 2$. The term with $k = 0$ is expected to be larger and has the form $(\mathbf{s}_1 \cdot \mathbf{s}_2)v_0(r_1, r_2)$. Matrix elements of this term are equal to products of radial integrals and matrix elements of $(\mathbf{s}_1 \cdot \mathbf{s}_2)$. In the present case, we denote by F_0 only the product of radial integrals. Instead of $\sum \mathbf{u}^{(k)}(i)$ in the tensor expansion of the interaction we use here $\sum \mathbf{s}_i$.

We now use (32.55) with $\sum_{\text{protons}} \mathbf{s}_i$ replacing $\sum \mathbf{u}^{(k)}(i)$ between core states and put $\sum \mathbf{F}^{(k)}(i) = \sum \boldsymbol{\mu}_i$ where the summation is over all protons and neutrons. We then use (32.58) and obtain for the reduced matrix element of the magnetic moment the result

$$(j\|\boldsymbol{\mu}^\nu\|j) + \frac{2}{3}\frac{F_0}{\Delta E}\left(J_c = 0\left\|\sum \mathbf{s}_i\right\|(l' + \tfrac{1}{2})^{-1}(l' - \tfrac{1}{2}), J_c = 1\right)^2$$
$$\times (j\|\mathbf{s}\|j)\frac{(l' + \frac{1}{2}\|\boldsymbol{\mu}^\pi\|l' - \frac{1}{2})}{(l' + \frac{1}{2}\|\mathbf{s}\|l' - \frac{1}{2})} \tag{32.80}$$

Since ΔE is negative we define

$$\boxed{\frac{2}{3}\frac{1}{\Delta E}\left(J_c = 0\left\|\sum \mathbf{s}_i\right\|(l' + \tfrac{1}{2})^{-1}(l' - \tfrac{1}{2}), J_c = 1\right)^2 = -\gamma^2} \tag{32.81}$$

We also note that since $\mathbf{j} = \mathbf{s} + \boldsymbol{l}$ is diagonal in j, the following relation holds

$$(l' + \tfrac{1}{2}\|g_l^\pi \boldsymbol{l} + g_s^\pi \mathbf{s}\|l' - \tfrac{1}{2}) = (l' + \tfrac{1}{2}\|g_l^\pi(\boldsymbol{l} + \mathbf{s}) + (g_s^\pi - g_l^\pi)\mathbf{s}\|l' - \tfrac{1}{2})$$
$$= (g_s^\pi - g_l^\pi)(l' + \tfrac{1}{2}\|\mathbf{s}\|l' - \tfrac{1}{2})$$

We can then bring (32.80) into the form

$$\boxed{\begin{aligned}&(j\|\boldsymbol{\mu}^\nu\|j) - (j\|\mathbf{s}\|j)\gamma^2 F_0(g_s^\pi - g_l^\pi)\\ &\qquad = (j\|g_l^\nu \boldsymbol{l} + g_s^\nu \mathbf{s} - \gamma^2 F_0(g_s^\pi - g_l^\pi)\mathbf{s}\|j)\end{aligned}} \tag{32.82}$$

We see from (32.82) that the first order corrections modify the g-factor of the intrinsic spin. The renormalized g_s^ν factor becomes

equal to

$$\boxed{g_s^\nu - \gamma^2 F_0(g_s^\pi - g_l^\pi)} \tag{32.83}$$

In order to see the change in g_s^ν due to the polarization of the core we must know the sign of F_0.

The interaction $(\mathbf{s}_1 \cdot \mathbf{s}_2)V(|\mathbf{r}_1 - \mathbf{r}_2|)$ between protons and neutrons should be attractive in space symmetric states with $T = 0$. Such states have spin $S = 1$ in which case the eigenvalue of $(\mathbf{s}_1 \cdot \mathbf{s}_2)$ is equal to 1/4 according to (10.65). Hence, the potential $V(|\mathbf{r}_1 - \mathbf{r}_2|)$ is attractive and the radial integral F_0 is *negative*. The modification of g_s^ν is thus given in nuclear magnetons by

$$g_s^\nu + \gamma^2|F_0|(g_s^\pi - g_l^\pi) = -2 \times 1.93 + \gamma^2|F_0|2 \times 2.29 \tag{32.84}$$

The absolute value of the g_s factor of the neutron is thus diminished. The same effect is obtained for the proton g_s factor which is expressed by

$$g_s^\pi + \gamma^2|F_0|(g_s^\nu - g_l^\nu) = 2 \times 2.79 - \gamma^2|F_0|2 \times 1.93 \text{ n.m.} \tag{32.85}$$

The corrections in (32.84) and (32.85) are in the right direction to obtain agreement with experimental data. They were introduced, along with other first order corrections by Arima and Horie (1954) and by Blin-Stoyle (1953).

We may consider in the proton-neutron interaction also the term $(\mathbf{s}_1 \cdot \mathbf{s}_2)(\mathbf{Y}_2(1) \cdot \mathbf{Y}_2(2))(4\pi/5)v_2(r_1, r_2)$. In view of (10.67) we obtain in the same way that (32.82) was obtained, a correction term to the magnetic moment proportional to $(j\|[\mathbf{s} \times \mathbf{Y}_2]^{(1)}\|j)$.

Let us briefly look at the second order corrections to energies of ground states due to core polarization as expressed by (32.51). These are given by

$$\sum_{\substack{j_1 j_2 \neq j \\ v'\alpha' J'}} \beta_k(j_1^{-1} j_2 \alpha J_\nu v' \alpha' J'_\nu J)^2 \Delta E_{j_1^{-1} j_2} \tag{32.86}$$

Substituting into (32.86) the values of the β-coefficients from (32.50), we obtain the expression

$$\sum_{v'\alpha'J'} \frac{(j^n v\alpha J_v \| \sum_i \mathbf{u}^{(k)}(i) \| j^n v'\alpha' J_v')^2}{2J_v+1} \sum_{j_1 j_2 \neq j} \frac{F_k(j_1 j j_2 j)^2}{\Delta E_{j_1^{-1} j_2}}$$
$$\times (J_c = 0 \| \sum_i \mathbf{u}^{(k)}(i) \| j_1^{-1} j_2 J_c = k)^2 \tag{32.87}$$

In view of (32.71) and the derivation of (32.74) from (32.73), we can express the change in energy of $v = 0$ and $v = 1$ states of j^n configurations, due to second order perturbation theory, by

$$C'n + \frac{n(n-1)}{2}\alpha' + \left[\frac{n}{2}\right]\beta' \tag{32.88}$$

In such cases, the contribution of core polarization to energies of ground states has the form (20.14). Its effect is just in renormalizing the coefficients of the binding energy formula of the seniority scheme.

Second order corrections due to excitation of a single nucleon do not, in general, assume the simple form (32.88). When considering single nucleon excitations from the j^n configuration, the matrix element (32.30) was obtained. The square of (32.30) is proportional to $n(n-1)^2$ which should be contrasted with the proportionality to $n(n-1)$ of the second order correction (32.18). The n dependence of the second order correction due to (32.30) is

$$n(n-1)^2 = n(n-1)(n-2) + n(n-1).$$

Whereas (32.18) is equivalent to an additional *two-body* interaction, the contribution of (32.30) includes also *three-body* terms (Racah and Stein 1967).

33

The Interacting Boson Model (IBA-1)

At the end of Section 24, the complexity of the shell model program for nuclei with valence protons and neutrons was demonstrated. In a typical nucleus, ^{154}Sm, valence protons and neutrons may couple to more than 4×10^{13} states with $J = 0$ and positive parity, more than 3×10^{14} $J = 2^+$ states and more than 5×10^{14} states with $J = 4^+$. To calculate eigenstates and eigenvalues of the shell model Hamiltonian, matrices of order 10^{14} should be constructed and diagonalized. To construct the Hamiltonian matrices in this case, 1192 matrix elements of two nucleon configurations and 11 single nucleon energies should be known. In simple cases, such matrix elements are determined from experimental data but this would be beyond hope in the present case. Diagonalization of gigantic matrices of order 10^{14} is beyond the capability of present day computers. Even if those difficulties could have been overcome, the results would have been disappointing. There is no way in which properties of a wave function with 10^{14} components could be studied. These difficulties are particularly frustrating in view of the fact that many nuclei for which the shell model program is so complex, exhibit remarkably simple regularities. The low lying levels of ^{154}Sm as well as ^{148}Sm in Fig. 33.1, may serve as a simple example. To understand these regularities a simple coupling scheme should be

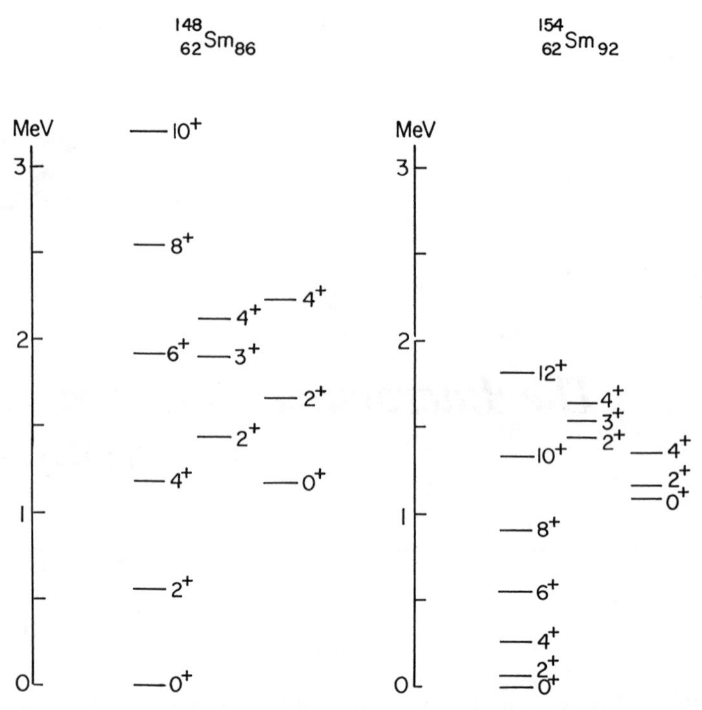

FIGURE 33.1. *Experimental low lying levels of* ^{148}Sm *and* ^{154}Sm.

found, truncating the huge shell model space and exhibiting the simple features of nuclear spectra.

Spectra of certain nuclei with many valence protons *and* neutrons exhibit simple regularities. Excited levels may be grouped into bands with strong intraband E2 transitions. In many nuclei states within these bands have energies whose dependence on J is given by $J(J+1)$. A typical spectrum is shown in Fig. 33.2. Such rotational bands have a very simple description in the collective model (Bohr and Mottelson 1953). The ground state of the nucleus in that model has a non-spherical, deformed shape described in terms of a few collective variables. Such a shape can rotate, more or less rigidly, giving rise to the ground state rotational band. Other bands may be due to rotations of that shape when it is in certain states of vibration. The problem of relating the collective variables to the (spherical) shell model is very complex and difficult. It is usually assumed that a deformed shape of the nucleus arises from the independent motion of

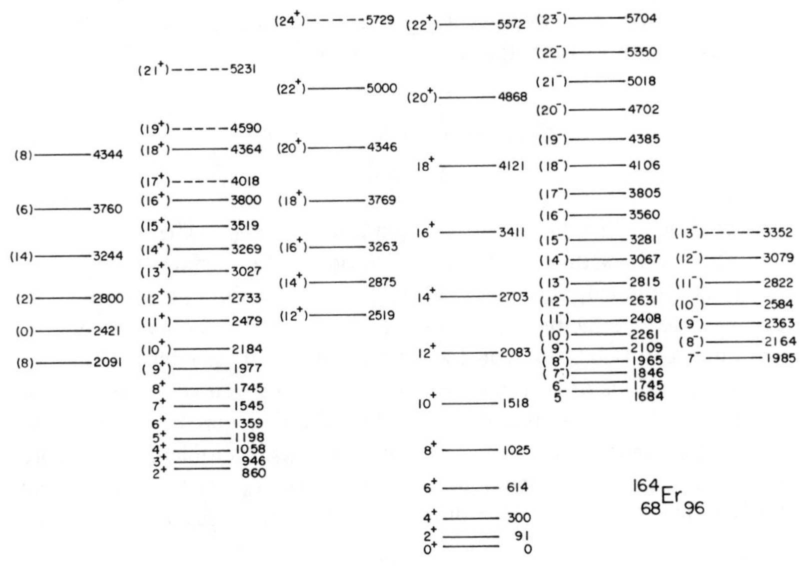

FIGURE 33.2. *Rotational bands in the experimental spectrum of* ^{164}Er.

nucleons in a *deformed potential well*. From wave functions of such a system, states with definite angular momentum J should be projected. This poses a very difficult problem and only rather crude approximations have been developed to solve it. In the case of odd-even nuclei, a single nucleon is usually coupled to the collective core. In the following sections only even-even nuclei will be considered. Only in Section 41 a simple model for a particle or particles coupled to a rotating core is presented.

An alternative approach to collective states in even-even nuclei has been offered by the *interacting boson model* (IBM or IBA for interacting boson approximation). This model will be presented in the following. As we shall see, it can be related, although to a certain approximation, to the spherical shell model. On the other hand, its description of collective states is essentially equivalent to that of the collective model. The collective model had tremendous success in describing collective states of nuclei. A very short description of its main features is given below. It is presented here only to establish the connection with the interacting boson model. A detailed description of the collective model is given by Bohr and Mottelson (1975).

The surface of an even-even nucleus is defined in terms of the distance of its points from the origin in every direction by

$$R(\theta,\phi) = R_0\left(1+\sum_{\mu}\alpha_\mu Y^*_{2\mu}(\theta,\phi)\right) \qquad (33.1)$$

The reality of $R(\theta,\phi)$ implies the condition $\alpha^*_\mu = (-1)^\mu \alpha_{-\mu}$. For non-vanishing values of α_μ, the deformation of the surface as given in (33.1) is a quadrupole deformation. The α_μ are the components of a quadrupole irreducible tensor (rank $k = 2$). In the collective model, the coefficients α_μ are considered as dynamical variables. Consider small oscillations around a spherical shape. For given values of surface tension and inertial parameters, a Hamiltonian can be constructed which is a quadratic function of the α_μ and their canonically conjugate momenta π_μ. It is convenient to express the α_μ variables by a real basis as in (28.12). The Hamiltonian can then be expressed by

$$\frac{1}{2}\pi_0^2 + \frac{1}{2}\sum_{\lambda=1,2}(\pi_\lambda^2+\pi_{\bar\lambda}^2) + \frac{\omega^2}{2}\alpha_0^2 + \frac{\omega^2}{2}\sum_{\lambda=1,2}(\alpha_\lambda^2+\alpha_{\bar\lambda}^2) \qquad (33.2)$$

where the hermitean operators $\pi_0, \pi_\lambda, \pi_{\bar\lambda}$ are cannonically conjugate to the hermitean operators $\alpha_0, \alpha_\lambda, \alpha_{\bar\lambda}$ respectively. This is the Hamiltonian of a five-dimensional harmonic oscillator. Creation and annihilation operators of the oscillator quanta may be defined, for $\lambda = 1,2$, by

$$d_0^+ = \frac{1}{\sqrt{2\hbar\omega}}(\pi_0 + i\omega\alpha_0)$$

$$d_\lambda^+ = \frac{1}{2\sqrt{\hbar\omega}}(\pi_\lambda + i\omega\alpha_\lambda + i\pi_{\bar\lambda} - \omega\alpha_{\bar\lambda})$$

$$d_{-\lambda}^+ = \frac{(-1)^\lambda}{2\sqrt{\hbar\omega}}(\pi_\lambda + i\omega\alpha_\lambda - i\pi_{\bar\lambda} + \omega\alpha_{\bar\lambda})$$

$$d_0 = \frac{1}{\sqrt{2\hbar\omega}}(\pi_0 - i\omega\alpha_0)$$

$$d_\lambda = \frac{1}{2\sqrt{\hbar\omega}}(\pi_\lambda - i\omega\alpha_\lambda - i\pi_{\bar\lambda} - \omega\alpha_{\bar\lambda})$$

$$d_{-\lambda} = \frac{(-1)^\lambda}{2\sqrt{\hbar\omega}}(\pi_\lambda - i\omega\alpha_\lambda + i\pi_{\bar\lambda} + \omega\alpha_{\bar\lambda})$$

Due to the cannonical commutation relations between the hermitean components of $\boldsymbol{\pi}$ and $\boldsymbol{\alpha}$, the operators defined above obey the following Bose commutation relations

$$[d_\mu, d^+_{\mu'}] = \delta_{\mu\mu'} \qquad [d_\mu, d_{\mu'}] = [d^+_\mu, d^+_{\mu'}] = 0 \qquad \mu = -2, -1, 0, 1, 2 \tag{33.3}$$

In terms of the spherical components of $\boldsymbol{\pi}$ and $\boldsymbol{\alpha}$ the creation and annihilation operators of the five-dimensional oscillator quanta are given by

$$d^+_\mu = \frac{1}{\sqrt{2\hbar\omega}}(\pi_\mu + i\omega\alpha_\mu) \qquad d_\mu = \frac{(-1)^\mu}{\sqrt{2\hbar\omega}}(\pi_{-\mu} - i\omega\alpha_{-\mu}) \tag{33.4}$$

The Hamiltonian (33.2) can then be expressed as

$$\boxed{\hbar\omega \sum_\mu d^+_\mu d_\mu + \tfrac{5}{2}\hbar\omega} \tag{33.5}$$

Thus, the collective Hamiltonian may be expressed in terms of creation and annihilation operators of $l = 2$ bosons—d-bosons. This is in analogy with the description of the three dimensional harmonic oscillator in terms of p-bosons.

In order to obtain a surface which is not spherical in the ground state, in which case $\langle\alpha_\mu\rangle \neq 0$, a more complicated potential energy must be added to the Hamiltonian (33.2). For this purpose, it is more convenient to use other variables, β and γ, defined by

$$\beta^2 = (\boldsymbol{\alpha}\cdot\boldsymbol{\alpha}) \qquad \beta^3 \cos 3\gamma = -\sqrt{\tfrac{7}{2}}(\boldsymbol{\alpha}\times\boldsymbol{\alpha})^{(2)}\cdot\boldsymbol{\alpha} \tag{33.6}$$

In addition to β and γ, three more dynamical variables are needed to express the 5 coordinates α_μ. They can be chosen as the three Euler angles which define the orientation of the deformed shape in space. If α'_μ are the components of the quadrupole tensor in a frame of reference defined by the system, we obtain from (5.23) the relation

$$\alpha_\mu = \sum_{\mu'} \alpha'_{\mu'} \mathcal{D}^{(2)}_{\mu'\mu}(\Omega) \tag{33.7}$$

where Ω stands for the three Euler angles. From (33.7) it is seen that only *two* of the five α' can be independent variables. The frame of reference determined by the dynamical variables of the system ("body-fixed") may be defined by the three principal axes of the deformed shape. In this frame of reference, due to (33.6), the dependence of α'_μ on β and γ is given by

$$\alpha'_0 = \beta\cos\gamma \qquad \alpha'_1 = \alpha'_{-1} = 0 \qquad \alpha'_2 = \alpha'_{-2} = \frac{1}{\sqrt{2}}\beta\sin\gamma \tag{33.8}$$

The surface is deformed if $\langle\beta\rangle \neq 0$ and $\langle\beta\rangle$ provides a measure for the extent of deformation. If $\beta > 0$ (and $\gamma = 0$) then $\alpha'_0 > 0$ and the surface is given, according to (33.1) by

$$R(\theta,\phi) = R_0(1+\alpha'_0 Y_{20}(\theta)) = R_0\left(1+\beta\sqrt{\frac{5}{16\pi}}(3\cos^2\theta - 1)\right)$$

Thus,

$$R(\theta=0) = R_0\left(1+2\beta\sqrt{\frac{5}{16\pi}}\right) > R_0\left(1-\beta\sqrt{\frac{5}{16\pi}}\right) = R(\theta=\pi/2)$$

and the shape is *prolate*. If $\beta < 0$ the shape is *oblate*. For $\gamma = 0$ the distance of the surface from the origin is independent of ϕ and thus the system has axial symmetry around the 3-axis. If $\gamma \neq 0$ there is no longer axial symmetry. If $\langle\gamma\rangle \neq 0$ the system is triaxial.

The case of axial symmetry is of special interest. In Section 30 it was mentioned that the projection of the total angular momentum on an axis of symmetry vanishes. This is due to the fact that a rotation around such an axis does not change the state of the system and hence the contribution of the infinitesimal part of the rotation is zero. As is well known, in addition to the operator $\mathbf{J}^2$, only one component, J_z say, may be brought to a diagonal form. This raises the question how in addition to J and M also the component along another axis may be sharply defined.

To answer this question let us consider three orthogonal axes defined by three unit vectors $\mathbf{e}_1$, $\mathbf{e}_2$ and $\mathbf{e}_3$ which are functions of the dynamical variables of the system. The components of $\mathbf{J}$ along these axes are then *scalar products*

$$J_1 = (\mathbf{J}\cdot\mathbf{e}_1) = (\mathbf{e}_1\cdot\mathbf{J}) \qquad J_2 = (\mathbf{J}\cdot\mathbf{e}_2) \qquad J_3 = (\mathbf{J}\cdot\mathbf{e}_3) \tag{33.9}$$

As scalar products, J_1, J_2 and J_3 *commute* with J_x, J_y and J_z which are the components in a *fixed* frame of reference. It is therefore possible for *one* of the J_i to have a sharply defined eigenvalue. Only one of the J_i may be chosen for that since they do not commute.

The commutation relations of the J_i may be directly calculated. Recalling the commutation relation (6.19) and those of **J** with components of any vector (6.22), we obtain

$$\begin{aligned}[J_1, J_2] &= [(\mathbf{J}\cdot\mathbf{e}_1), (\mathbf{J}\cdot\mathbf{e}_2)] \\ &= [J_x e_{1x} + J_y e_{1y} + J_z e_{1z}, J_x e_{2x} + J_y e_{2y} + J_z e_{2z}] \\ &= [J_x e_{1x}, J_y e_{2y} + J_z e_{2z}] + [J_y e_{1y}, J_x e_{2x} + J_z e_{2z}] \\ &\quad + [J_z e_{1z}, J_x e_{2x} + J_y e_{2y}] \end{aligned} \tag{33.10}$$

To obtain the final form for (33.10) we consider for instance the commutator

$$\begin{aligned}[J_x e_{1x}, J_y e_{2y} + J_z e_{2z}] &= J_x[e_{1x}, J_y e_{2y}] + [J_x, J_y e_{2y}]e_{1x} + J_x[e_{1x}, J_z e_{2z}] \\ &\quad + [J_x, J_z e_{2z}]e_{1x} = iJ_x e_{1z} e_{2y} + iJ_z e_{1x} e_{2y} + iJ_y e_{1x} e_{2z} \\ &\quad - iJ_x e_{1y} e_{2z} - iJ_y e_{1x} e_{2z} - iJ_z e_{1x} e_{2y} \\ &= -iJ_x(e_{1y} e_{2z} - e_{1z} e_{2y}) = -iJ_x e_{3x} \end{aligned} \tag{33.11}$$

The last equality follows from the fact that the vector product of the unit vectors $\mathbf{e}_1$ and $\mathbf{e}_2$ is equal to the unit vector $\mathbf{e}_3$. For the commutator (33.10) we thus obtain

$$[J_1, J_2] = -iJ_x e_{3x} - iJ_y e_{3y} - iJ_z e_{3z} = -iJ_3 \tag{33.12}$$

The other commutators are given by analogous expressions so that we obtain the following commutation relations

$$[J_1, J_2] = -iJ_3 \qquad [J_2, J_3] = -iJ_1 \qquad [J_3, J_1] = -iJ_2 \tag{33.13}$$

These relations are similar to those of J_x, J_y and J_z but there is a difference in sign. The eigenvalues of J_1, J_2 and J_3 follow from (33.13) and the relation

$$J_1^2 + J_2^2 + J_3^2 = J_x^2 + J_y^2 + J_z^2 = \mathbf{J}^2 \tag{33.14}$$

The equality (33.14) holds even though J_x does not commute with e_{iy} etc. Collecting all terms on the l.h.s. of (33.14) we obtain

$$J_x^2\left(\sum_{i=1}^{3} e_{ix}^2\right) + J_y^2\left(\sum_{i=1}^{3} e_{iy}^2\right) + J_z^2\left(\sum_{i=1}^{3} e_{iz}^2\right) \qquad (33.15)$$

plus terms proportional to $\sum_{i=1}^{3} e_{ix}e_{iy}$, $\sum_{i=1}^{3} e_{ix}e_{iz}$ and $\sum_{i=1}^{3} e_{iy}e_{iz}$. These latter terms vanish due to the orthogonality of the three unit vectors $\mathbf{e}_1, \mathbf{e}_2, \mathbf{e}_3$ and hence (33.15) is equal to $\mathbf{J}^2$.

To calculate the eigenvalues of J_1, J_2 or J_3 we define

$$J^{0\prime} = J_3 \qquad J^{+\prime} = J_1 + iJ_2 \qquad J^{-\prime} = J_1 - iJ_2 \qquad (33.16)$$

The commutation relations of these operators are

$$[J^{0\prime}, J^{+\prime}] = -J^{+\prime} \qquad [J^{0\prime}, J^{-\prime}] = J^{-\prime} \qquad [J^{+\prime}, J^{-\prime}] = -2J^{0\prime} \qquad (33.17)$$

From (33.17) follows that $J^{0\prime}$, $J^{+\prime}$ and $J^{-\prime}$ generate the same $SU(2)$ Lie algebra as the operators J_x, J_y, J_z. The only difference is that the roles of $J^{+\prime}$ and $J^{-\prime}$ are interchanged in comparison with J^+ and J^-. Hence the eigenvalues of $J_3 = J^{0\prime}$ say, are the same as those of J_x, J_y or J_z. Surely, the eigenvalues of the Casimir operator (33.14) are the same whether calculated by using J^+, J^-, J^0 or $J^{+\prime}, J^{-\prime}, J^{0\prime}$. Thus, the projection of $\mathbf{J}$ on the *intrinsic* axis $\mathbf{e}_3$, usually denoted by K is an integer (or half integer) if J is an integer (or half integer).

The Hamiltonian of the nucleus, in any model, is invariant under spatial rotations. We conclude from this fact that there are $2J+1$ states for any allowed value of J and the energy is independent of the projection M on an axis fixed in space. This is not the case for the projection K on a body-fixed axis. It is a quantum number associated with the dynamics of the system. Unlike the geometrical quantum number M, the value of K determines the internal state of the system. As we will see in the following, a state with given J and M is usually associated with one value of K.

Let us come back to the Hamiltonian of the collective model. The special Hamiltonian (33.2) or (33.4) can be immediately diagonalized. The operator $\sum_\mu d_\mu^+ d_\mu$ is the *number operator* of the d-bosons which are the quanta of the five-dimensional harmonic oscillator. The Hamiltonian is diagonal if that number is definite and the energy is accordingly proportional, apart from the constant $5\hbar\omega/2$, to N, the number of d-bosons. This situation is the *vibrational limit* of the collective model. We will discuss it in detail within the interacting boson model.

More realistic Hamiltonians which have eigenstates corresponding to rotational spectra, contain potential energy terms which depend on β. In the lowest states of such Hamiltonians, the value of $\langle\beta\rangle$ may be rather large. The exact calculation of eigenvalues and eigenstates is rather difficult. The calculation may be carried out either by solving the differential equation due to the collective Hamiltonian or by using the description of states in terms of d-boson states. Due to the definitions (33.4) and (33.6), the expression of β in terms of the oscillator d-bosons is proportional to $\sum d^+_\mu d_\mu + \frac{5}{2} - \sum(-1)^\mu(d^+_\mu d^+_{-\mu} + d_\mu d_{-\mu})$. For potential energies, other than the one in (33.2), eigenstates of the Hamiltonian need not have a definite number of d-bosons.

An approximate solution of the problem may be obtained in a certain limit. If the nuclear shape is strongly deformed but axially symmetric in the ground state then $\langle\beta\rangle = \beta_0 \neq 0$ and $\langle\gamma\rangle = 0$. The deformation may be rather rigid, as would be the case for a potential well for β which is deep and narrow at $\beta = \beta_0$. Then, with a large moment of inertia of the deformed shape, the system would behave adiabatically with respect to rotations. Like in the case of molecular spectra, the rotation may not affect very much the internal structure. For most nuclei it is assumed that the shape is invariant under rotations by π around an axis perpendicular to the symmetry axis. The lowest states form then a rotational band with spins $J = 0, 2, 4, \ldots$ and energies proportional to $(1/2\mathcal{I})J(J+1)$ where $\mathcal{I}$ is the moment of inertia which is rather constant. Each state has a definite projection of $\mathbf{J}$ on the symmetry axis given by $K = 0$.

Excited bands in this *rotational limit* arise from vibrational states of the deformed shape. In one kind of vibration, axial symmetry is preserved. This vibration is in the variable β and the lowest mode is called a *beta vibration*. Its lowest excitation energy is higher than some rotational levels of the ground state band. Rotations of this vibrating shape are also characterized by $K = 0$ and the β-band has states with $J = 0, 2, 4, \ldots$. Another type of vibration is associated with the variable γ and the lowest excitation of this kind is called a *gamma vibration*. In such a vibration, axial symmetry can no longer be strictly preserved. The vibration leads to a component of $\mathbf{J}$ along the major axis and the lowest band of this type is characterized by $K = 2$. The γ-band has states with $J = 2, 3, 4, \ldots$. In the absence of rotation-vibration interactions these bands are not coupled. The moment of inertia is constant and the bands continue to higher J-values in this approximation.

Another limit where the collective motion assumes a simple form is for γ-soft, or γ-unstable, nuclei. In the rotational limit the dependence of the potential energy on γ has a deep minimum at $\gamma = 0$. In the other limit, the potential energy is *independent of* γ. The nucleus is deformed since there is a deep minimum at $\beta_0 \neq 0$ but has no longer axial symmetry. The energy levels arising from rotations of such a vibrating surface turn out to have a simple expression first determined by Wilets and Jean (1956). We will describe this dependence within the interacting boson model.

It was mentioned above that finding the eigenvalues and eigenstates of the collective Hamiltonian (or Bohr Hamiltonian) can be carried out by solving a differential equation. In that equation differential operators with respect to β and γ appear, as well as the operators L_1, L_2, L_3 which may be expressed in terms of the Euler angles and their derivatives. An alternative approach is to use basis states of d-bosons in terms of which the collective coordinates and momenta may be expressed. In the vibrational limit this is a very simple matter and the Hamiltonian (33.2) is expressed as (33.4) which is diagonal in the number of d-bosons. If the potential energy is not $\frac{1}{2}\omega^2\beta^2$ as in (33.2) but has a deep minimum at $\beta_0 \neq 0$, its expression in boson creation and annihilation operators is very complicated. Still, the d-boson states furnish a complete scheme and this approach may be used as was done by Gneuss and Greiner (1971) and by other authors.

The states of d-bosons describing a rotating and vibrating deformed shape are very complicated. The number of d-bosons is not a good quantum number and the eigenstates are linear combinations of states with different numbers of bosons. In the case of simple Hamiltonians there is no upper limit to the number of d-bosons in any eigenstate. For practical reasons, dictated by the computer, a truncation of that number had to be introduced.

An elegant method of truncation was found by Janssen, Jolos and Dönau (1974). They constructed a Hamiltonian for d-bosons ("phonons") whose eigenstates were intended to describe collective states of nuclei. Those eigenstates belong to irreducible representations of the $U(6)$ Lie group. Each symmetric irreducible representation of $U(6)$ is uniquely characterized by an integer N. The basis states of those representations are linear combinations of states with different values of n_d, the number of d-bosons. The numbers of d-bosons are, however, limited by N

$$n_d \leq N \tag{33.18}$$

These authors diagonalized their Hamiltonian which contained only a few parameters to obtain eigenstates and eigenvalues. Later, they (Jolos, Dönau and Janssen, 1975) realized that $U(6)$ has $SU(3)$ as a subgroup and $SU(3)$ Hamiltonians yield rotational spectra as explained in Section 30.

Meanwhile, Arima and Iachello (1975) have independently introduced a very elegant boson model in which $U(6)$ symmetry is practically built in. In their interacting boson model, they introduced in addition to d-bosons also s-bosons (with $l = 0$) thereby obtaining a six-dimensional space in which $U(6)$ transformations are easy to apply. The quantum number N characterizing the irreducible representations of $U(6)$ is then simply the total number of s- and d-bosons. They also showed that their Hamiltonian is equivalent to that of Janssen, Jolos and Dönau (1974). We will now describe the interacting boson model in more detail.

The following description does not provide many details of the interacting boson model. Many important formulae may be found in the specific books by Iachello and Arima (1987) and by Bonatsos (1988). Some of the discussion concerning $O(5)$ symmetry in Section 35 is not included in these books. Comparison of experimental data with theoretical predictions may be found in Casten (1990) and in a book edited by Casten, Volume 6 in this series. Odd-even nuclei have been considered in terms of a nucleon coupled to an even-even core which is described in the boson model (Iachello and Scholten, 1979). This model will not be presented here, apart from a short reference in Section 41. A detailed description of this *interacting boson fermion model* (IBFM) is given by Iachello and van Isacker (1991).

In the interacting boson model IBA-1 (Arima and Iachello 1975), it is assumed that low lying collective states of even-even nuclei can be described by the states of a given (fixed) number N of bosons. The bosons can be only in two different states, one with angular momentum $l = 0$ (s-boson) and the other with $l = 2$ (d-boson). The exact nature of these bosons is not defined and hence, we will not use configuration space wave functions for describing their states. The formalism of second quantization is more convenient and we will use creation and annihilation operators of s-bosons and d-bosons.

The operator s^+ creates a boson in the $l = 0$ state. The operator d_μ^+ creates a boson in the $l = 2$ state with $l_z = \mu$. The corresponding annihilation operators are the hermitean conjugates of these, s and d_μ. The operators d_μ^+ transform under rotations like the components of an irreducible tensor with rank $l = 2$. The operator d_μ annihilates a boson with $l = 2, l_z = \mu$ and hence, apart from a phase factor, is the $-\mu$ component of an irreducible tensor of rank $l = 2$. The phase factor is given, in analogy with the fermion case (17.12), by

$$\boxed{\tilde{d}_\mu = (-1)^\mu d_{-\mu}} \tag{33.19}$$

The creation and annihilation operators satisfy the Bose commutation relations which guarantee that the particles they create are in fully symmetric states and obey the Bose-Einstein statistics. Any two creation operators or two annihilation operators commute, and the following relations hold

$$\boxed{\begin{array}{ll} [s, s^+] = 1 & [s^+, d_\mu] = [d_\mu^+, s] = 0 \\ [d_\mu, d_{\mu'}^+] = \delta_{\mu\mu'} & \end{array}} \tag{33.20}$$

The creation and annihilation operators considered here may be viewed as those of bosons of six independent one dimensional harmonic oscillators. Hence, the numbers of s-bosons are given by the (integral) eigenvalues of the operator

$$\boxed{\hat{n}_s = s^+ s} \tag{33.21}$$

The number of d-bosons in a state with $l_z = \mu$ is similarly given by the eigenvalues of $d_\mu^+ d_\mu$. The number operator of all d-bosons is thus equal to

$$\boxed{\hat{n}_d = \sum_\mu d_\mu^+ d_\mu = (d^+ \cdot \tilde{d})} \tag{33.22}$$

Single boson operators can be constructed, in analogy to the fermion case in Section 17, as

$$d_\mu^+ d_{\mu'}, \quad d_\mu^+ s, \quad s^+ d_\mu, \quad s^+ s \tag{33.23}$$

Altogether there are 36 such independent operators. To be able to use tensor algebra it is convenient to define single boson operators which are components of irreducible tensors. Taking linear combinations of the operators in (33.23) we obtain by using the definition (33.19), the following irreducible tensors

$$(d^+ \times \tilde{d})^{(k)}_\kappa = \sum_{\mu\mu'} (2\mu 2\mu' \mid 22k\kappa) d^+_\mu \tilde{d}_{\mu'} \qquad k = 0, 1, 2, 3, 4 \tag{33.24}$$

as well as the two hermitean combinations with rank $k = 2$

$$d^+_\mu s + s^+ \tilde{d}_\mu \qquad i(d^+_\mu s - s^+ \tilde{d}_\mu) \tag{33.25}$$

and the scalar ($k = 0$)

$$s^+ s \tag{33.26}$$

The $k = 1$ tensor in (33.24) is proportional to the angular momentum operator in the space of d-bosons. In view of (8.13) we obtain

$$J_\kappa = \sqrt{10}(d^+ \times \tilde{d})^{(1)}_\kappa \tag{33.27}$$

All operators (33.23), as well as any linear combination of them, commute with the number operator of s- and d-bosons given by

$$\hat{N} = \hat{n}_s + \hat{n}_d = s^+ s + \sum_\mu d^+_\mu d_\mu = s^+ s + (d^+ \cdot \tilde{d}) \tag{33.28}$$

A Hamiltonian of the boson system with single boson terms and boson-boson interactions may now be constructed. The most general rotationally invariant Hamiltonian of this kind is obtained as a linear combination of all scalar products of the irreducible tensors in (33.24), (33.25) and (33.26) to which the two single boson scalars, (33.26) and the $k = 0$ term in (33.24), may be added. The various terms obtained in this way however, are not independent. The requirement that the boson Hamiltonian be an hermitean operator also reduces the number of independent terms. Even the scalar products constructed from the irreducible tensors (33.24) are not independent. The situation is similar to that in the case of the fermion Hamiltonian (18.11). The various terms in that expansion are linearly independent only if antisymmetric as well as symmetric states are considered. For antisymmetric states only, there are less independent terms in the Hamiltonian as demonstrated by its expression as (18.2).

The only allowed states of d-bosons are fully symmetric. Hence, there are only three allowed states of two d-bosons, those with $J = 0, 2, 4$. Thus, there are only three independent terms among the five scalar products of the tensors (33.24). To make explicit use of this fact we transform all scalar products by changing the order of coupling. Due to the commutation relations (33.20), this yields single boson (scalar) terms in addition to real boson-boson interactions. The resulting Hamiltonian may be expressed in the following form

$$
\boxed{
\begin{aligned}
H = {} & \epsilon_s s^+ s + \epsilon_d (d^+ \cdot \tilde{d}) + \frac{1}{2} \sum_{J=0,2,4} c_J (d^+ \times d^+)^{(J)} \cdot (\tilde{d} \times \tilde{d})^{(J)} \\
& + \frac{1}{2} \tilde{v}_0 \{ (d^+ \times d^+)_0^{(0)} s^2 + (s^+)^2 (\tilde{d} \times \tilde{d})_0^{(0)} \} \\
& + \frac{1}{\sqrt{2}} \tilde{v}_2 \{ [(d^+ \times d^+)^{(2)} \times \tilde{d} s]_0^{(0)} + [s^+ d^+ \times (\tilde{d} \times \tilde{d})^{(2)}]_0^{(0)} \} \\
& + \frac{1}{2} u_0 (s^+)^2 s^2 + \frac{1}{\sqrt{2}} u_2 s^+ s (d^+ \cdot \tilde{d})
\end{aligned}
}
\tag{33.29}
$$

The Hamiltonian (33.29) contains 9 arbitrary parameters, two coefficients of single boson energies and 7 of boson-boson interactions. Their definition was made to conform with accepted standard notation. The Hamiltonian (33.29) commutes with the boson number operator (33.28) and its eigenvalues may be used to label the various eigenstates. For a given value of N, we can use it to eliminate the operator $s^+s = N - (d^+ \cdot \tilde{d})$ and obtain for the Hamiltonian (33.29) the equivalent expression

$$\boxed{\begin{aligned} H &= \epsilon_s N + \tfrac{1}{2}u_0 N(N-1) + \epsilon(d^+ \cdot \tilde{d}) \\ &+ \frac{1}{2}\sum_{J=0,2,4} c'_J (d^+ \times d^+)^{(J)} \cdot (\tilde{d} \times \tilde{d})^{(J)} \\ &+ \tfrac{1}{2}\tilde{v}_0\{(d^+ \times d^+)_0^{(0)} s^2 + (s^+)^2 (\tilde{d} \times \tilde{d})_0^{(0)}\} \\ &+ \frac{1}{\sqrt{2}}\tilde{v}_2\{[(d^+ \times d^+)^{(2)} \times \tilde{d}s]_0^{(0)} + [s^+ d^+ \times (\tilde{d} \times \tilde{d})^{(2)}]_0^{(0)}\} \end{aligned}} \tag{33.30}$$

The single d-boson energy ϵ in (33.30) is given by

$$\epsilon = \epsilon_d - \epsilon_s + (N-1)\left(\frac{u_2}{\sqrt{5}} - u_0\right) \tag{33.31}$$

The coefficients c'_J in (33.30) are defined by

$$c'_J = c_J + u_0 - \frac{2u_2}{\sqrt{5}} \tag{33.32}$$

In the expression (33.30) of the boson Hamiltonian, by using N, one parameter was absorbed into the N-dependence of ϵ according to (33.31). The first two terms in that expression contribute equally to all states with given N. Level spacings, for any value of N, are independent of the values of these terms. Thus, for given N, only 6 parameters completely determine level spacings and eigenstates of the most general boson Hamiltonian considered here. Five of these parameters are independent of N and only ϵ is a linear function of N given by (33.31).

Once values of these parameters have been fixed, the Hamiltonian may be diagonalized in the subspace of N bosons. The eigenvalues thus obtained should yield the energy levels of the nucleus considered. Rates of electromagnetic transitions between various states are determined by matrix elements of the transition operator between the various eigenstates. The most important transitions which reveal the nature of the collective states are electric quadrupole (E2) transitions. It is assumed in the interacting boson model that the E2 transition operator is a *single boson operator* for which the most general expression is

$$\boxed{T_\mu^{(2)} = \alpha_2(d_\mu^+ s + s^+ \tilde{d}_\mu) + \beta_2(d^+ \times \tilde{d})_\mu^{(2)}} \tag{33.33}$$

The E2 transition rates are thus determined by the coefficients α_2 and β_2 which may still depend on the value of N.

To obtain the matrix of the general boson Hamiltonian in the space of N bosons, it is necessary to define a basis in which matrix elements of the Hamiltonian may be calculated. The states of such a basis will be labeled by certain quantum members. These are eigenstates if the Hamiltonian is diagonal in that basis. We saw in preceding sections how the elegant and powerful methods of group theory help in constructing such bases. Group theory is much more important for the interacting boson model. For certain values of the parameters in (33.30) it is possible to calculate eigenvalues and transition probabilities in a compact analytic form. Such limiting situations turn out to be of great physical interest. At the same time, their systems of eigenstates form convenient bases for calculating matrix elements of the general boson Hamiltonian. We shall now proceed to review the group theory of the interacting boson model.

The single boson states $s^+|0\rangle$ and $d_\mu^+|0\rangle$ span a six-dimensional space. The set of single boson operators (33.23) transforms any single boson state into a single boson state. Hence, they may be considered as generators of a Lie group of linear transformations. Indeed, the commutator of any two single boson operators is a single boson operator as follows from the commutation relations (33.20). Since the operators (33.23) form a complete set, they obey the conditions of a Lie algebra. The group of transformations generated by single boson operators may be identified, as in the case of fermion operators, as the unitary group in six dimension $U(6)$. The number of generators of $U(n)$ is n^2 and there are 36 operators in the set (32.23).

The commutation relations of the generators (33.24), (33.35) and (33.36) may be obtained by using the relations (33.20). They are as follows

$$[(d^+ \times \tilde{d})^{(k')}_{\kappa'}, (d^+ \times \tilde{d})^{(k'')}_{\kappa''}] = \sum_{k\kappa} (-1)^k [1 - (-1)^{k+k'+k''}][(2k'+1)(2k''+1)]^{1/2} \times \begin{Bmatrix} 2 & 2 & k \\ k' & k'' & 2 \end{Bmatrix} (k'\kappa' k''\kappa'' \mid k'k''k\kappa)(d^+ \times \tilde{d})^{(k)}_{\kappa} \tag{33.34}$$

$$[d^+_\mu s, s^+ \tilde{d}_{\mu'}] = (-1)^{\mu+1} \delta_{\mu,-\mu'} s^+ s + \sum_{k\kappa} (2\mu 2\mu' \mid 22k\kappa)(d^+ \times \tilde{d})^{(k)}_{\kappa} \tag{33.35}$$

$$[d^+_\mu s, (d^+ \times \tilde{d})^{(k)}_{\kappa}] = (-1)^{\mu+1} \sum_{\mu'} (2\mu' 2, -\mu \mid 22k\kappa) d^+_{\mu'} s$$
$$[s^+ \tilde{d}_\mu, (d^+ \times \tilde{d})^{(k)}_{\kappa}] = (-1)^{k+\mu} \sum_{\mu'} (2\mu' 2, -\mu \mid 22k\kappa) s^+ \tilde{d}_{\mu'} \tag{33.36}$$

The other commutators are much simpler and there is no need to write them explicitly.

When acting on states with several bosons, the single boson operators induce linear transformations among them. We can consider sets of states which transform irreducibly among themselves. In each set, we can find a basis of linearly independent states in terms of which any state which belongs to the irreducible set may be expanded. Such basis sets form bases of irreducible representations of the $U(6)$ group of unitary transformations. By definition, generators of the group transform states which belong to an irreducible representation into themselves. This is also the case for the Hamiltonian (33.29) which is constructed of such $U(6)$ generators.

The matrix of the Hamiltonian, constructed in the set of states which are bases of irreducible representations, has a simple form. It has non-vanishing sub-matrices along the diagonal, each of which is defined by basis states of a certain irreducible representation. All matrix elements between states in different irreducible representations vanish. Hence, eigenstates of such a Hamiltonian may be classified according to the irreducible representation to which they belong. A set of states which belong to a given irreducible representation includes, in general, states with different values of J (and M) as well as different states with the same J (and M) values. Hence, the quantum numbers (labels) of the irreducible representation to which a state belongs do not, in general, uniquely characterize it.

The eigenvalues of the Hamiltonian which belong to the various eigenstates within the same irreducible representation need not be equal. It is only if the Hamiltonian commutes with *all* generators that it has the same eigenvalue in all such states (Schur's lemma). A simple example of such a situation, mentioned in preceding sections, is offered by rotationally invariant Hamiltonians. They commute with the three components of the total angular momentum which are the generators of the group of three dimensional rotations $O(3)$ or $SU(2)$. In such cases, all $2J+1$ eigenvalues of the eigenstates with given J (and $J \geq M \geq -J$) are indeed equal.

As explained in Section 26, irreducible representations of $U(n)$ groups are characterized by their symmetry properties. The commutation relations (33.20) guarantee the Bose statistics of the bosons from which follows that all states must be fully symmetric. For given N there is only one fully symmetric irreducible representation of $U(N)$. The Young diagram has one row of N squares and it is thus fully characterized by the number N of bosons. All results which will be derived below hold for fully symmetric boson states.

A Hamiltonian which commutes with all generators of $U(6)$ is any linear combination of the linear Casimir operator (33.28) and the quadratic Casimir operator defined by

$$\boxed{\begin{aligned} C_{U(6)} = &\sum_{k=0}^{4} (d^+ \times \tilde{d})^{(k)} \cdot (d^+ \times \tilde{d})^{(k)} + \tfrac{1}{2}(d^+ s + s^+ \tilde{d}) \cdot (d^+ s + s^+ \tilde{d}) \\ &- \tfrac{1}{2}(d^+ s - s^+ \tilde{d}) \cdot (d^+ s - s^+ \tilde{d}) + (s^+ s)(s^+ s) \end{aligned}} \tag{33.37}$$

It can be directly verified by using the commutation relations (33.34), (33.35), (33.36) that the operator (33.37) commutes with all the generators (33.23). The eigenvalues of the Casimir operator (33.37) will be now obtained. We consider first the summation on the r.h.s. of (33.37) and express it as

$$\sum_{k=0}^{4}(d^+ \times \tilde{d})^{(k)} \cdot (d^+ \times \tilde{d})^{(k)}$$

$$= \sum_{\substack{k,\kappa \\ \mu,\mu',\nu,\nu'}} (2\mu 2\mu' \mid 22k\kappa)(2\nu 2\nu' \mid 22k,-\kappa) \times (-1)^{\kappa+\mu'+\nu'} d_\mu^+ d_{-\mu'} d_\nu^+ d_{-\nu'}$$

$$= \sum_{k\kappa\mu\mu'\nu\nu'} (2\mu 2\mu' \mid 22k\kappa)(2,-\nu' 2,-\nu \mid 22k\kappa) \times (-1)^{\mu+\nu'} d_\mu^+ d_{-\mu'} d_\nu^+ d_{-\nu'} \tag{33.38}$$

The last equality was obtained by using the equality $\kappa = \mu + \mu'$ and the symmetry properties of the Clebsch-Gordan coefficients. We can now use their orthogonality relations to obtain for (33.38) the expression

$$\sum_{\mu,\mu'\nu,\nu'} (-1)^{\mu+\nu'} \delta_{\mu,-\nu'} \delta_{\mu',-\nu} d_\mu^+ d_{-\mu'} d_\nu^+ d_{-\nu'} = \sum_{\mu,\nu} d_\mu^+ d_\nu d_\nu^+ d_\mu \tag{33.39}$$

Using the commutation relations (33.20) we obtain from (33.39) the following expression for the operator (33.38)

$$-\sum_{\mu,\nu} d_\mu^+ d_\nu \delta_{\mu\nu} + \sum_{\mu,\nu} d_\mu^+ d_\nu d_\mu d_\nu^+ = -\sum_{\mu} d_\mu^+ d_\mu + \sum_{\mu,\nu} d_\mu^+ d_\mu d_\nu d_\nu^+$$

$$= -\sum_{\mu} d_\mu^+ d_\mu + \sum_{\mu,\nu} d_\mu^+ d_\mu d_\nu^+ d_\nu + 5\sum_{\mu} d_\mu^+ d_\mu = \hat{n}_d^2 + 4\hat{n}_d \tag{33.40}$$

The second and third term on the r.h.s. of (33.37) are equal to

$$(d^+ \cdot \tilde{d}) s s^+ + s^+ s (\tilde{d} \cdot d^+)$$

$$= (d^+ \cdot \tilde{d}) s^+ s + (d^+ \cdot \tilde{d}) + s^+ s (d^+ \cdot \tilde{d}) + 5 s^+ s$$

$$= 2\hat{n}_s \hat{n}_d + \hat{n}_d + 5\hat{n}_s \tag{33.41}$$

The equality in (33.41) was obtained by using the commutation relations (33.20). Combining now (33.40),(33.41) and the last term on the r.h.s. of (33.37) which is $\hat{n}_s^2$, we obtain that the Casimir operator (33.37) is equal to the operator

$$\boxed{\begin{aligned} C_{U(6)} &= \hat{n}_d^2 + 4\hat{n}_d + 2\hat{n}_s\hat{n}_d + \hat{n}_d + 5\hat{n}_s + \hat{n}_s^2 \\ &= (\hat{n}_d + \hat{n}_s)^2 + 5(\hat{n}_d + \hat{n}_s) = \hat{N}(\hat{N} + 5) \end{aligned}} \tag{33.42}$$

Hence, in the case of fully symmetric representations of $U(6)$, which are the only ones allowed by the Bose commutation relations (33.20), the quadratic Casimir operator (33.37) is equal to the simple linear and quadratic function (33.42) of the number operator $\hat{N}$. Its eigenvalues are then equal to $N(N + 5)$.

Let us now turn back to the general $U(6)$ Hamiltonian. As explained above, each of its eigenstates is characterized by the quantum number N which specifies the irreducible representation to which it belongs. This quantum number has the same value for all fully symmetric eigenstates of a given nucleus. There may be in a given irreducible representation of $U(6)$ several states with various values of J (and M). When going over from $U(6)$ to its subgroup $O(3)$, whose generators are given by (33.27), a $U(6)$ irreducible representation splits into several irreducible representations of $O(3)$. Each of the latter is characterized by a definite value of J and has a basis set of $2J + 1$ states. In most cases, however, N and J (and M) do not specify uniquely the states. There may be in a given irreducible representation of $U(6)$ several equivalent irreducible representations of $O(3)$. Other quantum numbers, in addition to N and J (and M) may be necessary to have a complete characterization of the boson system.

As in the case of nucleons, additional quantum numbers may be found by considering subgroups of $U(6)$ of which $O(3)$ is a subgroup. Two states with the same value of J (and M) which belong to the same $U(6)$ irreducible representation may belong to different irreducible representations of such a subgroup. These irreducible representations may be used as additional quantum numbers. If the Hamiltonian is constructed only from the generators of the subgroup, these additional quantum numbers specify *eigenstates* of the Hamiltonian—they are good quantum numbers. In other cases they may serve as labels for a set of states in which the matrix of the Hamiltonian may be constructed. This procedure will be described in the next section.

34

The $U(5)$ Scheme— the Vibrational Limit

As explained in Section 33, subgroups of $U(6)$ may furnish a classification scheme for states of s- and d-bosons. The simplest subgroup of $U(6)$ that we consider is the group $U(5)$ whose generators are the 25 operators in (33.24). More precisely, the subgroup we consider is $U(5) \otimes U(1)$ where the unitary transformations of $U(5)$ transform states of a d-boson into themselves and $U(1)$ trivially transforms the state of an s-boson into itself. Clearly, the generators (33.24) form a Lie algebra, as also follows formally from their commutation relations (33.34). A boson Hamiltonian constructed from only the generators of $U(5) \otimes U(1)$ is obtained from (33.29) by putting $\tilde{v}_0 = 0$ and $\tilde{v}_2 = 0$. The conservation of N was used to derive (33.30). We now consider Hamiltonians which are constructed only from $U(5)$ generators. Starting from (33.30) and putting $\tilde{v}_0 = \tilde{v}_2 = 0$ we obtain a Hamiltonian with $U(5)$ symmetry and will therefore refer simply to the $U(5)$ subgroup of $U(6)$.

The Hamiltonian (33.30), with $\tilde{v}_0 = \tilde{v}_2 = 0$, commutes with $\hat{n}_d = (d^+ \cdot \tilde{d})$. Hence, the number of d-bosons n_d is a good quantum number for such Hamiltonians (as well the number $n_s = N - n_d$). The fully symmetric irreducible representations of $U(5)$ are uniquely characterized by n_d. The corresponding Young diagrams have only one row with n_d squares in it. The eigenstates of the $U(5)$ boson Hamiltonian are given accordingly by the schematic expression

$$(d^+)^{n_d}(s^+)^{N-n_d}|0\rangle \tag{34.1}$$

They are characterized by N, n_d and by the total angular momentum J which is due to the d-bosons only (as expressed by (33.27)). Still, there may be several sets with the same value of J in one irreducible representation of $U(5)$. In addition to N, n_d, J (and M) other quantum numbers may be necessary to uniquely specify the eigenstates (34.1).

Such additional quantum numbers may be found if a subgroup of $U(5)$ itself is found which includes $O(3)$ as a subgroup. In the case of l^n configurations, considered in Section 28, we saw that $U(2l+1)$ has a subgroup $O(2l+1)$ which itself has $O(3)$ as a subgroup. Also here $O(5)$, the group of real orthogonal transformations in the five-dimensional space of a d-boson, is a subgroup of $U(5)$ and contains $O(3)$ as a subgroup. Its transformations leave invariant the symmetrical bilinear form which is the $J = 0$ state of two d-bosons, $(d^+ \times d^+)_0^{(0)}|0\rangle$. Its $5 \times 4/2 = 10$ generators are the components of

$$\boxed{(d^+ \times \tilde{d})_\kappa^{(3)} \qquad (d^+ \times \tilde{d})_\kappa^{(1)}} \tag{34.2}$$

Indeed, the generators (34.2) annihilate the state $(d^+ \times d^+)_0^{(0)}|0\rangle$. Due to the Wigner-Eckart theorem when the operator $(d^+ \times \tilde{d})_\kappa^{(k)}$ acts on such a state, the resulting state must have $J = k$, $M = \kappa$. The state $(d^+ \times d^+)_0^{(0)}|0\rangle$ is symmetric but a state $(d^+ \times d^+)_\kappa^{(k)}|0\rangle$ with k odd, is antisymmetric and hence vanishes due to the Bose commutation relations. From (34.2) we see that the $O(3)$ generators (33.27) form a sub algebra of the $O(5)$ Lie algebra and hence $O(3) \subset O(5)$. This follows directly from the definition of $O(5)$ since $O(3)$ transformations transform the M components of any J state among themselves and hence transform the state $(d^+ \times d^+)_0^{(0)}|0\rangle$ into itself.

In the case of l^n nucleon configurations, the quantum numbers due to the $O(2l+1)$ group were those of the seniority scheme. Also here,

$O(5)$ introduces the seniority scheme for d-bosons. In order to define the seniority quantum number we may use the Casimir operator of $O(5)$ whose eigenvalues uniquely characterize the $O(5)$ irreducible representations. Let us go one step back and look at the Casimir operator of $U(5)$.

The quadratic Casimir operator of $U(5)$ is simply the summation over k which is the first term on the r.h.s. of (33.37). From the derivation following (33.37), we obtain for the Casimir operator of $U(5)$ the form (33.40). Thus,

$$\boxed{C_{U(5)} = \sum_{k=0}^{4} (d^+ \times \tilde{d})^{(k)} \cdot (d^+ \times \tilde{d})^{(k)} = \hat{n}_d^2 + 4\hat{n}_d} \tag{34.3}$$

and its eigenvalues are given by $n_d(n_d + 4)$. As in the case of $U(6)$, in states which are fully symmetric, as prescribed by the commutation relations (33.20), the quadratic Casimir operator is a linear combination of $\hat{n}_d$ (which is the linear Casimir operator of $U(5)$) and its square.

The definition of the (quadratic) Casimir operator of $O(5)$ is

$$\boxed{C_{O(5)} = 2 \sum_{k=1,3} (d^+ \times \tilde{d})^{(k)} \cdot (d^+ \times \tilde{d})^{(k)}} \tag{34.4}$$

The operator (34.4) commutes with all $O(5)$ generators (34.2) as follows directly from the commutation relations (33.34). There is no linear Casimir operator of the $O(5)$ group since such an operator, commuting with the angular momentum components, must be a scalar. There is no scalar operator among those in (34.2). We shall now proceed to calculate the eigenvalues of (34.4).

As in the case of nucleons, the Casimir operator is closely related to a pairing interaction. This is a two body operator which has eigenvalue 5 in the $J = 0$ state and eigenvalues 0 in the $J = 2, 4$ states of two d-bosons. It can be simply expressed by $(d^+ \cdot d^+)(\tilde{d} \cdot \tilde{d})$. By using the change of coupling transformation (10.10) and observing the commutation relations, we may express it as

$$\begin{aligned}(d^+ \cdot d^+)(\tilde{d} \cdot \tilde{d}) &= \sum_{k=0}^{4} (-1)^k (d^+ \times \tilde{d})^{(k)} \cdot (d^+ \times \tilde{d})^{(k)} - (d^+ \cdot \tilde{d}) \\ &= \sum_{k=0}^{4} (d^+ \times \tilde{d})^{(k)} \cdot (d^+ \times \tilde{d})^{(k)} - (d^+ \cdot \tilde{d}) \\ &\quad - 2 \sum_{k=1,3} (d^+ \times \tilde{d})^{(k)} \cdot (d^+ \times \tilde{d})^{(k)} \\ &= \hat{n}_d^2 + 3\hat{n}_d - C_{O(5)}\end{aligned} \tag{34.5}$$

The first equality in (34.5) is the analog of the expansion (12.18) of the fermion pairing interaction. The last equality is due to (34.3). The irreducible representations of $O(5)$, which are included in the $U(5)$ irreducible representation characterized by n_d, have Young diagrams with one row, whose length τ is the boson seniority. States with maximum seniority are characterized by $\tau = n_d$. In general, states may have seniorities

$$\tau = n_d, \tau = n_d - 2, n_d - 4, \ldots, 1 \text{ or } 0$$

In states with $\tau = n_d$ there are no d-boson pairs coupled to $J = 0$. Recalling the general relation between pairing interactions and Casimir operators, we find by applying (34.5) to a state with $n_d = \tau$, $C_{O(5)}(\tau) - \tau(\tau + 3) = 0$, from which the eigenvalues of $C_{O(5)}$ are obtained as equal to $\tau(\tau + 3)$ (they depend only on τ and not on n_d). We shall derive this result more formally by following the procedure adopted in Section 19. There we used the $SU(2)$ algebra, of $J = 0$ pair creation and annihilation operators, to derive the eigenvalues of the pairing interaction.

In the present case we first evaluate the commutator

$$[\tilde{d} \cdot \tilde{d}, d^+ \cdot d^+] = 4 \sum_{\mu} d_\mu^+ d_\mu + 2 \times 5 = 4n_d + 2 \times 5 \tag{34.6}$$

We can then define

$$P^+ = \tfrac{1}{2}(d^+ \cdot d^+) \qquad P^- = \tfrac{1}{2}(\tilde{d} \cdot \tilde{d})$$

$$[P^-, P^+] = n_d + \tfrac{1}{2} \times 5 = 2P^0 \tag{34.7}$$

which satisfy the following commutation relations

$$[P^0, P^+] = P^+ \qquad [P^0, P^-] = -P^- \qquad [P^+, P^-] = -2P^0 \tag{34.8}$$

There is a difference in sign between the last commutator in (34.8) and the analogous commutator in (19.7) of the fermion case. This difference is due to the commutation relations obeyed by the boson creation and annihilation operators. The operators P^+, P^- and P^0 are elements of the Lie algebra of the $SU(1,1)$ group rather than $SU(2)$. We shall see in the following the different behavior of the irreducible representations in the two cases.

To calculate the eigenvalues of the pairing interaction (34.5), which is equal to $4P^+P^-$, we first observe that the Casimir operator in the present case is given by

$$\mathbf{P}^2 = (P^0)^2 - \tfrac{1}{2}P^+P^- - \tfrac{1}{2}P^-P^+ = (P^0)^2 - P^0 - P^+P^- \tag{34.9}$$

Due to the commutation relations (34.8), this operator $\mathbf{P}^2$ commutes with P^+ and P^- as well as with P^0. We can thus bring both $\mathbf{P}^2$ and P^0 into diagonal form. From (34.9) follows that the eigenvalue γ of $\mathbf{P}^2$ is *smaller* than or *equal* to the square of the eigenvalues m of P^0 (both P^+P^- and P^-P^+ are products of hermitean conjugate operators and thus have non-negative diagonal matrix elements). We shall consider only positive eigenvalues of P^0, $m > 0$ since only these are the possible eigenvalues of the specific P^0 operator in (34.7).

According to (34.8) when P^- acts on an eigenstate of P^0 with eigenvalue $m > 0$, it yields an eigenstate with eigenvalue $m - 1$ (it reduces n_d by two). This procedure may be continued, however, only a finite number of times. A state with $n_d = 1$ or $n_d = 0$ is annihilated by the annihilation operator P^-. Hence, there is a state with lowest value of positive m, to be denoted by m_0, which is annihilated by P^- as well as by P^+P^-. In that case we obtain by applying (34.9) to the state with m_0 the result

$$\gamma = m_0^2 - m_0 = m_0(m_0 - 1) \tag{34.10}$$

The value of n_d for which m_0 is reached is directly related to the seniority τ of that state. A state with seniority τ and $n_d = \tau$ has no pairs coupled to $J = 0$ and hence it is annihilated by P^-,

$$P^-|n_d = \tau, \tau JM\rangle = 0 \tag{34.11}$$

Applying to such a state the relation (34.9) we obtain, due to (34.7), an explicit expression for (34.10), namely

$$\gamma = m_0(m_0 - 1) = \tfrac{1}{4}(\tau + \tfrac{5}{2})(\tau + \tfrac{5}{2} - 2) \tag{34.12}$$

Substituting this expression for γ into (34.9) we obtain for the eigenvalues of the pairing interaction $4P^+P^-$ the expression

$$(n_d + \tfrac{5}{2})(n_d + \tfrac{5}{2} - 2) - (\tau + \tfrac{5}{2})(\tau + \tfrac{5}{2} - 2) = n_d(n_d + 3) - \tau(\tau + 3) \tag{34.13}$$

Comparing (34.13) with (34.5) we see that the eigenvalues of $C_{O(5)}$ are indeed equal to $\tau(\tau + 3)$. This is a special case of the general formula (28.25) for $l = 2$.

The eigenvalue γ in (34.12) may be expressed in analogy with the $SU(2)$ case, (19.9), as $p(p - 1)$. Choosing for p the positive value we obtain

$$p = \tfrac{1}{2}(\tau + \tfrac{5}{2}) \tag{34.14}$$

The eigenvalues of P^0 of states which belong to the irreducible representation of $SU(1,1)$ characterized by this value of p are

$$p = \tfrac{1}{2}(\tau + \tfrac{5}{2}), p + 1 = \tfrac{1}{2}(\tau + \tfrac{7}{2}), p + 2 = \tfrac{1}{2}(\tau + \tfrac{9}{2}), \ldots \tag{34.15}$$

The corresponding states are obtained by applying successively the pair creation operator P^+ on the state with $n_d = \tau$. There are infinitely many basis states in this irreducible representation. Another irreducible representation is determined by $-p$ and the eigenvalues of

P^0 are $-p, -p-1, -p-2, \ldots$. We shall not consider this case which does not arise from the specific generators (34.7) nor any other irreducible representations of $SU(1,1)$. The values of p (and of the eigenvalues of P^0) are *quarter* integers. Comparing this situation with the fermion case (19.10), we see that there, for half integral spin j, the number of components is even and $(2j+1)/2$ is an integer. In the case of bosons considered here, the spin k is integral and $(2k+1)/2$ is half integral yielding values of p which are quarter integral.

In the case of seniority of nucleons, an irreducible representation of $SU(2)$ characterized by v (or $s = \frac{1}{2}((2j+1/2) - v)$) contains a *finite* number of states, with given J and M, in which the nucleon number n was limited by $v \leq n \leq 2j+1-v$ (or $-s \leq S^0 \leq s$). The generators of $SU(2)$, e.g. S^+, transform one such state into another. The number n could not exceed $2j+1-v$ because of the Pauli principle. In the case of bosons considered here, the lowest value of the boson number is $n_d = \tau$ but it may assume the values $\tau+2, \tau+4, \ldots$ indefinitely. This is due to the Bose-Einstein statistics of the bosons and follows formally from the fact that the term $-P^+P^-$ in (34.9) is always negative (or zero). In the fermion case, the analogous operator, $S^+S^- = \mathbf{S}^2 - (S^0)^2 + S^0$, is non-negative which limits the values of S^0. The operator P^+ adding a pair of d-bosons coupled to $J=0$ to any state with given τ, makes the possible number of bosons unlimited. The irreducible representation of $SU(1,1)$ characterized by τ has accordingly infinitely many states. Unlike $SU(2)$, the group $SU(1,1)$ is *non-compact*.

In the case of fermions, generators of the quasi-spin group $SU(2)$ transform states with given seniority v, α J and M which have various values of n among themselves. Such a set of states is the basis of an irreducible representation of $SU(2)$, characterized by v, which has a finite number of states. On the other hand, generators of the $Sp(2j+1)$ group transform among themselves states with definite v and various α, J and M, all of which have the same number of particles n. The number n does not change since $Sp(2j+1)$ is a subgroup of $U(2j+1)$. In the case of bosons, $SU(1,1)$ plays the role corresponding to that of $SU(2)$ for fermions. Its generators tranform states with given τ, α, J and M which have various values of n_d among themselves. This set of states is the basis of an irreducible representation of $SU(1,1)$, characterized by τ, which has infinite dimension. On the other hand, generators of $O(5)$, which is a subgroup of $U(5)$, transform states with given v and given n_d and various values of α, J and M among themselves. In analogy to the fermion case, considered in Section 19, we will use the generators (34.7) of $SU(1,1)$ to establish relations between states and matrix elements with different values of n_d in the boson seniority scheme.

To define the angular momentum of the various states we consider the group $O(3)$ which is a subgroup of $O(5)$. The irreducible representations of $O(5)$ split into several irreducible representations of $O(3)$ with given values of J (and $M, -J \leq M \leq J$). We define in this way the following chain of groups

$$\text{I} \qquad U(6) \supset U(5) \supset O(5) \supset O(3) \tag{34.16}$$

We may now consider a linear combination of the Casimir operators of this chain of groups. By their definition, they all commute and thus can be diagonalized simultaneously. Altogether there are four Casimir operators, linear and quadratic ones of $U(5)$, $C_{O(5)}$ and $C_{O(3)} = \mathbf{J}^2$. They are also linearly independent as will be shown below. There are four independent operators in the $U(5)$ boson Hamiltonian (33.30) without the linear and quadratic terms in N and where we put $\tilde{v}_0 = \tilde{v}_2 = 0$. Hence, the most general $U(5)$ Hamiltonian may be expressed as a linear combination of the 4 Casimir operators. Since the latter contain also single boson terms, we subtract them from $C_{O(3)}$ and replace $C_{O(5)}$ by the pairing interaction (34.5). When these operators are diagonal we obtain the following closed expression for the eigenvalues of the most general $U(5)$ Hamiltonian

$$\boxed{\begin{aligned} &\epsilon n_d + \alpha n_d(n_d - 1)/2 + \beta[n_d(n_d + 3) - \tau(\tau + 3)] \\ &\quad + \gamma[J(J + 1) - 6n_d] \end{aligned}} \tag{34.17}$$

From (34.17) follows that any $U(5)$ Hamiltonian is diagonal in the seniority scheme. Its eigenstates have definite seniorities τ. The seniority scheme in this case is not just a convenient basis for labelling states but τ is a good quantum number for any Hamiltonian with interactions between two d-bosons. The reason is purely geometrical and is analogous to the situation of $j = 5/2$ identical fermions (Section 16). There are only three fully symmetric states of two d bosons, namely $(d^+ \times d^+)^{(J)}_M|0\rangle$ for $J = 0$, $\tau = 0$ and $J = 2, 4$, $\tau = 2$. Any two body interaction that reproduces the two-boson matrix elements in these states, the c'_J of (33.30), can be used as the exact boson boson interaction in any d^n configuration. We can thus replace any two-body interaction by a linear combination of 1 (a constant interaction), $2(\mathbf{J}_1 \cdot \mathbf{J}_2)$ and $4P^+P^-$—the pairing interaction (34.5). Hence, the most general two-body interaction of d-bosons is diagonal in the seniority scheme. It can always be expressed by (34.17) which also includes single boson energies ϵn_d, by adjusting the α, β and γ coefficients. It is possible to add to (34.17) terms linear and quadratic in N (due to the linear and quadratic Casimir operators of $U(6)$) which add the same amount to all eigenvalues of (34.17). The Hamiltonian will be still diagonal if to (34.17) we add the square of the scalar generator (33.26) and its product with the $k = 0$ operator in (33.24). Both are generators of $U(6)$, as well as of $U(5) \otimes U(1)$. This adds to (34.17) terms proportional to $\hat{n}_s^2$ and to $\hat{n}_s \hat{n}_d$. For given N, their contribution is in the linear and quadratic terms in N as well as by making ϵ depending on N as in (33.31).

The eigenstates of the $U(5)$ Hamiltonian are characterized by the quantum numbers N, n_d, τ and J (and M). As seen from (34.17) their eigenvalues are simple functions of those numbers. The characterization, however, is not complete. When irreducible representations of $O(5)$ are reduced to those of $O(3)$, the same $O(3)$ irreducible representation may appear more than once. In other words, there may be several states with seniority τ and the same value of J (and M). An additional quantum number is needed for a complete and unique characterization of states. There is no intermediate group between $O(5)$ and $O(3)$ so that the procedure described above cannot be followed. A way to distinguish between such states, with the same values of τ and J (and M) has been found. It is based on the amount of couplings of *three* d-boson into a state with $J = 0$. We shall not go here into the exact definition of the quantum number n_Δ counting such triplets, nor into the way that an orthogonal set of states is constructed with given τ and n_Δ. Its use becomes necessary only for values of $n_d \geq 6$. In any case, as evident from (34.17), the energy eigenvalues are independent of n_Δ. Any two states with the same values of N, n_d, τ and J are degenerate in the case of $U(5)$ Hamiltonians with only single boson terms and boson boson interactions.

The use of n_Δ is necessary for finding the allowed J-values of states with given seniority τ. For this purpose, the following procedure should be adopted. First, a number λ is defined by

$$\boxed{\tau = 3n_\Delta + \lambda} \tag{34.18}$$

The allowed values of J are then given by

$$\boxed{J = \lambda, \lambda+1, \lambda+2, \ldots, 2\lambda-2, 2\lambda} \tag{34.19}$$

The value $J = 2\lambda - 1$ does not appear among the states with given τ (even if it is equal to λ, i.e. $\lambda = 1$). For example, for $n_d = 3$, there are states with $\tau = 3$ and $\tau = 1$. The only $\tau = 1$ state is a state with $J = 2$ (like the state of a one d-boson). It contains no states in which the 3 d-bosons are coupled to $J = 0$, hence $n_\Delta = 0$. From (34.18) we find for this state the value $\lambda = 1$ and from (34.19) follows $J = 2$ but not $J = 2\lambda - 1 = 1$. The states with $\tau = 3$ and $n_\Delta = 0$ have $\lambda = 3$ and their J-values are $J = 3,4,6$. There is also one state with $\tau = 3$ where the three d-bosons are coupled to $J = 0$ and hence $n_\Delta = 1$ and $\lambda = 0$. For $n_d = 4$ there are states with $\tau = 0,2,4$. The $\tau = 0$ state has $J = 0$ in agreement with (34.18) from which follows $n_\Delta = 0$ and $\lambda = 0$. The $\tau = 2$ states have $J = 2,4$, $n_\Delta = 0$ and $\lambda = 2$. The $\tau = 4$ states with $n_\Delta = 0$ have $\lambda = 4$ and there are states with $J = 4,5,6,8$. There is one state which may have $n_\Delta = 1$ and it must have $J = 2$ since 3 of the 4 d-bosons are coupled to $J_0 = 0$. Indeed from (34.18) follows that $\lambda = 1$ and the only possible state has $J = 2$.

The reason for the absence of the state with $J = 2\lambda - 1$ is rather simple. According to (34.18) there are $n_d - \lambda$ d-bosons coupled to $J = 0$, either pairwise or in triples. The state with highest J value of the remaining λ d-bosons is obtained when all projections of d-boson spins on the z-axis are equal to 2. This is the state with $M = 2\lambda$ and $J = 2\lambda$. There is only *one* symmetric state with $M = 2\lambda - 1$ in which one projection is equal to 1 and all others to 2. This must be the state with $J = 2\lambda$ and projection $M = 2\lambda - 1$. Hence, there is no fully symmetric state with $J = 2\lambda - 1$. From this follows that there is no state with $J = 1$ in the whole spectrum of the IBA-1 boson model. According to (34.19) such a state may arise only for $\lambda = 1$ but then $J = 1$ is equal to $2\lambda - 1$ which spin is excluded. Actually, the spin of the state in which ($\lambda = 1$) is coupled to the spin $J_0 = 0$ of pairs and triples is equal to $J = 2$. This provides a simple argument for the formal result (34.19). This feature is seen here in the $U(5)$ basis but it is true in any basis of $U(6)$.

Let us now return to (34.17) and note that the situation leading to it is remarkable indeed. Once the $U(5)$ subgroup has been adopted, the Hamiltonian could be expanded in Casimir operators of the chain of subgroups (including $U(5)$) in (34.16). As a result, the eigenvalues are obtained as a simple algebraic function of the quantum numbers. Such a situation is referred to as *dynamical symmetry*. We saw such a case in Section 30 for the $SU(3)$ group. It may seem strange that restriction of $U(6)$ to a *subgroup* apparently implies a *stronger* symmetry. The reason is that the general boson Hamiltonian (33.30) *does not commute* with all $U(6)$ generators. It is merely constructed from $U(6)$ generators and hence, its eigenstates belong to definite irreducible representations of $U(6)$. If the Hamiltonian is constructed from generators of only a subgroup, it is more specific and has a higher symmetry.

From an alternative point of view, we may consider a Hamiltonian commuting with *all* generators of $U(6)$. It will then have the same eigenvalue in all states which belong to the same irreducible representation. Such a Hamiltonian may be expressed as a linear combination of the linear and quadratic Casimir operators of $U(6)$. If we add to such a Hamiltonian the Casimir operator(s) of a subgroup, the original dynamical symmetry will be broken. The resulting dynamical symmetry will not be as strong but certainly more interesting. We shall see in the following that this behavior appears also for other subgroups of $U(6)$. Once such a subgroup is chosen, the Hamiltonian has dynamical symmetry and can be expressed as a linear combination of Casimir operators of the groups in the given chain.

An example of the spectrum of a $U(5)$ Hamiltonian is shown in Fig. 34.1. It was calculated by taking $\epsilon > 0$, large compared to boson-boson interactions. The ground state has only s-bosons ($n_s = N$) and its spin is $J = 0$. The first excited state has $n_d = 1$ and $J = 2$. The next three states have $n_d = 2$ and $J = 0, 2, 4$ (the number n_d is written in parentheses above each level). They may be close in energy as is the case in certain nuclei (Fig. 34.2) but are not degenerate due to the boson-boson interactions. The s-bosons play no role in the spectrum and their contribution to the eigenstates is simply the presence of s-bosons whose number is $n_s = N - n_d$. The states are equivalent to the eigenstates of a five-dimensional harmonic oscillator. It is similar to the three-dimensional oscillator considered in Section 30 and the d-bosons correspond to the usual oscillator p-bosons. The states in the $U(5)$ case are thus equivalent to those of the vibrational limit mentioned in Section 33. The spectrum, however, need not be harmonic, the boson-boson interactions introduce some anharmonicities.

Simple selection rules on electric quadrupole transitions (E2) are obtained in the $U(5)$ limit. The term in (33.33) whose coefficient is

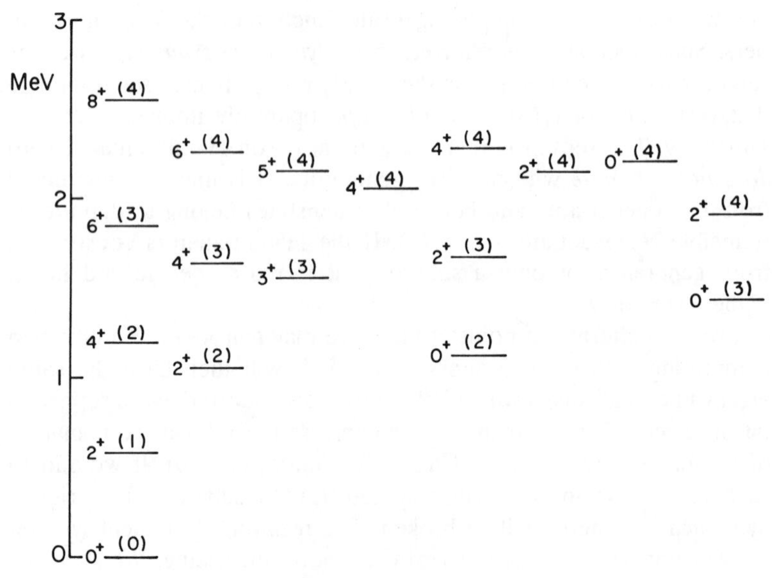

FIGURE 34.1. *Typical spectrum of a* $U(5)$ *Hamiltonian.*

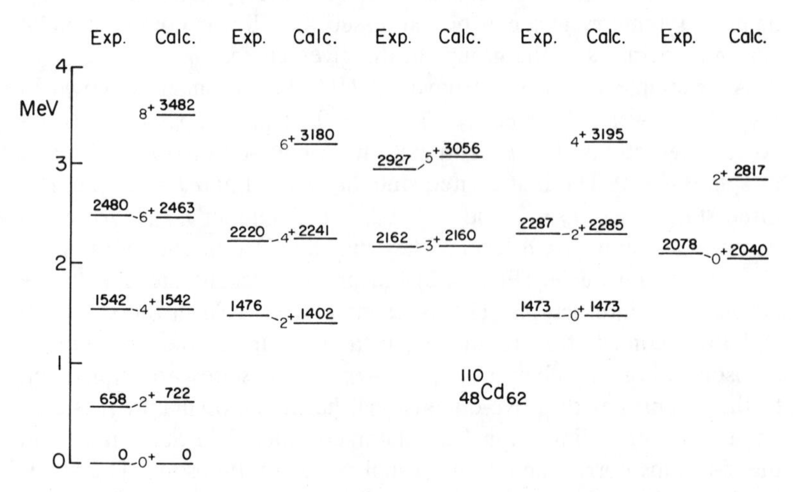

FIGURE 34.2. *Vibrational-like spectrum of* ^{110}Cd.

α_2 changes the number of d-bosons by one and has non-vanishing matrix elements only when $\Delta n_d = \pm 1$. Since it adds or eliminates one d-boson, the seniority τ must change by one unit, $\Delta\tau = \pm 1$. The other term in (33.33), multiplied by β_2 is a generator of $U(5)$. It does not change the number of d-bosons and hence its non-vanishing matrix elements satisfy $\Delta n_d = 0$. Since it is a single boson tensor operator of *even* rank ($k = 2$), it can change the seniority by at most two, $\Delta\tau = \pm 2, 0$. In particular, it has non-vanishing expectation value in the first excited state with $n_d = 1$, $J = 2$, i.e. a non-vanishing quadrupole moment. This result follows naturally from the interacting boson model. It is in better agreement with experiment than some earlier vibrational models which predicted vanishing quadrupole moments for the lowest $J = 2$ states in vibrational nuclei.

In order to calculate matrix elements of single boson operators and of boson-boson interactions, explicit expressions for the various states should be used. One way of constructing fully symmetric functions of d-bosons is by using coefficients of fractional parentage (c.f.p.). In analogy to the antisymmetric states of nucleons, considered in Section 15, we express a symmetric function of $n_d = n$ d-bosons as the linear combination

$$\boxed{\psi(d^n\alpha JM) = \sum_{\alpha_1 J_1} [d^{n-1}(\alpha_1 J_1)dJ|\}d^n\alpha J]\psi(d^{n-1}(\alpha_1 J_1)d_n JM)} \tag{34.20}$$

The wave functions on the r.h.s. of (34.20) are symmetric in the $n-1$ d-bosons with coordinates 1 to $n-1$. The n-th d-boson is coupled to the $\alpha_1 J_1$ states with Clebsch-Gordan coefficients to yield a state with given J (and M). The α and α_1 stand for additional quantum numbers which are necessary to uniquely specify states with the same values of $n = n_d, J$ (and M). As explained above, the additional quantum numbers we use are the seniority τ and n_Δ. In the following, we may not specify them explicitly and leave the more general notation α, α_1. The wave functions on the r.h.s. of (34.20) form a complete basis of states with given J (and M) which are symmetric in the first $n-1$ d-bosons. Hence, also a fully symmetric state may be expressed by the linear combination (34.20).

To calculate the c.f.p. for $n_d = 3$ we proceed as in the case of nucleons in Section 15. We start with a state of two d-bosons coupled to a symmetric state with a certain even spin J_0. To this state we couple, with Clebsch-Gordan coefficients the state of the third d-boson and then symmetrize it

$$\begin{aligned} \mathcal{S}\psi(d_{12}^2(J_0)d_3JM) &= \psi(d_{12}^2(J_0)d_3JM) + \psi(d_{13}^2(J_0)d_2JM) \\ &\quad + \psi(d_{32}^2(J_0)d_1JM) \\ &= \psi(d_{12}^2(J_0)d_3JM) + \psi(d_{13}^2(J_0)d_2JM) \\ &\quad + \psi(d_{23}^2(J_0)d_1JM) \end{aligned} \tag{34.21}$$

The last equality in (34.21) is due to the symmetry of the d^2J_0 state. We now apply to the r.h.s. of (34.21) the change of coupling transformations (10.11) obtaining

$$\begin{aligned} &\psi(d_{12}^2(J_0)d_3JM) + \sum_{J_1} \sqrt{(2J_0+1)(2J_1+1)} \begin{Bmatrix} J_0 & 2 & J \\ J_1 & 2 & 2 \end{Bmatrix} \\ &\quad \times \psi(d_{12}^2(J_1)d_3JM) + \sum_{J_1} (-1)^{J_1} \sqrt{(2J_0+1)(2J_1+1)} \begin{Bmatrix} J_0 & 2 & J \\ J_1 & 2 & 2 \end{Bmatrix} \\ &\quad \times \psi(d_{12}^2(J_1)d_3JM) = \psi(d_{12}^2(J_0)d_3JM) \\ &\quad + 2 \sum_{J_1 \text{ even}} \sqrt{(2J_0+1)(2J_1+1)} \begin{Bmatrix} J_0 & 2 & J \\ J_1 & 2 & 2 \end{Bmatrix} \psi(d_{12}^2(J_1)d_3JM) \end{aligned} \tag{34.22}$$

The phases $(-1)^{J_1}$ on the l.h.s. of (34.22) are due to interchange of d-bosons 1 and 2. The normalization of the wave function (34.22) may be carried out by the orthogonality relation (10.13) and sum rule(10.14) of $6j$-symbols. We thus obtain for the c.f.p. the values

$$\boxed{[d^2(J_1)dJ|\}d^3[J_0]J] = \frac{\delta_{J_0J_1} + 2\sqrt{(2J_0+1)(2J_1+1)} \begin{Bmatrix} J_0 & 2 & J \\ J_1 & 2 & 2 \end{Bmatrix}}{\sqrt{3 + 6(2J_0+1) \begin{Bmatrix} J_0 & 2 & J \\ J_0 & 2 & 2 \end{Bmatrix}}}} \tag{34.23}$$

The symbol $[J_0]$ in the c.f.p. (34.23) indicates that the state with J_0 is the principal parent of the state considered. This symbol is actually not necessary in the case $n = 3$ since then there is only one allowed (symmetric) state with given J (J may be equal to 0, 2, 3, 4 or 6).

The c.f.p. thus defined may be used in the same way as described in the case of nucleons. We will not repeat the various expressions derived in Section 15. Let us, however, establish the relation between c.f.p. and reduced matrix elements of creation operators, the analog of (17.28). From the definition (34.20) follows

$$[d^{n-1}(\alpha_1 J_1)dJ|\}d^n\alpha J] = \int \psi^*(d^n\alpha JM)\psi(d^{n-1}(\alpha_1 J_1)d_n JM)$$

$$= \int \frac{1}{n}(\mathcal{S}\psi^*(d^n\alpha JM))\psi(d^{n-1}(\alpha_1 J_1)d_n JM)$$

$$= \frac{1}{n}\int \psi^*(d^n\alpha JM)\mathcal{S}\psi(d^{n-1}(\alpha_1 J_1)d_n JM) \tag{34.24}$$

In (34.24) the operator $\mathcal{S}$ is a symmetrizer analogous to the antisymmetrizer $\mathcal{A}$ for nucleons (a special case of it appears in (34.21)). It creates a linear combination of the original functions and functions in which the n-th boson is exchanged with each of the first $n-1$ bosons. The r.h.s. of (34.24) is related to the matrix element $\langle 0|A(n\alpha JM)[A^+(n-1,\alpha_1 J_1)\times d^+]^{(J)}_M|0\rangle$ but normalizations should be examined. The normalization of any term $d^+_{\mu_1}d^+_{\mu_2}\cdots d^+_{\mu_n}|0\rangle$ is 1. On the other hand, $(1/n)\mathcal{S}\psi_{\mu_1}(1)\psi_{\mu_2}(2)\cdots\psi_{\mu_n}(n)$ is a sum of n orthogonal (and normalized) functions so that the square of its norm is $(1/n^2)n = 1/n$ and the normalization coefficient is $\sqrt{n}$. We thus obtain

$$[d^{n-1}(\alpha_1 J_1)dJ|\}d^n\alpha J] = \frac{1}{n}\int \psi^*(d^n\alpha JM)\mathcal{S}\psi(d^{n-1}(\alpha_1 J_1)d_n JM)$$

$$= \frac{1}{\sqrt{n}}\times\frac{\sqrt{n}}{n}\int \psi^*(d^n\alpha JM)\mathcal{S}\psi(d^{n_1}(\alpha_1 J_1)d_n JM)$$

$$= \frac{1}{\sqrt{n}}\langle 0|A(n\alpha JM)[A^+(n-1,\alpha_1 J_1)\times d^+]^{(J)}_M|0\rangle$$

$$= \frac{1}{\sqrt{n}}\sum_{\mu M_1}\langle d^n\alpha JM|d^+_\mu|d^{n-1}\alpha_1 J_1 M_1\rangle(J_1 M_1 2\mu\mid J_1 2JM) \tag{34.25}$$

Recalling the definition of reduced matrix elements and the relation (8.11) we obtain

$$\boxed{[d^{n-1}(\alpha_1 J_1)dJ|\}d^n\alpha J] = \frac{1}{\sqrt{n}}\frac{1}{\sqrt{2J+1}}(d^n\alpha J\|d^+\|d^{n-1}\alpha_1 J_1)} \tag{34.26}$$

The c.f.p. in (34.26) may be expressed also in terms of the reduced matrix element of the annihilation operator $\tilde{d}$. Taking the complex conjugate of the r.h.s. of (34.25) we obtain, by using $(d_\mu^+)^\dagger = d_\mu$ and (8.11), the equivalent expression

$$\boxed{[d^{n-1}(\alpha_1 J_1)dJ|\}d^n\alpha J] = (-1)^{J-J_1}\frac{1}{\sqrt{n}}\frac{1}{\sqrt{2J+1}}(d^{n-1}\alpha_1 J_1\|\tilde{d}\|d^n\alpha J)} \tag{34.27}$$

Let us now consider matrix elements of the quadrupole operator in the seniority scheme. First we consider matrix elements of the operator $s^+\tilde{d} + d^+s$ between states (34.1) with given numbers of d-bosons

$$\langle n_s n_d \tau JM|s^+\tilde{d}_\mu + d_\mu^+ s|n_s+1, n_d-1, \tau_1 J_1 M_1\rangle \tag{34.28}$$

In (34.28) and in the following expressions, states with the same values of n_d, τ and J (and M) should be distinguished by the additional quantum number n_Δ. Within the seniority scheme, matrix elements between states may be reduced to those between states with lower n_d numbers with the *same seniorities*. Thus, it is not necessary to specify the value of n_Δ at each stage of the derivation and we will omit it for the sake of conciseness. It is clear that only the d^+s term contributes to the matrix element (34.28). From the structure (34.1) of states it follows that (34.28) is equal to the product

$$\langle n_s|s|n_s+1\rangle\langle n_d\tau JM|d_\mu^+|n_d-1, \tau_1 J_1 M_1\rangle \tag{34.29}$$

The matrix element of s may be directly calculated by observing that the normalization factor for the state $(s^+)^{n_s}|0\rangle$ is $(n_s!)^{-1/2}$. This follows from calculating the expectation value

$$\langle 0|s^{n_s}(s^+)^{n_s}|0\rangle = \langle 0|s^{n_s-1}[s,(s^+)^{n_s}]|0\rangle$$
$$= n_s\langle 0|s^{n_s-1}(s^+)^{n_s-1}|0\rangle = \cdots = n_s! \qquad (34.30)$$

From this follows

$$\langle n_s|s|n_s+1\rangle = \frac{1}{\sqrt{n_s!(n_s+1)!}}\langle 0|s^{n_s}s(s^+)^{n_s+1}|0\rangle$$
$$= \frac{(n_s+1)!}{\sqrt{n_s!(n_s+1)!}} = \sqrt{n_s+1} \qquad (34.31)$$

Hence, we obtain for the matrix element (34.28) the result

$$\boxed{\begin{aligned}&\langle n_s n_d \tau JM|s^+\tilde{d}_\mu + d^+_\mu s|n_s+1, n_d-1, \tau_1 J_1 M_1\rangle \\ &\quad = \sqrt{N-n_d+1}\langle n_d\tau JM|d^+_\mu|n_d-1,\tau_1 J_1 M_1\rangle\end{aligned}} \qquad (34.32)$$

The matrix element in (34.32) may be expressed by using the Wigner-Eckart theorem and (34.26) in terms of the c.f.p. $[d^{n_d-1}(\tau_1 J_1)dJ|\}d^{n_d}\tau J]$. The transition probabilities are expressed in terms of the *reduced* matrix elements of the relevant operator according to (8.29).

The c.f.p. in the scheme of the seniority τ in configurations with $n_d > \tau$ may be simply expressed in terms of these in configurations with smaller values of n_d. The recursion formulae may be derived by methods used in de-Shalit and Talmi (1963). It is, however, simpler to derive recursion relations directly for the reduced matrix elements of d^+. To do that, we need the normalization of states in the d-boson seniority scheme. Let us calculate the normalization of the state with given n_d and τ obtained by applying $(d^+\cdot d^+)$ to the normalized state with n_d-2 and τ. Let us thus evaluate the expectation value

$$\langle n_d-2,\tau JM|(\tilde{d}\cdot\tilde{d})(d^+\cdot d^+)|n_d-2,\tau JM\rangle$$
$$= \langle n_d-2,\tau JM|[(\tilde{d}\cdot\tilde{d}),(d^+\cdot d^+)] + (d^+\cdot d^+)(\tilde{d}\cdot\tilde{d})|n_d-2,\tau JM\rangle$$
$$= 4(n_d-2)+10+(n_d-2)(n_d-2+3)-\tau(\tau+3)$$
$$= n_d(n_d+3)-\tau(\tau+3) \qquad (34.33)$$

The result (34.33) was obtained by using (34.6) and (34.13). Its r.h.s. is the eigenvalue of $(\tilde{d}\cdot\tilde{d})(d^+\cdot d^+)$ in the state $|n_d-2,\tau JM\rangle$ and this fact will be used in the following. The square root of its inverse is the normalization factor $\mathcal{N}_{n_d\tau}$ for the state $(d^+\cdot d^+)|n_d-2,\tau JM\rangle$. From the derivation of (34.33) it is clear that the normalization factor is the same for all states with given n_d and τ and is independent of J or n_Δ.

To reduce the matrix element of d^+_μ in (34.32) we first consider the case in which $\tau_1=\tau+1$. It is equal to

$$\begin{aligned}
&\langle n_d\tau JM|d^+_\mu|n_d-1,\tau+1,J_1M_1\rangle\\
&\quad=\mathcal{N}_{n_d\tau}\mathcal{N}_{n_d-1,\tau+1}\\
&\qquad\times\langle n_d-2,\tau JM|(\tilde{d}\cdot\tilde{d})d^+_\mu(d^+\cdot d^+)|n_d-3,\tau+1,J_1M_1\rangle\\
&\quad=\mathcal{N}_{n_d\tau}\mathcal{N}_{n_d-1,\tau+1}\\
&\qquad\times\langle n_d-2,\tau JM|\{(\tilde{d}\cdot\tilde{d})(d^+\cdot d^+)\}d^+_\mu|n_d-3,\tau+1,J_1M_1\rangle\\
&\quad=\mathcal{N}_{n_d\tau}\mathcal{N}_{n_d-1,\tau+1}\mathcal{N}^{-2}_{n_d\tau}\langle n_d-2,\tau JM|d^+_\mu|n_d-3,\tau+1,J_1M_1\rangle
\end{aligned}\tag{34.34}$$

The last equality in (34.34) follows from applying the operator in curly brackets to the state $\langle n_d-2,\tau JM|$ and recalling (34.33). Substituting in (34.34) the values of the normalization factors which follow from (34.33), we obtain a recursion relation which holds also for the *reduced* matrix elements of d^+

$$\begin{aligned}
&(n_d\tau J\|d^+\|n_d-1,\tau+1,J_1)\\
&\quad=(\mathcal{N}_{n_d-1,\tau+1}/\mathcal{N}_{n_d\tau})(n_d-2,\tau J\|d^+\|n_d-3,\tau+1,J_1)\\
&\quad=\sqrt{\frac{n_d(n_d+3)-\tau(\tau+3)}{(n_d-1)(n_d+2)-(\tau+1)(\tau+4)}}\\
&\qquad\times(n_d-2,\tau J\|d^+\|n_d-3,\tau+1,J_1)\\
&\quad=\sqrt{\frac{n_d-\tau}{n_d-2-\tau}}(n_d-2,\tau J\|d^+\|n_d-3,\tau+1,J_1)
\end{aligned}\tag{34.35}$$

This relation may be successively used on the r.h.s. of (34.35) to reduce it to a matrix element between states with lower n_d numbers. This may be carried on until the state with $n_d=\tau+1$ is reached. We thus obtain the explicit formula

$$\boxed{\begin{aligned}&(n_d \tau n_\Delta J \| d^+ \| n_d - 1, \tau + 1, n'_\Delta J') \\ &\quad = \sqrt{\frac{n_d - \tau}{2}} (\tau + 2, \tau n_\Delta J \| d^+ \| \tau + 1, \tau + 1, n'_\Delta, J')\end{aligned}}$$

(34.36)

The same procedure may be used to obtain recursion relations for matrix elements (34.32) for the case $\tau_1 = \tau - 1$. We obtain in this case, in analogy with (34.35), the following result

$$\begin{aligned}&(n_d \tau J \| d^+ \| n_d - 1, \tau - 1, J_1) \\ &\quad = (\mathcal{N}_{n_d - 1, \tau - 1} / \mathcal{N}_{n_d \tau})(n_d - 2, \tau J \| d^+ \| n_d - 3, \tau - 1, J_1) \\ &\quad = \sqrt{\frac{n_d(n_d + 3) - \tau(\tau + 3)}{(n_d - 1)(n_d + 2) - (\tau - 1)(\tau + 2)}} \\ &\qquad \times (n_d - 2, \tau J \| d^+ \| n_d - 3, \tau - 1, J_1) \\ &\quad = \sqrt{\frac{n_d + \tau + 3}{n_d + \tau + 1}} (n_d - 2, \tau J \| d^+ \| n_d - 3, \tau - 1, J_1)\end{aligned} \tag{34.37}$$

Iteration of the recursion relation (34.37) leads to the explicit formula

$$\boxed{\begin{aligned}&(n_d \tau n_\Delta J \| d^+ \| n_d - 1, \tau - 1, n'_\Delta J') \\ &\quad = \sqrt{\frac{n_d + \tau + 3}{2\tau + 3}} (\tau \tau n_\Delta J \| d^+ \| \tau - 1, \tau - 1, n'_\Delta J')\end{aligned}}$$

(34.38)

The reduction factors in (34.36) and (34.38) are analogous to (19.28) and (19.26) in the case of nucleons. There, they were obtained as the ratios between Clebsch-Gordan coefficients of the $SU(2)$ group. In the present case, these factors are ratios of Clebsch-Gordan coefficients of the $SU(1,1)$ group. Since we have not dealt with them here, it was necessary to use the derivation given above for calculating these ratios.

The reduced matrix element on the r.h.s. of (34.36) may be expressed in terms of the matrix element between states with $n_d = \tau$ and $n_d = \tau + 1$. We start with

$$\langle \tau+2, \tau JM | d_\mu^+ | \tau+1, \tau+1, J_1 M_1 \rangle$$
$$= \mathcal{N}_{\tau+2,\tau} \langle \tau\tau JM | (\tilde{d}\cdot\tilde{d}) d_\mu^+ | \tau+1, \tau+1, J_1 M_1 \rangle \quad (34.39)$$

and evaluate the commutator

$$[(\tilde{d}\cdot\tilde{d}), d_\mu^+] = \sum_\kappa (-1)^\kappa [d_\kappa d_{-\kappa}, d_\mu^+] = 2(-1)^\mu d_{-\mu} = 2\tilde{d}_\mu \quad (34.40)$$

Using the result (34.40) in (34.39) we obtain the following relation between the reduced matrix elements

$$(\tau+2, \tau n_\Delta J \| d^+ \| \tau+1, \tau+1, n'_\Delta J')$$
$$= \frac{2}{\sqrt{4\tau+10}} (\tau\tau n_\Delta J \| \tilde{d} \| \tau+1, \tau+1, n'_\Delta J')$$
$$= (-1)^{J+J'} \sqrt{\frac{2}{2\tau+5}} (\tau+1, \tau+1, n'_\Delta J' \| d^+ \| \tau\tau n_\Delta J) \quad (34.41)$$

The last equality is obtained by observing that matrix elements of d_μ^+ and $\tilde{d}_\mu$ are real and using the Wigner-Eckart theorem.

Finally, we may use the relations (34.36), (34.38) and (34.41) to obtain similar relations between c.f.p. in the seniority scheme of d-bosons. Using (34.26) we obtain from (34.36) the relation

$$[d^{n_d-1}(\tau+1, n'_\Delta J') dJ | \} d^{n_d} \tau n_\Delta J]$$
$$= \sqrt{\frac{(n_d-\tau)(\tau+2)}{2n_d}} [d^{\tau+1}(\tau+1, n'_\Delta J') dJ | \} d^{\tau+2} \tau n_\Delta J] \quad (34.42)$$

From (34.38) follows by using (34.26) the relation

$$[d^{n_d-1}(\tau-1, n'_\Delta J')dJ|\}d^{n_d}\tau n_\Delta J]$$
$$= \sqrt{\frac{(n_d+\tau+3)\tau}{n_d(2\tau+3)}}[d^{\tau-1}(\tau-1, n'_\Delta J')dJ|\}d^\tau \tau n_\Delta J] \tag{34.43}$$

These relations are the analogs of (19.28) and (19.26) in the case of nucleons. The analog of (19.32) is obtained by applying (34.26) to the relation (34.41) and is expressed by

$$[d^{\tau+1}(\tau+1, n'_\Delta J')dJ|\}d^{\tau+2}\tau n_\Delta J]$$
$$= (-1)^{J+J'}\sqrt{\frac{(2J'+1)(2\tau+2)}{(2J+1)(\tau+2)(2\tau+5)}}$$
$$\times [d^\tau(\tau n_\Delta J)dJ_1|\}d^{\tau+1}\tau+1, n'_\Delta J'] \tag{34.44}$$

Let us now consider the second term in the quadrupole operator (33.33) and calculate matrix elements of $(d^+\times\tilde{d})^{(2)}_\kappa$. Using the formula (10.28) we can express the reduced matrix elements by

$$(n_d\tau J\|(d^+\times\tilde{d})^{(2)}\|n_d\tau' J')$$
$$= (-1)^{J+J'}\sqrt{5}\sum_{n''_d\tau''J''}(n_d\tau J\|d^+\|n''_d\tau''J'')$$
$$\times (n''_d\tau''J''\|\tilde{d}\|n_d\tau' J')\begin{Bmatrix} 2 & 2 & 2 \\ J' & J & J'' \end{Bmatrix} \tag{34.45}$$

The seniority τ' may be equal to τ or differ from it by two units. The reduced matrix elements on the r.h.s. of (34.45) may be expressed by using (34.36) and (34.38), thereby leading to a reduction of the matrix element (34.45) to one in a smaller configuration. The value of n''_d on the r.h.s. of (34.45) is $n''_d = n_d - 1$ whereas the possible values of τ'' are $\tau+1$ and $\tau-1$. Let us first consider the case $\tau' = \tau-2$ which is equivalent to the case with $\tau' = \tau+2$.

In the case $\tau' = \tau - 2$, the value of τ'' is uniquely determined to be $\tau'' = \tau - 1$. We then express the products on the r.h.s. of (34.45) by

$$(n_d \tau J \| d^+ \| n_d - 1, \tau - 1, J'')(n_d - 1, \tau - 1, J'' \| \tilde{d} \| n_d, \tau - 2, J')$$

$$= (n_d \tau J \| d^+ \| n_d - 1, \tau - 1, J'')(-1)^{J' + J''}$$

$$\times (n_d, \tau - 2, J' \| d^+ \| n_d - 1, \tau - 1, J'') \tag{34.46}$$

We now make use of (34.38) for one of the reduced matrix elements on the r.h.s. of (34.46) and of (34.36) for the other, to obtain for (34.46) the expression

$$\sqrt{\frac{n_d + \tau + 3}{2\tau + 3}} (\tau \tau J \| d^+ \| \tau - 1, \tau - 1, J'')(-1)^{J' + J''}$$

$$\times \sqrt{\frac{n_d - (\tau - 2)}{2}} (\tau \tau - 2, J' \| d^+ \| \tau - 1, \tau - 1, J'')$$

$$= \sqrt{\frac{(n_d + \tau + 3)(n_d - \tau + 2)}{2(2\tau + 3)}} (\tau \tau J \| d^+ \| \tau - 1, \tau - 1, J'')$$

$$\times (\tau - 1, \tau - 1, J'' \| \tilde{d} \| \tau \tau - 2, J') \tag{34.47}$$

Substituting this expression in (34.45) we obtain the result

$$\boxed{\begin{aligned} &(n_d \tau n_\Delta J \| (d^+ \times \tilde{d})^{(2)} \| n_d \tau - 2, n'_\Delta J') \\ &= \sqrt{\frac{(n_d + \tau + 3)(n_d - \tau + 2)}{2(2\tau + 3)}} \\ &\times (\tau \tau n_\Delta J \| (d^+ \times \tilde{d})^{(2)} \| \tau, \tau - 2, n'_\Delta J') \end{aligned}} \tag{34.48}$$

which is the analog of (19.44) in the case of nucleons.

The case where $\tau' = \tau$ is more complicated since in that case τ'' may be either $\tau - 1$ or $\tau + 1$. The terms with $\tau'' = \tau - 1$ in (34.45) may be expressed in a similar way to (34.46) yielding the expression

$$(n_d\tau J\|d^+\|n_d-1,\tau-1,J'')(-1)^{J'+J''}(n_d\tau J'\|d^+\|n_d-1,\tau-1,J'')$$
$$=\frac{n_d+\tau+3}{2\tau+3}(\tau\tau J\|d^+\|\tau-1,\tau-1,J'')$$
$$\times(-1)^{J'+J''}(\tau\tau J'\|d^+\|\tau-1,\tau-1,J'')$$
$$=\frac{n_d+\tau+3}{2\tau+3}(\tau\tau J\|d^+\|\tau-1,\tau-1,J'')(\tau-1,\tau-1,J''\|\tilde{d}\|\tau\tau J') \tag{34.49}$$

The result (34.49) was obtained by using (34.38) on both reduced matrix elements on the l.h.s. of (34.49). For $n_d=\tau$, the reduced matrix elements on the r.h.s. of (34.49) are the only possible ones. We obtain by substituting them in (34.45) the result

$$\frac{n_d+\tau+3}{2\tau+3}(\tau\tau J\|(d^+\times\tilde{d})^{(2)}\|\tau\tau J) \tag{34.50}$$

To the result (34.49) we should add the contribution of the $\tau''=\tau+1$ terms which may be expressed as

$$(n_d\tau J\|d^+\|n_d-1,\tau+1,J'')(-1)^{J'+J''}(n_d\tau J'\|d^+\|n_d-1,\tau+1,J'')$$
$$=\frac{n_d-\tau}{2}(\tau+2,\tau J\|d^+\|\tau+1,\tau+1,J'')(-1)^{J'+J''}$$
$$\times(\tau+2,\tau J'\|d^+\|\tau+1,\tau+1,J'')$$
$$=\frac{n_d-\tau}{2}\frac{2}{2\tau+5}(\tau\tau J\|\tilde{d}\|\tau+1,\tau+1,J'')$$
$$\times(\tau+1,\tau+1,J''\|d^+\|\tau\tau J') \tag{34.51}$$

The equalities in (34.51) have been obtained by using first (34.36) and then (34.41) on both reduced matrix elements on the l.h.s. of (34.51). The r.h.s. of (34.51) contains terms obtained when the reduced matrix element of $(\tilde{d}\times d^+)^{(2)}$ is evaluated. Due to the commutation relations (33.20) and orthogonalilty of Clebsch-Gordan coefficients we obtain $(d^+\times\tilde{d})^{(k)}_\kappa=(-1)^k(\tilde{d}\times d^+)^{(k)}$ for any $k\neq 0$. Thus, for $k=2$ (or $k=4$) we may use (34.51) in the following derivations. The terms (34.51) do not yield the full matrix elements of $(\tilde{d}\times d^+)^{(2)}$ in the states with $n_d=\tau$. To obtain the full matrix elements we add to (34.51) the terms

$$\frac{n_d-\tau}{2\tau+5}(\tau\tau J\|\tilde{d}\|\tau+1,\tau-1,J'')(\tau+1,\tau-1,J''\|d^+\|\tau\tau J') \tag{34.52}$$

and then subtract them. The terms (34.51) and (34.52) substituted in (34.45) lead to the result

$$\frac{n_d-\tau}{2\tau+5}(\tau\tau J\|(\tilde{d}\times d^+)^{(2)}\|\tau\tau J') \tag{34.53}$$

Finally, we consider the terms (34.52) with minus sign which can be expressed, by using (34.41) in which $\tau-1$ replaces τ, on both reduced matrix elements in (34.52) as

$$-\frac{n_d-\tau}{2\tau+5}\frac{2}{2(\tau-1)+5}(\tau\tau J\|d^+\|\tau-1,\tau-1,J'')$$
$$\times(\tau-1,\tau-1,J''\|\tilde{d}\|\tau\tau J') \tag{34.54}$$

When the terms (34.54) are substituted in (34.45) we obtain

$$-\frac{n_d-\tau}{2\tau+5}\frac{2}{2\tau+3}(\tau\tau J\|(d^+\times\tilde{d})^{(2)}\|\tau\tau J') \tag{34.55}$$

Taking the sum of (34.50), (34.53) and (34.55) we obtain the relation

$$\boxed{\begin{aligned}&(n_d\tau n_\Delta J\|(d^+\times\tilde{d})^{(2)}\|n_d\tau n'_\Delta J')\\ &\qquad=\frac{2n_d+5}{2\tau+5}(\tau\tau n_\Delta J\|(d^+\times\tilde{d})^{(2)}\|\tau\tau n'_\Delta J')\end{aligned}} \tag{34.56}$$

This result is the analog of (19.40) in the case of fermions.

The derivations of (34.48) and (34.56) hold if we replace $(d^+\times\tilde{d})^{(2)}$ by the hexadecapole operator $(d^+\times\tilde{d})^{(4)}$. This is not the case for the operators $(d^+\times\tilde{d})^{(k)}$ with $k=1,3$. These operators are generators of the $O(5)$ group and hence they are diagonal in τ and their matrix elements are independent of n_d. The result (34.48) may be derived also for these operators but both sides of the equality vanish in that case. In the case $\tau'=\tau$, the final expression should be obtained by adding to (34.50) the terms (34.53) and (34.55) *with the opposite sign*. Instead of (34.56) we obtain

$$\boxed{\begin{aligned}&(n_d \tau n_\Delta J \| (d^+ \times \tilde{d})^{(k)} \| n_d \tau n'_\Delta J')\\ &\qquad = (\tau \tau n_\Delta J \| (d^+ \times \tilde{d})^{(k)} \| \tau \tau n'_\Delta J') \qquad k = 1,3\end{aligned}} \tag{34.57}$$

which expresses the statement made above.

Let us consider now several examples of matrix elements and E2 transition probabilities. Among the states with given n_d there is one state with maximum spin given by $J = 2n_d$. In such a "stretched state" all spins of the n_d bosons are coupled symmetrically and the state with $M = J = 2n_d$ is a product of single boson states each with $m = 2$. There are no pairs of d-bosons coupled to $J = 0$ in such states and hence in each of them $\tau = n_d$. These states may be considered as the "ground state band" with $J = 0, 2, 4, \ldots, 2N$. Reduced matrix elements of the quadrupole operator between such states are given by (34.32) and (34.26) as

$$\begin{aligned}&\sqrt{N - n_d + 1}\\ &\qquad \times (n_d, \tau = n_d, J = 2n_d \| d^+ \| n_d - 1, \tau_1 = n_d - 1, J_1 = 2n_d - 2)\\ &\quad = \sqrt{N - n_d + 1}\sqrt{n_d}\sqrt{2J + 1}\\ &\qquad \times [d^{n_d - 1}(\tau_1 = n_d - 1, J_1 = 2n_d - 2) dJ |\} d^{n_d} \tau = n_d, J = 2n_d]\end{aligned} \tag{34.58}$$

The c.f.p. on the r.h.s. of (34.58) is equal to 1 since the only parent of the state with $J = 2n_d$ is the state with $J_1 = 2n_d - 2$. Multiplying (34.58) by the coefficient α_2 in (33.33) and using (8.29) we obtain for these transitions the result

$$B(\mathrm{E2}; J \to J - 2) = \alpha_2^2 n_d (N - n_d + 1) = \frac{\alpha_2^2}{4} J(2N + 2 - J) \tag{34.59}$$

35

The $O(6)$ *Limit.* $O(5)$ *Symmetry*

We now consider another limit of the $U(6)$ Hamiltonian which leads to dynamical symmetry. In the $U(5)$ case discussed in the preceding section, we encountered the $O(5)$ group which is a subgroup of $U(6)$ and contains $O(3)$ as its subgroup. In that case, $U(5)$ is an intermediate group between $O(5)$ and $U(6)$. There is, however, another such intermediate group, namely $O(6)$—the group of real orthogonal transformations in the 6-dimensional space of s- and d-bosons (Arima and Iachello 1978). The number of generators of $O(5)$, given by (34.2) is 10 $(= 5 \times 4/2)$ whereas the number of $O(6)$ generators is $15(= 6 \times 5/2)$. The components of another quadrupole operator should be added to the generators (34.2) to obtain the $O(6)$ Lie algebra. Since the s-boson should be included, the only $k = 2$ tensor among the $U(6)$ generators is one of those in (33.25). It is customary to define the $O(6)$ Lie algebra by the generators (34.2), $(d^+ \times \tilde{d})^{(k)}_\kappa$, $k = 1,3$ and the components of

$$\boxed{\mathbf{Q}' = s^+\tilde{d} + d^+s} \tag{35.1}$$

From (33.36) follows that the commutators of $\mathbf{Q}'_\mu$ with the operators in (34.2) are components of $\mathbf{Q}'$. From (33.35) we obtain for the commutators of components of $\mathbf{Q}'$ the expression

$$[Q'_\mu, Q'_{\mu'}] = \sum_{k\kappa}(1-(-1)^k)(2\mu 2\mu' \mid 22k\kappa)(d^+ \times \tilde{d})^{(k)}_\kappa \tag{35.2}$$

Thus, the quadrupole operator (35.1) and the $k = 1$, $k = 3$ tensors in (33.24) form a Lie algebra. All these generators annihilate the symmetric $J = 0$ state

$$P'^+|0\rangle = \tfrac{1}{2}[(d^+ \cdot d^+) - (s^+)^2]|0\rangle \tag{35.3}$$

The minus sign on the r.h.s. of (35.3) is due to the choice of $\mathbf{Q}'$ in (35.1). A plus sign would be consistent with choosing the other quadrupole operator in (33.25) as can be verified by using the commutation relations (33.20). Thus, the group generated by this Lie algebra is the group $O(6)$ of real orthogonal transformations in the six dimensional space of s- and d-boson. The (quadratic) Casimir operator of $O(6)$ can be expressed as

$$C_{O(6)} = \mathbf{Q}' \cdot \mathbf{Q}' + 2\sum_{k=1,3}((d^+ \times \tilde{d})^{(k)} \cdot (d^+ \times \tilde{d})^{(k)}) = \mathbf{Q}' \cdot \mathbf{Q}' + C_{O(5)} \tag{35.4}$$

As may be verified by using the commutation relations (33.34), (33.35) and (33.6), it commutes with all generators of $O(6)$. The irreducible representations of $O(6)$ included in the fully symmetric irreducible representations of $U(6)$, belong to Young diagrams with *one* row whose length σ differs from N by an even number,

$$\sigma = N, N-2, \ldots, 1 \text{ or } 0.$$

The eigenvalues of the Casimir operator (35.4) should be functions of this σ which is the generalized seniority quantum number in the space of s- *and* d-bosons. It is the analog of the nucleon seniority in the quasi-spin scheme discussed in Section 22. We shall now proceed to calculate the eigenvalues of $C_{O(6)}$.

The Casimir operator (35.4) is closely related to a pairing interaction analogous to the one for the $O(5)$ case and the nucleon case considered in preceding sections. The pairing interaction may be expressed in terms of the operator P'^{+} in (35.3) and its hermitean conjugate $(P'^{+})^{\dagger} = P'^{-}$ as

$$\boxed{P'^{+}P'^{-} = \tfrac{1}{4}((d^{+}\cdot d^{+}) - (s^{+})^{2})((\tilde{d}\cdot\tilde{d}) - s^{2})} \tag{35.5}$$

To obtain another expression for (35.5) we make use of the result (34.5) derived above, to obtain

$$\begin{aligned} 4P'^{+}P'^{-} &= (d^{+}\cdot d^{+})(\tilde{d}\cdot\tilde{d}) - (s^{+})^{2}(\tilde{d}\cdot\tilde{d}) - (d^{+}\cdot d^{+})s^{2} + (s^{+})^{2}s^{2} \\ &= \hat{n}_d^2 + 3\hat{n}_d - C_{O(5)} - Q'\cdot Q' + s^{+}s(\tilde{d}\cdot d^{+}) \\ &\quad + ss^{+}(d^{+}\cdot\tilde{d}) + (s^{+})^{2}s^{2} \\ &= \hat{n}_d^2 + 3\hat{n}_d - C_{O(6)} + 5\hat{n}_s + \hat{n}_s\hat{n}_d + \hat{n}_d + \hat{n}_s\hat{n}_d + \hat{n}_s^2 - \hat{n}_s \\ &= N(N+4) - C_{O(6)} \end{aligned} \tag{35.6}$$

From the general relationship between Casimir operators and pairing interactions we can deduce from (35.6) that the eigenvalues of $C_{O(6)}$ are given by $\sigma(\sigma+4)$. To obtain this result formally we can repeat the derivation of (34.13) replacing P^{+} and P^{-} by P'^{+} and P'^{-}, n_d by N, τ by σ and the number of dimensions 5 by 6. We thus obtain for the eigenvalues of the pairing interaction (35.5) the result

$$\boxed{\begin{aligned} &\tfrac{1}{4}[(N+\tfrac{1}{2}\times 6)(N+\tfrac{1}{2}\times 6-2) - (\sigma+\tfrac{1}{2}\times 6-2)(\sigma+\tfrac{1}{2}\times 6)] \\ &\quad = \tfrac{1}{4}\left[N(N+4) - \sigma(\sigma+4)\right] \\ &\quad = \tfrac{1}{4}(N-\sigma)(N+\sigma+4) \end{aligned}} \tag{35.7}$$

From (35.7) and (35.6) the eigenvalues of $C_{O(6)}$ are obtained as $\sigma(\sigma+4)$.

The most general boson Hamiltonian constructed from $O(6)$ generators is a linear combination of the scalar products $\mathbf{Q}' \cdot \mathbf{Q}'$ and $(d^+ \times \tilde{d})^{(k)} \cdot (d^+ \times \tilde{d})^{(k)}$ for $k = 1,3$. Due to (35.4) and (34.4) we may express such Hamiltonian as a linear combination of $C_{O(6)}$, $C_{O(5)}$ and $C_{O(3)}$. Adding to that the linear and quadratic Casimir operator of $U(6)$ and using the pairing interaction (35.5) instead of $C_{O(6)}$, we obtain a linear combination of Casimir operators which may be simultaneously diagonalized. Thus, the most general $O(6)$ Hamiltonian has eigenvalues which are a linear combination of the eigenvalues of those Casimir operators. They are given by

$$\boxed{\begin{aligned} &Na + \tfrac{1}{2}N(N-1)b + \tfrac{1}{4}A(N-\sigma)(N+\sigma+4) \\ &\quad + \tfrac{1}{6}B\tau(\tau+3) + CJ(J+1) \end{aligned}} \tag{35.8}$$

The last two terms in (35.8) contain also single boson energies. In the state of a single s-boson ($N = \sigma = 1$), $\tau = 0$ and $J = 0$ so that their contribution vanishes. On the other hand, in the state of a single d-boson ($N = \sigma = 1$), $\tau = 1$ and $J = 2$, they contribute to its energy $2B/3 + 6C$. We also see that in the case of the $O(6)$ subgroup of $U(6)$, there is a dynamical symmetry characterized by the group chain

$$\text{III} \qquad U(6) \supset O(6) \supset O(5) \supset O(3) \tag{35.9}$$

The notation III in (35.9), indicating the *third* case of dynamical symmetry, has historical origin (Arima and Iachello 1978) and is used here to conform with the literature. The dynamical symmetry III describes nuclei which in the collective model are γ-unstable or γ-soft (Wilets and Jean 1956).

From (35.8) follows that *any* $O(6)$ Hamiltonian is diagonal in the $O(5)$ seniority scheme. The quantum number τ which determines the eigenvalues of $C_{O(5)}$ in (35.8) is the seniority of the d-bosons. The operator $\mathbf{Q}' \cdot \mathbf{Q}'$ in the $O(6)$ Hamiltonian admixes states with s-bosons into those of d-bosons. Hence n_d is no longer conserved. The mixing is, however, such that *pairs* of s bosons replace pairs of d-bosons coupled to $J = 0$. Such admixtures do not change the seniority of the d-bosons. The $O(5)$ group here is the *same* $O(5)$ group as in the $U(5)$ chains (34.16). Originally, the seniority of d-bosons in the $U(5)$ limit was denoted by v whereas in the $O(6)$ limit it was denoted by τ. This distinction is not justified and we will denote the $O(5)$ seniority by

τ in both limits as well as in other situations to be considered below. To uniquely characterize the states, another quantum number is needed. When going over from $O(5)$ to $O(3)$ several equivalent irreducible representations may appear. To distinguish between states with the same values of σ, τ, J (and M) we use also here the same n_Δ as in the $U(5)$ limit (in the case of $O(6)$ it was denoted originally by ν_Δ). The energy eigenvalues, however, are given by (35.8) and are independent of n_Δ. Any two states with the same values of N, σ, τ, J (and M) are degenerate in the case of the $O(6)$ Hamiltonian with single boson energies and boson-boson interactions.

As explained above, σ is the generalized seniority in the space of s- and d-bosons. A state with seniority σ in the case of $\sigma = n_s + n_d = N$ has no pairs coupled to the state (35.3) and is annihilated by the P'^- operator. Among the states with $\sigma = N$ there are states with $\tau = N$, which are annihilated also by the operator P^- and have $n_s = 0$, $n_d = N$. Other states, with $\sigma = N$ annihilated by P^-, as well as by P'^-, have $n_s = 1$, $n_d = N - 1$ and $\tau = N - 1 = \sigma - 1$. All other states with $\sigma = N$, have lower values of τ. Thus the allowed values of τ are given by

$$\boxed{\tau = \sigma,\ \sigma - 1,\ \sigma - 2, \ldots, 0}$$

A simple example is offered by $N = 2$. The state (35.3) is the state with $\sigma = N - 2 = 0$. It contains one coherent pair of s- and d-bosons coupled to $J = 0$. The other $J = 0$ state orthogonal to it, $[(d^+ \cdot d^+) + 5(s^+)^2)|0\rangle$, has $\sigma = 2$. It has one pair of d-bosons coupled to $J = 0$ and hence, it is not annihilated by P^- and has $\tau = 0$. Other $\sigma = 2$ states are $(d^+ \times d^+)^{(J)}|0\rangle$ with $J = 2,4$ and these have $\tau = 2$. There is one more $\sigma = 2$ state, namely $s^+ d^+_\mu|0\rangle$, and this state, containing *one* unpaired d-boson has $\tau = 1$.

A typical spectrum of a Hamiltonian with $O(6)$ dynamical symmetry is shown in Fig. 35.1 for $N = 6$. It was calculated by taking the coefficient $A(> 0)$ to be larger than B and B larger than C. The spectrum is divided into parts according to the values of σ and n_Δ which are written in parantheses above each part. Spin values are indicated and levels with the same value of τ are grouped together. There are several nuclei whose spectra are well described by this limit, like ^{196}Pt whose spectrum is shown in Fig. 35.2. The lowest states belong to the $O(6)$ irreducible representation with $\sigma = N$. The values of τ of those states range between 0 and $\tau = \sigma = N$. The allowed values of J for any given τ are determined by the irreducible representations of

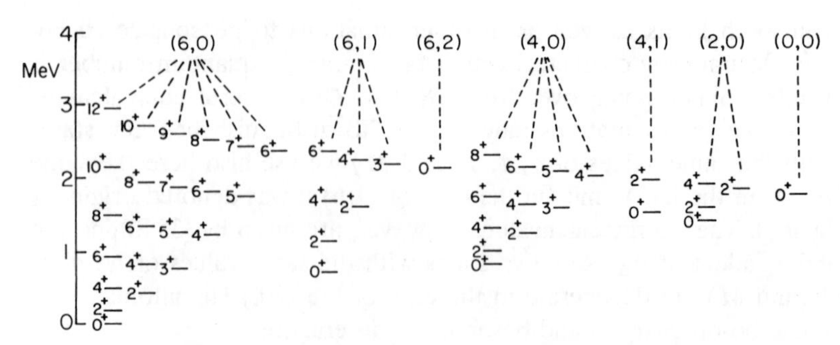

FIGURE 35.1. *Typical spectrum of a $O(6)$ Hamiltonian.*

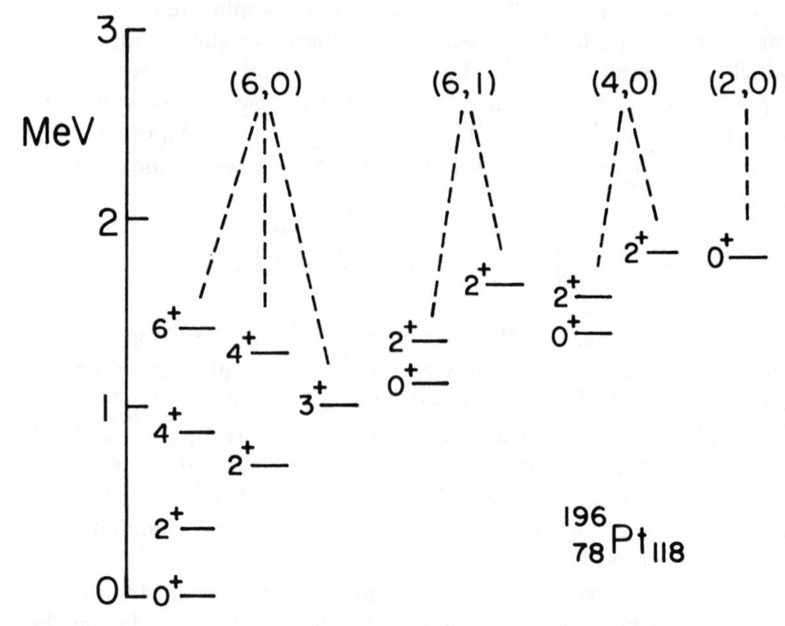

FIGURE 35.2. *$O(6)$-like spectrum of* ^{196}Pt.

$O(3)$ contained in the irreducible $O(5)$ representation with the given τ. These J values for given τ are determined by (34.18) and (34.19) exactly as in the $U(5)$ limit. Nuclei with $O(6)$ symmetry correspond to γ-unstable nuclei in the collective model.

The electric quadrupole transitions in the $O(6)$ limit depend on the values of α_2 and β_2 in the operator (33.33). If $\beta_2 = 0$ then (33.33) is proportional to the operator $\mathbf{Q}'$. The latter is a generator of $O(6)$ and hence has simple selection rules. It cannot have non-vanishing matrix elements between states in different irreducible representations, i.e. $\Delta\sigma = 0$. We recall the discussion following (25.24) and the derivation of (26.30) for generators of the group considered. Their non-vanishing matrix elements depend only on the irreducible representation to which the states belong and on their labels within it. These matrix elements are independent of the particular realization of the group and its irreducible representations. Hence, we obtain the following reduction relation

$$\boxed{\begin{aligned}&(N\sigma\tau n_\Delta J\|\mathbf{Q}'\|N\sigma'\tau' n'_\Delta J')\\ &\qquad = (N=\sigma,\sigma\tau n_\Delta J\|\mathbf{Q}'\|N=\sigma,\sigma\tau' n'_\Delta J')\delta_{\sigma\sigma'}\end{aligned}}$$

The operator $\mathbf{Q}'$ either adds or eliminates one d-boson, thereby changing the seniority τ by one unit, $\Delta\tau = \pm 1$. As a result, all diagonal matrix elements of $\mathbf{Q}'$ vanish. If the electric quadrupole operator is proportional to $\mathbf{Q}'$, all eigenstates in the $O(6)$ limit have vanishing quadrupole moments. If, however, the term with β_2 in (33.33) does not vanish, there may be non-vanishing quadrupole moments in certain states even in the $O(6)$ limit.

We shall not discuss transition probabilities in detail. We only compare the $U(5)$ result (34.59) with the $O(6)$ limit. We consider transitions between two states both with $\sigma = N$, one with $\tau, J = 2\tau$ and the other with $\tau - 1$, $J = 2\tau - 2$. In both cases all spins of the τ, or $\tau - 1$, d-bosons are parallel and thus, have $n_\Delta = 0$. In this sequence of levels with $J = 0, 2, 4, \ldots, 2N$ the transition probabilities were found to be given by

$$B(\mathrm{E2}, J \to J-2) = \alpha_2^2 \frac{J}{8(J+3)}(2N+2-J)(2N+6+J) \tag{35.10}$$

The transition probabilities in the present case, for fixed J, go roughly like N^2 whereas in the $U(5)$ limit (34.59) is only linear in N. The difference in N-dependence should be attributed to the coherent combinations of s- and d-boson states in the $O(6)$ wave functions.

The $O(5)$ classification of states leads to various reduction formulae for matrix elements of the transition operator (33.33). These results, which are not limited to the $O(6)$ scheme, are due to Leviatan et al. (1986). They are described below up to the end of this section. In both $U(5)$ and $O(6)$ limits, the group chains include the $O(5)$ group (and of course its $O(3)$ sub-group). This feature is not confined to these limits. The boson Hamiltonian has $O(5)$ symmetry whenever the coefficient $\tilde{v}_2$ in (33.30) vanishes. Such Hamiltonians contain 7 arbitrary coefficients and may be constructed in the following way. The boson Hamiltonian in the $U(5)$ limit contains 6 independent terms. In order to obtain the most general $U(6)$ boson Hamiltonian with $\tilde{v}_2 = 0$ we may add to the $U(5)$ Hamiltonian the Casimir operator of $O(6)$ multiplied by an arbitrary constant. Since $C_{O(6)}$ does not commute with the $U(5)$ generators (it does not commute with n_d) it is linearly independent of the Casimir operators in the $U(5)$ chain. The general boson Hamiltonians we shall now consider, are then given by

$$\epsilon_s' N + \tfrac{1}{2} u_0 N(N-1) + \epsilon \hat{n}_d + \alpha \hat{n}_d(\hat{n}_d - 1)/2 + \kappa C_{O(6)} + \frac{B}{6} C_{O(5)} + C C_{O(3)} \tag{35.11}$$

The single boson energy ϵ may depend linearly on N according to (33.31). The Hamiltonian (35.11) contains also single boson terms, arising from $C_{O(5)}$ and $C_{O(3)}$, which are equal to $(2B/3 + 6C)n_d$. The single boson term due to $C_{O(6)}$ has the same value for a d-boson and s-boson and hence, it is linear in N. Since $C_{O(5)}$ and certainly $C_{O(3)}$ commute with both $\hat{n}_d$ and $C_{O(6)}$, we may bring them to a diagonal form and express (35.11) by

$$\boxed{\epsilon_s' N + \tfrac{1}{2} u_0 N(N-1) + \epsilon \hat{n}_d + \alpha \hat{n}_d(\hat{n}_d - 1)/2 + \kappa C_{O(6)} + \tfrac{1}{6} B\tau(\tau+3) + CJ(J+1)} \tag{35.12}$$

Any such Hamiltonian is diagonal in the $O(5)$ seniority scheme independent of the relative sizes of ϵ, α and κ.

Since $\hat{n}_d$ and $C_{O(6)}$ do not commute, there is no complete dynamical symmetry of the general Hamiltonian (35.11). The eigenvalues are not characterized by simple quantum numbers like n_d for $\kappa = 0$ or σ for $\epsilon = 0$, $\alpha = 0$. Still, the operator $\epsilon \hat{n}_d + \alpha \hat{n}_d(\hat{n}_d - 1)/2 + \kappa C_{O(6)}$ commutes with all generators of $O(5)$ (it is an $O(5)$ scalar). Due to Schur's lemma all its eigenstates which belong to the same irreducible representation of $O(5)$ have equal eigenvalues. In any such $O(5)$ irreducible representations, the states with given τ, have J-values determined by (34.18) and (34.19). These states form a *τ-multiplet* of states and their energy differences are determined by the $CJ(J+1)$ term *independent* of the values of ϵ, α and κ in (35.11). The various multiplets with the same value of τ are distinguished by another label ("principal quantum number"). It characterizes the eigenstates (and eigenvalues) of the operator $\epsilon\hat{n}_d + \alpha\hat{n}_d(\hat{n}_d - 1)/2 + \kappa C_{O(6)}$, as does n_d in the $U(5)$ limit or σ in the $O(6)$ limit. In other cases, no simple quantum numbers may be obtained. Hence, we may assign each multiplet an integer ν whose range is $\nu = 1, 2, \ldots$ which indicates its relative position in the spectrum.

The relative positions of the various multiplets with the same value of τ, as well as those with different τ values, depend, in general, on the actual values of ϵ, α, κ as well as on N. This is also true of the structure of the eigenstates. Starting from the $U(5)$ basis, the operator $\hat{n}_d$ is diagonal and the non-diagonal elements are due to the $\mathbf{Q}' \cdot \mathbf{Q}'$ term in $C_{O(6)}$. Since $\mathbf{Q}' \cdot \mathbf{Q}'$ commutes with all $O(5)$ generators its diagonal elements within a τ-multplet are equal. Also its non-diagonal elements between states with given n_Δ, J in two τ-multiplets with different values of n_d are the same for all states and are thus independent of n_Δ and J (and M). When the Hamiltonian matrix with any values of ϵ, α and κ, for a given value of τ, is diagonalized, the various n_Δ, J states in a τ-multiplet are shifted by the same amount. This is in agreement with the statement made above that spacings within a τ-multiplet are given by $CJ(J+1)$ for any values of ϵ, α and κ. Moreover, it follows that for any values of ϵ, α and κ in (35.11), eigenstates of the Hamiltonian are linear combinations of $U(5)$ states with various values of n_d (differing by even numbers) whose amplitudes are independent of n_Δ and $J(M)$. The wave function of any state in the ν-th τ-multiplet can thus be expressed by

$$|N, \nu, \tau, n_\Delta, J, M\rangle = \sum_{n_d} a^{\nu}_{n_d \tau} |N, n_d, \tau, n_\Delta, J, M\rangle \tag{35.13}$$

The amplitudes $a^{\nu}_{n_d \tau}$ in (35.13) are obtained by diagonalization of (35.12) or just of the operator $\epsilon\hat{n}_d + \alpha\hat{n}_d(\hat{n}_d - 1)/2 + \kappa C_{O(6)}$. They depend, in general, on the actual values of ϵ, α, κ as well as on N. In fact, the expansion (35.13) follows directly by observing that states in each τ-multiplet transform among themselves in the same way under action of elements of the $O(5)$ group. These transformations are the same as those of states in τ-multiplets characterized by values of n_d. Hence, the expansion coefficients in (35.13) cannot depend on n_Δ or $J(M)$.

In the general case of a $O(5)$ boson Hamiltonian (35.11), the dependence of its eigenvalues on τ may be rather complicated. In the $O(6)$ limit (35.8) and in the $U(5)$ limit (34.17), the dependence of the eigenvalues on τ is given simply by the $\tau(\tau+3)$ term. There is a large class of $O(5)$ Hamiltonians for which this simple dependence on τ holds for the lowest τ-multiplets ($\nu = 1$) with $\tau = N, N-1, \ldots, 1, 0$. This is the same behavior as in the $O(6)$ limit for the levels with $\sigma = N$. To construct such Hamiltonians we generalize the pairing interaction (35.5) and consider the operator

$$\boxed{P_\beta^+ P_\beta^- = \tfrac{1}{4}((d^+ \cdot d^+) - \beta(s^+)^2)((\tilde{d} \cdot \tilde{d}) - \beta s^2)} \tag{35.14}$$

For $\beta = 1$, (35.14) coincides with (35.5) of the $O(6)$ limit. If, however, we put $\beta = 0$ the operator (35.14) is reduced to the pairing interaction (34.5) of the $U(5)$ limit. If we add to (35.14), in addition to the terms with N, also $\frac{1}{6}B\tau(\tau+3) + CJ(J+1)$, we obtain an $O(5)$ Hamiltonian which reduces in the $\beta = 1$ case to the most general $O(6)$ Hamiltonian. In the case of $\beta = 0$ it reduces to a special $U(5)$ Hamiltonian.

To find eigenstates and eigenvalues of the generalized pairing interaction (35.14), we construct a linear combination of states with different values of τ and J. If the eigenvalues of those are indeed equal, such a state could be an eigenstate of (35.14). We consider a coherent or an *intrinsic state* of the form

$$(d_0^+ + xs^+)^N|0\rangle \tag{35.15}$$

The state (35.15) has a zero projection of the angular momentum on a given axis which was conveniently chosen as the z-axis. From the state (35.15) we can project states with definite angular momenta whose values are $J = 0, 2, 4, \ldots, 2N$. This can be proved by induction, starting from (35.15) with $N = 1$ from which two states may be projected with $J = 0$ and $J = 2$. The binomial expansion of (35.15) contains the terms

$$\binom{N}{\nu} x^{N-\nu} \mid (s^+)^{N-\nu}(d_0^+)^\nu|0\rangle$$

Hence, it is only necessary to check the spins of states projected from $(d_0^+)^\nu|0\rangle$. If it has been shown that states projected from $(d_0^+)^{\nu-1}|0\rangle$ have spin values $J = 0, 2, \ldots, 2(\nu-1)$ we obtain

$$(d_0^+)^\nu|0\rangle = d_0^+\left(\sum_{J_0\,\text{even}}^{2(\nu-1)} a_{J_0}|d^{\nu-1}J_0M_0 = 0\rangle\right)$$

$$= \sum_J \sum_{J_0\,\text{even}}^{2(\nu-1)} a_{J_0}(20J_00 \mid 2J_0J0)|d^\nu JM = 0\rangle \qquad (35.16)$$

Due to their symmetry property (7.18), the Clebsch-Gordan coefficients on the r.h.s. of (35.16) vanish unless $(-1)^{J_0-J} = 1$. Thus, it follows that the states projected from $(d_0^+)^\nu|0\rangle$ have all even values of J, including $J = 2(\nu-1)+2 = 2\nu$. Since this property holds for $\nu = 1$, it holds by induction for any ν and N.

We now apply to (35.15) the pairing interaction (35.14) and recalling the lemma (23.8), we obtain

$$P_\beta^+P_\beta^-(d_0^+ + xs^+)^N|0\rangle = NP_\beta^+(d_0^+ + xs^+)^{N-1}[P_\beta^-,(d_0^+ + xs^+)]|0\rangle$$

$$+\tfrac{1}{2}N(N-1)P_\beta^+(d_0^+ + xs^+)^{N-2}$$

$$\times[[P_\beta^-,(d_0^+ + xs^+)],(d_0^+ + xs^+)]|0\rangle + \cdots \qquad (35.17)$$

The commutators in (35.17) can be directly evaluated. Since the operators d_μ for $\mu \neq 0$ commute with d_0^+ and s^+ we obtain

$$[P_\beta^-,(d_0^+ + xs^+)] = [(d_0^2 - \beta s^2), d_0^+ + xs^+] = 2(d_0 - \beta xs) \qquad (35.18)$$

$$[[P_\beta^-,(d_0^+ + s^+)], d_0^+ + xs^+] = 2[(d_0 - \beta xs),(d_0^+ + xs^+)] = 2(1-\beta x^2) \qquad (35.19)$$

Hence, the higher commutators in (35.17) vanish. Since (35.18) contains only annihilation operators, the first term on the r.h.s. of (35.17) vanishes. From (35.19) follows that if $1-\beta x^2 = 0$ then the state (35.15) is an eigenstate of (35.14) with eigenvalue zero. Since the operator (35.14) is a product of hermitean conjugate operators, its eigenvalues cannot be negative ($\langle|P_\beta^+ P_\beta^-|\rangle \geq 0$). Hence, the eigenvalue 0 of (35.14) is its lowest eigenvalue.

We can add now to the pairing interaction (35.14), multiplied by a constant $4A > 0$, the terms $\frac{1}{6}BC_{O(5)}$ and $CJ(J+1)$. We thus obtain an $O(5)$ Hamiltonian, like (35.11), in which

$$\epsilon_s' = 5\beta A \qquad u_0 = 2\beta^2 A \qquad \epsilon = A(4 + 2\beta(N-3) - 2\beta^2(N-1))$$

$$\alpha = 2A(1-\beta)^2 \qquad \kappa = -\beta A$$

and the coefficient multiplying $C_{O(5)}$ is equal to $\frac{1}{6}B - (1-\beta)A$. For $\beta = 0$, also $\kappa = 0$ and the Hamiltonian has $U(5)$ symmetry. For $\beta = 1$, both ϵ and α vanish and an $O(6)$ Hamiltonian is obtained. The lowest eigenvalues of the resulting $O(5)$ Hamiltonian are those for which (35.14) has vanishing eigenvalues and are then equal to

$$\frac{1}{6}B\tau(\tau+3) + CJ(J+1) \tag{35.20}$$

The intrinsic state $(d_0^+ + (1/\sqrt{\beta})s^+)^N|0\rangle$ is an eigenstate of the rotationally invariant Hamiltonian (35.14) with $O(5)$ symmetry. Hence it is a linear combination of states with definite values of τ and J, each of which is an eigenstate of (35.14) with eigenvalue 0. There may be other states which are orthogonal to projected states with the *same* values of τ and J. These are also eigenstates of the $O(5)$ Hamiltonian (35.14) but have higher eigenvalues. The τ-multiplets which are included in the intrinsic state have $\tau = N, N-1, \ldots, 0$. To see it, we may put $\beta = 1$ in which case we obtain the $O(6)$ limit and the lowest states have $\sigma = N, \tau = \sigma, \sigma - 1, \ldots, 0$. The states with given τ that can be projected from (35.15) are those with *even* values of J only. As shown above, states with odd values of J with $\tau \leq N$ cannot be projected from the state (35.15). They belong to τ-multiplets where states with even values of J are degenerate in the case of the Hamiltonian (35.14). Thus, the coefficient of $J(J+1)$ which determines spacings of states in the τ-multiplet vanishes in the case of (35.14) and also states with odd values of J have the eigenvalue 0 of the pairing interaction (35.14). In the case $\beta = 1$, the state (35.15) with $x = 1$ is an eigenvalue of the $\kappa = 0$ component of $\mathbf{Q}'$ of (35.1). To see it we apply Q_0' to (35.15) and make use of the lemma (23.8). We thus obtain

$$[s^+\tilde{d}_0 + sd_0^+, d_0^+ + xs^+] = s^+ + xd_0^+$$

and higher commutators vanish. Hence, for $x = 1$ follows the result

$$(s^+\tilde{d}_0 + sd_0^+)(d_0^+ + s^+)^N|0\rangle = N(d_0^+ + s^+)(d_0^+ + s^+)^{N-1}|0\rangle$$
$$= N(d_0^+ + s^+)^N|0\rangle \qquad (35.21)$$

If β is changed between 1 and 0, the transition region between the $O(6)$ and $U(5)$ limits is covered. The case $\beta = 0$ is a special $U(5)$ limit in which (35.15) reduces to one $J = 0$ state with $n_s = N$, $n_d = 0$. There is no need to limit the range of β to the interval between 0 and 1. If β becomes negative, the coefficient x of s^+ in (35.15) becomes imaginary. For $\beta = -1$, $x = i$, the Hamiltonian contains the Casimir operator of the orthogonal group $\overline{O(6)}$ whose generators are $(d^+ \times \tilde{d})_\kappa^{(k)}$, $k = 1,3$, and instead of $\mathbf{Q}'$ in (35.12), the quadrupole generator is the other quadrupole operator in (33.25).

For example in the case $N = 2$, the instrinsic state is

$$\left(d_0^+ + \frac{1}{\sqrt{\beta}}s^+\right)^2|0\rangle$$
$$= \left\{\sum_{J=2,4}(2020\,|\,22J0)(d^+ \times d^+)_0^{(J)} + \tfrac{1}{5}(d^+ \cdot d^+)\right.$$
$$\left. + \frac{2}{\sqrt{\beta}}s^+d_0^+ + \frac{1}{\beta}(s^+)^2\right\}|0\rangle$$

The projected states with $\tau = 2$ whose eigenvalues vanish are the $J = 2$ and $J = 4$ states $(d^+ \times d^+)_0^{(J)}|0\rangle$ and the $\tau = 1$ state is $s^+d_0^+|0\rangle$. Due to the rotational invariance of the Hamiltonian, also the states with $M \neq 0$ have the same eigenvalues as the corresponding $M = 0$ states. The state with eigenvalue 0 and $\tau = 0$ has $J = 0$ and is equal to $[\frac{1}{5}(d^+ \cdot d^+) + (1/\beta)(s^+)^2]|0\rangle$. The orthogonal linear combination is the other $\tau = 0$, $J = 0$ eigenstate of (35.14) with non-vanishing eigenvalue.

In the case $N = 3$ the intrinsic state is

$$\left(d_0^+ + \frac{1}{\sqrt{\beta}}s^+\right)^3|0\rangle = \left[(d_0^+)^3 + \frac{3}{\sqrt{\beta}}(d_0^+)^2s^+ + \frac{3}{\beta}d_0^+(s^+)^2\right.$$
$$\left. + \frac{1}{\beta\sqrt{\beta}}(s^+)^3\right]|0\rangle.$$

The $\tau = 3$ states may be projected only from the component with $n_d = 3$ and are thus the $J = 6$, $J = 4$ and $J = 0$ states projected from $(d_0^+)^3|0\rangle$. The $\tau = 3$, $J = 3$ cannot be projected from the intrinsic state. The $\tau = 2$ states with $J = 2,4$ are projected from the $(d_0^+)^2 s^+|0\rangle$ component. The $\tau = 1$ eigenstate with vanishing eigenvalue has $J = 2$ and is a linear combination of the $d_0^+(s^+)^2|0\rangle$ state and the $J = 2$ state projected from the $(d_0^+)^3|0\rangle$ state. The $\tau = 0$, $J = 0$ state with eigenvalue 0 is a linear combination of the $J = 0$ component of the $(d_0^+)^2 s^+|0\rangle$ state and the $(s^+)^3|0\rangle$ state.

Let us now turn to electromagnetic transition probabilities in the general case of $O(5)$ symmetry. The results derived below hold in particular in the $U(5)$ and $O(6)$ limits as well as in the transition region between them. The reduced matrix element of the quadrupole operator (33.33) between two states (35.13) is given by

$$\boxed{\begin{aligned}&(N\nu\tau n_\Delta J\|\mathbf{Q}\|N\nu'\tau' n'_\Delta J')\\ &\qquad = \sum_{n_d n'_d} a^{\nu}_{n_d\tau} a^{\nu'}_{n'_d\tau'} (N n_d \tau n_\Delta J\|\mathbf{Q}\|N n'_d \tau' n'_\Delta J')\end{aligned}} \tag{35.22}$$

Due to the different selection rules of the two terms in (33.33), those with coefficients α_2 and β_2, they cannot contribute to the same matrix element. If $\tau' = \tau \pm 1$, only the first term may contribute whereas for $\tau' = \tau$, $\tau \pm 2$ only the second term may give a non-vanishing contribution. We start with the former case.

If $\tau' = \tau + 1$ we obtain, due to (34.32) (and the similar result for the transposed matrix element), for (35.22) the expression

$$\begin{aligned}&\sum_{n_d n'_d} a^{\nu}_{n_d\tau} a^{\nu'}_{n'_d\tau+1} (N n_d \tau n_\Delta J\|s^+\tilde{d} + d^+ s\|N n'_d \tau+1, n'_\Delta J')\\ &\qquad = \sum_{n_d} a^{\nu}_{n_d\tau} a^{\nu'}_{n_d-1,\tau+1} \sqrt{N - n_d + 1}\\ &\qquad\quad \times (n_d \tau n_\Delta J\|d^+\|n_d - 1, \tau+1, n'_\Delta J')\\ &\qquad\quad + \sum_{n_d} a^{\nu}_{n_d\tau} a^{\nu'}_{n_d+1,\tau+1} \sqrt{N - n_d}\\ &\qquad\quad \times (n_d \tau n_\Delta J\|\tilde{d}\|n_d + 1, \tau+1, n'_\Delta J')\end{aligned}$$

$$
\begin{aligned}
&= \sum_{n_d} a^{\nu}_{n_d\tau} a^{\nu'}_{n_d-1,\tau+1} \sqrt{N-n_d+1} \\
&\quad \times (n_d \tau n_\Delta J \| d^+ \| n_d-1, \tau+1, n'_\Delta J') \\
&\quad + (-1)^{J+J'} \sum_{n_d} a^{\nu}_{n_d\tau} a^{\nu'}_{n_d+1,\tau+1} \sqrt{N-n_d} \\
&\quad \times (n_d+1, \tau+1, n'_\Delta J' \| d^+ \| n_d \tau n_\Delta J)
\end{aligned}
\tag{35.23}
$$

The last equality follows from the Wigner-Eckart theorem as was the case on the r.h.s. of (34.41). We now use (34.36) and (34.38) to reduce (35.23) to the form

$$
\begin{aligned}
&\sum_{n_d} a^{\nu}_{n_d\tau} a^{\nu'}_{n_d-1,\tau+1} \sqrt{N-n_d+1} \sqrt{\frac{n_d-\tau}{2}} \\
&\quad \times (\tau+2, \tau n_\Delta J \| d^+ \| \tau+1, \tau+1, n'_\Delta J') \\
&\quad + (-1)^{J+J'} \sum_{n_d} a^{\nu}_{n_d\tau} a^{\nu'}_{n_d+1,\tau+1} \sqrt{N-n_d} \sqrt{\frac{n_d+\tau+5}{2\tau+5}} \\
&\quad \times (\tau+1, \tau+1, n'_\Delta J' \| d^+ \| \tau \tau n_\Delta J)
\end{aligned}
\tag{35.24}
$$

The reduced matrix elements in the first summation in (35.24) may be now expressed in terms of those in the second summation with the help of (34.41). We thus obtain for the matrix element (35.22) the expression

$$
\begin{aligned}
&(-1)^{J+J'} (\tau+1, \tau+1, n'_\Delta J' \| d^+ \| \tau \tau n_\Delta J) \frac{1}{\sqrt{2\tau+5}} \\
&\quad \times \left\{ \sum_{n_d} a^{\nu}_{n_d\tau} a^{\nu'}_{n_d-1,\tau+1} \sqrt{(N-n_d+1)(n_d-\tau)} \right. \\
&\qquad \left. + \sum_{n_d} a^{\nu}_{n_d\tau} a^{\nu'}_{n_d+1,\tau+1} \sqrt{(N-n_d)(n_d+\tau+5)} \right\}
\end{aligned}
\tag{35.25}
$$

The reduced matrix element is thus *factorized* into a product of two terms. One factor is determined by the Hamiltonian and depends on N (both explicitly and also through the coefficients $a^{\nu}_{n_d\tau}$), ν and τ of the two states. The other term is the same for any Hamiltonian (35.11) and is completely determined by the $O(5)$ symmetry of such Hamiltonians. It is a function of only the seniority quantum numbers τ, n_Δ and J of both states.

To obtain the electric quadrupole transition probability from the reduced matrix element (35.22) or (35.25), the latter should be squared and then divided by $2J'+1$ if J' is the spin of the initial (or by $2J+1$ for the $J \to J'$ transition). Multiplying the result by α_2^2, we obtain from (35.25) the result

$$B(\mathrm{E2};\nu'\tau+1,n'_\Delta J' \to \nu\tau n_\Delta J) = \frac{\alpha_2^2}{2J'+1}F_N(\nu\nu'\tau)(\tau+1,\tau+1,n'_\Delta J'\|d^+\|\tau\tau n_\Delta J)^2 \tag{35.26}$$

where the factor $F_N(\nu\nu'\tau)$ can be explicitly obtained by comparing (35.26) and (35.25).

Before drawing the conclusions from the expression (35.26) of the B(E2), let us see that a similar factorization also occurs for transitions between states with $\tau' = \tau$, $\tau \pm 2$ and τ. From the general expression (35.22) we obtain the reduced matrix element of the operator, whose coefficient in (33.33) is β_2, in the form

$$(N\nu\tau n_\Delta J\|(d^+ \times \tilde{d})^{(2)}\|N\nu'\tau' n'_\Delta J') = \sum_{n_d} a^{\nu'}_{n_d\tau'} a^{\nu}_{n_d\tau}(n_d\tau n_\Delta J\|(d^+ \times \tilde{d})^{(2)}\|n_d\tau' n'_\Delta J') \tag{35.27}$$

The reduced matrix elements on the r.h.s. of (35.27) are given by (34.48) for $\tau' = \tau \pm 2$ and by (34.56) for $\tau' = \tau$. In the former case, the reduced matrix element can be expressed as

$$\boxed{\begin{aligned}&(\tau+2,\tau n_\Delta J\|(d^+\times\tilde{d})^{(2)}\|\tau+2,\tau+2,n'_\Delta J')\\&\qquad\times\sum_{n_d} a^{\nu'}_{n_d\tau+2}a^{\nu}_{n_d\tau}\sqrt{\frac{(n_d+\tau+5)(n_d-\tau)}{2(2\tau+5)}}\end{aligned}} \tag{35.28}$$

In the case where $\tau'=\tau$ we obtain from (34.56) the reduced matrix element as

$$\boxed{(\tau\tau n_\Delta J\|(d^+\times\tilde{d})^{(2)}\|\tau\tau n'_\Delta J')\sum_{n_d} a^{\nu'}_{n_d\tau}a^{\nu}_{n_d\tau}\frac{2n_d+5}{2\tau+5}} \tag{35.29}$$

In both (35.28) and (35.29) the reduced matrix element is factorized into a term which is due only to the $O(5)$ symmetry of the Hamiltonian, which depends on $\tau, n_\Delta, n'_\Delta, J, J'$, and a term which depends on ν and ν' as well as on τ. The latter term, which is independent of $n_\Delta, n'_\Delta, J, J'$, depends also on N since the coefficients $a^{\nu}_{n_d\tau}$ are determined by diagonalization of the Hamiltonian (35.11) for a given value of N. The B(E2) value obtained from either (35.28) or (35.29) has thus the general form as (35.26). It is a product of $\beta_2^2/(2J'+1)$ and two factors. One which depends only on τ, τ', $n_\Delta, n'_\Delta, J, J'$ and another which depends on ν, ν', τ, τ' and N.

Let us now consider E2 transitions from states in any $\nu'\tau'$ multiplet to states in any $\nu\tau$ multiplet. The number ν' may be different from ν or both states with τ' and τ could have the same ν. A direct consequence of (35.26), as well as of (35.28) and (35.29), is that the *ratios* of B(E2) values of such transitions depend only on τ, n_Δ, J and τ', n'_Δ, J'. These ratios are independent of ν, ν' and have thus the same value for transitions in the given nucleus. They are also independent of N and hence, should be the same for *all* nuclei with $O(5)$ symmetry. They are determined only by the $O(5)$ symmetry of the Hamiltonian and are independent of the actual values of its parameters ϵ, α and κ in (35.11) or (35.12). In particular, they should be the same throughout the $U(5)$–$O(6)$ region.

Another set of predictions follows from (35.26), (35.28) and (35.29) for E2 transitions from a state with given τ', n'_Δ, J' to a state with given τ, n_Δ, J either with the same value of ν, or for different ν', ν. Such transitions may be considered for different values of N and thus in different nuclei. In *ratios* of B(E2) values of such transitions, the factors which depend on $\tau', n'_\Delta, J', \tau, n_\Delta, J$ cancel and the ratio of the other factors is independent of $n'_\Delta, J', n_\Delta, J$. For examples, the ratio between the B(E2) of the transition from the state with $N_1\nu'_1\tau + 1, n'_\Delta, J'$ to the state $N_1\nu_1\tau n_\Delta J$ and the B(E2) of the transition from the state with $N_2\nu'_2\tau + 1, n'_\Delta J'$ to the state with $N_2\nu_2\tau n_\Delta J$ is given, according to (35.26) by

$$\alpha_2^2 F_{N_1}(\nu_1\nu'_1\tau)/\alpha_2'^2 F_{N_2}(\nu_2\nu'_2\tau) \tag{35.30}$$

The coefficients α_2 and α'_2 in (35.30) may depend on the values of N_1 and N_2 respectively. Hence, ratios of these B(E2) values should be the same for all transitions between the states, with various values of n_Δ and J, in the τ-multiplets considered.

The predictions derived above, about ratios of rates of E2 transitions, are based only on $O(5)$ symmetry. They hold, in particular, in the $O(6)$ and $U(5)$ limits and throughout the transition region between them. Experimental data which confirm these predictions indicate agreement with $O(5)$ symmetry. They should not be quoted as evidence for any particular limit, either $O(6)$ or $U(5)$.

As mentioned above, the reduction formulae (34.48) and (34.56) for matrix elements of the quadrupole operator $(d^+ \times \tilde{d})^{(2)}$ hold also for the operator $(d^+ \times \tilde{d})^{(4)}$. Therefore, the ratios of B(E4) and B(E2) for transitions between the same states are independent of ν', ν and N. In fact, this holds for the ratio of reduced matrix elements

$$(\nu\tau n_\Delta J\|(d^+ \times \tilde{d})^{(4)}\|\nu'\tau' n'_\Delta J')/(\nu\tau n_\Delta J\|(d^+ \times \tilde{d})^{(2)}\|\nu'\tau' n'_\Delta J') \tag{35.31}$$

if the denominator does not vanish.

We can use (35.29) in order to derive a relation between the quadrupole moment (or hexadecapole moment) of a state and the expectation value of $\hat{n}_d$ in that state. From (35.29) we see that matrix elements (35.27) for $\tau' = \tau$ are proportional to those between states with $n_d = \tau$. The proportionality factor, for $\nu' = \nu$, is given by (35.29) as

$$\boxed{\sum_{n_d}(a^\nu_{n_d\tau})^2\frac{2n_d+5}{2\tau+5} = \frac{2\bar{n}_d+5}{2\tau+5}} \tag{35.32}$$

Hence, in particular, for $k = 2$ or $k = 4$,

$$\boxed{\begin{aligned}&(N\nu\tau n_\Delta J\|(d^+ \times \tilde{d}^+)^{(k)}\|N\nu\tau n_\Delta J)\\ &\quad = \frac{2\bar{n}_d + 5}{2\tau + 5}(N n_d = \tau, \tau n_\Delta J\|(d^+ \times \tilde{d})^{(k)}\|N n_d = \tau, \tau n_\Delta J)\end{aligned}} \tag{35.33}$$

All these features of E2 moments and transitions follow directly from $O(5)$ symmetry. They hold exactly for any values of N, α, ϵ and κ in (35.11) or (35.12) and in particular through the whole $U(5)$–$O(6)$ region.

36

The $SU(3)$ Dynamical Symmetry— Rotational Spectra

Once the term with $\tilde{v}_2$ is present in (33.30) the $O(5)$ group can no longer be used to classify eigenstates of the boson Hamiltonian. This term admixes states whose numbers of s-bosons differ by *one*. For instance, the state $(d^+ \times d^+)^{(2)}_\mu|0\rangle(\tau = 2)$ is admixed with the $\tau = 1$ state $s^+d^+_\mu|0\rangle$. Hence, the seniority τ of d-bosons can no longer be a good quantum number. In that case, however, there is a possible dynamical symmetry associated with another subgroup of $U(6)$. It appears if $\tilde{v}_2 \neq 0$, $\tilde{v}_0$ is proportional to it and ϵ and the c'_L have certain values. The subgroup from whose generators such a Hamiltonian can be constructed is the $SU(3)$ group. In Section 30, the $SU(3)$ group was considered in the case of fermion states. Here it appears as a symmetry group of s- and d-bosons. As we recall from Sections 6 and 30, the $U(3)$ group has 9 generators, one of which commutes with all the others. Once it is removed, only 8 elements of the Lie algebra remain which generate transformations whose matrices have determinants equal to 1. The orthogonal group in 3 dimensions $O(3)$ is a subgroup of $U(3)$ as well as of $SU(3)$. Its generators yield infinitesimal three dimensional rotations and hence, in the space of s- and d-bosons they are given by the components of $\mathbf{J}$ in (33.27). The other

5 generators must be the components of a quadrupole operator which we will now consider.

The commutation relations of the components of **J** with those of any tensor operator are equal to components of the same tensor. The components of the quadrupole tensor needed here, should have their commutators equal to linear combinations of $SU(3)$ generators. A quadrupole tensor **Q** such that $\mathbf{Q}\cdot\mathbf{Q}$ yields the term multiplying $\bar{\nu}_2$ in (33.30) should have the general form

$$\tilde{\mathbf{Q}} = s^+\tilde{d} + d^+s + \chi(d^+ \times \tilde{d})^{(2)} = \mathbf{Q}' + \chi(d^+ \times \tilde{d})^{(2)} \qquad (36.1)$$

The commutator of Q_μ and $Q_{\mu'}$ can be evaluated by using (33.34), (33.35) as well as (35.2). We obtain

$$\begin{aligned}[\tilde{Q}_\mu, \tilde{Q}_{\mu'}] &= [Q'_\mu, Q'_{\mu'}] + \chi[Q'_\mu, (d^+ \times \tilde{d})^{(2)}_{\mu'}] \\ &\quad + \chi[(d^+ \times \tilde{d})^{(2)}_\mu, Q'_{\mu'}] + \chi^2[(d^+ \times \tilde{d})^{(2)}_\mu, (d^+ \times \tilde{d})^{(2)}_{\mu'}] \\ &= \sum_{k\kappa}(1-(-1)^k)(2\mu 2\mu' \mid 22k\kappa)(d^+ \times \tilde{d})^{(k)}_\kappa \\ &\quad - \chi^2 \sum_{k\kappa}(1-(-1)^k)5\begin{Bmatrix} 2 & 2 & k \\ 2 & 2 & 2 \end{Bmatrix} \\ &\quad \times (2\mu 2\mu' \mid 22k\kappa)(d^+ \times \tilde{d})^{(k)}_\kappa \end{aligned} \qquad (36.2)$$

The terms linear in χ vanish due to the commutation relations (33.36). The summation in (36.2) has non-vanishing terms for $k = 1$ and $k = 3$ only. To obtain the $SU(3)$ Lie algebra the term with $k = 3$ should vanish which implies the condition

$$\left(1 - \chi^2 5\begin{Bmatrix} 2 & 2 & 3 \\ 2 & 2 & 2 \end{Bmatrix}\right) = (1 - \chi^2 5 \cdot \tfrac{4}{35}) = 1 - \tfrac{4}{7}\chi^2 = 0 \qquad \chi^2 = \tfrac{7}{4}$$

Instead of the general quadrupole operator (36.1) we take for the $SU(3)$ generators the components of

$$\boxed{\mathbf{Q} = s^+\tilde{d} + d^+s - \frac{\sqrt{7}}{2}(d^+ \times \tilde{d})^{(2)}} \qquad (36.3)$$

An equally acceptable choice would have been $\chi = +\sqrt{7}/2$. We shall discuss later the physical meaning of the choice (36.3). The commutation relations of the components of (36.3) are given by (36.2) with $\chi^2 = \frac{7}{4}$ and $k = 1$, to be

$$\boxed{[Q_\mu, Q_{\mu'}] = \frac{15}{4} \sum_\kappa (2\mu 2\mu' \mid 221\kappa)(d^+ \times \tilde{d})^{(1)}_\kappa} \tag{36.4}$$

Apart from the N, N^2 terms, a rotationally invariant Hamiltonian constructed from the generators of $SU(3)$ has only two independent terms—the scalar products $\mathbf{Q} \cdot \mathbf{Q}$ and $\mathbf{J} \cdot \mathbf{J}$. Instead of using these scalars, any such Hamiltonian may be expressed as a linear combination of the quadratic Casimir operator of $SU(3)$ and the $O(3)$ Casimir operator. The $SU(3)$ Casimir operator is a linear combination of $\mathbf{Q} \cdot \mathbf{Q}$ and $\mathbf{J} \cdot \mathbf{J}$. These scalar products commute with all components of $\mathbf{J}$. A linear combination that commutes with all Q_μ is obtained by using the commutation relations (36.4) as well as those of the components of $\mathbf{J}$ with any $T^{(k)}_\kappa$. We thus define

$$\boxed{C_{SU(3)} = 2(\mathbf{Q} \cdot \mathbf{Q}) + \tfrac{3}{4}\mathbf{J} \cdot \mathbf{J}} \tag{36.5}$$

We can express the $SU(3)$ Hamiltonian (omitting the linear and quadratic terms in N) as the following combination

$$-\kappa \mathbf{Q} \cdot \mathbf{Q} - \kappa' \mathbf{J} \cdot \mathbf{J} = -\tfrac{1}{2}\kappa C_{SU(3)} + (\tfrac{3}{8}\kappa - \kappa')\mathbf{J} \cdot \mathbf{J} \tag{36.6}$$

The eigenvalues of the $SU(3)$ Hamiltonian (36.6) belong to irreducible representations of $SU(3)$. All states which belong to a given irreducible representation have the *same* eigenvalue of $C_{SU(3)}$ and their energies are thus proportional to $J(J+1)$. A rotational spectrum arises naturally in the $SU(3)$ limit of the boson model as was the case in the nucleon $SU(3)$ model described in Section 30.

The dynamical symmetry expressed by (36.6) was obtained by using the chain of groups

$$\text{II} \qquad U(6) \supset SU(3) \supset O(3) \tag{36.7}$$

The fact that this dynamical symmetry is denoted by II has historical reasons and conforms to the literature.

The irreducible representations of $SU(3)$ are characterized by the lengths of the two rows in the Young diagrams. As in Section 30, we define the number of squares in the top row by $\lambda + \mu$ and in the second row by μ. The total number of squares is thus $\lambda + 2\mu$. The eigenvalues of the Casimir operator $C_{SU(3)}$ in the (λ, μ) irreducible representation are given by (30.14) as

$$\boxed{C_{SU(3)}(\lambda, \mu) = \lambda^2 + \mu^2 + \lambda\mu + 3(\lambda + \mu)} \tag{36.8}$$

The overall normalization of (36.5) was chosen to yield the eigenvalues (36.8). The eigenvalues of the $SU(3)$ Hamiltonian (36.6) are thus given by

$$-\tfrac{1}{2}\kappa(\lambda^2 + \mu^2 + \lambda\mu + 3(\lambda + \mu)) + (\tfrac{3}{8}\kappa - \kappa')J(J+1) \tag{36.9}$$

Due to (36.9) the energy eigenvalues fall naturally into rotational bands as will be discussed below in more detail. The energy of each eigenstate depends only on (λ, μ) and J (in addition to N). Still, these quantum numbers do not furnish a complete scheme which uniquely characterizes the various states. Going from an irreducible representation of $SU(3)$ to those of $O(3)$, several equivalent $O(3)$ representations may appear. In other words, there may be several states with the same value of J (and M) in the set of states characterized by the same (λ, μ). Another quantum number is necessary as in the fermion case. It becomes necessary in all but the lowest irreducible representation. This number is denoted by a non-negative integer K.

The irreducible representation of $SU(3)$ to which the states of *one* boson belong has $\lambda = 2$, $\mu = 0$. As explained in Section 30, states in the third oscillator shell, with $l = 0$ and $l = 2$, are constructed from *two* oscillator p-bosons. They belong to the $SU(3)$ irreducible representation $\lambda = 2$, $\mu = 0$ in which the two p-bosons are coupled symmetrically to $J = 0$ or $J = 2$. In the system with N s- and d-bosons there is always an irreducible representation in which the $2N$ p-bosons are coupled symmetrically. The Young diagram for this representation has one row of length $\lambda = 2N$ ($\mu = 0$). This is the lowest representation if the quadrupole interaction is attractive, i.e. $\kappa > 0$, since for given N, $C_{SU(3)}$ has then its maximum value. Other irreducible representations which are fully symmetric in the s- and d-bosons, need not be fully symmetric in the oscillator p-bosons and may have $\mu \neq 0$. It is sufficient that the states are symmetric under exchange of *two* p-bosons which are symmetrically coupled. Hence, allowed irreducible representations may be obtained from the one with $(2N, 0)$ by removing *two* squares from the top row and putting them in the second row. Thus, allowed irreducible representations of $SU(3)$ are given by

$$(2N,0),(2N-4,2),(2N-8,4),\ldots$$

Representations like $(2N-2,1)$ in which μ is odd are not fully symmetric in the s- and d-bosons and are not allowed. Two squares may also be removed from the top row (if $\lambda > 2$) and placed in the third row (if $\mu \geq 2$). Since we consider irreducible representations of $SU(3)$ (and not of $U(3)$), the irreducible representation thus obtained is equivalent to the one in which both λ and μ are reduced by two. For example, if $N = 8$ all allowed irreducible representations for s- and d-bosons are given by

$$(16,0),(12,2),(8,4),(4,6),(0,8),(10,0),(6,2),(2,4),(4,0),(0,2)$$

When going over from the irreducible representations of $SU(3)$ to those of $O(3)$, states with definite J values are obtained. The possible J values in any irreducible representation of $SU(3)$ follow from the mathematical properties of those representations and their decompositions into irreducible representations of $O(3)$. Thus, they are not specific to the boson realization of $SU(3)$ considered here. Let us take the case that $\lambda \geq \mu$ (if $\mu \geq \lambda$ the roles of λ and μ in the following are interchanged). The allowed values of K are given by the same rules as in the fermion case (30.34), (30.35) and (30.36), to be

$$\boxed{K = \mu, \mu-2, \ldots, 0 \qquad \text{(for even values of } \mu)} \tag{36.10}$$

The allowed values of J for any given K are then

$$\boxed{\begin{array}{ll} J = K, K+1, K+2, \ldots, K+\lambda & \text{for} \quad K > 0 \\ J = 0,2,4,\ldots,\lambda & \text{for} \quad K = 0 \end{array}} \tag{36.11}$$

In the lowest $SU(3)$ irreducible representation $(2N,0)$ there is only one K value, $K = 0$ and the states form the *ground state band* with $J = 0,2,4,\ldots,2N$. In the next higher representation $(2N-4,2)$ there are two possible K values, $K = 0$ and $K = 2$. Accordingly, there is

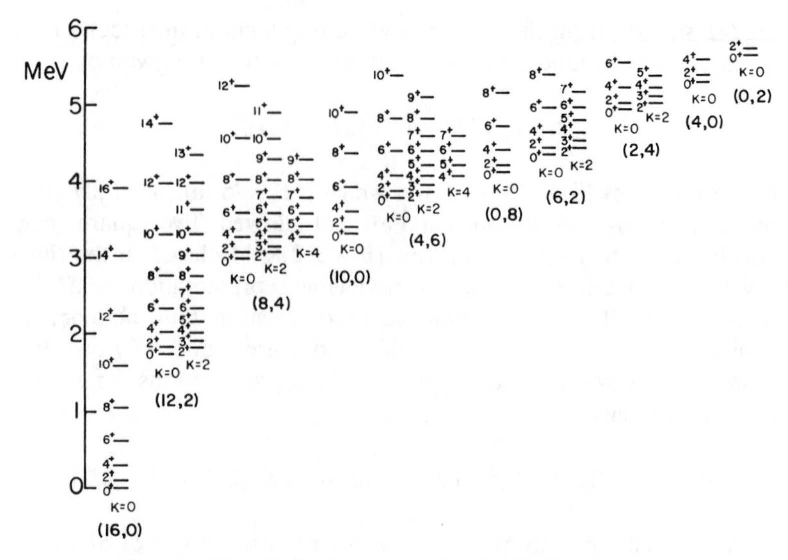

FIGURE 36.1. *Typical spectrum of a $SU(3)$ Hamiltonian.*

an excited band with $K = 0$ and $J = 0, 2, 4, \ldots, 2N - 4$ (which corresponds to the *β-band* of the collective model) and a $K = 2$ band with $J = 2, 3, 4, 5, \ldots, 2N - 2$ (corresponding to the *γ-band*). The energy eigenvalues (36.9) are independent of K and excited levels with $J = 2, 4, \ldots$ are degenerate. As explained in Section 30, the β-band is defined to also include the state with $J = 2N - 2$, whereas the highest J-value in the γ-band is equal to $2N - 3$. The dynamical symmetry II thus corresponds to a specific rotational limit of the collective model.

These features are seen in the $SU(3)$ spectrum in Fig. 36.1 calculated with $\kappa > 0$ and the coefficient of $J(J + 1)$ positive but smaller than κ. Such a spectrum agrees well with experimental spectra observed in the rare earth region, an example of which is given in Fig. 36.2. If the electric quadrupole transition operator is proportional to $\mathbf{Q}$ in (36.3), it is proportional to a generator of the $SU(3)$ group. Its selection rules are then very simple—it has non-vanishing matrix elements only between states which belong to the same irreducible representation. This leads to strong intra-band transitions and weak transitions between bands in different irreducible representations. There may be transitions between the $K = 0$ and $K = 2$ bands which both belong to the $(2N - 4, 2)$ irreducible representation.

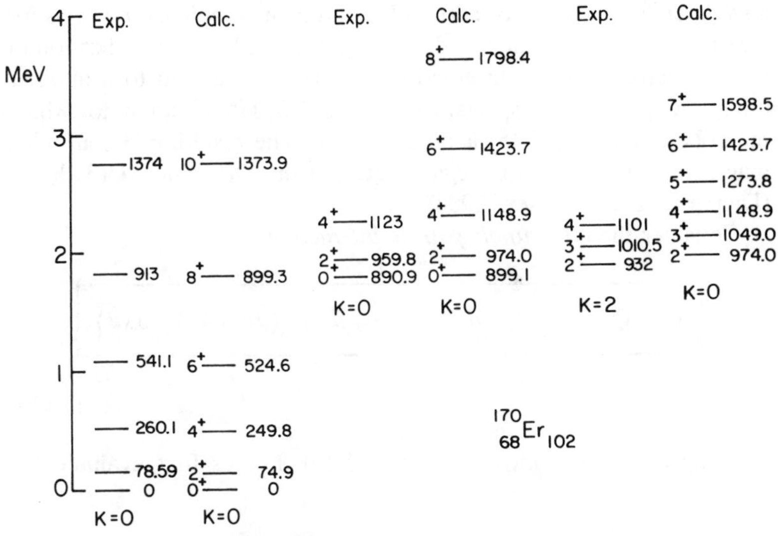

FIGURE 36.2. *Rotational spectrum of* ^{170}Er.

We shall not consider here the various expressions for strengths of allowed E2 transitions. We only quote the result for transitions within the ground state band, $\lambda = 2N$, $\mu = 0$. The strength of the transition from the state with J to that with $J - 2$ is given by

$$B(\mathrm{E2}, J \to J - 2) = \alpha_2^2 \frac{3J(J-1)}{4(2J-1)(2J+1)}(2N + 2 - J)(2N + 1 + J) \tag{36.12}$$

Also in this case the effects of coherence are seen in (36.12) as compared to the $U(5)$ case, (34.59). The coefficient β_2 in (33.33) was taken in the present case to be equal to $-\alpha_2\sqrt{7}/2$.

The boson Hamiltonians (35.11) or (35.12) contain 1 independent term less than the most general boson Hamiltonian (33.30). It is therefore possible to express any $U(6)$ Hamiltonian by adding to (35.11) the Casimir operator of $SU(3)$ multiplied by an arbitrary coefficient. The operator $C_{SU(3)}$ is certainly linearly independent of the other Casimir operators in (35.11) since it does not commute with $\hat{n}_d$, $C_{O(6)}$ or $C_{O(5)}$. The resulting Hamiltonian does not have dynamical symmetry since some Casimir operators do not commute. Still it is interesting to note that the general $U(6)$ Hamiltonian may be expressed as a linear combination of the Casimir operators in the various group chains. It is also possible to obtain for certain Hamiltonians simple expressions for a certain set of eigenstates and their eigenvalues by a generalization of the generalized pairing interaction (35.14). We can add to it another interaction, which is a special case of a $SU(3)$ Hamiltonian, for which the coherent state (35.15) is an eigenstate. The resulting Hamiltonian may describe a transition region between the $O(6)$ and $SU(3)$ limits (Dieperink and Scholten, 1980).

Consider the *quadrupole pairing interaction*

$$\boxed{\mathbf{Q}_{\alpha}^{+} \cdot \mathbf{Q}_{\alpha}^{-} = ((d^{+} \times d^{+})^{(2)} - \alpha s^{+} d^{+}) \cdot ((\tilde{d} \times \tilde{d})^{(2)} - \alpha s \tilde{d})} \tag{36.13}$$

and apply it to the intrinsic state (35.15). Let us first evaluate the following expression

$$((\tilde{d} \times \tilde{d})_{\kappa}^{(2)} - \alpha s \tilde{d}_{\kappa})(d_0^{+} + x s^{+})^{N} |0\rangle \tag{36.14}$$

In order to use the lemma (23.8) we calculate the commutators of $((\tilde{d} \times \tilde{d})_{\kappa}^{(2)} - \alpha s \tilde{d}_{\kappa})$ and $d_0^{+} + x s^{+}$. Since the operators d_{μ} for $\mu \neq 0$ commute with d_0^{+} and s^{+}, we obtain

$$\begin{aligned}
&[(\tilde{d} \times \tilde{d})_{\kappa}^{(2)} - \alpha s \tilde{d}_{\kappa}, d_0^{+} + x s^{+}] \\
&\quad = \left[2 \sum_{\kappa \neq 0} (-1)^{\kappa} (2\kappa 20 \mid 222\kappa) d_{-\kappa} d_0 \right. \\
&\qquad \left. + (2020 \mid 2220) d_0^2 - \alpha s d_0, d_0^{+} + x s^{+} \right] \\
&\quad = 2 \sum_{\kappa \neq 0} (-1)^{\kappa} (2\kappa 20 \mid 222\kappa) d_{-\kappa} + 2(2020 \mid 2220) d_0 - \alpha s - x \alpha d_0
\end{aligned} \tag{36.15}$$

The first commutator thus yields zero when applied to the vacuum state. The double commutator is obtained from (36.15) as

$$[[(\tilde{d} \times \tilde{d})_{\kappa}^{(2)} - \alpha s \tilde{d}_{\kappa}, d_0^{+} + x s^{+}], d_0^{+} + x s^{+}] = 2(2020 \mid 2220) - 2x\alpha \tag{36.16}$$

From (36.16) follows that all higher commutators vanish. Hence, if we put

$$\alpha = \frac{1}{x}(2020 \mid 2220) = -\frac{1}{x}\sqrt{\frac{2}{7}} \tag{36.17}$$

the expression (36.14) vanishes and hence $\mathbf{Q}^+_\alpha \cdot \mathbf{Q}^-_\alpha(d^+_0 + xs^+)^N|0\rangle = 0$. Thus, if the value of x is determined by the vanishing of (35.17) as $1/\sqrt{\beta}$ and $\alpha = -\sqrt{2\beta/7}$, the state (35.15) is an eigenstate also of $P^+_\beta P^-_\beta$ defined by (35.14), with $\beta = x^{-2}$, with eigenvalue 0. Any Hamiltonian which is a linear combination with arbitrary coefficients of $P^+_\beta P^-_\beta$ and $\mathbf{Q}^+_\alpha \cdot \mathbf{Q}^-_\alpha$ has in this case (35.15) as an eigenstate with a vanishing eigenvalue.

We can now see that a special linear combination of the (monopole) pairing interaction (35.14) and the quadrupole pairing interaction (36.13) is indeed equivalent to the Casimir operator $C_{SU(3)}$ given by (36.5). Let us start with the $\tilde{\mathbf{Q}} \cdot \tilde{\mathbf{Q}}$ interaction with the more general quadrupole operator (36.1) and write it explicitly as

$$\begin{aligned}
&(s^+\tilde{d} + d^+s + \chi(d^+ \times \tilde{d})^{(2)}) \cdot (s^+\tilde{d} + d^+s + \chi(d^+ \times \tilde{d})^{(2)}) \\
&\quad = (s^+)^2\tilde{d} \cdot \tilde{d} + (d^+ \cdot d^+)s^2 + s^+s(\tilde{d} \cdot d^+) \\
&\quad + ss^+(d^+ \cdot \tilde{d}) + \chi^2(d^+ \times \tilde{d})^{(2)} \cdot (d^+ \times \tilde{d})^{(2)} \\
&\quad + \chi\{s^+\tilde{d} \cdot (d^+ \times \tilde{d})^{(2)} + d^+ \cdot (d^+ \times \tilde{d})^{(2)}s \\
&\quad + (d^+ \times \tilde{d})^{(2)} \cdot d^+s + s^+(d^+ \times \tilde{d})^{(2)} \cdot \tilde{d}\}
\end{aligned} \tag{36.18}$$

The scalar products on the r.h.s. of (36.18) may now be transformed into a form which may be compared with the linear combination of $P^+_\beta P^-_\beta$ and $\mathbf{Q}_\alpha \cdot \mathbf{Q}_\alpha$. We first notice the following equalities

$$\tilde{d} \cdot (d^+ \times \tilde{d})^{(2)} = (d^+ \times \tilde{d})^{(2)} \cdot \tilde{d} \qquad (d^+ \times \tilde{d}) \cdot d^+ = d^+ \cdot (d^+ \times \tilde{d})^{(2)} \tag{36.19}$$

These may be proved by using the commutation relations (33.20) and the properties of Clebsch-Gordan coefficients. They may, however, be demonstrated by a simple argument. The commutators of components of $\tilde{d}$ and those of $(d^+ \times \tilde{d})^{(2)}$ are equal to terms with *one* operator $\tilde{d}_\mu$. Changing the order of couplings in the scalar $\tilde{d} \cdot (d^+ \times \tilde{d})^{(2)}$ we must obtain a scalar but $\tilde{d}$ is a rank $k = 2$ tensor operator and hence the coefficients of the $\tilde{d}_\mu$ components must vanish. We can now apply to the scalar products in (36.19) the change of coupling transformation (10.12) to obtain

$$(d^+ \times \tilde{d})^{(2)} \cdot \tilde{d} = d^+ \cdot (\tilde{d} \times \tilde{d})^{(2)} \qquad d^+ \cdot (d^+ \times \tilde{d})^{(2)} = (d^+ \times \tilde{d}^+)^{(2)} \cdot \tilde{d} \tag{36.20}$$

The next term in (36.18) to be transformed into the desired form has the coefficient χ^2. We apply to it the change of coupling transformation (10.10). Due to the commutation relations (33.20) we obtain also a (scalar) single boson term in addition to the boson-boson interactions and the result is

$$(d^+ \times \tilde{d})^{(2)} \cdot (d^+ \times \tilde{d})^{(2)}$$
$$= (d^+ \cdot \tilde{d}) + 5 \sum_L \begin{Bmatrix} 2 & 2 & 2 \\ 2 & 2 & L \end{Bmatrix} (d^+ \times d^+)^{(L)} \cdot (\tilde{d} \times \tilde{d})^{(L)} \tag{36.21}$$

Due to the Bose commutation relations, the summation in (36.21) is over even values of L only (terms with odd L values vanish). Terms with $L = 0$ and $L = 2$ may be directly compared with those in $P_\beta^+ P_\beta^-$ and $\mathbf{Q}_\alpha^+ \cdot \mathbf{Q}_\alpha^-$ but those interactions do not contain $L = 4$ terms. We recall, however, that these $L = 0, 2, 4$ terms appear in a simple relation. If the boson-boson interactions in the states with $L = 0$, $L = 2$ and $L = 4$ are all equal to 1, their sum is simply the total number of boson-boson interactions in the given state. We obtain thus the relation

$$\sum_L (d^+ \times d^+)^{(L)} \cdot (\tilde{d} \times \tilde{d})^{(L)} = n_d(n_d - 1) \tag{36.22}$$

The r.h.s. of (32.22) is the total number of interactions between d-bosons multiplied by 2. This factor is due to the missing normalization of $(2^{-1/2})^2 = \frac{1}{2}$ on the l.h.s. of (36.22). The relation (36.22) may be formally obtained in the same way as (36.21), by transforming $(d^+ \cdot \tilde{d})(d^+ \cdot \tilde{d})$. Using (36.22) we can express the $L = 4$ term in (36.21) by $L = 0$ and $L = 2$ terms. Inserting the values of the $6j$-symbols,

$$\begin{Bmatrix} 2 & 2 & 2 \\ 2 & 2 & 0 \end{Bmatrix} = \tfrac{1}{5}, \quad \begin{Bmatrix} 2 & 2 & 2 \\ 2 & 2 & 2 \end{Bmatrix} = -\tfrac{3}{70} \quad \text{and} \quad \begin{Bmatrix} 2 & 2 & 2 \\ 2 & 2 & 4 \end{Bmatrix} = \tfrac{2}{35},$$

we obtain for (36.21) the form

$$\tfrac{2}{7}n_d^2 + \tfrac{5}{7}n_d + \tfrac{1}{7}(d^+ \cdot d^+)(\tilde{d} \cdot \tilde{d}) - \tfrac{1}{2}(d^+ \times d^+)^{(2)} \cdot (\tilde{d} \times \tilde{d})^{(2)} \tag{36.23}$$

Collecting all terms we can express the $\tilde{\mathbf{Q}} \cdot \tilde{\mathbf{Q}}$ interaction (36.18) by

$$\begin{aligned}(s^+)^2(\tilde{d} \cdot \tilde{d}) &+ (d^+ \cdot d^+)s^2 + 2n_s n_d + 5n_s + n_d + 2\chi s^+ d^+ \cdot (\tilde{d} \times \tilde{d})^{(2)} \\ &+ 2\chi(d^+ \times d^+)^{(2)} \cdot \tilde{d}s + \chi^2 \tfrac{2}{7} n_d^2 + \chi^2 \tfrac{5}{7} n_d \\ &+ \chi^2 \tfrac{1}{7}(d^+ \cdot d^+)(\tilde{d} \cdot \tilde{d}) - \chi^2 \tfrac{1}{2}(d^+ \times d^+)^{(2)} \cdot (\tilde{d} \times \tilde{d})^{(2)}\end{aligned} \tag{36.24}$$

Since we are trying to construct an interaction by using $SU(3)$ generators which will have (35.15) as an eigenstate, the interaction must be degenerate in the states with $J = 0, 2, 4 \ldots$ projected from (35.15). Hence, we should now add to (36.24) the term $\tfrac{3}{8}(\mathbf{J} \cdot \mathbf{J})$ so that the result will be equal to $\tfrac{1}{2}C_{SU(3)}$ for $\chi = -\sqrt{7}/2$. Recalling (33.27) we evaluate the expression

$$\begin{aligned}\tfrac{3}{8} &\times 10(d^+ \times \tilde{d})^{(1)} \cdot (d^+ \times \tilde{d})^{(1)} \\ &= \tfrac{15}{4}\left\{\tfrac{3}{5}(d^+ \cdot \tilde{d}) + 3\sum_L \begin{Bmatrix} 2 & 2 & 1 \\ 2 & 2 & L \end{Bmatrix} (d^+ \times d^+)^{(L)} \cdot (\tilde{d} \times \tilde{d})^{(L)}\right\} \\ &= \tfrac{15}{4}\{\tfrac{3}{5}n_d - \tfrac{3}{25}(d^+ \cdot d^+)(\tilde{d} \cdot \tilde{d}) - \tfrac{3}{10}(d^+ \times d^+)^{(2)} \cdot (\tilde{d} \times \tilde{d})^{(2)} \\ &\qquad + \tfrac{2}{5}(n_d(n_d - 1) - \tfrac{1}{5}(d^+ \cdot d^+)(\tilde{d} \cdot \tilde{d}) \\ &\qquad - (d^+ \times d^+)^{(2)} \cdot (\tilde{d} \times \tilde{d})^{(2)})\} \\ &= \tfrac{3}{2}n_d^2 + \tfrac{3}{4}n_d - \tfrac{3}{4}(d^+ \cdot d^+)(\tilde{d} \cdot \tilde{d}) - \tfrac{21}{8}(d^+ \times d^+)^{(2)} \cdot (\tilde{d} \times \tilde{d})^{(2)}\end{aligned} \tag{36.25}$$

The r.h.s. of (36.25) was obtained by following the same steps as in evaluating (36.21) and by substituting the values

$$\begin{Bmatrix} 2 & 2 & 1 \\ 2 & 2 & 0 \end{Bmatrix} = -\tfrac{1}{5}, \quad \begin{Bmatrix} 2 & 2 & 1 \\ 2 & 2 & 2 \end{Bmatrix} = -\tfrac{1}{10} \quad \text{and} \quad \begin{Bmatrix} 2 & 2 & 1 \\ 2 & 2 & 4 \end{Bmatrix} = \tfrac{2}{15}$$

The sum of (36.24) and (36.25) may now be compared with the linear combination $AP_\beta^+ P_\beta^- + B(\mathbf{Q}_\alpha^+ \cdot \mathbf{Q}_\alpha^-)$. In the latter, however, there are no n_d^2 terms. This is no problem since we may subtract terms linear and quadratic in $N = n_s + n_d$ from $\frac{1}{2}C_{SU(3)}$ without any change in energy spacings or structure of eigenstates. Substituting in (36.24) $\chi = -\sqrt{7}/2$ we obtain for the terms with n_s, n_d in the sum of (36.24) (with $\chi = -\sqrt{7}/2$) and (36.25) the form

$$2n_d^2 + 3n_d + 5n_s + 2n_s n_d = 2N^2 + 3N + 2n_s - 2n_s^2 - 2n_s n_d \tag{36.26}$$

Subtracting $2N^2 + 3N$ from the sum of (36.24) with $\chi = -\sqrt{7}/2$ and (36.25) we obtain the expression

$$\begin{aligned}(s^+)^2(\tilde{d}\cdot\tilde{d}) + (d^+\cdot d^+)s^2 - \sqrt{7}s^+d^+\cdot(\tilde{d}\times\tilde{d})^{(2)} - \sqrt{7}(d^+\times d^+)^{(2)}\cdot\tilde{d}s \\ - \tfrac{1}{2}(d^+\cdot d^+)(\tilde{d}\cdot\tilde{d}) - \tfrac{7}{2}(d^+\times d^+)^{(2)}\cdot(\tilde{d}\times\tilde{d})^{(2)} + 2n_s - 2n_s^2 - 2n_s n_d\end{aligned} \tag{36.27}$$

This can now be compared with the linear combination

$$\begin{aligned}AP_\beta^+P_\beta^- + B(\mathbf{Q}_\alpha^+\cdot\mathbf{Q}_\alpha^-) &= A(d^+\cdot d^+)(\tilde{d}\cdot\tilde{d}) \\ &- A\beta(s^+)^2(\tilde{d}\cdot\tilde{d}) - A\beta(d^+\cdot d^+)s^2 \\ &+ A\beta^2(s^+)^2s^2 + B(d^+\times d^+)^{(2)}\cdot(\tilde{d}\times\tilde{d})^{(2)} - B\alpha s^+d^+\cdot(\tilde{d}\times\tilde{d})^{(2)} \\ &- B\alpha(d^+\times d^+)^{(2)}\cdot\tilde{d}s + B\alpha^2 n_s n_d\end{aligned} \tag{36.28}$$

Comparing (36.27) with (36.28) we see that they are equal, provided the coefficients obey the relations

$$\begin{aligned}A\beta &= -1 \qquad A = -\tfrac{1}{2} \qquad A\beta^2 = -2 \\ B\alpha &= \sqrt{7} \qquad B = -\tfrac{7}{2} \qquad B\alpha^2 = -2\end{aligned}$$

These relations may be satisfied if the coefficients are given by

$$\beta = 2 \qquad \alpha = -2/\sqrt{7} \qquad A = -\tfrac{1}{2} \qquad B = -\tfrac{7}{2} \tag{36.29}$$

Thus, any Hamiltonian (36.28) in which $\beta = 2$, $\alpha = -2/\sqrt{7}$ and $B = 7A$ is diagonal in the $SU(3)$ basis and has the intrinsic state (35.15) with $x = 1/\sqrt{2}$ as an eigenstate with eigenvalue 0.

The intrinsic state (35.15) with $x = 1/\sqrt{2}$, namely

$$\left(d_0^+ + \frac{1}{\sqrt{2}}s^+\right)^N |0\rangle \tag{36.30}$$

is an eigenstate with eigenvalue 0 of a Hamiltonian which is proportional to the Casimir operator of $SU(3)$. The state (36.30) is also an eigenstate of the $\kappa = 0$ component of the quadrupole operator $\mathbf{Q}$ in (36.3). Applying Q_0 to (36.30) we make use of the lemma (23.8) and first calculate

$$\left[s^+\tilde{d}_0 + sd_0^+ - \frac{\sqrt{7}}{2}(d^+ \times \tilde{d})_0^{(2)}, d_0^+ + \frac{1}{\sqrt{2}}s^+\right]$$
$$= s^+ + \frac{1}{\sqrt{2}}d_0^+ - \frac{\sqrt{7}}{2}[(d^+ \times \tilde{d})_0^{(2)}, d_0^+]$$
$$= s^+ + \frac{1}{\sqrt{2}}d_0^+ - \frac{\sqrt{7}}{2}(2020 \mid 2220)d_0^+$$

Substituting the value $(2020 \mid 2220) = -\sqrt{2/7}$, we obtain for the commutator the result $s^+ + \sqrt{2}d_0^+$. The higher commutators vanish and thus we obtain

$$\left(s^+\tilde{d}_0 + sd_0^+ - \frac{\sqrt{7}}{2}(d^+ \times \tilde{d})_0^{(2)}\right)\left(d_0^+ + \frac{1}{\sqrt{2}}s^+\right)^N |0\rangle$$
$$= N(s^+ + \sqrt{2}d_0^+)\left(d_0^+ + \frac{1}{\sqrt{2}}s^+\right)^{N-1} |0\rangle$$
$$= \sqrt{2}N\left(d_0^+ + \frac{1}{\sqrt{2}}s^+\right)^N |0\rangle \qquad (36.31)$$

It is worthwhile to note that the intrinsic state (36.30) is an eigenstate, with eigenvalue 0, of the Hamiltonian (36.28), with $\beta = 2$ and $\alpha = -2/\sqrt{7}$, for *any* values of A and B. All the states projected from (36.30) belong to the irreducible representation $(2N, 0)$ of $SU(3)$ even though the Hamiltonian (36.28) with arbitrary coefficients A, B does not have $SU(3)$ symmetry.

From this discussion it follows that the lowest $SU(3)$ states, those which belong to the $(2N, 0)$ irreducible representation may be obtained by projecting states with good angular momentum J from the intrinsic state (36.30). All these states are degenerate with eigenvalue 0. If an interaction proportional to $(\mathbf{J} \cdot \mathbf{J})$ is added, these states form a rotational band with energies proportional to $J(J+1)$. For example, the intrinsic state with $N = 2$ is given by

$$\left(d_0^+ + \frac{1}{\sqrt{2}}s^+\right)^2 |0\rangle = \left\{ \tfrac{1}{5}(d^+ \cdot d^+) + \sum_{J=2,4} (2020 \mid 22J0)(d^+ \times d^+)_0^{(J)} \right.$$

$$\left. + \sqrt{2}s^+ d_0^+ + \tfrac{1}{2}(s^+)^2 \right\} |0\rangle \qquad (36.32)$$

The state with $J = 4$ projected from (36.32) is $2^{-1/2}(d^+ \times d^+)_0^{(4)}|0\rangle$. The (unnormalized) state with $J = 2$ is given by $\{(2020 \mid 2220)(d^+ \cdot d^+)_0^{(2)} + \sqrt{2}s^+ d_0^+\}|0\rangle$ or, by using the value $(2020 \mid 2220) = -\sqrt{2/7}$, as simply $[(d^+ \times d^+)_0^{(2)} - \sqrt{7}s^+ d_0^+]|0\rangle$. The $J = 0$ state is given by $[2(d^+ \cdot d^+) + 5(s^+)^2]|0\rangle$. It can be verified that these states, which belong to the (4,0) irreducible representation of $SU(3)$, transform among themselves when operated by components of the quadrupole operator **Q** defined by (36.3). The states which belong to the other symmetric $SU(3)$ irreducible representation with $N = 2$, namely (0,2), are the states orthogonal to the (4,0) states. Thus, the normalized state with $J = 0$ is

$$\frac{1}{3\sqrt{2}}[-(d^+ \cdot d^+) + 2(s^+)^2]|0\rangle \qquad (36.33)$$

and the $J = 2$ normalized state with $J_z = M$ is

$$\frac{1}{3\sqrt{2}}[\sqrt{7}(d^+ \times d^+)_M^{(2)} + 2s^+ d_M^+]|0\rangle \qquad (36.34)$$

The (0,2) irreducible representation of $SU(3)$ has the same number of states as the (2,0) irreducible representation with $N = 1$. The transformation of the (0,2) states under $SU(3)$ generators is, however, different. These two representations are called conjugate representations. The single boson operators s^+ and d_μ^+ which create the basis of the (2,0) irreducible representation transform under Q_κ components according to

$$[Q_\kappa, s^+] = [d_\kappa^+ s, s^+] = d_\kappa^+$$

$$[Q_\kappa, d_\mu^+] = [s^+ \tilde{d}_\kappa, d_\mu^+] - \frac{\sqrt{7}}{2}[(d^+ \times \tilde{d})_\kappa^{(2)}, d_\mu^+]$$

$$= (-1)^\kappa s^+ \delta_{\kappa,-\mu} - \frac{\sqrt{7}}{2}\sum_{\mu'}(2\kappa 2\mu \mid 222\mu')d_{\mu'}^+ \qquad (36.35)$$

We can now calculate the commutation relations of Q_κ with the operators in (36.33) and (36.34) which create the basis of the (0,2) irreducible representation. We first obtain by using the commutation relations (33.20) and the symmetry properties of the Clebsch-Gordan coefficients the formula

$$[s^+\tilde{d}_\kappa + d_\kappa^+ s, (d^+ \times d^+)_\mu^{(J)}] = 2\sqrt{\frac{2J+1}{5}}(2\kappa J\mu \mid 2J2\mu')s^+ d_{\mu'}^+ \tag{36.36}$$

The other necessary formula, obtained by using the relation (10.24), is

$$[(d^+ \times \tilde{d})_\kappa^{(2)}, (d^+ \times d^+)_\mu^{(J)}] = 2\sqrt{5(2J+1)} \sum_{J' \text{ even}} \begin{Bmatrix} 2 & 2 & 2 \\ J & 2 & J' \end{Bmatrix} \times (2\kappa J\mu \mid 2JJ'\mu')(d^+ \times d^+)_{\mu'}^{(J')} \tag{36.37}$$

Using these formulae we obtain for the transformation of the operators in (36.33) the result

$$\left[Q_\kappa, \frac{1}{3\sqrt{2}}(-(d^+ \cdot d^+) + 2(s^+)^2)\right] = \frac{1}{3\sqrt{2}}(\sqrt{7}(d^+ \times d^+)_\kappa^{(2)} + 2s^+ d_\kappa^+) \tag{36.38}$$

This relation is the same as the corresponding relation in (36.35). The other commutator, with the operator in (36.34), is given by

$$\begin{aligned}
&[Q_\kappa, \sqrt{7}(d^+ \times d^+)_\mu^{(2)} + 2s^+ d_\mu^+)] \\
&\quad = -35 \sum_{J' \text{ even}} \begin{Bmatrix} 2 & 2 & 2 \\ 2 & 2 & J' \end{Bmatrix} (2\kappa 2\mu \mid 22J'\mu')(d^+ \times d^+)_{\mu'}^{(J')} \\
&\qquad + 2\sqrt{7}(2\kappa 2\mu \mid 222\mu')s^+ d_{\mu'}^+ + 2s^+ s(-1)^\kappa \delta_{\kappa,-\mu} + 2d_\kappa^+ d_\mu^+ \\
&\qquad - \frac{\sqrt{7}}{2} 2(2\kappa 2\mu \mid 222\mu')s^+ d_{\mu'}^+ \\
&\quad = \sum_{J' \text{ even}} (2\kappa 2\mu \mid 22J'\mu') \left[2 - 35 \begin{Bmatrix} 2 & 2 & 2 \\ 2 & 2 & J' \end{Bmatrix}\right] (d^+ \times d^+)_{\mu'}^{(J')} \\
&\qquad + \sqrt{7}(2\kappa 2\mu \mid 222\mu')s^+ d_{\mu'}^+ + 2(-1)^\kappa s^+ s \delta_{\kappa,-\mu}
\end{aligned} \tag{36.39}$$

Substituting into (36.39) the values

$$\begin{Bmatrix} 2 & 2 & 2 \\ 2 & 2 & 0 \end{Bmatrix} = \tfrac{1}{5}, \qquad \begin{Bmatrix} 2 & 2 & 2 \\ 2 & 2 & 2 \end{Bmatrix} = -\tfrac{3}{70}$$

$$\text{and} \qquad \begin{Bmatrix} 2 & 2 & 2 \\ 2 & 2 & 4 \end{Bmatrix} = \tfrac{2}{35}$$

we see that the term with $J' = 4$ vanishes and we obtain the result

$$\left[Q_\kappa, \frac{1}{3\sqrt{2}}(\sqrt{7}(d^+ \times d^+)^{(2)}_\mu + 2s^+ d^+_\mu) \right]$$
$$= (-1)^\kappa \delta_{\kappa,-\mu} \frac{1}{3\sqrt{2}}(-(d^+ \cdot d^+) + 2(s^+)^2)$$
$$+ \frac{\sqrt{7}}{2} \sum_{\mu'} (2\kappa 2\mu \mid 222\mu') \frac{1}{3\sqrt{2}}(\sqrt{7}(d^+ \times d^+)^{(2)}_{\mu'} + 2s^+ d^+_{\mu'}) \tag{36.40}$$

Comparing (36.40) with the corresponding relation in (36.35) we see a change in sign in front of the $J' = 2, \mu'$ term.

The states of (36.33) and (36.34) of the (0,2) irreducible representation may be viewed as hole states of a single s- or d-boson. In the Young diagrams of $SU(3)$, a column with three squares is equivalent to the (0,0) representation. Hence, the (0,2) diagram with two rows of two squares may be characterized by the two squares which augment it to the (0,0) representation. States of one boson added to the (0,2) states are included in the product

$$(0,2) \otimes (2,0) = (2,2) \oplus (1,1) \oplus (0,0)$$

Thus, states of one boson may be coupled to the (0,2) states to "fill the hole". As pointed out in Section 30, these hole states are not due to the Pauli principle that does certainly not apply to s- and d-bosons. They are due to the antisymmetry of the states in the p-particles whose coordinates appear in the same column of the Young diagram. There are only three orbital states for a p-particle and hence antisymmetric states of 3 p-particles are invariant under $SU(3)$ transformations.

Let us point out that the irreducible representation (0,2) may be obtained from single boson states provided we change the sign of χ in (36.10) from $-\sqrt{7}/2$ to $\sqrt{7}/2$. If we define

$$\boxed{Q_\kappa^{(+)} = s^+\tilde{d}_\kappa + sd_\kappa^+ + \frac{\sqrt{7}}{2}(d^+ \times d^+)_\kappa^{(2)}} \tag{36.41}$$

we obtain the following commutation relations

$$[Q_\kappa^{(+)}, s^+] = [d_\kappa^+ s, s^+] = d_\kappa^+$$

$$[Q_\kappa^{(+)}, d_\mu^+] = [s^+\tilde{d}_\kappa, d_\mu^+] + \frac{\sqrt{7}}{2}[(d^+ \times \tilde{d})_\kappa^{(2)}, d_\mu^+]$$

$$= (-1)^\kappa s^+ \delta_{\kappa,-\mu} + \frac{\sqrt{7}}{2}(2\kappa 2\mu \mid 22 2\mu')d_{\mu'}^+ \tag{36.42}$$

Comparing (36.42) to (36.35) we see that there is a change in sign in the second relation. The relations (36.42) correspond exactly to (36.38) and (36.40). We will make use of this fact in a subsequent section.

Let us return to the general Hamiltonian $AP_\beta^+ P_\beta^- + B(\mathbf{Q}_\alpha^+ \cdot \mathbf{Q}_\alpha^-)$. For $B = 0$, the Hamiltonian has $O(5)$ symmetry and in particular for $\beta = 1$ it is the generalized pairing interaction of the $O(6)$ limit. For $B = 7A$, $\alpha = -2/\sqrt{7}$ and $\beta = 2$ according to (36.29), the Hamiltonian is proportional to the Casimir operator $C_{SU(3)}$ of the $SU(3)$ group. By varying the values of these parameters, the transition region between the $O(6)$ and $SU(3)$ limits may be covered. The same transition region may be obtained also in a simpler way. We can start from the $SU(3)$ Hamiltonian (36.6) in which the quadrupole operator $\mathbf{Q}$ is given by (36.1) with $\chi = -\sqrt{7}/2$. If we now vary χ from the value $-\sqrt{7}/2$ to zero we obtain a smooth transition from the $SU(3)$ generator $\mathbf{Q}$ of (36.3) to the $O(6)$ generator $\mathbf{Q}'$ in (35.1). Hence, starting with the most general $SU(3)$ Hamiltonian we obtain a smooth transition to an $O(6)$ Hamiltonian in which, according to (35.4) the combination $C_{O(6)} - C_{O(5)}$ appears. In the general $O(6)$ Hamiltonian, $C_{O(6)}$ and $C_{O(5)}$ have independent coefficients. Still, the eigenstates of $C_{O(6)} - C_{O(5)}$ if characterized by definite values of σ and τ do not change if the coefficients of $C_{O(6)}$ and $C_{O(5)}$ are varied.

Eigenstates of the Hamiltonian $\kappa\tilde{\mathbf{Q}} \cdot \tilde{\mathbf{Q}} + \kappa'\mathbf{J} \cdot \mathbf{J}$ depend only on the coefficient χ in $\tilde{\mathbf{Q}}$. The operator $\mathbf{J}^2$ has eigenvalues $J(J+1)$ in any state with given J. The parameter κ determines only the strength of the quadrupole interaction and thus, eigenstates are determined by the value of the parameter χ. For $\chi = -\sqrt{7}/2$, we obtain eigenstates of the $SU(3)$ scheme, whereas for $\chi = 0$, the $O(6)$ states are eigenstates. As a result, matrix elements of the operator $\tilde{\mathbf{Q}}$, which appears in the Hamiltonian, between eigenstates also depend on χ. Electromagnetic E2 transition rates may be calculated by using the same $\tilde{\mathbf{Q}}$ operator. This way, structure of eigenstates and transition rates are determined by *one* parameter χ. Applications of such Hamiltonians to actual nuclei have been made by Casten and collaborators and have been called "consistent Q formalism". A detailed description of such applications is given in the book by Casten (1990).

We may approach the situation described above from a different angle by starting from an arbitrary intrinsic state (35.15) defined by

$$(d_0^+ + xs^+)^N|0\rangle \tag{36.43}$$

As shown above, the state (36.43) is an eigenstate, with eigenvalue 0, of the Hamiltonian

$$\boxed{\begin{aligned} AP_\beta^+ P_\beta + B(\mathbf{Q}_\alpha^+ \cdot \mathbf{Q}_\alpha^-) &= \frac{A}{4}((d^+ \cdot d^+) - \beta(s^+)^2)((\tilde{d} \cdot \tilde{d}) - \beta s^2) \\ &+ B((d^+ \times d^+)^{(2)} - \alpha s^+ d^+) \cdot ((\tilde{d} \times \tilde{d})^{(2)} - \alpha s d) \end{aligned}} \tag{36.44}$$

with $\beta = (1/x)^2$, $\alpha = -(1/x)\sqrt{\frac{2}{7}}$. The state (36.43) is a linear combination of states with spins $J = 0, 2, \ldots, 2N$ as shown by (35.16) and as may be conveniently verified in the $SU(3)$ limit (for $B = 7A$, $x = 1/\sqrt{2}$). All these states have vanishing eigenvalues of the Hamiltonian (36.44). Some of these states have definite values of the seniority τ (like the state with $J = 2N$ which has $\tau = N$). Others are linear combinations of states with different values of τ. If we add to (36.44) with $B = 0$, an operator with $O(5)$ symmetry like $\gamma C_{O(5)} + C(\mathbf{J} \cdot \mathbf{J})$, eigenstates of the resulting Hamiltonian have definite values of τ and eigenvalues equal to $\gamma\tau(\tau + 3) + CJ(J + 1)$. We may, instead, add to (36.44) with $A \neq 0$, $B \neq 0$, just the operator $C(\mathbf{J} \cdot \mathbf{J})$. Eigenstates of the combined Hamiltonian in this case are the linear combinations with definite values of J, as they appear in (36.43), with eigenvalues equal to $CJ(J+1)$. Thus, given *any* intrinsic state (36.43), a Hamiltonian may be constructed whose lowest eigenstates have spins $J = 0, 2, \ldots, 2N$ with eigenvalues given by $CJ(J+1)$. Such states may be described as members of the ground state rotational band.

37

Mapping States of Identical Fermions onto Boson States

The interacting boson model introduced in Section 33 has been described from the point of view of the collective model of the nucleus. The 5 components of a d-boson have been related to the 5 degrees of freedom of a shape, with constant volume, which may have quadrupole deformations. The s-boson has been added to simplify the construction of eigenstates which belong to irreducible representations of the $U(6)$ group, the fully symmetric ones being characterized by an integer N. We saw how simple Hamiltonians including single boson energies and boson boson interactions give very good description of various kinds of collective motion.

It is interesting to find out whether this boson model may be related to the shell model. An attractive feature of IBA-1 is that the s- and d-bosons are defined in a frame of reference fixed in space. This is the frame of reference in which nucleon states in the (spherical) shell model are defined. A shell model basis of the interacting boson model will furnish a shell model description of collective states of nuclei. Clearly, such a description is possible since the shell model wave functions form a *complete* set of states. The hope, however, is to find a *simple* prescription for obtaining states of collective bands without having to diagonalize Hamiltonian matrices of order 10^{15} as implied by a straightforward application of the shell model (see end of Section 24).

It is important to realize that IBA-1 *cannot* have a *direct* shell model basis. There are both protons and neutrons in nuclei and it is impossible to replace them by fermions of *one* kind. In the collective model it is usually assumed that the collective variables describe proton and neutron degrees of freedom in the same way. If, however, we consider wave functions of protons and neutrons we must observe the Pauli principle and resulting symmetry properties of the various states. The latter will be rather different for states with low values of isospin than for states with maximum value of T.

It is equally important to realize that IBA-1 may be considered as a phenomenological model, related to the collective model, which *does not need* a shell model foundation. Even the fact that the total number of bosons $N = n_s + n_d$ is conserved does not necessarily imply a fermion substructure of the bosons (in their original paper, Arima and Iachello (1975) did not attach any physical meaning to the parameter N). To demonstrate this point we shall briefly consider a boson model for diatomic molecules. This model, invented by Iachello (1981a), is very similar to IBA-1. Like the latter, the bosons are simply related to some collective coordinates. Those coordinates are components of the vector $\mathbf{r}$ connecting the centers of mass of the two atoms or ions. Thus, these coordinates have a clear and well defined physical meaning. In that model, the total number of bosons is a good quantum number characterizing eigenstates of a simple Hamiltonian. Yet, this model has definitely no fermion or "microscopic" basis.

Vibrations of the three components of $\mathbf{r}$ around zero may be described by wave functions of a three-dimensional harmonic oscillator or by states of p-bosons with odd parity. Low lying states of the molecule form rotational bands in which the equilibrium distance between the atoms $r = r_0$ is different from zero. The direction of the molecular axis in space is determined by the polar angles θ and ϕ. In higher states there are also vibrations of r along the axis. Such states can be expanded in terms of p-boson states in which their number n_p is not bound. An approximate simpler description is to consider states of fully symmetric irreducible representations of the $U(4)$ Lie algebra characterized by an integer N. The numbers of p-bosons in such states satisfy the restriction $n_p \leq N$. Finally, s-bosons are added in order to simplify the construction of generators and irreducible representations of the $U(4)$ group. Simple Hamiltonians are then constructed from the following 16 generators of $U(4)$

$$s^+s, \quad s^+p_\mu, \quad p_\mu^+s, \qquad (p^+ \times \tilde{p})_\kappa^{(k)} \quad k = 0,1,2 \tag{37.1}$$

The rotationally invariant Hamiltonian should be hermitean as well as parity conserving. Such a Hamiltonian can be constructed as a linear combination of the scalar operators in (37.1) and scalar products of other operators in (37.1). Not all scalar products are independent since there are only 4 symmetric states of two s,p-bosons. To further simplify the Hamiltonian, we can make use of the conservation of the total number of bosons, namely

$$s^+s + \sum_\mu p_\mu^+ p_\mu = s^+s + (p^+ \cdot \tilde{p}) = n_s + n_p = N \tag{37.2}$$

where we use the definition $\tilde{p}_\mu = (-1)^\mu p_{-\mu}$.

We will not consider here the general case of a $U(4)$ Hamiltonian. Instead, we consider special cases in which there is a dynamical symmetry. We consider subgroups of $U(4)$ which contain $O(3)$ as a subgroup. An obvious one is the $U(3)$ group whose generators are $(p^+ \times \tilde{p})_\kappa^{(k)}$, $k = 0,1,2$. Another such subgroup is $O(4)$ whose generators can be taken to be the 6 components of the two hermitean vectors

$$s^+\tilde{p} + p^+s \qquad (p^+ \times \tilde{p})^{(1)} \tag{37.3}$$

A different $O(4)$ group can be generated by choosing the other hermitean vector $i(s^+\tilde{p} - p^+s)$. The $O(3)$ Lie algebra is a subalgebra of the $O(4)$ one whose elements are listed in (37.3). Its generators are the components of angular momentum which are equal in this case to

$$L_\mu = \sqrt{2}(p^+ \times \tilde{p})_\mu^{(1)} \tag{37.4}$$

The most general Hamiltonian with dynamical symmetry, contains linear and quadratic terms in N which do not contribute to level spacings and the linear combination

$$a(s^+\tilde{p} + p^+s) \cdot (s^+\tilde{p} + p^+s) + b\mathbf{L} \cdot \mathbf{L} \tag{37.5}$$

The dynamical symmetry in this case is that of the chain of groups

$$U(4) \supset O(4) \supset O(3) \tag{37.6}$$

The Casimir generator of $O(4)$ is given by the linear combination (37.5) with coefficients $a = 1$ and $b = 1$. Hence, the Hamiltonian (37.5) may be expressed as a linear combination of the Casimir operators of $O(4)$ and $O(3)$

$$H = A' C_{O(4)} + B\mathbf{L}\cdot\mathbf{L} \tag{37.7}$$

The eigenvalues of $C_{O(4)}$ for fully symmetric irreducible representations may be calculated in the same way as in the $O(6)$ case (Section 35). Replacing in (35.5), (35.6) and (35.7) the operators d^{+}_{μ}, d_{μ} by p^{+}_{μ}, p_{μ} and the number of dimensions 6 by 4, we obtain the eigenvalues of $C_{O(4)}$ to be equal to $\sigma(\sigma+2)$ where $\sigma = N, N-2, \ldots, 1$ or 0. Instead of using $C_{O(4)}$ in (37.7), we may use the pairing interaction associated with it. The analog of (35.5) is

$$P'^{+}P'^{-} = \tfrac{1}{4}((p^{+}\cdot p^{+}) - (s^{+})^{2})((\tilde{p}\cdot\tilde{p}) - s^{2}) \tag{37.8}$$

and its eigenvalues, the analogs of (35.7), are given by

$$\tfrac{1}{4}(N-\sigma)(N+\sigma+2) \tag{37.9}$$

The eigenvalues of a Hamiltonian with $O(4)$ dynamical symmetry may be thus expressed as

$$\frac{A}{4}(N-\sigma)(N+\sigma+2) + BL(L+1) \tag{37.10}$$

The spectrum has rotational bands, each specified by the quantum number σ or by a vibrational quantum number $\nu = N - \sigma$. With $A > 0$, the lowest band has $\sigma = N$ or $\nu = 0$. Higher vibrational states have band heads ($L = 0$ states) whose energies are given by

$$\frac{A}{4}\nu(2N+2-\nu) = A_{1}\nu + A_{2}\nu^{2}$$

The spectrum thus obtained is the same as obtained from the (approximate) solution of the Schrödinger *radial* equation with the Morse potential.

Members of rotational bands whose energies are given by (37.10) have angular momenta $L = 0, 1, 2, \ldots$. Hence, they correspond to rotational states of diatomic molecules with different nuclei. If, however, the two atoms are identical, states with odd values of L and odd parity do not appear in these bands. Rotational bands with $L = 0, 2, 4, \ldots$

may be described by the other dynamical symmetry associated with the chain of groups

$$U(4) \supset U(3) \supset O(3)$$

The 9 generators of $U(3)$ are components of the tensors $(p^+ \times \tilde{p})^{(k)}$ with $k = 0,1,2$. As shown in Section 30 and Section 36, rotational bands of $U(3)$ (or $SU(3)$) with $K = 0$ include states with $L = 0,2,4,\ldots$.

It is worthwhile to mention that a spectrum like (37.10) with B independent of L cannot arise from a Schrödinger equation in which the potential is regular (Talmi 1981). A variational argument shows that if there are 3 levels with L_0, L_0+1, L_0+2 (or L_0, L_0+2, L_0+4) whose spacings are determined by $BL(L+1)$, then the rotational band does not terminate but continues to states with arbitrarily large values of $L > L_0$ and energies given by $BL(L+1)$. This argument may be applied also to the IBA-1 boson model. There is no collective Hamiltonian with a potential energy which is a regular function of β and γ, which reproduces *exactly* the $SU(3)$ spectrum (36.9) or the $O(6)$ spectrum (35.8).

Let us now return to the interacting boson model of nuclei and its relation to the shell model. To establish a correspondence, we should look in the shell model for "building blocks" with spins $J = 0$ and $J = 2$ and positive parity. The simplest of those are states of fermion pairs with these spins. Such a pair could be a particle hole pair in which a nucleon is raised from closed shells into a higher orbit. Such a $J = 0$ pair state would have a rather high excitation energy since a nucleon must be excited across *two* major shells to have the same spin and parity as the hole state. It is difficult to imagine that such high excitations play a dominant role in the structure of low lying states. There is also no reason why the number of such excitations should be constant for a given nucleus. On the other hand, the number of pairs of *nucleons* is fixed in a given nucleus which may be the basis of conservation of N in the boson model for nuclei.

We can now ask which nucleon pairs should be considered, proton-proton and neutron-neutron pairs or perhaps proton-neutron pairs. In most nuclei where collective bands appear, valence protons and neutrons occupy different major shells. It is then impossible to couple a valence proton and a valence neutron to a state with $J = 0$ and positive parity. Hence, we must consider pairs of valence identical nucleons coupled to $J = 0$ and $J = 2$ and see how they can be related to the s, d bosons. It is clear at this stage that we are naturally led into a possible boson description of proton states and a boson description of neutron states. This does not mean that we can ignore the proton-neutron interaction. This strong and attractive interaction plays a dominant role in the interacting boson model. Yet it will be introduced at the proper stage later on.

Pairs of identical nucleons with $J = 0$ have been considered in nuclear physics since the early days of the shell model. As we saw above, the concept of seniority is based on coupling of identical nucleons into $J = 0$ pairs. In a single j-orbit there is only one $J = 2$ pair state. When configurations based on several orbits are admixed there are several $J = 2$ states. We saw, however, that in the quasi-spin scheme of the surface delta interaction (Section 22) and certainly in the case of generalized seniority (Section 24) there is one $J = 2$ pair state with special properties. Thus, we can first see whether there is a boson description of states with generalized seniority $v = 0$ and $v = 2$.

Let us recall the eigenvalues (23.9) and (24.10) of the shell model Hamiltonian satisfying the conditions (23.5), (23.7), (24.3) and (24.9). We obtained there

$$H(S^+)^N|0\rangle = (NV_0 + \tfrac{1}{2}N(N-1)W)(S^+)^N|0\rangle \tag{37.11}$$

and

$$H(S^+)^{N-1}D^+_\mu|0\rangle = ((N-1)V_0 + V_2 + \tfrac{1}{2}N(N-1)W)(S^+)^{N-1}D^+_\mu|0\rangle \tag{37.12}$$

The *same* eigenvalues may be obtained from the *boson Hamiltonian*

$$\boxed{H^B = V_0 s^+ s + V_2(d^+ \cdot \tilde{d}) + \tfrac{1}{2}W(s^+)^2 s^2 + W s^+ s(d^+ \cdot \tilde{d})} \tag{37.13}$$

These eigenvalues are obtained from the eigenstates $(s^+)^N|0\rangle^B$ and $(s^+)^{N-1}d^+_\mu|0\rangle^B$ of H^B respectively.

The boson creation operators in the model described above *do not* create $J = 0$ and $J = 2$ pairs of fermions. They obey exactly the boson commutation relations with their hermitean conjugates. On the other hand, the S^+ and D_μ operators do not commute. The commutation relation between S and S^+ are given, due to (19.6) by

$$[S,S^+] = -2\sum_j \alpha_j^2 S_j^0 = \sum_j \alpha_j^2(2j+1) - \sum_j \alpha_j^2 n_j \qquad (37.14)$$

which is rather different from the boson commutator. Still, the commutator of S and S^+, if both are divided by $[\sum \alpha_j^2(2j+1)]^{1/2}$, is close to 1 as long as $\sum \alpha_j^2 n_j \ll \sum \alpha_j^2(2j+1)$. This has been used as an argument for replacing fermion pair operators by boson ones. However, the eigenvalues of the boson Hamiltonian (37.13) reproduce the eigenvalues (37.11) and (37.12) for *any value* of $N \leq \Omega$. What was established above, for any $N \leq \Omega$, is a *mapping* between certain fermion states and boson states as well as between fermion operators and boson operators. Derivation of the eigenvalues of the fermion Hamiltonian can be carried out equally well in the boson model. It is worthwhile to iterate the fact that the special eigenvalues in (37.11) and (37.12) obtained in the boson model are *exactly* equal to those obtained for fermion states where the Pauli principle is strictly obeyed.

The boson Hamiltonian (37.13) may replace the shell model fermion Hamiltonian only for a very special set of states. No simple prescription can be found for the other states. In very special cases it is possible to find a boson expression not only for a simple Hamiltonian but also for the pair creation and pair annihilation operators. Consider the operators S^+ and S with *equal* α_j *coefficients* denoted by $\mathcal{S}^+$ and $\mathcal{S}^-$ in the quasi-spin scheme. The boson operators

$$(\sqrt{\Omega - s^+s})s \qquad s^+\sqrt{\Omega - s^+s} \qquad (37.15)$$

obey the commutation relation

$$\begin{aligned}
&\left[s^+\sqrt{\Omega - s^+s}, \left(\sqrt{\Omega - s^+s}\right)s\right] \\
&\quad = s^+(\Omega - s^+s)s - \left(\sqrt{\Omega - s^+s}\right)ss^+\sqrt{\Omega - s^+s} \\
&\quad = s^+(\Omega - s^+s)s - \sqrt{\Omega - s^+s}[s,s^+]\sqrt{\Omega - s^+s} \\
&\qquad - \left(\sqrt{\Omega - s^+s}\right)s^+s\sqrt{\Omega - s^+s} \\
&\quad = s^+(\Omega - s^+s)s - (\Omega - s^+s) - s^+s(\Omega - s^+s) = 2s^+s - \Omega
\end{aligned} \qquad (37.16)$$

The boson operator which is one half the r.h.s. of (37.16) satisfies with the operators (37.15) the following commutation relations (note that

s^+s commutes with $\sqrt{\Omega - s^+s}$)

$$\left[s^+s - \tfrac{1}{2}\Omega, s^+\sqrt{\Omega - s^+s}\right] = s^+\sqrt{\Omega - s^+s}$$

$$\left[s^+s - \tfrac{1}{2}\Omega, (\sqrt{\Omega - s^+s})s\right] = -\left(\sqrt{\Omega - s^+s}\right)s$$

Thus, the operator s^+s is the boson expression for the number operator of fermion pairs. The operators (37.15) are the boson equivalents of $\mathcal{S}^-$ and $\mathcal{S}^+$ and the operator on the r.h.s. of (37.16) is the equivalent of $2\mathcal{S}^0$. The boson operators satisfy the commutation relations of the $SU(2)$ Lie algebra exactly like the corresponding fermion operators $\mathcal{S}^-$, $\mathcal{S}^+$ and $\mathcal{S}^0$. The pairing interaction $\mathcal{S}^+\mathcal{S}^-$ is equivalent to the boson operator

$$s^+\sqrt{\Omega - s^+s}\left(\sqrt{\Omega - s^+s}\right)s = s^+(\Omega - s^+s)s = \Omega s^+s - (s^+)^2s^2 \tag{37.17}$$

This has the same form as (37.13) without the $d^+, \tilde{d}$ operators and its eigenvalues are given by $\Omega N - N(N-1)$ like those of the $\mathcal{S}^+\mathcal{S}$ fermion Hamiltonian. This boson expression of the pairing interaction, (Holstein and Primakoff 1940, Pang et al. 1968) is based on the $SU(2)$ Lie algebra which is not sufficiently general as discussed in Section 23. It has a boson expression only for S-pairs but not for D-pairs. The expressions (37.15) vanish for $N = \Omega$ and do not describe fermion states if $N > \Omega$.

The boson model with the Hamiltonian (37.13) is less ambitious. It does not have boson expressions for pair fermion creation and annihilation operators S^+ and S. Still, its Hamiltonian has only single boson energies and boson boson interactions and its eigenvalues (for $N \leq \Omega$) are exactly equal to those of the shell model Hamiltonian in fully antisymmetric fermion states. The boson Hamiltonian (37.13) contains no terms which annihilate s-bosons and create d-bosons. This conforms to the fact that its eigenstates $(s^+)^N|0\rangle^B$ correspond to fermion eigenstates with generalized seniority $v = 0$. A boson Hamiltonian corresponding to a fermion Hamiltonian whose eigenstates have definite generalized seniorities could still contain, unlike (37.13), interactions between d-bosons. Let us now consider states with several d-bosons.

Complications begin already for fermion states corresponding to states of two d-bosons. The boson state $(d^+ \times d^+)^{(J)}_M|0\rangle^B$ with $J = 0$

is orthogonal to the state $(s^+)^2|0\rangle^B$ and the boson state with $J = 2$ is orthogonal to the $s^+d^+|0\rangle^B$ state. The fermion state $(D^+ \times D^+)^{(J)}_M|0\rangle$ with $J = 0$ is, in general, not orthogonal to the state $(S^+)^2|0\rangle$. Similarly, the fermion state with $(D^+ \times D^+)^{(J)}_M|0\rangle$, $J = 2$ is, in general, not orthogonal to the $S^+D^+_M|0\rangle$ state. In order to establish the correspondence with boson states, the fermion states must first be orthogonalized. Before going into it, let us make sure that any fermion state constructed from S^+ and D^+_M pair creation operators corresponds to a non-vanishing state similarly constructed from s^+ and d^+_M boson creation operators. The inverse of this statement is definitely not true as will be discussed in detail in the following.

To demonstrate the correspondence described above it is more convenient to use configuration space for the fermions. To construct the fermion states we first couple with Clebsch-Gordan coefficients pair states and then antisymmetrize the resulting wave function. The antisymmetrization may be carried out in two steps. We first exchange two nucleon coordinates in any pair state with two coordinates in any other pair state. Such permutations are added with plus signs and hence, if the wave function does not vanish, it is a fermion state which corresponds to a *symmetric* function of s- and d-bosons. In the second step all other permutations are applied to the fermion wave function. If the resulting antisymmetric function does not vanish, it corresponds to a symmetric state of s- and d-bosons. It should be realized, however, that two fermion states which correspond in this way to two orthogonal boson states may become, as a result of the complete antisymmetrization, non-orthogonal and even equal. We shall meet in the following such a case occurring already for $N = 2$.

If all α_j coefficients in the operator S^+ (23.1) are equal, the simpler quasi-spin scheme may be applied. In that case, the orthogonalization of states constructed from $S^+ \equiv \mathcal{S}^+$ and D^+_M operators may be carried out by using the seniority quantum number. Any state created by the action of n_d operators D^+_M may have components whose highest possible seniority is equal to

$$\boxed{v = 2n_d} \tag{37.18}$$

If n_s operators $\mathcal{S}^+$ act on such a state then the total number of pairs is $N = n_s + n_d = n_s + \frac{1}{2}v$. We may obtain from such a state a component with definite seniority v by projecting out of it components with seniorities smaller than v. If the result does not vanish we obtain a state with seniority v which may be mapped on a boson state obtained by acting on the (boson) vacuum by n_s operators s^+ and $n_d = \frac{1}{2}v$ operators d_M^+.

We can now present an example where a simple boson state does not correspond to a fermion state. The state $(D^+ \times D^+)_0^{(0)}|0\rangle$ where $D_M^+ = \sum(\frac{7}{2}m\frac{7}{2}m' \mid \frac{7}{2}\frac{7}{2}J = 0, M = 0)a^+_{(7/2)m}a^+_{(7/2)m'}$ which could correspond to the $(d^+ \times d^+)_0^{(0)}|0\rangle$ state cannot have seniority $v = 4$ since there is no state with $J = 0$, $v = 4$ in the $(\frac{7}{2})^4$ configuration. If the component with seniority $v = 0$ is projected from it, the rest simply vanishes. Thus, there is no fermion state in this case which corresponds to the boson state $(d^+ \times d^+)^{J=0}|0\rangle^B$. That boson state may be described as a spurious state. We will later consider such states in more detail.

The non-vanishing fermion states with definite seniorities v obtained in this way correspond to states obtained by applying $\frac{1}{2}v$ operators d_M^+ and $N - \frac{1}{2}v$ operators s^+ to the boson vacuum. We see now that the boson Hamiltonian (37.13) with $V_0 = \Omega$, $V_2 = 0$, $W = -2$, yields eigenvalues of the pairing interaction also for such states. Recalling the eigenvalues (22.5) of the pairing interaction in states with given $N = 2n$ and v we see that the Hamiltonian

$$\boxed{H^B = \Omega s^+ s - (s^+)^2 s^2 - 2s^+ s(d^+ \cdot \tilde{d})} \tag{37.19}$$

is indeed the boson operator corresponding to $\mathcal{S}^+\mathcal{S}^-$, not only for the states with $v = 0$ and $v = 2$ but also for states with higher values of v.

The correspondence described above, between fermion states with seniority v and boson states with $n_d = \frac{1}{2}v$, loses its meaning beyond the middle of the fermion shell. The seniority v cannot exceed the value of Ω whereas there is no limit on the number n_d of boson states. If we construct fermion states with higher numbers, $n_d > \Omega/2$ of pair creation operators D_M^+, they are linear combination of states with seniorities v smaller than Ω. Such states vanish when states with $v \leq \Omega$ are projected out. Since this is an important point let us recall that in a single j-orbit the relation between the quasi-spin value s and seniority v is given by (19.10) as $s = \frac{1}{2}((2j+1)/2 - v)$. Since $s \geq 0$, it follows that $v \leq (2j+1)/2$. If several orbits are admixed according to (22.1), then $(2j+1)/2$ is replaced by $\Omega = \frac{1}{2}\sum(2j+1)$ and hence $v \leq \Omega$. This fact holds also in generalized seniority where it assumes the following form—any state with $n > \Omega$ may be expressed as $S^+B^+|0\rangle$ where B^+ creates a state with $n-2$ (identical) fermions.

To prove this property we create the state with $n > \Omega$ identical fermions by acting on the state of the completely filled valence shell by an operator *annihilating* $2\Omega - n$ fermions. The closed valence shell may be expressed as in Section 24 by $(S^+)^\Omega|0\rangle$ and the annihilation operator which may be denoted by $B_{\gamma JM}(2\Omega - n)$, is a linear combination of products of $2\Omega - n$ single fermion annihilation operators. To prove our point we make use of (23.8) to expand the state considered as follows

$$B_{\gamma JM}(2\Omega - n)(S^+)^\Omega|0\rangle = (S^+)^\Omega B_{\gamma JM}(2\Omega - n)|0\rangle$$
$$+\sum_{\nu=1}^{\Omega}\binom{\Omega}{\nu}(S^+)^{\Omega-\nu}[\cdots[B_{\gamma JM}(2\Omega - n), S^+], \ldots, S^+]]|0\rangle \tag{37.20}$$

The multiple commutators in (37.20) contain ν operators S^+. Since $B_{\gamma JM}(2\Omega - n)$ contains products of $2\Omega - n$ fermion annihilation operators, the multiple commutators for $\nu > 2\Omega - n$ vanish provided $2\Omega - n < \Omega$. Thus, the last non-vanishing term on the r.h.s.of (37.20) has the factor $(S^+)^{\Omega-(2\Omega-n)} = (S^+)^{n-\Omega}$ which for $n > \Omega$ is a positive power of S^+. All other non-vanishing terms on the r.h.s. of (37.20) have higher powers of S^+ as factors. This implies that a factor S^+ is common to all non-vanishing terms on the r.h.s. of (37.20) and hence to the state on the l.h.s. of (37.20).

This difficulty, appearing when the middle of the shell is crossed, may be easily fixed. Beyond the middle of the shell we simply consider fermion *hole* pairs and make the correspondence between hole states with seniority v, constructed from hole pair states with $J = 0$ and $J = 2$, and states with $n_d = v/2$ d-bosons. In Section 24 we saw how states with generalized seniorities $v = 0$ and $v = 2$ may be constructed as hole states.

If the α_j coefficients in the S^+ operator (23.1) are not all equal, we may still use the orthogonalization procedure suggested by the seniority scheme. We may couple n_d operators D^+_μ consecutively to a state with total J (and M) and multiply them by the operator $(S^+)^{n_s}$. This yields a creation operator of a state of $n = 2N = 2(n_s + n_d)$ fermions coupled to the given J (and M). We can then project out of that state all components which have the form $(S^+)^{n_s+1}B^+_{JM}|0\rangle$ where B^+_{JM} creates a state with $n = 2n_d - 2$ fermions. The non-vanishing resulting states, after normalization, may be mapped onto boson states

$$(s^+)^{n_s}(d^+)^{n_d}_{\gamma JM}|0\rangle^B$$

where γ denotes the additional necessary quantum number which distinguishes between different orthogonal boson states. Also in this case, if $N > \Omega$ the resulting states vanish after the projection and beyond the middle of the shell, hole pairs are used for construction of fermion states.

A simpler process of orthogonalization can be carried out within the space of states created by S^+ and D^+_M operators. This space, which may be called *the* $S - D$ *space*, is a rather small subspace of the shell model subspace of the valence nucleons. When a state

$$[(S^+)^{n_s}(D^+)^{n_d}_{JM}]_\gamma|0\rangle$$

is created, we have to remove from it all components which can be expressed as $[(S^+)^{n_s+1}(D^+)^{n_d-1}_{JM}]_{\gamma'}|0\rangle$, $[(S^+)^{n_s+2}(D^+)^{n_d-2}_{JM}]_{\gamma''}|0\rangle$ etc. The $\gamma, \gamma', \gamma'' \ldots$ characterize different orthogonal fermion states. These components do not include all states of the form $(S^+)^{n_s+1}B^+_{JM}|0\rangle$ where the only condition on the operator B^+_{JM}, with given J and M, is that it creates a state with $2n_d - 2$ fermions. The states resulting from this procedure, if they do not vanish, are normalized and mapped onto $(s^+)^{n_s}(d^+)^{n_d}_{\gamma JM}|0\rangle^B$ boson states where the quantum number γ of boson states corresponds to that of fermion states. Using this procedure, it may happen that states with $n_d = N > \Omega$ will not vanish after projecting out states created by operators which have the S^+ operator as a factor. In practice, most of the wave functions of such states will be removed and they may be safely ignored. Also in this procedure, beyond the middle of the shell we use $S - D$ states constructed from fermion *hole* pairs with $J = 0$ and $J = 2$.

The correspondence between fermion and boson states in either of the two procedures described above, may not be completely defined. For d-boson states with $n_d \geq 4$ there may be several independent states with the same values of J (and M). In a sufficiently large shell model subspace there may be an equal number of fermion states (with seniority $v = 2n_d$) created by the action of n_d operators D^+. The set of additional quantum numbers γ which characterize orthogonal fermion states must be related to the corresponding quantum numbers of the boson states. We shall not discuss here this problem.

In Sections 23 and 24 we saw that states with generalized seniorities $v = 0$ and $v = 2$ are eigenstates of the shell model Hamiltonians of actual semi-magic nuclei. What can be said about the states described above which are constructed from *several* D^+_M operators? Are any of the two kinds of (orthogonal) states described above exact or approximate eigenstates of the shell model Hamiltonian? The experimental information on semi-magic nuclei is not sufficient to obtain an answer to these questions. As we shall see in the next section, however, the success of the boson model does not depend critically on the answer.

The correspondence between fermion and boson states may be carried out without any complications in certain models introduced by Ginocchio (1980). In those models, a fermion subspace of the shell model can be mapped onto a boson space and the mapping is both simple and exact. The various fermion states are orthogonal and to each of them corresponds a well-defined boson state. This situation is due to both fermion states and boson states being basis states of irreducible representations of the same Lie group. Such a Lie group may be $O(6)$ which was described in Section 35. In its realization in terms of bosons it has the generators (34.2) and (35.1). These are the odd tensor operators $(d^+ \times \tilde{d})^{(k)}_\kappa$, $k = 1,3$ and the quadrupole operator

$$Q'_\mu = s^+\tilde{d}_\mu + d^+_\mu s \tag{37.21}$$

The commutation relations of these generators with boson creation operators are

$$[(d^+ \times \tilde{d})^{(k)}_\kappa, s^+] = 0 \qquad k = 1,3 \qquad [s^+\tilde{d}_\mu + d^+_\mu s, s^+] = d^+_\mu \tag{37.22}$$

$$[(d^+ \times \tilde{d})^{(k)}_\kappa, d^+_\mu] = \sum_\nu (-1)^\mu (2\nu 2, -\mu | 22k\kappa) d^+_\nu \qquad k = 1,3,$$

$$[s^+\tilde{d}_\kappa + d^+_\kappa s, d^+_\mu] = (-1)^\kappa \delta_{\mu,-\kappa} s^+ \tag{37.23}$$

A corresponding set of generators satisfying these commutation relations with the fermion S^+ and D^+_μ operators may be found in certain shell model sub-spaces.

The simplest case is that of a single $j = \frac{3}{2}$ orbit. The single nucleon tensor operators

$$(a^+_{3/2} \times \tilde{a}_{3/2})^{(k)}_\kappa \qquad k = 0,1,2,3 \tag{37.24}$$

are generators of the $U(4)$ Lie algebra as explained in Section 25. The 15 operators with $k = 1,2,3$ are the elements of the $SU(4)$ Lie algebra which is the *same* as the $O(6)$ Lie algebra. The operators (37.24) satisfy the commutation relations (25.7) listed in Section 25. We compare now these commutation relations to those of the boson generators of $O(6)$, $(d^+ \times \tilde{d})_\kappa^{(k)}$, $k = 1,3$ and $s^+\tilde{d}_\kappa + d_\kappa^+ s$ given by (33.34), (33.35) and (33.36). We see that the same coefficients of generators appear in the expansion of the commutators if we define the fermion generators by

$$\frac{1}{\sqrt{2}}(a_{3/2}^+ \times \tilde{a}_{3/2})_\kappa^{(k)} \qquad k = 1,3, \qquad (a_{3/2}^+ \times \tilde{a}_{3/2})_\kappa^{(2)} \tag{37.25}$$

The coefficient $2^{-1/2}$ in front of the $k = 1,3$ tensors is the origin of the factor 2 in front of the $\sum_{k=1,3}(d^+ \times \tilde{d})^{(k)} \cdot (d^+ \times \tilde{d})^{(k)}$ term in the Casimir operator (35.4). The Casimir operator of the $SU(4)$ group is, according to (25.4) with $j = \frac{3}{2}$, the sum

$$\sum_{k=1,2,3} (a_{3/2}^+ \times \tilde{a}_{3/2})^{(k)} \cdot (a_{3/2}^+ \times \tilde{a}_{3/2})^{(k)}.$$

Hence, the Casimir operator in the boson model is given by (35.4) including the factor 2. In the present case, there is only one $J = 0$ pair creation operator $S^+ = \frac{1}{2}\sum(-1)^{3/2-m} a_{(3/2)m}^+ a_{(3/2)-m}^+$ and only one $J = 2$ operator

$$D_M^+ = \frac{1}{\sqrt{2}} \sum (\tfrac{3}{2}m\tfrac{3}{2}m' \mid \tfrac{3}{2}\tfrac{3}{2}2M) a_{(3/2)m}^+ a_{(3/2)m'}^+$$

The odd tensor operators in (37.25) with $k = 1,3$ commute with S^+ as shown in Section 19. The commutator of the quadrupole tensor ($k = 2$) in (37.24) with $(1/\sqrt{2})S^+$ was calculated in (19.36) to be equal to D_κ^+. The commutators of the operators in (37.25) with D_M^+ may be directly calculated.

We evaluate the commutator

$$\begin{aligned}
[(a_j^+ \times \tilde{a}_j)_\kappa^{(k)}, D_\mu^+] &= \left[(a_j^+ \times \tilde{a}_j)_\kappa^{(k)}, \frac{1}{\sqrt{2}}(a_j^+ \times a_j^+)_\mu^{(2)}\right] \\
&= \frac{1}{\sqrt{2}} \sum_{mm'\nu\nu'} (-1)^{j+m'}(jmjm' \mid jjk\kappa)(j\nu j\nu' \mid jj2\mu) \\
&\quad \times [a_{jm}^+ a_{j,-m'}, a_{j\nu}^+ a_{j\nu'}^+] \\
&= \sqrt{2} \sum_{m\nu\nu'} (-1)^{j-\nu}(jmj,-\nu \mid jjk\kappa)(j\nu j\nu' \mid jj2\mu) a_{jm}^+ a_{j\nu'}^+
\end{aligned}$$

Expressing the Clebsch-Gordan coefficients by $3j$-symbols and applying the transformation (10.24) we obtain

$$(-1)^{k+1}\sqrt{10}\sum_{JM}\sqrt{2J+1}\begin{Bmatrix} j & 2 & j \\ k & j & J \end{Bmatrix}$$

$$\times(-1)^{\mu}(JM2,-\mu\,|\,J2k\kappa)(a_j^+\times a_j^+)_M^{(J)} \qquad (37.26)$$

The pair creation operators in the summation vanish unless the value of J is even. If $k=1,3$ the only possible even value is $J=2$ and for $j=\frac{3}{2}$ we obtain

$$\left[\frac{1}{\sqrt{2}}(a_{3/2}^+\times\tilde{a}_{3/2})_\kappa^{(k)},D_\mu^+\right]=5\sqrt{2}\begin{Bmatrix} \frac{3}{2} & 2 & \frac{3}{2} \\ k & \frac{3}{2} & 2 \end{Bmatrix}$$

$$\times\sum_M(-1)^{\mu}(2M2,-\mu\,|\,22k\kappa)D_M^+$$

$$=\sum_M(-1)^{\mu}(2M2,-\mu\,|\,22k\kappa)D_M^+ \quad k=1,3 \qquad (37.27)$$

The last equality was obtained by substituting the numerical values of the $6j$-coefficients for $k=1$ or $k=3$. In the case of $k=2$ the possible even values of J are $J=2$ and $J=0$. Since, however,

$$\begin{Bmatrix} \frac{3}{2} & 2 & \frac{3}{2} \\ 2 & \frac{3}{2} & 2 \end{Bmatrix}=0$$

we obtain in this case by putting in (37.26) $k=2$, $j=\frac{3}{2}$, $J=0$, the result

$$[(a_{3/2}^+\times\tilde{a}_{3/2})_\kappa^{(2)},D_\mu^+]=-\sqrt{10}\begin{Bmatrix} \frac{3}{2} & 2 & \frac{3}{2} \\ 2 & \frac{3}{2} & 0 \end{Bmatrix}$$

$$\times(-1)^{\mu}(002-\mu\,|\,022\kappa)(a_{3/2}^+\times a_{3/2}^+)_0^{(0)}$$

$$=(-1)^{\kappa}\delta_{\mu,-\kappa}\frac{1}{\sqrt{2}}S^+ \qquad (37.28)$$

Thus, the commutators of the generators (37.25) with $(1/\sqrt{2})S^+$ and D_μ^+, given by (37.27) and (37.28), are the same as (37.22) and (37.23).

Hence $(1/\sqrt{2})S^+|0\rangle$ and $D_M^+|0\rangle$ form the basis of the same irreducible representation of $O(6)$ as the boson states $s^+|0\rangle^B$ and $d_\mu^+|0\rangle^B$.

The shell model sub-space of a single $j = \frac{3}{2}$ orbit is rather limited. It may, however, be generalized (Ginocchio 1980) by coupling the spin $i = \frac{3}{2}$ to an integral spin k obtaining several (degenerate) j-orbits with $j = k - \frac{3}{2}, k - \frac{1}{2}, k + \frac{1}{2}, k + \frac{3}{2}$ (for $k \geq 2$) or $j = \frac{1}{2}, \frac{3}{2}, \frac{5}{2}$ for $k = 1$. The description of states of nucleons occupying the given j-orbits in terms of states of the k and i spins is just a change of coupling transformation. The creation operators of a nucleon in states with given k, m_k, i, m_i are defined in terms of the a_{jm}^+ operators by

$$a_{im_ikm_k}^+ = \sum_{j=|k-i|}^{k+i} (im_ikm_k \mid ikjm)a_{jm}^+$$

States of the system should be fully antisymmetric, irrespective of the representation. For two (identical) nucleons there are 4 independent states with $J = 0$ (3 states if $k = 1$). The $J = 0$ state of interest in the model has the two k-spins coupled to $K = 0$ and the two i-spins to $I = 0$. The irreducible tensors (with ranks 1,2,3) in (37.25) should operate only on the variables of the i-spin ("pseudo spin") and not on those of k ("pseudo orbital angular momentum"). Then also in the $J = 2$ pair, considered in the model, the total K vanishes, $K = 0$, whereas the total I is equal to 2. In that case, only the i-spin is *active* and the states constructed with these $J = 0$ and $J = 2$ pair creation operators form irreducible representations of the $O(6)$ Lie algebra. It should be emphasized that there is no physical significance to the k-spin and i-spin vectors, beyond the specific model.

Let us consider the $J = 0$ state with $K = 0$ and $I = 0$. Its expansion in states with definite values of $\mathbf{j} = \mathbf{k} + \mathbf{i}$ is given according to (10.10) by

$$\begin{aligned}
&|ii(0)kk(0)J = 0\rangle \\
&= \sum_j \sqrt{2j+1}(-1)^{i+j+k} \begin{Bmatrix} k & i & j \\ i & k & 0 \end{Bmatrix} |ik(j)ik(j)J = 0\rangle \\
&= \frac{1}{\sqrt{(2i+1)(2k+1)}} \sum_j \sqrt{2j+1}|ik(j)ik(j)J = 0\rangle \qquad (37.29)
\end{aligned}$$

Recalling the normalization of S_j^+ we see that the state (37.29) is obtained by acting on the vacuum with the operator

$$S^+ = \sum_{j=k-3/2}^{j=k+3/2} S_j^+ \tag{37.30}$$

where all α_j coefficients are equal. The normalization coefficient for the state created by (37.30) is $(\sum(2j+1)/2)^{-1/2} = \Omega^{-1/2} = (2(2k+1))^{-1/2}$.

The operators in (37.24) acting only on the i-spin variables should also be expressed in terms of the j orbits. Making use of (9.27) and (10.9) we obtain for them the expressions

$$\boxed{\sum_{jj'}(-1)^{k+(3/2)+j+L}\sqrt{(2j+1)(2j'+1)} \times \begin{Bmatrix} \frac{3}{2} & j & k \\ j' & \frac{3}{2} & L \end{Bmatrix} (a_j^+ \times \tilde{a}_{j'})_M^{(L)} \qquad L = 1,2,3} \tag{37.31}$$

In (37.31) the rank of the irreducible tensors is denoted by L since it is customary to reserve the letter k for the pseudo-orbital angular momentum. The operators (37.31) act only on the variables of the i-spin. Hence, the commutation relations between these operators are the same as those between the operators (37.24) and thus, they generate the $O(6)$ group. This fact may be verified by a direct calculation in which the identity (10.17) between $6j$-symbols should be used.

The D_M^+ operator can now be obtained by taking the commutator of the quadrupole operator in (37.31) with S^+ in (37.30). We thus obtain

$$\boxed{D_M^+ = [Q_M, S^+] = \sum_{jj'}(-1)^{k+(3/2)+j}\sqrt{(2j+1)(2j'+1)} \times \begin{Bmatrix} \frac{3}{2} & j & k \\ j' & \frac{3}{2} & 2 \end{Bmatrix} (a_j^+ \times a_{j'}^+)_M^{(2)}} \tag{37.32}$$

It may be verified that the r.h.s. of (37.32) acting on the vacuum state, is indeed equal to the expansion (9.27) combined with (10.9) in terms of j and j' of the state $|ii(2)kk(0)J = 2\rangle$ multiplied by $\sqrt{2k+1}$. The overall normalization of the state $D_M^+|0\rangle$ is obtained by using the orthogonality properties of $6j$-symbols to be $(2k+1)^{-1/2} = (\Omega/2)^{-1/2}$. Since both D_M^+ and $(1/\sqrt{2})S^+$ operators leave the $K = 0$ state unchanged, their commutation relations with the operators (37.31) are the same as in the $j = \frac{3}{2}$ case.

Any state constructed from these $(1/\sqrt{2})S^+$ and D_M^+ operators which belongs to a certain irreducible representation of $O(6)$, uniquely corresponds to a boson state of the same irreducible representation. The inverse is also true. Any boson state which belongs to a certain $O(6)$ irreducible representation uniquely corresponds to a fermion state of that representation if the latter does not vanish. Ginocchio (1980) has shown that up to boson number N where $2N \leq \Omega = 4k + 2$ all $O(6)$ irreducible representations realized by boson states can be also obtained from fermion states. The correspondence between boson states and fermion states is simple (Ginocchio and Talmi 1980). Boson states are obtained by replacing in the corresponding fermion states the operators $(1/\sqrt{2})S^+, D_M^+$ by s^+, d_M^+ even though the state $(S^+)^2|0\rangle$ is *not* orthogonal to the $(D^+ \cdot \tilde{D}^+)|0\rangle$ state. This is due to the commutators of the $O(6)$ generators, the $k = 2$ tensor in (37.31) and the $k = 1,3$ tensors in (37.31) divided by $\sqrt{2}$, with $(1/\sqrt{2})S^+, D_M^+$ being the same as (37.22) and (37.23). Hence, also matrix elements of the generators (37.31) are numerically equal to those of the boson generators taken between corresponding states. This is particularly true of the eigenvalues of boson and fermion Hamiltonians constructed from the corresponding $O(6)$ generators.

The equality of matrix elements of generators stated above is a direct result of the concept of irreducible representations. The generators give rise to definite irreducible transformations among the orthogonal basis wave functions. The matrices of these transformations in a certain basis are determined up to a multiplicative constant. Once a set of generators was chosen which satisfies the Lie algebra commutation relations with given coefficients, the matrices of each irreducible representation are completely determined. This is due to the fact that the commutation relations are non-linear in the generators. The elements of these matrices may be equally well calculated with either boson or fermion sets of orthonormal wave functions. The normalization coefficients of corresponding boson and fermion states, however, may be very different. For example, for $N = 2$ the boson state with $J = 0$ and $\sigma = 0$ is given by (35.3) as $[(d^+ \cdot d^+) - (s^+)^2]|0\rangle^B$. Its normalization coefficient can be simply calculated since ${}^B\langle 0|s^2(d^+ \cdot d^+)|0\rangle^B = 0$ and is equal to $(12)^{-1/2}$. The corresponding fermion state is

$$[(D^+ \cdot D^+) - \tfrac{1}{2}(S^+)^2]|0\rangle \tag{37.33}$$

but its normalization factor is more complicated and is given by

$$\mathcal{N}^{-2} = \langle 0|(\tilde{D} \cdot \tilde{D})(D^+ \cdot D^+)|0\rangle + \tfrac{1}{4}\langle 0|S^2(S^+)^2|0\rangle - \langle 0|S^2(D^+ \cdot D^+)|0\rangle$$

Also the orthogonality of the fermion state (37.33) to the one with $\sigma = 2$, $\tau = 0$, $[(D^+ \cdot D^+) + \frac{5}{2}(S^+)^2]|0\rangle$ is more complicated than in the boson case. It is expressed by

$$\langle 0|(\tilde{D} \cdot \tilde{D})(D^+ \cdot D^+)|0\rangle - \tfrac{5}{4}\langle 0|S^2(S^+)^2|0\rangle + 2\langle 0|S^2(D^+ \cdot D^+)|0\rangle = 0$$

Another model of Ginocchio (1980) is based on the $SU(3)$ group. Instead of a model in which i is active, as described above, a model in which k is active and the single fermion operators act on its variables may be chosen. In order to have in the model only pair states created by S^+ and D_M^+ operators, the value of k must be limited to 1. The single fermion operators acting on the k-spin variables are irreducible tensors with ranks $L = 0, 1, 2$. These operators generate, as explained in Section 28, the $U(3)$ group. When the $L = 0$ term which is the number operator commuting with all others, is removed, the $SU(3)$ Lie algebra is obtained. As discussed in Section 30, the $SU(3)$ generators are the $L = 1$ vector and the quadrupole tensor with $L = 2$.

The k-spin is coupled to a half integral i-spin to yield (degenerate) orbits with eigenvalues of $\mathbf{j} = \mathbf{k} + \boldsymbol{i}$ equal to $i - 1, i, i + 1$ (for $i > \frac{1}{2}$) and $\frac{1}{2}$, $\frac{3}{2}$ for $i = \frac{1}{2}$. The $J = 0$ pair state of interest in the model, has $K = 0$, $I = 0$ and is given by (37.29). Thus, the $J = 0$ pair creation operator has equal α_j coefficients and is given by

$$\boxed{S^+ = \sum_{j=|i-1|}^{i+1} S_j^+} \tag{37.34}$$

The normalization coefficient for the state created by (37.34) is

$$\left(\sum(2j+1)/2\right)^{-1/2} = (\Omega)^{-1/2} = ((6i+3)/2)^{-1/2}$$

The single fermion operators are given in this case by

$$T_M^{(L)} = \sum_{j,j'=|i-1|}^{i+1} (-1)^{1+i+j'+L}\sqrt{(2j+1)(2j'+1)} \times \begin{Bmatrix} 1 & j & i \\ j' & 1 & L \end{Bmatrix} (a_j^\dagger \times \tilde{a}_{j'})_M^{(L)} \quad L = 1,2 \tag{37.35}$$

The operators (37.35) act only on the variables of the k-spin. Hence the commutation relations between them are the same as in (28.3) with $l = 1$ and they generate the $SU(3)$ group. This fact may be verified by a direct calculation in which the identity (10.17) of $6j$-symbols should be used. It should be remembered that in (28.3) unit tensors appear which are related to operators like (37.35) by the relation (17.25). In this way we obtain

$$[T_\mu^{(L)}, T_{\mu'}^{(L')}] = (-1)^J(1-(-1)^{L+L'+J})\sqrt{(2L+1)(2L'+1)} \times \begin{Bmatrix} L & L' & J \\ 1 & 1 & 1 \end{Bmatrix} (L\mu L\mu' \mid LL'JM) T_M^{(J)} \tag{37.36}$$

The expression (37.36) can be brought into the form (28.3), with $l = 1$, if we substitute in (37.36) the unit tensors $U_M^{(L)}$. The latter are related to $T_M^{(L)}$, according to (17.25) by $U_M^{(L)} = T_M^{(L)}/\sqrt{2L+1}$. The rank $k = 1$ tensor in (37.35) is proportional to the angular momentum operator in the shell model subspace. Putting in (37.36) $L = 1$ and $L' = 1,2$ and substituting the numerical value of the $6j$-symbol, we find the proportionality factor and obtain the relation

$$L_\mu = \sqrt{2} T_\mu^{(1)} \tag{37.37}$$

The factor $\sqrt{2}$ is the same as in the case of orbital angular momentum $k = 1$. We recall that $(l\|L\|l) = \sqrt{l(l+1)(2l+1)}$ and hence $\mathbf{L} = \sqrt{2 \times 3}\mathbf{U}^{(1)} = (\sqrt{2 \times 3}/\sqrt{3})\mathbf{T}^{(1)}$. When the rank $L = 2$ tensor Q_μ

in (37.31) acts on the $S^+|0\rangle$ state it creates a $J = 2$ state in which the pseudo-spin is $I = 0$ and pseudo-orbital angular momentum is $K = 2$. The commutator of Q_μ with $D^+_{\mu'}$, should be a linear combination of S^+ and $D^+_{\mu+\mu'}$. No creation operators of states with $J = 4$ may appear in this commutator since the maximum value of K in a two-fermion state is $K = 2$.

In the boson model, the $SU(3)$ generators are given by the angular momentum $\mathbf{L}$ and quadrupole operator (36.3) as

$$Q^B_\mu = s^+\tilde{d}_\mu + d^+_\mu s - \frac{\sqrt{7}}{2}(d^+ \times \tilde{d})^{(2)}_\mu$$
$$L^B_\mu = \sqrt{10}(d^+ \times \tilde{d})^{(1)}_\mu \qquad (37.38)$$

The commutation relation of the components of angular momentum L^B_μ with the operators Q^B_μ are the well-known ones. The commutation relations between the components Q^B_μ are given by (36.4). The commutators of L^B_μ and s^+, d^+_μ are the usual ones for angular momentum components. The commutators of Q^B_μ with s^+ and d^+_μ are given in (36.35). We can compare the commutation relation (36.4) with the relation (37.36) where we put $L = L' = 2$. The only possible value of J is $J = 1$ and we find

$$[T^{(2)}_\mu, T^{(2)}_{\mu'}] = \sqrt{5}(2\mu 2\mu' \mid 221M)T^{(1)}_\mu = \sqrt{\tfrac{5}{2}}(2\mu 2\mu' \mid 221M)L_M \qquad (37.39)$$

In the boson case we obtain from (36.4) and (33.27)

$$[Q^B_\mu, Q^B_{\mu'}] = \tfrac{3}{4}\sqrt{\tfrac{5}{2}}(2\mu 2\mu' \mid 221M)L^B_\mu$$

Comparing this expression with (37.39) we see that the boson generator $\mathbf{Q}^B$ corresponds to the fermion operator $\sqrt{3/4}\mathbf{T}^{(2)}$. This is the reason why the Casimir operator of $SU(3)$, given according to (28.24) by $\sum_{L=1,2}(\mathbf{T}^{(L)} \cdot \mathbf{T}^{(L)})$, is proportional to $\frac{1}{2}(\mathbf{L} \cdot \mathbf{L}) + \frac{4}{3}(\mathbf{Q} \cdot \mathbf{Q}) = \frac{2}{3}(2(\mathbf{Q} \cdot \mathbf{Q}) + \frac{3}{4}(L \cdot L))$ as in (36.5).

We therefore define the D_μ^+ operator by

$$D_\mu^+ = \left[\sqrt{\tfrac{3}{4}}T_\mu^{(2)}, S^+\right] = \sqrt{\tfrac{3}{4}} \sum_{j,j'=|i-1|}^{i+1} (-1)^{i+j'+1}\sqrt{(2j+1)(2j'+1)} \times \begin{Bmatrix} 1 & j & i \\ j' & 1 & 2 \end{Bmatrix} (a_j^+ \times a_{j'}^+)_\mu^{(2)} \tag{37.40}$$

The state created by the operator (37.40), $D_\mu^+|0\rangle$, is indeed equal to the expansion (9.27) in terms of j, j' of the state $|kk(2)ii(0)J = 2\rangle$ multiplied by $(3(2i+1)/4)^{1/2}$. Its normalization factor can be directly calculated by using the orthogonality relation (10.13) of the $6j$-symbols. It is equal to $[\frac{1}{2}\sum(2j+1)]^{-1/2} = (\Omega)^{-1/2} = [(6i+3)/2]^{-1/2}$ and thus is equal to the normalization factor of the S^+ operator in (37.34).

The commutators of the boson operators (37.38) with s^+ and d_μ^+ were calculated in Section 36. Those of L_κ^B with s^+ and d_μ^+ are the usual ones for components of angular momentum with rank 0 and 2 tensor operators. The commutators of Q_κ^B with s^+ and d_μ^+ are given in (36.35) by

$$[Q_\kappa^B, s^+] = d_\kappa^+$$
$$[Q_\kappa^B, d_\mu^+] = (-1)^\kappa \delta_{\kappa,-\mu} s^+ - \frac{\sqrt{7}}{2}\sum_M (2\kappa 2\mu \mid 222M) d_M^+ \tag{37.41}$$

In the fermion space considered here, (37.40) corresponds to the first equation in (37.41). We then evaluate by direct calculation, using the identity (10.17) of $6j$-symbols, the commutator

$$\left[\sqrt{\tfrac{3}{4}}T_\kappa^{(2)}, D_\mu^+\right] = (-1)^\kappa \delta_{\kappa,-\mu} S^+ + \frac{\sqrt{7}}{2}\sum_M (2\kappa 2\mu \mid 222M) D_M^+ \tag{37.42}$$

The result (37.42) is the same as (37.41) apart from the sign of the second term on its r.h.s. It will be exactly the same if we change the sign in the quadrupole operator Q_μ^B in (37.38) as in (36.41). It was explained in detail in Section 36 that this is a possible choice for generators of $SU(3)$. Hence states constructed from S^+ and D^+_μ operators can be classified according to the irreducible representations of $SU(3)$. If we define them by using Q^B_μ in (37.38), then, according to (36.40), the states $S^+|0\rangle$ and $D^+_\mu|0\rangle$ span the space of the irreducible representation (0,2). Any boson state constructed of s^+ and d^+_μ operators which belong to the (λ,μ) irreducible representation of $SU(3)$ corresponds to a fermion state which belongs to the (μ,λ) conjugate representation, provided the latter does not vanish. The fermion state may be constructed by replacing in the creation operator of the state, each s^+, d^+_μ by S^+, D^+_μ respectively. The normalization factors of the boson and fermion states may well be different and should be explicitly calculated. As explained in Section 30, all states of the (λ,μ) irrreducible representation have the same spins and other quantum numbers as those included in the conjugate (μ,λ) irreducible representation.

In the present case, however, there are boson states for which corresponding fermion states do not exist. This occurs not only beyond the middle of the shell, as in the $O(6)$ model, but for smaller fermion numbers. Let us consider boson states which belong to the lowest irreducible $SU(3)$ representation $(0,2N)$. One of these is the state with $J = 2N$. The $M = J = 2N$ component of the fermion state which should correspond to it is simply proportional to $(D^+_2)^N|0\rangle$. In such a state the projections of the k vectors of *all* fermions along the 3-axis must be $\kappa = +1$. Hence, according to the Pauli principle the fermions must be in states with different projections of their i-spins. There are exactly $2i+1$ such different states and hence, if $2N > 2i+1$ the antisymmetric states must vanish. Since the fermion state considered should belong to an *irreducible* representation, vanishing of one state implies that *all* states of that irreducible representation vanish for $N > (2i+1)/2 = \frac{1}{3}\Omega$. Beyond the middle of the shell, pairs of nucleon *holes* are coupled to S and D pairs. Thus, for nucleon numbers between $2\Omega/3$ and $4\Omega/3$ all states of the $(0,2N)$ irreducible representation vanish due to the Pauli principle.

In the Ginocchio $O(6)$ model fermion states of all fully symmetric irreducible representations of $O(6)$ with given N ($\leq \frac{1}{2}\Omega$) are present. Hence, all these states can be mapped on boson states which belong to fully symmetric $O(6)$ irreducible representations with given N (i.e. $\sigma = N, N-2, \ldots, 1$ or 0). In the $SU(3)$ model, however, the Pauli principle completely eliminates states of the $(0,2N)$ irreducible representations for $N > \frac{1}{3}\Omega$. In more realistic cases, the effect of the

Pauli principle may be more complicated. It is difficult to imagine that there are always fermion states constructed with S^+ and D^+_μ operators which are in one to one correspondence with all s, d boson states for $N \le \Omega/2$. On the other hand, it could be expected that lack of exact correspondence between fermion and boson states will show up only for rather large numbers of n_d pairs with $J = 2$.

It is worth while to point out that the fermion states with $SU(3)$ symmetry in the Ginocchio model are quite different from the states in the Elliott $SU(3)$ scheme described in Section 30. In the latter, the shell model subspace is that of an oscillator major shell and hence the orbits have spins $\frac{1}{2}, \frac{3}{2}, \ldots, j$. For $i > \frac{3}{2}$ the orbits in the $SU(3)$ Ginocchio model have spins $i-1, i, i+1$ only and for all of them $j > \frac{1}{2}$. In the p-shell where the $i = \frac{1}{2}$ spin may be identified with the real intrinsic spin and the $k = 1$ spin with the orbital angular momentum, both $SU(3)$ schemes coincide. In the case $i = \frac{3}{2}$ the orbits have the same spins, $\frac{1}{2}, \frac{3}{2}, \frac{5}{2}$, as the orbits in the $2s, 1d$ oscillator shell but, nevertheless, the two $SU(3)$ schemes are different. In both cases the nucleon states are eigenstates of the quadrupole interaction $\mathbf{Q} \cdot \mathbf{Q}$ but the quadrupole operators are different. The quadrupole operator in Elliott's $SU(3)$ scheme is $r^2 Y_{2m}(\theta\phi)$ whereas in the model described here it is given by (37.35). In the $J = 2$ state, $D^+|0\rangle$, of the Elliott scheme, the d^2 component is proportional to the ${}^1D_2$ state ($S = 0$, $L = 2$, $J = 2$) which, according to Table 9.1 is given by

$$(\sqrt{12}|(\tfrac{5}{2})^2 J = 2\rangle - \sqrt{6}|\tfrac{5}{2}\tfrac{3}{2} J = 2\rangle + \sqrt{7}|(\tfrac{3}{2})^2 J = 2\rangle)/5$$

On the other hand, in the Ginocchio $SU(3)$ scheme we obtain from (37.40) the expansion within the d^2 configuration to be

$$(\sqrt{21}|(\tfrac{5}{2})^2 J = 2\rangle + 3\sqrt{7}|\tfrac{5}{2}\tfrac{3}{2} J = 2\rangle - 4|(\tfrac{3}{2})^2 J = 2\rangle)/5$$

The Ginocchio models described above are simple and elegant. They furnish an example where a fermion subspace may be *exactly* mapped onto a boson space. Thus, such a mapping is possible in principle and does not violate any mathematical or physical laws. The application of these models, however, to proton or neutron configurations in semi-magic nuclei is highly questionable. This is due not only to the S^+ operators in these models including only certain orbits with the same parity and equal α_j coefficients. In Sections 23 and 24 we saw that a realistic S^+ operator, as well as a D^+_M operator, must include *all* orbits in a major shell, with both parities, and that the α_j coefficients cannot be all equal. The main objection against the actual use of these models is that their states which can be simply mapped onto boson states belong to irreducible representations of the $O(6)$ or $SU(3)$ groups. Such states may be eigenstates of Hamiltonians with $O(6)$ or $SU(3)$ symmetry. Such symmetries may be present in spectra of nuclei with both valence protons and neutrons but not in semi-magic nuclei. In the latter the coupling scheme is that of seniority or generalized seniority. Ground states of actual semi-magic nuclei are well described by $(S^+)^N|0\rangle$ states which are definitely not eigenstates of $O(6)$ or $SU(3)$ Hamiltonians. The appearance of rotational or $SU(3)$ spectra as well as $O(6)$ spectra is due to the proton-neutron interaction in the shell model Hamiltonian as will be discussed in the next section.

The mapping of fermion states with good seniority or generalized seniority onto boson states is simple only for $v = 0$ and $v = 2$ states. In other cases, as described above in detail, the fermion states must first be orthogonalized. The boson model which is obtained in this way has a $U(5)$ symmetry rather than the $O(6)$ or $SU(3)$ ones. The most general Hamiltonian with $U(5)$ symmetry is given by (33.29) with $\tilde{v}_0 = 0$ and $\tilde{v}_2 = 0$. We can make use of the conservation of N and express it as a generalization of (37.13) by

$$\boxed{\begin{aligned} &V_0N + \tfrac{1}{2}WN(N-1) + (V_2 - V_0)(d^+\cdot\tilde{d}) \\ &\quad + \tfrac{1}{2}C_0[(d^+\cdot\tilde{d})^2 - (d^+\cdot\tilde{d})] \\ &\quad + C_2(d^+\times d^+)^{(2)}\cdot(\tilde{d}\times\tilde{d})^{(2)} + C_4(d^+\times d^+)^{(4)}\cdot(\tilde{d}\times\tilde{d})^{(4)} \end{aligned}} \tag{37.43}$$

In the special case of the Hamiltonian (37.13), $C_2 = C_4 = 0$ and $C_0 = -W$. The eigenvalues of the boson Hamiltonian are equal to the eigenvalues of the shell model Hamiltonian of states with generalized seniorities $v = 0$ and $v = 2$. For states with $n_d > 2$ of actual Hamiltonians describing semi-magic nuclei, it is impossible to make any statements because not much is known about levels with higher (generalized) seniorities.

We will now consider matrix elements of single nucleon operators and corresponding boson operators. In the Ginocchio models the generators of $O(6)$ or $SU(3)$ are either single fermion operators, or single boson operators whose matrix elements are equal to those of the fermion operators taken between corresponding states. In the case of eigenstates in the seniority scheme or states with generalized seniority, the situation is more complicated. Let us first recall the recursion relations (19.37), (19.40) and (19.44) for matrix elements in the seniority scheme. As explained in Section 22, these relations hold for hermitean tensor operators also in the quasi-spin scheme provided $2j+1$ is replaced by $2\Omega = \sum(2j+1)$. These relations can be thus expressed as

$$\langle nv'J'M' \mid T_\kappa^{(k)}|nvJM\rangle = \delta_{vv'}\langle vvJ'M'|T_\kappa^{(k)}|vvJM\rangle \qquad k \text{ odd} \tag{37.44}$$

$$\langle nvJ'M'|T_\kappa^{(k)}|nvJM\rangle = \frac{\Omega-n}{\Omega-v}\langle vvJ'M'|T_\kappa^{(k)}|vvJM\rangle \qquad k \text{ even} \tag{37.45}$$

$$\begin{aligned}\langle nv+2,J'M'|T_\kappa^{(k)}|nvJM\rangle &= \sqrt{\frac{(n-v)(2\Omega-n-v)}{2(2\Omega-2-2v)}} \\ &\quad\times \langle v+2,v+2,J'M'|T_\kappa^{(k)}|v+2,vJM\rangle \\ &\qquad k \text{ even}\end{aligned} \tag{37.46}$$

We can now compare these expressions and their dependence on n with matrix elements of single boson operators. We look first at the generators of $U(5)$ namely

$$(d^+ \times \tilde{d})_\kappa^{(k)} \qquad k = 0,1,2,3,4 \tag{37.47}$$

By their structure, they do not have non-vanishing matrix elements between states with different n_d numbers. Moreover, since they do not include the operators s^+, s, their matrix elements are independent of n_s. This behavior is in agreement with (37.44) for $k = 1$ and $k = 3$ but not with (37.45) for $k = 2$ and $k = 4$. The $k = 0$ operator $(d^+ \cdot \tilde{d})$ has the eigenvalues n_d which correspond to $2v$. Although v is a good quantum number in the seniority scheme, there is no fermion operator with eigenvalues v. The definition of v in the fermion case is through the eigenvalues of the pairing operator as explained in Section 19.

Let us now calculate matrix elements of the only boson operator among the $U(6)$ generators with non-vanishing matrix elements between states with different n_d values. This is the operator

$$s^+\tilde{d}_\mu + d_\mu^+ s \tag{37.48}$$

A normalized state with given n_d and n_s can be expressed due to (34.30) by

$$\mathcal{N}_{n_d}(n_s!)^{-1/2}(s^+)^{n_s}(d^+)^{n_d}_{\gamma JM}|0\rangle \tag{37.49}$$

The matrix elements of (37.48) between the state (37.49) and a state with $n_d + 1$ are equal to

$$\begin{aligned}
&\mathcal{N}_{n_d+1}\mathcal{N}_{n_d}(n_s!)^{-1/2}((n_s-1)!)^{-1/2}\langle 0|d^{n_d+1}_{\gamma' J' M'}s^{n_s-1}d_\mu^+ s(s^+)^{n_s}(d^+)^{n_d}_{\gamma JM}|0\rangle \\
&\quad = (n_s!)^{-1/2}((n_s-1)!)^{-1/2}\langle 0|s^{n_s}(s^+)^{n_s}|0\rangle\mathcal{N}_{n_d+1}\mathcal{N}_{n_d} \\
&\quad\quad \times \langle 0|d^{n_d+1}_{\gamma' J' M'}d_\mu^+(d^+)^{n_d}_{\gamma JM}|0\rangle \\
&\quad = \sqrt{n_s}\mathcal{N}_{n_d+1}\mathcal{N}_{n_d}\langle 0|d^{n_d+1}_{\gamma' J' M'}d_\mu^+(d^+)^{n_d}_{\gamma JM}|0\rangle
\end{aligned} \tag{37.50}$$

Thus, the matrix element of (37.48) between states with n_s, n_d and $n_s - 1, n_d + 1$ is equal to the matrix element between such two states with $n_s = 1$ and $n_s = 0$ multiplied by

$$\sqrt{n_s} = \sqrt{N - n_d} = \sqrt{\frac{n-v}{2}} \tag{37.51}$$

The last equality in (37.51) is obtained by identifying n_d with $2v$. The r.h.s. of (37.51) is equal to the first factor on the r.h.s. of (37.46). When $n, v \ll 2\Omega$, (37.51) is approximately equal to the factor in (37.46) but due to the Pauli principle there is an additional factor

$$\sqrt{(2\Omega - n - v)/(2\Omega - 2 - 2v)} = \sqrt{(\Omega - N - n_d)/(\Omega - 1 - 2n_d)}$$

We see that the single boson operators (37.48) with $k = 2,4$ cannot reproduce the dependence on n of the fermion operators. Even between states with seniorities $v = 0$ and $v = 2$ where the correspondence is simple, the single boson operators must be modified. In view of (37.45) we must make the seniority conserving part of the fermion $k = 2$ and $k = 4$ operators correspond to the boson operators

$$\frac{\Omega - 2N}{\Omega - 2n_d}(d^+ \times \tilde{d})^{(k)}_\kappa \qquad k = 2,4 \tag{37.52}$$

The expression (37.52) holds for $N \leq \Omega/2$. Beyond the middle of the shell, the number N of *hole* pairs is equal to $(2\Omega - n)/2$. Hence, the coefficient of $(d^+ \times \tilde{d})^{(k)}_\kappa$ in (37.52) becomes equal to $(\Omega - n)/(\Omega - 2n_d) = (2N - \Omega)/(\Omega - 2n_d)$. Due to this coefficient, the matrix element of (37.52) changes its sign beyond the middle of the shell, as the fermion matrix element (37.45). The seniority changing part of the $k = 2$ single fermion operator must, in view of (37.46) and (37.51), correspond to the boson operator

$$\boxed{d^+_\mu s\sqrt{\frac{\Omega - N - \hat{n}_d}{\Omega - 1 - 2\hat{n}_d}} + \sqrt{\frac{\Omega - N - \hat{n}_d}{\Omega - 1 - 2\hat{n}_d}}s^+\tilde{d}_\mu} \tag{37.53}$$

The coefficient in (37.46) is symmetric between particles and holes. Hence, the coefficients in (37.53) are the same in the two halves of the shell.

The operators (37.52) and (37.53) are no longer single boson operators. The coefficients of the single boson terms are functions of the operator $\hat{n}_d = (d^+ \cdot \tilde{d})$. The dependence on N and n_d in (37.52) and (37.53) is simple. It is based, however, on the relations (37.45) and (37.46) which hold in the quasispin scheme with equal α_j coefficients in S^+. In the realistic case of those coefficients not being equal, the dependence of the corresponding boson operators on N and n_d could be much more complex. Some calculations (Pittel et al. 1982) demonstrate how complicated and different from the simple dependence of (37.45) and (37.46) the actual dependence on n could be, even for states with generalized seniority $v = 0$ and $v = 2$. In the next section we will mention some of the approximations that have been made for the corresponding operators in the boson model.

We started by looking for a shell model ("microscopic") basis of the interacting boson model. We were naturally led into a model in which certain states and operators of identical nucleons are mapped onto states and operators of s- and d-bosons. This implies a model with *two kinds* of bosons s_π and d_π proton bosons and s_ν and d_ν neutron bosons. This model is not very useful for semi-magic nuclei. The only well known states in those nuclei are the $J = 0$ ground states and $J = 2$ first excited states. The eigenvalues of those states were calculated in Section 23 and Section 24. The boson model may be used for calculating energies of higher excited states but very little is known experimentally about them. The boson description, however, turns out to be very useful for describing levels of nuclei where there are both valence protons and neutrons. Such nuclei are considered in the next section. The shell model Hamiltonian in these nuclei may be written as

$$H = H_\pi + H_\nu + V_{\pi\nu} \tag{37.54}$$

In (37.54), H_π is the Hamiltonian of the valence protons including single proton energies and two-body interactions and H_ν is that Hamiltonian of the valence neutrons. The eigenstates of $H_\pi + H_\nu$ are proton eigenstates of H_π multiplied by neutron eigenstates of H_ν. The proton-neutron interaction $V_{\pi\nu}$ admixes these products of proton and neutron eigenstates. In looking for the eigenstates and eigenvalues of H, including the proton-neutron interaction, the boson model for H_π and H_ν as well as for $V_{\pi\nu}$ becomes extremely useful as we will see in the following section.

38

Valence Protons and Neutrons. The S–D Space

The most complicated nuclear systems occur in nuclei with both valence protons and valence neutrons. In the interesting cases, those valence nucleons occupy several j-orbits. In actual cases, the number of two-nucleon matrix elements required to specify the effective interaction, becomes very large. The order of the interaction matrices to be diagonalized becomes so large that exact diagonalization by present day computers is practically impossible. At the end of Section 24 an example of such a situation was presented. In ${}^{154}_{62}Sm_{92}$ the 12 valence protons may occupy the 5 orbits in the 50–82 major shell and the 10 valence neutrons may occupy the 6 orbits in the next major shell, 82–126. To determine the effective interaction in this range of nuclei, 1307 two nucleon matrix elements are needed, 290 of which are diagonal. There is no consistent way to determine them from experimental data. The matrix to be diagonalized for obtaining the positions of $J = 0$ states with positive parity in ${}^{154}Sm$ is of order 41,654,193,517,797. The one for the $J = 2$ states with positive parity is even bigger, its order is 346,132,052,934,889. It is clear that the straightforward application of the shell model cannot be made in such cases. A very drastic truncation scheme must be used within the

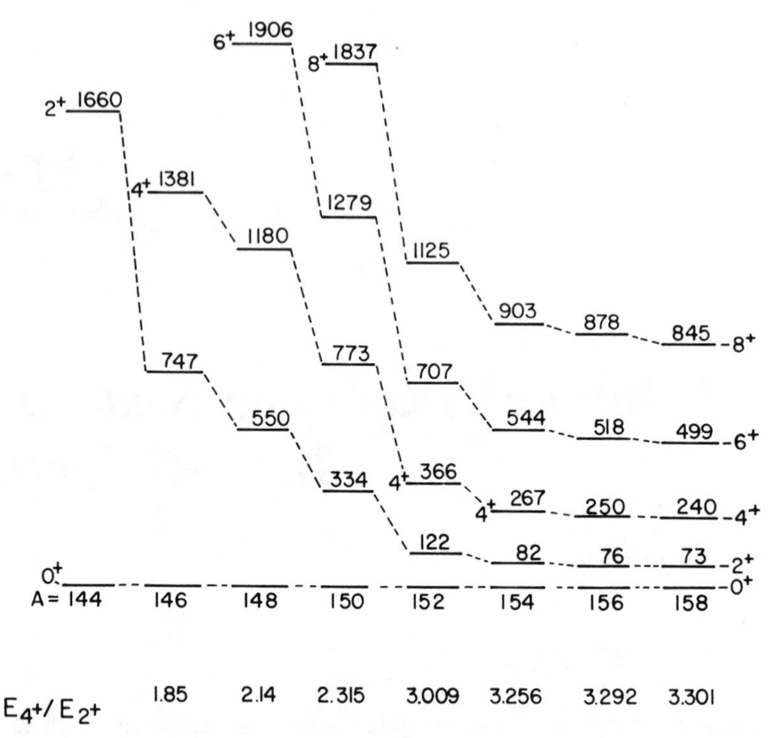

FIGURE 38.1. *Dependence of spectra of even* Sm *isotopes on the number of valence neutrons (energies in* keV*).*

framework of the shell model to reduce considerably the dimensions of the problem.

A quick glance at the positive parity low lying energy levels experimentally observed in ^{154}Sm indicates that the required truncation should not just lead to a numerically manageable problem. The observed levels of ^{154}Sm in Fig. 33.1 are grouped into a set of rather clear "rotational bands". These bands are simply described by the collective model which is very successful for such nuclei. These simple features of low lying levels suggest that there should be a simple scheme within the shell model that will reproduce them. All other 10^{14} higher states with $J = 0$ or $J = 2$ are practically of little interest. They cannot be determined experimentally and most of them lie so high that the description in terms of the valence shell becomes very doubtful. If we look at spectra of lighter Sm isotopes, Fig. 38.1, we see a gradual transition from the semi-magic ^{144}Sm, whose states are well described by generalized seniority, to ^{154}Sm with a rotational spectrum. A good truncation scheme should also be able to reproduce this transition as a function of the number of valence neutrons.

As remarked above in Section 24, the scheme of generalized seniority introduces a tremendous simplification of the shell model problem. Could this scheme be applied also to nuclei with valence protons and neutrons? To answer this question we may look at simple cases where both valence protons and neutrons are in the same j-orbit. This point was considered in Section 27. As mentioned there, if the Hamiltonian is diagonal in the seniority scheme, spacings of energy levels with the same values of isospin T and seniority quantum numbers v, t are independent of n and T. In the case of identical valence nucleons this feature is observed in many cases. We can try to see whether it holds for valence protons and neutrons. Let us look at the $1f_{7/2}$-orbit. We saw that $(1f_{7/2})^n$ configurations give only a fair description of nuclei with N or Z between 20 and 28. Nevertheless, let us compare the positions of the $T = 1$ levels with $J = 2,4,6$ in ^{42}Ca (or ^{42}Sc) and the $T = 3$ levels with $J = 2,4,6$ in $^{48}_{21}$Sc$_{27}$. Both sets of levels have the same seniority quantum numbers $v = 2$, $t = 1$. In ^{42}Sc, the $J = 0$ ground state has also $T = 1$ but in $^{48}_{21}$Sc$_{27}$ the $J = 0$ state is the analog of the ground state of $^{48}_{20}$Ca$_{28}$. It has isospin $T = 4$ and lies much higher than the $T = 3$ states. The comparison is made in Fig. 27.1 and it is seen that not only are spacings not equal, they are almost exactly reversed. This is rather general as seen also from Fig. 27.2 where spacings between $J = 2,4,6,8$ levels of $(1g_{9/2})^n$ configurations are presented. In ^{92}Mo these spacings are those of the $v = 2$, $t = 1$ levels of the $(1g_{9/2})^{12}$ configuration of protons and neutrons with $T = 4$. In $^{88}_{39}$Y$_{49}$ these $v = 2$, $t = 1$ levels belong to the $(1g_{9/2})^{10}$ configuration with isospin $T = 5$. Thus, seniority is badly broken in j^n configurations of protons and neutrons for states with $T < n/2$. Some of the states, like those considered above, have definite seniority quantum numbers, irrespective of the Hamiltonian. They are the only states with given T and J in the configuration considered. The fact that level spacings are far from being the same, proves that the Hamiltonian of the j^n configurations considered for $T < n$, is far from diagonal in the seniority scheme.

There is another indication that neither seniority nor generalized seniority give a good description of nuclei considered here. Hamiltonians which are diagonal in the seniority scheme yield for even even nuclei energy levels with $J = 0,2,\ldots$ whose spacings are independent of n. This feature holds also in the case of generalized seniority as demonstrated by constant 0–2 spacings in semi-magic nuclei. The significant decrease of this spacing as more nucleons are added, as in Fig. 38.1, is a clear signature of the transition to rotational spectra. In the collective model it is assumed that the rotation of the deformed shape is adiabatic. This implies that rotational levels lie lower than other excitations like vibrational or intrinsic ones. This is the case in spectra of diatomic molecules where rotational levels are much lower than vibrational and electronic excitations. Hence, the lower the $J = 2$ excitation, the better the description as a rotational spectrum. It seems that seniority breaking interactions must be responsible for the lowering of the $J = 2$ level and the transition to rotational spectra.

As long as there are valence nucleons of one kind only, the eigenstates of the shell model Hamiltonian have definite generalized seniority. Once valence nucleons of the other kind are added, the seniority breaking proton neutron interaction strongly affects the structure of levels. Whenever several valence protons and valence neutrons are present, the proton-neutron interaction dominates and leads in many cases to collective rotational spectra. *The whole rich range of nuclear level schemes is the result of the competition between interactions of identical nucleons and the seniority breaking interactions between protons and neutrons.*

In Section 31 we noted that the $T = 1$ part of the nuclear interaction, apart from the attractive pairing term, is repulsive on the average. It cannot create the central potential well in which nucleons move. On the other hand, the $T = 0$ part of the nuclear interaction is attractive and leads to an attractive central potential well. It is therefore not surprising that the $T = 0$ part appearing in the proton-neutron interaction, could lead to *deformation* of the central potential well thereby paving the way to rotational spectra.

There are many interactions which are not diagonal in the seniority scheme. As explained in Section 26, irreducible tensor operators with odd ranks k are diagonal in the seniority scheme of j^n configurations for all values of T. Hence, odd tensor interactions, as well as a monopole term with $k = 0$, are diagonal in the seniority scheme. Even tensor interactions, with $k > 0$, are usually not diagonal in the seniority scheme. The only linear combination of scalar products of even tensor operators which is diagonal in the seniority scheme for all values of T is (26.31). Thus, we expect that the effective interaction between j-nucleons which breaks seniority, contains in its expansion scalar products of even tensor operators. This expansion may be determined from the interaction energies in the two nucleon states

$V_J = V(j^2J) = \langle j^2JM|V|j^2JM\rangle$ with even values of J ($T = 1$) as well as with odd J ($T = 0$).

There are no cases where all these interaction energies could be determined in a consistent way. As remarked above, states with $T < n/2$ are not accurately described by jj-coupling wave functions. Yet, to obtain an idea of the tensor expansion of the effective interaction, let us consider energy levels in some nuclei, even though j^n configurations may be strongly perturbed. We then use the relation (12.13) for the j-orbit to obtain the expansion coefficients which are given by

$$F^k(jjjj) = (2k+1)\sum_J (-1)^{2j+J}(2J+1)\begin{Bmatrix} j & j & J \\ j & j & k \end{Bmatrix} V_J \tag{38.1}$$

The simplest case is in the $1f_{7/2}$ orbit. The values of V_J relative to V_0 may be taken from the excited states with spin J above the $J = 0$ ground state of ${}^{42}_{21}Sc_{21}$. The ground state energy V_0 may be determined from the following binding energy differences

$$\text{B.E.}(^{42}\text{Sc}) - \text{B.E.}(^{40}\text{Ca}) - (\text{B.E.}(^{41}\text{Sc}) - \text{B.E.}(^{40}\text{Ca}))$$
$$- (\text{B.E.}(^{41}\text{Ca}) - \text{B.E.}(^{40}\text{Ca}))$$

Reversing the sign of this energy we obtain $V_0 = -3.19$ MeV. The other V_J are then obtained by adding this V_0 to their excitation energies. The positions of the $T = 1$, $J = 4$ and $J = 6$ levels may be taken from the analog levels in ^{42}Ca. Substituting the values of the V_J thus obtained in (38.1) we obtain the values of the expansion coefficients F^k listed in Table 38.1 (column 2). We see that the F^k coefficients with even k values are large but their values are very far from following the $2k + 1$ behavior which is seen in (26.31). We notice that the coefficient of the attractive quadrupole term F^2 is the largest in absolute value and it is even larger than the coefficient of the monopole term F^0.

As indicated in Section 21, the $(1f_{7/2})^n$ configurations of either protons or neutrons ($T = n/2$) are strongly perturbed. This fact is even more pronounced in configurations with $T < n/2$. If level spacings taken from ^{42}Sc are adopted, the calculated levels of the particle-hole configuration in ${}^{48}_{21}Sc_{27}$ are only in rough agreement with experiment. Comparing the two nucleon levels of ^{42}Sc to the two hole

TABLE 38.1 *Coefficients of the tensor expansion of* $1f_{7/2}^2$ *effective interactions (in* MeV*)*

k	F^k		
	From ^{42}Sc	From *MBZ*	From ^{42}Sc and ^{48}Sc
0	−11.48	−9.54	−12.44
1	−1.27	−2.89	−3.59
2	−12.97	−14.93	−14.59
3	−5.34	−4.57	−5.18
4	−10.75	−10.94	−10.30
5	−8.31	−8.15	−6.64
6	−8.33	−4.70	−9.32
7	−3.09	+1.10	−5.64

spectrum of ^{54}Co we see rather large variations in the positions of $J = 2, 4, 6$ levels. Still, the main feature of many nuclei with $20 \leq Z$, $N \leq 28$ have been fairly reproduced by McCullen, Bayman and Zamick (1964) using an effective charge independent interaction in $(1f_{7/2})^n$ configurations.

A nice example of such a feature is the existence of a "spin gap" leading to an isomeric state with $T = \frac{1}{2}$ of $(1f_{7/2})^3$ configurations. The state with $T = \frac{1}{2}$, $J = \frac{19}{2}$ may be obtained from a principal parent with $T_0 = 1$, $J_0 = 6$. Its interaction energy may be calculated from (15.21) where the c.f.p. given by (15.34) should be used. We obtain the resulting interaction energy to be given by

$$\tfrac{3}{2}V(1f_{7/2}^2 T = 1,\ J = 6) + \tfrac{3}{2}V(1f_{7/2}^2 T = 0, J = 7)$$

The state with $T = \frac{1}{2}$, $J = \frac{17}{2}$ may be also obtained from the same principal parent. For that state, the calculated interaction energy is

$$1.5V(1f_{7/2}^2 T = 1, J = 6) + .7308V(1f_{7/2}^2 T = 0, J = 5)$$

$$+ .7692V(1f_{7/2}^2 T = 0, J = 7)$$

The value of $V(1f_{7/2}^2 T = 0, J = 7)$ is considerably more attractive than $V(1f_{7/2}^2 T = 0, J = 5)$. Hence, the state with $T = \frac{1}{2}$, $J = \frac{17}{2}$ is considerably higher than the $T = \frac{1}{2}$, $J = \frac{19}{2}$ state. The two nucleon matrix elements of McCullen, Bayman and Zamick put the former about 1 MeV above the latter. The calculated position of the $T = \frac{1}{2}$, $J = \frac{19}{2}$ level is at 3.64 MeV whereas the $T = \frac{1}{2}$, $J = \frac{17}{2}$ state lies at 4.62 MeV above the $T = \frac{1}{2}$, $J = \frac{7}{2}$ ground state. The interaction energies in the states with $T = \frac{1}{2}$ and $J = \frac{15}{2}, \frac{13}{2}$ contain also contributions of $V(1f_{7/2}^2 T = 1, J = 4)$ which is more attractive than $V(1f_{7/2}^2 T = 1, J = 6)$ but they contain a smaller amount of $V(1f_{7/2}^2 T = 0, J = 7)$ which makes them lie higher than the $T = \frac{1}{2}$, $J = \frac{19}{2}$ state. The calculated positions of the $T = \frac{1}{2}$, $J = \frac{15}{2}$ state is 3.71 MeV and of the $T = \frac{1}{2}$, $J = \frac{13}{2}$ state is 3.85 MeV above the ground state. Thus, a "spin gap" is obtained between the $J = \frac{19}{2}$ state and the $T = \frac{1}{2}$, $J = \frac{11}{2}$ state which is calculated to lie at 2.44 MeV above the ground state. This makes the $J = \frac{19}{2}$ state a long lived isomeric state. It is actually observed in the $(1f_{7/2})^{-3} \equiv (f_{7/2})^3$ configuration in ^{53}Fe (the experimental situation in ^{43}Sc is less clear cut). The $T = \frac{1}{2}$, $J = \frac{19}{2}$ level is observed in that nucleus at 3.04 MeV, the $J = \frac{17}{2}$ has not been found and the $J = \frac{15}{2}$ state and $J = \frac{13}{2}$ states are at 3.46 MeV and 3.18 MeV respectively. The experimentally observed $T = \frac{1}{2}$, $J = \frac{11}{2}$ level lies at 2.34 MeV above the $J = \frac{7}{2}$ ground state (the $T = \frac{1}{2}$, $J = \frac{9}{2}$ level calculated to lie at 1.68 MeV is observed at 1.33 MeV). The coefficients of the tensor expansion of the effective interaction of McCullen, Bayman and Zamick (MBZ) are given in the third column of Table 38.1. The F^k coefficients in the fourth column of Table 38.1 were obtained from an interaction which roughly fits the measured levels of both ^{42}Sc and ^{48}Sc. We see that these three interactions show essentially the same behavior of the expansion coefficients.

The only other case of valence protons and valence neutrons occupying the same orbit is $j \equiv 1g_{9/2}$. Levels of $(1g_{9/2})^n$ configurations of this kind are found in nuclei with $Z > 38$ and $N < 50$. The experimental information is not plentiful and, as explained in Section 21, the valence protons occupy both $1g_{9/2}$ and $2p_{1/2}$ orbits. These configurations were analyzed by Serduke et al. (1976) and by Gross and Frenkel (1976). The coefficients of the tensor expansion may be obtained from the effective interaction determined by these authors by using (38.1). The resulting values of the F^k are listed in Table 38.2. The interaction taken from Serduke et al. (1976) is the one denoted in their paper by "$T = 0$ fit". From Table 38.2 it is seen that the quadrupole-quadrupole interaction (with $k = 2$) is the most attractive one. It is larger in absolute value even than the monopole term ($k = 0$).

In many nuclei with rotational spectra, valence protons and neutrons are in different orbits. In that case the interaction between a

TABLE 38.2 *Coefficients of the tensor expansion F^k of $1g_{9/2}^2$ effective interactions (in* MeV)

$k =$	0	1	2	3	4	5	6	7	8	9
Serduke Lawson	−5.39	−1.56	−11.05	−3.27	−8.75	−4.90	−6.89	−5.65	−5.46	−1.78
Gross Frenkel	−5.90	−3.12	−12.03	−3.84	−8.63	−5.29	−6.34	−4.36	−4.66	−.23

TABLE 38.3 *Coefficients of the tensor expansion F^k of the $1d_{3/2}$ proton–$1f_{7/2}$ neutron effective interaction (in* MeV)

k	0	1	2	3
F^k	−5.89	1.51	−4.93	−.64

proton in a j-orbit and a neutron in a j'-orbit (the neutron j-orbit is completely filled) is given by (10.7) as

$$V(j_\pi j'_\nu J) = \tfrac{1}{2}[\langle jj'T = 1, J|V|jj'T = 1, J\rangle + \langle jj'T = 0, J|V|jj'T = 0, J\rangle] \tag{38.2}$$

The tensor expansion of this proton-neutron interaction has coefficients F^k given by (12.13) as

$$F^k(jj'jj') = (2k+1)\sum_J (-1)^{j+j'+J}(2J+1)\begin{Bmatrix} j & j' & J \\ j' & j & k \end{Bmatrix} V(j_\pi j'_\nu J) \tag{38.3}$$

The simplest example of such an interaction was given in Section 1 where $j_\pi \equiv 1d_{3/2}$ and $j'_\nu \equiv 1f_{7/2}$. Differences of the corresponding two body matrix elements of the effective interaction can be taken from the spectrum of ${}^{38}_{17}\mathrm{Cl}_{21}$. The interaction energy in the $J = 2$ ground state is given by binding energy differences as

$$\text{B.E.}({}^{38}_{17}\mathrm{Cl}_{21}) - \text{B.E.}({}^{36}_{16}\mathrm{S}_{20}) - [\text{B.E.}({}^{37}_{17}\mathrm{Cl}_{20}) - \text{B.E.}({}^{36}\mathrm{S})] - [\text{B.E.}({}^{37}_{16}\mathrm{S}_{21}) - \text{B.E.}({}^{36}\mathrm{S})]$$

and is equal to −1.8 MeV. The matrix elements substituted in (38.3) yield the coefficients F^k listed in Table 38.3. Also in this case the quadrupole term is large and attractive.

TABLE 38.4 *Coefficients of the tensor expansion* F^k *of the* $1g_{9/2}$ *proton–* $2d_{5/2}$ *neutron effective interaction (in* MeV*)*

k	0	1	2	3	4	5
F^k	−3.33	−1.03	−2.82	−.72	−1.38	−1.11

TABLE 38.5 *Coefficients of the tensor expansion* F^k *of the* $1h_{9/2}$ *proton–* $2g_{9/2}$ *neutron effective interaction (in* MeV*)*

k	0	1	2	3	4	5	6	7	8	9
F^k	−2.47	.18	−2.49	−.05	−1.88	−.51	−1.10	−.80	−.64	−1.20

Another example of a proton-neutron interaction is offered by protons in the $1g_{9/2}$ orbit and neutrons in the $2d_{5/2}$ orbit. This case was described in Section 31. The coefficients of the tensor expansion of the interaction, obtained by using (38.3) are listed in Table 38.4. The same feature is seen here as in Table 38.3. The quadrupole term is much more attractive than the term with $k = 4$ and other $k \neq 0$ terms. The predominant role of the quadrupole term in the proton-neutron interaction is seen also in another case discussed in Section 31. Differences of matrix elements in the $1h_{9/2}$ proton– $2g_{9/2}$ neutron configuration may be obtained from levels of ${}^{210}_{83}\mathrm{Bi}_{127}$. The interaction energy in the $J = 1$ ground state, taken from binding energy differences is −.668 MeV. The expansion coefficients, obtained by using (38.3) are listed in Table 38.5.

A very drastic simplification of the proton-neutron interaction emerges if we replace it by a quadrupole-quadrupole interaction (in addition to the monopole interaction). The arguments for this approximation are based on the larger size of the quadrupole term and the empirical observation that the quadrupole degree of freedom is the main ingredient of the collective model. The s,d interacting boson model shares this feature with the collective model. As we shall see in the following, taking for the proton-neutron interaction the quadrupole term leads in a natural way to the s,d boson model. If the valence protons and neutrons are in different orbits, j and j', the quadrupole interaction is assumed to replace only the $T = 0$ part in (38.2). We saw that $T = n/2$ states are well described by generalized seniority. This essentially implies that the matrix elements $\langle jj'T = 1J|V|jj'T = 1J\rangle$ may be reproduced by an odd tensor interaction and a monopole ($k = 0$) interaction.

If valence protons and neutrons are in the *same* j-orbit, matrix elements in both $T = 0$ (J odd) and $T = 1$ (J even) states are needed to uniquely determine the tensor expansion as in (38.1). As seen in Table 38.2 the tensor expansion of the j^2 interaction (for $j \equiv 1g_{9/2}$) contains a large quadrupole term and terms with $k = 4, 6, 8$ as well. The question may be raised whether there is an apparent contradiction between this feature and the fact that the interaction is diagonal in seniority in $(1g_{9/2})^n$ configurations of identical nucleons ($T = n/2$). The answer is rather simple. For $T = n/2$ states the seniority breaking matrix elements of the $k = 2$ term are exactly cancelled by similar matrix elements of the $k = 4, 6, 8$ terms in the interaction. Thus, the two-nucleon interaction, restricted to $j^2T = 1$ states, may be expressed by an odd tensor interaction and a monopole ($k = 0$) term. This is definitely not the case for the full interaction acting on states with $T < n/2$. The linear combination whose coefficients are given in Table 38.2 or Table 38.1 breaks seniority in $T < n/2$ states in a major way. The most important seniority breaking term is the $k = 2$ quadrupole quadrupole interaction. Hence, we keep for j^2, $T = 0$ states, with J even, the interaction which is diagonal in seniority. For j^2, $T = 0$ states, with J odd, we replace the interaction by a quadrupole interaction (and a monopole term).

A simple example may show the importance of even tensor interactions in states with $T < n/2$. Consider the $J = 0$ states of the j^4 configuration with two protons and two neutrons. In the seniority scheme there are two states with $v = 0$, $t = 0$, one with isospin $T = 0$ and the other with $T = 2$. We may expand these states in the complete scheme furnished by coupling proton states j^2J_π with neutron states j^2J_ν. As explained in preceding sections, it is most convenient to use the isospin formalism but it is not mandatory. The expansion coefficients do not vanish only for $J_\pi = J_\nu$ and, as in the case discussed at the end of Section 15, are proportional to certain $4 \rightarrow 2$ c.f.p. of the $T = 0$ and $T = 2$ states. The $n \rightarrow n-2$ c.f.p. are given by (15.33), generalized to include isospins. The formula is

$$
\begin{aligned}
&[j^{n-2}(\alpha_2 T_2 J_2) j^2 (T'J')J|\}j^n \alpha TJ] \\
&\quad = \sum_{\alpha_1 T_1 J_1} [j^{n-2}(\alpha_2 T_2 J_2) j T_1 J_1|\}j^{n-1}\alpha_1 T_1 J_1] \\
&\quad \times [j^{n-1}(\alpha_1 T_1 J_1) j TJ|\}j^n \alpha TJ] \\
&\quad \times (-1)^{T_2+T+1+J_2+J+2j} \sqrt{(2T_1+1)(2T'+1)(2J_1+1)(2J'+1)} \\
&\quad \times \begin{Bmatrix} T_2 & \frac{1}{2} & T_1 \\ \frac{1}{2} & T & T' \end{Bmatrix} \begin{Bmatrix} J_2 & j & J_1 \\ j & J & J' \end{Bmatrix}
\end{aligned}
\tag{38.4}
$$

The only states of the j^3 configuration which appear in non-vanishing c.f.p. are the states with $v = 1$, $t = \frac{1}{2}$, $J_1 = j$ and isospins $T_1 = \frac{1}{2}$ and $T_1 = \frac{3}{2}$ respectively. Hence, the corresponding c.f.p. are equal to 1. Putting in (38.4) $n = 4$, $J = 0$, $J_1 = j$ as well as $T_2 = 1$, $J' = J_2$, we obtain by substituting the actual values of the $6j$-symbols, the following results. The $4 \rightarrow 2$ c.f.p. for $T = 2$ ($T_1 = \frac{3}{2}$) are equal to the $3 \rightarrow 2$ c.f.p. with the same value of J_2. The 4–2 c.f.p. for $T = 0$ ($T_1 = \frac{1}{2}$) are equal to the $3 \rightarrow 2$ c.f.p. with the same value of J_2. The $3 \rightarrow 2$ c.f.p. are given by (15.11) for $T = 2$ and by (15.36) for $T = 0$.

If the expansion of the two-nucleon interaction contains only odd tensors and a monopole term ($k = 0$), the state $j_\pi^2(J_\pi = 0)j_\nu^2(J_\nu = 0)$ is an eigenstate of Hamiltonian. There are no non-vanishing matrix elements connecting it with any other state since J_π and J_ν are both even. Since that state is a linear combination of the $T = 0$ and $T = 2$ states, with $v = 0$, $t = 0$, the charge independent interaction considered must have the same eigenvalue for these states. This is contrary to the rather large symmetry energy observed in nuclei. It follows that any acceptable interaction must contain in its expansion scalar products of even tensors. If seniority is a good quantum number for states with any T, the coefficients F^k with $k > 0$, even, must be proportional to $2k + 1$ as in (26.31). By using (26.31), this even tensor interaction may be replaced by a sum of an odd tensor interaction, a $k = 0$ tensor interaction and a term proportional to $T(T + 1) - \frac{3}{4}n$ (de-Shalit and Talmi 1963).

At the end of last section, the Hamiltonian for a system of valence protons and valence neutrons was expressed as

$$H = H_\pi + H_\nu + V_{\pi\nu} \tag{38.5}$$

In (38.5), H_π contains the kinetic energy of the valence protons, their interactions with the closed shells and their mutual interactions. The same terms of the neutrons are contained in H_ν. The interaction between protons and neutrons is contained in $V_{\pi\nu}$. For the sake of simplicity, let us consider cases in which valence protons and valence neutrons occupy different major shells. As explained above, all states of such systems have unique isospin given by $T = \frac{1}{2}|N - Z|$. Thus, no simplification is introduced by the isospin formalism and we may use as basis states those constructed by coupling all valance proton states with all valence neutron states. Let us characterize all states of valence protons with given value of J_π by a label α_π and all valence neutron states by $\alpha_\nu J_\nu$. If all valence protons or neutrons occupy the

j-orbit, the labels α_π or α_ν have been considered in preceding sections and may include the seniority quantum number. In more complicated cases, no simple labels exist and the number of different α_π or α_ν labels may reach more than 10^5. Given such bases

$$|\alpha_\pi J_\pi M_\pi\rangle \qquad |\alpha_\nu J_\nu M_\nu\rangle \tag{38.6}$$

all states of the combined system may be expressed in terms of the following complete basis of states

$$\boxed{\begin{aligned}&|\alpha_\pi J_\pi \alpha_\nu J_\nu JM\rangle\\ &\quad = \sum_{M_\pi M_\nu} (J_\pi M_\pi J_\nu M_\nu \mid J_\pi J_\nu JM)|\alpha_\pi J_\pi M_\pi\rangle|\alpha_\nu J_\nu M_\nu\rangle\end{aligned}} \tag{38.7}$$

If the bases (38.6) are chosen as the sets of eigenstates of H_π and H_ν, all non-diagonal elements of the submatrix of the Hamiltonian in the shell model space considered are those of the proton-neutron interaction $V_{\pi\nu}$. Even if $V_{\pi\nu}$ contains only quadrupole-quadrupole interactions, those non-diagonal matrix elements are very complicated. They depend on the strengths of the quadrupole interaction in the various orbits and on the configuration mixings in the eigenstates (38.6). Matrix elements in the basis (38.7) are greatly simplified if the proton-neutron quadrupole interaction can be expressed as

$$V_{\pi\nu} = \kappa(\mathbf{Q}_\pi \cdot \mathbf{Q}_\nu) \tag{38.8}$$

where κ is a negative constant number. This expression may arise if the relation

$$F^{(2)}(j_\pi j'_\nu j''_\pi j'''_\nu) = F_\pi(jj'')F_\nu(j'j''')$$

holds for all j-orbits in the valence shell considered. In Section 22 we noticed this feature in the quasi-spin scheme where all F^k coefficients were equal. In Section 30, when considering an effective quadrupole interaction leading to $SU(3)$ symmetry, the relation (38.8) holds as an operator equation. In that case $\mathbf{Q}_\pi$ or $\mathbf{Q}_\nu$ are given by $\sum r_i^2\mathbf{Y}_2(i)$.

Matrix elements of (38.8) in the scheme of states (38.7) are given due to (10.27) by

$$\boxed{\begin{aligned}&\kappa\langle\alpha_\pi J_\pi\alpha_\nu J_\nu JM|\mathbf{Q}_\pi\cdot\mathbf{Q}_\nu|\alpha'_\pi J'_\pi\alpha'_\nu J'_\nu JM\rangle\\ &\quad=\kappa(-1)^{J_\nu+J'_\pi+J}\begin{Bmatrix}J_\pi & J_\nu & J\\ J'_\nu & J'_\pi & 2\end{Bmatrix}\\ &\qquad\times(\alpha_\pi J_\pi\|\mathbf{Q}_\pi\|\alpha'_\pi J'_\pi)(\alpha_\nu J_\nu\|\mathbf{Q}_\nu\|\alpha'_\nu J'_\nu)\end{aligned}} \tag{38.9}$$

The number of matrix elements (38.9) is the same as in the general case. The information needed to write down these matrix elements is, however, very much reduced. It is only necessary to have the (reduced) matrix elements of the quadrupole operator $\mathbf{Q}_\pi$ between all proton states and those of $\mathbf{Q}_\nu$ between all neutron states. Still, the order of the matrices to be diagonalized has not decreased and their eigenstates must be expressed as a linear combination of very many basis functions. Looking at (38.9) we see that the important matrix elements of the proton-neutron interaction are those in which there are large matrix elements of the proton quadrupole operator $\mathbf{Q}_\pi$ and of the neutron operator $\mathbf{Q}_\nu$. This suggests a criterion for useful truncation schemes for the lowest eigenvalues. The only states (38.7) to be included in the diagonalization should be those in which the proton and neutron states have large $(\alpha_\pi J_\pi\|\mathbf{Q}_\pi\|\alpha'_\pi J'_\pi)$ and large $(\alpha_\nu J_\nu\|\mathbf{Q}_\nu\|\alpha'_\nu J'_\nu)$ reduced matrix elements.

In some simple models described in preceding sections, matrix elements of certain quadrupole operators have non-vanishing matrix elements only within certain sets of states. In such cases the truncation suggested above is *exact*. In Section 30 we saw that $\mathbf{Q}_\pi$ (and $\mathbf{Q}_\nu$) which are proportional to generators of the $SU(3)$ algebra have non-vanishing matrix elements only between states which are in the same irreducible representation of $SU(3)$. If H_π and H_ν are diagonal in the $SU(3)$ scheme, their lowest eigenstates belong to certain (λ_π,μ_π) and (λ_ν,μ_ν) irreducible representations of $SU(3)$. These irreducible representations furnish labels α_π and α_ν and the only non-vanishing matrix elements (38.9) with these labels are those for which also $\alpha'_\pi\equiv(\lambda_\pi,\mu_\pi)$ and $\alpha'_\nu\equiv(\lambda_\nu,\mu_\nu)$. In other words, the matrix whose elements are given by (38.9) is reduced into submatrices along the diagonal characterized by $(\lambda_\pi,\mu_\pi),(\lambda_\nu,\mu_\nu)$. Each subspace which defines such a submatrix is *completely decoupled* from all other states.

Thus, there is a tremendous reduction in the size of the matrices to be diagonalized. If the valence protons are in a high harmonic oscillator shell and the valence neutrons in a higher shell, the total number of $J = 0$ or $J = 2$ states may be about 10^{14} or higher. The number of states (38.7) with α_π, α_ν given by irreducible representations of $SU(3)$ would be of order 10^2 or 10^3. The lowest eigenstates are determined by the relative importance of H_π, H_ν and $V_{\pi\nu}$. If H_π, H_ν are large compared to $V_{\pi\nu}$, the lowest eigenstates are obtained from the lowest eigenstates of H_π and H_ν. If $V_{\pi\nu}$ is much larger than H_π, H_ν, the lowest eigenstates are determined by those $SU(3)$ representations that maximize the matrix elements (38.9).

Other cases in which a small subspace is exactly decoupled from the huge space of states of valence protons and neutrons occur in the Ginocchio models. If $\mathbf{Q}_\pi$ and $\mathbf{Q}_\nu$ in (38.8) are proportional to generators of $O(6)$ or $SU(3)$ respectively, matrix elements (38.9) vanish between states with α_π, α_ν and α'_π, α'_ν which label states in different irreducible representations of the appropriate group. As explained in Section 37, the lowest eigenstates of H_π, H_ν constructed from either $O(6)$ or $SU(3)$ generators are constructed by coupling appropriate S_π, D_π pairs and S_ν, D_ν pairs respectively. In these models the reduction in size from the huge matrices to those considered here is even more impressive than in Elliott's $SU(3)$ model.

In the general case, the α_j coefficients in S^+_π and S^+_ν are not equal. We saw that the $J = 0$ and $J = 2$ lowest eigenstates of H_π and H_ν are given by generalized seniority. In the absence of the proton neutron interaction, the lowest eigenstates of $H_\pi + H_\nu$ are thus equal to

$$(S^+_\pi)^{N_\pi} (S^+_\nu)^{N_\nu} |0\rangle \tag{38.10}$$

The state (38.10) is obtained by coupling the lowest $J_\pi = 0$ state of the valence protons with the lowest $J_\nu = 0$ of the valence neutrons. The strong proton neutron interaction has non-vanishing matrix elements between the state (38.10) and other states. To obtain them we may act on (38.10) with $\kappa \mathbf{Q}_\pi \cdot \mathbf{Q}_\nu$ and see which state is obtained. Using the lemma (23.8) we obtain

$$\begin{aligned}(\mathbf{Q}_\pi \cdot \mathbf{Q}_\nu)(S^+_\pi)^{N_\pi} (S^+_\nu)^{N_\nu} |0\rangle &= \sum (-1)^\mu Q_{\pi\mu} Q_{\nu,-\mu} (S^+_\pi)^{N_\pi} (S^+_\nu)^{N_\nu} |0\rangle \\ &= \sum (-1)^\mu N_\pi N_\nu (S^+_\pi)^{N_\pi - 1} (S^+_\nu)^{N_\nu - 1} [Q_{\pi\mu}, S^+_\pi][Q_{\nu,-\mu}, S^+_\nu]|0\rangle\end{aligned} \tag{38.11}$$

The higher commutators of $\mathbf{Q}_\pi$ (or $\mathbf{Q}_\nu$) with S_π^+ (or S_ν^+) vanish since the commutators on the r.h.s. of (38.11) are linear combinations of products of two fermion creation operators.

To simplify (38.11) we make a rather *drastic approximation*. We assume that the operators $\mathbf{Q}_\pi$ and $\mathbf{Q}_\nu$ which enter the definition (38.8) are proportional to the quadrupole operators introduced in Section 24 and Section 37. This happens to be the case in the Ginocchio models as well as in Elliott's $SU(3)$ model. In the general case, for actual nuclei, this could only be an approximation. Let us still see the consequences of this assumption. Using (24.36) we obtain from (38.11) the result

$$(\mathbf{Q}_\pi \cdot \mathbf{Q}_\nu)(S_\pi^+)^{N_\pi}(S_\nu^+)^{N_\nu}|0\rangle = N_\pi N_\nu (S_\pi^+)^{N_\pi-1}(S_\nu^+)^{N_\nu-1}(D_\pi^+ \cdot D_\nu^+)|0\rangle \tag{38.12}$$

The state in (38.12) is obtained by coupling the lowest $J_\pi = 2$ state and the lowest $J_\nu = 2$ state to a state with a total $J = 0$. In the next step we apply $(\mathbf{Q}_\pi \cdot \mathbf{Q}_\nu)$ to the state (38.12). To evaluate it we must know the commutation relations between components of the quadrupole operator and pair creation operators with $J = 2$. Let us use, for the sake of conciseness, for either the protons or neutrons, the definition

$$[\mathbf{Q},\mathbf{D}^+]_M^{(L)} = \sum_{\mu,\mu'} (2\mu 2\mu' \mid 22LM)[Q_\mu, D_{\mu'}^+] \tag{38.13}$$

From (38.13) follows, due to the orthogonality properties of the Clebsch-Gordan coefficients, the relation

$$[Q_\mu, D_{\mu'}^+] = \sum_{LM} (2\mu 2\mu' \mid 22LM)[\mathbf{Q},\mathbf{D}^+]_M^{(L)} \tag{38.14}$$

With this definition we obtain

$$\begin{aligned}(\mathbf{Q}_\pi \cdot \mathbf{Q}_\nu)&(S_\pi^+)^{N_\pi-1}(S_\nu^+)^{N_\nu-1}(D_\pi^+ \cdot D_\nu^+)|0\rangle \\ &= (N_\pi - 1)(N_\nu - 1)(S_\pi^+)^{N_\pi-2}(S_\nu^+)^{N_\nu-2}(D_\pi^+ \cdot D_\nu^+)(D_\pi^+ \cdot D_\nu^+)|0\rangle \\ &+ (S_\pi^+)^{N_\pi-1}(S_\nu^+)^{N_\nu-1}([\mathbf{Q}_\pi,\mathbf{D}_\pi^+]^{(L)} \cdot [\mathbf{Q}_\nu,\mathbf{D}_\nu^+]^{(L)})|0\rangle \end{aligned} \tag{38.15}$$

In the case of the Ginocchio $O(6)$ model, $[\mathbf{Q},\mathbf{D}^+]_0^{(0)}$ is, according to (37.28), proportional to S^+ whereas $[\mathbf{Q},\mathbf{D}^+]_M^{(L)}$ for $L > 0$ vanish. In Ginocchio's $SU(3)$ model, it follows from (37.42) that $[\mathbf{Q},\mathbf{D}^+]_0^{(0)}$

is proportional to S^+ and $[\mathbf{Q},\mathbf{D}^+]^{(2)}_M$ is proportional to D^+_M whereas $[\mathbf{Q},\mathbf{D}^+]^{(L)}_M$ vanishes for other values of L.

The vanishing of $[\mathbf{Q},\mathbf{D}^+]^{(L)}_M$ for odd L values ($L=1$ or $L=3$) follows directly from the equality of the α_j coefficients in S^+ in both Ginocchio models. In that case, $\mathbf{Q}$ can be constructed explicitly and is given by (24.39) as $Q_\mu = \frac{1}{2}[D^+_\mu, S^-]$. Using this expression we obtain with the help of the Jacobi identity

$$\begin{aligned}[Q_\mu, D^+_{\mu'}] &= \tfrac{1}{2}[[D^+_\mu, S^-], D^+_{\mu'}] \\ &= -\tfrac{1}{2}[[D^+_{\mu'}, D^+_\mu], S^-] - \tfrac{1}{2}[[S^-, D^+_{\mu'}], D^+_\mu] \\ &= [Q_{\mu'}, D^+_\mu]\end{aligned} \tag{38.16}$$

From (38.16) and the symmetry properties of the Clebsch-Gordan coefficients in (38.13) we then obtain

$$[\mathbf{Q},\mathbf{D}^+]^{(L)}_M = (-1)^L[\mathbf{Q},\mathbf{D}^+]^{(L)}_M \tag{38.17}$$

which proves the vanishing $[\mathbf{Q},\mathbf{D}^+]^{(L)}_M$ for odd values of L. The interesting feature of the Ginocchio model is the vanishing of (38.13) for $L=4$. In Elliott's $SU(3)$ model the $L=4$ term does not vanish.

In the general case, with unequal α_j coefficients, the relation (38.17) need not hold. Similarly, the vanishing of (38.13) for $L=4$ is not guaranteed. Moreover, $[\mathbf{Q},\mathbf{D}^+]^{(0)}_0$ need not be proportional to S^+ nor $[\mathbf{Q},\mathbf{D}^+]^{(2)}_M$ need be proportional to D^+_M. We assume, however, that these features hold *approximately*. Under this assumption the state on the r.h.s. of (38.15) is a linear combination of the state $(S^+_\pi)^{N_\pi}(S^+_\nu)^{N_\nu}|0\rangle$, the state $(S^+_\pi)^{N_\pi-1}(S^+_\nu)^{N_\nu-1}(D^+_\pi\cdot D^+_\nu)|0\rangle$ and the state

$$\begin{aligned}&(S^+_\pi)^{N_\pi-2}(S^+_\nu)^{N_\nu-2}(D^+_\pi\cdot D^+_\nu)(D^+_\pi\cdot D^+_\nu)|0\rangle \\ &\quad = (S^+_\pi)^{N_\pi-2}(S^+_\nu)^{N_\nu-2}\left(\sum_J(-1)^J(D^+_\pi\times D^+_\pi)^{(J)}\cdot(D^+_\nu\times D^+_\nu)^{(J)}\right)|0\rangle\end{aligned} \tag{38.18}$$

The equality in (38.18) was obtained by the change of coupling transformation (10.10). From (38.12) and (38.18), taking into account the normalization of the various states, matrix elements of $\kappa(\mathbf{Q}_\pi\cdot\mathbf{Q}_\nu)$ can be obtained.

More matrix elements may be obtained by acting on the state in (38.15) by $(\mathbf{Q}_\pi \cdot \mathbf{Q}_\nu)$. With our assumptions, the states thus obtained are linear combinations of states of protons constructed from S_π^+ and D_π^+ operators which are coupled to neutron states constructed from S_ν^+ and D_ν^+ operators. The matrix of $\kappa(\mathbf{Q}_\pi \cdot \mathbf{Q}_\nu)$ has non-vanishing matrix elements between one such state and another one. These states form the *$S-D$ space*. The sub-matrix of the Hamiltonian (38.5) with $V_{\pi\nu}$ given by (38.8) is completely decoupled from the many other states in the huge shell model space of all states of valence protons and valence neutrons. This feature is obtained also for $J=2$ states if we start with the lowest eigenstates of $H_\pi + H_\nu$. These are the $J_\pi = 0$, $J_\nu = 2$ and $J_\pi = 2$, $J_\nu = 0$ states, $(S_\pi^+)^{N_\pi}(S_\nu^+)^{N_\nu-1}D_\nu^+|0\rangle$ and $(S_\pi^+)^{N_\pi-1}(S_\nu^+)^{N_\nu}D_\pi^+|0\rangle$. This feature holds also for any value of J if we start from a state constructed from the operators $S_\pi^+, D_\pi^+, S_\nu^+, D_\nu^+$ operating on the vacuum state. Thus, of all proton and neutron states (38.7) which appear in the matrix elements (38.9) we keep only those which are in the $S-D$ space. Surely, the total number of states in the $S-D$ space is a tiny fraction of the total number of shell model states of several valence protons and valence neutrons.

In the procedure described above, the Hamiltonian itself generates the subspace of states in which it should be diagonalized. In this respect it is similar to the Lanczos method. Instead of starting with an arbitrary linear combination of all shell model states, we start here from a state which is expected from physical considerations to be an important component of the exact eigenstate. When the matrix is generated we make the approximation of neglecting the contribution of states which have small amplitudes and thus small matrix elements between them and the important states which we keep.

In Section 37 it was pointed out that it is not known whether states constructed from more than one D^+ operator are eigenstates of the Hamiltonian of identical valence nucleons. We assume, nevertheless, that H_π (and H_ν) is diagonal in either one of the two sets of orthogonal states introduced in Section 37. Whenever there are several valence protons and several valence neutrons, the dominant part of the Hamiltonian is the proton-neutron interaction. The $S-D$ space is constructed to maximize its contribution in the matrix to be diagonalized. If $H_\pi + H_\nu$ are not exactly diagonal in the basis of the $S-D$ space they contribute some non-diagonal matrix elements which are, however, expected to be small compared with those of $V_{\pi\nu}$.

The list of approximations made above leads to a very drastic truncation of the shell model space of valence protons and neutrons. The submatrices of the shell model Hamiltonian in the $S-D$ space are of rather low order. Once they are given, they could be directly diagonalized. Still, the calculation of matrix elements (38.9) is rather complicated. It involves calculating matrix elements of the quadrupole operator between rather complicated fermion states constructed from S and D-pairs. At this stage it is also difficult to see how simple spectra, like rotational states, arise from diagonalization of the Hamiltonian matrix. It is here that the interacting boson model wields its great power. The states in the $S-D$ space can be mapped, as discussed in Section 37, onto boson states constructed by $s_\pi^+, d_{\pi\mu}^+, s_\nu^+, d_{\nu\mu}^+$ creation operators. It was discussed in that section how H_π and H_ν can then be replaced by equivalent boson operators. The problem which should be considered here is how to construct boson operators $\mathbf{Q}_\pi^B$ (and $\mathbf{Q}_\nu^B$) which will have between states of s and d-bosons the same matrix elements of the nucleon $\mathbf{Q}_\pi$ (and $\mathbf{Q}_\nu$) operators between corresponding fermion states.

In Section 37 we saw that in the Ginocchio models this problem is rather easy to solve. There, the fermion quadrupole operator ($\mathbf{Q}_\pi$ or $\mathbf{Q}_\nu$) with $L = 2$ in (37.31) is a generator of $O(6)$ and hence, the corresponding generator in the boson picture is (37.21). Similarly, (37.35) with $L = 2$ is the $SU(3)$ quadrupole generator in the fermion space and the corresponding boson generator is given in (37.38). Due to the group structure, matrix elements of corresponding generators between corresponding fermion and boson states are equal. In the general case, as discussed at the end of last section, the situation is not simple. The boson operators corresponding to fermion single quadrupole operators may not be just single boson operators. This is exhibited in a simple case by the expressions (37.52) and (37.53). From those expressions it may be concluded that if a single-boson quadrupole is adopted for $\mathbf{Q}_\pi$ (and $\mathbf{Q}_\nu$) its coefficients may well change as a function of proton (and neutron) number.

Actually, in the Ginocchio models, due to the group structure, calculating matrix elements of operators between nucleon states is as easy as calculating matrix elements of corresponding boson operators between boson states. Also odd-even nuclei can then be considered. In the Fermion Dynamical Symmetry Model (FDSM) it is assumed that nuclear states are eigenstates of the shell model Hamiltonians with $O(6)$ or $SU(3)$ dynamical symmetries. In low lying states nucleons in the orbit with opposite parity to other orbits in a major shell (like $1g_{9/2}$ below 50) are coupled to states with $v = 0$, $J = 0$. We shall not enter here a discussion of the advantages of this model nor into criticism of its difficulties. The interested reader may find a detailed description in the paper by Wu et al. (1987). We now return to the general case.

The result of the approximations made above is a boson model, IBA-2, with two kinds of s and d-bosons, proton s_π, d_π and neutron s_ν, d_ν bosons. This was actually the aim of the assumptions made so far. The choice of the proton-neutron interaction was motivated by the phenomenological success of the boson model. The truncation to the S–D space made it possible to map the problem onto a boson model. It is very difficult to estimate the accuracy of the approximations made above. As emphasized before, the effective interaction in the shell model is known only in some simple cases. We saw some evidence that the Hamiltonian of identical valence nucleons has lowest eigenstates with definite generalized seniority. Not all possible shell model Hamiltonians have this property. It is the experimentally determined two-body effective interaction in $T = 1$ states which is diagonal in states with generalized seniority 0 and 2. In the present case of valence protons and valence neutrons, the two-body effective interaction, including that in $T = 0$ states, is even harder to determine from experiment. Not all possible quadrupole-quadrupole interactions can be well approximated by the boson model. The fact that the s, d boson model can reproduce successfully many features of collective spectra may be taken as an indication that the chain of approximations made above is indeed justified.

In Section 37 the problem of the Pauli principle has been discussed in the context of mapping fermion states onto boson states. We saw examples of matrix elements of shell model Hamiltonians and other operators between certain fermion states which are equal to those of boson Hamiltonians and other operators between corresponding boson states. This equality is exactly satisfied in Ginocchio's models for all states of the $S - D$ space. This feature does not violate the Pauli principle at all. The fermion states mapped on (fully symmetric) boson states are fully antisymmetric, strictly obeying the Pauli principle. Difficulties occur since not all states of s and d-bosons have corresponding fermion states. This happens beyond the middle of shells but there it was fixed by considering valence holes instead of valence nucleons. It may also happen for smaller numbers of identical nucleons as demonstrated in Section 37 by several examples. Such cases, however, depend critically on the size and nature of the valence shell. It is expected that in shells with many j-orbits such difficulties occur only at rather high proton and neutron states. In the Ginocchio

$SU(3)$ model, the lowest states of the boson system for $N > \Omega/3$ do not correspond to fermion states whereas in the $O(6)$ model *all* boson states up to the middle of the shell ($N \leq \Omega/2$) correspond to non-vanishing fermion states. In the general case, it is expected that this kind of violation of the Pauli principle may inflict some of the high states of the protons or the neutrons. Such states are not expected to contribute much to the low lying eigenstates of the Hamiltonian of valence protons and neutrons. Once this problem has been settled for proton states and neutron states, introduction of the proton neutron interaction does not introduce any further restrictions on the correspondence between fermion and boson states.

39

The Proton Neutron Interacting Boson Model (IBA-2)

The considerations presented above lead to a model in which the determination of low lying eigenstates of the shell model Hamiltonian is replaced by diagonalization of a boson Hamiltonian. This boson Hamiltonian contains single boson terms and interactions between bosons. The boson space considered is that of N_π proton bosons, s_π, d_π, and N_ν neutron bosons s_ν, d_ν. According to Section 37, the number $N_\pi(N_\nu)$ is one half the number of valence protons (neutrons) in the first half of the major shell and one half of the number of holes between the middle and the end of the shell (Arima et al. 1977, Otsuka et al. 1978). The boson Hamiltonian which should have the same eigenvalues as the fermion Hamiltonian (38.4) with $V_{\pi\nu}$ as in (38.7), also has the form (38.4). Since from now on only boson operators will appear, no different notation will be introduced. The boson Hamiltonians H_π and H_ν should have the same properties as the fermion ones. The latter have eigenstates with definite generalized seniority which implies, as explained in Section 37, that the boson Hamiltonians should have $U(5)$ symmetry. The boson quadrupole interaction breaks this symmetry and the quadrupole boson operator is approximated by

$$Q_\mu = d_\mu^+ s + s^+ \tilde{d}_\mu + \chi (d^+ \times \tilde{d})_\mu^{(2)} \tag{39.1}$$

To reproduce the fermion matrix elements, the parameter χ, as well as κ in (35.7), may have to be functions of N_π and N_ν. The possible dependence of these coefficients on numbers of d-bosons is generally ignored.

The boson Hamiltonian of IBA-2 is rather simple. Still it includes the main ingredients of the nuclear interaction. These are interactions between identical nucleons leading to generalized seniority and seniority breaking proton-neutron interactions. The competition between these two kinds of interactions and the resulting structure of eigenstates in the boson model is much more transparent than in the fermion shell model.

Before entering a detailed discussion of IBA-2, let us see some general features of the model. The Hamiltonian of $s_\pi, d_\pi, s_\nu, d_\nu$ bosons may be constructed from generators of the $U(6)$ group for protons, $U_\pi(6)$, and the one for neutrons, $U_\nu(6)$. These generators are listed in (33.24), (33.25) and (33.26) and are given by

$$\begin{gathered} (d_\pi^+ \times \tilde{d}_\pi)_\kappa^{(k)} \qquad k = 0,1,2,3,4 \\ d_{\pi\kappa}^+ s_\pi + s_\pi^+ \tilde{d}_{\pi\kappa} \qquad i(d_{\pi\kappa}^+ s_\pi - s_\pi^+ \tilde{d}_{\pi\kappa}) \qquad s_\pi^+ s_\pi \end{gathered} \tag{39.2}$$

as well as by

$$\begin{gathered} (d_\nu^+ \times \tilde{d}_\nu)_\kappa^{(k)} \qquad k = 0,1,2,3,4 \\ d_{\nu\kappa}^+ s_\nu + s_\nu^+ \tilde{d}_{\nu\kappa} \qquad i(d_{\nu\kappa}^+ s_\nu - s_\nu^+ \tilde{d}_{\nu\kappa}) \qquad s_\nu^+ s_\nu \end{gathered} \tag{39.3}$$

The two sets of generators (39.2) and (39.3) act on different sets of states and hence generate the group $U_\pi(6) \otimes U_\nu(6)$.

The most general Hamiltonian constructed from the generators (39.2) and (39.3) commutes with the number operators $\hat{N}_\pi$ and $\hat{N}_\nu$. Such a Hamiltonian, containing single boson terms and boson-boson interactions, which is hermitean and rotationally invariant, is a linear combination of the following three terms. The first term is a Hamiltonian H_π constructed from only the proton boson generators (39.2). Such Hamiltonians were considered in Section 33 and are expressed by (33.30) in which proton boson operators appear. The second term is the Hamiltonian H_ν constructed from the neutron boson generators (39.3). It is also given by (33.30) expressed in terms of the s_ν^+, s_ν, $d_{\nu\mu}^+$, $\tilde{d}_{\nu\mu}$ (and N_ν) operators. The third term contains scalar products of proton boson generators (39.2) and neutron boson generators (39.3). The hermitean tensor operators in (39.2) commute with those in (39.3) and hence, all of those scalar products are hermitean. Altogether such $V_{\pi\nu}$ interaction may contain 16 independent terms. Using the conservation of N_π and of N_ν some of these terms may be simplified. The scalar operators in (39.2) are number operators n_{s_π} and n_{d_π} and those in (39.3) are n_{s_ν} and n_{d_ν}. Using the relations

$$\begin{aligned} n_{s_\pi} + n_{d_\pi} &= N_\pi \\ n_{s_\nu} + n_{d_\nu} &= N_\nu \end{aligned} \tag{39.4}$$

products of those scalar operators may be expressed as linear combinations of $N_\pi N_\nu$, $n_{d_\pi} N_\nu$, $n_{d_\nu} N_\pi$ and $n_{d_\pi} n_{d_\nu}$. The first of these terms contributes only to binding energies and the second and third may be added to H_π and H_ν, making ϵ_{d_π} depend on N_ν and ϵ_{d_ν} depend on N_π. Only the last term $n_{d_\pi} n_{d_\nu}$ is a genuine proton boson-neutron boson interaction. All these terms represent monopole ($k = 0$) interactions. If this interaction between proton bosons and neutron bosons is the same irrespective of their being s- or d-bosons the total interaction is proportional to $N_\pi N_\nu$ only. We shall not express explicitly the other scalar products.

Let us first consider an important special case. There may be Hamiltonians in which single boson energies and all interactions between bosons are independent of their being proton bosons or neutron bosons. In other words, the interactions between two proton bosons, two neutron bosons or a proton boson and neutron boson in the *same state* are equal. This is similar to charge independence of the interaction between nucleons but it is definitely different from it. This difference will be discussed in detail in the following. Boson Hamiltonians with this special symmetry are not in agreement with our knowledge of nuclear interactions. Still, they are of great theoretical interest. For such Hamiltonians it is convenient to introduce a formalism in which proton bosons and neutron bosons are considered as two states of the *same* s and d-bosons. There is a simple and elegant way to express this formalism.

Hamiltonians with this symmetry between proton and neutron bosons may be constructed from special linear combinations of the generators (39.2) and (39.3). These combinations are fully symmetric in proton and neutron bosons and are given by

$$(d_\pi^+ \times \tilde{d}_\pi)_\kappa^{(k)} + (d_\nu^+ \times \tilde{d}_\nu)_\kappa^{(k)} \qquad k = 0,1,2,3,4 \qquad s_\pi^+ s_\pi + s_\nu^+ s_\nu$$

$$d_{\pi\kappa}^+ s_\pi + s_\pi^+ \tilde{d}_{\pi\kappa} + d_{\nu\kappa}^+ s_\nu + s_\nu^+ \tilde{d}_{\nu\kappa} \tag{39.5}$$

$$i[d_{\pi\kappa}^+ s_\pi - s_\pi^+ \tilde{d}_{\pi\kappa} + d_{\nu\kappa}^+ s_\nu - s_\nu^+ \tilde{d}_{\nu\kappa}]$$

The operators (39.5) are generators of the Lie algebra of a $U(6)$ group which is a subgroup of $U_\pi(6) \otimes U_\nu(6)$ and may be denoted by $U_{\pi+\nu}(6)$. Hence, the eigenstates of such a Hamiltonian are characterized by the irreducible representations of the $U(6)$ group. Since the eigenstates are of two kinds of bosons, proton bosons and neutron bosons, they need not be fully symmetric. The eigenvalue which belongs to a given state is independent of the separate numbers N_π and N_ν ($N_\pi + N_\nu = N$) provided a state with this symmetry properties can be constructed with these numbers of proton bosons and neutron bosons. This situation is analogous to the case of charge independent nucleon Hamiltonians considered in Section 26. There, the states become fully antisymmetric by using the isospin formalism. The isospin T (with n) determines the symmetry properties of isospin states belonging to irreducible representations of the $SU_T(2)$ Lie group which multiply the space and spin functions to yield a fully antisymmetric state. A similar procedure may be adopted here by introducing an analogous spin named F-spin (Arima et al. 1977).

To obtain an appropriate $SU(2)$ Lie algebra we first add to the $36 + 36 = 72$ generators (39.2) and (39.3) other operators which change a proton boson into a neutron boson and others that induce the opposite change. These are 36 operators $s_\pi^+ s_\nu$, $d_{\pi\kappa}^+ s_\nu$, $s_\pi^+ d_{\nu\kappa}$ and $d_{\pi\kappa}^+ d_{\nu\kappa'}$ and the 36 hermitean conjugate operators. The 144 operators thus obtained are generators of the $U(2 \times 6) = U(12)$ Lie algebra operating on the 12-dimensional space of proton and neutron s and d-bosons. Among the new generators we choose only those that just change the charge of the boson but not its spatial state. Consider the operator

$$\boxed{F^+ = s_\pi^+ s_\nu + \sum_\kappa d_{\pi\kappa}^+ d_{\nu\kappa} = s_\pi^+ s_\nu + (d_\pi^+ \cdot \tilde{d}_\nu)} \tag{39.6}$$

and its hermitean conjugate

$$(F^+)^\dagger = F^- = s_\nu^+ s_\pi + \sum_\kappa d_{\nu\kappa}^+ d_{\pi\kappa} = s_\nu^+ s_\pi + (d_\nu^+ \cdot \tilde{d}_\pi) \tag{39.7}$$

The commutator of F^+ and F^- can be readily evaluated to give

$$\begin{aligned}[F^+, F^-] &= s_\pi^+ s_\pi - s_\nu^+ s_\nu + \sum_\kappa (d_{\pi\kappa}^+ d_{\pi\kappa} - d_{\nu\kappa}^+ d_{\nu\kappa}) \\ &= N_\pi - N_\nu = 2F^0\end{aligned} \tag{39.8}$$

The operator F^0 defined by (39.8) satisfies the following commutation relations

$$[F^0, F^+] = F^+ \qquad [F^0, F^-] = -F^- \tag{39.9}$$

The commutation relations (39.8) and (39.9) define generators of a $SU(2)$ Lie algebra which has been discussed on several occasions before. The generators F^+, F^- and F^0 are spherical components of a spin vector–F-spin.

Instead of the $U(12)$ group we now consider the subgroup

$$U(12) \supset U_F(2) \otimes U(6) \tag{39.10}$$

The eigenstates of the Hamiltonians considered here belong to irreducible representations of $U_{\pi+\nu}(6)$ with certain symmetry properties. The Hamiltonian is invariant under permutations of indices specifying the various bosons. Hence, when such a permutation is applied to an eigenstate, it leads to an eigenstate with the same eigenvalue. If the eigenstate is not fully symmetric or fully antisymmetric, the other eigenstate will be independent of the original one. Still, no new *physical* state is obtained. Since the Hamiltonian is invariant under such permutations, they only lead to assigning the bosons different numbers. To avoid this spurious degeneracy, we multiply each eigenstate by a F-spin state which belongs to an irreducible representation of $SU_F(2)$ with the same symmetry and take their sum. We obtain thereby states which belong to fully symmetric irreducible representations of $U_F(2) \otimes U(6)$. The symmetry type of the irreducible representation of $SU_F(2)$ is thus the *same* as that of $U(6)$. The former is completely determined by N and F which determines the eigenvalues of $\mathbf{F}^2$ as $F(F+1)$. The larger the F, the more symmetric the state. Each boson has F-spin equal to $\frac{1}{2}$ and, according to (39.8), $M_F = \frac{1}{2}$ for a proton boson and $M_F = -\frac{1}{2}$ for a neutron boson. The maximum value of F is thus $F = \frac{1}{2}N = \frac{1}{2}(N_\pi + N_\nu)$ and its minimum value, for given N_π, N_ν, due to $F \geq |M_F|$, is given by $\frac{1}{2}|N_\pi - N_\nu|$.

For any values of N_π and N_ν, the most symmetric states are those with $F = \frac{1}{2}(N_\pi + N_\nu) = \frac{1}{2}N$. The fully symmetric eigenstates have the same eigenvalues for given N irrespective of the values of N_π or N_ν. In particular, these same states appear in the case of identical bosons, say $N_\pi = N$. This clearly demonstrates the fact that such states are fully equivalent to states of IBA-1 with the same value of N. Thus, IBA-1 is a special case of IBA-2 for eigenstates with $F = \frac{1}{2}N = \frac{1}{2}(N_\pi + N_\nu)$ of Hamiltonians which are fully symmetric in proton and neutron bosons. Such Hamiltonians commute with the F-spin components (39.6), (39.7) and (39.8). They are constructed from scalars and scalar products of $U(6)$ generators given by (39.5). It can be directly verified that each operator in (39.5) commutes with F^+, F^- and F^0. We shall see in the following that some lowest eigenstates of more general IBA-2 Hamiltonians may still have well defined F-spins equal to $F = \frac{1}{2}(N_\pi + N_\nu)$. Such states uniquely correspond to IBA-1 states. In more interesting cases, in which there is good agreement with experiment, F-spin may no longer be a good quantum number. Still, IBA-1 may offer a fair approximation or a model of IBA-2.

The irreducible representations of $U(6)$ which have $F = \frac{1}{2}(N_\pi + N_\nu) = \frac{1}{2}N$ are fully symmetric. They are characterized by Young diagrams which have one row with N squares. In addition to these representations, which are equivalent to the IBA-1 states, there are other irreducible representations with mixed symmetry. The Young diagrams of these have two rows of lengths N_1 and N_2 ($N_2 \leq N_1$). In analogy with isospin T (Section 26) and spin S (Section 28) we can write down the relations

$$\boxed{N = N_1 + N_2 \qquad F = \tfrac{1}{2}(N_1 - N_2)} \tag{39.11}$$

The quadratic Casimir operator for the $U_{\pi+\nu}(6)$ group may be constructed as in (33.37) by replacing the generators there by the generators (39.5). Its eigenvalues for these irreducible representations may be calculated from a general expression for eigenvalues of the quadratic Casimir operator of $U(k)$ group for the irreducible representation defined by $f_1 \geq f_2 \geq \cdots \geq f_k$, $\sum f_i = N$. We quote the general formula for these eigenvalues as

$$\sum_{i=1}^{k} f_i(f_i - 2i + 1) + k \sum_{i=1}^{k} f_i \tag{39.12}$$

Putting in (39.12) $f_1 = N_1$, $f_2 = N_2$, $f_3 = \cdots = f_6 = 0$ we obtain the eigenvalues

$$\begin{aligned} &N_1(N_1 - 1) + N_2(N_2 - 3) + 6(N_1 + N_2) \\ &\quad = N_1(N_1 + 5) + N_2(N_2 + 3) \end{aligned} \tag{39.13}$$

The expressions (26.12) and (28.6) are special cases of (39.12) with $n = n_1 + n_2$, $T = (n_1 - n_2)/2$ and $S = (n_1 - n_2)/2$ respectively. There, n_1 and n_2 are the lengths of the two *columns* of the Young diagram characterizing the irreducible representations of $U(2j + 1)$ and $U(2l + 1)$ respectively. Hence, to apply (39.12) to those cases we should put

$$f_1 = \cdots = f_{n_2} = 2, \qquad f_{n_2+1} = \cdots = f_{n_1} = 1, \qquad f_{n_1+1} = \cdots = f_k = 0$$

where $k = 2j + 1$ in (26.12) and $k = 2l + 1$ in (28.6).

Substituting in (39.12) the values of N_1 and N_2 from (39.11) we obtain these eigenvalues to be equal to

$$\boxed{\tfrac{1}{2}[N(N + 8) + 4F(F + 1)]} \tag{39.14}$$

For $F = \frac{1}{2}N$ (39.13) reduces to the value $N(N + 5)$ as in (33.42). A Hamiltonian equal to the Casimir operator of $U_{\pi+\nu}(6)$ multiplied by a negative constant has lowest eigenstates which have the maximum value of F $(= \frac{1}{2}N)$. The eigenvalues (39.14) may be obtained directly from the definition of the Casimir operator. For that we first derive an expression for $\mathbf{F}^2 = F^+F^- + (F^0)^2 - F^0$.

From (39.6) and (39.7) we obtain after slight rearrangements

$$F^+F^- = \left(s_\pi^+ s_\nu + \sum_\kappa d_{\pi\kappa}^+ d_{\nu\kappa}\right)\left(s_\pi s_\nu^+ + \sum_\kappa d_{\pi\kappa} d_{\nu\kappa}^+\right)$$

$$= s_\pi^+ s_\pi s_\nu^+ s_\nu + s_\pi^+ s_\pi + s_\pi^+ s_\nu (d_\nu^+ \cdot \tilde{d}_\pi)$$

$$+ s_\pi s_\nu^+ (d_\pi^+ \cdot \tilde{d}_\nu) + (d_\pi^+ \cdot \tilde{d}_\nu)(\tilde{d}_\pi \cdot d_\nu^+)$$

The last term on the r.h.s. may be rewritten by using the change of coupling transformation (10.10). Due to the commutation relations between $d_{\pi\kappa}^+$ and $d_{\pi\kappa'}$ we obtain

$$(d_\pi^+ \cdot \tilde{d}_\nu)(\tilde{d}_\pi \cdot d_\nu^+) = (d_\pi^+ \cdot \tilde{d}_\nu)(d_\nu^+ \cdot \tilde{d}_\pi)$$

$$= \sum_{L=0}^{4} (-1)^L (d_\pi^+ \times d_\nu^+)^{(L)} \cdot (\tilde{d}_\nu \times \tilde{d}_\pi)^{(L)} + (d_\pi^+ \cdot \tilde{d}_\pi)$$

$$= \sum_{L=0}^{4} (d_\pi^+ \times d_\nu^+)^{(L)} \cdot (\tilde{d}_\pi \times \tilde{d}_\nu)^{(L)} + (d_\pi^+ \cdot \tilde{d}_\pi) \tag{39.15}$$

The last equality in (39.15) is due to the symmetry properties of the vector addition coefficients. We can use (10.10) to obtain

$$(d_\pi^+ \cdot \tilde{d}_\pi)(d_\nu^+ \cdot \tilde{d}_\nu) = \sum_{L=0}^{4} (-1)^L (d_\pi^+ \times d_\nu^+)^{(L)} \cdot (\tilde{d}_\pi \times \tilde{d}_\nu)^{(L)}$$

$$= \sum_{L=0}^{4} (d_\pi^+ \times d_\nu^+)^{(L)} \cdot (\tilde{d}_\pi \times \tilde{d}_\nu)^{(L)}$$

$$-2 \sum_{L=1,3} (d_\pi^+ \times d_\nu^+)^{(L)} \cdot (\tilde{d}_\pi \times \tilde{d}_\nu)^{(L)} \tag{39.16}$$

Using (39.16) and (39.15) we obtain for F^+F^- the expression

$$F^+F^- = s_\pi^+ s_\pi s_\nu^+ s_\nu + s_\pi^+ s_\pi + s_\pi^+ s_\nu (\tilde{d}_\pi \cdot d_\nu^+) + s_\pi s_\nu^+ (d_\pi^+ \cdot \tilde{d}_\nu)$$
$$+ (d_\pi^+ \cdot \tilde{d}_\pi)(d_\nu^+ \cdot \tilde{d}_\nu) + (d_\pi^+ \cdot \tilde{d}_\pi)$$
$$+ 2 \sum_{L=1,3} (d_\pi^+ \times d_\nu^+)^{(L)} \cdot (\tilde{d}_\pi \times \tilde{d}_\nu)^{(L)} \tag{39.17}$$

We add now to (39.17) the term

$$(F^0)^2 - F^0 = \tfrac{1}{4}(N_\pi - N_\nu)^2 - \tfrac{1}{2}(N_\pi - N_\nu)$$
$$= \tfrac{1}{4}(N_\pi + N_\nu)^2 - N_\pi N_\nu + \tfrac{1}{2}(N_\pi + N_\nu) - N_\pi$$
$$= \tfrac{1}{2}N(\tfrac{1}{2}N + 1) - (s_\pi^+ s_\pi + (d_\pi^+ \cdot \tilde{d}_\pi))(s_\nu^+ s_\nu + (d_\nu^+ \cdot \tilde{d}_\nu))$$
$$- (s_\pi^+ s_\pi + (d_\pi^+ \cdot \tilde{d}_\pi)) \tag{39.18}$$

The result is

$$\mathbf{F}^2 = F^+F^- + (F^0)^2 - F^0$$
$$= \tfrac{1}{2}N(\tfrac{1}{2}N + 1) + s_\pi^+ s_\nu (\tilde{d}_\pi \cdot d_\nu^+) + s_\pi s_\nu^+ (d_\pi^+ \cdot \tilde{d}_\nu)$$
$$- s_\pi^+ s_\pi (d_\nu^+ \cdot \tilde{d}_\nu) - (d_\pi^+ \cdot \tilde{d}_\pi) s_\nu^+ s_\nu$$
$$+ 2 \sum_{L=1,3} (d_\pi^+ \times d_\nu^+)^{(L)} \cdot (\tilde{d}_\pi \times \tilde{d}_\nu)^{(L)} \tag{39.19}$$

The eigenvalues $F(F+1)$ of (39.19) are related to the symmetry of eigenstates. The higher the value of F the higher the symmetry. In the case of two bosons, $N = 2$, the symmetric states have $F = 1$ and hence, the sum of the terms on the r.h.s of (39.19), apart from the $\frac{1}{2}N(\frac{1}{2}N + 1)$ term, vanishes. The antisymmetric states have $F = 0$ and that sum is then equal to -2. Thus, the operator $\mathbf{F}^2 - \frac{1}{2}N(\frac{1}{2}N + 1) + 1$ is analogous to the Majorana operator for nucleons defined in Section 12. This boson operator has been introduced by several authors as a term into IBA-2 Hamiltonians. The term Majorana operator has been used for the operator

$$\begin{aligned}\mathcal{M}_0 &= \tfrac{1}{2}N(\tfrac{1}{2}N+1) - \mathbf{F}^2 \\ &= -s_\pi^+ s_\nu (\tilde{d}_\pi \cdot d_\nu^+) - s_\pi s_\nu^+ (d_\pi^+ \cdot \tilde{d}_\nu) + s_\pi^+ s_\pi (d_\nu^+ \cdot \tilde{d}_\nu) \\ &\quad + (d_\pi^+ \cdot \tilde{d}_\pi) s_\nu^+ s_\nu - 2 \sum_{L=1,3} (d_\pi^+ \times d_\nu^+)^{(L)} \cdot (\tilde{d}_\pi \times \tilde{d}_\nu)^{(L)} \\ &= (s_\pi^+ d_\nu^+ - d_\pi^+ s_\nu^+) \cdot (s_\pi \tilde{d}_\nu - \tilde{d}_\pi s_\nu) \\ &\quad - 2 \sum_{L=1,3} (d_\pi^+ \times d_\nu^+)^{(L)} \cdot (\tilde{d}_\pi \times \tilde{d}_\nu)^{(L)}\end{aligned} \tag{39.20}$$

It may be directly verified that r.h.s. of (39.20) vanishes when applied to states with two identical bosons or to the other symmetric states $s_\pi^+ s_\nu^+ |0\rangle$, $(d_\pi^+ \cdot d_\nu^+)|0\rangle$, $(s_\pi^+ d_{\nu\kappa}^+ + d_{\pi\kappa}^+ s_\nu^+)|0\rangle$, $(d_\pi^+ \times d_\nu^+)_\kappa^{(2)}|0\rangle$, $(d_\pi^+ \times d_\nu^+)_\kappa^{(4)}|0\rangle$. The Majorana operator (39.20) has eigenvalue 2 when applied to antisymmetric states, i.e. $(s_\pi^+ d_{\nu\kappa}^+ - d_{\pi\kappa}^+ s_\nu^+)|0\rangle$ and $(d_\pi^+ \times d_\nu^+)_\kappa^{(L)}|0\rangle$ with $L = 1, 3$.

From the expression (39.20) of $\mathcal{M}_0$ follows that it has eigenvalue zero in any state of s_π and s_ν bosons. All such states are annihilated by the $\tilde{d}_{\pi\kappa}$ and $\tilde{d}_{\nu\kappa}$ operators on the r.h.s. of (39.20). In fact, apart from their charge, s_π and s_ν bosons are in the same $l = 0$ state and hence must alsways be coupled symmetrically (the F-spin of any such state is equal to $\frac{1}{2}N = \frac{1}{2}n_s$).

A generalization of (39.20) , also called Majorana operator, is defined by

$$\begin{aligned}\mathcal{M} &= \sum_{L=1,3} \xi_L (d_\pi^+ \times d_\nu^+)^{(L)} \cdot (\tilde{d}_\pi \times \tilde{d}_\nu)^{(L)} \\ &\quad + \xi_2 (s_\pi^+ d_\nu^+ - d_\pi^+ s_\nu^+) \cdot (s_\pi \tilde{d}_\nu - \tilde{d}_\pi s_\nu)\end{aligned} \tag{39.21}$$

with arbitrary coefficients ξ_1, ξ_2, ξ_3. For $\xi_1 = \xi_3 = -2$, $\xi_2 = 1$ the operator (39.21) reduces to (39.20). Each term in (39.21) commutes with the components of **F**. It is sufficient to show that the components of **F** commute with the creation operators $(d_\pi^+ \times d_\nu^+)^{(L)}_\kappa$, $L = 1,3$ and $s_\pi^+ d_{\nu\kappa}^+ - d_{\pi\kappa}^+ s_\nu^+$ since the latter are multiplied in (39.21) by their hermitean conjugates. These pair creation operators are antisymmetric in the proton and neutron bosons. Their commutators with F^+ or F^- yield creation operators for the same antisymmetric states of two proton bosons or two neutron bosons which must vanish. The operator (39.21) with any value of ξ_L is thus diagonal in F-spin. The eigenvalues, however, are no longer simple functions of F and N. Still, if $\xi_1, \xi_3 < 0$ and $\xi_2 > 0$, the lower the value of F, the higher the eigenvalue of $\mathcal{M}$. If such an operator with large absolute values of the coefficients ξ_L is added to an arbitrary IBA-2 Hamiltonian the lowest eigenstates of the resulting Hamiltonian, have larger components with $F = \frac{1}{2}N$. We shall return to this point in the following.

The Majorana operator (39.20) is simply related to the Casimir operator of $U_{\pi+\nu}(6)$. By its construction, the quadratic Casimir operator is made of three terms. One is constructed from proton bosons and its eigenvalues are equal according to (33.42) to $N_\pi(N_\pi + 5)$. The other part, made of neutron bosons, has eigenvalues $N_\nu(N_\nu + 5)$. The third part is bilinear in proton and neutron generators and is equal to

$$2\left\{ s_\pi^+ s_\pi s_\nu^+ s_\nu + \sum_{L=0}^{4} (d_\pi^+ \times \tilde{d}_\pi)^{(L)} \cdot (d_\nu^+ \times \tilde{d}_\nu)^{(L)} + \tfrac{1}{2}(s_\pi^+ \tilde{d}_\pi + d_\pi^+ s_\pi) \right.$$

$$\left. \cdot (s_\nu^+ \tilde{d}_\nu + d_\nu^+ s_\nu) - \tfrac{1}{2}(s_\pi^+ \tilde{d}_\pi - d_\pi^+ s_\pi) \cdot (s_\nu^+ \tilde{d}_\nu - d_\nu^+ s_\nu) \right\} \qquad (39.22)$$

Using the change of coupling transformation (10.10) and the sum rule (10.21) we obtain

$$\sum_{L=0}^{4} (d_\pi^+ \times \tilde{d}_\pi)^{(L)} \cdot (d_\nu^+ \times \tilde{d}_\nu)^{(L)}$$

$$= \sum_{L=0}^{4} (d_\pi^+ \times d_\nu^+)^{(L)} \cdot (\tilde{d}_\pi \times \tilde{d}_\nu)^{(L)}$$

$$= (d_\pi^+ \cdot \tilde{d}_\pi)(d_\nu^+ \cdot \tilde{d}_\nu) + 2 \sum_{L=1,3} (d_\pi^+ \times d_\nu^+)^{(L)} \cdot (\tilde{d}_\pi \times \tilde{d}_\nu)^{(L)} \qquad (39.23)$$

The last equality in (39.23) is due to (39.16). Substituting from (39.23) into (39.22) we obtain

$$2\left\{ s_\pi^+ s_\pi s_\nu^+ s_\nu + (d_\pi^+ \cdot \tilde{d}_\pi)(d_\nu^+ \cdot \tilde{d}_\nu) + 2 \sum_{L=1,3} (d_\pi^+ \times d_\nu^+)^{(L)} \cdot (\tilde{d}_\pi \times \tilde{d}_\nu)^{(L)} \right.$$

$$\left. + s_\pi^+ s_\nu (\tilde{d}_\pi \cdot d_\nu^+) + s_\pi s_\nu^+ (d_\pi^+ \cdot \tilde{d}_\nu) \right\} \tag{39.24}$$

Comparing this expression with (39.20) we find that (39.24) is equal to

$$2\mathbf{F}^2 - N(\tfrac{1}{2}N + 1) + 2s_\pi^+ s_\pi (d_\nu^+ \cdot \tilde{d}_\nu) + 2(d_\pi^+ \cdot \tilde{d}_\pi) s_\nu^+ s_\nu$$
$$+ 2s_\pi^+ s_\pi s_\nu^+ s_\nu + 2(d_\pi^+ \cdot \tilde{d}_\pi)(d_\nu^+ \cdot \tilde{d}_\nu) \tag{39.25}$$

Adding now to (39.25) the terms $N_\pi(N_\pi + 5) = (s_\pi^+ s_\pi + (d_\pi^+ \cdot \tilde{d}_\pi))(s_\pi^+ s_\pi + (d_\pi^+ \cdot \tilde{d}_\pi) + 5)$ and $N_\nu(N_\nu + 5) = (s_\nu^+ s_\nu + (d_\nu^+ \cdot \tilde{d}_\nu))(s_\nu^+ s_\nu + (d_\nu^+ \cdot \tilde{d}_\nu) + 5)$ we finally obtain the following expression for the Casimir operator of $U_{\pi+\nu}(6)$

$$\boxed{2\mathbf{F}^2 + \tfrac{1}{2}N(N + 8)} \tag{39.26}$$

It should be made clear that this expression of $C_{U(6)}$ holds only for irreducible representations allowed in IBA-2. Only such irreducible representations are uniquely characterized by F and N according to (39.11). The operator (39.26) has the eigenvalues $2F(F+1) + \frac{1}{2}N(N+8)$ which are equal to (39.14) given above.

The Hamiltonians considered above, whose eigenstates are characterized by definite values of F-spin, have $SU_F(2) \otimes U(6)$ symmetry. Dynamical symmetries of such IBA-2 Hamiltonians follow if the operators (39.5) used in their construction are restricted to generators of subgroups of $U(6)$. In this way the dynamical symmetries of $U(5)$, $O(6)$ and $SU(3)$ are obtained like in the case of IBA-1. In each case the Hamiltonian may be expressed as a linear combination of the Casimir operators of the appropriate chain of groups. The eigenvalues are thus linear combinations of the eigenvalues of those Casimir operators. The difference between the present case and IBA-1 is that for $F < \frac{1}{2}(N_\pi + N_\nu)$ the irreducible representations of the various groups are not fully symmetric in the proton and neutron s and d-bosons.

We first consider the dynamical symmetry due to the $U_{\pi+\nu}(5)$ subgroup. As in Section 34, for IBA-1, we shall express the Hamiltonian as a linear combination of Casimir operators of a chain of groups. The quadratic Casimir operator for identical bosons was defined in (34.3). In the present case it is constructed from generators (39.5) as

$$\begin{aligned}
C_{U_{\pi+\nu}(5)} &= \sum_{L=0}^{4}((d_\pi^+ \times \tilde{d}_\pi)^{(L)} + (d_\nu^+ \times \tilde{d}_\nu)^{(L)}) \\
&\quad \cdot ((d_\pi^+ \times \tilde{d}_\pi)^{(L)} + (d_\nu^+ \times \tilde{d}_\nu)^{(L)}) \\
&= \sum_{L=0}^{4}(d_\pi^+ \times \tilde{d}_\pi)^{(L)} \cdot (d_\pi^+ \times \tilde{d}_\pi)^{(L)} \\
&\quad + \sum_{L=0}^{4}(d_\nu^+ \times \tilde{d}_\nu)^{(L)} \cdot (d_\nu^+ \times \tilde{d}_\nu)^{(L)} \\
&\quad + 2\sum_{L=0}^{4}(d_\pi^+ \times \tilde{d}_\pi)^{(L)} \cdot (d_\nu^+ \times \tilde{d}_\nu)^{(L)} \\
&= (d_\pi^+ \cdot \tilde{d}_\pi)((d_\pi^+ \cdot \tilde{d}_\pi) + 4) + (d_\nu^+ \cdot \tilde{d}_\nu)((d_\nu^+ \cdot \tilde{d}_\nu) + 4) \\
&\quad + 2\sum_{L=0}^{4}(d_\pi^+ \times \tilde{d}_\pi)^{(L)} \cdot (d_\nu^+ \times \tilde{d}_\nu)^{(L)}
\end{aligned} \tag{39.27}$$

The last equality was obtained by using (34.3). The irreducible representations of $U_{\pi+\nu}(5)$ have Young diagrams with at most two rows of lengths n_{d_1} and $n_{d_2}(n_{d_2} \leq n_{d_1})$. The eigenvalues of the Casimir operator in terms of n_{d_1} and n_{d_2} may be obtained from (39.12). We may instead make use of the fact that the $U(5)$ states are those of d-bosons only. We may then express the last term on the r.h.s. of (39.27) by using (39.23) which may be expressed by using the part of (39.20) where creation and annihilation operators of only d-bosons appear. We may then use the F-spin vector $\mathbf{F}'$ operating in the space of d-bosons only. In terms of this part of $\mathbf{F}^2$ and replacing N in (39.20) by n_d, we obtain for the last term on the r.h.s. of (39.27) the expression

$$2\mathbf{F}'^2 - n_d(\tfrac{1}{2}n_d + 1) + 2(d_\pi^+ \cdot \tilde{d}_\pi)(d_\nu^+ \cdot \tilde{d}_\nu) \tag{39.28}$$

Adding to (39.28) the first two terms on the r.h.s. of (39.27),

$$n_{d_\pi}(n_{d_\pi} + 4) \qquad \text{and} \qquad n_{d_\nu}(n_{d_\nu} + 4),$$

we obtain the eigenvalues of the Casimir operator (39.27) in the form

$$2F'(F'+1)+\tfrac{1}{2}n_d(n_d+6)$$

In the case considered here $n_d = n_{d_1} + n_{d_2}$, $F' = \frac{1}{2}(n_{d_1} - n_{d_2})$ and the eigenvalues are given by

$$\begin{aligned} &n_{d_1}(n_{d_1}-1)+n_{d_2}(n_{d_2}-3)+5(n_{d_1}+n_{d_2}) \\ &\quad = n_{d_1}(n_{d_1}+4)+n_{d_2}(n_{d_2}+2) \end{aligned} \tag{39.29}$$

In the case of identical nucleons, or for $F = \frac{1}{2}(N_\pi + N_\nu)$, $n_{d_1} = n_d$, $n_{d_2} = 0$ and (39.29) is reduced to $n_d(n_d+4)$ as in (34.3).

To obtain an explicit expression for the eigenvalues of Hamiltonians with $U_{\pi+\nu}(5)$ dynamical symmetry we consider the subgroup $O_{\pi+\nu}(5)$ of $U_{\pi+\nu}(5)$. Its 10 generators are odd tensor operators given by

$$(d_\pi^+ \times \tilde{d}_\pi)^{(1)}_\kappa + (d_\nu^+ \times \tilde{d}_\nu)^{(1)}_\kappa \qquad (d_\pi^+ \times \tilde{d}_\pi)^{(3)}_\kappa + (d_\nu^+ \times \tilde{d}_\nu)^{(3)}_\kappa \tag{39.30}$$

The quadratic Casimir operator is constructed from these tensors, in analogy with (34.4), as

$$\boxed{\begin{aligned} C_{O_{\pi+\nu}(5)} = 2\sum_{L=1,3} &((d_\pi^+ \times \tilde{d}_\pi)^{(L)} + (d_\nu^+ \times \tilde{d}_\nu)^{(L)}) \\ &\cdot((d_\pi^+ \times \tilde{d}_\pi)^{(L)} + (d_\nu^+ \times \tilde{d}_\nu)^{(L)}) \end{aligned}} \tag{39.31}$$

The irreducible representations of $O(5)$ need not be fully symmetric as in the case of identical bosons. They are characterized in this case by Young diagrams with two rows of lengths τ_1 and τ_2 ($\tau_2 \le \tau_1$). The eigenvalues of the Casimir operator (39.31) are given in terms of these quantum numbers of the seniority of d_π- and d_ν-bosons by (28.25) as

$$\boxed{\tau_1(\tau_1+3)+\tau_2(\tau_2+1)} \tag{39.32}$$

In the case of identical bosons, or for $F = \frac{1}{2}(N_\pi + N_\nu)$, $\tau_1 = \tau$, $\tau_2 = 0$ and (39.32) is reduced to $\tau(\tau+3)$ as in (34.13).

The eigenvalues (39.32) may be obtained by using the F-spin in the same way as the spin S was used in Section 28 or the isospin T in Section 26. In fact, we may use the result (28.18) for $l = 2$ and apply it to d^n boson states. The phase $(-1)^L$ in (28.18) may be replaced in the present case by $(-1)^{F+1}$ since symmetric $F = 1$ d-boson states have even L values and $F = 0$ states have odd values of L. Replacing $(-1)^{F+1}$ by $(1 + 4(\mathbf{f}_1 \cdot \mathbf{f}_2))/2$ we obtain for the pairing interaction P, instead of (28.22)

$$P = \tfrac{1}{4}n_d(n_d - 1) + (\mathbf{F}^2 - \tfrac{3}{4}n_d) - \sum_{L\,\text{odd}} (2L + 1)(U^{(L)} \cdot U^{(L)}) + 2n_d \tag{39.33}$$

In the states with boson seniority τ and $n_d = \tau$, the l.h.s. of (39.33) vanishes. In such states, $\tau = \tau_1 + \tau_2$ and the value of F is $\frac{1}{2}(\tau_1 - \tau_2)$ which yields for the Casimir operator $2\sum_{L=1,3}(2L + 1)(U^{(L)} \cdot U^{(L)})$ the eigenvalues

$$\tfrac{1}{2}\tau(\tau - 1) + (\tau_1 - \tau_2)\left(\frac{\tau_1 - \tau_2}{2} + 1\right) - \tfrac{3}{2}\tau + 4\tau = \tau_1(\tau_1 + 3) + \tau_2(\tau_2 + 1)$$

which is the result (39.32) stated above.

The last subgroup in the chain of groups considered here is $O_{\pi+\nu}(3)$ whose generators are the components of $\mathbf{J}_\pi + \mathbf{J}_\nu$. Its Casimir operator is taken to be

$$(\mathbf{J}_\pi + \mathbf{J}_\nu) \cdot (\mathbf{J}_\pi + \mathbf{J}_\nu) = \mathbf{J}^2 \tag{39.34}$$

and its eigenvalues are $J(J + 1)$. The general Hamiltonian considered here may be expressed by a linear combination of the linear and quadratic Casimir operators of the groups in the chain

$$\text{I}' \qquad U_\pi(6) \otimes U_\nu(6) \supset U_{\pi+\nu}(6) \supset U_{\pi+\nu}(5) \supset O_{\pi+\nu}(5) \supset O_{\pi+\nu}(3) \tag{39.35}$$

with arbitrary coefficients. The eigenvalues of such a Hamiltonian are then the linear combination of eigenvalues of the Casimir operators and may be written as

$$\begin{aligned} &aN + b\tfrac{1}{2}N(N - 1) + cF(F + 1) + \epsilon(n_{d_\pi} + n_{d_\nu}) \\ &\quad + \alpha[n_{d_1}(n_{d_1} + 4) + n_{d_2}(n_{d_2} + 2)] \\ &\quad + \beta(\tau_1(\tau_1 + 3) + \tau_2(\tau_2 + 1)) + \gamma J(J + 1) \end{aligned} \tag{39.36}$$

We shall not discuss these eigenvalues in detail. In the case of a typical vibrational spectrum, $\epsilon > 0$ is considerably larger than the coefficients α, β, γ which determine the order of levels within a multiplet. In that case there is not much difference between states with $n_{d_2} = 0$ and $n_{d_2} = 1$. To avoid intruding of such states into the low-lying spectrum, the coefficient c in (39.36) should be taken to be large and repulsive.

The next dynamical symmetry is associated with the subgroup $O_{\pi+\nu}(6)$ of $U_{\pi+\nu}(6)$. Its 15 generators are given in the present case by the following operators taken from (39.5)

$$\begin{gathered}(d_\pi^+ \times \tilde{d}_\pi)^{(L)} + (d_\nu^+ \times \tilde{d}_\nu)^{(L)} \qquad L = 1,3 \\ d_\pi^+ s_\pi + s_\pi^+ \tilde{d}_\pi + d_\nu^+ s_\nu + s_\nu^+ \tilde{d}_\nu\end{gathered} \tag{39.37}$$

The quadratic Casimir operator is constructed from these generators to be

$$\begin{aligned}2\sum_{L=1,3} & ((d_\pi^+ \times \tilde{d}_\pi)^{(L)} + (d_\nu^+ \times \tilde{d}_\nu)^{(L)}) \\ & \cdot ((d_\pi^+ \times \tilde{d}_\pi)^{(L)} + (d_\nu^+ \times \tilde{d}_\nu)^{(L)}) \\ & + (d_\pi^+ s_\pi + s_\pi^+ \tilde{d}_\pi + d_\nu^+ s_\nu + s_\nu^+ \tilde{d}_\nu) \\ & \cdot (d_\pi^+ s_\pi + s_\pi^+ \tilde{d}_\pi + d_\nu^+ s_\nu + s_\nu^+ \tilde{d}_\nu) \\ = & \, C_{O_{\pi+\nu}(5)} + (\mathbf{Q}'_\pi + \mathbf{Q}'_\nu) \cdot (\mathbf{Q}'_\pi + \mathbf{Q}'_\nu)\end{aligned} \tag{39.38}$$

The quadrupole operators $\mathbf{Q}'$ for protons and neutrons were introduced in (35.1). The irreducible representations of $O_{\pi+\nu}(6)$ which can be multiplied by $SU_F(2)$ irreducible representations to yield fully symmetric states, have Young diagrams with at most two rows whose lengths are σ_1 and σ_2 ($\sigma_2 \leq \sigma_1$). All eigenvalues of the Casimir operator of the orthogonal group $O(2k)$ with an *even* number of dimensions are given by an expression similar to (28.25). The formula for the eigenvalues is given in terms of σ_i—the lengths of the k rows by

$$\sum_{i=1}^{k} \sigma_i(\sigma_i + 2k - 2i) = \sigma_1(\sigma_1 + 2k) + \sigma_2(\sigma_2 + 2k - 2) + \cdots + \sigma_k^2 \tag{39.39}$$

In the present case ($k = 3$) with only σ_1 and σ_2 not vanishing, the eigenvalues are $\sigma_1(\sigma_1 + 4) + \sigma_2(\sigma_2 + 2)$. In the case of identical bosons, or $F = \frac{1}{2}N$, the fully symmetric representations have $\sigma_2 = 0$ and $\sigma_1 = \sigma$ as in (35.7). As explained in Section 35, σ is the generalized seniority of s and d-bosons.

The eigenvalues $\sigma_1(\sigma_1 + 4) + \sigma_2(\sigma_2 + 2)$ of the Casimir operator of $O_{\pi+\nu}(6)$ may be obtained directly by using a similar procedure to the one leading to (39.33). Instead of using the pairing interaction for d-bosons we may use the generalized pairing interaction (35.5). The latter has non-vanishing eigenvalues in two boson states only in the $J = 0$ state $((d^+ \cdot d^+) - (s^+)^2)|0\rangle$. To the r.h.s. of (28.18) we add the contributions of $Q' \cdot Q'$, taking into account the normalization of the various tensors, obtaining

$$(-1)^L(2 - 2(2l + 1)\delta_{L0}) + 4l - 2\delta_{L0} + 2$$

For $l = 2$ we obtain instead of (39.33) the following expression for the generalized pairing interaction

$$\tfrac{1}{4}N(N-1) + (\mathbf{F}^2 - \tfrac{3}{4}N) - \tfrac{1}{2}C_{O(6)} + \tfrac{5}{2}N \qquad (39.40)$$

In states with highest generalized boson seniority, $\sigma = N = \sigma_1 + \sigma_2$, $F = \frac{1}{2}(\sigma_1 - \sigma_2)$, the generalized pairing interaction vanishes and we obtain from (39.40) the eigenvalues of $C_{O(6)}$ as

$$\tfrac{1}{2}N^2 + 3N + 2F(F+1) = \sigma_1(\sigma_1 + 4) + \sigma_2(\sigma_2 + 2) \qquad (39.41)$$

The expression (39.41) is identical with the one obtained from (39.39).

The Hamiltonian in this case of dynamical symmetry may be expressed as a linear combination of linear and quadratic Casimir operators of the groups in the chain

$$\text{III}' \quad U_\pi(6) \otimes U_\nu(6) \supset U_{\pi+\nu}(6) \supset O_{\pi+\nu}(6) \supset O_{\pi+\nu}(5) \supset O_{\pi+\nu}(3) \qquad (39.42)$$

with arbitrary coefficients. The eigenvalues of such Hamiltonians are equal to the linear combination of eigenvalues of these Casimir operators. They may be thus expressed by

$$\boxed{\begin{aligned} &aN + b\tfrac{1}{2}N(N-1) + cF(F+1) + A(\sigma_1(\sigma_1+4) + \sigma_2(\sigma_2+2)) \\ &\quad + B(\tau_1(\tau_1+3) + \tau_2(\tau_2+1)) + CJ(J+1) \end{aligned}} \tag{39.43}$$

In nuclei whose spectra were described by the $O(6)$ limit of IBA-1, the lowest levels have $\sigma = N$, the $\sigma = N-2$ levels being considerably higher. With values of the coefficients A which reproduce this feature, the IBA-2 states with $\sigma_1 = N-1$, $\sigma_2 = 1$ levels are considerably higher than those with $\sigma_1 = N$, $\sigma_2 = 0$. If in addition, the coefficient c is large and repulsive, the lowest IBA-2 levels will be the analogs of IBA-1 levels with $F = \frac{1}{2}N$.

Another dynamical symmetry of IBA-2 Hamiltonians which commute with components of the F-spin, is associated with the subgroup $SU_{\pi+\nu}(3)$. The allowed irreducible representations in the IBA-1 case were considered in Section 36. These fully symmetric irreducible representations in s and d bosons could have Young diagrams with two rows with lengths $\lambda + \mu$ and μ with the values of λ and μ restricted to *even* numbers. In the IBA-2 case considered here, this restriction is removed and states with $F < N/2$ may belong to irreducible representations with odd values of μ.

The generators of $SU_{\pi+\nu}(3)$ are components of the special quadrupole tensor

$$\begin{aligned} \mathbf{Q}_\pi + \mathbf{Q}_\mu &= s_\pi^+\tilde{d}_\pi + d_\pi^+ s_\pi - \frac{\sqrt{7}}{2}(d_\pi^+ \times \tilde{d}_\pi)^{(2)} \\ &\quad + s_\nu^+\tilde{d}_\nu + d_\nu^+ s_\nu - \frac{\sqrt{7}}{2}(d_\nu^+ \times \tilde{d}_\nu)^{(2)} \end{aligned} \tag{39.44}$$

and of the total angular momentum vector $\mathbf{J} = \mathbf{J}_\pi + \mathbf{J}_\nu$. The Casimir operator of $SU_{\pi+\nu}(3)$ is defined in the present case by

$$\boxed{C_{SU_{\pi+\nu}(3)} = 2(\mathbf{Q}_\pi + \mathbf{Q}_\nu)\cdot(\mathbf{Q}_\pi + \mathbf{Q}_\nu) + \tfrac{3}{4}(\mathbf{J}_\pi + \mathbf{J}_\nu)\cdot(\mathbf{J}_\pi + \mathbf{J}_\nu)} \tag{39.45}$$

with eigenvalues given by (36.8) as $\lambda^2 + \mu^2 + \lambda\mu + 3(\lambda + \mu)$. The chain of groups for this dynamical symmetry is given by

$$\text{II}' \qquad U_\pi(6) \otimes U_\nu(6) \supset U_{\pi+\nu}(6) \supset SU_{\pi+\nu}(3) \supset O_{\pi+\nu}(3) \tag{39.46}$$

The general Hamiltonian with this symmetry can be expressed as a linear combination of the Casimir operators of the groups in this chain. The eigenvalues are then obtained by the linear combination of eigenvalues of the Casimir operators. They may be written as

$$\boxed{\begin{aligned} &aN + b\tfrac{1}{2}N(N-1) + cF(F+1) \\ &\qquad + \kappa(\lambda^2 + \mu^2 + \lambda\mu + 3\lambda + 3\mu) + \kappa' J(J+1) \end{aligned}} \tag{39.47}$$

The choice of coefficients κ and κ' here is different from the definition (36.6).

The allowed states of IBA-1, discussed in Section 36, correspond to IBA-2 states with $F = \frac{1}{2}N = \frac{1}{2}(N_\pi + N_\nu)$. They belong to bases of irreducible representations of $SU(3)$ which are fully symmetric in s and d-bosons. As explained in Section 36, these irreducible representations have the following (λ, μ) quantum numbers

$$\begin{aligned} &(2N,0), (2N-4,2), (2N-8,4), \ldots \\ &\qquad\qquad (2N-6,0), (2N-10,2), \ldots \\ &\qquad\qquad\qquad\qquad (2N-12,0), \ldots \end{aligned}$$

In IBA-2 other irreducible representations may occur if $F < (N_\pi + N_\nu)/2$. For example, if $N_\pi = N-1, N_\nu = 1$, some of the allowed states belong to bases of irreducible representations included in the product $(2N-2,0) \otimes (2,0)$. These are $(2N,0), (2N-4,2)$ with $F = N/2$ *and* $(2N-2,1)$ whose states have $F = (N-2)/2$. Among the states which belong to the irreducible representation $(2N-2,1)$ there is a state with $J = 1$. It arises from $K = \mu = 1$, according to (36.10) and $J = K = 1$ which appears in (36.11). Such a collective $J = 1$ state is not included at all in IBA-1 states. For $F = N/2$, in fully symmetric irreducible representations, μ (as well as λ) is always an *even* number. If $K = 0$ then the rotational band includes, according to (36.11) states with only even values of J. The fact that IBA-1 states do not include one with $J = 1$ follows simply in the $SU(3)$ limit but it is true, of course, in any scheme of the symmetric irreducible representations of $U(6)$. In the $U(5)$ limit it is a consequence of (34.18) and (34.19) as explained above. The eigenvalue of the $SU(3)$ Casimir operator in $(2N-2,1)$ states is larger than that eigenvalue in the states of the $(2N-4,2)$ irreducible representation. Hence, for $\kappa<0$ in (38.47), that $J=1$ state should be *lower* than the states of the $K=0$ and $K=2$ bands of the $(2N-4,2)$ irreducible representation with $F=N/2$. Hence, to avoid this feature which contradicts experimental data, it is necessary to make the coefficient c in (39.47) sufficiently repulsive. Properties of this $J=1$ state will be considered in the next section.

The dynamical symmetries of IBA-2 described above are all in cases where all eigenstates of the Hamiltonian have definite values of F-spin. As emphasized above, assuming that the Hamiltonian is fully symmetric in proton and neutron bosons is not in agreement with the information we have on nuclear interactions. It is therefore important to notice that dynamical symmetries may occur also for Hamiltonians which do not commute with components of the F-spin vector. The first case we consider is a dynamical symmetry associated with the $SU(3)$ group.

The $SU(3)$ symmetry considered above was obtained by constructing Hamiltonians from the $U(6)$ generators (39.5) which are certain linear combinations of the $U_\pi(6)\otimes U_\nu(6)$ generators in (39.2) and (39.3). To go to the $SU(3)$ subgroup of $U(6)$ the choice of generators from (39.5) was restricted to the components of the quadrupole operator (39.44) and components of $\mathbf{J}=\mathbf{J}_\pi+\mathbf{J}_\nu$. The same $SU_{\pi+\nu}(3)$ group may be obtained by a different route (Dieperink and Bijker 1982, Dieperink and Talmi 1983). We may first choose from the generators (39.2) the 8 components of $\mathbf{Q}_\pi$ and $\mathbf{J}_\pi$ and from the set (39.3) the operators $\mathbf{Q}_\nu$ and $\mathbf{J}_\nu$. This choice yields the algebra of $SU_\pi(3)\otimes SU_\nu(3)$ which is a subalgebra of the $U_\pi(6)\otimes U_\nu(6)$ one. In the next step $SU_\pi(3)\otimes SU_\nu(3)$ is reduced into its subgroup $SU_{\pi+\nu}(3)$ whose generators are $\mathbf{Q}_\pi+\mathbf{Q}_\nu$ and $\mathbf{J}_\pi+\mathbf{J}_\nu$. The chain of groups in this case is

$$\text{II}'' \qquad U_\pi(6)\otimes U_\nu(6)\supset SU_\pi(3)\otimes SU_\nu(3)\supset SU_{\pi+\nu}(3)\supset O_{\pi+\nu}(3) \tag{39.48}$$

The chain (39.48) is similar to the one in (39.46) but it represents a wider class of dynamical symmetries.

The Hamiltonians with dynamical symmetry II″ may be expressed as a linear combination of the Casimir operators of the groups in the chain (39.48). As in the previous cases, we do not include $C_{U_\pi(6)}$ and $C_{U_\nu(6)}$ which just add linear and quadratic terms in N_π and in N_ν. We can thus express such Hamiltonians as

$$\boxed{\tfrac{1}{2}\kappa_\pi C_{SU_\pi(3)} + \tfrac{1}{2}\kappa_\nu C_{SU_\nu(3)} + \tfrac{1}{2}\kappa C_{SU_{\pi+\nu}(3)} + \kappa' \mathbf{J}^2} \tag{39.49}$$

The eigenstates of the Hamiltonian (39.49) are obtained by coupling proton boson states of the (λ_π, μ_π) irreducible representation to neutron boson states with (λ_ν, μ_ν) to yield states with a definite (λ, μ) irreducible representation of $SU(3)$. The Casimir operators in (39.49) commute and the eigenvalues of the Hamiltonian are expressed as the linear combination of eigenvalues of the Casimir operators. The eigenvalues in the states described above are thus given by

$$\boxed{\begin{aligned} &\tfrac{1}{2}\kappa_\pi(\lambda_\pi^2 + \mu_\pi^2 + \lambda_\pi\mu_\pi + 3\lambda_\pi + 3\mu_\pi) \\ &\quad + \tfrac{1}{2}\kappa_\nu(\lambda_\nu^2 + \mu_\nu^2 + \lambda_\nu\mu_\nu + 3\lambda_\nu + 3\mu_\nu) \\ &\quad + \tfrac{1}{2}\kappa(\lambda^2 + \mu^2 + \lambda\mu + 3\lambda + 3\mu) + \kappa' J(J+1) \end{aligned}} \tag{39.50}$$

The eigenvalues of a $SU(3)$ Hamiltonian which commutes with components of $\mathbf{F}$ are given as a special case of (39.50) with $\kappa_\pi = \kappa_\nu = 0$. If κ_π or κ_ν do not vanish, the Hamiltonian (39.49) is no longer fully symmetric in proton and neutron bosons. In contrast to (39.46), $U_{\pi+\nu}$ does not appear in the group chain (39.48). Hence, F-spin which defines symmetry of $U_{\pi+\nu}(6)$ states, need no longer be a good quantum number.

Other choices of the coefficients in (39.49) yield more interesting Hamiltonians. Recalling (39.45), we may express the Hamiltonian (39.49) as

$$\boxed{\begin{aligned} &2\kappa(\mathbf{Q}_\pi \cdot \mathbf{Q}_\nu) + (\tfrac{3}{4}\kappa + 2\kappa')(\mathbf{J}_\pi \cdot \mathbf{J}_\nu) \\ &\quad + (\kappa + \kappa_\pi)(\mathbf{Q}_\pi \cdot \mathbf{Q}_\pi) + (\kappa + \kappa_\nu)(\mathbf{Q}_\nu \cdot \mathbf{Q}_\nu) \\ &\quad + (\tfrac{3}{8}\kappa + \tfrac{3}{8}\kappa_\pi + \kappa')\mathbf{J}_\pi^2 + (\tfrac{3}{8}\kappa + \tfrac{3}{8}\kappa_\nu + \kappa')\mathbf{J}_\nu^2 \end{aligned}} \tag{39.51}$$

If we put $\kappa_\pi = \kappa_\nu = -\kappa$ and $\kappa' = 0$ in (39.51) we obtain a Hamiltonian in which there are no interactions between identical bosons. The Hamiltonian then is equal to interactions between proton-bosons and neutron bosons given by

$$H' = 2\kappa \mathbf{Q}_\pi \cdot \mathbf{Q}_\nu + \tfrac{3}{4}\kappa \mathbf{J}_\pi \cdot \mathbf{J}_\nu \tag{39.52}$$

If we choose $\kappa_\pi = \kappa_\nu = -\kappa$ and $\kappa' = -\frac{3}{8}\kappa$ we obtain

$$H'' = 2\kappa \mathbf{Q}_\pi \cdot \mathbf{Q}_\nu - \tfrac{3}{8}\kappa \mathbf{J}_\pi \cdot \mathbf{J}_\pi - \tfrac{3}{8}\kappa \mathbf{J}_\nu \cdot \mathbf{J}_\nu \tag{39.53}$$

In both H' and H'' there is no quadrupole interaction between identical bosons. In the case of (39.53) there is a dipole interaction between identical bosons for which n_{d_π} and n_{d_ν} are good quantum numbers. The dipole interaction is a boson model for an effective interaction between identical nucleons which is diagonal in the (nucleon) seniority scheme. These features are in qualitative agreement with properties of the nuclear interaction described in preceding sections. Quantitatively, however, neither Hamiltonian (39.52), (39.53) nor any of the form (39.49) can give a satisfactory description of rotational nuclear spectra. This will become clear when we calculate their eigenvalues.

The eigenstates of (39.49) for an attractive value of κ and values of κ_π and κ_ν which are not too repulsive, belong to the $(2N_\pi + 2N_\nu, 0)$ irreducible representation of $SU_{\pi+\nu}(3)$. Such states may be obtained only from proton boson and neutron boson states with $(\lambda_\pi, \mu_\pi) = (2N_\pi, 0)$ and $(\lambda_\nu, \mu_\nu) = (2N_\nu, 0)$ respectively. For these values, (39.50) becomes equal to

$$\boxed{\begin{aligned} &4\kappa N_\pi N_\nu + (\kappa + \kappa_\pi)(2N_\pi^2 + 3N_\pi) \\ &\qquad + (\kappa + \kappa_\nu)(2N_\nu^2 + 3N_\nu) + \kappa' J(J+1) \end{aligned}} \tag{39.54}$$

As noted above, the Hamiltonian (39.49) does not commute with components of the F-spin vector if $\kappa_\pi \neq 0$ or $\kappa_\nu \neq 0$. Eigenstates which belong to the irreducible representation $(2N_\pi + 2N_\nu, 0)$, however, all have definite values of F-spin equal to $F = (N_\pi + N_\nu)/2$. The reason is that this representation is included (only) in the product of the $(2N_\pi, 0)$ and $(2N_\nu, 0)$ irreducible representations. Hence, variation of the values of κ_π and κ_ν does not change the structure of these eigenstates. Still, the eigenvalues (39.54) depend (for $\kappa_\pi \neq 0$ or

$\kappa_\nu \neq 0$) explicitly on N_π and N_ν and therefore, on $N = N_\pi + N_\nu$ and $M_F = (N_\pi - N_\nu)/2$. Only for $\kappa_\pi = \kappa_\nu = 0$ do the eigenvalues (39.54) depend only on N.

Eigenstates in higher irreducible representations of $SU_{\pi+\nu}(3)$ do not in general have definite F-spins. For example, eigenstates within the $(2N_\pi + 2N_\nu - 4, 2)$ irreducible representation are obtained by reduction of the product of $(2N_\pi, 0)$ and $(2N_\nu, 0)$ irreducible representations. They are, however, also contained in the product of $(2N_\pi - 4, 2)$ and $(2N_\nu, 0)$ and in the product of $(2N_\pi, 0)$ and $(2N_\nu - 4, 2)$ representations. In the case $\kappa_\pi = \kappa_\nu = 0$, states (with the same value of J) of the three equivalent $(2N_\pi + 2N_\nu - 4, 2)$ irreducible representations are degenerate since the eigenvalues (39.54) depend then only on λ and μ. One linear combination of such states has $F = \frac{1}{2}(N_\pi + N_\nu)$ whereas the other orthogonal combinations have higher values of F-spin. If, however, κ_π or κ_ν do not vanish, states are no longer degenerate and eigenstates are linear combinations of states with different F-spin values. The $J = 1$ state mentioned above, which belongs to the $(2N_\pi + 2N_\nu - 2, 1)$ irreducible representation, can be obtained only by coupling states in the $(2N_\pi, 0)$ and $(2N_\nu, 0)$ irreducible representations. Hence, its structure is not affected by the values of κ_π and κ_ν and it has, as well as other states in that representation, definite F-spin. States in the irreducible representation $(2N - 2, 1)$ may occur for $N_\pi = N - 1$, $N_\nu = 1$ and hence, the F-spin of these states is $F = (N - 2)/2$.

Another dynamical symmetry which also leads to eigenstates with $SU(3)$ labels may be obtained in IBA-2. As in the group chain II″ in (39.48), we start with the generators (39.2) and (39.3) but choose different linear combinations. For proton bosons we choose from (39.2) the components of $\mathbf{J}_\pi$ and the components of

$$\boxed{\mathbf{Q}_\pi^{(-)} = s_\pi^+ \tilde{d}_\pi + d_\pi^+ s_\nu - \frac{\sqrt{7}}{2}(d_\pi^+ \times \tilde{d}_\pi)^{(2)}} \tag{39.55}$$

For the neutron bosons, along with components of $\mathbf{J}_\nu$ we take components of

$$\boxed{\mathbf{Q}_\nu^{(+)} = s_\nu^+ \tilde{d}_\nu + d_\nu^+ s_\nu + \frac{\sqrt{7}}{2}(d_\nu^+ \times \tilde{d}_\nu)^{(2)}} \tag{39.56}$$

As explained in Section 36, the components of *either* $\mathbf{Q}^{(-)}$ or $\mathbf{Q}^{(+)}$, together with the components of $\mathbf{J}$, generate the $SU(3)$ algebra. When this choice of generators is made, the $U_\pi(6) \otimes U_\nu(6)$ Lie algebra is limited to that of the subalgebra of the direct product of the proton $SU(3)$ and the neutron $SU(3)$ algebras. To distinguish between the $SU(3)$ algebra with generators $\mathbf{Q}^{(-)}$ and that with the $\mathbf{Q}^{(+)}$ generators, we denote the latter by $\overline{SU(3)}$. In the dynamical symmetries considered above, both $\mathbf{Q}_\pi$ and $\mathbf{Q}_\nu$ were taken to be equal to $\mathbf{Q}_\pi^{(-)}$ and $\mathbf{Q}_\nu^{(-)}$ respectively. Using instead $\mathbf{Q}_\pi^{(+)}$ and $\mathbf{Q}_\nu^{(+)}$ yields equivalent Hamiltonians with the same eigenvalues and same structure of eigenstates. If the interaction between proton bosons and neutron bosons is given by $\kappa \mathbf{Q}_\pi^{(-)} \cdot \mathbf{Q}_\nu^{(+)}$ it is essentially different from the interactions $\kappa \mathbf{Q}_\pi^{(+)} \cdot \mathbf{Q}_\nu^{(+)}$ or $\kappa \mathbf{Q}_\pi^{(-)} \cdot \mathbf{Q}_\nu^{(-)}$. The Hamiltonian containing it has *different* eigenvalues and eigenstates. We notice that the interaction $\kappa \mathbf{Q}_\pi^{(-)} \cdot \mathbf{Q}_\nu^{(+)}$ is equivalent to $\kappa \mathbf{Q}_\pi^{(+)} \cdot \mathbf{Q}_\nu^{(-)}$ in which the signs in the quadrupole operators were interchanged. A trivial (canonical) transformation $s_\pi \to -s_\pi$, $s_\pi^+ \to -s_\pi^+$, $s_\nu \to -s_\nu$, $s_\nu^+ \to -s_\nu^+$ transforms one interaction into the other.

The situation in which the signs in $\mathbf{Q}_\pi$ and $\mathbf{Q}_\nu$ are different can be traced to the fermion operators. In Section 37 we saw that the part of the quadrupole operator that changes seniority has a constant sign throughout the shell. On the other hand, the seniority conserving term may well change its sign when the shell is being filled. In the case of a single j-orbit or in the quasi-spin scheme it changes sign in the middle of the shell as in (37.45). In more complicated cases, the change in sign may occur for other nucleon numbers. The boson operator (37.52), which is the image of the seniority conserving part, must change its sign as well. Thus, the signs of the second term in $\mathbf{Q}_\pi$ and $\mathbf{Q}_\nu$ may depend on the occupation numbers of the protons and neutrons in the nucleus considered. In the geometrical picture of these states it would seem that proton states have *prolate* deformation whereas the neutron states have *oblate* deformation or *vice versa*.

In the chain of groups leading to dynamical symmetry, the restriction to the generators (39.55) and (39.56) as well as $\mathbf{J}_\pi$ and $\mathbf{J}_\nu$ amounts to going from

$$U_\pi(6) \otimes U_\nu(6)$$

to the subgroup

$$SU_\pi(3) \otimes \overline{SU_\nu(3)}$$

The next step is to choose as generators the linear combinations

$$\boxed{\mathbf{Q}_\pi^{(-)} + \mathbf{Q}_\nu^{(+)} \qquad \mathbf{J} = \mathbf{J}_\pi + \mathbf{J}_\nu} \tag{39.57}$$

The generators (39.57) form a Lie algebra of a $SU(3)$ group which is different from $SU_{\pi+\nu}(3)$ constructed in (39.48). We denote it by $SU^*_{\pi+\nu}(3)$. It has $O(3)$ as a subgroup generated by components of $\mathbf{J} = \mathbf{J}_\pi + \mathbf{J}_\nu$. Thus we obtain the chain of groups

$$\mathrm{II}''' \quad U_\pi(6) \otimes U_\nu(6) \supset SU_\pi(3) \otimes \overline{SU_\nu(3)} \supset SU^*_{\pi+\nu}(3) \supset O_{\pi+\nu}(3) \tag{39.58}$$

The Hamiltonian with this dynamical symmetry may be expressed as in (39.49) by

$$\tfrac{1}{2}\kappa_\pi C_{SU_\pi(3)} + \tfrac{1}{2}\kappa_\nu C_{\overline{SU_\nu(3)}} + \tfrac{1}{2}\kappa C_{SU^*_{\pi+\nu}(3)} + \kappa' \mathbf{J}^2 \tag{39.59}$$

At the end of Section 37 we saw that single boson states which transform under action of $\mathbf{Q}^{(-)}$ according to the irreducible representation $(2,0)$ transform under the action of $\mathbf{Q}^{(+)}$ according to the conjugate representation $(0,2)$. Since these states are the building blocks of all irreducible representations of $SU(3)$ in the space of s- and d-bosons, states of the (λ,μ) irreducible representation transform under action of $\mathbf{Q}^{(+)}$ according to the conjugate representation (μ,λ). Hence, the eigenvalues of $C_{\overline{SU_\nu(3)}}$ in eigenstates characterized by (λ_ν,μ_ν) are given by the same expression as for $C_{SU_\nu(3)}$. For $\kappa_\nu < 0$ the lowest eigenvalues of $\frac{1}{2}\kappa_\nu C_{\overline{SU_\nu(3)}}$ are for states in the irreducible representation $(0,2N_\nu)$, the next higher states are in the $(2,2N_\nu-4)$ irreducible representation and so on. The roles of λ and μ are simply reversed in these conjugate representations. For $\kappa < 0$ the lowest states of $\frac{1}{2}\kappa C_{SU^*_{\pi+\nu}(3)}$ are those with maximum values of λ and μ among the irreducible representations included in the product of $(2N_\pi,0)$ and $(0,2N_\nu)$ representations.

The lowest eigenvalues of the proton boson-neutron boson interaction are in eigenstates which are in the $(2N_\pi,2N_\nu)$ irreducible representation of $SU^*_{\pi+\nu}(3)$. If κ_π and κ_ν are not too repulsive, those are also the lowest eigenstates of the Hamiltonian (39.59). The eigenvalues of those states are given by the linear combination of eigenvalues of the Casimir operators in (39.59). The explicit expression is given, like in the $SU_{\pi+\nu}(3)$ case by (39.50). For $(\lambda,\mu)=(2N_\pi,2N_\nu)$ states, the eigenvalues are given by

$$2\kappa N_\pi N_\nu + (\kappa+\kappa_\pi)(2N_\pi^2+3N_\pi) + (\kappa+\kappa_\nu)(2N_\nu^2+3N_\nu) + \kappa' J(J+1) \tag{39.60}$$

The proton boson-neutron boson interaction in (39.60) is only half of that interaction in (39.54). The overlap between prolate and oblate "shapes" is not as good as in the prolate-prolate or oblate-oblate case. Other irreducible representations included in the product of $(2N_\pi,0)$ and $(0,2N_\nu)$ irreducible representations have (λ,μ) values of $(2N_\pi-k,2N_\nu-k)$ for any k satisfying $k\le \min(2N_\pi,2N_\nu)$.

In the case of $SU_{\pi+\nu}(3)$ dynamical symmetry, we saw that for $\kappa_\pi=0$, $\kappa_\nu=0$ there were degeneracies in states of higher irreducible representations like $(2N_\pi+2N_\nu-4,2)$. In the case of $SU^*_{\pi+\nu}(3)$ dynamical symmetry with $\kappa_\pi=-\kappa$, $\kappa_\nu=-\kappa$, there are rather large degeneracies even in the ground state of Hamiltonians like (39.52) or (39.53). The eigenvalues of (39.59) given by (39.60) with $\kappa_\pi=-\kappa$, $\kappa_\nu=-\kappa$ are equal to $2\kappa N_\pi N_\nu+\kappa' J(J+1)$ in all states characterized by

$$(\lambda_\pi,\mu_\pi)=(2N_\pi,0) \qquad (\lambda_\nu,\mu_\nu)=(2k,2N_\nu-4k)$$
$$(\lambda,\mu)=(2N_\pi+2k,2N_\nu-4k)$$

with $k\le\frac{1}{2}N_\nu$. These eigenvalues are obtained also for all states characterized by $(\lambda_\pi,\mu_\pi)=(2N_\pi-4k,2k)$, $(\lambda_\nu,\mu_\nu)=(0,2N_\nu)$ and (λ,μ) $=(2N_\pi-4k,2N_\nu+2k)$ with $k\le\frac{1}{2}N_\pi$. It seems that if the proton shape is prolate and the neutron shape is oblate, or *vice versa*, and one shape is in its ground state, the amount of quadrupole proton-neutron interaction is independent of the state of excitation of the other shape. This large degeneracy which does not correspond to the situation in any nucleus, can be removed in several ways. If $\kappa+\kappa_\pi<0$, $\kappa+\kappa_\nu<0$, the degeneracy of the ground state does not occur. This is true, in particular, for $\kappa_\pi=\kappa_\nu=0$ in which case, however, several excited states are degenerate. An obvious way to remove the degeneracy is to break the $SU^*_{\pi+\nu}(3)$ dynamical symmetry by adding to the Hamiltonian the physically meaningful terms $\epsilon_\pi(d_\pi^+\cdot\tilde{d}_\pi)$ and $\epsilon_\nu(d_\nu^+\cdot\tilde{d}_\nu)$.

As explained above, the dynamical symmetries in IBA-1 have simple geometrical interpretations. The $U(5)$ symmetry corresponds to the vibrational limit of the collective model, $SU(3)$ symmetry leads to rotational states of an axially symmetric deformed shape. The dynamical symmetry associated with $O(6)$ is that of deformed γ-unstable (or γ-soft) nuclei for which the potential energy does not depend on γ. No limit of IBA-1 Hamiltonians with only boson-boson interactions corresponds to triaxial deformation in the collective model. In IBA-2 there is much more freedom and triaxial shapes may emerge. The $SU^*(3)$ symmetry can lead to such cases. If the proton and neutron shapes, both deformed with axial symmetry, are not aligned, a triaxial shape is obtained.

Dynamical symmetries analogous to those corresponding to the chain of groups II″ in (39.48) can be obtained also for other subgroups of $U(6)$. Instead of the group chain I′ in (39.35) we may choose the following chain of subgroups

$$\text{I}'' \qquad U_\pi(6) \otimes U_\nu(6) \supset U_\pi(5) \otimes U_\nu(5) \supset U_{\pi+\nu}(5) \supset O_{\pi+\nu}(5) \supset O_{\pi+\nu}(3) \tag{39.61}$$

Hamiltonians with this kind of dynamical symmetry may be expressed as a linear combination of the Casimir operators in (39.61). We omit the Casimir operators of $U_\pi(6) \otimes U_\nu(6)$ which contribute linear and quadratic terms in N_π and in N_ν and obtain

$$\boxed{\begin{aligned} &\epsilon_\pi n_{d_\pi} + \epsilon_\nu n_{d_\nu} + \alpha_\pi n_{d_\pi}(n_{d_\pi} - 1) \\ &\quad + \alpha_\nu n_{d_\nu}(n_{d_\nu} - 1) + \alpha C_{U_{\pi+\nu}(5)} + \beta C_{O_{\pi+\nu}(5)} + \gamma \mathbf{J}^2 \end{aligned}} \tag{39.62}$$

The eigenvalues of the Hamiltonian (39.62) are then given by

$$\begin{aligned} &\epsilon_\pi n_{d_\pi} + \epsilon_\nu n_{d_\nu} + \alpha_\pi n_{d_\pi}(n_{d_\pi} - 1) + \alpha_\nu n_{d_\nu}(n_{d_\nu} - 1) \\ &\quad + \alpha[n_{d_1}(n_{d_1} + 4) + n_{d_2}(n_{d_2} + 2)] \\ &\quad + \beta(\tau_1(\tau_1 + 3) + \tau_2(\tau_2 + 1)) + \gamma J(J + 1) \end{aligned} \tag{39.63}$$

For the values $\epsilon_\pi = \epsilon_\nu = \epsilon$ and $\alpha_\pi = \alpha_\nu = 0$ the eigenvalues (39.63) are identical to those in (39.36) and the eigenstates have definite values of F-spins.

The dynamical symmetry associated with the $O(6)$ subgroup may be obtained in a similar way. Instead of the group chain III′ in (39.42), we may use the chain

$$\text{III}'' \qquad U_\pi(6)\otimes U_\nu(6)\supset O_\pi(6)\otimes O_\nu(6)\supset O_{\pi+\nu}(6)$$
$$\supset O_{\pi+\nu}(5)\supset O_{\pi+\nu}(3) \tag{39.64}$$

Hamiltonians with this dynamical symmetry do not necessarily commute with components of the F-spin. They may be expressed as

$$\boxed{A_\pi C_{O_\pi(6)} + A_\nu C_{O_\nu(6)} + AC_{O_{\pi+\nu}}(6) + BC_{O_{\pi+\nu}}(5) + C\mathbf{J}^2} \tag{39.65}$$

In (39.65), the linear and quadratic terms in N_π and N_ν due to $C_{U_\pi(6)}$ and $C_{U_\nu(6)}$ were omitted. The eigenvalues of the Hamiltonians (39.65) are given by

$$A_\pi\sigma_\pi(\sigma_\pi+4)+A_\nu\sigma_\nu(\sigma_\nu+4)+A[\sigma_1(\sigma_1+4)+\sigma_2(\sigma_2+4)]$$
$$+B(\tau_1(\tau_1+3)+\tau_2(\tau_2+1))+CJ(J+1) \tag{39.66}$$

For $A_\pi = A_\nu = 0$ the eigenvalues (39.66) coincide with those in (39.43) and the eigenstates of (39.65) have definite values of F-spin. The lowest eigenstates for $A_\pi < 0$, $A_\nu < 0$, $A < 0$ and smaller values of B and C in (39.65), are characterized by $\sigma_\pi = N_\pi$, $\sigma_\nu = N_\nu$ and $\sigma_1 = N_\pi + N_\nu$, $\sigma_2 = 0$. These states have definite F-spin equal to $F = N/2$ even if $A_\pi \neq 0$ or $A_\nu \neq 0$, since the irreducible representation of $O_{\pi+\nu}(6)$ to which they belong, can be obtained only from proton boson states with $\sigma_\pi = N_\pi$ and neutron boson states with $\sigma_\nu = N_\nu$. Hence, their structure is independent of the actual values of A_π and A_ν. We shall not discuss further these dynamical symmetries and will consider next more general IBA-2 Hamiltonians.

40

Simple IBA-2 Hamiltonians. Some Applications

It was mentioned above that certain Hamiltonians which do not commute with components of F-spin may still have eigenstates which have definite F-spins. The eigenvalues of such eigenstates, however, depend on M_F. Simple examples of such Hamiltonians are offered by adding terms, proportional to F^0 and $(F^0)^2$, to a Hamiltonian which is a scalar in F-spin. We saw more interesting examples in Section 39 in which the lowest eigenstates have definite values of F-spin, $F = N/2$. In general, eigenstates of IBA-2 Hamiltonians are not characterized by definite values of F-spin. Still, in some cases, the lowest eigenstates may have large amplitudes of $F = N/2$ states. In such cases it may be a good approximation to replace the IBA-2 Hamiltonian by another one whose eigenstates have definite F-spins. The latter is equivalent, for $F = N/2$ states, to an IBA-1 Hamiltonian whose coefficients depend explicitly on M_F (Scholten 1980, Harter et al. 1985).

The projection onto an equivalent IBA-1 Hamiltonian is accomplished by calculating the matrix of any given IBA-2 Hamiltonian in a scheme of states with definite values of $F = N/2$ and M_F. If nondiagonal matrix elements between these $F = N/2$ states and states with $F < N/2$ are put to zero, the matrix thus obtained defines in the space of $F = N/2$ states, a certain IBA-2 Hamiltonian. The latter has eigenstates which have definite F-spin values, $F = N/2$, and corresponding eigenvalues which depend on M_F. In order to obtain the M_F dependence, we use the Wigner-Eckart theorem for the $SU_F(2)$ group. We first express the IBA-2 Hamiltonian as a linear combination of irreducible tensors with respect to **F**. Since each creation and annihilation operator is a component of an irreducible F-spin tensor of rank $k = \frac{1}{2}$, IBA-2 Hamiltonians are linear combinations of irreducible tensors of ranks $k = 0$, $k = 1$ and $k = 2$.

We write accordingly

$$H = H_0^{(0)} + H_0^{(1)} + H_0^{(2)} \tag{40.1}$$

The fact that H does not change N_π and N_ν implies that the components of the F-spin tensors $H^{(1)}$ and $H^{(2)}$ in (40.1) both have $\kappa = 0$. Technically it follows from the transformation properties of boson creation and annihilation operators under action of components of **F**. From the definitions (39.6) and (39.7) we obtain for the non-vanishing commutators the results

$$\begin{aligned} &[F^-, s_\pi^+] = s_\nu^+ \qquad [F^-, d_{\pi\mu}^+] = d_{\nu\mu}^+ \\ &[F^+, s_\nu^+] = s_\pi^+ \qquad [F^+, d_{\nu_\mu}^+] = d_\pi^+ \end{aligned} \tag{40.2}$$

as well as

$$\begin{aligned} &[F^-, s_\nu] = -s_\pi \qquad [F^-, d_{\nu\mu}] = -d_{\pi\mu} \\ &[F^+, s_\pi] = -s_\nu \qquad [F^+, d_{\pi\mu}] = -d_{\nu\mu} \end{aligned} \tag{40.3}$$

Thus, proton creation operators and neutron annihilation operators have $\kappa = \frac{1}{2}$ whereas neutron creation operators and proton annihilation operators have $\kappa = -\frac{1}{2}$. IBA-2 Hamiltonians are linear combinations of products of equal numbers of proton creation and annihilation operators and neutron creation and annihilation operators. Hence, IBA-2 Hamiltonians are linear combinations of the zeroth components of irreducible F-spin tensors.

The decomposition may be obtained by a simple procedure. Taking the commutator of F^+ and H we obtain, due to the general relations (6.25),

$$[F^+, H] = [F^+, H_0^{(0)}] + [F^+, H_0^{(1)}] + [F^+, H_0^{(2)}] = \sqrt{2} H_1^{(1)} + \sqrt{6} H_1^{(2)} \tag{40.4}$$

Taking the commutator of F^+ and (40.4) we obtain

$$[F^+,[F^+,H]] = 2\sqrt{6}H_2^{(2)} \tag{40.5}$$

Commuting now F^- with (40.5) yields

$$[F^-,[F^+,[F^+,H]]] = 4\sqrt{6}H_1^{(2)} \tag{40.6}$$

and by commuting (40.6) with F^- we obtain $H_0^{(2)}$ in the form

$$[F^-,[F^-,[F^+,[F^+,H]]]] = 24H_0^{(2)} \tag{40.7}$$

From (40.4) and (40.5) $H_1^{(1)}$ is obtained as

$$\sqrt{2}H_1^{(1)} = [F^+,H] - \tfrac{1}{4}[F^-,[F^+,[F^+,H]]] \tag{40.8}$$

The commutator of F^- and (40.8) yields

$$2H_0^{(1)} = [F^-,[F^+,H]] - \tfrac{1}{4}[F^-,[F^-,[F^+,[F^+,H]]]] \tag{40.9}$$

Finally $H_0^{(0)}$ is obtained by subtracting from H the expressions for $H_0^{(1)}$ given by (40.9) and for $H_0^{(2)}$ given by (40.7). The commutators can be directly evaluated by using the relations (40.2) and (40.3).

Since in the commutation relations (40.2) and (40.3) the spatial state of the boson does not change, it is sufficient to carry out the decomposition (40.1) for some typical terms in the IBA-2 Hamiltonian. Let us start with the single boson term in the Hamiltonian and consider the terms $s_\pi^+ s_\pi$ and $s_\nu^+ s_\nu$. These include in their expansion only $k = 0$ and $k = 1$ terms. For these operators we obtain

$$\begin{gathered} [F^+,s_\pi^+ s_\pi] = [s_\pi^+ s_\nu, s_\pi^+ s_\pi] = -s_\pi^+ s_\nu \qquad [F^+,[F^+,s_\pi^+ s_\pi]] = 0 \\ [F^-,[F^+,s_\pi^+ s_\pi]] = [s_\nu^+ s_\pi, -s_\pi^+ s_\nu] = -s_\pi^+ s_\pi + s_\nu^+ s_\nu \end{gathered} \tag{40.10}$$

The commutators with $s_\nu^+ s_\nu$ give similar results

$$[F^+,s_\nu^+ s_\nu] = s_\pi^+ s_\nu \qquad [F^-[F^+,s_\nu^+ s_\nu]] = -s_\pi^+ s_\pi + s_\nu^+ s_\nu \tag{40.11}$$

The results for $(d_\pi^+ \cdot \tilde{d}_\pi)$ and $(d_\nu^+ \cdot \tilde{d}_\nu)$ are the direct analogs of (40.10) and (40.11). Hence, the scalar part, with respect to F-spin, of

$\epsilon_\pi(d_\pi^+ \cdot \tilde{d}_\pi) + \epsilon_\nu(d_\nu^+ \cdot \tilde{d}_\nu)$ is given by

$$\frac{\epsilon_\pi + \epsilon_\nu}{2}((d_\pi^+ \cdot \tilde{d}_\pi) + (d_\nu^+ \cdot \tilde{d}_\nu)) \tag{40.12}$$

whereas the $k = 1$ part is equal to

$$\frac{\epsilon_\pi - \epsilon_\nu}{2}((d_\pi^+ \cdot \tilde{d}_\pi) - (d_\nu^+ \cdot \tilde{d}_\nu)) \tag{40.13}$$

Products of single boson operators can be similarly decomposed. The term $s_\pi^+ s_\pi s_\pi^+ s_\pi$ is equal to

$$\begin{aligned} s_\pi^+ s_\pi s_\pi^+ s_\pi &= \tfrac{1}{3}[(s_\pi^+ s_\pi + s_\nu^+ s_\nu)(s_\pi^+ s_\pi + s_\nu^+ s_\nu) + \tfrac{1}{2}(s_\pi^+ s_\pi + s_\nu^+ s_\nu)] \\ &+ \tfrac{1}{2}[s_\pi^+ s_\pi s_\pi^+ s_\pi - s_\nu^+ s_\nu s_\nu^+ s_\nu] \\ &+ \tfrac{1}{6}[s_\pi^+ s_\pi s_\pi^+ s_\pi + s_\nu^+ s_\nu s_\nu^+ s_\nu - 4 s_\pi^+ s_\pi s_\nu^+ s_\nu - s_\pi^+ s_\pi - s_\nu^+ s_\nu] \end{aligned} \tag{40.14}$$

where the first term on the r.h.s. is the $k = 0$ part, the second is the $k = 1$ part and the third has rank $k = 2$ with respect to F-spin. The term $s_\nu^+ s_\nu s_\nu^+ s_\nu$ has a very similar expression with a change of sign of the $k = 1$ part. The $s_\pi^+ s_\pi s_\nu^+ s_\nu$ term has no $k = 1$ part (due to $(1010 \mid 1110) = 0$) and is equal to the following linear combination of $\kappa = 0$ components of $k = 0$ and $k = 2$ irreducible tensors

$$\begin{aligned} s_\pi^+ s_\pi s_\nu^+ s_\nu &= \tfrac{1}{6}[(s_\pi^+ s_\pi + s_\nu^+ s_\nu)(s_\pi^+ s_\pi + s_\nu^+ s_\nu) - s_\pi^+ s_\pi - s_\nu^+ s_\nu] \\ &- \tfrac{1}{6}[s_\pi^+ s_\pi s_\pi^+ s_\pi + s_\nu^+ s_\nu s_\nu^+ s_\nu - 4 s_\pi^+ s_\pi s_\nu^+ s_\nu - s_\pi^+ s_\pi - s_\nu^+ s_\nu] \end{aligned} \tag{40.15}$$

Similar equalities hold for other terms, like $d_{\pi\mu}^+ d_{\pi\mu'} d_{\nu\mu''}^+ d_{\nu\mu'''}$. The single boson terms $s_\pi^+ s_\pi + s_\nu^+ s_\nu$ in (40.14) and (40.15) are due to interchanging of non-commuting creation and annihilation operators. Hence, they do not appear in the decomposition of terms like

$$s_\pi^+ d_{\pi\mu} s_\nu^+ d_{\nu\mu'}$$

We obtain, for instance, the following expression for such products in terms of $\kappa = 0$ components of $k = 0$ and $k = 2$ F-spin tensors

$$s_\pi^+(\tilde{d}_\pi\cdot\tilde{d}_\nu)s_\nu^+ = \tfrac{1}{6}[(s_\pi^+\tilde{d}_\pi + s_\nu^+\tilde{d}_\nu)\cdot(s_\pi^+\tilde{d}_\pi + s_\nu^+\tilde{d}_\nu)]$$
$$-\tfrac{1}{6}[(s_\pi^+\tilde{d}_\pi\cdot s_\pi^+\tilde{d}_\pi) + (s_\nu^+\tilde{d}_\nu\cdot s_\nu^+\tilde{d}_\nu) - 4(s_\pi^+\tilde{d}_\pi\cdot s_\nu^+\tilde{d}_\nu)]$$

The decomposition of scalar products of irreducible tensors constructed of d-boson operators is more complicated. Making use of the orthogonality properties of $3j$-symbols and the identity (10.24), we obtain the following result

$$(d_\pi^+\times\tilde{d}_\pi)^{(L)}\cdot(d_\pi^+\times\tilde{d}_\pi)^{(L)}$$
$$= \frac{1}{3}\Bigg[(d_\pi^+\times\tilde{d}_\pi)^{(L)}\cdot(d_\pi^+\times\tilde{d}_\pi)^{(L)} + (d_\nu^+\times\tilde{d}_\nu)^{(L)}\cdot(d_\nu^+\times\tilde{d}_\nu)^{(L)}$$
$$+ (d_\pi^+\times\tilde{d}_\pi)^{(L)}\cdot(d_\nu^+\times\tilde{d}_\nu)^{(L)} + (2L+1)\sum_J(-1)^{L+J}$$
$$\times\begin{Bmatrix}2 & 2 & L\\ 2 & 2 & J\end{Bmatrix}(d_\pi^+\times\tilde{d}_\pi)^{(J)}\cdot(d_\nu^+\times\tilde{d}_\nu)^{(J)}$$
$$+\frac{1}{2}\frac{2L+1}{5}((d_\pi^+\cdot\tilde{d}_\pi) + (d_\nu^+\cdot\tilde{d}_\nu))\Bigg]$$
$$+\tfrac{1}{2}[(d_\pi^+\times\tilde{d}_\pi)^{(L)}\cdot(d_\pi^+\times\tilde{d}_\pi)^{(L)} - (d_\nu^+\times\tilde{d}_\nu)^{(L)}\cdot(d_\nu^+\times\tilde{d}_\nu)^{(L)}]$$
$$+\frac{1}{6}\Bigg[(d_\pi^+\times\tilde{d}_\pi)^{(L)}\cdot(d_\pi^+\times\tilde{d}_\pi)^{(L)} + (d_\nu^+\times\tilde{d}_\nu)^{(L)}\cdot(d_\nu^+\times\tilde{d}_\nu)^{(L)}$$
$$-2(d_\pi^+\times\tilde{d}_\pi)^{(L)}\cdot(d_\nu^+\times\tilde{d}_\nu)^{(L)} - 2(2L+1)\sum_J(-1)^{L+J}$$
$$\times\begin{Bmatrix}2 & 2 & L\\ 2 & 2 & J\end{Bmatrix}(d_\pi^+\times\tilde{d}_\pi)^{(J)}\cdot(d_\nu^+\times\tilde{d}_\nu)^{(J)}$$
$$-\frac{2L+1}{5}((d_\pi^+\cdot\tilde{d}_\pi) + (d_\nu^+\cdot\tilde{d}_\nu))\Bigg] \qquad (40.16)$$

In (40.16), as in (40.14), the first term on the r.h.s. is the $k=0$ part, the second term has $k=1$ and the third term has rank $k=2$ with respect to F-spin. The decomposition of $(d_\nu^+\times\tilde{d}_\nu)^{(L)}\cdot(d_\nu^+\times\tilde{d}_\nu)^{(L)}$ is given by (40.16) in which the sign of the $k=1$ term is changed.

Similarly, we obtain the following decomposition of the scalar product of a tensor constructed of proton d-boson operators and one of neutron d-bosons

$$
\begin{aligned}
&(d_\pi^+ \times \tilde{d}_\pi)^{(L)} \cdot (d_\nu^+ \times \tilde{d}_\nu)^{(L)} \\
&\quad = \tfrac{1}{6}\Bigg[(d_\pi^+ \times \tilde{d}_\pi)^{(L)} \cdot (d_\pi^+ \times \tilde{d}_\pi)^{(L)} + (d_\nu^+ \times \tilde{d}_\nu)^{(L)} \cdot (d_\nu^+ \times \tilde{d}_\nu)^{(L)} \\
&\qquad + 4(d_\pi^+ \times \tilde{d}_\pi)^{(L)} \cdot (d_\nu^+ \times \tilde{d}_\nu)^{(L)} - 2(2L+1)\sum_J (-1)^{L+J} \\
&\qquad \times \begin{Bmatrix} 2 & 2 & L \\ 2 & 2 & J \end{Bmatrix} (d_\pi^+ \times \tilde{d}_\pi)^{(J)} \cdot (d_\nu^+ \times \tilde{d}_\nu)^{(J)} \\
&\qquad - \frac{2L+1}{5}((d_\pi^+ \cdot \tilde{d}_\pi) + (d_\nu^+ \times \tilde{d}_\nu))\Bigg] \\
&\quad - \tfrac{1}{6}\Bigg[(d_\pi^+ \times \tilde{d}_\pi)^{(L)} \cdot (d_\pi^+ \times \tilde{d}_\pi)^{(L)} + (d_\nu^+ \times \tilde{d}_\nu)^{(L)} \cdot (d_\nu^+ \times \tilde{d}_\nu)^{(L)} \\
&\qquad - 2(d_\pi^+ \times \tilde{d}_\pi)^{(L)} \cdot (d_\nu^+ \times \tilde{d}_\nu)^{(L)} - 2(2L+1)\sum_J (-1)^{L+J} \\
&\qquad \times \begin{Bmatrix} 2 & 2 & L \\ 2 & 2 & J \end{Bmatrix} (d_\pi^+ \times \tilde{d}_\pi)^{(J)} \cdot (d_\nu^+ \times \tilde{d}_\nu)^{(J)} \\
&\qquad - \frac{2L+1}{5}((d_\pi^+ \cdot \tilde{d}_\pi) + (d_\nu^+ \times \tilde{d}_\nu))\Bigg] \qquad (40.17)
\end{aligned}
$$

In (40.17), as in (40.15) there are only $\kappa = 0$ components of $k = 0$ and $k = 2$ F-spin tensors.

We can now apply the Wigner-Eckart theorem for the $SU_F(2)$ group which may be written as

$$
\langle FM_F\alpha|H_0^{(k)}|FM_F\alpha'\rangle = (-1)^{F-M_F}\begin{pmatrix} F & k & F \\ -M_F & 0 & M_F \end{pmatrix}(F\alpha\|H^{(k)}\|F\alpha') \qquad (40.18)
$$

The labels α and α' in (40.18) stand for the quantum numbers characterizing the various states with given F, M_F like J, M and any additional necessary quantum numbers. To obtain the reduced matrix elements we can calculate the matrix elements (40.18) in states with

$M_F = F = N/2$. Thus, we can write

$$\langle FM_F\alpha|H_0^{(k)}|FM_F,\alpha'\rangle = \langle F,M_F = F,\alpha|H_0^{(k)}|F,M_F = F,\alpha'\rangle \times \frac{(-1)^{F-M_F}\begin{pmatrix} F & k & F \\ -M_F & 0 & M_F \end{pmatrix}}{\begin{pmatrix} F & k & F \\ -F & 0 & F \end{pmatrix}} \tag{40.19}$$

The states with $M_F = F = N/2$ have $N_\pi = N$, $N_\nu = 0$ and are states of only proton bosons. The matrix elements

$$\langle F = N/2, M_F = N/2, \alpha|H_0^{(k)}|F = N/2, M_F = N/2, \alpha'\rangle$$

can be calculated from the expressions of $H_0^{(k)}$ in which we keep only terms with proton boson creation and annihilation operators. Hence, for $M_F = F = N/2$, states and operators are those of identical bosons equivalent to IBA-1 states and operators. The IBA-1 operators obtained from $H_0^{(k)}$ by replacing $s_\nu^+, s_\nu, d_{\nu\mu}^+, d_{\nu\mu}$ by zeroes may be denoted by $\tilde{H}_0^{(k)}$. The subscript π may be dropped so that s^+, s, d_μ^+, d_μ replace $s_\pi^+, s_\pi, d_{\pi\mu}^+, d_{\pi\mu}$. We then obtain the IBA-1 Hamiltonian projected from (39.1) for each value of M_F in the form

$$\sum_{k=0,1,2} \tilde{H}_0^{(k)}(-1)^{N_\nu} \frac{\begin{pmatrix} N/2 & k & N/2 \\ -\frac{1}{2}(N_\pi - N_\nu) & 0 & \frac{1}{2}(N_\pi - N_\nu \end{pmatrix}}{\begin{pmatrix} N/2 & k & N/2 \\ -N/2 & 0 & N/2 \end{pmatrix}} \tag{40.20}$$

An equivalent IBA-2 Hamiltonian may be obtained from (40.20) if in $\tilde{H}_0^{(k)}$ single boson operators of identical bosons are replaced by sums of single proton boson operators and single neutron boson operators (like writing $(d_\pi^+ \times \tilde{d}_\pi)^{(L)} + (d_\nu^+ \times \tilde{d}_\nu)^{(L)}$ instead of $(d^+ \times \tilde{d})^{(L)}$). That IBA-2 Hamiltonian commutes with components of F-spin and its eigenvalues for $F = N/2, M_F$ eigenstates are equal to the expectation values in states with $F = N/2, M_F$ of the original IBA-2 Hamiltonian.

Let us first apply the formalism described above to the case of a single boson operator like those in (40.12) and (40.13). For $k = 1$ we obtain by using the expressions for $3j$-symbols, the formula

$$(-1)^{N_\nu}\begin{pmatrix} N/2 & 1 & N/2 \\ -\frac{1}{2}(N_\pi - N_\nu) & 0 & \frac{1}{2}(N_\pi - N_\nu) \end{pmatrix} \Big/ \begin{pmatrix} N/2 & 1 & N/2 \\ -N/2 & 0 & N/2 \end{pmatrix}$$
$$= (N_\pi - N_\nu)/N$$

Substituting this value into (40.20) we obtain, in view of (40.12) and (40.13), that the matrix of $(d_\pi^+ \times \tilde{d}_\pi)_M^{(L)}$ constructed from states with $F = N/2$, can be calculated from the IBA-1 operator

$$\tfrac{1}{2}(d^+ \times \tilde{d})_M^{(L)} + \tfrac{1}{2}(d^+ \times \tilde{d})_M^{(L)} \frac{N_\pi - N_\nu}{N} = \frac{N_\pi}{N}(d^+ \times \tilde{d})_M^{(L)} \tag{40.21}$$

The dependence on N_π/N in (40.21) holds also for the operator $s_\pi^+ \tilde{d}_\pi$ or $d_\pi^+ s_\pi$. The projection of $(d_\nu^+ \times \tilde{d}_\nu)^{(L)}$ on states with definite F-spin equal to $N/2$ is similarly given by

$$\tfrac{1}{2}(d^+ \times \tilde{d})_M^{(L)} - \tfrac{1}{2}(d^+ \times \tilde{d})_M^{(L)} \frac{N_\pi - N_\nu}{N} = \frac{N_\nu}{N}(d^+ \times \tilde{d})_M^{(L)} \tag{40.22}$$

In the special case of the single nucleon term $\epsilon_\pi(d_\pi^+ \cdot \tilde{d}_\pi) + \epsilon_\nu(d_\nu^+ \cdot \tilde{d}_\nu)$ of the IBA-2 Hamiltonian, the projected IBA-1 operator is

$$\boxed{\frac{\epsilon_\pi N_\pi + \epsilon_\nu N_\nu}{N}(d^+ \cdot \tilde{d})} \tag{40.23}$$

The next operators in the IBA-2 Hamiltonian which we consider may be expressed as

$$\sum_{L=0}^{4}\{a_L^\pi(d_\pi^+ \times \tilde{d}_\pi)^{(L)} \cdot (d_\pi^+ \times \tilde{d}_\pi)^{(L)} + a_L^\nu(d_\nu^+ \times \tilde{d}_\nu)^{(L)} \cdot (d_\nu^+ \times \tilde{d}_\nu)^{(L)}$$
$$+ a_L(d_\pi^+ \times \tilde{d}_\pi)^{(L)} \cdot (d_\nu^+ \times \tilde{d}_\nu)^{(L)}\} \tag{40.24}$$

We note that in Section 33 it was explained that not all scalar products whose coefficients are a_L^π (or a_L^ν) are independent. We may choose only three independent scalar products but this does not affect the following derivations. For the first two terms under the summation sign in (40.24) we use the expansion like (40.16) whereas for the last term we use the expansion (40.17). For the $k = 2$, $\kappa = 0$ component we use the formula

$$(-1)^{N_\nu} \begin{pmatrix} N/2 & 2 & N/2 \\ -\frac{1}{2}(N_\pi - N_\nu) & 0 & \frac{1}{2}(N_\pi - N_\nu) \end{pmatrix} \Big/ \begin{pmatrix} N/2 & 2 & N/2 \\ -N/2 & 0 & N/2 \end{pmatrix}$$
$$= (\tfrac{3}{4}(N_\pi - N_\nu)^2 - \tfrac{1}{2}N(\tfrac{1}{2}N + 1))/\tfrac{1}{2}N(N-1)$$

Substituting this value as well as the result for $k = 1$, into (40.20) we obtain the projected IBA-1 operator in the form

$$\sum_{L=0}^{4} \Big\{ a_L^\pi (d^+ \times \tilde{d})^{(L)} \cdot (d^+ \times \tilde{d})^{(L)}$$
$$\times \left(\frac{1}{3} + \frac{1}{2}\frac{N_\pi - N_\nu}{N} + \frac{1}{6}\frac{N_\pi(N_\pi - 1) + N_\nu(N_\nu - 1) - 4N_\pi N_\nu)}{N(N-1)} \right)$$
$$+ a_L^\pi \frac{2L+1}{5}(d^+ \cdot \tilde{d}) \left(\frac{1}{6} - \frac{1}{6}\frac{N_\pi(N_\pi - 1) + N_\nu(N_\nu - 1) - 4N_\pi N_\nu}{N(N-1)} \right)$$
$$+ a_L^\nu (d^+ \times \tilde{d})^{(L)} \cdot (d^+ \times \tilde{d})^{(L)}$$
$$\times \left(\frac{1}{3} - \frac{1}{2}\frac{N_\pi - N_\nu}{N} + \frac{1}{6}\frac{N_\pi(N_\pi - 1) + N_\nu(N_\nu - 1) - 4N_\pi N_\nu}{N(N-1)} \right)$$
$$+ a_L^\nu \frac{2L+1}{5}(d^+ \cdot \tilde{d}) \left(\frac{1}{6} - \frac{1}{6}\frac{N_\pi(N_\pi - 1) + N_\nu(N_\nu - 1) - 4N_\pi N_\nu}{N(N-1)} \right)$$
$$+ a_L (d^+ \times \tilde{d})^{(L)} \cdot (d^+ \times \tilde{d})^{(L)}$$
$$\times \left(\frac{1}{6} - \frac{1}{6}\frac{N_\pi(N_\pi - 1) + N_\nu(N_\nu - 1) - 4N_\pi N_\nu}{N(N-1)} \right)$$
$$+ a_L \frac{2L+1}{5}(d^+ \cdot \tilde{d})$$
$$\times \left(-\frac{1}{6} + \frac{1}{6}\frac{N_\pi(N_\pi - 1) + N_\nu(N_\nu - 1) - 4N_\pi N_\nu}{N(N-1)} \right) \Big\}$$
$$= \sum_{L=0}^{4} \Big\{ \left(a_L^\pi \frac{N_\pi(N_\pi - 1)}{N(N-1)} + a_L^\nu \frac{N_\nu(N_\nu - 1)}{N(N-1)} + a_L \frac{N_\pi N_\nu}{N(N-1)} \right)$$
$$\times (d^+ \times \tilde{d})^{(L)} \cdot (d^+ \times \tilde{d})^{(L)}$$
$$+ (a_L^\pi + a_L^\nu - a_L)\frac{2L+1}{5}\frac{N_\pi N_\nu}{N(N-1)}(d^+ \cdot \tilde{d}) \Big\} \tag{40.25}$$

Finally we consider the quadrupole-quadrupole interaction. As in preceding sections we use the definitions

$$\begin{aligned}\tilde{\mathbf{Q}}_\pi &= s_\pi^+\tilde{d}_\pi + d_\pi^+ s_\pi + \chi_\pi(d_\pi^+ \times \tilde{d}_\pi)^{(2)} \\ \tilde{\mathbf{Q}}_\nu &= s_\nu^+\tilde{d}_\nu + d_\nu^+ s_\nu + \chi_\nu(d_\nu^+ \times \tilde{d}_\nu)^{(2)}\end{aligned} \tag{40.26}$$

and consider the quadrupole interaction

$$\begin{aligned}&\kappa_\pi(\tilde{\mathbf{Q}}_\pi \cdot \tilde{\mathbf{Q}}_\pi) + \kappa_\nu(\tilde{\mathbf{Q}}_\nu \cdot \tilde{\mathbf{Q}}_\nu) + \kappa(\tilde{\mathbf{Q}}_\pi + \tilde{\mathbf{Q}}_\nu)\cdot(\tilde{\mathbf{Q}}_\pi + \tilde{\mathbf{Q}}_\nu) \\ &\qquad = (\kappa_\pi + \kappa)(\tilde{\mathbf{Q}}_\pi \cdot \tilde{\mathbf{Q}}_\pi) + (\kappa_\nu + \kappa)(\tilde{\mathbf{Q}}_\nu \cdot \tilde{\mathbf{Q}}_\nu) + 2\kappa(\tilde{\mathbf{Q}}_\pi \cdot \tilde{\mathbf{Q}}_\nu)\end{aligned} \tag{40.27}$$

The projection onto states with $F = N/2$ of the scalar products in (40.27) can be carried out in a procedure similar to the one applied to (40.24). The only difference is in the single boson terms which appear in this procedure. Among the several terms in the scalar product of two quadrupole operators some give rise to scalar products $(d^+ \cdot \tilde{d})$ and s^+s and others like $(s_\pi^+\tilde{d}_\pi \cdot s_\pi^+\tilde{d}_\pi)$ or $s_\pi^+\tilde{d}_\pi \cdot (d_\pi^+ \times \tilde{d}_\pi)^{(2)}$ do not. The IBA-1 operator projected from (40.27) is thus calculated to be

$$\begin{aligned}&(\kappa_\pi + \kappa)(s^+\tilde{d} + d^+s + \chi_\pi(d^+ \times \tilde{d})^{(2)}) \cdot (s^+\tilde{d} + d^+s + \chi_\pi(d^+ \times \tilde{d})^{(2)}) \\ &\quad \times \frac{N_\pi(N_\pi - 1)}{N(N-1)} + (\kappa_\pi + \kappa)(5s^+s + (d^+ \cdot \tilde{d}) + \chi_\pi^2(d^+ \cdot \tilde{d}))\frac{N_\pi N_\nu}{N(N-1)} \\ &\quad + (\kappa_\nu + \kappa)(s^+\tilde{d} + \tilde{d}^+s + \chi_\nu(d^+ \times \tilde{d})^{(2)}) \\ &\quad \cdot (s^+\tilde{d} + d^+s + \chi_\nu(d^+ \times \tilde{d})^{(2)})\frac{N_\nu(N_\nu - 1)}{N(N-1)} \\ &\quad + (\kappa_\nu + \kappa)(5s^+s + (d^+ \cdot \tilde{d}) + \chi_\nu^2(d^+ \cdot \tilde{d}))\frac{N_\pi N_\nu}{N(N-1)} \\ &\quad + 2\kappa[(s^+\tilde{d} + d^+s)\cdot(s^+d + d^+s) \\ &\quad + (\chi_\pi + \chi_\nu)(s^+\tilde{d} + d^+s)\cdot(d^+ \times \tilde{d})^{(2)} \\ &\quad + \chi_\pi\chi_\nu(d^+ \times \tilde{d})^{(2)} \cdot (d^+ \times \tilde{d})^{(2)} - 5s^+s - d^+ \cdot \tilde{d} - \chi_\pi\chi_\nu(d^+ \cdot \tilde{d})] \\ &\quad \times \frac{N_\pi N_\nu}{N(N-1)}\end{aligned} \tag{40.28}$$

The last term in (40.28) may be simplified by introducing the notation

$$\chi = \tfrac{1}{2}(\chi_\pi + \chi_\nu) \qquad \chi' = \tfrac{1}{2}(\chi_\pi - \chi_\nu)$$
$$\mathbf{Q}(\chi) = s^+\tilde{d} + d^+s + \chi(d^+ \times \tilde{d})^{(2)} \tag{40.29}$$

and may then be expressed as

$$\boxed{\begin{aligned}&2\kappa[\mathbf{Q}(\chi)\cdot\mathbf{Q}(\chi) - \chi'^2(d^+ \times \tilde{d})^{(2)}\cdot(d^+ \times \tilde{d})^{(2)} \\ &\quad - 5s^+s - (d^+\cdot\tilde{d}) - \chi_\pi\chi_\nu(d^+\cdot\tilde{d})]\frac{N_\pi N_\nu}{N(N-1)}\end{aligned}} \tag{40.30}$$

In the case of no quadrupole interaction between identical bosons, $\kappa + \kappa_\pi = \kappa + \kappa_\nu = 0$, (40.30) is the only quadrupole term and it represents the proton boson-neutron boson interaction. It contains the quadrupole interaction $\mathbf{Q}(\chi)\cdot\mathbf{Q}(\chi)$ but also a quadrupole interaction between d-bosons. It contains also a single boson term which may be written, due to $s^+s = N - (d^+\cdot\tilde{d})$, as

$$\boxed{\begin{aligned}&2\kappa(-5s^+s - (d^+\cdot\tilde{d}) - \chi_\pi\chi_\nu(d^+\cdot\tilde{d}))\frac{N_\pi N_\nu}{N(N-1)} \\ &\quad = -\frac{10\kappa N_\pi N_\nu}{N(N-1)}N + 2\kappa(4 - \chi_\pi\chi_\nu)\frac{N_\pi N_\nu}{N(N-1)}(d^+\cdot\tilde{d})\end{aligned}} \tag{40.31}$$

As explained in Section 33, the quadrupole interaction $(d^+ \times \tilde{d})^{(2)}\cdot(d^+ \times \tilde{d})^{(2)}$ may be replaced by a linear combination of scalar products $(d^+ \times \tilde{d})^{(L)}\cdot(d^+ \times \tilde{d})^{(L)}$ with $L = 0$, $L = 1$ and $L = 3$.

It may be worthwhile to point out that in the case of the $SU^*_{\pi+\nu}(3)$ dynamical symmetry given by (39.57), the coefficient χ in (40.29) vanishes whereas $\chi' = \pm\sqrt{7}/2$. The resulting IBA-1 Hamiltonian (40.30) contains the Casimir operator of $O(6)$ along with the term $\frac{7}{4}(d^+ \times \tilde{d})^{(2)}\cdot(d^+ \times \tilde{d})^{(2)}$ which is the scalar product of two $U(5)$ generators. It also contains single d-boson energies which are absent in the exact $O(6)$ limit of IBA-1. In any case, since $SU^*_{\pi+\nu}(3)$ eigenstates do not have well defined F-spins, the resulting IBA-1 Hamiltonian is not expected to be a good approximation even for the lowest eigenstates.

Let us now consider some IBA-2 Hamiltonians which are constructed to reproduce level schemes in actual nuclei. We use the definition (38.4) for the boson Hamiltonian, namely

$$H = H_\pi + H_\nu + V_{\pi\nu} \tag{40.32}$$

and recall the implications for the various terms due to our knowledge of nuclear interactions. As repeatedly emphasized above, the Hamiltonian of identical valence nucleons has eigenstates with definite generalized seniorities. Hence, the boson Hamiltonians H_π and H_ν should be constructed from $U(5)$ generators only. The strong attractive interaction between valence protons and valence neutrons breaks generalized seniority in a major way. As a result, the interaction $V_{\pi\nu}$ should contain a strong quadrupole interaction of the form

$$\begin{aligned}\kappa\tilde{\mathbf{Q}}_\pi \cdot \tilde{\mathbf{Q}}_\nu = \kappa(s_\pi^+\tilde{d}_\pi + d_\pi^+ s_\pi + \chi_\pi(d_\pi^+ \times \tilde{d}_\pi)^{(2)}) \\ \cdot (s_\nu^+\tilde{d}_\nu + d_\nu^+ s_\nu + \chi_\nu(d_\nu^+ \times \tilde{d}_\nu)^{(2)})\end{aligned} \tag{40.33}$$

The proton boson-neutron boson interaction may well include also other terms which are scalar products of other irreducible tensors

$$\sum_{L=0}^{4} a_L (d_\pi^+ \times \tilde{d}_\pi)^{(L)} \cdot (d_\nu^+ \times \tilde{d}_\nu)^{(L)} \tag{40.34}$$

The terms in (40.34), however, commute with both n_{d_π} and n_{d_ν}. States between which they have non-vanishing matrix elements correspond to nucleon states with the same proton seniority and neutron seniority. In the s, d boson model it is only the quadrupole interaction that corresponds to seniority breaking interactions between valence protons and neutrons. It gives rise to coherent combinations of states with different numbers of s and d bosons which can reproduce states of collective or rotational bands.

Hamiltonians of identical bosons with $U(5)$ dynamical symmetry have been described in Section 34. They are diagonal in the scheme defined by eigenstates of n_d, boson seniority τ and total spin J. The most general $U(5)$ Hamiltonian is a linear combination of Casimir operators of the groups in the chain (34.16). The eigenvalues are then given by (34.17). As explained in Section 37, the only reliable information on semi-magic nuclei with identical valence nucleons is about

$J = 0$ ground states and first excited $J = 2$ states. This information may be used in the boson model to determine the parameter ϵ_π multiplying $(d_\pi^+ \cdot \tilde{d}_\pi)$ in H_π or ϵ_ν—the coefficient of $(d_\nu^+ \cdot \tilde{d}_\nu)$ in H_ν. This single d-boson energy may be put equal to $V_2 - V_0$ which is the excitation energy of the $J = 2$ state in semi-magic nuclei. There is no experimental information about values of the other coefficients in H_π and H_ν. To simplify matters, it has been often assumed that the other coefficients are small compared to ϵ and can be neglected. The simplest IBA-2 Hamiltonian consistent with our knowledge about nuclear interactions can thus be expressed as

$$\boxed{\epsilon_\pi(d_\pi^+ \cdot \tilde{d}_\pi) + \epsilon_\nu(d_\nu^+ \cdot \tilde{d}_\nu) + \kappa\tilde{\mathbf{Q}}_\pi \cdot \tilde{\mathbf{Q}}_\nu} \tag{40.35}$$

with the quadrupole operators defined by (40.26). There are only rough estimates about the value of κ and not even that for the coefficients χ_π and χ_ν. In applications, they were taken as free parameters to be determined by the fit to experimental data.

The basic Hamiltonian (40.35) does not commute with components of $\mathbf{F}$ even if $\epsilon_\pi = \epsilon_\nu$ and $\chi_\pi = \chi_\nu$. Its eigenstates do not have definite values of F and in particular, the lowest eigenstates are not pure $F = N/2$ states. To increase the amplitude of $F = N/2$ components in the lowest eigenstates in some applications of IBA-2 to nuclear spectra, the Majorana operator (39.21) was added to the Hamiltonian (40.35). As remarked above, if the coefficient ξ_2 is large and positive and $\xi_1 < 0$, $\xi_3 < 0$, the eigenvalues for $F < N/2$ states of the operator $\mathcal{M}$ are much higher than the vanishing ones for $F = N/2$ states. As a result, the amplitude of $F = N/2$ states in the lowest eigenstates is increased and they can be better approximated by IBA-1 states. This practice raises the question whether there is a shell model basis for the operator $\mathcal{M}$. The first two terms in (39.21) have a simple physical meaning. It can be directly verified that $-\xi_L$ is the eigenvalue of $\mathcal{M}$ in the state $(d_\pi^+ \times d_\nu^+)_M^{(L)}|0\rangle$ for $L = 1,3$. Such eigenvalues may arise from the interaction (40.34) and hence ξ_1, ξ_3 cannot be considered as independent parameters. On the other hand, if the general interaction (40.34) is not included in the Hamiltonian, it is consistent to include the first two terms of $\mathcal{M}$ in the IBA-2 Hamiltonian. No reliable experimental information is available to determine the size and even the sign of the coefficients ξ_1 and ξ_3. They are coefficients of boson interactions which correspond to simple interactions between protons and neutrons.

The situation is rather different for the third term, with coefficient ξ_2, in (39.21). Acting on states of one proton boson and one neutron boson that term has eigenvalue $2\xi_2$ in the antisymmetric state $(d_\pi^+ s_\nu^+ - d_\nu^+ s_\pi^+)|0\rangle$. All other eigenvalues of this quadrupole-quadrupole term vanish. Let us return for a moment to the corresponding four fermion system of two valence protons and two valence neutrons. The states which correspond to single boson states are the $J_\pi = 0$ and $J_\pi = 2$ of protons and $J_\nu = 0$, $J_\nu = 2$ states of valence neutrons. Let us consider the case in which the 0-2 spacings of the protons is equal to that of the neutrons. The combined system has two states with $J = 0$ obtained by coupling $J_\pi = 0$, $J_\nu = 0$ and $J_\pi = 2$, $J_\nu = 2$. There are three $J = 2$ states obtained by coupling $J_\pi = 0$, $J_\nu = 2$, and $J_\pi = 2$, $J_\nu = 0$ as well as $J_\pi = 2$, $J_\nu = 2$. Any proton-neutron interaction which pushes apart the two degenerate levels, with $J_\pi = 0$, $J_\nu = 2$ and $J_\pi = 2$, $J_\nu = 0$, must contain a quadrupole term $\kappa\tilde{\mathbf{Q}}_\pi \cdot \tilde{\mathbf{Q}}_\nu$. In that case, according to (10.27) the non-diagonal matrix element assumes the form

$$V = \kappa\langle J_\pi = 0, J_\nu = 2, J = 2, M|\tilde{\mathbf{Q}}_\pi \cdot \tilde{\mathbf{Q}}_\nu|J_\pi = 2, J_\nu = 0, J = 2, M\rangle$$

$$= \kappa \begin{Bmatrix} 0 & 2 & 2 \\ 0 & 2 & 2 \end{Bmatrix} (J_\pi = 0\|\tilde{\mathbf{Q}}_\pi\|J_\pi = 2)(J_\nu = 2\|\tilde{\mathbf{Q}}_\nu\|J_\nu = 0) \tag{40.36}$$

If the interaction is attractive, $\kappa < 0$, the symmetric combination of the two $J = 2$ states is pushed *down* by the amount $|V|$ and the antisymmetric one is pushed up by $|V|$. The same quadrupole interaction must also affect the $J = 0$ state obtained by coupling $J_\pi = 0$, $J_\nu = 0$. There is a non-diagonal matrix-element connecting it to the other $J = 0$ state with $J_\pi = 2$, $J_\nu = 2$. It is given by

$$V' = \kappa\langle J_\pi = 0, J_\nu = 0, J = 0, M = 0|\tilde{\mathbf{Q}}_\pi \cdot \tilde{\mathbf{Q}}_\nu$$

$$|J_\nu = 2, J_\nu = 2, J = 0, M = 0\rangle$$

$$= \kappa \begin{Bmatrix} 0 & 0 & 0 \\ 2 & 2 & 2 \end{Bmatrix} (J_\pi = 0\|\tilde{\mathbf{Q}}_\pi\|J_\pi = 2)(J_\nu = 0\|\tilde{\mathbf{Q}}_\nu\|J_\nu = 2) \tag{40.37}$$

Since the interaction $\kappa(\tilde{\mathbf{Q}}_\pi \cdot \tilde{\mathbf{Q}}_\nu)$ is hermitean and real, $\tilde{\mathbf{Q}}_\pi$ and $\tilde{\mathbf{Q}}_\nu$ must both be hermitean or anti-hermitean. Substituting the numerical values of the $6j$-symbols we obtain $V' = \pm\sqrt{5}V$. Thus, any interaction appearing in (40.36) that contributes to the position of the symmetric or antisymmetric combination of the $J = 2$ states must have non-vanishing non-diagonal element between the $|J_\pi = 0, J_\nu = 0, J = 0\rangle$ state and the $|J_\pi = 2, J_\nu = 2, J = 0\rangle$ state. Adding a monopole interaction between two protons in $J_\pi = 0$ state and two neutrons in $J_\nu = 2$ state as well as between $J_\pi = 2$ protons and $J_\nu = 0$ neutrons may push up both $J = 2$ states so that the symmetric state is in the original position. Still, the non-diagonal element between the two $J = 0$ states will not be eliminated. Coming back to the boson model we see that the interaction $\kappa\tilde{\mathbf{Q}}_\pi \cdot \tilde{\mathbf{Q}}_\nu$ in (40.33) has indeed non-vanishing matrix elements between the $s_\pi^+ s_\nu^+|0\rangle$ and $(d_\pi^+ \cdot d_\nu^+)|0\rangle$ states. On the other hand, the Majorana operator $\mathcal{M}$ in (39.21) has no such matrix element. We conclude that the quadrupole part of the Majorana operator cannot be the boson image of a two-nucleon (hermitean) interaction.

We should realize that the argument made above does not strictly exclude the Majorana operator, even with $\xi_2 \neq 0$, from IBA-2 Hamiltonians. As emphasized above, the restriction to quadrupole interaction, which is the basis of the s, d boson model, is a very drastic approximation. The truncation which was carried out in Section 38 ignores the contributions to energies from a huge number of states. Although each contribution may be very small the total effect may be considerable. Such contributions, even if they do not invalidate the model, may still strongly affect or renormalize the values of the parameters. The quadrupole part of the Majorana operator may arise in the model due to such contributions. Another parameter of the model which may be strongly renormalized is the single d-boson energy ϵ_π or ϵ_ν. As shown above, it may be directly related to the 0–2 spacings, $V_2 - V_0$, in semi-magic nuclei. It is nice if good agreement with experiment can be obtained by adopting these values for ϵ_π and ϵ_ν. There are indeed such cases but in others it may be necessary to use renormalized values. The basic Hamiltonian (40.35) is the simplest IBA-2 Hamiltonian which is based on the correct ingredients of the nuclear interaction. It strongly couples valence protons and valence neutrons leading to collective states. Still, there may be other interactions which enhance these features. Their effects may be mimicked by the quadrupole term in the Majorana operator which forces proton and neutron boson states to be more symmetrically coupled. Such effects may also reduce the values of ϵ_π and ϵ_ν thereby accelerating the transition to rotational spectra.

A common feature of all IBA-2 Hamiltonians which reproduce experimental data is their not commuting with components of the **F** vector. A Hamiltonian which does commute with **F** has the same eigenvalues in nuclei with any number of proton bosons and neutron bosons if their sum is a given number $N = N_\pi + N_\nu$. As we saw in several examples, if $N_\pi = N$ or $N_\nu = N$, low lying levels of semi-magic nuclei are characterized by generalized seniority. The corresponding boson states have $U(5)$ symmetry. Once there are several valence protons and valence neutrons, the spectrum becomes more collective and may become rotational. In the boson model, such a transition is exhibited by a change from $U(5)$ symmetry in the direction of $SU(3)$ or $O(6)$ symmetry. This behavior is demonstrated by eigenstates of the basic Hamiltonian (40.35) even if its coefficients have fixed values. To obtain quantitative agreement with experimental data, values of these coefficients may have to change. These facts have clear implications for the possible use of F-spin as a good quantum number. Even if the lowest eigenstates for given M_F can be well approximated by $F = N/2$ states, they cannot be obtained from states with $M_F' = M_F - 1$, say, by applying to them the operator F^+. Consider, for instance, the ground state of a semi-magic nucleus with $2N$ valence neutrons. In the boson model it is described by the $(s_\nu^+)^N|0\rangle$ state with $F = N/2$. Applying successively to this state the operator F^- we obtain the states $(s_\pi^+)^{N_\pi}(s_\nu^+)^{N_\nu}|0\rangle$ with $N_\pi + N_\nu = N$. All of them have $F = N/2$ but they belong to $U(5)$ symmetry and are not eigenstates of a realistic IBA-2 Hamiltonian. None of them can describe the ground state of a nucleus with collective features like rotational spectra. A simple consequence of this feature is that the coefficients of an IBA-1 Hamiltonian projected from a realistic IBA-2 Hamiltonian must depend explicitly on M_F (or N_π, N_ν) as in the examples shown above. Unlike the situation of isospin T-multiplets in nuclei, there are no such F-spin multiplets in actual nuclei. The distinction between F-spin and isospin as seen in this behavior should be discussed in some detail.

Charge independence of the nuclear interaction of protons and neutrons is one of the best established facts of nuclear physics. It is possible to exploit this symmetry by using the isospin variables and formalism. Eigenstates of the nuclear Hamiltonian (without the Coulomb interaction between protons) belong to definite eigenvalues of the total isospin T. For states in which valence protons and neutrons occupy the same (major) shell, the isospin classification yields a great simplification. The Hamiltonian matrix in the shell model space considered, reduces to a form in which it has sub-matrices along the diagonal. Each of these is characterized by a definite value of T and all elements connecting states with different values of T vanish.

Isospin symmetry is not of much help in configurations where valence protons are in a set of j-orbits which are fully occupied by neutrons (valence neutrons occupying a set of higher j'-orbits). As stated and restated above, all states of such configurations have the same value of isospin which is one half the difference between the total numbers of neutrons and protons $T = (N - Z)/2$. In such cases the isospin formalism is redundant–valence nucleons in j-orbits are protons and nucleons in j'-orbits are neutrons. The use of proton and neutron states rather than states of the isospin formalism is both justified and useful. Also in configurations with valence protons and neutrons in the same set of orbits it is allowed to use the proton neutron states for the calculation of energy levels. This is allowed also for transitions in which the numbers of protons and neutrons do not change. The only requirement is that the Hamiltonian will be charge independent. This means that the nuclear interactions between two neutrons, two protons and a proton-neutron pair should be the same when they are in the same state. Thus, no special care should be taken when valence protons and neutrons are in different shells. States of a valence proton and a valence neutron are different from all states of a valence proton pair or a valence neutron pair. More care should be applied to the case of protons and neutrons in the same valence shell.

As mentioned above, *all* states of configurations in which the neutrons fill completely the shell of valence protons and valence neutrons occupy orbits in a higher shell, have the *same* isospin $T = (N - Z)/2$. In the boson description of such ground state configurations, some of the states corresponding to nucleon states in the $S - D$ space have the maximum value of F-spin, $F = (N_\pi + N_\nu)/2$. If $N_\pi \neq 0$, $N_\nu \neq 0$, however, there are other $S - D$ states with lower values of F. Thus, boson states with different values of F may correspond to nucleon states with the same value of isospin.

Another feature of the relation between F-spin and isospin may be seen in the following example. If $N > Z$ in the ground state configuration, then $T = -M_T = (N - Z)/2$. Operating on states of this configuration with the operator T^- yields zero since $-(N - Z)/2$ is the lowest value of M_T with the given value of T. In the boson model, however, these states have $F = (N_\pi + N_\nu)/2$ and $M_F = (N_\pi - N_\nu)/2$. As long as $N_\pi < \Omega_\pi$, $N_\nu < \Omega_\nu$, the operator F^- acting on such states yields states with the same value of F and $M'_F = M_F - 1$. In the latter states, there are $N_\pi - 1$ proton bosons and $N_\nu + 1$ neutron bosons. Such states correspond to nucleon states with $N + 2$ neutrons and $Z - 2$ protons and hence must have isospins $T' \geq (N - Z)/2 + 2$. Thus, boson states with the same value of F-spin and different values of M_F may correspond to fermion states (in different nuclei) with different values of isospin.

Similar behavior may be observed when we consider an even-even nucleus whose ground state configuration has $T = -M_T = (N - Z)/2$. Operating twice with T^+ on its lowest states we obtain (for $N \geq Z + 4$) states with the same value of isospin T and $M_T' = -(N - Z)/2 + 2$. Due to the symmetry energy, such states lie several MeV above the ground state of the nucleus whose isospin is $T' = T - 2$. In the boson model, however, the state of the original nucleus has $F = (N_\pi + N_\nu)/2$, $M_F = (N_\pi - N_\nu)/2$. Acting on it by F^+ leads to the state with the same F and $M_F = (N_\pi - N_\nu)/2 + 1$ (we consider $N_\nu > 0$). This state should be a major component of boson states describing the low lying states of the nucleus with $Z + 2$ protons and $N - 2$ neutrons. These latter states, however, have isospins $T' = T - 2$.

A simple example can show what happens to the correspondence between nucleon states and boson states. For the sake of simplicity, consider a nucleus with the same closed shells of protons and neutrons and 4 valence neutrons in the j-orbit. The ground state has $J = 0$, $T = 2$, $M_T = -2$ and is given by $(S_\nu^+)^2|0\rangle$ where S_ν^+ creates a $J = 0$ state of two j neutrons outside the closed shells whose state is denoted by $|0\rangle$. The boson image of this state is $(s_\nu^+)^2|0\rangle$ and it has $F = 1$, $M_F = -1$. Operating on this boson state by F^+ we obtain the state $s_\pi^+ s_\nu^+|0\rangle$ which has also $F = 1$ but $M_F = 0$. As emphasized above, the proton neutron interaction does not allow the $s_\pi^+ s_\nu^+|0\rangle$ state to be an eigenstate of the boson Hamiltonian which is based on experiment. A linear combination of this state and the state $(d_\pi^+ \cdot d_\nu^+)|0\rangle$ may, however, be an eigenstate which adequately describes the ground state of the nucleus with two protons and two neutrons in the j-orbit.

We may now look at the nucleon state corresponding to $s_\pi^+ s_\nu^+|0\rangle$. It should be the $S_\pi^+ S_\nu^+|0\rangle$ state where S_π^+ creates the $J = 0$ state of two j-protons. A linear combination of this state and the state $(D_\pi^+ \cdot D_\nu^+)|0\rangle$ may be an approximate eigenstate of the shell model Hamiltonian and give adequate description of the ground state. This ground state, however, has isospin $T = 0$, $M_T = 0$. The much higher $J = 0$ state with $T = 2$, $M_T = 0$ of this nucleus may be expressed as a linear combination of states in which the two protons are coupled to $J_\pi = J'$, the two neutrons to $J_\nu = J'$ and then J_π and J_ν are coupled to $J = 0$. Recalling the discussion following (38.4) and the c.f.p. (15.11), the j^4 state with $T = 2$, $M_T = 0$ is expressed as

$$\sqrt{\frac{2j-1}{3(2j+1)}}|j^2 J_\pi = 0, j^2 J_\nu = 0, J = 0\rangle - \frac{2}{\sqrt{3(2j+1)(2j-1)}}$$

$$\times \sum_{J'>0,\text{even}} \sqrt{2J'+1}|j^2 J_\pi = J', j^2 J_\nu = J', J = 0\rangle \qquad (40.38)$$

Thus, the state described above with $T = 2$ is a clear $S - D$ state of the nucleus with $M_T = -2$. For $M_T = 0$, however, it is the linear combination of states (40.38) which have rather small amplitudes of $S - D$ states. Pair states with $J = 0$ have rather small weights (less than $\frac{1}{3}$). Pairs with $J = 2$ have even smaller weights which are the smallest among those of pairs with $J > 0$, even. If the closed shells contain more neutrons than protons, the situation is more complicated but these features are still present. Hence, the boson model may offer a good description of nucleon states which have the *lowest* isospin given by $T = (N - Z)/2$ but may be definitely inadequate for states in the same nucleus with higher isospins.

There is an elegant way to avoid the difficulty with charge independence in cases where valence protons and neutrons occupy the same set of orbits. States of proton bosons, which correspond to states of two protons, are assigned isospin $T = 1$ and $M_T = 1$. States of neutron bosons are assigned $T = 1$, $M_T = -1$ and another kind of bosons which correspond to proton-neutron pairs coupled to $J = 0$ and $J = 2$ are introduced with $T = 1$ and $M_T = 0$. The isospin of the bosons, which are the basic building blocks of the model, is $T = 1$ and there are 3 states with $M_T = 1, 0, -1$. The resulting model is based on the $SU_T(3) \otimes U(6)$ group and is called IBA-3. Applications of IBA-3 are rather limited since in most nuclei where it is necessary to replace the shell model by the interacting boson model, valence protons and neutrons occupy different major shells. A detailed description of IBA-3 is presented by Elliott (1985).

One of the important features of IBA-2 is the existence of collective states with $J = 1$ in its spectrum. As explained in Sections 34 and 39, no such collective states appear in IBA-1. In IBA-2, there are states with $J = 1$ which are obtained by coupling of d_π and d_ν-bosons. If the IBA-2 states are characterized by F-spin values, the $J = 1$ states cannot be fully symmetric and have $F = N/2$. The lowest F-spin for which such a state appears is $F = (N-2)/2$, as shown in the last section. A $J=1$ state of IBA-2 may decay to the $J=0$ ground state by $M1$ electromagnetic radiation. Let us calculate the rate of such transitions.

The $M1$ operator is proportional to

$$\boxed{\boldsymbol{\mu} = g_\pi \mathbf{J}_\pi + g_\nu \mathbf{J}_\nu = \tfrac{1}{2}(g_\pi + g_\nu)(\mathbf{J}_\pi + \mathbf{J}_\nu) + \tfrac{1}{2}(g_\pi - g_\nu)(\mathbf{J}_\pi - \mathbf{J}_\nu)} \tag{40.39}$$

where g_π is the g-factor of a d_π-boson and g_ν is that of a d_ν-boson. The first term on the r.h.s. of (40.39) has vanishing matrix elements between any two orthogonal states and only the second term contributes to $M1$ transitions. In the $SU(3)$ limit of IBA-2, the $J=0$ ground state belongs to the lowest fully symmetric representation $(2N_\pi + 2N_\nu, 0)$ and has F-spin equal to $F = (N_\pi + N_\nu)/2 = N/2$. The lowest $J=1$ state belongs to the irreducible representation $(2N_\pi + 2N_\nu - 2, 1)$ and, as explained in the last section, has $F = (N_\pi + N_\nu - 2)/2 = (N-2)/2$. In this case, the $J=0$ ground state has non-vanishing $M1$ matrix element leading from it only to this lowest $J=1$ state which is the only $J=1$ state with $F = (N-2)/2$. This is due to the fact that, according to (40.13), the operator (40.39) cannot change the F-spin value by more than 1. In other cases, there may be several $J=1$ states which can be reached from the $J=0$ ground state by $M1$ transitions.

Let us calculate the sum of squares of matrix elements of the hermitean operator (40.39) taken between the $J=0$ ground state and all $J=1$ states. This sum over α, specifying the various $J=1$ states, is give by

$$\sum_{\alpha M \kappa} |\langle \alpha J = 1M | \mu_\kappa | J = 0, g.s. \rangle|^2$$

$$= \sum_{\alpha M \kappa} \langle J = 0, g.s. | (-1)^\kappa \tfrac{1}{2}(g_\pi - g_\nu)(J_\pi - J_\nu)_{-\kappa} | \alpha J = 1, M \rangle$$

$$\times \langle \alpha J = 1, M | \tfrac{1}{2}(g_\pi - g_\nu)(J_\pi - J_\nu)_\kappa | J = 0, g.s. \rangle$$

$$= \langle J = 0, g.s. | \tfrac{1}{4}(g_\pi - g_\nu)^2 (\mathbf{J}_\pi - \mathbf{J}_\nu)^2 | J = 0, g.s. \rangle \tag{40.40}$$

The operator whose expectation value in the $J = 0$ ground state appears on the r.h.s. of (40.40) is equal to

$$\tfrac{1}{4}(g_\pi - g_\nu)^2(\mathbf{J}_\pi - \mathbf{J}_\nu)^2 = \tfrac{1}{4}(g_\pi - g_\nu)^2((\mathbf{J}_\pi + \mathbf{J}_\nu)^2 - 4\mathbf{J}_\pi \cdot \mathbf{J}_\nu) \tag{40.41}$$

The expectation value of $(\mathbf{J}_\pi + \mathbf{J}_\nu)^2 = \mathbf{J}^2$ vanishes in $J = 0$ states. The operator whose expectation value should be evaluated is thus equal to

$$-(g_\pi - g_\nu)^2\mathbf{J}_\pi \cdot \mathbf{J}_\nu = -10(g_\pi - g_\nu)^2(d_\pi^+ \times \tilde{d}_\pi)^{(1)} \cdot (d_\nu^+ \times \tilde{d}_\nu)^{(1)} \tag{40.42}$$

The combined rates of all $M1$ transitions between the $J = 0$ ground state and the various $J = 1$ states are given by

$$\sum B(M1) = -\frac{3}{4\pi}(g_\pi - g_\nu)^2\langle J = 0, g.s.|\mathbf{J}_\pi \cdot \mathbf{J}_\nu|J = 0, g.s.\rangle \tag{40.43}$$

The expression (40.43) can be further evaluated in the case where the IBA-2 ground state has definite F-spin, $F = (N_\pi + N_\nu)/2 = N/2$. The projection of the operator (40.42) on states with definite F-spin is obtained by putting, for $L = 1$, $a_L^\pi = 0$, $a_L^\nu = 0$ and $a_L = -10(g_\pi - g_\nu)^2$ in (40.25). Using this projection the result is

$$-\frac{3}{4\pi}(g_\pi - g_\nu)^2\frac{N_\pi N_\nu}{N(N-1)}\langle J = 0, g.s.|\mathbf{J}^2 - 6(d^+ \cdot \tilde{d})|J = 0, g.s.\rangle \tag{40.44}$$

where the state $|J = 0, g.s.\rangle$, with $F = N/2$, is constructed from one kind of s-and d-bosons. The operator $\mathbf{J}^2$ in (40.44) has a vanishing expectation value and we obtain the sum rule (Ginocchio 1991)

$$\boxed{\sum B(M1) = \frac{9}{2\pi}(g_\pi - g_\nu)^2\frac{N_\pi N_\nu}{N(N-1)}\langle J = 0, g.s.|\hat{n}_d|J = 0, g.s.\rangle} \tag{40.45}$$

The combined $M1$ rates may determine in this way the expectation value of the number operator of d-bosons in the ground state considered. In the case of $SU(3)$ mentioned above, the average number of d-bosons in the ground state is $4N(N-1)/3(2N-1)$ and the rate of the $J=0$ to $J=1$ transition is equal to $(6/\pi)(g_\pi - g_\nu)^2 N_\pi N_\nu/(2N-1)$. The measured rates of $M1$ transitions are rather large in agreement with the prediction of IBA-2 (Iachello 1981b).

Energy levels of various nuclei have been calculated in the IBA-2 model. The number of parameters in this model is rather high (20 or more) and hence the Hamiltonians used in these calculations were chosen to have only a small number of adjustable parameters. Successful calculations were carried out by several authors (e.g. Iachello et al. 1979, Bijker et al. 1980, Puddu et al. 1980, Duval and Barrett 1981). The examples to be presented below have one attractive feature. The single proton d-boson energy is higher than the single proton s-boson energy by the value $V_2 - V_0$ of the semi-magic nucleus obtained by removing all valence neutrons. The energy difference between a single neutron d-boson and single neutron s-boson is equal to $V_0 - V_2$ of the semi-magic nucleus with no valence protons. The nuclei described in the following pages were described in terms of the $O(6)$ limit of IBA-1. Detailed descriptions of some cases can be found in the book by Casten (1990) and results of IBA-2 calculations are presented in Volume 6 of this series edited by R. F. Casten.

The Hamiltonian chosen to reproduce the data is very simple indeed (Novoselsky and Talmi, 1986). It is defined by

$$\boxed{H = \epsilon_\pi(d_\pi^+ \cdot \tilde{d}_\pi) + \epsilon_\nu(d_\nu^+ \cdot \tilde{d}_\nu) + \kappa\tilde{\mathbf{Q}}_\pi(\chi_\pi)\cdot\tilde{\mathbf{Q}}_\nu(\chi_\nu) + \lambda\mathbf{L}_\pi\cdot\mathbf{L}_\nu} \tag{40.46}$$

To simplify the Hamiltonian even further, ϵ_π was taken to be equal to $\epsilon_\nu, \epsilon_\pi = \epsilon_\nu = \epsilon$. The value of ϵ was taken for Xe isotopes to be equal to the 0-2 spacing of the semi-magic ${}^{136}_{54}\mathrm{Xe}_{82}$ nucleus (1.313 MeV), for Ba isotopes to the 0–2 spacing in ${}^{138}_{56}\mathrm{Ba}_{82}$ (1.436 MeV) and for Pt isotopes, ϵ was taken to be 1.068 MeV (the 0-2 spacing in ${}^{206}_{80}\mathrm{Hg}_{126}$). Of all possible interactions between proton bosons and neutron bosons only the dipole interaction was added to the dominant quadrupole-quadrupole interaction. The parameters κ, χ_π, χ_ν and λ were determined, for each nucleus, by the best fit to the experimental data.

As explained in Section 37, it is not possible to determine these parameters without very detailed knowledge of the nucleon configurations which are approximated by s- and d-bosons. Once the co-

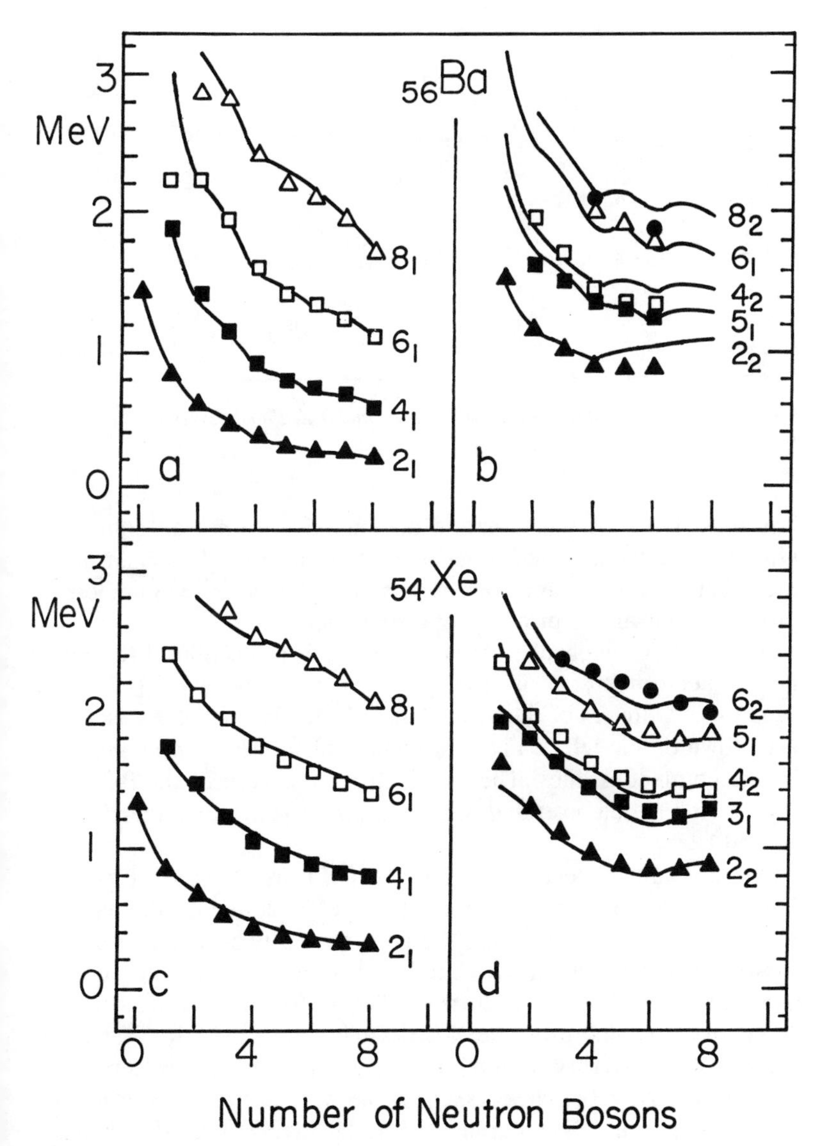

FIGURE 40.1. *Experimental and calculated levels of even* Xe *and* Ba *isotopes (number of neutron bosons is* $(82 - N)/2$*).*

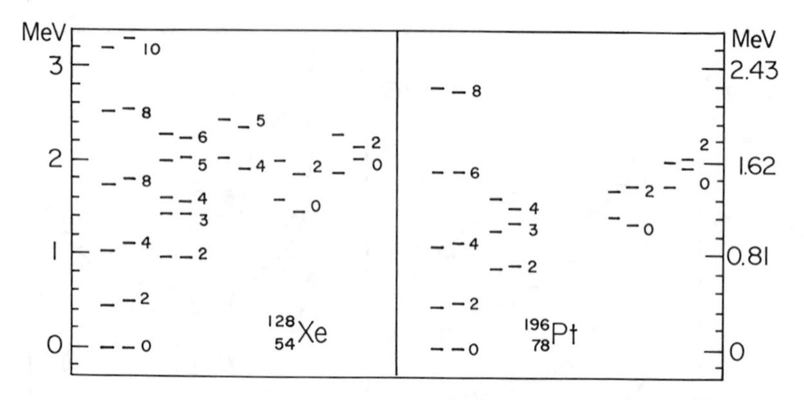

FIGURE 40.2. *Corresponding experimental and calculated levels of* ^{128}Xe *and* ^{196}Pt.

efficients α_j in the pair creation operator S^+ are not all equal, the effects of the Pauli principle become rather involved. It is clear, however, that these parameters are expected to change as the numbers of proton bosons and neutron bosons are varied.

In Fig. 40.1 the change of level spacings as a function of the number of neutron bosons is displayed. Experimental levels (triangles, squares and circles) of the ground state bands in Ba and Xe isotopes are shown on the left of Fig. 40.1 and calculated levels are connected by the continuous lines. The rather strong decrease of the 0-2 spacings is clearly demonstrated. Other levels are shown on the right of Fig. 40.1.

More detailed spectra of ^{128}Xe and ^{196}Pt are shown in Fig. 40.2. For each value of J, the experimental level is plotted on the left and the calculated one to its right. The great similarity between spectra of these two nuclei is clearly evident. In ^{128}Xe there are 2 proton bosons (4 protons outside the closed shells of $Z = 50$) and 4 neutron bosons (8 neutrons missing from the closed shells of $N = 82$). In ^{196}Pt there are 2 proton bosons (4 proton holes) and 8 neutron bosons (8 neutron holes in the $N = 126$ closed shells). The ratio between the single d-boson energy ϵ in ^{128}Xe and in ^{196}Pt, $1.313/1.068 = 1.23$ is fairly close to the ratio of the parameters κ and λ for these two nuclei (the values for ^{196}Pt are $\kappa = -.50$ MeV and $\lambda = -.24$ MeV). The values of χ_π and χ_ν for ^{196}Pt are taken to be equal to those for ^{128}Xe. The fact that in both cases χ_π and χ_ν have opposite signs ($\chi_\pi = \pm 1.462$, $\chi_\nu = \mp 1.656$) is inconsistent with equal α_j coefficients. It may well be reproduced in more realistic cases (Pittel et al. 1982, Talmi 1982).

In addition to the good agreement obtained for energy levels, IBA-2 calculations yield fairly good agreement between experimental and calculated E2 transition probabilities. The quadrupole operator is taken to be

$$e_\pi \tilde{\mathbf{Q}}_\pi(\chi_\pi) + e_\nu \tilde{\mathbf{Q}}_\nu(\chi_\nu) \tag{40.47}$$

with χ_π and χ_ν values equal to those in the Hamiltonian (40.46). For simplicity, the effective "charges" e_π and e_ν were assumed to be equal, $e_\pi = e_\nu = e$. The value of e was determined from the $B(\mathrm{E2}, 0_1 \to 2_1)$ in the semi-magic nucleus with only N_π proton bosons. In this case, the rate of transition between the $(s_\pi^+)^{N_\pi}|0\rangle$ and $(s_\pi^+)^{N_\pi-1}d_\pi^+|0\rangle$ states is given by (34.59) with $N = N_\pi$, $J = 2$, $\alpha_2 = e$ multiplied by 5 ((34.59) gives the rate of the $2 \to 0$ transition), i.e. $5e^2N_\pi$. The value of e determined from ^{136}Xe is 12 $e_p fm^2$ (e_p is the charge of a single proton) and the value from ^{138}Ba is 13 $e_p fm^2$. A partricularly interesting feature of the calculated transition rates is the reproduction of the rise in $B(\mathrm{E2})$ values as neutrons are added. The agreement between experimental rates and calculated values (on the continuous line) is demonstrated in Fig. 40.3 (where b stands for 100 fm^2). The rise in the calculated $B(\mathrm{E2})$ values is due to the coherent admixtures of proton and neutron s- and d-bosons.

The effective "charge" determined from E2 transitions in ^{196}Pt is 16 $e_p fm^2$. The quadrupole moment of the first excited $J = 2$ level in ^{196}Pt is then calculated to be 97 $e_p fm^2$. This value compares favorably with measured values 78 ± 6 $e_p fm^2$ or 70 ± 12 $e_p fm^2$. This rather large quadrupole moment is inconsistent with the $O(6)$ limit of IBA-1, which has been used to describe these nuclei, if the quadrupole operator is $\mathbf{Q}'$.

Attempts have been made to obtain good descritpion of strongly deformed nuclei by using Hamiltonians like (40.46) with ϵ determined from 0-2 spacings in semi-magic nuclei. Although good fits to energy levels were obtained, the calculated $B(\mathrm{E2})$ values were in rather poor agreement with the experimental ones. The situation did not improve much even if e_π and e_ν as well as χ_π and χ_ν in the quadrupole operator (40.47) were taken as free parameters to be determined by the experimental $B(\mathrm{E2})$ values. Much better agreement between calculated and experimental transition rates was obtained when the value of ϵ was taken to be considerably smaller than the 0–2 spacings of corresponding semi-magic nuclei. The resulting IBA-2 Hamiltonians are much closer to the $SU(3)$ limit (Novoselsky 1988).

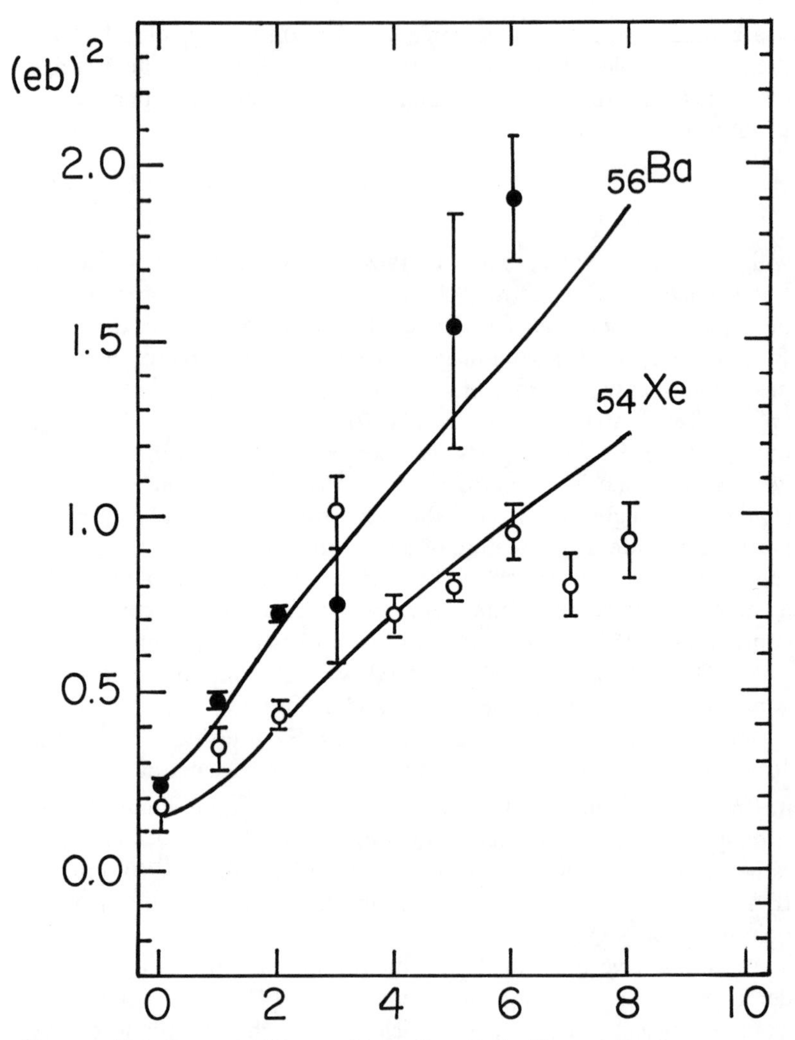

FIGURE 40.3. *Rates of* E2 *transitions* $(0 \rightarrow 2)$ *in* Xe *and* Ba *isotopes as a function of the number of neutron bosons.*

There were many attempts to improve IBA-1 and IBA-2 descriptions of strongly deformed nuclei by introducing in addition to s- and d-bosons also g-bosons (with $J = 4$). Such an extension of the boson model introduces many additional parameters and, as a result, the agreement with experiment may improve. At the same time, the simplicity of the model and its predictive power are greatly hampered.

It is remarkable that a simple model with only s- and d-bosons can reproduce all kinds of collective motion. This is due to the fact that it includes the two main ingredients of nuclear interactions. These are the $T = 1$ interactions between identical nucleons which lead to states with definite seniority (or generated seniority) and the strong attractive $T = 0$ quadrupole interaction.

41

Particles Coupled to a Rotating Core

In Section 33 a brief survey of the collective model was presented. In that model a deformed shape is defined by a set of collective variables. The spectrum due to this model is that of rotations and vibrations of the deformed shape. It was explained in Section 33 that, in a state with definite values of J and M, the projection of the total angular momentum on the axis of symmetry, defined by the dynamical variables, may have a definite value K. Some valence nucleons may be considered whose coordinates do not take place in the collective variables of the deformed even-even core. Such nucleons may be strongly coupled to the deformed core in which case the projections of their spins on the symmetry axis are good quantum numbers. In the collective model, properties of the combined system are deduced by constructing the nucleon wave functions in the body fixed frame of reference and then transforming states of the combined system to a frame of reference fixed in space. Such an approach is outside the framework of the spherical shell model and will not be followed here. It turns out, however, that it is possible to deal with the problem of nucleons strongly coupled to a rotating core by the usual methods of spectroscopy. It is therefore of some interest to derive the results of the collective model for this case by standing on the firmer ground of a frame of reference fixed in space.

These results obtained below turn out to be approximately valid also for cores described by various $SU(3)$ models. Such even-even cores may occur in the description of odd nuclei in the Ginocchio $SU(3)$ model (or FDSM). They may also occur in the interacting boson fermion model where a valence nucleon is coupled to the system of s, d bosons describing the even-even core (Iachello and Scholten 1979). This model is described in detail in a recently published book by Iachello and van Isacker (1991).

In the collective model, valence nucleons occupy orbits in a *deformed* potential well whose axis of symmetry is along one of the axes of the body-fixed frame of reference. In the model considered here, we start with valence nucleons occupying orbits in a *spherical* potential well. This potential well is determined by the *monopole* part of the interaction between valence nucleons and the core. The nucleon-core quadrupole interaction is introduced and then, as shown in detail below, single nucleon orbits become linear combinations of orbits in a spherical potential well. In the following considerations it is assumed that core nucleons occupy orbits which are orthogonal to those of the valence nucleons.

Let us first consider a single nucleon which may occupy the $j, j', \ldots$ orbits. The states of the axially symmetric core which will be considered here, are those of the ground state band characterized by $L = 0, 2, 4, \ldots$. The restriction to even values of L follows from the invariance of the core wave function under rotations by angle π around an axis perpendicular to the axis of symmetry. All states of the system are obtained by coupling states of the single nucleon with all states of the core. This basis of states is referred to as a weak coupling basis. Eigenstates of nucleons strongly coupled to the core may be expanded in states of this basis. To obtain such eigenstates, Brink et al. (1987) introduced a quadrupole interaction between a nucleon and the core and showed how to diagonalize it as follows. The interaction is taken to be proportional to the scalar product of the quadrupole operator of a single nucleon and the quadrupole operator of the core. Components of the latter are proportional to the spherical harmonics of rank $k = 2$ which are functions of the angles of a collective coordinate $\mathbf{R}$. The interaction can thus be expressed as

$$V = C' \mathbf{Y}_2(\mathbf{R}) \cdot \mathbf{q}(\mathbf{r}) \tag{41.1}$$

In the scheme defined by j, L and total spin J, matrix elements of (41.1) are given according to (10.27) and (8.15) by

$$\begin{aligned}
&\langle jLJM|V|j'L'JM\rangle \\
&= C'(-1)^{L+j'+J}(j\|\mathbf{q}\|j')(L\|\mathbf{Y}_2\|L')\begin{Bmatrix} j & L & J \\ L' & j' & 2 \end{Bmatrix} \\
&= C'(j\|\mathbf{q}\|j')(-1)^{J+j'}\sqrt{\frac{5}{4\pi}}\sqrt{(2L+1)(2L'+1)} \\
&\times \begin{pmatrix} L & L' & 2 \\ 0 & 0 & 0 \end{pmatrix}\begin{Bmatrix} j & L & J \\ L' & j' & 2 \end{Bmatrix}
\end{aligned} \tag{41.2}$$

In obtaining (41.2) we assumed that the core angular momentum is defined by

$$\mathbf{L} = \mathbf{R} \times \mathbf{P} \tag{41.3}$$

where $\mathbf{P}$ is the (collective) momentum conjugate to $\mathbf{R}$. It is also assumed that the wave functions which completely determine the rotational core states, are spherical harmonics of order L of the collective coordinate. It is worth while to note that the effect of the core is entirely determined by the values of the reduced matrix elements $(L\|\mathbf{Y}_2\|L')$.

Using the identity (10.23), the r.h.s. of (41.2) may be transformed into the expression

$$\begin{aligned}
&C'\sqrt{\frac{5}{4\pi}}\sum_m(-1)^{j-m}\begin{pmatrix} j & 2 & j' \\ -m & 0 & m \end{pmatrix}(j\|\mathbf{q}\|j') \\
&\quad\times\sqrt{2L+1}\begin{pmatrix} L & j & J \\ 0 & m & -m \end{pmatrix}\sqrt{2L'+1}\begin{pmatrix} L' & j & J \\ 0 & m & -m \end{pmatrix} \\
&= C\sum_m\langle jm|q_0|j'm\rangle\sqrt{2L+1}\begin{pmatrix} L & j & J \\ 0 & m & -m \end{pmatrix} \\
&\quad\times\sqrt{2L'+1}\begin{pmatrix} L' & j' & J \\ 0 & m & -m \end{pmatrix}
\end{aligned} \tag{41.4}$$

The interaction whose matrix elements are given by (41.4) is a sum of *separable* interactions. First we consider states in which the nucleon is in a fixed j-orbit and we put $j' = j$ in (41.4). The matrix (41.4) may then be diagonalized in a straightforward manner. The Hamiltonian of the system includes in addition to (41.1) also the single nucleon energy ϵ_j and the (rotational) energy of the core. The eigenstates of the axially symmetric core considered here have values of L equal to $L = 0, 2, 4, \ldots$. The corresponding energy eigenvalues are given by $L(L+1)/2\mathcal{I}$ and if differences between diagonal elements $\{L(L+1) - L'(L'+1)\}/2\mathcal{I}$ are much smaller than the non-diagonal elements of V, then (41.4) may be diagonalized first and the core energies be treated as a perturbation. This point will be discussed further in the following.

For each term with given value of m in (41.4) we obtain the separable matrix

$$C\langle jm_0|q_o|jm_0\rangle\sqrt{2L+1}\begin{pmatrix} L & j & J \\ 0 & m_0 & -m_0 \end{pmatrix} \times \sqrt{2L'+1}\begin{pmatrix} L' & j & J \\ 0 & m_0 & -m_0 \end{pmatrix} \tag{41.5}$$

The L values in (41.5) are all even since (41.2) vanishes unless L, L' are both even or odd. The matrix (41.5) has rank 1 so that only one of its eigenvalues does not vanish. The eigenvector which belongs to this eigenvalue has the components

$$\sqrt{2L+1}\begin{pmatrix} L & j & J \\ 0 & m_0 & -m_0 \end{pmatrix} \qquad L \text{ even} \tag{41.6}$$

The corresponding eigenvalue is equal to

$$C\langle jm_0|q_0|jm_0\rangle \tag{41.7}$$

The projection m_0 is indeed a good quantum number. The vector (41.6) is also an eigenvector of the sum over m in (41.4) with the same eigenvalue (41.7). To see it we derive a special orthogonality relation. From (7.26) we obtain

$$\sum_L (2L+1)\begin{pmatrix} L & j & J \\ 0 & m_0 & -m_0 \end{pmatrix}\begin{pmatrix} L & j & J \\ 0 & m & -m \end{pmatrix} = \delta_{mm_0} \tag{41.8}$$

as well as

$$\sum_{L}(2L+1)\begin{pmatrix} L & j & J \\ 0 & m_0 & -m_0 \end{pmatrix}\begin{pmatrix} L & j & J \\ 0 & -m & m \end{pmatrix} = \delta_{m,-m_0} \tag{41.9}$$

Due to the symmetry relation (7.28) of $3j$-symbols, (41.9) may be expressed as

$$\sum_{L}(-1)^L(2L+1)\begin{pmatrix} L & j & J \\ 0 & m_0 & -m_0 \end{pmatrix}\begin{pmatrix} L & j & J \\ 0 & m & -m \end{pmatrix} = (-1)^{j+J}\delta_{m,-m_0} \tag{41.10}$$

By adding (41.10) and (41.8) we obtain

$$\sum_{L\,\text{even}}(2L+1)\begin{pmatrix} L & j & J \\ 0 & m_0 & -m_0 \end{pmatrix}\begin{pmatrix} L & j & J \\ 0 & m & -m \end{pmatrix} = \tfrac{1}{2}(\delta_{mm_0} + (-1)^{j+J}\delta_{m,-m_0}) \tag{41.11}$$

We note that the expectation value of q_0 in the state with m is equal to that in the state with $-m$

$$\langle jm|q_0|jm\rangle = \langle j,-m|q_0|j,-m\rangle$$

Hence, the matrix (41.4) may be written as

$$\langle jLJM|V|jL'JM\rangle = 2C\sum_{m>0}\langle jm|q_0|jm\rangle\sqrt{2L+1}\begin{pmatrix} L & j & J \\ 0 & m & -m \end{pmatrix} \times \sqrt{2L'+1}\begin{pmatrix} L' & j & J \\ 0 & m & -m \end{pmatrix} \tag{41.12}$$

From (41.11) follows that the vector (41.6) with $m_0 > 0$ is orthogonal to all other vectors

$$(2L+1)^{1/2}\begin{pmatrix} L & j & J \\ 0 & m & -m \end{pmatrix}, \qquad m_0 \neq m > 0$$

From this orthogonality follows that the vector (41.6) is an eigenvector of (41.12) with the eigenvalue (41.7). The value of m_0 (>0) is a good quantum number which uniquely characterizes the eigenstates of V. From (41.11) follows the normalization of these eigenstates which may be written as

$$|m_0 jJM\rangle = \sqrt{2}\sum_{L\,\text{even}} \sqrt{2L+1}\begin{pmatrix} L & j & J \\ 0 & m_0 & -m_0 \end{pmatrix} |jLJM\rangle \qquad m_0 > 0 \tag{41.13}$$

The states (41.13) with $m_0 > 0$ form an orthogonal and *complete* basis for states obtained by coupling j with the various even values of L to yield states with given value of J. If in (41.13) we put $-m_0$ instead of m_0, we obtain, due to the symmetry properties of $3j$-symbols, the *same* state multiplied by the phase $(-1)^{j+J}$.

The quantum number m_0 appears as a projection of $\mathbf{j}$ on a symmetry axis of a "body fixed" frame of reference. The projection of $\mathbf{L}$ vanishes and that of $\mathbf{J}$ is equal to m_0. To see explicitly the nature of m_0, we show how it is related to the eigenvalue of $(\mathbf{j}\cdot\hat{\mathbf{R}})$ where $\hat{\mathbf{R}}$ is the unit vector in the direction of the collective coordinate $\mathbf{R}$. We first notice that if instead of (41.1) we start from any scalar product $(\mathbf{Y}_k(\hat{\mathbf{R}})\cdot\mathbf{q}^{(k)}(\mathbf{r}))$ we obtain the *same* expression (41.4) with q_0 replaced by $q_0^{(k)}$ and C replaced by $\sqrt{(2k+1)/4\pi}$. The spherical components of $\hat{\mathbf{R}}$ are equal to $\sqrt{4\pi/3}Y_{1\kappa}(\hat{\mathbf{R}})$. In this case, due to the symmetry properties of $3j$-symbols, it follows from (41.2) that $L' = L \pm 1$ in the non-vanishing terms. For the present discussion, let us consider cores with odd parity states and odd L-values in addition to even parity states with even values of L. Hence we obtain in this case

$$\begin{aligned}\langle jLJM|(\mathbf{j}\cdot\hat{\mathbf{R}})|jL'JM\rangle &= \sum_m \langle jm|j_0|jm\rangle\sqrt{2L+1}\begin{pmatrix} L & j & J \\ 0 & m & -m \end{pmatrix} \\ &\quad\times\sqrt{2L'+1}\begin{pmatrix} L' & j & J \\ 0 & m & -m \end{pmatrix}\end{aligned} \tag{41.14}$$

Hence, the eigenstates of $(\mathbf{j}\cdot\bar{\mathbf{R}})$ are characterized by m (ranging from $-j$ to $+j$) and their normalized expression is

$$|m_0 jJM\rangle' = \sum_L \sqrt{2L+1}\begin{pmatrix} L & j & J \\ 0 & m_0 & -m_0 \end{pmatrix}|jLJM\rangle \tag{41.15}$$

The summation in (41.15) is over all values of L. The corresponding eigenvalues of $(\mathbf{j}\cdot\hat{\mathbf{R}})$ are given by

$$\langle jm_0|j_0|jm_0\rangle = m_0 \tag{41.16}$$

It is worthwhile to point out that once $\mathbf{J}\cdot\hat{\mathbf{R}} = \mathbf{j}\cdot\hat{\mathbf{R}}$ is chosen to be diagonal, along with $\mathbf{j}^2$, $\mathbf{J}^2$ and $\mathbf{J}_z$, then $\mathbf{L}^2$ defined by (41.3) cannot be diagonal as seen in (41.15). If L completely characterizes the state of the core, then there is only *one* independent state with definite values of L, j, J and M. In this case no more quantum numbers may be specified. If $(\mathbf{J}\cdot\hat{\mathbf{R}})$ is diagonal and $j \neq 0$, $\mathbf{L}^2$ can no longer have a definite eigenvalue in that state. Indeed, $(\mathbf{J}\cdot\hat{\mathbf{R}})$ is a scalar with respect to the components of $\mathbf{J} = \mathbf{j} + \mathbf{L}$ and commutes with $\mathbf{J}^2$ and J_z as well as with J_x and J_y. Since $(\mathbf{J}\cdot\hat{\mathbf{R}}) = (\mathbf{j}\cdot\hat{\mathbf{R}})$ and components of $\mathbf{j}$ commute with components of $\hat{\mathbf{R}}, (\mathbf{j}\cdot\hat{\mathbf{R}})$ behaves as a component of $\mathbf{j}$ and commutes with $\mathbf{j}^2$. Hence, the operator

$$\mathbf{L}^2 = (\mathbf{J} - \mathbf{j})^2 = \mathbf{J}^2 + \mathbf{j}^2 - 2(\mathbf{J}\cdot\mathbf{j}) \tag{41.17}$$

does not commute with $(\mathbf{J}\cdot\hat{\mathbf{R}})$ since the latter does not commute with $(\mathbf{J}\cdot\mathbf{j})$. Components of $\mathbf{J}$ in $(\mathbf{J}\cdot\mathbf{j})$ commute with $(\mathbf{J}\cdot\hat{\mathbf{R}})$ but j_x, j_y and j_z do not commute with it. In the case $j = 0$, which may occur for several valence nucleons, $J = L$ and $m_0 = 0$. In this case, states of the system are completely specified by L and M (and other necessary quantum numbers specifying the $j = 0$ state of the valence nucleons).

Coming back to the eigenstates (41.13) we see that they are not eigenstates of j_0 but are equal to the linear combination of the states (41.15) with m_0 and $-m_0$

$$\frac{1}{\sqrt{2}}\{|m_0 jJM\rangle' + (-1)^{j+J}|-m_0 jJM\rangle'\} \tag{41.18}$$

The states (41.18) are, by their construction, eigenstates with definite values of m_0^2. As explained above, we take only the eigenstates with $m_0 > 0$ which define an orthogonal and *complete* set of states. It may be directly verified that the states (41.13) are eigenstates of $(\mathbf{J}\cdot\hat{\mathbf{R}})^2$ by using the general expression (41.4). If states of the core are limited to those with even values of L, the operator $(\mathbf{J}\cdot\hat{\mathbf{R}})$, for $j > 0$, cannot be diagonalized. This operator commutes with J_z (also with J_x and J_y) but it is not invariant under the rotation by π around an axis (in the body fixed frame) which is perpendicular to the symmetry axis. The states (41.13) are invariant under the rotation in which m_0 is transformed into $-m_0$ and *vice versa*.

The eigenvalue (41.7) which belongs to (41.13) is equal to C multiplying the expectation value of the quadrupole operator q_0 in a state with $j_z = m_0$. It may be viewed as the expectation value of the interaction between the nucleon quadrupole operator and a quadrupole field in a "body fixed" frame of reference. That is the frame of reference in which the projection of $\mathbf{j}$ on $\mathbf{R}$ is equal to m_0. Still no explicit use of that frame of reference has been made above.

Let us now turn to the energy of the core which was neglected so far. The results obtained above for eigenstates of the system, hold if the core energy may be included to first order in perturbation theory. Otherwise, no particular form for this energy has been assumed. In order to obtain explicit results for energies of the combined system we should specify the core energy. In the following we assume that the core has a fixed moment of inertia $\mathcal{I}$ and the Hamiltonian of the system is defined by

$$\boxed{H = \frac{1}{2\mathcal{I}}\mathbf{L}^2 + \epsilon_j + V} \tag{41.19}$$

where V is the quadrupole interaction (41.1), ϵ_j is the single nucleon energy and $\mathbf{L}^2/2\mathcal{I}$ is the rotational energy of the core. In the absence of the j-nucleon, eigenstates of the core are characterized by $L = 0, 2, 4, \ldots$ and the corresponding eigenvalues are equal to $L(L+1)/2\mathcal{I}$. In the strong coupling limit it is assumed that the core rotational energy is small compared to the interaction energy. If the rotational energy is ignored, the eigenvalues of the Hamiltonian are thus given by diagonalization of V. This leads to the eigenstates (41.13) and the eigenvalues of H are equal in this limit to

$$\epsilon_j + C\langle jm_0|q_0|jm_0\rangle = \epsilon_j + C(-1)^{j-m_0}\begin{pmatrix} j & 2 & j \\ -m_0 & 0 & m_0 \end{pmatrix}(j\|\mathbf{q}\|j)$$

$$= \epsilon_j + C(j\|\mathbf{q}\|j)\frac{3m_0^2 - j(j+1)}{\sqrt{j(j+1)(2j+1)(2j-1)(2j+3)}} \qquad (41.20)$$

The reduced matrix element of the quadrupole operator of a single j-nucleon is given by (10.46) and (10.47). It follows that this reduced matrix element is *negative*. Hence, if the quadrupole interaction is attractive, $C < 0$, the lowest eigenvalue (41.20) is for $m_0 = \frac{1}{2}$. All states with various values of J with this quantum number are degenerate. These are states with $J = m_0$, $J = m_0 + 1$, $J = m_0 + 2, \ldots$. The next higher group of degenerate states, for $j > \frac{1}{2}$ is given by $m_0 = \frac{3}{2}$ and so on. For a repulsive quadrupole interaction, $C > 0$, the lowest group of degenerate states has $m_0 = j$, the next has $m_0 = j - 1$ etc.

The degeneracy described above is removed when we take into account the energy of the core in the Hamiltonian (41.19). If it is considered as a small perturbation, its contribution to the energy is equal to its expectation values in the states (41.13). The expectation value of $\mathbf{L}^2$ is given by

$$\langle m_0 jJM|\mathbf{L}^2|m_0 jJM\rangle = 2\sum_{L\,\text{even}} L(L+1)(2L+1)\begin{pmatrix} L & j & J \\ 0 & m_0 & -m_0 \end{pmatrix}^2 \qquad (41.21)$$

To evaluate the sum we make use of an identity of $3j$-symbols (Zemel 1975) which will now be derived.

Instead of coupling $\mathbf{j}$ and $\mathbf{L}$ to $\mathbf{J}$ we consider the coupling of $\mathbf{j}$ and $\mathbf{J}$ to $\mathbf{L}$. Due to the symmetries of the vector addition coefficients the outcome will be applicable also to (41.21). Thus, we start with

$$\mathbf{L}^2 = (\mathbf{J} + \mathbf{j})^2 = \mathbf{J}^2 + \mathbf{j}^2 + 2(\mathbf{j}\cdot\mathbf{J}) \qquad (41.22)$$

from which follows

$$-2(\mathbf{j}\cdot\mathbf{J}) = \mathbf{J}^2 + \mathbf{j}^2 - \mathbf{L}^2 \qquad (41.23)$$

We now evaluate matrix elements of (41.23) in the scheme $jmJM$. The r.h.s. of (41.23) is diagonal in the $jJLM_L$ scheme. We use Clebsch-Gordan coefficients which transform one scheme to the other and obtain

$$\begin{aligned}
&-\langle jmJM|2(\mathbf{j}\cdot\mathbf{J})|jm'JM'\rangle \\
&\quad=\sum_{LM_L}[J(J+1)+j(j+1)-L(L+1)] \\
&\qquad\times(jmJM\mid jJLM_L)(jm'JM'\mid jJLM_L) \\
&\quad=[J(J+1)+j(j+1)]\delta_{mm'}\delta_{MM'} \\
&\qquad-\sum_{L}L(L+1)(jmJM\mid jJLM_L)(jm'JM'\mid jJLM_L)
\end{aligned} \tag{41.24}$$

To obtain the summation over even values of L we write the matrix element of (41.23) between the states $jmJM$ and $j,-m'J,-M'$ (for $m\neq -m'$, $M\neq -M'$) obtaining

$$\begin{aligned}
&-\langle jmJM|2(\mathbf{j}\cdot\mathbf{J})|j,-m'J,-M'\rangle \\
&\quad=-\sum L(L+1)(jmJM\mid jJLM_L)(j,-m'J,-M'\mid jJLM_L) \\
&\quad=(-1)^{j+J+1}\sum_{L}(-1)^{L}L(L+1) \\
&\qquad\times(jmJM\mid jJLM_L)(jm'JM'\mid jJL,-M_L)
\end{aligned} \tag{41.25}$$

For the value $M_L=0$ we can combine (41.24) and (41.25) to obtain

$$\begin{aligned}
&2\sum_{L\,\text{even}}L(L+1)(jmJM\mid jJL0)(jm'JM'\mid jJL0) \\
&\quad=[J(J+1)+j(j+1)]\,\delta_{mm'}\,\delta_{MM'} \\
&\qquad+[\langle jmJM|2(\mathbf{j}\cdot\mathbf{J})|jm'JM'\rangle \\
&\qquad\quad+(-1)^{j+J}\langle jmJM|2(\mathbf{j}\cdot\mathbf{J})|j,-m'J,-M'\rangle]
\end{aligned} \tag{41.26}$$

The matrix elements on the r.h.s. of (41.26) can be directly evaluated by using (6.20). We then obtain

$$\begin{aligned}\langle jmJM|2(\mathbf{j}\cdot\mathbf{J})|jm'JM'\rangle &= 2mM\,\delta_{mm'}\,\delta_{MM'}\\ &+\sqrt{J(J+1)-M'(M'+1)}\sqrt{j(j+1)-m'(m'-1)}\,\delta_{m,m'-1}\,\delta_{M,M'+1}\\ &+\sqrt{J(J+1)-M'(M'-1)}\sqrt{j(j+1)-m'(m'+1)}\,\delta_{m,m'+1}\,\delta_{M,M'-1}\end{aligned} \tag{41.27}$$

We now put in (41.26) $m' = m = m_0$, $M' = M = -m_0$ and obtain, due to (41.27) and $m_0 > 0$, the r.h.s. of (41.26), equal to the following sum of non-vanishing terms

$$\begin{aligned}&J(J+1)+j(j+1)-2m_0^2+(-1)^{j+J}\sqrt{J(J+1)-m_0(m_0-1)}\\ &\quad\times\sqrt{j(j+1)-m_0(m_0-1)}\,\delta_{m_0,-m_0+1}\,\delta_{-m_0,m_0-1}\end{aligned} \tag{41.28}$$

The only value of $m_0 > 0$ for which $\delta_{m_0,-m_0+1}$ does not vanish is $m_0 = \frac{1}{2}$. Hence, we obtain, expressing the Clebsch-Gordan coefficients in (41.26) by $3j$-symbols, the result

$$\boxed{\begin{aligned}&2\sum_{L\,\text{even}} L(L+1)(2L+1)\begin{pmatrix} L & j & J\\ 0 & m_0 & -m_0\end{pmatrix}\begin{pmatrix} L & j & J\\ 0 & m_0 & -m_0\end{pmatrix}\\ &\qquad = J(J+1)+j(j+1)-2m_0^2+\delta_{m_0,1/2}(-1)^{j+J}(J+\tfrac{1}{2})(j+\tfrac{1}{2})\end{aligned}} \tag{41.29}$$

Equation (41.29) is a precise mathematical identity of $3j$-symbols. The L-values on the l.h.s. of (4.29) are limited, due to their appearance in non-vanishing $3j$-symbols, to $|j-J| \leq L \leq j+J$.

We can now apply (41.29) and (41.21) to obtain the expectation value of the energy of the core in the states (41.13) which is given by

$$\frac{1}{2\mathcal{I}}J(J+1)+\delta_{m_0,1/2}\frac{1}{2\mathcal{I}}(-1)^{j+J}(J+\tfrac{1}{2})(j+\tfrac{1}{2})+\frac{1}{2\mathcal{I}}[j(j+1)-2m_0^2] \tag{41.30}$$

The expression (41.30) is the well known formula for the rotational energy of an odd even nucleus. For $m_0 > \frac{1}{2}$ the dependence of (41.30) on J is simply given by $J(J+1)$. For $m_0 = \frac{1}{2}$ (m_0 is usually denoted by K), there is also another term which is called the *decoupling* term. States in a given rotational band have spins $J = m_0, J = m_0 + 1, J = m_0 + 2, \ldots$. The result (41.30) is correct only to first order perturbation theory. Due to (41.27), (41.26) and (41.21), the kinetic energy of the core has non-vanishing non-diagonal matrix elements between states with different values of m_0. These arise from the term $2(\mathbf{j} \cdot \mathbf{J})$ which is called the Coriolis coupling term. This name arises naturally if the body fixed frame of reference is adopted.

The results obtained above may be generalized to the case in which several j-orbits are available for a single nucleon. The mixing of these orbits is caused by the quadrupole interaction. We may consider also several nucleons in a j-orbit or in several orbits. In that case j, j' characterize the various antisymmetric states of the valence nucleons in which their mutual interactions have been diagonalised. For even nucleon numbers j, j' will assume integral values. The matrix of the quadrupole interaction is given by the r.h.s. of (41.4) in which the summation is extended over all j, j' values of the orbits (or states) considered. We have to consider states which are linear combinations of the states (41.13) with the various j-values

$$|m_0 JM\rangle = \sum_j \alpha_j \sqrt{2} \sum_{L\,\text{even}} \sqrt{2L+1} \begin{pmatrix} L & j & J \\ 0 & m_0 & -m_0 \end{pmatrix} |jLJM\rangle$$

$$\sum_j \alpha_j^2 = 1$$

(41.31)

The expansion coefficients in (41.31) were derived by Bohr and Mottelson (1975). The coefficients α_j in (41.31) have no relation to the α_j introduced in Section 23. We now multiply the interaction matrix

$$\langle jLJM|V|j'L'JM\rangle = C \sum_{jj'm} \langle jm|q_0|j'm\rangle \times \sqrt{2L+1} \begin{pmatrix} L & j & J \\ 0 & m & -m \end{pmatrix} \sqrt{2L'+1} \begin{pmatrix} L' & j' & J \\ 0 & m & -m \end{pmatrix} \tag{41.32}$$

whose rows and columns are characterized by j, L, by the column vector whose components are the coefficients of $|jLJM\rangle$ in (41.31). We obtain

$$\sqrt{2}C\sum_{j'm}\langle jm|q_0|j'm\rangle\alpha_{j'}\sqrt{2L+1}\begin{pmatrix} L & j & J \\ 0 & m & -m \end{pmatrix}$$

$$\times \sum_{L' \text{ even}}(2L'+1)\begin{pmatrix} L' & j' & J \\ 0 & m & -m \end{pmatrix}\begin{pmatrix} L' & j' & J \\ 0 & m_0 & -m_0 \end{pmatrix}$$

$$= \sqrt{2}C\sum_{j'm}\langle jm|q_0|j'm\rangle\alpha_{j'}\sqrt{2L+1}\begin{pmatrix} L & j & J \\ 0 & m & -m \end{pmatrix}$$

$$\times \tfrac{1}{2}(\delta_{mm_0} + (-1)^{j+J}\delta_{m,-m_0}) \tag{41.33}$$

The equality in (41.33) is due to the orthogonality relation (41.11). The r.h.s. of (41.33) may be expressed as

$$\sqrt{2}C\sum_{j'}\alpha_{j'}\sqrt{2L+1}\tfrac{1}{2}\Bigg[\langle jm_0|q_0|j'm_0\rangle\begin{pmatrix} L & j & J \\ 0 & m_0 & -m_0 \end{pmatrix}$$

$$+ (-1)^{j+J}\langle j,-m_0|q_0|j',-m_0\rangle\begin{pmatrix} L & j & J \\ 0 & -m_0 & m_0 \end{pmatrix}\Bigg]$$

$$= \sqrt{2}C\sqrt{2L+1}\tfrac{1}{2}(1+(-1)^L)\begin{pmatrix} L & j & J \\ 0 & m_0 & -m_0 \end{pmatrix}$$

$$\times \sum_{j'}\langle jm_0|q_0|j'm_0\rangle\alpha_{j'} \tag{41.34}$$

The r.h.s. of (41.34) does not vanish only for even values of L. Hence, the state (41.31) is an eigenstate of (41.32) provided the coefficients of α_j form an eigenvector of the matrix $\langle jm_0|q_0|j'm_0\rangle$. Recalling that the Hamiltonian of the system contains also non-degenerate single nucleon energies, we take the α_j in (41.31) to satisfy the eigenvalue equation

$$\sum_{j'}(\epsilon_j\delta_{jj'} + \langle jm_0|q_0|j'm_0\rangle)\alpha_{j'} = Q_{m_0}\alpha_j \tag{41.35}$$

With these coefficients the state (41.31) is an eigenstate of the Hamiltonian with m_0 as a good quantum number, when the core rotational energy is neglected. If the states with j and j' are those of several nucleons in one orbit, ϵ_j denotes their single nucleon energy, as well as their mutual interaction energy and q_0 should be replaced by the sum of $\kappa = 0$ components of their quadrupole operators.

The α_j coefficients in (41.35) define an eigenstate of a Hamiltonian, including the interaction with the core, in the "body fixed" frame of reference. If we replace the ϵ_j by a single nucleon harmonic oscillator Hamiltonian and put $\mathbf{q} = r^2\mathbf{Y}_2(\mathbf{r})$, these eigenstates are Nilsson orbitals. There are several independent solutions of (41.35), their number being equal to the number of j-orbits, $j \geq m_0$, included in the summations. In such states, the angular momentum of the nucleon (or nucleons) is no longer diagonal. The lowest states of the system are those in which the single nucleon occupies the lowest eigenstates among the various eigenstates of (41.35) for all values of m_0. In the case of several nucleons, if their mutual interactions are much smaller than the single nucleon energies, the states of the system are simple. Each of the lowest single nucleon orbits is filled by at most two identical nucleons. In the following we will consider only the lowest among the eigenstates of (41.35) with given m_0 and hence, will not introduce for it a special notation.

Let us now introduce the rotational energy of the core as a perturbation. The expectation value of $\mathbf{L}^2/2\mathcal{I}$ in the states (41.31) can be directly evaluated. The operator $\mathbf{L}^2$ commutes with components of $\mathbf{j}$ and, hence, its expectation value is given by

$$2\sum_{L\,\text{even},j} \alpha_j^2 L(L+1)(2L+1)\begin{pmatrix} L & j & J \\ 0 & m_0 & -m_0 \end{pmatrix}^2$$

$$= J(J+1) - 2m_0^2 + \sum_j \alpha_j^2 (j+1)$$

$$+ \delta_{m_0,1/2}\sum_j \alpha_j^2(-1)^{j+J}(J+\tfrac{1}{2})(j+\tfrac{1}{2}) \qquad (41.36)$$

Dividing the r.h.s. of (41.36) by $2\mathcal{I}$ we obtain the contribution in first order of perturbation theory to the energy of the system. To express (41.36) in conventional form we introduce the state

$$\boxed{|m_0\rangle = \sum_j \alpha_j |j, m = m_0\rangle} \qquad (41.37)$$

To specify uniquely this state we should indicate which set of α_j of the independent solutions of (41.35) is chosen. We will suppress this label and add it only in cases where we consider two such states. The state (41.37) may be interpreted as an "intrinsic state" in the body fixed frame of reference. Using this well defined state, we may express one summation on the r.h.s. of (41.36) as

$$\langle m_0|\mathbf{j}^2|m_0\rangle = \sum_j \alpha_j^2 j(j+1) \tag{41.38}$$

To obtain a similar expression for the last term on the r.h.s. of (41.36) we introduce another state. We first rewrite (41.31) as

$$\begin{aligned}
&\sqrt{2}\sum_{L\,\text{even},j} \alpha_j\sqrt{2L+1}\begin{pmatrix} L & j & J \\ 0 & m_0 & -m_0 \end{pmatrix}|jLJM\rangle \\
&= \sqrt{2}\sum_{L\,\text{even},j} \alpha_j\sqrt{2L+1}(-1)^{j+J}\begin{pmatrix} L & j & J \\ 0 & -m_0 & m_0 \end{pmatrix}|jLJM\rangle \\
&= (-1)^{J+1/2}\sqrt{2}\sum_{L\,\text{even},j} \alpha_j(-1)^{j-1/2}\sqrt{2L+1} \\
&\quad\times \begin{pmatrix} L & j & J \\ 0 & -m_0 & m_0 \end{pmatrix}|jLJM\rangle
\end{aligned} \tag{41.39}$$

This leads to the definition of another intrinsic state as

$$\boxed{|\overline{m_0}\rangle = \sum_j (-1)^{j+1/2}\alpha_j|j, m = -m_0\rangle} \tag{41.40}$$

The last term on the r.h.s. of (41.36) may be now expressed as

$$-\delta_{m_0,1/2}(-1)^{J+1/2}(J+\tfrac{1}{2})\langle\tfrac{1}{2}|j^+|\overline{\tfrac{1}{2}}\rangle = \delta_{m_0,1/2}(-1)^{J+1/2}(J+\tfrac{1}{2})a \tag{41.41}$$

In (41.41) the conventional notation is given for the decoupling parameter a of the rotational energy.

As mentioned above, the expression of rotational energies in the case of nucleons coupled to a core is usually derived by using states of nucleons strongly coupled to a rotating frame of reference. They were derived here in a frame of reference fixed in space using the formalism described in this book. In constructing the states (41.31) or (41.13), only a limited number of L-values are used. These are the ones which may be coupled to the various j-values to obtain the total spin J. These finite numbers of L-values are not due to any approximation. More important, the assumption that the core rotational energy may be taken as a perturbation has a precise meaning. Energy differences between rotational states of the core with $L+2$ and L are equal to $(2L+3)/\mathcal{I}$. They should be small compared to the non-diagonal element of the nucleon-core quadrupole interaction between these states of the core coupled to the nucleon state. This non-diagonal element is given, in the case of a single j-orbit, by (41.4) in terms of $C\langle jm_0|q_0|jm_0\rangle$ and $3j$-symbols. A similar condition may be obtained if the non-diagonal matrix elements of the core energy between states with different quantum numbers m_0 should be small compared to energy differences between these m_0-states. Matrix elements of $\mathbf{L}^2/2\mathcal{I}$ between states with m_0 and m_0+1 can be obtained from (41.26) by using (41.27) with $m=m_0$, $m'=m_0+1$. Hence, the strong coupling limit may be valid only up to a certain value of L which is determined by the parameters of the system.

The expression (41.29) of the expectation value of $\mathbf{L}^2$ in the state (41.13) may be applied to calculating the magnetic moment of such a state. According to (8.24) the g-factor of a j-nucleon coupled to the core in the state with given L is equal to

$$
\begin{aligned}
g(jLJ) &= \frac{1}{J(J+1)}(g_c(\mathbf{L}\cdot\mathbf{J}) + g_j(\mathbf{j}\cdot\mathbf{J})) \\
&= \frac{1}{2J(J+1)}[g_c\{J(J+1)+L(L+1)-j(j+1)\} \\
&\qquad + g_j\{J(J+1)+j(j+1)-L(L+1)\}] \\
&= \frac{1}{2J(J+1)}[(g_j+g_c)J(J+1)+(g_j-g_c)j(j+1) \\
&\qquad -(g_j-g_c)L(L+1)]
\end{aligned} \tag{41.42}
$$

In (41.42), g_j is the g-factor of the nucleon and g_c is the core g-factor. The g-factor of the state $|m_0 jJM\rangle$ is obtained by multiplying (41.42) by the amplitude of $|jLJM\rangle$ in (41.13) and summing over L. Using the identity (41.29) we obtain

$$\boxed{\begin{aligned} g = {} & \frac{1}{2J(J+1)}[(g_j+g_c)J(J+1)+(g_j-g_c)j(j+1) \\ & -(g_j-g_c)\{J(J+1)+j(j+1) \\ & -2m_0^2+\delta_{m_0,1/2}(-1)^{j+J}(J+\tfrac{1}{2})(j+\tfrac{1}{2})\}] \\ = {} & g_c+(g_j-g_c)\frac{1}{J(J+1)} \\ & \times [m_0^2-\tfrac{1}{2}\delta_{m_0,1/2}(-1)^{j+J}(J+\tfrac{1}{2})(j+\tfrac{1}{2})] \end{aligned}} \tag{41.43}$$

If there are several nucleons in the same orbit, the g-factor of any of their states is equal to that of a single nucleon g_j. The magnetic moment of the eigenstate (41.31) is then given by the g-factor

$$\begin{aligned} g_c+(g_j-g_c)\frac{1}{J(J+1)}[m_0^2+\tfrac{1}{2}\delta_{m_0,1/2}(-1)^{J+1/2}(J+\tfrac{1}{2}) \\ \times \sum_j \alpha_j^2(-1)^{j+1/2}(j+\tfrac{1}{2})] \end{aligned} \tag{41.44}$$

The situation is more complicated if the state (41.31) is that of a single nucleon coupled to the core and the various j-orbits have different g-factors. In that case, the magnetic moment operator of the nucleons is no longer proportional to $\mathbf{j}$. It has non-vanishing matrix elements between states with $j=l+\frac{1}{2}$ and $j'=l-\frac{1}{2}$ (provided their radial functions are not orthogonal). The magnetic moment operator of the combined system is

$$\begin{aligned}\boldsymbol{\mu} &= g_c\mathbf{L} + \sum_i g_s^{(i)}\mathbf{s}_i + \sum_i g_l^{(i)}\boldsymbol{l}_i \\ &= g_c\mathbf{J} + \sum_i (g_s^{(i)} - g_c)\mathbf{s}_i + \sum_i (g_l^{(i)} - g_c)\boldsymbol{l}_i = g_c\mathbf{J} + \boldsymbol{\mu}'\end{aligned} \tag{41.45}$$

In (41.45) the summation is over all orbits with non-vanishing coefficients α_j in (41.31). The $\mathbf{s}_i$ and $\boldsymbol{l}_i$ are the (second quantized) operators of intrinsic spin and orbital spin of nucleons in the i-th orbit and $g_s^{(i)}$ and $g_l^{(i)}$ are the corresponding spin and orbital g-factors. The diagonal elements of the $\kappa = 0$ component of (41.45) in states with given J and $M = J$ are the magnetic moments. The non-diagonal elements determine the rates of M1 transitions. The first term on the r.h.s. of (41.45) has only non-vanishing diagonal elements which contribute $g_c J$ to the magnetic moment.

The operator $\boldsymbol{\mu}'$ in (41.45) acts only on the nucleon coordinates and hence is diagonal in L. Its reduced matrix element between a state (41.31) with $m_0 J$ and coefficients α_j and a state with m_0', J' and coefficients β_j is given, according to (10.42) by

$$\begin{aligned}&2\sum_{L\,\text{even},j,j'} \alpha_j\beta_{j'}(j\|\boldsymbol{\mu}'\|j')(-1)^{j+L+J'+1}\sqrt{(2J+1)(2J'+1)} \\ &\times \begin{Bmatrix} j & J & L \\ J' & j' & 1 \end{Bmatrix}(2L+1) \\ &\times \begin{pmatrix} L & j & J \\ 0 & m_0 & -m_0 \end{pmatrix}\begin{pmatrix} L & j' & J' \\ 0 & m_0' & -m_0' \end{pmatrix}\end{aligned} \tag{41.46}$$

The sum of products of the $6j$-symbol and two $3j$-symbols may be evaluated by using the identity (10.24). First we obtain

$$\sum_L (-1)^{L+1+m_0+m_0'}(2L+1)\begin{pmatrix} j & J & L \\ m_0 & -m_0 & 0 \end{pmatrix}$$

$$\times \begin{pmatrix} J' & j' & L \\ m_0' & -m_0' & 0 \end{pmatrix}\begin{Bmatrix} j & J & L \\ J' & j' & 1 \end{Bmatrix}$$

$$= \sum_\kappa \begin{pmatrix} j & j' & 1 \\ m_0 & -m_0' & -\kappa \end{pmatrix}\begin{pmatrix} J' & J & 1 \\ m_0' & -m_0 & \kappa \end{pmatrix} \quad (41.47)$$

To obtain a similar summation over *even* values of L we write (41.47) where m_0' is replaced by $-m_0'$ and then change the signs of the lower arguments in

$$\begin{pmatrix} J' & j' & L \\ -m_0' & m_0' & 0 \end{pmatrix}$$

obtaining

$$\sum (-1)^{m_0-m_0'+j'+J'+1}(2L+1)$$

$$\times \begin{pmatrix} j & J & L \\ m_0 & -m_0 & 0 \end{pmatrix}\begin{pmatrix} J' & j' & L \\ m_0' & -m_0' & 0 \end{pmatrix}\begin{Bmatrix} j & J & L \\ J' & j' & 1 \end{Bmatrix}$$

$$= \sum_\kappa \begin{pmatrix} j & j' & 1 \\ m_0 & m_0' & -\kappa \end{pmatrix}\begin{pmatrix} J' & J & 1 \\ -m_0' & -m_0 & \kappa \end{pmatrix} \quad (41.48)$$

Combining (41.47) with (41.48) we obtain the relation

$$2\sum_{L\,\text{even}} (2L+1)\begin{pmatrix} L & j & J \\ 0 & m_0 & -m_0 \end{pmatrix}\begin{pmatrix} L & j' & J' \\ 0 & m_0' & -_0' \end{pmatrix}\begin{Bmatrix} j & J & L \\ J' & j' & 1 \end{Bmatrix}$$

$$= (-1)^{m_0+m_0'+1}\left[\begin{pmatrix} j & j' & 1 \\ m_0 & -m_0' & \kappa \end{pmatrix}\begin{pmatrix} J' & J & 1 \\ m_0' & -m_0 & \kappa \end{pmatrix}\right.$$

$$\left. + (-1)^{j'+J'-2m_0'}\begin{pmatrix} j & j' & 1 \\ m_0 & m_0' & -\kappa' \end{pmatrix}\begin{pmatrix} J' & J & 1 \\ -m_0' & -m_0 & \kappa' \end{pmatrix}\right]$$

Making use of this relation we can bring the reduced matrix element (41.46) to the form

$$2\sum_{j,j'}\alpha_j\beta_{j'}(j\|\boldsymbol{\mu}'\|j')(-1)^{j+J'+m_0+m_0'}\sqrt{(2J+1)(2J'+1)}$$
$$\times\frac{1}{2}\left[\begin{pmatrix} j & j' & 1 \\ m_0 & -m_0' & -\kappa \end{pmatrix}\begin{pmatrix} J' & J & 1 \\ m_0' & -m_0 & \kappa \end{pmatrix}\right.$$
$$\left.+(-1)^{j'+J'-2m_0'}\begin{pmatrix} j & j' & 1 \\ m_0 & m_0' & -\kappa' \end{pmatrix}\begin{pmatrix} J' & J & 1 \\ -m_0' & -m_0 & \kappa' \end{pmatrix}\right] \tag{41.49}$$

Since $m_0 > 0$ and $m_0' > 0$, the second term in (41.49) may be different from zero only if $m_0' = m_0 = \frac{1}{2}$ and $\kappa' = 1$. Any other values of $m_0 > 0$ and $m_0' > 0$ lead to $\kappa' > 1$ in which case the $3j$-symbol vanishes.

The result (41.49) may be further simplified by applying the relation (11.29) between $3j$-symbols to

$$\begin{pmatrix} J' & J & 1 \\ -\frac{1}{2} & -\frac{1}{2} & 1 \end{pmatrix}$$

We thus obtain

$$\sum_{j,j'}\alpha_j\beta_{j'}(j\|\boldsymbol{\mu}'\|j')\sqrt{(2J+1)(2J'+1)}(-1)^{J-m_0}\begin{pmatrix} J & 1 & J' \\ -m_0 & \kappa & m_0' \end{pmatrix}$$
$$\times\left[(-1)^{J'-J+j-m_0+\kappa}\begin{pmatrix} j & 1 & j' \\ -m_0 & \kappa & m_0' \end{pmatrix}+\delta_{m_0,1/2}\delta_{m_0',1/2}\right.$$
$$\times(-1)^{j-1/2}\begin{pmatrix} j & 1 & j' \\ -\frac{1}{2} & 1 & -\frac{1}{2} \end{pmatrix}(-1)^{j'+J}\frac{1}{2\sqrt{2}}$$
$$\left.\times\{(2J+1)(-1)^{J+J'+1}+2J'+1\}\right]$$
$$=\sum_{j,j'}\alpha_j\beta_{j'}\sqrt{(2J+1)(2J'+1)}(-1)^{J-m_0}\begin{pmatrix} J & 1 & J' \\ -m_0 & \kappa & m_0' \end{pmatrix}$$
$$\times\left[(-1)^{J'-J+\kappa}\langle jm_0|\mu_\kappa'|j'm_0'\rangle+\delta_{m_0,1/2}\delta_{m_0',1/2}\langle j,\tfrac{1}{2}|\mu_1'|j',-\tfrac{1}{2}\rangle\right.$$
$$\left.\times\frac{(-1)^{j'+J}}{2\sqrt{2}}\{(2J+1)(-1)^{J+J'+1}+2J'+1\}\right] \tag{41.50}$$

Recalling the definition of the intrinsic states $|m_0\rangle$ in (41.37) and $|\overline{m_0}\rangle$ in (41.40), we can further simplify the expression (41.50). We define $|m_0\rangle$ and $|\overline{m_0}\rangle$ by the coefficients α_j and $|m_0'\rangle$ and $|\overline{m_0'}\rangle$ by the coefficients β_j. We then obtain the reduced matrix element of the operator $\boldsymbol{\mu}'$ (in (41.45)) in the form

$$\boxed{\begin{aligned}&\sqrt{(2J+1)2J'+1)}(-1)^{J-m_0}\begin{pmatrix} J & 1 & J' \\ -m_0 & \kappa & m_0' \end{pmatrix}\\ &\times\Big[(-1)^{J'-J+\kappa}\langle m_0|\mu_\kappa'|m_0'\rangle\\ &\quad-\delta_{m_0,1/2}\delta_{m_0',1/2}(-1)^{J+1/2}\frac{1}{2\sqrt{2}}\langle\tfrac{1}{2}|\mu_1'|\overline{\tfrac{1}{2}}'\rangle\\ &\quad\times\{(2J+1)(-1)^{J+J'+1}+(2J'+1)\}\Big]\end{aligned}}$$

(41.51)

To obtain the value of $B(M1)$ of the transition from the state with J' in the rotational band characterized by m_0' and the set β_j, to the state with J in the band with m_0 and the set α_j, the square of (41.51) should be divided by $2J'+1$ and multiplied by $(3/4\pi)(e\hbar/2Mc)^2$. To obtain the magnetic moment in a given state we should put $J'=J$, $m_0'=m_0$, $\beta_j=\alpha_j$ in the reduced matrix element (41.51) for $\kappa=0$, multiply it by

$$\begin{pmatrix} J & 1 & J \\ -J & 0 & J \end{pmatrix}$$

and add it to g_cJ. Substituting the actual expressions of the $3j$-symbols we obtain the contribution of $\boldsymbol{\mu}'$ to the magnetic moment given by

$$J\frac{m_0}{J(J+1)}\left[\langle m_0|\mu_0'|m_0\rangle-\delta_{m_0,1/2}(-1)^{J+1/2}\frac{2J+1}{\sqrt{2}}\langle\tfrac{1}{2}|\mu_1'|\overline{\tfrac{1}{2}}\rangle\right]$$

(41.52)

To conform with conventional notation we define g_{m_0} and b by

$$\begin{aligned} \langle m_0|\mu_0'|m_0\rangle &= m_0(g_{m_0} - g_c) \\ -\sqrt{2}\langle\tfrac{1}{2}|\mu_1'|\bar{\tfrac{1}{2}}\rangle &= b(g_{m_0} - g_c) \end{aligned} \tag{41.53}$$

Adding g_cJ to (41.52) we obtain the magnetic moment of the system of nucleon(s) coupled to the core as $\mu = gJ$ and the g-factor given by

$$\boxed{g = g_c + (g_{m_0} - g_c)\frac{m_0^2}{J(J+1)}[1 + \delta_{m_0,1/2}(-1)^{J+1/2}(2J+1)b]} \tag{41.54}$$

In the case of a single nucleon in the j-orbit, (41.54) is reduced to (41.43).

As already mentioned, all results derived above for given J and j, or set of j-values, may be valid if there are states of the rotational core with $L \leq J + j$. The properties of the core states which we use are that their energies are proportional to $L(L+1)$ and that the reduced matrix elements of the quadrupole operator of the core in its interaction with the coupled nucleon(s) are proportional to

$$\sqrt{(2L+1)(2L'+1)}\begin{pmatrix} L & L' & 2 \\ 0 & 0 & 0 \end{pmatrix}$$

In view of this, we can examine cores described by $SU(3)$ limits of either the interacting boson model or fermion models like the Ginocchio model (or FDSM). As we saw in preceding sections, the ground state band in these models is defined by the lowest $SU(3)$ irreducible representation $(\lambda, 0)$ where λ is an even number. The states within this irreducible representation have angular momenta $L = 0, 2, 4, \ldots, \lambda$ and their energies are proportional to $L(L+1)$. The quadrupole operator of the core is a generator of $SU(3)$ and hence, its matrix elements between states with L, L' in that band depend only on L, L' and λ. They do not depend on the particular realization of the group.

The matrix elements of the quadrupole operator between states which belong to the basis of the $(\lambda, 0)$ irreducible representation are

equal to

$$
\begin{aligned}
(L\|\mathbf{Q}\|L) &= (2L+1)\begin{pmatrix} L & L & 2 \\ 0 & 0 & 0 \end{pmatrix}(2\lambda+3) \\
(L' = L+2\|\mathbf{Q}\|L) &= \sqrt{(2L+1)(2L'+1)} \\
&\quad \times \begin{pmatrix} L & L' & 2 \\ 0 & 0 & 0 \end{pmatrix} 2\sqrt{(\lambda-L)(\lambda+L+3)}
\end{aligned}
\tag{41.55}
$$

In the interacting boson model, the lowest $SU(3)$ representation is $(2N,0)$ where N is the total number of s and d-bosons in IBA-1 and the total number of s_π, d_π, s_ν and d_ν bosons in IBA-2. In the Ginocchio model (or FDSM) λ is equal to the number of fermions in the orbits with $j = i-1, i, i+1$ for half integer i. In both models, λ becomes large for large numbers of bosons or nucleons in the core. For L values much smaller than λ the matrix elements (41.55) become proportional to the reduced matrix elements of $\mathbf{Y}_2(\mathbf{R})$ as they are given in (41.2) by

$$
(L\|\mathbf{Y}_2\|L') = \sqrt{\frac{5}{4\pi}}\sqrt{(2L+1)(2L'+1)}\begin{pmatrix} L & L' & 2 \\ 0 & 0 & 0 \end{pmatrix}
\tag{41.56}
$$

For such L-values, cores of $SU(3)$ models coupled to nucleon(s) are equivalent to the model of particle(s) coupled to a rotational core described above. In all models it is assumed that the nucleons coupled to the core are in j-orbits unoccupied by core nucleons. Otherwise, there are corrections due to the Pauli principle which complicate very much the simple picture presented above.

Let us now consider briefly the *weak coupling limit* of the Hamiltonian (41.19). In this case, which is the opposite of the strong coupling limit considered above, the energy of the core is taken to be large compared to the particle-core interaction V. The kinetic energy should then be diagonalized which implies that L^2 is diagonal and equal to $L(L+1)$. Due to the Coriolis coupling terms, which were ignored above, m_0 is no longer a good quantum number. In the present case, along with J and j, L has a definite value. In this limit, all states with $|L-j| \leq J \leq L+j$ are degenerate with the energy $L(L+1)/2\mathcal{I}$.

The lowest eigenvalue of the core energy is in the state with $L = 0$ and hence, $J = j$. The next higher degenerate states are $L = 2$ states with $J - j - 2,\ j-1,\ j,\ j+1,\ j+2$ (for $j > 2$), the next group of degenerate states has $L = 4$, etc. In this limit, the spacings between groups of degenerate states are determined by L and hence, are equal to the spacings of levels of the even–even core.

The Yrast states (lowest states for given values of J) have spins $J = j + L$ for $L = 0,2,4,\ldots$. A state with given $J - L + j$ can decay by an E2 electromagnetic transition only to the state with $J' = L - 2 + j$. All other lower states have spins J' which are *smaller* than $J - 2$. Hence, Yrast states, with $J = j + L$, decay to lower Yrast states by a cascade of E2 transitions. Such a set of states is called a *decoupled band* since the spin j of the single nucleon is decoupled from the rotational motion of the core. The core energy in these Yrast states is equal to

$$\frac{1}{2\mathcal{I}}L(L+1) = \frac{1}{2\mathcal{I}}(J-j)(J-j+1). \tag{41.57}$$

States in a decoupled band have spins $J, J+2, J+4, \ldots$ unlike the spin sequence in a normal band which is $J, J+1, J+2, \ldots$. The weak coupling limit is a schematic model (Vogel 1970) of decoupled bands which were experimentally observed. In certain cases, the simple features of the weak coupling limit are obtained also if the nucleon-core interaction is taken into account (Stephens et al. 1972).

The E2 transition between the states with $J = L + 2 + j$ and $J = L + j$ is due only to the quadrupole operator of the core. The single nucleon quadrupole operator has vanishing matrix elements between states with $L + 2$ and L. Rate of this transition is given, according to (10.48), by

$$B(\text{E2}) = \frac{1}{2L+2j+5}(L+2, j, J = L+j+2\|\mathbf{T}^{(2)}\|L, j, J' = L+j)^2 \tag{41.58}$$

where

$$\mathbf{T}^{(2)} = \sum_i \frac{e_i}{e} r_i^L \mathbf{Y}_L(\theta_i \phi_i)$$

and the summation is over core nucleons only. Due to (10.42), we express (41.58) by

$$\frac{1}{2L+2j+5}(2L+2j+5)(2L+2j+1)(L+2\|\mathbf{T}^{(2)}\|L)^2 \times \begin{Bmatrix} L+2 & L+j+2 & j \\ L+j & L & 2 \end{Bmatrix}^2 \tag{41.59}$$

The value of the $6j$-symbol in (41.59) is given in the Appendix. It may be obtained also by a simple argument. This $6j$-symbol appears in the change-of-coupling transformation (10.11) and hence

$$\langle L,L+2(2)L+j,L+j+2 \mid L,L+j(j)L+2,L+j+2\rangle^2 = 5(2j+1)\begin{Bmatrix} L+2 & L & 2 \\ L+j & L+j+2 & j \end{Bmatrix}^2 \tag{41.60}$$

The Clebsch-Gordan coefficient for coupling 2 and $L+j$ to $L+j+2$, $M=L+j+2$, as well as for coupling j to $L+2$ to $L+j+2$, $M=L+j+2$, is equal to 1. The other coefficients which appear in (41.60), for the state with $M=L+j+2$, are obtained by using (7.24) to be equal to

$$\begin{aligned}
&(L+2,M=L+2,L,M_1=-L \mid L+2,L,2,M_2=2)^2 \\
&= 5\begin{pmatrix} L+2 & L & 2 \\ L+2 & -L & -2 \end{pmatrix}^2 \\
&= \frac{5}{2L+5}(L,M_1=L,2,M_2=2 \mid L,2,L+2,M=L+2)^2 \\
&= \frac{5}{2L+5}
\end{aligned}$$

$$\begin{aligned}
&(L+j,M=L+j,L,M_1=-L \mid L+j,L,j,m=j)^2 \\
&= (2j+1)\begin{pmatrix} L+j & L & j \\ L+j & -L & -j \end{pmatrix}^2 \\
&= \frac{2j+1}{2L+2j+1}(L,M_1=L,j,m=j \mid L,j,L+j,M=L+j)^2 \\
&= \frac{2j+1}{2L+2j+1}
\end{aligned} \tag{41.62}$$

Substituting the values (41.61) and (41.62) into the l.h.s. of (41.60) we obtain

$$\begin{Bmatrix} L+j+2 & L+2 & j \\ L & L+j & 2 \end{Bmatrix} = \frac{1}{(2L+5)(2L+2j+1)} \tag{41.63}$$

Using the value (41.63) of the $6j$-symbol in (41.59) we obtain

$$\frac{1}{2L+2j+5}(L+2,j,J=L+j+2\|\mathbf{T}^{(2)}\|L,j,J'=L+j)^2$$

$$= \frac{1}{2L+5}(L+2\|\mathbf{T}^{(2)}\|L)^2 \tag{41.64}$$

The equality in (41.64) has a simple physical meaning. Since L and j, as well as $L+2$ and j, are completely aligned, the presence of the spin j does not affect the probability of the transition between core states with $L+2$ and L.

The weak coupling limit may be considered in the case where an even number of valence j-nucleons is present. If the interaction between valence nucleons and the core is ignored, eigenvalues of the Hamiltonian are equal to the sum of the eigenvalues of the core energy and the energy of the j^n configuration, $E(j^n, J_0)$, in the state with spin J_0. The latter is equal to the interaction energy of the j-nucleons to which single nucleon energies, $n\epsilon_j$, should be added. The core energy in states with $J = L + J_0$ is given by (41.57) in which j is replaced by J_0. The lowest state of the valence nucleons has $J_0 = 0$ and the core energy in states of the ground state band is equal to $J(J+1)/2\mathcal{I}$. States of valence nucleons with higher J_0 values have higher energies $E(j^n J_0)$. All states with given values of J_0 and L are degenerate in the weak coupling limit. Among all states with $|L - J_0| \leq J \leq L + J_0$, the one with highest spin has $J = L + J_0$. Energies of these latter states are equal to

$$E(j^n J_0) + \frac{1}{2\mathcal{I}}L(L+1) = E(j^n J_0) + \frac{1}{2\mathcal{I}}(J-J_0)(J-J_0+1) \tag{41.65}$$

From (41.65) it follows that the core energies in states with $J_0 > 0$ are lower than in corresponding states with $J_0 = 0$. Although $E(j^n J_0)$ is higher than $E(j^n J_0 = 0)$, the energy (41.65) becomes lower for high values of J. The value of J for which this happens is determined by the inequality

$$\frac{1}{2\mathcal{I}}(2J - J_0 + 1)J_0 > E(j^n J_0) - E(j^n J_0 = 0) \tag{41.66}$$

At this value of J, the band with J_0 crosses the ground state band and becomes the Yrast band. Differences between energies in the ground state band of states with $J+2$ and J are equal to $(2J+3)/\mathcal{I}$. These differences form a linear function of J with the slope $2/\mathcal{I}$. Beyond the band crossing, described above, these differences are equal to $(2J - 2J_0 + 3)/\mathcal{I}$. The slope is the same as in the ground state band but the differences are shifted down by $2J_0/\mathcal{I}$. In an ideal situation the value of J_0 may be deduced from this shift.

The weak coupling limit provides a schematic description (Stephens and Simon 1972) of band crossings which are experimentally observed, exhibiting an effect called *back-bending*. To obtain agreement with experiment, the interaction between the core and the valence nucleons is taken into account. A detailed description of the situation may be found in Stephens (1975).

Appendix

SINGLE PARTICLE IN A CENTRAL POTENTIAL WELL

Hamiltonian

$$H_0 = -\frac{\hbar^2}{2m}\left(\frac{\partial^2}{\partial r^2} + \frac{2}{r}\frac{\partial}{\partial r} - \frac{1}{r^2}\boldsymbol{l}^2\right) + V(r)$$

$$\boldsymbol{l}^2 = -\frac{1}{\sin\theta}\frac{\partial}{\partial\theta}\sin\theta\frac{\partial}{\partial\theta} - \frac{1}{\sin^2\theta}\frac{\partial^2}{\partial\phi^2}$$

Wave functions

$$\phi_{nlm}(r,\theta,\phi) = \frac{1}{r}R_{nl}(r)Y_{lm}(\theta,\phi)$$

Radial equation

$$\frac{d^2R_{nl}(r)}{dr^2} - \frac{l(l+1)}{r^2}R_{nl}(r) + \frac{2m}{\hbar^2}(E_{nl} - V(r))R_{nl}(r) = 0$$

Spherical harmonics

$$\boldsymbol{l}^2 Y_{lm}(\theta,\phi) = l(l+1)Y_{lm}(\theta,\phi)$$

$$l_z Y_{lm}(\theta,\phi) = mY_{lm}(\theta,\phi)$$

$$(l_x \pm il_y)Y_{lm}(\theta,\phi) = \sqrt{l(l+1) - m(m\pm 1)}\,Y_{lm\pm 1}(\theta,\phi)$$

Parity

$$Y_{lm}(\pi - \theta, \phi + \pi) = (-1)^l Y_{lm}(\theta,\phi)$$

TABLE *Some spherical harmonics*

$Y_{lm}(\theta,\phi)$ \ m	0	± 1	± 2	± 3
$l = 0$	$\frac{1}{\sqrt{4\pi}}$			
$l = 1$	$\sqrt{\frac{3}{4\pi}}\cos\theta$	$\mp\sqrt{\frac{3}{8\pi}}\sin\theta e^{\pm i\phi}$		
$l = 2$	$\sqrt{\frac{5}{16\pi}}(3\cos^2\theta - 1)$	$\mp\sqrt{\frac{15}{8\pi}}\cos\theta\sin\theta e^{\pm i\phi}$	$\sqrt{\frac{15}{32\pi}}\sin^2\theta e^{\pm 2i\phi}$	
$l = 3$	$\sqrt{\frac{7}{16\pi}}(5\cos^3\theta - 3\cos\theta)$	$\mp\sqrt{\frac{21}{64\pi}}(4\cos^2\theta\sin\theta - \sin^3\theta)e^{\pm i\phi}$	$\sqrt{\frac{105}{32\pi}}\cos\theta\sin^2\theta\, e^{\pm 2i\phi}$	$\mp\sqrt{\frac{35}{64\pi}}\sin^3\theta e^{\pm 3i\phi}$

Addition theorem

$$\frac{4\pi}{2l+1}\sum_m Y_{lm}^*(\hat{\mathbf{r}}_1)Y_{lm}(\hat{\mathbf{r}}_2) = P_l(\hat{\mathbf{r}}_1\cdot\hat{\mathbf{r}}_2) \qquad \hat{\mathbf{r}} = \mathbf{r}/r$$

$$\sum_m Y_{lm}^*(\theta,\phi)Y_{lm}(\theta,\phi) = \frac{2l+1}{4\pi}$$

Harmonic oscillator potential

$$H_0 = \frac{1}{2m}\mathbf{p}^2 + \frac{1}{2}m\omega^2\mathbf{r}^2$$

$$E_{nl} = \hbar\omega(2(n-1) + l + \tfrac{3}{2}) = \hbar\omega(N + \tfrac{3}{2})$$

$$R_{nl}(r) = \sqrt{\frac{(2\nu)^{l+3/2}2^{n+l+1}(n-1)!}{\sqrt{\pi}(2n+2l-1)!!}}\, r^{l+1}e^{-\nu r^2}L_{n-1}^{l+1/2}(2\nu r^2) \qquad \nu = \frac{m\omega}{2\hbar}$$

$$= \sqrt{\frac{2^{l-n+3}(2\nu)^{l+3/2}(2l+2n-1)!!}{\sqrt{\pi}(n-1)!((2l+1)!!)^2}}\, r^{l+1}e^{-\nu r^2}$$

$$\times \sum_{k=0}^{n-1}(-1)^k 2^k \binom{n-1}{k} \frac{(2l+1)!!}{(2l+1+2k)!!}(2\nu r^2)^k$$

$$R_{n=1,l}(r) = \sqrt{\frac{2^{l+2}(2\nu)^{l+3/2}}{\sqrt{\pi}(2l+1)!!}}\, r^{l+1}e^{-\nu r^2}$$

$$\langle Nl|\mathbf{r}^2|Nl\rangle = \frac{\hbar}{m\omega}(N + \tfrac{3}{2}) = \frac{1}{2\nu}(N + \tfrac{3}{2})$$

Spin-orbit interaction

$$H = H_0 + V_{SO}(r)(\mathbf{s} \cdot \boldsymbol{l})$$

$$\mathbf{j} = \mathbf{s} + \boldsymbol{l}$$

Radial equation

$$\frac{d^2 R_{nlj}(r)}{dr^2} - \frac{l(l+1)}{r^2} R_{nlj}(r) + \frac{2m}{\hbar^2}\left(E_{nlj} - V(r) - \frac{l}{2} V_{SO}(r)\right) R_{nlj}(r) = 0 \qquad \text{for} \quad j = l + \tfrac{1}{2}$$

$$\frac{d^2 R_{nlj}(r)}{dr^2} - \frac{l(l+1)}{r^2} R_{nlj}(r) + \frac{2m}{\hbar^2}\left(E_{nlj} - V(r) + \frac{l+1}{2} V_{SO}(r)\right) R_{nlj}(r) = 0 \qquad \text{for} \quad j = l - \tfrac{1}{2}$$

If $V_{SO}(r)$ is a small perturbation $\int_0^\infty R_{nl}^2(r) V_{SO}(r)\, dr = 2a_{nl}$

Eigenvalues

$$E_{nlj} = E_{nl} + a_{nl} l \qquad \text{for} \quad j = l + \tfrac{1}{2}$$

$$E_{nlj} = E_{nl} - a_{nl}(l+1) \qquad \text{for} \quad j = l - \tfrac{1}{2}$$

Eigenstates

$$\psi_{nljm} = \begin{pmatrix} \sqrt{\frac{l+\frac{1}{2}+m}{2l+1}}\phi_{nlm-1/2} \\ \sqrt{\frac{l+\frac{1}{2}-m}{2l+1}}\phi_{nlm+1/2} \end{pmatrix} = \sqrt{\frac{l+\frac{1}{2}+m}{2l+1}}\phi_{nlm-1/2}\chi_{1/2} + \sqrt{\frac{l+\frac{1}{2}-m}{2l+1}}\phi_{nlm+1/2}\chi_{-1/2} \qquad \text{for} \quad j = l+\tfrac{1}{2}$$

$$\psi_{nljm} = \begin{pmatrix} \sqrt{\frac{l+\frac{1}{2}-m}{2l+1}}\phi_{nlm-1/2} \\ -\sqrt{\frac{l+\frac{1}{2}+m}{2l+1}}\phi_{nlm+1/2} \end{pmatrix} = \sqrt{\frac{l+\frac{1}{2}-m}{2l+1}}\phi_{nlm-1/2}\chi_{1/2} - \sqrt{\frac{l+\frac{1}{2}+m}{2l+1}}\phi_{nlm+1/2}\chi_{-1/2} \qquad \text{for} \quad j = l-\tfrac{1}{2}$$

$$j^{+}\psi_{nljm} = (j_x + i j_y)\psi_{nljm} = \sqrt{j(j+1) - m(m+1)}\,\psi_{nljm+1}$$

$$j^{-}\psi_{nljm} = (j_x - i j_y)\psi_{nljm} = \sqrt{j(j+1) - m(m-1)}\,\psi_{nljm-1}$$

$$j^{0}\psi_{nljm} = j_z\psi_{nljm} = m\psi_{nljm}$$

$$\mathbf{J}^2\psi_{nljm} = j(j+1)\psi_{nljm}$$

Single particle magnetic moments

$$\mu = g_e l + \tfrac{1}{2}g_s \qquad j = l + \tfrac{1}{2}$$

$$\mu = \frac{j}{j+1}\left(g_e(l+1) - \tfrac{1}{2}g_s\right) \qquad j = l - \tfrac{1}{2}$$

Quadrupole moment

$$Q = -e\langle \mathbf{r}^2 \rangle \frac{2j-1}{2j+1}$$

COUPLING OF ANGULAR MOMENTA

Clebsch-Gordan coefficients

Definition

$$\psi_{j_1 j_2 j m}(1,2) = \sum_{m_1 m_2} \psi_{j_1 m_1}(1)\psi_{j_2 m_2}(2)(j_1 m_1 j_2 m_2 \mid j_1 j_2 j m)$$

$$\psi_{j_1 m_1}(1)\psi_{j_2 m_2}(2) = \sum_{jm} (j_1 m_1 j_2 m_2 \mid j_1 j_2 j m)\psi_{j_1 j_2 j m}(1,2)$$

Orthogonality

$$\sum_{m_1 m_2} (j_1 m_1 j_2 m_2 \mid j_1 j_2 j m)(j_1 m_1 j_2 m_2 \mid j_1 j_2 j' m') = \delta_{jj'}\delta_{mm'}$$

$$\sum_{jm} (j_1 m_1 j_2 m_2 \mid j_1 j_2 j m)(j_1 m_1' j_2 m_2' \mid j_1 j_2 j m) = \delta_{m_1 m_2'}\delta_{m_2 m_2'}$$

Symmetry properties

$$(j_2 m_2 j_1 m_1 \mid j_2 j_1 j m) = (-1)^{j_1 + j_2 - j}(j_1 m_1 j_2 m_2 \mid j_1 j_2 j m)$$

$$(j_1, -m_1 j_2, -m_2 \mid j_1 j_2 j, -m) = (-1)^{j_1 + j_2 - j}(j_1 m_1 j_2 m_2 \mid j_1 j_2 j m)$$

For $j_1 = j_2$

$$\sum_{jm}(1 \pm (-1)^{2j_1-j})(j_1 m_1 j_1 m_1' \mid j_1 j_1 j m)(j_1 m_2 j_1 m_2' \mid j_1 j_1 j m) = \delta_{m_1 m_2}\delta_{m_1' m_2'} \pm \delta_{m_1 m_2'}\delta_{m_1' m_2}$$

$3j$-symbols

Definition

$$\begin{pmatrix} j_1 & j_2 & j_3 \\ m_1 & m_2 & m_3 \end{pmatrix} = \frac{(-1)^{j_1-j_2-m_3}}{\sqrt{2j_3+1}}(j_1 m_1 j_2 m_2 \mid j_1 j_2 j_3, -m_3)$$

Symmetry properties

$$\begin{pmatrix} j_1 & j_2 & j_3 \\ m_1 & m_2 & m_3 \end{pmatrix} = \begin{pmatrix} j_2 & j_3 & j_1 \\ m_2 & m_3 & m_1 \end{pmatrix} = \begin{pmatrix} j_3 & j_1 & j_2 \\ m_3 & m_1 & m_2 \end{pmatrix}$$

$$= (-1)^{j_1+j_2+j_3}\begin{pmatrix} j_2 & j_1 & j_3 \\ m_2 & m_1 & m_3 \end{pmatrix} = (-1)^{j_1+j_2+j_3}\begin{pmatrix} j_3 & j_2 & j_1 \\ m_3 & m_2 & m_1 \end{pmatrix}$$

$$= (-1)^{j_1+j_2+j_3}\begin{pmatrix} j_1 & j_3 & j_2 \\ m_1 & m_3 & m_2 \end{pmatrix}$$

$$\begin{pmatrix} j_1 & j_2 & j_3 \\ -m_1 & -m_2 & -m_3 \end{pmatrix} = (-1)^{j_1+j_2+j_3}\begin{pmatrix} j_1 & j_2 & j_3 \\ m_1 & m_2 & m_3 \end{pmatrix}$$

Orthogonality

$$\sum_{m_1 m_2} \begin{pmatrix} j_1 & j_2 & j_3 \\ m_1 & m_2 & m_3 \end{pmatrix} \begin{pmatrix} j_1 & j_2 & j_3' \\ m_1 & m_2 & m_3' \end{pmatrix} = \frac{1}{2j_3+1} \delta_{j_3 j_3'} \delta_{m_3 m_3'}$$

$$\sum_{j_3 m_3} (2j_3+1) \begin{pmatrix} j_1 & j_2 & j_3 \\ m_1 & m_2 & m_3 \end{pmatrix} \begin{pmatrix} j_1 & j_2 & j_3 \\ m_1' & m_2' & m_3 \end{pmatrix} = \delta_{m_1 m_1'} \delta_{m_2 m_2'}$$

$$\sum_{m_1 m_2 m_3} \begin{pmatrix} j_1 & j_2 & j_3 \\ m_1 & m_2 & m_3 \end{pmatrix} \begin{pmatrix} j_1 & j_2 & j_3 \\ m_1 & m_2 & m_3 \end{pmatrix} = 1$$

Simple relation

$$\begin{pmatrix} j' & j'' & j \\ \frac{1}{2} & \frac{1}{2} & -1 \end{pmatrix} = -\frac{1}{2\sqrt{j(j+1)}} \{(2j'+1)(-1)^{j'+j''+j} + (2j''+1)\} \begin{pmatrix} j' & j'' & j \\ \frac{1}{2} & -\frac{1}{2} & 0 \end{pmatrix}$$

Coupling of states of the same system

$$C^l_{l'l''} Y_{lm}(\hat{\mathbf{r}}) = \sum_{m'm''} Y_{l'm'}(\hat{\mathbf{r}}) Y_{l''m''}(\hat{\mathbf{r}}) (l'm'l''m'' \mid l'l''lm)$$

$$C^l_{l'l''} = (-1)^l \sqrt{\frac{(2l'+1)(2l''+1)}{4\pi}} \begin{pmatrix} l & l' & l'' \\ 0 & 0 & 0 \end{pmatrix}$$

Some simple expressions

$$\begin{pmatrix} j_1 & j_2 & j_3 \\ 0 & 0 & 0 \end{pmatrix} = \tfrac{1}{2}(1 + (-1)^{j_1+j_2+j_3})(-1)^g \sqrt{\frac{(2g-2j_1)!(2g-2j_2)!(2g-2j_2)!}{(2g+1)!}} \frac{g!}{(g-j_1)!(g-j_2)!(g-j_3)!}$$

$$2g = j_1 + j_2 + j_3$$

$$\begin{pmatrix} J & J+\frac{1}{2} & \frac{1}{2} \\ M & -M-\frac{1}{2} & \frac{1}{2} \end{pmatrix} = (-1)^{J-M+1} \left[\frac{J+M+1}{(2J+1)(2J+2)} \right]^{1/2}$$

$$\begin{pmatrix} J & J+1 & 1 \\ M & -M-1 & 1 \end{pmatrix} = (-1)^{J-M} \left[\frac{(J+M+1)(J+M+2)}{(2J+1)(2J+2)(2J+3)} \right]^{1/2}$$

$$\begin{pmatrix} J & J+1 & 1 \\ M & -M & 0 \end{pmatrix} = (-1)^{J-M+1} \left[\frac{2(J+M+1)(J-M+1)}{(2J+1)(2J+2)(2J+3)} \right]^{1/2}$$

$$\begin{pmatrix} J & J & 1 \\ M & -M-1 & 1 \end{pmatrix} = (-1)^{J-M} \left[\frac{2(J-M)(J+M+1)}{2J(2J+1)(2J+2)} \right]^{1/2}$$

$$\begin{pmatrix} J & J & 1 \\ M & -M & 0 \end{pmatrix} = (-1)^{J-M} \frac{M}{[J(J+1)(2J+1)]^{1/2}}$$

$$\begin{pmatrix} J & J+\frac{3}{2} & \frac{3}{2} \\ M & -M-\frac{3}{2} & \frac{3}{2} \end{pmatrix} = (-1)^{J-M+1} \left[\frac{(J+M+1)(J+M+2)(J+M+3)}{(2J+1)(2J+2)(2J+3)(2J+4)} \right]^{1/2}$$

$$\begin{pmatrix} J & J+\frac{3}{2} & \frac{3}{2} \\ M & -M-\frac{1}{2} & \frac{1}{2} \end{pmatrix} = (-1)^{J-M} \left[\frac{3(J+M+1)(J+M+2)(J-M+1)}{(2J+1)(2J+2)(2J+3)(2J+4)} \right]^{1/2}$$

$$\begin{pmatrix} J & J+\frac{1}{2} & \frac{3}{2} \\ M & -M-\frac{3}{2} & \frac{3}{2} \end{pmatrix} = (-1)^{J-M+1} \left[\frac{3(J+M+1)(J+M+2)(J-M)}{2J(2J+1)(2J+2)(2J+3)} \right]^{1/2}$$

$$\begin{pmatrix} J & J+\frac{1}{2} & \frac{3}{2} \\ M & -M-\frac{1}{2} & \frac{1}{2} \end{pmatrix} = (-1)^{J-M} \left[\frac{(J+M+1)}{2J(2J+1)(2J+2)(2J+3)} \right]^{1/2} (J-3M)$$

$$\begin{pmatrix} J & J+2 & 2 \\ M & -M-2 & 2 \end{pmatrix} = (-1)^{J-M} \left[\frac{(J+M+1)(J+M+2)(J+M+3)(J+M+4)}{(2J+1)(2J+2)(2J+3)(2J+4)(2J+5)} \right]^{1/2}$$

$$\begin{pmatrix} J & J+2 & 2 \\ M & -M-1 & 1 \end{pmatrix} = (-1)^{J-M+1} 2 \left[\frac{(J+M+1)(J+M+2)(J+M+3)(J-M+1)}{(2J+1)(2J+2)(2J+3)(2J+4)(2J+5)} \right]^{1/2}$$

$$\begin{pmatrix} J & J+2 & 2 \\ M & -M & 0 \end{pmatrix} = (-1)^{J-M} \left[\frac{6(J+M+1)(J+M+2)(J-M+1)(J-M+2)}{(2J+1)(2J+2)(2J+3)(2J+4)(2J+5)} \right]^{1/2}$$

$$\begin{pmatrix} J & J+1 & 2 \\ M & -M-2 & 2 \end{pmatrix} = (-1)^{J-M} 2 \left[\frac{(J+M+1)(J+M+2)(J+M+3)(J-M)}{2J(2J+1)(2J+2)(2J+3)(2J+4)} \right]^{1/2}$$

$$\begin{pmatrix} J & J+1 & 2 \\ M & -M-1 & 1 \end{pmatrix} = (-1)^{J-M+1} \left[\frac{(J+M+1)(J+M+2)}{2J(2J+1)(2J+2)(2J+3)(2J+4)} \right]^{1/2} 2(J-2M)$$

$$\begin{pmatrix} J & J+1 & 2 \\ M & -M & 0 \end{pmatrix} = (-1)^{J-M+1} \left[\frac{6(J+M+1)(J-M+1)}{2J(2J+1)(2J+2)(2J+3)(2J+4)} \right]^{1/2} 2M$$

$$\begin{pmatrix} J & J & 2 \\ M & -M-2 & 2 \end{pmatrix} = (-1)^{J-M} \left[\frac{6(J-M-1)(J-M)(J+M+1)(J+M+2)}{(2J-1)(2J)(2J+1)(2J+2)(2J+3)} \right]^{1/2}$$

$$\begin{pmatrix} J & J & 2 \\ M & -M-1 & 1 \end{pmatrix} = (-1)^{J-M} \left[\frac{6(J+M+1)(J-M)}{(2J-1)(2J)(2J+1)(2J+2)(2J+3)} \right]^{1/2} (2M+1)$$

$$\begin{pmatrix} J & J & 2 \\ M & -M & 0 \end{pmatrix} = (-1)^{J-M} \frac{2[3M^2 - J(J+1)]}{[(2J-1)(2J)(2J+1)(2J+2)(2J+3)]^{1/2}}$$

CHANGE OF COUPLING. $9j$-SYMBOLS

Definition

$$\psi(J_1J_3(J_{13})J_2J_4(J_{24})JM) = \sum_{J_{12}J_{34}} \sqrt{(2J_{13}+1)(2J_{24}+1)(2J_{12}+1)(2J_{34}+1)} \begin{Bmatrix} J_1 & J_2 & J_{12} \\ J_3 & J_4 & J_{34} \\ J_{13} & J_{24} & J \end{Bmatrix}$$

$$\times \psi(J_1J_2(J_{12})J_3J_4J_{34})JM)$$

Expansion in $3j$-symbols

$$\begin{Bmatrix} J_{11} & J_{12} & J_{13} \\ J_{21} & J_{22} & J_{23} \\ J_{31} & J_{32} & J_{33} \end{Bmatrix} = \sum_{\text{all } M_{ij}} \begin{pmatrix} J_{11} & J_{12} & J_{13} \\ M_{11} & M_{12} & M_{13} \end{pmatrix} \begin{pmatrix} J_{21} & J_{22} & J_{23} \\ M_{21} & M_{22} & M_{23} \end{pmatrix} \begin{pmatrix} J_{31} & J_{32} & J_{33} \\ M_{31} & M_{32} & M_{33} \end{pmatrix}$$

$$\times \begin{pmatrix} J_{11} & J_{21} & J_{31} \\ M_{11} & M_{21} & M_{31} \end{pmatrix} \begin{pmatrix} J_{12} & J_{22} & J_{32} \\ M_{12} & M_{22} & M_{32} \end{pmatrix} \begin{pmatrix} J_{13} & J_{23} & J_{33} \\ M_{13} & M_{23} & M_{33} \end{pmatrix}$$

Symmetry properties

$$\begin{Bmatrix} J_{11} & J_{21} & J_{31} \\ J_{12} & J_{22} & J_{32} \\ J_{13} & J_{23} & J_{33} \end{Bmatrix} = \begin{Bmatrix} J_{11} & J_{12} & J_{13} \\ J_{21} & J_{22} & J_{23} \\ J_{31} & J_{32} & J_{33} \end{Bmatrix}$$

$$\begin{Bmatrix} J_{21} & J_{22} & J_{23} \\ J_{11} & J_{12} & J_{13} \\ J_{31} & J_{32} & J_{33} \end{Bmatrix} = \begin{Bmatrix} J_{11} & J_{12} & J_{13} \\ J_{31} & J_{32} & J_{33} \\ J_{21} & J_{22} & J_{23} \end{Bmatrix}$$

$$= \begin{Bmatrix} J_{31} & J_{32} & J_{33} \\ J_{21} & J_{22} & J_{23} \\ J_{11} & J_{12} & J_{13} \end{Bmatrix} = (-1)^{s} \begin{Bmatrix} J_{11} & J_{12} & J_{13} \\ J_{21} & J_{22} & J_{23} \\ J_{31} & J_{32} & J_{33} \end{Bmatrix}$$

$$S = J_{11} + J_{12} + J_{13} + J_{21} + J_{22} + J_{23} + J_{31} + J_{32} + J_{33}$$

Orthogonality

$$\sum_{J_{13}J_{24}}(2J_{13}+1)(2J_{24}+1)\begin{Bmatrix} J_1 & J_2 & J_{12} \\ J_3 & J_4 & J_{34} \\ J_{13} & J_{24} & J \end{Bmatrix}\begin{Bmatrix} J_1 & J_2 & J'_{12} \\ J_3 & J_4 & J'_{34} \\ J_{13} & J_{24} & J \end{Bmatrix} = \frac{\delta(J_{12},J'_{12})\delta(J_{34},J'_{34})}{(2J_{12}+1)(2J_{34}+1)}$$

Combined change-of-coupling transformation

$$\sum_{J_{13}J_{24}}(-1)^{J_2+J_4+J_{24}+J_2+J_3+J_{23}}(2J_{13}+1)(2J_{24}+1)$$
$$\times\begin{Bmatrix} J_1 & J_3 & J_{13} \\ J_2 & J_4 & J_{24} \\ J_{12} & J_{34} & J \end{Bmatrix}\begin{Bmatrix} J_1 & J_4 & J_{14} \\ J_3 & J_2 & J_{23} \\ J_{13} & J_{24} & J \end{Bmatrix} = (-1)^{J_3+J_4+J_{34}}\begin{Bmatrix} J_1 & J_4 & J_{14} \\ J_2 & J_3 & J_{23} \\ J_{12} & J_{34} & J \end{Bmatrix}$$

6j-SYMBOLS (RACAH COEFFICIENTS)

Definition

$$\begin{Bmatrix} J_1 & J_2 & J \\ J_4 & J_3 & J' \end{Bmatrix} = (-1)^{J_2+J+J_3+J'}\sqrt{(2J+1)(2J'+1)}\begin{Bmatrix} J_1 & J_2 & J \\ J_3 & J_4 & J \\ J' & J' & 0 \end{Bmatrix}$$

Change of coupling transformations

$$\psi(J_1J_3(J')J_2J_4(J')0) = \sum_{J}(-1)^{J_2+J_3+J+J'}\sqrt{(2J+1)(2J'+1)}\begin{Bmatrix} J_1 & J_2 & J \\ J_4 & J_3 & J' \end{Bmatrix}\psi(J_1J_2(J)J_3J_4(J)0)$$

$$\psi(J_1J_3(J_{13})J_2JM) = \sum_{J_{12}}(-1)^{J_2+J_3+J_{12}+J_{13}}\sqrt{(2J_{12}+1)(2J_{13}+1)}\begin{Bmatrix} J_1 & J_2 & J_{12} \\ J & J_3 & J_{13} \end{Bmatrix}\psi(J_1J_2(J_{12})J_3JM)$$

$$\psi(J_1,J_2J_3(J_{23})J) = \sum_{J_{12}}(-1)^{J_1+J_2+J_3+J}\sqrt{(2J_{12}+1)(2J_{23}+1)}$$
$$\times\begin{Bmatrix} J_1 & J_2 & J_{12} \\ J_3 & J & J_{23} \end{Bmatrix}\psi(J_1J_2(J_{12})J_3JM)$$

Symmetry properties

$$\begin{Bmatrix} J_1 & J_2 & J_3 \\ J_4 & J_5 & J_6 \end{Bmatrix} = \begin{Bmatrix} J_2 & J_1 & J_3 \\ J_5 & J_4 & J_6 \end{Bmatrix} = \begin{Bmatrix} J_3 & J_2 & J_1 \\ J_6 & J_5 & J_4 \end{Bmatrix} = \begin{Bmatrix} J_1 & J_3 & J_2 \\ J_4 & J_6 & J_5 \end{Bmatrix} = \begin{Bmatrix} J_4 & J_5 & J_3 \\ J_1 & J_2 & J_6 \end{Bmatrix}$$

Special value

$$\begin{Bmatrix} J_1 & J_2 & J_3 \\ J_4 & J_5 & 0 \end{Bmatrix} = \frac{(-1)^{J_1+J_2+J_3}}{\sqrt{(2J_1+1)(2J_2+1)}}\delta_{J_1J_5}\,\delta_{J_2J_4}$$

Orthogonality

$$\sum_J (2J+1) \begin{Bmatrix} J_1 & J_2 & J \\ J_3 & J_4 & J' \end{Bmatrix} \begin{Bmatrix} J_1 & J_2 & J \\ J_3 & J_4 & J'' \end{Bmatrix} = \frac{\delta_{J'J''}}{2J'+1}$$

Combined change-of-coupling transformations

$$\sum_J (-1)^{J+J'+J''} (2J+1) \begin{Bmatrix} J_1 & J_2 & J' \\ J_4 & J_3 & J \end{Bmatrix} \begin{Bmatrix} J_1 & J_4 & J'' \\ J_2 & J_3 & J \end{Bmatrix} = \begin{Bmatrix} J_1 & J_2 & J' \\ J_3 & J_4 & J'' \end{Bmatrix}$$

$$\begin{Bmatrix} J_1 & J_2 & J_3 \\ J_4 & J_5 & J_6 \end{Bmatrix} \begin{Bmatrix} J_1 & J_2 & J_3 \\ J_7 & J_8 & J_9 \end{Bmatrix}$$

$$= \sum_J (-1)^{J_1+J_2+J_3+J_4+J_5+J_6+J_7+J_8+J_9+J} (2J+1) \begin{Bmatrix} J_1 & J_5 & J_6 \\ J & J_9 & J_8 \end{Bmatrix} \begin{Bmatrix} J_4 & J_2 & J_6 \\ J_9 & J & J_7 \end{Bmatrix} \begin{Bmatrix} J_4 & J_5 & J_3 \\ J_8 & J_7 & J \end{Bmatrix}$$

(Biedenharn-Elliott identity)

Sum rules

$$\sum_J (-1)^{J_1+J_2+J} (2J+1) \begin{Bmatrix} J_1 & J_1 & J' \\ J_2 & J_2 & J \end{Bmatrix} = \sqrt{(2J_1+1)(2J_2+1)}\,\delta_{J'0}$$

$$\sum_J (2J+1) \begin{Bmatrix} J_1 & J_2 & J' \\ J_1 & J_2 & J \end{Bmatrix} = (-1)^{2(J_1+J_2)}$$

for $J_1 = J_2$

$$\sum_J (1 \pm (-1)^{2J_1+J})(2J+1) \begin{Bmatrix} J_1 & J_1 & J' \\ J_1 & J_1 & J \end{Bmatrix} = 1 \pm (2J_1+1)\delta_{J'0}$$

Relations with $3j$-symbols

$$\begin{pmatrix} J_1 & J_2 & J_3 \\ M_1 & M_2 & M_3 \end{pmatrix} \begin{Bmatrix} J_1 & J_2 & J_3 \\ J_4 & J_5 & J_6 \end{Bmatrix} = \sum_{M_4 M_5 M_6} (-1)^{J_4+J_5+J_6+M_4+M_5+M_6}$$

$$\times \begin{pmatrix} J_1 & J_5 & J_6 \\ M_1 & M_5 & -M_6 \end{pmatrix} \begin{pmatrix} J_4 & J_2 & J_6 \\ -M_4 & M_2 & M_6 \end{pmatrix} \begin{pmatrix} J_4 & J_5 & J_3 \\ M_4 & -M_5 & M_3 \end{pmatrix}$$

$$\sum_{M_3} \begin{pmatrix} J_1 & J_2 & J_3 \\ M_1 & M_2 & M_3 \end{pmatrix} \begin{pmatrix} J_4 & J_5 & J_3 \\ M_4 & M_5 & -M_3 \end{pmatrix} = \sum_{J_6 M_6} (-1)^{J_3+J_6+M_1+M_4} (2J_6+1)$$

$$\times \begin{Bmatrix} J_1 & J_2 & J_3 \\ J_4 & J_5 & J_6 \end{Bmatrix} \begin{pmatrix} J_4 & J_2 & J_6 \\ M_4 & M_2 & M_6 \end{pmatrix} \begin{pmatrix} J_1 & J_5 & J_6 \\ M_1 & M_5 & -M_6 \end{pmatrix}$$

$$\begin{pmatrix} j & j' & J \\ \frac{1}{2} & -\frac{1}{2} & 0 \end{pmatrix} = -\sqrt{(2l+1)(2l'+1)} \begin{pmatrix} l & l' & J \\ 0 & 0 & 0 \end{pmatrix} \begin{Bmatrix} j & j' & J \\ l' & l & \frac{1}{2} \end{Bmatrix}$$

$l = j \pm \frac{1}{2}$, $l' = j' \pm \frac{1}{2}$ so that $(-1)^{l+l'+J} = +1$

Relations with 9j-symbols

$$\begin{Bmatrix} j_1 & j_2 & J_{12} \\ j_3 & j_4 & J_{34} \\ J_{13} & J_{24} & J \end{Bmatrix} = \sum_{J'} (-1)^{2J'}(2J'+1) \begin{Bmatrix} j_1 & j_3 & J_{13} \\ J_{24} & J & J' \end{Bmatrix} \begin{Bmatrix} j_2 & j_4 & J_{24} \\ j_3 & J' & J_{34} \end{Bmatrix} \begin{Bmatrix} J_{12} & J_{34} & J \\ J' & j_1 & j_2 \end{Bmatrix}$$

$$\sum_{J'} (2J'+1) \begin{Bmatrix} J_{12} & J_{34} & J' \\ J_{13} & J_{24} & J \end{Bmatrix} \begin{Bmatrix} j_1 & j_2 & J_{12} \\ j_3 & j_4 & J_{34} \\ J_{13} & J_{24} & J' \end{Bmatrix} = (-1)^{2J} \begin{Bmatrix} j_1 & j_3 & J_{13} \\ J_{34} & J & j_4 \end{Bmatrix} \begin{Bmatrix} j_2 & j_4 & J_{24} \\ J & J_{12} & j_1 \end{Bmatrix}$$

Special value

$$\begin{Bmatrix} j_1 & j_2 & S \\ j_3 & j_4 & S \\ L & L & 1 \end{Bmatrix} = (-1)^{j_1+j_4+S+L} \frac{j_1(j_1+1) - j_2(j_2+1) - j_3(j_3+1) + j_4(j_4+1)}{2\sqrt{S(S+1)(2S+1)L(L+1)(2L+1)}} \begin{Bmatrix} j_1 & j_2 & S \\ j_4 & j_3 & L \end{Bmatrix}$$

Some simple expressions

$$\begin{Bmatrix} j_1 & j_2 & j_3 \\ j_2+\frac{1}{2} & j_1-\frac{1}{2} & \frac{1}{2} \end{Bmatrix} = (-1)^{j_1+j_2+j_3} \left[\frac{(j_1+j_3-j_2)(j_2+j_3-j_1+1)}{2j_1(2j_1+1)(2j_2+1)(2j_2+2)} \right]^{1/2}$$

$$\begin{Bmatrix} j_1 & j_2 & j_3 \\ j_2-\frac{1}{2} & j_3-\frac{1}{2} & \frac{1}{2} \end{Bmatrix} = (-1)^{j_1+j_2+j_3} \left[\frac{(j_1+j_2+j_3+1)(j_1+j_2-j_3)}{2j_1(2j_1+1)2j_2(2j_2+1)} \right]^{1/2}$$

$$\begin{Bmatrix} j_1 & j_2 & j_3 \\ j_2-1 & j_1-1 & 1 \end{Bmatrix} = (-1)^{j_1+j_2+j_3}\left[\frac{(j_1+j_2+j_3)(j_1+j_2+j_3+1)(j_1+j_2-j_3)(j_1+j_2-j_3-1)}{(2j_1-1)2j_1(2j_1+1)2j_2(2j_2-1)(2j_2+1)}\right]^{1/2}$$

$$\begin{Bmatrix} j_1 & j_2 & j_3 \\ j_2 & j_1-1 & 1 \end{Bmatrix} = (-1)^{j_1+j_2+j_3}\left[\frac{2(j_1+j_2+j_3+1)(j_1+j_2-j_3)(j_1+j_3-j_2)(j_2+j_3-j_1+1)}{(2j_1-1)2j_1(2j_1+1)2j_2(2j_2+1)(2j_2+2)}\right]^{1/2}$$

$$\begin{Bmatrix} j_1 & j_2 & j_3 \\ j_2+1 & j_1-1 & 1 \end{Bmatrix} = (-1)^{j_1+j_2+j_3}\left[\frac{(j_1+j_3-j_2)(j_1+j_3-j_2-1)(j_2+j_3-j_1+1)(j_2+j_3-j_1+2)}{(2j_1-1)2j_1(2j_1+1)(2j_2+1)(2j_2+2)(2j_3+3)}\right]^{1/2}$$

$$\begin{Bmatrix} j_1 & j_2 & j_3 \\ j_2 & j_1 & 1 \end{Bmatrix} = (-1)^{j_1+j_2+j_3+1}\frac{[j_1(j_1+1)+j_2(j_2+1)-j_3(j_3+1)]}{2[j_1(j_1+1)(2j_1+1)j_2(j_2+1)(2j_2+1)]^{1/2}}$$

$$\begin{Bmatrix} j_1 & j_2 & j_3 \\ j_2-\frac{3}{2} & j_1-\frac{3}{2} & \frac{3}{2} \end{Bmatrix} = (-1)^{j_1+j_2+j_3}$$

$$\times\left[\frac{(j_1+j_2+j_3-1)(j_1+j_2+j_3)(j_1+j_2+j_3+1)(j_1+j_2-j_3-2)(j_1+j_2-j_3-1)(j_1+j_2-j_3)}{(2j_1-2)(2j_1-1)2j_1(2j_1+1)(2j_2-2)(2j_2-1)2j_2(2j_2+1)}\right]^{1/2}$$

$$\begin{Bmatrix} j_1 & j_2 & j_3 \\ j_2-\frac{1}{2} & j_1-\frac{3}{2} & \frac{3}{2} \end{Bmatrix} = (-1)^{j_1+j_2+j_3}$$

$$\times\left[\frac{3(j_1+j_2+j_3)(j_1+j_2+j_3+1)(j_1+j_2+j_3-1)(j_1+j_2-j_3)(j_1+j_3-j_2)(j_1+j_2-j_3+1)}{(2j_1-2)(2j_1-1)2j_1(2j_1+1)(2j_2-1)2j_2(2j_2+1)(2j_2+2)}\right]^{1/2}$$

$$\begin{Bmatrix} j_1 & j_2 & j_3 \\ j_2+\frac{1}{2} & j_1-\frac{3}{2} & \frac{3}{2} \end{Bmatrix} = (-1)^{j_1+j_2+j_3}$$

$$\times \left[\frac{3(j_1+j_2+j_3+1)(j_1+j_2-j_3)(j_1+j_3-j_2-1)(j_1+j_3-j_2)(j_2+j_3-j_1+1)(j_2+j_3-j_1+2)}{(2j_1-2)(2j_1-1)2j_1(2j_1+1)2j_2(2j_2+1)(2j_2+2)(2j_2+3)}\right]^{1/2}$$

$$\begin{Bmatrix} j_1 & j_2 & j_3 \\ j_2+\frac{3}{2} & j_1-\frac{3}{2} & \frac{3}{2} \end{Bmatrix} = (-1)^{j_1+j_2+j_3}$$

$$\times \left[\frac{(j_1+j_3+j_2-2)(j_1+j_3-j_2-1)(j_1+j_3-j_2)(j_2+j_3-j_1+1)(j_2+j_3-j_1+2)(j_2+j_3-j_1+3)}{(2j_1-2)(2j_1-1)2j_1(2j_1+1)(2j_2+1)(2j_2+2)(2j_2+3)(2j_2+4)}\right]^{1/2}$$

$$\begin{Bmatrix} j_1 & j_2 & j_3 \\ j_2-\frac{1}{2} & j_1-\frac{1}{2} & \frac{3}{2} \end{Bmatrix} = (-1)^{j_1+j_2+j_3}$$

$$\times \frac{[2(j_2+j_3-j_1)(j_1+j_3-j_2)-(j_1+j_2+j_3+2)(j_1+j_2-j_3-1)][(j_1+j_2+j_3+1)(j_1+j_2-j_3)]^{1/2}}{4[(2j_1-1)j_1(j_1+1)(2j_1+1)(2j_2-1)j_2(j_2+1)(2j_2+1)]^{1/2}}$$

$$\begin{Bmatrix} j_1 & j_2 & j_3 \\ j_2+\frac{1}{2} & j_1-\frac{1}{2} & \frac{3}{2} \end{Bmatrix} = (-1)^{j_1+j_2+j_3}$$

$$\times \frac{[(j_1+j_3-j_2-1)(j_2+j_3-j_1)-2(j_1+j_2+j_3+2)(j_1+j_2-j_3)][(j_1+j_3-j_2)(j_2+j_3-j_1+1)]^{1/2}}{4[(2j_1-1)j_1(j_1+1)(2j_1+1)j_2(j_2+1)(2j_2+1)(2j_2+3)]^{1/2}}$$

$$\begin{Bmatrix} j_1 & j_2 & j_3 \\ j_2-2 & j_1-2 & 2 \end{Bmatrix} = (-1)^{j_1+j_2+j_3} \left[\frac{(j_1+j_2+j_3-2)(j_1+j_2+j_3-1)(j_1+j_2+j_3)(j_1+j_2+j_3+1) \times(j_1+j_2-j_3-3)(j_1+j_2-j_3-2)(j_1+j_2-j_3-1)(j_1+j_2-j_3)}{(2j_1-3)(2j_1-2)(2j_1-1)2j_1(2j_1+1)(2j_2-3)(2j_2-2)(2j_2-1)2j_2(2j_2+1)} \right]^{1/2}$$

$$\begin{Bmatrix} j_1 & j_2 & j_3 \\ j_2-1 & j_1-2 & 2 \end{Bmatrix} = (-1)^{j_1+j_2+j_3} 2 \left[\frac{(j_1+j_2+j_3-1)(j_1+j_2+j_3)(j_1+j_2+j_3+1)(j_1+j_2-j_3-2) \times(j_1+j_2-j_3-1)(j_1+j_2-j_3)(j_1+j_3-j_2)(j_2+j_3-j_1+1)}{(2j_1-3)(2j_1-2)(2j_1-1)2j_1(2j_1+1)(2j_2-2)(2j_2-1)2j_2(2j_2+1)(2j_2+2)} \right]^{1/2}$$

$$\begin{Bmatrix} j_1 & j_2 & j_3 \\ j_2 & j_1-2 & 2 \end{Bmatrix} = (-1)^{j_1+j_2+j_3} \left[\frac{6(j_1+j_2+j_3)(j_1+j_2+j_3+1)(j_1+j_2-j_3)(j_1+j_2-j_3-1) \times(j_1+j_3-j_2-1)(j_1+j_3-j_2)(j_2+j_3-j_1+1)(j_2+j_3-j_1+2)}{(2j_1-3)(2j_1-2)(2j_1-1)2j_1(2j_1+1)(2j_2-1)2j_2(2j_2+1)(2j_2+2)(2j_2+3)} \right]^{1/2}$$

$$\begin{Bmatrix} j_1 & j_2 & j_3 \\ j_2+1 & j_1-2 & 2 \end{Bmatrix} = (-1)^{j_1+j_2+j_3} 2 \left[\frac{(j_1+j_2+j_3+1)(j_1+j_2-j_3)(j_1+j_3-j_2-2)(j_1+j_3-j_2-1) \times(j_1+j_3-j_2)(j_2+j_3-j_1+1)(j_2+j_3-j_1+2)(j_2+j_3-j_1+3)}{(2j_1-3)(2j_1-2)(2j_1-1)2j_1(2j_1+1)2j_2(2j_2+1)(2j_2+2)(2j_2+3)(2j_2+4)} \right]^{1/2}$$

$$\begin{Bmatrix} j_1 & j_2 & j_3 \\ j_2+2 & j_1-2 & 2 \end{Bmatrix} = (-1)^{j_1+j_2+j_3}$$

$$\times \left[\frac{(j_1+j_3-j_2-3)(j_1+j_3-j_2-2)(j_1+j_3-j_2-1)(j_1+j_3-j_2) \times (j_2+j_3-j_1+1)(j_2+j_3-j_1+2)(j_2+j_3-j_1+3)(j_2+j_3-j_1+4)}{(2j_1-3)(2j_1-2)(2j_1-1)2j_1(2j_1+1)(2j_2+1)(2j_2+2)(2j_2+3)(2j_2+4)(2j_2+5)} \right]^{1/2}$$

$$\begin{Bmatrix} j_1 & j_2 & j_3 \\ j_2-1 & j_1-1 & 2 \end{Bmatrix} = (-1)^{j_1+j_2+j_3}$$

$$\times \frac{(j_3(j_3+1)-j_1^2-j_2^2+j_1j_2+1)[(j_1+j_2+j_3)(j_1+j_2+j_3+1)(j_1+j_2-j_3-1)(j_1+j_2-j_3)]^{1/2}}{[(2j_1-2)(2j_1-1)j_1(j_1+1)(2j_1+1)(2j_2-2)(2j_2-1)j_2(j_2+1)(2j_2+1)]^{1/2}}$$

$$\begin{Bmatrix} j_1 & j_2 & j_3 \\ j_2 & j_1-1 & 2 \end{Bmatrix} = (-1)^{j_1+j_2+j_3}$$

$$\times \frac{[(j_3+j_2+1)(j_3-j_2)-j_1^2+1][6(j_1+j_2+j_3+1)(j_1+j_2-j_3)(j_1+j_3-j_2)(j_2+j_3-j_1+1)]^{1/2}}{2[(2j_1-2)(2j_1-1)j_1(j_1+1)(2j_1+1)(2j_2-1)j_2(j_2+1)(2j_2+1)(2j_2+3)]^{1/2}}$$

$$\begin{Bmatrix} j_1 & j_2 & j_3 \\ j_2+1 & j_1-1 & 2 \end{Bmatrix} = (-1)^{j_1+j_2+j_3} \frac{[(j_3+j_2+2)(j_3-j_2-1)-(j_1-1)(j_1+j_2+2)] \times [(j_1+j_3-j_2-1)(j_1+j_3-j_2)(j_2+j_3-j_1+1)(j_2+j_3-j_1+2)]^{1/2}}{[(2j_1-2)(2j_1-1)j_1(j_1+1)(2j_1+1)j_2(j_2+1)(2j_2+1)(2j_2+3)(2j_2+4)]^{1/2}}$$

$$\begin{Bmatrix} j_1 & j_2 & j_3 \\ j_2 & j_1 & 2 \end{Bmatrix} = (-1)^{j_1+j_2+j_3} \frac{[3X(X-1)-4j_1(j_1+1)j_2(j_2+1)]}{2[(2j_1-1)j_1(j_1+1)(2j_1+1)(2j_1+3)(2j_2-1)j_2(j_2+1)(2j_2+1)(2j_2+3)]^{1/2}}$$

$$X = j_1(j_1+1) + j_2(j_2+1) - j_3(j_3+1)$$

IRREDUCIBLE TENSOR OPERATORS

Definition

$$[J^0, T_\kappa^{(k)}] = \kappa T_\kappa^{(k)} \qquad [J^\pm, T_\kappa^{(k)}] = \sqrt{k(k+1) - \kappa(\kappa \pm 1)}\, T_{\kappa \pm 1}^{(k)}$$

Wigner-Eckart theorem

$$\langle \alpha J M | T_\kappa^{(k)} | \alpha' J' M' \rangle = (-1)^{J-M} \begin{pmatrix} J & k & J' \\ -M & \kappa & M' \end{pmatrix} (\alpha J \| \mathbf{T}^{(k)} \| \alpha' J')$$

$$(\alpha J \| \mathbf{T}^{(k)} \| \alpha' J') = (2k+1) \sum_{MM'} (-1)^{J-M} \begin{pmatrix} J & k & J' \\ -M & \kappa & M' \end{pmatrix} \langle \alpha J M | T_\kappa^{(k)} | \alpha' J' M' \rangle$$

$$(T_\kappa^{(k)})^\dagger = (-1)^\kappa T_{-\kappa}^{(k)} \qquad \text{for hermitean tensor operator}$$

then $(\alpha' J' \| \mathbf{T}^{(k)} \| \alpha J)^* = (-1)^{J-J'} (\alpha J \| \mathbf{T}^{(k)} \| \alpha' J')$

Simple reduced matrix elements

$$(\alpha J \| 1 \| \alpha' J') = \sqrt{2J+1}\, \delta_{\alpha\alpha'} \delta_{JJ'}$$

$$(\alpha J \| \mathbf{J} \| \alpha' J') = \sqrt{J(J+1)(2J+1)}\, \delta_{\alpha\alpha'} \delta_{JJ'}$$

$$(l\|\mathbf{Y}_k\|l') = (-1)^l \sqrt{\frac{(2l+1)(2k+1)(2l'+1)}{4\pi}} \begin{pmatrix} l & k & l' \\ 0 & 0 & 0 \end{pmatrix}$$

$$(\tfrac{1}{2}l\,j\|\mathbf{Y}_k\|\tfrac{1}{2}l'\,j') = (-1)^{j-1/2} \sqrt{\frac{(2j+1)(2k+1)(2j'+1)}{4\pi}} \begin{pmatrix} j & k & j' \\ -\frac{1}{2} & 0 & \frac{1}{2} \end{pmatrix} \tfrac{1}{2}[1 + (-1)^{l+l'+k}]$$

Tensor products

$$[\mathbf{T}^{(k_1)} \times \mathbf{T}^{(k_2)}]^{(k)}_{\kappa} = \sum T^{(k_1)}_{\kappa_1} T^{(k_2)}_{\kappa_2} (k_1\kappa_1 k_2\kappa_2 \mid k_1 k_2 k \kappa)$$

Scalar product

$$(\mathbf{T}^{(k)}_1 \cdot \mathbf{T}^{(k)}_2) = (-1)^k \sqrt{2k+1}[\mathbf{T}^{(k)}_1 \times \mathbf{T}^{(k)}_2]^{(0)}_0 = \sum_{\kappa} (-1)^{\kappa} T^{(k)}_{1\kappa} T^{(k)}_{2,-\kappa}$$

$$(\mathbf{T}^{(k')}_1 \cdot \mathbf{T}^{(k')}_2)(\mathbf{U}^{(k'')}_1 \cdot \mathbf{U}^{(k'')}_2) = \sum_{r} (-1)^{k'+k''+k} [\mathbf{T}^{(k')}_1 \times \mathbf{U}^{(k')}_1]^{(k)} \cdot [\mathbf{T}^{(k')}_2 \times \mathbf{U}^{(k'')}_2]^{(k)}$$

Magnetic moments

$$\boldsymbol{\mu} = g\mathbf{J} = g_1\mathbf{J}_1 + g_2\mathbf{J}_2 \qquad g\mathbf{J}^2 = g_1(\mathbf{J}_1 \cdot \mathbf{J}) + g_2(\mathbf{J}_2 \cdot \mathbf{J})$$

$$\mu = gJ = \frac{1}{2(J+1)}[g_1\{J(J+1) + J_1(J_1+1) - J_2(J_2+1)\} + g_2\{J(J+1) + J_2(J_2+1) - J_1(J_1+1)\}]$$

Matrix elements

$$(\alpha_1 J_1 \alpha_2 J_2 J \| [\mathbf{T}^{k_1)}(1) \times \mathbf{T}^{(k_2)}(2)]^{(k)} \| \alpha_1' J_1' \alpha_2' J_2' J')$$

$$= \sqrt{(2J+1)(2k+1)(2J'+1)} \begin{Bmatrix} J_1 & J_2 & J \\ J_1' & J_2' & J' \\ k_1 & k_2 & k \end{Bmatrix} (\alpha_1 J_1 \| \mathbf{T}^{(k_1)}(1) \| \alpha_1' J_1')(\alpha_2 J_2 \| \mathbf{T}^{(k_2)}(2) \| \alpha_2' J_2')$$

$$\langle \alpha_1 J_1 \alpha_2 J_2 J M | (\mathbf{T}^{(k)}(1) \cdot \mathbf{T}^{(k)}(2)) | \alpha_1' J_1' \alpha_2' J_2' J M \rangle$$

$$= (-1)^{J_2 + J + J_1'} \begin{Bmatrix} J_1 & J_2 & J \\ J_2' & J_1' & k \end{Bmatrix} (\alpha_1 J_1 \| \mathbf{T}^{(k)}(1) \| \alpha_1' J_1')(\alpha_2 J_2 \| \mathbf{T}^{(k)}(2) \| \alpha_2' J_2')$$

$$(\alpha_1 J_1 \alpha_2 J_2 J \| \mathbf{T}^{(k)}(1) \| \alpha_1' J_1' \alpha_2' J_2' J') = (-1)^{J_1 + J_2 + J' + k}$$

$$\times \sqrt{(2J+1)(2J'+1)} \begin{Bmatrix} J_1 & J & J_2 \\ J' & J_1' & k \end{Bmatrix} (\alpha_1 J_1 \| \mathbf{T}^{(k)}(1) \| \alpha_1' J_1') \delta_{\alpha_2 \alpha_2'} \delta_{J_2 J_2'}$$

$$(\alpha_1 J_1 \alpha_2 J_2 J \| \mathbf{T}^{(k)}(2) \| \alpha_1' J_1' \alpha_2' J_2' J') = (-1)^{J_1 + J_2' + J + k}$$

$$\times \sqrt{(2J+1)(2J'+1)} \begin{Bmatrix} J_2 & J & J_1 \\ J' & J_2' & k \end{Bmatrix} (\alpha_2 J_2 \| \mathbf{T}^{(k)}(2) \| \alpha_2' J_2') \delta_{\alpha_1 \alpha_1'} \delta_{J_1 J_1'}$$

If $\mathbf{T}^{(k_1)}$ and $\mathbf{T}^{(k_2)}$ act on the *same* system

$$(\alpha J \| [\mathbf{T}^{(k_1)}(1) \times \mathbf{T}^{(k_2)}(1)]^{(k)} \| \alpha' J') = (-1)^{J+k+J'} \sqrt{2k+1}$$

$$\times \sum_{\alpha'' J''} (\alpha J \| \mathbf{T}^{(k_1)}(1) \| \alpha'' J'')(\alpha'' J'' \| \mathbf{T}^{(k_2)}(1) \| \alpha' J') \begin{Bmatrix} k_1 & k_2 & k \\ J' & J & J'' \end{Bmatrix}$$

$$\langle \alpha J M | (\mathbf{T}^{(k)}(1) \cdot \mathbf{U}^{(k)}(1) | \alpha' J M \rangle = \frac{1}{2J+1} \sum_{\alpha'' J''} (\alpha J \| \mathbf{T}^{(k)}(1) \| \alpha'' J'')(\alpha' J \| \mathbf{U}^{(k)}(1) \| \alpha'' J'')^*$$

Rates of transitions due to operator $\mathbf{T}^{(k)}$ are proportional to

$$\sum_{\kappa M_f} |\langle \alpha_f J_f M_f | T_\kappa^{(k)} | \alpha_i J_i M_i \rangle|^2 = \frac{1}{2J_i+1} |(\alpha_f J_f \| \mathbf{T}^{(k)} \| \alpha_i J_i)|^2$$

Transition probability per unit time of electromagnetic transitions of 2^L multipole

$$T(L) = 8\pi c \frac{e^2}{\hbar c} \frac{L+1}{L[(2L+1)!!]^2} k^{2L+1} B(L) \qquad k = \frac{2\pi}{\lambda} \ll R$$

For electric multipoles the reduced transition rate is

$$B(EL) = \frac{1}{2J_i+1} \left| \left(J_f \left\| \sum_i \frac{e_i}{e} r_i^L \mathbf{Y}_L(\theta_i, \phi_i) \right\| J_i \right) \right|^2$$

For a single particle

$$B_{s.p.}(EL) = \frac{1}{4\pi}(2L+1)(2j_f+1)\begin{pmatrix} j_i & j_f & L \\ \frac{1}{2} & -\frac{1}{2} & 0 \end{pmatrix}^2 \langle r^L \rangle^2$$

$$\langle r^L \rangle = \int_0^\infty R_{n_i l_i j_i}(r) R_{n_f l_f j_f}(r) r^L \, dr$$

For magnetic multipoles

$$B(ML) = \frac{1}{2J_i+1}\left(\frac{\hbar}{mc}\right)^2 L(2L+1)\left|\left(J_f\left\|\sum_i r_i^{L-1}\left[\mathbf{Y}_{L-1}(\theta_i\phi_i)\times\left(\frac{g_{l_i}}{L+1}\mathbf{l}_i+\frac{1}{2}g_{s_i}\mathbf{s}_i\right)\right]^{(L)}\right\|J_i\right)\right|^2$$

$$\begin{aligned} B_{s.p.}(ML) = \left(\frac{\hbar}{mc}\right)^2 \frac{(2L+1)(2j_f+1)}{16\pi}\Bigg[&\frac{g_l}{L+1}\begin{pmatrix} j_i & j_f & L-1 \\ \frac{1}{2} & -\frac{1}{2} & 0 \end{pmatrix} \\ &\times[(j_i+j_f+L+1)(j_i+L-j_f)(j_f+L-j_i)(j_i+j_f-L+1)]^{1/2} \\ &+\left(\frac{1}{2}g_s-\frac{g_l}{L+1}\right)\begin{pmatrix} j_i & j_f & L \\ \frac{1}{2} & -\frac{1}{2} & 0 \end{pmatrix} \\ &\times\left(L+\tfrac{1}{2}\{(-1)^{(1/2)+l_i-j_i}(2j_i+1)+(-1)^{(1/2)+l_f-j_f}(2j_f+1)\}\right)\Bigg]^2 \langle r^{L-1}\rangle^2 \end{aligned}$$

In case $L = |j_i - j_f|$

$$B_{s.p.}(ML) = \left(\frac{\hbar}{2mc}\right)^2 \left(\frac{1}{2}g_s - \frac{g_l}{L+1}\right)^2 \frac{4L^2(2L+1)}{4\pi}(2j_f+1)\begin{pmatrix} j_i & j_f & L \\ \frac{1}{2} & -\frac{1}{2} & 0 \end{pmatrix}^2 \langle r^{L-1}\rangle^2$$

Beta decay rates for Fermi transitions are proportional to

$$T(T+1) - M_T^{(i)} M_T^{(f)}$$

Sum rule for Gamow-Teller transitions

$$\frac{1}{4}\sum_f \left|\langle f|\sum_k \tau^+(k)\sigma(k)|i\rangle\right|^2 - \frac{1}{4}\sum_f \left|\langle f|\sum_k \tau^-(k)\sigma(k)|i\rangle\right|^2 = 3(N-Z)$$

Rates for single nucleon transitions are proportional to

$$\frac{1}{2j_i+1}(j_f\|\sigma\|j_i)^2 = 6(2j_f+1)\begin{Bmatrix} \frac{1}{2} & j_f & l \\ j_i & \frac{1}{2} & 1 \end{Bmatrix}^2$$

MATRIX ELEMENTS OF TWO-BODY INTERACTIONS

Two nucleon wave functions

$$\psi_{j_1 j_2 TJM} = \eta_{T,M_T} \frac{1}{\sqrt{2(1+\delta_{j_1 j_2})}} \sum_{m_1 m_2} (j_1 m_1 j_2 m_2 \mid j_1 j_2 JM)$$

$$\times [\psi_{j_1 m_1}(1)\psi_{j_2 m_2}(2) + (-1)^T \psi_{j_1 m_1}(2)\psi_{j_2 m_2}(1)]$$

For $j_1 = j_2$

$$\psi_{j^2 TJM} = \eta_{T,M_T} \sum_{mm'} (jmjm' \mid jjJM)\psi_{jm}(1)\psi_{jm'}(2) \qquad (-1)^{T+J} = -1$$

For a j-proton j'-neutron state

$$\psi_{j_\pi j'_\nu JM} = \sum (jmj'm' \mid jj'JM)\psi_{jm}(1_\pi)\psi_{j'm'}(2_\nu)$$

Slater expansion

$$V(|\mathbf{r}_1 - \mathbf{r}_2|) = V\left(\sqrt{r_1^2 + r_2^2 - 2r_1 r_2 \cos\omega_{12}}\right) = \sum_{k=0}^{\infty} v_k(r_1, r_2) P_k(\cos\omega_{12})$$

For

$$V(|\mathbf{r}_1 - \mathbf{r}_2|) = \frac{1}{|\mathbf{r}_1 - \mathbf{r}_2|}, \qquad v_k(r_1, v_2) = \frac{r_<^k}{r_>^{k+1}}$$

$$v_k(r_1, r_2) = \frac{2k+1}{2} \int_{-1}^{+1} V(|\mathbf{r}_1 - \mathbf{r}_2|) P_k(\cos\omega_{12}) d\cos\omega_{12}$$

$$P_k(\cos\omega_{12}) = \frac{4\pi}{2k+1} \sum_{\kappa} (-1)^{\kappa} Y_{k\kappa}(\theta_1\phi_1) Y_{k,-\kappa}(\theta_2\phi_2)$$

Radial integrals with angular coefficients

$$F^k(j_1 j_2 j_1' j_2') = \frac{4\pi}{2k+1} (j_1 \| \mathbf{Y}_k \| j_1')(j_2 \| \mathbf{Y}_k \| j_2')$$

$$\times \int R_{n_1 l_1 j_1}(r_1) R_{n_1' l_1' j_1'}(r_1) R_{n_2 l_2 j_2}(r_2) R_{n_2' l_2' j_2'}(r_2) v_k(r_1, r_2)\, dr_1\, dr_2$$

$$G^k(j_1 j_2 j_1' j_2') = \frac{4\pi}{2k+1} (j_1 \| \mathbf{Y}_k \| j_2')(j_1' \| \mathbf{Y}_k \| j_2)$$

$$\times \int R_{n_1 l_1 j_1}(r_1) R_{n_2' l_2' j_2'}(r_1) R_{n_1' l_1' j_1'}(r_2) R_{n_2 l_2 j_2}(r_2) v_k(r_1, r_2)\, dr_1\, dr_2$$

Matrix elements between antisymmetric states (for $j_1 \neq j_2, j_1' \neq j_2'$)

$$\begin{aligned}
&\langle j_1 j_2 TJM|V|j_1' j_2' TJM\rangle_a \\
&\quad = \sum_k (-1)^{j_2+j_1'+J} \begin{Bmatrix} j_1 & j_2 & J \\ j_2' & j_1' & k \end{Bmatrix} F^k + (-1)^T \sum_k \begin{Bmatrix} j_1 & j_2 & J \\ j_1' & j_2' & k \end{Bmatrix} G^k \\
&\quad = \sum_k (F^k + (-1)^T F'^k)(-1)^{j_2+j_1'+J} \begin{Bmatrix} j_1 & j_2 & J \\ j_2' & j_1' & k \end{Bmatrix} \\
&\quad = \sum_k (F^k + (-1)^T F'^k) \langle j_1 j_2 JM|\mathbf{u}^{(k)}(1) \cdot \mathbf{u}^{(k)}(2)|j_1' j_2' JM\rangle
\end{aligned}$$

$$F'^k = \sum_t (-1)^{j_2+j_1'+k+t}(2k+1) \begin{Bmatrix} j_1 & j_2' & t \\ j_2 & j_1' & k \end{Bmatrix} G^t$$

$$(j\|\mathbf{u}^{(k)}(1)\|j') = \delta_{jj_1}\delta_{j'j_1'} \qquad (j\|\mathbf{u}^{(k)}(2)\|j') = \delta_{jj_2}\delta_{j'j_2'}$$

For $j_1 = j_2$, $j_1' \neq j_2'$

$$\begin{aligned}
\langle j^2 TJM|V|j'j''TJM\rangle_a &= \sqrt{2}\langle j^2 TJM|V|j'j''TJM\rangle = \sqrt{2}\sum_k (-1)^{j+j'+J} \begin{Bmatrix} j & j & J \\ j'' & j' & k \end{Bmatrix} F^k \\
&= \sqrt{2}\sum_k F^k \langle j^2 JM|\mathbf{u}^{(k)}(1) \cdot \mathbf{u}^{(k)}(2)|j'j''JM\rangle \qquad (-1)^{T+J} = -1
\end{aligned}$$

For $j_1 = j_2,\ j_1' = j_2'$

$$\langle j^2 TJM|V|j'^2 TJM\rangle = \sum_k (-1)^{j+j'+J} \begin{Bmatrix} j & j & J \\ j' & j' & k \end{Bmatrix} F^k$$
$$= \sum_k F^k \langle j^2 JM|\mathbf{u}^{(k)}(1)\cdot\mathbf{u}^{(k)}(2)|j'^2 JM\rangle \qquad (-1)^{T+J} = -1$$

Define $F_k = F^k + (-1)^T F'^k$ for $j_1 \neq j_2, j_1' \neq j_2'$, $F_k = \sqrt{2}F^k$ for $j_1 = j_2, j_1' \neq j_2'$ or $j_1 \neq j_2,\ j_1' = j_2'$ and $F_k = F^k$ for $j_1 = j_2,\ j_1' = j_2'$, then in general

$$\langle j_1 j_2 TJM|V|j_1' j_2' TJM\rangle_a = V_{TJ}(j_1 j_2 j_1' j_2') = \sum_k F_k (-1)^{j_2+j_1'+J} \begin{Bmatrix} j_1 & j_2 & J \\ j_2' & j_1' & k \end{Bmatrix}$$
$$= \sum_k F_k \langle j_1 j_2 JM|\mathbf{u}^{(k)}(1)\cdot\mathbf{u}^{(k)}(2)|j_1' j_2' JM\rangle$$

For $j_1 \neq j_2,\ j_1' \neq j_2'$

$$F_k(j_1 j_2 j_1' j_2') = (2k+1)\sum_J (-1)^{j_2+j_1'+J}(2J+1) \begin{Bmatrix} j_1 & j_2 & J \\ j_2' & j_1' & k \end{Bmatrix} V_{TJ}(j_1 j_2 j_1' j_2')$$

for either $T = 0$ or $T = 1$.

If $j_1 = j_2$ or $j_1' = j_2'$

$$F_k(j_1 j_2 j_1' j_2') = (2k+1)\sum_J (-1)^{j_2+j_1'+J}(2J+1) \begin{Bmatrix} j_1 & j_2 & J \\ j_2' & j_1' & k \end{Bmatrix} V_{TJ}(j_1 j_2 j_1' j_2') \qquad (-1)^{T+J} = -1$$

For proton j, j'' orbits and neutron j', j''' orbits

$$\langle j_\pi j'_\nu JM|V|j''_\pi j'''_\nu JM\rangle = \sum_k F_k(jj'j''j''')(-1)^{j'+j''+J}\begin{Bmatrix} j & j' & J \\ j''' & j'' & k \end{Bmatrix}$$

$$= \tfrac{1}{2}[\langle jj'T=0, JM|V|j''j'''T=0, JM\rangle + \langle jj'T=1, JM|V|j''j'''T=1, JM\rangle]$$

For spin dependent interactions the expansion includes also the following terms

$$[\mathbf{s}_1 \times \mathbf{Y}_k(1)]^{(r)} \cdot [\mathbf{s}_2 \times \mathbf{Y}_{k'}(2)]^{(r)}$$

For example, $(\mathbf{s}_1 \cdot \mathbf{s}_2)V(|\mathbf{r}_1 - \mathbf{r}_2|)$ yields terms

$$(\mathbf{s}_1 \cdot \mathbf{s}_2)(\mathbf{Y}_k(1) \cdot \mathbf{Y}_k(2)) = \sum_r [\mathbf{s}_1 \times \mathbf{Y}_k(1)]^{(r)} \cdot [\mathbf{s}_2 \times \mathbf{Y}_k(2)]^{(r)}(-1)^{k+r+1}$$

Instead of $(\tfrac{1}{2}lj\|\mathbf{Y}_k\|\tfrac{1}{2}l'j')$ terms, there are now

$$(\tfrac{1}{2}lj\|[\mathbf{s} \times \mathbf{Y}_k]^{(r)}\|\tfrac{1}{2}l'j') = \sqrt{(2j+1)(2r+1)(2j'+1)}\begin{Bmatrix} \frac{1}{2} & l & j \\ \frac{1}{2} & l' & j' \\ 1 & k & r \end{Bmatrix}(\tfrac{1}{2}\|\mathbf{s}\|\tfrac{1}{2})(l\|\mathbf{Y}_k\|l')$$

In LS-coupling

$$\phi_{ll'LM_L} = \frac{1}{\sqrt{2(1+\delta_{ll'})}} \sum_{m_l m_l'} (l m_l l' m_l' \mid l l' L M_L)[\phi_{l m_l}(\mathbf{r}_1)\phi_{l' m_l'}(\mathbf{r}_2) - (-1)^{S+T}\phi_{l m_l}(\mathbf{r}_2)\phi_{l' m_l'}(\mathbf{r}_1)]$$

$$\psi_{ll'SLJM} = \sum_{M_S M_L} (S M_S L M_L \mid SLJM)\chi_{S M_S}\phi_{ll'LM_L}$$

Transformation to jj-coupling

$$\psi_{ll'jj'JM} = \sum_{SL} \sqrt{(2j+1)(2j'+1)(2S+1)(2L+1)} \begin{Bmatrix} \frac{1}{2} & \frac{1}{2} & S \\ l & l' & L \\ j & j' & J \end{Bmatrix} \psi_{ll'SLJM}$$

If the interaction is a scalar product of $\mathcal{T}^{(k)}$ of spins and $\mathcal{W}^{(k)}$ of space coordinates ($k = 0$ for central interactions, $k = 2$ for tensor forces) its matrix elements are given by

$$\langle SLJM|\mathcal{T}^{(k)} \cdot \mathcal{W}^{(k)}|S'L'JM\rangle = (-1)^{L+S'+J} \begin{Bmatrix} S & L & J \\ L' & S' & k \end{Bmatrix} (S\|\mathcal{T}^{(k)}\|S')(L\|\mathcal{W}^{(k)}\|L')$$

Matrix elements of a potential interaction

$$\langle ll'LM_L|V|ll'LM_L\rangle - (-1)^{S+T+l+l'-L}\langle ll'LM_L|V|l'lLM_L\rangle$$

$$= \sum_k \frac{4\pi}{2k+1}(l\|\mathbf{Y}_k\|l)(l'\|\mathbf{Y}_k\|l')(-1)^{l+l'+L}\begin{Bmatrix} l & l' & L \\ l' & l & k \end{Bmatrix}$$

$$\times \int R_{nl}^2(r_1)R_{n'l'}^2(r_2)v_k(r_1,r_2)\,dr_1\,dr_2 - (-1)^{S+T}\sum_k \frac{4\pi}{2k+1}(l\|\mathbf{Y}_k\|l')^2\begin{Bmatrix} l & l' & L \\ l & l' & k \end{Bmatrix}$$

$$\times \int R_{nl}(r_1)R_{n'l'}(r_1)R_{nl}(r_2)R_{n'l'}(r_1)v_k(r_1,r_2)\,dr_1\,dr_2$$

Two particles in a harmonic oscillator potential Hamiltonian

$$\frac{1}{2m}\mathbf{p}_1^2 + \frac{1}{2}m\omega^2\mathbf{r}_1^2 + \frac{1}{2m}\mathbf{p}_2^2 + \frac{1}{2}m\omega^2\mathbf{r}_2^2 = \frac{1}{4m}\mathbf{P}^2 + m\omega^2\mathbf{R}^2 + \frac{1}{m}\mathbf{p}^2 + \frac{m}{4}\omega^2\mathbf{r}^2$$

$$\mathbf{r} = \mathbf{r}_2 - \mathbf{r}_1 \qquad \mathbf{p} = \tfrac{1}{2}(\mathbf{p}_2 - \mathbf{p}_1) \qquad \mathbf{R} = \tfrac{1}{2}(\mathbf{r}_1 + \mathbf{r}_2) \qquad \mathbf{P} = \mathbf{p}_1 + \mathbf{p}_2$$

Wave functions

$$\phi_{N_1l_1N_2l_2LM_L}(\mathbf{r}_1,\mathbf{r}_2) = \sum_{\substack{\mathcal{N}\mathcal{L}nl \\ \mathcal{M}m}} a^{N_1l_1N_2l_2}_{\mathcal{N}\mathcal{L}nlL}(\mathcal{L}\mathcal{M}lm \mid \mathcal{L}lLM_L)\phi_{\mathcal{N}\mathcal{L}\mathcal{M}}(\mathbf{R})\phi_{nlm}(\mathbf{r})$$

Matrix elements

$$\langle N_1 l_1 N_2 l_2 L M_L | \mathcal{W}_\kappa^{(k)} | N_1' l_1' N_2' l_2' L' M_L' \rangle$$

$$= \sum_{\mathcal{N}\mathcal{L}\mathcal{M}} a_{\mathcal{N}\mathcal{L}nlL}^{N_1 l_1 N_2 l_2} a_{\mathcal{N}\mathcal{L}n'l'L'}^{N_1' l_1' N_2' l_2'} (\mathcal{L}\mathcal{M} l m | \mathcal{L} l L M_L)(\mathcal{L}\mathcal{M} l' m' | \mathcal{L} l' L' M_L') \langle n l m | \mathcal{W}_\kappa^{(k)} | n' l' m' \rangle$$

For $k = 0$ (central interaction)

$$\langle N_1 l_1 N_2 l_2 L M_L | V(|\mathbf{r}_1 - \mathbf{r}_2|) | N_1' l_1' N_2' l_2' L' M_L' \rangle$$

$$= \delta_{LL'} \delta_{MM'} \sum_{\mathcal{N}\mathcal{L}nn'l} a_{\mathcal{N}\mathcal{L}nlL}^{N_1 l_1 N_2 l_2} a_{\mathcal{N}\mathcal{L}n'lL}^{N_1' l_1' N_2' l_2'} \int R_{nl}(r) R_{n'l}(r) V(r)\, dr$$

Radial integral is a linear combination of

$$I_l = \int R_{n=1,l}^2(r) V(r) dr = \frac{2^{l+2} \nu^{l+\frac{3}{2}}}{\sqrt{\pi}(2l+1)!!} \int e^{-\nu r^2} r^{2l+2} V(r)\, dr \qquad \nu = \frac{m\omega}{2\hbar}$$

For $V(r) = e^2/r$

$$I_l = e^2 \int e^{-\nu r^2} r^{2l+1} dr = \frac{2^{l+1} l!}{(2l+1)!!} e^2 \sqrt{\frac{\nu}{\pi}}$$

THE δ-POTENTIAL INTERACTION

Matrix elements between antisymmetric states of nucleons in l and l'-orbits ($l \neq l'$) in LS-coupling

$$\Delta E_{ll'TSL} = (1-(-1)^{S+T})(2l+1)(2l'+1)\begin{pmatrix} l & l' & L \\ 0 & 0 & 0 \end{pmatrix}^2 \frac{1}{4\pi}\int R_{nl}^2(r)R_{n'l'}^2(r)\frac{dr}{r^2}$$

$$= (1-(-1)^{S+T})C_0(2l+1)(2l'+1)\begin{pmatrix} l & l' & L \\ 0 & 0 & 0 \end{pmatrix}^2$$

For $l = l'$ $(n = n')$

$$\Delta E_{l^2L} = \tfrac{1}{2}(1+(-1)^L)C_0(2l+1)^2\begin{pmatrix} l & l & L \\ 0 & 0 & 0 \end{pmatrix}^2 \qquad (-1)^{S+T+L} = -1$$

$$\Delta E_{l^2L=0} = (2l+1)C_0$$

In jj-coupling, for $T = 1$, $\Delta E_{ll'jj'J} = 0$ unless $(-1)^{l+l'+J} = 1$

For $j \neq j'$

$$\Delta E_{jj',T=1,J} = C_0(2j+1)(2j'+1)\begin{pmatrix} j & j' & J \\ \frac{1}{2} & -\frac{1}{2} & 0 \end{pmatrix}^2 \qquad (-1)^{l+l'+J} = 1$$

For $j = j'$, $(-1)^J = 1$ and $T = 1$

$$\Delta E_{j^2 J} = \tfrac{1}{2} C_0 (2j+1)^2 \begin{pmatrix} j & j & J \\ \frac{1}{2} & -\frac{1}{2} & 0 \end{pmatrix}^2 \qquad (-1)^J = 1$$

$$\Delta E_{j^2 J=0} = \tfrac{1}{2}(2j+1) C_0$$

For $T = 0$ states

$$\Delta E_{jj'T=0,J} = C_0 (2j+1)(2j'+1) \begin{pmatrix} j & j' & J \\ \frac{1}{2} & -\frac{1}{2} & 0 \end{pmatrix}^2 \frac{\{(2j+1) + (-1)^{j+j'+J}(2j'+1)\}^2}{4J(J+1)} \qquad (-1)^{l+l'+J} = 1$$

$$\Delta E_{jj'T=0,J} = C_0 (2j+1)(2j'+1) \begin{pmatrix} j & j' & J \\ \frac{1}{2} & -\frac{1}{2} & 0 \end{pmatrix}^2 \left[1 + \frac{\{(2j+1) + (-1)^{j+j'+J}(2j'+1)\}^2}{4J(J+1)} \right] \qquad (-1)^{l+l'+J} = -1$$

For $j' = j (l' = l, n' = n)$

$$\Delta E_{j^2 J} = \tfrac{1}{2}(2j+1)^2 \begin{pmatrix} j & j & J \\ \frac{1}{2} & -\frac{1}{2} & 0 \end{pmatrix}^2 \left[1 + \frac{(2j+1)^2}{J(J+1)} \right] C_0 \qquad T = 0,\ J \text{ odd}$$

For a j-proton and j'-neutron state (any J)

$$\Delta E_{j_\pi j'_\nu J} = \tfrac{1}{2}(2j+1)(2j'+1)\begin{pmatrix} j & j' & J \\ \tfrac{1}{2} & -\tfrac{1}{2} & 0 \end{pmatrix}^2 \left[1 + \frac{\{(2j+1)+(-1)^{j+j'+J}(2j'+1)\}^2}{4J(J+1)}\right] C_0$$

The interaction $4(\mathbf{s}_1 \cdot \mathbf{s}_2)\delta(|\mathbf{r}_1 - \mathbf{r}_2|)$ is equivalent to

$$-3\delta(|\mathbf{r}_1 - \mathbf{r}_2|) \quad \text{for} \quad T = 1 \text{ states} \qquad \delta(|\mathbf{r}_1 - \mathbf{r}_2|) \quad \text{for} \quad T = 0 \text{ states}$$

$$\Delta E_{j_\pi j'_\nu J} = \tfrac{1}{2}(2j+1)(2j'+1)\begin{pmatrix} j & j' & J \\ \tfrac{1}{2} & -\tfrac{1}{2} & 0 \end{pmatrix}^2 \left[-3 + \frac{\{(2j+1)+(-1)^{j+j'+J}(2j'+1)\}^2}{4J(J+1)}\right] C_0 \quad \text{for} \quad (-1)^{l+l'+J} = 1$$

$$\Delta E_{j_\pi j'_\nu J} = \tfrac{1}{2}(2j+1)(2j'+1)\begin{pmatrix} j & j' & J \\ \tfrac{1}{2} & -\tfrac{1}{2} & 0 \end{pmatrix}^2 \left[1 + \frac{\{(2j+1)+(-1)^{j+j'+J}(2j'+1)\}^2}{4J(J+1)}\right] C_0 \quad \text{for} \quad (-1)^{l+l'+J} = -1$$

Matrix elements between space symmetric ll' and $l''l'''$ $T = 1$ states

$$\frac{2(-1)^{l+l''}}{\sqrt{(1+\delta_{ll'})(1+\delta_{l''l'''})}}\sqrt{(2l+1)(2l'+1)(2l''+1)(2l'''+1)}\begin{pmatrix} l & l' & L \\ 0 & 0 & 0 \end{pmatrix}\begin{pmatrix} l'' & l''' & L \\ 0 & 0 & 0 \end{pmatrix} C_0(ll'l''l''')$$

$$C_0(ll'l''l''') = \frac{1}{4\pi}\int R_{nl}(r)R_{n'l'}(r)R_{n''l''}(r)R_{n'''l'''}(r)\frac{dr}{r^2}$$

In the **Surface Delta Interaction (SDI)** *all* $C_0(ll'l''l''')$ are equal

Non-vanishing eigenvalue

$$C_0 \sum_{l,l'} (2l+1)(2l'+1) \begin{pmatrix} l & l' & L \\ 0 & 0 & 0 \end{pmatrix}^2$$

Corresponding eigenstate (not normalized)

$$\sum_{l \leq l'} (-1)^l \sqrt{(2l+1)(2l'+1)} \begin{pmatrix} l & l' & L \\ 0 & 0 & 0 \end{pmatrix} \phi^s_{ll'LM} \qquad (\phi^s \text{ is symmetric and normalized})$$

Matrix elements in jj-coupling between jj' and $j''j'''$ $T = 1$ states

$$(-1)^{j+j''+1+l'+l'''} \sqrt{\frac{(2j+1)(2j'+1)(2j''+1)(2j'''+1)}{(1+\delta_{jj'})(1+\delta_{j''j'''})}}$$
$$\times \begin{pmatrix} j & j' & J \\ \frac{1}{2} & -\frac{1}{2} & 0 \end{pmatrix} \begin{pmatrix} j'' & j''' & J \\ \frac{1}{2} & -\frac{1}{2} & 0 \end{pmatrix} C_0 \qquad \text{for} \quad (-1)^{l+l'+J} = (-1)^{l''+l'''+J} = 1$$

In SDI, non-vanishing eigenvalue

$$\tfrac{1}{2} C_0 \sum_{j,j'} (2j+1)(2j'+1) \begin{pmatrix} j & j' & J \\ \frac{1}{2} & -\frac{1}{2} & 0 \end{pmatrix}^2 \qquad (-1)^{l+l'+J} = 1$$

Corresponding eigenstate (not normalized)

$$\sum_{j\le j'}(-1)^{j-(1/2)+l'}\sqrt{(2j+1)(2j'+1)}\begin{pmatrix} j & j' & J \\ \frac{1}{2} & -\frac{1}{2} & 0\end{pmatrix}\psi^a_{jj'JM}$$

(ψ^a is antisymmetric and normalized)

For $J = 0$, the eigenstate is

$$\sum_j (-1)^l\sqrt{2j+1}\psi_{j^2 J=0, M=0}$$

and the eigenvalue

$$\frac{1}{2}C_0\sum_j(2j+1) = C_0\sum_j \Omega_j = C_0\Omega$$

CLOSED SHELLS. PARTICLES AND HOLES

Interaction energy in the closed (fully occupied) j-orbit of identical nucleons

$$\sum_{J\,\text{even},M}\langle j^2JM|V|j^2JM\rangle = \sum_{J\,\text{even}}(2J+1)\langle j^2JM_0|V|j^2JM_0\rangle = \frac{1}{2}(2j+1)F_0 - \frac{1}{2}\sum_k F_k$$

Interaction energy of a j-proton (neutron) with the closed neutron (proton) j-orbit

$$\frac{1}{2j+1}\sum_{\text{all } J}(2J+1)\langle j^2 JM|JM|V|j^2 JM\rangle = F_0$$

Interaction energy of a j-nucleon with the same kind of nucleons in the closed j'-orbit

$$\frac{1}{2j+1}\sum_{|j-j'|}^{j+j'}(2J+1)\langle jj'T=1, JM|V|jj'T=1, JM\rangle = \sqrt{\frac{2j'+1}{2j+1}}F_0 = (2j'+1)\bar{V}(jj'T=1)$$

Interaction energy of a j-proton (neutron) with the closed neutron (proton) j'-orbit

$$\frac{1}{2j+1}\sum_{|j-j'|}^{j+j'}(2J+1)\tfrac{1}{2}[\langle jj'T=1, JM|V|jj'T=1, JM\rangle$$

$$+\langle jj'T=0, JM|V|jj'T=0, JM\rangle] = (2j'+1)\bar{V}(j_\pi j'_\nu)$$

Interaction energy in a state of a single j-hole (j^{2j} configuration of identical nucleons)

$$\sum_{J\text{ even}}(2J+1)\langle j^2 JM|V|j^2 JM\rangle - \frac{2}{2j+1}\sum_{J\text{ even}}(2J+1)\langle j^2 JM|V|j^2 JM\rangle$$

Interaction energy in a state of a particle and a hole (Pandya relation).

For a proton j-hole and a j'-neutron (the neutron j-orbit is closed)

$$\langle j_\pi^{2j}(j)j'_\nu JM|\sum_{i=1}^{2j} V(0,i)|j_\pi^{2j}(j)j'_\nu JM\rangle = \frac{1}{2j'+1}\sum_{|j-j'|}^{j+j'}(2J'+1)\langle j_\pi j'_\nu J'M'|V|j_\pi j'_\nu J'M'\rangle$$
$$-\sum_{J'}(2J'+1)\begin{Bmatrix} j & j' & J \\ j & j' & J' \end{Bmatrix}\langle j_\pi j'_\nu J'M'|V|j_\pi j'_\nu J'M'\rangle$$

For a neutron (proton) j-hole and a j'-neutron (proton) (no protons (neutrons) in the j, j'-orbits)

$$\langle j^{2j}(j)j'JM|\sum_{i=1}^{2j} V(0,i)|j^{2j}(j)j'JM\rangle = \frac{1}{2j'+1}\sum_{|j-j'|}^{j+j'}(2J'+1)\langle jj'T=1,J'M'|V|jj'T=1,J'M'\rangle$$
$$-\sum_{J'}(2J'+1)\begin{Bmatrix} j & j' & J \\ j & j' & J' \end{Bmatrix}\langle jj'T=1,J'M'|V|jj'T=1,J'M'\rangle$$

For excitation of one nucleon from closed j-orbit ($T = 0$) into the empty j'-orbit

$$\langle j^{2(2j+1)-1}(j)j'TJM|\sum_{j=1}^{4j+1} V(0,i)|j^{2(2j+1)-1}(j)j'TJM\rangle = \frac{1}{2(2j'+1)}\sum_{T'J'}(2J'+1)(2T'+1)V_{T'J'}(jj'jj')$$
$$-\sum_{T'J'}(2J'+1)(2T'+1)\begin{Bmatrix} j & j' & J \\ j & j' & J' \end{Bmatrix}\begin{Bmatrix} \frac{1}{2} & \frac{1}{2} & T \\ \frac{1}{2} & \frac{1}{2} & T' \end{Bmatrix}V_{T'J'}(jj'jj')$$

The second term is equal for $T = 1$ to

$$-\sum_{J'}(2J'+1)\begin{Bmatrix} j & j' & J \\ j & j' & J' \end{Bmatrix}\tfrac{1}{2}[\langle jj'T=1,J'M'|V|jj'T=1,J'M'\rangle + \langle jj'T=0,J'M'|V|jj'T=0,J'M'\rangle]$$

For $T = 0$ it is equal to

$$-\sum_{J'}(2J'+1)\begin{Bmatrix} j & j' & J \\ j & j' & J' \end{Bmatrix}[\tfrac{3}{2}\langle jj'T=1,J'M'|V|jj'T=1,J'M'\rangle - \tfrac{1}{2}\langle jj'T=0,J'M'|V|jj'T=0,J'M'\rangle]$$

COEFFICIENTS OF FRACTIONAL PARENTAGE (C.F.P.)

Definition (identical nucleons)

$$\psi(j^n\alpha JM) = \sum_{\alpha_1 J_1}[j^{n-1}(\alpha_1 J_1)jJ|\}j^n\alpha J]\psi(j^{n-1}(\alpha_1 J_1)j_n JM)$$

For $n = 3$

$$[j^2(J_1)jJ|\}j^3[J_0]J] = \left[\delta_{J_1J_0} + 2\sqrt{(2J_0+1)(2J_1+1)}\begin{Bmatrix} j & j & J_0 \\ J & j & J_1 \end{Bmatrix}\right]\left[3 + 6(2J_0+1)\begin{Bmatrix} j & j & J_0 \\ J & j & J_0 \end{Bmatrix}\right]^{-1/2}$$

For $J_0 = 0$

$$[j^2(0)jJ = j|\}j^3[J_0 = 0]J = j] = \sqrt{\frac{2j-1}{3(2j+1)}}$$

$$[j^2(J_1)jJ = j|\}j^3[J_0 = 0]J = j] = -\frac{2\sqrt{2J_1+1}}{\sqrt{3(2j+1)(2j-1)}} \qquad J_1 > 0 \text{ even}$$

Applications

$$(j^n\alpha J\|\sum_{i=1}^{n}\mathbf{f}^{(k)}(i)\|j^n\alpha' J') = n\sum_{\alpha_1 J_1}[j^{n-1}(\alpha_1 J_1)jJ|\}j^n\alpha J][j^{n-1}(\alpha_1 J_1)jJ'|\}j^n\alpha' J']$$
$$\times(-1)^{J_1+j+J+k}\sqrt{(2J+1)(2J'+1)}\begin{Bmatrix} j & J & J_1 \\ J' & j & k \end{Bmatrix}(j\|\mathbf{f}^{(k)}\|j)$$

$$\langle j^n\alpha JM|\sum_{i<k}^{n}\mathbf{g}(i,k)|j^n\alpha' JM\rangle = \frac{n}{n-2}\sum_{\alpha_1\alpha_1' J_1}[j^{n-1}(\alpha_1 J_1)jJ|\}j^n\alpha J][j^{n-1}(\alpha_1' J_1)jJ|\}j^n\alpha' J]$$
$$\times\langle j^{n-1}\alpha_1 J_1 M_1|\sum_{i<k}^{n-1}\mathbf{g}(i,k)|j^{n-1}\alpha_1' J_1 M_1\rangle$$

Recursion relation

$$n[j^{n-1}(\alpha_0 J_0)jJ|\}j^{n-1}[\alpha_0 J_0]J][j^{n-1}(\alpha_1 J_1)jJ|\}j^n[\alpha_0 J_0]J]$$
$$= \delta_{\alpha_1\alpha_0}\delta_{J_1J_0} + (n-1)\sum_{\alpha_2 J_2}(-1)^{J_0+J_1}\sqrt{(2J_0+1)(2J_1+1)}\begin{Bmatrix} J_2 & j & J_1 \\ J & j & J_0 \end{Bmatrix}$$
$$\times [j^{n-2}(\alpha_2 J_2)jJ_0|\}j^{n-1}\alpha_0 J_0][j^{n-2}(\alpha_2 J_2)jJ_1|\}j^{n-1}\alpha_1 J_1]$$

$n \to n-2$ coefficients

$$\psi(j^n\alpha JM) = \sum_{\alpha_2 J_2 J'}[j^{n-2}(\alpha_2 J_2)j^2(J')J|\}j^n\alpha J]\psi(j^{n-2}(\alpha_2 J_2)j^2_{n-1,n}(J')JM)$$

Application

$$\langle j^n\alpha JM|\sum_{i<k}\mathbf{g}(i,k)|j^n\alpha' JM\rangle = \frac{n(n-1)}{2}\sum_{\alpha_2 J_2 J'}[j^{n-2}(\alpha_2 J_2)j^2(J')J|\}j^n\alpha J][j^{n-2}(\alpha_2 J_2)j^2(J')J|\}j^n\alpha' J]$$
$$\times \langle j^2 J'M'|\mathbf{g}(n-1,n)|j^2 J'M'\rangle$$

Expression in terms of $n \to n-1$ c.f.p.

$$[j^{n-2}(\alpha_2 J_2)j^2(J')J|\}j^n\alpha J] = \sum_{\alpha_1 J_1}[j^{n-2}(\alpha_2 J_2)jJ_1|\}j^{n-1}\alpha_1 J_1][j^{n-1}(\alpha_1 J_1)jJ|\}j^n\alpha J]$$
$$\times (-1)^{J_2+J+2j}\sqrt{(2J_1+1)(2J'+1)}\begin{Bmatrix} J_2 & j & J_1 \\ j & J & J' \end{Bmatrix}$$

$n \to n-1$ c.f.p. with isospin

$$\psi(j^n \alpha T M_T J M) = \sum_{\alpha_1 T_1 J_1} [j^{n-1}(\alpha_1 T_1 J_1) j T J | \} j^n \alpha T J] \psi(j^{n-1}(\alpha_1 T_1 J_1) j_n T M_T J M)$$

For $n = 3$

$$\begin{aligned} &[j^2(T_1 J_1) j T J | \} j^3 [T_0 J_0] T J] \\ &\quad = \left[\delta_{T_1 T_0} \delta_{J_1 J_0} - 2\sqrt{(2T_0+1)(2T_1+1)(2J_0+1)(2J_1+1)} \begin{Bmatrix} \frac{1}{2} & \frac{1}{2} & T_1 \\ T & \frac{1}{2} & T_0 \end{Bmatrix} \begin{Bmatrix} j & j & J_1 \\ J & j & J_0 \end{Bmatrix} \right] \\ &\quad \times \left[3 - 6(2T_0+1)(2J_0+1) \begin{Bmatrix} \frac{1}{2} & \frac{1}{2} & T_0 \\ T & \frac{1}{2} & T_0 \end{Bmatrix} \begin{Bmatrix} j & j & J_0 \\ J & j & J_0 \end{Bmatrix} \right]^{-1/2} \end{aligned}$$

For $T_0 = 1,\ J_0 = 0$

$$[j^2(T_1 J_1) j T = \tfrac{1}{2}, J = j | \} j^3 [10] T = \tfrac{1}{2}, J = j] = \begin{cases} \sqrt{\dfrac{2(j+1)}{3(2j+1)}} & T_1 = 1,\ J_1 = 0 \\ \dfrac{\sqrt{2J_1+1}}{\sqrt{6(j+1)(2j+1)}} & T_1 = 1,\ J_1 > 0 \text{ even} \\ -\dfrac{\sqrt{2J_1+1}}{\sqrt{2(j+1)(2j+1)}} & T_1 = 0,\ J_1 \text{ odd} \end{cases}$$

Application

$$(j^n\alpha TJ\|\sum_i^n \mathbf{h}^{(k')}(i)\mathbf{f}^{(k)}(i)\|j^n\alpha'T'J') = n\sum_{\alpha_1T_1J_1}[j^{n-1}(\alpha_1T_1J_1)jTJ|\}j^n\alpha TJ][j^{n-1}(\alpha_1T_1J_1)jT'J'|\}j^n\alpha'T'J']$$
$$\times(-1)^{T_1+(1/2)+T+k'+J_1+j+J+k}\sqrt{(2T+1)(2T'+1)(2J+1)(2J'+1)}$$
$$\times\begin{Bmatrix}\frac{1}{2} & T & T_1\\ T' & \frac{1}{2} & k'\end{Bmatrix}\begin{Bmatrix}j & J & J_1\\ J' & j & k\end{Bmatrix}(\tfrac{1}{2}\|\mathbf{h}^{(k')}\|\tfrac{1}{2})(j\|\mathbf{f}^{(k)}\|j)$$

$n \to n-1$ c.f.p. in LS-coupling

$$\psi(l^n\alpha SLJM) = \sum_{\alpha_1S_1L_1}[l^{n-1}(\alpha_1S_1L_1)lSL|\}l^n\alpha SL]\psi(l^{n-1}(\alpha_1S_1L_1)l_nSLJM)$$

For $n = 3$

$$[l^2(S_1L_1)lSL|\}l^3[S_0L_0]SL] = \left[\delta_{S_1S_0}\delta_{L_1L_0} + 2\sqrt{(2S_0+1)(2S_1+1)(2L_0+1)(2L_1+1)}\begin{Bmatrix}\frac{1}{2} & \frac{1}{2} & S_1\\ S & \frac{1}{2} & S_0\end{Bmatrix}\begin{Bmatrix}l & l & L_1\\ L & l & L_0\end{Bmatrix}\right]$$
$$\times\left[3 + 6(2S_0+1)(2L_0+1)\begin{Bmatrix}\frac{1}{2} & \frac{1}{2} & S_0\\ S & \frac{1}{2} & S_0\end{Bmatrix}\begin{Bmatrix}l & l & L_0\\ L & l & L_0\end{Bmatrix}\right]^{-1/2}$$

IN THE FORMALISM OF SECOND QUANTIZATION

Single nucleon irreducible tensor operators

$$\mathbf{T}_{\kappa}^{(k)} = \sum_{jmj'm'} \langle jm|\mathbf{T}_{\kappa}^{(k)}|j'm'\rangle a_{jm}^{+} a_{j'm'}$$

$$= \sum_{jmj'm'} (-1)^{j-m} \begin{pmatrix} j & k & j' \\ -m & \kappa & m' \end{pmatrix} (j\|\mathbf{T}^{(k)}\|j') a_{jm}^{+} a_{j'm'}$$

$$= \frac{1}{\sqrt{2k+1}} \sum_{jj'} (j\|\mathbf{T}^{(k)}\|j') [a_{j}^{+} \times \tilde{a}_{j'}]_{\kappa}^{(k)}$$

$$\tilde{a}_{jm} = (-1)^{j+m} a_{j,-m}$$

The number operator

$$\hat{n} = \sum_{j} \hat{n}_j = \sum_{jm} a_{jm}^{+} a_{jm} = \sum_{jm} (-1)^{j-m} a_{jm}^{+} \tilde{a}_{j,-m}$$

$$[\hat{n}, a_{jm}^{+}] = a_{jm}^{+} \qquad [\hat{n}, a_{jm}] = -a_{jm}$$

Angular momentum

$$\mathbf{J} = \frac{1}{\sqrt{3}} \sum_j \sqrt{j(j+1)(2j+1)} [a_j^+ \times \tilde{a}_j]^{(1)}$$

$$[J^0, a_{jm}^+] = m a_{jm}^+ \qquad [J^\pm, a_{jm}^+] = \sqrt{j(j+1) - m(m \pm 1)}\, a_{jm\pm 1}^+$$

$$[J_0, \tilde{a}_{jm}] = m \tilde{a}_{jm} \qquad [J^\pm, \tilde{a}_{jm}] = \sqrt{j(j+1) - m(m \pm 1)}\, \tilde{a}_{jm\pm 1}$$

Creation operator of a normalized two nucleon state

$$A^+(j_1 j_2 J M) = (1 + \delta_{j_1 j_2})^{-\frac{1}{2}} (a_{j_1}^+ \times a_{j_2}^+)_M^{(J)}$$

$$[A^+(j_1 j_2 J M)]^\dagger = (1 + \delta_{j_1 j_2})^{-1/2} \sum (j_1 m_1 j_2 m_2 \mid j_1 j_2 J M) a_{j_2 m_2} a_{j_1 m_1} = A(j_1 j_2 J M)$$

Relation with c.f.p.

$$[j^{n-1}(\alpha_1 J_1) j J | \} j^n \alpha J] = \frac{(-1)^n}{\sqrt{n}} \frac{1}{\sqrt{2J+1}} (j^n \alpha J \| a_j^+ \| j^{n-1} \alpha_1 J_1)$$

Two-body interaction

$$\mathbf{G} = \frac{1}{4} \sum_{\substack{j_1 j_2 j_1' j_2' \\ JM}} (1 + \delta_{j_1 j_2})(1 + \delta_{j_1' j_2'}) V_J(j_1 j_2 j_1' j_2') A^+(j_1 j_2 JM) A(j_1' j_2' JM)$$

$$= \frac{1}{4} \sum_{j_1 j_2 j_1' j_2'} \sum_{m_1 m_2 m_1' m_2'} \langle j_1 m_1 j_2 m_2 | G | j_1' m_1' j_2' m_2' \rangle_a a_{j_1 m_1}^+ a_{j_2 m_2}^+ a_{j_2' m_2'} a_{j_1' m_1'}$$

$$= \sum_{j_1 j_2 j_1' j_2' k} [(1 + \delta_{j_1 j_2})(1 + \delta_{j_1' j_2'})]^{1/2} F_k(j_1 j_2 j_1' j_2')(a_{j_1}^+ \times \tilde{a}_{j_1'})^{(k)} \cdot (a_{j_2}^+ \times \tilde{a}_{j_2'})^{(k)} - \frac{1}{2} \sum_{jk} \frac{1}{2j+1} F_k(jjjj) \hat{n}_j$$

In the isospin formalism

$$T^0 = \sum_{jm\mu} \mu a_{jm\mu}^+ a_{jm\mu} = \frac{1}{2} \sum_{jm} (a_{jm1/2}^+ a_{jm1/2} - a_{jm,-1/2}^+ a_{jm,-1/2})$$

$$T^+ = \sum_{jm} a_{jm1/2}^+ a_{jm,-1/2} \qquad T^- = \sum_{jm_1} a_{jm,-1/2}^+ a_{jm1/2}$$

$$A^+(jj'TM_TJM) = (1 + \delta_{jj'})^{-1/2} \sum_{mm'\mu\mu'} (jmj'm' \mid jj'JM)(\tfrac{1}{2}\mu\tfrac{1}{2}\mu' \mid \tfrac{1}{2}\tfrac{1}{2}TM_T) a_{jm\mu}^+ a_{j'm'\mu'}^+$$

$$\mathbf{G} = \frac{1}{4} \sum_{\substack{j_1 j_2 j_1' j_2' \\ TM_TJM}} (1 + \delta_{j_1 j_2})(1 + \delta_{j_1' j_2'}) V_{TJ}(j_1 j_2 j_1' j_2') A^+(j_1 j_2 TM_TJM) A(j_1' j_2' TM_TJM)$$

THE SENIORITY SCHEME

Quasi-spin operators

$$S_j^+ = \sqrt{\frac{2j+1}{2}}A^+(j^2J=0, M=0) = \frac{1}{2}\sum_m (-1)^{j-m} a_{jm}^+ a_{j,-m}^+ \qquad S_j^- = (S_j^+)^\dagger$$

$$[S_j^+, S_j^-] = \sum_m a_{jm}^+ a_{jm} - \frac{2j+1}{2} = 2S_j^0$$

$$[S_j^0, S_j^+] = S_j^+ \qquad [S_j^0, S_j^-] = -S_j^-$$

Pairing interaction

$$P = 2S_j^+ S_j^- \qquad \langle j^2JM|P|j^2JM\rangle = (2j+1)\delta_{J0}\delta_{M0}$$

$$S_j^+ S_j^- = \mathbf{S}_j^2 - (S_j^0)^2 + S_j^0 = \mathbf{S}_j^2 - \frac{1}{4}\left(\frac{2j+1}{2} - n\right)\left(\frac{2j+5}{2} - n\right)$$

Eigenvalues

$$s(s+1) - \frac{1}{4}\left(\frac{2j+1}{2} - n\right)\left(\frac{2j+5}{2} - n\right)$$

Seniority v

$$S_j^- |j^v vJM\rangle = 0$$

$$s = \frac{1}{2}\left(\frac{2j+1}{2} - v\right)$$

$$P(n,v) = \frac{n-v}{2}(2j+3-n-v) = \frac{n-v}{2}(2j+2) - \frac{n(n-1)}{2} + \frac{v(v-1)}{2}$$

$$(S_j^+)^{n/2}|0\rangle \qquad v = 0,\ J = 0$$

$$(S_j^+)^{(n-2)/2} A^+(j^2JM)|0\rangle \qquad v = 2,\ J = 2,4,\ldots,2j-1$$

$$(S_j^+)^{(n-1)/2} a_{jm}^+|0\rangle \qquad v = 1,\ J = j$$

Normalized state

$$\mathcal{N}_{n,v}(S_j^+)^{(n-v)/2}|j^v v\alpha JM\rangle$$

$$\mathcal{N}_{n,v}^{-2} = \frac{\left(\frac{n-v}{2}\right)!\left(\frac{2j+1}{2} - v\right)!}{\left(\frac{2j+1-n-v}{2}\right)!}$$

Quasi-spin tensors

$$[S_j^0, T_\sigma^{(s)}] = \sigma T_\sigma^{(s)} \qquad [S_j^\pm, T_\sigma^{(s)}] = \sqrt{s(s+1) - \sigma(\sigma \pm 1)}\, T_{\sigma \pm 1}^{(s)}$$

$$[S_j^0, a_{jm}^+] = \tfrac{1}{2} a_{jm}^+ \qquad [S_j^+, a_{jm}^+] = 0 \qquad [S_j^-, a_{jm}^+] = (-1)^{j-m} a_{j,-m} = -\tilde{a}_{jm}$$

$$[S_j^0, -\tilde{a}_{jm}] = -\tfrac{1}{2}(-\tilde{a}_{jm}) \qquad [S_j^-, -\tilde{a}_{jm}] = 0 \qquad [S_j^+, -\tilde{a}_{jm}] = a_{jm}^+$$

Recursion relations of c.f.p.

$$[j^{n-1}(v-1, \alpha_1 J_1) jJ|\}j^n v\alpha J] = \sqrt{\frac{v(2j+3-n-v)}{n(2j+3-2v)}}[j^{v-1}(v-1, \alpha_1 J_1) jJ|\}j^v v\alpha J]$$

$$[j^{n-1}(v+1, \alpha_1 J_1) jJ|\}j^n v\alpha J] = \sqrt{\frac{(v+2)(n-v)}{2n}}[j^{v+1}(v+1, \alpha_1 J_1) jJ|\}j^{v+2} v\alpha J]$$

$$[j^{v+1}(v+1, \alpha_1 J_1) jJ|\}j^{v+2} v\alpha J] = (-1)^{J+j-J_1}\sqrt{\frac{2(2J_1+1)(v+1)}{(2J+1)(v+2)(2j+1-2v)}}[j^v(v\alpha J) jJ_1|\}j^{v+1} v+1, \alpha_1 J_1]$$

Special values

$$[j^{n-1}(v_1 = 1, J_1 = j)jJ = 0|\}j^n v = C, J = 0] = 1$$

$$[j^{n-1}(v_1 = 0, J_1 = 0)jJ = j|\}j^n v = 1, J = j] = \sqrt{\frac{2j+2-n}{n(2j+1)}}$$

$$[j^{n-1}(v_1 = 2, J_1)jJ = j|\}j^n v = 1, J = j] = -\sqrt{\frac{2(n-1)(2J_1+1)}{n(2j+1)(2j-1)}} \qquad J_1 > 0 \text{ even}$$

$$[j^{n-2}(vJ)j^2(0)J|\}j^n vJ]^2 = \frac{(n-v)(2j+3-n-v)}{n(n-1)(2j+1)}$$

Special orthogonality relations

$$\sum_{\alpha J}(2J+1)[j^{n-1}(v+1, \alpha_1 J_1)jJ|\}j^n v\alpha J][j^{n-1}(v+1, \alpha_1' J_1)jJ|\}j^n v\alpha J] = \frac{(n-v)(v+1)}{n(2j+1-2v)}(2J_1+1)\,\delta_{\alpha_1\alpha_1'}$$

$$\sum_{\alpha J}(2J+1)[j^{n-1}(v-1, \alpha_1 J_1)jJ|\}j^n v\alpha J][j^{n-1}(v-1, \alpha_1' J_1)jJ|\}j^n v\alpha J] = \frac{(2j+3-n-v)(2j+4-v)}{n(2j+5-2v)}(2J_1+1)\,\delta_{\alpha_1\alpha_1'}$$

$$\sum_{\alpha J}(2J+1)[j^{n-1}(v-1, \alpha_1 J_1)jJ|\}j^n v\alpha J][j^{n-1}(v+1, \alpha_1' J_1)jJ|\}j^n v\alpha J] = 0$$

Matrix elements of single nucleon operators

$$\langle j^n v\alpha JM|T_\kappa^{(k)}|j^n v'\alpha' J'M'\rangle = \langle j^v v\alpha JM|T_\kappa^{(k)}|j^v v\alpha' J'M'\rangle \delta_{vv'} \qquad k \text{ odd}$$

$$\langle j^n v\alpha JM|T_\kappa^{(k)}|j^n v\alpha' J'M'\rangle = \frac{2j+1-2n}{2j+1-2v}\langle j^v v\alpha JM|T_\kappa^{(k)}|j^v v\alpha' J'M\rangle \qquad k > 0 \text{ even}$$

$$\langle j^n v\alpha JM|T_\kappa^{(k)}|j^n v-2,\alpha' J'M'\rangle = \sqrt{\frac{(n-v+2)(2j+3-n-v)}{2(2j+3-2v)}}\langle j^v v\alpha JM|T_\kappa^{(k)}|j^n v-2,\alpha' J'M'\rangle \qquad k > 0 \text{ even}$$

Matrix elements of two-body interactions

$$\langle j^n v\alpha JM|V|j^n v-4,\alpha' JM\rangle = \sqrt{\frac{(n-v+2)(n-v+4)(2j+3-n-v)(2j+5-n-v)}{8(2j+3-2v)(2j+5-2v)}} \times \langle j^v v\alpha JM|V|j^v v-4,\alpha' JM\rangle$$

$$\langle j^n v\alpha JM|V|j^n v-2,\alpha' JM\rangle = \frac{2j+1-2n}{2j+1-2v}\sqrt{\frac{(n-v+2)(2j+3-n-v)}{2(2j+3-2v)}}\langle j^v v\alpha JM|V|j^v v-2,\alpha' JM\rangle$$

$$\langle j^n v\alpha JM|V|j^n v\alpha' JM\rangle = \langle j^v v\alpha JM|V|j^v v\alpha' JM\rangle + E_0\frac{n-v}{2}\delta_{\alpha\alpha'} + \frac{(n-v)(2j+1-n-v)}{2(2j+1-2v)} \times [\langle j^{v+2} v\alpha JM|V|j^{v+2} v\alpha' JM\rangle - \langle j^v v\alpha JM|V|j^v v\alpha' JM\rangle - E_0\delta_{\alpha\alpha'}]$$

$$E_0 = \frac{2}{2j+1}\sum_{J' \text{ even}}(2J'+1)V_{J'} = F_0 - \frac{1}{2j+1}\sum_k F_k$$

The pairing property of odd tensor interactions

$$\langle j^n v\alpha JM|V^{\text{odd}}|j^n v'\alpha' JM\rangle = \left[\langle j^v v\alpha JM|V|j^v v\alpha' JM\rangle + \frac{n-v}{2}V_0\delta_{\alpha\alpha'}\right]\delta_{vv'}$$

Any interaction with single nucleon and two-body terms which is diagonal in the seniority scheme is equal to

$$V = V^{(s=0)} + xS_j^0 + y(S_j^0)^2 = V^{(s=0)} + \frac{n(n-1)}{2}\alpha + \frac{n}{2}\beta$$

$$\langle j^n v\alpha JM|V|j^n v\alpha JM\rangle = \langle j^v v\alpha JM|V^{(s=0)}|j^v v\alpha JM\rangle + \frac{n(n-1)}{2}\alpha + \frac{n}{2}\beta$$

$$= \langle j^v v\alpha JM|V|j^v v,\alpha JM\rangle + \frac{n(n-1)}{2}\alpha + \frac{n}{2}\beta - \frac{v(v-1)}{2}\alpha - \frac{v}{2}\beta$$

Binding energies

$$\text{B.E.}(j^n g.s.) = \text{B.E.}(n=0) + n\epsilon_j + \frac{n(n-1)}{2}\alpha + \left[\frac{n}{2}\right]\beta$$

$$\left[\frac{n}{2}\right] = \frac{1}{2}(n - \tfrac{1}{2}(1-(-1)^n)$$

$$\alpha = \frac{2(j+1)\bar{V}_2 - V_0}{2j+1} \qquad \beta = \frac{2(j+1)}{2j+1}(V_0 - \bar{V}_2)$$

$$\bar{V}_2 = \sum_{J>0\,\text{even}}(2J+1)V_J \Big/ \sum_{J>0\,\text{even}}(2J+1) = \frac{1}{(j+1)(2j+1)}\sum_{J>0\,\text{even}}(2J+1)V_J$$

Separation energies

$$\text{B.E.}(j^n g.s.) - \text{B.E.}(j^{n-1} g.s.) = \epsilon_j + (n-1)\alpha + \frac{1+(-1)^n}{2}\beta$$

Necessary and sufficient (but not independent) conditions for a two-body interaction to be diagonal in the seniority scheme are

$$V_J - \bar{V}_2 + 2 \sum_{k>0\,\text{even}} (2k+1) \begin{Bmatrix} j & j & k \\ j & j & J \end{Bmatrix} (V_k - \bar{V}_2) = 0 \qquad J > 0 \text{ even}$$

Matrix element between the $v = 1$, $J = j$ state and the $v = 3$, $J = j$ state of the j^3 configuration obtained from J_0

$$-2 \left[V_{J_0} - \bar{V}_2 + 2 \sum_{J_1>0\,\text{even}} (2J_1+1) \begin{Bmatrix} j & j & J_1 \\ j & j & J_0 \end{Bmatrix} (V_{J_1} - \bar{V}_2) \right] \sqrt{\frac{2J_0+1}{(2j+1)(2j-1)}}$$

$$\times \left[1 + 2(2J_0+1) \begin{Bmatrix} j & j & J_0 \\ j & j & J_0 \end{Bmatrix} - \frac{4(2J_0+1)}{(2j+1)(2j-1)} \right]^{-1/2}$$

GENERALIZED SENIORITY

The quasi-spin scheme

$$\mathcal{S}^+ = \sum_j S_j^+ \qquad \mathcal{S}^- = \sum_j S_j^-$$

$$[\mathcal{S}^+, \mathcal{S}^-] = \sum_{jm} a_{jm}^+ a_{jm} - \frac{1}{2}\sum_j (2j+1) = \sum_j \hat{n}_j - \sum_j \Omega_j = \hat{n} - \Omega = 2\mathcal{S}^0$$

$$[\mathcal{S}^0, \mathcal{S}^+] = \mathcal{S}^+ \qquad [\mathcal{S}^0, \mathcal{S}^-] = -\mathcal{S}^-$$

The pairing interaction

$$2\mathcal{S}^+\mathcal{S}^- = 2\mathcal{S}^2 - 2\mathcal{S}^0(\mathcal{S}^0 - 1)$$

Eigenvalues

$$2s(s+1) - \tfrac{1}{2}(\Omega - n)(\Omega + 2 - n) = \frac{n - v}{2}(2\Omega + 2 - n - v)$$

$$T_\kappa^{(k)} = \sum_{j,j'} \frac{(j\|\mathbf{T}^{(k)}\|j')}{\sqrt{2k+1)}}(a_j^+ \times \tilde{a}_{j'})_\kappa^{(k)}$$

$$[\mathcal{S}^+, T_\kappa^{(k)}] = 0 \qquad k \text{ odd for hermitean } \mathbf{T}^{(k)}$$

$$[\mathcal{S}^+, T_\kappa^{(k)}] = 0 \qquad k \text{ even for antihermitean } \mathbf{T}^{(k)}$$

For SDI

$$\mathcal{S}_1^+ = \sum_j (-1)^{l_j} S_j^+ \qquad \mathcal{S}_1^- = \sum_j (-1)^{l_j} S_j^- \qquad \mathcal{S}_1^0 = \mathcal{S}^0 = \sum_j S_j^0$$

$$\left[\mathcal{S}_1^+, \sum_{jj'} (\tfrac{1}{2} l j \| [\mathbf{s} \times \mathbf{Y}_k]^{(r)} \| \tfrac{1}{2} l' j')(a_j^+ \times \tilde{a}_{j'})_\rho^{(r)} \right] = 0 \qquad \text{for any } k, r$$

$$H_{SDI} = H^{(s=0)} + C_0 \Omega \sum_j \left(S_j^0 + \frac{2j+1}{4} \right) = H^{(s=0)} + \tfrac{1}{2} C_0 \Omega \hat{n}$$

$$H_{SDI} \mathcal{S}_1^+ |0\rangle = C_0 \Omega \mathcal{S}_1^+ |0\rangle$$

$$H_{SDI} (\mathcal{S}_1^+)^N |0\rangle = N C_0 \Omega (\mathcal{S}_1^+)^N |0\rangle = \tfrac{1}{2} n C_0 \Omega (\mathcal{S}_1^+)^N |0\rangle$$

$$H_{SDI} (\mathcal{S}_1^+)^N a_{jm}^+ |0\rangle = N C_0 \Omega (\mathcal{S}_1^+)^N a_{jm} |0\rangle = \tfrac{1}{2} (n-1) C_0 \Omega (\mathcal{S}_1^+)^N a_{jm}^+ |0\rangle$$

Generalized seniority, ground states

$$(S^+)^N |0\rangle \qquad S^+ = \sum \alpha_j S_j^+$$

$$[H, (S^+)^N] = \sum_{\nu=1}^{N} \binom{N}{\nu} (S^+)^{N-\nu} [\cdots [H, S^+], \ldots, S^+], S^+]$$

If

$$HS^+|0\rangle = [H,S^+]|0\rangle = V_0S^+|0\rangle \qquad H|0\rangle = 0$$

and

$$[[H,S^+],S^+] = W(S^+)^2$$

then for any N

$$H(S^+)^N|0\rangle = E_N(S^+)^N|0\rangle = \left(NV_0 + \frac{N(N-1)}{2}W\right)(S^+)^N|0\rangle$$

Pair separation energies

$$\text{B.E.}(N) - \text{B.E.}(N-1) = V_0 + (N-1)W$$

Conditions on Hamiltonians with $(S^+)^N|0\rangle$ eigenstates

$$2(\alpha_j^2 + \alpha_{j'}^2)\frac{V(j^2j'^2J=0)}{\sqrt{(2j+1)(2j'+1)}} = \alpha_j\alpha_{j'}\frac{4}{(2j+1)(2j'+1)}\sum_{J'}(2J'+1)V(jj'jj'J') - \alpha_j\alpha_{j'}W = \alpha_j\alpha_{j'}(4\bar{V}_{jj'} - W)$$

$$(\alpha_j^2 + \alpha_{j'}^2)V(j^2 j'^2 J) - 2\alpha_j\alpha_{j'} \sum_{J'} (2J'+1)(-1)^{j+j'-J'} \begin{Bmatrix} j & j' & J' \\ j' & j & J \end{Bmatrix} V(jj'jj'J') = 0 \qquad J > 0 \text{ even}$$

$$\alpha_j(\alpha_j + \alpha_{j'}) \left[V(j^2 jj'J) + 2 \sum_{J' \text{even}} (2J'+1) \begin{Bmatrix} j & j & J' \\ j & j' & J \end{Bmatrix} V(j^2 jj'J') \right] = 0 \qquad J \text{ even}$$

$$(\alpha_{j_1}^2 + \alpha_{j_2}\alpha_{j_3})V(j_1 j_1 j_2 j_3 J) + \sqrt{2}\alpha_{j_1}(\alpha_{j_2} + \alpha_{j_3}) \sum_{J'} (2J'+1) \begin{Bmatrix} j_1 & j_1 & J \\ j_2 & j_3 & J' \end{Bmatrix} V(j_1 j_3 j_2 j_1 J') = 0$$

$$(\alpha_{j_1}\alpha_{j_2} + \alpha_{j_3}\alpha_{j_4})V(j_1 j_2 j_3 j_4 J) + (\alpha_{j_1}\alpha_{j_4} + \alpha_{j_2}\alpha_{j_3}) \sum_{J'} (2J'+1) \begin{Bmatrix} j_1 & j_2 & J \\ j_3 & j_4 & J' \end{Bmatrix} V(j_1 j_4 j_3 j_2 J')$$

$$- (-1)^{j_3+j_4-J}(\alpha_{j_1}\alpha_{j_3} + \alpha_{j_2}\alpha_{j_4}) \sum_{J'} (2J'+1) \begin{Bmatrix} j_1 & j_2 & J \\ j_4 & j_3 & J' \end{Bmatrix} V(j_1 j_3 j_4 j_2 J') = 0$$

First excited states with $J = 2$

$$D_M^+ = \sum_{j \le j'} \beta_{jj'} A^+(jj'J = 2, M) = \sum_{j \le j'} \beta_{jj'}(1 + \delta_{jj'})^{-1/2} \sum_{mm'} (jmj'm' \mid jj'2, M) a_{jm}^+ a_{j'm'}^+$$

If

$$HD_M^+|0\rangle = V_2 D_M^+|0\rangle$$

and

$$[[H,S^+],D_M^+] = WS^+D_M^+$$

then for any N

$$H(S^+)^{N-1}D_M^+|0\rangle = (N-1)V_0 + V_2 + \tfrac{1}{2}N(N-1)W)(S^+)^{N-1}D_M^+|0\rangle = (E_N + V_2 - V_0)(S^+)^{N-1}D_M^+|0\rangle$$

Single nucleon quadrupole operator $\mathbf{Q}$ defined by

$$[Q_M, S^+] = D_M^+$$

$$Q_M(S^+)^N|0\rangle = N(S^+)^{N-1}[Q_M,S^+]|0\rangle = N(S^+)^{N-1}D_M^+|0\rangle$$

In the quasi-spin scheme

$$Q_M = \tfrac{1}{2}[D_M, S^-]$$

In general,

$$Q_M = \sum_{jj'} \gamma_{jj'}(a_j^+ \times \tilde{a}_{j'})_M^{(2)} \qquad \gamma_{j'j} = (-1)^{j-j'}\gamma_{jj'}$$

$$\gamma_{jj}\alpha_j = \frac{1}{\sqrt{2}}\beta_{jj} \qquad \gamma_{jj'}(\alpha_j + \alpha_{j'}) = \beta_{jj'}$$

For hole states

$$S' = \sum_j \frac{1}{\alpha_j} S_j^-$$

$$S'(S^+)^\Omega|0\rangle = \Omega(S^+)^{\Omega-1}|0\rangle$$

$$\tilde{D}'_M = [Q_M, S'] = \sum_{j\le j'} \beta'_{jj'}(1+\delta_{jj'})^{-1/2}(\tilde{a}_j \times \tilde{a}_{j'})^{(2)}_M$$

$$\beta'_{jj} = \sqrt{2}\frac{\gamma_{jj}}{\alpha_j} = \frac{\beta_{jj}}{\alpha_j^2}$$

$$\beta'_{jj'} = \gamma_{jj'}\left(\frac{1}{\alpha_j} + \frac{1}{\alpha_{j'}}\right) = \frac{\gamma_{jj'}(\alpha_j + \alpha_{j'})}{\alpha_j \alpha_{j'}} = \frac{\beta_{jj'}}{\alpha_j \alpha_{j'}}$$

$$\tilde{D}'_M (S^+)^N|0\rangle = N(N-1)(S^+)^{N-2} D^+_M|0\rangle$$

SENIORITY AND THE $Sp(2j+1)$ GROUP

Generators of $U(2j+1)$

$$[(a_j^\dagger \times \tilde{a}_j)^{(k')}_{\kappa'}, (a_j^\dagger \times \tilde{a}_{j'})^{(k'')}_{\kappa''}] = \sqrt{(2k'+1)(2k''+1)} \sum_{k\kappa} (-1)^{k+1}(1-(-1)^{k'+k''+k})$$
$$\times \begin{Bmatrix} k' & k'' & k \\ j & j & j \end{Bmatrix} (k'\kappa' k''\kappa'' \mid k'k''k\kappa)(a_j^\dagger \times \tilde{a}_j)^{(k)}_\kappa$$

Casmir operator of $U(2j+1)$

$$\sum_k (a_j^\dagger \times \tilde{a}_j)^{(k)} \cdot (a_j^\dagger \times \tilde{a}_j)^{(k)} = (2j+1)\hat{n} - \hat{n}(\hat{n}-1)$$

Generators of $Sp(2j+1)$

$$(a_j^\dagger \times \tilde{a}_j)^{(k)}_\kappa |j^2 J = 0, M = 0\rangle \qquad k \text{ odd}$$

Casimir operator

$$C_{Sp(2j+1)} = 2 \sum_{k\,\text{odd}} (a_j^\dagger \times \tilde{a}_j)^{(k)} \cdot (a_j^\dagger \times \tilde{a}_j)^{(k)}$$

Eigenvalues

$$\frac{v}{2}(2j+3-v)$$

Average interaction energies

$$\overline{\langle j^n JM | \sum_{i<k} V_{ik} | j^n JM \rangle} = \alpha \frac{n(n-1)}{2}$$

$$\alpha = \sum_{J \text{ even } M} \langle j^2 JM|V|j^2 JM\rangle \Big/ \sum_{J \text{ even}} (2J+1) = \frac{1}{j(2j+1)} \sum_{J \text{ even}} (2J+1)V_J$$

$$\overline{\langle j^n vJM|\sum_{i<k} V_{ik}|j^n vJM\rangle} = \frac{n(n-1)}{2}\bar{V}_2 + \frac{n-v}{2}(2j+3-n-v)\frac{V_0 - \bar{V}_2}{2j+1}$$

$$= \frac{n(n-1)}{2}\frac{(2j+2)\bar{V}_2 - V_0}{2j+1} + \frac{n-v}{2}\frac{2j+2}{2j+1}(V_0 - \bar{V}_2) + \frac{v(v-1)(V_0 - \bar{V}_2)}{2j+1}$$

$$\bar{V}_2 = \frac{1}{(j+1)(2j-1)} \sum_{J>0 \text{ even}} (2J+1)V_J$$

With isospin

generators of $U(2j+1)$

$$[U_{\kappa'}^{(k')}, U_{\kappa''}^{(k'')}] = \sum_{k\kappa} \sqrt{2k+1}(-1)^{k+1}(1-(-1)^{k'+k''+k}) \begin{Bmatrix} k' & k'' & k \\ j & j & j \end{Bmatrix} (k'\kappa' k''\kappa'' \mid k'k''k\kappa) U_\kappa^{(k)}$$

Casimir operator of $U(2j+1)$

$$\sum_k (2k+1)(\mathbf{U}^{(k)} \cdot \mathbf{U}^{(k)})$$

Eigenvalues

$$(2j+1)n - \frac{n(n-1)}{2} - 2[T(T+1) - \tfrac{3}{4}n]$$

Generators of $Sp(2j+1)$

$$U_\kappa^{(k)}|j^2 J=0, M=0\rangle = 0 \qquad k \text{ odd}$$

Casimir operator

$$C_{Sp(2j+1)} = 2\sum_{k\,\text{odd}} (2k+1)(\mathbf{U}^{(k)}\cdot\mathbf{U}^{(k)})$$

Eigenvalues

$$C_{Sp(2j+1)}(v,t) = \frac{-v(v-1)}{2} + 2v(j+1) - 2(t(t+1) - \tfrac{3}{4}v) = \frac{v}{2}(4j+8-v) - 2t(t+1)$$

Eigenvalues of the pairing interaction

$$P(n,T,v,t) = \tfrac{1}{2}[C_{Sp(2j+1)}(n,T) - C_{Sp(2j+1)}(v,t)] = \frac{n-v}{4}(4j+8-n-v) - T(T+1) + t(t+1)$$

C.f.p. of states with seniorities $v=0$, $t=0$

$$[j^{n-1}(v_1=1, t_1=\tfrac{1}{2}, T_1=T+\tfrac{1}{2}, J=j)jTJ=0|\}j^n v=0, t=0, T, J=0]^2 = \frac{(n-2T)(T+1)}{n(2T+1)}$$

$$[j^{n-1}(v_1=1, t_1=\tfrac{1}{2}, T_1=T-\tfrac{1}{2}, J=j)jTJ=0|\}j^n v=0, t=0, T, J=0]^2 = \frac{(n+2T+2)T}{n(2T+1)}$$

for $T = n/2, (n/2)-2, (n/2)-4, \ldots, 1$ or 0.

C.f.p. with seniorities $v = 1$, $t = \frac{1}{2}$

Case I $\quad T = n/2, (n/2) - 2, (n/2) - 4, \dots$

$$[j^{n-1}(v_1 = 0, t_1 = 0, T_1 = T - \tfrac{1}{2}, J_1 = 0) jTJ = j|\} j^n v = 1, t = \tfrac{1}{2}, T, J = j]^2 = \frac{4(j+1) - n - 2T}{2n(2j+1)}$$

$$[j^{n-1}(v_1 = 2, t_1 = 0, T_1 = T + \tfrac{1}{2}, J_1) jTJ = j|\} j^n v = 1, t = \tfrac{1}{2}, T, J = j]^2$$

$$= \frac{(n - 2T)}{4n(j+1)(2j+1)}(2J_1 + 1) \qquad J_1 \text{ odd}$$

$$[j^{n-1}(v_1 = 2, t_1 = 1, T_1 = T + \tfrac{1}{2}, J_1) jTJ = j|\} j^n v = 1, t = \tfrac{1}{2}, T, J = j]^2$$

$$= \frac{(n - 2T)(T + 3)}{4n(j+1)(2j-1)(2T+1)}(2J_1 + 1) \qquad J_1 > 0 \text{ even}$$

$$[j^{n-1}(v_1 = 2, t_1 = 1, T_1 = T - \tfrac{1}{2}, J_1) jTJ = j|\} j^n v = 1, t = \tfrac{1}{2}, T, J = j]^2$$

$$= \frac{4(j+1)(2T^2 + nT - 1) + n - 2T}{2n(j+1)(2j-1)(2j+1)(2T+1)}(2J_1 + 1) \qquad J_1 > 0 \text{ even}$$

Case II $T = (n/2)-1, (n/2)-3, (n/2)-5, \ldots$

$$[j^{n-1}(v_1 = 0, t_1 = 0, T_1 = T+\tfrac{1}{2}, J_1 = 0)jTJ = j|\}j^n v = 1, t = \tfrac{1}{2}, T, J = j]^2 = \frac{4j+6-n+2T}{2n(2j+1)}$$

$$[j^{n-1}(v_1 = 2, t_1 = 0, T_1 = T-\tfrac{1}{2}, J_1)jTJ = j|\}j^n v = 1, t = \tfrac{1}{2}, T, J = j]^2$$

$$= \frac{n+2T+2}{4n(j+1)(2j+1)}(2J_1+1) \qquad J_1 \text{ odd}$$

$$[j^{n-1}(v_1 = 2, t_1 = 1, T_1 = T+\tfrac{1}{2}, J_1)jTJ = j|\}j^n v = 1, t = \tfrac{1}{2}, T, J = j]^2$$

$$= \frac{4(j+1)[(T+1)(n-2T-2)+1]-(n+2T+2)}{2n(j+1)(2j-1)(2j+1)(2T+1)}(2J_1+1) \qquad J_1 > 0 \text{ even}$$

$$[j^{n-1}(v_1 = 2, t_1 = 1, T_1 = T-\tfrac{1}{2}, J_1)jTJ = j|\}j^n v = 1, t = \tfrac{1}{2}, T, J = j]^2$$

$$= \frac{(n+2T+2)(2T-1)}{4n(j+1)(2j-1)(2T+1)}(2J_1+1) \qquad J_1 > 0 \text{ even}$$

Magnetic moments of odd proton nuclei (Z' protons and N' neutrons in the j-orbit)

Case I $Z' > N'$

$$\mu = \mu_\pi \left(1 - \frac{N'}{4(j+1)(T+1)}\right) + \mu_\nu \frac{N'}{4(j+1)(T+1)}$$

Case II $Z' < N'$

$$\mu = \mu_\pi \left(1 - \frac{2j+1-N'}{4(j+1)(T+1)}\right) + \mu_\nu \frac{2j+1-N'}{4(j+1)(T+1)}$$

Average interaction energies

$$\overline{\langle j^n T M_T J M | \sum_{i<k} V_{ik} | j^n T M_T J M \rangle} = \alpha \frac{n(n-1)}{2} + \beta[T(T+1) - \tfrac{3}{4}n]$$

$$\alpha = \frac{3}{4j(2j+1)} \sum_{J\,\text{even}} (2J+1)V_J + \frac{1}{4(j+1)(2j+1)} \sum_{J\,\text{odd}} (2J+1)V_J$$

$$\beta = \frac{1}{2j(2j+1)} \sum_{J\,\text{even}} (2J+1)V_J - \frac{1}{2(j+1)(2j+1)} \sum_{J\,\text{odd}} (2J+1)V_J$$

$$\overline{\langle j^n v t T M_T J M | \sum_{i<k} V_{ik} | j^n v t T M_T J M \rangle}$$

$$= \alpha \frac{n(n-1)}{2} + \beta[T(T+1) - \tfrac{3}{4}n] + \gamma \frac{n-v}{2} + \frac{\gamma}{2j+2}\left[\frac{v(v-1)}{4} + t(t+1) - \tfrac{3}{4}v\right]$$

$$\alpha = \frac{1}{4(2j+1)}[(6j+5)\bar{V}_2 + (2j+1)\bar{V}_1 - 2V_0]$$

$$\beta = \frac{1}{2(2j+1)}[(2j+3)\bar{V}_2 - (2j+1)\bar{V}_1 - 2V_0]$$

$$\gamma = \frac{2(j+1)}{2j+1}(V_0 - \bar{V}_2)$$

$$\bar{V}_2 = \frac{1}{(j+1)(2j-1)} \sum_{J>0\,\text{even}} (2J+1)V_J \qquad \bar{V}_1 = \frac{1}{(j+1)(2j+1)} \sum_{J\,\text{odd}} (2J+1)V_J$$

Seniority scheme mass formula

$$\text{B.E.}(j^n g.s.) = \text{B.E.}(n=0) + nC + \frac{n(n-1)}{2}\alpha + \beta[T(T+1) - \tfrac{3}{4}n]$$
$$+ \gamma \left[\frac{n}{2}\right] + \text{Coulomb energy}$$

SENIORITY IN l^n CONFIGURATIONS

Generators of the $U(2l+1)$ group

$$[U^{(k')}_{\kappa'}, U^{(k'')}_{\kappa''}] = \sum_{k\kappa} (-1)^k (1 - (-1)^{k'+k''+k}) \begin{Bmatrix} k' & k'' & k \\ l & l & l \end{Bmatrix} (k'\kappa' k''\kappa'' \mid k'k''k\kappa) U^{(k)}_{\kappa}$$

Casimir operator of $U(2l+1)$

$$\sum_k (2k+1)(\mathbf{U}^{(k)} \cdot \mathbf{U}^{(k)})$$

Eigenvalues

$$(2l+1)n - \frac{n(n-1)}{2} - 2[S(S+1) - \tfrac{3}{4}n]$$

Generators of $O(2l+1)$

$$U_\kappa^{(k)}|l^2 L=0, M=0\rangle = 0 \qquad k \text{ odd}$$

Casimir operator

$$C_{O(2l+1)} = 2\sum_{k \text{ odd}}(2k+1)(\mathbf{U}^{(k)}\cdot\mathbf{U}^{(k)})$$

Eigenvalues

$$C_{O(2l+1)}(v,S) = \frac{v}{2}(4l+4-v) - 2S(S+1)$$

Pairing interaction

$$\langle l^2 LM|V|l^2 LM\rangle = (2l+1)\,\delta_{L0}\,\delta_{M0}$$

Eigenvalues

$$P(n,v) = \tfrac{1}{2}[C_{O(2l+1)}(n,S) - C_{O(2l+1)}(v,S)] = \frac{n-v}{4}(4l+4-n-v)$$

IN THE $SU(4)$ SCHEME

Eigenvalues of the Casimir operator $C_{SU(4)}$

$$4[P(P+4) + P'(P'+2) + P''^2]$$

Eigenvalues of the Majorana space exchange operator P^x

$$\frac{n(16-n)}{8} - \frac{1}{2}[P(P+4) + P'(P'+2) + P''^2]$$

Margenau-Wigner magnetic moments of odd-even nuclei

$$\mu = \tfrac{1}{2}L + \tfrac{1}{2}g_s \qquad \text{for} \quad J = L + \tfrac{1}{2}$$

$$\mu = \frac{J}{J+1}(\tfrac{1}{2}(L+1) - \tfrac{1}{2}g_s) \qquad \text{for} \quad J + L - \tfrac{1}{2}$$

$g_s = g_s^{\pi}$ for odd proton nuclei and $g_s = g_s^{\nu}$ for odd neutron nuclei

Average interaction energies

$$\bar{V}(PP'P'') = a\frac{n(n-1)}{2} + bP^x$$

$$= \left(a - \frac{b}{4}\right)\frac{n(n-1)}{2} + \frac{15}{8}bn - \frac{1}{2}b[P(P+4) + P'(P'+2) + P''^2]$$

$$a = \tfrac{1}{2}(\bar{V}_s + \bar{V}_a) \qquad b = \tfrac{1}{2}(\bar{V}_s - \bar{V}_a)$$

Casimir operator of the $U(2l+1)$ group

$$\sum_k (2k+1)(\mathbf{U}^{(k)} \cdot \mathbf{U}^{(k)}) = 2\sum_{i<j}^{n} P_{ij}^x + (2l+1)\hat{n}$$

Generators of $O(2l+1)$

$$U_\kappa^{(k)}|l^2 L = 0, M = 0\rangle = 0 \qquad k \text{ odd}$$

Casimir operator

$$C_{O(2l+1)} = 2\sum_{k \text{ odd}} (2k+1)(\mathbf{U}^{(k)} \cdot \mathbf{U}^{(k)})$$

Interaction energy in ground states

$$\frac{n(n-1)}{2}a + b\left[\frac{n(16-n)}{8} - \frac{1}{2}T(T+4)\right] - nlc + \frac{1-(-1)^n}{2}\left[-\frac{3}{4}b + lc\right]$$

$$a = \tfrac{1}{2}(\bar{V}_s' + \bar{V}_a) \qquad b = \frac{1}{2(2l+1)}\left[(2l-1)\bar{V}_s' + V_0 - (2l+1)\bar{V}_a\right] \qquad c = \frac{1}{2l+1}(\bar{V}_s' - V_0)$$

THE $SU(3)$ SCHEME

Harmonic oscillator Hamiltonian

$$H_0 = \tfrac{1}{2}(\mathbf{p}^2 + \mathbf{r}^2) = \tfrac{1}{2}(\mathbf{p} + i\mathbf{r}) \cdot (\mathbf{p} - i\mathbf{r}) + \tfrac{3}{2}\hbar = \hbar(\mathbf{A}^+ \cdot \mathbf{A} + \tfrac{3}{2})$$

Generators of $SU(3)$

$$\mathbf{L} = \sum l_i = \frac{1}{\hbar}\sum_i (\mathbf{r}_i \times \mathbf{p}_i) \qquad Q_\mu^{(2)} = \sum_i q_i^{(2)} = \frac{\sqrt{6}}{2\hbar}\sum_i ((\mathbf{r}_i \times \mathbf{r}_i)_\mu^{(2)} + (\mathbf{p}_i \times \mathbf{p}_i)_\mu^{(2)})$$

Casimir operator

$$C_{SU(3)} = \tfrac{1}{4}(\mathbf{Q}^{(2)} \cdot \mathbf{Q}^{(2)}) + \tfrac{3}{4}(\mathbf{L} \cdot \mathbf{L})$$

Eigenvalues

$$C(\lambda, \mu) = \lambda^2 + \lambda\mu + \mu^2 + 3(\lambda + \mu)$$

Dynamical symmetry

$$H = H_0 + \tfrac{1}{2}a(\mathbf{Q}^{(2)} \cdot \mathbf{Q}^{(2)}) + \tfrac{1}{2}b(\mathbf{L} \cdot \mathbf{L}) = H_0 + 2aC_{SU(3)} + \tfrac{1}{2}(b - 3a)\mathbf{L}^2$$

Eigenvalues

$$n(N + \tfrac{3}{2}\hbar) + 2a(\lambda^2 + \lambda\mu + \mu^2 + 3(\lambda + \mu)) + \tfrac{1}{2}(b - 3a)L(L + 1)$$

L states in $(\lambda\mu)$ irreducible representation $(\lambda \geq \mu)$

$$K = \mu, \mu - 2, \ldots, 1 \text{ or } 0$$

For $K \neq 0$ $\quad L = K, K + 1, K + 2, \ldots, K + \lambda$

For $K = 0$ $\quad L = \lambda, \lambda - 2, \lambda - 4, \ldots, 1 \text{ or } 0$

PROTONS AND NEUTRONS IN DIFFERENT ORBITS

Single nucleon operators

$$\left([j^{n-1}(\alpha_1 T_1 J_1)j' M'_T J']_a \| \sum_{i=1}^{n} h(i)\mathbf{f}^{(k)} \| j^n \alpha T M_T J\right) = \sqrt{n}[j^{n-1}(\alpha_1 T_1 J_1) jTJ|\} j^n \alpha TJ] \langle T_1 \tfrac{1}{2} T' M'_T | h(n) | T_1 \tfrac{1}{2} T M_T \rangle$$

$$\times (-1)^{J_1+j+J'+k} \sqrt{(2J+1)(2J'+1)} \begin{Bmatrix} j' & J' & J_1 \\ J & j & k \end{Bmatrix} (j' \| \mathbf{f}^{(k)} \| j)$$

$$\left([j_1^{n_1}(\alpha_1 T_1 J_1) j_2^{n_2}(\alpha_2 T_2 J_2) TJ]_a \left\| \sum_{i=1}^{n_1+n_2} \mathbf{h}^{(k')}(i)\mathbf{f}^{(k)}(i) \right\| [j_1^{n_1-1}(\alpha'_1 T'_1 J'_1) j_2^{n_2+1}(\alpha'_2 T'_2 J'_2) T' J']_a\right)$$

$$= (-1)^{n_2} \sqrt{n_1(n_2+1)} [j_1^{n_1-1}(\alpha'_1 T'_1 J'_1) j_1 T_1 J_1 |\} j_1^{n_1} \alpha_1 T_1 J_1]$$

$$\times [j_2^{n_2}(\alpha_2 T_2 J_2) j_2 T'_2 J'_2 |\} j_2^{n_2+1} \alpha'_2 T'_2 J'_2] \times (-1)^{T_2 - T'_2 + (1/2) + J_2 - J'_2 + j_2}$$

$$\times \sqrt{(2T_1+1)(2T'_2+1)(2T+1)(2T'+1)(2J_1+1)(2J'_2+1)(2J+1)(2J'+1)}$$

$$\times \begin{Bmatrix} T_1 & T_2 & T \\ T'_1 & T'_2 & T' \\ \tfrac{1}{2} & \tfrac{1}{2} & k' \end{Bmatrix} \begin{Bmatrix} J_1 & J_2 & J \\ J'_1 & J'_2 & J' \\ j_1 & j_2 & k \end{Bmatrix} (\tfrac{1}{2} \| \mathbf{h}^{(k')} \| \tfrac{1}{2})(j_1 \| \mathbf{f}^{(k)} \| j_2)$$

Special cases (maximum isospins)

$$\left([j_1^{n_1}(v_1 = 1, J_1 = j_1) j_2^{n_2}(v_2 = 0, J_2 = 0) J = j_1]_a \left\| \sum_{i+1}^{n_1+n_2} \mathbf{f}^{(k)}(i) \right\| [j_1^{n_1-1}(v_1' = 0, J_1' = 0) j_2^{n_2+1}(v_2' = 1, J_2' = j_2) J' = j_2]_a \right)$$

$$= \sqrt{n_1(n_2+1)} \sqrt{\frac{(2j_1+2-n_1)(2j_2+1-n_2)}{(2j_1+1)(2j_2+1)}} (j_1 \| \mathbf{f}^{(k)} \| j_2)$$

$$\left([j_1^{n_1}(v_1 = 0, J_0) j_2^{n_2}(v_2 = 1, J_2 = j_2)]_a J = j_2 \left\| \sum_{i}^{n_1+n_2} \mathbf{f}^{(k)}(i) \right\| [j_1^{n_1-1}(v_1' = 1, J_1' = j_1) j_2^{n_2+1}(v_2' = 0, J_2' = 0)]_a J' = j_1 \right)$$

$$= (-1)^{k+1} \frac{\sqrt{n_1(n_2+1)}}{\sqrt{(2j_1+1)(2j_2+1)}} (j_2 \| \mathbf{f}^{(k)} \| j_1)$$

Two-body interactions

$$\langle [j_1^{n_1}(\alpha_1 J_1) j_2^{n_2}(\alpha_2 J_2) JM]_a | \sum_{i=1}^{n_1} \sum_{h=n_1+1}^{n_1+n_2} V_{ih} | [j_1^{n_1}(\alpha_1 J_1) j_2^{n_2}(\alpha_2 J_2) JM]_a \rangle$$

$$= n_1 n_2 \sum_{\alpha_{11} J_{11} \alpha_{22} J_{22}} [j_1^{n_1-1}(\alpha_{11} J_{11}) j_1 J_1 |\} j_1^{n_1} \alpha_1 J_1]^2 [j_2^{n_2-1}(\alpha_{22} J_{22}) j_2 J_2 |\} j_2^{n_2} \alpha_2 J_2]^2$$

$$\times (2J_1+1)(2J_2+1) \sum_{J_1=|j_1-j_2|}^{j_1+j_2} (2J'+1) V_{J'}(j_1 j_2) \sum_{J_0} (2J_0+1) \begin{Bmatrix} J_{11} & j_1 & J_1 \\ J_{22} & j_2 & J_2 \\ J_0 & J' & J \end{Bmatrix}^2$$

Average interaction energy (center of mass theorem)

$$\frac{1}{(2J_1+1)(2J_2+1)}\sum_{J=|J_1-J_2|}^{J_1+J_2}(2J+1)\langle[j_1^{n_1}(\alpha_1 J_1)j_2^{n_2}(\alpha_2 J_2)JM]_a|\sum_{i=1}^{n_1}\sum_{h=n_1+1}^{n_1+n_2}V_{ih}|[j_1^{n_1}(\alpha_1 J_1)j_2^{n_2}(\alpha_2 J_2)JM]_a\rangle$$

$$= n_1 n_2 \frac{1}{(2j_1+1)(2j_2+1)}\sum_{J'=|j_1-j_2|}^{j_1+j_2}(2J'+1)V_{J'}(j_1 j_2)$$

$$= n_1 n_2 \bar{V}(j_1 j_2) = \frac{n_1 n_2}{\sqrt{(2j_1+1)(2j_2+1)}}F_0(j_1 j_2 j_1 j_2)$$

Interaction energy for $J_1 = 0$ or $J_2 = 0$

$$n_1 n_2 \bar{V}(j_1 j_2) = \frac{n_1 n_2}{\sqrt{(2j_1+1)(2j_2+1)}}F_0(j_1 j_2 j_1 j_2)$$

General matrix elements

$$
\begin{aligned}
&\langle[j_1^{n_1}(\alpha_1 J_1) j_2^{n_2}(\alpha_2 J_2) JM]_a | \sum_{i=1}^{n_1} \sum_{h=n_1+1}^{n_1+n_2} V_{ih} | [j_1^{n_1}(\alpha_1' J_1') j_2^{n_2}(\alpha_2' J_2') JM]_a \rangle \\
&= n_1 n_2 \sum_{\alpha_{11} J_{11} \alpha_{22} J_{22}} [j_1^{n_1-1}(\alpha_{11} J_{11}) j_1 J_1 |\} j_1^{n_1} \alpha_1 J_1][j_1^{n_1-1}(\alpha_{11} J_{11}) j_1 J_1' |\} j_1^{n_1} \alpha_1' J_1'] \\
&\quad \times [j_2^{n_2-1}(\alpha_{22} J_{22}) j_2 J_2 |\} j_2^{n_2} \alpha_2 J_2][j_2^{n_2-1}(\alpha_{22} J_{22}) j_2 J_2' |\} j_2^{n_2} \alpha_2' J_2'] \\
&\quad \times \sqrt{(2J_1+1)(2J_2+1)(2J_1'+1)(2J_2'+1)} \sum_{J'=|j_1-j_2|}^{j_1+j_2} (2J'+1) V_{J'}(j_1 j_2) \\
&\quad \times \sum_{J_0} (2J_0+1) \begin{Bmatrix} J_{11} & j_1 & J_1 \\ J_{22} & j_2 & J_2 \\ J_0 & J' & J \end{Bmatrix} \begin{Bmatrix} J_{11} & j_1 & J_1' \\ J_{22} & j_2 & J_2' \\ J_0 & J' & J \end{Bmatrix} \\
&= n_1 n_2 [j_1^{n_1-1}(\alpha_{11} J_{11}) j_1 J_1 |\} j_1^{n_1} \alpha_1 J_1][j_1^{n_1-1}(\alpha_{11} J_{11}) j_1 J_1' |\} j_1^{n_2-1} \alpha_1' J_1'] \\
&\quad \times [j_2^{n_2-1}(\alpha_{22} J_{22}) j_2 J_2 |\} j_2^{n_2} \alpha_2 J_2][j_2^{n_2-1}(\alpha_{22} J_{22}) j_2 J_2' |\} j_2^{n_2} \alpha_2' J_2'] \\
&\quad \times (-1)^{j_1+j_2+J_{11}+J_{22}+J_1+2J_2+J_1'+J} \sqrt{(2J_1+1)(2J_1'+1)(2J_2+1)(2J_2'+1)} \\
&\quad \times \sum_k F_k(j_1 j_2 j_1 j_2) \begin{Bmatrix} J_1 & J_2 & J \\ J_2' & J_1' & k \end{Bmatrix} \begin{Bmatrix} j_1 & J_1 & J_{11} \\ J_1' & j_1 & k \end{Bmatrix} \begin{Bmatrix} j_2 & J_2 & J_{22} \\ J_2' & j_2 & k \end{Bmatrix}
\end{aligned}
$$

Special case

$$\langle [j' j^n(\alpha_1 J_1) JM]_a | \sum_{h=2}^{n+1} V_{1h} | [j' j^n(\alpha_1' J_1') JM]_a \rangle$$

$$= n \sum_{\alpha_{11} J_{11}} [j^{n-1}(\alpha_{11} J_{11}) j J_1 | \} j^n \alpha_1 J_1][j^{n-1}(\alpha_{11} J_{11}) j J_1' | \} j^n \alpha_1' J_1'] \sqrt{(2J_1+1)(2J_1'+1)}$$

$$\times \sum_{J'=|j-j'|}^{j+j'} (2J'+1) V_{J'}(j'j) \begin{Bmatrix} J_{11} & j & J_1 \\ j' & J & J' \end{Bmatrix} \begin{Bmatrix} J_{11} & j & J_1' \\ j' & J & J' \end{Bmatrix}$$

Special case $j' = \frac{1}{2}$

$$\langle [j' = \tfrac{1}{2}, j^n(\alpha_0 J_0) JM]_a | \sum_{h=2}^{n+1} V_{1h} | [j' = \tfrac{1}{2}, j^n(\alpha_0' J_0') JM]_a \rangle = \delta_{\alpha_0 \alpha_0'} \delta_{J_0 J_0'} [na + b(J(J+1) - J_0(J_0+1) - \tfrac{3}{4})]$$

$$a = \frac{1}{2j+1} [(j+1) V_{J=j+1/2}(j\tfrac{1}{2}) + j V_{J=j-1/2}(j\tfrac{1}{2})]$$

$$b = \frac{1}{2j+1} [V_{J=j+1/2}(j\tfrac{1}{2}) - V_{J=j-1/2}(j\tfrac{1}{2})]$$

General case with isospin

$$
\begin{aligned}
&\langle [j_1^{n_1}(\alpha_1 T_1 J_1) j_2^{n_2}(\alpha_2 T_2 J_2) T M_T J M]_a | \sum_{i=1}^{n_1} \sum_{h=n_1+1}^{n_1+n_2} V_{ih} | [j_1^{n_1}(\alpha_1' T_1' J_1') J_2^{n_2}(\alpha_2' T_2' J_2') T M_T J M]_a \rangle \\
&= n_1 n_2 \sum_{\substack{\alpha_{11} T_{11} J_{11} \\ \alpha_{22} T_{22} J_{22}}} [j_1^{n_1-1}(\alpha_{11} T_{11} J_{11}) j_1 T_1 J_1 |\} j_1^{n_1} \alpha_1 T_1 J_1] \\
&\quad \times [j_1^{n_1-1}(\alpha_{11} T_{11} J_{11}) j_1 T_1' J_1' |\} j_1^{n_1} \alpha_1' T_1' J_1'][j_2^{n_2-1}(\alpha_{22} T_{22} J_{22}) j_2 T_2 J_2 |\} j_2^{n_2} \alpha_2 T_2 J_2] \\
&\quad \times [j_2^{n_2-1}(\alpha_{22} T_{22} J_{22}) j_2 T_2' J_2' |\} j_2^{n_2} \alpha_2' T_2' J_2'] \\
&\quad \times \sqrt{(2T_1+1)(2T_2+1)(2T_1'+1)(2T_2'+1)(2J_1+1)(2J_2+1)(2J_1'+1)(2J_2'+1)} \\
&\quad \times \sum_{T'J'} (2T'+1)(2J'+1) V_{T'J'}(j_1 j_2) \sum_{T_0} (2T_0+1) \begin{Bmatrix} T_{11} & \frac{1}{2} & T_1 \\ T_{22} & \frac{1}{2} & T_2 \\ T_0 & T' & T \end{Bmatrix} \begin{Bmatrix} T_{11} & \frac{1}{2} & T_1' \\ T_{22} & \frac{1}{2} & T_2' \\ T_0 & T' & T \end{Bmatrix} \\
&\quad \times \sum_{J_0} (2J_0+1) \begin{Bmatrix} J_{11} & j_1 & J_1 \\ J_{22} & j_2 & J_2 \\ J_0 & J' & J \end{Bmatrix} \begin{Bmatrix} J_{11} & j_1 & J_1' \\ J_{22} & j_2 & J_2' \\ J_0 & J' & J \end{Bmatrix}
\end{aligned}
$$

MATRIX ELEMENTS BETWEEN DIFFERENT CONFIGURATIONS

Two nucleon excitations

$$\langle [j^{n-2}(\alpha_1 T_1 J_1) j' j''(T'J')TM_T JM]_a | \sum_{i<h}^{n} V_{ih} | j^n \alpha T M_T J M\rangle$$

$$= \sqrt{\frac{n(n-1)}{2}} [j^{n-2}(\alpha_1 T_1 J_1) j^2 (T'J')TJ|\} j^n \alpha T J] \langle [j' j'' T' J']_a | V | j^2 T' J'\rangle$$

$$\langle [j_1^{n-2}(\alpha_1 T_1 J_1) j_2^{n_2+2}(\alpha_2 T_2 J_2) T J]_a | \sum_{i<h}^{n_1+n_2} V_{ih} | [j_1^{n_1}(\alpha_1' T_1' J_1') j_1^{n_2}(\alpha_2' T_2' J_2') T J]_a\rangle$$

$$= \sqrt{\frac{n_1(n_1-1)(n_2+1)(n_2+2)}{4}} \sum_{T'J'} [j_1^{n_1-2}(\alpha_1 T_1 J_1) j^2 (T'J') T_1' J_1' |\} j_1^{n_1} \alpha_1' T_1' J_1']$$

$$\times [j_2^{n_2}(\alpha_2' T_2' J_2') j_2^2 (T'J') T_2 J_2 |\} j_2^{n_2+2} \alpha_2 T_2 J_2]$$

$$\times (-1)^{2J_2' + J_1 - J_2 + J + 2T_2' + T_1 - T_2 + T} \sqrt{(2T_1'+1)(2T_2+1)(2J_1'+1)(2J_2+1)}$$

$$\times \begin{Bmatrix} T_1 & T' & T_1' \\ T_2' & T & T_2 \end{Bmatrix} \begin{Bmatrix} J_1 & J' & J_1' \\ J_2' & J & J_2 \end{Bmatrix} \langle j_2^2 T' J' | V | j_1^2 T' J'\rangle$$

$$\langle [j_1^{n_1-2}(\alpha_1 T_1 J_1) j_2^{n_2+1}(a_2 T_2 J_2) j_3^{n_3+1}(\alpha_3 T_3 J_3)(T_{23} J_{23}) T J]_a | \sum_{i<h}^{n_1+n_2+n_3} V_{ih} | [j_1^{n_1}(\alpha_1' T_1' J_1') j_2^{n_2}(\alpha_2' T_2' J_2') j_3^{n_3}(\alpha_3' T_3' J_3')(T_{23}' J_{23}') T J]_a \rangle$$

$$= \sqrt{n_1(n_1-1)(n_2+1)(n_3+1)/2} \sum_{T'J'} [j_1^{n_1-2}(\alpha_1 T_1 J_1) j_1^2 (T'J') T_1' J_1' | \} j_1^{n_1} \alpha_1' T_1' J_1']$$

$$\times [j_2^{n_2}(\alpha_2' T_2' J_2') j_2 T_2 J_2 | \} j_2^{n_2+1} \alpha_2 T_2 J_2][j_3^{n_3}(\alpha_3' T_3' J_3') j_3 T_3 J_3 | \} j_3^{n_3+1} \alpha_3 T_3 J_3](-1)^{2T_{23}'-T_{23}+T_1+T+2J_{23}'-J_{23}+J_1+J}$$

$$\times \sqrt{(2T_1'+1)(2T_2+1)(2T_3+1)(2T_{23}+1)(2T_{23}'+1)(2T'+1)}$$

$$\times \sqrt{(2J_1'+1)(2J_2+1)(2J_3+1)(2J_{23}+1)(2J_{23}'+1)(2J'+1)}$$

$$\times \begin{Bmatrix} T_2' & \frac{1}{2} & T_2 \\ T_3' & \frac{1}{2} & T_3 \\ T_{23}' & T' & T_{23} \end{Bmatrix} \begin{Bmatrix} T_1 & T' & T_1' \\ T_{23}' & T & T_{23} \end{Bmatrix} \begin{Bmatrix} J_2' & j_2 & J_2 \\ J_3' & j_3 & J_3 \\ J_{23}' & J' & J_{23} \end{Bmatrix} \begin{Bmatrix} J_1 & J' & J_1' \\ J_{23}' & J & J_{23} \end{Bmatrix} \langle [j_2 j_3 T' J']_a | V | j_1^2 T' J' \rangle$$

$$\langle [j_1^{n_1-1}(\alpha_1 T_1 J_1) j_2^{n_2-1}(\alpha_2 T_2 J_2)(T_{12} J_{12}) j_1' j_2' (T'J') T J]_a | \sum_{i<h}^{n_1+n_2} V_{ih} | [j_1^{n_1}(\alpha_1' T_1' J_1') j_1^{n_2}(\alpha_2' T_2' J_2') T J]_a \rangle$$

$$= \sqrt{n_1 n_2} \sum [j_1^{n_1-1}(\alpha_1 T_1 J_1) j_1 T_1' J_1' | \} j_1^{n_1} \alpha_1' T_1' J_1'][j_2^{n_2-1}(\alpha_2 T_2 J_2) j_2 T_2' J_2' | \} j_2^{n_2} \alpha_2' T_2' J_2']$$

$$\times [(2T_1'+1)(T_2'+1)(2T_{12}+1)(2T'+1)(2J_1'+1)(2J_2'+1)(2J_{12}+1)(2J'+1)]$$

$$\times \begin{Bmatrix} T_1 & \frac{1}{2} & T_1' \\ T_2 & \frac{1}{2} & T_2' \\ T_{12} & T' & T \end{Bmatrix} \begin{Bmatrix} J_1 & j_1 & J_1' \\ J_2 & j_2 & J_2' \\ J_{12} & J' & J \end{Bmatrix} \langle [j_1' j_2' T' J']_a | V | [j_1 j_2 T' J']_a \rangle$$

$$
\begin{aligned}
&\langle [j_1^{n_1-1}(\alpha_1 T_1 J_1) j_2^{n_2-1}(\alpha_2 T_2 J_2)(T_{12} J_{12}) j_3^{n_3+2}(\alpha_3 T_3 J_3) T J]_a | \sum_{1<h}^{n_1+n_2+n_3} V_{ih} \\
&\quad | [j_1^{n_1}(\alpha_1' T_1' J_1') j_2^{n_2}(\alpha_2' T_2' J_2')(T_{12}' J_{12}') j_3^{n_3}(\alpha_3' T_3' J_3') T J]_a \rangle = \sqrt{n_1 n_2 (n_3+2)/2} \\
&\quad \times \sum_{T'J'} [j_1^{n_1-1}(\alpha_1 T_1 J_1) j_1 T_1' J_1' |\} j_1^{n_1} \alpha_1' T_1' J_1'][j_2^{n_2-1}(\alpha_2 T_2 J_2) j_2 T_2' J_2' |\} j_2^{n_2} \alpha_2' T_2' J_2'] \\
&\qquad \times [j_3^{n_3}(\alpha_3' T_3' J_3') j_3^2 (T'J') T_3 J_3 |\} j_3^{n_3+2} \alpha_3 T_3 J_3](-1)^{T_{12}+T_3'+T'+T+J_{12}+J_3'+J'+J} \\
&\quad \times [(2T_1'+1)(2T_2'+1)(2T_{12}+1)(2T_{12}'+1)(2T_3+1)(2T'+1) \\
&\qquad \times (2J_1'+1)(2J_2'+1)(2J_{12}+1)(2J_{12}'+1)(2J_3+1)(2J'+1)]^{1/2} \\
&\quad \times \begin{Bmatrix} T_1 & \frac{1}{2} & T_1' \\ T_2 & \frac{1}{2} & T_2' \\ T_{12} & T' & T \end{Bmatrix} \begin{Bmatrix} T_{12} & T' & T_{12}' \\ T_3' & T & T_3 \end{Bmatrix} \begin{Bmatrix} J_1 & j_1 & J' \\ J_2 & j_2 & J_2' \\ J_{12} & J' & J \end{Bmatrix} \begin{Bmatrix} J_{12} & J' & J_{12}' \\ J_3' & J & J_3 \end{Bmatrix} \langle j_3^2 T' J' | V | [j_1 j_2 T' J']_a \rangle
\end{aligned}
$$

$$\langle [j_1^{n_1-1}(\alpha_1 T_1 J_1) j_2^{n_2-1}(\alpha_2 T_2 J_2)(T_{12} J_{12}) j_3^{n_3+1}(\alpha_3 T_3 J_3) j_4^{n_4+1}(\alpha_4 T_4 J_4)(T_{34} J_{34}) T J]_a |$$

$$\sum_{i<h}^{n_1+n_2+n_3+n_4} V_{ih} | [j_1^{n_1}(\alpha_1' T_1' J_1') j_2^{n_2}(\alpha_2' T_2' J_2')(T_{12}' J_{12}') j_3^{n_3}(\alpha_3' T_3' J_3') j_4^{n_4}(\alpha_4' T_4' J_4')(T_{34}' J_{34}') T J]_a \rangle$$

$$= \sqrt{n_1 n_2 (n_3+1)(n_4+1)} \sum [j_1^{n_1-1}(\alpha_1 T_1 J_1) j_1 T_1' J_1' |\} j_1^{n_1} \alpha_1' T_1' J_1']$$

$$\times [j_2^{n_2-1}(\alpha_2 T_2 J_2) j_2 T_2' J_2' |\} j_2^{n_2} \alpha_2' T_2' J_2'][j_3^{n_3}(\alpha_3' T_3' J_3') j_3 T_3 J_3 |\} j_3^{n_3+1} \alpha_3 T_3 J_3]$$

$$\times [j_4^{n_4}(\alpha_4' T_4' J_4') j_4 T_4 J_4 |\} j_4^{n_4+1} \alpha_4 T_4 J_4] \times (-1)^{T_{12}+T_{34}+T+J_{12}+J_{34}+J}$$

$$\times [2T_1'+1)(2T_2'+1)(2T_{12}+1)(2T_{12}'+1)(2T_3+1)(2T_4+1)(2T_{34}+1)(2T_{34}'+1)]^{1/2}(2T'+1)$$

$$\times [(2J_1'+1)(2J_2'+1)(2J_{12}+1)(2J_{12}'+1)(2J_3+1)(2J_4+1)(2J_{34}+1)(2J_{34}'+1)]^{1/2}(2J'+1)$$

$$\times \begin{Bmatrix} T_1 & \frac{1}{2} & T_1' \\ T_2 & \frac{1}{2} & T_2' \\ T_{12} & T' & T_{12}' \end{Bmatrix} \begin{Bmatrix} T_3' & \frac{1}{2} & T_3 \\ T_4' & \frac{1}{2} & T_4 \\ T_{34}' & T' & T_{34} \end{Bmatrix} \begin{Bmatrix} T_{12} & T' & T_{12}' \\ T_{34}' & T & T_{34} \end{Bmatrix}$$

$$\times \begin{Bmatrix} J_1 & j_1 & J_1' \\ J_2 & j_2 & J_2' \\ J_{12} & J' & J_{12}' \end{Bmatrix} \begin{Bmatrix} J_3' & j_3 & J_3 \\ J_4' & j_4 & J_4 \\ J_{34}' & J' & J_{34} \end{Bmatrix} \begin{Bmatrix} J_{12} & J' & J_{12}' \\ J_{34}' & J & J_{34} \end{Bmatrix} \langle [j_3 j_4 T' J']_a | V | [j_1 j_2 T' J']_a \rangle$$

Single nucleon excitations

$$\left\langle [j^{n-1}(\alpha_1 T_1 J_1) j' TJ]_a \right| \sum_{i<h}^{n} V_{ih} \left| j^n \alpha TJ \right\rangle = (n-1)\sqrt{n}$$

$$\times \sum_{\alpha_2 T_2 J_2 T' J'} [j^{n-2}(\alpha_2 T_2 J_2) j^2 (T'J') T_1 J_1 |\} j^n \alpha_1 T_1 J_1][j^{n-2}(\alpha_2 T_2 J_2) j^2 (T'J') TJ |\} j^n \alpha TJ]$$

$$\times (-1)^{T_2+(1/2)+(1/2)+T+J_2+j+j'+J} \sqrt{(2T_1+1)(2T'+1)(2J_1+1)(2J'+1)}$$

$$\times \begin{Bmatrix} T_2 & \frac{1}{2} & T_1 \\ \frac{1}{2} & T & T' \end{Bmatrix} \begin{Bmatrix} J_2 & j & J_1 \\ j' & J & J' \end{Bmatrix} \frac{1}{\sqrt{2}} \langle [j j' T' J']_a | V | j^2 T' J' \rangle$$

THE INTERACTING BOSON MODEL IBA-1

Generators of the $U(6)$ group

$$(d^+ \times \tilde{d})^{(k)}_\kappa = \sum_{\mu\mu'} (2\mu 2\mu' \mid 22k\kappa) d^+_\mu \tilde{d}_{\mu'} \qquad \tilde{d}_\mu = (-1)^\mu d_{-\mu} \qquad k = 0,1,2,3,4$$

$$d^+_\mu s + s^+ \tilde{d}_\mu \qquad i(d^+_\mu s - s^+ \tilde{d}_\mu)$$

$$s^+ s$$

$$J_\kappa = \sqrt{10}(d^+ \times \tilde{d})_\kappa \qquad \hat{N} = \hat{n}_s + \hat{n}_d = s^+ s + \sum d^+_\mu d_\mu = s^+ s + (d^+ \cdot \tilde{d})$$

$$[(d^+ \times \tilde{d})^{(k')}_{\kappa'}, (d^+ \times \tilde{d})^{(k'')}_{\kappa''}] = \sum_{k\kappa} (-1)^k (1 - (-1)^{k+k'+k''}) \sqrt{(2k'+1)(2k''+1)}$$

$$\times \begin{Bmatrix} 2 & 2 & k \\ k' & k'' & 2 \end{Bmatrix} (k'\kappa' k''\kappa'' \mid k'k''k\kappa)(d^+ \times \tilde{d})^{(k)}_{\kappa}$$

$$[d^+_\mu s, s^+ \tilde{d}_{\mu'}] = (-1)^{\mu+1} \delta_{\mu,-\mu'} s^+ s + \sum_{k\kappa} (2\mu 2\mu' \mid 22k\kappa)(d^+ \times \tilde{d})^{(k)}_{\kappa}$$

$$[d^+_\mu s, (d^+ \times \tilde{d})^{(k)}_{\kappa}] = (-1)^{\mu+1} \sum_{\mu'} (2\mu' 2, -\mu \mid 22k\kappa) d^+_{\mu'} s$$

$$[s^+ \tilde{d}_{\mu,} (d^+ \times \tilde{d})^{(k)}_{\kappa}] = (-1)^{k+\mu} \sum_{\mu'} (2\mu' 2, -\mu \mid 22k\kappa) s^+ \tilde{d}_{\mu'}$$

General $U(6)$ Hamiltonian

$$H = \epsilon_s s^+ s + \epsilon_d (d^+ \cdot \tilde{d}) + \frac{1}{2} \sum_{J=0,2,4} c_J (d^+ \times d^+)^{(J)} \cdot (\tilde{d} \times \tilde{d})^{(J)} + \tfrac{1}{2} \tilde{v}_0 \{(d^+ \times d^+)^{(0)}_0 s^2 + (s^+)(\tilde{d} \times \tilde{d})^{(0)}_0\}$$

$$+ \frac{1}{\sqrt{2}} \tilde{v}_2 \{[(d^+ \times d^+)^{(2)} \times \tilde{d} s]^{(0)}_0 + [s^+ d^+ \times (\tilde{d} \times \tilde{d})^{(2)}]^{(0)}_0\} + \tfrac{1}{2} u_0 (s^+)^2 s^2 + \frac{1}{\sqrt{2}} u_2 s^+ s (d^+ \cdot \tilde{d})$$

Using $\hat{n}_s + \hat{n}_d = N$

$$H = \epsilon_s N + \tfrac{1}{2} u_0 N(N-1) + \epsilon(d^+ \cdot \tilde{d}) + \tfrac{1}{2} \sum_{J=0,2,4} c'_J (d^+ \times d^+)^{(J)} \cdot (\tilde{d} \times \tilde{d})^{(J)} + \tfrac{1}{2} \tilde{v}_0 \{(d^+ \times d^+)_0^{(0)} s^2 + (s^+)(\tilde{d} \times \tilde{d})_0^{(0)}\}$$

$$+ \frac{1}{\sqrt{2}} \tilde{v}_2 \{[(d^+ \times d^+)^{(2)} \times \tilde{d} s]_0^{(0)} + [s^+ d^+ \times (\tilde{d} \times \tilde{d})^{(2)}]_0^{(0)}\}$$

$$\epsilon = \epsilon_d - \epsilon_s + (N-1)\left(\frac{u_2}{\sqrt{5}} - u_0\right) \qquad c'_J = c_J + u_0 - \frac{2u_2}{\sqrt{5}}$$

Single boson quadrupole operator

$$T_\mu^{(2)} = \alpha_2 (d_\mu^+ s + s^+ \tilde{d}_\mu) + \beta_2 (d^+ \times \tilde{d})_\mu^{(2)}$$

Casimir operator of $U(6)$

$$C_{U(6)} = \sum_{k=0}^{4} (d^+ \times \tilde{d})^{(k)} \cdot (d^+ \times \tilde{d})^{(k)} + \tfrac{1}{2}(d^+ s + s^+ \tilde{d}) \cdot (d^+ s + s^+ \tilde{d}) - \tfrac{1}{2}(d^+ s - s^+ \tilde{d}) \cdot (d^+ s - s^+ \tilde{d}) + (s^+ s)(s^+ s)$$

$$= (\hat{n}_d + \hat{n}_s)^2 + 5(\hat{n}_d + \hat{n}_s) = \hat{N}(\hat{N} + 5)$$

Generators of the $U(5)$ group

$$(d^{+} \times \tilde{d})^{(k)}_{\kappa} \qquad k = 0, 1, 2, 3, 4$$

Casimir operator of $U(5)$

$$C_{U(5)} = \sum_{k=0}^{4} (d^{+} \times \tilde{d})^{(k)} \cdot (d^{+} \times \tilde{d})^{(k)} = \hat{n}_d^2 + 4\hat{n}_d$$

Generators of the $O(5)$ group

$$(d^{+} \times \tilde{d})^{(1)}_{\kappa} \qquad (d^{+} \times \tilde{d})^{(3)}_{\kappa}$$

Casimir operator of $O(5)$

$$C_{O(5)} = 2 \sum_{k=1,3} (d^{+} \times \tilde{d})^{(k)} \cdot (d^{+} \times \tilde{d})^{(k)}$$

Eigenvalues

$$\tau(\tau + 3) \qquad \tau = n_d, n_d - 2, \ldots, 1 \text{ or } 0$$

Pairing interaction

$$(d^{+} \cdot d^{+})(\tilde{d} \cdot \tilde{d}) = \hat{n}_d^2 + 3\hat{n}_d - C_{O(5)}$$

Eigenvalues

$$n_d(n_d + 3) - \tau(\tau + 3)$$

Eigenvalues of general $U(5)$ Hamiltonian

$$\epsilon_s N + \tfrac{1}{2}u_0 N(N-1) + \epsilon n_d + \alpha n_d(n_d - 1)/2 + \beta[n_d(n_d + 3) - \tau(\tau + 3)] + \gamma[J(J+1) - 6n_d]$$

Allowed J values

$$J = \lambda, \lambda + 1, \lambda + 2, \ldots, 2\lambda - 2, 2\lambda \qquad \tau = 3n_\Delta + \lambda$$

Coefficients of fractional parentage

$$\psi(d^n \alpha J M) = \sum_{\alpha_1 J_1} [d^{n-1}(\alpha_1 J_1) dJ |\} d^n \alpha J] \psi(d^{n-1}(\alpha_1 J_1) d_n J M)$$

$$[d^{n-1}(\alpha_1 J_1) dJ |\} d^n \alpha J] = \frac{1}{\sqrt{n}} \frac{1}{\sqrt{2J+1}} (d^n \alpha J \| d^+ \| d^{n-1} \alpha_1 J_1)$$

$$= (-1)^{J-J_1} \frac{1}{\sqrt{n}} \frac{1}{\sqrt{2J+1}} (d^{n-1} \alpha_1 J_1 \| \tilde{d} \| d^n \alpha J)$$

Special case

$$[d^2(J_1)dJ|\}d^3J] = \frac{\delta_{J_1J_0} + 2\sqrt{(2J_0+1)(2J_1+1)}\begin{Bmatrix} J_0 & 2 & J \\ J_1 & 2 & 2 \end{Bmatrix}}{\sqrt{3 + 6(2J_0+1)\begin{Bmatrix} J_0 & 2 & J \\ J_0 & 2 & 2 \end{Bmatrix}}}$$

Matrix elements of the quadrupole operator

$$\langle n_s n_d \tau J M | s^+ \tilde{d}_\mu + d_\mu^+ s | n_s + 1, n_d - 1, \tau_1 J_1 M_1 \rangle = \sqrt{N - n_d + 1}\langle n_d \tau J M | d_\mu^+ | n_d - 1, \tau_1 J_1 M_1 \rangle$$

Reduction formulae

$$(n_d \tau n_\Delta J \| d^+ \| n_d - 1, \tau + 1, n'_\Delta J') = \sqrt{\frac{n_d - \tau}{2}}(\tau + 2, \tau n_\Delta J \| d^+ \| \tau + 1, \tau + 1, n'_\Delta J')$$

$$(n_d \tau n_\Delta J \| d^+ \| n_d - 1, \tau - 1, n'_\Delta J') = \sqrt{\frac{n_d + \tau + 3}{2\tau + 3}}(\tau \tau n_\Delta J \| d^+ \| \tau - 1, \tau - 1, n'_\Delta J')$$

$$(\tau + 2, \tau n_\Delta J \| d^+ \| \tau + 1, \tau + 1, n'_\Delta J') = (-1)^{J+J'}\sqrt{\frac{2}{2\tau + 5}}(\tau + 1, \tau + 1, n'_\Delta J' \| d^+ \| \tau \tau n_\Delta J)$$

Recursion formulae for c.f.p.

$$[d^{n_d-1}(\tau+1,n'_\Delta J')dJ|\}d^{n_d}\tau n_\Delta J] = \sqrt{\frac{(n_d-\tau)(\tau+2)}{2n_d}}[d^{\tau+1}(\tau+1,n'_\Delta J')dJ|\}d^{\tau+2}\tau n_\Delta J]$$

$$[d^{n_d-1}(\tau-1,n'_\Delta J')dJ|\}d^{n_d}\tau n_\Delta J] = \sqrt{\frac{\tau(n_d+\tau+3)}{n_d(2\tau+3)}}[d^{\tau-1}(\tau-1,n'_\Delta J')dJ|\}d^{\tau}\tau n_\Delta J]$$

$$[d^{\tau+1}(\tau+1,n'_\Delta J')dJ|\}d^{\tau+2}\tau n_\Delta J] = (-1)^{J+J'}\sqrt{\frac{(2J'+1)(2\tau+2)}{2n_d(2J+1)(\tau+2)(2\tau+5)}}[d^{\tau}(\tau n_\Delta J)dJ'|\}d^{\tau+1}\tau+1,n'_\Delta J']$$

Reduction formulae for the quadrupole matrix elements

$$(n_d\tau n_\Delta J\|(d^+\times\tilde{d})^{(2)}\|n_d\tau-2,n'_\Delta J') = \sqrt{\frac{(n_d+\tau+3)(n_d-\tau+2)}{2(2\tau+3)}}(\tau\tau n_\Delta J\|(d^+\times\tilde{d})^{(2)}\|\tau,\tau-2,n'_\Delta J')$$

$$(n_d\tau n_\Delta J\|(d^+\times\tilde{d})^{(2)}\|n_d\tau n'_\Delta J') = \frac{2n_d+5}{2\tau+5}(\tau\tau n_\Delta J\|(d^+\times\tilde{d})^{(2)}\|\tau\tau n'_\Delta J')$$

$$(n_d\tau n_\Delta J\|(d^+\times\tilde{d})^{(k)}\|n_d\tau' n'_\Delta J') = (\tau\tau n_\Delta J\|(d^+\times\tilde{d})^{(k)}\|\tau\tau n'_\Delta J')\delta_{\tau\tau'} \qquad k=1,3$$

Reduced transition probability from state with n_d, $J=2n_d$ to state with n_d-1, $J=2(n_d-1)$

$$B(\mathrm{E2};J\to J-2) = \alpha_2^2 n_d(N-n_d+1) = \frac{\alpha_2^2}{4}J(2N+2-J)$$

Generators of the $O(6)$ group

$$(d^+ \times \tilde{d})^{(k)}_\kappa \qquad k = 1,3, \qquad Q'_\kappa = s^+\tilde{d}_\kappa + d^+_\kappa s$$

$$[Q'_\mu, Q'_{\mu'}] = \sum_{k\kappa}(1-(-1)^k)(2\mu 2\mu' \mid 22k\kappa)(d^+ \times \tilde{d})^{(k)}_\kappa$$

Casimir operator of $O(6)$

$$C_{O(6)} = \mathbf{Q}' \cdot \mathbf{Q}' + 2\sum_{k=1,3}(d^+ \times \tilde{d})^{(k)} \cdot (d^+ \times \tilde{d})^{(k)} = \mathbf{Q}' \cdot \mathbf{Q}' + C_{O(5)}$$

Eigenvalues

$$\sigma(\sigma+4) \qquad \sigma = N, N-2, \ldots, 1 \text{ or } 0$$

Pairing interaction

$$P'^+P' = \tfrac{1}{4}((d^+ \cdot d^+) - (s^+)^2)((\tilde{d} \cdot \tilde{d}) - s^2) = \tfrac{1}{4}[\hat{N}(\hat{N}+4) - C_{O(6)}]$$

Eigenvalues

$$\tfrac{1}{4}[N(N+4) - \sigma(\sigma+4)] = \tfrac{1}{4}(N-\sigma)(N+\sigma+4)$$

Eigenvalues of general $O(6)$ Hamiltonians

$$Na + \tfrac{1}{2}N(N-1)b + \tfrac{1}{4}A(N-\sigma)(N+\sigma+4) + \tfrac{1}{6}B\tau(\tau+3) + CJ(J+1)$$

$$\tau = \sigma, \sigma-1, \sigma-2, \ldots, 0$$

Reduced transition probability between $N = \sigma$ states with $\tau, J = 2\tau$ and $\tau - 1$, $J = 2\tau - 2$

$$B(\mathrm{E2}, J \to J - 2) = \alpha_2^2 \frac{J}{8(J+3)}(2N + 2 - J)(2N + 6 - J)$$

Most general $O(5)$ Hamiltonian

$$\epsilon_s' \hat{N} + \tfrac{1}{2} u_0 \hat{N}(\hat{N} - 1) + \epsilon \hat{n}_d + \alpha \hat{n}_d(\hat{n}_d - 1)/2 + \kappa C_{O(6)} + \frac{B}{6} C_{O(5)} + C C_{O(3)}$$

In partially diagonal form

$$\epsilon_s N + \tfrac{1}{2} u_0 N(N - 1) + \epsilon \hat{n}_d + \alpha \hat{n}_d(\hat{n}_d - 1)/2 + \kappa C_{O(6)} + \frac{B}{6}\tau(\tau + 3) + CJ(J + 1)$$

Eigenstates

$$|N, \nu, \tau, n_\Delta, JM\rangle = \sum_{n_d} a^{\nu}_{n_d \tau} |N, n_d, \tau, n_\Delta, JM\rangle$$

Generalized pairing interaction

$$P_\beta^+ P_\beta^- = \tfrac{1}{4}((d^+ \cdot d^+) - \beta(s^+)^2)((\tilde{d} \cdot \tilde{d}) - \beta s^2)$$

Special eigenstate

$$P_\beta^+ P_\beta^- (d_0^+ + x s^+)^N |0\rangle = 0 \qquad \text{for} \quad \beta x^2 = 1$$

Matrix elements of the quadrupole operator

$$(N\nu\tau n_\Delta J\|\mathbf{T}^{(2)}\|N\nu'\tau+1,n'_\Delta J') = \alpha_2(-1)^{J+J'}(\tau+1,\tau+1,n'_\Delta J'\|d^+\|\tau\tau n_\Delta J)\frac{1}{\sqrt{2\tau+5}}$$

$$\times\left\{\sum_{n_d} a^{\nu}_{n_d\tau}a^{\nu'}_{n_d-1,\tau+1}\sqrt{(N-n_d+1)(n_d-\tau)} + \sum_{n_d} a^{\nu}_{n_d\tau}a^{\nu'}_{n_d+1,\tau+1}\sqrt{(N-n_d)(n_d+\tau+5)}\right\}$$

$$(N\nu\tau n_\Delta J\|\mathbf{T}^{(2)}\|N\nu'\tau+2,n'_\Delta J') = \beta_2(\tau+2,\tau,n_\Delta J\|(d^+\times\tilde{d})^{(2)}\|\tau+2,\tau+2,n'_\Delta J')$$

$$\times\sum_{n_d} a^{\nu}_{n_d\tau}a^{\nu'}_{n_d\tau+2}\sqrt{\frac{(n_d+\tau+5)(n_d-\tau)}{2(2\tau+5)}}$$

$$(N\nu\tau n_\Delta J\|\mathbf{T}^{(2)}\|N\nu'\tau n'_\Delta J') = \beta_2(\tau\tau n_\Delta J\|(d^+\times\tilde{d}^{(2)}\|\tau\tau n'_\Delta J')\sum_{n_d} a^{\nu}_{n_d\tau}a^{\nu'}_{n_d\tau}\frac{2n_d+5}{2\tau+5}$$

Special relation for $k = 2,4$

$$(N\nu\tau n_\Delta J\|(d^+\times\tilde{d})^{(k)}\|N\nu\tau n_\Delta J) = \frac{2\bar{n}_d+5}{2\tau+5}(Nn_d=\tau,\tau n_\Delta J\|(d^+\times\tilde{d})^{(k)}\|Nn_d=\tau,\tau n_\Delta J)$$

Generators of the $SU(3)$ group

$$J_\mu, \qquad Q_\mu = s^+\tilde{d}_\mu + d^+_\mu s - \frac{\sqrt{7}}{2}(d^+ \times \tilde{d})^{(2)}_\mu = Q'_\mu - \frac{\sqrt{7}}{2}(d^+ \times \tilde{d})^{(2)}_\mu$$

$$[Q_\mu, Q_{\mu'}] = \frac{15}{4}\sum_\kappa (2\mu 2\mu' \mid 221\kappa)(d^+ \times \tilde{d})^{(1)}_\kappa$$

Casimir operator of $SU(3)$

$$C_{SU(3)} = 2(\mathbf{Q}\cdot\mathbf{Q}) + \tfrac{3}{4}(\mathbf{J}\cdot\mathbf{J})$$

Eigenvalues

$$C_{SU(3)}(\lambda, \mu) = \lambda^2 + \mu^2 + \lambda\mu + 3(\lambda + \mu)$$

General $SU(3)$ Hamiltonian

$$-\kappa(\mathbf{Q}\cdot\mathbf{Q}) - \kappa'(\mathbf{J}\cdot\mathbf{J}) = -\tfrac{1}{2}\kappa C_{SU(3)} + (\tfrac{3}{8}\kappa - \kappa')(\mathbf{J}\cdot\mathbf{J})$$

Eigenvalues

$$-\tfrac{1}{2}\kappa(\lambda^2 + \mu^2 + \lambda\mu + 3(\lambda + \mu)) + (\tfrac{3}{8}\kappa - \kappa')J(J + 1)$$

Allowed states ($\lambda \geq \mu, \mu$ even)

$$K = \mu, \mu - 2, \ldots, 0$$

$$J = K, K+1, K+2, \ldots, K+\lambda \qquad \text{for} \quad K > 0$$

$$J = 0, 2, 4, \ldots, 2\lambda \qquad \text{for} \quad K = 0$$

Reduced transition probability in the ground state band

$$B(\mathrm{E2}; J \to J-2) = \alpha_2^2 \frac{3J(J-1)}{4(2J-1)(2J+1)}(2N+2-J)(2N+1+J)$$

A generalized quadrupole pairing interaction

$$\mathbf{Q}_\alpha^+ \cdot \mathbf{Q}_\alpha^- = ((d^+ \times d^+)^{(2)} - \alpha s^+ d^+) \cdot ((\tilde{d} \times \tilde{d})^{(2)} - \alpha s \tilde{d})$$

Special eigenstate

$$(\mathbf{Q}_\alpha^+ \cdot \mathbf{Q}_\alpha^-)(d_0^+ + x s^+)^N |0\rangle = 0 \qquad \text{for} \quad \alpha x = -\sqrt{\tfrac{2}{7}}$$

A generalized pairing interaction and a quadrupole pairing interaction

$$[A P_\beta^+ P_\beta^- + B(\mathbf{Q}_\alpha^+ \cdot \mathbf{Q}_\alpha^-)](d_0^+ + x s^+)^N |0\rangle = 0 \qquad \text{for} \quad x = \frac{1}{\sqrt{\beta}}, \quad \alpha = -\sqrt{\frac{2\beta}{7}}$$

For $\beta = 2$, $\beta = 7A$ it is diagonal in the $SU(3)$ scheme, for $\beta = 0$, $\beta = 1$ it is a special $O(6)$ Hamiltonian

GINOCCHIO MODELS

$$a^{+}_{im_i k m_k} = \sum_{j=|k-i|}^{k+i} (i m_i k m_k \mid i k j m) a^{+}_{jm}$$

$$|ii(0)kk(0)J=0\rangle = \frac{1}{\sqrt{(2i+1)(2k+1)}} \sum_{j=|k-i|}^{k+i} \sqrt{2j+1}\,|ik(j)ik(j)J=0\rangle$$

The $O(6)$ model $(i=\frac{3}{2})$

$$S^{+} = \sum_{j=|k-3/2|}^{k+3/2} S^{+}_{j}$$

$$T^{(L)}_{M} = \sum_{jj'} (-1)^{k+(3/2)+j+L} \sqrt{(2j+1)(2j'+1)} \begin{Bmatrix} \frac{3}{2} & j & k \\ j' & \frac{3}{2} & L \end{Bmatrix} (a^{+}_{j} \times \tilde{a}_{j'})^{(L)}_{M}$$

$$D^{+}_{M} = [T^{(2)}_{M}, S^{+}] = \sum_{jj'} (-1)^{k+(3/2)+j} \sqrt{(2j+1)(2j'+1)} \begin{Bmatrix} \frac{3}{2} & j & k \\ j' & \frac{3}{2} & 2 \end{Bmatrix} (a^{+}_{j} \times a^{+}_{j'})^{(2)}_{M}$$

The $SU(3)$ model ($k = 1$)

$$S^+ = \sum_{J=|i-1|}^{i+1} S_j^+$$

$$T_M^{(L)} = \sum_{jj'} (-1)^{1+i+j'+L} \sqrt{(2j+1)(2j'+1)} \begin{Bmatrix} 1 & j & i \\ j' & 1 & L \end{Bmatrix} (a_j^+ \times \tilde{a}_{j'})_M^{(L)}$$

$$D_M^+ = [\sqrt{\tfrac{3}{4}} T_M^{(2)}, S^+] = \sqrt{\tfrac{3}{4}} \sum_{jj'} (-1)^{i+j'+1} \sqrt{(2j+1)(2j'+1)} \begin{Bmatrix} 1 & j & i \\ j' & 1 & 2 \end{Bmatrix} (a_j^+ \times a_j^+)_M^{(2)}$$

THE PROTON NEUTRON INTERACTING BOSON MODEL (IBA-2)

Generators of $U_\pi(6) \otimes U_\nu(6)$

$$(d_\pi^+ \times \tilde{d}_\pi)_\kappa^{(k)} \quad k = 0,1,2,3,4, \quad d_{\pi\kappa}^+ s_\pi + s_\pi^+ \tilde{d}_{\pi\kappa}, \quad i(d_{\pi\kappa}^+ s_\pi - s_\pi^+ \tilde{d}_{\pi\kappa}), \quad s_\pi^+ s_\pi$$

$$(d_\nu^+ \times \tilde{d}_\nu)_\kappa^{(k)} \quad k = 0,1,2,3,4, \quad d_{\nu\kappa}^+ s_\nu + s_\nu^+ \tilde{d}_{\nu\kappa}, \quad i(d_{\nu\kappa}^+ s_\nu - s_\nu^+ \tilde{d}_{\nu\kappa}), \quad s_\nu^+ s_\nu$$

Generators of $U_{\pi+\nu}(6)$

$$(d_\pi^+ \times \tilde{d}_\pi)_\kappa^{(k)} + (d_\nu^+ \times \tilde{d}_\nu)_\kappa^{(k)}, \quad k = 0,1,2,3,4, \quad s_\pi^+ s_\pi + s_\nu^+ s_\nu$$

$$d_{\pi\kappa}^+ s_\pi + s_\pi^+ \tilde{d}_{\pi\kappa} + d_{\nu\kappa}^+ s_\nu + s_\nu^+ \tilde{d}_{\nu\kappa}, \quad i(d_{\pi\kappa}^+ s_\pi - s_\pi^+ \tilde{d}_{\pi\kappa} + d_{\nu\kappa}^+ s_\nu - s_\nu^+ \tilde{d}_{\nu\kappa})$$

F-spin operators

$$F^+ = s_\pi^+ s_\nu + \sum_\kappa d_{\pi\kappa}^+ d_{\nu\kappa} = s_\pi^+ s_\nu + (d_\pi^+ \cdot \tilde{d}_\nu)$$

$$F^- = (F^+)^\dagger = s_\nu^+ s_\pi + \sum_\kappa d_{\nu\kappa}^+ d_{\pi\kappa} = s_\nu^+ s_\pi + (d_\nu^+ \cdot \tilde{d}_\pi)$$

$$[F^+, F^-] = s_\pi^+ s_\pi - s_\nu^+ s_\nu + \sum_\kappa (d_{\pi\kappa}^+ d_{\pi\kappa} - d_{\nu\kappa}^+ d_{\nu\kappa}) = N_\pi - N_\nu = 2F^0$$

$$[F^0, F^+] = F^+ \qquad [F^0, F^-] = -F^-$$

The Casimir operator of $U_{\pi+\nu}(6)$

$$C_{U_{\pi+\nu}(6)} = 2\mathbf{F}^2 + \tfrac{1}{2}\hat{N}(\hat{N} + 8)$$

Eigenvalues

$$2F(F+1) + \tfrac{1}{2}N(N+8)$$

The Majorana operator

$$\mathcal{M}_0 = (s_\pi^+ d_\nu^+ - d_\pi^+ s_\nu^+) \cdot (s_\pi \tilde{d}_\nu - \tilde{d}_\pi s_\nu) - 2 \sum_{L=1,3} (d_\pi^+ \times d_\nu^+)^{(L)} \cdot (\tilde{d}_\pi \times \tilde{d}_\nu)^{(L)} = \tfrac{1}{2} N (\tfrac{1}{2} N + 1) - \mathbf{F}^2$$

A generalization of the Majorana operator

$$\mathcal{M} = \sum_{L=1,3} \xi_L (d_\pi^+ \times d_\nu^+)^{(L)} \cdot (\tilde{d}_\pi \times \tilde{d}_\nu)^{(L)} + \xi_2 (s_\pi^+ d_\nu^+ - d_\pi^+ s_\nu^+) \cdot (s_\pi \tilde{d}_\nu - \tilde{d}_\pi s_\nu)$$

Generators of the $SU_{\pi+\nu}(3)$ group

$$\mathbf{Q}_\pi + \mathbf{Q}_\nu = s_\pi^+ \tilde{d}_\pi + d_\pi^+ s_\pi - \frac{\sqrt{7}}{2} (d_\pi^+ \times \tilde{d}_\pi)^{(2)} + s_\nu^+ \tilde{d}_\nu + d_\nu^+ s_\nu - \frac{\sqrt{7}}{2} (d_\nu^+ \times \tilde{d}_\nu)^{(2)}$$

$$\mathbf{J} = \mathbf{J}_\pi + \mathbf{J}_\nu = \sqrt{10} ((d_\pi^+ \times \tilde{d}_\pi)^{(1)} + (d_\nu^+ \times \tilde{d}_\nu)^{(1)})$$

Casimir operator of $SU_{\pi+\nu}(3)$

$$C_{SU_{\pi+\nu}(3)} = 2(\mathbf{Q}_\pi + \mathbf{Q}_\nu) \cdot (\mathbf{Q}_\pi + \mathbf{Q}_\nu) + \tfrac{3}{4} (\mathbf{J}_\pi + \mathbf{J}_\nu) \cdot (\mathbf{J}_\pi + \mathbf{J}_\nu)$$

Eigenvalues of the general $SU(3)$ Hamiltonian

$$aN + b\tfrac{1}{2}N(N-1) + cF(F+1) + \kappa\big(\lambda^2 + \mu^2 + \lambda\mu + 3(\lambda+\mu)\big) + \kappa' J(J+1)$$

Dynamical symmetry withough definite F-spin

$$H + \tfrac{1}{2}\kappa_\pi C_{SU_\pi(3)} + \tfrac{1}{2}\kappa_\nu C_{SU_\nu(3)} + \tfrac{1}{2}\kappa C_{SU_{\pi+\nu}(3)} + \kappa' \mathbf{J}^2$$

Eigenvalues

$$\tfrac{1}{2}\kappa_\pi(\lambda_\pi^2 + \mu_\pi^2 + \lambda_\pi\mu_\pi + 3\lambda_\pi + 3\mu_\pi) + \tfrac{1}{2}\kappa_\nu(\lambda_\nu^2 + \mu_\nu^2 + \lambda_\nu\mu_\nu + 3\lambda_\nu + 3\mu_\nu)$$

$$+ \tfrac{1}{2}\kappa(\lambda^2 + \mu^2 + \lambda\mu + 3\lambda + 3\mu) + \kappa' J(J+1)$$

Special cases

$$H' = 2\kappa \mathbf{Q}_\pi \cdot \mathbf{Q}_\nu + \tfrac{3}{4}\kappa \mathbf{J}_\pi \cdot \mathbf{J}_\nu$$

$$H'' = 2\kappa \mathbf{Q}_\pi \cdot \mathbf{Q}_\nu - \tfrac{3}{8}\kappa \mathbf{J}_\pi \cdot \mathbf{J}_\pi - \tfrac{3}{8}\kappa \mathbf{J}_\nu \cdot \mathbf{J}_\nu$$

Generators of $\overline{SU(3)}$

$$\mathbf{Q}^{(+)} = s^+\tilde{d} + d^+ s + \frac{\sqrt{7}}{2}(d^+ \times \tilde{d})^{(2)}, \quad \mathbf{J}$$

Generators of $SU^*_{\pi+\nu}(3) \subset SU_\pi(3) \otimes \overline{SU}_\nu(3)$

$$\mathbf{Q}_\pi^{(-)} + \mathbf{Q}_\nu^{(+)} = s_\pi^+ \tilde{d}_\pi + d_\pi^+ s_\pi - \frac{\sqrt{7}}{2}(d_\pi^+ \times \tilde{d}_\pi)^{(2)} + s_\nu^+ \tilde{d}_\nu + d_\nu^+ s_\nu + \frac{\sqrt{7}}{2}(d_\nu^+ \times \tilde{d}_\nu)^{(2)}, \qquad \mathbf{J}_\pi + \mathbf{J}_\nu$$

Dynamical symmetry

$$\tfrac{1}{2}\kappa_\pi C_{SU_\pi(3)} + \tfrac{1}{2}\kappa_\nu C_{\overline{SU_\nu(3)}} + \tfrac{1}{2}\kappa C_{SU^*_{\pi+\nu}(3)} + \kappa' \mathbf{J}^2$$

Projection onto states with exact F-spin, $F = N/2$

$$a_\pi(d_\pi^+ \times \tilde{d}_\pi)^{(L)}_\mu + a_\nu(d_\nu^+ \times \tilde{d}_\nu)^{(L)}_M \to \frac{a_\pi N_\pi + a_\nu N_\nu}{N_\pi + N_\nu}(d^+ \times \tilde{d})^{(L)}_M$$

$$\sum_{L=0}^{4}\{a_L^\pi(d_\pi^+ \times \tilde{d}_\pi)^{(L)} \cdot (d_\pi^+ \times \tilde{d}_\pi)^{(L)} + a_L^\nu(d_\nu^+ \times \tilde{d}_\nu)^{(L)} \cdot (d_\nu^+ \times \tilde{d}_\nu)^{(L)} + a_L(d_\pi^+ \times \tilde{d}_\pi) \cdot (d_\nu^+ \times \tilde{d}_\nu)^{(L)}\}$$

$$\to \sum_{L=0}^{4}\left\{\left(a_L^\pi \frac{N_\pi(N_\pi - 1)}{N(N-1)} + a_L^\nu \frac{N_\nu(N_\nu - 1)}{N(N-1)} + a_L \frac{N_\pi N_\nu}{N(N-1)}\right)(d^+ \times \tilde{d})^{(L)} \cdot (d^+ \times \tilde{d})^{(L)} + \right.$$

$$\left. + (a_L^\pi + a_L^\nu - a_L)\frac{2L+1}{5}\frac{N_\pi N_\nu}{N(N-1)}(d^+ \cdot \tilde{d})\right\}$$

$$\tilde{\mathbf{Q}}_\pi = s_\pi^+ \tilde{d}_\pi + d_\pi^+ s_\pi + \chi_\pi(d_\pi^+ \times \tilde{d}_\pi)^{(2)}, \qquad \tilde{\mathbf{Q}}_\nu = s_\nu^+ \tilde{d}_\nu + d_\nu^+ s_\nu + \chi_\nu(d_\nu^+ \times \tilde{d}_\nu)^{(2)}$$

$$\kappa_\pi(\tilde{\mathbf{Q}}_\pi \cdot \tilde{\mathbf{Q}}_\pi) + \kappa_\nu(\tilde{\mathbf{Q}}_\nu \cdot \tilde{\mathbf{Q}}_\nu) + \kappa(\tilde{\mathbf{Q}}_\pi + \tilde{\mathbf{Q}}_\nu) \cdot (\tilde{\mathbf{Q}}_\pi + \tilde{\mathbf{Q}}_\nu)$$

$$\rightarrow (\kappa_\pi + \kappa)\frac{N_\pi(N_\pi - 1)}{N(N-1)}(s^+\tilde{d} + d^+s + \chi_\pi(d^+ \times \tilde{d})^{(2)}) \cdot (s^+\tilde{d} + d^+s + \chi_\pi(d^+ \times \tilde{d})^{(2)}$$

$$+ (\kappa_\pi + \kappa)\frac{N_\pi N_\nu}{N(N-1)}(5s^+s + (d^+ \cdot \tilde{d}) + \chi_\pi^2(d^+ \cdot \tilde{d}))$$

$$+ (\kappa_\nu + \kappa)\frac{N_\nu(N_\nu - 1)}{N(N-1)}(s^+\tilde{d} + d^+s + \chi_\nu(d^+ \times \tilde{d})^{(2)}) \cdot (s^+\tilde{d} + d^+s + \chi_\nu(d^+ \times d)^{(2)}$$

$$+ (\kappa_\nu + \kappa)\frac{N_\pi N_\nu}{N(N-1)}(5s^+s + (d^+ \cdot \tilde{d}) + \chi_\nu^2(d^+ \cdot \tilde{d}))$$

$$+ 2\kappa\frac{N_\pi N_\nu}{N(N-1)}[\mathbf{Q}(\chi) \cdot \mathbf{Q}(\chi) - \chi'^2(d^+ \times \tilde{d})^{(2)} \cdot (d^+ \times \tilde{d})^{(2)} - 5s^+s - (d^+ \cdot \tilde{d}) - \chi_\pi\chi_\nu(d^+ \cdot \tilde{d})]$$

$$\chi = \tfrac{1}{2}(\chi_\pi + \chi_\nu) \qquad \chi' = \tfrac{1}{2}(\chi_\pi - \chi_\nu) \qquad \mathbf{Q}(\chi) = s^+\tilde{d} + d^+s + \chi(d^+ \times \tilde{d})^{(2)}$$

Sum rule for $M1$ strength to the $J = 0$ ground state with $F = N/2$

$$\sum B(M1) = \frac{9}{2\pi}(g_\pi - g_\nu)^2\frac{N_\pi N_\nu}{N(N-1)}\langle J = 0, g.s.|\hat{n}_d|J = 0, g.s.\rangle$$

PARTICLES COUPLED TO AN AXIALLY SYMMETRIC CORE

Eigenstates with $\pm m_0$ projection of spin of a single j-nucleon

$$|m_0 jJM\rangle = \sqrt{2}\sum_{L\,\text{even}} \sqrt{2L+1}\begin{pmatrix} L & j & J \\ 0 & m_0 & -m_0 \end{pmatrix}|jLJM\rangle$$

Expectation value of $\mathbf{L}^2$

$$\langle m_0 jJM|\mathbf{L}^2|m_0 jJM\rangle = 2\sum_{L\,\text{even}} L(L+1)(2L+1)\begin{pmatrix} L & j & J \\ 0 & m_0 & -m_0 \end{pmatrix}^2$$

$$= J(J+1) + j(j+1) - 2m_0^2 + \delta_{m_0 1/2}(-1)^{j+J}(J+\tfrac{1}{2})(j+\tfrac{1}{2})$$

Eigenstates of a nucleon or nucleons in states with $j, j', \ldots$

$$|m_0 JM\rangle = \sum_j \alpha_j \sqrt{2}\sum_{L\,\text{even}} \sqrt{2L+1}\begin{pmatrix} L & j & J \\ 0 & m_0 & -m_0 \end{pmatrix}|jLJM\rangle \qquad \sum_j \alpha_j^2 = 1$$

where

$$\sum_{j'}(\epsilon_j \delta_{jj'} + \langle jm_0|q_0|j'm_0\rangle)\alpha_{j'} = Q_{m_0}\alpha_j$$

q_0 is the $\kappa = 0$ component of the particles quadrupole moment

Energy of the core in first order perturbation theory

$$\frac{1}{2\mathcal{I}}2\sum_{L\,\text{even}\,j}\alpha_j^2L(L+1)(2L+1)\begin{pmatrix}L & j & J\\ 0 & m_0 & -m_0\end{pmatrix}^2$$

$$=\frac{1}{2\mathcal{I}}\left[J(J+1)-2m_0^2+\sum_j\alpha_j^2j(j+1)+\delta_{m_01/2}\sum_j\alpha_j^2(-1)^{j+J}(J+\tfrac{1}{2})(j+\tfrac{1}{2})\right]$$

$$=\frac{1}{2\mathcal{I}}[J(J+1)-2m_0^2+\langle m_0|\mathbf{j}^2|m_0\rangle+\delta_{m_01/2}(-1)^{J+1/2}(J+\tfrac{1}{2})\langle\tfrac{1}{2}|j^+|\bar{\tfrac{1}{2}}\rangle]$$

$$=\frac{1}{2\mathcal{I}}[J(J+1)-2m_0^2+\langle m_0|\mathbf{j}^2|m_0\rangle-\delta_{m_01/2}(-1)^{J+1/2}(J+\tfrac{1}{2})a]$$

$$|m_0\rangle=\sum_j\alpha_j|jm=m_0\rangle\qquad|\overline{m_0}\rangle=\sum_j(-1)^{j+1/2}\alpha_j|jm=-m_0\rangle$$

Magnetic moment in case of a single j-nucleon

$$\mu=g_cJ+\frac{m_0^2}{J+1}(g_j-g_c)[1-\delta_{m_01/2}(-1)^{j+J}(2J+1)(j+\tfrac{1}{2})]$$

Magnetic moment in the general case of a single nucleon

$$\mu = g_c J + \frac{m_0^2}{J+1}(g_{m_0} - g_c)[1 + \delta_{m_0 1/2}(-1)^{J+1/2}(2J+1)b]$$

$$\boldsymbol{\mu}' = \sum_i [(g_{s_i} - g_c)\mathbf{s}_i + (g_{l_i} - g_c)\boldsymbol{l}_i]$$

$$\langle m_0|\mu_0'|m_0\rangle = m_0(g_{m_0} - g_c), \qquad \sqrt{2}\langle \tfrac{1}{2}|\mu_1'|\bar{\tfrac{1}{2}}\rangle = b(g_{m_0} - g_c)$$

REFERENCES*

Amit, D. and Katz, A. (1964), Nucl. Phys. **58**, 388 (see p. 17).
Arima, A., Harvey, M. and Shimizu, K. (1969), Phys. Lett. **B30**, 519 (see p. 445).
Arima, A. and Horie, H. (1954), Prog. Theor. Phys. **11**, 509; **12**, 623 (see p. 729).
Arima, A., Horie, H. and Tanabe, Y. (1954), Prog. Theor. Phys. **11**, 143 (see p. 138).
Arima, A. and Iachello, I. (1975), Phys. Rev. Lett. **35**, 1069 (see pp. 741, 818).
——— (1978), Phys. Rev. Lett. **40**, 385 (see p. 780).
Arima, A., Otsuka, T., Iachello, F. and Talmi, I. (1977), Phys. Lett. **B66**, 205 (see pp. 867, 870).
Arima, A. and Terasawa, T. (1960), Prog. Theor. Phys. **23**, 115 (see p. 236).
Arvieu, R. and Moszkowski, S. A. (1966), Phys. Rev. **145**, 830 (see p. 442).
Auerbach, N. (1966), Nucl. Phys. **76**, 321 and Phys. Rev. **163** (1967), 1203 (see pp. 463, 478, 480, 483).
Auerbach, N. and Talmi, I. (1964a), Phys. Lett. **9**, 153 (see p. 656).
——— (1964b), Phys. Lett. **10**, 297 (see pp. 659, 662).
——— (1965), Nucl. Phys. **64**, 458 (see pp. 410, 658).
Bacher, R. F. and Goudsmit, S. (1934), Phys. Rev. **46**, 948 (see p. 265).
Balashov, V. V. and Eltekov, V. A. (1960), Nucl. Phys. **16**, 423 (see p. 236).

*Page numbers in parentheses which appear at the end of each reference indicate the locations of citations within the text.

Ball, J. B., McGrory, J. B. and Larsen, J. S. (1972), Phys. Lett. **B41**, 581 (see p. 658).

Bardeen, J., Cooper, L. N. and Schrieffer, J. R. (1957), Phys. Rev. **108**, 1175 (see pp. 219, 448).

Barrett, B. R., Hewitt, R. G. L. and McCarthy, R. J. (1971), Phys. Rev. **C3**, 1137 (see p. 236).

Bartlett, J. H. (1932a), Nature **130**, 165 (see p. 21).

——— (1932b), Phys. Rev. **41**, 370 (see p. 21).

Bayman, B. F. and Lande, A. (1966), Nucl. Phys. **77**, 1 (see p. 266).

Bayman, B. F., Reiner, A. S. and Sheline, R. K. (1959), Phys. Rev. **115**, 1627 (see p. 410).

Bertsch, G. F. (1972), *The Practitioner's Shell Model*, North-Holland, Amsterdam (see p. 258).

Bethe, H. A. and Bacher, R. F. (1936), Rev. Mod. Phys. **8**, 82 (see pp. 23, 31).

Bethe, H. A. and Rose, M. E. (1937), Phys. Rev. **51**, 283 (see p. 52).

Biedenharn, L. C. (1953), J. Math Phys. **31**, 289 (see p. 152).

Biedenharn, L. C. and Louck, J. D. (1981), *Encyclopedia of Mathematics and Its Applications*, Vol. 8, *Angular Momentum in Quantum Physics*, Vol. 9, *The Racah–Wigner Algebra in Quantum Theory*, Addison-Wesley, Reading, MA (see p. 18).

Biedenharn, L. C. and van Dam, H. (1965), eds., *Quantum Theory of Angular Momentum*, Academic Press, New York, NY (see p. 18).

Bijker, R., Dieperink, A. E. L., Scholten, O. and Spanhoff, R. (1980), Nucl. Phys. **A344**, 207 (see p. 916).

Blin-Stoyle, R. J. (1953), Proc. Phys. Soc. (London) **A66**, 1158; Blin-Stoyle, R. J. and Perks, M. A., Proc. Phys. Soc. (London) **A67** (1954), 885 (see p. 729).

Blomqvist, J. (1984), in *International Review of Nuclear Physics*, World Press, Singapore, Vol. 2, p. 1 (see p. 420).

Bohr, A. and Mottelson, B. R. (1953), Dan. Vid. Selsk., Mat.-Fys. Medd. No. 16 (see p. 732).

——— (1969), *Nuclear Structure* Vol. I, Benjamin, Reading, MA (see pp. 165, 169).

——— (1975), *Nuclear Structure* Vol. II, Benjamin, Reading, MA (see pp. 733, 934).

Bohr, N. (1936), Nature **137**, 344 (see p. 25).

Bonatsos, D. (1988), *Interacting Boson Models of Nuclear Structure*, Oxford University Press (see pp. 19, 741).

Brink, D. M., Buck, B., Huby, R., Nagarajan, M. A. and Rowley, N. (1987), J. Phys. G. **13**, 629 (see p. 924).

Brody, T. A. and Moshinsky, M. (1960), *Tables of Transformation Brackets*, Monografias del Instituto de Fisica, Mexico (see p. 236).

Brussaard, P. J. and Glaudemans, P. W. M. (1977), *Shell Model Applications in Nuclear Spectroscopy*, North Holland, Amsterdam (see p. 227).

Casten, R. F. (1990), *Nuclear Structure from a Simple Perspective*, Oxford University Press (see pp. 741, 814, 916).

Cohen, S. and Kurath, D. (1965), Nucl. Phys. **73**, 1 (see p. 17).

Cohen, S., Lawson, R. D., MacFarlane, M. H. and Soga, M. (1964), Phys. Lett. **10**, 195 (see p. 410).

Cohen, S., Lawson, R. D., MacFarlane, M. H., Pandya, S. P. and Soga, M. (1967), Phys. Rev. **160**, 903 (see pp. 463, 478, 480, 483).

Cohen, S., Lawson, R. D. and Soper, J. M. (1966), Phys. Lett. **21**, 306 (see p. 462).

Condon, E. U. and Shortley, G. H. (1935), *The Theory of Atomic Spectra*, Cambridge University Press (reprinted 1953) (see p. 297).

Dawson, J. F., Talmi, I. and Walecka, J. D. (1962), Ann. Phys. (NY) **18**, 339 (see p. 236).

de-Shalit, A. (1953), Phys. Rev. **91**, 1479 (see pp. 83, 188, 190, 191).

de-Shalit, A. and Feshbach, H. (1974), *Theoretical Nuclear Physics*, John Wiley and Sons, New York, NY (see pp. 165, 169).

de-Shalit, A. and Talmi, I. (1963), *Nuclear Shell Theory*, Academic Press, New York, NY (see pp. 160, 165, 169, 258, 273, 293, 557, 678, 767).

Dieperink, A. E. L. and Bijker, R. (1982), Phys. Lett. **B116**, 77 (see p. 886).

Dieperink, A. E. L. and Scholten, O. (1980), Nucl. Phys. **A346**, 125 (see p. 804).

Dieperink, A. E. L. and Talmi, I. (1983), Phys. Lett. **B131**, 1 (see p. 886).

Duval, P. D. and Barrett, B. R. (1981), Phys. Rev. **C23**, 492 (see p. 916).

Eckart, C. (1930), Rev. Mod. Phys. **2**, 305 (see p. 109).

Edmonds, A. R. (1957), *Angular Momentum in Quantum Mechanics*, Princeton University Press, Princeton, NJ (see p. 18, 63).

Elliott, J. P. (1953), Proc. Roy. Soc. (London) **A218**, 345 (see p. 152).

——— (1958), Proc. Roy. Soc. (London) **A245**, 128; 562 (see pp. 621, 625).

——— (1985), Rep. Prog. Phys. **48**, 171, where earlier references are listed (see p. 913).

Elliott, J. P. and Skyrme, T. H. R. (1955), Proc. Roy. Soc. (London), **A232**, 561 (see p. 53).

Elsasser, W. M. (1933), J. de Phys. et Rad. **4**, 549 (see p. 22).

——— (1934a), J. de Phys. et Rad. **5**, 389 (see p. 22).

——— (1934b), J. de Phys. et Rad. **5**, 635 (see p. 23).

——— (1935), J. de Phys. et Rad. **6**, 473 (see p. 23).

Fano, U. and Racah, G. (1959), *Irreducible Tensorial Sets*, Academic Press, New York, NY (see p. 18, 157).

Feenberg, E. (1949), Phys. Rev. **75**, 320 (see p. 26).

Feenberg, E. and Hammack, K. C. (1949), Phys. Rev. **75**, 1877 (see p. 26).

Feenberg, E., Hammack, K. C. and Nordheim, L. (1949), Phys. Rev. **75**, 1968 (see p. 31).

Flowers, B. H. (1952), Proc. Roy. Soc. (London) **A212**, 248 (see pp. 350, 541).

Ford, K. W. (1955), Phys. Rev. **98**, 1516 (see p. 410).

Gaarde, C., Larsen, J. S., Harakeh, M. K., van der Werf, S. Y., Igarashi, M. and Müller-Arnke, A. (1980), Nucl. Phys. **A334**, 248 (see p. 130).

Gal, A. (1968), Ann. Phys. (NY) **49**, 341 (see p. 236).

Gambhir, Y. K., Rimini, A. and Weber, T. (1969), Phys. Rev. **188**, 1513 (see p. 448).

Garvey, G. T. and Kelson, I. (1966), Phys. Rev. Lett. **16**, 197 (see p. 573).

Garvey, G. T., Gerace, W. J., Jaffe, R. L., Talmi, I. and Kelson, I. (1969), Rev. Mod. Phys. **41**, S1 (see p. 573).

Ginocchio, J. N. (1980), Ann. Phys. (NY) **126**, 234 where earlier references are given (see pp. 829, 832, 835).

——— (1991), Phys. Lett. **B265**, 6 (see p. 915).

Ginocchio, J. N. and Haxton, W. C. (1993), in *Symmetries in Science VI*, B. Gruber and M. Ramek eds., Plenum, New York (see p. 398).

Ginocchio, J. N. and Talmi, I. (1980), Nucl. Phys. **A337**, 431 (see p. 834).

Glaubman, M. J. (1953), Phys. Rev. **90**, 1000 (see p. 183).

Gloeckner, D. H. and Serduke, F. J. D. (1974), Nucl. Phys. **A220**, 477 (see pp. 410, 413, 658, 679).

Gneuss, G. and Greiner, W. (1971), Nucl. Phys. **171**, 449 (see p. 740).

Goldstein, S. and Talmi, I. (1956), Phys. Rev. **102**, 589 (see p. 13).

Green, I. M. and Moszkowski, S. A. (1965), Phys. Rev. **139B**, 790 (see p. 220).

Gross, R. and Frenkel, A. (1976), Nucl. Phys. **A267**, 85 (see p. 658).

Harter, H., Gelberg, A. and von Brentano, P. (1985), Phys. Lett. **B157**, 1 (see p. 895).

Harvey, M. (1968), Advances in Nuclear Physics Vol. 1, 67 (see p. 643).

Haxel, O., Jensen, J. H. D. and Suess, H. E. (1949a), Die Naturwissenschaften **35**, 376 (see p. 31).

——— (1949b), Phys. Rev. **75**, 1766 (see p. 31).

——— (1950), Zeits. f. Physik **128**, 1295 (see p. 31).

Hecht, K. T. and Adler, A. (1969), Nucl. Phys. **A137**, 129 (see p. 445).

Heisenberg, W. (1932), Zeits. f. Physik **77**, 1 (see p. 21).

Holstein, T. and Primakoff, H. (1940), Phys. Rev. **58**, 1098 (see p. 824).

Hund, F. (1937), Zeits. f. Physik **105**, 202 (see p. 599).

Iachello, F. (1981a), Chem. Phys. Lett. **78**, 581 (see p. 818).

——— (1981b), Nucl. Phys. **A358**, 89c (see p. 916).

Iachello, F. and Arima, A. (1987), *The Interacting Boson Model*, Cambridge University Press (see p. 19, 741).

Iachello, F., Puddu, G. Scholten, O., Arima, A. and Otsuka, T. (1979), Phys. Lett. **B89**, 1 (see p. 916).

Iachello, F. and Scholten, O. (1979), Phys. Rev. Lett. **43**, 679 (see pp. 741, 924).

Iachello, F. and van Isacker, P. (1991), *The Interacting Boson Fermion Model*, Cambridge University Press (see pp. 741, 924).

Janssen, D., Jolos, R. V. and Dönau, F. (1974), Nucl. Phys. **A224**, 93 (see pp. 740, 741).

Jensen, J. H. D., Suess, H. and Haxel, O. (1949), Die Naturwissenschaften **36**, 155 (see p. 31).

Jolos, R. V., Dönau, F. and Janssen, D. (1975), Yad. Fiz. **22**, 965 (see p. 741).

Kerman, A. K. (1961), Ann. Phys. (NY) **12**, 300 (see p. 350, 431, 432).

Kleinheinz, P., Lunardi, S., Ogawa, M. and Maier, M. R. (1978), Zeits. f. Physik **A284**, 351. P. Kleinheinz, R. Broda, P. J. Daly, S. Lunardi, M. Ogawa and J. Blomqvist, Zeits. f. Physik **A290** (1979), 279 (see p. 419).

Kumar, K. (1966), J. Math. Phys. **7**, 671 (see p. 236).

——— (1967), Aust. J. Phys. **20**, 205 (see p. 236).

Kurath, D. (1956), Phys. Rev. **101**, 216 (see p. 10).

——— (1957), Phys. Rev. **106**, 975 (see p. 10).

Kurath, D. and Pičman, L. (1959), Nucl. Phys. **10**, 313 (see p. 706).

Lande, A. (1933), Phys. Rev. **44**, 1028 (see pp. 27, 44).

——— (1934), Phys. Rev. **46**, 477 (see pp. 27, 44, 613).

Lawson, R. D. (1981), Zeits, f. Physik **A303**, 51 (see p. 420).

Lawson, R. D. and Goeppert-Mayer, M. (1960), Phys. Rev. **117**, 174 (see p. 236).

Lawson, R. D. and Uretsky, J. L. (1957a), Phys. Rev. **106**, 1369 (see p. 402).

——— (1957b), Phys. Rev. **108**, 1300 (see p. 650).

Lederer, C. M. and Shirley, V. S. (1978), eds. *Table of Isotopes*, John Wiley and Sons, New York, NY (see p. 14).

Leviatan, A., Novoselsky, A. and Talmi, I. (1986), Phys. Lett. **B172**, 144 (see p. 784).

Lorazo, B. (1970), Nucl. Phys. **A153**, 255 (see p. 448).

Magnus, W. and Oberhettinger, F. (1949), *Formulas and Theorems for the Functions of Mathematical Physics*, Chelsea, New York (see p. 46).

Margenau, H. and Wigner, E. P. (1940), Phys. Rev. **58**, 103 (see pp. 28, 612).

Mayer, M. G. (1948), Phys. Rev. **74**, 235 (see p. 26).

——— (1949), Phys. Rev. **75**, 1969 (see pp. 30, 31).

——— (1950a), Phys. Rev. **78**, 16 (see p. 30).

——— (1950b), Phys. Rev. **78**, 22 (see p. 30).

McCullen, J. D., Bayman, B. F. and Zamick, L. (1964), Phys. Rev. **134**, B515 (see pp. 852, 853).

McNeill, J. H., Blomqvist, J., Chishti, A. A., Daly, P. J., Gelletly, W., Hotchkis, M. A. C., Piiparinen, M., Varley, B. J. and Woods, P. J. (1989), Phys. Rev. Lett. **63**, 860 (see p. 423).

Moshinsky, M. (1959), Nucl. Phys. **13**, 104 (see p. 236).

Mottelson, B. R. (1958), in *The Many Body Problem*, Lecture Notes of the Les Houches Summer School, Dunod, Paris (1959), p. 283 (see p. 448).

Nolte, E., Colombo, G., Gui, S. Z., Korschinek, G., Schollmeier, W., Kubik, P., Anstavsson, S., Geier, R. and Morinaga, H. (1982), Zeits. f. Physik **A306**, 223; E. Nolte, G. Korschinek and Ch. Setzensack (1982), Zeits. f. Physik **A309**, 33 (see p. 426).

Nordheim, L. (1949), Phys. Rev. **75**, 1894 (see pp. 27, 28).

Novoselsky, A. (1988), Nucl. Phys. **A483**, 282 (see p. 919).

Novoselsky, A. and Talmi, I. (1986), Phys. Lett. **B172**, 139 (see p. 916).

Otsuka, T., Arima, A., Iachello, F. and Talmi, I. (1978), Phys. Lett. **B76**, 139 (see p. 867).

Pandya, S. P. (1956), Phys. Rev. **103**, 956 (see pp. 13, 258).

Pang, S. C., Klein, A. and Dreizler, R. M. (1968), Ann. Phys. (NY) **49**, 477 (see p. 824).

Pittel, S., Duval, P. D. and Barrett, B. R. (1982), Phys. Rev. **C25**, 2834 (see p. 844).

Post, H. R. (1953), Proc. Phys. Soc. (London) **A66**, 649 (see p. 54).

Pryce, M. H. L. (1952), Proc. Phys. Soc. (London) **A65**, 773 (see p. 181).

Puddu, G., Scholten, O. and Otsuka, T. (1980), Nucl. Phys. **A348**, 109 (see p. 916).

Racah, G. (1942), Phys. Rev. **62**, 438 (see pp. 18, 104, 149, 157).

——— (1943), Phys. Rev. **63**, 367 (see pp. 18, 265, 350, 351, 401, 575, 587).

——— (1952), in *L. Farkas Memorial Volume*, Research council of Israel, Jerusalem, p. 294 (see pp. 289, 294, 350).

——— (1964), J. Quant. Spect. Rad. Transfer **4**, 617 (see pp. 297, 298).

——— (1965), *Group Theory and Spectroscopy*, Erg. d. Exakten Wissenschaften **37**, 27, Springer, Berlin. Based on mimeographed lecture notes, Princeton 1951, taken by E. Merzbacher and D. Park (see p. 503).

Racah, G. and Stein, J. (1967), Phys. Rev. **156**, 58 (see p. 730).

Racah, G. and Talmi, I. (1952), Physica **18**, 1097 (see p. 211).

Redlich, M. G. (1958), Phys. Rev. **110**, 468 (see p. 706).

Redmond, P. J. (1954), Proc. Roy. Soc. (London) **A222**, 84 (see pp. 265, 274).

Ring, P. and Schuck, P. (1980), *The Nuclear Many-Body Problem*, Springer, New York (see p. 682).

Rosenfeld, L. (1948), *Nuclear Forces*, North-Holland, Amsterdam (see p. 26).

Rotenberg, M., Bivins, R., Metropolis, N. and Wooten, J. K., Jr. (1959), *The 3j- and 6j-Symbols*, Technology Press, MIT, Cambridge, MA (see p. 104).

Schmidt, Th. (1937), Zeits, f. Physik **106**, 358 (see pp. 28, 613).

Scholten, O. (1980), Ph.D. Thesis, University of Groningen, Netherlands (unpublished) (see p. 895).

Schwartz, C. and de-Shalit, A. (1954), Phys. Rev. **94**, 1257 (see pp. 265, 276).

Schwinger, J. (1952), unpublished. Reproduced in Biedenharn and van Dam (1965) (see p. 138).

Serduke, F. J. D., Lawson, R. D. and Gloeckner, D. H. (1976), Nucl. Phys. **A256**, 45 (see p. 853).

Shlomo, S. and Talmi, I. (1972), Nucl. Phys. **A198**, 81 (see pp. 468, 489).

Smirnov, Yu. F. (1961), Nucl. Phys. **27**, 177 (see p. 236).

——— (1962), Nucl. Phys. **39**, 346 (see p. 236).

Stephens, F. S. (1975), Rev. Mod. Phys. **47**, 43 (see p. 949).

Stephens, F. S., Diamond, R. M., Leigh, J. R., Kammuri, T. and Nakai, K. (1972), Phys. Rev. Lett. **29**, 438 (see p. 946).

Stephens, F. S. and Simon, R. S. (1972), Nucl. Phys. **A183**, 257 (see p. 949).

Suess, H. E., Haxel, O. and Jensen, J. H. D. (1949), Die Naturwissenschaften **36**, 153 (see p. 31).

Talmi, I. (1952), Helv. Phys. Acta **25**, 185 (see pp. 232, 236).

——— (1953), Phys. Rev. **90**, 1001 (see pp. 181, 183).

——— (1957), Phys. Rev. **107**, 326 (see p. 402).

——— (1960), in *Proc. Varenna Summer School*, Rendiconti S. I. F. XV, Bologna (1962) (see p. 258).

——— (1962a), Phys. Rev. **126**, 2116 (see pp. 286, 653).

——— (1962b), Phys. Rev. **126**, 1096 (see p. 408).

——— (1962c), Rev. Mod. Phys. **34**, 704 (see pp. 619, 667).

——— (1969), in *Proc. Mendeleevian Conf. Torino and Rome*, Academia delle Scienze, Torino (1971), p. 275 (see p. 429).

——— (1971), Nucl. Phys. **A172**, 1 (see pp. 448, 468, 476).

——— (1975), Phys. Lett. **B55**, 255 (see p. 489).

——— (1981), Phys. Lett. **B103**, 177 (see p. 821).

——— (1982), Phys. Rev. **C25**, 3189 (see p. 495).

——— (1983), in *Electromagnetic Properties of Atomic Nuclei*, Tokyo Institute of Technology, p. 4 (see p. 619).

——— (1984), Nucl. Phys. **A423**, 189 (see pp. 725, 727).

——— (1985), in *Proc. Varenna Summer School*, Rendiconti S.I.F. **XCI**, Bologna (see p. 429).

Talmi, I. and Thieberger, R. (1956), Phys. Rev. **103**, 923 (see p. 572).

Talmi, I. and Unna, I. (1960a), Ann. Rev. Nucl. Sci. **10**, 353 (see p. 120).

——— (1960b), Nucl. Phys. **19**, 225 (see p. 410).

——— (1960c), Phys. Rev. Lett. **4**, 469 (see p. 664).

Teitelbaum, P. (1954), Bull. Res. Counc. Israel **III**, 299 (see p. 555).

Varshalovich, D. A., Moskalev, A. N. and Khersonskii, V. K. (1988), *Quantum Theory of Angular Momentum*, World Scientific, Singapore (see p. 18).

Vergados, J. D. (1968), Nucl. Phys. **A111**, 681 (see p. 638).

Vichniac, G. (1972), M. Sc. Thesis, The Weizmann Institute of Science (unpublished) (see p. 655).

Vogel, P. (1970), Phys. Lett. **B33**, 400 (see p. 946).

Wapstra, A. H. and Audi, G. (1985), Nucl. Phys. **A432**, 1 (see p. 14).

Wapstra, A. H., Audi, G. and Hoekstra, R. (1988), Atomic Data and Nuclear Data Tables **39**, 281 (see p. 15).

Wigner, E. P. (1931), *Group Theory and Its Application to the Quantum Mechanics of Atomic Spectra*, translated from the German edition by J. J. Griffin, Academic Press, New York (1959) (see p. 109).

——— (1937), Phys. Rev. **51**, 106; 947 (see pp. 599, 604).

——— (1940), unpublished, reproduced in Biedenharn and van Dam (1965) (see pp. 18, 138, 149).

Wildenthal, B. H. (1984), Prog. Part. Nucl. Phys. **11**, 5, where earlier references are given. B. A. Brown and B. H. Wildenthal (1988), Ann. Rev. Nucl. Sci. **38**, 29 (see pp. 17, 564, 573).

Wilets, L. and Jean, M. (1956), Phys. Rev. **102**, 788 (see p. 740).

Wu, C. L., Feng, D. H., Chen, X. G., Chen, J. Q. and Guidry, M. W. (1987), Phys. Rev. **C36**, 1157, where earlier references are given (see p. 865).

Zamick, L. (1971), Ann. Phys. (NY) **66**, 784 (see p. 725).

Zemel, A. (1975), M.Sc. Thesis, The Weizmann Institute of Science, unpublished. Presented by G. Goldring in *Heavy Ion Collisions*, R. Bock, Ed., Vol. 3, North Holland, Amsterdam (1982), p. 509 (see p. 931).

INDEX